COMPREHENSIVE INSECT PHYSIOLOGY BIOCHEMISTRY AND PHARMACOLOGY

Volume 7

ENDOCRINOLOGY I

COMPREHENSIVE INSECT PHYSIOLOGY BIOCHEMISTRY AND PHARMACOLOGY

Volume 7

ENDOCRINOLOGY I

Executive Editors

G. A. KERKUT
Department of Neurophysiology,
University of Southampton, UK

L. I. GILBERT
Department of Biology,
University of North Carolina, USA

PERGAMON PRESS

OXFORD · NEW YORK · TORONTO · SYDNEY · PARIS · FRANKFURT

UK	Pergamon Press Ltd., Headington Hill Hall, Oxford OX3 0BW, England
USA	Pergamon Press Inc., Maxwell House, Fairview Park, Elmsford, New York 10523, USA
CANADA	Pergamon Press Canada Ltd., Suite 104, 150 Consumers Road, Willowdale, Ontario M2J 1P9, Canada
AUSTRALIA	Pergamon Press (Aust.) Pty. Ltd., P.O. Box 544. Potts Point, N.S.W. 2011, Australia
FRANCE	Pergamon Press SARL, 24 rue des Ecoles, 75240 Paris, Cedex 05, France
FEDERAL REPUBLIC OF GERMANY	Pergamon Press GmbH, Hammerweg 6, D-6242 Kronberg-Taunus, Federal Republic of Germany

First edition 1985

Library of Congress Cataloging in Publication Data

Main entry under title:
Comprehensive insect physiology, biochemistry, and pharmacology.

Contents: v. 1. Embryogenesis and reproduction.
1. Insects—Physiology—Collected works. I. Kerkut, G. A.
II. Gilbert, Lawrence I. (Lawrence Irwin), 1929–

QL495.C64 1984 595.7′01 83–25743

British Library Cataloguing in Publication Data

Comprehensive insect physiology, biochemistry and pharmacology.
1. Insects
I. Kerkut, G. A. II. Gilbert, Lawrence I.
595.7 QL463

ISBN 0–08–030808–2 (volume 7)
ISBN 0–08–026850–1 (set)

Filmset by Filmtype Services Ltd., Scarborough
Printed in Great Britain by A. Wheaton & Co. Ltd., Exeter

Contents

Foreword

Aristotle was enchanted by the phenomenon of insect metamorphosis and the early microscopists such as Robert Hook, Marcello Malpighi, Anton van Leeuwenhoek, René de Réaumur and Pieter Lyonet were fascinated by the structure and function of the different parts of insects and made some of the first important contributions to our knowledge of insect physiology. More detailed functional studies were made by Borelli in his book "De Motu Animalium", published in 1680, and his interpretation of insect walking patterns remained in our textbooks until 1955.

In general, the 18th and 19th century research workers were more concerned with the morphology and classification of insects, though physiologists such as Claude Bernard and naturalists such as John Lubbock and Henri Fabre were always interested in the functional analysis of insects.

One of the milestones in the study of insect physiology was the publication by Wigglesworth of his small book on insect physiology in 1934. This was stimulated by an appreciation of the way in which studies on the basic physiology of insects were necessary before one could understand and ultimately control the activity of insect pests of man and crops.

Wigglesworth initially studied medicine and then carried out research at the London School of Hygiene and Tropical Medicine. His innate gift for planning simple but fundamental experiments on *Rhodnius* led rapidly to an increase in our knowledge about moulting, the control of larval and adult stages, and provided the foundation for insect endocrinology. Furthermore he inspired a group of co-workers who later played a key role in the application of modern techniques to solve the problems of insect physiology and biochemistry.

Wigglesworth's "Insect Physiology" was followed by a more detailed and full-sized textbook, "Principles of Insect Physiology", which was published in 1947 and is now in its 7th edition (1972).

The three-volume edition of "Physiology of Insecta", a multi-authored work edited by Morris Rockstein, was published in 1964 and a new edition in six volumes followed in 1973.

The study of insect biochemistry developed more slowly, partly because there was no special distinction between physiology and biochemistry; the investigator just used the methods available for his studies. David Keilin started his studies working on insects: "From 1919 onwards I had been actively engaged in the study of the anatomy of the respiratory system, respiratory adaptation and respiration of dipterous larvae and pupae. Among the vast amount of material I was investigating, special attention was given to the larvae of *Gasterophilus intestinalis*". For these studies Keilin developed a method for the spectroscopic analysis of respiratory pigments of insect pupae under the microscope, which ultimately led to the discovery of the cytochromes.

The pteridines were discovered in insect pigments, and the one gene–one enzyme hypothesis of Beadle and Tatum, which was the cornerstone of molecular biology, was a result of biochemical and genetic analysis of *Drosophila*.

The rapid expansion of biochemistry after 1945 led to many more workers studying insect biochemistry and the first textbook on the subject by Darcy Gilmore was published in 1961. This was followed by the multi-authored "Biochemistry of Insects", edited by Morris Rockstein, in 1978.

The first evidence that a steroid hormone acts at the level of the gene came from the studies of Clever and Karlson in the 1960s on the puffing by the polytene chromosomes of *Chironomus*.

Though insect physiologists and biochemists initially published their papers in journals such as *Biological Bulletin*, *Journal of Biological Chemistry*, *Biochemical Journal*, *Journal of Physiology*, *Journal of Experimental Zoology*, *Journal of Experimental Biology*, *Roux' Archiv für Entwicklungsmechanik*, and *Zeitschrift für vergleichende Physiologie*, the great expansion of insect physiology and biochemistry from 1945 onwards led to the establishment of journals and other periodicals specialising in insects, such as the *Journal of Insect Physiology*, *Insect Biochemistry*, *Annual Review of Entomology*, and *Advances in Insect Physiology*.

It is also fitting to mention the work of other pioneers in the study of insect physiology and biochemistry, such as Autrum, Bounhiol, Bodenstein, Butenandt, Chadwick, Dethier, Fraenkel, Fukuda, Joly, Lees, Karlson, Kopec, Piepho, Richards, Roeder, Berta Scharrer, Snodgrass and Williams; these and many others laid the foundations of the subject and all following research workers have stood on the shoulders of these giants.

In July 1980 a meeting was held at Pergamon Press in Oxford to discuss the possibility of publishing a series of volumes on insect physiology, biochemistry and pharmacology. The idea was to produce 12 volumes that would provide an up-to-date summary and orientation on the physiology, biochemistry, pharmacology, behaviour and control of insects that would be of value to research workers, teachers and students. The volumes should provide the reader with the classical background to the literature and include all the important basic material. In addition, special attention would be given to the literature from 1950 to the present day. Emphasis would be given to illustrations, graphs, EM pictures and tabular summaries of data.

We were asked to act as Executive Editors and by December 1980 we had produced a 27-page booklet giving details of the aims and objectives of the project, details of the proposed volumes and chapters, suggested plans within the chapters, abbreviations, preparation of diagrams and tables, and journal citations to ensure uniformity of presentation as far as possible. This booklet was sent to authors of the chapters and their comments invited. By the middle of 1981 most of the chapters had been assigned to authors and the project was under way. The details of the volumes and the chapters they contain are given on the following pages so that the reader can see the contents of each of the other volumes.

In addition, there is a final volume, Volume 13, which is the Index Volume. Although each volume will contain its own subject index, species index and author index, Volume 13 will contain the combined subject, species and author indexes for all 12 text volumes so that any material in these volumes can be rapidly located.

All references in the volumes are given with full titles of papers, journal, volume, and first and last pages. The references to the authors in the text are given with their initials so that it is clear that the text refers to D. Smith and not, say, to A. Smith. There are more than 50,000 references to the literature, more than 10,000 species of insect referred to, and all should be readily found in the 12 different volumes.

There are 240 authors of the 200 chapters in the volumes and they have produced a series of very readable, up-to-date, and critical summaries of the literature. In addition, they have considered the problems associated with their subject, indicated the present state of the subject and suggested its developmental pathway over the next decade.

We are very grateful to our colleagues for the efficient way that they have met the challenge and the deadlines in spite of their many other commitments.

This series of volumes will be very useful to libraries, but an important case can be made that the books should be considered as research instruments. A set of volumes should also be available in the laboratory for constant reference. They will provide the research worker with an account of the literature and will always be instantly available for consultation. For this reason they should be considered as research equipment equally important as microscopes, oscilloscopes or spectrophotometers.

The volumes should save research workers many weeks of time each year in that not only will they provide an awareness of the literature and the background, and so save valuable research time, but the full index to authors, subject and species, and the full literature references, should also make it much easier to write reports and papers on their own new research work.

It is hoped that these volumes will do much to strengthen the case for insects as a source of research material, not only because insects are important medical and agricultural pests (over 200 million people at present have malaria: insects eat or destroy about 20% of planted food crops), but also because in many cases insects are the ideal unique research material for studying and solving fundamental biological problems.

G. A. KERKUT
Southampton

L. I. GILBERT
Chapel Hill

Preface to Volume 7

During the early 20th century, research in insect biology was characterized by a fascination with form and diversity, but with the birth of insect physiology in the 1920's and 1930's function and causality became the main objects of investigation. By creating novel experimental paradigms that combined the now classic techniques of ligation, transplantation and parabiosis, the early students of insect development identified the endocrine glands that control development – the brain, corpora allata and prothoracic glands – and they named the humoral products of these glands the brain hormone, juvenile hormone and ecdysone, respectively. Finally, with great ingenuity and some serendipity, they used these same paradigms to predict, and then prove, how these endocrine glands interact to direct insect metamorphosis.

By the 1950's and 1960's, the technology of chemical analysis had been improved and refined, and a second generation of insect endocrinologists employed these techniques to elucidate the chemistry of the growth and development hormones. The establishment of the structures of juvenile hormone and ecdysone fueled efforts to isolate and identify other hormones of these types, and to begin to unravel their biosynthetic pathways. This intensified search lead to the discovery that several homologs of juvenile hormone existed as well as different ecdysteroids, and that these varied among different insect species and at different stages of the life cycle of the same species. Furthermore, ecdysteroids were shown to be synthesized by the prothoracic gland as well as other tissues, but via apparently similar biosynthetic pathways. From these findings, it became apparent that both the juvenile hormones and the ecdysteroids comprised two families of molecules with structural and functional homologies, a conclusion which would profoundly affect the direction of subsequent research in the field. The delay in similar research on the brain hormone, or the prothoracicotropic hormone (PTTH) as it is now termed, reflects a general lag in the growth of the neuropeptide field in insects. The chemistry of this particular neurohormone has proven rather elusive, and it is only now, some 60 years after its discovery, that its structure is finally being determined. Not surprisingly, mounting evidence suggests that PTTH may also exist as a family of functionally, and perhaps structurally, related hormones.

A third generation of insect endocrinologists has come of age in the past decade; following the example of their predecessors, their emergence coincided with technological developments in immunological, biochemical, and chemical methodologies. With the use of these formidable tools, essential information has been obtained which provides a basis for the expansion of this field into new and exciting directions. This information includes the titers of the hormones during development, a more critical definition of their endocrinological functions during the insect's life, determinations of the evolutionary diversity of the hormones, elucidation of their biosynthetic pathways and of the mechanisms regulating their synthesis and titers, and most recently knowledge of the mechanisms of their action on target tissues. In the course of these studies, many parallels with vertebrate endocrine systems were discovered, indicating the tremendous potential of insect endocrinology to contribute substantially to a general understanding of the functioning of eukaryotic neuroendocrine and endocrine systems. The basis of this potential derives from the fact that insect endocrine systems have a functional complexity analogous to vertebrate systems, but in an evolutionary sense have retained a remarkable degree of structural simplicity. With the application of genetics and molecular biological techniques to investigations of the function of these structurally "simple" systems, the insect endocrine field is in a position to make significant contributions to understanding the endocrinological basis of development, growth, reproduction and homeostasis. For example, a number of vertebrate peptide hormones have been identified in insects by immunological procedures; if these hormones are shown to function in insects as they do in vertebrates, there would be available a comparatively simple system with which to elucidate the basis of that function and the mechanisms by which function is controlled. It is problems such as these which will undoubtedly be the focus of the work of the fourth generation of insect endocrinologists.

This volume details the origins and history of insect endocrinology, from its remarkably intuitive beginnings, through the occasionally circuitous, but ultimately successful, investigations by several generations of endocrinologists who have progressively transformed the field to one characterized by vigor, enthusiasm, and scientific excellence. From the opening chapter by the founder of the field, Sir Vincent Wigglesworth, documenting the origins of this science, to the chapter defining the new investigative area of comparative vertebrate–invertebrate endocrinology, this volume, augmented by volume 8 of this series, presents the evolution of insect endocrinology to the most recent discoveries with a level of detail that makes it an unsurpassed resource for today's endocrinologist. Perhaps more importantly, however, this volume will stand as the main reference for future generations of insect endocrinologists, serving as their intellectual window to the laboratories and minds of their scientific predecessors.

W. E. Bollenbacher
Chapel Hill

Contributors to Volume 7

Baker, F. C.
Zoecon Corporation, 975 California Avenue, PO Box 10975, Palo Alto, CA 94304, USA

Bergamasco, R.
Division of Applied Organic Chemistry, CSIRO, GPO Box 4331, Melbourne, Victoria 3001, Australia

Bollenbacher, W. E.
Department of Biology, University of North Carolina, Chapel Hill, NC 27514, USA

Chang, E. S.
Bodega Marine Laboratory, University of California, PO Box 247, Bodega Bay, Davis, CA 94923, USA

Feyereisen, R.
Department of Entomology and Agricultural Chemistry, Oregon State University, Corvallis, OR 97331, USA

Granger, N. A.
Department of Anatomy, University of North Carolina, Chapel Hill, NC 27514, USA

Goodman, W. G.
Department of Entomology, University of Wisconsin, Madison, WI 53706, USA

Hammock, B. D.
Department of Entomology and Environmental Toxicology, University of California, Davis, CA 95616, USA

Horn, D. H. S.
Division of Applied Organic Chemistry, CSIRO, GPO Box 4331, Melbourne, Victoria 3001, Australia

Karlson, P.
Institutsgruppe Lahnberge der Medizinischen, Facultät der Philipps-Universität, Institut für Physiologische Chemie, D-3550 Marburg, Federal Republic of Germany

Koolman, J.
Institutsgruppe Lahnberge der Medizinischen, Facultät der Philipps-Universität, Institut für Physiologische Chemie, D-3550 Marburg, Federal Republic of Germany

Kramer, K. J.
US Grain Marketing Research Laboratory, US Department of Agriculture, 1515 College Avenue, Manhattan, KS 66506, USA

Loughton, B. G.
Department of Biology, York University, 4700 Keele Street, Downsview, Ontario M3J 1P3, Canada

Mordue, W.
Department of Zoology, University of Aberdeen, Tillydrone Avenue, Aberdeen AB9 2TN, UK

Morgan, P. J.
Department of Zoology, University of Aberdeen, Tillydrone Avenue, Aberdeen AB9 2TN, UK

Orchard I.
Department of Biology, York University, 4700 Keele Street, Downsview, Ontario M3J 1P3, Canada

Osterbur, D. L.
Department of Genetics, University of California, Berkeley, CA 94720, USA

Rees, H. H.
Department of Biochemistry, University of Liverpool, PO Box 147, Liverpool L69 3BX, UK

Schooley, D. A.
Zoecon Corporation, 975 California Avenue, PO Box 10975, Palo Alto, CA 94304, USA

Sedlak, B. J.
Developmental Biology Center, University of California, Irvine, CA 92717, USA

Smith, S. L.
Department of Biological Sciences, Bowling Green State University, Bowling Green, OH 43403, USA

Wigglesworth, V. B.
Department of Zoology, University of Cambridge, Downing Street, Cambridge, CB2 3EJ, UK

Yund, M. A.
Department of Genetics, University of California, Berkeley, CA 94720, USA

Contents of All Volumes

Volume 13 — CUMULATIVE INDEXES

1 Historical Perspectives

SIR VINCENT B. WIGGLESWORTH, FRS

University of Cambridge, Cambridge, UK

INTRODUCTION

This chapter aims to cover the period up to the early 1960s, the time when the explosive expansion in the number of workers in the field of insect endocrinology and the consequent increase in the rate of publication took place. But I have set no hard and fast line at which to end, and have occasionally included a word or two about recent work where this seemed desirable.

1 THE CONCEPTS OF "INTERNAL SECRETION" AND "HORMONES"

The effects of castration on the secondary sexual characters of vertebrates have been familiar to agriculturalists from ancient times. John Hunter experimented with the influence of the testes on the comb of barn door fowls; but Berthold, in the middle of the nineteenth century, was the first to show that the masculinizing effect of the testis in birds is due to a substance liberated into the circulation. The term "internal secretion" was proposed by Claude Bernard in 1859 for this "endocrine" activity which was a secondary function of the testis.

In the second half of the nineteenth century, largely as the result of clinical observations, a series of primary glands of internal secretion were recognized, notably the thyroid, pituitary and adrenal glands, and the secondary endocrine functions of the ovary, pancreas and other organs were uncovered. It had long been recognized that the entry of the acid contents from the stomach into the duodenum induced the flow of pancreatic juice. The work of Bayliss and Starling, in 1904, proved that the agent concerned, "secretin", produced by the duodenum, was transported to the pancreas by the blood stream – that is, as an internal secretion. Bayliss and Starling thought of secretin, in common with other internal secretions, as a "chemical messenger" and they sought a general term which would convey this meaning. Eventually they adopted the name "hormone" proposed by my former tutor W.B. Hardy. This is the Greek word for "I excite" but has been generally adopted to mean a chemical messenger whatever may be the nature of the message.

The absence of any effect from castration, with or without the transfer of gonads of the opposite sex in such insects as *Lymantria dispar* (Oudemans, J. 1899; Meisenheimer, J. 1910), *Bombyx mori* (Kellogg, V. 1904) and *Gryllus domesticus* (Regen, J. 1910) seemed to exclude any endocrine function for the sex glands of insects. Coupled with the absence of any consistent effects from vertebrate hormones when applied to insects, these results gave rise to the belief that insects do not use hormones. This dogma may well have delayed research. It was sometimes said that metamorphosis might be induced by a hormone – were it not for the fact that insects did not secrete hormones!

(It should be pointed out that in more recent times some exceptions to the negative results on hormonal effects of the gonads in insects have been reported, notably in *Orgyia*, in the glow-worm *Lampyris* and in *Chironomus*; and it is uncertain whether hormones are involved in the phenomena of parasitic reversal of sex in insects).

2 HORMONES AND THE CONTROL OF GROWTH; THE MOULTING HORMONE

The word "moulting" in insects, like the German "Häutung", has been used for many years with two meanings: it may refer to the more or less prolonged period of growth which leads up to the casting of the old cuticle, or it may refer to the final act of shedding this cuticle. "Ecdysis" is the synonymous with "moulting" and is likewise used in the two senses; but it is sometimes convenient to use ecdysis for the final act alone. Jenkin, P. and Hinton, H (1966) proposed the name "apolysis" for the initial separation of the epidermis from the cuticle; and Hinton, H. (1958) following Snodgrass, R. (1935) regarded apolysis as the turning point of growth, marking the boundary between one instar and the next; he recommended the abolition of the word "moulting" and reserved "ecdysis" for the final act. This had the inconvenient result of leaving no word for the prolonged process of moulting – which is in fact the crucial period of active growth and development in the insect.

The hormonal control of growth in this sense was first demonstrated by Kopeč, S. (1917, 1922) when, in the course of extensive studies on the physiology of the central nervous system in *Lymantria dispar*, he discovered that during the last larval stage there is a critical period of growth and that if the caterpillar is divided in the middle by a tight ligature before this period, the anterior half goes on to form the pupa but the posterior half remains unchanged. If the brain of a larva was excised before the critical period, further development was arrested; whereas cutting the nerve cord just behind the head had no effect and pupation took place. Kopeč concluded that the brain was secreting a hormone necessary for "metamorphosis" – but by metamorphosis he meant continued development to pupa and imago, what we are here calling "moulting". He was not concerned with the *change in form* which we shall consider later under the heading of metamorphosis.

This important discovery attracted very little attention. That may well have been because Kopeč was regarding the brain as a source of internal secretion and at the time that was difficult to accept. Some 12 years later (Wigglesworth, V. 1934) I found that moulting in the larva of the bloodsucking bug *Rhodnius prolixus* was likewise arrested by decapitation before a critical period, a week or so after the great meal of blood. This was a chance observation made while testing whether the swallowing of air was necessary for rupturing the old cuticle at ecdysis. But the presence of a circulating hormone was proved when a larva, decapitated immediately after feeding, was induced to moult when connected by a capillary tube to a moulting larva.

As a possible source of this hormone I suggested the corpus allatum. Since the time of Lyonet, P. (1762) the corpus allatum (together with the corpus cardiacum) had been regarded as nervous structures. The corpus allatum was so named by Janet, C. (1899) who cast doubt on its nervous character. The detailed study of Nabert, A. (1913) showed clearly that it was indeed a glandular organ and he suggested that it might be an endocrine gland. I failed to get experimental support for the suggestion that it secreted the moulting hormone, but this view made the results acceptable to the reader in a way that Kopeč's brain secretion was not. The brain, however, was considered to play an essential contributory stimulating role, because stretching of the abdominal wall of the *Rhodnius* larva above a certain degree was necessary to initiate secretion of the hormone; and this effect was eliminated if the nerve cord was cut behind the head.

I had first learned of Kopeč's work (in conversation with Julian Huxley) when I was writing up these experiments — which were therefore described as confirmation of Kopeč's findings. About the same time Fraenkel, G. (1935a) obtained similar results on puparium formation in *Calliphora erythrocephala*, but he made no claims about the source of the hormone.

2.1 Role of epithelial continuity

In the parabiosis experiments on *Rhodnius* larvae, when the two insects were brought into direct contact with one another they finally moulted simultaneously and the epithelium had become continuous between them. The same applied to union between larvae of *Rhodnius* and *Cimex* which belong to different families. The same conjunction and simultaneous moulting had been observed by Crampton, H. (1899) when saturniid pupae were joined in parabiosis, and by Bodenstein, D. (1933) when appendages of vanessid larvae were transplanted. On the other hand larvae connected by a capillary tube moulted independently, and the question arises whether these experiments indicate that communication is maintained by way of the continuous epidermis or whether a circulating hormone is responsible for signalling the final shedding of the cuticle — such as the "eclosion hormone" described in recent years by Truman in Lepidoptera (see chapter 12, volume 8).

When the epidermis of an insect is interrupted by an excision it shows an inborn capacity for maintaining existing continuity and spreading in a thin sheet until general continuity is restored. Excised segments of a limb implanted into another insect show the cells from the two ends migrating over the outer surface until they meet and re-establish continuity. In an excised segment from a trachea the same migration takes place within the lumen. Restoration of continuity inhibits migration and growth (like "contact inhibition" among cells in tissue culture). The role of diffusible chemical factors in this process is not known.

2.2 Neurosecretory cells and moulting

After a lecture by myself in Copenhagen in 1937, in which I was still supporting the corpus allatum as a possible source of the moulting hormone, Hanström suggested that there might be neurosecretory cells in the brain which could be responsible. I supplied him with material and he duly demonstrated neurosecretory cells in the dorsum of the cerebrum in *Rhodnius* (Hanström, B. 1938) and when this area of the brain was excised and implanted into decapitated larvae of *Rhodnius* these were induced to moult (Wigglesworth, V. 1940b). This was the first demonstration of a function, and a hormonal function, for the neurosecretory cells which Ernst and Berta Scharrer had been describing in the central nervous system of vertebrate and invertebrate animals (see chapter 3, this volume). It appeared that Kopeč had been right after all and that secretory cells in the brain were the source of the moulting hormone. Meanwhile several authors had reported that after excision of the corpora allata in Lepidoptera the larvae continued to moult (Bounhiol, J. 1938; Plagge, E. 1938).

2.3 Prothoracic glands and moulting

In 1931 Hachlow, V. (confirmed by Bodenstein, D., 1938) had already questioned Kopeč's findings; because in various Lepidoptera he had found that if the pupae were sectioned transversely and the cut ends sealed against a glass plate, only those anterior or posterior fragments which included the thorax proceeded to develop. But the nature of this thoracic factor was not defined.

In following up the work of Fraenkel on Diptera, Burtt, E. (1938) and Hadorn, E., and Neel, J. (1938) showed that Weismann's ring in the larvae of Diptera is made up of the corpus allatum, the corpus cardiacum (Scharrer, B. and Hadorn, E. 1938) and large lateral cells sometimes called the peritracheal gland (Possompès, B. 1953). Then Fukuda, S. (1940) found in very convincing experiments that the immediate source of the moulting hormone in both larvae and pupae of *Bombyx mori* was the homologous gland which he called the "prothoracic gland". This gland had been described and figured by Lyonet, P. (1762) in his classic publication on the larva of *Cossus*, under the name of the "granulated vessels" (of "unknown function"). This gland was observed by Verson, E. and Bisson, E. (1891) in the silkworm and called the "hypostigmatic gland"; and its origin in the embryo as an ingrowth from the second maxillary segment was described by Toyama, K. (1902).

It was soon realized that the prothoracic gland was probably homologous with the lateral cells of Weismann's ring in Diptera which Thomsen, E. (1942) and Vogt, M. (1943) held responsible for puparium formation, and with the glands variably situated in the back of the head ("ventral glands") or in the thorax ("prothoracic", "intersegmental", "peritracheal" or "pericardial") glands in various insects (Williams, C. 1948; Pflugfelder, O. 1949). These glands reside primarily in the back of the head but in many insects they have been carried backwards into the thorax, often in association with the salivary glands (Wells, M. 1954). They were found to be maintained by the juvenile hormone and to degenerate and disappear soon after, or shortly before, the completion of the moult to the adult stage (Wigglesworth, V. 1955a) — save in the Thysanura which continue to moult in adult life (Gabe, M. 1953).

There was a brief period during which there appeared to be a dichotomy between insects induced to moult by secretion from the neurosecretory cells in the brain and insects with moulting hormone produced by the prothoracic glands or their equivalent. It had been noticed by Plagge, E. (1938) that although the implanted brain could cause pupation in *Deilephila* larvae from which the brain had been removed, it could *not* induce pupation in the posterior fragments of ligated larvae. Piepho, H. (1942) suggested that the neurosecretion served to activate the prothoracic gland in *Galleria mellonella* which then secreted the moulting hormone; and this conception was proved experimentally by Williams in the resting pupa of *Hyalophora cecropia* in which an active (chilled) brain implanted into the isolated abdomen of the pupa will induce renewed development only if prothoracic glands are implanted at the same time (Williams, C. 1947). This experiment was soon repeated in *Periplaneta* (Bodenstein, D. 1953), *Calliphora* (Possompès, B. 1953) and *Rhodnius* (Wigglesworth, V. 1952a).

The hormone from the neurosecretory cells of the dorsum of the brain is generally called the "activation hormone" or "prothoracicotropic hormone" (see chapter 4, this volume), the hormone from the prothoracic gland is called the "moulting hormone", the active substance being ecdysone (see chapters 6, 7 and 8, this volume).

3 CHANGE IN FORM AT METAMORPHOSIS: THE JUVENILE HORMONE

In the course of the early experiments on the control of moulting in *Rhodnius* (Wigglesworth, V. 1934, 1936) it was observed that when fourth instar larvae were decapitated around the time of the "critical period", some failed to moult, some moulted into normal fifth instar larvae, but a small number showed partial development of adult characters — in the wings, genitalia, cuticle structure, pigmentation, etc. They had undergone a partial "metamorphosis". It was evident that some factor from the head was inhibiting metamorphosis. By cutting through the head at different levels it was demonstrated that the corpus allatum was the source of this factor, which was called the "inhibitory hormone". It was made clear in the description that this hormone was considered to be inhibiting the *change in form* which is the criterion of metamorphosis. However, many readers persisted in regarding it as inhibiting the moulting hormone, in other words, using "metamorphosis" in the same sense as Kopeč to signify the continuation of moulting, leading automatically to pupal and adult development. As late as the Paris Symposium of 1947 Bounhiol for example was still adopting this interpretation.

It was readily shown that in *Rhodnius* the "inhibitory hormone" was operating from the earliest stages of larval growth (indeed probably already within the egg) because first instar larvae exposed to the moulting hormone in the absence of the corpus allatum developed adult characters. In the fifth-instar larva the corpus allatum no longer secreted the inhibitory hormone and thus allowed metamorphosis to proceed, whereas a fifth-stage larva with corpora allata from fourth-stage or other young larvae implanted, developed into giant sixth- or even seventh-stage larvae.

In the late 1930s when the *Rhodnius* experiments were in progress, Bounhiol, J. (1938) was studying metamorphosis in the silkworm. In view of the results on *Rhodnius* Avel, M., suggested that he should test the effects of the corpus allatum on this material. Bounhiol devised most elegant operative procedures which showed that the same results could be obtained; silkworm larvae, even in the second instar, if deprived of the corpus allatum, underwent metamorphosis to diminutive adult moths which were capable of developing eggs. Confirmation came also in *Carausius* (Pflugfelder, O. 1937), *Tenebrio* (Radtke, A. 1942), *Melanoplus* (Pfeiffer, I. 1945) and others.

A few years later (Wigglesworth, V. 1939, 1940a, b) it was found that the adult *Rhodnius*, exposed to the moulting hormone by parabiosis with fifth-stage larvae, could be induced to form a new cuticle of normal adult type beneath the existing cuticle. But if such a moulting adult was provided with abundant "inhibitory hormone" by implantation of corpora allata from young larvae, or by parabiosis, the new cuticle showed a partial recovery of larval characters. This so-called "reversal of metamorphosis" was far from complete, but despite the smallness of the effect it was highly significant in principle, for it showed that the corpus allatum is not a mere "inhibitor" but is exerting a positive influence on the manifestation of larval characters. It was therefore proposed to change the name from "inhibitory hormone" to "juvenile hormone" (or in Greek form "neotenin"). A similar reversal was obtained by Piepho, H. (1939) in contemporary experiments in which fragments of pupal integument of *Galleria* implanted into young larvae produced cuticle of larval type. A striking confirmation was obtained by Ozeki, K. (1959) in adult earwigs (*Anisolabis*), which provided their own juvenile hormone, when they were induced to moult again by implantation of active ventral glands: they developed once more the ecdysial line on the thorax, absolutely characteristic of the larva.

3.1 Mode of action of the moulting hormone

As early as 1905 Dewitz, J. put forward the view that the "oxidase" which becomes active at the time of pupation was actually initiating metamorphosis. Williams, C. (1951), working on the larva and pupa of *Hyalophora cecropia*, stressed the importance of the build-up of the cytochrome system in the initiation of insect growth. Whether the synthesis of the complete cytochrome system was indeed the prime result of the action of the prothoracic gland hormone or merely one consequence among many of the renewed growth which the hormone brings about remained uncertain.

With the generous co-operation of Prof. P. Karlson, after the chemical isolation of ecdysone in 1954, I was able to inject known doses of the purified hormone in a series of experiments aimed at defining its primary action. I communicated the results to the Society for Experimental Biology Symposium on the Biological Action of Growth Substances in 1956 (Wigglesworth, V. 1957, 1963a) and at the Physiology of Insect Development Conference of the Paul Weiss series of symposia in the same year (Campbell, F. (ed.), 1959). I developed the idea that the characteristic defect in the tissues of the resting or dormant insect, that is, in the absence of the moulting hormone, is a failure of protein synthesis. In the presence of the hormone in *Rhodnius* active synthesis of RNA and protein became evident in the epidermal cells within 6 h after injection. Associated with this renewed activity there was a great enlargement of the nucleolus, and the mitochondria became enlarged and began to multiply. The characteristic U-shaped curve of respiration during the moulting cycle of the insect was interpreted as a curve of protein synthesis, which is a strongly endothermic process with a high oxygen requirement (Zwicky, K. and Wigglesworth, V. 1956). It was shown that labelled glycine is incorporated into protein at a far higher rate in developing pupae than it is in dormant pupae of *Sphinx ligustri* (Bricteux-Grégoire S. *et al.* 1957) and *Hyalophora cecropia* (Telfer, W. and Williams, C. 1960). The histological changes in the epidermal cells were identical with those seen in response to local injury (Wigglesworth, V. 1937) and it was suggested that the hormone might be influencing intracellular permeability relations so that enzymes necessary for these syntheses gained access to their substrates in those cells responsible for growth; that is, particularly the epidermal cells.

These ideas were followed within a few years by the observations of Clever, U. (1961) and Becker, H. (1962) on the effect of ecdysone in inducing "puffing" of chromosomal bands in the polytene chromosomes of the salivary glands and other tissues of *Chironomus* and *Drosophila* (see chapter 8, volume 2). These results fitted well with the belief that the hormone was promoting nucleoprotein synthesis in a most direct manner; they gave rise to a vast corpus of research into the analysis of protein synthesis during growth and moulting. For example, according to the ideas developed by Karlson and his colleagues (Karlson, P. 1967; Karlson, P. and Sekeris, C. 1966) the enzymes involved in the sclerotization of the puparium in *Calliphora* are synthetic products of mRNA attached to ribosomes in the cytoplasm, and this mRNA is derived from a specific region of the chromosomes in the nucleus, where a specific "RNA-polymerase" is formed at a fixed site on the chromosomal DNA — this site being one of the chromosomal regions activated and "puffing" under the action of ecdysone. The pursuit of these ideas has been highly productive of information about the chemistry of sclerotization (see chapters 1–3, volume 3). The detailed links between the moulting hormone and these syntheses are still *sub judice*.

3.2 Mode of action of the juvenile hormone

The control of metamorphosis, like all other changes in the form and pattern of the body, was attributed by Goldschmidt, R. (1923) to regulation of the relative rates of competing processes of growth; a theory which has the great advantage (or disadvantage) of being able to explain anything. After the demonstration of the role of the corpus allatum secretion I suggested, on the basis of Goldschmidt's hypothesis, that the two competing processes in metamorphosis were: (a) the rate of growth and differentiation of the tissues; and (b) the time of deposition of the new cuticle (Wigglesworth, V. 1936). Such an effect was clearly demonstrated when the moulting of a larva of *Rhodnius* was accelerated by parabiosis with another larva at a more advanced stage of moulting. The partial arrest of metamorphosis which resulted was certainly due to accelerated deposition of the new cuticle, but this was caused not by the juvenile hormone, as I supposed, but by the moulting hormone. Indeed at that time the question was still open as to whether the "inhibitory" effect of the corpus allatum was due to a higher concentration of a single moulting hormone or to a separate factor. When moulting hormone became available Williams (see Karlson, P. 1956) showed that large doses of pure ecdysone cause rapid and atypical development in pupae of *Hyalophora*, the new cuticle being secreted before bristles, scales and primordia of the genitalia were fully developed. The same effect was obtained in *Locusta* by implantation of additional ventral glands (Halbwachs, M. *et al.*, 1957).

By 1940, when the reversal of metamorphosis in the epidermal cells had been demonstrated, such an inhibitory or *status quo* interpretation could provide at most only a partial explanation of juvenile hormone activity. It was clear that there were latent systems within the epidermal cells which reacted to the hormonal environment by producing larval or adult structures, as the case might be. These systems were pictured as dormant enzyme systems activated perhaps by cofactors (Wigglesworth, V. 1940a). At that time it was not universally customary to consider genes as being concerned in controlling the normal body form. The turning point came with a paper by Avery, O. *et al.* (1944) on the "transformation principle" (DNA) which could induce permanent inheritable changes in pneumococci. The changes in cell determination induced by the juvenile hormone in *Rhodnius* were compared with this genetic phenomenon (Wigglesworth, V. 1954a).

A year or two later the larva, pupa and adult in the insect were regarded as different forms of a polymorphic organism, the manifestation of these forms being dependent upon the changed synthetic activity of the epidermal cells. It was believed that the juvenile hormone reacts with the gene-controlled enzyme system in these cells in such a way that the amount and timing of hormone secretion controls the realization of these alternative morphological forms. In other words, gene activity was at the root of the matter — but the point of action of the juvenile hormone was not known (Wigglesworth, V. 1957). "Receptor proteins" lay far in the future (see chapter 15, this volume).

Since Aristotle, and in more recent times William Harvey, metamorphosis had been regarded essentially as a reversion to embryonic development. This idea was fostered by the discovery of the imaginal discs by Weismann in the middle of the last century. But hormonal studies on metamorphosis soon led to the realization that imaginal discs are a speciality of those insects whose larvae live very different lives from the adult and are in no way necessary for metamorphosis. They serve to free certain ectodermal cells from the task of forming functional cuticle so that they can prepare for the formation of the specialized structures of the future adult. In general, all the epidermal cells have the triple capacity to form larval, pupal or adult characters.

By implanting a variable number of corpora allata into the last-stage larva or the pupa of *Galleria*, or by making the implants on progressively later days, Piepho, H. (1942, 1951) was able to obtain a graded series of intermediates between larva and pupa, and between pupa and imago. He concluded that before a larval moult there is much juvenile hormone in the body, before the pupal moult much less and before the imaginal moult less still. How the constant morphology at each developmental stage is normally maintained remained unexplained and the need for a delicate balance between the timing of growth and differentiation of the tissues and the deposition of the cuticle still remains (see chapter 2, volume 8). In many Lepidoptera (e.g. *Hyalophora*) the larvae have distinctive "morphs" for each instar. Attempts to show whether these forms are controlled by the level of juvenile hormone have given indefinite results (Staal, G. 1967b).

Parabiosis experiments between the larval stages of *Rhodnius*, with and without retention of the corpus allatum, did point to the degree of differentiation of the external genitalia during the larval stages being under the control of the corpus allatum (Wigglesworth, V. 1936) but these results can be explained by varying activity of the moulting hormone. Adult males of *Cimex* with genitalia showing intermediate grades between larva and adult were obtained when a fifth-stage larva of *Cimex* was decapitated and joined in parabiosis to a third-stage larva of *Rhodnius* (Wigglesworth, V. 1936). This result, which was attributed to the "inhibitory hormone" (juvenile hormone) of the corpus allatum, was probably due to accelerated moulting brought about by the moulting hormone. The intermediate structures agree exactly in form with the successive stages observed in the normal epidermal rudiments by Christophers, S. and Cragg, F. (1922).

3.3 Juvenile hormone and behaviour

The juvenile hormone has definite effects on behaviour, but the interpretation of these effects is uncertain. It can be argued that the juvenile hormone controls the development of the nerve centres and changed behaviour is the result of changed morphology in the brain. But in the sphingid *Mimas tiliae*, if the juvenile hormone is above a threshold concentration, the larval behaviour is that which is normal before a larval moult; if the hormone is below this threshold the larva migrates down the tree trunk to the soil as before the pupal moult (Piepho, H. *et al.*, 1960). This looks like a direct effect on behaviour. Likewise in *Galleria*, the larva spins a slight flat web before a larval moult; before moulting to the pupa the full-grown larva spins a tough cocoon; and if active corpora allata are implanted into the last-stage larva, so that it moults again into a larva, it will spin a larval web or an intermediate type (Piepho, H. 1950). But it has been shown by Bounhiol, J. (1953) that throughout larval life the silk gland of *Bombyx mori* is restrained by the corpora allata as though it were an imaginal organ; so an indirect action by way of the silk gland is possible.

3.4 Supposed antagonism between juvenile hormone and moulting hormone

The supposed antagonism between the juvenile hormone and moulting hormone which had been claimed by many authors (notably Bounhiol, J. 1938) does not exist. The two hormones are doing two different things: the moulting hormone inducing growth and moulting, and the juvenile hormone controlling the body form. But we have already seen how an excessive supply of ecdysone can accelerate the moulting process in such a way as to reduce the differentiation of adult characters and thus to mimic the effect of the juvenile hormone.

Another interesting interaction was demonstrated many years ago by Bückmann, D. (1959) when he showed that the red–brown colour change in the larva of *Cerura vinula*, which is assumed at the wandering stage before pupation, can be induced experimentally by the injection of a small dose of ecdysone, amounting to 66–330 "*Calliphora* units" (section 8.2). This produces the brown colour (due to the formation of the red ommochrome pigment in the fat body and gut contents) but does not induce moulting, whereas a large dose of ecdysone (3300–6600 "*Calliphora* units") will induce direct moulting of the larva without any change in colour. This observation has become of great interest in recent years when it has been discovered that it is a common feature in many insects at the time of pupation to have first a very small peak of ecdysone secretion followed after an interval by a large peak which initiates moulting; and it has been established that the presence of juvenile hormone in the circulating haemolymph can suppress the secretion of ecdysone by the prothoracic gland. When the juvenile hormone level is reduced the prothoracic gland is activated and ecdysone appears at a high level in the circulating blood. We shall return to this factor in considering diapause (section 5).

3.5 Prothetely and metathetely

It had been known, at least from early in the last century, that silkworms exposed to high temperature might develop enlarged antennae or mealworm larvae diminutive wing pads (Pruthi, H. 1924). This precocious appearance of imaginal characters was termed "prothetely" by Kolbe, J. (1903). Other abnormalities in which adult insects retained some pupal characters, or pupae retained larval characters, were termed "metathetely" (as described in *Simulium* parasitized by the nematode *Mermis* by Strickland, E. (1911), a condition equivalent to "neoteny". Goldschmidt, R. (1923) regarded such abnormalities as resulting from an upset in the time relations between metamorphosis and evagination of the imaginal discs. Von Lengerken, H. (1924) explained them as due to a disturbance in the enzymes which regulate metamorphosis. Indeed the results of the early hormone experiments on *Rhodnius* were regarded as examples of experimental prothetely and metathetely (Wigglesworth, V. 1934) the incidence of which in the normal insect could be mimicked in a small degree by abnormal temperatures, (Wigglesworth, V. 1952b).

It had long been known that the imbalance which produced these effects resulted from abnormal temperatures, infestation by parasites, and cross-matings between species (Cockayne, E. 1941). From his observations on *Simulium* parasitized by *Mermis* Strickland, E. (1911) suggested that "there may be in the insect body two sets of enzymes which we may term larval enzymes and imaginal enzymes, and that these may be accelerated or retarded by different factors".

4 HORMONES AND POLYMORPHISM

In winding up a symposium on embryonic differentiation in 1934, Morgan, T. hazarded the suggestion that the initial differentiation of cells in the germ band of the early embryo might be the result of localized chemical factors which evoked specific genetic activities in those cell nuclei which came within their sphere of influence. This idea that local chemical differences or gradients of one sort or another call forth potential elements or capabilities in the chromosomal blueprint still remains the most satisfactory hypothesis to describe the first and greatest polymorphism to which the insect is subject: differentiation in the egg.

The juvenile hormone was one of the first unequivocally morphogenetic hormones to be recognized. It not only ensured the development of larval characters but, at a lower concentration, pupal characters; and it exerted these effects by an action upon the gene system in the cells, whether in the nucleus or in the cytoplasm. It was noted that the juvenile hormone acts upon some characters in the cell long after mitosis in the epidermis has ceased, and in the absence of DNA synthesis. The site of action was judged to be some element in the gene system that persists in the cytoplasm (Wigglesworth, V. 1940b, 1954a).

The adult pattern exists in the young larva of *Rhodnius* in invisible form. The proof was that when epidermal cells migrate during the repair of a burn in a young insect, the pattern appearing in the cuticle at the next moult is distorted by this migration. And when the same larva undergoes metamorphosis to the adult, the totally different adult pattern also is shown to have been distorted by the same early migration. The adult characters remain latent so long as the juvenile hormone is present (Wigglesworth, V. 1940a).

Such experiments emphasized the similarities of these processes: (1) differentiation of the parts of the body during embryonic development; (2) the successive differences of the body form in larva, pupa and imago; and (3) the polymorphism of individuals in a single species. Lubbock, J. (1883) clearly recognized that the diversity in form at successive stages of the life cycle was closely comparable with the polymorphism that may exist between different individuals of a species. But he was more influenced by the theory of recapitulation, so fashionable at that time, and adopted the old Aristotelian theory of the nature of metamorphosis: that insect growth is a kind of embryonic succession; and he advanced the view, already adumbrated by William Harvey (1651) of metamorphosis being the result of precocious emergence from the egg.

In many insects which occur naturally in more than one form, the characters are controlled by genetic factors. The different forms exist side by side in the natural environment, the relative abundance of each form depending upon the pressure of natural selection to which each is exposed. This phenomenon is termed "balanced polymorphism" (Ford, E. 1953). But there are other insects in which the different forms occur in individuals of uniform genetic constitution (Kennedy, J. 1961). These forms likewise, it has been argued, are produced by the action of genes, but the gene function is "switched", or latent genes are brought into action, under the impact of some change in the environment (Wigglesworth, V. 1961a).

Most genes are "autonomous", their action being confined to the cell in which they occur. That is the source of the "mosaics" which appear when clones of cells in one individual differ in genetic constitution. "Vermilion" eye colour in *Drosophila* is an exception; it fails to develop in a vermilion mosaic eye if the other eye is of wild type. The circulating factor was referred to, somewhat unsuitably as a "gene hormone". The classic investigations of Ephrussi, B. and Beadle, G. (1942) on *Drosophila* and of Kühn, Butenandt and others on *Ephestia* (in which an homologous gene is responsible for the usual black colour of the eye) led to the explanation that the mutant insects are deficient in the enzyme which converts tryptophan into kynurenine which then by way of hydroxykynurenine forms the brown ommochrome pigment of the eye (Kühn, A. 1936). This was one of the clearest early examples of the "one gene, one enzyme" hypothesis. This metabolic cycle is used by the insect for the elimination of excess tryptophan (Linzen, B. and Bückmann, D. 1961) and, as we have seen its activation can figure among the subsidiary effects of ecdysone (Bückmann, D. 1959).

The arrest of hormone secretion which results in diapause is brought about by certain signals from the environment; and at the same time a "gene switch" may cause striking changes in structure or pigmentation (Müller, H. 1955, following Danilyevsky).

Poulton, E. (1903) recognized that the ability of certain caterpillars, such as *Biston betularius*, to adapt their coloration and pattern to resemble that of the twigs on which they happen to rest, was due to the existence of genetic types (evolved by natural selection) whose action is evoked by gene switching in accordance with the visual impression received by the insect. Many grasshoppers and other insects respond in the same way (Kennedy, J. 1961) — presumably by the use of hormonal agents.

Likewise in the striking dimorphism of the egg parasitoid *Trichogramma semblidis*, the adult was shown by Salt, G. (1937) to be winged when reared in eggs of Lepidoptera, apterous and with other striking morphological differences when reared in those of *Sialis*. Similarly the ichneumonid *Gelis* occurs in two genetic forms determined by some limiting factor in the diet, in that the form produced depends on the size of the host in which the larva matures (Salt, G. 1952).

These variants approximate to the many examples described by geneticists in which the penetrance of genes is changed by environmental factors. For instance tetraptera (Astauroff, B. 1930) and aristopedia (Villee, C. 1942) in *Drosophila* in which expression is influenced by changes in temperature; or antennaless in *Drosophila* which is suppressed if the larvae receive a diet rich in riboflavine (Gordon, C. and Sang, J. 1941).

In certain examples of environmental polymorphism it was suggested that the juvenile hormone itself was involved; such as the green colour of solitary *Locusta* (Joly, P. 1956); perhaps in the production of neotenic apterous forms of aphids (Lees, A. 1961); and in the production of soldiers in the primitive termite *Kalotermes* (Lüscher, M. 1958). This last example is reminiscent of the graded concentration of the juvenile hormone which determines the stages of metamorphosis in Lepidoptera, for they are termites already, held as larvae by the presence of juvenile hormone, which are induced to transform into soldiers by an excessive amount of this agent — an example of the principle emphasized by Child, C. (1941) that qualitative differences in form can be induced by quantitative differences in a single inductor substance.

Light, S. (1942) concluded that the genetic mechanisms for all the female castes of ants exist in all female-producing eggs, and that some environmental factor, particularly nutrition, determined their realization. Since deficiency of juvenile hormone can lead in *Formica rufa* to the laying of eggs poor in yolk which tend to give rise to workers, a change in corpus allatum secretion could well play an indirect part in caste production (Bier, K. 1954).

Light, S. (1942) concluded that in termites also the castes are controlled by different genetic patterns present in all individuals and that the actual fate of the individual depends upon which of several genetic mechanisms comes into final play, under the influence of feeding, "ectohormones", the "pheromones" of Karlson, P. and Lüscher, M. (1959) or other environmental factors. The effect of the juvenile hormone on soldier formation is established; the extent to which juvenile hormone secretion or corpus allatum inhibitors are involved has been subject to prolonged debate. In the honey bee it has long been known that the production of the queen is controlled by nutrition. Only in very recent years has it been proved that the juvenile hormone plays an essential role, and that royal jelly provides a nutritional stimulus to juvenile hormone secretion.

5 HORMONES AND DIAPAUSE

The term "diapause" was originally coined by Wheeler, W. (1893) for the stage of development at which the embryo ceases to move tail foremost over the yolk and remains stationary for a time before blastokinesis. But it was later extended by Henneguy, L. (1904) to dormancy at any stage of development, including the arrest of reproduction in the adult female (see chapter 11, volume 8).

It has long been recognized that such arrest may be due to deficiencies in nutrition in general, or of salts or vitamins in particular, to lack of oxygen or of water, or to reduced temperature. Arrest which continued when all such limiting factors were removed was called diapause. It was commonly regarded as the result of some inhibitory constituent accumulating in the body; for example Roubaud, E. (1919) regarded diapause as a state of developmental fatigue ("asthenobiosis") attributed to the build-up of an excess of excretory products. A less specific inhibitor was referred to as "Latenzstoff" (Goldschmidt, R. 1927) which was regarded by Bytinski-Salz, H. (1933) as in Roubaud's hypothosis as a metabolic product accumulating early in metamorphosis.

There are many forms of diapause in which the metabolic cause remains unknown; notably in the early stages of embryogenesis. But when the presence of a moulting hormone was demonstrated in *Rhodnius* it was at once evident that the state of the unfed larva, or of the larva decapitated soon after feeding, in which the moulting hormone was absent, was in effect a state of diapause; and it was suggested that natural diapause in insects might well result from "a temporary failure of growth-promoting hormones, due sometimes perhaps to an inborn rhythm, sometimes perhaps to the indirect effect of environmental factors" (Wigglesworth, V.

1934). This idea was confirmed by the clear-cut experiments of Williams, C. (1946) described above in which it was shown that in the overwintering pupa of *Hyalophora* in diapause the neurosecretory cells of the brain were inactive. They became activated after a period at a low temperature and their secretion initiated the activity of the prothoracic gland in secreting the moulting hormone. The same inactivity of the neurosecretory cells in the brain was found in *Lucilia* larvae (Fraser, A. 1960), in the larva of the rice stem borer *Chilo* (Fukaya, M. 1955), in the pupa of *Luehdorfia japonica* (Ichikawa, M. and Nishiitsutsuji, J. 1955) and in the overwintering larva of the wheat-stem sawfly *Cephus cinctus* (Church, N. 1955).

There was uncertainty about the nature of the arrest of activity in the neuroendocrine system. Diapause in *Lucilia* larvae induced by desiccation could be terminated and the larvae caused to pupate by placing them in empty glass vials. The ring gland showed signs of renewed activity and the effect was attributed by Mellanby, K. (1938) to a nervous stimulus to hormone secretion. In the European corn borer *Ostrinia nubilalis* the neurosecretory cells contained stored secretion; the nervous control of release appeared to prevent its discharge (Cloutier, E. *et al.*, 1962); and Schoonhoven, L. (1962) arrived at similar conclusions on the diapausing pupa of *Bupalus piniarius*, whereas in the dormant pupa of *Hyalophora* far-reaching enzyme changes in the brain were described by Van der Kloot, W. (1955), notably an extreme deficiency in cholinesterase. At one stage it was proposed that specialized cells in the hind-gut of *Ostrinia* set free a hormone ("proctodone") which served to activate the neurosecretory cells of the brain (Beck, S. and Alexander, N. 1964).

It was recognized that during diapause the arrest of development is not complete. Changes with a low optimum temperature (often about 5°) are taking place during the cold rest. This process, which is essential if growth is to be renewed when the environmental temperature rises, was named by Andrewartha, H. (1952) "diapause development". The nature of this latent development, and any role of hormones during the process, are unknown.

It was noted by Rahm, U. (1952) that the prothoracic gland in *Sialis* larvae was already activated by the brain in the autumn, but secretion remained arrested during the winter. This was attributed to the corpus allatum secretion suppressing liberation of the moulting hormone from the prothoracic gland — but this suggestion was based only on the supposed antagonism between these two hormones. Later Fukaya, M. and Mitsuhashi, J. (1961) and other Japanese workers produced experimental evidence that in diapausing larvae of *Chilo* the corpus allatum secretion keeps the brain and thoracic gland inactive; and Waku, Y. (1960) noted that in *Plodia* larvae the corpus allatum becomes enlarged during diapause. At that time I did not find the evidence for inhibition of the prothoracic gland by the juvenile hormone entirely convincing; but in recent years further evidence by Japanese workers has been incontrovertible and work on diapause in *Diatraea* has demonstrated a variant of this effect; and in a wide range of other insects which feed continuously during the last larval instar it has become established that there is a control in the liberation of ecdysone by a regulated concentration of the juvenile hormone (see chapter 1, volume 8).

In the adult insect, diapause takes the form of a reproductive arrest. The general physiological activities of the adult female appear normal, but reproduction fails because the eggs fail to ripen — as in a starved insect. This normally results from a failure of juvenile hormone secretion; it was early studied in *Leptinotarsa* in which, after extirpation of the corpora allata, the adult female enters the diapause state with a reduced level of oxygen consumption and a cessation of egg production. The beetle burrows into the soil and settles down, its wing muscles degenerate and it is incapable of flight; yolk formation ceases and the ovaries become dormant. Implantation of active corpora allata reverse all these changes and reproduction is rapidly restored. Here the corpora allata proved all-important in vitellogenesis (de Wilde, J. 1954).

In such insects as mosquitoes, in which neurosecretion is of prime importance in yolk formation, the onset and termination of diapause in the hibernating adult are under the control of neurosecretion from the brain — but this relation has only been firmly established in recent years.

As already pointed out, the control of diapause in the insect egg did not find a hormonal explanation. One exception is diapause in the egg of the silkworm. It was discovered by Duclaux, E. (1869) that the overwintering eggs of *Bombyx mori* will not develop if kept warm. They will not even hatch in the spring unless they have been exposed to a temperature around 0° for several months. The incidence of this diapause is dependent upon genetic factors; the "voltinism", or the number of generations in the year, is hereditary; but as shown in the classic work of Kogure, M. (1933) it is influenced by the temperature and by the length of day at which the eggs of the preceding generation were incubated or to which the young larvae were exposed. Thus the voltinism of a given brood of silkworms is determined by some influence from the somatic cells of the mother. It is here that the hormonal control comes in (Umeya, Y. 1926).

As was first demonstrated independently by Fukuda, S. (1951) and Hasegawa, K. (1951) the suboesophageal ganglion (which contains a pair of large neurosecretory cells) secretes a "diapause hormone" which leads to the production of diapause eggs. But the secretion of this factor is under the control of the brain, acting by way of the oesophageal commissures. The brain normally inhibits the secretion of the diapause hormone, but if the brain is removed immediately after pupation in moths that would have laid non-diapause eggs, they proceed to lay eggs which enter diapause. The biochemical nature and the action of this hormone have been studied in recent years (see chapter 11, volume 8).

6 HORMONES AND REPRODUCTION

6.1 Role of the corpus allatum

Bytinski-Salz, H. (1933) found that if the sphingids *Celerio gallii* males and *Celerio euphorbiae* females were cross-mated the resultant females failed to develop into imagines. But if the ovaries of these hybrid females were transplanted into hybrid males (which develop normally) full-grown eggs are matured within them. Bytinski-Salz suggested that this effect was due to a hormonal influence from the new host.

A year or two later it was observed that in the adult female of *Rhodnius* the cells of the corpus allatum are shrunken immediately after moulting or after fasting for some 4 weeks; but that the cells are swollen with dense cytoplasm at a week after moulting or at 6 days after feeding in the long-starved insects. These observations suggested that the corpus allatum might be involved in reproduction (Wigglesworth, V. 1936). They agreed with the observations of Ito, H. (1918) that in Lepidoptera which mature their eggs during the pupal stage the corpora allata show signs of maximum activity in the pupa; and of Holmgren, N. (1909) that in queen termites the corpora allata attain a colossal size — regarded by him as a hypertrophic degeneration.

Decapitation with removal of the corpus allatum prevented the formation of ripe eggs in newly moulted adults of *Rhodnius* or in starved adults newly fed, whereas removal of the brain with the corpus allatum left intact, allowed complete maturation of the eggs. Parabiosis experiments showed that a circulating hormone was responsible and that the hormone was non-specific: the corpus allatum of *Triatoma* induced development of eggs in *Rhodnius* (Wigglesworth, V. 1936).

In the female *Rhodnius* deprived of the corpus allatum, development of the oocyte proceeded normally until the stage at which it loses connection with the nurse cells by means of the nutritive cords; by that time it had attained a length of about 0.45 mm (the ripe egg is about 1.5 mm long). At this point its development was arrested; the oocyte died, and the follicular cells, instead of producing yolk, seemed to proliferate amitotically and absorb the dead oocyte. And so the process continued with each successive oocyte. The same changes in the ovary were seen in the starved adult with corpus allatum intact (Wigglesworth, V. 1936).

The role of the corpus allatum in egg production (see chapter 6, volume 8) was confirmed in the grasshopper *Melanoplus*, where it was also needed for the secretory activity of the lower part of the oviduct which produces the ootheca (Pfeiffer, I. 1939); and in *Cimex* after fertilization (Mellanby, K. 1939), in *Leptinotarsa* (de Wilde, J. 1954), in the earwig *Anisolabis* (Ozeki, K. 1958) and in mosquitos (Larsen, J. and Bodenstein, D. 1959). In *Rhodnius* and most other insects it seemed to be the final stages of oocyte development, including the deposition of yolk, for which the corpus allatum secretion was necessary; but in *Dytiscus*, Joly, P. (1945) showed that the corpus allatum was needed at quite early stages of oocyte development.

The role of the gland was complicated by observations on Lepidoptera in many of which egg development takes place during pupal–adult development. In *Bombyx mori* removal of the corpus allatum in the larva did not prevent the development of eggs (Bounhiol, J. 1938) and eggs were produced in the isolated abdomen of *Hyalophora* into which prothoracic glands had been implanted (Williams, C. 1952). On the other hand, in *Galleria* the production of eggs was greatly reduced by excising the corpus allatum in the late larval stage (Röller, H. 1962). It was early recognized, of course, that the ripening of the oocytes is highly dependent upon nutrition. This was evident in the unfed *Rhodnius* in which each oocyte in turn dies when it reaches the stage when yolk formation should occur. It was seen in the underfed pupa of *Bombyx mori* (Hasegawa, K. 1947). It was pointed out by Johansson, A. (1955) that in *Oncopeltus* and in *Leucophaea* the activity of the corpora allata depends on nutrition. Indeed during starvation the corpus allatum secretion appeared to be inhibited by the brain (Johansson, A. 1958).

It was originally assumed that the gonadotropic hormone of the corpus allatum was distinct from the juvenile hormone (Wigglesworth, V. 1936); but Pfeiffer, I. (1945) pointed out that in *Melanoplus* the corpus allatum of the adult continues to secrete the juvenile hormone. She suggested that the same hormone might be concerned in yolk formation. Indeed it proved that the corpus allatum of young *Rhodnius* larvae would induce yolk formation (Wigglesworth, V. 1948) and later that farnesol had both a juvenile hormone effect and a yolk-promoting effect in *Rhodnius* (Wigglesworth, V. 1961b). In the parthenogenetic *Carausius*, where feeding and egg production are continuous processes in the adult female, there seems to be no call for hormonal regulation, and removal of the corpora allata does not influence yolk formation (Pflugfelder, O. 1937).

In the adult male of *Rhodnius* the development of spermatozoa was found not to be affected by decapitation: the vesiculae became distended with spermatozoa as usual; but the secretion in the accessory glands was wanting and these glands remained attenuated. Retention of the corpus allatum ensures their normal development; the corpus allatum of either sex being equally effective (Wigglesworth, V. 1936). The accessory glands of male insects secrete the substance of the spermatophore, the packet in which the mass of spermatozoa is transferred to the female. Spermatophores are proteinaceous structures which are produced (notably by male Lepidoptera) in very great numbers (Norris, M. 1932).

The corpus allatum seemed to be more active in male Lepidoptera than in the female. That was very evident in *Ephestia kuehniella* (Rehm, M. 1951). This sexual difference in the size of the corpus allatum was very evident in *Hyalophora* in which the gland in the adult female weighs about 65 μg and about 400 μg in the adult male; and this difference was reflected in the extractable juvenile hormone which was eight-fold greater in concentration in the lipid extracted from the male; and since the male contains five times as much lipid the total content in the male was about 40 times that in the female (Gilbert, L. and Schneiderman, H. 1961). This high content was markedly reduced if ovaries were implanted into the male pupa. The developing ovaries evidently consumed juvenile hormone.

6.2 Role of neurosecretion and nervous control

Already by 1942 it had been found by E. Thomsen that the corpus allatum secretion plays a part in the deposition of yolk in the oocytes of *Calliphora*; but that occasional females developed the eggs fully even when the corpus allatum had been removed. That was confirmed by Possompès, B. (1953) after extirpation of the corpus allatum in *Calliphora* larvae. In 1952 Thomsen, E., proved that the source of the hormone concerned was the medial group of neurosecretory cells in the brain. If these were excised the egg chambers failed to grow beyond 0.17 mm. Reimplanted neurosecretory cells led to complete development. Corpora allata induced only slight growth; but the corpus cardiacum served as a store for the neurosecretory product and could thus replace the implanted neurosecretory cells. The role of neurosecretory cells in egg development was confirmed by Clements, A. (1956) in *Culex* and notably by Gillett, J. (1958) in *Aedes aegypti* (see chapter 7, volume 8).

In these insects there was therefore good evidence that the neurosecretory cells in the brain were the chief source of the hormone necessary for yolk formation. In some other insects the results were equivocal: in *Rhodnius* there was no reduction in the number of eggs produced when the brain was removed, provided that the corpus allatum remained (Wigglesworth, V. 1963b). In the stick insects *Carausius* and *Clitumnus* eggs are produced in the absence of the medial neurosecretory cells but in smaller numbers (Dupont-Raabe, M. 1952) and in *Oncopeltus* the essential source of the yolk-forming hormone was the corpus allatum; but this gland depends on connection with the neurosecretory cells for its full activity, and these cells in turn are dependent upon nutrition (Johansson, A. 1958).

It was thus realized that the ovary is dependent upon a complex system comprising neurosecretory cells in the protocerebrum and suboesophageal ganglion, the corpus allatum and corpus cardiacum, and that these components influence one another by nervous stimuli, by neurosecretion carried along nerve pathways and by humoral factors in the circulating blood.

These interactions were closely studied in Orthoptera and Diploptera. In *Leucophaea maderae* the brain is the final regulator of the corpus allatum at all stages of reproduction. Secretion is arrested during pregnancy in the ovoviviparous species by an inhibitory nervous influence from the brain; but the corpus allatum cannot continue secretion if all nervous connection with the brain and suboesophageal ganglion is severed (Lüscher, M. and Engelmann, F. 1955). The developing eggs in the brood sac induce inhibition by the brain — perhaps by a nervous stimulus from mechanical stretching, perhaps by a humoral stimulus from the eggs (Engelmann, F. 1957). In *Blattella* and other cockroaches the mechanical stretching of the brood chamber by the ootheca was found to maintain a neural inhibition of secretion (Roth, L. and Stay, B. 1961).

There are other nervous functions which were early recognized as being important in reproduction. In *Schistocerca* copulation is necessary if a full complement of eggs is to be laid. It seems to provide the normal stimulus for the liberation of the neurosecretory material (Norris, M. 1954). In *Cimex* the corpus allatum does not become active until the sperm, after passing through the body cavity, have reached the appropriate part of the female genital tract (Mellanby, K. 1939). In *Culex* and *Aedes* prolonged distension of the gut can lead to egg development even in the absence of a blood meal (Larsen, J. and Bodenstein, D. 1959).

The colour changes in the maturing male of *Schistocerca*, notably the bright yellow pigment in the epidermis, are brought about by secretion from the corpora allata. These changes are associated with the liberation of a volatile lipophil secretion which serves as a pheromone (Loher, W. 1960) that may lower the sensory threshold of the female and so favour discharge of the neurosecretory hormone (Highnam, K. and Lusis, O. 1962).

In summary, it appeared that the neurosecretory cells, the corpus cardiacum and corpus allatum acted as an endocrine unit, the relative importance of the components varying in different insects, and that the neurosecretory system provided a link between the nervous and endocrine moieties of the co-ordinated mechanism (Highnam, K. 1962). The essential requirements are that the oocytes must be kept alive and they must be furnished with the protein and other reserves needed for yolk formation. In recent years it has been shown, notably in mosquitoes, that ecdysone, produced in the ovary, plays an important part in inducing the synthesis by the fat body of the yolk protein (vitellogenin) needed for the late stages of egg production (chapter 7, volume 8).

6.3 Hormonal control of ovulation and oviposition

It was early pointed out that humoral factors, probably derived from neurosecretory cells in the brain, are involved in ovulation and oviposition. The haemolymph of fecundated *Bombyx mori* will induce oviposition when injected into unfertilized females (Mokia, G. 1941) and it is at the time of oviposition that the neurosecretory material in the adult female *Bombyx* is most actively discharged along the axons (Arvy, L. *et al.*, 1953). In *Dermestes*, mating was found to act as a stimulus to oviposition and appeared to hasten final maturation of the eggs (Roth, L. and Willis, E. 1950); in *Aedes aegypti* females ovulation also seemed to be provoked by some hormonal factor liberated when the spermathecae are filled with sperm (Gillett, J. 1956).

In the pyrrhocorid bug *Iphita limbata* the neurosecretory cells of the pars intercerebralis seem to be concerned in oviposition. Haemolymph from a laying female will induce oviposition when injected into a female not yet gravid; whereas haemolymph from earlier stages, or from a female which has started mating again, does not have this effect. Implanted neurosecretory cells likewise were active (Nayar, K. 1958). Nayar suggested that an "ovarian hormone" could be concerned in the oviposition response, for injection of a watery extract of the ripe ovary would likewise stimulate the neurosecretory cells to discharge their secretion and induce immediate oviposition.

More recent work has been concerned in differentiating hormonal from nervous stimuli in the control of oviposition by mating and by the presence of spermatozoa or by the secretion of male accessory glands.

7 HORMONES IN THE GENERAL FUNCTIONS OF THE BODY

7.1 Neurosecretion and neurohormones

It is interesting to reflect that in the classic physiology of Galen the nerves were depicted as conveying the "animal spirits" generated in the brain to all parts of the body, inducing movement in muscles and sensation in organs of sense. More than a thousand years later Vesalius, the great renaissance anatomist, in considering the animal spirits, posed the question whether these were in the nature of a substance transported along fine tubes or a "quality" transmitted like sound or light along the solid substance of the nerve. In the course of the next three centuries the idea of the nerves conveying material from the brain to the periphery became almost wholly discredited and nerves were believed to transport only electrical disturbances or impulses to and from the central nervous system.

In the 1930s Ernst Scharrer developed the conception of "neurosecretion" which was in effect a revival of the classical idea (see chapter 3, this volume). Scharrer's theory was based on histology; the discovery in the central nervous system of vertebrates and invertebrates of cells containing apparently secretory products. The discovery that neurosecretory cells in the dorsum of the brain in insects are the source of the hormone that initiates moulting was a striking confirmation of the general idea of neurosecretion — which was thus linked with endocrinology (Gabe, M. 1966).

Neurosecretory cells were commonly recognized by their containing an acidophil product staining with acid fuchsin or with phloxin; a product which became basophil after permanganate oxidation and then staining a deep blue-black with chrome haematoxylin. The product also stained strongly with the paraldehyde fuchsin of Gomori, G. (1941) (Gabe, M. 1953) — usually to give a purple colour (Type A), sometimes a green colour (Type B) (Nayar, K. 1955; Gabe, M. 1966). Many other histological types were defined, but this classification soon became equivocal. Hanström, B. (1940, 1953) traced the evolutionary origin of the neurosecretory cells in the dorsum of the insect brain from the lateral frontal organ of the lower Crustacea, regarding them as derivatives of ordinary neurones. In Thysanura (e.g. in *Petrobius*) they form two groups of cells on the surface of the protocerebrum, lying within a connective tissue sheath; in Pterygota the cells have become embedded in the brain. Hanström, B. (1940) showed that the secretion in the cells in the dorsum of the brain in *Petrobius* passes down the axons to the corpus cardiacum (see chapter 2, this volume). In contrast, Clark, R. (1956) suggested, from observations on annellids, that glandular cells of the epidermis were incorporated in the most primitive parts of the brain to form the neurosecretory cells.

In their general morphology they are essentially identical with ordinary neurones, with axon processes which will convey nervous impulses (Gosbee, J. *et al.*, 1968). As was first shown by Thomsen, E. (1954) the cells and the axons coming from them in *Calliphora erythrocephala*, when examined in the fresh state under dark ground illumination, have a luminous blue appearance — clearly a "physical colour" (Tyndall blue) indicating the presence of fairly uniform minute particles capable of scattering the shorter wavelengths of light. This was confirmed with the electron microscope (originally by Bargmann in vertebrates) which showed that the material was always made up of minute spheres ranging from 100 to 300 nm in diameter (Scharrer, B. 1962) which pass down the axons (Thomsen, E. 1954) and accumulate in bulbous extremities in the corpus cardiacum (Meyer, G. and Pflugfelder, O. 1958; Scharrer, B. 1952; Arvy, L. and Gabe, M.

1953). After each burst of secretion from the neurosecretory cells there was commonly an increase in volume of the corpora cardiaca (e.g. in *Ephestia kühniella* (Rehm, M. 1951) and *Bombyx mori* (Arvy, L. *et al.*, 1953). This transport was made very evident by Scharrer, B. (1952) in *Leucophaea maderae*: after cutting the nerve to the corpus cardiacum on one side, there was a striking accumulation of the blue-staining material above the cut and it disappeared from the corpus cardiacum on that side. Thus it became apparent that the secretion was not liberated directly from the neurosecretory cells, but only after transport to the corpus cardiacum. From the observations of Rehm, M. (1951, 1955) in Lepidoptera during moulting, by Arvy, L. and Gabe, M. (1954) in *Leptinotarsa* and by other authors, it became evident that the periods when the neurosecretory cells are laden with secretion are those in which the hormone is *not* being liberated. The same proved true during reproduction in *Schistocerca* (Highnam, K. 1962) and in *Ostrinia* coming out of diapause (Cloutier, E. *et al.*, 1962).

No-one could fail to be struck by the superficial resemblance between the corpus cardiacum/corpus allatum system in insects and the pituitary gland system in vertebrates (Wigglesworth, V. 1934) but this similarity became highly significant after the detailed studies of Hanström, B. (1941) and Scharrer, B. and Scharrer, E. (1944) which showed that in both structures the secretory material can be traced from neurosecretory cells (in hypothalamus and protocerebrum) to the neurohypophysis and corpus cardiacum respectively, whereas the corpus allatum can be compared with the glandular lobe of the hypophysis. It was Knowles, F., when studying Crustacea, who proposed that the structures of nervous origin which form the sites at which the products of the neurosecretory cells are liberated into the circulation should be called "neurohaemal organs".

Neurosecretory cells in insects were not for long confined to the two groups, medial and lateral, in the dorsum of the protocerebrum which sent two nerves to the corpora cardiaca. They were soon found also in the deutocerebrum and tritocerebrum (Dupont-Raabe, M. 1954) and in the suboesophageal ganglion, in the fused ganglia of *Rhodnius* and in the ganglion chain of many other insects (Dupont-Raabe, M. 1965). Moreover the corpus cardiacum, besides containing the swollen terminations of the neurosecretory neurones, was also found to contain the cell bodies of intrinsic neurosecretory cells (Scharrer, B. 1963).

It was observed by Newport, G. (1836) during his studies of the circulation of Lepidoptera (*Sphinx ligustri*) that there were fusiform enlargements on the side branches of the so-called ventral sympathetic nerve; and these were figured as ganglion-like swellings by Leydig, F. (1864). These and other closely related structures ("perisympathetic organs") were described by Dupont-Raabe, M. (1965) and shown to be the neurohaemal organs which serve the numerous neurosecretory cells of the ventral ganglia. But not all neurosecretory cells discharge their product by way of compact neurohaemal organs. Neurosecretory cells in the corpus cardiacum send their axons into the nerves of the stomatogastric system to supply the gut and the dorsal aorta as in aphids and *Periplaneta* (Bowers, B. and Johnson, B. 1966), and the axons from the neurosecretory cells in the fused ganglia of *Rhodnius* enter nerves which are freely exposed in the abdominal cavity and they discharge their products through the axon walls in these nerves (Maddrell, S. 1966).

Most of our knowledge of the functions of these diverse neurosecretory cells is of very recent date, or is still in process of being acquired, and will not be included in this historical account. But one of the earliest indications of possible hormonal activity was the observation by Brown, F. and Meglitsch, A. (1940) and Thomsen, M. (1943) that an active substance (soluble in water and resistant to boiling) from the corpus cardiacum of *Periplaneta* or *Calliphora* will bring about contraction of the chromatophores of Crustacea (cf. Hanström, B. 1940). Then Steele, J. (1963) showed that these extracts cause breakdown of glycogen with an increase in blood trehalose ("hyperglycaemic hormone") and more recently an "adipokinetic hormone" (Mayer, R. and Candy, D. 1969) has been recognized (see chapter 3, volume 8). The diuretic hormone of *Rhodnius* which is liberated immediately after the large meal of blood comes mainly from the neurosecretory cells in the fused ventral ganglia as described above (Maddrell, S. 1966). In recent years similar diuretic hormones and antidiuretic hormones have been found in many other insects coming from the abdominal ganglia, and evidence is growing that all the neurosecretory products are small peptides.

The function and mode of action of the neurosecretion which is carried directly to the tissues by the axons — supplying, for example, the aorta, the intestine and the rectal glands are only now beginning to be understood. But it was early proposed that they could be acting upon intermediary metabolism (as the "hyperglycaemic hormone" or the "adipokinetic hormone" described above). Activation of protein synthesis was early suggested by Thomsen, E. and Möller, I. (1959) as the prime effect of the neurosecretion in adult *Calliphora erythrocephala* — both in yolk formation and in intestinal digestion.

Diel rhythms of colour change, oviposition, enzyme secretion and locomotary activity are well known to be under the control of environmental stimuli. In a number of instances it was suggested that a hormonal link was provided by neurosecretory cells in the central nervous system. The relative importance of electrical pacemakers situated in the central nervous system and neurosecretory centres has been a subject of controversy. Neurosecretory cells in the deuto- and tritocerebrum of Phasmids secrete the hormonal factor controlling colour change (Dupont-Raabe, M. 1957). The neurosecretory cells in the suboesophageal ganglion of *Periplaneta* were regarded by Harker, J. (1960) as the source of a hormone controlling the daily rhythm of locomotor activity. It was shown by Fraenkel, G. (1935b) that Muscid flies defer the hardening of the cuticle so long as they are burrowing their way to the surface of the soil. In *Calliphora* and other insects this delay results from the absence of a peptide hormone released into the blood by the brain, which acts on the epidermal cells and leads to the normal darkening and hardening of the cuticle (Cottrell, C. 1960, 1962). The source of this hormone proved to be the neurosecretory cells of the pars intercerebralis of the brain (Fraenkel, G. and Hsiao, C. 1962). Because the sclerotization process is sometimes equivalent to the tanning of leather the hormone was named "bursicon" (see chapters 9, 10, volume 8). Other ganglia, including the terminal abdominal ganglion, were found to be sources in *Periplaneta* (Mills, R. 1967). The absorption of moulting fluid from the new tracheae and tracheoles after hatching or moulting in mosquito larvae was another process which appeared to be hormonally controlled (Wigglesworth, V. 1938), probably by bursicon itself.

Colour changes in insects are of two types: (1) morphological changes depending upon permanent deposition of pigments. Such changes are commonly dependent on genetic differences or on metabolic changes occurring during the action of growth hormones. The colour changes involving approximation to the background colour were regarded as nervous responses to sensory impressions resulting in changes in gene activity. Only in recent years has the neurosecretory involvement been defined (see chapter 13, volume 8).

On the other hand: (2) the so-called physiological changes in colour, dependent upon the movements of pigment granules within the epidermal cells, have long been known to be hormonally controlled. In *Carausius morosus* and in mantids the darkening at night and the pale coloration by day are induced indirectly. The nervous stimulus reaches a centre in the head and leads to the production of a hormone which causes pigment movement in the cells (Atzler, M. 1930; Giersberg, H. 1928). It was shown by Janda, V. (1936) that pieces of cuticle and epidermis transplanted from one insect to another change colour simultaneously with the new host. As to the source of the hormone involved, in *Carausius* several factors appeared to be involved: (a) from the corpus cardiacum (probably derived from neurosecretory cells of the deuto- or tritocerebrum); (b) from the tritocerebral region of the brain where the centre can be localized in many insects. *Carausius morosus*, rendered light in colour after extirpation of the brain, became very dark if the brains of dark-adapted individuals were implanted. The injection of extracts from neurosecretory cells in the tritocerebrum had the same effect (Dupont-Raabe, M. 1951, 1959). The expansion of the chromatophores on the air sacs of *Corethra* larvae was found to be similarly controlled (Dupont-Raabe, M. 1957; Gersch, M. 1956; Hadorn, E. and Frizzi, W. 1949; Kopenec, A. 1949).

7.2 Neurohumours

Neurohumours are pharmacologically active substances present in extracts from the nervous system or circulating in the haemolymph and presumably derived from the nervous system. When they are normal physiological components of the body they are commonly concerned in providing a chemical stimulus for the postsynaptic neuron; or on liberation at the nerve ending they may serve to stimulate the muscle or gland that is being supplied.

Adrenaline and noradrenaline are the classic neurohumours in vertebrates. Adrenaline is extractable from insects and experimentally applied will produce detectable effects — but until very recently there was no good evidence of what functional role it may play in insects. Other catecholamines, notably 5-hydroxytyramine (dopamine) are plentiful in insects, and there is recent evidence for their importance in normal physiology.

Acetylcholine is plentiful in the central nervous system of insects. It is doubtless important in synaptic transmission. Abundant cholinesterase ensures that it does not escape from the synaptic junctions. Unlike vertebrates, insects do not employ acetylcholine as a neuro-muscular transmitter (Wigglesworth, V. 1958b).

Serotonin (5-hydroxytryptamine) is readily extractable from the nervous system and corpus cardiacum of insects (Colhoun, E. 1963; Treherne, J. 1966) and it has many pharmacological effects on insects; but whether these are related with normal physiology has been uncertain.

Glutamate may well play a part as a neuromuscular stimulant in motor nerve endings; it is very doubtful whether it is active in the circulating blood.

8 BEGINNINGS OF HORMONE CHEMISTRY IN INSECTS

8.1 Activation hormone and other neurosecretory products

The nature of the activation (prothoracicotropic) hormone secreted by the neurosecretory cells of the brain has presented many difficulties (see chapter 4, this volume). The usual method of study was to remove the brain from the pupa of the silkworm *Bombyx mori* or other Lepidoptera, and thus to produce a persistent or "dauer" pupa in which the prothoracic glands remain inactive and further development does not take place. It was then found that lipid extracts from the brain would activate the prothoracic glands and cause renewed growth; even cholesterol alone was effective (Kobayashi, M. and Kirimura, J. 1958; Kobayashi, M. and Yamasaki, M. 1966; Kirimura, J. *et al.*, 1962), and oily extracts with juvenile hormone activity have the effect (Williams, C. 1959). But in this material the prothoracic gland secretion seems to be only temporarily suppressed; after some months the gland may become active spontaneously. It is therefore possible that non-specific agents might activate the gland and thus be mistaken for the activation hormone.

In most vertebrate animals the active product of the neurosecretory cells is generally proving to be a small polypeptide, and highly active aqueous extracts were obtained by Ichikawa, M. and Ishizaki, H. (1967) from the brain of *Bombyx mori*. The partially purified product seemed to be a mixture of peptides with molecular weight ranging from 9,000 to 31,000 (chapter 4, this volume).

The active principle in the hormone controlling diapause in the eggs of *Bombyx mori* has been recognized in recent years as a lipid-soluble peptide of about 14 amino acids (see chapter 12, volume 1); and current work is pointing to similar conclusions for the various neurohormones involved in insect metabolism (chapter 5, this volume; chapter 4, volume 3).

8.2 Prothoracic gland hormone

It was early recognized that the sclerotization of the integument of muscid larvae, as originally described by Fraenkel, G. (1935a) would make *Calliphora* ideal material for the extraction of the hormone and for its bioassay during the perfection of an isolation procedure. Becker, E. and Plagge, E. (1939) devised the "*Calliphora* test" which consisted of injecting an extract into the posterior fragment of *Calliphora* larvae in which the anterior segments, containing the brain and Weismann's ring, had been separated by a tight ligature and had hardened, while the posterior fragment had retained its soft white larval character. The quantity of active substance which induced 50–70% pupation among the injected specimens, within 24 h, was defined as a "*Calliphora* unit". A "*Calliphora* factory" was set up by Becker, E. (1941) in Berlin–Dahlem in order to set about concentrating and isolating the active principle. After the death of Becker this problem was taken up by Butenandt and Karlson. Working with kilogram quantities of puparia they had been able to define a satisfactory scheme for isolation; but in 1946 it was dicovered that active extracts could be obtained from *Bombyx mori* pupae which were obtainable from the silk industry, and they continued the work on 50 kg batches of pupae. But final purification was possible only after processing 500 kg in the factory of Hoffmann-La Roche (Basel) in a series of countercurrent distribution steps which yielded in the end 25 mg of the crystalline hormone, ecdysone (Butenandt, A. and Karlson, P. 1954; chapter 6, this volume).

At symposia in those days many of us questioned Karlson as to whether the hormone might not be a steroid. But that was excluded by the molecular weight; moreover all known steroids were lipid-soluble. At that time the elementary formula was believed to be $C_{18}H_{30}O_4$; but eventually the correct molecular weight was found to be 464 and the empirical formula $C_{27}H_{44}O_6$. Application of the novel X-ray crystallography methods by W. Hoppe to crystals of pure ecdysone showed that the fundamental structure was indeed that of a steroid. The full structure was finally elucidated by Huber, R. and Hoppe, W. (1965) and it was later duly synthesized by a number of chemists using different pathways (Karlson, P. 1967).

It came as a surprise (not least to Butenandt who was a leading authority on steroids) that ecdysone was a new sort of highly hydroxylated steroid; it has five hydroxyl groups as opposed to the one hydroxyl group of cholesterol. When ecdysone was first isolated a second, more polar substance with higher activity had been isolated (β-ecdysone). This proved to have a sixth hydroxyl group and is commonly termed "ecdysterone" (20-hydroxyecdysone). It was the solubility in water which had misled the chemists. But during recent years it has been found that steroids of this type are abundant in certain plants, notably certain ferns, where again they had been overlooked. A long list of Japanese plants yielded a wide variety of new "ecdysones" (Nakanishi, K. *et al.*, 1966). Indeed the foliage of yew (*Taxus*) is such a rich source that 1 g of dry leaves yields 500 μg of pure ecdysterone and 70 g will yield as much of the active substance as 1000 kg of *Bombyx* pupae (Staal, G. 1967a). The relation between ecdysone and cholesterol suggested that the hormone might be derived from cholesterol in the diet, and Karlson, P. and Hoffmeister, H. (1963) showed that tritiated cholesterol fed to *Calliphora* larvae led to the appearance of labelled ecdysone (chapter 7, this volume).

In recent years it has been realized that ecdysone is normally converted into 20-hydroxyecdysone by a wide range of tissues and that this product is the active principle in the living insect (chapter 8, this volume). But the precise physiological nature of this relation is still *sub judice*.

The quantitative requirement of ecdysone for the completion of moulting is about 10 $\mu g\ g^{-1}$ of body weight as estimated in *Rhodnius* (Wigglesworth, V. 1955b) and *Hyalophora* (Williams, C.: see Butenandt, A. and Karlson, P. 1954). For the colour changes in the full-grown caterpillar of *Cerura* not more than one-tenth of this dose is required (Bückmann, D. 1959).

8.3 Juvenile hormone

It was recognized early that the juvenile hormone effect in insects was brought about by agents which were effective in a wide range of insects: the corpora allata of *Periplaneta* producing giant larvae in *Oncopeltus* (Novak, V. 1951) and in *Rhodnius*; the ring gland from *Calliphora* larvae inducing the local formation of larval cuticle in *Rhodnius* (Wigglesworth, V. 1954b) and so forth. But active extracts were first obtained by Williams, C. (1956) from the abdomen of the adult male of *Hyalophora cecropia* — and this still remains the most active source known. A simple ethanol–ethyl ether extract of the abdomen yielded a deep orange-coloured oil which shows strong juvenile hormone activity (chapter 10, this volume). Similar extracts from other insects proved inactive without suitable fractionation. With the use of this active extract it was possible to devise simple assays: by application to the integument of *Rhodnius* fifth-stage larvae or to punctures in the pupal cuticle of *Tenebrio* (Wigglesworth, V. 1958a). The hormone was rapidly broken down in metabolism if injected in emulsified form (Gilbert, L. and Schneiderman, H. 1960). It was effective if applied in a mixture with paraffin wax (the "wax test" of Gilbert, L. and Schneiderman, H.) or in a drop of oil containing a dilute solution (Wigglesworth, V. 1969).

By the use of countercurrent separation in suitable solvents the active substance was carried along with the non-saponifiable fraction in the oily extracts; and material with juvenile hormone activity was thus extracted from many insects and other invertebrates (Gilbert, L and Schneiderman, H. 1958) as well as the suprarenal glands of cattle and many other organs in vertebrates, including man and even in some plants, bacteria and yeasts.

Active material was first isolated in pure form from the excreta of *Tenebrio* and from yeast by Schmialek, P. (1961) and in this case was shown to be a mixture of the open chain terpene alcohol farnesol and its aldehyde farnesal. Farnesol was quite active in *Rhodnius* as well as *Tenebrio* but far less so in Lepidoptera (Wigglesworth: in Schmialek, P. 1961). The less polar derivative of farnesol, farnesyl methyl ether, was highly active in *Rhodnius* (in fact it was later shown to have an activity in *Rhodnius* of 25% that of the natural juvenile hormone (Wigglesworth, V. 1969)). These results at once suggested that the natural hormone might well be a related isoprenoid compound. Some chemists seemed positively to eschew this lead but Bowers, W. *et al.* (1965) followed it up by preparing the methyl ester of farnesenic acid (in place of the ether used by Schmialek) and introducing an epoxy ring in the 10 — 11 position (a common feature in natural isoprenoids) and obtained a compound with a very much wider range of juvenile hormone activity. Bowers had in fact synthesized what is now recognized as JH III, the principal juvenile hormone of most insects. Benefiting from the experience of Bowers, Röller extracted and identified the major component of the active material extracted from *Hyalophora* as a compound closely similar to that synthesized by Bowers *et al.* except that in place of the methyl groups at C_7 and C_{11} usual in isoprenoids it had two ethyl groups (Röller, H. and Bjerke, J. 1965; Röller, H. and Dahm, K. 1969). This substance came to be known as JH I; and a very similar compound isolated by Meyer, A. *et al.* (1968) from the same material, which has a methyl group at C-7 and an ethyl group at C-11, and is somewhat less active, is known as JH II.

Of the eight geometric isomers of the JH I compound, the *trans*, *trans*, *cis* isomer (which is the natural substance) proved to be by far the most active (Röller, H. and Dahm, K. 1968). It is this substance which has four times the activity of farnesyl methyl ether in restraining metamorphosis in *Rhodnius* — but in most insects the difference in activity is far greater than this. There is one asymmetric carbon atom at C_{11} in the molecule of JH I so that there is a D and an L series of isomers. As tested on *Rhodnius* the natural, optically active, compound proved considerably more active as a hormone than the synthetic racemic product (Wigglesworth, V. 1973).

REFERENCES

Andrewartha, H. G. (1952). Diapause in relation to the ecology of insects. *Biol. Rev. 27*, 50–107.

Arvy, L. and Gabe, M. (1953). Données histophysiologiques sur la neurosécrétion chez quelques Ephéméroptères. *Cellule 55*, 203–222.

Arvy, L. and Gabe, M. (1954). Modifications de la neurosécrétion protocérébrale et des glandes endocrines céphaliques de *Leptinotarsa decemlineata* Say au cours de la métamorphose. 79ᵉ *Congr. Soc. savantes*, 189–196.

Arvy, L., Bounhiol, J. J. and Gabe, M. (1953). Déroulement de la neurosécrétion protocérébrale chez *Bombyx mori* L au cours du développement post-embryonaire. *C.R. Acad. Sci., Paris 236*, 627–629.

Astauroff, B. L. (1930). Analyse der erblichen Störungsfälle der bilateralen Symmetrie im Zusammenhang mit der selbständigen Variabilität ähnlicher Strukturen. *Z. Indukt. Abstamm.-u Vererb-Lehre, 55*, 183–262.

Atzler, M. (1930). Untersuchungen über morphologischen und physiologischen Farbwechsel von *Dixippus morosus*. *Z. vergl. Physiol. 13*, 505–533.

Avery, O. T., MacLeod, C. M. and McCarty, M. (1944). Studies on the chemical nature of the substance inducing transformation of pneumoccal types. *J. Exp. Med. 79.*, 137–158.

Beck, S. D. and Alexander, N. (1964). Hormonal activation of the insect brain. *Science 143*, 478–479.

Becker, E. (1941). Uber Versuche zur Anreicherung und physiologische Charakterisierung des Wirkstoffes der Puparisierung. *Biol. Zbl. 61*, 360–388.

Becker, E. and Plagge, E. (1939). Uber das die Pupariumbildung auslösende Hormon der Fliegan. *Biol. Zbl. 59*, 326–341.

Becker, H. J. (1962). Die Puffs der Speicheldrüsen-chromosomen von *Drosophila melanogaster*. *Chromosoma 13*, 341–384.

Bier, K. (1954). Ueber den Saisondimorphismus der Oogenesis von *Formica rufa* und dessen Bedeutung für die Kastendetermination. *Biol. Zbl. 73*, 170–190.

Bodenstein, D. (1933). Beintransplantationen an Lepidopteren raupen. I. Transplantationen zur Analyse der Raupen- und Puppenhäutung. *Wilhelm Roux Archiv. Entw.Mech. 128*, 564–583.

Bodenstein, D. (1938). Untersuchungen zum Metamorphose problem. II. Entwicklungsrelationen in verschmolzenen Puppenteilen. *Wilhelm Roux Archiv. Entw.Mech. 137*, 636–660.

Bodenstein, D. (1953). Studies on the humoral mechanisms in growth and metamorphosis of the cockroach *Periplaneta americana*. II. The function of the prothoracic gland and the corpus cardiacum. *J. Exp. Zool. 123*, 413–433.

Bounhiol, J. J.(1938). Recherches expérimentales sur le déterminisme de la métamorphose chez les Lépidoptères. *Bull. Biol.* Suppl. *24*, 1–199.

Bounhiol, J. J. (1953). Rôle du corpus allatum dans la métamorphose des insectes. *Trans. IXth Int. Congr. Ent. 2*, 63–72.

Bowers, B. and Johnson, B. (1966). An electron microscope study of the corpora cardiaca and secretory neurons in the Aphid, *Myzus persicae* (Sulz.). *Gen. Comp. Endocr. 6*, 213–230.

BOWERS, W. S., THOMPSON, M. J. and UEBEL, E. C. (1965). Juvenile and gonadotropic hormone activity of 10, 11–epoxyfarnesenic acid methyl ester. *Life Sci. 4*, 2323–2331.

BRICTEUX-GRÉGOIRE, S., VERLY, W. G. and FLORKIN, M. (1957). Protein synthesis in *Sphinx ligustri* pupae. *Nature, Lond. 179*, 678–679.

BROWN, F. A. and MEGLITSCH, A. (1940). Comparison of the chromatophorotropic activity of insect corpora cardiaca with that of Crustacean sinus glands. *Biol. Bull. 79*, 409–418.

BÜCKMANN, D. (1959). Die Auslösung der Umfärbung durch das Häutungshormon bei *Cerura vinula* L. (Lepidoptera, Notodontidae). *J. Insect Physiol. 3*, 159–189.

BURTT, E. T. (1938). On the corpora allata of dipterous insects II. *Proc. Roy. Soc. B. 126*, 210–223.

BUTENANDT, A. and KARLSON, P. (1954). Ueber die Isoleirung eines Metamorphose-Hormons der Insekten in kristallisierter Form. *Z. Naturf. 9b*, 389–391.

BYTINSKI-SALZ, H. (1933). Untersuchungen an Lepidopterenhybriden II. Entwicklungsphysiologische Experimente über die Wirkung der disharmonischen Chromosomenkombinationen. *Wilhelm Roux Arch. Entw. Mech. 129*, 356–378.

CAMPBELL, F. L. (ed.) (1959). *Physiology of Insect Development.* Chicago University Press.

CHILD, C. M. (1941). *Patterns and Problems of Development.* University of Chicago Press, Illinois.

CHRISTOPHERS, S. R. and CRAGG, F. W. (1922). On the so-called "penis" of the bed-bug (*Cimex lectularius* L.) and on the homologies generally of the male and female genitalia of this insect. *Indian J. Med. Res. 9*, 445–463.

CHURCH, N. S. (1955). Hormones and the termination and reinduction of diapause in *Cephus cinctus* Nort. (Hymenoptera: Cephidae). *Canad. J. Zool. 33*, 339–369.

CLARK, R. B. (1956). On the origin of neurosecretory cells. *Ann. Sci. Nat. Zool.* (11) *18*, 199–207.

CLEMENTS, A. N. (1956). Hormonal control of ovary development in mosquitoes. *J. Exp. Biol. 33*, 211–223.

CLEVER, U. (1961). Genaktivitäten in den Riesenchromosomen von *Chironomus tentans*, und ihre Beziehugen zur Entwicklung. I. Genaktivierung durch Ecdyson, *Chromosoma 12*, 607–675.

CLOUTIER, E. J., BECK, S. D., MCLEOD, D. G. R. and SILHACEK, D. L. (1962). Neural transplants and insect diapause. *Nature, 195*, 1222–1224.

COCKAYNE, E. A. (1941). Prothetely in a larva of *Smerinthus* hybr. *hybridus* Stephens (Lepidoptera). *Proc. Roy. Ent. Soc. Lond. A, 16*, 55–59.

COLHOUN, E. H. (1963). The physiological significance of acetylcholine in insects and observations upon other pharmacologically active substances. *Adv. Insect Physiol. 1*, 1–46.

COTTRELL, C. B. (1960). The control of cuticle hardening and darkening and the mechanism of expansion at imaginal ecdysis of Cychorrhapha Diptera. Thesis, Cambridge.

COTTRELL, C. B. (1962). The imaginal ecdysis of blowflies. The control of cuticular hardening and darkening. *J. Exp. Biol. 39*, 35–430.

CRAMPTON, H. E. (1899). An experimental study upon Lepidoptera. *Wilhelm Roux Archiv. EntwMech. 9*, 293–318.

DEWITZ, J. (1905). Untersuchungen über die Verwandlung der Insektenlarven. II. *Arch. Anat. Physiol., Lpz., Abt. Physiol.*, pp. 389–415.

DUCLAUX, E. (1869). De l'influence du froid de l'hiver sur le développement de l'embryon du ver à soie et sur l'éclosion de la graine. *C.R. Acad. Sci. Paris. 69*, 1021–1022.

DUPONT-RAABE, M. (1951). Etude morphologique et cytologique du cerveau de quelques Phasmides. *Bull. Soc. Zool. Fr. 76*, 386–397.

DUPONT-RAABE, M. (1952). Contribution à l'étude du role endocrine du cerveau et notamment de la pars intercerebralis chez les phasmides. *Archiv. Zool. Exp. Gén. 89*, 128–138.

DUPONT-RAABE, M. (1954). Répartition des activités chromatiques dans le ganglion susoesophagien des Phasmides; mise en evidence d'une région sécrétiore dans la partie deuto et tritocérébrale. *C.R. Acad. Sci. Paris 238*, 950–951.

DUPONT-RAABE, M. (1957). Les mécanismes de l'adaptation chromatique chez les insectes. *Archiv. Zool. Exp. Gén. 94*, 61–293.

DUPONT-RAABE, M. (1959). Neurohormones chez les insectes. *Bull. Soc. Zool. Fr. 84*, 272–316.

DUPONT-RAABE, M. (1965). Etude des phénomènes de neurosécretion au niveau de la chaine nerveuse des Phasmides. *Bull. Soc. Zool. Fr. 90*, 631–654.

ENGELMANN, F. (1957). Die Steuerung der Ovarfunktion bei der ovoviviparen Schabe *Leucophaea maderae* (Fabr.). *J. Insect Physiol. 1*, 257–278.

EPHRUSSI, B. and BEADLE, G. W. (1942). Chemistry of "eye color hormones" of *Drosophila. Quart. Rev. Biol. 17*, 326–338.

FORD, E. B. (1953). The genetics of polymorphism in the Lepidoptera. *Adv. Genet. 5*, 43–87.

FRAENKEL, G. (1935a). A hormone causing pupation in the blowfly *Calliphora erythrocephala. Proc. Roy. Soc. B. 118*, 1–12.

FRAENKEL, G. (1935b). Observations and experiments on the blowfly (*Calliphora erythrocephala*) during the first day after emergence. *Proc. Zool. Soc. Lond.* pp. 893–904.

FRAENKEL, G. and HSIAO, C. (1962). Hormonal and nervous control of tanning in the fly. *Science 138*, 27–29.

FRASER, A. (1960). Humoral control of metamorphosis and diapause in the larvae of certain Calliphoridae (Diptera: Cyclorrhapha). *Proc. Roy. Soc. Edinb. B, 67*, 127–140.

FUKAYA, M. (1955). The role of the brain in the metamorphosis of the rice stem borer, *Chilo suppressalis, Jap. J. Appl. Zool. 20*, 179–183.

FUKAYA, M. and MITSUHASHI, J. (1961). Larval diapause in the rice stem borer with special reference to its hormonal mechanisms. *Bull. Nat. Inst. Agric. Sci. (Japan)* (C) No. 13, 1–32.

FUKUDA, S. (1940). Induction of pupation in silkworm by transplanting the prothoracic gland. *Proc. Imp. Acad. Tokyo 16*, 414–416.

FUKUDA, S. (1951). The production of the diapause eggs by transplanting the suboesophageal ganglion in the silkworm. *Proc. Imp. Acad. Japan, 27*, 672–677.

GABE, M. (1953). Sur quelques applications de la coloration par la fuchsine-paraldehyde. *Bull. Mic. Appl. 3*, 153–162.

GABE, M. (1966). *Neurosecretion.* Pergamon Press, Oxford.

GERSCH, M. (1956). Untersuchungen zur Frage der hormonalen Beeinflussung der Melanophoren bei der *Corethra*-Larve. *Z. vergl. Physiol. 39*, 190–208.

GIERSBERG, H. (1928). Ueber de morphologischen und physiologischen Farbwechsel der Stabheuschrecke *Dixippus morosus. Z. vergl. Physiol. 7*, 657–695.

GILBERT, L. I. and SCHNEIDERMAN, H. A. (1958). Occurrence of substances with juvenile hormone activity in adrenal cortex of vertebrates. *Science 128*, 844.

GILBERT, L. I. and SCHNEIDERMAN, H. A. (1960). The development of a bioassay for the juvenile hormone of insects. *Trans. Amer. Mic. Soc. 79*, 38–67.

GILBERT, L. I. and SCHNEIDERMAN, H. A. (1961). The content of juvenile hormone and lipid in Lepidoptera: sexual differences and developmental changes. *Gen. Comp. Endocr. 1*, 453–472.

GILLETT, J. D. (1956). Initiation and promotion of ovarian development in the mosquito *Aëdes (Stegomyia) aegypti* (Linnaeus). *Ann. Trop. Med. Parasit. 50*, 375–380.

GILLETT, J. D. (1958). Induced ovarian development in decapitated mosquitoes by transfusion of haemolymph. *J. Exp. Biol. 35*, 685–693.

GOLDSCHMIDT, R. (1923). Einige Materialien zur Theorie der abgestimmten Reaktionsgeschwindigkeiten. *Archiv. Mik. Anat. 98*, 292–313.

GOLDSCHMIDT, R. (1927). *Physiologische Theorie der Vererbung.* Berlin, Springer.

GOMORI, G. (1941). Observations with differential stains on human islets of Langerhans. *Amer. J. Path. 17*, 395–406.

GORDON, C. and SANG, J. H. (1941). The relation between nutrition and exhibition of the gene Antennaless *(Drosophila melanogaster). Proc. Roy. Soc.* B. *130*, 151–184.

GOSBEE, J. L., MILLIGAN, J. V. and SMALLMAN, B. N. (1968). Neural properties of the protocerebral neurosecretory cells of the adult cockroach *Periplaneta americana. J. Insect Physiol. 14*, 1785–1792.

HACHLOW, V. (1931). Zur Entwicklungsmechanik der Schmetterlinge. *Wilhelm Roux Archiv. EntwMech. 125*, 26–49.

HADORN, E. and FRIZZI, W. (1949). Experimentelle Untersuchungen zur Melanophoren-Reaktion von *Corethra. Rev. Suisse Zool. 56*, 306–316.

HADORN, E. and NEEL, J. (1938). Der hormonale Einfluss der Ringdrüse (corpus allatum) auf die Puparium-bildung bei Fliegen. *Wilhelm Roux Archiv. EntwMech. 138*, 281–304.

HALBWACHS, M. C., JOLY, L. and JOLY, P. (1957). Résultats d'implantations de "glandes ventrales" à *Locusta migratoria* L. *J. Insect Physiol. 1*, 143–149.

HANSTRÖM, B. (1938). Zwei Probleme betreffs der hormonalen Localization im Insektenkopf. *Acta Univ. Lund.* N.F., Avd. 2, *39*, 1–17.

HANSTRÖM, B. (1940). Inkretorische Organe, Sinnesorgane und Nervensystem des Kopfes einiger niederen Insektenordnungen. *K. Svenska Vetens Akad. Handl. 18*, 3–265.

HANSTRÖM, B. (1941). Einige Parallelen im Bau und in der Herkunft der inkretorischen Organe der Arthropoden und der Vertebraten. *Acta Univ. Lund.*, N.F. Avd.2, *37*, 1–19.

HANSTRÖM, B. (1953). Neurosecretory pathways in the head of crustaceans, insects and vertebrates. *Nature, 171*, 72–73.

HARKER, J. E. (1960). The effect of perturbations in the environmental cycle on the diurnal rhythm of activity of *Periplaneta americana*. L. *J. Exp. Biol.*, *37*, 154–163.

HARVEY, W. (1651). *Generatione Animalium*. exerc. XLV and L. Amsterdam.

HASEGAWA, K. (1947). Studies on the development of the ova in the silkworm, *Bombyx mori*. (ii) On the degenerated ovum. *Bull. Seric. Exp. Sta. 12*, 481–492.

HASEGAWA, K. (1951). Studies on the voltinism in the silkworm *Bombyx mori* L. with special reference to the organs concerning determination of voltinism. *Proc. Japan. Acad. 27*, 557.

HENNEGUY, L. F. (1904). *Les Insectes*, Paris.

HIGHNAM, K. C. (1962). Neurosecretory control of ovarian development in *Schistocerca gregaria*. *Quart. J. Mic. Sci. 103*, 57–72.

HIGHNAM, K. C. and LUSIS, O. (1962). The influence of mature males on the neurosecretory control of ovarian development in the desert locust. *Quart. J. Mic. Sci. 103*, 73–83.

HINTON, H. E. (1958). Concealed phases in the metamorphosis of insects. *Science Progress, 182*, 260–275.

HOLMGREN, N. (1909). Termitenstudien, 1. *K. Svenska VetensAkad. Handl. 44*, 1–215.

HUBER, R. and HOPPE, W. (1965). Die Kristall- und Molekülstrukturanalyse des Insektenverpuppungshormons Ecdyson mit der automatisierten Faltmolekülmethode. *Chem. Ber. 98*, 2403–2424.

ICHIKAWA, M. and NISHIITSUTSUJI, J. (1955). Studies on insect metamorphosis. (v) Implantation of larval brains into the pupae of *Luehdorfia japonica*. *Mem. Coll. Sci. Univ. Kyoto* B. *22*, 11–15.

ISHIZAKI, H. and ICHIKAWA, M. (1967). Purification of the brain hormone of the silkworm *Bombyx mori*. *Biol. Bull. 133*, 355–368.

ITO, H. (1918). On the glandular nature of the corpora allata of the Lepidoptera. *Bull. Imp. Tokyo Sericultural Coll. 1*, 63–103.

JANDA, V. (1936). Ueber den Farbwechsel transplantierter Hautstücke und Küntslich verbundener Körperfragmente bei *Dixippus morosus*. *Zool. Anz. 115*, 177–185.

JANET, C. (1899). Sur les nerfs céphaliques, les corpora allata et le tentorium de la fourmi *Myrmica rubra*. *Mém. Soc. Zool. Fr. 12*, 295–335.

JENKIN, P. M. and HINTON, H. E. (1966). Apolysis in arthropod moulting cycles. *Nature, 211*, 871.

JOHANSSON, A. S. (1955). The relationship between corpora allata and reproductive organs in starved female *Leucophaea maderae* (Blattaria). *Biol. Bull.*, *108*, 40–44.

JOHANSSON, A. S. (1958). Relation of nutrition to endocrine-reproductive functions in the milkweed bug *Oncopeltus fasciatus* (Dallas) (Heteroptera: Lygaeidae). *Nyt. Mag. Zool. 7*, 5–132.

JOLY, P. (1945). La fonction ovarienne et son contrôle humoral chez les Dytiscidés. *Arch. Zool. Exp. Gén. 84*, 49–164.

JOLY, P. (1956). Croissance et indices de gregarisation chez *Locusta migratoria* (L). *Insectes sociaux 3*, 17–24.

KARLSON, P. (1956). Biochemical studies on insect hormones. *Vitam. Horm. 14*, 227–266.

KARLSON, P. (1967). The chemistry of insect hormones and insect pheromones. *Pure Appl. Chem. 14*, 75–87.

KARLSON, P. and HOFFMEISTER, H. (1963). Zur Biogenese des Ecdysons. i. Umwandlung von Cholesterin in Ecdyson. *Hoppe-Seylers. Z. Physiol. Chem. 331*, 298–300.

KARLSON, P. and LÜSCHER, M. (1959). "Pheromones": a new term for a class of biologically active substances. *Nature, 183*, 55.

KARLSON, P. and SEKERIS, C. E. (1966). Biochemical mechanisms of hormone action. *Acta Endocr. 53*, 505–518.

KELLOGG, V. L. (1904). Influence of the primary reproduction organs on the secondary sexual characters. *J. Exp. Zool. 1*, 601–605.

KENNEDY, J. S. (1961). Ed. *Insect Polymorphism. Symp. R. Ent. Soc. Lond. 1*.

KIRIMURA, J., SAITO, M. and KOBAYASHI, M. (1962). Steroid hormone in an insect *Bombyx mori*. *Nature, 195*, 729–730.

KOBAYASHI, M. and KIRIMURA, J. (1958). The brain hormone in the silkworm, *Bombyx mori* L. *Nature, 181*, 1217.

KOBAYASHI, M. and YAMASAKI, M. (1966). The proteinic brain hormone in an insect *Bombyx mori* L. *Appl. Ent. Zool. 1*, 53–60.

KOGURE, M. (1933). The influence of light and temperature on certain characters of the silkworm *Bombyx mori*. *J. Dep. Agric. Kyushu Univ. 4*, 1–93.

KOLBE, J. (1903) Ueber vorschnelle Entwicklung (Prothetelie) von Puppen- und Imago-Organen bei Lepidopteren- und Coleopteren-Larven. *Allg. Zeitschr. Ent. 8*, 1–9, 25–39.

KOPEČ, S. (1917). Experiments on metamorphosis of insects. *Bull. Int. Acad. Cracovie* B. 57–60.

KOPEČ, S. (1922). Studies on the necessity of the brain for the inception of insect metamorphosis. *Biol. Bull.*, *42*, 322–342.

KOPENETZ, A. (1949). Farbwechsel der Larve von *Corethra plumicornis*. *Z. vergl. Physiol. 31*, 490–505.

KÜHN, A. (1936). Versuche über die Wirkungsweise der Erbanlagen. *Naturwissenschaften 24*, 1–10.

LARSEN, J. R. and BODENSTEIN, D. (1959). The humoral control of egg maturation in the mosquito. *J. Exp. Zool. 140*, 343–381.

LEES, A. D. (1961). Clonal polymorphism in aphids. *Symp. R. Ent. Soc. Lond. 1*, 68–79.

LENGERKEN, H. VON (1924). Prothetelie bei Coleopterenlarven. *Zool. Anz. 58*, 179–185.

LEYDIG, F. (1864). *Tafeln zur Vergleichenden Anatomie*. Tübingen.

LIGHT, S. F. (1942). The determination of the castes of social insects. *Quart. Rev. Biol. 17*, 312–326; *18*, 46–63.

LINZEN, B. and BÜCKMANN, D. (1961). Biochemische und histologische Untersuchungen zur Umfärbrung der Raupe von *Cerura vinula*. *Z. Naturf. 16b*, 6–17.

LOHER, W. (1960). The chemical acceleration of the maturation process and its hormonal control in the male of the desert locust. *Proc. Roy. Soc.* B, *153*, 380–397.

LUBBOCK, J. (1883). *The Origin and Metamorphosis of Insects*. London: Macmillan.

LÜSCHER, M. (1958). Uber die Entstehung der Soldaten bei Termitten. *Rev. Suisse Zool. 65*, 372–377.

LÜSCHER, M. and ENGELMANN, F. (1955). Uber die Steurerung der Corpora allata-Funktion bei der Schabe *Leucophaea maderae*. *Rev. Suisse Zool. 62*, 649–657.

LYONET, P. (1762). *Traité anatomique de la chenille qui ronge le bois de saule*. La Haye.

MADDRELL, S. H. P. (1966). The site of release of the diuretic hormone in *Rhodnius* — a new neurohaemal system in insects. *J. Exp. Biol. 45*, 499–508.

MAYER, R. J. and CANDY, D. J. (1969). Control of haemolymph lipid concentration during locust flight: an adipokinetic hormone from the corpora cardiaca. *J. Insect Physiol. 15*, 611–620.

MEISENHEIMER, J. (1910). Zur Ovarialtransplantation bei Schmetterlingen. *Zool. Anz. 35*, 446–450.

MELLANBY, K. (1938). Diapause and metamorphosis of the blowfly *Lucilia sericata* Meig. *Parasitology 30*, 392–402.

MELLANBY, K. (1939). Fertilization and egg production in the bed-bug, *Cimex lectularius* L. *Parasitology 31*, 193–199.

MEYER, A. S., SCHNEIDERMAN, H. A., HANZMANN, E. and KO, J. H. (1968). Two juvenile hormones from the cecropia silkmoth. *Proc. Nat. Acad. Sci. U.S.A. 60*, 853–860.

MEYER, G. F. and PFLUGFELDER, O. (1958). Elektronenmikroskopische Untersuchungen an den Corpora cardiaca von *Carausius morosus* Br. *Z. Zellforsch. 48*, 556–564.

MILLS, R. R. (1967). Control of cuticular tanning in the cockroach: bursicon release by nervous stimulation. *J. Insect Physiol. 13*, 815–820.

MOKIA, G. G. (1941). Contribution to the study of hormones in insects. *C.R. Acad. Sci. U.R.S.S. 30*, 371–373.

MÜLLER, H. J. (1955). Die Saisonformenbildung von *Arachnia levana*; ein photoperiodisch gesteuerter Diapause-Effekt. *Naturwissenschaften 42*, 134–135.

NABERT, A. (1913). Die *Corpora allata* der Insekten. *Z. Wiss. Zool. 104*, 181–358.
NAKANISHI, K., KOREEDA, M., SASAKI, S., CHANG, M. L. and HSU, H. Y. (1966). The structure of ponasterone A, an insect moulting hormone from the leaves of *Podocarpus nakaii*, Hay. *J. Chem. Soc. Chem. Commun.*, 915–917.
NAYAR, K. K. (1955). Studies on the neurosecretory system of *Iphita limbata* Stål. (i) Distribution and structure of the neurosecretory cells of the nerve ring. *Biol. Bull., 108*, 296–307.
NAYAR, K. K. (1958). Studies on the neurosecretory system of *Iphita limbata* Stål. (v) Probable endocrine basis of oviposition in the female insect. *Proc. Indian Acad. Sci. 67*, 233–249.
NEWPORT, G. (1836). On the respiration of insects. *Phil. Trans. Roy. Soc. Lond., 1836*, 529–566.
NORRIS, M. J. (1932). Contributions towards the study of insect fertility. (i) The structure and operation of the reproductive organs of the genera *Ephestia* and *Plodia* (Lepidoptera, Phycitidae). *Proc. Zool. Soc. Lond.* 1932, 595–611.
NORRIS, M. J. (1954). Sexual maturation in the desert locust *Schistocerca gregaria* Forskål. *Anti-locust Bull. 18*, 1–44.
NOVAK, V. J. A. (1951). New aspects of the metamorphosis of insects. *Nature, 167*, 132–133.
OUDEMANS, J. T. (1899). Falter aus castrierten Raupen. *Zool. Jahrb., Syst. 12*, 71–85.
OZEKI, K. (1958). Neurosecretion of the earwig, *Anisolabis maritima. Sci. Papers Coll. Gen. Educ. Univ. Tokyo 8*, 85–92.
OZEKI, K. (1959). Secretion of molting hormone from the ventral glands. *Sci. Papers Coll. Gen. Educ. Univ. Tokyo 9*, 256–262.
PFEIFFER, I. W. (1939). Experimental study of the function of the corpora allata in the grasshopper, *Melanoplus differentialis. J. Exp. Zool. 82*, 439–461.
PFEIFFER, I. W. (1945). The influence of the corpora allata over the development of nymphal characters in the grasshopper *Melanoplus differentialis. Trans. Conn. Acad. Arts Sci. 36*, 489–515.
PFLUGFELDER, O. (1937). Bau, Entwicklung und Funktion der Corpora allata und cardiaca von *Dixippus morosus* Br. *Z. Wiss. Zool. 149*, 477–512.
PFLUGFELDER, O. (1949). Die Funktion der Pericardialdrüsen der Insekten. *Zool. Anz. 14*, suppl. 169–173.
PIEPHO, H. (1939). Raupenhäutungen bereits verpuppter Hautstücke bei der wachsmotte *Galleria mellonella* L. *Naturwissenschaften 27*, 301–302.
PIEPHO, H. (1942). Untersuchungen zur Entwicklungsphysiologie der Insektenmetamorphose. Uber die Puppenhäutung der Wachsmotte *Galleria mellonella* L. *Wilhelm Roux Archiv. EntwMech. 141*, 500–583.
PIEPHO, H. (1950). Hormonale Grundlagen der Spinntätigkeit bei Schmetterlingsraupen. *Z. Tierpsychol. 7*, 424–434.
PIEPHO, H. (1951). Uber die Lenkung der Insektenmetamorphose durch Hormone. *Verh. Dtsch. Zool. Ges.* Wilhelmshaven, 1951, 62–75.
PIEPHO, H., BÖDEN, E. and HOLZ, I. (1960). Uber die Hormonabhängigkeit des Verhaltens von Schwärmerraupen vor den Häutungen. *Z. Tierpsychol. 17*, 261–269.
PLAGGE, E. (1938). Weitere Untersuchungen über das Verpuppungshormon bei Schmetterlingen. *Biol. Zbl. 58*, 1–12.
POSSOMPÈS, B. (1953). Recherches expérimentales sur le déterminisme de la métamorphose de *Calliphora erythrocephala* Meig. *Archiv. Zool. Exp. Gen. 89*, 203–364.
POULTON, E. B. (1903). Experiments in 1893, 1894 and 1896 upon the colour relation between lepidopterous larvae and their surroundings. *Trans. Ent. Soc. Lond.* 1903, 311–368.
PRUTHI, H. S. (1924). Studies in insect metamorphosis. I. Prothetely in mealworms *(Tenebrio molitor)* and other insects. *Biol. Rev. 1.* 139–147.
RAABE, M. *See* DUPONT-RAABE.
RADTKE, A. (1942). Hemmung der Verpuppung beim Mehlkäfer *Tenebrio molitor* L. *Naturwissenschaften, 30*, 451–452.
RAHM, U. H. (1952). Die innersekretorische Steuerung der postembyonalen Entwicklung von *Sialis lutaria* L. (Megaloptera). *Rev. Suisse Zool. 59*, 173–237.
REGEN, J. (1910). Kastration und ihre Folgeerscheinungen bei *Gryllus campestris* L. *Zool. Anz. 35*, 427–432.
REHM, M. (1951). Die zeitliche Folge der Tätigkeitsrhythmen inkretorischer Organe von *Ephestia kühniella* wahrend der Metamorphose und des Imaginal-lebens. *Wilhelm Roux Arch. EntwMech. 145*, 205–248.
REHM, M. (1955). Morphologische und histologische Untersuchungen an Neurosekretorischen Zellen von Schmetterlingen. *Z. Zellforsch. 42*, 19–58.
RÖLLER, H. (1962). Ueber den Einfluss der Corpora allata auf den Stoffwechsel der Wachsmotte. *Naturwissenschaften 49*, 524.
RÖLLER, H. and BJERKE, J. S. (1965). Purification and isolation of juvenile hormone and its action in lepidopteran larvae. *Life Sci. 4*, 1617–1624.
RÖLLER, H. and DAHM, K. H. (1968). The chemistry and biology of juvenile hormone. *Recent Prog. Horm. Res. 24*, 651–680.
ROTH, L. M. and STAY, B. (1961). Oöcyte development in *Diploptera punctata* (Eschscholtz) (Blattaria). *J. Insect Physiol. 7*, 186–202.
ROTH, L. M. and WILLIS, E. R. (1950). The oviposition of *Dermestes ater* Degeer, with notes on bionomics under laboratory conditions. *Amer. Midl. Nat. 44*, 427–447.
ROUBAUD, E. (1919). Etudes sur le sommeil d'hiver préimaginal des muscides. *Bull. Biol. Fr. Belg. 56*, 455–544.
SALT, G. (1937). The egg-parasite of *Sialis lutaria:* a study of the influence of the host upon a dimorphic parasite. *Parasitology, 29*, 539–553.
SALT, G. (1952). Trimorphism in the ichneumonid parasite *Gelis. Quart. J. Mic. Sci. 93*, 453–474.
SCHARRER, B. (1952). Neurosecretion. XI. The effects of nerve section on the intercerebralis-cardiacum-allatum system of the insect *Leucophaea maderae. Biol. Bull., 102*, 261–272.
SCHARRER, B. (1963). Neurosecretion. XIII. The ultrastructure of the corpus cardiacum of the insect *Leucophaea maderae. Z. Zellforsch. 60*, 761–796.
SCHARRER, B. and HADORN, E. (1938). The structure of the ring gland (corpus allatum) in normal and lethal larvae of *Drosophila. Proc. Nat. Acad. Sci. Wash. 24*, 236–247.
SCHARRER, B. and SCHARRER, E. (1944). Neurosecretion. VI. A comparison between the intercerebralis-cardiacum-allatum system of insects and the hypothalamo-hypophysical system of vertebrates. *Biol. Bull. 87*, 242–251.
SCHMIALEK, P. (1961). Die Identifizierung zweier im Tenebriokot und in Hefe vorkommender Substanzen mit Juvenilhormonewirkung. *Z. Naturf. 16b*, 461–464.
SCHOONHOVEN, L. M. (1962). Diapause and the physiology of host-parasite synchronization in *Bupalus piniarius* L. (Geometridae) and *Eucarcelia rutilla* Vill. (Tachinidae). *Arch. Néer. Zool. 15*, 111–174.
SNODGRASS, R. E. (1935). *Principles of Insect Morphology*. New York.
STAAL, G. B. (1967a). Plants as a source of insect hormones. *Koninkl. Nederl. Akad. Wetensch. 70*, 409–418.
STAAL, G. B. (1967b). Endocrine aspects of larval development in insects. *J. Endocr. 37*, xiii–xiv.
STEELE, J. E. (1963). The site of action of insect hyperglycemic hormone. *Gen. Comp. Endocr. 3*, 46–52.
STRICKLAND, E. H. (1911). Some parasites of *Simulium* larvae and their effects on the development of the host. *Biol. Bull., 21*, 302–338.
TELFER, W. H. and WILLIAMS, C. M. (1960). The effects of diapause, development, and injury on the incorporation of radioactive glycine into the blood proteins of the cecropia silkworm, *J. Insect Physiol. 5*, 61–72.
THOMSEN, E. (1942). An experimental and antomical study of the corpus allatum in the blowfly *Calliphora erythrocephala* Meig. *Vidensk. Medd. Dansk. Naturh. Foren. Kbh. 106*, 319–405.
THOMSEN, E. (1952). Functional significance of the neurosecretory brain cells and the corpus cardiacum in the female blowfly *Calliphora erythrocephala* Meig. *J. Exp. Biol. 29*, 137–172.
THOMSEN, E. (1954). Studies on the transport of neurosecretory material in *Calliphora erythrocephala* by means of ligaturing experiments. *J. Exp. Biol., 31*, 322–330.
THOMSEN, E. and MÖLLER, I. (1959). Neurosecretion and intestinal protease activity in an insect, *Calliphora erythrocephala* Meig. *Nature, 183*, 1401–1402.
THOMSEN, M. (1943). Effect of corpus cardiacum and other insect organs on the colour change of the shrimp *Leander adspersus. Biol. Medd. Kbh. 19*, 1–38.
TOYAMA, K. (1902). Contributions to the study of silkworms. i. On the embryology of the silkworm, *Bull, Coll. Agric. Tokyo, 5*, 73–117.
TREHERNE, J. E. (1966). The *Neurochemistry of Arthropods*. Cambridge University Press.

Umeya, Y. (1926). Experiments on ovarian transplantations and blood transfusion in silkworm with special reference to the alternation of voltinism. *Bull. Ser. Exp. Sta.* Chosen, *1*, 1–26.

Van der Kloot, W. G. (1955). The control of neurosecretion and diapause by physiological changes in the brain of the cecropia silkworm, *Biol. Bull.*, *109*, 276–294.

Verson, E. and Bisson, E. (1891). Cellule glandulari ipostigmatiche nel *Bombyx mori. Bull. Soc. Ent. Ital. 23*, 3–20.

Villee, C. A. (1942). The phenomenon of homoeosis. *Amer. Nat. 76*, 494–506.

Vogt, M. (1943). Zur Produktion und Bedeutung metamorphosefördernder Hormone während der Larvenentwicklung von *Drosophila. Biol. Zbl. 63*, 395–446.

Waku, Y. (1960). Studies on the hibernation and diapause in insects. *Sci. Rep. Tôhoku Univ. 26*, 327–352.

Wells, M. J. (1954). The thoracic glands of Hemiptera Heteroptera, *Quart. J. Mic. Sci.*, *95*, 231–244.

Wheeler, W. M. (1893). A contribution to insect embryology. *J. Morph. 8*, 1–160.

Wigglesworth, V. B. (1934). The physiology of ecdysis in *Rhodnius prolixus* (Hemiptera). II. Factors controlling moulting and "metamorphosis". *Quart. J. Mic. Sci.*, *77*, 191–222.

Wigglesworth, V. B. (1936). The function of the corpus allatum in the growth and reproduction of *Rhodnius prolixus* (Hemiptera). *Quart. J. Mic. Sci. 79*, 91–121.

Wigglesworth, V. B. (1937). Wound healing in an insect (*Rhodnius prolixus* Hemiptera). *J. Exp. Biol. 14*, 364–381.

Wigglesworth, V. B. (1938). The absorption of fluid from the tracheal system of mosquito larvae at hatching and moulting. *J. Exp. Biol. 15*, 248–254.

Wigglesworth, V. B. (1939). Häutung bei Imagines von Wanzen. *Naturwissenschaften, 27*, 301.

Wigglesworth, V. B. (1940a). Local and general factors in the development of "pattern" in *Rhodnius prolixus* (Hemiptera). *J. Exp. Biol. 17*, 180–200.

Wigglesworth, V. B. (1940b). The determination of characters at metamorphosis in *Rhodnius prolixus* (Hemiptera). *J. Exp. Biol. 17*, 201–222.

Wigglesworth, V. B. (1948). The function of the corpus allatum in *Rhodnius prolixus* (Hemiptera). *J. Exp. Biol. 25*, 1-14.

Wigglesworth, V. B. (1952a). The thoracic gland in *Rhodnius prolixus* (Hemiptera) and its role in moulting. *J. Exp. Biol. 29*, 561–570.

Wigglesworth, V. B. (1952b). Hormone balance and the control of metamorphosis in *Rhodnius prolixus* (Hemiptera). *J. Exp. Biol. 29*, 620–631.

Wigglesworth, V. B. (1954a). The *Physiology of Insect Metamorphosis*. Cambridge University Press.

Wigglesworth, V. B. (1954b). Secretion of juvenile hormone by the corpus allatum of *Calliphora. Nature. 174*, 556.

Wigglesworth, V. B. (1955a). The breakdown of the thoracic gland in the adult insect, *Rhodnius prolixus. J. Exp. Biol. 32*, 485–491.

Wigglesworth, V. B. (1955b). The role of the haemocytes in the growth and moulting of an insect, *Rhodnius prolixus* (Hemiptera). *J. Exp. Biol. 32*, 649–663.

Wigglesworth, V. B. (1957). The action of growth hormones in insects. *Symp. Soc. Exp. Biol. 11*, 204–227.

Wigglesworth, V. B. (1958a). Some methods for assaying extracts of the juvenile hormone in insects. *J. Insect Physiol. 2*, 73–84.

Wigglesworth, V. B. (1958b). The distribution of esterase in the nervous system and other tissues of the insect *Rhodnius prolixus. Quart. J. Mic. Sci. 99*, 441–450.

Wigglesworth, V. B. (1961a). Insect Polymorphism — a tentative synthesis. *Symp. R. Ent. Soc. Lond. 1*, 104–113.

Wigglesworth, V. B. (1961b). Some observations on the juvenile hormone effect of farnesol in *Rhodnius prolixus* Stål. (Hemiptera). *J. Insect Physiol. 7*, 73–78.

Wigglesworth, V. B. (1963a). The action of moulting hormone and juvenile hormone at the cellular level in *Rhodnius prolixus. J. Exp. Biol. 40*, 231–245.

Wigglesworth, V. B. (1963b). The juvenile hormone effect of farnesol and some related compounds: quantitative experiments. *J. Insect Physiol. 9*, 105–119.

Wigglesworth, V. B. (1969). Chemical structure and juvenile hormone activity: comparative tests on *Rhodnius prolixus. J. Insect Physiol. 15*, 73–94.

Wigglesworth, V. B. (1973). Assays on *Rhodnius* for juvenile hormone activity. *J. Insect Physiol. 19*, 205–211.

Wilde, J. de (1954). Aspects of diapause in adult insects, with special regard to the Colorado beetle, *Leptinotarsa decemlineata* Say. *Arch.-Néerl. Zool. 10*, 375–385.

Williams, C. M. (1946). Physiology of insect diapause: I. the role of the brain in the production and termination of pupal dormancy in the giant silkworm *Platysamia cecropia. Biol. Bull.*, *90*, 234–243.

Williams, C. M. (1947). Physiology of insect diapause. II. Interaction between the pupal brain and prothoracic glands in the metamorphosis of the giant silkworm, *Platysamia cecropia. Biol. Bull.*, *93*, 89–98.

Williams, C. M. (1948). Physiology of insect diapause. III. The prothoracic glands in the cecropia silkworm, with special reference to their significance in embryonic and postembryonic development. *Biol. Bull.*, *94*, 60–65.

Williams, C. M. (1951). Biochemical mechanisms in insect growth and metamorphosis. *Fed. Proc. 10*, 546–552.

Williams, C. M. (1952). Physiology of insect diapause. IV. The brain and prothoracic glands as an endocrine system in the cecropia silkworm. *Biol. Bull.*, *103*, 120–138.

Williams, C. M. (1956). The juvenile hormone of insects. *Nature, 178*, 212–213.

Williams, C. M. (1959). The juvenile hormone. I. Endocrine activity of the corpora allata of the adult *Cecropia* silkworm. *Biol. Bull.*, *116*, 323–338.

Zwicky, K. and Wigglesworth, V. B. (1956). The course of oxygen consumption during the moulting cycle of *Rhodnius prolixus* Stål. (Hemiptera). *Proc. R. Ent. Lond.* A. *31*, 153–160.

2 Structure of Endocrine Glands

BONNIE JOY SEDLAK

University of California, Irvine, California, USA

INTRODUCTION

The cellular morphology of insect glands has been studied with the intention of learning the structural basis of function. Specifically, the goal of most studies has been to visually define the organelles involved with the production and/or secretion of hormones.

This review will present the ultrastructure of the corpora cardiaca (CC), corpora allata (CA) and the prothoracic glands (PG). The corpus cardiacum is both a neurohemal storage site for numerous brain hormones and the site of production of its own product(s) from its intrinsic cells. Among the hormones reportedly contained within these glands are a cardioaccelerator hormone, hyperglycemic factor and adipokinetic hormone (Goldsworthy, G. and Mordue, W. 1974). The corpus allatum is the site of juvenile hormone synthesis and release, and in at least one insect the CA are the storage sites of the brain peptide, prothoracicotropic hormone (PTTH). PTTH is the hormone which stimulates the PG to synthesize its steroid products. Although the PG may be the sole or major source of the prohormone (ecdysone) other cells have also been implicated in ecdysone production (e.g. oneocytes).

This review will concentrate on the cellular aspects of gland structure since previous reviews have very adequately summarized the gross morphology and histology of the glands (e.g. see Herman, W. 1967; Cassier, P. 1979).

1 CORPORA CARDIACA

1.1 General structure

The CC are composed of neural, glandular and glial elements. The extrinsic cells (neural) store and secrete neurohormones which are produced either in the brain, or in ganglionic cell bodies. The nervi corporis cardiaci (NCC I and NCC II) extend from the brain to the CC and carry the hormonal products of neurosecretory cells (NSC) to the extrinsic cells of the CC and to target areas beyond the CC (e.g. corpora allata; aorta). The intrinsic cells (glandular) have their cell bodies located within the CC itself and produce their own secretions.

Experimental evidence supporting the idea that the CC are neurohemal organs came from severing the nervi corporis cardiaci of *Leucophaea maderae* (Scharrer, B. 1952). Following a unilateral cut at the level where the NCC nerve enters the CC, there was an accumulation of neurosecretory material proximal to the cut; distal to the cut (within the CC) the neurosecretory material was depleted. This showed that NSC material is transported down the NCC nerves from the brain and stored within the CC.

At the cellular level the predominant feature of the CC are the NSC granules contained within the intrinsic cell bodies and axons of brain NSC. Neurosecretory neurons are nerve cells which are capable of generating an action potential but in addition produce specific chemical mediators (see chapter 3, this volume). In the NSC it is the glandular function which stands above all other neuronal attributes. The chemical nature and staining characteristics of the products carried within NSC axons identify the fibers as either A or B types (Highnam, K. 1961). The A type fibers are peptidergic in nature and carry NSC granules of 100–300 nm size ranges. The B type fibers, on the other hand, are aminergic and produce monoamines (such as noradrenaline, serotonin, and dopamine) stored within smaller granules. The proteinaceous NSC material of A type axons stains with aldehyde fuchsin and identifies these axons as belonging to the classical neurosecretory neuron (Highnam, K. 1961).

1.2 Extrinsic cells

The extrinsic cells of the CC usually contain axons with NSC granules of various sizes and electron densities (Fig. 1). Neither the size of the granule nor the degree to which it stains with heavy metals (i.e. lead and/or uranium) have been particularly helpful in identifying whether a particular axon carries a specific hormone because the electron density of NSC granules can depend on the fixation procedures employed. Nordmann, J. (1977) showed that granules have a variety of electron densities if fixed at pH 7.0 but if the same tissue is fixed at pH 5.0 or 6.0 all the granules appear electron-dense. In spite of this variability in granule morphology, the number of axon types have been identified on the basis of the size and staining density of NSC granules which are carried within each fiber. Only one axon type has been found in the following CC: *Rhodnius prolixus*, where the granules are 150–200 nm in diameter and appear electron-dense (Beaulaton, J. 1967); *Calliphora stygia* with electron-dense granules 60–140 nm in diameter (Johnson, B. 1966) and *Calliphora erythrocephala* with 100–300 nm granules.

There are two axon types in both the blatterian species studied; in *Leucophaea maderae* the granules range from 90 to 150 nm and from 200 to 280 nm (Scharrer, B. 1963); in *Periplaneta americana* both types of axons contain electron-dense granules of either 150 nm or 300 nm (Scharrer, B. and Kater, S. 1969). Two axon types were also found in *Carausius morosus*; one contained 100–300 nm granules which were electron-lucent, the other contained 100–150 nm electron-opaque granules. The latter class of axons were much less numerous than the electron-lucent variety (Smith, U. and Smith, D. 1966). In *Hypera postica* the two axon types contained 60–220 nm granules which varied from electron-dense to opaque in their staining intensity and another class of clear granules 10–20 nm (Tombes, A. and Smith, D. 1970).

Three axon types are found in the orthopterans studied. *Leptinotarsa decemlineata* have 120 nm opaque granules, 150 nm gray granules and small pale granules (Schooneveld, H. 1979). In *Schistocerca gregaria* two types of granules are electron-dense and either 200 nm or 300 nm in size; the third type is found infrequently and contains electron-dense 100–140 nm granules (Prentø, P. 1972). There are also three axon types found in *Myzus persicae*; with either 140–180 nm electron-dense granules, or 90–110 nm electron-dense and 200–300 nm electron-lucent granules associated with synaptic-like vesicles (Bowers, B. and Johnson, B. 1966). In *Tyria jacobeae* the three types of axon are classified as small and electron-dense, large and electron-dense, and large semi-opaque granules; the sizes of granule varied from 75 to 260 nm (Philogene, B. and Massalski, A. 1976).

There is a discrepancy reported for *Locusta migratoria migratoriodes*. Cassier, P. and Fain-Maurel, M. (1970) report three classes of axons: (1) 250 nm electron-opaque granules; (2) a mixed axon with both 250 nm electron-opaque plus 300 nm electron-dense granules within the same axon and (3) another mixed axon with 250–300 nm spherical, 200–650 nm ovoid and 350 nm electron-dense granules. In contrast Cazal, M. *et al*., (1971) report four classes of NSC granules, three of which are electron-dense and vary in size (60–90 nm; 100–200 nm; 200–350 nm) the fourth axon has 280 nm electron-lucent granules.

In *Aedes sollicitans* there are five types of axons: (1) 150–250 nm with mixed electron-dense and electron-lucent granules, (2) 50–100 nm granules with the same variation in staining morphology with a clear cytoplasm, and (3) 100–200 nm granules also in both electron-dense and lucent staining but with a dark cytoplasm. Two other classes are 100–200 nm electron-lucent with either a dark or an electron-lucent cytoplasm (Meola, S. and Lea, A. 1972).

1.3 Intrinsic cells

Intrinsic cells are seen in most CC studied and are distinguished from the extrinsic cells by the presence of such organelles as rough endoplasmic reticulum, Golgi complexes and dense bodies. These cells also contain NSC granules presumably synthesized within these cells (rather than being synthesized within the brain).

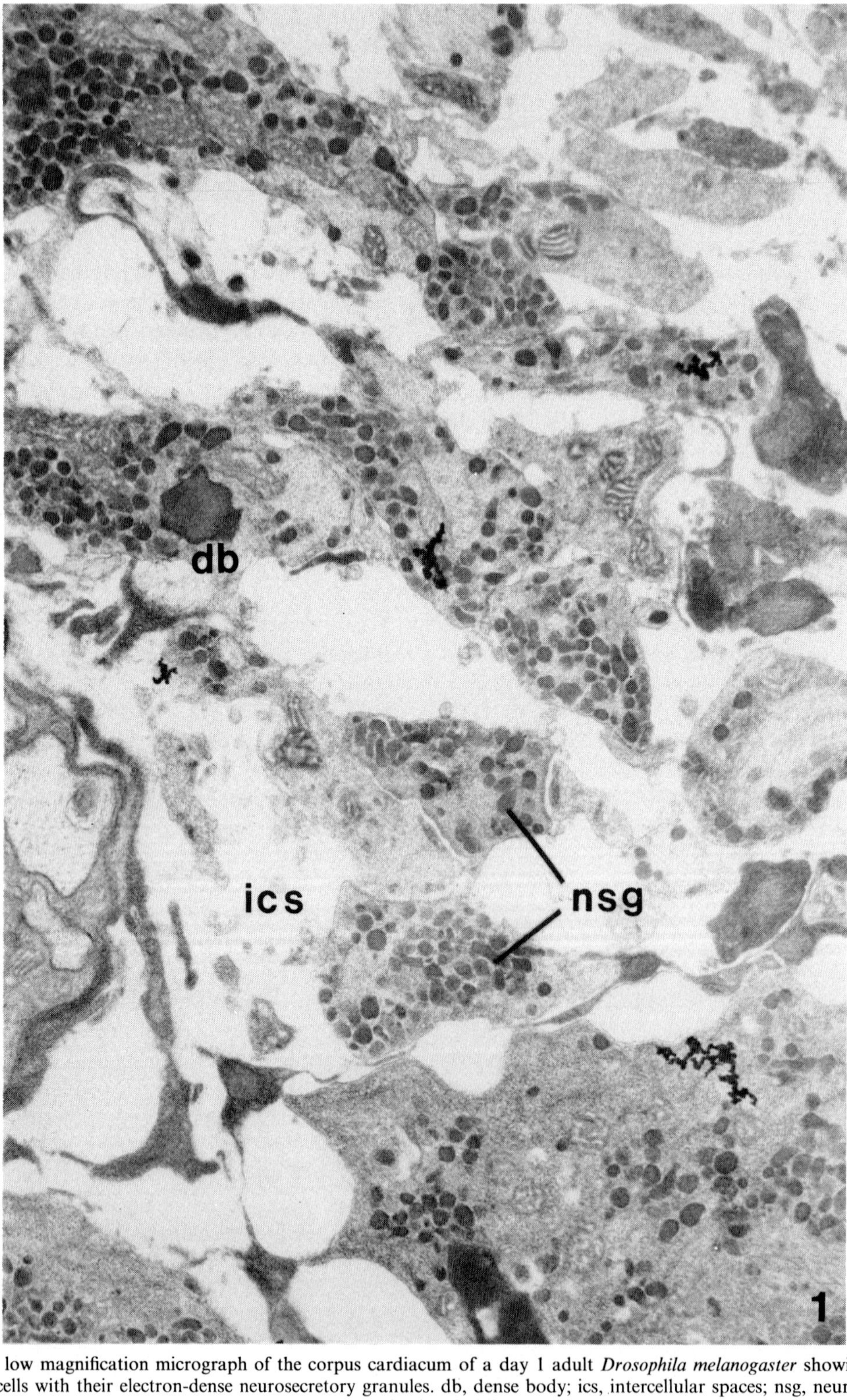

FIG. 1. A low magnification micrograph of the corpus cardiacum of a day 1 adult *Drosophila melanogaster* showing several extrinsic cells with their electron-dense neurosecretory granules. db, dense body; ics, intercellular spaces; nsg, neurosecretory granules (× 16,000).

There are lateral intrinsic cells found within *Drosophila melanogaster* CC (Aggarwal, S. and King, R. 1971). The granules within these cells were reported to be electron-lucent in this study but appear to be electron-dense in Fig. 2 which shows an adult CC of the same species (Sedlak, B. unpublished observations). The differences in fixation may account for the differences in staining intensity of these granules. In the Aggarwal and King study the glands were dissected in Ringers and then fixed in OsO_4 plus $K_2Cr_2O_7$, whereas Sedlak fixed glands directly in glutaraldehyde followed by postfixation in OsO_4. It is interesting that the CC of the female sterile mutant *fes* has very swollen rough endoplasmic reticulum in contrast to the wild type glands, indicating there was change in the exportable protein manufactured by these genetically altered cells (Aggarwal, S. and King, R. 1971).

Intrinsic cells were found in *Myzus persicae* which contain 150 nm granules (Bowers, B. and Johnson, B. 1966); *Carasuis morosus* have electron-opaque granules 100–300 nm (Smith, U. and Smith, D. 1966); *Calliphora stygia* have 100–300 nm granules (Johnson, B. 1966), and the cells of *Calliphora erythrocephala* have 100–300 nm granules (Norman, T. 1965); *Rhodnius prolixus* have 150–200 nm granules (Beaulaton, J. 1967); *Locusta migratoria migratoriodes* have 600–800 nm granules (Cassier, P. and Fain-Maurel, M. 1970; Cazal, M. *et al.*, 1971); *Periplaneta americana* have 300 nm granules (Willey, R. and Chapman, G. 1960); *Hypera postica* have 400 nm granules (Tombes, A. and Smith, D. 1970).

Raina, A. and Borg, T. (1980) studied *Pectinophora gossypiella* CC during larval diapause and found changes in the NSC granule population. Developing larvae have electron-dense (100 ± 17 nm) plus electron-lucent (125 ± 15 nm) granules and the electron-dense granules increased to 140 ± 30 nm during diapause; some electron-lucent granules (105 ± 20 nm) were also seen. At the termination of diapause the electron-dense granules increased in number along with organelles such as rough endoplasmic reticulum, Golgi and lipid, indicating an overall increase in synthesis at this time.

1.4 Characterization of neurosecretory granules

It would be useful if individual axons could be classified morphologically to identify the hormonal product carried within. Unfortunately (as mentioned above) the great variability seen in both the sizes of NSC granules and in their staining intensity has made such identification impossible. The most critical technique available for identification is antibody localization using cytological immunochemistry. Such localization has been used for vertebrate neuropeptides within the anterior pituitary (see Moriarty, G. 1973) but the techniques have not yet been applied to the insect's CC. Nonetheless, fluorescent histochemistry and autoradiography have been used with some success to probe structure–function relationships in the CC of *Schistocerca gregaria* (Lafon-Cazal, M. and Arluison, M. 1976).

Monoamines were localized within the CC of both *Schistocerca gregaria* and *Locusta migratoria migratoriodes* using a fluorescent histochemical labeling technique. Fluoresence was found within the neural (extrinsic) but not the glandular (intrinsic) region of the CC and differences in the color of the stain indicate that two neurons are present, one containing catecholamines (staining green) and the other indolalkylamines (staining yellow).

In addition [^{3}H]serotonin (5-hydroxytryptamine, 5HT) was introduced to CC both *in vivo* and *in vitro* and glands were processed for electron microscopic autoradiography. In the *in vivo* study grains were found within two types of axons in the neural lobe, one axon did not contain granules whereas the other axon had 100 nm electron-dense granules. There was also weak uptake of 5HT into 100 nm dense granules of the glandular lobe. In contrast to this, the *in vitro* experiments showed that only one type of axon containing labeled 100 nm dense granules, was located within the neural lobe. The grains were distributed equally over all cell components including the mitochondria, microtubules, 50 nm dense granules and 100 nm dense granules. It is argued that the 100 nm vesicles are the site of storage for 5HT and thus the presence of labeled 5HT distinguished the axons containing such vesicles as serotoninergic. It must be remembered, however, that the label was also found outside these membrane-bound areas (i.e. within the axonal cytoplasm). In addition, the 100 nm vesicles in the glandular axons were labeled but were not considered serotoninergic since the axons in this region

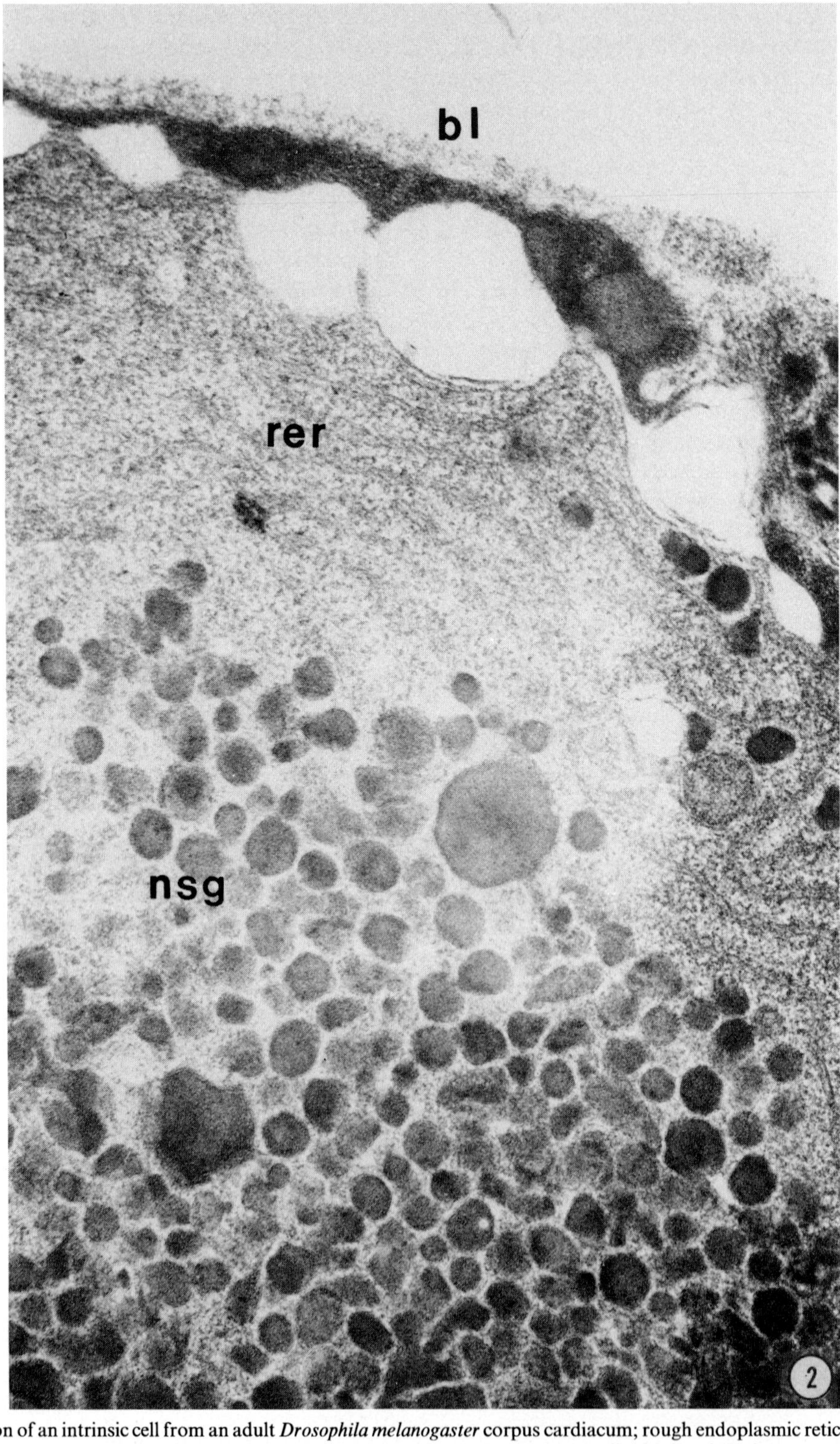

FIG. 2 Portion of an intrinsic cell from an adult *Drosophila melanogaster* corpus cardiacum; rough endoplasmic reticulum is found in the peripheral region of the cells and neurosecretory granules are clustered centrally. bl, basal lamina; nsg, neurosecretory granules; rer, rough endoplasmic reticulum (× 45,000).

did not fluoresce to indicate the presence of catecholamines.

1.5 Mode of release

The mode of release of NSC material has been given considerable attention. The two most commonly recorded features associated with release at the cellular level are exocytosis figures and the clusters of electron-lucent synaptic-like vesicles which are called synaptoids.

1.5.1 Exocytosis

Areas of the plasma membrane in CC cells invaginate in a characteristic omega shape which has been interpreted as the morphological indicator of a recently exocytosed NSC granule; very few granules are actually caught in the process of exocytosis. The first evidence for exocytosis came from studies of CC in *Calliphora erythrocephala* (Norman, T. 1965) and *Carasius morosus* (Smith, U. Smith, D. 1966).

Stimulation increases the number of exocytosis figures seen in the CC of several species. For example, electrical stimulation of *Calliphora erythrocephala* CC resulted in a significantly greater number of exocytosis figures when compared with unstimulated glands (Norman, T. 1969). Similarly, the omega figures increased in electrically stimulated CC of *Schistocerca gregaria* versus control CC. In the same species acetylcholine also stimulated exocytosis of NSC granules, but the exocytosis figures were not seen as frequently as with electrical stimulation (Krogh, I. and Normann, T. 1977). When CC of *Leucophaea maderae* were stimulated by serotonin the exocytosis of material took place within a very short time (within 3 min; Fig. 3). The exocytosis figures seen were actually smaller than those in unstimulated glands. Since there were very few omega figures seen in unstimulated CC, it is unlikely that release of all material occurs via this process.

Tannic acid plus glutaraldehyde were used to enhance the contrast of material in the process of exocytosis. The combined fixation is thought to complex with basic proteins, glycoproteins and polypeptides and then chelate heavy metal ions such as OsO_4 and lead. Although the method successfully enhanced contrast, and thus the ability to easily recognize exocytosis figures in freshwater snail and rat neurosecretory axons, similar exocytosis figures were not readily seen in *Locusta migratoria migratoriodes* CC axons (Roubos, E. and Vander Wal-Divendal, R. 1980). The authors of this report agree with Scharrer, B. and Wurzelmann, S. (1974, 1978) who proposed that exocytosis was not seen because the process occurred too rapidly to visualize with consistency.

1.5.2 Synaptoids

Synaptoids are presumed to be the release site within NSC axons. These areas resemble presynaptic regions of non-NSC axons since they contain clusters of small (30–50 nm) electron-lucent vesicles and often a local electron-dense thickening of the plasma membrane (Scharrer, B. 1968). Such synaptoids occur near the circulatory system and in areas where the axon abuts on its target tissue (e.g. the corpora allata; Scharrer, B. 1963). Figure 4 shows synaptoids within an axonal ending at the CA of *Manduca sexta* (Sedlak, B. 1981).

After electrical stimulation of *Periplaneta americana* CC there was not a significant decrease in the amount of NSC material contained within the axons of the glands (Scharrer, B. and Kater, S. 1969). It was assumed that the gland stored vast excesses of NSC material so that a depletion of product was not obvious. There was, however, an increase in the synaptoids associated with cardioaccelerator hormone release. In order to preserve the ultrastructural integrity for this study, the glands were not directly stimulated. Instead the nerves which had previously been determined to be the efferent stimulatory pathway for cardioaccelerator hormone were electrically stimulated (Kater, S. 1968).

Scharrer, B. and Wurzelmann, S. (1974) proposed that the microvesicles represent the endocytotic recapture of membrane exocytosed along with NSC granules. Retrieval of membrane was demonstrated via the uptake of an exogenously applied tracer (horseradish peroxidase) at the synaptoid site. Although a mixture of zinc-iodide–OsO_4, which stains synaptic vesicles, was also found to stain synaptoids this method was not considered to give unequivocal evidence that the synaptoid sites contained neurotransmitters.

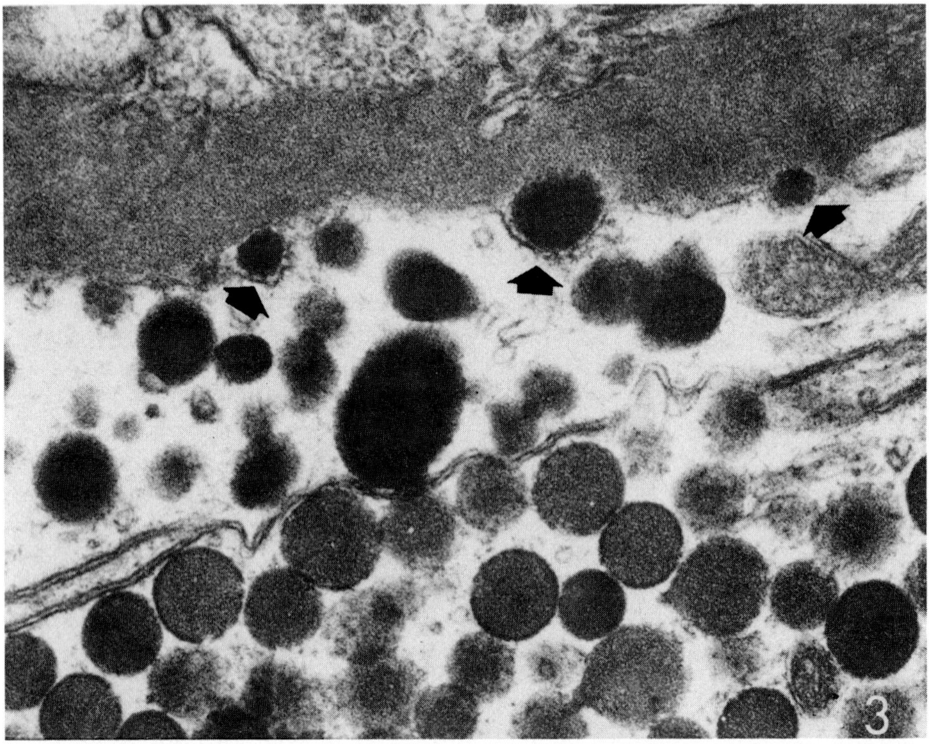

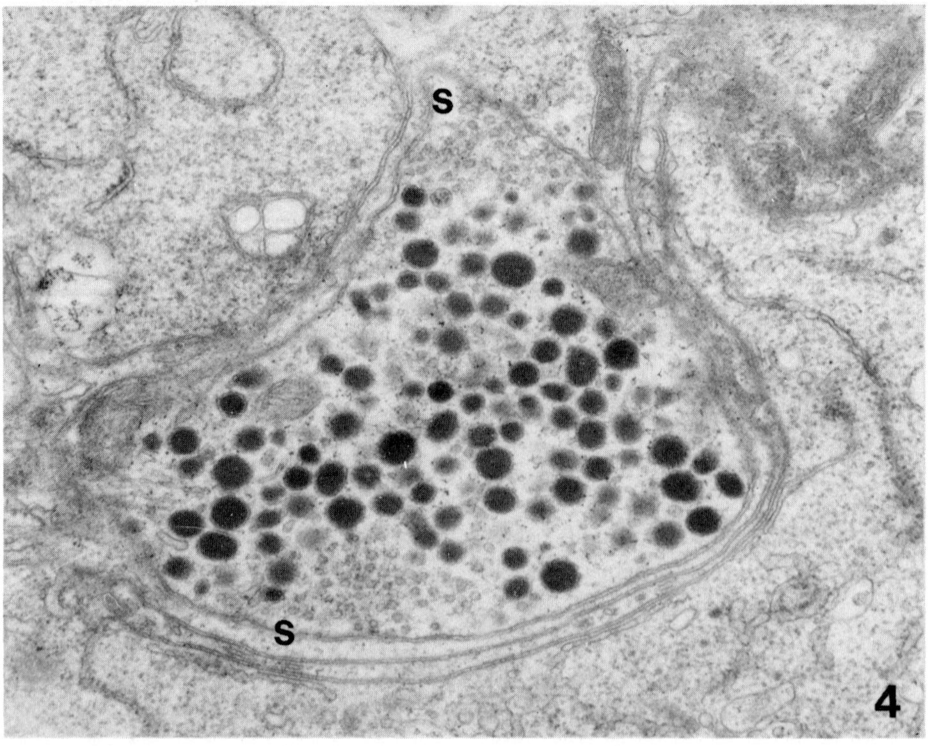

FIG. 3. Part of neurosecretory axon in corpus cardiacum of *Leucophaea maderae* showing three exocytotic figures (arrows) in tandem, demonstrating release of peptidergic granules induced by the administration of serotonin. From Scharrer, B. and Wurzelman, S. (1978). FIG. 4. Clusters of 30–55 nm electron-lucent vesicles mark two synaptoid release sites. This particular axon terminates within the corpus allatum of *Manduca sexta*. S, synaptolid. From Sedlak, B. (1981).

Thus, it has been accepted that synaptoids indicate sites of release of neurohormone, but it has not yet been established whether these microvesicles are morphological indicators of the retrieval of plasma membrane exocytosed along with its granule or whether the vesicles contain a specific substance such as a neurotransmitter. Norman, T. (1969) proposed that the microvesicles formed are only one part of the overall exocytosis mechanism since the microvesicles formed following the release of a granule represent the fragmentation of membrane which once constituted the omega figure. These vesicles appear only transiently and may eventually be autodigested by the lysosomal system.

1.6 Summary of corpora cardiaca ultrastructure

The CC have two main cellular components which contain NSC granules: the extrinsic cells storing brain hormones and the intrinsic cells storing products of their own making. Axon types have been classified on the basis of the size and staining density of these granules even though the morphology of granules may differ with the fixative used (Nordmann, J. 1977). Secretory activity within the CC has been visualized as the exocytosis of the granule and by the presence of synaptoids. Whether or not these microvesicles represent membrane retrieval following the release of the granule has not been determined with certainty, but it can be stated conservatively that the synaptoids do represent release sites of a hormone.

2 CORPORA ALLATA

2.1 General structure

Although the cellular structure of the CA is seen to be highly variable if comparisons are made between species, some generalizations can be made about gland composition. The glands are made up of closely interdigitated cells, and in many species the gland region is surrounded by a stromal sheath. This sheath area is of a variable diameter ranging from a thin basal lamina to a lamelliated layer within which may be strewn tracheoles and NSC axons. These NSC axons ending within the CA have their cell bodies residing within the brain and the axons traverse the CC before reaching the CA.

CA are usually rich in mitochondria and have a predominance of smooth endoplasmic reticulum, rough endoplasmic reticulum, or both. Other organelles present: nucleus, ribosomes, Golgi complexes, lysosomes and various inclusions, all appear to vary in their presence and individual morphology for each species. Each of these organelles will be discussed in detail but I will first consider the problems of determining gland activity since the ultimate goal of most studies is to define the structures which are present at times of high versus low levels of secretory activity.

Problems in correlating ultrastructural morphology with the production and release of JH arise because it is difficult to assess physiological activity in many insects. The most successful attempts in establishing the structural basis for endocrine activity came from studies of adult CA. In the adults of many species the JH serves as a reproductive hormone necessary for the uptake of vitellogenin (yolk protein) into maturing occytes (see chapter 6, volume 8). In females which undergo cyclic egg production it is possible to demonstrate peaks of gland activity via an ovarian bioassay. That is, the CA are considered to be actively secreting JH when the oocytes are increasing in volume via vitellogenic uptake. The CA is thought to be inactive during quiescent periods of the reproductive cycle, e.g. virgin females or pregnant females (Scharrer, B. 1964a). Alternatively, the activity of some CA were assessed by comparing reproductive female glands with male glands (Fain-Maurel, M. and Cassier, P. 1969). Another assessment of activity was made between the CA of wild type and female sterile mutants of *Drosophila melanogaster* (King, R. *et al.*, 1966).

Assessing activity at developmental stages other than the adult is more difficult because of the absence of a criteria for active hormone production. Exceptions to this generalization are those studies which compare glands from animals in diapause or aged adults (when the JH titer is presumed to be clearly at a minimal level) to other stages such as larval or young adults (which should have a higher titer of hormone) (Waku, Y. and Gilbert, L. 1964; Panov, A. and Bassurmanova, O. 1970; Takeda, N. 1977; Yin, C. and Chippendale, G. 1979). Additional exceptions are the studies which report the ultrastructure of developing larval glands in the context of the actual JH titers as established by radioimmunoassay, e.g. *Diploptera punctata* (Szibbo, C. and Tobe, S. 1981), or mass spectroscopy, e.g. *Manduca sexta* (Sedlak, B. *et al.*, 1983). In some studies the nutritional state has been used to assess activity of the glands; Thomsen, E. and Thomsen, M. (1970) showed that feeding sugar to adult females inactivated the CA.

Before reviewing the status of each organelle, a few general morphological observations will be discussed. Specifically the size changes in glands and the presence of two populations of cells (light versus dark) will be considered from the point of view of secretory activity.

2.2 Volume changes

Many authors have studied the changes in CA volume and attempted to correlate such alterations in volume with fluctuations in glandular activity. The volume of the CA increases in glands assumed to be active on the basis of developmental age, e.g. adult *Nebria brevicollis* (Ganagarajah, G. 1965), larval *Drosophila melanogaster* (Aggarwal, S. and King, R. 1969), prediapause *Monema flavescens* (Takeda, N. 1977), and late larval *Choleva augustata* (Deleurance, S. and Charpin, S. 1978). Large glands have been assumed to be active on the basis of reproductive activity in *Leucophaea maderae* (Scharrer, B. 1964a), (where CA increase in size following gonadectomy; Scharrer, B. and von Harnack, M. 1961), *Galleria tanaceti* (Siew, Y. 1965), *Schistocerca gregaria* (Odhiambo, T. 1966a), *Eurygaster intergryceps* (Panov, A. and Bassurmanova, O. 1970) and *Leptinotarsa decemlineata* (Schooneveld, H. 1979).

Similarly decreases in gland volume occur in glands assumed to be inactive where a lack of secretory activity is based on developmental age, e.g. diapause *Hypera postica* (Tombes, A. and Smith, D. 1970) have smaller glands versus the glands of reproductive females of this species. If *Caliphora erythrocephala* are fed a sugar diet the animal is considered to be physiologically inactive (Thomsen, E. and Thomsen, M. 1970); such enforced inactivation results in the presence of small glands. However, Lea, A. and Thomsen, E. (1969) report that there was no size correlation with secretory activity in the glands of the same species. In *Periplaneta americana* the CA volume decreases when NSC granules reach their peak accumulation (Khan, T. *et al.*, 1978).

At the cellular level CA volume changes are manifest as differences in the interdigitation between cells. Large cells rich in cytoplasm have a smoother contour in comparison with cells in which the cytoplasm is reduced (Fig. 5). The plasma membranes are consequently highly convoluted in the smaller cells. Convoluted plasma membranes are seen in adult *Aphis craccivora* (Elliot, H. 1976), diapausing females of *Hypera postica* (Tombes, A. and Smith, D. 1970), diapausing *Pectinophora gossypiella* (Raina, A. and Borg, T. 1980), and in the pupal versus larval stages of *Manduca sexta* (Sedlak, B. *et al.*, 1983). The CA in all of these stages are considered to be relatively inactive and in *Manduca sexta* where the JH titer has actually been determined (Bergot, B. *et al.*, 1984), the pupal stages indeed have very little JH. Similarly, the CA volume increases in preovulatory females of *Leptinotarsa decemlineata* and decreases in prediapause animals (Schooneveld, H. 1979); these changes correlate with alterations in JH synthesis (Kramer, S. 1978). In *Nauphoeta cinerea* the CA volume increases at the same time as there is an increase in size of oocytes, and at the time the JH titer rises (as assayed via the *Galleria* wax test; Lanzrein, B. *et al.*, 1978).

It is not clear, however, that volume changes accurately reflect hormone synthesis in all insects. For example in *Schistocerca gregaria* (Odhiambo, T. 1966a) the CA increases in adult males but no cyclic volume changes occur in the glands of reproductive females where fluctuations in JH would be expected. Mordue, W. (1965) reports that volume changes are not a criterion of activity during the pupal to adult development of *Tenebrio molitor*. Similarly in larval *Locusta migratoria migratoriodes* (Johnson, R. and Hill, S. 1973) the increase in gland volume does not correlate with JH activity.

A significant contribution to understanding how changes in volume reflect gland activity is the study by Szibbo, C. and Tobe, S. (1981). The authors demonstrated in *Diploptera punctata* that the increase in CA volume occurs at the time of peak gonadotropic activity and at a time when the JH titer is estimated to be at its maximum. Using autoradiography Szibbo and Tobe showed that the

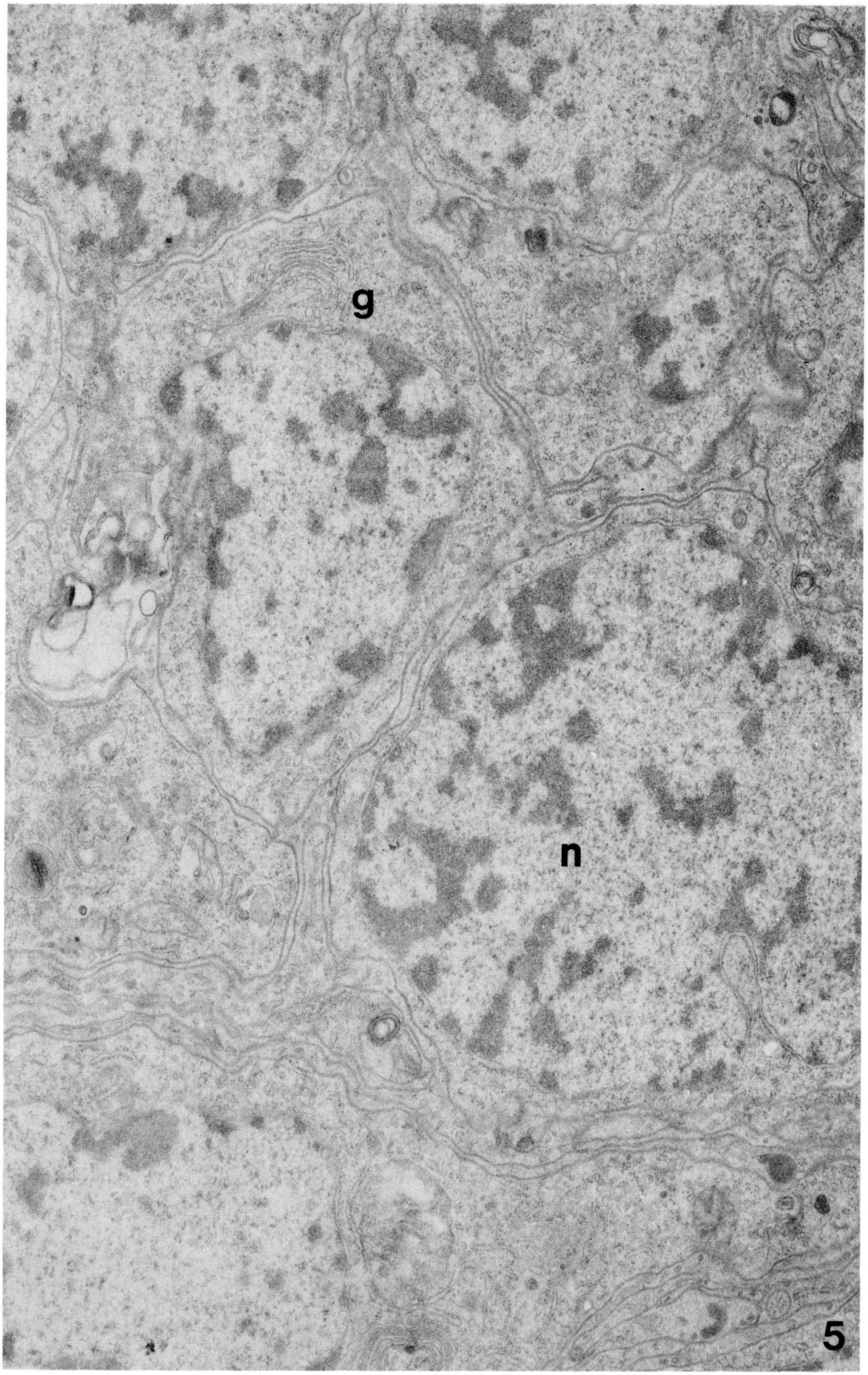

FIG. 5. A low magnification micrograph of corpus allatum cells from adult *Oncopeltus fasciatus* shows that adjacent cells interdigitate. g, Golgi complex; n, nucleus (× 8000).

CA volume increases as a result of cell divisions which temporally precede the attainment of maximal gland volume. Scharrer, B. and von Harnack, M. (1958) also reported mitoses in the adult CA of *Leucophaea maderae*.

2.3 Light versus dark cells

Light and dark cells have been seen in the adult CA of *Calliphora erythrocephala* (Thomsen, E. and Thomsen, M. 1970), *Oncopeltus fasciatus* (Dorn, A. 1973; Liechty, L. and Sedlak, B. 1978), *Locusta migratoria migratoriodes* (Schooneveld, H. 1979), *Choleva augustata* (Deleurance, S. and Charpin, S. 1978). This apparent difference in density of the cells actually reflects a difference in cellular content in *Oncopeltus*. Figure 6 shows that the light cells in this insect have more vesicles, and dark cells have a denser packing of rough endoplasmic reticulum and free ribosomes (Liechty, L. and Sedlak, B. 1978). Dorn, A. (1973) suggests that the light cells of *Oncopeltus* are active, based upon the fact that the light cells have more cytoplasm and larger nuclei; the dark cells are, by the process of elimination, the inactive cell population. On the other hand, Liechty, L. and Sedlak, B. (1978) propose that the light cells are inactive since these predominate in drug-inactivated glands (see discussion of precocene-treated CA). Whether this difference in electron-density between cells is physiologically significant, or whether the difference is only the result of fixation artifacts, is not entirely clear since the epidermis and other epithelial tissues also stain differentially in many insects.

2.4 Nucleus

Some authors report changes in the nuclei of CA. In the CA of *Diatraea grandiosella* (Yin, C. and Chippendale, G. 1979) the nuclear profile alters from a smooth contour in active glands (pre- to mid-diapausing insects) to a convoluted outline in inactive glands (late diapausing and non-diapausing insects) (see chapter 11, volume 8).

In *Hyphantrea cunea* (Melinkova, F. and Panov, A. 1975) the perinuclear space increases in active glands. In both *Schistocerca gregaria* (Odhiambo, T. 1966a) and *Drosophila melanogaster* (King, R. *et al.*, 1966) there is an apparent increase in nuclear volume with increasing age of the animal; such a nuclear increase does not occur in the female sterile mutant (*fes*) in which the CA are assumed to be inactivated by a genetic mutation.

Nucleoli are noted to fragment in the inactive adult glands of *Aphis craccivora* (Elliot, H. 1976) and change from compact nucleoli (in diapausing females) to "hollow" nucleoli (in active females) of *Eurygaster intergriceps* (Panov, A. and Bassurmanova, O. 1970). There is chromosome degeneration seen in the senescent adults of *Hyalophora cecropia* (Waku, Y. and Gilbert, L. 1964). These degenerative changes in the nucleoli and chromatin presumably represent a stage in gland degeneration.

There is not sufficient data or consistency of data between species to warrant any generalizations about the meaning of these changes in nuclear morphology.

2.5 Endoplasmic reticulum

The greatest variation in any cellular structure is found in the endoplasmic reticulum. When the presence of this membranous organelle has been interpreted in the context of secretory activity it becomes apparent that both the smooth endoplasmic reticulum (SER) and the rough endoplasmic reticulum (RER) are found in glands assumed to be active. Thus, the mere presence of either form of endoplasmic reticulum is still no indication of gland activity.

2.5.1 Rough endoplasmic reticulum

RER is specifically involved in producing exportable protein and the participation of RER in JH synthesis could be indirect, i.e. to provide, for example, a carrier protein for the hormone.

RER increases in active glands in some insects; e.g. preovulatory females of *Leucophaea maderae* (Scharrer, B. 1964a), active females of *Eurygaster intergriceps* (Panov, A. and Bassurmanova, O. 1970), adult *Rhodnius prolixus* (Baehr, J. *et al.*, 1973), fifth instar *Oncopeltus fasciatus* (Dorn, A. 1973) and previtellogenic females of *Folsomia candida* (Palevody, C. and Grimal, A. 1976). In larval *Choleva augustata* (Charpin, P. 1975), in the light cells (presumed to be active) of adult *Oncopeltus fasciatus* (Dorn, A. 1973), and female adults of

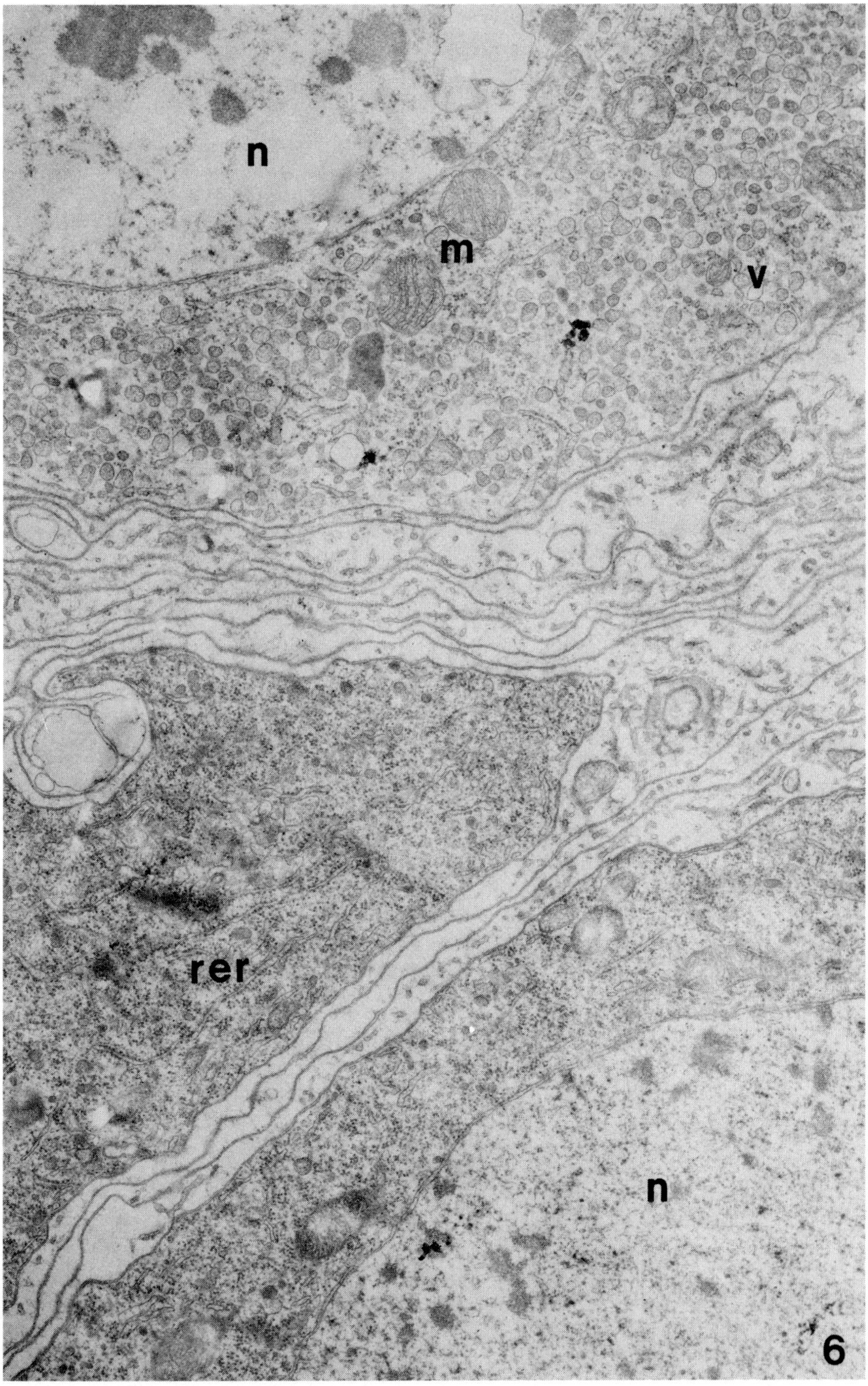

FIG. 6. Two dark and one light cell from a corpus allatum of adult *Oncopeltus fasciatus*. This illustration shows the differences in organelle content in each cell type. Dark cells are rich in free ribosomes and rough endoplasmic reticulum, whereas abundant vesicles are predominant in the light cells. n, nucleus; m, mitochondrion; rer, rough endoplasmic reticulum; v, vesicle (× 18,000).

Leucophaea maderae (Scharrer, B. 1964a) the RER is arranged in a whorled configuration, with parallel cisternae stacked concentrically. Decreases in the quantity of RER are observed in senescent adult *Hyalophora cecropia* (Waku, Y. and Gilbert, L. 1964) at the time when all metabolism would cease.

The polysome population which is responsible for synthesizing protein (e.g. enzymes) for internal use, also varies in some species and an increase in polysomes could reflect either an increase in general metabolism or specifically in hormone metabolism. Polyribosomes decrease with increasing age in adult *Drosophila melanogaster* (King, R. *et al.*, 1966) but increase in adult *Schistocerca gregaria* (Odhiambo, T. 1966).

2.5.2 Smooth Endoplasmic Reticulum

Since SER has been found as the predominant organelle in vertebrate steroid-secreting glands, i.e. testis (Christensen, A. 1965) and ovaries (Bjersing, L. 1967), by analogy this membranous organelle is the most likely candidate to be associated directly with production of the sesquiterpenoid juvenile hormone (chapter 10, this volume). Consistent with this prediction is the fact that SER is found in CA during the times when many insects are thought to be secreting JH. For example, smooth endoplasmic reticulum is found in the CA of larval *Drosophila melanogaster* (Aggarwal, S. and King, R. 1969), reproductive females of *Hypera postica* (Tombes, A. and Smith, D. 1970), previtellogenic females of *Folsomia candida* (Palevody, C. and Grimal, A. 1976), and diapausing *Monema flavescens* (Takeda, N. 1977). The SER increases at the periphery of glands in sexually mature *Schistocerca gregaria* (Odhiambo, T. 1966) and there is also an increase in the amount of SER of castrated females of *Leucophaea maderae* (Scharrer, B. 1975). Since castration in female cockroaches presumably removes a negative feedback regulation of the CA, these glands remain active. The SER which is found

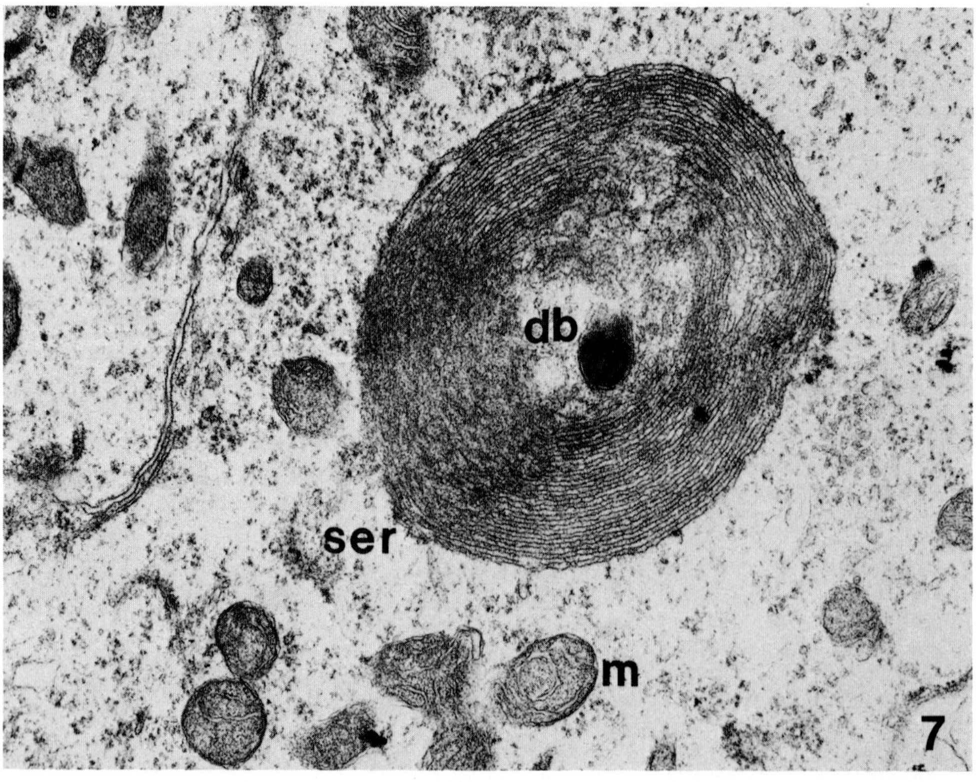

Fig. 7. The smooth endoplasmic reticulum in larval *Manduca sexta* corpus allatum is organized into concentric stacks of cisternae; a dense body is found in the central area of this stack. db, dense body; m, mitochondria; SER, smooth endoplasmic reticulum (× 32,000).

in the CA of non-diapausing *Pectinophora gossypiella* is associated with mitochondria at the termination of diapause. The SER breaks down at this time (Raina, A. and Borg, T. 1980).

SER is found arranged in distinctive whorls of concentric cisternae in active glands of *Diatraea grandiosella* (Yin, C. and Chippendale, G. 1979) larval *Choleva augustata* (Deleurance, S. and Charpin, S. 1978) and larval *Manduca sexta* (Sedlak, B. *et al.*, 1983). Figure 7 shows the concentric cisternae of SER in *Manduca sexta*. In many insects the SER decreases when the animal becomes physiologically inactive; presumably at a time when the hormone titer has decreased.

In *Choleva augustata* the SER fluctuates throughout larval life, being present in first and second instars and during the third instar until the time when the animal ceases feeding (the alimentation period). At this time when the insect is preparing for pupation the quantity of SER decreases. Deleurance, S. and Charpin, S. (1978) correlated the time of SER appearance with hormonal activity; it was assumed that the SER is the biosynthetic site of JH synthesis and that the SER fluctuations represent cycles of secretory activity in the CA. The SER whorls decrease in sugar-inactivated *Calliphora erythrocephala* (Thomsen, E. and Thomsen, M. 1970). "Flat stacks" of SER are considered to be the resting configuration in inactive *Hyphantrea cunea* (Melnikova, F. and Panov, A. 1975) and the "allatum body" is thought to represent a resting state of SER in *Drosophila melanogaster* (King, R. *et al.*, 1966) since the structure is found in glands of sexually immature or aged female adults. The SER in *Manduca sexta* is present throughout the fourth and fifth larval instars (when the JH titer is highest) and only decreases at the end of the instar when the JH titer begins to fall preceding pupation (Sedlak, B. *et al.*, 1983). SER is completely absent from pupal stages when JH is at its minimal value. This is the only insect for which the JH titer has been determined with precision by mass spectroscopy (Bergot, B. *et al.*, 1984) and the coincidence of the presence of SER in glands actively secreting JH argues strongly that the SER is involved in JH biosynthesis.

2.6 Golgi complexes

Golgi complexes vary from rarely being seen in the CA of both adult *Aphis craccivora* (Elliot, H. 1976) and senescent adult *Hyalophora cecropia* (Waku, Y. and Gilbert, L. 1964), to being large and well developed in adult glands of *Locusta migratoria migratoriodes* (Schooneveld, H. 1979). In *Schistocerca gregaria* (Odhiambo, T. 1966a) the Golgi increase prior to maturation of the oocyte (Odhiambo, T. 1966b) and also increase in previtellogenic *Folsomia candida* (Palevody, C. and Grimal, A. 1976). The Golgi cisternae increase both in number and in the electron-density of their contents in the active glands of castrated females of *Leucophaea maderae* (Scharrer, B. 1978) and since the Golgi increase in these reproductively active females the complex may be an indicator of hormone processing.

In larval *Manduca sexta* (Sedlak, B. *et al.*, 1983) the number of Golgi complexes per cell increases significantly in pupal versus larval glands. Since the hormone titer is low in pupae it is likely that the Golgi are associated not with JH production but with biosynthesis of a non-hormone product. In fact, numerous dense bodies arise during early pupal development and the Golgi may be involved in the production of these lysosome-like structures.

2.7 Mitochondria

It has been noted that the mitochondrial number increases in the CA of several species: adult *Rhodnius prolixus* (Baehr, J. *et al.*, 1973), inactive glands of *Diatraea grandiosella* (Yin, C. and Chippendale, G. 1979), diapausing *Monema flavescens* (Takeda, N. 1977), and larval *Drosophila melanogaster* (King, R. *et al.*, 1966).

In *Leucophaea maderae* there was a change in the population of mitochondria; these organelles were more numerous in reproductively active female adults and their number decreased in inactive cells (Scharrer, B. 1964a). The decrease was presumably due to autophagic destruction of the mitochondria since fragments of the organelles were found within lysosome-like bodies. Similarly in *Locusta migratoria migratoriodes* Fain-Maurel, M. and Cassier, P. (1969) note that in males the mitochondrial population is homogeneous whereas the morphology undergoes cyclic variation in reproductively active female glands.

Swollen mitochondria were noted in a female sterile mutant of *Drosophila melanogaster* (King, R.

et al., 1966), in senescent adults of *Hyalophora cecropia* (Waku, Y. and Gilbert, L. 1964) and mitochondrial degeneration occurred in adult *Choleva augustata* (Deleurance, S. and Charpin, S. 1978).

2.8 Inclusions

In the hope of finding unique structures within the CA which could be associated with JH synthesis and/or storage, many inclusions have been presented as possible candidates for such sites of JH. Arguing against the possibility of finding any JH site of storage are the studies by Stay, B. and Tobe, S. (1977) which convincingly demonstrate the continuous production of JH from glands *in vitro*. If all insects secrete their hormone as do *Diploptera punctata*, then the possibility of finding inclusions containing the hormone is unlikely.

2.8.1 Lysosomes

If glands do not store JH within these unique inclusions, what then is their role in the cell? The structural resemblance of some of these inclusions to lysosomes suggests that the dense bodies might play a role in the degradation of JH or the enzymes necessary to synthesize the hormone. The temporal appearance of dense bodies with periods of senescence or aging adds credibility to such a possible role for dense bodies.

Dense bodies, or structures assumed to be lysosomes on the basis of their cellular morphology, were found to increase in the CA of senescent adults of *Hyalophora cecropia* (Waku, Y. and Gilbert, L. 1964), adult *Schistocerca gregaria* (Odhiambo, T. 1966a), aged *Drosophila melanogaster* (King, R. *et al.*, 1966), inactive *Diatraea grandiosella* (Yin, C. and Chippendale, G. 1979), at the termination of diapause in *Pectinophora gossypiella* (Raina, A. and Borg, T. 1980), and in pupal *Manduca sexta* (Sedlak, B. *et al.*, 1983).

In vertebrates, dense bodies have been associated with hormone degradation in cells of the anterior pituitary (Smith, R. and Farquahar, M. 1966) and this seems a likely explanation for the presence of lysosomes in pupal CA of *Manduca sexta* when the JH titer is minimal. Lysosome-like structures increase in vitellogenic females of *Folsomia candida* (Palevody, C. and Grimal, A. 1976) and in castrated females of *Leucophaea maderae* (Scharrer, B. 1978) where the role of this organelle might be in the turnover of metabolites in these presumably active glands.

Dense multivesicular bodies (MVB), which are one type of lysosome, were noted in adult *Oncopeltus fasciatus* (Dorn, A. 1973) and pleomorphic dense bodies were also noted (Sedlak, B. and Liechty, L. unpublished observations; see Fig. 8). Both MVB and lysosomes increased at the end of the third larval instar of *Choleva augustata* (Deleurance, S. and Charpin, S. 1978) when this animal ceases feeding and begins a wandering type of behavior in preparation for pupation.

2.8.2 Lipid

Because of the lipid solubility of JH, the lipid content within the CA has attracted some interest. It is, however, questionable whether JH is stored within glands or simply diffuses out through the lipid bilayer as soon as it is synthesized (Stay, B. and Tobe, S. 1977). Lipid droplets were found as inclusions in the glands of adult *Oncopeltus fasciatus* (Dorn, A. 1973), adult *Folsomia candida* (Palevody, C. and Grimal, A. 1976), and non-diapausing *Pectinophora gossypiella* (Raina, A. and Borg, T. 1980). The amount of lipid decreases in the CA of sugar-inactivated *Calliphora erythrocephala* (Thomsen, E. and Thomsen, M. 1970) and in glands of the senescent adults of *Hyalophora cecropia* (Waku, Y. and Gilbert, L. 1964).

The evidence for lipid as a storage site for JH in *Hyphantrea cunea* lacks substantiation. In the report by Melnikova, F. and Panov, A. (1975) the figure showing an area designated to be lipid is probably a dense body, i.e. lipid characteristically has a smooth circular profile in cross-section whereas the structures in the micrograph have an irregular outline typically seen in dense bodies.

2.8.3 Glycogen

Other inclusions that have been noted are the energy storage form of glucose, glycogen. Glycogen is found in glands of adult *Oncopeltus fasicatus* (Dorn, A. 1973), in pupal *Manduca sexta* (Sedlak, B. *et al.*, 1983) and "paracrystalline glycogen" was

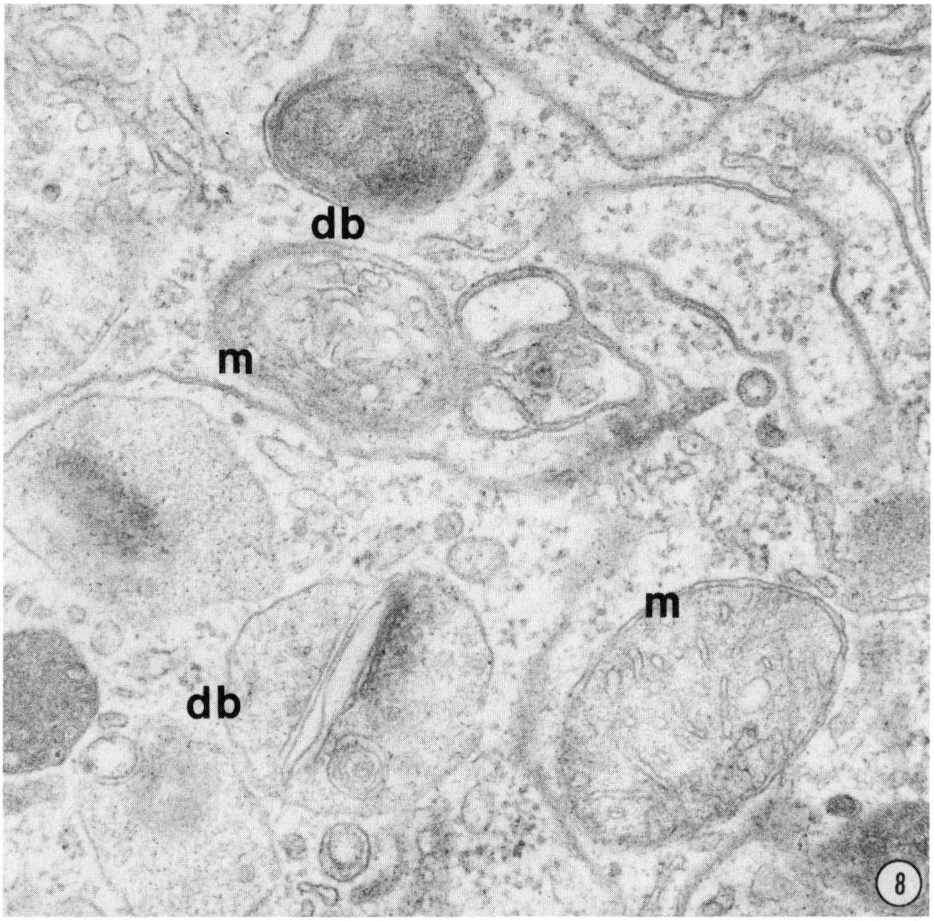

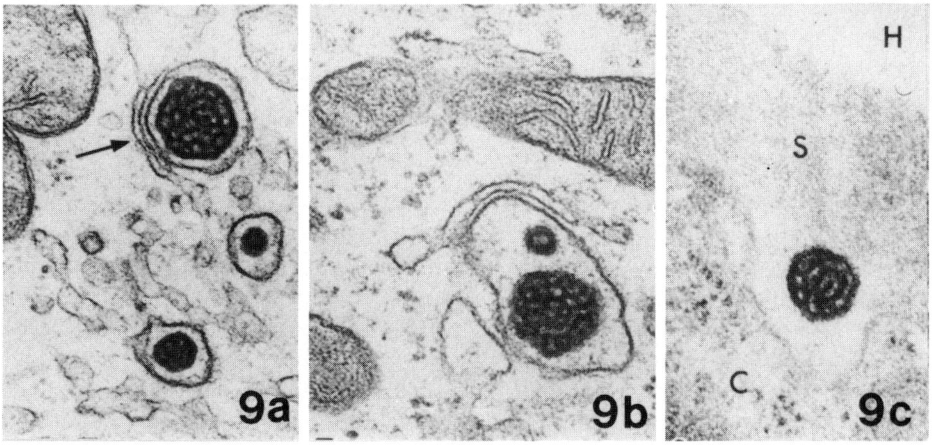

FIG. 8. The dense bodies in adult corpus allatum of *Oncopeltus fasciatus* vary in their morphology; some of these lysosome-like structures have a dense amorphous core or a membranous interior plus vesicles. db, dense bodies; m, mitochrondrion (× 52,000).

FIG. 9. Membrane-enclosed "C" bodies in corpus allatum of *Leucophaea*. (**a**) Three intracellular membrane-enclosed bodies showing spatial relationship to Golgi element (arrow) (× 50,000). (**b**) Two "C" bodies in the same spacious compartment (× 50,000). (**c**) "C" body, presumably released from corpus allatum cell, (C) by exocytosis facing stromal sheath (S) which provides access to hemocoele (H) (× 38,400). From Scharrer, B. (1971).

reported for larval glands of *Hyphantrea cunea* (Melnikova, F. and Panov, A. 1975). The amount of glycogen deposits apparently increases in the CA of castrated females of *Leucophaea maderae* (Scharrer, B. 1971).

Unique inclusions Figures 9a, b, and c show a unique inclusion which Scharrer, B. (1971) reported in *Leucophaea maderae*; this "C" body is a dense body with a characteristic internal morphology. This inclusion is prevalent at times of low glandular activity, e.g. late in development and in starved individuals. The inclusion is found both within the cells and within the thick extracellular stroma. Scharrer suggests that this "C" body accumulates when the system is "sluggish" in terms of its endocrine physiology. She also suggests that the lamellated body reported by Waku, Y. and Gilbert, L. (1964) for *Hyalophora cecropia* CA may be an inclusion similar to the "C" body found in *Leucophaea*. Whether or not this "C" body is a cellular site of JH remains unresolved.

2.9 Corpora allata regulation

One method of regulating the JH titer in insects is by controllng synthesis and release of the hormone from the CA (chapter 11, this volume; chapter 1, volume 8). It has been established experimentally (by nerve section and allatectomy) that the brain can exert control by either non-NSC axons (Fig. 10) or by NSC axons (Fig. 11). The control by either axon can be positive or negative, although neurosecretory control is usually positive (i.e. the activation of the gland and production of its hormone) whereas in many adult insects non-NSC control is inhibitory (deKort, C. and Granger, N. 1981).

There is morphological evidence for both NSC and non-NSC axons within the CA of many insects. The structural evidence for NSC control is the presence of NSC axons ending within the CA itself. NSC axons have characteristic synaptoids which have been interpreted as release sites for these axons. These synaptoids are seen as microvesicles clustered at the axonal plasma membrane. It has been postulated that these vesicles represent membrane recycling and their presence is taken as an indicator of the release of the neurosecretory granules. In addition to synaptoids, evidence of exocytosis of NSC granules has been visualized as typical omega-shaped plasma membranes, e.g. adult female *Drosophila melanogaster* (see Fig. 12; Sedlak, B. unpublished observation).

Neurosecretory neurons were reported in the CA for adult *Schistocerca gregaria* (Odhiambo, T. 1966a), larval *Drosophila melanogaster* (Aggarwal, S. and King, R. 1969), female adult *Hypera postica* (Tombes, A. and Smith, D. 1970), adult *Oncopeltus fasciatus* (Dorn, A. 1973; Unnithan, G. *et al.*, 1971) and adult *Labidura riparia* (Caussanel, C. *et al.*, 1979). In some insects such as *Aphis craccivora* (Elliot, H. 1976) the NSC were correlated with gland activity, i.e. the NSC axons varied in diameter, increasing in active larval stages and decreasing in the adult. Such a variation in NSC axon diameter was also correlated with an increase in JH titer in larval *Manduca sexta* (Sedlak, B. 1981; Sedlak, B. *et al.*, 1983), i.e. large-diameter NSC axons were found in mid-instar glands immediately preceding the rise in JH titer (Fig. 11) whereas much smaller axons were evident in early larval stages and early pupal stages (when the JH titer is low).

In *Locusta migratoria migratoriodes* (Goltzene, F. and Porte, A. 1978) the NSC are needed for vitellogenic uptake in the ovary and Mordue, W. (1965) reports an increase in NSC granule staining concurrent with egg production. The increase in the electron-density of the NSC granules occurred in *Schistocerca gregaria* (Odhiambo, T. 1966b) prior to maturation of the ovary and presumably reflected an active state of the glands. The electron-density of NSC granules may not, however, be considered a critical criterion for determining secretory activity since many glands contain NSC axons with granules of both variable size and electron-density. In *Manduca sexta*, for example, granules vary in size from 80 to 350 nm, and these may be electron-lucent or dense; such NSC granules are found during all stages of larval and pupal development (Sedlak, B. 1981). Variablity of NSC granule electron-opacity may also depend upon the pH used during fixation (Nordmann, J. 1977).

Strong, L. (1965) reported that when NSC were cauterized in *Schistocerca gregaria* there was a subsequent decrease in CA volume. An increase in NSC granules occurred in *Periplaneta americana* at a time when the gland volume decreased (Khan, T. *et al.*, 1978). If a large gland represents secretory

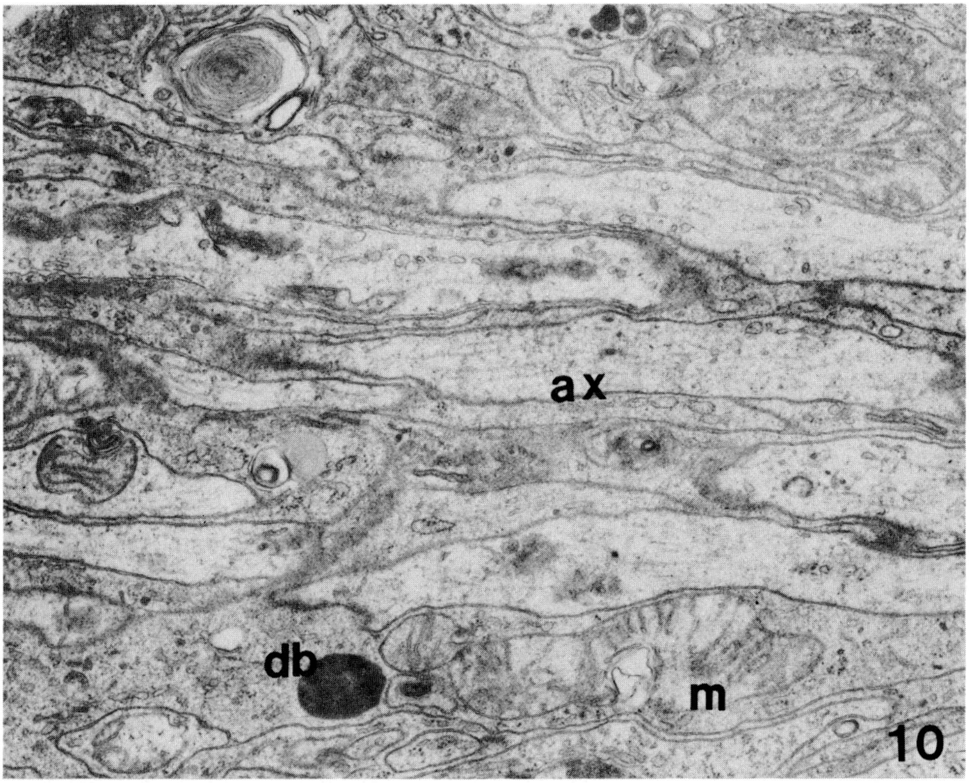

FIG. 10. The axonal tracts found within this corpus allatum cell of an adult *Oncopeltus fasciatus* appear to be non-neurosecretory; they are devoid of ns granules and are rich in microtubules. ax, axon; db, dense body; m, mitochondrion (× 17,000).

activity, then in *Schistocerca gregaria* removal of the NSC would remove a positive influence on the gland leading to a decrease in gland size. On the other hand, in *Periplaneta americana* the increase in NSC granules resulted in the inactivation of the gland (i.e. decrease in gland volume). It must be noted, however, that the size of the CA is not necessarily an accurate indicator of secretory activity.

As noted previously, innervation of the CA is commonly via two nerves (NCA I and NCA II). The NCA I nerve fibers, consisting of conventional and NSC axons, have their cells bodies within the protocerebrum and the nerve fibers traverse the CC prior to entering the CA region. The NCA II fibers found in primitive insects (see Cassier, P. 1979) arise from cells within the suboesophageal ganglion.

Nervous inhibition was reported for pregnant females of *Leucophaea maderae* (Engelmann, F. and Lüscher, M. 1956), fasting females of *Oncopeltus fasciatus* (Johansson, A. 1958), female *Rhodnius prolixus* (Baehr, J. 1973), virgin females of *Dyndimus veriscolor* (Dargo, A. 1976) and diapausing females of *Pyrrhocoris apterus* (Hodkova, G. 1976). Figure 10 shows several non-NSC axons in female adult *Oncopeltus fasciatus* (Sedlak, B. and Liechty, L. unpublished observations).

2.10 Experimental basis of gland activity

Experimental manipulation of the CA has provided some criteria to assess the basis of active versus inactive glands. Three examples of such experimentation will be discussed in some detail: (1) castration of female adults; (2) treatment with the drug precocene and (3) autoradiography. All three of the studies which will be described provide evidence that the SER is probably the site of JH biosynthesis.

2.10.1 Castration of female adults

The structural basis for endocrine activity in the CA

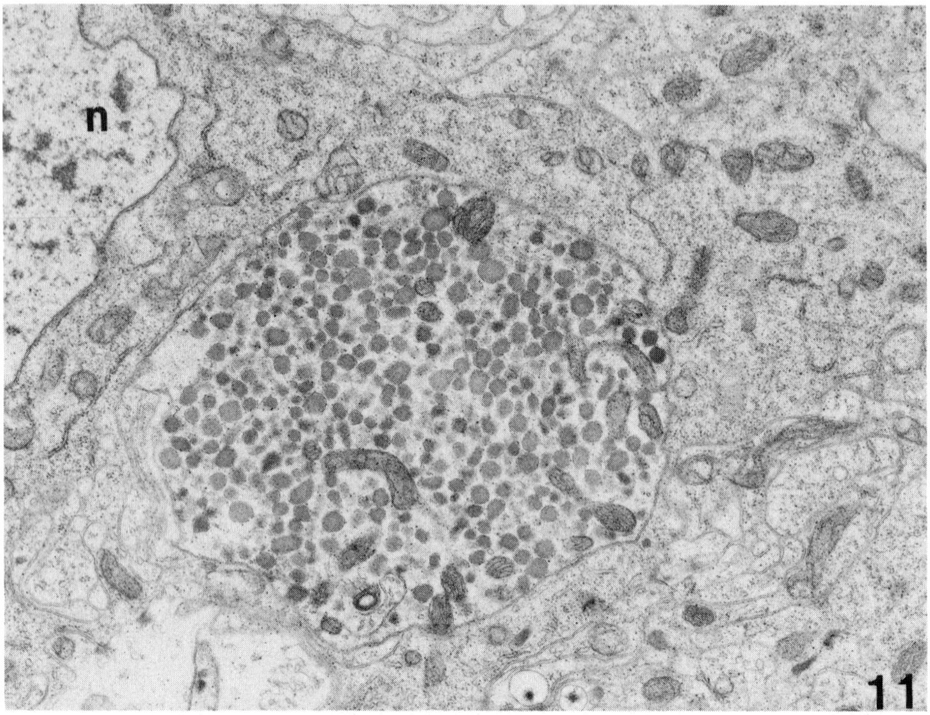

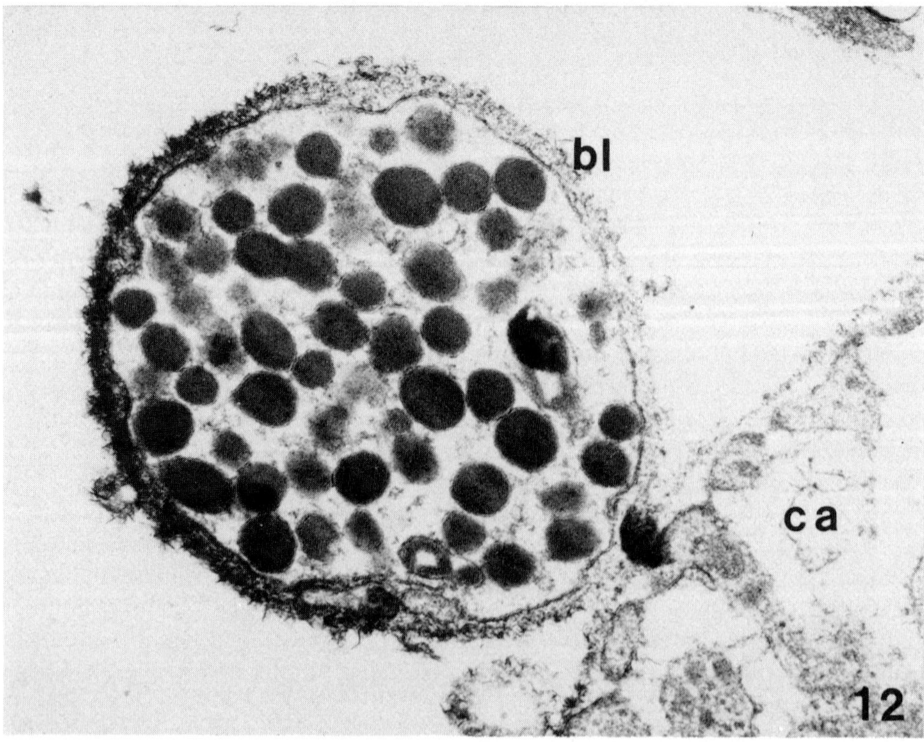

FIG. 11. A neurosecretory cell axon from a corpus allatum cell; although this axon contains granules which are fairly electron-dense many axons have granules which are very electron-lucent. n, nucleus (× 15,000). FIG. 12. An electron-dense granule is seen in the process of exocytosis from a neurosecretory cell axon which is located adjacent to the corpus allatum of a female adult *Drosophila melanogaster*. bl, basal lamina; ca, corpus allatum (× 62,000).

of *Leucophaea maderae* was studied in female adult insects at various stages of their reproductive cycle. Glands were presumed to be inactive in pregnant females or in females at the beginning of their reproductive cycle; glands were considered to be active in hormone secretion if the terminal oocyte had reached its maximal length (Scharrer, B. 1964a). This ovarian assay is based on the assumption that JH, acting as a gonadotropic homone in adults, is necessary for the uptake of yolk protein into developing oocytes (chapter 6, volume 8).

Using the ovarian assay to determine secretory activity of the CA, the active glands showed an increase in both volume and in SER. When the CA of castrated females were examined it was found that, on the basis of an increase in the SER, these glands closely resembled active female glands (Scharrer, B. 1978). The interpretation was that removal of an ovary in effect removed a negative feedback control on the gland, thus increasing CA activity.

2.10.2 Precocene

Probing the structural basis for gland activity has also been approached by studying CA which have been treated with precocene. Precocenes are naturally occurring compounds extracted from plants which have been reported to have an anti-JH effect on some insects (Bowers, W. *et al.*, 1976; chapter 15, volume 8). The precocenes induce physiological responses which resemble those in which there is a JH deficiency, such as the precocious metamorphosis of immature stages and the sterilization of reproductive female adults. In *Oncopeltus fasciatus*, nymphs can be induced to metamorphose to adults from as early as the first larval instar; the female adults of the species fail to incorporate yolk protein into their developing oocytes. Since these JH effects could be reversed by treatment with JH III or JH analogues it was thought that the primary precocene effect was on the CA itself. Indeed the CA volume decreases when animals are treated with precocene (Bowers, W. and Martinez-Pardo, R. 1977). Precocene also decreases the stainability of material in the NSC cells of the brain, suggesting that the chemical has its effect at the level of brain control of the CA (Unnithan, G. *et al.*, 1977). If, however, the nerve connection between brain and CA is severed, precocene still has an effect on the CA (Masner, P. *et al.*, 1978).

The ultrastructure of precocene-treated glands did show that, with a long exposure to precocene, the CA underwent irreversible degeneration. The cells were filled with lysosome-like bodies, autophagic vacuoles, and multivesicular bodies (Unnithan, G. *et al.*, 1971). However, Liechty, L. and Sedlak, B. (1978) showed that short-term exposure to precocene resulted in a more specific change in the CA, i.e. the accumulation of vesicles usually found to predominate in only the light cell population of non-treated glands. These authors report that the light cell population predominates on the day of eclosion, whereas the dark cells are more numerous within 5 days, i.e. the day the gland can be considered to be active on the basis of the presence of mature, yolky oocytes. In the precocene-treated animals, the light cell population was dominant, suggesting that the vesicle accumulation within light cells was actually a structural definition of gland inactivation.

Feyereisen, R. *et al.* (1981) have shown that the short-term exposure of precocene II to *Diploptera punctata* results in the aggregation (and eventual degeneration) of SER. These authors suggest that the vesicles seen by Liechty, L. and Sedlak, B. (1978) are actually vesiculated SER. Furthermore, Feyereisen *et al.* point out that the differences seen in all of the precocene studies might be due to: (1) differences in drug dosage and drug exposure, and (2) differential sensitivity of a mixed population of cells. The latter sensitivity is seen in the morphological differences between adjacent cells (i.e. the light and dark cell-population) and is suggested by the fact that in the study by Feyereisen *et al.* not all cells responded to the drug in an identical manner.

Short-term exposure to precocene caused gland degeneration in *Locusta migratoria migratoriodes* (Schooneveld, H. 1979) in which the volume of these CA decreased, electron-dense cells accumulated, and then hemocytes invaded the CA to phagocytose the dead cells. In *Schistocerca gregaria* (Unnithan, G. *et al.*, 1980) degeneration of the gland was observed at the histological level.

2.10.3 Autoradiography

An autoradiographic study was done by Tobe, S. and Saleuddin, A. (1977) in an attempt to demonstrate the cellular sites of JH biosynthesis in *Schistocerca gregaria*. A labeled precursor of JH

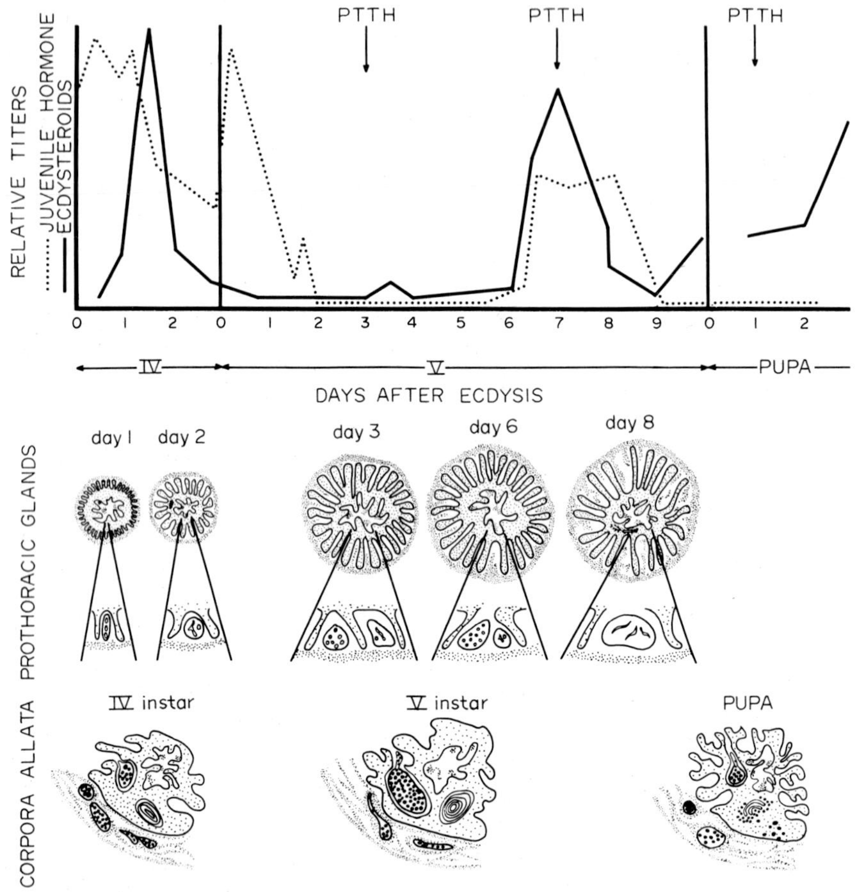

FIG. 13. Diagrammatic summary of hormonal fluctuations and temporal changes in both prothoracic gland and corpus allatum cell structure. Ecdysteroid titer (after Bollenbacher, W. *et al.*, 1981) and juvenile hormone titer (based on summed JH I and JH II titers; Bergot, B. *et al.*, 1984) are represented as relative titers. The prothoracicotropic hormone (PTTH) cycles of activity are indicated by arrows at their peaks of activity (Agui, N. *et al.*, 1980). Drawings of cell structure emphasize developmental changes. The prothoracic gland cell alters most significantly from early to late in each larval instar, whereas the most notable changes in corpora allatal cell structure occur between larval and pupal development. (From Sedlake, B. *et al.*, 1983.)

([^{3}H]faresenic acid) was incubated with CA and after a time, to allow its uptake and conversion to two final products in JH synthesis ([^{3}H]methyl farnesoate and [^{3}H]C_{16}JH), the CA were processed for autoradiography. The figures show that grains were distributed throughout the gland cells with a reportedly higher distribution of label over the two forms of SER, i.e. the concentric SER and the vesiculated SER. Grains were also found over most other areas of the cell including the mitochondria and Golgi complexes, indicating that many of these organelles may be involved in JH biosynthesis.

This study is inconclusive since: (1) the label was not localized specifically over any particular organelle, and (2) it is not at all clear that the grains represent the actual hormone.

2.10.4 Corpus allatum structure and juvenile hormone titer

To date, the most critical method for identification and measurement of JH titer is by mass spectrometery. The JH titers of *Manduca sexta* have been determined by mass spectrometry (Bergot, B. *et al.*, 1984) and have been correlated with the structural changes in the CA to assess the basis of CA gland activity. Using a colony of precisely timed *Manduca sexta* it was possible to study gland activity during larval–pupal development and to correlate JH titers with the ultrastructural changes in the CA. It was

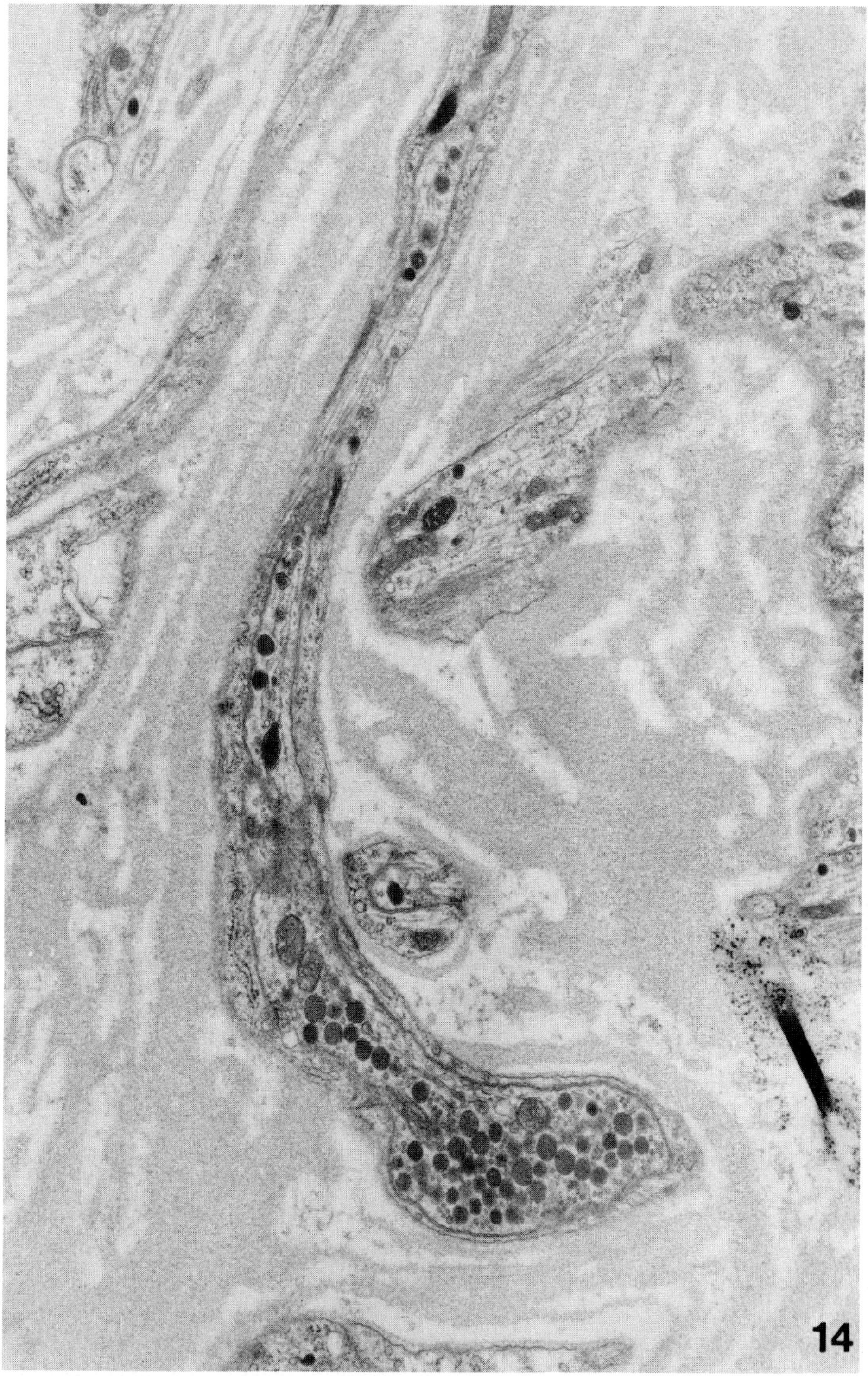

FIG. 14. The neurosecretory cell axon residing within layers of basal lamina at the periphery of a *Manduca sexta* corpus allatum is thought to be the release site for prothoracicotropic hormone. From Sedlak, B. (1981) (× 17,000).

further possible to determine the structural basis for synthesis versus secretion as separate events (Sedlak, B. *et al.*, 1983). Figure 13 is a diagrammatic summary of the changes in CA structure and the relative JH titer.

In *Manduca*, whorls of SER predominate in the fourth larval instar and until late in the fifth larval instar but are absent from the pupal cells. The presence of this organelle indicates that the gland is capable of synthesizing hormone continuously during larval life. The actual secretion of JH may be under NSC control. This assumption is based on the observation that NSC axon diameters reach a maximum size just prior to the late JH surge. These axonal diameters are much smaller earlier in larval life and again later during pupal life. Thus, in *Manduca* an interpretation of changing cell structure in the context of the actual juvenile hormone titer suggests that synthesis (via the SER) and secretion (stimulated by NSC) are separate events which can be visualized at the cell level.

2.11 Prothoracicotropic hormone (PTTH) and the corpora allata

PTTH activity has been found within the CA of *Manduca sexta* (Agui, N. *et al.*, 1980). To determine when activation of the prothoracic glands occurred after stimulation by the tropic hormone, PTTH, the glands were incubated *in vitro* with brains and ecdysone was measured via radioimmunoassay; (Bollenbacher, W. *et al.*, 1980; chapter 4, this volume). In a subsequent study the CA and CC were used as the source of tropic hormone and it was found that the CA was the release site for PTTH (Agui, N. *et al.*, 1980). This result was not predicted on the basis of past assumptions that the CC was the major (perhaps the only) neurohemal organ in the endocrine axis. Yet it has been demonstrated ultrastructurally that the NSC axons are located both within the gland proper and within the stromal sheath surrounding the CA (Fig. 13; Sedlak, B. 1981). It has been suggested that those NSC axons within the sheath are the PTTH release sites. Indeed if the sheath actually represents an extension of the CC which envelops the CA (Scharrer, personal communication), the CC technically remain the sole neurohemal organ and the CA is relegated to the role of release site for PTTH.

2.12 Summary of corpora allata ultrastructure

The great structural variability seen when comparing CA cells between species may be the result of real differences in the developmental ages and/or physiological states of these glands, or may result from artifactual differences in tissue preparation of the glands for microscopy.

The numerous differences preclude any meaningful generalizations about a universal basis for structural activity. In spite of this obvious limitation in interpretation, those individual studies which carefully determine hormonal activity based on, for example, reproductive ability of female adults (Scharrer, B. 1964a), or drug-induced inhibition of the CA (Liechty, L. and Sedlak, B. 1978; Feyereisen, R. *et al.*, 1981), autoradiographic localization of JH (Tobe, S. and Salueddin, A. 1977) and comparisons of the structural changes during larval development with the actual JH titer (Sedlak, B. *et al.*, 1983), have helped to define structural activity for the CA within particular species. The SER remains the most likely candidate for the structural site of JH synthesis according to these studies.

3 PROTHORACIC GLANDS

3.1 General structure

In contrast to the CA the PG show a rather consistent cellular morphology when glands of different species are compared. In general, each gland cell has very few organelles and inclusions in the cytoplasm. The mere presence of a specific organelle does not define functional activity, however, and in some early studies it was difficult to establish a structural basis for secretory activity. With the current availablity of RIA titers for ecdysteroids and some experimental manipulation, our understanding of the relationship between structure and function has become clearer.

3.2 Intercellular spaces

The most commonly recorded characteristics of PG cells are the peripheral intercellular spaces (ICS) which can vary in their depth during development

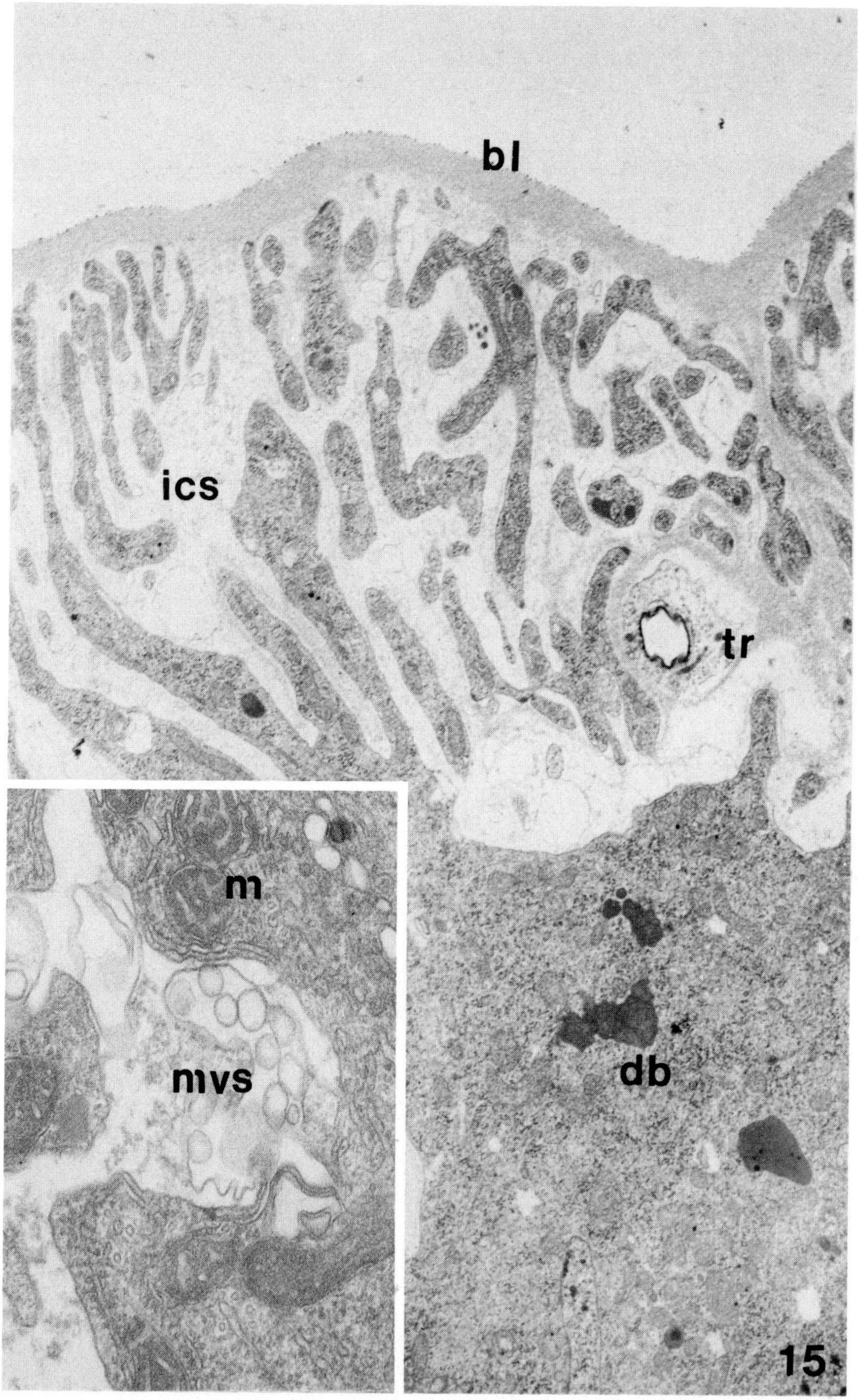

FIG. 15. The peripheral area of each prothoracic gland cell of *Manduca sexta* is characterized by a channel system formed via the invagination of the plasma membrane. bl, basal lamina; db, dense body; ics, intercellular space; tr, tracheole (× 8000). Inset. Within the intercellular spaces of *Manduca sexta* prothoracic glands are distinctive membranous structures; the multivesicular sac is a cluster of vesicles bounded by an outer membrane. m, mitochondrion; mvs, multivesicular sac (× 20,000).

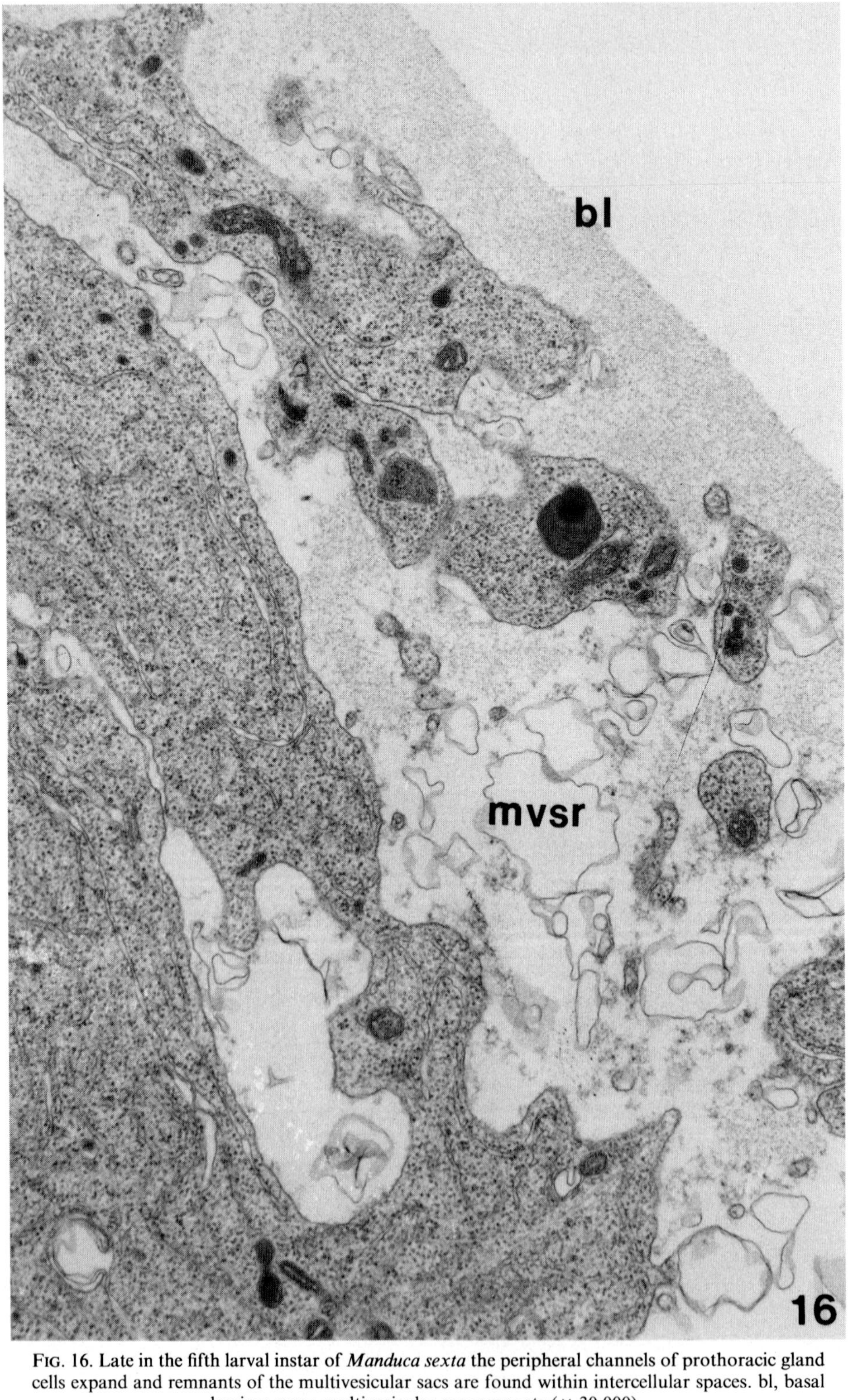

FIG. 16. Late in the fifth larval instar of *Manduca sexta* the peripheral channels of prothoracic gland cells expand and remnants of the multivesicular sacs are found within intercellular spaces. bl, basal lamina; mvsr, multivesicular sac remnants (× 30,000).

(Figs 15–16). These intercellular spaces have been named lacunae and channels, and are formed by the invagination of the plasma membrane. The surface area of each cell is increased by such infolds. ICS are found in most orders of insects (Herman, W. 1967) but there are exceptions, e.g. *Gryllus bimaculatus* (Romer, F. 1974) have no folds and in *Musca domestica*, Herman, W. (1967) reports that the lacunae are not formed by plasma membrane infolds.

In some insects when ICS are present, the increased surface area has been correlated with fluctuations in the molting hormone; e.g. for *Drosophila melanogaster* (Aggarwal, S. and King, R. 1969), *Galleria mellonella* (Gersch, M. *et a.*, 1975; Blazsek, I. *et al.*, 1975), *Tenebrio molitor* (Romer, F. 1971; Glitho, I. *et al.*, 1979) and *Manduca sexta* (Sedlak, B. *et al.*, 1983). In all of these insects the increase in ICS occurs at the time when the later surge in ecdysteroids is detected by the RIA (Richards, G. 1981; Bollenbacher, W. *et al.*, 1978; Delbecque, J. *et al.*, 1978; Bollenbacher, W. *et al.*, 1981). Thus, the increase in surface area which occurs concurrent with the appearance of hemolymph ecdysteroids reflects the release of hormone from the glands.

3.3 Microvesicles

The exact cellular mechanism of release has not been defined but the microvesicles, which are present in the intercellular spaces of several species, may be involved in the release of hormone from PG cells. Such vesicles were found in *Leucophaea maderae*, *Blaberus craniifer* (Scharrer, B. 1964) and *Hyalophora cecropia* (McDaniel, C. *et al.*, 1976). Scharrer, B. (1964b) suggested that the microvesicles were involved with the transport of material between the gland and the hemolymph.

In *Galleria mellonella* (Blazsek, I. and Mala, J. 1978) the microvesicles were associated with the localization of sterols when PG were initially fixed in glutaraldehyde then incubated with digitonin. The digitonin complexes with all sterols having a free $-OH$ in position three, e.g. cholesterol, ecdysone and 20-hydroxyecdysone. In this study the electron-dense digitonin crystals were localized preferentially at the site of microvesicles. During the larval growth phase, digitonin crystals were seen at the base of caveolae (thought to be structures associated with pinocytosis), whereas at cocoon formation and during early prepupal stages the crystals were seen at microvesicles within intercellular spaces. When microvesicles were no longer evident (i.e. at the end of the prepupal stage) the digitonin staining was also missing. These results were interpreted to mean that first there was endocytotic uptake of cholesterol at the time the larva is feeding and growing logarithmically. Later, during prepupal stages, the labeled microvesicles were thought to represent "reverse micropinocytosis" (i.e. exocytosis) of ecdysone. These conclusions were based on the indirect evidence that a late ecdysteroid peak is found in many insects at the time when digitonin stained the exocytotic vesicles.

In *Manduca sexta* microvesicles, called multivesicular sacs, were found throughout the fourth and fifth larval instars but these sacs changed in morphology following cycles of PTTH activity (Fig. 15, inset). The change was from membrane-bound vesicles to an empty sac in which the inner membranes were no longer present. The change to empty sacs preceded the major increase in hemolymph ecdysteroids, suggesting that release of a gland product occurred in response to the tropic hormone. The adrenal cortex may be the vertebrate endocrine organ analogous to the PG. The adrenal cortex secretes its corticosteroid product in response to a tropic peptide ACTH (adrenocorticosteroid tropic hormone). In the adrenal cortex the steroid hormones are presumably dissolved in the lipid and there is structural evidence that the lipid is exocytosed from the gland cells (Rhodin, J. 1971). Thus, exocytosis of hormone may be a general mechanism of hormone release.

3.4 Mitochondria

Mitochondrial morphology in the PG varies in some species; this organelle is thinner during intermolt phases of *Leucophaea maderae*, *Blaberus craniifer* (Scharrer, B. 1964b) and *Drosophila melanogaster* (Aggarwal, S. *et al.*, 1969) than at other developmental phases in these insects. Gland mitochondria reportedly increase in size in the last larval instar of *Galleria mellonella* (Gersch, M. *et al.*, 1975), and Blazsek, I. *et al.* (1975) found mitochondrial degeneration in the penultimate instar. These organelles swell in *Antheraea pernyi*

(Herman, W. 1967) from early to late postmolt; such swelling might reflect organelle degeneration.

Pleomorphic mitochondria have been noted in PG of several species, e.g. in the last larval instar of *Galleria mellonella* (Blazsek, I. *et al.*, 1975), there are three types of mitochondria; (1) elongated, (2) spherical with tubulovesicular cristae, and (3) spherical with a crystalline center. In *Manduca sexta* the mitochondria have a much more electron-dense matrix during the fifth larval instar in comparison with those in the fourth larval instar (Sedlak, B. *et al.*, 1983).

Beaulaton, J. (1968) examined the mitochondrial population in the last instar larvae of *Antheraea pernyi* and *Bombyx mori* and correlated changes in mitochondria with the presence of the molting hormone; molting hormone was bioassayed only for *Bombyx mori* and expressed in *Calliphora* units (see chapter 1, this volume). During both larval instars the organelles progressively changed from rod-like structures with electron-dense matrices to swollen macromitochondria lacking the typical inner membranes. Intermediate stages showing dilated mitochondria with tubular cristae, preceded a stage with central core tubules. The changes were considered to be physiologically significant since intermediate morphologies were consistently recognized even when the fixation was varied. Macromitochondria were associated with SER and in *Bombyx mori* the transformation to this final morphology occurred concurrent with a rise in the molting hormone titer in both larval instars. These two observations were interpreted to indicate that the mitochondria participated with the SER in the production of ecdysone. In the adrenal cortex, there are "cytoplasmic droplets", which are vacuoles with a low electron-opacity, associated with mitochondria in a relationship believed to be involved with the transfer of cholesterol from the droplet to the mitochondrion during steroid hormone production (Giacomelli, F. *et al.*, 1965).

Beaulaton, J. and Gras, R. (1980) studied organelle changes in PG cells of *Rhodnius prolixus* with different tissue preparations at times of secretory activity and inactivity. PG cells were assumed to be inactive during the period of fast following each ecdysis and active during the middle portion of a molting cycle. Thirteen different aldehyde fixatives were used and the osmotic pressure was adjusted with sucrose during the various times studied. The morphological changes seen in the mitochondria, and both RER and SER were thought to occur because the osmotic pressure in the gland and hemolymph were not consistent during each developmental stage. It was concluded that the changes observed reflected physiological alterations in the gland rather than artifacts of fixation.

All of the developmental changes recorded for mitochondria are consistent with the tenet that mitochondrial enzymes participate in ecdysone biosynthesis.

3.5 Endoplasmic reticulum

Since SER predominates in the endocrine components of vertebrate reproductive organs, i.e. testis (Christensen, A. 1965) and ovary (Bjersing, L. 1967), the presence of SER might be expected in the PG of insects. It is not clear, however, that these steroid-secreting sex glands are analogous to the PG; in fact the steroid-secreting adrenal cortex may be more appropriately compared with the PG (see discussion above).

Nonetheless SER has been found in some glands, e.g. SER associated with mitochondria is present in *Antheraea pernyi* and *Bombyx mori* (Beaulaton, J. 1968); tubules thought to be SER were reported for *Musca domestica* (Herman, W. 1967), and in *Gryllus bimaculatus* (Romer, F. 1974) the SER was initially abundant but decreased post-ecdysis.

RER has also been found in glands of several species, e.g. *Musca domestica* (Herman, W. 1967), *Tenebrio molitor* (Romer, F. 1971), and *Oncopeltus fasciatus* (Dorn, A. and Romer, F. 1976). The RER was arranged as whorls in PG of cocoon stages in *Galleria mellonella* (Blazsek, I. *et al.*, 1975) and flat stacks of RER were seen in *Leucophaea maderae* and *Blaberus craniifer* (Scharrer, B. 1964b). Since the RER produces exportable protein, the presence of this organelle cannot be directly involved with the synthesis of the steroid hormone ecdysone. McDaniel, C. *et al.* (1976) report that in *Hyalophora cecropia* the RER appears by the fourth day of pupal–adult life whereas prior to this stage free ribosomes were predominant and following this stage there is a decrease in RER. Since the RER is found at a time when the gland is thought

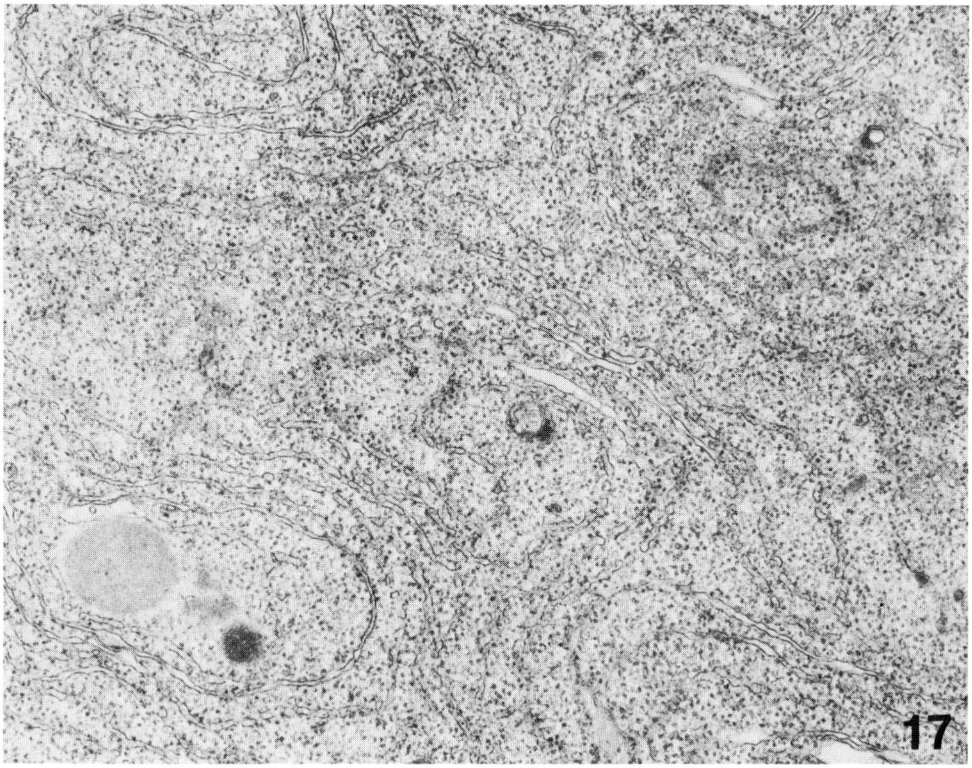

FIG. 17. The rough endoplasmic reticulum in prothoracic gland cells of *Manduca sexta* is plentiful late in the last larval instar (× 30,000).

to be secreting ecdysone the authors conclude that the RER is synthesizing a carrier protein for ecdysone.

The RER stacks which Sedlak, B. *et al.* (1983) found in late fourth larval instars and the RER whorls found in late fifth larval instars of *Manduca sexta* appear coincidentally with each major surge of hemolymph ecdysteroids (Fig. 17). These authors suggest that the RER is producing exportable protein in response to the molting hormone signal but it is unlikely that a hormone carrier is being synthesized because the ecdysteroids are already present in the circulation before the time RER whorls are seen. The protein could contribute to the reconstitution of the plasma membrane in the pupal stage; i.e. larval stages have deeply invaginated channels whereas pupal cells lack such extensive ICS.

Finally, both SER and RER have been found in PG cells of *Galleria mellonella* (Blazsek, I. *et al.*, 1975) and *Oncopletus fasciatus* (Dorn, A. and Romer, F. 1976).

3.6 Oenocytes as an alternative source for ecdysone

It has been proposed that the oenocytes are a source of ecdysone in some insects (Locke, M. 1969; Romer, F. 1974; Dorn, A. and Romer, F. 1976). The proposition is mainly based on the fact that oenocyte ultrastructure is similar to that of vertebrate steroid producing glands. The testicular interstitial cells (Christensen, A. 1965) and the ovarian corpus luteum (Bjersing, L. 1967) are rich in SER, as are the oenocytes of *Calpodes ethlius* (Locke, M. 1969); *Gryllus binaculatus* (Romer, F. 1974) and *Oncopeltus fasciatus* (Dorn, A. and Romer, F. 1976). In *Gryllus* the SER fluctuates during the molting cycle, i.e. the SER is scarce after hatching but increases during the next 20–50 h. Shortly before molting there is autolysis of the SER. Based on the fact that SER is maximally present at 38–54 h post-ecdysis, these cells are considered to be at their peak of secretory activity. It is not, however, known when the molting hormone titer falls and rises in this insect. Thus the conclusion that hormone secretion is reflected by structural changes in oenocytes remains tentative.

In *Calpodes* where neither the SER morphology nor quantity changes during the molting cycle there are at least two peaks of ecdysteroids detected (Dean, R. *et al.*, 1980). In this case the SER does not change with hormone secretion. It is, of course possible, although unlikely, that biosynthesis of the hormone is continuous at the SER but that the pulses of ecdysone occur because secretion is a separately controlled event.

In *Oncopeltus* the PG break down in the adult yet there are detectable quantities of molting hormone present (as determined by the *Calliphora* bioassay) in both males and females. Thus the bioassay indicates that there is another endocrine organ which secretes this molting hormone in the adult and the oenocytes are suggested as a prime candidate based on structural criteria, i.e. the presence of SER.

The role of oenocytes in ecdysone production remains unresolved. There may indeed be another endocrine organ supplying ecdysone after the PG degenerate, but the sole presence of SER in cells does not necessarily indicate that these cells are the alternative source for the hormone. They are, however, a viable candidate worthy of further investigation.

In addition to the presence of SER, both *Calpodes* and *Gryllus* oenocytes have a membrane network which increases the cell surface area. These membrane invaginations are reminiscent of those in the PG cells and may similarly reflect secretion of cell product from oenocytes. In *Calpodes* the plasma membrane reticular system, a tubular network found at the periphery of each cell, fluctuates during development. It disappears before molting and before pupation (Locke, M. 1969) at times following a rise in ecdysteroid titer (Dean, R. *et al.*, 1980). In *Gryllus*, clefts, which are fissures in the cytoplasm, appear during the premolt stages and increase 54 h later at a time of presumed secretory activity. It is not known, however, whether or not these clefts connect with the plasma membrane and thus may not be involved with the export of a cell product at the cell surface (Romer, F. 1974).

The ovary has also been proposed as an alternative source for ecdysone production in mosquitoes (Hagedorn, H. *et al.*, 1975; chapter 7, volume 8) but there are no structural studies of the ovary which would indicate the organelle basis for ecdysone activity in this tissue.

3.7 Degeneration of glands and lysosomes

Degeneration of the PG is common. In *Periplaneta americana* (Bodenstein, D. 1953); *Rhodnius prolixus* (Wigglesworth, V. 1955); *Leucophaea maderae* (Scharrer, B. 1964b, 1966), and *Hyalophora cecropia* (Herman, W. and Gilbert, L. 1966) degeneration occurs in the adult. Pupal glands degenerate in *Galleria mellonella* (Blazsek, I. *et al.*, 1975) and *Tenebrio molitor* (Srivastava, U. 1958).

Lysosomes which indicate cellular death were present in the PG of early instar larvae of *Drosophila melanogaster* (King, R. *et al.*, 1966), *Musca domestica* (Herman, W. 1967), penultimate instar larvae of *Galleria mellonella* (Blazsek, I. *et al.*, 1975) but were rarely found in the last larval instar larvae of the same species (Gersch, M. *et al.*, 1975). In *Leucophaea maderae* and *Blaberus craniifer* PG regression takes place within a few days of the final molt. The nuclei become pycnotic and pleomorphic autophagic vacuoles appear. Eventually hemocytes invade the area to phagocytose the cellular debris (Scharrer, B. 1966b).

At the end of larval life and until pupal ecdysis the PG cells of *Tenebrio molitor* are reduced in size (Glitho, I. *et al.*, 1979). The decrease in volume reflects the loss of organelles, e.g. endoplasmic reticulum, Golgi and mitochondria disappear from the cytoplasm and at the same time autophagic vacuoles (cytolysosomes) containing endoplasmic reticulum, mitochondria and pycnotic nuclei increase in number. This autolysis occurs at the time when the ecdysteroid titer falls. In this study, ecdysteroid titers were determined by radioimmunoassay for the same animals which were used in the structural analysis. By the second day after ecdysis to the pupal stage no PG cells, not even degenerating cells, were found. Yet, there was a considerable quantity of ecdysteroids detected in three animals where no PG cells could be found. The absence of PG in the presence of hormone suggests that there is an alternative endocrine organ producing ecdysone in pupal *Tenebrio*.

3.8 Inclusions

The most commonly reported inclusion is glycogen which was found in *Leucophaea maderae* (Scharrer, B. 1964b), *Oncopeltus fasciatus* (Herman, W. 1967), *Gryllus bimaculatus* (Romer, F. 1974), *Tenebrio molitor* (Glitho, I. *et al.*, 1979), and *Hyalophora cecropia* (McDaniel, C. *et al.*, 1976). The glycogen increased post-molt in *Oncopeltus fasciatus* (Herman, W. 1967). Rosettes of glycogen were found in *Musca domestica* (Herman, W. 1967) and glycogen converting to vacuoles was reported by Romer, F. (1971) in *Tenebrio molitor*.

3.9 Prothoracic gland regulation

Axons with neurosecretory granules have been observed in the PG of *Calliphora erythrocephala* (Norman, T. 1965), *Leucophaea maderae* (Scharrer, B. 1964b), *Drosophila melanogaster* (Aggarwal, S. and King, R. 1971), *Tenebrio molitor* (Romer, F. 1971) and *Galleria mellonella* (Blazsek, I. *et al.*, 1975; Gersch, M. *et al.*, 1975). In *Galleria*, Benedeczky, I. *et al.* (1980) report the nerve bundles contain electron-dense granules of 120–150 nm diameter, 50 nm electron-lucent granules and some rarely found 100–150 nm granules with an electron-dense core. There were, however, no synaptoids seen at the endocrine PG junction with these nerves, even though synaptoids were found at non-endocrine cells. The neurosecretory neurons do not have synaptoids but exocytosis figures were reported. The authors conclude that only peptidergic NSC control the PG cells.

3.10 Immunocytochemistry

The potential use of immunocytochemical techniques to localize hormones at the sites of biosynthesis and/or release is promising but problematic. Some vertebrate peptide hormones have been localized in specific areas of the adenohypophysis (Moriarty, G. 1973) indicating the specificity and sensitivity of antibodies to probe the position of each hormone (antigen). This technique utilizes an indirect staining procedure, i.e. the primary antigen (the hormone) is linked to a non-specific antibody which can be visualized in the electron microscope because this antibody carries an electron-dense tracer such as ferritin or horseradish peroxidase. Thus, the electron-dense marker indirectly localizes the primary antigen (the hormone).

There have not yet been any studies localizing steroid hormones by this procedure in either vertebrates or invertebrates. The steroids can easily pass into and out of cells by dissolving in the plasma membrane lipid bilayer and the hormones can just as readily be dissolved out of cells during the alcoholic dehydration required during tissue processing for electron microscopy.

The only immunocytochemical study attempting to localize ecdysteroids in PG is that of Birkenbeil, H. *et al.* (1979) who used saponin to increase the permeability of *Galleria mellonella* PG cells. Since the method purposefully destroyed membranes with the intent of allowing access of the antibody, the ultrastructure was not adequate enough to identify any membranous organelles (such as RER and SER) with the exception of the mitochondria which seemed unaffected by the saponin treatment. The antibody linked to peroxidase was seen to be associated with "hyaloplasm" and microtubules. The microtubules were thought to be involved in the transport of the hormone within the cells. This study is very inconclusive due to the poor preservation of cells; the potential of the immunocytochemical technique still awaits technical advances before any meaningful answers can be obtained about the site of synthesis and/or secretion of ecdysone.

3.11 Structural changes in relationship to ecdysteroids and juvenile hormone

3.11.1 Prothoracic glands and ecdysteroid titers

Glitho, I. *et al.* (1979) studied changes in PG cells during larval–pupal development in *Tenebrio molitor* in the context of radioimmunoassay (RIA) analysis of ecdysteroid titers. The RIA data of Delbecque, J. *et al.* (1978) were used and in addition RIA titers were determined for the same animals used for the morphological study. Early in the larval instar, at a time of low molting hormone titer, the PG contained abundant glycogen, and vesicles (thought to be lipid) were associated with RER and mitochondria. In contrast, late in the same instar when the ecdysteroid titer was high, the plasma membrane infolds decreased, vesicles were rare and glycogen absent. The size of both PG cells and intercellular channels continued to decrease until pupal ecdysis. Concurrent with the cell size decrease, which is thought to reflect an inactive gland, there was an increase in cytolysosomes indicating PG cell autolysis. Interestingly, the gland began to degenerate prior to the pupal–adult transformation. Since the degeneration took place prior to the time the ecdysteroid peak was detected, it was assumed that there was another endocrine organ, rather than the PG, which provides the ecdysone in the adult.

In *Manduca sexta*, the titers of Bollenbacher, W. *et al.* (1981) were correlated with structural changes in PG gland cells during larval development to determine the morphological basis for secretory activity (see Fig. 13). Although the same animals were not used for collecting RIA and morphological data, the precise timing of the animals provided confidence in the correlation between structure and function. In addition, the cycles of PTTH activity are known for this animal and provide a means of assessing the structural basis for gland stimulation by PTTH. The most significant changes in gland cell structure were found in the intercellular spaces and microvesicles. The ICS increased at the time of peak ecdysteroid secretion and the microvesicles (multivesicular sacs — MVS) changed from membrane-bound vesicles to a remnant with an outer membrane devoid of the typical inner vesicles. The change to MVS remnants occurred at the time of the second PTTH cycle of activity which precedes the major late ecdysteroid peak. Thus, it was suggested that the MVS releases a gland cell product in response to stimulation by the tropic hormone. The RER whorls, which appear late in each instar following detection of the major ecdysteroid peak, were thought to occur in response to a positive stimulation of the PG by its own hormone. Positive feedback of the PG was suggested by Agui, N. and Hiruma, K. (1977) and others (chapter 1, volume 8).

3.11.2 Prothoracic glands and juvenile hormone

JH is though to maintain the PG (Gilbert, L. 1962) and consequently modulate ecdysone secretion in many insects. There is also structural evidence that specific morphological changes occur in response to this hormone.

Yin, C. and Chippendale, G. (1979) suggest that the changes in PG ultrastructure seen in *Diatraea grandiosella* occur in response to JH, i.e. in the cells of non-diapausing insects there is a cytoplasm richer in organelles such as mitochondria, ribosomes, and inclusions (glycogen) in comparison with cells from inactive diapausing insects. JH is assumed to both initiate and maintain diapause and, by inference, the JH controls PG cell secretory activity, perhaps indirectly via the brain. Since the PG of *Galleria mellonella* (Blazsek, I. *et al.*, 1975) involute when the JH titer falls, JH presumably maintains the gland in this insect also.

Friedel, T. *et al.* (1980) suggested that the molting hormone has a negative feedback control on the neurosecretory brain cells to ultimately influence JH production (chapter 1, volume 8). In *Manduca sexta* there is, in fact, a decrease in the amount of smooth endoplasmic reticulum which occurs concurrently with a rise in ecdysteroid titer (Sedlak, B. *et al.*, 1983).

3.12 Summary of prothoracic gland ultrastructure

The availability of ecdysteroid titers has made it possible to interpret changes in PG cell structure as representing cycles of hormone synthesis and release. For example, increases in the intercellular channels occur coincidentally with a rise in ecdysteroid titer in several insects (e.g. in *Manduca sexta* where the titer was directly compared with cell structure, Sedlak, B. *et al.*, 1983). The cellular features most often suggested to be involved with hormone release are the microvesicles, located within these intercellular spaces. These vesicles were stained with digitonin to show the presence of sterols such as ecdysone in *Galleria mellonella* (Blazsek, I. and Mala, J. 1978) and in *Manduca* the microvesicles undergo a change concurrently with the stimulation of PG by its tropic hormone PTTH suggesting that the structures export a cell product (Sedlak, B. 1981). Finally, since the mitochondria in *Bombyx mori* changed in response to a rise in molting hormone these organelles were also implicated in ecdysone synthesis (Beaulaton, J. 1968). The critical determination of the sites of synthesis and release of ecdysone still awaits more studies such as the immunocytochemical localization of the hormone.

ACKNOWLEDGEMENTS

This work was supported by grant number AG 01979 from the National Institutes of Health.

REFERENCES

AGGARWAL, S. K. and KING, R. C. (1969). A comparative study on the ring glands from wild type and l(2)gl mutant *Drosophilia melanogaster*. *J. Morph. 129*, 171–200.

AGGARWAL, S. K. and KING, R. C. (1971). An electron microscopic study of the corpus cardiacum of adult *Drosophila melanogaster* and its afferent nerves. *J. Morph. 134*, 437–445.

AGUI, N., BOLLENBACHER, W. E., GRANGER, N. A. and GILBERT, L. I. (1980). Corpus allatum is release site for insect prothoracicotropic hormone. *Nature 285*, 669–670.

AGUI, N. and HIRUMA, K. (1977). Ecdysone as a feedback regulator for the neurosecretory brain cells in *Mamestra brassicae*. *J. Insect Physiol. 23*, 1393–1396.

BAEHR, J. C., CASSIER, P. and FAIN-MAUREL, M. A. (1973). Contribution expérimentale et infrastructurale à l'etude de la dynamique du corpus allatum de *Rhodnius prolixus* Stal. Influence de la nutrition, de Ì activité ovarienne, de la pars intercerebralis et de ses connections. *Arch. Zool. Exp. Gen. 114*, 611–626.

BEAULATON, J. (1967). Sur la localization ultrastructurale d'une activite cholinesterasique dans le corps cardiaque de *Rhodnius prolixus* Stal. aux quatrieme et cinquieme stades larvaires. *J. Microscopie 6*, 65–80.

BEAULATON, J. (1968). Modifications ultrastructurales des cellules secrétricés de la glande prothoracique de vers à soie au cours des deux derniers âges larvaire. I: Le chondriome et ses relations avec le réticulum agranulaire. *J. Cell Biol. 39*, 501–525.

BEAULATON, J. and GRAS, R. (1980). Influence of aldehyde fixatives on the ultrastructure of the prothotracic gland cells in *Rhodnius prolixus* (Insecta Heteroptera) with special reference to their osmotic effects. *Microsc. Acta 82*, 351–366.

BENEDECZKY, I., MALA, J. and SEHNAL, F. (1980). Ultrastructural study on the innervation of prothoracic glands in *Galleria mellonella*. *Gen. Comp. Endocr. 41*, 400–407.

BERGOT, B. J., HALL, M. S., FURRER, A. A. and SCHOOLEY, D. A. (1984). Juvenile hormone titers in *Manduca sexta* throughout the life cycle. *J. Insect Physiol.* (Submitted).

BIRKENBEIL, H., ECKERT, M. and GERSCH, M. (1979). Electron microscopical immunocytochemical evidence of ecdysteroids in the prothoracic gland of *Galleria mellonella*. *Cell Tiss. Res. 200*, 285–290.

BJERSING, L. (1967). On the ultrastructure of granulosa lutein cells in porcine corpus luteum. With special reference to endoplasmic reticulum and steroid hormone synthesis. *Z. Zellforsch. Mikrosk. Anat. 82*, 187–211.

BLAZSEK, I. and MALA, J. (1978). Steroid transport through the surface of the prothoracic gland cells in *Galleria mellonella* L. *Cell Tiss. Res. 187*, 507–513.

BLAZSEK, I., BALAZ, A., NOVAK, V. J. A. and MALA, J. (1975). Ultrastructural study of prothoracic glands of *Galleria mellonella L.* in the penultimate, last larval, and pupal stages. *Cell Tiss. Res. 158*, 269–280.

BODENSTEIN, D. (1953). Studies on the humoral mechanisms in growth and metamorphosis of the cockroach, *Periplaneta americana*. II. The function of the prothoracic gland and the corpus cardiacum. *J. Exp. Zool. 123*, 413–433.

BOLLENBACHER, W. E., ZVENKO, H., KUMARAN, A. K. and GILBERT, L. I. (1978). Changes in ecdysone content during post-embryonic development of the wax moth, *Galleria mellonella*: the role of the ovary. *Gen. Comp. Endocr. 34*, 169–179.

BOLLENBACHER, W. E., AGUI, N., GRANGER, N. A. and GILBERT, L. I. (1979). *In vitro* activation of insect prothoracic glands by the prothoracicotropic hormone. *Proc. Nat. Acad. Sci. 76*, 5148–5152.

BOLLENBACHER, W. E., SMITH, S. L., GOODMAN, W. and GILBERT, L. I. (1981). Ecdysteroid titer during larval–pupal–adult development of the tobacco hornworm, *Manduca sexta*. *Gen. Comp. Endocr. 44*, 302–306.

BOWERS, B. and JOHNSON, B. (1966). An electron microscopic study of the corpora cardiaca and secretory neurons of the aphid, *Myzus persicae* (Sulz.). *Gen. Comp. Endocr. 6*, 213–230.

BOWERS, W. S. and MARTINEZ-PARDO, R. (1977). Antiallatotropins: inhibition of corpus allatum development. *Science 197*, 1369–1371.

BOWERS, W. S., OHTA, T., CLEERE, J. S. and MARSELLA, P. A. (1976). Discovery of insect antijuvenile hormones in plants. *Science 193*, 542–547.

CASSIER, P. (1979). The corpora allata of insects. *Int. Rev. Cytol. 57*, 1–73.

CASSIER, P. and FAIN-MAUREL, M. A. (1970). Contribution a l'étude intrastructurale du systéme neurosecrétéur rétrocérébral chez *Locusta migratoria migratoriodes* (R. et. F.) I. Les corpora cardiaca. *Z. Zellforsch. Mikrosk. Anat. 111*, 471–482.

CASSIER, P. and FAIN-MAUREL, M. A. (1970). Contribution a l'étude intrastructurale du systéme nerosecrétéur rétrocérébral chez *Locusta migratoria migratoriodes* (R, et. F.) II. le transit des neurosécrétions. *Z. Zellporsch. Mikrosk. Anat 111*, 483–492.

CAUSSANEL, C., BAEHR, J. C., CASSIER, P., DRAY, P. and PORCHERON, P. (1979). Correlations humorales et ultrastructurales au cours de la vitellogenese et de la periode de soins aus oeufs chez *Labidura riparia*. *C.R. Acad. Sci. D. 288*, 513–516.

CAZAL, M., JOLY, L. and PORTE, A. (1971). Etude intrastructurale des corpora cardiaca et de quelques formations annexes chez *Locusta migratoria* L. *Z. Zellforsch. Mikrosk. Anat. 114*, 61–72.

CHARPIN, P. (1975). Evolution ultrastructurale des corps allatum au cours du dernier stade larvaire chez *Choleva augustata* Fab. (Coléopterès Catopidae de la sous-famille des Catopuae). *C.R. Acad. Sci.* (Paris) *280D*, 1997–1999.

CHRISTENSEN, A. K. (1965). The fine structure of testicular interstitial cells in guinea pigs. *J. Cell Biol. 26*, 911–934.

DARJO, A. (1976). Activité des corpora allata et controle photoperiodique de la maturation ovarienne chez *Locusta migratoria*. *J. Insect Physiol. 22*, 347–355.

DEAN, R. L., BOLLENBACHER, W. E., LOCKE, M., SMITH, S. L. and GILBERT, L. I. (1980). Haemolymph ecdysteroid levels and cellular events in the intermoult/moult sequence of *Calpodes ethlius*. *J. Insect Physiol. 26*, 267–280.

DELEURANCE, S. and CHARPIN, S. (1978). Ultrastructural dynamics of the corpus allatum of *Choleva augustata* Fab. (Coleoptera, Catopidae). *Cell Tiss. Res. 191*, 151–160.

DELBECQUE, J. P., HIRN, M., DELACHAMBRE, J. and DEREGGI, M. (1978). Cuticular cycle and molting hormone levels during the metamorphosis of *Tenebrio molitor* (Insecta: Coleoptera). *Devel. Biol. 64*, 11–30.

DORN, A. (1973). Electron microscopic study on the larval and adult corpus allatum of *Oncopeltus facsiatus* Dallas. (Insecta, heteroptera), *Z. Zellforsch. Mikrosk. Anat. 145*, 447–458.

DORN, A. and ROMER, F. (1976). Structure and function of prothoracic glands and oenocytes in embryos and last larval instars of *Oncopeltus fasciatus* (Insecta, Heteroptera). *Cell Tiss. Res. 171*, 331–350.

DEKORT, C. A. D. and GRANGER, N. A. (1981). Regulation of the juvenile hormone titer. *Ann. Rev. Ent. 26*, 1–28.

ELLIOT, H. J. (1976). Structural analysis of the corpus allatum of an aphid *Aphis craccivora*. *J. Insect Physiol. 22*, 1275–1279.

ENGELMANN, F. and LÜSCHER, M. (1956). Zur Frage der Auslösung der Metamorphose bei Insecten. *Naturwissenschaften 43*, 43–44.

FAIN-MAUREL, M. A. and CASSIER, P. (1969). Pléomorphisme mitochondrial dans les corpora allata de *Locusta migratoria migratorioides* au cours de la vie imaginale. *Z. Zellforsch. Mikrosk. Anat. 102*, 543–553.

FEYEREISEN, R., JOHNSON, G., KOENER, J., STAY, B. and TOBE, S. S. (1981). Precocenes as pro-allatocidins in adult female *Diploptera punctata*: a functional and ultrastructural study. *J. Insect Physiol. 27*, 855–868.

FRIEDEL, T., FEYEREISEN, R., MEINDALL, E. C. and TOBE, S. S. (1980). The allatostatic effect of 20-hydroxyecdysone on the adult viviparous cockroach, *Diploptera punctata*. *J. Insect Physiol. 26*, 665–670.

GANAGARAJAH, G. (1965). The neuroendocrine complex of adult *Nebria brevicollis* (F.) and its relation to reproduction. *J. Insect Physiol. 11*, 1377–1387.

GERSCH, M., BIRKENBEIL, H. and UDE, J. (1975). Ultrastructure of the prothoracic gland cells of the last instar of *Galleria mellonella* in relation to the state of development. *Cell Tiss. Res. 160*, 389–397.

GIACOMELLI, F., WIENER, J. and SPIRO, D. (1965). Cytological alterations related to stimulation of the zona glomerulosa of the adrenal gland. *J. Cell Biol. 26*, 499–522.

GILBERT, L. I. (1962). Maintenance of the prothoracic gland by the juvenile hormone in insects. *Nature 193*, 1205–1207.

GLITHO, I., DELBECQUE, J. P. and DELACHAMBRE, J. (1979). Prothoracic gland involution related to molting hormone levels during the metamorphosis of *Tenebrio molitor L. J. Insect Physiol. 25*, 187–192.

GOLDSWORTHY, G. J. and MORDUE, W. (1974). Neurosecretory hormones in insects. *J. Endocr. 60*, 529–558.

GOLTZENE, F. and PORTE, A. (1978). Endocrine control by neurosecretory cells of the pars intercerebralis and the corpora allata during the earlier phases of vitellogenesis in *Locusta migratoria migratorioides* R. and F. (Orthoptera). *Gen. Comp. Endocr. 35*, 35–45.

HAGEDORN, H. H., O'CONNOR, J. D., FUCHS, M. S., SAGE, B., SCHLAEGER, D. A. and BOHM, M. K. (1975). The ovary as a source of α-ecdysone in an adult mosquito. *Proc. Nat. Acad. Sci. 72*, 3255–3259.

HERMAN, W. S. (1967). The ecdysial glands of arthropods. *Int. Rev. Cytol. 22*, 269–347.

HERMAN, W. S. and GILBERT, L. I. (1966). The neuroendocrine system of *Hyalophora ceropia* (L) (Lepidoptera: Saturniidae). I. The anatomy and histology of the ecdysial glands. *Gen. Comp. Endocr. 7*, 275–291.

HIGHNAM, K. C. (1961). The histology of the neurosecretroy system of the adult female desert locust, *Schistocerca gregaria. Quart. J. Mic. Sci. 102*, 27–38.

HODKOVÁ, G. (1976). Nervous inhibition of corpora allata by photoperiod in *Pyrrhocoris apterus. Nature 263*, 521–523.

JOHANSSON, A. S. (1958). Relation of nutrition to endocrine-reproductive functions in the milkweed bug *Oncopeltus fasciatus* (Dallas) (Heteroptera: Lygaeidae). *Nytt. Mag. Zool. 7*, 5–132.

JOHNSON, B. (1966). Ultrastructure of probable sites of release of neurosecretory materials in an insect *Calliphora stygia* Fabr. (Diptera). *Gen. Comp. Endocr. 6*, 99–108.

JOHNSON, R. A. and HILL, L. (1973). The activity of the corpora allata in the fourth and fifth larval instars of the migratory locust. *J. Insect Physiol. 19*, 1921–1932.

KATER, S. B. (1968). Cardioaccelerator release in *Periplaneta americana* (L). *Science 160*, 765–767.

KHAN, T. R., SINGH, S. B., SINGH, R. K. and SINGH, T. K. (1978). Neurosecretory control of corpora allata activity in cockroach *Periplaneta americana. Experientia 34*, 49–51.

KING, R. C., AGGARWAL, S. K. and BODENSTEIN, D. (1966). The comparative submicroscopic cytology of the corpus allatum–corpus cardiacum complex of wild type and *fes* adult female *Drosophila melanogaster. J. Exp. Zool. 161*, 151–175.

KRAMER, S. J. (1978). Age-dependent changes in corpus allatum activity *in vitro* in the adult Colorado potato beetle, *Leptinotarsa decemlineata. J. Insect Physiol. 24*, 461–464.

KROGH, I. M. and NORMANN, T. C. (1977). The corpus cardiacum and neurosecretory cells of *Schistocerca gregaraia.* Electron microscopy of resting and secreting cells. *Acta Zool. (Stockh). 58*, 69–78.

LAFON-CAZAL, M. and ARLUISON, M. (1976). Localization of monoanimes in the corpora cardiaca and the hypocerebral ganglion of locusts. *Cell Tiss. Res. 172*, 517–527.

LANZREIN, B., GENTINETTA, V., FEHR, R. and LÜSCHER, M. (1978). Correlation between haemolymph juvenile hormone titre, corpus allatum volume and corpus allatum *in vivo* and *in vitro* activity during oocyte maturation in a cockroach (*Nauphoeta cinerea*). *Gen. Comp. Endocr. 36*, 339–345.

LEA, A. O. and THOMSEN, E. (1969). Size independent secretion by the corpus allatum of *Calliphora erythrocephala. J. Insect Physiol. 15*, 477–482.

LIECHTY, L. and SEDLAK, B. J. (1978). The ultrastructure of precocene-induced effects on the corpora allata of the adult milkweed bug, *Oncopeltus fasciatus. Gen. Comp. Endocr. 36*, 433–436.

LOCKE, M. (1969). The ultrastructure of the oenocytes in the molt/intermolt cycle of an insect. *Tissue Cell 1*, 103–154.

MASNER, P., KAELIN, M. and HUEHLE, T. (1978). Effect of a substance with antiallatotropic activity on the neuroendocrine system of *Oncopeltus fasciatus* Dall. (Lygaeidae Heteroptera). *Gen. Comp. Endocr. 34*, 106.

MCDANIEL, C. N., JOHNSON, E., SAUM, T. and BERRY, S. J. (1976). Ultrastructure of active and inhibited prothoracic glands. *J. Insect Physiol. 22*, 473–481.

MELNIKOVA, F. J. and PANOV, A. A. (1975). Ultrastructure of the larval corpus allatum of *Hyphantrea cunea.* Drury (Insecta, Lepidoptera) *Cell Tiss. Res. 162*, 395–410.

MEOLA, S. M. and LEA, A. O. (1972). The ultrastructure of the corpus cardiacum of *Aedes sollicitans* and the histology of the cerbral neurosecretory system of mosquitoes. *Gen. Comp. Endocr. 18*, 210–234.

MORDUE, W. (1965). Studies on oocyte production and associated histological changes in the neuro-endocrine system in *Tenebrio molitor* L. *J. Insect Physiol. 11*, 493–503.

MORIARTY, G. C. (1973). Adenohypophysis: ultrastructural cytochemistry. *J. Histochem. Cytochem. 21*, 855–894.

NORDMANN, J. J. (1977). Ultrastructural appearance of neurosecretory granules in the sinus gland of the crab after different fixation procedures. *Cell Tiss. Res. 185*, 557–563.

NORMAN, T. C. (1965). The neurosecretory system of the adult *Calliphora erythrocephala.* I. The fine structure of the corpus cardiacum with some observations on adjacent organs. *Z. Zellforsch. Mikr. Anat. 67*, 461–501.

NORMAN, T. C. (1969). Experimentally induced exocytosis of neurosecretory granules. *Exp. Cell Res. 55*, 285–287.

ODHIAMBO, T. R. (1966a). Morphometric changes and the hormonal activity of the corpus allatum in the adult male of the desert locust. *J. Insect Physiol. 12*, 655–664.

ODHIAMBO, T. R. (1966b). The fine structure of the corpus allatum of the sexually inactive male of the desert locust. *J. Insect Physiol. 12*, 819–828.

PALEVODY, C. and GRIMAL, A. (1976). Variations cytologiques des corps allates au cours de cycle reproducteur de collembole *Folsomia candida. J. Insect Physiol. 22*, 63–72.

PANOV, A. A. and BASSURMANOVA, O. K. (1970). Fine structure of the glands in inactive and active corpus allatum of the bug, *Eurygaster intergriceps. J. Insect Physiol. 16*, 1265–1281.

PHILOGENE, B. J. R. and MASSALSKI, A. (1976). Neurosecretory system of the brain corpus cardiacum complex in diapausing *Tyria jacobaeae*, L. (Lepidoptera Arctiidae). *Int. J. Insect Morphol. Embryol 5*, 289–299.

PRENTØ, P. (1972). Histochemistry of neurosecetion of the pars intercerebrales–corpus cardiacum system of the desert locust *Schistocerca gregaria. Gen. Comp. Endocr. 18*, 482–500.

RAINA, A. K. and BORG, T. K. (1980). Corpora cardiaca-allata complex of the larvae of the pink bollworm *Pectinophora gossypiella*: an ultrastructural study in relation to diapause. *Acta Zool. (Stockh). 61*, 65–77.

RHODIN, J. A. G. (1971). Ultrastructure of the adrenal cortex of the rat under normal and experimental conditions. *J. Ultrastruct. Res. 34*, 23–71.

RICHARDS, G. (1981). The radioimmune assay of ecdysteroid titres in *Drosophila melanogaster. Mol. Cell. Endocr. 21*, 181–197.

ROMER, F. (1971). Die prothorakaldrusen der Larvae von *Tenebrio molitor* L. (Tenebrionidae, Coleoptera) und ihre Veranderungen warhend eines Hautungsklus. *Z. Zellforsch. Mikrosk. Anat. 122*, 425–455.

ROMER, F. (1974). Ultrastructural changes of the oenocytes of *Gryllus bimaculatus* DEG (Saltatoria, Insecta) during the molting cycle. *Cell Tiss. Res. 151*, 27–46.

ROMER, F., EMMERICH, H. and NOWOCK, J. (1974). Biosynthesis of ecdysones with isolated prothoracic glands and oenocytes of *Tenebrio molitor in vitro. J. Insect Physiol. 20*, 1975–1987.

ROUBOS, E. W. and VAN DER WAL-DIVENDAL, R. M. (1980). Ultrastructural analysis of peptide hormone release by exocytosis. *Cell Tiss. Res. 207*, 267–275.

SCHARRER, B. (1952). Neurosecretion. XI. The effects of nerve section on the intercerebralis-cardiacum-allatum system of the insect *Leucophaea maderae. Biol. Bull. Woods Hole 102*, 261–272.

SCHARRER, B. (1963). Neurosecretion. XIII. The ultrastructure of the corpus cardiacum of the insect *Leucophaea maderae. Z. Zellforsch. Mikrosk. Anat. 60*, 761–796.

SCHARRER, B. (1964a). Histophysiological studies on the corpus allatum of *Leucophaea maderae*. IV. Ultrastructure during normal activity cycle. *Z. Zellforsch. Mikrosk. Anat. 62*, 126–148.

SCHARRER, B. (1964b). The fine structure of *Blattrian* prothoracic glands. *Z. Zellforsch. Mikrosk. Anat. 64*, 301–326.

SCHARRER, B. (1966). Ultrastructural study of the regressing prothoracic glands of blattarian insects. *Z. Zellforsch. Mikrosk. Anat. 69*, 1–21.

SCHARRER, B. (1968). Neurosecretion. XIV. Ultrastructural study of sites of release of neurosecretory material in Blatterian insects. *Z. Zellforsch. Mikrosk. Anat. 89*, 1–16.

SCHARRER, B. (1971). Histophysiological studies on the corpus allatum of *Leucophaea maderae*. V. Ultrastructure of sites of origin and release of distinctive cellular product. *Z. Zellforsch. Mikrosk. Anat. 120*, 1–16.

SCHARRER, B. (1978). Histophysiological studies on the corpus allatum of *Leucophaea maderae*. VI. Ultrastructural characteristics in gonadectomized females. *Cell Tiss. Res. 194*, 533–545.

SCHARRER, B. and VON HARNACK, M. (1961). Histophysiological studies on the corpus allatum of *Leuophaea maderae*. IV. The effect of castration. *Biol. Bull. 121*, 193–208.

SCHARRER, B. and KATER, S. B. (1969). Neurosecretion. XV. An electron microscope study of the corpora cardiaca of *Periplaneta americana* after experimentally induced hormone release. *Z. Zellforsch. Mikrosk. Anat. 95*, 177–186.

SCHARRER, B. and WURZELMANN, S. (1974). Observations on synaptoid vesicles in insect neurons. *Zool. Jb. Physiol. 78*, 387–396.

SCHARRER, B. and WURZELMANN, S. (1978). Neurosecretion. XVII. Experimentally induced release of neurosecretory material by exocytosis in the insect *Leucophaea maderae*. *Cell Tiss. Res. 190*, 173–180.

SCHOONEVELD, H. (1979). Precocene induced necrosis and haemocyte mediated breakdown of corpora allata in nymphs of the locust *Locusta migratoria*. *Cell Tiss. Res. 203*, 25–33.

SEDLAK, B. J. (1981). An ultrastructural study of neurosecretroy fibers within the corpora allata of *Manduca sexta*. *Gen. Comp Endocr. 44*, 207–218.

SEDLAK, B. J., MARCHIONE, L., DEVOKIN, B. and DAVINO, R. (1983). Correlations between endocrine gland ultrastructure and hormone titers in the fifth larval instar of *Manduca sexta*. *Gen. Comp. Endocr.* (In press).

SIEW, Y. C. (1965). The endocrine control of adult reproductive diapause in the crysomelid beetle, *Galeruca tanaceti* L.-II. *J. Insect Physiol. 11*, 463–479.

SMITH, R. E. and FARQUAHAR, M. G. (1966). Lysosome function in the regulation of the secretory process in cells of the anterior pituitary gland. *J. Cell Biol. 31*, 319–347.

SMITH, U. and SMITH, D. S. (1966). Observations on the secretory processes in the corpus cardiacum of the stick insect *Carausius morosus*. *J. Cell Sci. 1*, 59–66.

SRIVASTAVA, U. S. (1958). Prothoracic glands in *Tenebrio molitor* L. (Coleoptera, Tenebrionidae). *Nature, Lond. 181*, 1668.

STAY, B. and TOBE, S. S. (1977). Control of juvenile hormone biosynthesis during the reproductive cycle of a viviparous cockroach. I. Activation and inhibition of corpora allata. *Gen. Comp. Endocr. 33*, 531–540.

STRONG, L. (1965). The relationships between the brain, corpora allata, and oocyte growth in the Central American locust, *Schistocerca* sp. I. The cerebral neurosecretory system, the corpora allata, and oocyte growth. *J. Insect Physiol. 11*, 135–146.

SZIBBO, C. M. and TOBE, S. S. (1981). Cellular and volumetric changes in relation to the activity cycle in the corpora allata of *Diploptera punctata*. *J. Insect Physiol. 27*, 655–665.

TAKEDA, N. (1977). Histo-physiological studies on the corpus allatum during prepupal dispause in *Monema falvescens Lepidoptera*. *J. Morph. 153*, 245–262.

THOMSEN, E. and THOMSEN, M. (1970). Fine structure of the corpus allatum of the female blowfly, *Calliphora erythrocephala*. *Z. Zellforsch. Mikrosk. Anat. 110*, 40–60.

TOBE, S. S. and SALEUDDIN, A. S. M. (1977). Ultrastructural localization of juvenile hormone biosynthesis by insect corpora allata. *Cell Tiss. Res. 183*, 25–32.

TOMBES, A. S. and SMITH, D. S. (1970). Ultrastructural studies on the corpora cardica-allata complex of the adult alfalfa weevil, *Hypera postica*. *J. Morph. 132*, 137–148.

UNNITHAN, G. C., BERN, H. A. and NAYAR, K. K. (1971). Ultrastructural analysis of the neuroendocrine apparatus of *Oncopeltus fasciatus* (Heteroptera). *Acta. Zool.* (Stockh.) *52*, 117–143.

UNNITHAN, G. C., NAIR, K. K. and BOWERS, W. S. (1977). Precocene-induced degeneration of the corpus allatum of adult females of the bug, *Oncopeltus fasciatus*. *J. Insect Physiol. 23*, 1081–1094.

UNNITHAN, G. C., NAIR, K. K. and SYED, A. (1980). Precocene-induced metamorphosis in the desert locust, *Schistocerca gregaria*. *Experientia 36*, 135–138.

WAKU, Y. and GILBERT, L. I. (1964). The corpora allata of the silkmoth, *Hyalophora cecropia*: an ultrastructural study. *J. Morph. 115*, 69–96.

WILLEY, R. B. and CHAPMAN, G. B. (1960). The ultrastructure of certain components of the corpora cardiaca in orthopteroid insects. *J. Ultrastruct. Res. 4*, 1–14.

WIGGLESWORTH, V. (1955). The breakdown of the thoracic gland in the adult insect, *Rhodnius prolixus*. *J. Exp. Biol. 32*, 485–491.

YIN, C. M. and CHIPPENDALE, G. M. (1979). Ultrastructural characteristics of insect corpora allata in relation to larval diapause. *Cell Tiss. Res. 197*, 453–461.

ADDENDUM

1. Corpora cardiaca

Two books with relevant chapters on the structure of neurohemal organs, the characterization of neurosecretory granules and the mode of release from neurohemal organs have been published.

In her book, Raabe, M. (1982) reviews the synthesis, storage and release of neurohormones and includes a comprehensive tabulation of histochemical staining of neurosecretory products and a size classification of neurosecretory granules in neurosecretory cells and the CC.

Gupta, A. P. (1983) edited a book in which individual authors discuss the structure of insect neurohemal organs. Included is a collection of elegant diagrams plus micrographs at both the light and electron microscope levels for the orders of Apterygota, Odonata, Isoptera, Orthoptera, Lepidoptera (with emphasis on *Manduca sexta*), Hymenoptera, Coeloptera (with emphasis on *Leptinotarsa decemlineata*), and adult Diptera. There is also a useful chapter in which Dr. Berta Scharrer summarizes the mechanisms of release of neurosecretory products. She discusses the role of exocytosis and microvesicles in neurohormone release and points to evidence from studies with the mammalian neurohypophysis that the microvesicles seen at neurosecretory terminals may be the site of calcium ion sequestration. Nordmann, J. and Chevalier, J. (1980), Shaw, F. and Morris, J. (1980) propose that hormone release is triggered by a calcium ion influx into the cell which must subsequently be removed to reestablish the basal level of cellular calcium, Dr. Scharrer cautions that "it remains to be seen to what extent these conclusions drawn for mammalian NS systems also apply to the microvesicles in the neurohemal organs of arthropods and other invertebrates".

2. CA volume changes

Several studies on the relationship between CA volume and CA activity were published in 1982.

Baehr, J. *et al.* (1982) report that during vitellogenesis when there is an increase in JH, there is a size increase in adult *Labidura riparia* CA and SER is found in the glands at this time. If the neurosecretory cells within the CC are cauterized in this insect, ovarian activity declines and the CA volume decreases; ribosomes and residual bodies predominate in the CA gland cells indicating degeneration.

In contrast Szibbo, C. *et al.* (1982) showed that there was an increase in CA volume and nuclear number during the last two instars of *Diploptera punctata.* At these stages the CA activity as determined by *in vitro* radiochemical assay for JH, was low and the increase in gland volume and cell number was attributed to somatic growth rather than hormone production.

Finally Khan, M. *et al.* (1982) measured CA activity *in vitro* (via JH radiochemical assay) and simultaneously measured CA volume in adult *Leptinotarsa decemlineata.* The authors report that there was no correlation between spontaneous JH secretion and volume changes, although higher production of JH was found in larger versus smaller glands. Khan points to a similar lack of correlation between CA volume and CA activity in *Schistocerca gregaria* (Tobe, S. and Pratt, G. 1975) where again there is a relationship between maximal gland activity and CA volume.

ADDITIONAL REFERENCES

Baehr, J. C., Cassier, P., Caussanel, C., and Porcheron, P., (1982). Activity of corpora allata, endocrine balance and reproduction in female *Labidura riparia* (Dermaptera). *Cell Tissue Res. 225*, 267–282.

Gupta, A. P. (1983). *Neurohemal Organs of Arthropods, Their Development, Evolution, Structures and Functions.* Charles C. Thomas, Springfield, Illinois.

Khan, M. A., Doderer, A., Koopmanschap, A. B., and de Kort, C. A. D. (1982). Improved assay conditions for measurement of corpus allatum activity *in vitro* in the adult Colorado potato beetle, *Leptinotarsa decemlineata. J. Insect Physiol, 28*, 279–284.

Nordmann, J. J., and Chevalier, J. (1980). The role of microvesicles in buffering (Ca^{2+}) in neurohypophysis. *Nature* (Lond.) *287*, 1357–1360.

Raabe, M. (1982). *Insect Neurohormones*, Plenum Press, N.Y.

Shaw, F. D. and Morris, J. F. (1980). Calcium localization in the rat neurohypophysis. *Nature* (Lond.) *287*, 56–58.

Szibbo, C. M., Rotin, D. M., Feyereisen, R. and Tobe, S. S. (1982). Synthesis and degradation of C_{16} juvenile hormone (JH III) during the final two stadia of the cockroach, *Diploptera punctata. Gen. Comp. Endocrinol. 48*, 25–32.

Tobe, S. S. and Pratt, G. E. (1975). The synthetic activity and glandular volume of the corpus allatum during ovarian maturation in the desert locust *Schistocerca gregaria. Life Sci. 17*, 417–422.

3 Neurosecretion

IAN ORCHARD and BARRY G. LOUGHTON

York University, Downsview, Ontario, Canada

1 INTRODUCTION

Insects have provided favourable preparations for the study of neurosecretory phenomena, leading to a substantial increase in our understanding of neurosecretion. The morphological work of Hanström and experimental physiology of Kopeč and Wigglesworth (for references see Gabe, M. 1966 and chapter 1, this volume) demonstrated the importance of neurosecretion in insects in the control of morphogenesis at the onset of research in this field. Since that time neurosecretory cells have been implicated in controlling a plethora of activities (Table 1). Since the pioneering work of the Scharrers and other investigators (see Gabe, M. 1966) it has become accepted that neurosecretory cells play a key role in regulation throughout the animal kingdom. Within the vertebrates, the neurosecretory cells function largely as an intermediary between the nervous and endocrine systems, generally not producing the final hormone (which is the product of non-neural tissue) but providing the "final common pathway" (Scharrer, B. 1967) between the hypothalamus and the anterior pituitary. Exceptions are the classical neurohormones vasopressin and oxytocin which are released from the posterior pituitary and act directly on target issues other than endocrine cells.

Insects rely more heavily on the neurosecretory cell to produce the final hormone, and most putative insect hormones are the product of neurosecretory cells. Exceptions to this rule include ecdysone and juvenile hormone (JH) which are produced from non-neural tissue. The control of synthesis and release of these two hormones from the prothoracic gland and corpora allata however, are under the influence of neurohormones (Wigglesworth, V. 1940; Bollenbacher, W. *et al.*, 1979; Williams, C. 1948; Bhaskaran, G. *et al.*, 1980; Tobe, S. and Stay, B. 1980; see volume 8, chapter 1) thereby reaffirming the principle of the "final common pathway" in insects.

The historical aspect of the study of the neurosecretory cells and their hormones has been dealt with in chapter 1 of this volume. It is the purpose of this chapter to present an overview of the morphology, biochemistry and physiology of neurosecretory cells in insects and thereby lay a foundation for other chapters on peptide hormones (chapters 4 and 5 this volume). We shall draw upon other invertebrates and vertebrates where pertinent to emphasize the gaps in the data on insects. We shall examine how the neurosecretory cell functions, and how it is integrated within the animal as a whole.

1.1 The concept of neurosecretion

The concept of neurosecretion was developed more than 60 years ago with the description by Speidel of giant neurons in the caudal spinal cord of selachians which had the attributes of glandular cells (Speidel, 1919, quoted in Gabe, M. 1966). The Scharrers subsequently made extensive observations on a wide variety of both invertebrates and vertebrates, and confirmed the apparent universal existence of nerve cells which appeared to manufacture a secretory product. Since that time, the term neurosecretory cell has been confined to those "nerve cells that specialize in the manufacture of chemical mediators to a degree greatly surpassing that of the more conventional neurons, and that have a secretory activity comparable to that of gland cells" (Scharrer, B. 1977). The present concept of neurosecretion includes physiological parameters not included in the earlier descriptions, in that these chemical mediators are released into the circulatory system and have a physiological effect upon the animal (see Gabe, M. 1966).

It is probably true to state that all neurons are secretory and that any aspect associated with secretion may be specialized in an individual neuron (e.g. the biochemical nature of the secretory product; synthesis; release; quantity of product formed, etc.) There are, therefore, differences between neurons and an obvious continuum of activities is evident. This has led to difficulties in terminology attested to by the numerous and varied articles on the subject (e.g. Dismukes, R. 1979 and subsequent commentaries; Scharrer, B. 1969; Gabe, M. 1966). Only at the extremes of the spectrum of activities may there be unambiguity. At one extreme are neurons (so-called conventional or ordinary neurons) which send axons directly to their target site for communication at specialized contact areas called synapses (see volume 5). The presynaptic terminal quickly releases precisely metered small amounts of neurotransmitter which travel across a very short distance (15–50 nm) to the adjacent (postsynaptic) cell and are subsequently removed almost as quickly. They are involved in the rapid communication of short messages, controlling immediate physiological reactions. Among examples within this group are the motor nerve cells releasing acetylcholine at the neuromuscular junction of vertebrates.

At the other extreme, neurosecretory neurons simply pour relatively large amounts of neurohormone into the circulatory system. These neurohormones are capable of regulating multiple and distant target cells. Compared to neurotransmitters, physiologically active titres of neurohormone are much less rapidly achieved, persist for some considerable time (before finally being destroyed) and are long-lasting in their effect. Neurohormones thereby carry relatively general, persistent messages. The classical neurosecretions of vertebrates are peptides such as oxytocin and vasopressin which regulate milk ejection and diuresis, amongst other physiological responses.

Similarly, the classical neurosecretions of invertebrates are largely peptides, hence the term "peptidergic neurosecretory cell" (Bern, H. and Knowles, F. 1966), although more recently in both vertebrates and invertebrates it has been recognized that biogenic amines may also act as neurohormones (Silverberg, A. *et al.*, 1978; Orchard, I. 1982a).

The borderline between "ordinary neurons" and "neurosecretory neurons" is less sharp than originally thought, and difficulties arise in cataloguing neurons which show characteristics of both. For example, some neurons are ultrastructurally similar to neurosecretory cells but appear to make direct contact with endocrine and non-endocrine tissue. They do not, however, make synaptic contact, but merely release their product in the immediate vicinity of adjacent cells. They have been described as releasing "local neurohormone" or "non-synaptic transmitter" (Dismukes, R. 1979). Examples in insects include neurons innervating epidermis (Maddrell, S. 1965), Malpighian tubules (Maddrell, S. and Gee, J. 1974), hindgut and reproductive tissue (Johnson, B. 1963; Tombes, A. 1976), skeletal muscle (Anwyl, R. and Finlayson, L. 1973; Hoyle, G. 1975) and prothoracic glands (Benedeczky, I. *et al.*, 1980; Finlayson, L. 1980). Unfortunately, we know little of the physiology of the vast majority of these examples, and the fragmentary nature of the data makes it impossible to catalogue them with any confidence. Two identified substances which may fall into this intermediate category in insects are proctolin and octopamine. Proctolin is a myotropic substance with potent activity on a number of invertebrate visceral, skeletal and heart muscle preparations (Brown, B. 1975; Piek, T. and Mantel, P. 1977; Miller, T. 1979; Cook, B. and Holman, G. 1979; Bishop, C. *et al.*, 1981). It has been found within identified neurons in the cockroach (O'Shea, M. and Adams, M. 1981; Bishop, C. *et al.*, 1981). Proctolin is a pentapeptide which has been proposed to be a neurotransmitter at the cockroach proctodeal muscles (Brown, B. 1975) in a similar fashion to some vertebrate peptidergic neurotransmitters (Hokfelt, T. *et al.*, 1980). However, its mode of action is unknown, and there is controversy as to whether it is a neurotransmitter or whether it modulates release of another transmitter such as glutamate (Cook, B. and Holman, G. 1979) or acetylcholine (Penzlin, H. *et al.*, 1981). Furthermore, proctolin has profound effects on tissues apparently not innervated by proctolin-containing cells and hence it has been considered a neurohormone, though it has yet to be detected in the haemolymph. Conversely recent observations on the distribution of the classical vertebrate neurohormones have suggested even they may have secondary roles as neurotransmitters or neuromodulators (Hokfelt, T. *et al.*, 1980).

Octopamine is another example of a multifunctional active factor in that it can be a neurotransmitter and neurohormone within insects, and can also modulate the activity expressed by other motoneurons (see Evans, P. 1980; Orchard, I. 1982a).

The possession of different mechanisms of nervous communication utilizing neurotransmitters, neurohormones or hard-to-define intermediaries such as local neurohormones, non-synaptic transmitters, or neuromodulators clearly makes good physiological sense. There is a need for complexity in communication; a flexibility in degree of privacy and speed of delivery, duration of message, and time of action leads to this variability. It is not surprising, therefore, to find that neurons have evolved to meet the demands of individual systems. One should not allow one's appreciation of phenomena to be limited by the inadequacy of definitions based on earlier more primitive concepts.

1.2 Identification of neurosecretory cells

Neurosecretory cells were originally described because of certain histological and histochemical features, and these features are still used as a first step in their identification. However, there are a number of dangers in attributing neurosecretory function to a neuron on histological grounds (Bern, H. 1966; Bern, H. and Knowles, F. 1966) and several authors have presented rigorous criteria including morphological, biochemical and physiological evidence which need to be satisfied before ascribing neurosecretory function to a cell (see Berlind, A. 1977). It is hardly surprising to note that within insects, indeed within any invertebrate, these criteria have been satisfied for only a few putative neurosecretory cells. However, one should not be discouraged since the neurosecretory cells which do

*Table 1: Examples of processes in insects for which there is believed to be neurohormonal control**

Function	Hormone	Insect	Tissue or cell	Reference
Control of:				
Ecdysone synthesis	Prothoracicotropic hormone	*Rhodnius*	MNC	Wigglesworth, V. 1940
		Manduca	LNC CA	Agui, N. *et al.*, 1979
		Hyalophora	MNC LNC	Williams, C. 1948
		Megoura	MNC	Steel, C. 1978
	Egg development neurohormone	*Aedes*	Brain	Hagedorn, H. *et al.*, 1979
Juvenile hormone synthesis	Allatotropin	*Manduca*	?	Bhaskaran, G. *et al.*, 1980
		Diploptera	?	Tobe, S. and Stay, B. 1980
Polymorphism of offspring	Virginoparae inducing hormone	*Megoura*	Brain Group I Cells	Steel, C. and Lees, A. 1977
Diapause in eggs	Diapause hormone	*Bombyx*	SOG	Kubota, Y. *et al.*, 1977
Pupariation	Anterior retraction factor	*Sarcophaga*	CNS	Zdarek, J. *et al.*, 1981
	Puparium immibolization factor	*Sarcophaga*	CNS	Zdarek, J. and Fraenkel, G. 1972
Cuticular tanning	Bursicon	*Tenebrio*	Thoracic g.	Grillot, J. *et al.*, 1976
		Manduca	Abd. g.	Truman, J. 1973
		Locusta	Abd. g.	Vincent, J. 1972
		Periplaneta	Terminal g.	Mills, R. *et al.*, 1965
	Puparium tanning factor	*Sarcophaga*	CNS	Fraenkel, G. *et al.*, 1972
Eclosion behaviour	Eclosion hormone	*Hyalophora*	Brain, CC and/or VNC	Truman, J. *et al.*, 1981
		Manduca		Truman, J. *et al.*, 1981
Reproductive behaviour	Calling hormone	*Hyalophora*	CC	Riddiford, L. 1977
Ovulation	Oviposition inducing hormone	*Locusta*	CC	Lange (personal communication)
	Myotropic ovulation hormone	*Rhodnius*	MNC	Kriger, F. (1980)
Haemolymph sugar levels	Hyperglycaemic hormone (trehalagon)	*Periplaneta*	CC	Steele, J. 1961
		Locusta	CC	Goldsworthy, G. 1969
				Orchard, I. and Loughton, B. 1981
		Calliphora	CC	Normann, T. and Duve, H. 1969
	Hypoglycaemic hormone	*Calliphora*	MNC	Duve, H. 1978
Haemolymph lipid levels	Hyperlipaemic hormone (Adipokinetic hormone)	*Locusta*	CC (glandular lobe)	Stone, J. *et al.*, 1975
	Hypolipaemic hormone	*Locusta*	CC (storage lobe)	Orchard, I. and Loughton, B. 1980
Protein synthesis in fat body	Protein synthesis inhibitor	*Locusta*	CC	Carlisle, J. and Loughton, B. 1979
	Protein synthesis stimulator	*Locusta*	CC	Carlisle, J. (personal communication)
Fluid secretion by Malpighian tubules	Diuretic hormone	*Rhodnius*	VGM	Maddrell, S. 1963
		Glossina	VGM	Maddrell, S. and Gee, J. 1974
		Lepidoptera spp.	Brain and CC/CA	Nicolson, S. 1980
		Carausius	CNS	Pilcher, D. 1970
Fluid secretion by Malpighian tubules and rectum	Antidiuretic hormone	*Periplaneta*		Steele, J. and Tolman, J. 1980
		Schistocerca		Mordue, W. 1970
		Locusta	CC	Cazal, M. and Girardie, A. 1968

Table 1: *Examples of processes in insects for which there is believed to be neurohormonal control** — Continued

Function	Hormone	Insect	Tissue or cell	Reference
Absorption from rectum	Chloride-transport stimulating hormone	*Schistocerca*	CC	Phillips, J. *et al.*, 1980
Heartbeat	Cardioaccelerator	*Periplaneta*	Brain/CC	Gersch, M. *et al.*, 1970
		Schistocerca	CC	Mordue, W. and Goldsworthy, G. 1969
		Locusta	CC	

Abd. g., Abdominal ganglia; CA, corpus allatum; CC, corpus cardiacum, CNS, central nervous system; MNC medial neurosecretory cells; LNC, lateral neurosecretory cells; SOG, sub-oesophageal ganglion; Terminal g., terminal ganglion; VGM, ventral ganglionic mass; VNC ventral nerve cord.

* Some cases are included for which the evidence of neurohormonal control is not yet complete. The reference list is meant only to be representative and is not complete.

satisfy the criteria indicate that there is indeed a good correlation between the morphological characterization and the cells being neurosecretory in a functional sense. Consequently it is generally accepted that the morphological identification of neurosecretory cells has some validity (Rowell, H. 1976). However, considerable caution should still be exercised since many of the characteristic features of neurosecretory cells are due to the peptidergic nature of their hormones, and as such are features shared by the recently discovered neurons which may use peptides as their neurotransmitter (Brown, B. 1975; O'Shea, M. and Adams, M. 1981; Hokfelt, T. *et al.*, 1980).

Historically, neurosecretory cells were recognized by the use of a number of light microscopic stains such as chrome–haematoxylin–phloxine, paraldehyde fuchsin, azan (review by Rowell, H. 1976) and this technique has been used extensively in insects (see Raabe, M. *et al.*, 1974; Rowell, H. 1976; Panov, A. 1980). The stains almost certainly do not stain the hormone *per se* but rather the carrier protein associated with them (see Friedel, T. and Loughton, B. 1982, and section 3.3.2). The number of cells identified with these techniques depends upon the variety of stains used and with the physiological state of the cell (complete review by Rowell, H. 1976; Panov, A. 1980). If used carefully (e.g. paraldehyde fuchsin stains connective tissue) these stains can be used as a first indication that a cell may be neurosecretory, and, importantly from the definition of neurosecretion, permit the tracing of their processes to a likely release site (neurohaemal area) closely opposed to the haemolymph or circulatory fluid.

Electron microscopic examination of cells which stain with these light microscopic stains has revealed a common substructure. The cells all have large numbers of so-called "elementary neurosecretory granules" which are membrane-bound vesicles usually about 100–300 nm in diameter and characteristically electron-dense (e.g. in insects, Finlayson, L. and Osborne, M. 1968, 1975; Morris, G. and Steel, C. 1975; 1977a,b; Fifield, S. and Finlayson, L. 1978; Scharrer, B. and Kater, S. 1969; Scharrer, B. 1968; chapter 2, this volume). It is a characteristic of neurosecretory cells to have an opalescent blue–white appearance under reflected light that allows them to be visualized in living preparations (chapter 1, this volume). This feature is a result of the scattering of light by the neurosecretory granules and is termed the Tyndall-blue effect. The feature has great practical value to the physiologist since it may be used to identify cells for impalement with microelectrodes (see section 3.1), dissecting free for bioassay (Agui, N. *et al.*, 1979; Berlind, A. and Maddrell, S. 1979) or ablation (Steel, C. and Lees, A. 1977).

While a characteristic feature of neurosecretory cells was originally believed to be the proteinaceous nature of the secretory products (i.e. peptide hormone) it has become apparent that amines may also act as neurohormones in that they can be released into the circulatory system and have a widespread action. There is particularly good evidence that octopamine is released into the haemolymph of the crustacean *Homarus* (Livingstone, M. *et al.*, 1980) and the insects *Locusta* and *Schistocerca* (Goosey, M. and Candy, D. 1980; David, J. and Lafon-Cazal, M. 1979; Orchard, I. *et al.*, 1981b). Physiological roles for circulating octopamine have been described in *Periplaneta* (Gole, J. and Downer, R. 1979; Downer, R. 1979) and *Locusta* (Orchard, I. *et al.*, 1981b, 1982). Within vertebrates norepinephrine has been suggested as serving a hormonal function (Silverberg, A. *et al.*, 1978). Aminergic cells have their own characteristic histochemistry (Falck-Hillarp technique; Klemm, N., 1980), and ultrastructurally possess large electron-dense granules, although of a lower range of diameters than peptidergic cells, being some 60–100 nm diameter (for a complete review of biogenic amines in insects see Evans, P. 1980). The precise position of aminergic cells in the overall scheme of animal co-ordinations has been obscure for some considerable time. The recent discovery that in insects octopamine (Orchard, I. 1982a) and in humans, norepinephrine (Silverberg, A. *et al.*, 1978), may act as neurohormones illustrates that aminergic cells may be classified as neurosecretory, although it must be remembered that they also appear to act as neurotransmitters and neuromodulators (Evans, P. 1980; Orchard, I. 1982a; see also Barrington, E. 1975).

Even though cells showing histochemical and ultrastructural evidence of neurosecretion have been extensively described in insects, the number of cases in which they are actually *known* to release hormones is still extremely low. This, however, is primarily due to the lack of suitable techniques for the recognition of hormonal activity, and the functional significance of many of these cells will no doubt become evident as techniques improve.

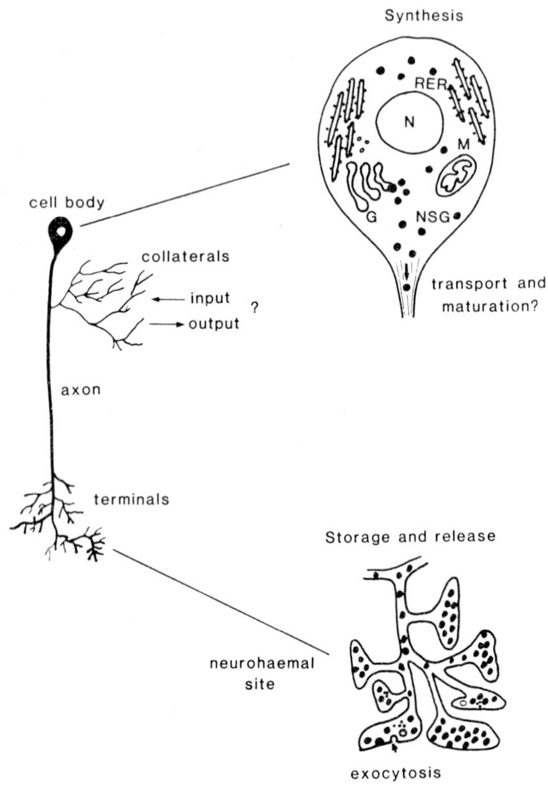

FIG. 1. Diagrammatic representation of a "typical" insect neurosecretory cell found within the central nervous system. The cell body is the site of synthesis of neurosecretory products. The material synthesized by the rough endoplasmic reticulum (RER) is packaged into membrane-bound neurosecretory granules (NSG) by the golgi apparati (G). These granules are transported along the axon, during which time they may undergo maturation. Storage of granules occurs at the swollen terminals which constitute a neurohaemal site. It is here that release of neurohormone occurs by exocytosis. Collaterals are evident branching from the axon close to the cell body. These collaterals may be the site of integration constituting both input and output. M, mitochondrion; N, nucleus.

2 MORPHOLOGY

2.1 Anatomy of neurosecretory cells

In gross anatomy neurosecretory cells are no different from other neurons. A variety of processes can emanate from a cell body which contains the nucleus, resulting in a cell which may be monopolar or multipolar. Monopolar neurosecretory cells are more typical of the central nervous system, while multipolar neurosecretory cells are common in the peripheral nervous system.

In the central nervous system the monopolar neurosecretory cell has an axon passing some considerable distance to arrive at a release site (Fig. 1). Soon after leaving the cell body, small processes emanate from the axon and pass into the neuropile resulting in extensive arborizations. It has been presumed that these processes are responsible for integration (see section 3.4; also Pipa, R. 1982). It has yet to be shown if peripheral neurosecretory cells have processes passing into the central nervous system although one would predict they should.

The specialization of neurosecretory cells arises at their terminals since this is the site of hormone release into the circulatory system. The axons divide repeatedly into fine branches each of which terminates in a swollen ending which may contain large numbers of neurosecretory granules (Fig. 1). The resulting increase in surface area enables large quantities of hormone to be released into the haemolymph. These branched terminals must lie outside the so called "blood–brain" barrier to allow the hormone easy access into the haemolymph. Consequently, the axon terminals penetrate the perineurium surrounding the central nervous system and are therefore only separated from the haemolymph by an acellular basement membrane (Brady, J. and Maddrell, S. 1967; Maddrell, S. 1974; Maddrell, S. and Nordmann, J. 1979; Normann, T. 1965; Finlayson, L. and Osborne, M. 1968). These terminals constitute regions of both storage and release of neurohormone and are called *neurohaemal areas*. Often, though not always, many different neurosecretory cells will send their axons to the same area which therefore becomes swollen with numerous branched endings. These larger areas of storage and release are termed *neurohaemal organs* and examples in insects are the corpora cardiaca and the perisympathetic organs.

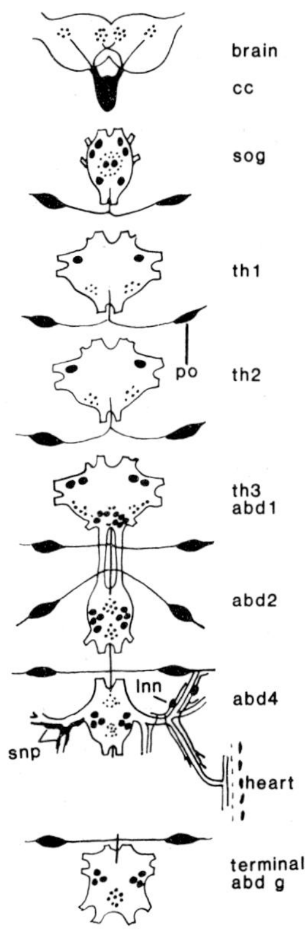

FIG. 2. Distribution of neurosecretory cells (dots) and neurohaemal organs (shown as black swellings) associated with the central and peripheral nervous system of a "typical" insect. abd, abdominal ganglion; cc, corpus cardiacum; lnn, link nerve neurosecretory neuron; po, perisympathetic organ; snp, segmental nerve neurosecretory process; sog, sub-oesophageal ganglion; th, thoracic ganglion. Compiled from Raabe, M. 1971a; Fifield, S. and Finlayson, L. 1978; Orchard, I. and Finlayson, L. 1977b; Orchard, I. 1975.

2.2 Distribution within the nervous system

The distribution of putative neurosecretory cells in the insect nervous system has been extensively reviewed in the literature in recent years (Maddrell, S. 1974; Maddrell, S. and Nordmann, J. 1979; Rowell, H. 1976; Raabe, M. 1971a,b; Finlayson, L. 1980; Orchard, I. 1982b). Figure 2 presents a generalized diagram of the distribution of putative neurosecretory cells and neurohaemal organs. However, it must be borne in mind that many of these cells have been described as neurosecretory only by histochemical or ultrastructural studies. Whether or not they all have axons passing to neurohaemal areas and release hormone into the haemolymph has not been determined. Some of these cells may use a peptide or amine as a neurotransmitter or as a neuromodulator.

Clearly, neurons which show cytological evidence of being neurosecretory are found in all parts of the nervous system. Indeed, neurosecretory cells have been described in every ganglion of the central nervous system (Raabe, M. 1971a,b; Raabe, M. *et al.*, 1971) and also within the peripheral nervous system (Finlayson, L. and Osborne, M. 1968; Fifield, S. and Finlayson, L. 1978; Hinks, C. 1975; Orchard, I. and Finlayson, L. 1977a,b; Wasserthal, W. and Wasserthal, L. 1980). Many neurosecretory axons end in storage and release sites — neurohaemal organs, while others may run directly to a target organ (see section 1.1).

The corpora cardiaca are the classical neurohaemal organs in insects where neurohormones which are produced essentially in neurosecretory cells of the brain are stored and released. The typical pattern to emerge for brain neurosecretory cells contributing axons to the corpora cardiaca, reveals them to be present in the pars intercerebralis (medial neurosecretory cells), the pars lateralis (lateral neurosecretory cells) and

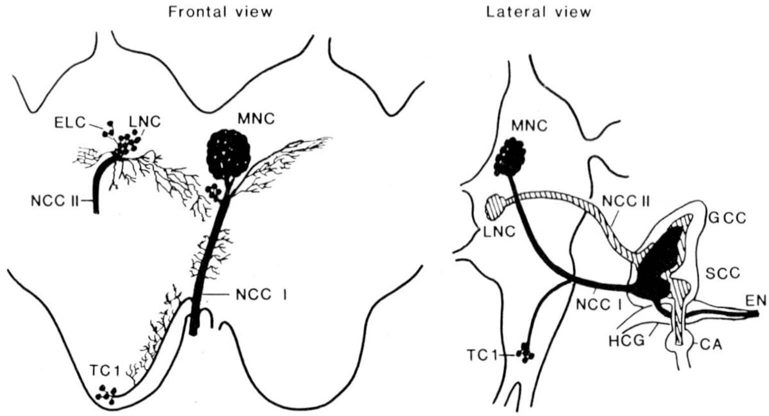

FIG. 3. Diagrammatic representation of nerve projections of the neurosecretory cells in the brain of an insect as typified by the locust. CA, corpus allatum; ELC, extralateral cells; EN, oesophageal nerve; GCC, glandular lobe of the corpus cardiacum; HCG, hypocerebral ganglion; LNC, lateral neurosecretory cells; MNC, medial neurosecretory cells; NCC I and NCC II, nervus corpus cardiacum I and II; SCC, storage lobe of the corpus cardiacum; TC, tritocerebral cells 1. Redrawn from figures based on cobalt backfillings in Koontz, M. and Edwards, J. 1980; Mason, C. 1973; Rademakers, L. 1977.

possibly in the tritocerebrum (tritocerebral cells) (Fig. 3). The number of cells present in these groups varies between species. For example, there may be as many as 500 medial neurosecretory cells, as in *Schistocerca*, or as few as 4–5, as in *Aphis* (complete review in Rowell, H. 1976).

These groups of neurosecretory cells provide the axons which constitute the nervi corporis cardiaci (NCC) passing to the corpora cardiaca (see Fig. 3). In some insects there are a pair of nerves, the NCC I and NCC II, passing to each side of the corpus cardiacum. In others these nerves are fused so that only one nerve, the NCC, passes into the corpus cardiacum on each side. In insects with separate NCC I and NCC II it has been shown that the medial neurosecretory cells contribute axons to NCC I, while lateral neurosecretory cells contribute axons to NCC II. The medial neurosecretory cell axons decussate and pass out of the contralateral NCC I while lateral neurosecretory cells have axons passing out of the ipsilateral NCC II. Exceptions of course are evident and, for example, Buys, C. and Gibbs, D. (1981) have described lateral neurosecretory cells in *Manduca* whose axons decussate and pass out through the contralateral NCC. Retrograde axonal transport of cobalt ions has been used extensively to map out axons contributing to the corpora cardiaca, and has aided our understanding of the neuroanatomy of this system considerably. Excellent examples can be found for *Periplaneta* (Pipa, R. 1978; Gundel, M. and Penzlin, H. 1980); *Schistocerca* (Mason, C. 1973); *Locusta* (Rademakers, L. 1977b) and *Manduca* (Buys, C. and Gibbs, D. 1981). Orthograde filling of NCC I and NCC II with cobalt can provide a clear picture of the fibre pathways within the corpora cardiaca (Fig. 3). In *Schistocerca* (Mason, C. 1973) fibres in NCC I project and end mainly in the neurohaemal storage lobe of the corpus cardiacum but are also found in the posterior glandular lobe. Others continue into the hypocerebral ganglion and oesophageal nerves. Some fibres in NCC II also end in the neurohaemal storage lobe and posterior glandular lobe, but others enter the corpora allata via the nervi corporis allati I (NCA I).

While it is clear that the corpus cardiacum is the major neurohaemal area for cerebral neurosecretory cells, other areas may also exist. Thus it has recently been shown that the peripheral regions of the corpus allatum in *Manduca* constitute a neurohaemal area, part of whose function is likely to be the storage and release site for those lateral neurosecretory cells which produce prothoracicotropic hormone (Gilbert, L. *et al.*, 1981). The corpus cardiacum may also produce neurohormones from intrinsic neurosecretory cells (Chapter 5, this volume; chapter 7, volume 8). These cells have been described in numerous insects (e.g. Smith, U. and Smith, D. 1966; Normann, T. 1965, 1973; Scharrer, B. and Kater, S. 1969; Bowers, G.

and Johnson, B. 1966; Highnam, K. 1961a; for complete review see Goldsworthy, G. and Mordue, W. 1974; and chapter 2, this volume). Intrinsic cells may exist as integral members of the neurohaemal area, or as a separate lobe to the corpus cardiacum as in *Schistocerca* and *Locusta* where they form the glandular lobe.

It was once believed that the brain and corpus cardiacum were the only sites of neurosecretory cells and that the corpus cardiacum was the only neurohaemal organ in insects. Raabe, M. (1965) was the first to describe neurosecretory cells in the ventral ganglia, and to locate neurohaemal organs on the median nervous system of phasmids, the so-called "organes périsympathiques". These perisympathetic organs are structurally identical to the corpus cardiacum and have since been found in at least 15 orders of insects (Raabe, M. *et al.*, 1971; Grillot, J. 1972). (Extensive electron microscopic studies may be found in Fifield, S. and Finlayson, L. 1978; Finlayson, L. and Osborne, M. 1968; Brady, J. and Maddrell, S. 1967; Chalaye, D. 1967; 1974a,b.) The neurosecretory cells within the ventral ganglia have neurohaemal areas with a very widespread distribution (see Fifield, S. and Finlayson, L. 1978). Many processes terminate along the median (unpaired) nervous system producing such a network of branches that the nerve is swollen, thereby producing the perisympathetic organ. Others form terminals along the peripheral segmental (paired) nerves (Figs 2 and 4). Diffuse neurohaemal terminations on segmental nerves in insects have been described in *Rhodnius* (Maddrell, S. 1967, 1970; Anwyl, R. and Finlayson, L. 1973; Davey, K. and Kuster, J. 1981) *Glossina* (Maddrell, S. and Gee, J. 1974) and *Carausius* (Finlayson, L. and Osborne, M. 1968; Fifield, S. and Finlayson, L. 1978).

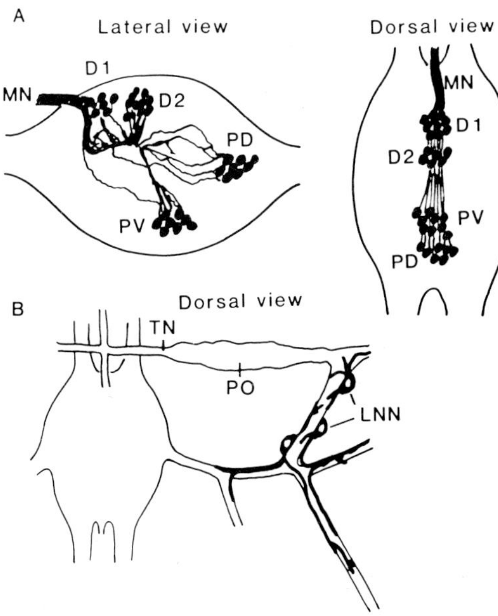

FIG. 4. **(A)** Cobalt backfilling of the median nerve (MN) in *Periplaneta* reveals four groups of neurons which may be associated with the perisympathetic organs associated with the transverse nerve of an abdominal ganglion. The four groups are located at different levels associated with the mid-saggital plane and are accordingly named: dorsal 1 (D1); dorsal 2 (D2); posterior dorsal (PD) and posterior ventral (PV). Redrawn from Ali, Z. and Pipa, R. (1978). **(B)** Projections of link nerve neurosecretory neurons (LNN) in *Carausius* as revealed by intracellular injection of procion yellow. PO, perisympathetic organ; TN, transverse nerve. Redrawn from Orchard, I. (1975).

Cobalt backfilling has also been used to trace neurons which project into the perisympathetic organs of *Periplaneta* (Ali, Z. and Pipa, R. 1978; see Fig. 4), *Schistocerca* (Evans, P. 1980) and *Carausius* (Orchard, I. 1982b). Groups of filled cell bodies occur at various levels adjacent to the mid-sagittal plane, and also in the lateral aspects of the ganglia. It would appear most likely, though not proven, that the filled cells represent in part the neurosecretory cells which terminate in the perisympathetic organs.

In addition to neurosecretory cells occupying sites within the central nervous system, there are numerous reports of neurosecretory cell bodies (so-called peripheral neurosecretory cells) lying on the peripheral nerves (Bowers, G. and Johnson, B. 1966; Gelperin, A. 1971; Hinks, C. 1975; Finlayson, L. and Orchard, I. 1978; see Finlayson, L. 1980). The most thorough investigation of these cells has been performed in the stick insect, *Carausius morosus* (Finlayson, L. and Osborne, M. 1968; Fifield, S. and Finlayson, L. 1978; Orchard, I. and Finlayson, L. 1976, 1977a,b,c; Orchard, I. 1976). These peripheral cells lie on the nerve linking the median nervous system with the somatic nervous system and are termed link nerve neurosecretory cells. They are multipolar with many axons running along and terminating on the surface of peripheral nerves (Fig. 4). Whether or not they are connected to the central nervous system has yet to be determined.

Neurosecretory cells and neurohaemal sites have also been described on the lateral cardiac nerves

(Johnson, B. 1966; Miller, T. and Thomson, W. 1968; Miller, T. and Usherwood, P. 1971; Fifield, S. and Finlayson, L. 1978; Chapter 9, volume 3). Some of these may control heart beat (Miller, T. and Usherwood, P. 1971) while others may use the heart as a convenient release site from which their products can be circulated to the rest of the body.

2.3 Light and electron microscopy

Light microscopy, coupled with so-called "specific" histochemical stains for neurosecretion (see Panov, A. 1980 for fuller account) has been used extensively to locate putative neurosecretory cells, their pathways and their neurohaemal areas. In addition, changes in staining intensity have been used as a means of assessing secretory cycles (Steel, C. and Harmsen, R. 1971), since these have been assumed to bear a relationship to the synthesis, transport and release of neurohormone. The nature of the neurosecretory material which stains, however, is far from clear (see section 3.2a) as is its exact relationship to neurohormone.

A characteristic feature of neurosecretory cells, revealed by ultrastructural analysis, is the presence of large numbers of electron-dense granules (Figs 5 and 6). These elementary neurosecretory granules are reported by many authors to originate within the golgi apparatus (Normann, T. 1965; Bassurmanova, O. and Panov, A. 1967; Finlayson, L. and Osborne, M. 1968; Beattie, T. 1971; Morris, G. and Steel, O. 1975, 1977a,b; Fifield, S. and Finlayson, L. 1978), located within the cell body (Figs 1 and 6). Material synthesized by the rough endoplasmic reticulum (RER) is transferred to the golgi apparatus for packaging into membrane-bound granules. These elementary neurosecretory granules then bud off from the saccules of the golgi and move away into the cytoplasm for transport to storage and release sites (see Berlind, A. 1977; Finlayson, L. and Osborne, M. 1975; Maddrell, S. and Nordmann, J. 1979). The exact nature of the material synthesized on the RER and packaged in these granules has not been established in insects, but by analogy with vertebrates and some other invertebrates the granules probably contain precursors of carrier protein (neurophysin?) and neurohormone (see sections 3.2 and 3.3).

In vertebrates and some invertebrates the proteins within the granules are then processed as they pass along the axon to the terminals. This has been studied most extensively using biochemical techniques in the hypothalamo-neurohypophysial system of vertebrates and in a variety of neurosecretory cells of molluscs (see section 3.3a). Ultrastructurally, the appearance of neurosecretory granules changes between the cell bodies of neurosecretory cells in the paraventricular and supraoptic nuclei and their processes in the neural lobe of the pituitary gland (Cannata, M. and Morris, J. 1973; Morris, J. and Dyball, R. 1974). The differences lie in the size and electron-density of granules after specific treatments. At a pH of 8·0, a significantly greater proportion of granules retain their electron-dense core within the perikarya and adjacent processes and are also of smaller diameter than within the neural lobe where only 5% of all granules retained their dense core. This is felt to represent the "maturation" of granules, with those in the neural lobe being more mature than those in the cell body—maturation being the processing of protein from a precursor into hormone plus neurophysin.

A similar phenomenon appears to occur within insects (Morris, G. and Steel, C. 1975; 1977c). In some neurosecretory cells of *Rhodnius*, the granules within the immediate vicinity of Golgi complexes have a prominent "bristle-coat" which is not seen in granules within the cytoplasm. The changes in granule diameter and apparent granule fusion observed in these cells suggests that considerable modifications in granule structure occur following release from the golgi.

This change in appearance of neurosecretory granules associated with their maturation illustrates a danger in attempting to assign neurosecretory terminals to their cell bodies purely on the basis of granule morphology. Whether this change implies that processing is occuring within neurosecretory granules in insects is not known. It has been shown that cell bodies of the identified diuretic hormone-producing cells in *Rhodnius* do contain physiologically active material (Berlind, A. and Maddrell, S. 1979). Furthermore, proctolin has been identified within the cell body of identified cells (O'Shea, M. and Adams, M. 1981) indicating that peptides may be present in an active form within the cell body.

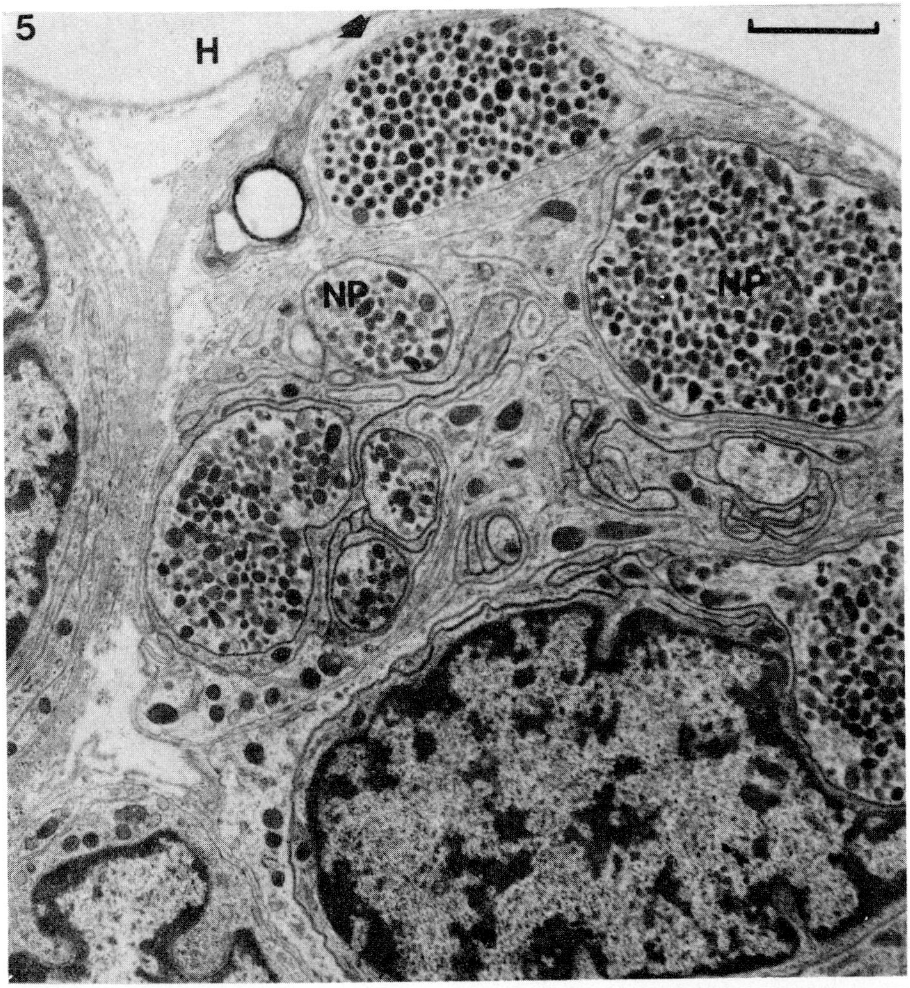

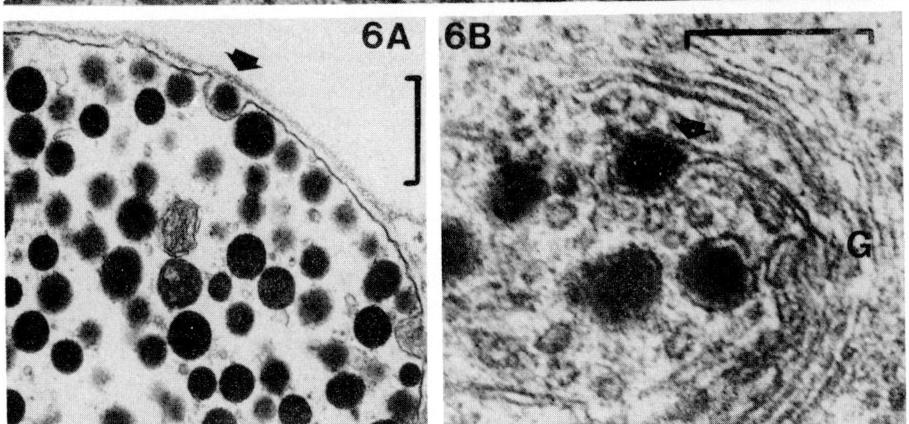

FIG. 5. Electron micrograph of neurohaemal tissue lying on the transverse nerve in *Carausius*. Neurosecretory processes (NP) are filled with neurosecretory granules. Some of these processes are devoid of a glial cell sheath and lie in close proximity (arrow) to the haemolymph space (H). Scale = 2 μm (courtesy of Dr. M. P. Osborne). FIG. 6. **(A)** Exocytosis (arrow) of a neurosecretory granule from a process of link nerve neuron in *Carausius*. Scale = 0.5 μm (courtesy of Dr. M. P. Osborne). **(B)** Electron micrograph shows a golgi body (G) producing granules. Note the continuity of golgi membranes with the limiting membrane of the granule (arrow). Link nerve neuron in *Carausius*. Scale = 0.5 μm (courtesy of Dr. M. P. Osborne).

The idea that the neurohormone is located within the granules was originally based on the fact that synthetic material was packaged into granules and the content of these granules was expelled at the terminals. Recently, however, direct evidence has been obtained which confirms this hypothesis. Stone, J. and Mordue, W. (1979, 1980) fractionated homogenates of glandular lobes of the corpus cardiacum of *Schistocerca* by differential centrifugation. Adipokinetic hormone is synthesized within, and released from, the glandular lobes (see chapter 5, this volume). Both total and specific adipokinetic activity was found in the subcellular fraction which, when viewed with the electron microscope, contained the largest proportion of neurosecretory granules and very little other material. Similar results were obtained from differential centrifugation of homogenized neurohypophysis whereby hormone was found within the granule fraction (Russell, J. and Holtz, R. 1981).

Following synthesis and packaging into elementary neurosecretory granules, the granules are then transported out of the cell body to the storage and release sites. The passage of granules in a proximal–distal direction was illustrated by early experiments in which nerve tracts were ligated or severed and the accumulation of neurosecretory material was observed to lie on the proximal side of the block (chapter 1, this volume). Furthermore, the transfer of material from the cell body to the axons can also be observed in a number of neurosecretory cells following their activation (Thomsen, M. 1965; Steel, C. and Harmsen, R. 1971; Morris, G. and Steel, C. 1975, 1977c). Unfortunately, the mechanism whereby the granules are transported is poorly understood in insects (see section 3.3a).

The transport of granules from the cell body typically results in a much larger concentration of granules at the terminals in neurohaemal areas (Figs 1 and 5). These terminals are usually swollen, filled with granules, and generally lie outside the blood–brain barrier (characteristics, in fact, used to identify neurohaemal areas). The sites of storage and release are usually the same, but in some systems it is apparent that there may be axonal swellings *en route* to the release site where storage may occur (Steel, C. 1977). It has also been suggested that storage sites may be found in the extensive arborizations of neurosecretory cells found within the neuropile (Highnam, K. and West, M. 1971).

It is generally accepted that neurosecretory cells release their granule contents by the mechanism of exocytosis. During this process the neurosecretory granule fuses with the plasma membrane producing a characteristic omega profile (Fig. 6; chapter 2, this volume). Evidence for this process in insects has accumulated from a number of electron-microscopic studies of different neurohaemal organs in which omega profiles were observed, and more importantly shown to increase at times when hormonal release was believed to be occurring (Maddrell, S. 1966; Normann, T. 1969; Brady, J. and Maddrell, S. 1967; Finlayson, L. and Osborne, M. 1968; Fifield, S. and Finlayson, L. 1978; Krogh, I. and Normann, T. 1977; Morris, G. and Steel, C. 1977a,b; Rademakers, L. 1977a; Rademakers, L. and Beenakkers, A. 1977). In particular, the frequency of omega profiles increases in the glandular cells of the corpus cardiacum of locusts which have been flown, a period when adipokinetic hormone is released from these cells (Rademakers, L. and Beenakkers, A. 1977; Rademakers, L. 1977a). Also, Krogh, I. and Normann, T. (1977) found that omega profiles were more frequently observed in corpora cardiaca of *Schistocerca* which had been stimulated either electrically or with acetylcholine; and a similar observation was made by Norman, T. (1965, 1970) in *Calliphora* using electrical stimulation or depolarization with high potassium saline. An increase in exocytosis is also evident in the corpus cardiacum of *Rhodnius* at times when the cells are electrically active and believed to be releasing hormone (Steel, C. 1978a; Orchard, I. and Steel, C. 1980).

The process of exocytosis will continually add membrane to the plasma membrane. Clearly, there has to be a mechanism for removing this excess membrane or the cell must suffer an enlargement of surface area. To emphasize the importance of this, Normann, T. (1970) calculated that exocytotic release of the granules at a 1 μm diameter ending would be capable of increasing the area of the plasma membrane by at least a factor of 10. Similarly, Maddrell, S. and Nordmann, J. (1979) calculated that under maximal stimulation the area of the membrane of the terminal in the rat neurohypophysis would double in an hour if membrane retrieval did not occur. Clearly, membrane retrieval is essential. One possible retrieval mechanism is microvesiculation, in which excess membrane is pinched off into the cytoplasm to form microvesicles of 30–40 nm diameter. These microvesicles are then presumed to be broken down to be used later in the manufacture of the neurosecretory granules in the cell body. In support of the microvesiculation hypothesis are those electron micrographs showing numerous vesicle clusters in the cytoplasm of neurosecretory endings during and following secretory activity (Rademakers, L. 1977a; Rademakers, L. and Beenakkers, A. 1977; Scharrer, B. 1968; Scharrer, B. and Wurzelmann, S. 1978; Normann, T. 1969, 1970; Krogh, I. and Normann, T. 1977; Smith, A. 1971). These microvesicles can be seen to be derived from the membrane by the process of pinocytosis, and molecular markers, such as ferritin added to extracellular fluid, are found in the small vesicles (Smith, A. 1971). Nordmann, J. (1979), however, argues that quantitative estimates have not been used to show an overall increase in microvesicle population after release. Rather, there are merely changes in the microvesicle clustering within terminals, with no concomitant increase in number of microvesicles. In a number of publications concerned with both rat neurohypophysis and crab sinus gland (Lescure, H. and Nordmann, J. 1980; Nordmann, J. 1979; Nordmann, J. and Morris, J. 1976, 1980) it has been suggested that membrane retrieval involves reclosure and retrieval of the empty granules as membranous sacs. These bud off from the membrane and form electron-lucent vacuoles similar in size to the neurosecretory granules. Such vacuoles have also been described in the glandular cells of the corpus cardiacum of *Schistocerca* (Krogh, I. and Normann, T. 1977).

Evidence for the release of neurosecretion by exocytosis is exceptionally good when one considers the numerous reports in vertebrates and invertebrates. However, it should be pointed out that this is not the only theory put forward for a mechanism of release. Other theories include the liberation of granule contents intracellularly into the cytoplasm from whence they diffuse out of the cell; the release of intact neurosecretory granules from the axon endings: and finally the breaking up of granules into microvesicles which are subsequently released by exocytosis. Comprehensive reviews of these hypotheses may be found in Finlayson, L. and Osborne, M. (1975); Berlind, A. (1977); and Maddrell, S. and Nordmann, J. (1979).

3 PHYSIOLOGY AND BIOCHEMISTRY

3.1 Electrical activity of neurosecretory cells

3.1.1 Electrical properties

Since it is implicit in the definition of neurosecretory cells that they are neuronal, one may expect to find they share characteristics with "conventional neurons" such as the generation and conduction of action potentials. Intracellular recordings (Fig. 7; Tables 2 and 3) established beyond doubt many years ago that this was true for neurosecretory cells in the vertebrates and invertebrates (reviews by Finlayson, L. and Osborne, M. 1975; Yagi, K. and Iwasaki, S. 1977; Orchard, I. 1982b).

The Tyndall blue–white effect shown by neurosecretory cells has been used by insect physiologists to identify the cells in living preparations and a number of intracellular recordings have been made from neurosecretory cells in the central nervous system identified in this way (Gosbee, J. *et al.*, 1968; Cook, D. and Milligan, J. 1972; Wilkens, J. and Mote, M. 1970; Miyazaki, S. 1980). Unfortunately, identification of the neurosecretory cells was not confirmed by intracellular staining techniques, leaving some doubt as to the validity of these original recordings. More recently, recordings have been made from medial neurosecretory cells in the protocerebrum of *Periplaneta* (Krauthamer, V. 1980). These cells were marked with horseradish peroxidase and subsequently examined at the ultrastructural level. Sections through the filled cell revealed the characteristic inclusions of the medial neurosecretory cells, thereby identifying the cells as neurosecretory.

Intracellular recordings have also been made from the link nerve neurosecretory cells in *Carausius*. These cells are clearly visible under the compound microscope (Orchard, I. 1976; Orchard, I. and Finlayson, L. 1977a,b), and the insertion of microelectrodes was monitored visually, providing no doubt as to the intrasomatic nature of the recordings. Final confirmation was obtained by

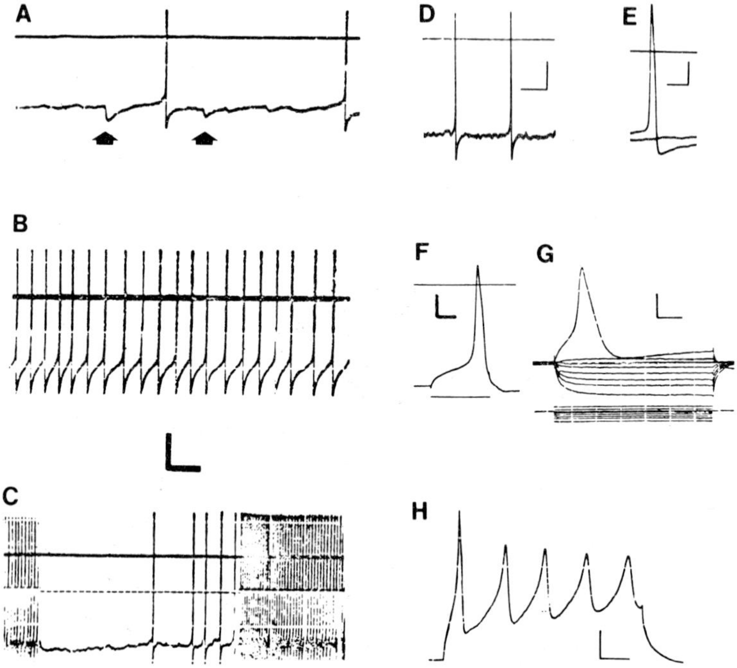

FIG. 7. Intracellular recordings from the cell bodies of neurosecretory cells in *Periplaneta* and *Carausius*. **A–C,** medial neurosecretory cells in *Periplaneta*. **A,** spontaneous overshooting action potentials and inhibitory postsynaptic potentials (arrows). **B, C,** spontaneous action potentials which show a decrease in frequency with hyperpolarizing current. Scale bars: **A,** 20 mV, 200 ms; **B,** 20 mV, 1 s; **C,** 20 mV, 5 s. Redrawn from Krauthamer (1980). **D–H,** link nerve neurosecretory neurons in *Carausius*. **D, E,** spontaneous overshooting action potentials; **F, G,** response to depolarizing and hyerpolarizing pulses; **H,** long-duration depolarizing current elicits a burst of action potentials which show a reduction in amplitude and an increase in duration. Scale bars: **D,** 10 mV, 2 s; **E,** 10 mV, 200 ms; **F,** 10 mV, 50 ms; **G,** 20 mV, 50 ms; **H,** 10 mV, 50 ms. Redrawn from Orchard, I. and Finlayson, L. (1977a) with permission of Pergamon Press. The composite figure is reproduced from Orchard, I. (1982b) with permission of Alan R. Liss Inc., New York.

filling the cells with procion yellow (Orchard, I. and Finlayson, L. 1977a).

In addition, intracellular recordings have been made from identified octopaminergic neurons (e.g. Heitler, W. and Goodman, C. 1978; Hoyle, G. and Dagan, D. 1978) and from identified proctolin-containing cells in the central nervous system of insects (O'Shea, M. and Adams, M. 1981). Whilst there is debate as to the neurosecretory nature of these cells, the properties of these recordings are reported here for completeness. Certainly octopamine is a neurohormone, but it is not known whether the identified cells from which recordings were made are those that release octopamine into the haemolymph. Similarly, proctolin may be a peptide transmitter or neuromodulator at the hind-gut musculature of cockroaches, but it may also be found to function as a neurohormone and so the properties of the identified proctolin cells may be similar to other classical neurosecretory cells.

Table 2 sets out the information available for the electrical properties of cell bodies of putative insect neurosecretory cells including octopamine and proctolin-containing neurons. It is clear that in spite of being specialized for glandular activity, insect neurosecretory cells have maintained all of the electrical properties of conventional neurons in that they demonstrate resting potentials, propagate action potentials, receive synaptic input and possess similar passive membrane properties. Despite the prolonged duration of the action potentials which have been shown to be 2–20 times greater than in conventional neurons (see Yagi, K. and Iwasaki, S. 1977) their electrical properties do not appear unique to the neurosecretory cell.

The cell bodies of neurosecretory cells in *Periplaneta* (Cook, D. and Milligan, J. 1972; Krauthamer, V. 1980), *Bombyx* (Miyazaki, S. 1980), *Carausius* (Orchard, I. and Finlayson, L. 1977a,b) and the cell bodies of octopaminergic

*Table 2: Electrical properties of insect neurosecretory cells**

Order	Insect	Neuron	Spontaneously active	Synaptic potential	Resting potential (mV)	Action potential (mV)	Over-shoot (mV)	Duration of action potential (ms)	Specific membrane resistance Ω cm²	Specific membrane capacity μF cm⁻²	Time constant (ms)	Reference
Dictyoptera	*Periplaneta americana*	MNC	Yes	—	20–40	2–3	—	5–12	—	—	—	Gosbee, J. *et al.*, 1968
		MNC	Yes	epsp/ipsp	55	60	5–10	5	60			Cook, D. and Milligan, J. 1972
		MNC	Yes	ipsp	44 ± 4.1	70 ± 2.3	26 ± 2.3	1.3 ± 0.42†	33,500	2.9	98.2	Krauthamer, V. 1980
		LW	—	epsp	60–80	70–90	10	—	—	—	—	O'Shea, M. and Adams, M. 1981
Diptera	*Sarcophaga bullata*	MNC	Yes	ipsp	2–40	1–40	—	2–20	—	—	—	Wilkens, J. and Mote, M. 1970
		CC terminals	No	—	40–60	0.5–1	—	10–20	—	—	—	Wilkens, J. and Mote, M. 1970
	Calliphora erythoce-phala	MNC	Yes	epsp	20–51	—	—	3–6	—	—	—	Normann, T. 1973
		CNC	Yes	epsp	15–65	0.5–35	—	3–7	1000	—	—	Normann, T. 1973
		Axons of CC	Yes	—	64	66.5–71	2.5–7	—	—	—	—	Normann, T. 1973
Phasmida	*Carausius morosus*	LNN	Yes	—	30–62	40–70	10	7–20†	3000	3.3	10	Orchard, I. and Finlayson, L. 1977a
Lepidoptera	*Bombyx mori*	MNC	Yes	ipsp	42 ± 7	67 ± 4	25 ± 4	39 ± 9†	—	—	—	Miyazaki, S. 1980
		DFC	Yes	—	42 ± 6	65 ± 4	23 ± 4	30 ± 13†	—	—	—	Miyazaki, S. 1980
Orthoptera	*Schistocerca gregaria*	DUM	Yes	epsp	56 ± 6	87 ± 7	15–20	4.8–40	—	—	—	Hoyle, G. and Dagan, D. 1978

* Including proctolin (LW) and octopamine cells (DUM).
† Measured at 50% spike heights; CC, corpora cardiaca; CNC, corpora cardiaca intrinsic neurosecretory cells; CC terminals, terminals of axons in corpora cardiaca; DFC, diapause factor neurosecretory cells; DUM, dorsal unpaired median cells (including DUMETi); LNN, link nerve neurons; LW, lateral white; MNC, medial neurosecretory cells.

and proctolin-containing neurons all have membranes capable of supporting overshooting action potentials and thus they appear to be electrically excitable. Other workers (Wilkens, J. and Mote, M. 1970; Normann, T. 1973) have described neurosecretory cells which do not support overshooting action potentials. However, these latter reports also describe low resting potentials and, therefore, probably represent recordings from damaged cells. Similarly the amplitude of action potentials reported by Gosbee, J. *et al.* (1968) are clearly too low in view of the fact that Cook, D. and Milligan, J. (1972) reported overshooting action potentials from the same group of cells. Overshooting action potentials are not recorded from the cell bodies of conventional neurons such as motor neurons or interneurons in the central nervous system of insects. Thus the presence of overshooting action potentials may turn out to be a characteristic feature of neurosecretory cells. The significance of this feature is unknown and has not been examined. Krauthamer, V. (1980) has suggested it may be related to the endogenous nature of firing which neurosecretory cells seem to demonstrate (see later). The somatic action potential may be part of the mechanism for generating this endogenous pattern. One possibility is that an overshooting action potential may be accompanied by an appreciable influx of calcium which in turn alters a potassium conductance. As we shall see in section 3.1a, a calcium component in the somatic action potential is a characteristic of invertebrate neurosecretory cells.

Intracellular recordings have shown that the neurosecretory cells tend to be "spontaneously

Table 3: Insect neurosecretory cells and organs in which electrophysiological studies have been reported

Order	Species	Structure	Method		Reference
Dictyoptera	*Periplaneta americana*	MNC	intracellular		Cook, D. and Milligan, J. 1972
		MNC	intracellular		Gosbee, J. *et al.*, 1968
		MNC (culture)		extracellular	Seshan, K. *et al.*, 1974
		MNC	intracellular		Krauthamer, V. 1980
		LCNC		extracellular	Miller, T. and Usherwood, P. 1971
		*LW	intracellular		O'Shea, M. and Adams, M. 1981
Diptera	*Sarcophaga bullata*	MNC	intracellular	extracellular	Wilkens, J. and Mote, M. 1970
		MNC		extracellular	Bruce, I. and Wilkens, J. 1976
	Calliphora erythrocephala	MNC	intracellular		Normann, T. 1973
		CNC	intracellular		Normann, T. 1973
		CC	intracellular		Normann, T. 1973
	Phormia terrae-nova	PNC		extracellular	Gelperin, A. 1971
Lepidoptera	*Antheraea pernyi*	NCC		extracellular	Mandel'shtam, Y. and Tyshchenko, V. 1967
	Dendrolimus pini	NCC		extracellular	Mandel'shtam, Y. and Tyshchenko, V. 1967
	Bombyx mori	MNC	intracellular		Miyazaki, S. 1980
		DFC	intracellular		Miyazaki, S. 1980
Heteroptera	*Rhodnius prolixus*	NCC		extracellular	Orchard, T. and Steel, C. 1980
		PNN		extracellular	Anwyl, R. and Finlayson, L. 1974
Orthoptera	*Locusta migratoria*	CC		extracellular	Orchard, I. and Loughton, B. 1981a,b
	Gryllus bimaculatas	*LW	intracellular		O'Shea, M. and Adams, M. 1981
	Schistocerca gregaria	DUMETi	intracellular	extracellular	Hoyle, G. 1974; Hoyle, G. and Dagan, D. 1978
Phasmida	*Carausius morosus*	PSO		extracellular	Finlayson, L. and Osborne, M. 1970
		PSO		extracellular	Orchard, I. and Osborne, M. 1977
		LNN	intracellular	extracellular	Orchard, I. and Finlayson, L. 1976, 1977a,b Orchard, I. 1976

CC, axons in corpora cardiaca; CNC, corpora cardiaca intrinsic neurosecretory cells; DFC, diapause factor neurosecretory cells; LCNC, lateral cardiac nerve neurosecretory cells; LNN, link nerve neurosecretory cells; LW, lateral white cell; MNC, medial neurosecretory cells; NCC, nervi corporis cardiaci; PNN, peripheral nerve neurosecretory axons; PSO, peri-sympathetic organ.
* The proctolin cell LW and octopamine cell DUMETi have been included for completeness. It is not clear whether they are indeed neurosecretory cells.

active" (Figs 7A, D and E) and there is evidence for both excitatory and inhibitory postsynaptic potentials (Table 2). The "spontaneous" action potentials are usually of a low frequency ($1-2\,s^{-1}$). Furthermore, neurosecretory cells can only be driven at low frequencies with electrical stimulation. The link nerve neurosecretory neurons in *Carausius* can only be driven on a one-to-one basis by short-duration stimuli with a frequency of $2\,s^{-1}$ (Orchard, I. and Finlayson, L. 1977a). Under prolonged steady depolarization, the response varies between preparations, but an initial increased firing rate rapidly declines to be replaced by abortive spiking (Fig. 7H) which may quickly cease altogether. This rapid adaptation may be due to accommodation or to the refractory properties of the membrane. Accommodation certainly appears to occur since a burst of action potentials can be initiated by the termination of long hyperpolarizing pulses. Similar properties have been observed for octopamine cells in locusts and grasshoppers (Hoyle, G. and Dagan, D. 1978).

Krauthamer, V. (1980) could find no evidence for excitatory postsynaptic potentials associated with the action potentials of the medial neurosecretory cells in *Periplaneta* and considered the action potentials to be endogenous since there was a decrease in frequency with slight hyperpolarization (Fig. 7c) and an increase in frequency in a calcium-reduced saline (possibly suggesting a calcium-activated potassium conductance). Further evidence for the endogenous nature of this activity was shown by Seshan, K. *et al*, (1974) who found spontaneously occurring action potentials in medial neurosecretory cells of *Periplaneta, in vitro*. Furthermore, excitatory postsynaptic potentials were not associated with action potentials in *Sarcophaga* (Wilkens, J. and Mote, M. 1970) or in *Carausius* (Orchard, I. and Finlayson, L. 1977a,b) indeed, the *Carausius* cells generate action potentials when they are isolated from the rest of the nervous system (Orchard, I. and Finlayson, L. 1976).

More detailed information concerning spontaneous activity of neurosecretory cells may be found within molluscs. For example, cell R15 in *Aplysia* and cell 11 in *Otala* have been shown to produce spontaneous activity which undergoes bursting activity. This activity is apparently endogenous, since the pattern of activity continues to occur in completely isolated cell bodies, can be reset by depolarizing and hyperpolarizing pulses and persists in high magnesium medium (Alving, B. 1968; Gainer, H. 1972). The burst pattern of R15 exhibits a diurnal cycle and a seasonal modulation (see Berlind, A. 1977). Cell 11 is quiescent during aestivation. It has been shown that calcium is involved in the seasonal modulation of bursting pacemaker potentials in *Otala* (Barker, J. and Gainer, H. 1975).

The source of the excitatory or inhibitory postsynaptic potentials onto the protocerebral neurosecretory cells is essentially unknown, although Bruce, I. and Wilkens, J. (1976) found evidence for optic, ocellar, antennal and thoracico-abdominal pathways contributing inputs to neurosecretory cells in *Sarcophaga*.

In addition to intracellular recordings, a number of reports have claimed extracellular recordings from insect neurosecretory systems. These reports are summarized in Table 3. Some of these recordings are from neurohaemal tissue or nerves leading to neurohaemal tissue. While the majority of axons in these preparations are probably neurosecretory, the preparations are a mixture of neurosecretory and conventional axons, and so some caution should be used in interpreting data from such recordings without additional information (e.g. Anwyl, R. and Finlayson, L. 1973; Orchard, I. and Steel, C. 1980; Orchard, I. and Finlayson, L. 1976).

An interesting phenomenon observed from extracellular recordings has been the discovery in *Carausius* and *Rhodnius* of action potentials which are initiated in peripheral nerves devoid of their cell bodies (Finlayson, L. and Osborne, M. 1968; Anwyl, R. and Finlayson, L. 1974; Orchard, I. and Osborne, M. 1977; Orchard, I. and Finlayson, L. 1977b). The current hypothesis suggests that these may result from a generator potential located in the unsheathed region of the neurosecretory endings which give rise to action potentials passing towards the central nervous system. The function of these action potentials has not been determined, but it is an interesting possibility that their frequency may reflect changes occurring in the haemolymph and thereby provide feedback information to the central nervous system. Since action potentials in neurosecretory cells induce hormone release, presumably these changes could also stimulate a concomitant release of hormone. While further clarification of the nature of these peripherally generated action potentials is required, it is worth mentioning that a similar initiation of action potentials at neurosecretory terminals has been described in neurosecretory bag cells in *Aplysia* (Dudek, F. and Blankenship, J. 1977) which results in the release of egg-laying hormone. Clearly though, this does not appear to be a characteristic feature of insect neurosecretory cells since other neurosecretory cells in *Carausius* and *Rhodnius* propagate action potentials towards their terminals (Orchard, I. and Finlayson, L. 1976; Orchard, I. and Steel, C. 1980; Ruegg, R. *et al.*, 1981).

The peripheral neurosecretory cells of *Carausius* are multipolar and each process propagates action potentials away from the cell body towards the terminals. The conduction velocity of these action potentials is about 3 times slower than either motor or sensory action potentials (Orchard, I. and Finlayson, L. 1976), which is in keeping with the few other measurements made of this parameter in other invertebrate neurosecretory cells.

3.1.1 Ionic mechanism(s) of the action potentials

The resting potential of insect neurosecretory cells may be conventional, since in *Carausius* link nerve neurosecretory neurons it is determined mainly by the concentration of potassium ions in the external medium (Orchard, I. 1976). Furthermore, TEA (tetraethylammonium salts) prolong the action potential (Orchard, unpublished). However, studies on the ionic mechanism of action potentials indicate that neurosecretory cells may show differences from other neuron types. In invertebrate neurosecretory cells it has been found to be a common occurrence that the somatic action potential involves a movement of both sodium and calcium ions during the rising phase (review by Yagi, K. and Iwasaki, S. 1977). The axonal action potentials are apparently brought about by movement of sodium ions, while the terminals again may revert to both sodium and calcium components (Cooke, I. 1977). The studies examining this problem in insects are summarized in Table 4. Of some considerable interest is the data available for *Carausius*. This is a phytophagous insect and has a haemolymph with a very low sodium ion concentration (15 mM). The link nerve neurosecretory neurons are exposed to this haemolymph because of the absence of any "blood–brain" barrier, and so must generate action potentials in a hostile environment. Since there is insufficient sodium to support sodium-dependent action potentials, these cells have evolved exclusively calcium-dependent ones in the cell bodies and apparently the axons (Orchard, I. 1976). Whilst no recordings have been made from terminals one would predict that they must also have calcium-dependent action potentials, as is suggested by extracellular recordings from the perisympathetic organs (Orchard, I. and Osborne, M. 1977). *Bombyx* also has an environment hostile to sodium-dependent action potentials. However, the medial neurosecretory cell bodies, and presumably much of

Table 4: Ionic regulation of action potentials in insect neurosecretory cells. Concentration of sodium in haemolymph and duration of somatic action potential are provided for comparison.

Order	Species	Haemolymph sodium (mM)	Neurosecretory cell	Duration of action potential*	Ionic species carrying inward current for action potential: Cell body	Axon	Reference
Phasmid	*Carausius morosus*	15	LNN	7–20	calcium	calcium	Orchard, I. 1976
Lepidoptera	*Bombyx mori*	15	MNC	39 ± 9	calcium	calcium and sodium	Miyazaki, S. 1980
			DFC	39 ± 13	calcium	calcium and sodium	Miyazaki, S. 1980
Dictyoptera	*Periplaneta americana*	150	MNC	1.3 ± 0.4	sodium some calcium?	sodium	Krauthamer, V. 1980

* Measured at 50% spike height; DFC, diapause factor neuorsecretory cell; LNN, link nerve neuron; MNC, medial neurosecretory cell.

their axons, lie within a blood–brain barrier and are therefore presumably exposed to a controlled ionic environment. In spite of this protective covering, these cells also have a calcium-dependent somatic action potential which has an additional small sodium component (Miyazaki, S. 1980). Furthermore, axonal action potentials generated close to the cell body show both sodium- and calcium-dependency. Here again, in view of the low sodium levels in the haemolymph, one would predict that the exposed terminals would have to rely on a calcium-dependent action potential. *Periplaneta*, on the other hand, has haemolymph with high sodium-ion concentration. In the medial neurosecretory cells of *Periplaneta* the action potentials appear to be dominated by a sodium component, although Krauthamer, V. (1980) does show evidence for a small calcium component.

Extracellular recordings from axons of link nerve neurosecretory neurons in *Carausius* reveal that they too are calcium-dependent (Orchard, I. 1976). However, the axons entering the corpus cardiacum of *Rhodnius* (Orchard, I. and Steel, C. 1980) and *Locusta* (Orchard, I. *et al.*, 1981a) have sodium-dependent action potentials, although some recordings of compound action potentials in the corpus cardiacum of *Locusta* (possibly from the terminals) do indicate there may be a calcium component (Orchard, I. *et al.*, 1981a).

Action potentials dependent upon both sodium and calcium ions as charge carriers are typical of some invertebrate neurosecretory cells (Yagi, K. and Iwasaki, S. 1977). Within insects the proportion of sodium/calcium entry appears to be variable, and may be dependent upon the location of the cell and the blood–brain barrier, and also on the ionic composition of the haemolymph. The relative proportion of sodium/calcium entry may also be correlated with the duration of the action potentials since the kinetics of the calcium channel are considerably slower than that of the sodium channel (Kostyuk, P. 1980). Hence, in *Carausius* and *Bombyx* the somatic action potentials have a very long duration and are calcium-dependent whereas in *Periplaneta*, which has a sodium-dominated action potential, they are of short duration. The physiological function of this calcium-dependency in the cell bodies of neurosecretory cells has not been examined. Possibilities which spring to mind include an involvement in the generation of endogenous electrical activity, an involvement in the regulation of synthesis and/or transport of neurosecretory material, or an involvement in release from the cell body (Takeda, N. 1976).

3.2 Biochemical nature of neurosecretory cells

3.2.1 Histochemistry

Nerve cells which appeared to possess the histochemical properties of glandular cells were described by Weyer, F. (1935) in the brain of the honey bee. Since that time the histochemical properties of neurosecretory cells have frequently provided the initial evidence of their presence. The differential staining of the classical neurosecretory neurons is based on their high concentration of protein granules which confer an enhanced affinity for protein stains. It is clear (Rowell, H. 1976) that not all neurosecretory cells stain equally with a given stain. Unfortunately, the chemical nature of the stainable material can be inferred in relatively few cases. Thus the so-called "A" cells, present in the pars intercerebralis of the protocerebrum and elsewhere in the central nervous system, stain intensely with paraldehyde fuchsin (PAF), performic acid–Alcian blue (PAB), performic acid–Victoria blue (PAVB), and chrome alum haematoxylin (CH) after suitable fixation, indicating the presence of sulphur-rich proteins. The neurosecretory cells of the vertebrate neurohypophysis stain in a similar fashion (Adams, C. and Sloper, J. 1955). It was originally held that this staining property of vertebrate neurosecretory cells was due to the high cysteine content of vasopressin and oxytocin (Block, R. and Van Dyke, H. 1952; Du Vigneaud, V. *et al.*, 1953a,b). However, it has now been established that the neurophysin carrier molecule is similarly rich in cysteine (Walter, R. *et al.*, 1971). The high proportion of sulphur-containing amino acids in the "A" cells was first proposed by Sloper, J. (1957) and subsequently confirmed by a succession of authors (Panov, A. 1980; Rowell, H. 1976). Recently Steel, C. and Morris, G. (1975) measured the sulphur content of identified neurosecretory cells in *Rhodnius* by X-ray microanalysis and demonstrated directly their enriched sulphur content compared to that of adjacent neurons. In addition, they determined that the neurosecretory granules were exceptionally high in sulphur. These findings were confirmed by Normann, T. and Hall, T. (1978) who used the same technique to investigate the neurosecretory cells of *Schistocerca*, *Calliphora* and the posterior pituitary of *Rana*. In addition to possessing a high concentration of sulphur, the neurosecretory granules were also enriched in calcium. No function for this calcium was determined. Though Hinks, C. (1971) found no evidence of cysteine in the majority of the "A" cells in the lepidopteran *Triphaena*, sulphur-rich proteins or peptides seem to be characteristic of many invertebrate neurosecretory cells since PAF-positive neurons have been observed in virtually all the major invertebrate phyla (see Gabe, M. 1966). The reason for the high frequency of sulphur-rich protein products in invertebrate neurosecretory systems is not clear. Since many different physiological functions appear to be controlled by cells with similar staining properties (e.g. "A" cells of the pars intercerebralis of insects have been implicated in the control of oogenesis, ovulation, protein synthesis, diuresis and colour change, and the sinus gland in the eyestalk of decapod crustacea has been shown to release, among others, red pigment concentrating hormone, distal retinal pigment hormone and hyperglycaemic hormone; (see Berlind, A. 1977; Kleinholtz, 1976), it seems evident that the staining properties of the cells are most probably due to the characteristics of a carrier molecule or common precursor molecule rather than the hormones *per se*. More direct histochemical evidence for a common neurosecretory molecule comes from the immunohistochemical analysis of neurosecretory proteins in *Periplaneta* (Eckert, M. 1973). Eckert prepared antisera against three different fractions derived from the corpus cardiacum. Each antiserum stained different cells of the pars intercerebralis, though all these cells were PAF-positive and thus apparently rich in cysteine. Thus a molecule common to this type of cells seems a reasonable hypothesis. The presence of such a molecule was postulated by Gabe, M. (1966) on the basis of the histochemical properties of neurosecretory material.

Additional evidence for the heterogeneity of function of the PAF-positive (Type "A" cells) in the pars intercerebralis comes from the immunocytochemical localization of insulin-like material in the brain of *Calliphora* (Duve, H. and Thorpe, A. 1979). Though approximately 26 cells in the pars are PAF-positive, only 6–8 of them show insulin-like immune reactivity.

The staining properties of the so-called "B" neurosecretory cells cannot be convincingly related to the characteristics of the proteins reacting with the stain except that they react with anionic dyes. The most commonly used stain for "B" cells is phloxine (see Rowell, H. 1976). Prentó, P. (1972) infers that the products of "B" cells of *Schistocerca* are rich in tyrosine and in basic amino acids, a claim that is confirmed by Hinks, C. (1971) for the "B" cells of *T. pronuba*. Rehm, M. (1955) claimed that the "B" cells of *Galleria* and *Ephestia* contained a high concentration of histidine and arginine. Though the staining properties of neurosecretory cells have been observed to change during development there seems general agreement that the "A" and "B" cells represent different classes of neurosecretory neuron rather than phases of activity of the same cell (but see Berlind, A. 1977).

The "C" neurosecretory cells stain with general protein stains other than those specific to "A" and "B" cells, and are recognized histologically after staining with azan or azan-containing preparations (Rowell, H. 1976).

Though the histochemical staining properties of supposed neurosecretory cells may differ, ultrastructural analysis has demonstrated the presence of elementary granules in all classes of neurosecretory cells. Unfortunately, classifications of neurosecretory cells based on ultrastructure frequently conflict with those based on histochemical staining (Maddrell, S. 1967; Schooneveld, H. 1974). Chalaye, D. (1975) has compared the histochemistry and ultrastructure of neurosecretory cells in the suboesophageal ganglion of *Locusta*. He found that the type "C" cells (defined by staining properties with azan) comprised a variety of ultrastructurally different cells. Thus it is possible that at least some of the "C" cells may be aminergic and even possibly octopaminergic since the "DUM" neurones emanating from the ventral ganglia were originally described as azan-positive cells (Evans, P. 1980). Once again it is necessary to have adjacent sections taken from within the same cell, one stained and the next examined at the ultrastructural level to be sure that all azan-positive cells are peptidergic.

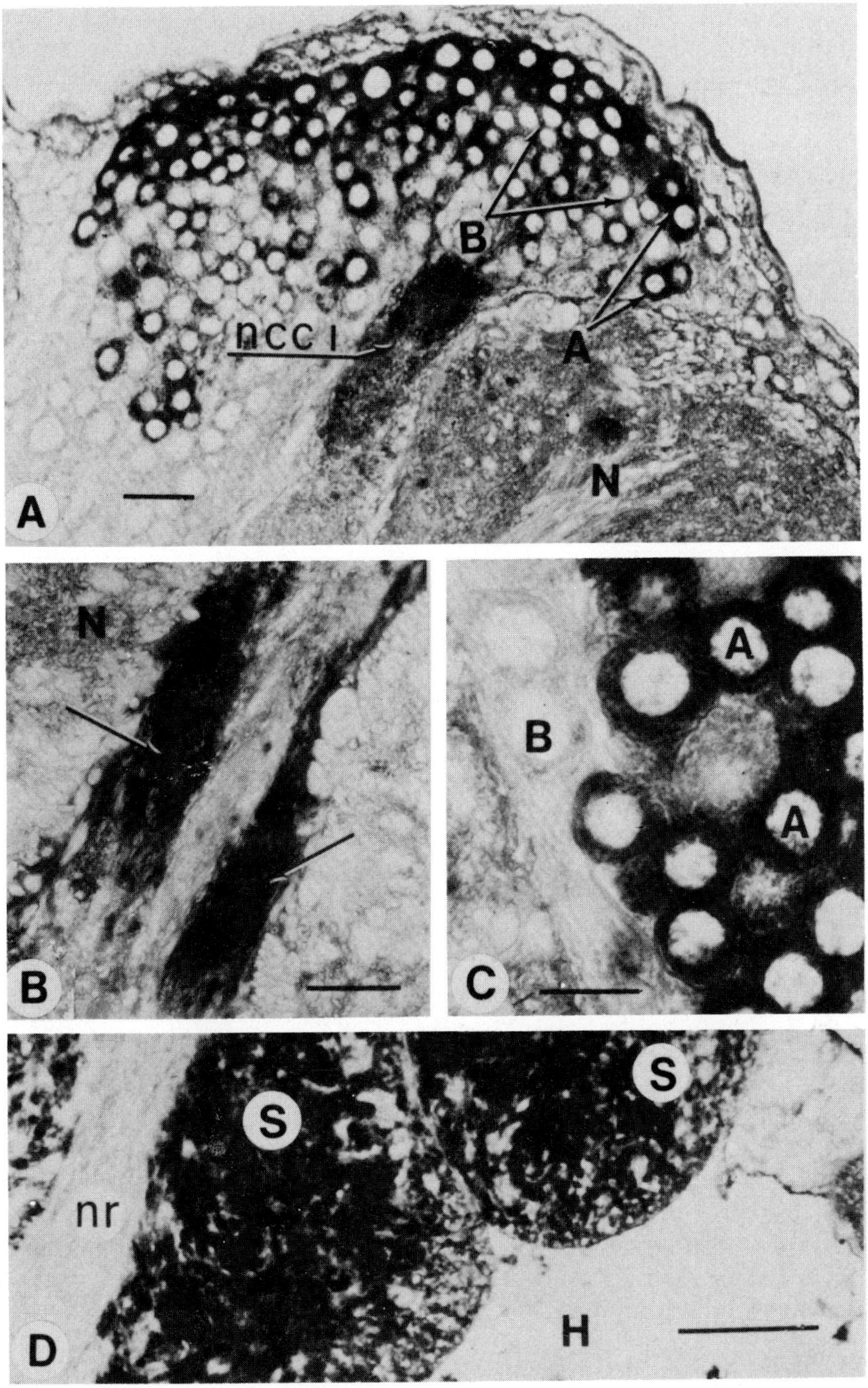

FIG. 8. Localization of locust neurosecretory protein by immuno-histochemistry. **A:** Immunoreactivity associated with the "A" cells (A) and axons of the nervus corpus cardiacum I (ncc I) in a section through the pars intercerebralis of *Locusta*. No immunoreactivity is present in the "B" cells (B) or neuropile (N). Scale bar: 50μM. **B:** Intense immunoreactivity associated with axons of ncc I (arrows) with no stain in the neuropile (N). Scale bar: 25μM. **C:** "A" and "B" cells in the pars intercerebralis. Scale bar: 20μM. **D:** Section through the storage lobe(s) of the corpus cardiacum. nr, recurrent nerve, H, haemocoel. Scale bar: 50 μM.

Finally, the value of histochemical and ultrastructural studies continues to be demonstrated by virtue of the range of endocrine functions now ascribable to cells whose neurosecretory function was previously based solely on histochemical and morphological grounds.

3.2.2 Chemistry

Though the histochemical properties of insect neurosecretory cells have received considerable attention, there have been few studies aimed directly at an understanding of the biochemical nature of their products. In this regard, only the products of the "A" cells have been studied. Highnam, K. and Mordue, A. (1970) used autoradiography following injections of [^{35}S]cysteine to measure incorporation into neurosecretory protein. They found that label appeared quickly in the neurosecretory cells and corpus cardiacum at levels up to twice that of other neurons. They also found circumstantial evidence of non-specific binding of label to neurosecretory material. This adsorption was highest in the corpus cardiacum (see also Mordue, A. and Highnam, K. 1973). Girardie, J. and Girardie, A. (1972) showed that the "A" cells of *Locusta migratoria* incorporated higher levels of [^{35}S]cysteine than "B" cells. However, unless one knows the physiological state of the two groups of cells, it is not possible to determine whether this is due to differences in the product or differences in the activity of the two cell types.

The cysteine-rich nature of the neurosecretory material of the "A" cells was used by Friedel, T. *et al.* (1980) as a means of recognizing neurosecretory molecules from the brain and corpus cardiacum of *Locusta*. Extracts of brains and corpora cardiaca were subjected to ion exchange chromatography following a prior injection of [^{3}H]cysteine. Gel chromatography of the [^{3}H]cysteine-labelled fractions yielded two molecules of high specific radioactivity with apparent molecular weights of 23,000 and 11,000. Gel electrophoresis of either fraction showed a single protein of molecular weight between 10,000 and 12,000 and an isoelectric point of 6·3–6·5. The molecules were soluble in 70% alcohol like the neurophysins of vertebrates. Gel electrophoresis of alcohol extracts of the storage lobe of the corpus cardiacum yielded two protein fractions. Isolation of either of these electrophoretic fractions followed by re-electrophoresis once again yielded both electrophoretic fractions. We believe that the two cysteine-rich molecules detected after gel filtration and the two electrophoretic fractions represent the monomer and dimer of the neurosecretory protein. Dimerization of vertebrate neurophysins has been reported by Nicolas, P. *et al.* (1976) and Tellam, R. and Winzor, D. (1980). Amino acid analysis of the 12,000 MW molecule (Table 5) showed a protein rich in lysine (16%) and cysteine (9%) with a calculated molecular weight of 11,200. Though less material was available, a similar amino acid composition was obtained after hydrolysis of the 23,000 MW fraction. Other than the considerable amount of lysine in the molecule the proportions of its constituent amino acids were similar to those of vertebrate neurophysins (North, W. and Valtin, H. 1977). Hinks, C. (1971) argued that the "A" cells of *Triphaena pronuba* contained material rich in lysine on the basis of its susceptibility to digestion with trypsin.

The neurosecretory nature of the cysteine-rich protein isolated by Friedel, T. *et al.* (1980) was confirmed by immunocytochemical localization. Antibodies against the molecules specifically stained the "A" cells, the axons and the storage lobe of the corpus cardiacum (Fig. 8). No stain was evident in the glandular lobe of the corpus cardiacum. Staining of electron micrographs of the corpus cardiacum demonstrated immune precipitate over the neurosecretory granules.

3.2.3 Molecular nature

Investigations of neurosecretory products have focused primarily on material with biological activity. The most extensive studies have concentrated on the isolation of the prothoracicotropic hormone (PTTH). Though the peptidergic nature of PTTH is now generally agreed upon, the size of the active molecules has been subject to considerable debate (Chapter 4, this volume). Ishizaki, H. and Ichikawa, M. (1967) resolved active fractions from the brain of *Bombyx* pupae with molecular weights of 9000, 12,000 and 31,000. Yamazaki, M. and Kobayashi, M. (1969) isolated an active protein molecule with an apparent molecular weight of 20,000 from the same source. Nagasawa, H. *et al.* (1980) reported a molecular weight of 4400 for the active principle in adult *Bombyx*. The cytological origin of these protein molecules was not known. The PTTH isolated from larval *Manduca* has a molecular weight of 25,000 (Kingan, T. 1981). Gilbert, L. *et al.* (1981) report an apparent molecular weight of 22,000 for a highly active form of the hormone extracted from the pupal brain of *Manduca*. A second less active fraction with a molecular weight of 7,000 was also detected. Only the "big" form was detected in the haemolymph. Activation factor II (Gersch, M. and Sturzbecher, J. 1970) isolated from the brain of *Periplaneta* was a heterogeneous protein fraction with molecular weights of 10,000 and 20,000. Even though several different species of insects were used in these investigations the wide discrepancies in results suggests that certain authors may have isolated precursors (see 3.3a) or polymers of PTTH. The cells producing PTTH have been identified in *Manduca* (Agui, N. *et al.*, 1979) and correspond to one of two PAF-positive cells located on each side of the brain previously observed by Nijhout, H. (1975). These cells apparently stain in a manner similar to the "A" cells. It will be interesting to discover if PTTH is rich in cysteine or whether the staining properties of the cell must be attributed to some other protein moiety. If the active molecule has a molecular weight of 5000 or greater, a carrier (neurophysin) molecule may not be required. Neurophysins are thought to preserve the integrity of the peptide hormone and prevent it from leaking from the neurosecretory granule.

Though PTTH has been most extensively studied, the only insect neurohormone to be fully characterized is adipokinetic hormone (AKH) (Stone, J. *et al.*, 1976; Chapter 5, this volume) (Fig. 9). This hormone was isolated from the glandular lobe of the corpus cardiacum of locusts and following elucidation of its structure has subsequently been synthesized and is now commercially available. The primary effect of the hormone appears to be the elevation of haemolymph lipid in response to the exigencies of flight (Mayer, R. and Candy, D. 1969; Beenakkers, A. *et al.*, 1981; chapter 5, this volume; chapter 12, volume 10). In addition, it has been found to inhibit protein synthesis (Carlisle, J. and Loughton, B. 1979) and (in relatively high doses) activate fat body glycogen phosphorylase (Gade, G. 1981; chapter 4, volume 8). In cockroaches it causes an elevation of haemolymph carbohydrate (Jones, J. *et al.*, 1977). More recently a second peptide has been isolated from the glandular lobe of the locust corpus cardiacum (Carlsen, J. *et al.*, 1979). This peptide appears somewhat smaller, but has many amino acids in common with AKH (see Fig. 9). In fact, the amino acid composition differs by only one amino acid from the erythrophore pigment concentrating hormone (RPCH) isolated (see Fig. 9) from crustacean sinus glands. The similarity between RPCH and AKH had been noted previously (Mordue, W. and Stone, J. 1977). RPCH had been found to stimulate release of fat body lipid into the haemolymph and AKH had been shown to cause "blanching" in *Palaeononetes*. The second peptide, termed AKH II by Carlsen, J. *et al.* (1979), causes release of lipid from locust fat body, but also causes the release of carbohydrate (Loughton, B. and Orchard, I. 1981) in neck-ligated insects. AKH does not cause this release in neck-ligated locusts. The similarity in apparent structure of AKH, RPCH and AKH II is reminiscent of the similarities between vertebrate vasopressin and oxytocin, and suggests an evolutionary pathway involving an original single precursor molecule followed by either mutations resulting in alteration of constituent amino acids in the hormone itself or alterations in the adjacent amino acids which altered the processing of the precursor and hence caused additions or deletions from the hormone (cf. Archer, R. 1981). It would be instructive if the synthetic pathway for AKH, RPCH and AKH II could be elucidated.

Eckert, M. and Gersch, M. (1978) separated four proteins from the corpora cardiaca of *Periplaneta* following their adsorption to an antiserum prepared against neurosecretory protein. The molecular weights of these proteins were 100,000, 48,000, 9000 and 7000. All these molecules were PAF-positive and antisera raised in rabbits against them stained neurosecretory granules (Eckert, M. 1980). Several workers have used antisera against

AKH I : PCA-Leu-Asn-Phe-Thr-Pro-Asn-Trp-Gly-Thr-NH_2

AKH II: PCA-Leu-Asn-Phe-Thr-Ser-Gly-Trp-NH_2

RPCH : PCA-Leu-Asn-Phe-Ser-Pro-Gly-Trp-NH_2

FIG. 9. The structure of locust adipokinetic hormone (AKH), Carlsen's peptide (AKH II) and crustacean erythrophore pigment concentrating hormone (RPCH). Sequence for AKH II is only suggested.

vertebrate peptide hormones to search for similar hormones in insects (see chapter 16, this volume). Remy, C. *et al.* (1979) and Remy, C. and Girardie, J. (1980) found two neurosecretory cells (PAF-positive) in the suboesophageal ganglion of *Locusta* which were stained immunocytochemically with anti-vasopressin and Remy, C. *et al.* (1977) had shown a similar circumstance in the phasmid *Clitumnus extradentatus*. Proux, J. and Rougon-Rapuzzi, G. (1980) demonstrated the presence of an immunoreactive vasopressin-like molecule in the suboesophageal ganglion which could also be detected in the haemolymph and which caused diuresis when injected into locusts. Recently Camier, M. *et al.* (1980), isolated a 10,000 MW protein from extracts of locust suboesophageal ganglion which reacted immunopositively with antibovine neurophysin II. The isoelectric point of these proteins was between 4·5 and 4·75. Vertebrate neurophysins have isoelectric points of 4·8 for vasopressin neurophysin and 4·6 for the neurophysin of oxytocin. The experiments of Remy, C. *et al.* (1979) and Remy, C. and Girardie, J. (1980) together with those of Camier, M. *et al.* (1980) and Proux, J. and Rougon-Rapuzzi, G. (1980) give the first direct evidence of the coexistence of a neurophysin-like molecule and a hormone in the same insect neurosecretory cells.

Evidence of an association between a peptidergic hormone and a neurophysin-like molecule comes from studies on the corpus cardiacum of *Locusta* (Loughton, B. *et al.*, unpublished). Alcohol extracts of the whole gland yielded three major bands after gel electrophoresis. Extracts of the storage lobe yielded two slowly migrating bands whereas glandular lobe extracts demonstrated two very similar fast-migrating bands. SDS gel electrophoresis of both glandular and storage lobe extracts yielded one band with a molecular weight of approximately 10,000.

Gel filtration of glandular lobe extracts yielded a 10,000 MW protein which migrated in the fashion of the fast-migrating, alcohol-soluble proteins upon gel electrophoresis. This peak also possessed adipokinetic activity. A much more active hyperlipaemic fraction was eluted much later from the column, corresponding to a molecular weight of 1000. The 10,000 MW fraction was coupled to

Table 5: A comparison of the amino acid composition of locust "neurophysin" with those of rat, pig, sheep and man (data from Friedel, T. et al, *1980 and North, W. and Valtin, H. 1977).*

	Locust	Rat		Pig	Bovine		Sheep	Man
		I	II		I	II		
Asp	0	5	8	5	7	5	3	7
Thre	7	3	3	2	2	2	2	2
Ser	7	7	7	7	5	6	6	4
Glu	10	13	10	13	10	14	13	10
Pro	7	8	10	7	9	8	8	9
Gly	8	14	14	14	14	16	15	12
Ala	4	9	10	7	8	6	7	9
1/2 Cys	8	14	14	14	14	14	14	14
Val	5	1	1	2	3	4	3	3
Met	1	1	1	1	0	1	1	0
Ile	4	2	2	2	2	2	4	1
Leu	5	7	6	7	6	6	6	7
Tyr	2	1	1	1	1	1	1	1
Phe	2	4	3	3	3	3	3	3
Trp	0	0	0	0	0	0	0	0
His	0	0	0	0	1	0	0	1
Lys	14	2	3	2	2	2	2	3
Arg	4	6	5	5	4	7	6	4
Number of residues	88	97	100	92	91	97	94	90
Calculated molecular weight	11,902	10,126	10,121	11,108	10,723	11,776	11,349	10,712

Sepharose beads and incubated with known quantities of synthetic adipokinetic hormone (AKH). The supernatant was subsequently tested for AKH activity. No AKH activity was detected in the supernatant. When AKH was incubated with control beads, coupled to bovine serum albumin or neurosecretory protein from the storage lobe of the corpus cardiacum the hyperlipaemic activity remained in the supernatant. Hence, it appears that the AKH selectively bound to the neurosecretory protein from the glandular lobe. Thus the glandular lobe of the locust corpus cardiacum has a protein which is soluble in alcohol, has a molecular weight of approximately 10,000 and which can bind AKH. In addition, this protein can be detected in the medium together with AKH activity after treatment of glandular lobes with "high potassium saline" (see 3.3) indicating its release from the neurosecretory granules together with a hormone (cf. vertebrate neurophysins; Seif, S. and Robinson, A. 1978).

3.2.4 One cell one product? The dilemma of multiple cell response

The activity of the neurosecretory cells of the pars intercerebralis frequently provides the investigators with an apparent paradox. Thus, Highnam, K. (1961a,b) reported that mating and oviposition brought about histological changes consistent with release of neurosecretory material from the brain of the female locust. Starvation in *Schistocerca* (Highnam, K. *et al.*, 1966; Highnam, K. and Mordue, A. 1974) resulted in an accumulation of stainable neurosecretory material in the median neurosecretory cells and in the corpus cardiacum. Feeding the starved animals caused an immediate depletion of the corpus cardiacum followed 4 h later by a similar depletion of the median neurosecretory cells. The histological data have been confirmed by autoradiographic experiments following the injection of [^{35}S]cysteine. A massive release of neurosecretion from the periphery of the corpus cardiacum could be detected within 10 min of feeding. A similar pattern of activity was reported by Dogra, G. and Gillott, C. (1971) in starved and fed *Melanoplus*. In the beetle *Galeruca tanaceti* (Siew, Y. 1965) and the grasshopper *Anacridium aegypti* (Geldiay, S. 1970) oogenesis was paralleled by a period of intense activity in both the "A" and "B" cells of the pars intercerebralis. Reproductive "diapause" brought about a cessation of this activity. Highnam, K. and Mordue, A. (1970) report that rapidly developing female *Schistocerca* have histologically "empty" neurosecretory systems and exhibit a rapid turnover of [^{35}S]cysteine. They argue that this suggests a continuing synthesis and release of neurosecretion. These observations were confirmed by Girardie, J. and Girardie, A. (1972) who point out that the neurosecretory cells in the subocellar zone do not alter their metabolic activity in the same manner as the median neurosecretory cells.

McCaffery, A. and Highnam, K. (1975a,b) report a general release of neurosecretory material from the median neurosecretory cells of *Locusta* following injection of juvenile hormone (JH) analogues or implantation of additional corpora allata.

The intriguing aspect of all these results is the apparent absence of differential activity within the median neurosecretory cells. In a similar vein Loughton, B. and Friedel, T. (1980) observed a massive fluctuation in the incorporation of [^{3}H]cysteine into the neurosecretory protein (localized in the "A" cells) of larval *Locusta*. The extent of this change was such that it was hard to believe only a few of the 1500 cells in the pars intercerebralis were involved.

Thus, though it is quite clear that a variety of neurosecretory cell types with separate functions can be distinguished in the pars intercerebralis (see also Raabe, M. *et al.*, 1979) these cells frequently appear to function as one in response to a variety of stimuli. Exceptions to these observations suggest that there are differences in activity though they require careful examination to be discerned. Thus, Steel, C. (1973) could distinguish cycles of activity in specific PAF-positive cells in the pars intercerebralis of *Rhodnius prolixus* at specified stages of development, and Grossman, M. and Davey, K. (1978) detected variations in the activity of certain PAF-positive cells in the brain of the maturing male tsetse fly.

Nevertheless, many cells of the pars intercerebralis appear to undergo cycles of activity in concert. Whether this is an illusion brought about by the gross level at which investigations have been conducted, or a result of a real need for the simultaneous release of a number of neurosecretory products, must await the identification of specific

"A" cells with a specific hormone product and the subsequent monitoring of the cells' activity.

The discovery of precursor neurosecretory protein molecules in vertebrates that contain the amino acid sequences for several hormones suggests another possibility (Archer, R. 1981). It is conceivable that the synthetic activity of many cells all producing the same precursor could be triggered whether or not the final product is the same. The ultimate release of a given hormone would depend on the processing of the precursor as it progressed to the axon terminal. In cells not producing the required hormone lysosomes would degrade the precursor in a non-specific fashion. Such a process seems profligate, but could explain the "all-or-none" activity frequently reported for the neurosecretory cell population of the pars intercerebralis. Alternatively, cells might alter their processing of the neurosecretory precursor to produce the required final product. In addition, it is possible for a cell to produce more than one peptide. Cells of the medulla oblongata apparently contain both substance P and thyrotropin releasing hormone-like peptides, and these two peptides appear to be in serotonin neurons. Other immunocytochemical studies on vertebrates have demonstrated the coexistence of neurosecretory peptides and classical neurotransmitters within neurons (see review by Hokfelt, T. *et al.*, 1980). Certain of these cells were termed APUD neurons by Pearse, A. (1969) (for amine content and/or precursor uptake and decarboxylation system). In the adrenal gland there is strong evidence for enkephalin-like peptides in catecholamine cells and vasointestinal-like peptide has been localized in cholinergic cells. As a consequence of these potentially multifunctional neurons, it is possible that many neurons may in fact be active together.

Finally, it appears that some peptide hormones may have more than one activity. Consequently a number of cells might be expected to respond to several discrete stimuli since the correct response of the insect endocrine system may require one or other of their hormone's activities.

3.3 Synthesis, transport and release of neurosecretion

3.3.1 Synthesis, processing and transport

The lack of any identifiable product has, until recently, severely hampered investigators of the synthesis of neurosecretion in insects. As a result most authors have relied on histochemical, autoradiographic or ultrastructural evidence of synthetic activity in neurosecretory cells to infer neurosecretory activity. Histochemical analysis must be treated with circumspection for, as Highnam, K. (1965) has pointed out, the observation of a "full" neurosecretory cell does not give any information as to its activity. It could be essentially quiescent, having accumulated neurosecretion, or it could be highly active, synthesizing and releasing hormone at a constant rate. Similarly an apparently "empty" cell could be synthesizing and releasing hormone at a maximum rate. Only when fluctuations in the staining intensity of neurosecretory cells are evident may one infer that they are actually changing their activity. Even when such changes are evident the precise nature of the change in the neurosecretory cell activity may be difficult to assess. Nevertheless, changes in the staining intensity of neurosecretory cells have been correlated with such varied physiological stimuli as oogenesis in the locust (Highnam, K. 1961a,b) feeding in *Glossina* (Grossman, M. and Davey, K. 1978), day length in *Dytiscus marginalis* (Berlind, A. 1977) and diurnal rhythm (Cymborowski, B. and Dutkowski, A. 1970). Alterations in staining intensity and in nuclear diameter were used by McCaffery, A. and Highnam, K. (1975a,b) to assess the effect of JH and certain of its mimics on the activity of the "A" cells of the pars intercerebralis of *Locusta*. Injections of JH caused massive transport and disappearance of stainable neurosecretory material from the "A" cells and their terminals in the corpus cardiacum. In addition, the nuclear volume increased in JH-injected insects. Increases in nuclear and nucleolar volume were observed by Thomsen, E. and Lea, A. (1968) to be correlated with increases in the number of neurosecretory granules in the median neurosecretory cells of *Calliphora erythrocephala* fed on a meat diet. No changes were observed in sugar-fed flies. Bradley, J. and Simpson, T. (1981) noted that PAF-positive neurosecretory cells in ovariectomized crickets contained less neurosecretory material than normal females and that this condition could be reversed by the injection of ecdysterone.

Cytological studies of neurosecretory synthetic activity are more convincing at the ultrastructural level. Since the neurosecretory material is peptidergic in nature the neurosecretory material of insects appears to be synthesized and packaged in the same manner as that of vertebrates. Though the cellular protein synthetic apparatus has been well known for some time (see Palade, G. 1975) it is only recently that direct evidence of the involvement of the golgi apparatus in the synthesis of neurophysin in rats has been forthcoming (Broadwell, R. and Oliver, C. 1980). Girardie, J. and Girardie, A. (1972) used EM autoradiography to follow the intracellular sequence of incorporation of [^{35}S]cysteine in the "A" cells of the pars intercerebralis. Label appeared first in the rough endoplasmic reticulum (RER), then over the golgi apparatus and finally over the neurosecretory granules. Labelled granules were subsequently found in the axons. Labelling and subsequent transport were also observed in "C" cells, but the time frame was extended.

The synthetic activity of neurosecretory cells has been assessed by examination of the extent of their protein synthetic apparatus at specific stages of development or after experimental manipulation. Bloch, B. *et al.*, (1966) found clear evidence of active protein synthesis in meat-fed adult *Calliphora*, whereas in sugar-fed flies the RER was limited and the number of golgi apparati declined. In starved *Rhodnius* larvae, Morris, G. and Steel, C. (1975) observed that lysosomal granules and multivesiculate bodies became prominent with a concomitant reduction in RER apparati and the golgi apparatus. The lysosomal granules could not be detected after feeding and the cells quickly assumed the appearance of active protein synthesis.

A further characteristic of inactive neurosecretory cells appears to be an increase (with time) of neurosecretory granules in the perikaryon (Scharrer, E. and Brown, S. 1961). This observation is supported by the increase in staining intensity in inactive neurosecretory systems (see above). The most comprehensive autoradiographic analysis of the synthetic activity of neurosecretory cells is that of Highnam, K. and Mordue, A. (1970, 1974). They showed that in isolated female *Schistocerca* the turnover of [^{35}S]cysteine in the neurosecretory cells of the pars intercerebralis was slow. In females reared together with males, turnover in these cells was rapid. They correlated these observations with oogenesis in that the isolated females matured more slowly than those reared with males. Highnam, K. and Mordue, A. (1974) also followed the incorporation of [^{35}S]cysteine to the neurosecretory system of adult *Schistocerca* before and after feeding. In starved females the incorporation of [^{35}S]cysteine into the products of the median neurosecretory cells was low, and transport and release of neurosecretory material appeared minimal. The neurosecretory system of fed insects showed an immediate change in activity. A high proportion of the material in the periphery of the corpus cardiacum was released within 10 min of feeding. The rate of synthesis of new material and the amount transported also appeared to increase. The authors also claimed that the rate of transport increased. However, this interpretation assumes that the amount transported stayed constant. It is perhaps more likely that the rate of movement stays constant and the amount of neurosecretory product transported is doubled. Girardie, J. and Girardie, A. (1972) confirmed the elevated synthesis of neurosecretory material in locust females with rapidly maturing oocytes. Electrical stimulation of the pars intercerebralis (Girardie, J. *et al.*, 1975) following injection of [^{35}S]cysteine caused a disappearance of label from the cells of the pars and from the corpus cardiacum. However, labelling of the whole neurosecretory system was observed if [^{35}S]cysteine was injected after electrical stimulation, suggesting not only an increased synthesis of neurosecretory material, but also its transport into the corpus cardiacum. It is interesting to note that the alteration in the labelling of cells, consistent with the release of sulphur-containing neurosecretory material, was followed by accelerated maturation in both male and female adults.

Girardie, J. *et al.* (1976) estimated the synthesis and release of neurosecretory material after electrical stimulation of cell bodies in the pars intercerebralis and after stimulating the storage lobe of the corpus cardiacum. They found that electrostimulation of the intact system resulted in a decrease in the amount of label in the brain. In insects which had been previously injected with [^{35}S]cysteine, stimulation of the isolated corpus cardiacum caused a reduction in the amount of label present. Stimulation of the cell bodies in the brain following section of NCC I resulted in an increase in the amount of label in the pars intercerebralis. In a second experiment they estimated label in the protein of the brain and corpus cardiacum following injection of [^{35}S]cysteine and electrostimulation. Their results essentially confirmed the earlier experiment (Girardie, J. and Girardie, A. 1977b). Electrostimulation of the cell bodies brought about an increase in incorporation into the neurosecretory cells of the pars intercerebralis when NCC I was sectioned. Stimulation of the isolated corpus cardiacum caused a 25% reduction in the amount of label in the gland that could be recovered from the medium as $HClO_3$ precipitable material. Gonadotrophic activity was also present in the corpus cardiacum perfusate.

Synthesis transport and release of neurosecretory protein in the adult locust and in the larva were investigated by Friedel, T. and Loughton, B. (1980a) and Loughton, B. and Friedel, T. (1980) using antiserum generated against a neurosecretory protein (see section 3.2b). Starved male locusts incorporated as much labelled cysteine into neurosecretory protein as fed controls. However, transport of the newly synthesized material to the corpus cardiacum was approximately half that of controls. Within 1 h of feeding the incorporation of [^{3}H]cysteine into neurosecretory protein in the brain more than doubled, and the amount of labelled neurosecretory protein reaching the corpus cardiacum tripled. The level of incorporation into non-neurosecretory proteins was unaltered. Since the absolute amount of neurosecretory protein in the corpus cardiacum did not alter appreciably the authors argued that release of neurosecretory material from the corpus cardiacum must have taken place. The continued production of neurosecretory material in the absence of transport to the corpus cardiacum in the starved locust suggests a continual degradation and hence turnover of neurosecretory granules. Morris, G. and Steel, C. (1977c) proposed that lysosomal degradation of neurosecretory granules occurs on a continuing basis in the starved *Rhodnius*, and have detected PAF-positive material in the lysosomes of neurosecretory cells. In a second study Loughton, B. and Friedel, T. (1980) followed the synthesis of neurosecretory protein in the brain and corpus cardiacum during the last larval instar of *Locusta*. Two major peaks of neurosecretory protein synthesis were detected during the stadium. Both of these events could be correlated with increases in the specific activity of neurosecretory protein in the storage lobe of the corpus cardiacum. Once again the absolute amount of neurosecretory protein in the brain and corpus cardiacum changed little by comparison with the changes in synthetic activity, and consequently it appears neurosecretory material must be released into the haemolymph during or immediately after synthesis. Incorporation of [^{3}H]cysteine into proteins of the glandular lobe of the corpus cardiacum varied in a manner independent of the "A" cells of the neurosecretory system. This second paper demonstrates the limitations of quantitative histochemistry in estimating neurosecretory activity since Clarke, K. and Langley, P. (1963) found no evidence of fluctuations in the staining properties of the cells of the pars intercerebralis during the fifth instar of *Locusta* (an observation confirmed in our laboratory). By contrast, Berlind, A. and Maddrell, S. (1979) observed "empty" neurosecretory cells in the ventral ganglion of *Rhodnius* after feeding. Since these cells are the source of diuretic hormone they assayed the hormone content of isolated neurosecretory cells from the ventral ganglia of *Rhodnius* at timed intervals after feeding (and hence diuresis). The amount of assayable hormone in the cells was almost halved within 1 h after feeding. At 4 h after feeding, when diuretic hormone was still being released, the cells contained two-thirds of that present in the unfed larva. This level persisted for at least 10 days after feeding. This suggests that the replenishment of diuretic hormone is a prolonged process. In this regard it should be noted that normally the *Rhodnius* larva will feed only once in the stadium. The day before ecdysis (D17) the neurosecretory cells contained almost as much hormone as they did in the unfed larva. This system appears to contrast markedly with the condition in the neurosecretory cells of the pars intercerebralis in which the level of neurosecretory product in the cells appears to be kept almost constant.

The mode of release of diuretic hormone in the neurosecretory cells of the ventral ganglion of *Rhodnius* has been investigated by Berlind, A. (1981). The amount of hormone available for release by artificial depolarization was increased after feeding. The replenishment of available hormone at the release site appeared to be reduced by treatment with vinblastine and colchicine, suggesting axonal transport was mediated at least in part by microtubules (see below). The decrease in the amount of hormone released after prolonged depolarization in 100 mM K^+ could be reversed by treatment with a Ca^{2+} ionophore. This indicates that the factor limiting release was a closing of membrane Ca^{2+} channels rather than depletion of the hormone.

Berry, R. and Arch, S. (1981) followed the incorporation of [^{3}H]leucine into the bag cells of *Aplysia* following their depolarization with 100 mM K^+. The virtue of this system lies in the fact that the neurosecretory products and their precursor molecules could be analysed. Depolarization caused a 25% increase in the synthesis of precursor molecules. The authors also demonstrated that depolarization (and hence activation) did not alter the rate of degradation or turnover of the egg laying hormone. An elevated rate of synthesis persisted for at least 8 h after activation.

Ligation of the sciatic nerve of the rat caused an accumulation of somatostatin proximal to the obstruction. Using the rate of accumulation as an index of transport, they estimate that the "rapidly transported" fraction of the somatostatin (19.2% of that produced) moves at a rate of 16.7 mm h^{-1} (Rasool *et al.*, personal communication in Reichlin, S. (1980). Estimates of the rate of movement of neurosecretory material in insects vary considerably. Friedel, T. and Loughton, B. (1980a) report 21 mm h^{-1} in *Locusta* for the "A" cells of the pars intercerebralis, but point out this is a minimum value. In *Schistocerca*, Highnam, K. and Mordue, A. (1974) revised their earlier estimate of 0.8–1.6 mm h^{-1} upward to 11 mm h^{-1} and Girardie, J. and Girardie, A. (1977) estimated a transport rate of 12 mm h^{-1} for these same cells. Steel, C. and Harmsen, R. (1971) calculated a rate of movement of 4.8 mm h^{-1} for the median neurosecretory cells of *Rhodnius*.

These values fall within the levels generally cited for fast axonal transport in vertebrates. Several authors have claimed increases in rate of transport following stimulation of the neurosecretory cells in some way. However, these increases were based on the observation that more neurosecretory material arrived at the axon terminals, consequently the absolute rate of movement may not have altered only the amount of material moved in a given length of time. Studies of axonal transport in vertebrate nerves have recognized a number of different rates of movement from 0.02 mm h^{-1} to a maximum of 21 mm h^{-1}. Generally rates above 8 mm h^{-1} are regarded as characteristic of "fast" axonal transport which is an active, energy-dependent process. The mechanism of active axonal transport is currently a matter of dispute. Application of the mitotic inhibitor colchicine prevents fast axonal transport and causes a build-up of neurosecretory granules in the perikaryon. These experiments have been taken to mean that microtubules are involved in the transport of neurosecretory granules. In fact, some authors have described connections between neurosecretory granules and microtubules (see Maddrell, S. and Nordmann, J. 1979). The mode of action of the drug has been questioned by Brady, S. *et al.* (1980) who argue that, since intact microtubules are not necessary for fast axonal transport in their experiments, colchicine may act at some site other than the microtubules.

The process leading to the production of active neuropeptides appears to conform to the pattern of synthesis previously reported for other peptide hormones (see Herbert, E. 1981). Sachs, H. and Takabatake, Y. (1964) first argued for the concomitant synthesis of neurophysin and hormone in the hypothalamus of vertebrates. A series of studies has culminated in the report of Brownstein, M. *et al.* (1980) who demonstrated the presence of 20,000 MW precursor molecules for both vasopressin and oxytocin which are synthesized in the perikaryon and packaged in neurosecretory granules. These precursor molecules apparently contain a glycopeptide moiety at the amino terminal which is separated from the hormone-containing sequence by a region rich in arginine and lysine. The hormone in turn is separated from the neurophysin sequence by another region rich in arginine and lysine residues. The precursor molecule is cleaved in the cell body and during the passage of the neurosecretory granule from the perikaryon of the cell to its axon terminus. It should be noted that this scheme requires the presence of a proteolytic enzyme in the granule. Russell, J. *et al.* (1979) have shown that treatment of the precursor molecule with trypsin yields a 10,000 MW function and an arginine–

vasopressin-like peptide. In addition, the hormone-containing sequence must be trimmed by a carboxypeptidase and then amidated. Thus, at least two other enzymes should be present in the neurosecretory granule. Once cleaved, the integrity of the neurohormone–neurophysin complex in the granule is maintained by the intravesicular pH (5.8). Optimum binding of hormone and neurophysin occurs at pH 5.5 (Russell, J. and Holz, R. 1981). At the present time all the known vertebrate neuropeptides appear to be synthesized first as larger precursor molecules (Acher, R. 1981). The final product may not always be the same since Nordmann, J. and Labouesse, J. (1981) have demonstrated the production of a different set of peptide products at regions of the neuron other than the neurohaemal release site in the neurohypophysis of rats. Several neuropeptides ACTH, β-endorphin, MSH, appear to enjoy a common precursor (pro-opiomelanocortin) which contains paired lysine and arginine, residues between the physiologically active sequences (Nakanishi, S. *et al.*, 1979) suggesting that tryptic cleavages are involved in the production of the hormones from their precursor molecule. The processing of the precursor occurs in a highly ordered manner even though the points of cleavage contain the same amino acid residues and the final product in the given cell may differ from that of another cell. How these processes are performed is not understood. One method for the differential production of final products from a common precursor would be to use different endopeptidases whose specificity for given amino acid sequences differed. Thus trypsin-like and pepsin-like enzymes would be expected to produce different end products from the same precursor (see Acher, R. 1981).

Processing of a precursor molecule has been described by Berry, R. (1981) and Berry, R. *et al*, (1981) in the bag cells of *Aplysia*. The precursor molecule is a 29,000 MW protein (PI 7.7) which gives rise to a 6000 MW fraction and a number of molecules 16,000–20,000 MW. This second larger molecule is finally processed to yield a molecule of 13,000 MW. Interestingly enough the 13,000 MW fraction is not transported out of the perikaryon. The heterogeneous 6000 MW fraction is cleaved to yield the egg laying hormone (4385 MW) or an acidic protein (4500 MW). Tryptic digestion of the 13,000 MW protein that remains in the cell body revealed that it contains an additional acidic protein sequence. No evidence of a neurophysin molecule was adduced despite the recognition of a molecule derived from the precursor with a molecular weight roughly corresponding to that of vertebrate neurophysins. Once again perhaps it should be recognized that egg-laying hormone is a relatively large hormone (4385 MW) which might not require the stabilizing capacity of a neurophysin.

Andrew, R. and Saleuddin, A. (1979) used two-dimensional gel electrophoresis to analyse the processing of neurosecretory protein synthesized by the crustacean X organ and released at the sinus gland. After incubation of crayfish eyestalks in medium containing [^{3}H]leucine, extracts of the X organ and sinus gland were analysed for the presence of labelled peptides by the method of O'Farrell, P. Z. *et al.* (1977). The X organ contained labelled peptides with molecular weights of 19,000–21,000, whereas in the sinus gland three peptides with molecular weights between 4000 and 10,000 were observed. Since the source of sinus gland peptides is the X organ, it was assumed that the smaller sinus gland peptides were derived from the larger molecules labelled in the X organ. Despite the fact that the neurosecretory cells of the brain and the storage lobe of the corpus cardiacum represent a similar anatomically attractive system no similar studies have been performed on insects to date.

3.3.2 Release

The mechanism of release of neurosecretory granules was outlined in section 2.3. In this section we shall examine the factors which control and regulate the release at neurohaemal organs.

In conventional neurons, action potentials passing into the terminal regions result in an influx of calcium ions which trigger the release of neurotransmitter. It has been possible to show this because of the ability to assay the release of neurotransmitter by virtue of recording the resultant postsynaptic response. Since the neuronal nature of neurosecretory cells has been confirmed, it has been argued that in neurosecretory cells as well, the action potentials must control the release of neurohormone. It has been more difficult to prove this, however, since there is effectively no postsynaptic cell to be used as a convenient bioassay for release of active material. Furthermore, in the majority of cases the function and therefore target site for the neurohormone is unknown. As a result it has not been possible to correlate action potentials in an individual neurosecretory cell with release of assayable quantities of neurohormone. Consequently, it has been necessary to study large groups of neurosecretory cells which release more than one hormone, are mixed with conventional axons, and generally have to be artificially stimulated *en masse*. With the advent of direct assays for the hormone itself (e.g. RIAs, for which none currently exist for insect neurohormones) it may eventually be possible to show that action potentials in one neurosecretory cell result in the release of neurohormone. The excellent preparations available to the neurobiologist examining release of neurotransmitter at synapses between axons or synapses (neuromuscular junction) between axon and muscle are not available to the neurobiologist examining release of neurohormone from neurosecretory cells. As a result there are numerous publications attempting to show a correlation between electrical activity and release of neurohormone — all of them indirect, all of them subject to different criticisms. However, taken together, they all provide evidence which cannot be ignored, that action potentials in neurosecretory cells result in the release of neurohormone. Furthermore, the mechanism of release conforms to the calcium hypothesis for neurotransmitter release.

In insects, electrical stimulation of neurosecretory cell groups or neurosecretory tracts induces histological and ultrastructural changes indicative of release. Hodgson, E. and Geldiay, S. (1959) and Gosbée, J. *et al*, (1968) described a decrease in stainability of the corpus cardiacum following electrical stimulation of the brain. However, light microscope stains do not stain for neurohormone but rather for carrier protein (neurophysin?), and so the precise relationship between the changes observed and release of neurohormone is not clear. Ultrastructural evidence for release — as indicated, for example, by exocytotic profiles and increased frequency of clear vesicles (see section 2.3) — can provide a clearer picture, since there is good evidence to support the idea that the neurohormone is located within the neurosecretory granules. Comparisons of unstimulated versus stimulated cells have shown a correlation between electrical stimulation and release (Scharrer, B. and Kater, S. 1969; Normann, T. 1970, 1974; Krogh, I. and Normann, T. 1977; see section 2.3).

More direct evidence has been obtained by bioassaying for hormone release following electrical stimulation. The electrical stimulation of NCC I leading to the corpus cardiacum in *Periplaneta* induced ultrastructural changes associated with release and a concomitant release of bioassayable neurohormone in the form of a cardioaccelerator (Scharrer, B. and Kater, S. 1969). Electrical stimulation of the brain resulted in the release of the cardioaccelerator so long as the NCC I were intact (Kater, S. 1968) indicating that information passed along these nerves to control release. Stimulation of NCC I of *Periplaneta* released cardioaccelerator activity while NCC II released both a hyperglycaemic and an ecdysiotropic factor (Gersch, M. 1972; Gersch, M. *et al.*, 1970). Girardie, J. and Girardie, A. (1977b, 1978) induced release of a gonadotropin from the corpus cardiacum of *Locusta* by stimulating the area containing the medial neurosecretory cells in the brain. Electrical stimulation of the brain of *Calliphora* using electrodes inserted through the posterior surface of the head resulted in hypertrehalosaemia of the haemolymph (Normann, T. and Duve, H. 1969). This response was only observed if the nervous connections between the corpora cardiaca and brain were intact, again indicating that hypertrehalosaemic hormone was released from the corpora cardiaca following the passage of information along the nervous connections.

A criticism of these approaches lies in the fact that there was no evidence to show that the electrical stimulation brought about normal physiological responses in the stimulated cells. Much of the release may have been the result of excessive currents passing through the nervous system. A better technique relies on recording the compound action potential initiated by electrical stimulation, thus allowing the use of current intensities which are sufficiently large to cause a just maximal compound action potential. This approach was used recently in examining the release of neurohormones and neurosecretory protein from the corpus cardiacum of *Locusta* (Friedel, T. *et al.*, 1981; Orchard, I. *et al.*, 1981a; Orchard, I. and Loughton, B. 1981a,b). The isolated corpus cardiacum of *Locusta* released factors which brought about diuresis, and which altered protein synthetic activity of the fat body following electrical stimulation of NCC I (Orchard, I. *et al.*, 1981a). Stimulation of NCC II resulted in the release of hypolipaemic hormone from the storage lobe of corpus cardiacum (Orchard, I. and Loughton, B. 1980) and hyperlipaemic and hyperglycaemic hormones from the glandular lobe (Orchard, I. and Loughton, B. 1981a,b; Loughton, B. and Orchard, I. 1981). The release of the latter hormones was dependent upon the axons of NCC II activating the glandular cells via an aminergic (octopaminergic?) synapse (Orchard, I. and Loughton, B. 1981b). Control experiments using tetrodotoxin, sodium-free saline and sub-threshold stimulation confirmed that the release of hyperlipaemic hormone from the corpus cardiacum was dependent upon the propagation of compound action potentials along the nerves and not an artifact caused by electrical stimulation.

A more recent approach used to correlate release with electrical stimulation has made use of the fact that the secretory products of many insect neurosecretory cells are rich in sulphur and thereby incorporate labelled cysteine or cystine. The use of label has therefore been used as a tool to monitor synthetic and secretory activity (see section 3.3.1). Girardie, J. and Girardie, A. (1977b, 1978) measured the release of [^{35}S]cysteine-labelled protein from the corpus cardiacum of *Locusta* following electrical stimulation of the brain and terminals. They found that a [^{35}S]cysteine-labelled protein (which in fact may be a neurosecretory carrier protein) was released following stimulation of the area containing the cell bodies of the medial neurosecretory cells. Using a similar approach, Friedel, T. *et al.* (1980) isolated a neurosecretory protein showing similarities to the neurophysins of vertebrates, from the brain and corpora cardiaca of *Locusta*. Using an antiserum against this protein Friedel, T. *et al.* (1980) localized the protein, which was labelled with [^{35}S]cystine, within the neurosecretory A cells of the pars intercerebralis, the NCC I and the storage lobe of *Locusta* (Fig. 8). Immunoreactive deposits were predominantly associated with the neurosecretory granules with little reaction within the cytoplasm. Electrical stimulation of NCC I resulted in the release of this neurosecretory protein (estimated following precipitation with the antiserum) provided a compound action potential was propagated along the nerve (Orchard, I. *et al.*, 1981a; Friedel, T. *et al.*, 1981). The release was prevented by tetrodotoxin and sodium-free saline (see Fig. 10).

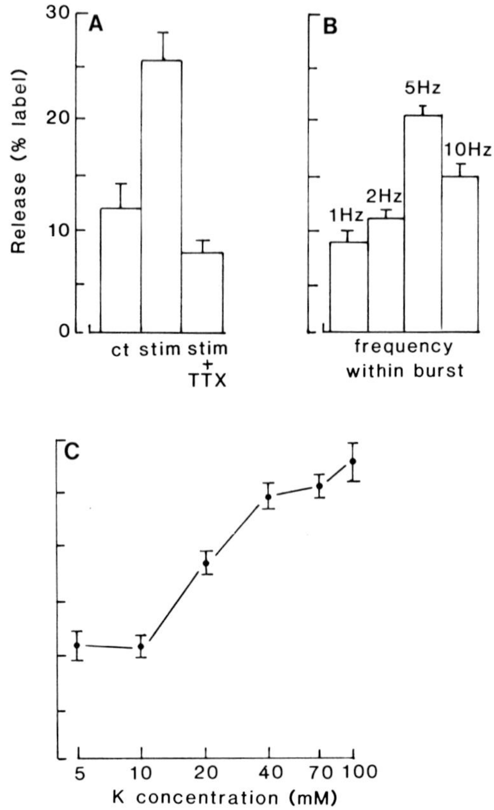

FIG. 10. Release of labelled neurosecretory protein from the storage lobe of the corpora cardiaca of *Locusta* **A:** Electrical stimulation of NCC II at sub-threshold current intensities (ct), at current intensities sufficient to induce a compound action potential (stim) and at the same current intensities in the presence of 10^{-6} M tetrodotoxin (stim + TTX). **B:** The release of neurosecretory protein is dependent upon the burst patterning of compound action potentials. A total of 1800 compound action potentials were delivered in 30 min at either 1 Hz; 2 Hz for 1 s every 2 s; 5 Hz for 1 s every 5 s; 10 Hz for 1 s every 10 s. **C:** The effect of external potassium (K) concentration on release of neurosecretory protein. Redrawn from Orchard, I. *et al.* (1981a).

Direct depolarization of secretory cell terminals using high potassium saline led to hormone release from insect abdominal neurohaemal sites (Maddrell, S. and Gee, J. 1974; Berlind, A. 1981) and from corpora cardiaca (Normann, T. 1974; Orchard, I. and Loughton, B. 1981b) and resulted in the release of labelled neurosecretory protein from *Locusta* corpora cardiaca (Girardie, J. and Girardie, A. 1977b, 1978; Friedel, T. *et al.*, 1981; Orchard, I. *et al.*, 1981a). The release of neurohormone (Maddrell, S. and Gee, J. 1974) and neurosecretory protein (Orchard, I. *et al.*, 1981a) was dependent upon the concentration of potassium ions in the bathing medium (Fig. 10).

The calcium hypothesis for release of neurohormone has been examined in many of the systems in which release has been demonstrated. Reduction or total removal of calcium from the incubation medium resulted in the inhibition of release of neurohormone following electrical stimulation (Orchard, I. *et al.*, 1981a; Orchard, I. and Loughton, B. 1981a; Gersch, M. *et al.*, 1970) or application of high potassium saline (Normann, T. 1974; Maddrell, S. and Gee, J. 1974; Friedel, T. *et al.*, 1981; Orchard, I. *et al.*, 1981a). A slightly different approach has made use of the calcium ionophores X537A and A23187. These antibiotic ionophores can make cations soluble in a lipid phase and are thus able to transport them across cell membranes (Pressman, B. 1976). X537A can transport mono- and divalent cations (and possibly catecholamines) through membranes while A23187 is thought to be more specific for divalent cations. As a result, they have been used as tools for investigating mechanisms of release, and of providing evidence for the calcium-dependency of the release since the ionophores result in an elevation of intracellular free calcium. Both ionophores have been shown to induce release in a variety of different tissues, but in particular from insect neurohaemal tissue (Orchard, I. *et al.*, 1979), crayfish sinus gland (Andrew, R. and Shivers, R. 1976) and rat neurohypophysis (Nakazato, Y. and Douglas, W. 1974), thereby implicating calcium in the control of release of neurohormone.

A further interesting facet in the release of neurosecretions is that during secretion it is the contents of the most recently synthesized granules which are released first, followed by the contents of the older granules. This "last in, first out" hypothesis was first postulated in mammals (e.g. Sachs, H. *et al.*, 1969), but there is now evidence that the same may be true in insects. In locusts it is possible to label the neurosecretory protein contained within the medial neurosecretory cells using [^{35}S] labelled cysteine (Girardie, J. and Girardie, A. 1977b, 1978). Upon electrical stimulation of the cell bodies a larger proportion of radiolabel was released after 2 h than was released after 10 h or 72 h, illustrating that the longer the neurosecretory material remained in the corpus cardiacum the less available it was for release. The theory, based essentially on the vertebrate work, is that the neurosecretory granules which arrive in the axon terminals and which are not soon released move away from the terminal and accumulate in large numbers in dilations of the axon. These swellings contain lysosomes which presumably act to destroy the granules.

A criticism of most of the experiments concerned with release of neurohormone described above is that neurosecretory tracts and neurohaemal organs contain a sizeable number of conventional neurons and axons. It is therefore impossible to state categorically that non-neurosecretory elements do not play a role following electrical stimulation or perfusion with high-potassium saline. For example, these elements may synapse with neurosecretory terminals and cause release. However, the consensus of opinion, based upon the knowledge gained from invertebrate and vertebrate preparations, is that action potentials in neurosecretory cells must result in the release of neurohormone and neurophysin, although it may not be possible to prove this directly.

A characteristic feature of electrical activity of neurosecretory cells lies in the patterning of electrical activity observed at times when hormone release is believed to be occurring. A number of recordings from neurosecretory cells in invertebrates and vertebrates have shown that they possess a bursting (phasic) pattern of activity (Gainer, H. 1978) and such a phenomenon has been described in insects (Finlayson, L. and Osborne, M. 1970; Orchard, I. and Finlayson, L. 1976; Bruce, I. and Wilkens, J. 1976; Orchard, I. and Steel, C. 1980; Krauthamer, V. 1980; Seshan, K. *et al.*, 1974). Since the production of action potentials is so closely associated with release of hormone it has been suggested that bursting may greatly increase the amount of hormone released per action potential (Gainer, H. 1978). This hypothesis has recently been investigated in both vertebrates (Dutton, A. and Dyball, R. 1979) and invertebrates (Orchard, I. *et al.*, 1981a). It was found that in the isolated rat neural lobe, electrical stimulation led to a tetrodotoxin-sensitive release of vasopressin (assayed by RIA) (Dutton, A. and Dyball, R. 1979). Stimulation with pulse patterns (which were derived from tape-recordings of phasically active cells in the supraoptic nucleus of dehydrated rats) released more vasopressin than the same number of pulses regularly spaced in time. Vasopressin release was related to pulse frequency within the bursts and the number of short (< 100 ms) interpulse intervals within the range of 2–8 pulses s^{-1}. As a result, it is apparent that phasic firing of vasopressin-releasing neurosecretory cells enhances release. In a somewhat similar fashion, the release of labelled neurosecretory protein from the storage lobe of the corpora cardiaca of *Locusta* was found to be dependent upon frequency and patterning of pulses with a 5 Hz pattern releasing 229% more than a 2 Hz pattern (Orchard, I. *et al.*, 1981a) (Fig. 10).

The mechanism for this potentiation has not been examined in neurosecretory cells, but by analogy with synaptic transmission and release of neurotransmitter it may result from an increase in calcium influx per action potential and hence an increase in the amount of hormone released. The wide duration of action potentials in cell bodies and terminals of neurosecretory cells may allow for a greater influx of calcium ions, and in addition the duration of each consecutive action potential in a burst may also increase (Orchard, I. and Finlayson, L. 1977a; Cooke, I. 1977; Gainer, H. 1978). Thus, the grouping of action potentials may potentiate the influx of calcium ions and thereby release more hormone. This is clearly of great benefit to neurosecretory cells in view of the substantial quantity of hormone required in the haemolymph.

The correlations between electrical activity and hormone release, coupled with the patterning of activity, suggests that an examination of the patterning and intensity of electrical activity of neurosecretory cells may enable the timing of hormone release to be determined. This approach was used to examine the cerebral neurosecretory cells controlling growth and moulting in the bug *Rhodnius prolixus* (Orchard, I. and Steel, C. 1980). The advantage of this preparation lay with the ease in which the neurosecretory system could be activated by a natural stimulus, a blood meal. The naturally occurring electrical activity can be recorded *in vivo* from the axons of neurosecretory cells from the brain of the fifth instar *Rhodnius*. In the unfed fifth the frequency and pattern of electrical activity varied little from preparation to preparation, and so enabled differences between treatments to be observed.

Electrical activity was low in unfed *Rhodnius* but quickly altered minutes after the insect took a blood meal. Bursting activity, and increased frequency of action potentials, reached a peak within 1–2 h after feeding. The bursting activity declined after 2 h and remained low for 5 days, but was then followed by a resurgence in the bursting pattern which lasted a few hours. Both these peaks of bursting activity were closely coupled with increases in the haemolymph ecdysteroid titre, and it may be inferred that release of PTTH occurred on these two occasions. The continuous firing components initiated at feeding maintained a high level for 5 days, perhaps indicating the release of other brain neurosecretions. Bursting activity and high-frequency firing resumed shortly before ecdysis and continued until 12 h after (for full interpretation of these results see Orchard, I. and Steel, C. 1980).

In adult *Rhodnius* the intensity and patterning of activity of the axons from the medial neurosecretory cells of mated and virgin females is different. These differences may be correlated with the release of a myotropic hormone involved in ovulation in the mated animals (Ruegg, R. *et al.*, 1981; see section 3.3.4). The link nerve neurosecretory cells in *Carausius* showed a higher frequency of firing during the dark phase of a diel cycle and it has been inferred this is representative of the release of hormone during this phase (Finlayson, L. 1980).

4 INTEGRATION AND REGULATION

A comprehensive review of neuroendocrine integration and regulation in insects may be found in chapter 1, volume 8. It is the purpose of the present section to examine the principles involved in neuroendocrine regulation and integration, concentrating on a small number of model examples. Neurosecretory cells must clearly integrate with the rest of the nervous system. They must be turned on and off, and their activities must be regulated in such a way that critical titres of hormone are maintained at appropriate times. Methods of integration may involve either excitatory or inhibitory input and in either case the source of this regulation must be nervous and/or hormonal.

4.1 Synaptic and hormonal activation

Neurosecretory cells receive synaptic input. Excitatory and inhibitory postsynaptic potentials have been recorded from insect neurosecretory cells (see section 3.1a) and ultrastructural evidence for synaptic input has been described on a number of occasions. The elaborate dendritic trees branching into the neuropile from neurosecretory cells indicate that there is plenty of opportunity for synaptic input (for complete discussion see Pipa, R. 1982). Unfortunately, there is little information concerning the nervous pathways involved in these integrative areas, or the identities of neurons providing the input. A number of internal and external conditions are known to result in the release of neurohormone (Maddrell, S. 1974; Goldsworthy, G. and Mordue, W. 1974; Maddrell, S. and Nordmann, J. 1979) but by and large the locations and nature of receptors, the control pathways or the transmitters used for most systems, are unknown.

One system for which there is a large body of information regarding regulation of release of neurohormone concerns that of adipokinetic hormone (AKH) in locusts (Fig. 11); see also chapter 12, volume 10). Within about 15 min of flight, AKH is released from the glandular lobe of the corpus cardiacum of locusts and serves as a hyperlipaemic hormone acting upon the fat body to stimulate the release of lipid and thereby an increase in the haemolymph lipid. Clearly, the initial stimulus for activation of the glandular cells is associated with flight, although the precise mechanism is not known. It is possible that a lowering of haemolymph trehalose levels during the first few minutes of flight may stimulate the release of hormone since injection of trehalose prevented the rise in lipid levels normally induced by flight (Houben, N. and Beenakkers, A. 1973). However, the source and nature of receptors for this stimulus are not known. The ultimate nervous control of release of AKH is via NCC I and NCC II. Axons in NCC II arising from lateral cells in the brain synapse with the glandular cells (Rademakers, L. 1977b) and electrical stimulation of NCC II resulted in the release of hyperlipaemic hormone (Orchard, I. and Loughton, B. 1981a). Axons in NCC I appeared to modulate the effect of NCC II, but did not in themselves cause release of hormone (Orchard, I. and Loughton, B. 1981a). The pharmacology of the synapse between axons in NCC II and the glandular cells has recently been characterized (Orchard, I. and Loughton, B. 1981b) and shown to be aminergic. The available evidence points to the transmitter being octopamine. This is the first description of the pharmacological nature of identified synapses impinging on insect neurosecretory cells. The octopaminergic nature of these synapses is particularly interesting because another feature of flight in locusts is a short-term elevation in haemolymph octopamine levels (Goosey, M. and Candy, D. 1980). Octopamine can act as a neurohormone (see Orchard, I. 1982a) and in locusts has been shown to act upon the fat body to release lipid (Orchard, I. *et al.*, 1981b; 1982). Furthermore, the levels of octopamine in the haemolymph may be sufficiently large to potentiate the neurally evoked release of hormone from the glandular cells (Singh, G. *et al.*, 1981). The possible role of octopamine in this system is shown in Fig. 11. Octopamine may stimulate the release of lipid within the first 10 min of flight, but AKH soon takes over the long-term maintenance of haemolymph lipid. Octopamine may also act upon the glandular lobe to potentiate the octopamine-stimulated release of AKH at this time. Here then there may be dual transmitter/hormonal control of glandular cells.

The medial neurosecretory cells in *Rhodnius* larvae are activated when the insect takes a blood meal. This stimulus precipitated structural changes in the 17 medial neurosecretory cells within minutes of feeding, which were indicative of the sudden stimulation of synthesis, transport and release of neurosecretion (Morris, G. and Steel, C. 1977a,b; Steel, C. and Harmsen, R. 1971; Steel, C. 1978a). Concomitantly, there was a change in the electrical activity recorded from near the terminals, with the background spike frequency increasing four-fold within an hour and high-frequency bursts of spikes (up to 200 Hz) evident within minutes of the bug completing a blood meal. Precisely what aspect of

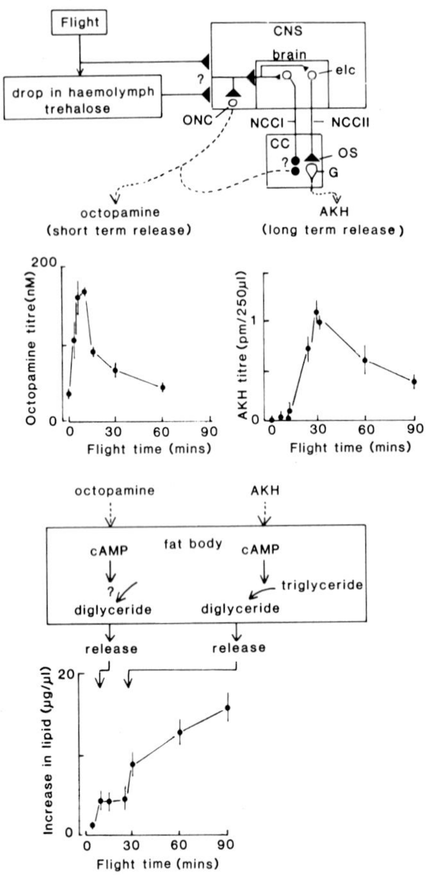

FIG. 11. Neurohormonal integration of haemolymph lipid during locust flight. Explanation in text. ——◄, excitatory input; ——●, neuromodulation; — — — — →, neurohormone; ONC, octopamine neurosecretory cell; OS, octopamine synapse; NCC I and NCC II, nervus corpus cardiacum I and II; CC, corpus cardiacum; G, glandular cells; AKH, adipokinetic hormone; elc, extralateral cells. Octopamine titre redrawn from Goosey, M. and Candy, D. (1980); AKH and haemolymph lipid titre redrawn from Cheeseman, P. and Goldsworthy, G. (1979).

feeding activated the neurosecretory cells is essentially unknown, although Wigglesworth, V. (1964) postulated that abdominal stretch receptors were activated by distension caused by the blood meal and that information of this stretch either directly or indirectly activated the medial neurosecretory cells. The nervous pathways used by these receptors within the central nervous system are not known.

Steel, C. and Lees, A. (1977) examined the location of the photoperiodic mechanism controlling the production of the sexual and parthenogenetic morphs by apterous parents in *Megoura*. Lesions which destroyed the Group I neurosecretory cells in the protocerebrum abolished the response to changed day length, while destruction of other neurosecretory cells, compound eyes and optic lobes was without effect. These results suggested that Group I cells were the source of a virginopara-promoting factor in whose absence oviparae were produced. If areas slighty lateral to the Group I cells were damaged, then the response to long day length was abolished. These experiments indicated that the areas lateral to the Group I neurosecretory cells may be the site of a neuronal photoperiodic clock which regulates the release of neurosecretory material from the Group I cells. The neuronal pathway connecting these two areas has not been determined.

Hormones could conceivably feed back either directly or indirectly via interneurons onto neurosecretory cells to alter their activity. Indeed, it has been postulated that the neurohormone from a cell may feed back to control its own release. This would seem logical since the amount of hormone released must in some way be regulated. Actual evidence for hormonal regulation of neurosecretory cells comes from a number of reports. We have already seen that octopamine may act as a neurohormone on the glandular cells. Other systems include the following. The link nerve neurosecretory cells in *Carausius* showed a higher frequency of action potentials during the dark phase of a diel cycle, and a secretory cycle, as judged by acridine orange staining, corresponding to the electrical cycle (Finlayson, L. and Orchard, I., 1978). The frequency of action potentials generated by these cells was directly influenced by the haemolymph (see Finlayson, L. and Orchard, I. 1978; Finlayson, L. 1980) to such an extent that haemolymph from a light phase insect decreased the rate of firing of link nerve neurosecretory cells from a dark phase insect. The suggestion from these and other results (Finlayson, L. 1980) was that an inhibitor may be present in the haemolymph of the insects during the light phase.

There have been a number of reports concerning the role of ecdysteroids in regulating neurosecretory activity (e.g. Steel, C. 1973; Agui, N. and Hiruma, K. 1977). The evidence for 20-hydroxyecdysone as a regulator is particularly strong in adult *Rhodnius* females (Ruegg, R. *et al.*, 1981, 1982) and a model for this system is shown in Fig. 12. The electrical activity of the corpus cardiacum of mated females 6 days after feeding was high (Ruegg, R. *et al.*, 1982), with evidence of bursting activity. This electrical activity corresponds to the timing of release of a myotropic ovulation hormone from the medial neurosecretory cells, whose activity was detected in the haemolymph (Ruegg, R. *et al.*, 1981). Both electrical activity and haemolymph myotropic activity were reduced or absent in virgin females. Clearly mating was necessary for the expression of these activities. The haemolymph ecdysteroid titre in mated females peaked 5 days after a blood meal. Ovariectomy prevented this peak (Ruegg, R. *et al.*, 1981) since the ovaries are the origin of ecdysone (Ruegg and Kriger, personal communication). Ovariectomy also prevented the appearance of bursting activity in the corpus cardiacum and prevented the release of the myotropic ovulation hormone (Ruegg, R. *et al.*, 1981, 1982). However, injection of 20-hydroxyecdysone into ovariectomized mated females restored the electrical activity and haemolymph myotropic activity back to the unoperated level of a mated insect (Ruegg, R. *et al.*, 1981, 1982). It was deduced from this work that 20-hydroxyecdysone must feed back onto the medial neurosecretory cells and stimulate the release of the myotropic ovulation hormone by the activation of a distinct pattern of electrical activity. The isolated brain and corpus cardiacum taken from ovariectomized mated females had a low level of electrical activity in the corpus cardiacum (Ruegg, R. *et al.*, 1982) and also did not release myotropic ovulation hormone into the medium (Kriger, personal communication). Perfusion with 20-hydroxyecdysone restored the electrical activity to that of an intact mated insect, and resulted in the release of the hormone. Clearly, the ecdysteroid was acting on the brain/corpus cardiacum complex to control the activity of the medial neurosecretory cells and thereby the release of myotropic ovulation hormone. Preliminary results suggest that this was not a direct action of 20-hydroxyecdysone, but may involve interneurons which use amine as their transmitter.

Steel, C. (1973, 1978) also examined the possibility that 20-hydroxyecdysone might act on the medial neurosecretory cells in *Rhodnius* larvae. He found that parabiosis of a decapitated insect 8 days after a blood meal (a time when ecdysteroid titres are rising) to an intact insect 1 day after a blood meal, induced changes in the medial neurosecretory cells of the younger insect indicative of the older insect. These changes were in the direction whereby the haemolymph from the older insect appeared to inhibit release of material from the cells. Steel, C. (1973) considered the factor causing these changes was probably 20-hydroxyecdysone and found that injection of 1 μg 20-hydroxyecdysone could mimic these effects. However, the site of action of 20-hydroxyecdysone could not be determined in these experiments.

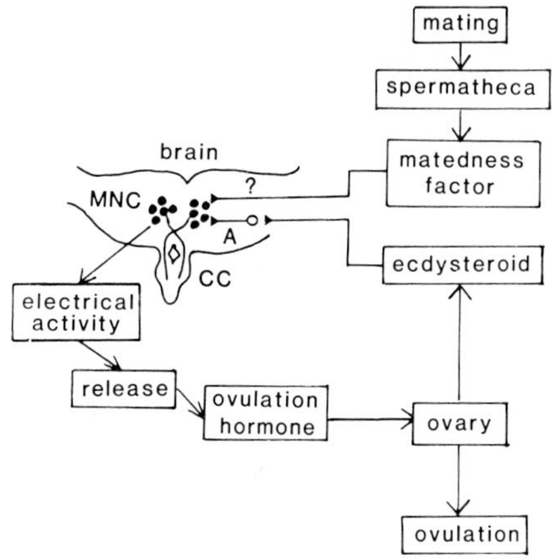

FIG. 12. Proposed model for control of release of ovulation hormone in *Rhodnius*. Ecdysteroid from the ovary and a matedness factor from the spermatheca provide excitatory input (———◀) onto the medial neurosecretory cells (MNC). The ecdysteroids act via an aminergic interneurone (A). The mechanism of action of the matedness factor is not understood. The resultant alteration in frequency and patterning of electrical activity within the corpus cardiacum (CC) results in the release of ovulation hormone which acts on the muscles of the ovarian sheath to induce ovulation. Redrawn from Ruegg (1981). Based upon evidence in Ruegg, R. *et al.* (1981, 1982); Orchard, Ruegg, Kriger and Davey, unpublished.

It has also been postulated that JH may regulate the activities of neurosecretory cells. Allatectomy of adult locusts led to an apparent decrease in synthetic activity of the medial neurosecretory cells and to a slight increase in material in the storage sites (McCaffery, A. and Highnam, K. 1975a). Implantation of active corpora allata or application of JH or the JH analogue farnesyl methyl ether (FME) increased the synthetic activity and the transport and release of neurosecretory material. The site of action of JH is unknown. Pratt, G. and Davey, K. (1972) showed that application of FME accelerated egg development in *Rhodnius*, but this did not occur if the cerebral neurosecretory cells were removed, possibly linking JH with the release of neurohormone. Removal of the corpora allata in *Calliphora* resulted in a reduction in the apparent synthetic activity of the medial neurosecretory cells (Thomsen, E. and Lea, A. 1968). Implantation of corpora allata into allatectomized insects stimulated the secretory activity, again implicating JH with the activity of the neurosecretory cells, although again whether this is a direct or indirect action is not known.

Friedel, T. and Gillott, C. (1976) have postulated the control of release of an ovulation hormone from the brain of female *Melanoplus* by factors released from the male accessory gland deposited during mating. Interestingly enough, Angela Lange (personal communication) has been able to demonstrate the release of a neurosecretory protein from the isolated corpus cardiacum of *Locusta* using extracts of mature male accessory glands. Whether or not factors are released which control oviposition remain to be seen, as too is the site of action of the accessory gland factor.

4.2 Nervous system as a target site for neurohormones

Neurohormones can act upon the nervous system to either initiate activity or modify existing activity (review by Truman, J. 1978; chapter 12, 13, volume 8). For example, the eclosion of the adult insect from the pupal case is accomplished by a stereotyped programme of movements that are triggered by a peptide hormone. Injection of this "eclosion hormone" can cause the release of the eclosion programme and application of the hormone to the isolated abdominal nerve cord induced a stereotyped motor programme which correlates well with the stereotyped movements (Truman, J. 1971, 1976, 1978).

The intrinsic cells of the corpora cardiaca of saturniid moths are believed to release a hormone which acts upon the abdominal nerve cord to induce the calling posture associated with calling behaviour (Riddiford, L. and Williams, C. 1971; Riddiford, L. 1977). Thus, abdomens isolated from virgin females of *Hyalophora cecropia* could be induced to assume the calling posture by injection of haemolymph from a "calling" female or injection of corpora cardiaca extracts (Riddiford, L. 1977).

These effects have, by and large, been assumed to be the major target for these hormones. However, it is becoming increasingly apparent that single hormones may have multiple target sites within the central nervous system as well as within non-nervous peripheral tissue. Once again, the bag cell neurons of *Aplysia* are proving to be a model system for studying such neuroendocrine integration (see review by Blankenship, J. 1980). The bag cells release a peptide hormone which acts upon the ovotestis to induce the release of oocytes. Since this was believed to be the function of the bag cells, the hormone was termed egg-laying hormone.

While *in vitro* studies have shown that egg-laying hormone has a direct non-neurally mediated effect upon the ovotestis, it is also known that this hormone can directly influence neural activity. Numerous identified cells within the parietovisceral ganglion are affected when the bag cells of the same ganglion are releasing their product (Mayeri, E. *et al.*, 1979a,b); when bag cell extract is applied to the ganglion or when purified bag cell peptide is applied (Branton, W. *et al.*, 1978). The spontaneously bursting neuron R15 shows an increased amplitude of its pacemaker potential and an intensification of bursting activity within 1 min of exposure to bag cell peptide, which may persist for 3 h. Neurons L2–L6 exhibit a slow hyperpolarization and reduction in activity 5–14 s after exposure to peptide, and other neurons are also affected by being either excited or inhibited for varying duration. None of these effects are mediated by conventional synaptic mechanisms but rather show a much slower, long-duration action of great variation.

Several neurons in the buccal and genital ganglia are also affected by bag cell peptide. Since these ganglia contain neurons associated with feeding and reproductive activity, and it is known that these behaviours are modified during egg-laying, it is possible that these neurons are associated with these behaviours (Stuart, D. and Strumwasser, F. 1980).

The notion put forward by these studies is that if there is a hormone controlling a physiological process, e.g. egg-laying, then maybe this hormone also controls other behaviours associated with this process. With this in mind, it is interesting to re-evaluate the function of the extensive arborizations of insect neurosecretory cells found within the neuropile. The view has long been held that these projections represent areas of integration, with the neurosecretory cell receiving information, i.e. being postsynaptic. However, the question remains whether all of these processes are postsynaptic or whether some of them may in fact be used to transmit information. Electron micrographs reveal apparent release sites within the neuropile (Schooneveld, H. 1974), but unfortunately we have no information as to the identity of these cells or their physiology. If neurosecretory cells do release within the neuropile, it will be of great importance to discover the nature of the substance released, and the type of message being sent. In view of the recent discovery of the importance of peptides in neural communication it will come as no surprise (though still important) if the neurohormone itself is released within the neuropile. However, it may turn out that a conventional transmitter is released within the neuropile while the neurohormone is released at the neurohaemal site. The coexistence of peptides and amines has been postulated for some time, as has the coexistence of several transmitters. If a neurosecretory cell can communicate using two different messages at different sites, then it would prove a most interesting problem to examine the physiological mechanisms underlying such an event.

5 CONCLUSION

Neurosecretory cells occupy a central position in the co-ordination of insects. At present our appreciation of their activities is limited, though certain patterns are apparent. As in vertebrates, neurosecretory cells provide the interface between the nervous system and endocrine systems through the secretion of peptidergic hormones. In addition, other insect neurosecretory hormones act directly to bring about a variety of endocrine-mediated events. Ideally one should have anatomical, electrophysiological, biochemical and integrative data for a single system. At present, such an ideal has yet to be realized; however, by a synthesis of data from different systems it becomes possible to prepare a general picture of insect neurosecretory cells. Two major types of neurosecretory cells emerge: those that are peptidergic and those that are aminergic. Within each category the product may be released into the haemolymph or immediately adjacent to the target organ. Neurosecretory cells may lie within the central nervous system or in the peripheral nervous system. They present a conventional neuronal form and a characteristic pattern of electrical activity with prolonged overshooting action potentials that occur in bursts. The release of hormone is mediated through the influx of Ca^{2+} ions following the arrival of action potentials at the neurosecretory terminals. The effects of their products appear to be associated with long-term responses.

Insect neurosecretory cells appear to have much in common with those of vertebrates and the impact of the recent surge of information on peptidergic neurones in the vertebrates has yet to reach entomology. In addition, the advent of new techniques should facilitate the separation, characterization and localization of many neurohormones which have been known hitherto only by function. The next few years should see a burgeoning of information on insect neurosecretory systems; the organisms provide a uniquely convenient system for manipulation and analysis and the opportunity awaits only the imagination and skill of the investigator.

ACKNOWLEDGEMENTS

We are indebted to the number of authors who gave permission for us to redraw figures from their papers. We are also grateful to Alan R. Liss Inc., New York for the reproduction of Figure 7. We would like to thank Krystyna Tarkowski and Angela Lange for their expert assistance in preparing the manuscript. Work originating in our laboratories was supported by the Natural Sciences and Engineering Research Council of Canada.

REFERENCES

Acher, R. (1981). Evolution of neuropeptides. *Trends Neurosci. 4*, 225–229.

Adams, C. W. M. and Sloper, J. C. (1965). Techniques for demonstrating neurosecretory material in the human hypothalamus. *Lancet* (26 March), 651–652.

Agui, N. and Hiruma, K. (1977). Ecdysone as a feedback regulator for the neurosecretory brain cells in *Mamestra brassicae. J. Insect Physiol. 23*, 1393–1396.

Agui, N., Granger, N. A., Gilbert, L. I. and Bollenbacher, W. E. (1979). Cellular localization of the insect prothoracicotropic hormone: *in vitro* assay of a single neurosecretory cell. *Proc. Nat. Acad Sci. USA. 76*, 5694–5698.

Ali, Z. I. and Pipa, R. (1978). The abdominal perisympathetic neurohaemal organs of the cockroach *Periplaneta americana*: innervation revealed by cobalt chloride diffusion. *Gen. Comp. Endocr. 36*, 396–401.

Alving, B. O. (1968). Spontaneous activity in isolated somata of *Aplysia* pacemaker neurons *J. Gen. Physiol. 51*, 29–45.

Andrew, R. D. and Saleuddin, A. S. M. (1979). Two dimensional gel electrophoresis of neurosecretory polypeptides in crustacean eyestalk. *J. Comp. Physiol. 134*, 303–313.

Andrew, R. D. and Shivers, R. R. (1976). Ultrastructure of neurosecretory granule exocytosis by crayfish sinus gland induced with ionic manipulations. *J. Morphol. 150*, 253–278.

Anwyl, R. and Finlayson, L. H. (1973). The ultrastructure of neurons with both a motor and a neurosecretory function in the insect, *Rhodnius prolixus*. *Z. Zellforsch. 146*, 367–374.

Anwyl, R. and Finlayson, L. H. (1974). Peripherally and centrally generated action potentials in neurons with both a motor and a neurosecretory function in the insect *Rhodnius prolixus*. *J. Comp. Physiol. 91*, 135–145.

Barker, J. L. and Gainer, H. (1975). Studies on bursting pacemaker potential activity in molluscan neurons. II. Regulation by divalent cations. *Brain Res. 84*, 479–500.

Barrington, E. J. W. (1975). *General and Comparative Endocrinology*, p.61. Clarendon Press, Oxford.

Bassurmanova, O. K. and Panov, A. A. (1967). Structure of the neurosecretory system in Lepidoptera. Light and electron microscopy of type A′-neurosecretory cells in the brain of normal and starved larvae of the silkworm *Bombyx mori*. *Gen. Comp. Endocr. 9*, 245–262.

Beattie, T. M. (1971). Histology, histochemistry and ultrastructure of neurosecretory cells in the optic lobe of the cockroach, *Periplaneta americana*. *J. Insect. Physiol. 17*, 1843–1855.

Beenakkers, A. M. Th., Van der Horst, D. J. and Van Marrewijk, W. J. A. (1981). Metabolism during locust flight. *Comp. Biochem. Physiol. 69B*, 315–321.

Benedeczky, I., Mala, J. and Sehnal, F. (1980). Ultrastructural study on the innervation of prothoracic glands in *Galleria mellonella*. *Gen. Comp. Endocrinol. 41*, 400–407.

Berlind, A. (1977). Cellular dynamics in invertebrate neurosecretory systems. *Int. Rev. Cytol. 49*, 171–251.

Berlind, A. (1981). Mobilization of a peptide neurohormone for release during a physiological secretion cycle. *Gen. Comp. Endocrinol. 44*, 444–453.

Berlind, A. and Maddrell, S. H. P. (1979). Changes in hormone activity of single neurosecretory cell bodies during a physiological secretion cycle. *Brain Res. 161*, 459–467.

Bern, H. A. (1966). On the production of hormones and the role of neurosecretion in neuroendocrine mechanisms. *Symp. Soc. Exptl. Biol. 20*, 325–344.

Bern, H. A. and Knowles, F. G. W. (1966). Neurosecretion. In *Neuroendocrinology*, pp.139–186. Edited by L. Martini and W. F. Ganong. Academic Press, New York.

Berry, R. W. (1981). Proteolytic processing in the biogenesis of the neurosecretory egg-laying hormone in *Aplysia*. I. Precursors, intermediates and products. *Biochemistry 20*, 6200–6205.

Berry, R. W. and Arch, S. (1981). Activation of neurosecretory cells enhances their synthesis of secretory protein. *Brain Res. 215*, 115–123.

Berry, R. W., Trump, M. J. and Baylen, J. T. (1981). Proteolytic processing in the biogenesis of the neurosecretory egg-laying hormone in *Aplysia*. II. Analysis of tryptic fragments. *Biochemistry. 20*, 6206–6211.

Bhaskaran, G., Jones, G. and Jones, D. (1980). Neuroendocrine regulation of corpus allatum activity in *Manduca sexta*: sequential neurohormonal and nervous inhibition in the last instar larva. *Proc. Nat. Acad. Sci. USA 77*, 4407–4411.

Bishop, C. A., O'Shea, M. and Miller, R. J. (1981). Neuropeptide proctolin (H–Arg–Tyr–Leu–Pro–Thr–OH), immunological detection and neuronal localization in insect central nervous system. *Proc. Nat. Acad. Sci. USA 78*, 5899–5902.

Blankenship, J. E. (1980). Physiological properties of peptide-secreting neuroendocrine cells in the marine mollusc *Aplysia*. In *The Role of Peptides in Neuronal Function*, pp.160–187. Edited by J. L. Barker and T. G. Smith, Marcel Dekker, Inc., New York and Basel.

Bloch, B., Thomsen, E. and Thomsen, M. (1966). The neurosecretory system of the adult *Calliphora erythrocephala* III. Electron microscopy of the medial neurosecretory cells of the brain and some adjacent cells. *Z. Zellforsch. Mik. Anat. 70*, 185–208.

Block, R. J. and Van Dyke, H. B. (1952). Amino acids in posterior pituitary protein. *Arch. Biochem. Biophys. 36*, 1–4.

Bollenbacher, W. E., Agui, N., Granger, N. A. and Gilbert, L. I. (1979). *In vitro* activation of insect prothoracic glands by the prothoracicotropic hormone. *Proc. Nat. Acad. Sci. USA 76*, 5148–5152.

Bowers, B. and Johnson, B. (1966). An electron microscope study of the corpora cardiaca and secretory neurons in the aphid *Myzus persicae* (Sulz). *Gen. Comp. Endocr. 6*, 213–230.

Bradley, J. T. and Simpson, T. A. (1981). Brain neurosecretion during ovarian development and after ovariectomy in adult *Acheta domesticus* L. *Gen. Comp. Endocr. 44*, 117–127.

Brady, J. and Maddrell, S. H. P. (1967). Neurohaemal organs in the medial nervous system of insects. *Z. Zellforsch. Mik. Anat. 76*, 389–404.

Brady, S. T., Crothers, S. D., Nosal, C. and McClure, W. O. (1980). Fast axonal transport in the presence of high Ca^{2+}: Evidence that microtubules are not required. *Proc. Nat. Acad. Sci. USA. 77*, 5909–5913.

Branton, W. D., Arch, S., Smock, T. and Mayeri, E. (1978). Evidence for mediation of a neuronal interaction by a behaviourally active peptide. *Proc. Nat. Acad. Sci. USA 75*, 5732–5736.

Broadwell, R. D. and Oliver, C. (1980). Morphological basis for the synthesis and packaging of neuronal peptides. In *The Role of Peptides in Neuronal Function*, pp.21–48. Edited by J. L. Barker, and T. G. Smith, Marcel Dekker Inc., New York and Basel.

Brown, B. E. (1975). Proctolin: a peptide transmitter candidate in insects. *Life Sci. 17*, 1241–1252.

Brownstein, M. J., Russell, J. T. and Gainer, H. (1980). Synthesis, transport and release of posterior pituitary hormones. *Science 207*, 373–378.

Bruce, I. C. and Wilkens, J. L. (1976). Neuronal activity in neurosecretory cells of the fleshfly, *Sarcophaga bullata*. *J. Comp. Physiol. 112*, 109–122.

Buys, C. M. and Gibbs, D. (1981). The anatomy of neurons projecting to the corpus cardiacum from the larval brain of the tobacco hornworm, *Manduca sexta* (L.). *Cell. Tiss. Res. 215*, 505–513.

Camier, M., Girardie, J., Remy, C., Girardie, A. and Cohen, P. (1980). Identification of immunoreactive neurophysin-like proteins in the central nervous system of an insect, *Locusta migratoria*. *Biochem. Biophys. Res. Comm. 93*, 792–796.

Cannata, M. A. and Morris, J. F. (1973). Changes in the appearance of hypothalamus-neurohypophysial neurosecretory granules associated with their maturation. *J. Endocrinol. 57*, 531–538.

Carlisle, J. and Loughton, B. G. (1979). Adipokinetic hormone inhibits protein synthesis in *Locusta*. *Nature 282*, 420–421.

Carlsen, J., Herman, W. S., Christensen, M. and Josefsson, L. (1979). Characterisation of a second peptide with adipokinetic and red-pigment-concentrating activity from the locust corpora cardiaca. *Insect. Biochem. 9*, 497–501.

Cazal, M. and Girardie, A. (1968). Controle humoral de l'equilibre hydrique chez *Locusta migratoria migratorioides*. *J. Insect Physiol. 14*, 655–668.

Chalaye, D. (1967). Neurosecretion au niveau de la chaine nerveuse ventrale de *Locusta migratoria migratorioides*. (R. et F.) (Orthoptere acridien). *Bull. Soc. Zool. Fr. 92*, 87–108.

Chalaye, D. (1975). Etude ultrastructurale des cellules neurosecretrices du ganglion sous-oesophagien de *Locusta migratoria migratorioides* (R. et F.) (Orthopteres). *Acrida. 4*, 19–32.

Chalaye, D. (1974a). Ultrastructure de la mass ganglionnaire metathoracique de *Locusta migratoria migratorioides* (R. et. F.) (Orthoptere): les cellules neurosecretrices et leur prolongements dans le neuropile. *Acrida. 3*, 19–34.

Chalaye, D. (1974b). Ultrastructure de la mass ganglionnaire metathoracique de *Locusta migratoria migratorioides* (R. et F.) (Orthoptere). II. Les organes perisympathiques abdominaux et thoraciques. *Acrida. 3*, 35–46.

Cheeseman, P. and Goldsworthy, G. J. (1979). The release of adipokinetic hormone during flight and starvation in *Locusta*. *Gen. Comp. Endocr. 37*, 35–43.

CLARKE, K. U. and LANGLEY, P. A. (1963). Studies on the initiation of growth and moulting in *Locusta migratoria migratorioides* R. and F. IV. The relationship between the stomatogastric nervous system and neurosecretion. *J. Insect Physiol. 9*, 423–430.

COOK, B. J. and HOLMAN, G. M. (1979). The action of proctolin and L-glutamic acid on the visceral muscles of the hindgut of the cockroach *Leucophaea maderae. Comp. Biochem. Physiol. 64C*, 21–28.

COOK, D. J. and MILLIGAN, J. V. (1972). Electrophysiology and histology of the medial neurosecretory cells in adult male cockroaches *Periplaneta americana. J. Insect. Physiol. 18*, 1197–1214.

COOKE, I. M. (1977). Electrical activity of neurosecretory terminals and control of peptide hormone release. In *Peptides in Neurobiology*, pp.345–374. Edited by H. Gainer. Plenum Press, New York.

CYMBOROWSKI, B. and DUTKOWSKI, A. (1970). Circadian changes in protein synthesis in the neurosecretory cells of the central nervous system of *Acheta domesticus J. Insect. Physiol. 16*, 341–348.

DAVEY, K. G. and KUSTER, J. E. (1981). The source of an anti-gonadotropin in the female of *Rhodnius prolixus* Stål. *Can. J. Zool. 59*, 761–764.

DAVID, J.-C., and LAFON-CAZAL, M. (1979). Octopamine distribution in the *Locusta migratoria* nervous and non-nervous systems. *Comp. Biochem. Physiol. 64C*, 161–164.

DISMUKES, R. K. (1979). New concepts of molecular communication among neurons. *Behav. Brain Sci. 2*, 409–448.

DOGRA, G. S. and GILLOTT, C. (1971). Neurosecretory activity and protease synthesis in relation to feeding in *Melanoplus sanguinipes J. Exp. Zool. 177*, 41–50.

DOWNER, R. G. H. (1979). Induction of hypertrehalosemia by excitation in *Periplaneta americana. J. Insect. Physiol. 25*, 59–63.

DUDEK, F. E. and BLANKENSHIP, J. E. (1977). Neuroendocrine cells of *Aplysia brasiliana*. I. Bag cell action potentials and after discharge. *J. Neurophysiol. 40*, 1301–1324.

DUTTON, A. and DYBALL, R. E. J. (1979). Phasic firing enhances vasopressin release from the rat neurohypophysis. *J. Physiol. 290*, 433–440.

DUVE, H. and THORPE, A. (1979). Immunofluorescent localization of insulin-like material in the median neurosecretory cells of the blowfly *Calliphora vomitoria* (Diptera). *Cell Tiss. Res. 200*, 187–191.

DUVE, H. (1978). The presence of a hypoglucemic and hypotrehalocemic hormone in the neurosecretory system of the blowfly, *Calliphora erythrocephala. Gen. Comp. Endocr. 36*, 102–110.

DU VIGNEAUD, V., LAWLER, H. C. and POPENOE, E. A. (1953a). Enzymatic cleavage of glycinamide from vasopressin and a proposed structure for this pressor-antidiuretic hormone of the posterior pituitary. *J. Amer. Chem. Soc. 75*, 4880–4881.

DU VIGNEAUD, V., RISSLER, C. and TRIPPET, S. (1953b). The sequence of amino acids in oxytocin with a proposal for the structure of oxytocin. *J. Biol. Chem. 205*, 949–957.

ECKERT, M. (1977). Immunologische Untersuchungen des neuroendokrinen Systems von Insekten. IV. Differenzierte immunohistochemische Darstellung von Neurosekreten des Gehirns und der Corpora Cardiaca von der Schabe *Periplaneta americana. Zool. Jb. Physiol. 81*, 25–41.

ECKERT, M. (1980). Enzymmarkierte Antikorper in der Immunhistochemie (Enzyme-labelled antibodies in immunocytochemistry). *Acta. Histochem. Suppl. XXII*, 55–65.

ECKERT, M. (1973). Immunologische Untersuchungen des neuroendokrinen Systems von Insekten. III. Immunohistochemische Markierung des neuroendokrinen Systems von *Periplaneta americana* durch fraktionierung von Retrocerebralkomplexextrakten gewonnenen Anti-Seren. *Zool. Jb. Physiol. 76*, 29–35.

ECKERT, M. and GERSCH, M. (1978). Immunochemical investigations on neuroendocrine system of the cockroach, *Periplaneta americana* L. In *Neurosecretion and Neuroendocrine Activity, Evolution, Structure and Function*, pp.365–369. Edited by A. Bargman, A. Oksche, A. Polenov, and B. Scharrer. Springer Verlag, Berlin.

ECKERT, M., GOETZE, H. and GERSCH, M. (1970). Immunologische Untersuchungen des neuroendokrinen Systems von Insekten. I. Kennzeichnung der Antigene der Corpora cardiaca und der corpora allata von *Periplaneta americana* durch Immunodiffusion and Immunoelektrophorese. *Zool. Jb. Physiol. 75*, 134–140.

EVANS, P. D. (1980). Biogenic amines in the insect nervous system. *Adv. Insect Physiol. 15*, 317–473.

FIFIELD, S. and FINLAYSON, L. H. (1978). Peripheral neurons and peripheral neurosecretion in the stick insect, *Carausius morosus. Proc. R. Soc. Lond. B. 200*, 63–85.

FINLAYSON, L. H. (1980). Distribution and function of neurosecretory cells. In *Insect Neurobiology and Pesticide Action (Neurotox 79)*, pp.297–304. Society of Chemical Industry, London.

FINLAYSON, L. H. and ORCHARD, I. (1978). Neurosecretory and electrical activity of extraganglionic neurons in the stick insect. In *Comparative Endocrinology*, pp.323–326. Edited by P. J. Gaillard, and H. H. Boer. Elsevier/North Holland Biomedical Press, Amsterdam.

FINLAYSON, L. H. and OSBORNE, M. P. (1968). Peripheral neurosecretory cells in the stick insect (*Carausius morosus*) and the blowfly larva (*Phormia terrae-novae*). *J. Insect. Physiol. 14*, 1793–1801.

FINLAYSON, L. H. and OSBORNE, M. P. (1970). Electrical activity of neurohaemal tissue in the stick insect, *Carausius morosus. J. Insect Physiol. 16*, 791–800.

FINLAYSON, L. H. and OSBORNE, M. P. (1975). Secretory activity of neurons and related electrical activity. *Adv. Comp. Physiol. Biochem. 6*, 165–258.

FRAENKEL, G., ZDAREK, J. and SIVASUBRAMANIAN, P. (1972). Hormonal factors in the central nervous system and haemolymph of pupariating fly larvae which accelerate puparium formation and tanning. *Biol. Bull., Woods Hole, 143*, 127–139.

FRIEDEL, T. and GILLOTT, C. (1976). Male accessory gland substance of *Melanoplus sanguinipes*: an oviposition stimulant under the control of the corpus allatum. *J. Insect Physiol. 22*, 489–495.

FRIEDEL, T. and LOUGHTON, B. G. (1980a). Neurosecretory protein synthesis in relation to starvation and feeding in adult male *Locusta migratoria. J. Insect Physiol. 26*, 33–37.

FRIEDEL, T. and LOUGHTON, B. G. (1982). Insect neurophysins. In *Insect Endocrinology*. Edited by R. G. H. Downer and H. Laufer. Alan R. Liss Inc, New York.

FRIEDEL, T., LOUGHTON, B. G. and ANDREW, R. D. (1980). A neurosecretory protein from *Locusta migratoria. Gen. Comp. Endocr. 41*, 487–498.

FRIEDEL, T., ORCHARD, I. and LOUGHTON, B. G. (1981). Release of neurosecretory protein from insect neurohaemal tissue following electrical stimulation. *Brain Res. 208*, 451–455.

GABE, M. (1966). Neurosecretion. *Int. Ser. Monog. Biol. 28*, Pergamon Press, Oxford.

GADE, G. (1981). Activation of fat body glycogen phosphorylase in *Locusta migratoria* by corpus cardiacum extract and synthetic adipokinetic hormone. *J. Insect Physiol. 27*, 155–161.

GAINER, H. (1972). Effects of experimentally induced diapause on the electrophysiology and protein synthesis of identified molluscan neurons. *Brain Res. 39*, 387–402.

GAINER, H. (1978). Input-output relations of neurosecretory cells. In: *Comparative Endocrinology*, pp.293–304. Edited by P. J. Gaillard and H. H. Boer. Elsevier/North Holland Biomedical Press, Amsterdam.

GELDIAY, S. (1970). Photoperiodic control of neurosecretory cells in the brain of the Egyptian grasshopper, *Anacridium aegyptium* L. *Gen. Comp. Endocr. 14*, 35–42.

GELPERIN, A. (1971). Abdominal sensory neurons providing negative feedback to the feeding behaviour of the blowfly. *Z. Vergl. Physiol. 72*, 17–31.

GERSCH, M. (1972). Experimentelle untersuchungen zum freisetzungsmechanismus von neurohormonen nach elektrischer reizung der corpora cardiaca von *Periplaneta americana in vitro. J. Insect. Physiol. 18*, 2425–2439.

GERSCH, M. and STURZBECHER, J. (1970). Experimentelle Stimulierung der zellulären Aktivität der Prothorakaldrüsen von *Periplaneta americana* durch den Aktivationsfaktor. *J. Insect Physiol. 16*, 1813–1826.

GERSCH, M., ECKERT, M., BAUMANN, E. and BIRKENBEIL, H. (1977). Immunological investigations of the neuroendocrine system of insects. V. Characterization of activation factor II by immunological biochemical and physiological methods. *Zool. Jb. Physiol. Bd. 81*, 153–164.

GERSCH, M., RICHTER, K., BÖHM, G. A. and STÜRZBECHER, J. (1970). Selektive Ausschuttung von Neurohormonen nach elektrischer Reizung der Corpora Cardiaca von *Periplaneta americana in vitro. J. Insect Physiol. 16*, 1991–2013.

Gilbert, L. I., Bollenbacher, W. E., Agui, N., Granger, N. A., Sedlak, B. J., Gibbs, D. and Buys, C. M. (1981). The prothoracicotropes: source of the prothoracicotropic hormone. *Am. Zool. 21*, 641–653.

Girardie, J. and Girardie, A. (1972). Evolution de la radioactivité des cellules neurosécrétrices de la pars intercerebralis chez *Locusta migratoria migratorioides* (Insecte Orthoptère) après injection de cysteine S^{35}. Etude autoradiographique aux microscopes optique et électronique. *Z. Zellforsch. Mik. Anat. 128*, 212–226.

Girardie, J. and Girardie, A. (1977a). Neurosécrétion médiane après section des nerfs hypocérébrocardiaques chez le criquet migrateur. *C.R. Acad. Sci. Ser. D. 284*, 1433–1436.

Girardie, J. and Girardie, A. (1977b). Liberation provoquée *in vitro* du produit de neurosecrétion des cellules protocérébrales medianes chèz le criquet migrateur. *J. Physiol. Paris. 73*, 707–721.

Girardie, J. and Girardie, A. (1978). "Last in, First out" in locust neurosecretory lobes of the corpora cardiaca. *Gen. Comp. Endocr. 34*, 104–105 (Abstract, 112.).

Girardie, J., Girardie, A. and Moulins, A. (1975). Preuves radiochimiques et physiologiques d'une activation par électrostimulation des cellules neurosécrétrices de la pars intercerebralis chez *Locusta migratoria* (Insecte Orthoptère). *Gen. Comp. Endocr. 25*, 416–424.

Girardie, J., Girardie, A. and Moulins, M. (1976). Etude radiochimique après électrostimulation de la dynamique fonctionelle des cellules neurosécrétrices protecérébrales médianes de *Locusta migratoria* (Insecte Orthoptère). *Gen. Comp. Endocr. 30*, 410–418.

Goldsworthy, G. J. (1969). Hyperglycaemic factors from the corpus cardiacum of *Locusta migratoria. J. Insect. Physiol. 15*, 2131–2140.

Goldsworthy, G. J., Cheeseman, P. (1978). Comparative aspects of endocrine control of energy metabolism. In: *Comparative Endocrinology*, pp.423–436. Edited by P. J. Gaillard and H. H. Boer. Elsevier/North Holland Biomedical Press, Amsterdam.

Goldsworthy, G. J. and Mordue, W. (1974). Neurosecretory hormones in insects. *J. Endocr. 60*, 529–558.

Gole, J. W. D. and Downer, R. G. H. (1979). Elevation of adenosine 3′,5′-monophosphate by octopamine in fat body of the american cockroach, *Periplaneta americana* L. *Comp. Biochem. Physiol. 64C*, 223–226.

Goosey, M. W. and Candy, D. J. (1980). The D-octopamine content of the haemolymph of the locust, *Schistocerca americana gregaria* and its elevation during flight. *Insect Biochem. 10*, 393–397.

Gosbee, J. L., Milligan, J. V. and Smallman, B. (1968). Neural properties of the protocerebral neurosecretory cells of the adult cockroach, *Periplaneta americana. J. Insect Physiol. 14*, 1785–1792.

Grillot, J.-P. (1972). Les organes périsympathetique lieux de stockage et de diffusion de produits de neurosécrétion chez les insectes. *Ann. Univ. Brazzaville. 8(C)*, 107–112.

Grossman, M. and Davey, K. G. (1978). The effect of feeding on the cerebral neurosecretory system of the adult male tsetse, *Glossina austeni* Neurt. *Can. J. Zool. 56*, 1988–1992.

Gundel, M. and Penzlin, H. (1980). Identification of neuronal pathways between the stomatogastric nervous system and the retrocerebral complex of the cockroach *Periplaneta americana* (L.). *Cell. Tiss. Res. 208*, 283–297.

Hagedorn, H. H., Shapiro, J. P. and Hanaoka, K. (1979). Ovarian ecdysone secretion is controlled by a brain hormone in an adult mosquito. *Nature. 282*, 92–94.

Heitler, W. J. and Goodman, C. S. (1978). Multiple sites of spike initiation in a bifurcating locust neurone. *J. Exp. Biol. 76*, 63–84.

Herbert, E. (1981). Discovery of pro-opiomelanocortin — a cellular polyprotein. *Trends Biochem. Sci. 6*, 184–188.

Highnam, K. C. (1961a). The histology of the neurosecretory system of the adult female desert locust, *Schistocerca gregaria. Q. J. Mic. Sci. 102*, 27–38.

Highnam, K. C. (1961b). Induced changes in the amount of material in the neurosecretory system of the desert locust. *Nature. 191*, 199–200.

Highnam, K. C. (1962). Neurosecretory control of ovarian development in the desert locust. *Mem. Soc. Endocr. 12*, 379–390.

Highnam, K. C. (1965). Some aspects of neurosecretion in arthropods. *Zool. Zb. (Physiol). 71*, 558–582.

Highnam, K. C. (1971). Estimates of neurosecretory activity during maturation in female locusts. In *Insect Endocrines. Proc. Int. Symp. Insect Endocr., Bono, 1966, Supplementa Acta Entomologica Bohemoslovaca*, pp.81–90. Edited by V. J. A. Novak and K. Slama. Academia, Prague.

Highnam, K. C. and Mordue, A. J. (Luntz), (1974). Induced changes in neurosecretory activity of adult female *Schistocerca gregaria* in relation to feeding. *Gen. Comp. Endocr. 22*, 519–525.

Highnam, K. C. and Mordue, A. J. (Luntz), (1970). Estimates of neurosecretory activity by an autoradiographic method in adult female *Schistocerca gregaria* (Forsk). *Gen. Comp. Endocr. 15*, 31–38.

Highnam, K. C. and West, M. W. (1971). The neuropilar neurosecretory reservoir of *Locusta migratoria migratorioides. Gen. Comp. Endocrinol. 16*, 574–585.

Highnam, K. C., Hill, L. and Mordue, W. (1966). The endocrine system and oöcyte growth in *Schistocerca* in relation to starvation and frontal ganglionectomy. *J. Insect Physiol. 12*, 977–994.

Hinks, C. F. (1975). Peripheral neurosecretory cells in some Lepidoptera. *Can. J. Zool. 53*, 1035–1038.

Hinks, C. F. (1971). Histochemical analysis of the neurosecretory cells in the adult brain of *Triphaena pronuba* (Lepidoptera: Noctuidae). *Can. Entomol. 103*, 1639–1648.

Hodgson, E. S. and Geldiay, S. (1959). Experimentally induced release of neurosecretory materials from roach corpora cardiaca. *Biol. Bull. Mar. Biol. Lab., Woods Hole. 117*, 275–283.

Hokfelt, T., Johansson, O., Ljungdahl, A., Lundberg, J. M. and Schultzberg, M. (1980). Peptidergic neurons. *Nature. 284*, 515–521.

Houben, N. M. D. and Beenakkers, A. M. T. (1973). Regulation of diglyceride release from the fatbody of the locust during flight. *J. Endocr. 57*, 1 iv–1 v.

Hoyle, G. (1975). Dorsal unpaired median insect neurons make neurosecretory endings on skeletal muscle. *J. Exp. Zool. 187*, 159–165.

Hoyle, G. and Dagan, D. (1978). Physiological characteristics and reflex activation of DUM (octopaminergic) neurons of locust metathoracic ganglion. *J. Neurobiol. 9*, 59–79.

Ishizaki, H. and Ichikawa, M. (1967). Purification of the brain hormone of the silkworm, *Bombyx mori. Biol. Bull. Mar. Biol. Lab., Woods Hole. 133*, 355–368.

Ishizaki, H. and Suzuki, A. (1980). Prothoracotropic hormone. In *Neurohormonal Techniques in Insects*, pp.244–276. Edited by T. A. Miller. Springer-Verlag, New York, Heidelberg, Berlin.

Johnson, B. (1963). A histological study of neurosecretion in aphids. *J. Insect Physiol. 9*, 727–739.

Johnson, B. G. (1966). Fine structure of lateral cardiac nerves of the cockroach, *Periplaneta americana* (L.). *J. Insect Physiol. 12*, 645–653.

Jones, J., Stone, J. V. and Mordue, W. (1977). The hyperglycaemic activity of locust adipokinetic hormone. *Physiol. Entomol. 2*, 185–187.

Kater, S. B. (1968). Cardioaccelerator release in *Periplaneta americana* (L). *Science. 160*, 765–767.

Kingan, T. G. (1981). Purification of the prothoracicotropic hormone from the tobacco hornworm, *Manduca sexta. Life Sci. 28*, 2585–2594.

Kleinholtz, L. H. (1976). Crustacean neurosecretory hormones and physiological specificity. *Amer. Zool. 16*, 151–166.

Klemm, N. (1980). Histochemical demonstration of biogenic monoamines (Falck-Hillarp Method) in the insect nervous system. In *Neuroanatomical Techniques*, pp.51–76. Edited by N. J. Strausfeld and T. A. Miller. Springer-Verlag, New York, Heidelberg, Berlin.

Koontz, M. and Edwards, J. S. (1980). The projection of neuroendocrine fibers (NCC I and II) in the brains of three Orthopteroid insects. *J. Morph. 165*, 285–299.

Kostyuk, P. G. (1980). Calcium ion channels in electrically excitable membrane. *Neuroscience. 5*, 945–959.

Krauthamer, V. (1980). Electrophysiology and morphology of two neuron types in the pars intercerebralis of the cockroach, *Periplaneta americana* (L). Thesis, State University of New York at Buffalo, Buffalo, USA.

Kriger, F. L. (1981). Neuroendocrine regulation of ovulation of an insect, *Rhodnius prolixus* Stål. PhD Thesis, York University, Toronto, Canada.

KROGH, I. M. and NORMANN, T. C. (1977). The corpus cardiacum neurosecretory cells of *Schistocerca gregaria*. Electron microscopy of resting and secreting cells. *Acta Zool. (Stockh). 58*, 69–79.
KUBOTA, L., ISOBE, M., GOTO, T. and HASEGAWA, K. (1976). Molecular size of the diapause hormone of the silkworm, *Bombyx mori. Z. Naturf. 31C*, 132–134.
LESCURE, H. AND NORDMANN, J. J. (1980). Neurosecretory granule release and endocytosis during prolonged stimulation of the rat neurohypophysis *in vitro. Neuroscience. 5*, 651–659.
LIVINGSTONE, M. S., HARRIS-WARRICK, R. M. and KRAVITZ, E. A. (1980). Serotonin and octopamine produce opposite postures in lobster. *Science 208*, 76–79.
LOUGHTON, B. G. and FRIEDEL, T. (1980). The synthesis of neurosecretory protein during the fifth instar of *Locusta migratoria migratorioides. Gen. Comp. Endocr. 40*, 261–267.
LOUGHTON, B. G. and ORCHARD, I. (1981). The nature of the hyperglycaemic factor from the glandular lobe of the corpus cardiacum of *Locusta migratoria. J. Insect Physiol. 27*, 383–385.
MADDRELL, S. H. P. (1963). Excretion in the blood-sucking bug, *Rhodnius prolixus* Stål. I. The control of diuresis. *J. Exp. Biol. 40*, 247–256.
MADDRELL, S. H. P. (1965). Neurosecretory supply to the epidermis of an insect. *Science, 150*, 1033–1034.
MADDRELL, S. H. P. (1966). The site of release of the diuretic hormone in *Rhodnius*—a new neurohaemal system in insects. *J. Exp. Biol. 45*, 499–508.
MADDRELL, S. H. P. (1967). Neurosecretion in insects. In *Insects and Physiology*, pp.103–118. Edited by J. W. L. Beament and J. E. Treherne, Oliver & Boyd, Edinburgh.
MADDRELL, S. H. P. (1970). Neurosecretory control systems in insects. In *Insect Ultrastructure* pp.101–116. Edited by A. C. Neville, Blackell, Oxford and Edinburgh.
MADDRELL, S. H. P. (1974). Neurosecretion. In *Insect Neurobiology*, pp.307–357. Edited by J. E. Treherne. North Holland Publishing, Amsterdam.
MADDRELL, S. H. P. and GEE, J. D. (1974). Potassium-induced release of the diuretic hormones of *Rhodnius prolixus* and *Glossina austeni*: Ca dependence, time course and localization of neurohaemal areas. *J. Exp. Biol. 61*, 155–171.
MADDRELL, S. H. P. and NORDMANN, J. J. (1979). *Neurosecretion*. Blackie, Glasgow and London.
MANDEL'SHTAM, Y. E. and TYSHCHENKO, V. P. (1968). Action potentials and structure of the nervi corporis cardiaci of pine moth (*Dendrolinus pini* L.) and china oak silkworm moth (*Antheraea perny* Guer.) larvae. *Akad. Sci. Dokl. Biol. Sci. 179*, 209–212.
MASON, C. A. (1973). New features of the brain-retrocerebral neuroendocrine complex of the locust *Schistocerca vaga* (Scudder). *Z. Zellforsch. Mik. Anat. 141*, 19–32.
MAYER, R. J. and CANDY, D. J. (1969). Control of haemolymph lipid concentration during locust flight: an adipokinetic hormone from the corpora cardiaca. *J. Insect Physiol. 15*, 611–620.
MAYERI, E., BROWNELL, P., BRANTON, W. D. and SIMON, S. B. (1979a). Multiple, prolonged actions of neuroendocrine bag cells on neurons in *Aplysia*. I. Effects on bursting pacemaker neurons. *J. Neurophysiol. 42*, 1165–1184.
MAYERI, E., BROWNELL, P. and BRANTON, W. D. (1979b). Multiple, prolonged actions of neuroendocrine bag cells on neurons in *Aplysia*. II. Effects on beating pacemaker and silent neurons. *J. Neurophysiol. 42*, 1185–1197.
MCCAFFERY, A. R. and HIGHNAM, K. C. (1975a). Effects of corpora allata on the activity of the cerebral neurosecretory system of *Locusta migratoria migratorioides*, R. & F. *Gen. Comp. Endocr. 25*, 358–372.
MCCAFFERY, A. R. and HIGHNAM, K. C. (1975b). Effects of corpus allatum hormone and its mimics on the activity of the cerebral neurosecretory system of *Locusta migratoria migratorioides* R. & F. *Gen. Comp. Endocr. 25*, 373–386.
MILLER, T. (1979). Nervous versus neurohormonal control of insect heartbeat. *Amer. Zool. 19*, 77–86.
MILLER, T. and THOMSON, W. W. (1968). Ultrastructure of cockroach cardiac innervation. *J. Insect Physiol. 14*, 1099–1104.
MILLER, T. and USHERWOOD, P. N. R. (1971). Studies of cardio-regulation in the cockroach, *Periplaneta americana. J. Exp. Biol. 54*, 329–348.
MILLS, R. R., MATHUR, R. B. and GUERRA, A. A. (1965). Studies on the hormonal control of tanning in the American cockroach. I. Release of an activation factor from the terminal abdominal ganglion. *J. Insect Physiol. 11*, 1047–1053.
MIYAZAKI, S. (1980). The ionic mechanism of action potentials in neurosecretory cells and non-neurosecretory cells of the silkworm. *J. Comp. Physiol. 140*, 43–52.
MORDUE, W. (1970). Evidence for the existence of diuretic and antidiuretic hormones in locusts. *J. Endocrinol. 46*, 119–120.
MORDUE, W. and GOLDSWORTHY, G. J. (1969). The physiological effects of corpus cardiacum extracts in locusts. *Gen. Comp. Endocrinol. 12*, 360–369.
MORDUE, W. and STONE, J. V. (1977). Relative potencies of locust adipokinetic hormone and prawn red pigment-concentrating hormone in insect and crustacean systems. *Gen. Comp. Physiol. 33*, 103–108.
MORDUE, A. J. (LUNTZ) and HIGHNAM, K. C. (1973). Incorporation of cysteine into the cerebral neurosecretory system of adult desert locusts. *Gen. Comp. Endocr. 20*, 351–357.
MORRIS, G. P. and STEEL, C. G. H. (1975). Ultrastructure of neurosecretory cells in the pars intercerebralis of *Rhodnius prolixus* (Hemiptera). *Tissue Cell, 7*, 73–90.
MORRIS, G. P. and STEEL, C. G. H. (1977a). Ultrastructural changes in the medial neurosecretory cells of *Rhodnius prolixus* following activation by feeding. *Amer. Zool. 17*, 929.
MORRIS, G. P. and STEEL, C. G. H. (1977b). Sequence of ultrastructural changes induced by activation in the posterior neurosecretory cells in the brain of *Rhodnius prolixus* with special reference to the role of lysosomes. *Tissue Cell. 9*, 547–561.
MORRIS, J. F. and DYBALL, R. E. J. (1974). A quantitative study of the ultrastructural changes in the hypothalamo-neurohypophysial system during and after experimentally induced hypersecretion. *Cell Tiss. Res. 149*, 525–535.
NAGASAWA, H., FU, G., XIANGCHEN, X., BANGYING, X., ZONGSHUN, W., XUJIA, Q., DINGYI, W., EYING, C., JINGZE, W., SUZUKI, A., ISOGAI, A., HOI, Y., TAMWAN, S. and ISHIZAKI, H. (1980). Large scale purification of prothoracicotropic hormone of the silkworm *Bombyx mori. Scientia Sinica 23*, 1053–1060.
NAKANISHI, S., INOUE, A., KITA, J., NAKAMURA, M., CHANG, A. C. Y., COHEN, S. N. and NUMAN, S. (1979). Nucleotide sequence of cloned cDNA for bovine caticotropin-′potropin precursor. *Nature (Lond). 278*, 423–427.
NAKAZATO, Y. and DOUGLAS, W. W. (1974). Vasopressin release from the isolated neurohypophysis induced by a calcium ionophore, X-537A. *Nature 249*, 479–481.
NICOLAS, P., CAMIER, M., DESSEN, P. and COHEN, P. (1976). Interactions of oxytocin and vasopressin with bovine neurophysins I and II. Effects of hormone binding on the protein quaternary structure: a simple model. *J. Biol. Chem. 251*, 3965–3971.
NICHOLSON, S. W. (1980). Diuresis and its hormonal control in butterflies. *J. Insect Physiol. 26*, 841–846.
NIJHOUT, H. F. (1975). Axonal pathways in the brain-rectrocerebral neuroendocrine complex of *Manduca sexta* (L) (Lepidoptera: Spingidae). *Int. J. Insect Morph. Embryol. 4*, 529–538.
NORDMANN, J. J. (1979). Hormone release and membrane retrieval in neurosecretion. In *Biologie cellulaire des processus neurosecretoires hypothalamiques*, No. 280, *Colloques Internationaus du CNRS*, Chapter 37, pp.619–636. Edited by J. D. Vincent and C. Kordon.
NORDMANN, J. J. and LABOUESSE, J. (1981). Neurosecretory granules: Evidence for aging process within the neurohypophysis. *Science 211*, 595–597.
NORDMANN, J. J. and MORRIS, J. F. (1976). Membrane retrieval in neurosecretory axon endings. *Nature (Lond). 261*, 723–725.
NORDMANN, J. and MORRIS, J. F. (1980). Depletion of neurosecretory granules and membrane retrieval in the sinus gland of the crab. *Cell. Tiss. Res. 205*, 31–42.
NORMANN, T. C. (1965). Neurosecretory system of an adult *Calliphora erythrocephala*. I. The fine structure of the corpus cardiacum, with some observations on adjacent organs. *Z. Zellforsch. Mikr. Anat. 67*, 461–501.
NORMANN, T. C. (1969). Experimentally induced exocytosis of neurosecretory granules. *Exp. Cell Res. 55*, 285–287
NORMANN, T. C. (1970). The mechanism of hormone release from neurosecretory axon endings in the insect *Calliphora erythrocephala*. In *Aspects of Neuroendocrinology*, pp.30–42. Edited by W. Bargmann and B. Scharrer. Springer-Verlag, Berlin, Heidelberg, New York.

NORMANN, T. C. (1973). Membrane potential of the corpus cardiacum neurosecretory cells of the blowfly, *Calliphora erythrocephala. J. Insect Physiol. 19,* 303–318.

NORMANN, T. C. (1974). Calcium-dependence of neurosecretion by exocytosis. *J. Exp. Biol. 61,* 401–409.

NORMANN, T. C. and DUVE, H. (1969). Experimentally induced release of a neurohormone influencing haemolymph trehalose level in *Calliphora erythrocephala* (Diptera). *Gen. Comp. Endocr. 12,* 449–459.

NORMANN, T. C. and HALL, T. H. (1978). Calcium and sulphur in neurosecretory granules and calcium in mitochondria as determined by electron microscope X-ray microanalysis. *Cell. Tiss. Res. 186,* 453–463.

NORTH, W. G. and VALTIN, H. (1977). The purification of rat neurophysins by a method of preparative polyacrylamide gel electrophoresis. *Anal. Biochem. 78,* 436–450.

O'FARRELL, P. Z., GOODMAN, H. M. and O'FARRELL, P. H. (1977). High resolution two dimensional electrophoresis of basic as well as acidic proteins. *Cell 12,* 1133–1142.

ORCHARD, I. (1975). The physiology of peripheral neurons in the stick insect, *Carausius morosus.* Thesis, University of Birmingham, England.

ORCHARD, I. (1976). Calcium dependent action potentials in a peripheral neurosecretory cell of the stick insect. *J. Comp. Physiol. 112,* 95–102.

ORCHARD, I. (1982a). Octopamine in insects. Neurotransmitter, neurohormone and neuromodulator. *Can. J. Zool. 60,* 659–669.

ORCHARD, I. (1982b). Neurosecretion: morphology and physiology. In *Insect Endocrinology,* (in press). Edited by R. G. H. Downer and H. Laufer, Alan R. Liss, Inc., New York. (In press).

ORCHARD, I. and FINLAYSON, L. H. (1976). The electrical activity of mechanoreceptive and neurosecretory neurons in the stick insect *Carausius morosus. J. Comp. Physiol. 107,* 327–338.

ORCHARD, I. and FINLAYSON, L. H. (1977a). Electrical properties of identified neurosecretory cells in the stick insect. *Comp. Biochem. Physiol. 58A,* 87–91.

ORCHARD, I. and FINLAYSON, L. H. (1977b). Studies on peripheral neurons and neurohaemal tissue in the thorax of the stick insect (*Carausius morosus*). *Experientia. 33,* 1440–1442.

ORCHARD, I. AND FINLAYSON, L. H. (1977c). Electrically excitable neurosecretory cell bodies in the periphery of the stick insect, *Carausius morosus. Experientia. 33,* 226–228.

ORCHARD, I. and LOUGHTON, B. G. (1980). A hypolipaemic factor from the corpus cardiacum of *Locusta migratoria. Nature. 286,* 494–496.

ORCHARD, I. and LOUGHTON, B. G. (1981a). The neural control of release of hyperlipaemic hormone from the corpus cardiacum of *Locusta migratoria. Comp. Biochem. Physiol. 68A,* 25–30.

ORCHARD, I. and LOUGHTON, B. G. (1981b). Is octopamine a transmitter mediating hormone release in insects? *J. Neurobiol. 12,* 143–153.

ORCHARD, I. and OSBORNE, M. P. (1977). The effects of cations upon the action potentials recorded from neurohaemal tissue of the stick insect. *J. Comp. Physiol. 118,* 1–12.

ORCHARD, I. and STEEL, C. G. H. (1980). Electrical activity of neurosecretory axons from the brain of *Rhodnius prolixus*: relation of changes in the pattern of activity to endocrine events during the moulting cycle. *Brain Res. 191,* 53–65.

ORCHARD, I., CARLISLE, J. A., LOUGHTON, B. G., GOLE, J. W. D. and DOWNER, R. G. H. (1982). *In vitro* studies on the effects of octopamine on locust fat body. *Gen. Comp. Endocr. 48,* 7–13.

ORCHARD, I., FRIEDEL, T. and LOUGHTON, B. G. (1981a). Release of neurosecretory protein from the corpora cardiaca of *Locusta migratoria* induced by high potassium saline and compound action potentials. *J. Insect Physiol. 27,* 297–304.

ORCHARD, I., LOUGHTON, B. G. and WEBB, R. A. (1981b). Octopamine and short-term hyperlipaemia in the locust. *Gen. Comp. Endocr. 45,* 175–180.

ORCHARD, I., OSBORNE, M. P. and FINLAYSON, L. H. (1979). Ultrastructural changes in neurohaemal tissue of the stick insect *Carausius morosus,* induced by the ionophores Br-X-537A and A-23187. *Cell. Tiss. Res. 197,* 227–234.

O'SHEA, M. and ADAMS, M. E. (1981). Pentapeptide (proctolin) associated with an identified neuron. *Science 213,* 567–569.

PALADE, G. (1975). Intracellular aspects of the process of protein synthesis. *Science. 189,* 347–358.

PANOV, A. A. (1980). Demonstration of neurosecretory cells in the insect central nervous system. In *Neuroanatomical Techniques,* pp.26–51. Edited by N. J. Strausfeld and T. A. Miller. Springer-Verlag, New York, Heidelberg, Berlin.

PEARSE, A. G. E. (1969). The cytochemistry and ultrastructure of polypeptide hormone-producing cells of the APUD series and the embryologic, physiologic and pathologic implications of the concept. *J. Histochem. Cytochem. 17,* 303–313.

PENZLIN, H., AGRICOLA, H., ECKERT, M. and KUSCH, T. (1981). Distribution of proctolin in the sixth abdominal ganglion of *Periplaneta americana* L. and the effect of protolin on the ileum of mammals. In *Neurotransmitters in Invertebrates.* Edited by K. A. Rozsa. *Adv. Physiol. Sci. 22,* 525–535.

PHILLIPS, J. E., MORDUE, W., MEREDITH, J. and SPRING, J. (1980). Purification and characteristics of the chloride transport stimulating factor from locust corpora cardiaca: a new peptide. *Canad. J. Zool. 58,* 1851–1860.

PIEK, T. and MANTEL, P. (1977). Myogenic contractions in locust muscle induced by proctolin and by wasp, *Philanthus triangulum,* venom. *J. Insect Physiol. 23,* 321–325.

PILCHER, D. E. M. (1970). Hormonal control of the malpighian tubules of the stick insect, *Carausius morosus. J. Exp. Biol. 52,* 653–665.

PIPA, R. L. (1978). Locations and central projections of neurons associated with the retrocerebral neuroendocrine complex of the cockroach *Periplaneta americana* (L). *Cell Tiss. Res. 193,* 443–455.

PIPA, R. L. (1982). Morphological considerations in the integration of nervous and endocrine systems. In *Insect Endocrinology.* (In press). Edited by R. G. H. Downer and H. Laufer, Alan R. Liss, New York. (In press).

PRATT, G. E. and DAVEY, K. G. (1972). The corpus allatum and oogenesis in *Rhodnius prolixus,* (Stål). III. The effects of mating. *J. Exp. Biol. 56,* 223–237.

PRENTØ, P. (1972). Histochemistry of neurosecretion in the pars intercerebralis-corpus cardiacum system of the desert locust *Schistocerca gregaria. Gen. Comp. Endocr. 18,* 482–500.

PRESSMAN, B. C. (1976). Biological applications of ionophores. *Ann. Rev. Biochem. 45,* 501–530.

PROUX, J. and ROUGON-RAPUZZI, G. (1980). Evidence for vasopressin-like molecule in migratory locust. Radioimmunological measurements in different tissues: correlation with various states of hydration. *Gen. Comp. Endocr. 42,* 378–383.

RAABE, M. (1965). Recherches sur la neurosécrétion dans la chaîne nerveuse ventrale du Phasme, *Clitumnus extradentatus*: Les elements neurosécréteurs. *C.R. Acad. Sci. Paris. 260,* 6710–6713.

RAABE, M. (1971a). Neurosecretion in the ventral nerve cord of insects. In *Insect Endocrines,* pp.105–114. Edited by J. A. Novak and K. Slama. Academia, Prague.

RAABE, M. (1971b). Neurosécrétion dans la chaîne nerveuse ventrale des insectes et organes neurohémaux metamerique. *Arch. Zool. Exp. Gen. 112,* 679–694.

RAABE, M., BAUDRY, N., GRILLOT J.-P. and PROVANSAL, A. (1971). Les organes périsympathiques des Insects Ptérygotes. Distribution. Caractères généraux. *C.R. Acad. Sci. Paris. 273,* 2324–2327.

RAABE, M., BAUDRY, N., GRILLOT, J.-P. and PROVANSAL, A. (1974). The perisympathetic organs of insects. In *Neurosecretion — The Final Common Pathway,* pp.59–71. Edited by F. Knowle and L. Vollrath. Springer-Verlag, Berlin.

RAABE, M., PANOV, A. A., DAVYDOVA, E. D. and CHERVIN, D. (1979). Neurosecretory products diversity in the pars intercerebralis of insects. *Experientia 35,* 404–405.

RADEMAKERS, L. H. P. M. (1977a). Effects of isolation and transplantation of the corpus cardiacum on hormone release from its glandular cells after flight in *Locusta migratoria* A quantitative electron microscopical study. *Cell. Tiss. Res. 184,* 213–224.

RADEMAKERS, L. H. P. M. (1977b). Identification of a secretomotor centre in the brain of *Locusta migratoria,* controlling the secretory activity of the adipokinetic hormone producing cells of the corpus cardiacum. *Cell. Tiss. Res. 184,* 381–395.

RADEMAKERS, L. H. P. M. and BEENAKKERS, A. M. T. (1977). Changes in the secretory activity of the glandular lobe of the corpus cardiacum of *Locusta migratoria* induced by flight. A quantitative electron microscopical study. *Cell. Tiss. Res. 180,* 155–171.

REHM, M. (1955). Morphologische und histochemische Untersuchungen an neurosecretionsche Zellen von Schmelterlingen. *Z. Zellforsch. Mik. Anat. 42*, 19–58.

REICHLIN, S. (1980). Peptides in neuroendocrine regulation. In *Peptides: Integrators of Cell and Tissue Function*, pp.235–250. Edited by F. E. Bloom. Raven Press, New York.

REMY, C. and GIRARDIE, J. (1980). Anatomical organisation of two vasopressin-neurophysin-like neurosecretory cells throughout the control nervous system of the migratory locust. *Gen. Comp. Endocr. 40*, 27–35.

REMY, C., GIRARDIE, J. and DUBOIS, M. P. (1977). Exploration immunocytologique des ganglions cerebroides et sous-oesophagien du phasme *Clitumnus extradentatus*: éxistence d'une neurosécrétion apparantée à la vasopressine-neurophysine. *C.R. Acad. Sci. Paris. D. 285*, 1495–1497.

REMY, C., GIRARDIE, J. and DUBOIS, M. P. (1979). Vertebrate neuro peptide-like substances in the suboesophageal ganglion of two insects: *Locusta migratoria* R. and F. (Orthoptera) and *Bombyx mori* L. (Lepidoptera) immunocytological investigation. *Gen. Comp. Endocr. 37*, 93– 100.

REYNOLDS, S. E. and TRUMAN, J. W. (1980). Eclosion hormones. In *Neurohormonal Techniques in Insects*, pp.196–215. Edited by T. A. Miller. Springer-Verlag, New York, Heidelberg, Berlin.

RIDDIFORD, L. M. (1977). The role of the corpora cardiaca in sex pheromone release by female saturniid moths. In *Insect Chemoreception*, vol. 3, pp.9–16. Edited by A. Skirkevicius. Lithuanian Academy of Sciences.

RIDDIFORD, L. M. and WILLIAMS, C. M. (1971). Role of the corpora cardiaca in the behaviour of saturniid moths. I. Release of sex pheromone. *Biol. Bull. Mar. Biol. Lab., Woods Hole. 140*, 1–7.

ROWELL, H. F. (1976). The cells of the insect neurosecretory system: constancy, variability and the concept of the unique identifiable neuron. *Adv. Insect Physiol. 12*, 63–123.

RUEGG, R. P. (1981). Factors influencing reproduction in *Rhodnius prolixus* (Insecta: Hemiptera) PhD Thesis, York University, Canada.

RUEGG, R. P., KRIGER, F. L., DAVEY, K. G. and STEEL, C. G. H. (1981). Ovarian ecdysone elicits release of a myotropic ovulation hormone in *Rhodnius* (Insecta: Hemiptera). *Int. J. Invert. Repr. 3*, 357–361.

RUEGG, R. P., ORCHARD, I. and DAVEY, K. G. (1982). 20-Hydroxy-Ecdysone as a modulator of electrical activity in neurosecretory cells of *Rhodnius prolixus*. *J. Insect Physiol. 28*, 243–248.

RUSSELL, J. T. and HOLTZ, R. W. (1981). Measurement of ΔpH and membrane potential in isolated neurosecretory vesicles from bovine neurohypophyses. *J. Biol. Chem. 256*, 5950–5953.

RUSSELL, J. T., BROWNSTEIN, M. J. and GAINER, H. (1979). Trypsin liberates an arginine vasopressin-like peptide and neurophysin from a M_r 20,000 putative common precursor. *Proc. Nat. Acad. Sci. USA. 76*, 6086–6090.

SACHS, H. and TAKABATAKE, Y. (1964). Evidence for a precursor in vasopressin biosynthesis. *Endocrinology. 75*, 943–948.

SACHS, H., FAWCETT, P., TAKABATAKE, Y. and PORTANOVA, R. (1969). Biosynthesis and release of vasopressin and neurophysin. *Rec. Prog. Horm. Res. 25*, 447–491.

SCHARRER, B. (1967). The neurosecretory neuron in neuroendocrine regulatory mechanisms. *Am. Zool. 7*, 161–169.

SCHARRER, B. (1968). Neurosecretion. XIV. Ultrastructural study of sites of release of neurosecretory material in blattarian insects. *Z. Zellforsch. Mik. Anat. 89*, 1–16.

SCHARRER, B. (1969). Neurohumors and neurohormones: definitions and terminology. *J. Neuro-Visceral Relations, Suppl. IX*, 1–20.

SCHARRER, B. (1977). Peptides in neurobiology Historical introduction. In *Peptides in Neurobiology*, pp.1–7. Edited by H. Gainer. Plenum Press, New York.

SCHARRER, B. and KATER, S. B. (1969). Neurosecretion. XV. An electron microscopic study of the corpora cardiaca of *Periplaneta americana* after experimentally induced hormone release. *Z. Zellforsch. Mik. Anat. 95*, 177–186.

SCHARRER, B. and WURZELMANN, S. (1978). Neurosecretion. XVII. Experimentally induced release of neurosecretory material by exocytosis in the insect *Leucophaea maderae*. *Cell. Tiss. Res. 190*, 173–180.

SCHARRER, E. and BROWN, S. (1961). Neurosecretion. XII. The formation of neurosecretory granules in the earthworm *Lumbricus terrestris*. *Z. Zellforsch. Mik. Anat. 54*, 530–540.

SCHOONEVELD, H. (1974). Ultrastructure of the neurosecretory system of the Colorado potato beetle, *Leptinotarsa decemlineata* (Say). II. Pathways of axonal secretion transport and innervation of neurosecretory cells. *Cell. Tiss. Res. 154*, 289–301.

SCOTT, A. P., RATCLIFFE, J. G., REES, L. H., LANDON, J., BENNETT, H. P. J., LOWRY, P. J. and MCMARTIN, C. (1973). Pituitary peptide. *Nature New Biol. 224*, 65–67.

SEIF, S. M. and ROBINSON, A. G. (1978). Localization and release of neurophysins. *Ann. Rev. Physiol. 40*, 345–376.

SESHAN, K. R., PROVINE, R. R. and LEVI-MONTALCINI, R. (1974). Structural and electrophysiological properties of nymphal and adult insect medial neurosecretory cells: an *in vitro* analysis. *Brain Res. 78*, 359–376.

SIEW, Y. C. (1965). The endocrine control of adult reproductive diapause in the chrysomelid beetle *Galeruca tanaceti* (L). *J. Insect Physiol. 11*, 973–981.

SILVERBERG, A. B., SHAH, S. D., HAYMOND, M. W. and CRYER, P. E. (1978). Norepinephrine: hormone and neurotransmitter in man. *Am. J. Physiol. 234*, E252–E256.

SINGH, G. J. P., ORCHARD, I. and LOUGHTON, B. G. (1981). Octopamine-like action of formamidines on hormone release in the locust, *Locusta migratoria*. *Pest Biochem. Physiol. 16*, 249–255.

SLOPER, J. C. (1957). Presence of a substance rich in protein-bound cystine or cysteine in the neurosecretory system of an insect. *Nature. 179*, 148.

SMITH, A. D. (1971). Summing up: some implications of the neuron as a secreting cell. *Phil. Trans. Roy. Soc. Lond. B. 261*, 423–427.

SMITH, U. and SMITH, D. S. (1966). Observations on the secretory processes in the corpus cardiacum of the stick insect, *Carausius morosus*. *J. Cell. Sci. 1*, 59–66.

STAY, B. and TOBE, S. (1981). Control of the corpora allata during a reproductive cycle in a viviparous cockroach. *Am. Zool. 21*, 663–674.

STEEL, C. G. H. (1973). Humoral regulation of the cerebral neurosecretory system of *Rhodnius prolixus* (Stål) during growth and moulting. *J. Exp. Biol. 58*, 177–187.

STEEL, C. G. H. (1975). A neuroendocrine feedback mechanism in the insect moulting cycle. *Nature. 253*, 267–269.

STEEL, C. G. H. (1977). The neurosecretory system in the aphid *Megoura viciae* with reference to unusual features associated with long distance transport of neurosecretion. *Gen. Comp. Endocr. 31*, 307–322.

STEEL, C. G. H. (1978b). Some functions of identified neurosecretory cells in the brain of the aphid, *Megoura viciae*. *Gen. Comp. Endocr. 34*, 219–228.

STEEL, C. G. H. (1978a). Nervous and hormonal regulation of neurosecretory cells in the insect brain. In *Comparative Endocrinology*, pp.327–330. Edited by P. J. Gaillard and H. H. Boer. Elsevier/North Holland Biomedical Press, Amsterdam.

STEEL, C. G. H. and HARMSEN, R. (1971). Dynamics of the neurosecretory system in the brain of an insect, *Rhodnius prolixus*, during growth and molting. *Gen. Comp. Endocr. 17*, 125–141.

STEEL, C. G. H. and LEES, A. D. (1977). The role of neurosecretion in the photoperiodic control of polymorphism in the aphid *Megoura viciae*. *J. Exp. Biol. 67*, 117–135.

STEEL, C. G. H. and MORRIS, G. P. (1975). Evidence from X-ray microanalysis of sulphur-rich protein in insect neurosecretion: A new approach to the estimation of neurosecretory activity. *Gen. Comp. Endocr. 26*, 517–524.

STEELE, J. E. (1961). Occurence of a hyperglycaemic factor in the corpus cardiacum of an insect. *Nature. 192*, 680–681.

STEELE, J. E. (1976). Hormonal control of metabolism in insects. *Adv. Insect Physiol. 12*, 239–323.

STEELE, J. E. and TOLMAN, J. H. (1980). Regulation of water transport in the cockroach rectum by the corpora cardiaca-corpora allata system: the requirement for Na^+. *J. Comp. Physiol. 138*, 357–365.

STONE, J. V. and MORDUE, W. (1979). Isolation of granules containing adipokinetic hormone from locust corpora cardiaca by differential centrifugation. *Gen. Comp. Endocr. 39*, 543–547.

STONE, J. V. and MORDUE, W. (1980). Adipokinetic Hormone. In *Neurohormonal Techniques in Insects*, pp.31–76. Edited by T.A. Miller. Springer-Verlag, Berlin.

STONE, J. V., MORDUE, W., BATLEY, K. E. and MORRIS, H. R. (1976). Structure of locust adipokinetic hormone, a neurohormone that regulates lipid utilisation during flight. *Nature. 263*, 207–211.

STUART, D. K. and STRUMWASSER, F. (1980). Neuronal sites of action of a neurosecretory peptide, egg-laying hormone, in *Aplysia californica*. *J. Neurophysiol. 43*, 499–519.

TAKEDA, N. (1976). The direct release of neurosecretory material from the cell in the pars intercerebralis of *Monema flavescens* (Lepidoptera:Heterogeneidae). *Appl. Ent. Zool. 11*, 143–153.

TELLAM, R. and WINZOR, D. J. (1980). Effect of phosphate on the macromolecular state of bovine neurophysin. *Arch. Biochem. Biophys. 201*, 20–24.

THOMSEN, M. (1965). The neurosecretory system of the adult *Calliphora erythrocephola*. II. Histology of the neurosecretory cells of the brain and some related structures. *Z. Zellforsch. Mikr. Anat. 67*, 693–717.

THOMSEN, E. and LEA, A. O. (1968). Control of the medial neurosecretory cells by the corpus allatum in *Calliphora erythryocephala*. *Gen. Comp. Endocr. 12*, 51–57.

TOBE, S. S. and STAY, B. (1980). Control of juvenile hormone biosynthesis during the reproductive cycle of a viviparous cockroach. *Gen. Comp. Endocrinol. 40*, 89–98.

TOMBES, A. S. (1976). Myoneural junctions and neurosecretory endings on spermathecal muscle fibers of two weevils, *Sitophilus granarius* and *Hypera postica*. *J. Insect Physiol.* 22, 1573–1580.

TRUMAN, J. W. (1971). Physiology of insect ecdysis. I. The eclosion behavior of saturniid moths and its hormonal release. *J. Exp. Biol. 54*, 805–814.

TRUMAN, J. W. (1973). Physiology of insect ecdysis. II. The assay and occurence of the eclosion hormone in the Chinese oak silkmoth, *Antheraea pernyi*. *Biol. Bull. Mar. Biol. Lab., Woods Hole. 144*, 200–211.

TRUMAN, J. W. (1976). Development and hormonal release of adult behavior patterns in silkmoths. *J. Comp. Physiol. 107*, 39–48.

TRUMAN, J. W. (1978). Hormonal release of stereotyped motor programmes from the isolated nervous system of the ceropia silkmoth. *J. Exp. Biol. 74*, 151–173.

TRUMAN, J. W., TAGHERT, P. H., COPENHAVER, P. F., TUBLITZ, N. J. and SCHWARTZ, L. M. (1981). Eclosion hormone may control all ecdyses in insects. *Nature. 291*, 70–71.

VINCENT, J.F. V. (1972). The dynamics of release and the possible identity of bursicon in *Locusta migratoria migratorioides*. *J. Insect Physiol. 18*, 757–780.

WALTER, R., SCHESINGER, D. H., SCHWARTZ, I. L. and CAPRA, J. D. (1971). Complete amino acid sequence of bovine neurophysin II. *Biochem. Biophys. Res. Commun. 44*, 293–298.

WASSERTHAL, W. and WASSERTHAL, L. T. (1980). Multinucleate neurons with neurohaemal and synapsing axons at the heart and alary muscles of the butterfly *Caligo beltrao* Illiger (Lepidoptera). *Cell. Tiss. Res. 212*, 351–362.

WEYER, F. (1935). Uber drusenartige Nervenzellen in Gehirn der Honigbiene, *Apis mellifica* L. *Zool. Anz. 112*, 137–141.

WIGGLESWORTH, V. B. (1940). The determination of characters at metamorphosis in *Rhodnius prolixus* (Hemiptera). *J. Exp. Biol. 17*, 201–222.

WIGGLESWORTH, V. B. (1964). The hormonal regulation of growth and reproduction in insects. *Adv. Insect Physiol. 2*, 247–336.

WILKENS, J. L. and MOTE, M. I. (1970). Neuronal properties of the neurosecretory cells in the fly *Sarcophaga bullata*. *Experientia. 26*, 275–276.

WILLIAMS, C. M. (1948). Physiology of insect diapause. III. The prothoracic glands in the Cercropia silkworm with special reference to their significance in embryonic and postembryonic development. *Biol. Bull. Mar. Biol. Lab. Woods Hole. 94*, 60–65.

YAGI, K. and IWASAKI, S. (1977). Electrophysiology of the neurosecretory cell. *Int. Rev. Cytol. 48*, 141–186.

YAMAZAKI, M. and KOBAYASHI, M. (1969). Purification of the proteinic brain hormone of the silkworm *Bombyx mori*. *J. Insect Physiol. 15*, 1981–1990.

ZDAREK, J. and FRAENKEL, G. (1972). The mechanism of puparium formation in flies. *J. Exp. Zool. 179*, 315–324.

ZDAREK, J., ROHLF, R., BLECHL, J. and FRAENKEL, G. (1981). A hormone effecting immobilization in pupariating fly larvae. *J. Exp. Biol. 93*, 51–63.

ADDENDUM I

The release of adipokinetic hormones from the glandular lobe of the CC of *Locusta* is under the synaptic control of axons within NCC II. Stimulation of NCC II has now been shown to produce an elevation in cyclic AMP in the glandular lobe. Application of octopamine mimics the effect of nerve stimulation. It is believed that the increase in cyclic AMP is coupled to the process of synaptic transmission (Orchard, I. *et al.*, 1983a).

NCC II controls the release of both AKH I and AKH II in *Locusta* (Orchard, I. and Lange, A., 1983a) and both hormones are released during flight (Orchard, I. and Lange, A., 1983b). The titre of AKH I is elevated 10–15 mins after flight commences while that of AKH II is elevated between 15–30 mins. The titre of adipokinetic activity demonstrated by Cheeseman, P. and Goldsworthy, G. (1979) (see Fig. 11) probably results from the combined activities of AKH I and II. The biphasic elevation of lipid during flight (see Fig. 11) has recently been confirmed in *Locusta* (see Orchard, I. and Lange, A., 1983b).

Recent results have confirmed that the action of 20-hydroxyecdysone on the medial neurosecretory cells of *Rhodnius* is indirect (see Fig. 12) and involves aminergic interneurons (Orchard, I. *et al.*, 1983b).

ORCHARD, I., and LANGE, A. B. (1983a) Release of identified adipokinetic hormones during flight and following neural stimulation in *Locusta migratoria*. *J. Insect Physiol. 29*, 425–429.

ORCHARD, I. and LANGE, A. B. (1983b) The hormonal control of haemolymph lipid during flight in *Locusta migratoria*. *J. Insect Physiol. 29*, 639–642.

ORCHARD, I., RUEGG, R. P. and DAVEY, K. G. (1983a) The role of central aminergic neurons in the action of 20-hydroxy-ecdysone on neurosecretory cells of *Rhodnius prolixus*. *J. Insect Physiol. 29*, 387–391.

ORCHARD, I., LOUGHTON, B. G., GOLE, J. W. D. and DOWNER, R. G. H. (1983b) Synaptic transmission elevates adenosine 3′,5′-monophosphate (cyclic AMP) in locust neurosecretory cells. *Brain Res. 258*, 152–155.

ADDENDUM II

The complexity of information that can be transmitted by a neuroendocrine system has been discussed by Håkanson, R. and Sundler, F (1983). They point out that the presence of neuropeptides and conventional neurotransmitters (in varying proportions) in the same cell and even in the same granule together with the potential for differential degradation of any given polypeptide precursor molecule enables a cell to transmit different messages at different times. Target cell response may vary according to the initial population of receptors and to the length of exposure to the neurohormone. Down-regulation of receptors could result in target cells becoming refractory to one message but not to another transmitted by the same cell.

HÅKANSON, R. and SUNDLER, F. (1983) The design of the neuroendocrine system: a unifying concept and its consequences. Trends in Pharmacol. *Sci 4*, 41–44.

4 Endocrinology of the Prothoracicotropic Hormone

WALTER E. BOLLENBACHER and NOELLE A. GRANGER

University of North Carolina, Chapel Hill, North Carolina, USA

1 HISTORICAL PERSPECTIVE

Over six decades ago a Polish biologist interested in the development of insects performed a series of simple experiments that were designed to understand how the process of insect metamorphosis occurred (Kopeć, S., 1918, 1922). At the time, and for many years after, the results of these experiments were generally ignored. With the accumulated hindsight of the intervening years, however, it has been acknowledged that these creative experiments and their insightful interpretations constituted the first evidence of a hormonal function for the brain of any animal. The neurohormone in evidence in these experiments was the insect brain hormone, the prothoracicotropic hormone or PTTH.

In his initial study of this process in the gypsy moth *Lymantria dispar*, Kopeć did not believe that the nervous system was important for insect metamorphosis. This conclusion was based upon his repeat of experiments initially performed by Conte, A. and Vaney, C. (1911), in which fully grown larvae of *Lymantria* were ligated between the head and body, and his results confirmed that separation of the head had no effect on pupation and adult development (Kopeć, S., 1912). However, when he utilized younger last-instar larvae and ligated them midway along the body, the anterior half pupated as observed before, but the posterior half remained larval. These results suggested that the head was essential for metamorphosis, and Kopeć, S. (1917) concluded that his former interpretation was incorrect and that something from the anterior end of the larva, possibly the brain, was essential for metamorphosis to occur. Proof of this idea came from a subsequent experiment where he assessed the effect of brain removal at different days after the last larval molt on the incidence of pupation of female *Lymantria* larvae. Of larvae ligated on the seventh day, 80% of the animals failed to pupate; while of those ligated on the 10th day, the time at which Conte and Vaney had apparently performed their ligations, almost all pupated. On the basis of these results, Kopeć concluded that metamorphosis of the insect depended on the presence of the brain at a certain time during development (a critical period), and that since removal of the subesophageal ganglion had no effect on metamorphosis, the influence of the brain was not nervous. From these experiments came his remarkable conclusion that "... the brain ought to be considered as a gland of internal secretion" (Kopeć, S., 1922). At this time there was no precedent that the brain of an animal could function as an endocrine gland, and thus the acceptance of this insightful conclusion was prejudiced. It was not until Ernst Scharrer (1928) demonstrated an endocrine function for certain hypothalamic neurons in the vertebrate brain that Kopeć's idea was confirmed and subsequently acknowledged as the first demonstration of an endocrine function for the brain of an animal.

During the time when the process of neurosecretion was being demonstrated as a general endocrine occurrence, two other series of experiments conducted on insects were designed to demonstrate that metamorphosis was hormonally controlled (von Buddenbrock, W., 1930; Bodenstein, D., 1933a; see Wigglesworth, V. this volume). In the first, von Buddenbrock, W. (1930) injected hemolymph of prepupal larvae into younger larvae of the same species and found that the injected larvae appeared to molt earlier than non-injected controls, an observation suggesting the existence of a hormonal factor that accelerated molting. However, it must be noted that a few years later, Fraenkel, G. (1935) found that the difference between the injected larvae and the controls was statistically within the limits of experimental error. The second study, by Bodenstein, D. (1933a) employed a slightly different approach, but more clearly demonstrated that molting was controlled by a blood-borne substance. When the legs of young larvae of the lepidopterans *Vanessa urticae* and *Vanessa io* were transplanted into older larvae of the same instar, the implants always molted simultaneously with the hosts. Furthermore, the legs of last-instar larvae transplanted into earlier-instar larvae underwent extralarval molts as the host larvae molted into the next larval instar. In apparent ignorance of Kopeć's studies, Bodenstein, D. (1933b) concluded that hormones were probably involved in the process of insect metamorphosis. Here, however, both von Buddenbrock and Bodenstein were probably assessing the effect of an endocrine event downstream from PTTH release, i.e. the effect of molting hormone, 20-hydroxyecdysone (see vol. 8).

Thereafter, a number of investigations went about unravelling the exact nature of the endocrine regulation of insect metamorphosis, particularly the function of the brain in this process. These investigations can be loosely organized into two groups: those confirming and elucidating the humoral effect of the brain, and those elucidating other endocrine centers involved in molting. *In toto*, these studies provided the majority of the basic information of the endocrinology of the prothoracicotrope–prothoracic gland endocrine axis, which surprisingly remained relatively unchanged until the 1970s.

In the first group is found the classical experiments of Sir Vincent Wigglesworth (1934), who, using the bloodsucking bug *Rhodnius prolixus*, successfully repeated, refined and expanded upon Kopeć's experiments. In *Rhodnius*, molting occurs in response to the intake of a bloodmeal, rather than

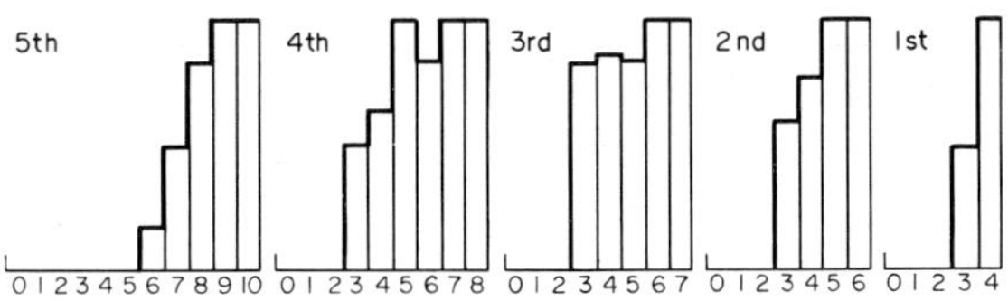

FIG. 1. Chart showing the head critical periods for first- to fifth-stage *Rhodnius prolixus* larvae, determined from the proportion of larvae which molted out of batches decapitated at different times after feeding. Figures on baseline represent days after feeding. The uppermost level represents 100% molting. (From Wigglesworth, V., 1934.)

as the result of a temporally predictable, environmentally cued, neuroendocrine rhythm as in *Lymantria*. In a series of decapitation experiments, Wigglesworth demonstrated that fourth (penultimate) instar *Rhodnius* larvae which had had their heads removed up to 4 days following a bloodmeal failed to undergo a molt, although these headless larvae could survive for months (Fig. 1). However, when the larvae were decapitated at least 5 days after feeding, the expected larval molts occurred, but the animals were unable to shed their old larval cuticles. By transfusing hemolymph from a larva older than 5 days to a larva decapitated shortly after feeding, Wigglesworth was able to demonstrate that this hemolymph could induce the younger larva to undergo a larval molt. The head thus presumably played a critical role in conditioning the hemolymph of the older larva to induce a molt in the younger larva.

At about the same time, in ligation studies on larvae of the flesh-fly, *Calliphora erythrocephala*, Fraenkel, G. (1935) demonstrated the presence of a hormone in the region of the anterior, central ganglion which was responsible for puparium formation. As had been found by Wigglesworth for *Rhodnius*, this hormone was secreted at a precise time during development, about 16 h before pupariation. Furthermore, it was a humoral factor, since injection of hemolymph from the anterior half of a ligated *Calliphora* larva into the non-developing posterior half of a larva ligated early in the last larval instar induced puparium formation. Fraenkel concluded that the mechanism involved in eliciting the molt was endocrinological and that it appeared to be similar, if not identical, in three different orders of insects (Lepidoptera, Diptera, Hymenoptera). Additional analogous studies were conducted in other Lepidoptera (Caspari, E. and Plagge, E., 1935; Plagge, E., 1938; Kühn, A. and Piepho, H., 1938; Bounhiol, J., 1938), as well as in other Diptera (Bodenstein, D., 1936) and in Hymenoptera (Schmeider, R., 1942), and the results of these studies consistently verified the presence and necessity of a hormonal factor from the region of the brain for stimulating molting and/or pupal metamorphosis.

Although Kopeć had originally suggested that the source of the factor responsible for the molt was the brain, the precise location or origin of the active principle had not been defined. Based on a histological correlation between the changing structure of the corpus allatum (CA) (Wigglesworth, V., 1934), which presumably reflected varying CA activity, and the time of the critical period for secretion of the head factor, Wigglesworth concluded that in *Rhodnius* the CA was the source of the "moulting hormone". An apparent confirmation of this hypothesis came 2 years later, when Wigglesworth, V. (1936) observed that fourth-instar nymphs decapitated posterior to the head, but cephalic to the CA, provided a better source of the molting hormone than larvae decapitated through the prothorax. However, it was subsequently found for two species of Orthoptera (Weed, I., 1936; Pflugfelder, O., 1937), as well as for various Lepidoptera (Bounhiol, J., 1938), that this was not the case. Furthermore, when brains were implanted into abdomens of *Ephestia kühniella* (Kühn, A. and Peipho, H., 1936) and of various sphingid larvae (Caspari, E. and Plagge, E., 1935; Plagge, E., 1936) which had been debrained before the head critical period, a statistically significant number of the implanted larvae underwent pupation, these results lending further support to the idea that the brain was the source of the humoral agent. A few years later, using a similar experimental design with *Rhodnius*, Wigglesworth, V. (1940a) showed that when brain–subesophageal ganglion or corpus allatum–sympathetic ganglion were implanted into abdomens of fourth-instar nymphs decapitated 24 h after feeding, only the nymphs implanted with brains were induced to molt (84%). On the basis of these results and those of other workers, Wigglesworth re-evaluated his original proposition and concluded that the brain, and not the CA, was the source of the head molting factor. The results of the earlier experiments (Wigglesworth, V., 1936) which utilized larvae decapitated cephalic to the CA that led to Wigglesworth's misinterpretation, were presumably due to the inclusion in the remaining thorax of the larva of a small piece of the posterior dorsal region of the brain. This piece, which had thus not been removed by ligation, was providing molting activity to the decapitated larvae.

It was not until 1940 that the first experiments directed at localizing the source of this "moulting hormone" within the brain were reported, and these experiments were conducted by Wigglesworth on *Rhodnius* (Wigglesworth, V., 1940b). When optic lobes, subesophageal ganglion and dorsal and ventral halves of the central portion of the brain were implanted separately into suitable hosts, only one tissue was found to be capable of eliciting a molt, and this tissue was the dorsal half of the central region of the brain. Wigglesworth then subdivided the brain into operational regions: (1) a small dorsal fragment of the protocerebral lobes and the region between the lobes; and (2) a much larger fragment consisting of the rest of the brain plus subesophageal ganglion. It was not by chance that Wigglesworth chose to divide the dorsal protocerebrum into these regions for his implantation studies. In 1938, Hanström noted the presence of some histologically peculiar nerve cells in the dorsal region of the protocerebrum of the brain of adult *Rhodnius*, and he suggested that since the cytoplasm of these cells contained rounded masses which stained with acid fuchsin, the cells were probably secretory in nature (Hanström, B., 1938). Wigglesworth confirmed the location and appearance of these cells in the dorsal half of the *Rhodnius* larval brain (Fig. 2) and divided the dorsal half in such a way that of the two anatomical regions for implantation, the small dorsal fragment contained these large cells. In response to these implantations 67% of the decapitated nymphs receiving the small dorsal fragment molted, while none of the nymphs implanted with the other fragment molted. Wigglesworth concluded that the "moulting hormone" was from the region of the brain containing these cells, but "... whether these cells are really the source of the hormone must be left undecided". In this way the central role of the brain in the control of insect postembryonic development was established.

During the time when the existence of a "moulting hormone" from the brain was being proven, there were several studies indicating that another center in the insect was necessary for molting to occur. In the first of these studies, Hachlow, V. (1931) transected the bodies of several species of lepidopteran pupae at different locations and observed to what extent the various isolated body fragments developed. On the basis of the development of the individual fragment(s), he inferred the contribution of the various portions of the body to the development of the pupa as a whole. The study revealed that only those fragments including the thorax were able to develop to the adult, a finding which suggested the presence of some factor in the thorax essential for metamorphosis. The subsequent discoveries by Burtt, E. (1938) and Hadorn, E. and Neel, J. (1938) that excision of the ring glands from the thoraces of the larval blowfly, *Calliphora erythrocephala* and the fruit-fly, *Drosophila melanogaster*, respectively, prevented pupariation, corroborated Hachlow's results that the thorax possessed a factor important for development. Furthermore, the factor probably originated from the ring gland. Shortly after these studies, Fukuda, S. (1940a,b, 1941) identified the prothoracic glands (PG) as the source of the hormonal factor in the thorax that elicited pupation in the commercial silkmoth *Bombyx mori*.

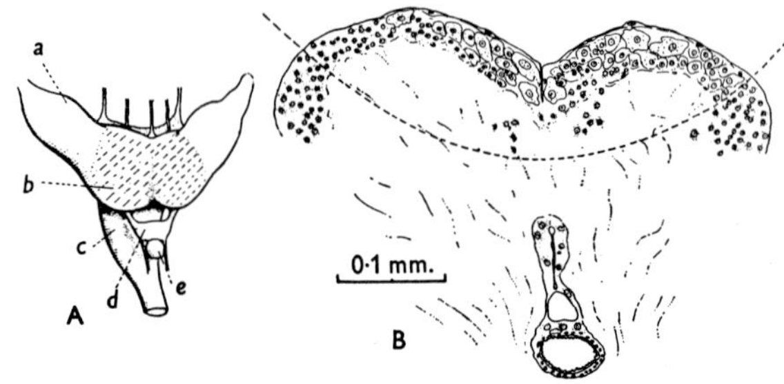

FIG. 2. **A**: Brain of a fifth-stage *Rhodnius prolixus* larva: a, optic lobes; b, protocerebrum; c, subesophageal ganglion; d, corpus cardiacum; e, corpus allatum. The shaded area of the protocerebrum shows the approximate extent of the region excised and implanted. **B**: Vertical section through the posterior part of the protocerebrum showing neurosecretory cells in the central region. The broken line marks off approximately the region implanted. (From Wigglesworth, V., 1940b.)

These early investigations established the endocrine importance of the brain and the PG in molting and metamorphosis, and set the stage for the identification of the brain hormone (PTTH) and molting hormone, as well as for the discovery of their functional relationship to one another and their precise roles in the endocrine control of molting and metamorphosis.

Despite the fact that humoral products of both the brain and the PG were now clearly shown to be necessary for molting, the way in which these moieties achieved essentially the same goal had to be resolved. Several years before Fukuda's studies, Plagge, E. (1938) had shown that the molting factor produced by the brain interacted in some way with the center in the thorax to elicit molting. He observed that when brains were implanted into debrained larvae of *Celerio lineata*, pupation would occur, but that when brains were implanted into isolated abdomens of the larvae, pupation did not result. These results suggested that the brain had to interact with something in the thorax to drive metamorphosis. The first formal suggestion of this relationship was made by Piepho, H. (1942), who proposed that a secretion from the PG might be produced in response to a secretion from the brain, and that there might be not one but a succession of active principles responsible for molting. Confirmation of this novel idea was finally provided about 5 years later by Williams, C. (1947, 1948b) from his studies of the pupal diapause of the silkworm *Hyalophora cecropia*. In an initial series of experiments, Williams corroborated the now numerous studies demonstrating that adult metamorphosis was initiated in response to a factor originating from the brain. This was accomplished by the demonstration that the permanent diapause of debrained diapausing *H. cecropia* pupae was terminated only when brains of chilled pupae were implanted into these animals. He then set about determining the level at which the implanted brain evoked this response by preparing isolated sections of debrained pupae and testing the developmental response of these sections to implanted chilled brains. When transected anterior and posterior halves of pupae were each implanted with a brain, only the anterior half developed normally, while the posterior half remained pupal. Refinement of these transection experiments revealed that the thorax was the so-called "differentiation center", since only body sections containing the thorax could develop when implanted with a brain. Furthermore, the effect of the thoracic center could be exerted even when a plastic tube was placed between an anterior section of a debrained pupa containing the thorax and an isolated abdomen. In this case a chilled brain implanted into either section elicited the development of both.

Aided by S. Fukuda's (1940a,b) identification of the endocrine function of the prothoracic glands in *Bombyx*, Williams, C. (1947, 1948a) was able to demonstrate that the PG in *H. cecropia* were the source of the endocrine factor in the thorax of this insect as well. He accomplished this with the demonstration that an isolated abdomen of a diapausing pupa, when implanted with a chilled brain plus two pairs of prothoracic glands,

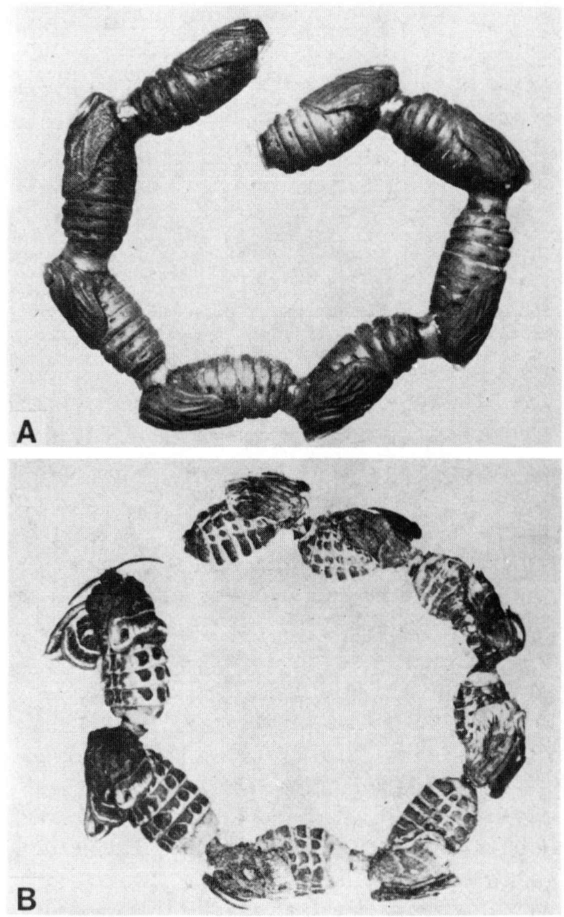

FIG. 3. **A**: A chain of eight brainless diapausing pupae established with a single chilled pupal brain implanted under the facial window of the anterior-most individual. (From Williams, C., 1952.) **B**: The preparation in figure 3A after $7\frac{1}{2}$ weeks. The single brain has caused all the pupae to undergo adult development. (From Williams, C., 1952.)

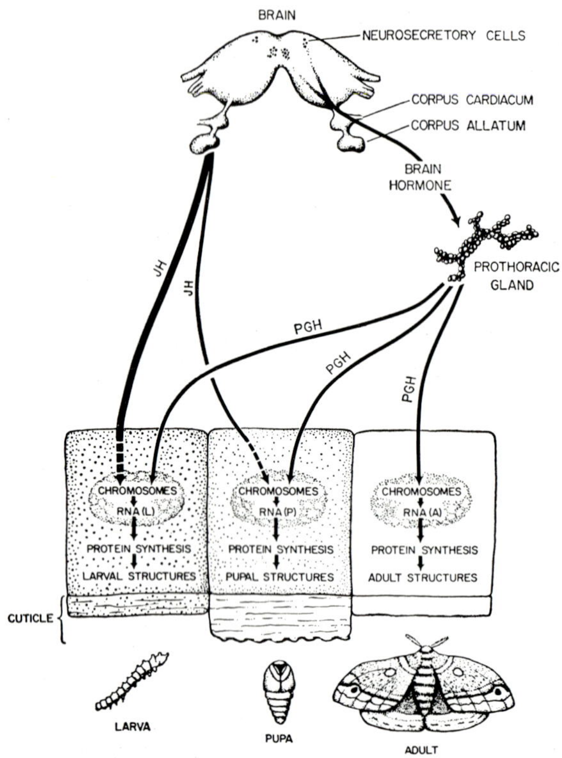

FIG. 4. Schematic diagram of the principal endocrine organs of the *Hyalophora cecropia* silkworm and the site of action of their hormones. The larva of this insect molts four (or occasionally five) times. These larval molts appear to be initiated by the brain hormone, which stimulates the prothoracic glands to secrete the prothoracic gland hormone. At the same time the corpora allata secrete juvenile hormone, which favors larval syntheses, so that when the larva molts in response to prothoracic gland hormone it molts into a larva. At the end of larval life the corpora allata cease secreting and thus the mature larva is left with a low concentration of juvenile hormone. At the next molt the epidermal cells respond to prothoracic gland hormone and to the low concentration of juvenile hormone by secreting pupal cuticle, and in response to these hormonal conditions other tissues within the insect either break down or transform into pupal structures. At the final molt juvenile hormone is absent and the epidermal cells respond to prothoracic gland hormone alone by secreting adult cuticle; the other pupal tissues either break down or develop into adult structures. (From Schneiderman, H. and Gilbert, L., 1964.)

developed into an adult abdomen. From these studies Williams concluded that the termination of pupal diapause and subsequent adult development required both brain and prothoracic gland factors, and that the brain factor exerted a "controlling action" on the prothoracic glands. In addition, he proposed that the only function of the brain factor was to affect the prothoracic glands and their secretory factor. The basis for this last conclusion was the fact that when an isolated abdomen was grafted to a brainless diapausing pupa still containing its prothoracic glands, and was then implanted with a chilled brain, the brainless pupa initiated development 1 day before the brain-containing abdomen.

Four years later, after a series of studies on the endocrine control of molting and pupation in larvae of *H. cecropia*, Williams, C. (1952) concluded that the same hormonal system was acting here as well (Fig. 3). Thus, by the mid-1950s the role of the brain and the prothoracic glands in insect postembryonic development and metamorphosis had been defined and, with the findings of comparable studies on the CA, gave rise to the so-called "classical" scheme for the endocrine control of insect postembryonic development (Fig. 4).

Ironically, although the brain hormone, as it was then called, or prothoracicotropic hormone (PTTH), as it is now called, was the first insect hormone discovered, it has until recently been the least understood both chemically and physiologically of the principal hormones that control insect development. The lack of such knowledge of the endocrinology of PTTH was inextricably bound to the types of assays that were developed to detect the presence of this moiety. In recent years new methods for assaying PTTH activity have been developed which promise to facilitate an elucidation of much of the endocrinology of this important neurohormone. The central role assays for PTTH have had in the progress of this field makes a discussion of these assays both necessary and important to an historical perspective on the work on PTTH since the early 1950s.

2 *IN VIVO* ASSAYS FOR PTTH

Since the time of its discovery, most of the research on PTTH has been conducted with the utilization of whole-animal bioassays as a means of assessing hormone activity. By definition, such bioassays measure the presence of the hormone indirectly, usually by scoring the degree to which a morphological response is evoked downstream. In the case of PTTH, this response has been a molt to either the pupa or the adult, which in both cases is a measure of a secondary endocrine response, the synthesis of ecdysone by the PG. Although it has been acknowledged for many years that a direct and specific means of assaying PTTH activity was needed, it has only been within the last 10 years that adequate biochemical and analytical methods, and essential endocrinological information, have become available to permit the development of new methods for assaying PTTH.

With the elaboration of the classical scheme for the endocrine control of insect metamorphosis came the first bioassays for PTTH. The work of Caspari, E. and Plagge, E. (1935), and Kühn, A. and Piepho, H. (1936) involved the first of such bioassays in their assessment of the activation of larval PG by implanted brains. Unfortunately, the response to the implanted brains in both cases was quite low: pupal cuticle secretion was induced in less than 5% of the debrained host larvae of the meal moth, *Ephestia kühniella* (Kühn, A. and Peipho, H., 1936), and with larvae of the two sphingid moths, *Sphinx ligustri* and *Deilephila euphorbiae*, ecdysis occurred in only about 20% of the animals (Caspari, E. and Plagge, E., 1935). Similar difficulties were encountered in obtaining a significant response to brain implantations of decapitated larvae of *Bombyx mori* (Fukuda, S., 1944), and more recently of *Galleria mellonella* (Beck, S., 1970; Hetru, C., 1976). By contrast, non-developing pupae of Lepidoptera have proven to be quite sensitive to the presence of PTTH.

2.1 *Hyalophora* pupal assay

The bioassay developed by Williams, C. (1947, 1952) in his studies defining brain–PG interaction in the termination of pupal diapause employed pupae of the *H. cecropia* moth. Debrained, diapausing pupae were used in this assay and the breaking of diapause, i.e. the initiation of pupal–adult development, was scored as a positive response to the presence of PTTH. Despite the success of the early experiments which utilized this assay to test the PTTH activity of fresh larval brains, developing pupal brains, or brains from chilled, diapausing pupae, it appeared that this assay would not be suitable for the chemical characterization of PTTH. This was due to the fact that implantation of up to 50 killed brains, e.g. frozen then thawed, failed to elicit the same effect (Williams, C., 1956). Thus, the *H. cecropia* bioassay and/or brains did not seem to be very well suited for probing the endocrinology of PTTH in this species. Other factors also indicated that *H. cecropia* would not be a good test animal for a PTTH assay. If diapausing pupae were debrained after chilling for up to 10 weeks, the resulting dauer pupae, that is pupae in a state of developmental arrest due to brain removal before the head critical period, were unresponsive to brains or brain extracts (Williams, C., 1956, 1968a). However, if the pupae were chilled for longer than 10 weeks but no longer than 35 weeks prior to debraining, and if they were then used immediately for assay, a discrete positive response to PTTH could be obtained (Williams, C., 1968a). Thus, there appeared to be a critical chilling time necessary to get a viable assay animal. If chilling exceeded 35 weeks, debrained pupae would respond non-specifically to substances such as zinc acetate, copper sulfate, cobalt chloride and ferrous sulfate and would initiate adult development as if they had been exposed to PTTH. Williams speculated that the window of time within which the chilled, diapausing pupae could be used represented the time when the endogenous brain was slowly renewing its secretion of PTTH to a subthreshold titer. This titer would then support adult development in the debrained pupae if another brain were immediately implanted. Beyond 35 weeks, the endogenous titer of the neurohormone became sufficiently high to permit non-specific stimuli to trigger adult development. He further surmised that once the brain was removed, the subthreshold, endogenous titer of PTTH would decay to a point where the animal was no longer sufficiently sensitive to respond to implanted brains or an extract of brains.

Despite the use of the *H. cecropia* assay under these stringent conditions, Williams was unable to demonstrate the presence of the hormone in extracts of brains of chilled diapausing *H. cecropia* pupae, nor from the brains of another large sphingid moth, *Samia cynthia* (Williams, C., 1968a). However, extracts of *Antheraea pernyi* brains very effectively elicited adult development in the debrained *H. cecropia* pupae. A positive response could be scored as early as 3–8 days after injection by monitoring the time of epidermal retraction (Williams, C., 1968a). The lower limit of sensitivity of the *H. cecropia* assay with *Antheraea* brains as the source of PTTH was 1–2 brain equivalents. That *Antheraea* brains could elicit a positive response in the *H. cecropia* assay was an interesting observation in that it implied interspecific cross-reactivity of PTTHs which was similar to that being observed with the PTTH bioassays used by the Japanese laboratories. Overall, its various shortcomings made the *H. cecropia* bioassay for PTTH not very useful and, because of this, it was not used routinely by other laboratories.

2.2 *Bombyx* pupal assay

Although Fukuda, S. (1944) had successfully utilized *Bombyx* in his studies identifying the role of the PG, the use of this insect as a bioassay animal for PTTH proved to be problematic. Fukuda was the first to note that pupal–adult development in *Bombyx* could not be curtailed by brain extirpation immediately following larval–pupal ecdysis, an observation later confirmed by Bounhiol, J. (1952a,b), and apparently due to the fact that the head critical period for PTTH release in the races of *Bombyx* used in these studies occurred before pupation.

However, Kobayashi, M. (1955) did succeed in creating "dauer pupae" of *Bombyx* suitable for use in a PTTH bioassay. He accomplished this by utilizing a specific race of *Bombyx* in which the critical period for PTTH release occurred after pupation. In this assay, which has been modified with use (Ishizaki, H., 1972; Nishiitsutsuji-Uwo, J. and Nishimura, M., 1972), *Bombyx* pupae within 10 min of pupation are chilled to 1–2°, maintained at this temperature for 8–16 h, debrained and then returned to 25°. Following surgery, the dauer pupae are monitored for 4 weeks, at which time all developing pupae are discarded, and the remaining dauer pupae are used for assaying PTTH. Only male pupae have been used in this assay because of the tendency of female pupae to undergo spontaneous adult development (Nishiitsutsuji-Uwo, J. and Nishimura, M., 1972). In addition, suitable dauer pupae must be used within 4 weeks, since both injected and non-injected pupae begin to develop spontaneously about 7 weeks after debraining. Despite this rigid protocol for obtaining suitable assay animals, the yield of developmentally arrested, brainless pupae for assay was still found to vary greatly with nutritional status, seasonal changes and parental history (Kobayashi, M., 1957; Ishizaki, H., 1972; Ishizaki, H. and Suzuki, A., 1980). The sensitivity of the *Bombyx* dauer pupae to PTTH in this assay is approximately that of the *Samia cynthia* pupal assay (see subsequent section).

Although the *Bombyx* assay proved very sensitive to PTTH under precisely defined conditions, it still has a significant drawback: its apparent sensitivity to various non-specific stimuli. For example, magnesium, iron, copper, and zinc salts, and even water and buffers, were found to induce a large percentage of debrained *Bombyx* pupae to undergo adult development (Nishiitsutsuji-Uwo, J. and Nishimura, M., 1972). This was clearly a major shortcoming of this assay, as concluded by Nishiitsutsuji-Uwo, J. and Nishimura, M. (1972), and as a consequence it was never used extensively in studies on PTTH, even though *Bombyx* served as the source of PTTH for most of the purification studies.

2.3 *Samia* pupal assay

Samia cynthia ricini is an obligatorily non-diapausing lepidopteran that becomes a dauer pupa if debrained within 20 h after pupation. The incidence of arrested development in the debrained pupae after 3 months was found to be 90%. Within the 30-day maximal scoring period for a positive PTTH response, only 7.7% of the pupae underwent adult development (Ishizaki, H. and Ichikawa, M., 1967; Ishizaki, H. and Suzuki, A., 1980). The head critical period for PTTH release for adult development in this insect thus occurs well after pupal ecdysis, making *Samia* a more suitable animal for assaying PTTH activity. One problem encountered in the original *Samia* assay was the occasional failure of injected PTTH to elicit adult development in freshly prepared brainless pupae (Ishizaki, H. and Ichikawa, M., 1980). Aging all debrained pupae for at least a month, but no longer than 10 months, prior to injection circumvented this problem. Beyond 10 months, dauer pupae begin to respond non-specifically to various agents, e.g. water. The finding that dauer pupae of *Samia* could be maintained for prolonged periods of time was a major advantage of this assay, since thousands of test animals can be prepared and then stored until needed. The specificity of the *Samia* PTTH assays was far greater than that of the *Bombyx* and *H. cecropia* assays, e.g. injections of water, magnesium, iron and copper salts had no effect (Nishiitsutsuji-Uwo, J. and Nishimura, M., 1972). However, fairly high concentrations of organic solvents such as acetone, ethyl alcohol and ether, often used in the experimental manipulation of the insects, could elicit a significant level of adult emergence (Nishiitsutsuji-Uwo, J. and Nishimura, M., 1975).

Although the *Samia* assay required about 20 days to score complete adult development (Ishizaki, H. and Ichikawa, M., 1967), the assay could be shortened to only 4–6 days if epidermal retraction were scored as had been done by Williams, C. (1968a) in the *H. cecropia* assay. A positive response is scored in this manner when the trachea in an excised portion of wing epidermis can be moved independently of the cuticle.

Since the *Samia* pupal assay is fairly specific and can be scored in a moderate amount of time, it has been the assay of choice and the most successfully employed pupal assay for PTTH (Ishizaki, H. and Suzuki, A., 1980). In addition, this assay was the first used to quantify PTTH activity. A unit of PTTH activity in this assay is expressed as one *Samia* unit, and this is defined as the amount of PTTH that initiates adult development in a brainless *Samia* pupa. This value is based on the PTTH activity present in a crude extract of a single *Samia* head, which possesses 10 *Samia* units of "crude PTTH" (Ishizaki, H. and Suzuki, A., 1980).

2.4 *Manduca* larval assay

It is only within the past decade that the endocrinology of the tobacco hornworm, *Manduca sexta*, has been investigated intensively. Because of its size, ease of rearing and synchrony of development, this lepidopteran has quickly become the preferred insect for investigations of the endocrine systems regulating insect growth and development. Although there are decided advantages for using *Manduca* in such studies, the pupae of this insect do not appear to be well-suited for a dauer pupal bioassay for PTTH, since *Manduca* pupae appear to be capable of undergoing metamorphosis spontaneously without a brain. In fact, Judy, K. (1972) reported that debrained, last-instar larvae of *Manduca* are capable not only of pupation but also of development to adults. This remarkable feat demonstrates the adaptability, or perhaps more appropriately the plasticity, of this insect's endocrine system. This experimental observation serves to illustrate the non-obligatory nature of the *Manduca* neuroendocrine system to its development, and thus the potential limitation of the use of this animal in a traditional PTTH bioassay.

Although a *Manduca* pupal assay for PTTH does not exist, a clever larval assay for this hormone was developed by Gibbs, D. and Riddiford, L. (1977) that circumvented many of the problems associated with a pupal assay. The basis of this assay is the fact that PTTH release in fourth-instar larvae occurs during a specified gate, which is established by a specific photoperiod (Truman, J., 1972). The gate is open during the scotophase of each day, and the release of the neurohormone (head critical period) on a given night seems to be linked to the developmental state of the larvae. Truman, J. (1972) found that a larva neck-ligated early in the head critical period did not molt into a fifth-instar larva but instead metamorphosed precociously to a pupa. The opening of this gate for PTTH release was precisely defined by Fain, M. and Riddiford, L. (1976), and they further demonstrated that neck-ligated larvae would molt to fifth-instar larvae in 2–3 days if injected with ecdysone at the time of ligation. They proposed that this response was due to the quantity of JH present at the time a threshold ecdysteroid titer for initiating a molt was reached, i.e., high JH dictating a larval molt and low JH permitting pupal metamorphosis. On the basis of this ecdysone response in the fourth-instar larvae, Gibbs, D. and Riddiford, L. (1977) developed a PTTH bioassay in which fourth-instar larvae ligated just before the head critical period were injected with brain extract containing PTTH. Since PTTH would immediately activate the PG to synthesize ecdysone in the presence of an elevated JH titer, a larval molt occurred. The process of spiracle apolysis, the first overt sign of larval molting in *Manduca* (Truman, J. *et al.*, 1973), was scored as a positive response in this assay, and this was observed within 48 h after PTTH injection. The percentage of injected larvae exhibiting spiracle apolysis at 48 h increased linearly with increasing concentrations of brain homogenate injected. The dose response range for this

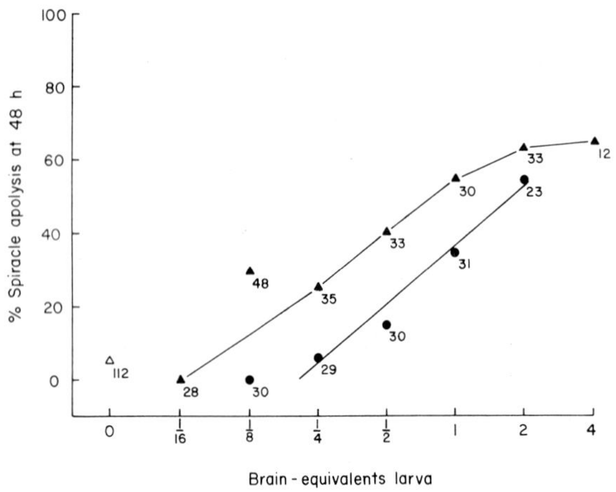

FIG. 5. Dose–response curves for injections of pupal brain homogenates or extracts into neck-ligated fourth-instar larvae. ▲, Fresh homogenates prepared from brains removed 1 day after pupation and stored frozen until homogenization; ●, dilutions of a single preparation of supernatant of boiled (5 min) homogenate from day 1 brains plus brains after 15–33 days in pupal diapause. Dilutions stored frozen at −20° until injection. (From Gibbs, D. and Riddiford, L., 1977.)

assay was 0.06–1.0 brain equivalents per animal (Fig. 5), making this the most sensitive of all the PTTH bioassays. The value of this assay was made apparent in Gibbs and Riddiford's study on PTTH in *Manduca* in which they tentatively identified a group of lateral protocerebral neurosecretory cells (NSC) as the prothoracicotropes and implied that the CA was the neurohemal organ for PTTH. Subsequent uses of this assay by Kingan, T. (1981) and Bollenbacher and Gilbert (1981, unpublished) in studies of the chemistry of the *Manduca* larval and pupal PTTHs, respectively, have proven its value in these types of studies as well.

From this discussion of the bioassays for PTTH, it becomes apparent that there are a number of drawbacks to these assays that have effectively restricted their use in research on PTTH, and thus have impeded progress in this field. The most obvious problem is that these assays are labor-intensive. A considerable amount of time is needed for rearing, timing, and debraining or ligating the large numbers of animals needed for the assays. These assays are not particularly sensitive, requiring at least 0.1 brain equivalents to evoke a positive response. In addition, a large population of test animals is required for each assay and thus, because of the lack of sensitivity of the assays, a large amount of test material is needed. Another problem is the time required to run these assays, since several days to weeks are required to score a response. Perhaps the most critical drawback, however, is that these assays, particularly the dauer pupal bioassays, have the reputation of being non-specific, as evidenced by the positive response elicited in the *Samia* and *Bombyx* assays by salts, buffers, and organic solvents.

In toto, the whole-animal bioassays have not proven to be adequate for elucidating the endocrinology of PTTH, but in the last 8 years with the development of *in vitro* assays for PTTH these problems are being circumvented.

3 *IN VITRO* ASSAYS FOR PTTH

The rationale for developing any *in vitro* assay for PTTH was based on the premise that PTTH activates the PG directly (Gilbert, L. and King, D., 1973; Goldsworthy, G. and Mordue, W., 1974). If this were the case, then ecdysone biosynthesis should be stimulated when the PG are incubated with PTTH *in vitro*.

The evolution of *in vitro* assays for PTTH was essentially a by-product of the new methods and techniques being developed to investigate the endocrinology of insect development, particularly the production of ecdysone by the prothoracic glands, the chemistries of ecdysone and 20-hydroxyecdysone and their relationship to each other (Horn, D. and Bergamasco, R., this volume). These methods and approaches included the establishment of *in vitro* organ culture methods, the development of a radioimmunoassay for ecdysteroids (Borst, D. and O'Connor, J., 1972) and the identification of ecdysone as the product of the PG (King, D. *et al.*, 1974).

3.1 Spermatogenesis assay

Kambysellis, M. and Williams, C. (1971a,b) were the first to demonstrate an *in vitro* effect of the brain on the PG. In a study of the role of ecdysone in spermatogenesis of saturniid silkworms, they were able to activate PG from diapausing *Antheraea pernyi* pupae by co-culturing them *in vitro* with active brains from chilled, diapausing *A. pernyi* pupae. This activation was monitored indirectly by the initiation of spermatogenesis in testes that were included in the PG–brain cultures. Unfortunately, the degree of activation of the PG assessed by spermatogenesis was not quantifiable. Although this assay was never pursued as a means of assessing PTTH activity, in retrospect it alluded to the feasibility of using an *in vitro* system to assay PTTH, albeit indirectly.

3.2 *Chilo* integument assay

A similar approach using integument to assess PG activation was employed by Agui, N. (1975, 1976), and in this case the assay proved somewhat quantitative. In this assay, brains from diapausing and non-diapausing *Mamestra brassicae* pupae were cultured in groups of five for 3 days, at which time five PG from diapausing pupae were added to the cultures and the resulting culture incubated for an additional 3 days. After this time, five fragments of diapausing *Chilo suppressalis* larval integument were co-cultured with the brains and PGs for an additional 4 days to assay the quantity of ecdysone produced by the PG in response to co-culture with brains (PTTH release). The assay for ecdysone, scored daily, was a determination of the rate of ecdysis of the *Chilo* integument, the rate reflecting the degree of PG activation by the brain (PTTH). When PG from freshly ecdysed diapausing pupae were cultured with non-diapausing brains, 100% activation of the glands occurred. As the age of the diapause PG increased, the number of ecdysing integument fragments decreased, indirectly indicating that the level of activation of the PG was decreasing, i.e. that the PG were becoming progressively refractory to PTTH as diapause progressed. Although this *in vitro* assay could effectively demonstrate that the brain was activating the PG, it had a number of shortcomings as an assay for PTTH: (1) it was very tedious and time-consuming to run; (2) it was scored subjectively; (3) it was indirect; and (4) it was not particularly quantitative.

3.3 Radio-synthesis assay

The first *in vitro* system incorporating a direct means of measuring ecdysone was reported by Gersch, M. and Stürzebecher, J. (1967) and Braüer, R. *et al.* (1977), in which [^{3}H]- or [^{14}C]-cholesterol was used to radiolabel the product of the PG *in vitro*. This product was then separated and identified by thin-layer chromatography. The ecdysteroid nature of the labeled product was verified by testing for molting hormone activity in the *Musca* bioassay for ecdysone. Activation factor I, one of two moieties from the corpora cardiaca of cockroaches which have prothoracicotropic activity (see section 4.2) in the *Antheraea* dauer pupal assay for PTTH (Gersch, M. and Stürzebecher, J., 1968), had a stimulatory effect on the *in vitro* incorporation of labeled cholesterol into what appeared to be ecdysone by inactive PG. Thus, it appeared that this system was monitoring directly, and in quantitative terms, the response of the PG to an activating factor. Although a significant step in the direction of an *in vitro* assay for PTTH, this assay was still tedious to run. Furthermore, in some insects metabolism of cholesterol to ecdysone by the PG has not been demonstrable (Gilbert, L. *et al.*, 1977), and this fact effectively precluded use of the radiochemical assay for studying PTTH.

3.4 *Manduca* PG assay

Clearly, the feature of any assay for PTTH most critical to its usefulness in endocrinological studies is the rapidity with which the assay can be run and results obtained. An *in vitro* PG assay developed recently for PTTH offers this feature, and circumvents many of the other problems previously mentioned, by combining *in vitro* organ incubation, not organ culture, and a radioimmunoassay (RIA) for ecdysone to quantify the ecdysone synthesis by the PG (Bollenbacher, W. *et al.*, 1979).

Because so much is known about the developmental endocrinology of *Manduca* relative to other insects, this animal was used in the development of this assay. The assay is basically an incubation of a PG with or without PTTH, and activation of the gland by the neurohormone is determined by an increase in the rate of ecdysone synthesis by the treated PG. Paired PG are used for the assay, with one member of the pair serving as the control and the other as the experimental. This paired-gland approach was possible because basal rates of ecdysone synthesis by members of a pair of glands from an animal were equivalent. Thus, this assay measures quite precisely a change in the rate of ecdysone synthesis between one gland treated with PTTH and the contralateral, untreated gland. Activation of the PG by PTTH is expressed as an activation ratio (A_r), which represents the quantity of ecdysone synthesized by the experimental gland divided by the amount synthesized by the control gland during the incubation period.

Since this assay was originally developed (Bollenbacher, W. *et al.*, 1979), it has been modified such that the PG incubations last only 2 h, and the ecdysone synthesized and quantitatively released by the PG into the medium is quantified directly by a "macro" RIA (standard concentration range of 0.25–32 ng ecdysone) of an aliquot of the incubation medium (Bollenbacher, W. *et al.*, 1983). The PG routinely used in the assay are excised from day 0 pupae, since glands at this stage exhibit a relatively low basal rate of synthesis from which activation can be clearly demonstrated. In addition, they contain sufficient substrate to support a substantial amount of synthesis when activated. Day 0 pupal PG were also selected for the assay because they could be obtained in large quantity, because they exhibited moderately constant basal activities and, most importantly, because recent studies have shown that they are not easily activated non-specifically. The latter is becoming progressively more important in light of recent findings that PG from other developmental stages, e.g. day 3 larval glands, appear capable of being activated non-specifically (Bollenbacher, W. *et al.*, 1983; Bollenbacher and Gilbert, unpublished observations).

Activation of the pupal PG by PTTH occurs in a characteristic dose-dependent manner over a very narrow dynamic range (2–3-fold) of PTTH concentration (Fig. 6). The maximum activation ratio (A_{max}) obtained in a dose–response typically ranges from 3 to 8, while the concentration of neurohormone (ED_{50}) necessary to half maximally activate the gland (A_{50}) remains relatively constant for a given extract. Because of this consistency in the ED_{50}, it is possible to use this value to quantify the amount of PTTH present in a test sample and between test samples. This is possible since the different amounts of PTTH in different extracts are expressed by the quantity of the extract needed to obtain an ED_{50}. Once an A_{max} for a test sample has been obtained, it is a simple matter to compute an ED_{50} from the dose–response curve. Reciprocals of the ED_{50} for the response curves of different extracts can then be compared, and the amount of PTTH present in each extract quantified in relative terms. The basic unit for quantifying PTTH activity in this assay is a *Manduca* PTTH unit, which is the amount of PTTH present in a day 1 pupal brain of *Manduca*. In an incubation volume of 0.025 ml, the sensitivity of this assay can approximate 0.05 PTTH units. If the incubation volume is further reduced, it is possible to increase the sensitivity to about 0.025 units. This is the greatest level of sensitivity of all the PTTH assays that have been developed, and since only a few pairs of glands are needed to assess the presence of PTTH, far less material is needed overall to get a positive result.

Recently, Carrow, G. *et al.* (1981) reported the use of day 3 larval PG of *Manduca* in an *in vitro* PG assay for PTTH that is a modification of the Bollenbacher, W. *et al.* (1979, 1983) assay. Although this larval PG assay is based on the pupal PG assay (Bollenbacher, W. *et al.*, 1979, 1983), the activation kinetics of the larval glands and their dose–responses of activation by PTTH are quite different than those reported for pupal glands. Bollenbacher, W. *et al.* (1983) have attempted to discover the basis for these apparent differences, which could potentially affect the use of the assay. They concluded that the use of larval glands in the assay could be the basis for the differences observed, since their use can give rise to non-specific activation that is not

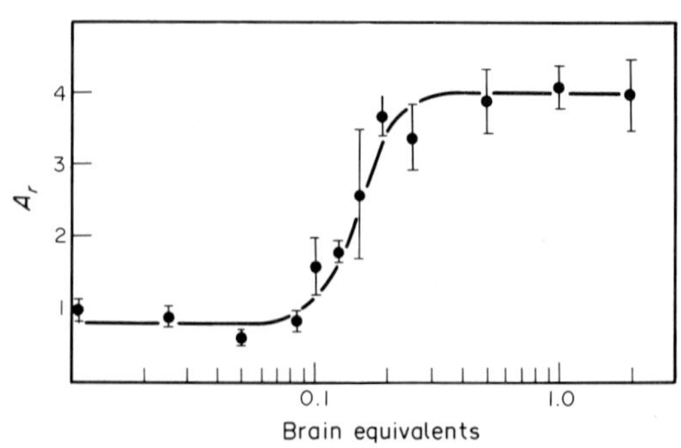

FIG. 6. Dose–response of activation of PGs by homogenates of day 6 last-instar larval *Manduca sexta* brains. The homogenate was defined as the supernatant after centrifugation of a crude homogenate at 10,000 ***g*** for 10 min. Each datum point represents the mean (±SEM) of the A_rs for three to six separate activation assays. (From Bollenbacher, W. *et al.*, 1979.)

representative of PTTH. The expression of this non-specific activation would vary, depending on how the glands were manipulated in the execution of the assay.

Irrespective of such differences, the *in vitro* PG assay for PTTH has provided a more critical means by which the endocrinology of PTTH can be elucidated, and this is best illustrated by the use of the assay in recent studies on the PTTH neuro-endocrine system during larval–pupal development of *Manduca sexta* (Agui, N. *et al.*, 1979, 1980; Bollenbacher, W. *et al.*, 1980; Bollenbacher, W. and Gilbert, L., 1981; Gilbert, L. *et al.*, 1981; Carrow, G. *et al.*, 1981; Bollenbacher, W. *et al.*, 1983; Agui, N. *et al.*, 1983; Bowen, M. *et al.*, 1983).

Although the *in vitro* assay for PTTH will now undoubtedly be the method of choice for probing the endocrinology of this neurohormone, it must be kept in mind that this type of assay is still biological, quantifying PTTH activity in terms of the hormone it stimulates; and because of this it has its own particular limitations. Hopefully, however, the use of an *in vitro* assay for PTTH will permit the purification of the hormone and the resolution of its chemistry, which would, in turn, make possible the generation of antibodies. With these antibodies to PTTH a radioimmunoassay could be then developed. This accomplishment would immeasurably facilitate a comprehensive investigation of the endocrinology of PTTH.

4 CHEMISTRY OF PTTH

Research into the chemistry of PTTH was not initiated in earnest until nearly 40 years after its discovery. Although the lack of adequate assays for the neurohormone was definitely a contributing factor to this lag time, more likely factors are the conceptual differences in research priorities during those years. Until the late 1950s most research on insect endocrine systems was descriptive in nature, dealing principally with various aspects of the now classical scheme for the endocrine regulation of insect development (see Granger, N. and Bollenbacher, W., 1981). Until this scheme was established, the chemical characterization of PTTH would have been premature, and perhaps impossible, since peptide neurohormone chemistry was still developing as a research discipline and techniques for purifying trace amounts of neurohormones were not yet developed. Thus, there was little reason to study the chemistry of PTTH.

The first study attempting to elucidate the chemistry of PTTH was reported in 1958 (Kobayashi, M. and Kirimura, J., 1958). From that time until the late 1970s, episodic and nominal progress was made by the few laboratories working on the problem. Furthermore, results obtained from these studies on the general physical nature (lipid, protein, mucopolysaccharide, etc.) of the molecule with PTTH activity were often completely contradictory.

In the past 5 years, however, substantial progress has been made on the purification of PTTH(s), in part due to the development and use of new bioassays and purification methods for detecting, separating, and quantifying PTTH activity (Bollenbacher, W. and Gilbert, L., 1981). On the basis of these studies, it appears that the endocrinology of PTTH may be far more intricate than initially thought, possibly rivalling comparable vertebrate systems in its chemical and functional complexity.

Because research on the chemistry of PTTH has taken such an unpredictable route during the past 25 years, and because the progress has been episodic, it is most clearly discussed in an evolutionary format. To present this topic in an historical context will also serve to illustrate the circuitous paths often taken in the solution of a scientific problem.

4.1 Pioneering studies

In 1958 Kobayashi, M. and Kirimura, J. made the first attempt to obtain "brain hormone" from the brains of the silkworm *Bombyx mori* L. For this study brains from approximately 8500 day 0–1 pupae of *Bombyx* were extracted in methanol and the extract partitioned with ethyl ether. The evaporated ethyl ether extract was assayed for PTTH activity by the *Bombyx* dauer pupal bioassay, and the test pupae were injected with the extract at different times after brain removal. Of 13 animals that were injected with varying concentrations of extract at the different times, 62% underwent adult development while none of the 19 control animals developed. However, even though adult development was elicited in response to the injected brain extract, 400 brain equivalents were required for the response, and the response occurred in a mean time of about 35 days. Thus, both the dose and the response time were far in excess of the number of implanted brains (one) needed to achieve a response in a normal period of time (approximately 15 days). The fact that the injected extract evoked the expected bioassay response led the authors to conclude that PTTH was an ether-soluble moiety, probably a non-polar lipid. However, as has already been discussed in this review, the *Bombyx* dauer pupal bioassay for PTTH can yield false-positives due to a non-specific response to injected material. This lack of specificity may have been the reason for the results obtained in this study and thus the conclusions drawn.

Three years following this study, Ichikawa, M. and Ishizaki, H. (1961) reported on the successful extraction of PTTH activity from the brains of pharate adult *Bombyx* utilizing the *Samia* pupal bioassay to follow activity. Initially, extracts of "devitalized" brains, i.e. brains either desiccated, heat-treated (100° for 90 min) or frozen (−18°), were injected into debrained pupae and were found to elicit adult development at a dose of only five brains per recipient. Several hundred brains of the same stage were then extracted with methanol, the methanol extract evaporated, and the residue suspended in water. This aqueous solution was then partitioned with ether and the PTTH activity of the ether and aqueous phases tested by bioassay. PTTH activity was detected only in the aqueous phase, and 12 brain equivalents of the aqueous material elicited adult development in approximately 95% of the test animals in a period of time comparable to the time for normal pupal–adult development in *Samia*. The PTTH activity could also be extracted in 2% NaCl, and 12 brain equivalents of this extract yielded the same percentage response as the methanol extraction. The partitioning properties of the pharate adult brain PTTH activity thus suggested that the material was a polar, water-soluble moiety, in direct contrast to the lipoidal moiety indicated by the results of Kobayashi, M. and Kirimura, J. (1958).

Subsequent to these initial, contradictory reports, these same two laboratories continued to publish mutually exclusive results in a series of papers during the 1960s. In 1962 Kobayashi, M. *et al.* attempted to crystallize the pupal "brain hormone" using material obtained from a methanol extract of 500,000 brains of day 0–2 *Bombyx* pupae. This extract was subjected to countercurrent distribution, and the active fraction from this procedure was crystallized in acetone and ethanol. When the crystalline material was subsequently assayed for PTTH activity with the *Bombyx* dauer pupal assay, only 0.02 μg evoked adult development. However, this amount represented hundreds of pupal brain equivalents, and the period of time necessary to score a positive response was about twice that of normal development. Furthermore, mention was made that other fractions generated by this purification protocol possessed brain hormone activity; this could possibly be the first indication of multiple moieties with PTTH activity.

Shortly thereafter, Kirimura, J. *et al.* (1962) presented data identifying this crystallized lipoidal material as cholesterol. Since the identification was based on infra-red spectroscopy, gas–liquid chromatography, a positive Liebermann–Burchard reaction and a melting point determination, the bulk of the material isolated was clearly cholesterol, but how cholesterol could function as the brain hormone was puzzling. In the same study an attempt was made to obtain a dose–response of development to the material in the bioassay, but all of the concentrations in the dose range assayed — 2 μg to 0.02 μg animal^{-1} — yielded a maximum response. These data indicated that the biological potency was far greater than previously thought; a peculiar finding since cholesterol is generally not considered to be a biologically active substance. Furthermore, the fact that cholesterol exists in insects at levels far in excess of the dose of PTTH sufficient in this study for a positive bioassay response argued against it being the brain hormone (Schneiderman, H. and Gilbert, L., 1964).

At about the same time, Ichikawa, M. and Ishizaki, H. (1963) proposed that PTTH was a protein, basing their conclusion on the results of a straightforward probe of the chemistry of the water-soluble fraction with brain hormone activity which had been obtained from pharate adult *Bombyx* brains. Using as an operational hypothesis that PTTH is a protein and the extraction protocol they had established a few years earlier (Ichikawa, M. and Ishizaki, H., 1961), Ichikawa and Ishizaki found the PTTH activity in the pharate adult brain to be non-dialysable and heat-stable at near neutral pH. However, under acidic (pH 4.8) and basic (pH 8.8) conditions, PTTH activity was heat-labile. The activity was also found to be precipitable by methods selective for protein, i.e. ammonium sulfate, acetone and trichloroacetic acid precipitation. Lastly, the PTTH activity could be hydrolyzed with proteases, i.e. activity was lost following treatment with Nagase (*Bacillus subtilis* subtilisin) and pronase (*Streptomyces griseus* proteinase), as evidenced by a negative dauer pupal bioassay response. Pepsin and trypsin, however, did not destroy the activity. This study thus demonstrated that the PTTH activity in the aqueous extract of pharate adult *Bombyx* brains could be attributable to a peptide, a finding consistent with the well-established proteinaceous nature of cerebral neurohormones produced by neurosecretory cells in vertebrates. However, the results of this study presented insect endocrinologists with a dilemma: the possible existence in brains of the same insect of two factors with an identical prothoracicotropic function. Kobayashi, M. (1963), in fact, gave consideration to the brain hormone activity observed in non-lipoidal fractions generated during the purification of the "cholesterol" brain hormone, speculating that these activities were associated with protein.

During the time of these studies of the *Bombyx mori* PTTH activities, Gersch, M. (1961) reported on the isolation from the protocerebrum of *Periplaneta americana* of a factor which he termed activation hormone (neurohormone D_1) and which elicited a positive bioassay response in a modified "Calliphora test" (ecdysone bioassay; see Wigglesworth, V., this volume) for this moiety. This putative PTTH was partially purified by the combined procedures of aqueous and/or alcoholic extraction, paper chromatography, repeated precipitation with alcohol to remove contaminating protein, and finally crystallization from water–methanol (Unger, H., 1957; Gersch, M. *et al.*, 1957). Gersch and his co-workers determined that partially purified neurohormone D_1 was heat-stable and that it was soluble in both saline and absolute ethanol, thus exhibiting properties of both of the *Bombyx* factors. Unfortunately, even the isolation and characterization of a PTTH-like moiety from the brain of an insect of a different order could not provide the critical information about its chemical structure, which might have resolved the conflicting evidence obtained on the chemical nature of the *Bombyx* PTTH.

At this point, for reasons less empirical than theoretical, the hypothesis that PTTH was a peptide began to gain acceptance, and a study by Schneiderman, H. and Gilbert, L. (1964) finally closed the door on the "sterol" brain hormone hypothesis. In this study the authors attempted to verify the studies of Kobayashi and his colleagues (Kobayashi, M. and Kirimura, J., 1958; Kobayashi, M. *et al.*, 1962) by assaying the brain hormone activity of cholesterol and several other sterols with the *Samia* pupal bioassay. Their own words provide the best summary of their results:

> "although it has been possible to confirm the effectiveness of cholesterol and other sterols as mimics of the brain hormone, the results ... are difficult to understand, for there appears to be no clear correlation between sterol concentration and brain hormone activity. The data suggest that some factor other than chemical structure may be responsible for the brain hormone activity of these alcoholic preparations. Perhaps surface activity or some other properties of the sterol solutions and suspensions are involved."

An adequate explanation of the data suggesting that cholesterol is PTTH has yet to be made. The simplest explanation is that proposed by Schneiderman and Gilbert, but equally likely is the possibility that the conditions under which the bioassay was used gave rise to non-specific responses. A somewhat less likely explanation for the cholesterol–PTTH hypothesis is that the cholesterol fraction was contaminated with trace amounts of the actual neurohormone. Regardless of the reasons for the results, the possibility that cholesterol was the brain hormone gradually came to be disregarded as a viable alternative to the hypothesis that it was a protein.

The gradual establishment of PTTH as a protein represented a significant advance in the field, but the subsequent chemical definition of this peptide neurohormone(s) has proven to be as difficult a problem as the identification of its chemical nature.

4.2 Purification of the *Bombyx* PTTH

Following the establishment of PTTH as a protein, a concerted effort was made by several Japanese laboratories to purify the neurohormone. However, these new attempts at purification proceeded along the same lines as the previous studies, yielding disparate results from the different groups and even a degree of controversy over the chromatographic and physical properties of the neurohormone, as evidenced by a series of three papers during the late 1960s (Kobayashi, M. and Yamazaki, M., 1966; Ishizaki, M. and Ichikawa, M., 1967; Yamazaki, M. and Kobayashi, M., 1969).

The initial report by Kobayashi, M. and Yamazaki, M. (1966) detailed a relatively mild extraction protocol for PTTH, which involved acetone drying of *Bombyx* pupal brains, followed by extraction with phosphate buffer. The resulting extract containing PTTH activity was then subjected to heat treatment (100° for 20 min) at different pHs, ammonium sulfate precipitation and enzyme hydrolysis. The activity was temperature-stable irrespective of pH, precipitable with ammonium sulfate (45–55% saturated), and pronase– and Nagase–sensitive, but resistant to trypsin and chymotrysin digestion. The stability of the neurohormone with regard to temperature and proteases thus confirmed the results of Ichikawa, M. and Ishizaki, M. (1963), except that pH did not appear to affect the thermal stability of the hormone. Gel filtration indicated that the active moiety had an apparent molecular weight (M_r) of 20 kd. It is significant that the use of this moderately conservative extraction protocol finally enabled Kobayashi and Yamazaki to determine an M_r for the peptide PTTH.

Most of the PTTH activity extracted by this protocol did not bind to DEAE cellulose but bound effectively to CM cellulose. Most significantly, the active fraction eluted from the CM cellulose ion exchange column exhibited a dose-dependent response in the *Bombyx* assay for PTTH, a result never before demonstrated. In addition, elution of the PTTH activity from a DEAE cellulose column revealed that at least some activity stayed bound to the resin, and thus suggested the possible existence of more than one peptide having PTTH activity. The existence of a cholesterol PTTH was passingly addressed in this study, but there were no further attempts by Kobayashi and Yamazaki to promote this sterol as PTTH. At this point it seemed clear they were convinced that the brain hormone was a peptide.

Ishizaki, H. and Ichikawa, M. (1967) also used a moderately conservative extraction protocol, similar to that developed by Kobayashi, M. and Yamazaki, M. (1966), to partially purify PTTH from 100,000 pupal–pharate adult *Bombyx* brains. Nevertheless, they again obtained results which apparently contradicted those of Kobayashi, M. and Yamazaki, M. (1966). The PTTH active extract obtained by their procedure was heat-stable, as was that of Kobayashi, M. and Yamazaki, M. (1966), and the activity was also precipitable by ammonium sulfate (30–70% saturation). However, when the dialyzed extract was subjected to gel filtration, bioassay of eluted fractions revealed the presence of two, and possibly three, peaks of PTTH activity having M_r of 31 kd, 12 kd and 9 kd. Furthermore, the PTTH activity in the extract appeared to bind to DEAE cellulose but did not bind to CM cellulose, in marked contrast to the ion exchange binding properties of the PTTH activity in the previous study. Since the extraction and chromatographic conditions employed by these two laboratories were similar and the extract exhibited similar physical properties (e.g. temperature stability, precipitation by ammonium sulfate, and susceptibility to proteolytic digestion), it is possible the different results were caused by sub-optimal column conditions or by the fact that the bioassays used to follow activity were different in the two studies. The differing specificities of the two bioassays, or perhaps a selective cross-reactivity of the PTTH of one species in a bioassay using another species, might have been a causal factor. Lastly, since the sources of PTTH for the two studies were different, i.e. early pupal *vs.* pharate adult brains, the results may have simply reflected the presence of different moieties with PTTH activity in *Bombyx mori* brains at these two stages. Despite their differences, both of these studies indicated that PTTH was a fairly large molecular weight peptide. In the case of the study by Ishizaki, H. and Ichikawa, M. (1967), the results suggested that the hormone might exist as several peptides, and the authors in fact suggested that the molecular weight heterogeneity of the PTTH activity could be explained if the neurohormone existed in multiple molecular forms. This was an intriguing and insightful interpretation of the data, but unfortunately, it was pursued no further.

In 1969, Yamazaki, M. and Kobayashi, M.

reported the results of further purification studies of the 20 kd PTTH. The hormone appeared to be a basic protein (pI • 8.5) and on the basis of preliminary analyses for sugar content, they suggested that it might be a glycopeptide. To obtain the 20 kd moiety for this study, they extracted 100,000 pupal brains according to their previously established extraction protocol. Gel filtration of this extract yielded two proteins, and 0.02 μg of the principal 20 kd moiety elicited a positive response in the *Bombyx* bioassay. In addition, this fraction was found to be heat-stable, irrespective of pH, and susceptible to proteolytic hydrolysis by pronase and Nagase, as had been the PTTH activity isolated from pharate adult brains by Ichikawa, M. and Ishizaki, H. (1963). The PTTH activity could also be hydrolyzed by trypsin, which had not been effective in their 1966 study. Since the methods used by these two laboratories to purify hormone were similar, the physical dissimilarities of the moieties isolated by the two groups suggest that these substances were indeed different. The possibility remains, however, that each laboratory may have selectively purified one of two or more molecules while discarding the others. Since, as reported by Kobayashi, M. and Yamazaki, M., in 1966, approximately 20% of the PTTH activity was absorbed by DEAE cellulose but not by CM cellulose, and since two proteins were clearly resolved from gel filtration of the extract, the second of these moieties may have been the material isolated by Ishizaki, H. and Ichikawa, M. (1967).

While the Japanese laboratories were pursuing the elusive *Bombyx* PTTH, Williams, C. (1968a) used a similar extraction protocol to isolate PTTH from pupal brains of *Antheraea.* Employing the dauer pupal assay, he was able to demonstrate that PTTH activity was present in brain extracts and that this activity had physical properties similar to those of the active moieties in *Bombyx* brains. The activity was also hydrolyzable by pronase and Nagase. Based on these preliminary data, Williams concluded that the *Antheraea* PTTH was a mucopolysaccharide. The rationale for this conclusion, which contrasts markedly with those of the other studies, evolved from a unique interpretation of the data, which were obtained with experimental protocols essentially the same as those of the Japanese workers. The possibility that *Antheraea* PTTH is a mucopolysaccharide was never pursued further by Williams, nor anyone else, and was never seriously considered to be a valid alternative chemical basis for a neurohormone.

In toto, and despite continued controversy over the exact physical properties of PTTH, the results of all these studies generated a general consensus that PTTH was a peptide. With this tenet it would have been a simple matter to define and standardize methods for the purification of what appeared to be several molecular forms of PTTH in *Bombyx* pupal brains, and then to assess their potential endocrinological roles in the regulation of post-embryonic development. However, the investigative directions of any scientific field often take inexplicable turns, and the purification of PTTH at this time took another course, involving a change in the tissue source of the hormone and the adoption of different and rather harsh methods for its purification.

In a study of the titer of PTTH in developing adult brains of *Bombyx*, Ishizaki, H. (1969) observed that the head of an adult *Bombyx* moth contained approximately 50 *Samia* units of activity, in contrast to a pupal brain content of about 10–15 units. Since these results suggested a better source for the large quantities of PTTH needed for its purification, subsequent studies by the Japanese laboratories on the purification of the *Bombyx* PTTH utilized adult heads, without any attempt to verify that the activity in the adult head was the same as that in pupal brains.

The endocrinological significance of such a large amount of PTTH activity in adult heads is perplexing in view of the classical scheme for the endocrine control of insect post-embryonic development, according to which PTTH activates the PG. In the adult, PTTH is no longer needed to activate the PG, since the PG have degenerated. Thus, it was suggested that PTTH may have an additional neuroendocrinological role in the adult, possibly in regulating reproduction (Ishizaki, H., 1969).

In the first study of the chemistry of PTTH from the heads of adult *Bombyx*, Nishiitsutsuji-Uwo, J. (1972) extracted approximately 250,000 adult heads with a considerably different and more lengthy extraction protocol than previously used and followed activity using both the *Samia* and *Bombyx* bioassays. The extraction protocol followed that reported by Ishizaki, H. and Ichikawa, M. (1967) through the heat treatment step. At that point a series of solvent partitions with phenol, ether, chloroform and methanol were employed to remove lipids, nucleic acids, sugars, etc. Picric acid and rivanol were then used at the end of the protocol for additional clean-up and removal of "unnecessary components". This extraction protocol was thus considerably harsher than previous ones. Since the PTTH activity obtained from adult heads by this protocol had an M_r of only 5 kd, it seems likely that some of the "unnecessary components" removed from the extract were the large molecular weight moieties with PTTH activity that had been isolated previously (Kobayashi, M. and Yamazaki, M., 1966; Ishizaki, H. and Ichikawa, M., 1967). The 5 kd material was sensitive to pronase and Nagase and, under certain conditions, to trypsin as well. However, these data on protease sensitivity, as well as those on ion exchange resin (DEAE and CM cellulose) binding, varied considerably, depending on which bioassay was used to assess hormone activity. Thus they precluded both a clear interpretation of the results, and any meaningful comparison with data obtained by less harsh extraction methods on the properties of the PTTH activity obtained from pupal brains of *Bombyx*.

As with the observed differences in the chemical properties of the pupal PTTHs, the difference between the pupal and adult hormones could be explained simply on the basis of the different extraction procedures and/or the different sources of the hormone. Unfortunately, neither of these possibilities were explored. A few years later Ishizaki's laboratory (Suzuki, A. *et al.*, 1975) attempted the partial purification of PTTH from adult heads of *Bombyx* using a purification protocol similar to that of Nishiitsutsuji-Uwo, J. (1972), rather than the protocol previously developed by that laboratory for the purification of pupal PTTH. Gel filtration chromatography revealed that the M_r of the adult head hormonal activity extracted by this procedure was ≥ 5 kd. Anion (DEAE) and cation (SP) exchange chromatography identified at least two moieties with PTTH-like activity. Although these results were similar to the earlier findings with pupal brains, the molecular weights of these moieties were much smaller. An interesting and important piece of information can thus be derived from this study: as with the pupal brain extracts, several zones of PTTH activity are resolved by ion exchange chromatography. Regrettably, the authors failed to recognize the potential significance of this result and chose one activity as being PTTH.

The stability of the crude PTTH brain extract was analyzed prior to column chromatography with specific proteases, glycosidases and reducing and oxidizing agents (Ishizaki, H. *et al.*, 1977), and the results revealed: (1) the moiety with PTTH activity probably did not contain sugar; (2) it was a peptide, as evidenced by its sensitivity to trypsin and chymotrypsin; (3) the N and C termini of this peptide were blocked; (4) it possessed at least one disulfide bond which was necessary for biological activity; and (5) it did not contain a sulfhydryl group, at least not in the region of the molecule necessary for biological activity.

In 1979, Nagasawa, H. *et al.*, using anion and cation exchange chromatographic methods substantially different from those previously reported (Ishizaki, H. and Ichikawa, M., 1967), achieved approximately a 2.8×10^4-fold purification of crude PTTH; this highly purified moiety had a M_r of 4.4 kd and was active in the *Samia* bioassay at an effective dose of 7 ng (Fig. 7). The chromatographic methods employed in this purification utilized different buffer systems and column conditions, as well as a step gradient protocol. It is apparently for this reason that only one active fraction was obtained, in contrast to the several active fractions routinely obtained by the different ion exchange chromatographic methods previously employed. This single active fraction had a M_r of 4.4 kd and exhibited enzymatic and chemical stabilities similar to the previously isolated small molecule (Suzuki, A. *et al.*, 1975), and thus probably represents the same moiety. The results of the studies from the 1960s suggesting the existence of several large PTTHs were not considered by the authors of this work, who concluded that "... the main principle of PTTH seemed to have a smaller molecular weight than that previously reported". The same laboratory has since purified this PTTH on a large scale (Nagasawa, H. *et al.*, 1980) using the same protocol (Nagasawa, H. *et al.*, 1979) and have confirmed both the molecular weight of 4.4 kd and the bioassay effective dose of 7 ng.

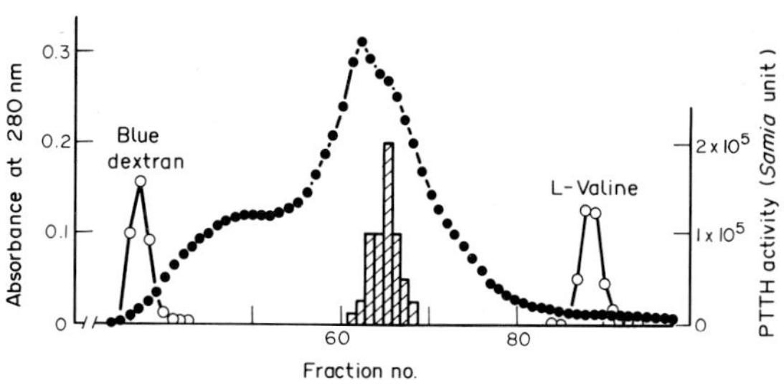

FIG. 7. Gel-filtration on Sephadex G-50 (superfine) of the SP-Sephadex fraction of PTTH activity from adult *Bombyx* heads. 0.2 M ammonium acetate was used as solvent. (From Nagasawa, H. *et al.*, 1979). ●, Absorbance of sample at 280 nm; ○, standards for ascertaining exclusion and inclusion columns; ▨, PTTH activity.

Recently, Suzuki, A. *et al.* (1982) conducted N-terminal end group analysis and determined the amino acid composition of the adult *Bombyx* small PTTH purified by HPLC. DEAE-ion exchange chromatography was used in the purification procedure, similar to the approach taken many years ago, and not surprisingly, two active fractions were obtained. One of these two fractions was selected for HPLC and chemical analysis. Since the second fraction with PTTH activity was not analyzed, it is not known whether the activity represents the existence of a completely different molecule. The activity chosen for analysis has an effective dose of 0.1 ng in the bioassay, and is thus apparently 70 times more potent that that purified by Nagasawa, H. *et al.* (1980). Enzymatic analysis of this moiety with N-terminal amino peptidases revealed that the N-terminal was not blocked as previously reported (Ishizaki, H. *et al.*, 1977); end group analysis revealed that glycine was the N-terminal amino acid. Since this was the only amino acid detected, the moiety was apparently pure. The amino acid composition of this peptide revealed integral ratios of amino acid residues and permitted a calculation of an M_r of 4.33–4.74 kd, consistent with the M_r of 4.4 kd determined by gel filtration.

This, then, is the present status of the research on adult *Bombyx* PTTH. Obviously, there remain a number of questions to be answered. First, it must be determined whether the other fractions having PTTH activity are different forms of the neurohormone. Second, it must be demonstrated that the purified activity is indeed a PTTH. This requires its isolation from the hemolymph of the insect at a time when release of the neurohormone is known to occur *in situ*, i.e. at a head critical period. Third, it should be established that the PTTH activity isolated from adult heads of *Bombyx* is the same as that isolated from *Bombyx* pupal brains. In fact, on the basis of the studies thus far conducted, the moieties isolated from these two sources appear to be different. Furthermore, there is no apparent endocrinological need for PTTH during adult life of *Bombyx*, and as mentioned earlier, the 4.4 kd moiety from adult heads may be a reproductive neuropeptide comparable to the mosquito EDNH (Fuchs, M. and Kang, S.-H., 1981; Gilbert, L. *et al.*, 1981). Its activity in the bioassay may simply mean that it is sufficiently similar to pupal brain PTTH to give a positive response. Finally, a more direct and specific assay is needed to monitor the 4.4 kd *Bombyx* PTTH activity. One similar to the *in vitro* PG assay used to study the *Manduca sexta* PTTH (Bollenbacher, W. *et al.*, 1979) apparently is now being developed (Suzuki, A. *et al.*, 1982).

A recent observation which sheds some light on the question of whether the 4.4 kd moiety from adult *Bombyx* is really a PTTH is that this material is inactive in both the *Manduca* larval and pupal bioassays for PTTH at doses up to 100 *Samia* units (Safranek, L. and Williams, C., 1980). Thus, either the PTTHs of these two species of Lepidoptera are dramatically different or the PTTH-like moiety isolated from *Bombyx* heads is not PTTH. The former explanation does not appear to be the case, since the PTTH activity present in day 1 pupal brains of *Bombyx* can quite effectively activate *Manduca* PG to synthesize ecdysone *in vitro*; conversely, *Manduca* pupal brain PTTH can activate *Bombyx* PG *in vitro* (Agui, N. *et al.*, 1983). The degree of interspecific cross-reactivity of these PTTHs therefore suggests, albeit indirectly, that the *Bombyx* adult head factor may not be PTTH. Research initiated a few years ago on the chemistry of the PTTH present in the pupal brain of *Manduca* supports this possibility. The *Manduca* PTTH is a very different peptide from that isolated from *Bombyx* adult heads, and in fact, appears to be similar to the putative larger molecular weight PTTHs isolated initially from *Bombyx* pupal brains.

4.3 Purification of the *Manduca* PTTH

Besides *Bombyx* and *Periplaneta*, the only other insect in which PTTH has been investigated to any degree is *Manduca sexta*. The first study of *Manduca* PTTH involved the development of the *in situ* larval bioassay for the hormone (Gibbs, D. and Riddiford, L., 1977), which permitted an assessment of the chemical nature and temperature stability of the *Manduca* PTTH. The activity detected by the larval assay in crude extracts of day 1 pupal brains was heat-stable (100° for 10 min) and was a peptide, since activity was lost following hydrolysis with pronase and trypsin.

Two groups have subsequently pursued the purification of *Manduca* PTTH, one investigating the chemistry of the hormone present in heads of *Manduca* fifth-instar larvae and following activity with the larval bioassay (Kingan, T., 1981), the other utilizing the *in vitro Manduca* PG assay for PTTH to investigate the PTTH(s) present in day 1 pupal brains of this insect (Bollenbacher, W. and Gilbert, L., 1981; Gilbert, L. *et al.*, 1981; Bollenbacher, W. and Bowen, M., 1983). The data resulting from these two studies have revitalized the idea of a large PTTH, and the latter study has again raised the possibility of several structural forms for PTTH.

Both of these studies reported the presence of a large molecular weight PTTH moiety in *Manduca*, similar in size to the PTTHs initially described in pupal brains of *Bombyx*. In addition, Bollenbacher, W. and Gilbert, L. (1981) reported the existence of a second factor with PTTH activity but of a lower molecular weight, slightly larger than the *Bombyx* adult head PTTH.

The PTTH activity in larval heads of *Manduca* (Kingan, T., 1981) was extracted by a protocol similar to that recently used for *Bombyx* heads (Suzuki, A. *et al.*, 1982), involving an acetone wash of the heads, extraction with 2% NaCl, heat treatment (83° for 5 min), ammonium sulfate precipitation, and acetone precipitation. This activity was purified by gel filtration, DEAE ion exchange chromatography, hydoxyapatite chromatography and by gel electrophoresis. A single electrophoretic band was obtained by gel electrophoresis with an M_r of 25 kd. This isolation procedure resulted in a 4400-fold purification of this moiety. If this peptide is indeed PTTH, and if it has been purified to homogeneity, then its concentration in the *Manduca* larval head is considerably greater than that of the moiety with PTTH activity in adult *Bombyx* heads (Suzuki, A. *et al.*, 1982). This apparently substantial difference between species in relative concentration (molarity) of the neurohormone may be related to: (1) the concentration necessary for activation of the PG in each species; (2) the nature of the purification protocol; (3) the stage of development from which the hormone was obtained; or (4) the possibility that one of these moieties is not a PTTH.

In their purification studies of the PTTH in the day 1 *Manduca* pupal brain, Bollenbacher and his colleagues (Bollenbacher, W. and Gilbert, L., 1981; Bollenbacher, W. and Bowen, M., 1983) found a large molecular weight peptide with PTTH activity similar to that reported by Kingan, T. (1981) but with some differences in actual molecular weight and chromatographic properties. The most significant difference between these two studies, however, was the finding by Bollenbacher and his colleagues that *Manduca* pupal brains possessed a second peptide with PTTH activity which had a much smaller molecular weight. For the purpose of subsequent discussion of the results of these studies, these two factors are considered to be neurohormones. However, their identification as neurohormones awaits chemical characterization, verification of activity in bioassays, and, perhaps most importantly, the demonstration of their release *in situ* at that time of development when PTTH activation of the PG is known to occur.

The purification procedure apparently yielding two PTTHs from the day 1 pupal brain was less complex than those used by other laboratories, in that it involved far fewer steps, and more importantly, was less harsh. The choice of a conservative method for extracting PTTH from pupal brains was critical in that it was felt this method would yield maximal PTTH activity from a tissue, would prevent or minimize the loss of potential multiple molecular forms of this neurohormone, and would interface with the *in vitro* assay for the hormone. Thus, samples could be assayed with minimal changes in buffer molarity, ionic strength, pH, etc. Of the various methods investigated for this study, which included those previously used by other laboratories, extraction in 25% Grace's medium or in distilled water at 4° was found to yield the best results. Sonication did not substantially change the efficiency of extraction. Following centrifugation of the extract, the supernatant was heated for 2 min at

100°, concentrated by ultrafiltration and lyophilization, and the PTTH content quantified by the *in vitro* assay using a dose-response protocol (Bollenbacher, W. *et al.*, 1979, 1983). This procedure yielded approximately a 10-fold purification with nearly quantitative recovery of activity. Gel filtration of the lyophilized material revealed two peaks of PTTH activity with mean M_rs of ~ 28 kd and ~ 7 kd (Fig. 8). These moieties were termed "big" and "small" PTTH, respectively and their M_r determinations were confirmed both by conventional size separation chromatography with several gel types and several different buffers at both low and high salt molarity and by HPLC with a series of Waters I protein columns (Bollenbacher, W. *et al.*, in preparation).

The potential significance of the gel filtration data was underscored by the subsequently observed differential stability of the two forms of PTTH in the presence of guanidine · HCl (Gn·HCl). The activity of big PTTH was found to be irreversibly denatured

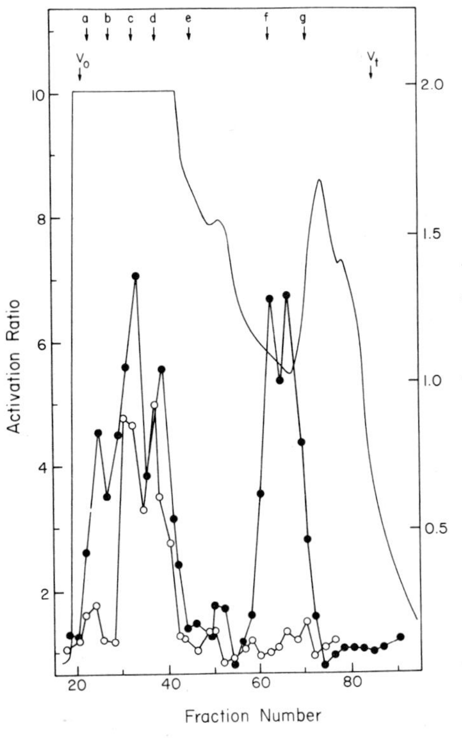

FIG. 8. Molecular weight determination by gel filtration (G-25) of the PTTH activity present in an extract of day 1 pupal brains of *Manduca sexta*. Fractions were screened for PTTH activity with day 0 pupal (○–○) and day 3 last larval instar (●–●) PG.

by Gn·HCl treatment, presumably due to the greater susceptibility of this larger peptide to degradation by harsh salt denaturants. However, small PTTH could be renatured after Gn·HCl treatment, and under these conditions it had a M_r of ~ 8 kd, a value consistent with the M_r derived by conventional gel filtration methods. Furthermore, the differential stabilities of the two putative PTTHs indicated that the big PTTH was neither an aggregate of the small form nor composed of subunits of the small form, and thus provided additional evidence for the existence of a big and small form of the hormone.

SDS gel electrophoresis of the big and small PTTH followed by the *in vitro* assay for activity yielded essentially the same results as Gn·HCl chromatography: The biological activity of the big PTTH was irreversibly lost upon SDS denaturation, while activity of the small form was recoverable. The M_r of the small form determined by this procedure was approximately 8 kd.

The big PTTH isolated by gel filtration was purified by DEAE ion exchange chromatography and subsequent isoelectric focusing (IEF). This procedure yielded two activities having pIs of 5.35 and 5.15. About 80% of the recovered extract was the 5.35 fraction. When the IEF purified material (both pIs) was subjected to SDS gel electrophoresis, both moieties were found to have an M_r of 28.5 kd, in close agreement to that determined by conventional gel filtration and HPLC. The two pIs may either reflect slight changes in the charge of the peptide during purification or may be an artifact of the IEF method itself. Once focused, big PTTH was found to be quite labile. This lability may be related to the reduction of disulfide bonds, although this has not yet been demonstrated. The difference between the M_r determined for this moiety and that of the PTTH isolated by Kingan may not be significant. This apparent difference could simply be a reflection of differences in technique. Further chemical analysis of the big PTTH activity will be required to resolve this issue.

Purification of the small PTTH identified in these studies (Bollenbacher, W. and Gilbert, L., 1981; Bollenbacher, W. *et al.*, unpublished) has not been completed, but preliminary analyses have revealed that it does not bind to ion exchangers as does big PTTH, despite the fact that it appears to have an acidic pI. In addition to differences in their physical properties, it was observed that the big and small factors affected different, developmentally specific responses from the PG (Bollenbacher, W. and Gilbert, L., 1981; Bollenbacher, W., unpublished). Each moiety was shown to activate the PG *in vitro* in a dose-dependent manner similar to crude PTTH, but the amount of each form of PTTH necessary to activate the PG varied, depending on the developmental stage of the glands. For example, the dose–responses of activation and ED_{50} obtained for big PTTH with day 3 larval and day 0 pupal PG were the same (ED_{50}s of 0.075 and 0.08, respectively), but the ED_{50}s of small PTTH obtained with pupal glands was approximately 20 times that found with larval glands (ED_{50}s of 0.16 and 3.8, respectively). Thus it appears that the small PTTH more effectively activates day 3 larval glands than pupal glands. The differential effect of small PTTH in activating larval and pupal PG was used by this laboratory during the purification of big and small PTTH to verify which of the two moieties was present.

Based upon these data, it appears that the large factor having PTTH activity is a neurohormone, and similar evidence is being accumulated to demonstrate that the small factor is also a neurohormone. The possibility that there may be two prothoracicotropins in *Manduca* is not that surprising, since if the early research by the Japanese and German laboratories on the chemistry of the *Bombyx* and *Periplaneta* PTTH activities is critically examined, there are tantalizing indications that there may exist more than one moiety with PTTH activity. Perhaps the reason that the first studies on the chemistry of PTTH overlooked the possibility of multiple molecular forms of the neurohormone was the apparent need for a simple endocrine model to describe the control of molting and metamorphosis. In the case of PTTH, this model would involve one neuropeptide produced by neurosecretory cells (NSC) and having a single function. Unfortunately, it appears that this is probably not the case, and although the more complex model certainly makes the problem of understanding the function of PTTH more difficult, it also makes the system much more interesting to study.

Confirmation of the two-form–two-function hypothesis of Bollenbacher, W. and Gilbert, L. (1981) and Bollenbacher, W. *et al.* (1983) is currently being attempted by a demonstration of the presence of these molecules in the hemolymph of *Manduca* at specific head critical periods (HCP) and at concentrations sufficient to activate the PG. The PTTH released into the hemolymph during the fourth larval instar HCP and first HCP of the fifth larval instar has been isolated and identified and characterized by a number of the same criteria used to characterize the day 1 pupal brain PTTHs. The hormone released into the hemolymph during the fourth larval instar HCP is released in a single burst over several hours, eliciting a large increase in the ecdysteroid titer and a subsequent larval molt. This hormone has been identified as big PTTH, as judged by its M_r, pI, and biological activity. By the same criteria, the PTTH activity found in the hemolymph during the first HCP of the fifth larval instar has provisionally been identified as small PTTH. This moiety is released in a pulsatile manner (three bursts) over approximately 18 h which results in pupal commitment. It appears to have a M_r of ~ 9 kd and a pI of 5.5. Substantial additional work is necessary before these data can support the two-form–two-function hypothesis.

If, with contemporary methods, it is ultimately shown unequivocally that there is more than one PTTH, the PTTH axis would be a unique system in that, at least in *Manduca*, several neurohormones would be released in different modes, and at selective times during development, to evoke a common response from the same target gland and possibly unique responses from other tissues. It cannot be assumed, however, that the mechanisms by which these different moieties activate the glands are the same, since they evoke different types of developmental responses, i.e. larval molting in the fourth instar, and pupal commitment for the first HCP of the fifth larval instar. Thus, the PTTH system would be the first for which discrete physiological functions have been demonstrated for multiple molecular forms of a neurohormone.

From this discussion of what is known of the chemistry of PTTH, it should be evident that considerably more knowledge is needed of this system, especially in insects representing several different orders, before a consensus can be reached on the chemical nature of the hormone. This is particularly true in regard to the question of the conservation of its structure among insects. With the use of the new methods for quantifying PTTH directly, and with the development and application of contemporary methods analogous to those used to assess the chemistry and function of vertebrate neurohormones, substantial progress should be made in this field over the next few years.

5 SITE OF PTTH SYNTHESIS AND RELEASE

5.1 The prothoracicotrope

Since V. Wigglesworth's (1940) pioneering study localizing the source of PTTH in the brain of *Rhodnius prolixus*, numerous similar attempts have been made to identify the PTTH NSC in the insect cerebrum (Wigglesworth, V., this volume; Orchard, I. and Loughton, B., this volume). Most of these studies have taken an indirect approach employing strategies that fall into three categories: (1) assessment by *in situ* bioassay of PTTH activity in implanted brains or brain fragments containing NSC; (2) correlation of morphological and histological changes in the cerebral NSC with PTTH-mediated developmental events; and (3) assessment of the effect of microcauterization of cerebral NSC on developmental events mediated by PTTH. More recently, however, additional and considerably different approaches have been taken entailing either electrophysiological stimulation of NSC or the direct measurement of PTTH activity by an *in vitro* assay, which demonstrates the presence of the hormone in individual NSC both directly and quantitatively (Steel, C. and Davey, K., vol. 8). Thus, a discussion of this topic will be structured around the approaches taken and methods employed over the years to identify the prothoracicotropes.

In overview, the results of numerous studies utilizing indirect approaches to identify the PTTH NSC in a wide variety of insects have implicated various groups of cerebral NSC as the source of PTTH. Since these studies took an indirect approach, very few have yielded precise results, and thus it is not possible to conclude whether the disparate findings are a function of the different approaches taken, or whether the prothoracicotropes are actually located in different areas of the brain in different insects.

The first approach employed to identify the site of PTTH synthesis utilized *in vivo* PTTH bioassays to detect the presence of neurohormone in a brain, a fragment of a brain or a group of NSC. This was accomplished either by implantation of the tissue or by injection of extracts of the tissue into a suitable test animal. Good examples of this type of study are those on *Rhodnius* (Wigglesworth, V., 1940b), *Hyalophora cecropia* (Williams, C., 1948b), *Samia* (Ichikawa, M. and Yashika, K., 1958) and *Manduca sexta* (Gibbs, D. and Riddiford, L., 1977).

Based on the observation that decapitated fourth-stage nymphs molted to the fifth instar in response to implantations of a small dorsal fragment of the brain which included the region between the lobes containing the medial NSC (Hanström, B., 1938), Wigglesworth, V. (1940b) concluded that the medial area of the brain contained the PTTH NSC. The fact that other areas of the brain failed to elicit a molt supported his conclusion. In *H. cecropia*, implantation of fragments of pupal brains into debrained, diapausing pupae revealed that both the lateral and the medial areas of the brain, both containing NSC, had PTTH activity. This finding led Williams, C. (1948b) to speculate that the PTTH NSC were located in both regions of the brain. By contrast, only the medial NSC of the pars intercerebralis of *Samia* elicited adult metamorphosis in the *Samia* assay (Ichikawa, M. and Yashika, K., 1958). Using the *Manduca* bioassay for PTTH, Gibbs, D. and Riddiford, L. (1977) localized PTTH activity in the *Manduca* pupal brain to the lateral region, which contains two groups of lateral NSC (Nijhout, H., 1975a). Thus by *in situ* bioassay, it appears that the prothoracicotropes are located in different areas of the brain in different insects. This is not a surprising discovery for insects of different orders, but for different species of the same order and family such a substantial positional variation of these cells is somewhat unexpected. However, since *in situ* bioassays were used in these studies, it is possible that their inconsistent results are in part a function of the assays. Regardless of the basis of these differences, these studies illustrate that localization of the prothoracicotropes by such methods only provides a rough approximation of a site of PTTH synthesis.

The method most frequently used to identify the PTTH NSC, and certainly the most equivocal, has been the correlation of histological changes in NSC appearance with PTTH-mediated developmental events. These studies have usually entailed an assessment in different NSC of the relative abundance and turnover of stainable neurosecretory material. Of these studies, those on *Mamestra*, *Pieris rapae* and *Rhodnius* best demonstrate the use of this approach and the extent to which the data can be interpreted.

In an initial histological study of the cerebral NSC in diapausing and non-diapausing *Mamestra*, Hiruma, K. and Agui, N. (1977) observed that the type II NSC of the medial NSC cluster in the diapausing pupal brain possessed a greater amount of paraldehyde–fuschin stainable material than did these same NSC in non-diapausing pupae. The accumulation of neurosecretory material in a brain programmed not to release its PTTH led them to speculate that these cells might be the prothoracicotropes. Subsequent experiments utilizing ablation and implantation to assess the function of the different NSC groups revealed that both the medial and lateral NSC possessed PTTH activity, the medial NSC having the greater amount of activity. Since the quantity of PTTH activity was higher in the type II medial NSC, these cells were further implicated as the source of the hormone. Agui, N. and Hiruma, K. (1977a,b) then used an *in vitro* approach to investigate the effect of 20-hydroxyecdysone on the accumulation of stainable neurosecretory material by the NSC in the diapausing *Mamestra* pupal brain. *In vitro* culture itself appeared to enhance the intensity of the PAF staining of the cerebral NSC, but when 20-hydroxyecdysone was added to the cultures a further intensification of PAF staining occurred specifically in the type II medial NSC. The increase in stainable material in these NSC implied a possible feedback regulation of the synthesis of PTTH by 20-hydroxyecdysone, the hormone whose synthesis PTTH regulates.

To demonstrate more directly that the type II cells were the PTTH NSC, Hiruma, K. and Agui, N. (1977) used the *in vitro Chilo* integument assay for PTTH (Agui, N., 1975) to show that an increase in PTTH activity occurred in brains cultured with 20-hydroxyecdysone. They found that when brains were co-cultured with 20-hydroxyecdysone, their PTTH activity increased 4–8-fold, which roughly correlated with the accumulation of PAF-positive material in the type II medial NSC. Thus, they concluded that synthesis of PTTH in *Mamestra* was regulated by 20-hydroxyecdysone and, although it was not clearly stated, implicated the type II medial NSC as the prothoracicotropes.

A detailed, strictly histological study on the ultrastructure of the cerebral NSC of *Pieris rapae crucivora* during larval–pupal development was conducted by Kono, Y. (1973). He found that the type II medial NSC in diapause and non-diapause-destined pharate pupae, which are possibly equivalent to the cells described by Agui, exhibited profound ultrastructural differences in the location and number of secretory granules. Daily ultrastructural observations of the type II medial NSC during the fifth (last) instar (Kono, Y., 1975) revealed specific responses to photoperiod in the secretory cycle of these cells, as evidenced by changing profiles of mitochondria, abundance of Golgi, the arrangement of rough endoplasmic reticulum and the numbers of secretory granules. On the basis of this study and his earlier work, Kono considered these NSC to be the source of PTTH. However, since the type IV cells in the lateral region of the brain also changed morphologically in response to photoperiod, there is some question about the interpretation that the medial cells are the prothoracicotropes.

The morphological studies that have thus far most convincingly shown a correlation between NSC structure and PTTH activity are those on *Rhodnius prolixus* by Steel and his colleagues (Steel, C. and Harmsen, R., 1971; Steel, C., 1973; Morris, G. and Steel, C., 1977a,b). The advantage of using *Rhodnius* for an identification of the prothoracicotrope by light and electron microscopic means is that substantial information already exists on the PTTH neuroendocrine system in this insect, thus enabling the interpretation of morphological data in a physiological context. The first study of the cerebral NSC of *Rhodnius* by Steel, C. and Harmsen, R. (1971) was concerned with the histological appearance of these cells during a molting cycle. Using a variety of staining techniques, they observed a developmental sequence of morphological changes which occurred in all of the medial NSC and which correlated with the physiological events associated with a bloodmeal. These changes in staining patterns were expressed at an ultrastructural level as well (Morris, G. and Steel, C.,

1977a,b), the ultrastructural changes in the cells suggesting the occurrence of rapid increases in transport, release and synthesis of material in response to stimulation by stretch receptors.

When Steel, C. (1973, 1975) probed potential regulation of the medial NSC by ecdysteroids during the molting cycle, he observed a specific accumulation of stainable material in the type II medial cells. This occurred either at the time of rising ecdysteroid titers *in situ* or in response to the injection of 20-hydroxyecdysone. On the basis of these findings he suggested that the type II medial NSC might be producing PTTH. Although the cumulative evidence was derived by indirect means, the conclusion drawn was strengthened by the earlier studies of Wigglesworth, V. (1940b). Despite the thoroughness and attention to detail of these descriptive studies of NSC morphology, histology and ultrastructure, the data generated are not sufficient proof of the function of a NSC (Highnam, K., 1965; Bern, H., 1966; Berlind, A., 1977). There are a number of reasons why this is so, of which the most significant are: (1) the conventional stains used to identify neurosecretory material may not in fact be reacting with the active factor to which the stain intensity is being correlated; (2) intense staining does not always correlate either with accumulated neurosecretory granules (Girardie, J., 1973) or with an extractable and quantifiable active factor (Highnam, K. and Goldsworthy, G., 1972; Meola, R. and Lea, A., 1971); (3) correlations of temporal changes in NSC cytology with physiological events are not adequate for an unambiguous assessment of secretory activity (see Highnam, K., 1965), e.g. the accumulation of material in the perikaryon of a cell may represent increased synthesis, decreased degradation, decreased transport, cessation of release or any combination thereof; and most importantly (4) the cell(s) in question has not been shown to possess neurohormonal activity. Similar criticisms apply to studies at the ultrastructural level.

The third *in situ* approach employed to localize the prothoracicotropes utilized microcautery, with which discrete regions of the brain containing certain groups of NSC are ablated, and the effect of this injury on the development of the animal then assessed. The earliest notable use of cauterization to identify the prothoracicotropes was by Van der Kloot, W. (1961), who employed microcautery to destroy specific clusters of NSC in the *in situ* brain of *Hyalophora cecropia* pupae and then observed the effect of the cautery on the adult development of the pupae. It was found that at least one lateral and one medial group of intact NSC were needed for adult development to occur, a result consistent with C. Williams' (1948b) earlier demonstration with transplantation studies that both the medial and lateral NSC served as sources of PTTH in *H. cecropia*.

A few years later Girardie, A. (1967) used this technique to demonstrate that destruction of the pars intercerebralis of the *Locusta* brain at the end of the instar prevented ecdysis but did not affect apolysis or formation of a new cuticle. Thus it appeared that in *Locusta*, the NSC of the pars intercerebralis were the source of PTTH, but not of eclosion hormone nor bursicon. However, the lateral NSC were later found to have a prothoracicotropic effect as well, although much weaker (Girardie, A., 1974). Using this same method, Charlet, M. and Schaller, F. (1976) observed that two pairs of anterior NSC, located just outside of the pars intercerebralis, control ecdysis in *Aeshna cyanea*.

Although these studies yielded results indicating that development could be arrested by ablation of regions of the brain containing groups of particular NSC, they were not sufficiently precise to define the location of the PTTH NSC. Even the most precise lesions generated by microcautery are somewhat diffuse, involving a number of NSC, axon tracts, arborizations, glial cells, and ordinary neurons in the cauterized area. Thus, the effect of the cautery could be a result of destruction of the prothoracicotropes, of axons of the PTTH NSC as they traverse the ablated area, or of other neural elements. Possibly those studies indicating the involvement of both medial and lateral areas of the brain as sources of PTTH are in fact demonstrating the existence of these various alternatives, only one area being the site of the actual prothoracicotropes.

A more refined method of electrocautery, which has potentially overcome most of the drawbacks of this technique, was used by Steel, C. (1978) to locate the PTTH NSC in the brain of the aphid *Megoura viciae*. The cuticle on the dorsum of the head of this insect is lightly pigmented, making recognition of the NSC in the brain possible without any manipulation (Fig. 9). Since there are only five widely separated NSC groups in the brain of *Megoura*, with only a few cells per group (Steel, C., 1977), and since the cuticle overlying the brain is sufficiently soft to permit the insertion of an electrode directly through it into a designated area of the brain, it was possible to cauterize selected NSC without significantly disturbing the physiology of the animal. By using reference maps of the head to make reproducible lesions, Steel was able to ablate areas as small as 30 μm in diameter with a maximum error

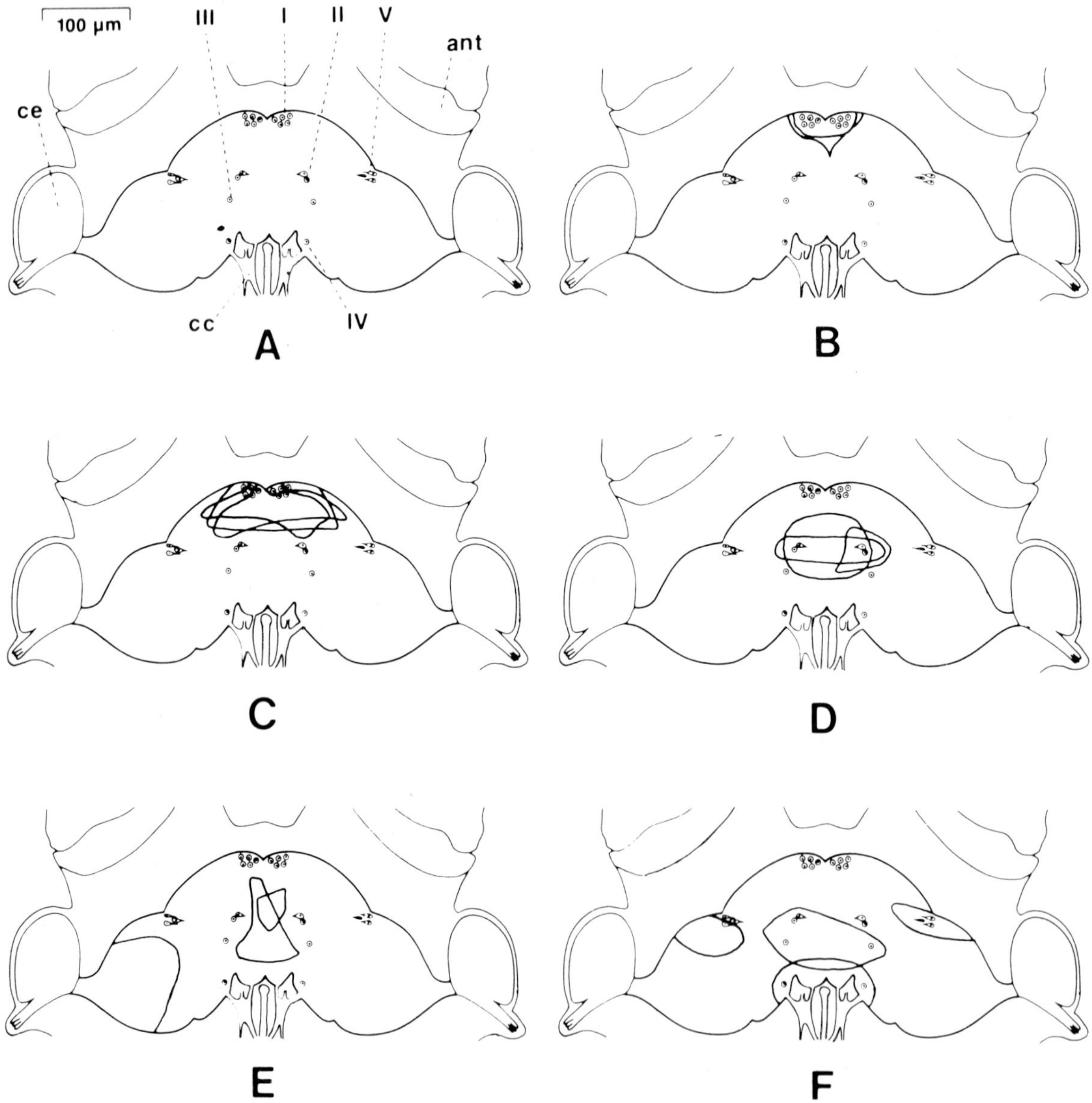

FIG. 9. Composite drawing showing the locations of various brain lesions, grouped according to their effects. **A**: Dorsal view of brain and adjacent structures to show the positions of NSC Groups I–V: ant, antenna; cc, corpus cardiacum; ce, compound eye; **B**: lesions which destroyed all group I NSC and only a few other neurons, produced by insertion of electrode through the cuticle between the antennae; **C**: larger lesions around Group I NSC which produced abnormal locomotor patterns — note that some Group I NSC remain intact in most cases; **D**: lesions which destroyed Group II NSC and various numbers of adjacent neurons, leading to interference with molting and tanning; **E**: lesions in similar locations to those in **D** and **F** but no NSC were destroyed and no abnormality was detected in the animal; **F**: lesions which destroyed NSC Groups III, IV, and V. (From Steel, C., 1978.)

in electrode placement of only about 20 μm. Lesions of the two cytologically distinct cells comprising the group II NSC, which are found in the mid-dorsal protocerebrum and which have axonal projections going to the corpora cardiaca, affected both the degree of cuticular tanning and the incidence of molting. Based on the assumption that a single NSC produces only one physiologically active principle, Steel concluded from these results that only one of the two group II NSC was the prothoracicotrope.

Although this approach was a significant improvement over previous attempts to localize the prothoracicotropes by cauterization, or by any other means, the fact that the presence of PTTH in the cell had not actually been demonstrated weakened the conclusions reached. Thus, the results of Steel's cautery study could also be interpreted as demonstrating the ablation of a cell that itself controlled the prothoracicotrope, possibly at the levels of synthesis, release, etc. Despite this possibility, the ability to demonstrate that the two group II NSC controlled molting in *Megoura* was a laudable accomplishment, and it underscores the potential usefulness of this method in future investigations of the control of PTTH release in general.

In recent years, refinement of neurophysiological techniques has produced another approach to identifying the location of the prothoracicotropes, in which groups of NSC are stimulated electrically (Girardie, J. *et al.*, 1976; Girardie, J. and Girardie, A., 1977), and the effects of this stimulation on development then assessed. The use of this technique and an RIA (De Reggi, M. *et al.*, 1975) to measure the ecdysone titers in the stimulated animals has permitted an evaluation of the function of medial and lateral NSC in the control of molting in *Locusta* (Girardie, J. and De Reggi, M., 1978). The measurement of ecdysone titers is essentially just one step upstream from scoring the overt biological response — a molt — to PTTH, and the results of this study corroborated those obtained previously with microcautery (Girardie, A., 1974). Electrical stimulation of the medial NSC of the *Locusta* brain on day 1 of the fifth instar resulted in an increased ecdysone titer on day 4, which was approximately four times larger, and 1 day earlier, than that occurring in normal or sham-stimulated animals. Subsequent peaks in the ecdysone titer of the stimulated animals also occurred earlier and were larger as well. A similar stimulation of the lateral NSC resulted in a temporal shift in the ecdysteroid titer like that observed with the medial NSC, but an increase in the size of the ecdysteroid peak did not occur. Thus control of PG in *Locusta* appears to occur via a dual mechanism involving both of these groups of NSC, similar to the type of control by NSC suggested to occur during *H. cecropia* development (Williams, C., 1948; Van der Kloot, W., 1961).

The only direct and unequivocal approach to the identification of the cells which synthesize PTTH is through the identification of their product, and the means by which to do this did not exist until the development of the *in vitro* PTTH assay (Bollenbacher, W. *et al.*, 1979). The sensitivity and quantitative nature of this assay permitted direct quantification of PTTH activity in groups of NSC and even individual NSC (Agui, N. *et al.*, 1979). A morphological basis for the identification study was provided by a histological characterization of the NSC of the *Manduca* brain (Nijhout, H., 1975a; Agui, N. *et al.*, 1978), which demonstrated the presence of medial and lateral clusters in each hemisphere (Fig. 10). The lateral clusters were found to contain two distinct groups, one composed of five relatively small NSC (L-NSC-II, ~ 15 μm) and the other of two larger cells (L-NSC-III, ~ 30 μm). To localize the prothoracicotropes, Agui, N. *et al.* (1979) quantified, by the *in vitro* assay, the PTTH activity in isolated regions of the brain containing either the medial or lateral NSC clusters. The lateral region of the brain containing the L-NSC possessed most of the PTTH activity found in the whole brain. Thus these results suggested that the lateral NSC were the source of PTTH. Low but significant levels of PTTH activity were also detected in the medial region of the brain, probably reflecting the presence of PTTH in the axon tracts of the lateral NSC as they traverse this region *en route* to the retrocerebral complex (Nijhout, H., 1975a), from which PTTH is released (Agui, N. *et al.*, 1980). Thus these findings confirmed the results of a previous localization study (Gibbs, D. and Ridiford, L., 1977), which used the *in situ Manduca* bioassay for PTTH to measure activity.

Microsurgical techniques were then used to separate for *in vitro* assay the two groups of NSC (L-NSC-II and L-NSC-III) in the lateral region of

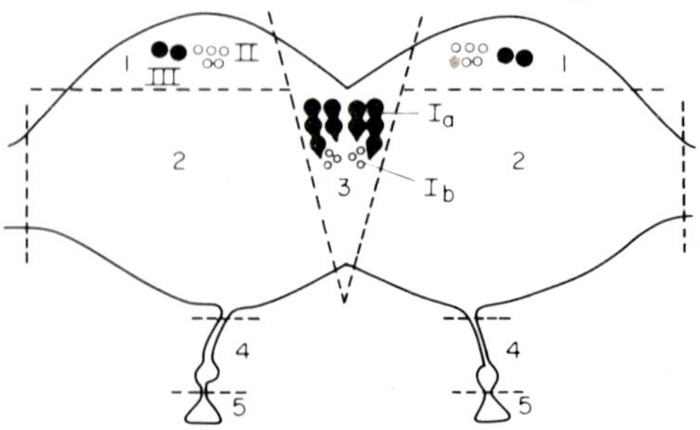

FIG. 10. Topographical diagram of the different groups of NSC in the protocerebrum of day 0 *Manduca sexta* pupae. The different groups of NSC are M-NSC-I_a (filled teardrops) and M-NSC-I_b (open circles), and L-NSC-II (open circles) and L-NSC-III (filled circles). For the distribution studies of PTTH activity, the pupal neuroendocrine system was divided into five operational regions: 1, 2, and 3 = the brain; 4 and 5 = the retrocerebral complex, CC and CA respectively. (From Agui, N. *et al.*, 1979.)

the brain. When their PTTH activity was quantified the L-NSC-III were found to contain all of the activity in this region of the brain. In an attempt to demonstrate that one or both of these cells were the prothoracicotopes, the PTTH content of the individual L-NSC-III was then quantified. PTTH was detected in only one of the two cells, at a level per cell comparable to that calculated on a per cell basis from a dose–response of activation by groups of L-NSC-III cells. Thus, this study revealed that a single NSC in the L-NSC-III group was the source of PTTH in the *Manduca* brain. Unfortunately, Agui and his colleagues were not able to determine which of the L-NSC-III was the prothoracicotrope, since the two cells comprising the group could not be distinguished from each other during their isolation. Nevertheless, the identification of the prothoracicotrope in *Manduca* by the assay of single neurosecretory cells marked the first demonstration of the cellular origin of a known neurohormone in an insect brain.

In summary, the prothoracicotropes have been unequivocally identified in one insect species and their location inferred with varying degrees of precision in a number of other species. Thus far, it appears that there is considerable phylogenetic variation in the site of PTTH synthesis, e.g. in *Manduca, Rhodnius* and *Locusta*. However, without a definitive localization, such as that accomplished in *Manduca*, it is not possible to conclude whether or not the prothoracicotropes are generally found in one region of the brain, and thus whether or not the site of synthesis has been conserved evolutionarily. Since some conservation of the structure of the active site of the PTTH molecule appears to have occurred (Agui, N. *et al.*, 1983), conservation of the site of synthesis is still a distinct possibility.

With the technological progress made in this field in recent years, precise and direct localizations of the PTTH NSC can now be attempted in other insects to answer this important evolutionary question. Even though the dissection and assay of individual cells may not be technically feasible in many species, it is reasonable to project the future use of antibodies to PTTH for immunocytochemical localizations in these species.

5.2 Neurohemal organ

In comparison to the efforts to identify the site of PTTH synthesis, the identification of the neurohemal organ for this neurohormone, i.e. its site of release, has received much less attention. Historically, the corpus cardiacum (CC) has been considered the neurohemal organ for PTTH. The basis for this conclusion consists mainly of histological evidence demonstrating that the CC had a structure fitting the definition of a classical neurohemal organ (Scharrer, B., 1952; Wigglesworth, V., 1964; Goldsworthy, G. and Mordue, W., 1974; Berlind, A., 1977) (Fig. 11). Since PTTH is a neurohormone, and since axons of the cerebral NSC terminate in the CC, it was concluded by extension that the CC was the site of PTTH release.

Paradoxically, a number of physiological studies over the years have implicated the CA, and not the CC, as the site of release. For example, Ichikawa, M. and Nishiitsutsuji-Uwo, J. (1959) noted that CA implanted into dauer pupae elicited adult development, and similar experiments by Kobayashi, M. and Yamashita, Y. (1959) and Morohoshi, S. and Shimada, J. (1975) on *Bombyx mori* yielded comparable results. In addition to the effect of implanted CA on development, extracts of *Bombyx* (Ishizaki, H., 1969) and *Manduca sexta* (Gibbs, D. and Riddiford, L., 1977) CA exhibited significant PTTH activity, as measured by the *Samia* and *Manduca* bioassays for PTTH, respectively. Surprisingly, other studies on PTTH release have indicated that the neurohormone may also be released directly from the brain (Nishiitsutsuji-Uwo, J., 1961; Takeda, N., 1976), possibly from the perikaryon of the prothoracicotrope itself or from processes of these cells projecting to the dorsal surface of the brain. Recently, this ability of the brain to release PTTH has been observed by Okajima (cited by Ishizaki, H. and Suzuki, A., 1980), who detected PTTH activity in medium of *in vitro* cultures of *Bombyx* brain–corpus cardiacum–corpus allatum complexes (BR–CC–CA). In these cultures the brain was physically separated from the CC–CA by a bridge of liquid paraffin, the only connection being the nerves running from the brain to the retrocerebral complex. The culture media bathing the brain and the CC–CA were assayed separately with the *Bombyx* assay and were both found to possess PTTH activity. This finding implied the existence of multiple sites of PTTH release, and as evidence to support this possibility the authors mentioned the widespread distribution of what appear cytologically to be neurohemal sites along the axons of many insect NSC (Johnson, B., 1963; Maddrell, S., 1966; Maddrell, S. and Gee, J., 1974; Steel, C., 1977).

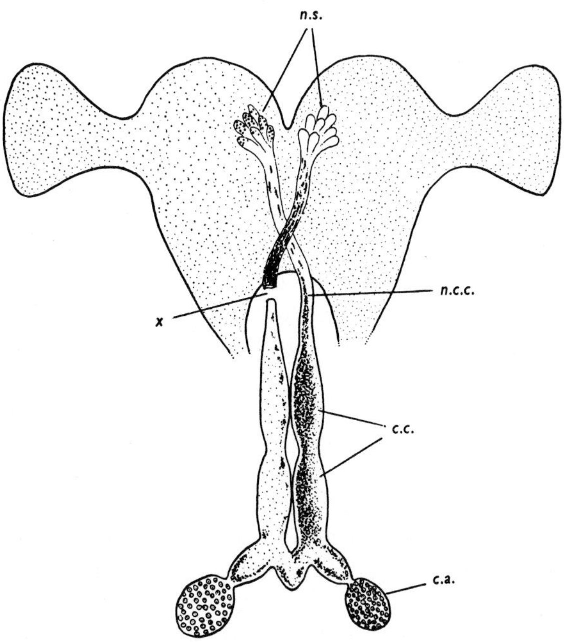

FIG. 11. Diagram of intercerebralis–cardiacum-allatum system in *Leucophaea maderae*. On the left side the nerve to the corpus cardiacum has been cut. Neurosecretory material has accumulated above the point of section and has almost disappeared below. On the operated side the corpus cardiacum has slightly decreased in size and the corpus allatum is enlarged. c.a., Corpus allatum; c.c., corpus cardiacum; n.c.c., medial nerve to the corpus cardiacum; n.s., neurosecretory cells; x, point of section of nerve. (From Scharrer, B., 1952.)

In the past few years the site of PTTH release has been probed in *Manduca* with the use of the *in vitro* PG assay for the hormone. The first investigation by Agui, N. *et al.* (1980) was based on the identification of the prothoracicotropes in the *Manduca* pupal brain (Agui, N. *et al.*, 1979) and the previous demonstration that the axons of these NSC traversed the CC and terminated in the CA (Nijhout, H., 1975a). To demonstrate in a quantitative manner that the CA was the neurohemal organ, Agui and his colleagues titered PTTH activity in the BR, CC, and CA during larval–pupal development, acting on the supposition that the neurohemal site would have a substantial level of PTTH activity and that temporal changes in the level of this activity would reflect those of the activity in the prothoracicotropes. The titers revealed that the PTTH activity in the CA was always significantly greater than that in the CC, and that fluctuations in the CA titer paralleled developmental changes in the brain titer of PTTH, thus confirming the experimental hypothesis. A recent ultrastructural study of the *Manduca* CA during this period of postembryonic development (Sedlak, B., 1981) provides indirect evidence supporting the conclusion that the CA is the neurohemal organ. Numerous axon terminals packed with neurosecretory granules were observed in the peripheral sheath of the CA, a logical site for the termination of axons from which neurohormone is released into the hemolymph. Thus, the preponderance of evidence points to the CA as the site of PTTH release in *Manduca*.

By contrast, Carrow, G. *et al.* (1981), using an *in vitro* culture system of larval BR–CC–CA similar to that of Okajima (see Ishizaki, H. and Suzuki, A., 1980), have proposed that the CC and the CA are both physiological sites of PTTH release in *Manduca* larvae. When they assessed spontaneous and potassium-evoked release of PTTH from BR–CC–CA, CC–CA complexes, and isolated BR, CC and CA *in vitro* with an *in vitro* assay for PTTH, they observed spontaneous release of significant and comparable amounts of PTTH from intact BR–CC–CA and isolated CA. By contrast, levels of PTTH activity barely above background for the assay appeared to be released spontaneously from

BR–CC and CC alone. In the presence of high potassium, PTTH release from isolated CA was stimulated 5-fold, and from isolated CC, 4-fold. Although potassium evoked an increased release of PTTH activity from the CC, the increase was barely equal to that released spontaneously by the CA. From these data the authors concluded that both the CC and CA are neurohemal organs for PTTH in *Manduca*. While multiple-release sites have been suggested for PTTH and have been found for some vertebrate hormones (Patel, Y. *et al.*, 1977), the substantial quantitative differences (5-fold) in either the spontaneous or evoked releases of PTTH from the CA and the CC suggests that release from the CC may simply be an artifact of the *in vitro* system rather than representing release from a neurohemal organ. The axons of the prothoracicotropes traverse the CC *en route* to the CA, and thus a certain amount of PTTH would be present in this organ (Agui, N. *et al.*, 1979). When incubated *in vitro*, the CC might then release enough of this hormone nonspecifically to yield activity in the medium. With the methods currently available for assessing PTTH levels, little more can be done at this time to resolve this issue of alternative sites of release for PTTH. However, with the development of immunocytochemical methods this issue should finally be resolved.

In conclusion, when the cumulative evidence on the location of the site of PTTH release in Lepidoptera is interpreted conservatively, the data indicate that the CA functions as the major site of PTTH release. However, there could be secondary sites of release that may be endocrinologically important. The question of the conservation of the site of PTTH release during insect evolution thus becomes one which, by its answer, may explain the varied results thus far obtained on this topic.

6 TITERS OF PTTH

From the knowledge gained in recent years from titer determinations of ecdysteroids during insect development, it is apparent that information about the titers of PTTH will be pivotal to an understanding of the mechanisms involved in the regulation of this neurohormone, particularly those concerned with synthesis, release, etc. Of the small amount of information currently available on PTTH titers, most has been derived from bioassay analyses of activity in brains and extracts of brains, or from head ligation studies that define the head critical period (HCP) for PTTH release. Only one study has actually focused on titering this hormone in the hemolymph and has attempted to correlate the results with levels of PTTH in the brain (Bollenbacher, W. and Gilbert, L., 1981; Bollenbacher, W., unpublished). As with the other areas of PTTH research, lack of progress in this area has primarily been a function of the bioassays available. HCP determinations are a notable exception in this regard, since they can precisely define the time of PTTH release into the hemolymph without relying on the PTTH bioassays to do so. However, this technique lacks a quantitative dimension.

In the first attempt to titer PTTH in an insect, Ishizaki, H. (1969) quantified with the *Samia* assay the PTTH activity present in different tissues of several different races of *Bombyx* throughout postembryonic development. For these titers of PTTH activity the hormone was extracted according to the procedure of Ishizaki, H. and Ichikawa, M. (1967), and serial dilutions of each of the extracts bioassayed. During the larval and pupal stages PTTH activity was consistently found in brains and heads at levels which remained relatively constant, heads possessing twice the activity of brains alone. The fact that the PTTH levels did not change during this period of development, even at the times when release was presumably occurring, led Ishizaki to conclude that release, and not synthesis, was the point at which PTTH secretion was regulated. Another interesting finding from this study was the substantially greater amount of PTTH activity (3–7 times) in adult heads than in larval and pupal heads. The physiological significance of this high level of activity in adult heads is unclear, since the PG during pupal–adult development degenerate, leaving the adult with no endocrinological need for the hormone. At this time, Ishizaki proposed that the PTTH activity in adult *Bombyx* might be important for reproduction, and in light of the increasing role the neuroendocrine system appears to have in regulating reproductive processes in adult insects, e.g. EDNH in the mosquito (Fuchs, M. and Kang, S.-H., 1981; H. Hagedorn, vol. 8), this may have been a very insightful interpretation of these data.

An important consequence of the discovery of such a large quantity of PTTH in adult heads was the immediate use of these heads as the source of PTTH for its purification. Another unexpected result of this study was the detection of considerable PTTH activity in the pharate adult thorax–abdomen of females of selected races of *Bombyx*. Since this activity would be extracerebral in origin, there is no easy explanation for its purpose, unless it is produced in this portion of the body for a function there. It has subsequently been found that there is a dramatic increase in ovary and oocyte ecdysteroid levels during the pharate adult development of *Bombyx mori*, which is coincidental with oocyte maturation (Legay, J.-M. *et al.*, 1976). If PTTH is involved in regulating this ecdysteroid titer, it might be expected that a high level of PTTH activity would be found. However, this does not explain why this activity would be localized in the thorax–abdomen, nor does it explain why it is specific to individual races of *Bombyx*. Regardless of these rather provocative findings, Ishizaki's study was a considerable accomplishment considering the methods available at the time for this research.

Some years later, Nishiitsutsuji-Uwo, J. (1975) used essentially the same experimental design to confirm Ishizaki's titer results and to identify the source of the large amount of PTTH activity present in the thorax–abdomen of *Bombyx*. In measuring the PTTH activity in the head and thorax–abdomen of several races of *Bombyx* other than those used by Ishizaki, she observed that the PTTH activity in the head remained high from the time of pupation through adulthood. These findings are in marked contrast to those of Ishizaki in that the measured levels of activity overall were 10–50 times higher, and that the activity in the adult head was not significantly higher than that in pupal and pharate adult heads. There also appeared to be a sexual dimorphism in the amount of activity present in the head, females possessing a greater amount than males. Even more striking than the differences found in PTTH levels in the head was the presence of large quantities of activity in the thorax–abdomen throughout pupal and pharate adult development of both females and males and in adult females. The removal of the brain in the early pupal period did not appear to change significantly the level in the adult thorax–abdomen, and ovariectomy may have actually increased the level. The reason(s) for the dramatic quantitative differences in these data relative to those of Ishizaki is not readily apparent, but they do not appear to result from the use of different races of *Bombyx*, different extraction protocols, and different bioassays, etc., unless the error introduced by each of these aspects of the studies combined to yield significant differences. The PTTH activity in the thorax–abdomen of *Bombyx* was not pursued beyond these initial studies, and because it has never been chemically identified as PTTH, there is no certainty that the activity was actually due to the hormone. If this activity were not PTTH, it would be far easier to explain the effects of debraining and gonadectomy on the level of this activity in the adult female thorax–abdomen.

In toto, these two studies raised many questions about PTTH titers, possibilities of alternative sites of synthesis, sequestration of the hormone, its function in insect post-embryonic development, and a possible function in reproduction as well. Unfortunately, these questions will probably remain unanswered until new, more precise and specific methods for monitoring and quantifying PTTH are developed.

Recently, Bollenbacher and his colleagues (Bollenbacher, W. and Gilbert, L., 1981; Gilbert, L. *et al.*, 1981; Bollenbacher, W. and Bowen, M., 1983; Agui, N. *et al.*, 1980) have used the *in vitro* PG assay to generate titers of PTTH in the brain, CA, CC and hemolymph of *Manduca*. Initially, they sought an estimate of the amount of PTTH present in the brain in order to assess the role of synthesis, and perhaps storage, in the control of the PTTH titer (Fig. 12). Thus a brain titer of the hormone during larval–larval, larval–pupal and early pupal–adult development was generated. The titer determination revealed that the levels of the neurohormone in the brain increased in a stepwise manner at each molt, with the highest amount measured on days 1–2 of pupal life. Dramatic increases and decreases in the titer were not observed before and after the HCP of the fourth larval instar and the commitment HCP of the last larval instar, suggesting, as Ishizaki, H. (1969) had speculated, that release was the level at which the PTTH titer is controlled.

A titer of PTTH in the CA during larval–pupal development revealed fluctuations generally paralleling those in the brain, but slight increases and

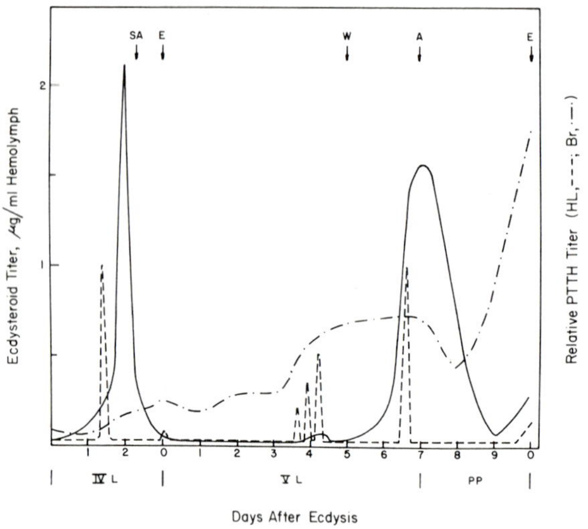

FIG. 12. Ecdysteroid and PTTH titers during larval and larval–pupal development of *Manduca sexta*. The ecdysteroid titer was determined for hemolymph and is expressed in 20-hydroxyecdysone equivalents. The PTTH titers were determined for both hemolymph and the brain and are expressed relative to the highest level of hormone activity found. The symbols SA, E, W, and A during the penultimate (IV L), last larval (V L), instar, and pharate pupal (PP) life signify spiracle apolysis, ecdysis, wandering, and apolysis, respectively. (From Bollenbacher, W. and Gilbert, L., 1981.)

decreases in the CA titer were also noted prior to and after each HCP. These changes in the CA titer around the time of the HCP provided additional support for the idea that release is the level at which the hemolymph titer is modulated.

The most important and potentially most useful PTTH titer data generated for *Manduca sexta* have been those relating to hormone titers in the hemolymph during larval and larval–pupal development (Bollenbacher, W. and Gilbert, L., 1981; Bollenbacher, W., unpublished). This was accomplished with the development of a procedure for extracting PTTH from the hemolymph, thus permitting its assay. The extraction procedure included a partial purification of the hemolymph by heat denaturation, followed by column chromatography of the remaining soluble hemolymph constituents to remove low molecular weight moieties, including ecdysteroids that might otherwise interfere with the *in vitro* PG assay.

On the basis of HCP data there is one period of PTTH release (4 h duration) in the penultimate larval instar and two in the fifth larval instar (18 and 6 h in duration, respectively) (Bollenbacher, W. and Gilbert, L., 1981; Bollenbacher, W., unpublished). From the hemolymph titer of PTTH it was found that during the HCP in the fourth instar, there was a single peak in the PTTH titer lasting for only a few hours, with a similar peak occurring during the second of the two HCP in the fifth instar. In contrast, there were three releases of PTTH during the first HCP of the fifth instar spread over a period of 18–20 h, with each lasting only a few hours, similar to that in the fourth instar. Thus the titer data revealed what appear to be two different modes of release for PTTH in *Manduca*. Because of the potential importance of differential modes of PTTH release to the control of the ecdysteroid titer, and, therefore, to insect development, it became necessary to define the chemical nature of the PTTH activity released. This would ensure that it was PTTH, and would permit a determination of its identity as the big and/or small PTTHs isolated from the brain of *Manduca*.

The PTTHs released into the hemolymph during the fourth instar HCP and first HCP of the fifth instar have been partially identified and characterized by the same protocols used for the characterization of the brain PTTHs (see section 4.3). The hormone released in a single burst in the fourth instar to elicit a large increase in the ecdysteroid titer and the subsequent larval molt was found to be big PTTH, as judged by its molecular weight, isoelectric point, biological activity and kinetics of PG activation *in vitro*. By the same criteria, the form released in a pulsatile manner to elicit the subtle increase in the ecdysteroid titer which results in pupal commitment tentatively appears to be small PTTH. A considerable amount of work is still needed to confirm these preliminary findings.

If these findings can be substantiated they would verify the results of the purification of the brain PTTHs in *Manduca* and would suggest that there are distinct functions for each of these PTTHs. Furthermore, these findings on the chemistry of the PTTHs in *Manduca* would then suggest that there are multiple molecular forms of this neurohormone, a common observation with vertebrate neuropeptide hormones. Each of these forms may also have distinct endocrine functions, based on the developmental response of the fourth and fifth instar larvae to the different HCP. It is entirely plausible that the two different PTTHs could be released in different modes to elicit different degrees of PG activation, differential increases in the ecdysteroid titer, and different developmental responses.

The possible *raison d'être* for a neuroendocrine system composed of two hormones whose basic action on their common target gland is the same, but whose developmental effect is entirely different, may become more clear once the chemistries of the PTTHs in other insect species have been elucidated. A two-form–two-function system, and the intervening cascade of events, may possibly have occurred in response to, or as an effect of, the evolution of homometabolous development. If there are indeed multiple forms of PTTH in *Manduca*, each with a unique function, this would be the first demonstration in insects, and possibly in any organism, of multiple functions for different structural forms of a neurohormone.

7 CONTROL OF PTTH RELEASE

The mechanism by which PTTH release is controlled appears to vary among the different insects, apparently depending on the evolution of the insect and the ecological niche it occupies. From the insects studied in an effort to understand how the control of PTTH release occurs, it is the general consensus that both extrinsic and intrinsic stimuli are involved and that the mediators of these stimuli can be either neural and/or endocrine. In general, there are three basic means by which control is expressed: (1) proprioceptive or mechanoreceptive stimuli; (2) photoperiodic stimuli, both circadian and seasonal; and (3) temperature.

7.1 Proprioceptive and mechanoreceptive stimuli

Investigations of the mechanisms by which nervous stimuli could control PTTH release first required knowledge of the time of release of the neurohormone *in situ*. With this knowledge it was then possible to determine whether factors such as diet, photoperiod, temperature, developmental stage, etc., were temporally correlated with release and thus possibly involved in the control of PTTH release. Head ligation, one of the classical experimental methods, proved to be the key to establishing when PTTH release occurred in an insect. An HCP, as defined earlier, is the time during which the presence of the brain becomes unnecessary for a subsequent molt; it can be defined more precisely as the time during which PTTH release from the brain occurs. Not surprisingly, the first insect for which HCPs were determined, and in which the stimulus for this release of PTTH was investigated, was *Rhodnius prolixus* (Wigglesworth, V., 1933). Molting in *Rhodnius* can only be initiated by a large bloodmeal, a succession of small meals being insufficient to elicit a molt. This observation led Wigglesworth, V. (1933) to conclude that distension, rather than nutritional state alone, was the principal stimulus for the molt. Additional support for this idea was provided when he demonstrated that occlusion of the anus of *Rhodnius* with a paraffin plug immediately after feeding prevented the normal rapid elimination of fluid. In doing so the size of the blood meal was artificially exaggerated, resulting in a 76% reduction in the size of the bloodmeal needed to elicit a molt (Wigglesworth, V., 1934). From these data it appeared that proprioceptive stimuli controlled PTTH release in *Rhodnius*.

Since proprioception is a nervous phenomenon, the possibility existed that the effect of stretching on the release of PTTH from the brain was mediated via the sensory nervous system. The involvement of neural elements in this process was confirmed by the failure of blood-fed animals to molt when their ventral nerve cords were sectioned immediately following the meal (Wigglesworth, V., 1934). Indirect electrophysiological proof of the role of stretch receptors in evoking PTTH release came from studies by Van der Kloot, W. (1960, 1961). He demonstrated that the two stretch receptors located in each abdominal segment of *Rhodnius*, one on either side, continued to discharge without adaptation as long as the abdomen was distended. Furthermore, electrical stimulation of the ventral nerve cord by stretching of the abdomen elicited action potentials in the nerves running from the brain to the corpora cardiaca. This finding was consistent with the idea that axons of the NSC previously identified as the likely source of PTTH passed through these nerves (Hanström, B., 1938; Wigglesworth, V., 1939, 1940b). Recently, Orchard, I. and Steel, C. (1980) found that the stimulus provided by a stretch receptor initiated bursting activity in the medial NSC of *Rhodnius*, and that this occurred simultaneously with structural changes in these cells indicative of a sudden onset of transport and release of neurosecretory material (Morris, G. and Steel, C., 1977a,b; Steel, C. and Harmsen, R., 1971). Within a few hours of this bursting activity the hemolymph titer of ecdysteroid increased subtly, presumably in response to PTTH stimulation of the PG (Steel, C. *et al.*, 1982). Together, these data on *Rhodnius* suggest that release of PTTH in this bug is exerted via neural channels which affect the prothoracicotropes themselves, and that the NSC become electrically active and transport neurosecretory material for release. About 2 h after feeding the bursting activity of these NSC was observed to subside and remain low for the first 5 days of the instar, as did the ecdysteroid titer (Orchard, I. and Steel, C., 1980; Steel, C. *et al.*, 1982). At about the fifth day these cells underwent a second period of bursting activity, which was followed temporally by a second and dramatic increase in the hemolymph ecdysteroid titer. This second period of electrical activity is in temporal agreement with the HCP for *Rhodnius*, i.e. at 28° decapitation 5–6 days after feeding fails to prevent a molt (Orchard, I. and Steel, C., 1980; Steel, C. *et al.*, 1982). From these studies it appears that there are two discrete periods of PTTH release in *Rhodnius*, each evoking a characteristic peak in the ecdysteroid titer, presumably via PG activation. Both periods of release are apparently required for molting to occur. These findings seem to conflict with those of Wigglesworth, V. (1934), who concluded that the HCP for *Rhodnius* consisted of a period of 5 days of continuous PTTH secretion from the time of feeding. However, Wigglesworth's conclusion that the brain was needed at least for the first 5 days was appropriate, given that the experimental methods employed could not resolve discrete periods of release, release being monitored downstream and indirectly by the molt.

Although the control of PTTH release in *Rhodnius* is beginning to be defined in more specific terms, little is known of the cascade of events between stimulation of the stretch receptor and activation of the prothoracicotropes. The means by which this proprioceptive cue is transduced neurophysiologically and neurochemically is certainly more complicated than initially proposed: (1) the signal is expressed both immediately upon gut distension with a bloodmeal and then a second time some days later; and (2) with each nymphal instar the period of time between the bloodmeal and the molt progressively increases, meaning that the time of the second release relative to the first is somehow stored and programmed into the development of the animal, to be cued for later expression at the critical time.

In the milkweed bug, *Oncopeltus fasciatus*, Nijhout, H. (1979) has recently demonstrated that abdominal stretch receptors are apparently involved in triggering the molt. For this insect a molting cycle is initiated once the nymph reaches a critical weight, a situation similar to that proposed for *Manduca* (see subsequent discussion). However, distension of the abdominal wall of starved, subcritical weight nymphs by saline or air also initiates a molt. Thus it appears that the critical weight is merely an index of the size at which a sufficient degree of abdominal stretching occurs to signal molting. If this is the case, then *Oncopeltus* would be another example of stretch-stimulated molting mediated by proprioception.

Stretch receptors, and thus proprioception, have also been implicated in the control of the release of PTTH in *Locusta migratoria* (Clarke, K. and Langley, P., 1963). Excision of the frontal ganglion of this orthopteran early in a nymphal instar, or section of both connectives of the frontal ganglion to the brain, arrests growth and molting. Since in *Locusta* two pairs of anterolateral nerves run from stretch receptors in the pharyngeal wall to the frontal ganglion, Clarke and Langley suggested that the receptors are stimulated during feeding, providing a continuous stimulus to the brain which triggers the release of PTTH.

Proprioceptive input may regulate PTTH release in other insects, but in these instances it appears to occur in response to external cues via other types of sensory pathways. For example, diapausing larvae of the wheat-stem sawfly, *Cephus cinctus*, do not proceed through development if they are removed from wheat straw (Church, N., 1955), and larvae of honeybees need to be in contact with their silk cocoon, or some surrounding cell, to initiate pupation (Fyg, W., 1956). When penultimate- and last-instar larvae of *Galleria mellonella* are immobilized by mechanical restraint, development is curtailed (Edwards, J., 1966; Sehnal, F. and Edwards, J., 1969). However, in this insect there is a critical time during the last instar after which restraint no longer has an effect. This time is temporally close to that of the second hemolymph ecdysteroid peak (Bollenbacher, W. *et al.*, 1978), the presumed time of PTTH release. Thus, the larva apparently needs to receive certain proprioceptive input from the periphery for PTTH release to occur and development to proceed. The ventral nerve cord seems to be the route by which this proprioceptive signal gets to the brain, since severance of the ventral nerve cord during the time when immobilization can affect development also inhibits or delays pupation. The more anterior the lesions occur along the nerve cord, the greater the reduction in the incidence of metamorphosis, and as with mechanical restraint, severance of the nerve cord after the critical developmental period during the last instar has little or no effect on pupation (Edwards, J., 1966).

In *Manduca sexta* (Nijhout, H., 1975b) and other lepidopterans such as *Galleria* (Beck, S., 1970; Sehnal, F., 1966), proprioceptive control of PTTH release may also involve a form of allometry; that is, for the release of PTTH to occur the brain must receive information from the body that a certain critical weight has been attained (Nijhout, H. and Williams, C., 1974a,b; Nijhout, H., 1975b). In *Manduca,* allometry may be controlling PTTH release by inactivation of the CA (Nijhout, H. and Williams, C., 1974b). Starvation of a last-instar larva prior to its attainment of a critical weight of 5 g was observed to prolong the activity of the CA and delay pupation. The involvement of JH in this process was shown by the effect of injecting JH into normal larvae weighing less than 5 g. Pupation of the JH-injected larvae was effectively delayed, presumably via the curtailment of PTTH release. Hence it appears that the effect of proprioceptive input on the release of PTTH from the prothoracicotropes in *Manduca* occurs indirectly via the regulation of JH titers rather than by directly affecting the response. Regarding the role of allometry in this process, it must be emphasized that, as with *Oncopeltus*, the attainment of a critical weight by the *Manduca* larva may only be an event that correlates temporally with some other neural and/or endocrine process actually regulating PTTH release.

Injury is another form of proprioceptive input which appears to effect PTTH release and which has not been evaluated critically. Some 30 years ago O'Farrell, A. and Stock, A. (1953) and O'Farrell, A. *et al.* (1956) noted that removal of a limb from the larva of the cockroach *Blattella germanica* delayed molting until the limb had regenerated. Wigglesworth, V. (1964) suggested that this effect might involve a mechanism similar to that apparently causing an inhibition of molting in *Rhodnius prolixus* in response to the injection of particulate matter into the hemolymph (Wigglesworth, V., 1955; see Gilbert, L. and Schneiderman, H., 1961). The extirpation of imaginal discs from lepidopteran larvae late in the last instar (Pohley, H., 1960; Madhavan, K. and Schneiderman, H., 1969; Krishnakumaran, A., 1972) has also been observed to delay the onset of molting. In this instance the delay was most pronounced when the molting cycle had already been initiated, and it was proportional to the magnitude of the injury. Apparently, injury was in some way either effecting the normal release of PTTH from the brain or affecting the ability of the PG to respond to the neurohormone.

Exactly how injury could effect PTTH release, if it actually does, is unclear, but it is possible that injury may result in activation of the CA. There exists histological and physiological evidence to support this idea (Norris, M., 1954; Scharrer, E. and von Harnack, M., 1958; Lüscher, M. and Engelmann, F., 1960; Ozeki, K., 1962). If this is the case, then the proposed mechanism for JH inhibition of PTTH release in larval *Manduca* (Nijhout, H. and Williams, C., 1974b) could be functioning in the response to injury as well.

In addition to its effect in delaying development, injury has been observed to give rise to extra larval molts, as evidenced by the response to injury of day 1 last-instar larvae of *Galleria* (Krishnakumaran, A., 1972). This occurrence would imply a more complex endocrine response to injury, one involving changes in CA activity as well as the release of PTTH and PG activation. Injury has also been observed to trigger adult development in dauer pupae of saturniid moths, an observation again suggesting a release of PTTH, but in this case in an animal in which there is no cerebral source for the hormone (McDaniel, C. and Berry, S., 1967). These results obviously contradict those presented earlier, and thus serve to illustrate that the effect of injury on the animal and its endocrine system is undoubtedly quite complex, involving a number of dramatic physiological changes. Recent work on the effects of injury has tended to substantiate this conclusion, indicating a complex response involving changes in patterns of protein and RNA synthesis in many tissues as well as an inducible immune system (Boman, H. *et al.*, 1981). Thus a physiological interpretation of a developmental response to injury in an insect must be made with caution. The importance of this point is underscored by the fact that much of the research on PTTH has been conducted on animals that have been manipulated surgically, i.e. injured.

7.2 Photoperiodic stimuli

Photoperiodicity is a major source of external control of insect development (D. Denlinger, vol. 8), and its apparently direct effect on the release of PTTH can be expressed in either a circadian or seasonal manner. Photoperiodic stimuli can be detected by the brain of the insect by "extraretinal" photoreceptors (Truman, J., 1976). In *Manduca*, and in many other Lepidoptera (Fujishita, M. and Ishizaki, M., 1981), PTTH release during normal development is governed by a circadian clock which confines release to a certain period of a 24 h light–dark cycle referred to as a gate. In *Manduca*, release of PTTH is photosensitive during larval development up to the second HCP of the last larval instar (Truman, J., 1972; Truman, J. and Riddiford, L., 1974; Bollenbacher, W., unpublished) and is temperature-compensated.

Light acts as a zeitgeber to entrain the circadian pacemaker (Aschoff, J., 1960; Truman, J., 1972, 1976), and thus the gate for PTTH release can be shifted by alterations in the photoperiod (Truman, J., 1972; Truman, J. and Riddiford, L., 1974; Fujishita, M. and Ishizaki, M., 1981). The reception of these circadian photoperiodic stimuli in *Manduca* appears to be extraretinal, since cauterization of the ocelli, the light-receptive organs in larvae, does not abolish the ability of the larvae to perceive photoperiod (Truman, J., 1972). These larvae exhibit a normal HCP, i.e. time of PTTH release, and ecdyse to the next instar at the same time as untreated larvae. It must be emphasized that photoperiod does not directly control PTTH release, but that it acts as a primary oscillator establishing a gate, or time of the day, when release can occur. The control of release itself is probably under intrinsic control involving secondary oscillators which may sense the acquisition of a particular developmental competency that is a prerequisite for PTTH release during the open gate.

The effect of seasonal photoperiodism on PTTH release is a well-established phenomenon (Hardie, J. and Lees, A.; Pener, M., vol. 8), and this stimulus apparently overrides the input of other environmental stimuli to cue the induction and duration of diapause in insects, i.e. a period of arrested development brought about partially by the seasonal photoperiodic signal preventing the release of PTTH (Tauber, M. and Tauber, C., 1976; Saunders, D., 1976; Beck, S., 1980; Bowen, M. *et al.*, 1983). In *Antheraea* photoperiod controls the onset of pupal diapause and the termination of this condition (Tanaka, Y., 1950; Williams, C. and Adkisson, P., 1964). A short day of 12 h light or less initiates diapause, while 17 h light or more triggers adult development. Furthermore, as in *Manduca*, the larval ocelli in *Antheraea* are not involved in the reception of the photoperiodic stimuli (Tanaka, Y., 1950). The photoperiodic receptor appears to be brain-centered. This was shown by Williams', C. and Adkisson's, P. demonstration (1964) that transfer of the brain from the head of a pupa to the tip of the abdomen resulted in a transfer of sensitivity to photoperiod to the hind end. Williams, C. (1969) then demonstrated by surgical excision that the area of the brain sensitive to photoperiod was just lateral to the medial NSC, in an area containing seven lateral NSC. Tetrodotoxin, a puffer-fish toxin which effectively blocks the generation of action potentials in the central nervous system, failed to affect either the maintenance of diapause in LD 12 : 12 or the termination of diapause in LD 17 : 7, indicating that nerve conduction probably does not play a role in the photoperiodic mechanism affecting PTTH activity. Thus, the possibility exists that the photoreceptor and the photoperiodic mechanism in general may reside in the lateral NSC, and, therefore, possibly in the prothoracicotropes themselves (see Bollenbacher, W. and Bowen, M., 1983). Analysis of the titers of PTTH in brains of diapausing *Antheraea* (Williams, C., 1967) suggested that the arrest of development in this species probably occurred at the level of release, since the levels of

PTTH in short-day brains remained high during diapause.

A similar seasonal photoperiodic control of PTTH release was found to occur in *Manduca*, where exposure of young larvae to short-day (LD 12:12) photoperiods induced a pupal diapause (Raabe, M., 1965). As in *Antheraea*, the induction of diapause appeared to be at the level of PTTH release (Safranek, L. and Williams, C., 1980; Bowen, M. *et al.*, 1983). The levels of PTTH activity in the brains of freshly pupated long- and short-day pupae are comparable. However, several days after pupation, activity begins to accumulate in the CA of the diapausing pupae (M.F. Bowen, personal communication), the neurohemal site for PTTH release in *Manduca*. Thus it appears that in *Manduca*, synthesis of PTTH continues in the diapausing pupal brain, and the synthesized hormone is transported to, and stored in, the CA. The intriguing question to be resolved is how the information provided by a short-day photoperiod, which is seen by the larva early in its development, is stored, presumably in the brain, and at a much later stage of development transduced into a signal that curtails PTTH release.

7.3 Temperature stimuli

In a circadian rhythm, such as the one that dictates the time of PTTH release, the gated release of the hormone is temperature-compensated (Pittendrigh, C., 1964; Wilkens, M., 1965), i.e. temperature has no effect on the gated release of PTTH (Truman, J., 1972). However, there are HCP during the development of insects when PTTH release is not photosensitive and not gated, e.g. the second release in the last larval instar of *Manduca* (Bollenbacher, W., unpublished observations). This type of release is apparently temperature-dependent. Therefore, temperature is also an important external cue that can regulate PTTH release, and this effect can be expressed either in a circadian or seasonal context. Temperature can act either alone or in concert with a particular photoperiod, and can affect the rate of development and the induction, reversal or termination of diapause (Danilevsky, A., 1965; Saunders, D., 1976; Tauber, M. and Tauber, C., 1976; Beck, S., 1980).

The effects of temperature in potentiating the termination of pupal diapause, and thus the release of PTTH, appear to be more ubiquitous than the effects of photoperiod. In the silkworms *Samia* and *Hyalophora cecropia*, for example, diapause termination occurs in response to extended exposure to temperatures less than 20° (Williams, C., 1956). In *Samia*, adult development never occurs if the diapausing pupa is maintained at temperatures higher than 20°. In *H. cecropia*, a lowered temperature apparently potentiates the recovery of PTTH activity in the brain, since the brain of a diapausing *H. cecropia* pupa chilled for a number of months accumulates substantial PTTH activity (Williams, C., 1956). Thus in this instance, chilling is a signal for the synthesis of PTTH in preparation for diapause termination in the spring. The actual release of the hormone in *H. cecropia*, as indicated by subsequent adult development, occurs in response to elevated temperature from 20 to 25°.

In *Manduca sexta* pupae, a somewhat different relationship of temperature to diapause termination appears to exist. Chilling of diapausing pupae is not necessary for adult development and only a few days of elevated temperature is needed to initiate development (Rasul, C., 1972). Brains of diapausing *Manduca* pupae contain at the time of pupation as much PTTH as brains of non-diapausing pupae (Bowen, M. *et al.*, 1983); thus it appears that in *Manduca*, unlike in *H. cecropia*, synthesis of PTTH is not the limiting, temperature-inducible factor in diapause termination. In *Manduca* only the release of PTTH is curtailed, and elevated temperatures probably evoke the release of previously stored hormone.

Temperature also has an effect on diapause induction in *Manduca*. Short-day photoperiods are ineffective in inducing diapause unless the ambient temperature during larval development is between 26 and 30° (Bell, R. *et al.*, 1975). Furthermore, the relationship of temperature cycles to photoperiod is also critical to diapause induction. If the cryophase coincides with the dark period, the incidence of diapause, and hence the inhibition of PTTH release, increases. Conversely, if the thermophase occurs during the dark period, the incidence of diapause decreases (Thurston, R., 1976).

Temperature can not only regulate PTTH release during the inductive phase or in the termination of pupal diapause, it can also affect a reversal of diapause after a photoperiodic signal has programmed its occurrence. For example, in *Sarcophaga* (Denlinger, D., 1972; Gibbs, D., 1975; Saunders, D., 1976) temperature step-ups after puparium formation decrease the percentage of pupal diapause. The size of the temperature increase, and the time after puparium formation that this increase occurs, both affect the incidence of diapause (Gibbs, D., 1975). By contrast, temperature step-downs with larvae reared under inductive short-day conditions decrease the incidence of diapause, presumably by promoting the release of PTTH. Therefore, variations in temperature can regulate development by overriding photoperiodic information and its expression.

This effect of temperature on development, i.e. PTTH release, occurs not only in holometabolous insects but in hemimetabolous insects as well. If a *Rhodnius prolixus* larva is exposed to high temperatures, either before or after feeding, molting is inhibited and development delayed, an effect apparently due to a general inhibition of protein synthesis and cell division (Okasha, A., 1968a,b). The extent to which temperature can modulate insect development, and whether this occurs at the level of PTTH release, is unclear, as the above studies indicate. It is clear, however, that temperature fluxes can dramatically affect development, and since it is an environmental cue that the insect encounters continually during its life cycle, the mechanism by which this stimulus exerts its effect, and particularly how it interacts with other environmental cues to control insect development, becomes an important point.

7.4 Endocrine regulators

The last aspect of the control of PTTH release pertinent to this discussion concerns the endocrine systems themselves, which in effect can function as the chemical mediators for the environmental stimuli controlling development. The regulation of PTTH release by humoral stimuli has already been indicated by the studies on critical body weight and JH inhibition of PTTH release in larval molting of *Manduca* (Nijhout, H. and Williams, C., 1974a,b; Nijhout, H., 1975b). This same form of humoral regulation by JH appears to occur during larval–pupal development of *Mamestra brassicae* as well (Hiruma, K. *et al.*, 1978a), based on the observed effects of JH on cerebral neurosecretion (Hiruma, K. *et al.*, 1978b; McCaffery, A. and Highnam, K., 1975a,b; Thomsen, E. and Lea, A., 1968). However, the most compelling evidence that JH regulates PTTH release comes from studies of larval diapause (see D. Denlinger, vol. 8). In several insect species, the foremost being the rice stem-borer, *Chilo suppressalis* and the southwestern corn-borer, *Diatraea grandiosella* (Fukaya, M. and Mitsuhashi, J., 1961; Fukaya, M. and Kobayashi, M., 1966; Yagi, S. and Fukaya, M., 1974), JH appears to be a feedback inhibitor of PTTH release, inducing and/or maintaining larval diapause (Yin, C.-M. and Chippendale, G., 1973; Chippendale, G. and Turunen, S., 1981; Takeda, N., 1978; Sieber, R. and Benz, G., 1980; Chippendale, G. and Yin, C.-M., 1979). The mechanism by which it exerts this effect is complex, since diapausing larvae of a number of species undergo larval–larval (stationary) molts. The occurrence of stationary molts indicates that the PG can be activated periodically, presumably by PTTH, but that for some reason this does not result in metamorphosis to the pupa. One proposal for the endocrine mechanism involved suggests that an intermediate titer of JH inhibits the brain–PG axis, while higher and lower titers activate it (Chippendale, G., 1977). The target for this JH effect appears to be located in the head, possibly the brain, since topical application of JH induces stationary molts in diapausing *Diatraea grandiosella* larvae, but not if these larvae are first neck-ligated (Chippendale, G. and Yin, C.-M., 1976).

Based on observations of the hormonal effects of 20-hydroxyecdysone on cerebral NSC, this hormone may also regulate the release and/or synthesis of PTTH (Marks, E. *et al.*, 1972; Steel, C., 1973, 1975; Agui, N. and Hiruma, K., 1977a,b); but no information is available as to when, and precisely how, this regulation might occur in the normal physiology of the insect. Considerable additional research is needed before the role of hormones in the control of PTTH release is understood.

In considering the evidence for the humoral regulation of PTTH release it must be remembered that any hormone found to effect a response, whether it is JH, 20-hydroxyecdysone, etc., is itself controlled by a mechanism integrating intrinsic and/or extrinsic cues. Thus, the regulation of development has as its ultimate basis the dynamic interaction of the insect with its constantly changing environment, an interaction that has evolved and become highly specialized for each insect species to ensure its survival.

From this discussion it should be apparent that the control of PTTH release is the cumulative effect of a complex and interrelated series of environmental and proprioreceptive–mechanoreceptive stimuli acting via the nervous and endocrine systems. An effective example of the complexity and diversity of cues and stimuli that can affect the development of an insect can be drawn from the studies on *Manduca*. In this species photoperiod, acting via a circadian pacemaker, sets the gate within which the prothoracicotropes can release their hormone. Whether or not release occurs is determined by proprioceptive input, apparently mediated by the feedback of JH on the PTTH cells. At certain times during development the normal release of PTTH is overridden by seasonal photoperiodic control, which is finely tuned by temperature. The unraveling of the physiological pathways involved in these events promises to be a challenging undertaking.

8 PROSPECTS

After a lengthy period of development it appears that the field of insect neuroendocrinology as it concerns the regulation of post-embryonic development has finally come of age. The next decade promises to be one of phenomenal growth in our understanding of: (1) the neuroendocrine role of PTTH in this process; and (2) the interaction of the insect's environment with its central nervous system to regulate this neuroendocrine function. Already it can be seen that the availability of technology has led to a renewed interest and different approaches to this problem. With these as driving forces, the next decade promises to be a provocative and exciting one for researchers in the field.

ACKNOWLEDGEMENTS

The authors thank Ms J. Hightower and Mr T. McMillan for their assistance in the preparation of this manuscript. This work was supported by NIH grants No. NS 18791 and AM 31642.

REFERENCES

AGUI, N. (1975). Activation of prothoracic glands by brains *in vitro*. *J. Insect Physiol. 21*, 903–913.

AGUI, N. (1976). Attempts to establish insect endocrine system *in vitro*. In *Invertebrate Tissue Culture, Research Applications*. Edited by K. Maramorosch. Pages 133–160. Academic Press, New York.

AGUI, N. and HIRUMA, K. (1977a). *In vitro* activation of neurosecretory brain cells in *Mamestra brassicae* by β-ecdysone. *Gen. Comp. Endocr. 33*, 467–472.

AGUI, N. and HIRUMA, K. (1977b). Ecdysone as a feedback regulator for the neurosecretory brain cells in *Mamestra brassicae*. *J. Insect Physiol. 23*, 1393–1396.

AGUI, N., BOLLENBACHER, W. E. and GILBERT, L. I. (1983). *In vitro* analysis of prothoracicotropic hormone specificity and prothoracic gland sensitivity in Lepidoptera. *Experientia 39*, 984–988.

AGUI, N., BOLLENBACHER, W. E., GRANGER, N. A. and GILBERT, L. I. (1980). Corpus allatum is release site for insect prothoracicotropic hormone. *Nature 285*, 669–670.

AGUI, N., GRANGER, N. A., GILBERT, L. I. and BOLLENBACHER, W. E. (1979). Cellular localization of the insect prothoracicotropic hormone: *In vitro* assay of a single neurosecretory cell. *Proc. Nat. Acad. Sci. 76*, 5694–5698.

ASCHOFF, J. (1960). Exogenous and endogenous components in circadian rhythms. *Cold Spring Harb. Symp. Quant. Biol. 25*, 11–28.

BECK, S. D. (1970). Neural and hormonal control of pupation in *Galleria mellonella* (Lepidoptera: Galleriidae). *Ann. Ent. Soc. Amer. 63*, 144–149.

BECK, S. D. (1980). *Insect Photoperiodism*. 2nd edition. Academic Press, New York.

BELL, R. A., RASUL, C. G. and JOACHIM, F. G. (1975). Photoperiodic induction of the pupal diapause in the tobacco hornworm, *Manduca sexta*. *J. Insect Physiol. 21*, 1471–1480.

BERLIND, A. (1977). Cellular dynamics in invertebrate neurosecretory systems. *Int. Rev. Cytol. 49*, 171–251.

BERN, H. A. (1966). On the production of hormones by neurones and the role of neurosecretion in neuroendocrine mechanisms, *Symp. Soc. Exp. Biol. 20*, 325–344.

BODENSTEIN, D. (1933a). Beintransplantationen an Lepidopteren Raupen. I. Transplantationen zur analyse der Raupen und Puppenhautung. *Wilhelm Roux Archiv. Ent. Mech. 128*, 564–583.

BODENSTEIN, D. (1933b). Zur Frage der Bedeutung hormoneller Beziehungen bei der Insektenmetamorphose. *Naturwiss. 21*, 861–863.

BODENSTEIN, D. (1936). Das Determinationsgeschehen bei Insekten mit Ausschluss der früernbryonalen Determination. *Ergeb. Biol. 13*, 174–234.

BOLLENBACHER, W. E. and GILBERT, L. I. (1981). Neuroendocrine control of postembryonic development in insects, the prothoracicotropic hormone. In *Neurosecretion: Molecules, Cells, Systems*. Edited by D. S. Farner and K. Lederis. Pages 361–370. Plenum, New York.

BOLLENBACHER, W. E. and BOWEN, M. F. (1983). The prothoracicotropic hormone. In *Endocrinology of Insects*. Edited by H. Laufer and R. G. H. Downer. Pages 89–99. A. R. Liss, New York.

BOLLENBACHER, W. E., AGUI, N., GRANGER, N. A. and GILBERT, L. I. (1979). *In vitro* activation of insect prothoracic glands by the prothoracicotropic hormone. *Proc. Nat. Acad. Sci. 76*, 5148–5152.

BOLLENBACHER, W. E., AGUI, N., GRANGER, N. A. and GILBERT, L. I. (1980). Insect prothoracic glands *in vitro*: a system for studying the prothoracicotropic hormone. In *Invertebrate Systems in Vitro*. Edited by E. Kurstak, K. Maramorosch, and A. Dubendorfer. Pages 253–271. Elsevier North Holland, Amsterdam.

BOLLENBACHER, W. E., O'BRIEN, M. A., KATAHIRA, E. J. and GILBERT, L. I. (1983). A kinetic analysis of the action of the insect prothoracicotropic hormone. *Mol. Cell. Endocr. 32*, 27–46.

BOLLENBACHER, W. E., SMITH, S. L., GOODMAN, W. and GILBERT, L. I. (1981). Ecdysteroid titer during larval–pupal–adult development of the tobacco hornworm, *Manduca sexta*. *Gen. Comp. Endocr. 44*, 302–306.

BOLLENBACHER, W. E., ZVENKO, H., KUMARAN, A. K. and GILBERT, L. I. (1978). Changes in ecdysone content during postembryonic development of the wax moth, *Galleria mellonella*: the role of the ovary. *Gen. Comp. Endocr. 34*, 169–179.

BOMAN, H. G., BOMAN, A. and PIGON, A. (1981). Immune and injury responses in *Cecropia* pupae — RNA isolation and comparison of protein synthesis *in vivo* and *in vitro*. *Insect Biochem.* *11*, 33–42.

BORST, D. W. and O'CONNOR, J. D. (1972). Trace analysis of ecdysones by gas–liquid chromatography, radioimmunoassay and bioassay. *Steriods 24*, 637–656.

BOUNHIOL, J. J. (1938). Recherches expérimentales sur le déterminisme de la métamorphose chez les Lépidoptères. *Bull. Biol., Suppl. 24*, 1–199.

BOUNHIOL, J. J. (1952a). L'achèvement de la métamorphose et la mue imaginale seraient commandés par la cerveau à la fin de la vie larvaire chez *Bombyx mori* L. *C. R. Acad. Sci., Paris 235*, 671–672.

BOUNHIOL, J. J. (1952b). Nature probablement sécrétoire du facteur cérébral conditionnant la mue imaginale de *Bombyx mori* L. *C. R. Acad. Sci., Paris 235*, 747–748.

BOWEN, M. F., BOLLENBACHER, W. E. and GILBERT, L. I. (1984). *In vitro* studies on the role of the brain and prothoracic glands in the pupal diapause of *Manduca sexta*. *J. Exp. Biol. 108*, 9–24.

BRAÜER, R., GERSCH, M., BÖHM, G.-A. and BAUMANN, E. (1977). Stimulation by activation factor I of the synthesis of α-ecdysone in the prothoracic glands of *Periplaneta americana in vitro*. *Zool. Jb. Physiol. Bd. 1*, 1–12.

BURTT, E. T. (1938). On the corpora allata of dipterous insects II. *Proc. R.. Soc. Lond. B 126*, 210–223.

CARROW, G., CALABRESE, R. L. and WILLIAMS, C. M. (1981). Spontaneous and evoked release of prothoracicotropin from multiple neurohemal organs of the tobacco hornworm. *Proc. Nat. Acad. Sci. 78*, 5866–5870.

CASPARI, E. and PLAGGE, E. (1935). Versuche zur Physiologie der Verpuppung von Schmetterlingsraupen. *Naturwiss. 23*, 751–752.

CHARLET, M. and SCHALLER, F. (1976). Blocage de l'exuviation chez la larve d'*Aeshna cyanea* (insecte Odonate) après électrocoagulation d'un centre neurosécréteur du protocérébron antérieur. *C.R. Acad. Sci., Paris 283*, 1539–1541.

CHIPPENDALE, G. M. (1977). Hormonal regulation of larval diapause. *Ann. Rev. Ent. 22*, 121–138.

CHIPPENDALE, G. M. and TURUNEN, S. (1981). Hormonal and metabolic aspects of the larval diapause of the southwestern corn borer, *Diatraea grandiosella* (Lepidoptera: Pyralidae). *Ent. Gen. 7*, 223–231.

CHIPPENDALE, G. M. and YIN, C.-M. (1976). Diapause of the southwestern corn borer, *Diatraea grandiosella* Dyar (Lepidoptera: Pyralidae): effects of a juvenile hormone mimic. *Bull. Ent. Res. 66*, 75–79.

CHIPPENDALE, G. M. and YIN, C.-M. (1979). Larval diapause of the European corn borer, *Ostrinia nubilalis*: further experiments examining its hormonal control. *J. Insect Physiol. 25*, 53–58.

CHURCH, N. S. (1955). Hormones and the termination and reinduction of diapause in *Cephus cinctus* Nort. (Hymenoptera: Cephidae). *Canad. J. Zool. 33*, 339–369.

CLARKE, K. U. and LANGLEY, P. A. (1963). Studies on the initiation of growth and moulting in *Locusta migratoria migratorioides* R. & F. IV. The relationship between the stomatogastric nervous system and neurosecretion. *J. Insect Physiol. 9*, 423–430.

CONTE, A. and VANEY, C. (1911). Production expérimentale des Lépidoptères acéphales. *C. R. Acad. Sci., Paris 152*. 404–406.

DANILENSKY, A. S. (1965). *Photoperiodism and Seasonal Development of Insects*. Oliver & Boyd, London.

DENLINGER, D. L. (1972). Induction and termination of pupal diapause in *Sarcophaga* (Diptera: Sarcophagidae). *Biol. Bull. 142*, 11–24.

DE REGGI, M. L., HIRN, M. H. and DELAAGE, M. A. (1975). Radioimmunoassay of ecdysone. An application to *Drosophila* larvae and pupae. *Biochem. Biophys. Res. Commun. 66*, 1307–1315.

EDWARDS, J. S. (1966). Neural control of metamorphosis in *Galleria mellonella* (Lepidoptera). *J. Insect Physiol. 12*, 1423–1433.

FAIN, M. J. and RIDDIFORD, L. M. (1976). Reassessment of the critical periods for prothoracicotropic hormone and juvenile hormone secretion in the larval molt of the tobacco hornworm *Manduca sexta*. *Gen. Comp. Endocr. 30*, 131–141.

FRAENKEL, G. (1935). A hormone causing pupation in the blowfly *Calliphora erythrocephala*. *Proc. R. Soc. Lond. B. 118*, 1–12.

FUCHS, M. S. and KANGS, S.-H. (1981). Ecdysone and mosquito vitellogenesis: a critical appraisal. *Insect Biochem. 11*, 627–633.

FUJISHITA, M. and ISHIZAKI, M. (1981). Circadian clock and prothoracicotropic hormone secretion in relation to the larval-larval ecdysis rhythm of the Saturniid *Samia cynthia ricini*. *J. Insect Physiol. 27*, 121–128.

FUKAYA, M. and KOBAYASHI, M. (1966). Some inhibitory actions of corpora allata in diapausing larvae of the rice stem borer, *Chilo suppressalis* Walker (Lepidoptera: Pyralidae). *Appl. Ent. Zool. 1*, 125–129.

FUKAYA, M. and MITSUHASHI, J. (1961). Larval diapause in the rice stem borer with special reference to its hormonal mechanism. *Bull. Nat. Inst. Agri. Sc. 13*, 1–30.

FUKUDA, S. (1940a). Induction of pupation in silkworm by transplanting the prothoracic gland. *Proc. Imp. Acad. Jpn. 16*, 414–416.

FUKUDA, S. (1940b). Hormonal control of moulting and pupation in the silkworm. *Proc. Imp. Acad. Jpn. 16*, 417–420.

FUKUDA, S. (1941). Role of the prothoracic gland in differentiation of the imaginal characters in the silkworm pupa. *Annot. Zool. Jap. 20*, 9–13.

FUKUDA, S. (1944). The hormonal mechanism of larval molting and metamorphosis in the silkworm. *J. Fac. Sci. 6*, 447–532.

FYG, W. (1956). Experimentelle Untersuchungen über die Entwicklung der Honigbiene. *Mitt. Schweiz. Ent. Oes. 29*, 404–416.

GERSCH, M. (1961). Insect metamorphosis and the activation hormone. *Amer. Zool. 1*, 53–57.

GERSCH, M. and STÜRZEBECHER, J. (1967). Zur Frage der Identität und des Vorkommens von Neurohormon D in verschiedenen Bereichen des Zentralnervensystems von *Periplaneta americana*. *Z. Naturf. Teil C. 226*, 563.

GERSCH, M. and STÜRZEBECHER, J. (1968). Weitere Untersuchungen zur Kennzeichnung des Aktivationshormons der Insektenhäutung. *J. Insect Physiol. 14*, 87–96.

GERSCH, M. and STÜRZEBECHER, J. (1970). Experimentelle Stimulierung der Zellulären Aktivität der Prothorakaldrüsen von *Periplaneta americana* durch den Aktivationsfaktor. *J. Insect Physiol. 16*, 1813–1826.

GERSCH, M., UNGER, H. and FISCHER, F. (1957). Die Isolierung eines Neurohormons aus dem Nervensystem von *Periplaneta americana* und einige biologische Testverfahren. *Wiss. Z. Friedrich-Schiller Univ. Jena 6*, 125–129.

GIBBS, D. (1975). Reversal of pupal diapause in *Sarcophaga argyrostoma* by temperature shifts after puparium formation. *J. Insect Physiol. 21*, 1179–1186.

GIBBS, D. and RIDDIFORD, L. M. (1977). Prothoracicotropic hormone in *Manduca sexta*: Localization by a larval assay. *J. Exp. Biol. 66*, 255–266.

GILBERT, L. I. and KING, D. S. (1973). Physiology of growth and development: endocrine aspects. In *The Physiology of Insecta*. Edited by M. Rockstein. Vol. 1, 2nd edition, pages 249–370. Academic Press, New York.

GILBERT, L. I. and SCHNEIDERMAN, H. A. (1961). Some biochemical aspects of insect metamorphosis. *Amer. Zool. 1*, 11–51.

GILBERT, L. I., BOLLENBACHER, W. E., AGUI, N., GRANGER, N. A., SEDLAK, B. J., GIBBS, D. and BUYS, C. M. (1981). The prothoracicotropes: Source of the prothoracicotropic hormone. *Amer. Zool. 21*, 641–653.

GILBERT, L. I., GOODMAN, W. and BOLLENBACHER, W. E. (1977). Biochemistry of regulatory lipids and sterols in insects. In *Biochemistry of Lipids II*. Edited by T. W. Goodwin. Pages 1–50. University Park Press, Baltimore.

GIRARDIE, A. (1967). La pars intercerebralis chez *Locusta migratoria* L. (Orthoptère) et son rôle dans le development. These, Strasbourg.

GIRARDIE, A. (1974). Rechèrches sur le rôle physiologique des cellules neurosécrétrices latérales du protocerebron de *Locusta migratoria migratorioides*. *Zool. Jb. Physiol. 178*, 310–326.

GIRARDIE, A. and DE REGGI, M. (1978). Moulting and ecdysone release in response to electrical stimulation of protocerebral neurosecretory cells in *Locusta migratoria*. *J. Insect Physiol. 24*, 797–802.

GIRARDIE, J. (1973). Aspects histologique, histochimique et ultrastructural des péricaryones neurosécréteurs lateraux du protocérébron de *Locusta migratoria migratorioides* (Insecte: Orthoptère). *Z. Zellforsch. Mik. Anat. 141*, 75–91.

GIRARDIE, J. and GIRARDIE, A. (1977). Libération provoquée *in vitro* due produit de neurosécrétion des cellules protocérébrales médianes chez le criquet migrateur. *J. Physiol. Paris 73*, 707–721.

GIRARDIE, J., GIRARDIE, A. and MOULINS, M. (1976). Etude radiochimique après électrostimulation de la dynamique fonctionnelle des cellules neurosécrétrices protocérébrales médianes de *Locusta migratoria* (Insecte: Orthoptère). *Gen. Comp. Endocr. 30*, 410–418.

GOLDSWORTHY, G. J. and MORDUE, W. (1974). Neurosecretory hormones in insects. *J. Endocr. 60*, 529–558.

GRANGER, N. A. and BOLLENBACHER, W. E. (1981). Hormonal control of insect metamorphosis. In *Metamorphosis: A Problem in Developmental Biology*. Edited by L. I. Gilbert and E. Frieden. 2nd edition, pages 105–137. Plenum, New York.

HACHLOW, V. (1931). Zur Entwicklungsmechanik der Schmetterlinge. *Wilhelm Roux Archiv. Ent. mech. 125*, 26–49.

HADORN, E. and NEEL, J. (1938). Der hormonale Einfluss der Ringdrüse (corpus allatum) auf die Pupariumbildung bei Fliegen. *Wilhelm Roux Archiv. Ent. mech. 138*, 281–304.

HANSTRÖM, B. (1938). Zwei Probleme betreffs der hormonalen Lokalization im Insektenkopf. *Acta Univ. Lund., N.F., Avd. 2, 39*, 1–17.

HETRU, C. (1976). Rétablissement de la mue par implantation de cerveau chez des larves permanentes de *Galleria mellonella. C.R. Acad. Sci., Paris (D) 282*, 81–83.

HIGHNAM, K. C. (1965). Some aspects of neurosecretion in arthropods. *Zool. Jb. (Physiol.) 71*, 558–582.

HIGHNAM, K. C. and GOLDSWORTHY, G. J. (1972). Regenerated corpora cardiaca and hyperglycemic factor in *Locusta migratoria. Gen. Comp. Endocr. 18*, 83–88.

HIRUMA, K. and AGUI, N. (1977). Relationship between histological changes and functions of the neurosecretory cells in the brain of the cabbage armyworm, *Mamestra brassicae* L. *Appl. Ent. Zool. 12*, 42–49.

HIRUMA, K., SHIMADA, H. and YAGI, S. (1978a). Activation of the prothoracic gland by juvenile hormone and prothoracicotropic hormone in *Mamestra brassicae. J. Insect Physiol. 24*, 215–220.

HIRUMA, K., YAGI, S. and AGUI, N. (1978b). Action of juvenile hormone on the cerebral neurosecretory cells of *Mamestra brassicae in vivo* and *in vitro. Appl. Ent. Zool. 13*, 149–157.

ICHIKAWA, M. and ISHIZAKI, H. (1961). Brain hormone of the silkworm, *Bombyx mori. Nature 191*, 933–934.

ICHIKAWA, M. and ISHIZAKI, H. (1963). Protein nature of the brain hormone of insects. *Nature 198*, 308–309.

ICHIKAWA, M. and NISHIITSUTSUJI-UWO, J. (1959). Studies on the role of the corpus allatum in the Eri-silkworm, *Philosamia cynthia ricini. Biol. Bull. 116*, 88–94.

ICHIKAWA, M. and YASHIKA, K. (1958). Neurosecretory cells as a source of the brain hormone in the phenomenon of insect metamorphosis. *Proc. 10th Int. Congr. Entomol. 2*, 269–270.

ISHIZAKI, H. (1969). Changes in titer of the brain hormone during development of the silkworm, *Bombyx mori. Devel. Growth Diff. 11*, 1–7.

ISHIZAKI, H. (1972). Arrest of adult development in debrained pupae of the silkworm, *Bombyx mori. J. Insect Physiol. 18*, 1621–1627.

ISHIZAKI, H. and ICHIKAWA, M. (1967). Purification of the brain hormone of the silkworm *Bombyx mori. Biol. Bull. 133*, 355–368.

ISHIZAKI, H. and SUZUKI, A. (1980). Prothoracicotropic hormone. In *Neurohormonal Techniques in Insects*. Edited by T. A. Miller. Pages 244–276. Springer-Verlag, New York.

ISHIZAKI, H., SUZUKI, A., ISOGAI, A., NAGASAWA, H. and TAMURA, S. (1977). Enzymatic and chemical inactivation of partially purified prothoracicotropic (brain) hormone of the silkworm, *Bombyx mori. J. Insect Physiol. 23*, 1219–1222.

JOHNSON, B. (1963). A histological study of neurosecretion in aphids. *J. Insect Physiol. 9*, 727–739.

JUDY, K. J. (1972). Diapause termination and metamorphosis in brainless tobacco hornworms (Lepidoptera). *Life Sci. 11*, 605–611.

KAMBYSELLIS, M. P. and WILLIAMS, C. M. (1971a). *In vitro* development of insect tissues. I. A macro-molecular factor prerequisite for silkworm spermatogenesis. *Biol. Bull. 141*, 527–540.

KAMBYSELLIS, M. P. and WILLIAMS, C. M. (1971b). *In vitro* development of insect tissues. II. The role of ecdysone in the spermatogenesis of silkworms. *Biol. Bull. 141*, 541–552.

KING, D. S., BOLLENBACHER, W. E., BORST, D. W., VEDECKIS, W. V., O'CONNOR, J. D., ITTYCHERIAH, P. I. and GILBERT, L. I. (1974). The secretion of α-ecdysone by the prothoracic glands of *Manduca sexta in vitro. Proc. Nat. Acad. Sci. 71*, 793–796.

KINGAN, T. G. (1981). Purification of the prothoracicotropic hormone from the tobacco hornworm *Manduca sexta. Life Sci. 28*, 2585–2594.

KIRIMURA, J., SAITO, M. and KOBAYASHI, M. (1962). Steroid hormone in an insect, *Bombyx mori. Nature 195*, 4842.

KOBAYASHI, M. (1955). Relationship between the brain hormone and the imaginal differentiation of silkworm, *Bombyx mori. J. Sericult. Sci. Jpn. 24*, 389–392.

KOBAYASHI, M. (1957). Studies on the neurosecretion in the silkworm, *Bombyx mori* L. *Bull. Seric. Exp. Stn. (Tokyo) 15*, 181–273.

KOBAYASHI, M. (1963). The chemistry and physiology of the brain hormone. *Proc. XVI Int. Congress Zool. 4*, 226–233.

KOBAYASHI, M. and KIRIMURA, J. (1958). The 'brain' hormone in the silkworm, *Bombyx mori* L. *Nature 181*, 1217.

KOBAYASHI, M. and YAMASHITA, Y. (1959). A function of corpus allatum in neurosecretory system in the silkworm *Bombyx mori. J. Seric. Sci., Tokyo 28*, 335–339.

KOBAYASHI, M. and YAMAZAKI, M. (1966). The proteinic brain hormone in an insect, *Bombyx mori* L. (Lepidoptera: Bombycidae). *Appl. Ent. Zool. 12*, 53–60.

KOBAYASHI, M., KIRIMURA, J. and SAITO, M. (1962). Crystallization of the 'brain' hormone of an insect. *Nature. 195*, 515–516.

KONO, Y. (1973). Light and electron microscopic studies on the neurosecretory control of diapause incidence in *Pieris rapae crucivora. J. Insect Physiol. 19*, 255–272.

KONO, Y. (1975). Daily changes of neurosecretory type-II cell structure of *Pieris* larvae entrained by short and long days. *J. Insect Physiol. 21*, 249–264.

KOPEĆ, S. (1912). Ueber die Functionen des Nervensystems der Schmetterlinge während der sukzessiven stadien ihrer Metamorphose. *Zool. Anz. 40*, 353–360.

KOPEĆ, S. (1917). Experiments on metamorphosis of insects. *Bull. Int. Acad. Sci. Cracovie BB*, 57–60.

KOPEĆ, S. (1918). Localisationsversuche an zentralen Nervensystem der Raupen und Falter. *Zool. Jahrb., Alot. F. allg. Zool. und Physiol. Bd. 36*, 453–502.

KOPEĆ, S. (1922). Studies on the necessity of the brain for the inception of insect metamorphosis. *Biol. Bull. 42*, 323–342.

KRISHNAKUMARAN, A. (1972). Injury induced molting in *Galleria mellonella* larvae. *Biol. Bull. 142*, 281–292.

KÜHN, A. and PIEPHO, H. (1936). Über hormonale Wirkungen bei der Verpuppung der Schmetterlinge. *Ges. Wiss. Gottingen Nachr. Biol. 2*, 141–154.

KÜHN, A. and PIEPHO, H. (1938). Die Reaktionen der Hypodermis und der Versonschen Drusen auf das Verpuppungshormon bei *Ephestia kühniella. Biol. Zbl. 58*, 12–51.

LEGAY, J.-M., CALVEZ, B., HIRN, M. and DE REGGI, M. L. (1976). Ecdysone and oocyte morphogenesis in *Bombyx mori. Nature 262*, 489–490.

LÜSCHER, M. and ENGELMANN, F. (1960). Histologische und experimentelle Untersuchungen über die Auslösung der Metamorphose bei *Leucophaea maderae* (Orthoptera). *J. Insect Physiol. 5*, 240–258.

MADDRELL, S. H. P. (1966). The site of release of the diuretic hormone in *Rhodnius* — a new neurohaemal system in insects. *J. Exp. Biol. 45*, 499–508.

MADDRELL, S. H. P. and GEE, J. D. (1974). Potassium-induced release of the diuretic hormones of *Rhodnius prolixus* and *Glossina austeni*: Ca dependence, time course and localization of neurohaemal areas. *J. Exp. Biol. 61*, 155–172.

MADHAVAN, K. and SCHNEIDERMAN, H. A. (1969). Hormonal control of imaginal disc regeneration in *Galleria mellonella* (Lepidoptera). *Biol. Bull. 137*, 321–331.

MARKS, E. P., ITTYCHERIAH, P. I. and LELOUP, A. M. (1972). The effect of β-ecdysone on insect neurosecretion *in vitro. J. Insect Physiol. 18*, 847–850.

MCCAFFERY, A. R. and HIGHNAM, K. C. (1975a). Effects of corpora allata on the activity of the cerebral neurosecretory system of *Locusta migratoria migratoriades* R. and F. *Gen. Comp. Endocr. 25*, 358–372.

MCCAFFERY, A. R. and HIGHNAM, K. C. (1975b). Effects of corpus allatum homone and its mimics on the activity on the cerebral neurosecretory system of *Locusta migratoria migratorioides* R. & F. *Gen. Comp. Endocr. 25*, 373–386.

MCDANIEL, C. N. and BERRY, S. J. (1967). Activation of the prothoracic glands of *Antheraea polyphemus. Nature 214*, 1032–1034.

MEOLA, R. and LEA, A. O. (1971). Independence of paraldehyde-fuchsin staining of the corpus cardiacum and the presence of the neurosecretory hormone required for egg development in the mosquito. *Gen. Comp. Endocr. 16*, 105–111.

Morohoshi, S. and Shimada, J. (1975). The control of growth and development in *Bombyx mori*. XXVII. Release of the brain hormone from the corpora allata through nerve axons from brain neurosecretory cells. *Proc. Jpn. Acad. 51*, 744–748.

Morris, G. P. and Steel, C. G. H. (1977a). Ultrastructural changes in the medial neurosecretory cells of *Rhodnius prolixus* following activation by feeding. *Amer. Zool. 17*, 929.

Morris, G. P. and Steel, C. G. H. (1977b). Sequence of ultrastructural changes induced by activation in the posterior neurosecretory cells in the brain of *Rhodnius prolixus* with special reference to the role of lysosomes. *Tissue Cell 9*, 547–561.

Nagasawa, H., Fu, G., Xiangchen, Z., Bangying, X., Zongshun, W., Xujia, Q., Dingyi, W., Eying, C., Jingze, W., Suzuki, A., Isogai, A., Hori, Y., Tamura, S. and Ishizaki, H. (1980). Large-scale purification of prothoracicotropic hormone of the silkworm (*Bombyx mori*). *Scientia Sinica 23*, 1053–1060.

Nagasawa, H., Isogai, A., Suzuki, A., Tamura, S. and Ishizaki, H. (1979). Purification and properties of the prothoracicotropic hormone of the silkworm, *Bombyx mori*. *Devel. Growth Diff. 21*, 29–38.

Nijhout, H. F. (1975a). Axonal pathways in the brain-retrocerebral neuroendocrine complex of *Manduca sexta* (L.) (Lepidoptera: Sphingidae). *Int. J. Insect Morph. Embryol. 4*, 529–538.

Nijhout, H. F. (1975b). A threshold size for metamorphosis in the tobacco hornworm, *Manduca sexta* (L.). *Biol. Bull 149*, 214–225.

Nijhout, H. F. (1979). Stretch-induced molting in *Oncopeltus fasciatus*. *J. Insect Physiol. 25*, 277–281.

Nijhout, H. F. (1981). Physiological control of molting in insects. *Amer. Zool. 21*, 631–640.

Nijhout, H. F. and Williams, C. M. (1974a). Control of moulting and metamorphosis in the tobacco hornworm, *Manduca sexta* (L.): Growth of the last-instar larva and the decision to pupate. *J. Exp. Biol. 61*, 481–491.

Nijhout, H. F. and Williams, C. M. (1974b). Control of moulting and metamorphosis in the tobacco hornworm, *Manduca sexta* (L.): Cessation of juvenile hormone secretion as a trigger for pupation. *J. Exp. Biol. 61*, 493–501.

Nishiitsutsuji-Uwo, J. (1961). Electron microscopic studies on the neurosecretory system in Lepidoptera. *Z. Zellforsh. Mik. Anat. 54*, 613–630.

Nishiitsutsuji-Uwo, J. (1972). Purification and some properties of insect brain hormone extracted from silkworm heads. *Botyu-Kagaku 37*, 93–111.

Nishiitsutsuji-Uwo, J. and Nishimura, M. S. (1972). Adult development induced by the injection of non-hormonal agents into brainless pupae of silkworms. *Appl. Ent. Zool. 7*, 207–216.

Nishiitsutsuji-Uwo, J. and Nishimura, M. S. (1975). Brain hormone contained in the silkworm body of *Bombyx mori*. *Experientia 31*, 1105–1107.

Nishiitsutsuji-Uwo, J. and Nishimura, M. S. (1975). Further studies on non-hormonal agents which mimic insect brain hormone. *Appl. Ent. Zool. 10*, 52–57.

Norris, M. J. (1954). Sexual maturation in the desert locust *Schistocerca gregaria*. *Forskal. Anti-locust Bull. 18*, 1–44.

O'Farrell, A. F. and Stock, A. (1953). Regeneration and the moulting cycle in *Blattella germanica* L. *Aust. J. Biol. Sci. 6*, 485–500.

O'Farrell, A. F., Stock, A. and Morgan, J. (1956). Regeneration and the moulting cycle in *Blattella germanica* L. iv. Single and repeated regeneration and metamorphosis. *Aust. J. Biol. Sci. 9*, 406–422.

Okasha, A. Y. K. (1968a). Effects of sub-lethal high temperature on an insect, *Rhodnius prolixus* (Stal.) II. Mechanism of cessation and delay of moulting. *J. Exp. Biol. 48*, 465–473.

Okasha, A. Y. K. (1968b). Effects of sub-lethal high temperature on an insect, *Rhodnius prolixus* (Stal.) III. Metabolic changes and their bearing on the cessation and delay of moulting. *J. Exp. Biol. 48*, 475–486.

Orchard, I. and Steel, C. G. H. (1980). Electrical activity of neurosecretory axons from the brain of *Rhodnius prolixus*: relation of changes in the pattern of activity to endocrine events during the moulting cycle. *Brain Res. 191*, 53–65.

Ozeki, K. (1962). Studies on the secretion of the juvenile hormone in the earwig, *Anisolabis maritima*. *Sci. Pap. Coll. Gen. Educ. Univ. Tokyo (Biol. Part) 12*, 65–72.

Patel, Y. C., Zingg, H. H. and Dreifuss, J. J. (1977). Calcium-dependent somatostatin secretion from rat neurohypophysis *in vitro*. *Nature 267*, 852–853.

Pflugfelder, O. (1937). Bau, Entwicklung und Funktion der Corpora Allata und Cardiaca von *Dixippus morosus* Br. *Z. Wiss. Zool. 149*, 477–512.

Piepho, H. (1942). Untersuchungen zur Entwicklungsphysiologie der Insektenmetamorphose. Über die Puppenhäutung der Wachsmotte *Galleria mellonella* L. *Wilhelm Roux Archiv. Ent. mech. 141*, 500–583.

Pittendrigh, C. S. (1964). On temperature independence in the clock system controlling emergence time in *Drosophila*. *Proc. Nat. Acad. Sci. 40*, 1018–1029.

Plagge, E. (1938). Weitere Untersuchungen über das Verpuppungshormon bei Schmetterlingen. *Biol. Zbl. 58*, 1–12.

Pohley, H. J. (1960). Experimentelle Untersuchungen über die Steuerung des Häutungsrhythmus bei der Mehlmotte *Ephestia kühniella*. *Zeller. Wilhem. Roux. Archiv. Ent. mech. 152*, 182–203.

Raabe, M. (1965). Étude des phénomènes de neurosécrétion au niveau de la chaîne nerveuse ventrale des Phasmides. *Bull. Soc. Zool. Fr. 90*, 631–654.

Rasul, C. G. (1972). Factors affecting induction and termination of the pupal diapause in the tobacco hornworm, *Manduca sexta* (L.). PhD. thesis, Fargo, North Dakota.

Safranek, L. and Williams, C. M. (1980). Studies of the prothoracicotropic hormone in the tobacco hornworm, *Manduca sexta*. *Biol. Bull. 158*, 141–153.

Saunders, D. S. (1976). *Insect Clocks*. Pergamon Press, New York.

Scharrer, B. (1952). Neurosecretion. XI. The effects of nerve section on the intercerebralis-cardiacum-allatum system of the insect *Leucophaea maderae*. *Biol. Bull. 102*, 261–272.

Scharrer, B. and von Harnack, M. (1958). Histophysiological studies on the corpus allatum of *Leucophaea maderae*. I. Normal life cycle in male and female adults. *Biol. Bull. 115*, 508–520.

Scharrer, E. (1928). Die lichtempfindlichkeit blinder Elritzen (Untersuchungen über das Zwischenhirn der Fische. I.) *Z. vergl. Physiol. 7*, 1–38.

Schmeider, R. G. (1942). The control of metamorphosis in Hymenoptera. *Anat. Rec. 84*, 514.

Schneiderman, H. A. and Gilbert, L. I. (1964). Control of growth and development in insects. *Science 143*, 325–333.

Sedlak, B. J. (1981). An ultrastructural study of neurosecretory fibers within the corpora allata of *Manduca sexta*. *Gen. Comp. Endocr. 44*, 207–218.

Sehnal, F. (1966). Kritisches Stadium der Bionomie und biometrik der in verschiedenen Lebensbedingungen gezüchteten Wachsmotte, *Galleria mellonella* L. (Lepidoptera). *Z. Wiss. Zool. 174*, 53–82.

Sehnal, F. and Edwards, J. S. (1969). Body constraint and developmental arrest in *Galleria mellonella* L.: further studies. *Biol. Bull. 137*, 352–358.

Sieber, R. and Benz, G. (1980). The hormonal regulation of the larval diapause in the codling moth, *Laspeyresia pomonella* (Lep. Tortricidae). *J. Insect Physiol. 26*, 213–218.

Steel, C. G. H. (1973). Humoral regulation of the cerebral neurosecretory system of *Rhodnius prolixus* (Stal.) during growth and molting. *J. Exp. Biol. 58*, 177–187.

Steel, C. G. H. (1975). A neuroendocrine feedback mechanism in the insect moulting cycle. *Nature 253*, 267–269.

Steel, C. G. H. (1977). The neurosecretory system in the aphid *Megoura viciae*, with reference to unusual features associated with long distance transport of neurosecretion. *Gen. Comp. Endocr. 31*, 307–322.

Steel, C. G. H. (1978). Some functions of identified neurosecretory cells in the brain of the aphid, *Megoura viciae*. *Gen. Comp. Endocr. 34*, 219–228.

Steel, C. G. H. and Harmsen, R. (1971). Dynamics of the neurosecretory system in the brain of an insect, *Rhodnius prolixus*, during growth and molting. *Gen. Comp. Endocr. 17*, 125–141.

Steel, C. G. H., Bollenbacher, W. E., Smith, S. L. and Gilbert, L. I. (1982). Haemolymph ecdysteroid titres during larval-adult development in *Rhodnius prolixus*: Correlations with moulting hormone action and brain neurosecretory cell activity. *J. Insect Physiol. 28*, 519–525.

Suzuki, A., Isogai, A., Hori, Y., Ishizaki, H. and Tamura, S. (1975). A simple procedure for partial purification of silkworm brain hormone. *Agr. Biol. Chem. 39*, 2157–2162.

SUZUKI, A., NAGASAWA, H., KATAOKA, H., HORI, Y., ISOGAI, A., TAMURA, S., GUO, F., ZHONG, X., ISHIZAKI, H., FUJISHITA, M. and MIZOGUCHI, A. (1982). Isolation and characterization of prothoracicotropic hormone from silkworm, *Bombyx mori*. *Agr. Biol. Chem. 46*, 1107–1109.

TAKEDA, N. (1976). The direct release of neurosecretory material from the cell in the pars intercerebralis of *Monema flavescens* (Lepidoptera: Heterogeneidae). *Appl. Ent. Zool. 11*, 143–153.

TAKEDA, N. (1978). Hormonal control of prepupal diapause in *Monema flavescens* (Lepidoptera). *Gen. Comp. Endocr. 34*, 123–131.

TANAKA, Y. (1950). Studies on hibernation with special reference to photoperiodicity and breeding of the Chinese Tussar silkworm I. *J. Sericult. Sci. Jpn. 19*, 358–371.

TAUBER, M. J. and TAUBER, C. A. (1976). Insect seasonality: diapause maintenance, termination, and post-diapause development. *Ann. Devel. Ent. 21*, 81–107.

THOMSEN, E. and LEA, A. O. (1968). Control of the medial neurosecretory cells by the corpus allatum in *Calliphora erythrocephella*. *Gen. Comp. Endocrinol. 12*, 51–57.

THURSTON, R. (1976). Diapause inhibition in *Manduca sexta*, and *Apanteles congregatus* by high scotophase temperatures. *Environ. Ent. 5*, 626–627.

TRUMAN, J. W. (1972). Physiology of insect rhythms I. Circadian organization of the endocrine events underlying the moulting cycle of larval tobacco hornworms. *J. Exp. Biol. 57*, 805–820.

TRUMAN, J. W. (1976). Extraretinal photoreception in insects. *Photochem. Photobiol. 23*, 215–225.

TRUMAN, J. W. and RIDDIFORD, L. M. (1974). Physiology of insect rhythms. III. The temporal organization of the endocrine events underlying pupation of the tobacco hornworm. *J. Exp. Biol. 60*, 371–382.

TRUMAN, J. W., RIDDIFORD, L. M. and SAFRANEK, L. (1973). Hormonal control of cuticle coloration in the tobacco hornworm, *Manduca sexta*: Basis of an ultrasensitive bioassay for juvenile hormone. *J. Insect Physiol. 19*, 195–203.

UNGER, H. (1957). Untersuchungen zur neurohormonalen Steuerung der Herztätigkeit bei Insekten. *Biol. Zbl. 76*, 204–225.

VAN DER KLOOT, W. G. (1960). Neurosecretion in insects. *Ann. Rev. Ent. 5*, 35–52.

VAN DER KLOOT, W. G. (1961). Insect metamorphosis and its endocrine control. *Amer. Zool. 1*, 3–9.

VON BUDDENBROCK, W. (1930). Beitrag zur Histologie und Physiologie der Rauperhäutung mit besonderer Berücksichtigung der Versonschen Drüsen. *Z. Morph. Ökol. Tiere 18*, 700–725.

WEED, I. G. (1936). Experimental study of moulting in the grasshopper *Melanoplus differentialis*. *Proc. Soc. Exp. Biol. Med. (Amer.) 34*, 885–886.

WIGGLESWORTH, V. B. (1933). The physiology of the cuticle and of ecdysis in *Rhodnius prolixus* (Triatomidae, Hemiptera); with special reference to the function of the oenocytes and of the dermal glands. *Quart. J. Mic. Sci. 76*, 296–318.

WIGGLESWORTH, V. B. (1934). The physiology of ecdysis in *Rhodnius prolixus* (Hempitera). II. Factors controlling moulting and metamorphosis. *Quart. J. Mic. Sci. 77*, 191–222.

WIGGLESWORTH, V. B. (1936). The function of the corpus allatum in the growth and reproduction of *Rhodnius prolixus* (Hemiptera). *Quart. J. Mic. Sci. 79*, 91–121.

WIGGLESWORTH, V. B. (1939). Hautung bei Imagines von Wanzen. *Naturwiss. 27*, 301.

WIGGLESWORTH, V. B. (1940a). Local and general factors in the development of "pattern" in *Rhodnius prolixus* (Hemiptera). *J. Exp. Biol. 17*, 180–200.

WIGGLESWORTH, V. B. (1940b). The determination of characters at metamorphosis in *Rhodnius prolixus* (Hemiptera). *J. Exp. Biol. 17*, 201–222.

WIGGLESWORTH, V. B. (1955). The role of the haemocytes in the growth and moulting of an insect, *Rhodnius prolixus* (Hemiptera). *J. Exp. Biol. 32*, 649–663.

WIGGLESWORTH, V. B. (1964). The hormonal regulation of growth and reproduction in insects. *Adv. Insect Physiol. 2*, 247–336.

WILKENS, M. B. (1965). The influence of temperature and temperature changes on biological clocks. In *Circadian Clocks*. Edited by J. Aschoff. Pages 146–163. North-Holland, Amsterdam.

WILLIAMS, C. M. (1947). Physiology of insect diapause. II. Interaction between the pupal brain and prothoracic glands in the metamorphosis of the giant silkworm, *Platysamia cecropia*. *Biol. Bull. 93*, 89–98.

WILLIAMS, C. M. (1948a). Physiology of insect diapause. III. The prothoracic glands in the *cecropia* silkworm, with special reference to their significance in embryonic and postembryonic development. *Biol. Bull. 94*, 60–65.

WILLIAMS, C. M. (1948b). Extrinsic control of morphogenesis as illustrated in the metamorphosis of insects. *Growth Symp. 126*, 61–74.

WILLIAMS, C. M. (1952). Physiology of insect diapause. IV. The brain and prothoracic glands as an endocrine system in the *cecropia* silkworm. *Biol. Bull. 103*, 120–138.

WILLIAMS, C. M. (1956). Physiology of insect diapause. X. An endocrine mechanism for the influence of temperature on the diapausing pupa of the *cecropia* silkworm. *Biol. Bull. 110*, 201–218.

WILLIAMS, C. M. (1968a). The present status of the brain hormone. In *Insects and Physiology*. Edited by J. W. L. Beament and J. E. Treherne. Pages 133–139. American Elsevier, New York.

WILLIAMS, C. M. (1968b). The effects of temperature gradients on the pupal-adult transformation of silkworms. *Biol. Bull. 134*, 347–355.

WILLIAMS, C. M. (1969). Photoperiodism and the endocrine aspects of insect diapause. *Symp. Soc. Exp. Biol. 23*, 285–300.

WILLIAMS, C. M. and ADKISSON, P. L. (1964). Physiology of insect diapause. XIV. An endocrine mechanism for the photoperiodic control of pupal diapause in the oak silkworm, *Antheraea pernyi*. *Biol. Bull. 127*, 511–525.

YAGI, S. and FUKAYA, M. (1974). Juvenile hormone as a key factor regulating larval diapause of the rice stem borer, *Chilo suppressalis* (Lepidoptera: Pyralidae). *Appl. Ent. Zool. 9*, 247–255.

YAMAZAKI, M. and KOBAYASHI, M. (1969). Purification of the proteinic brain hormone of the silkworm, *Bombyx mori*. *J. Insect Physiol. 15*, 1981–1990.

YIN, C.-M. and CHIPPENDALE, G. M. (1973). Juvenile hormone regulation of the larval diapause of the southwestern corn borer, *Diatraea grandiosella*. *J. Insect Physiol. 19*, 2403–2420.

5 Chemistry of Peptide Hormones

W. MORDUE and P. J. MORGAN

University of Aberdeen, Aberdeen, UK

1 CHEMISTRY OF PEPTIDE HORMONES

Insect peptide hormones are involved in the control of a wide range of physiological, biochemical and developmental functions, including water balance, lipid and carbohydrate metabolism, muscle contraction, reproduction, and growth and development (see Steele, vol. 8). The existence of some of these hormones and the functions which they control have been known for some time, yet the structures or the amino acid sequences of some of these peptides have been elucidated only in recent years. Our knowledge of the chemical structures of insect peptides, until the last few years, was limited by the insensitive, slow, low-resolution chromatographic techniques which were available. These problems were compounded by the lack of sufficient starting material. Therefore the structural characterization of the first two neurohormones, proctolin and adipokinetic hormone, by these methods, were substantial achievements.

However, more recently the development of high performance liquid chromatography (HPLC), especially in the reverse-phase has revolutionized the field of peptide chromatography. HPLC offers the facility to rapidly separate small amounts of material with high resolution and recovery. The advances in column and detector technology have made it possible to perform amino acid analysis using reverse-phase HPLC and fluorometric detection using femta-mole levels of material. Yet in more routine amino acid analysis of purified biological material the limit of reliable detection for small peptides is nearer 1–10 picomoles of peptide. Therefore, although HPLC should not be heralded as the panacea for peptide isolation, undoubtedly it has made a major impact in this field of research, and hopefully in the next few years our knowledge of the structural chemistry of insect peptide hormones will be greatly advanced.

Such new-found knowledge would open up numerous new avenues of research, and enable us to answer questions which we hitherto have been unable to answer, such as the existence of vertebrate-like hormones in insects (Kramer, this volume), the role of certain peptides as neurotransmitters/neuromodulators (vol. 11), and the synergistic interplay between hormones (Steel and Davey, vol. 8). Therefore peptides are a most important class of compounds and knowledge of their chemistry is fundamental to our greater understanding of their role(s). In this chapter the chemistry of only a few of the many peptides known to exist has been discussed. This is for two reasons; firstly, there is still very little information about the chemistry of some insect peptide hormones, and secondly, other peptides will have been covered in some detail in other chapters and volumes of this series.

1.1 Diuretic hormone(s) (DH)

Diuretic hormones(s) (DH) is the name given to an as yet uncharacterized neurosecretory factor(s), that is involved in the regulation of insect water balance. The presence of a diuretic hormone has been shown or implicated in a wide range of insects (see Table 1); yet study of its chemical nature has been limited to only a few insects. Readers who may wish to know more of the background physiology and actions of DH are directed to reviews by

Table 1: Occurrence of diuretic hormone(s) in insects

Insect	Source(s)	Reference
Apis	Corpus allatum	Altmann, G. (1956)
Anisotarsus	Median neurosecretory cells/corpus cardiacum	Nunez, J. (1956)
Rhodnius	Mesothoracic ganglionic mass (brain, sub-oesophageal ganglion, prothoracic ganglion)	Maddrell, S. (1963)
Carausius	Median neurosecretory cells/corpus cardiacum (suboesophageal ganglion)	Unger, H. (1965); Vietinghoff, U. (1967); Pilcher, D. (1970)
Dysdercus	Median neurosecretory cells	Berridge, M. (1966)
Periplaneta	Terminal abdominal ganglion	Mills, R. (1967)
Corethra	Abdominal ganglia (1–3)	Gersch, M. (1967)
Locusta	Median neurosecretory cells — corpus cardiacum (brain, sub-oesophageal and thoracic ganglia)	Cazal, M. and Girardie, A. (1968); Mordue, W. (1972); Morgan (unpublished)
Schistocerca	Median neurosecretory cells/corpus cardiacum	Mordue, W. (1969)
Glossina	Thoracic ganglia (and head)	Gee, J. (1975)
Pieris	Brain — corpus cardiacum	Nicolson, S. (1976)
Anopheles	Head and thoracic ganglia	Nijhout, H. and Carrow, G. (1978)
Danaus	Brain — corpus cardiacum	Dores, R. *et al.* (1979); Nicolson, S. (1980)
Papilio	Brain — corpus cardiacum	Nicolson, S. (1980)
Acraea	Brain — corpus cardiacum	Nicolson, S. (1980)

Maddrell, S. (1971), Goldsworthy, G. and Mordue, W. (1974), and Gee, J. (1977).

1.1.1 THE PEPTIDERGIC NATURE OF DIURETIC HORMONE

By the selective use of proteolytic enzymes it has been established for a number of insects that DH is a peptide or polypeptide. Trypsin, which cleaves specifically peptide bonds in which the carbonyl group is donated by either lysine or arginine, has been found to destroy the diuretic activity of hormone extracts from *Rhodnius prolixus* (Aston, R., 1979), *Danaus plexippus* (Dores, R. *et al.*, 1979), *Calliphora* (Schwartz, L. and Reynolds, S., 1979), *Glossina austeni* (Gee, J., 1975) and *Locusta migratoria* (Morgan, unpublished). This suggests that the DH from insects might contain either lysine or arginine, or both.

Chymotrypsin, which hydrolyses peptide bonds in which phenylalanine, tryptophan or tyrosine donate the carbonyl group, has been found to destroy the diuretic activity in hormone extracts from *Rhodnius* (Aston, R., 1979), *Glossina* (Gee, J., 1975), *Schistocerca gregaria* (Mordue, W. and Goldsworthy, J., 1969) and *Locusta* (Morgan, unpublished). This implies that at least one of the above amino acid residues is present in the diuretic hormone molecule.

A variety of other enzymes, including neuraminidase, α-amylase, DNase, RNase and lipase, were tested for their effects on the diuretic activity of DH from *Glossina* (Gee, J., 1975). Incubation with neuraminidase, which cleaves the bonds between *N*-acetyl neuraminic acid and sugar residues, and α-amylase which hydrolyses glycosidic linkages in starch chains, were both found to destroy the DH activity; none of the other enzymes had any effect. On the basis of these experiments Gee, J. (1975) postulated that DH of *Glossina* is a sialoglycoprotein. However, when neuraminidase, α-amylase, leucine aminopeptidase and lipase were tested for their effects on DH from *Rhodnius*, they were without effect on the diuretic activity. The lack of effect of leucine aminopeptidase, which hydrolyses the peptide bonds of *N*-terminal amino acids, has also been observed in experiments on *Locusta* DH (Morgan, unpublished). Aston, R. and Hughes, L. (1980) have interpreted this finding as implying that either DH has no N-terminal amino acid (i.e. is a cyclic peptide) or it is protected from hydrolysis, or even that the loss of the N-terminal amino acid does not affect the biological activity of the DH molecule. However, the latter explanation would seem unlikely, as there seems no reason why after the removal of a terminal residue the enzyme does not proceed to attack the new N-terminal residue.

One of the cautions which must be observed in the interpretation of enzyme digestion experiments such as those described above is that of specificity of the enzymes (see also Aston, R. and Hughes, L., 1980). The presence of contaminating enzymes in "pure" enzyme preparations may produce misleading results; yet it does seem reasonable to postulate that DH is a peptide. However, when considering the nature of DH one must be aware of the mimic effects of pharmacological stimulants; 5-hydroxytryptamine, which has been identified in the nervous system of a number of insect species (Welsh, J. and Moorhead, M., 1960; Berridge, M. and Patel, N., 1968; Hiripi, L. and S-Rozsa, K., 1973), is a potent stimulant of fluid secretion by the Malpighian tubules of both *Rhodnius* and *Carausius* at concentrations as low as 3×10^{-8} M (Maddrell, S. *et al.*, 1969). However, Maddrell, S. *et al.* (1971) demonstrated that 5-HT is not the DH in these insects; therefore, as DH also seems extractable from all parts of the insect CNS, care must be taken when isolating DH that pharmacological effects are not confused with the effect of the natural hormone.

1.1.2 Hormone stability

A major problem in the isolation of DH has been its lability; it is only in *Danaus plexippus* that the hormone has been shown to be stable in aqueous extracts at room temperature (Dores, R. *et al.*, 1979). In other insects, such as *Rhodnius prolixus* (Aston, R. and White, A., 1974), *Glossina austeni* (Gee, J., 1975), *Locusta* (Rafaeli, A. and Mordue, W., 1982; Morgan, P. and Mordue, W., 1983), *Anopheles freeborni* (Nijhout, H. and Carrow, G., 1978) and *Calliphora* (Schwartz, L. and Reynolds, S., 1979), the hormone has been found to be unstable at room temperature. DH of *Glossina, Anopheles* and *Calliphora* can be stabilized if the extracts are heated briefly at 100°. This could indicate that the loss of diuretic activity is due to degradation by proteolytic enzymes released along with DH during the extraction procedure. However, similar heat treatments of DH extracts from *Rhodnius* were found to destroy all diuretic activity, indicating a molecular instability of the hormone or an alternative mechanism for the loss of activity (Aston, R. and White, A., 1974; Aston, R., 1979). *Rhodnius* DH has a tendency to adhere to charged surfaces such as glass, chromatography paper and cellulose acetate (Maddrell, S. *et al.*, 1971); indeed, it has been shown that the half-life of *Rhodnius* DH is related to the exposed surface area of glass beads (Aston, R., 1979). However, such loss of DH activity due to physical adsorption can be reduced by prior treatment of potential adsorption sites with DH or by increasing the concentration of DH (Aston, R., 1979; Hughes, L., 1979). Alternatively it has been found that the half-life of DH can be increased by including 0.1% bovine serum albumen (BSA) in the hormone extracts. It is suggested that the losses of DH are reduced through non-specific hormone binding and also perhaps by competition for potential binding sites on charged surfaces (Aston, R. and Hughes, L., 1980).

In *Locusta*, boiling DH in glass test tubes for 10 min has been found to destroy diuretic activity (Morgan, unpublished). Such a loss of activity could have been due to adherence of DH to charged sites on the glass. However, hormone extracts heated in polypropylene vials at 70° for 60 min lost most of their activity, whereas those heated for only 10 min retained their diuretic potency, see Table 2 (Morgan, P. and Mordue, W., 1983). This result suggests that short periods of heating might have destroyed enzymes which might degrade the hormone, whereas the hormone is sensitive to, and is destroyed by, longer periods of heating. To test this, the extracts heated for 10 min were left at room temperature (*ca.* 23°) and tested for activity after 24 h and 4 days. After 24 h considerable diuretic activity was still detectable, contrasting with non-heated extracts. However, after 4 days the extracts had lost their diuretic activity completely. This implies that short periods of heating can partly

Table 2: The effect on diuretic hormone activity of incubating aqueous extracts of corpus cardiacum storage lobe at 70° (after Morgan, P. and Mordue, W., 1983)

	Δ rate (nl/min)	SE	*n*	*t*
(1) Control (immediate assay)	3.04	0.35	17	8.8
(2) After 60 min at 70°	0.86	0.5	18	1.7
(3) After 10 min at 70°				
(a) Immediate assay	2.5	0.25	17	10
(b) Assayed after 24 h at 22–23°	2.6	0.43	8	6.1
(c) Assayed after 4 days at 22–23°	0.23	0.34	18	0.97

stabilize diuretic activity by denaturing degradative enzymes in the homogenate. Yet either not all the enzymes were destroyed, or the hormone has an inherent molecular instability, or it sticks to vacant charged sites in the polypropylene vials, so that activity is lost after 4 days (Morgan, P. and Mordue, W., 1983).

The lability of DH has complicated the separation and chemical characterization of DH. However, besides the use of 0.1% BSA to suppress the loss of DH through adsorption, it has been shown that DH activity can be preserved for long periods if extracts are kept at −20° or below (Aston, R. and White, A., 1974; Hughes, L., 1979; Dores, R. *et al.*, 1979; Morgan, P. and Mordue, W., 1983).

1.1.3 Solubility in organic solvents

The solubility of putative hormones in different organic solvents can be useful. For example, small peptides are soluble in methanol, and can be separated readily from large molecular weight proteins such as degradative enzymes which will be precipitated out in this solvent. Furthermore, solubility in a volatile solvent, without loss of biological activity, can reduce drastically the time needed for evaporative drying. However, as for many of the properties of DH, the solubilities of DH from different insects in different organic solvents show inconsistencies. In *Rhodnius prolixus*, DH was found to be poorly soluble in a variety of organic solvents, including methanol (Aston, R. and White, A., 1974; Aston, R., 1979), which suggested that *Rhodnius* DH is a polar molecule (Aston, R., 1979). Similarly, DH of *Danaus plexippus* was found to be poorly soluble in methanol and chloroform, yet soluble in acetic acid (Dores, R. *et al.*, 1979). By contrast *Calliphora* DH appears to be soluble to some extent in ethanol, methanol, acetone and acetic acid (Schwartz, L. and Reynolds, S., 1979), although the authors suggested that greater solubility in these solvents might have been disguised by losses during evaporative drying. The DHs of both *Glossina austeni* and *Locusta migratoria* are soluble in methanol (Gee, J., 1975; Rafaeli, A. and Mordue, W., 1982), although in the case of *Glossina* there is no indication as to whether any activity is lost compared to aqueous extracts. In *Locusta* there is negligible loss of activity (Rafaeli, A. and Mordue, W., 1982; Morgan, unpublished). The variation in solubilities of the different insect DHs probably reflects subtle differences in their structure.

1.2 Isolation and purification of diuretic hormone

The lability of DH has contributed to the lack of success of a number of attempts to separate DH from crude extracts. Gee, J. (1975) using *Glossina austeni* DH, was unable to detect diuretic activity in the eluant from a Bio-Gel P-6 column, and the loss of activity during lyophilization has hampered efforts to characterize DH of *Calliphora* further (Schwartz, L. and Reynolds, S., 1979). However, some progress has been made in the isolation of diuretic hormone in *Rhodnius prolixus*, *Locusta* and *Periplaneta americana* (Aston, R. and White, A., 1974; Aston, R., 1979; Hughes, L., 1977, 1979; Morgan, P. and Mordue, W., 1983; Goldbard, G. *et al.*, 1970). The isolation procedure, and the information they provide about the chemical nature of each DH, will be considered separately.

1.2.1 *Rhodnius*

Using a crude aqueous extract prepared by sonication of metathoracic ganglionic masses (MTGM), the source of DH in *Rhodnius prolixus* (Maddrell, S., 1963), differential centrifugation at 100,000 ***g*** has been used as the first stage of the purification procedure (Aston, R. and White, A., 1974; Aston, R., 1979; Hughes, L., 1979). Aston, R. and White, A. (1974) found that 80% of the diuretic activity sedimented with the insoluble debris, leaving only 20% activity in the supernatant. It now appears that most of the diuretic activity can be solubilized by more efficient extraction procedures (Hughes, L., 1977, 1979).

Initial attempts to isolate DH were made using gel filtration chromatography, utilizing polyacrylamide gels. When samples of the 100,000 ***g*** supernatant were applied to P-30, P-60 and P-2 columns diuretic activity was found to elute as two distinct zones (Aston, R. and White, A., 1974). Most of the diuretic activity was found in the void volume, suggesting a molecular weight of greater than 60,000. However, some diuretic activity was found in the second zone with an apparent molecular weight of less than 2000. The diuretic factor extracted from the haemolymph of *Rhodnius* is comparable to the low molecular weight hormone, and it is suggested that this is the form in which DH is released into the haemolymph following a blood meal (Aston, R. and White, A., 1974).

The currently accepted concept for the synthesis and release of vertebrate peptide hormones is that the biologically active hormones are synthesized as parts of larger precursor proteins (prohormones), which themselves have no biological activity (Schwyzer, R., 1982). The physiologically active hormone is then released from the prohormone by the action of specific enzymes. Hughes, L. (1979) has implied that the high molecular weight form of *Rhodnius* DH might be a prohormone or possibly a preprohormone, although it would be unusual for a prohormone to possess biological activity. It is possible, however, that the high molecular weight form is biologically inactive but is converted into the physiologically active smaller hormone during the assay procedure (Hughes, L., 1979).

It is known that Bio-Gel matrices exhibit cation exchange properties in distilled water, due to the presence of carboxylic acid groups (Williams, K., 1972). In consequence Aston, R. (1979) has suggested that the faster elution time of DH from a Bio-Gel P-2 column using distilled water indicates the anionic (acidic) nature of *Rhodnius* DH. Further evidence for such anionic properties of DH is claimed from electrophoretic mobilities; high molecular weight DH when electrophoresed for 1 h at pH 7.1 remained at the origin, whereas at pH 9.0 the diuretic activity was detectable in two adjacent fractions which migrated towards the anode (Aston, R., 1979). However, the low molecular weight DH could not be detected when run on such a system. Therefore the anionic nature of the high molecular weight factor may be due to components other than the "physiological" hormone.

It must be stressed that the low molecular weight diuretic hormone is distinguished clearly from 5-HT; the chromatographic characteristics of the two compounds are different (Aston, R. and White, A., 1974).

1.2.2 *LOCUSTA*

Gel filtration chromatography has also been applied successfully to the separation of DH from the storage lobes of the corpus cardiacum, the source of DH in the locust (Morgan, P. and Mordue, W., 1983). DH was separated on a Bio-Gel P-6 column with distilled water as eluant; the elution characteristics suggest that *Locusta migratoria* DH is a small molecular weight compound, probably with a molecular weight below 2000 (Morgan, P. and Mordue, W., 1983). The diuretic hormone isolated from *Locusta* is similar to the low molecular weight hormone of *Rhodnius prolixus*. There is no evidence for the existence of a high molecular weight diuretic hormone in *Locusta* (Morgan, P. and Mordue, W., 1983).

The recoveries of *Locusta* DH from the Bio-Gel P-6 column were poor; considerable loss of *Rhodnius* DH, especially of the low molecular weight hormone, has been noted during purification (Aston, R., 1979). This loss of activity on the Bio-Gel columns may have been due to the free carboxyl groups on the gels (Williams, K., 1972). However, this would suggest a cationic property for the *Locusta* DH, rather than an anionic property as suggested for *Rhodnius* DH (Aston, R., 1979). A buffer of high ionic strength can be used to suppress the charges on the gels; however, when 0.025 M ammonium acetate was used as the solvent, the resolution of the separations and recoveries were not improved (Morgan, unpublished).

The use of reverse-phase high performance liquid chromatography (RP-HPLC) has been found to be a powerful technique for the separation of peptides (Hearn, M., 1980). RP-HPLC has been used to separate *Locusta* DH from methanol extracts of locust storage lobes, and the diuretic activity is separated into at least two fractions, DH I and DH II (see Fig. 1) (Morgan, P. and Mordue, W., 1983). RP-HPLC separates peptides according to relative hydrophobicities of the molecules (Meek, J. and Rossetti, Z., 1981; Hearn, M., 1980), so that the more hydrophobic (non-polar) the molecule, the longer it is retained on the column; more polar compounds have shorter retention times. The resolving power of RP-HPLC is such that only minor changes in the structure of the molecule, such as an amino acid replacement or deletion, will alter the relative hydrophobicity of the molecule and this will affect the retention time (Hearn, M., 1980). Therefore any difference in structure between DH I and DH II may be quite subtle. It is worth noting that other

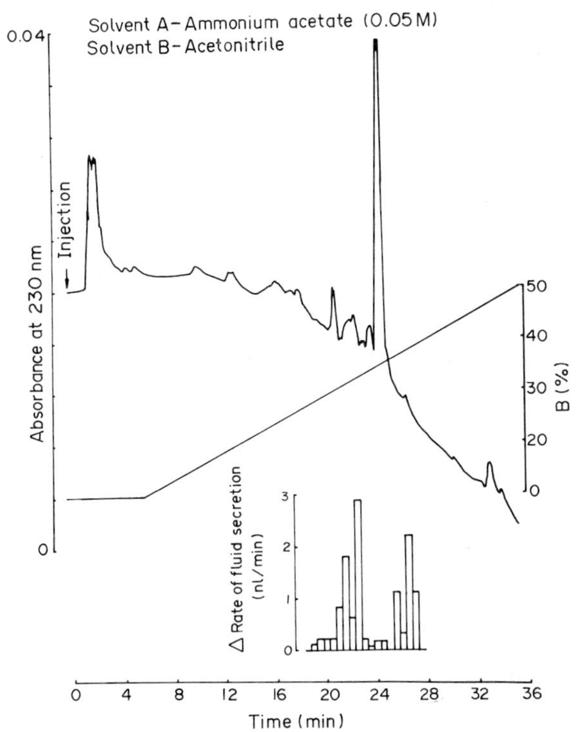

FIG. 1. Separation of 20 storage lobes from *Locusta migratoria* by reverse-phase HPLC on a Shandon Hypersil ODS column, using a linear gradient elution. Absorbance was monitored at 230 nm at 0.04 absorbance units full scale. Flow rate — 1.5 ml/min; chart speed — 0.5 cm/min; solvent A — 0.05 M ammonium acetate buffer (ph 5.6); solvent B — acetonitrile. Inset shows fractions showing diuretic activity.

insect peptide hormones have been isolated in more than one form; for example, AKH I and AKH II (Carlsen, J. *et al.*, 1979), diapause hormone A and B (Isobe, M. and Goto, T., 1980; Yamashita and Hasegawa, vol. 1), and the prothoracicotropic hormone (see Bollenbacher and Granger, this volume).

The retention times of the two fractions with diuretic activity are similar to AKH, and as AKH is a non-charged (non-polar) molecule (Stone, J. *et al.*, 1976), this implies that both DH I and II are relatively hydrophobic (i.e. non-polar) molecules, although this disagrees with the cationic property inferred above. If DH of *Locusta* is less polar than *Rhodnius* DH, this might explain why *Locusta* DH is soluble in methanol, whereas *Rhodnius* DH is not.

In the separation shown in Fig. 1, the lowest wavelength at which absorbance could be monitored was 230 nm, due to restrictions imposed by the absorbance of the solvent system. Yet when the absorbance was measured at 280 nm, a wavelength at which tryptophan and tyrosine absorb, no absorbance peaks corresponding to DH were observed. AKH, which contains a tryptophan residue in its structure, was found to absorb strongly at 280 nm (Morgan, P. and Mordue, W., 1983). This result may reflect the absence of these two amino acid residues from the DH molecule of *Locusta*; yet experiments with chymotrypsin (see above) would suggest the presence of a tyrosine, tryptophan or phenyalanine residue. In comparison to AKH, the absorbance peaks for DH monitored at 230 nm (Fig. 1) are small. This reflects the high concentration of AKH relative to DH, and suggests that the concentration of DH may be too low to be detectable at 280 nm. Alternatively, contaminating enzymes (such as trypsin) may have been present with the chymotrypsin, making its proteolytic action less specific, despite the use of chymotrypsin treated with 1-chloro-3-tosylanido-7-amino-L-2-heptanone (TLCK), to inactivate any trypsin.

Clearly the results from HPLC have been encouraging, and hopefully future work using both reverse-phase HPLC and high performance gel filtration chromatography should allow the chemical nature of DH to be fully characterized.

1.2.3 PERIPLANETA

The source of DH in *Periplaneta americana* is the terminal abdominal ganglion (Mills, R., 1967). Using a Bio-Gel P-30 column extracts of the terminal abdominal ganglia were eluted in 0.1 M sodium phosphate buffer. Diuretic activity was found in the void volume, suggesting a diuretic factor with a molecular weight of greater than 30,000. This factor may be similar to the high molecular weight form found in *Rhodnius prolixus* (Aston, R. and White, A., 1974); no low molecular weight hormone was observed. It has also been suggested that the *Periplaneta* DH might be analogous to the tanning hormone, bursicon (Mills, R. and Whitehead, D., 1970; Reynolds, vol. 8).

It is pertinent at this point to comment on the target tissues of DH. There may be two or three target tissues where a DH could exert its effect. The first is the Malpighian tubules, where DH promotes elevated fluid secretion, and this action forms the basis of an *in vitro* bioassay system used in most studies to detect diuretic activity. However, a second site for DH action might also be the rectum (Mordue, W., 1972), where it could act by reducing rectal resorption; indeed this is the basis of the bioassay used in the isolation of a DH in *Periplaneta* (Goldbard, G. *et al.*, 1970). Maddrell, S. (1972) has also postulated that DH may control the rate of water absorption by the midgut. Therefore, in view of the three potential sites of DH action it is possible, if not likely, that the same DH acts at each site, providing a neatly integrated control mechanism. However, until the DH has been chemically characterized care should be taken not to exclude the possibility of chemically distinct hormones acting at different sites.

1.3 Is insect diuretic hormone a vasopressin-like molecule?

Recently, as the result of immunochemical and physiological studies, it has been suggested that DH of *Locusta migratoria* is a vasopressin-like molecule (Proux, J. and Rougon-Rapuzzi, G., 1980; Proux, J. *et al.*, 1980, 1982). This immunoreactive material has been identified throughout the nervous system from the brain to the abdominal ganglia, but with the highest concentration in the sub-oesophageal ganglion (SOG) (Proux, J. and Rougon-Rapuzzi, G., 1980). The localization of vasopressin-like material in the nervous system of a number of insects has been reported (Remy, C. *et al.*, 1977, 1979; Strambi, C. *et al.*, 1978, 1979; Remy, C. and Girardie, J., 1980; see Kramer, this volume).

1.3.1 THE CHEMICAL CHARACTERISTICS OF THE VASOPRESSIN-LIKE MOLECULE

Using a Sephadex G-25 column, eluted in 0.1 N HCl, an acidic extract of SOG was filtered, and tested for vasopressin-like immunoreactivity. The latter was found to elute as a peak in a position similar to arginine vasopressin (Proux, J. and Rougon-Rapuzzi, G., 1980), indicating similar molecular weights. The structure of arginine vasopressin (AVP) is shown in Fig. 2. As the curves for the displacement of [^{125}I]AVP from antibodies by locust vasopressin-like material and synthetic vasopressin were superimposable over a wide range of concentrations, it was concluded that synthetic AVP and locust vasopressin-like material have very

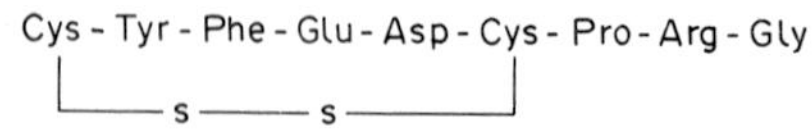

M.W. 1083

FIG. 2. The structure of arginine vasopressin (MW 1083).

similar structures (Proux, J. and Rougon-Rapuzzi, G., 1980). However, as injection of synthetic vasopressin, even at pharmacological doses, was found not to affect dye excretion, it has more recently been suggested that any structural similarity is restricted to its antigenic determinant (Proux, J. *et al.*, 1982).

Proux, J. and Rougon-Rapuzzi, G. (1980) showed that the locust vasopressin-like material is stable to neutral pH, freezing, thawing and room temperature, although there is no indication as to how long the extracts were left at room temperature. If the vasopressin-like material is stable at room temperature then this contrasts with the marked instability of DH from most other insects (Aston, R. and White, A., 1974; Gee, J., 1975; Nijhout, H. and Carrow, G., 1978; Schwartz, L. and Reynolds, S., 1979), and also with the stability of DH extracted from the storage lobes of *Locusta* (Rafaeli, A. and Mordue, W., 1982; Morgan, P. and Mordue, W., 1983).

As no vasopressin-like material was reported to be found in the corpus cardiacum of *Locusta* (Proux, J. and Rougon-Rapuzzi, G., 1980), this may indicate that the vasopressin-like hormone is chemically distinct from DH found in the storage lobes (Mordue, W., 1969; Morgan, P. and Mordue, W., 1983; see above). However, Proux, J. *et al.* (1982) discount the corpora cardiaca storage lobes as a source of diuretic hormone, as they were unable to affect either the *in vivo* rate of dye elimination or the rate of *in vitro* urine excretion, contrary to the well-established findings of Mordue, W. (1969, 1972), and Mordue, W. and Goldsworthy, G. (1969). In our laboratory, in addition to the diuretic activity found in the corpus cardiacum, diuretic activity has also been noted in extracts of the sub-oesophageal ganglion (Morgan, unpublished observations). Currently, work is in progress to determine whether the diuretic activity extracted from both the corpus cardiacum and the sub-oesophageal ganglion is due to the same hormone or structurally different factors.

2 CHLORIDE TRANSPORT STIMULATING HORMONE (CTSH)

The insect rectum is an important organ involved in the regulation of both ion and water balance (see Phillips, J., 1980; Bradley, vol. 4). In the locust, *Schistocerca gregaria*, it has been shown that transport of Cl^- from the rectal lumen into the haemolymph is under the control of a hormone from the corpus cardiacum (Spring, J. and Phillips, J., 1980a,b). As the hormone stimulates increased transport of chloride ions it has been named chloride transport stimulating hormone (CTSH) (Spring, J. and Phillips, J., 1980b).

2.1 The chemical properties of CTSH

CTSH activity from both crude or partially purified extracts was found to be destroyed by trypsin, indicating that the CTSH molecule contains peptide linkages in its structure (Phillips, J. *et al.*, 1980). CTSH activity can be extracted in both aqueous solvents and 80% methanol with only trace activity being left in the protein precipitate. In crude aqueous extracts left at room temperature, CTSH activity has a half-life of about 20 h (Phillips, J. *et al.*, 1980); yet little loss in activity occurs if the homogenates are stored frozen (Phillips, J. *et al.*, 1980).

Crude extracts exposed to buffers with pH values below 5, showed a loss of CTSH activity within 12 h at 4°, whereas at higher pH (7–10) CTSH was stable for as long as 8 months at −20°. Separation of CTSH from crude extracts by electrophoresis on cellulose acetate strips at pH values above 5 showed that CTSH migrated towards the anode in a single band (Phillips, J. *et al.*, 1980). This result indicates that not only does the CTSH molecule have a net negative charge at neutral to alkaline pH, but also the isoelectric point must occur below pH 5.

The anionic nature of the CTSH molecule at alkaline pH is further confirmed by the fact that CTSH activity does not elute from DEAE-Sephadex A-50 anion exchange column eluted in Tris buffer (pH 9.3); yet running a NaCl gradient from 15 to 150 mM displaced the hormone from the column giving a 17-fold purification of the crude homogenate (Phillips, J. *et al.*, 1980).

2.2 Separation of CTSH by gel filtration chromatography

Separation of corpus cardiacum homogenates on a Bio-Gel P-6 column showed all the CTSH activity to be eluted in the void volume, indicating a molecular weight in excess of 6000 (Phillips, J. *et al.*, 1980). This result was confirmed by dialysis, and by using a Sephadex G-75 column; both methanol and saline extracts were eluted in a single peak, slightly later than the elution volume for cytochrome C. Similarly, 96% of CTSH activity was found to elute shortly after cytochrome C on a Bio-Gel P-30 column with trace activity (4%) in the void volume. From the elution volumes of molecular weight marker proteins, the molecular weight of CTSH in the main peak was estimated to be about 8000 daltons, and the CTSH in the void volume >30,000 (Phillips, J. *et al.*, 1980). Phillips, J. *et al.* (1981) concluded that it is not yet clear whether the high molecular form is a "prehormone" or an artefact due to the non-specific absorption of CTSH to higher molecular weight protein. Recovery of CTSH from the Bio-Gel P-30 column was reported to vary from less than 35% to greater than 85%; it is possible that this variability reflects the instability of the purified CTSH (Phillips, J. *et al.*, 1980). However, relative to the protein concentration in the fraction containing the low molecular weight CTSH, it is estimated that at least 100-fold purification is achieved (Phillips, J. *et al.*, 1980).

2.2.1 Separation of CTSH by thin layer chromatography

Methanol extracts of either storage lobes or glandular lobes chromatographed on thin layer cellulose plates (eluted in isopropanol:acetate:water) showed all the CTSH activity to remain at the origin ($R_F = 0$). As adipokinetic hormone (AKH), antidiuretic hormone (ADH) and diuretic hormone (DH) from locust corpus cardiacum are reported to have R_F values greater than 0.5 (Mordue, W. and Goldsworthy, G., 1969; Stone, J. *et al.*, 1976), this result shows that CTSH is structurally distinct from three other peptides extractable from locust corpora cardiaca (Phillips, J. *et al.*, 1980). This was further confirmed by reciprocal bioassay tests.

3 NEUROHORMONE D

3.1 Neurohormone D and insect heartbeat

Neurohormone D was the name given to the material isolated from the corpus cardiacum of *Carausius morosus* by paper chromatography some years ago (Unger, H., 1957). Neurohormone D corresponded to a group of compounds, which appeared to affect a variety of physiological functions, including increasing the frequency and amplitude of heartbeat (Unger, H., 1957; Gersch, M. *et al.*, 1960; Miller, T., vol. 3), stimulating colour change in *Carausius* (Mothes, G. 1960), induction of nervous activity in the phallus nerves of *Periplaneta americana* (Gersch, M. and Sturzebecher, J., 1971) and increasing the rate of dye excretion by the Malpighian tubules (Unger, H., 1957, 1965). In support of the osmoregulatory function of neurohormone D, Gersch, M. *et al.* (1969) showed that neurohormone D isolated from the corpora cardiaca of *Periplaneta* is able to increase the flux of water across the toad bladder. However, Vietinghoff, U. (1967) proposed that neurohormone D had an opposite, antidiuretic role in insects.

Although neurohormone D has been shown to have a number of diverse effects, it is still not clear whether these effects can be attributed to a single hormone, or if each is due to a different compound. Recently, neurohormone D has been isolated and purified from the corpora cardiaca of *Periplaneta americana* (Baumann, E. and Gersch, M., 1982), heartbeat being used to monitor hormonal activity. This suggests that neurohormone D is a factor involved in the regulation of heartbeat. However, some authors have been critical of the physiological importance of neurohormonal factors regulating heartbeat rate, where, as in the case of the American cockroach, the heart is extensively innervated (Miller, Th., 1979; vol. 3). Support for this contention is provided by the fact that injections of corpus cardiacum extracts into intact animals were found not to alter the rate of insect heartbeat (Mordue, W. and Goldsworthy, G., 1969; Roussel, J. and Cazal, M., 1969; Normann, T., 1972). It has been suggested that the effect of corpus cardiacum extracts on the beating of isolated hearts may be due to pharmacological stimulation by peptides, which exert their primary effects on other target tissues (Goldsworthy, G. and Mordue, W., 1974). An example of this is AKH, whose primary role is the mobilization of lipid in the fat body, but which also exerts a pharmacological effect on the rate of heartbeat (Stone, J. and Mordue, W., 1980a).

Clearly the control of heartbeat is complex, involving both nervous and neurohormonal regulation. It must be borne in mind that a number of compounds are known to stimulate heartbeat, such as acetylcholine, 5-hydroxytryptamine, octopamine, dopamine and proctolin, and all except perhaps dopamine are excitatory at relatively low concentrations. Acetylcholine acts by stimulating the lateral cardiac nerve cord, whereas the others act directly on the myocardium (Miller, Th., 1979). Therefore, although neurohormone D has been isolated in pure form, and identified as a heart-accelerating peptide, its physiological significance remains obscure. It will be of interest to know what effects purified neurohormone D might have on other bioassay systems (e.g. fluid secretion), to clarify whether heart-acceleration is its primary role or a pharmacological event.

3.2 Isolation of neurohormone D and its preliminary chemical characteristics

Gersch, M. and Sturzebecher, J. (1971) prepared extracts of both brain and ventral nerve cord from 5000 donor *Periplaneta americana*, and separated these extracts by paper chromatography, developed in a mixture of propanol–conc. ammonia (22:18). The area corresponding to neurohormone D activity was chromatographed twice more by paper chromatography in a less polar solvent mixture (propanol–chloroform–conc. ammonia, 4:2:1) to free the "biologically active zone" from contaminating amino acids (i.e. they were reduced to below the threshold level of the ninhydrin reaction). Using this experimental procedure 2.5 mg of purified neurohormone D were isolated from both the ventral nerve cord and brains of 5000 cockroaches (Gersch, M. and Sturzebecher, J., 1971). The R_F values of neurohormone D separated using different solvent systems varied only slightly, ranging between R_F 0.20 to 0.25. An R_F value of 0.33 obtained in an earlier report (Gersch, M. *et al.*, 1960), was considered erroneous, as positive staining by ninhydrin has been used as a characteristic of neurohormone D. However, Gersch, M. and Sturzebecher, J. (1971) were able to show that more purified neurohormone D does not react with ninhydrin. The reaction to ninhydrin reported in their earlier study (Gersch, M. *et al.*, 1960) was explained by the presence of contaminating amino acids, chiefly alanine, which was found to have an R_F value of 0.33 (Gersch, M. and Sturzebecher, J., 1971).

Although the purified neurohormone D prepared by triple chromatography was insensitive to ninhydrin, colour formation was obtained after chlorination with Erlich's reaction mixture (*O*-dimethyl-amino benzaldehyde-HCl) and with KI-σ-toluidine. The latter reaction with KI-σ-toluidine after chlorination is specific to NH-groups, which indicates the presence of peptide bonds and therefore the peptide nature of neurohormone D. Erlich's reaction forms coloured condensation products with both primary amines and indole derivatives; yet, as ninhydrin which reacts with primary amines showed no reaction, it was concluded that one of the amino acids contained in neurohormone D must contain a heterocyclic ring. Further support for the peptide nature of neurohormone D was provided by the fact that after acid hydrolysis (in 6 N HCl for 24 h) the hydrolysed neurohormone D showed a ninhydrin-positive reaction, indicating free amino acids (Gersch, M. and Sturzebecher, J., 1971). These authors also used infra-red spectral analysis to indicate C — N, C = O and N — H bonds, which are characteristic of peptide bonds. The peptide character was further confirmed in a later study, in which trypsin was shown to destroy the heart-accelerating of neurohormone D (Baumann, E. and Gersch, M., 1973). As trypsin hydrolyses peptide bonds, where the carbonyl group is donated by either the basic amino acids lysine or arginine, Baumann, E. and Gersch, M. (1973) suggested that one or both of these amino acids might be present in the neurohormone D molecule. However, in their recent work, in which the amino acid composition of neurohormone D was revealed, neither amino acid was found to be present (see Table 3; Baumann, E. and Gersch, M., 1982). This underlines the limitations of enzyme digestion experiments in revealing information about the chemical structure of peptides or proteins. Purified trypsin may still contain chymotrypsin as a contaminant, unless treated with an inhibitor of chymotrypsin activity such as TPLCK (1-chloro-3-tosylamido-7-amino-L-2-heptanone). Chymotrypsin hydrolyses peptide bonds, where the carbonyl group is donated by one of the aromatic amino acids, tyrosine, tryptophan or phenylalanine. It is possible that the digestion of neurohormone D may have been due to contaminating chymotrypsin rather than the action of trypsin *per se*.

Using a Sephadex G-75 gel filtration chromatography column, eluted in water, Gersch, M. and Sturzebecher, J. (1971) were able to show that neurohormone D activity elutes in fractions between the molecular markers cytochrome *c* (13,000) and oxytocin (1000), with a suggested molecular weight of 2000. From the preliminary work it was possible to conclude that neurohormone D is a small molecular weight peptide extractable from all parts of the nervous system, which is heat labile (Gersch, M. and Sturzebecher, J., 1971; Baumann, E. and Gersch, M., 1973).

Although neurohormone D has been known for some time, and a large amount of work on the hormone has been performed, it is only relatively recently that the amino acid composition has been established (Baumann, E. and Gersch, M., 1982). The first stage in the isolation and purification of neurohormone D was to separate the hormone from aqueous ethanol extracts of corpora cardiaca of *Periplaneta americana* by gel filtration on a Bio-Gel P-6 column. Neurohormone D was found to be weakly retained by the P-6 column and to elute in a similar position to tyrosine. The fractions containing neurohormone D were pooled and applied to thin-layer silica gel chromatography plates and developed in ethanol/water (9:1 v/v). Only one UV absorbing spot showed the characteristics of neurohormone D; this had an R_F value of 0.2. This biologically active material was then applied to cellulose-acetate strips and subjected to electrophoresis. The neurohormone D activity remained at the origin, indicating no net charge on the molecule at the pH of the buffer, pH 1.9.

Finally, the electrophoresed neurohormone D was rechromatographed on a Bio-Gel P-6 column, eluted in distilled water, from which 20 μg of purified material was recovered from an initial starting material of 1000 corpora cardiaca (Baumann, E. and Gersch, M., 1982).

Table 3: Amino acid composition of neurohormone D isolated from corpora cardiaca of Periplaneta *(after Baumann, E. and Gersch, M., 1982)*

Asx	2	Pro	1
Glu	1	Ser	1
Phe	1	Val	1

3.3 Amino acid composition of neurohormone D

The amino acid composition was determined by both dansylation of the purified peptide followed by microchromatography, and by using a conventional automatic analyser. Both methods were in agreement, showing neurohormone D to possess Asx, Glu, Phe, Pro, Ser and Val (see Table 3), although it was not possible to establish clearly whether the aspartic acid and glutamic acid residues exist as asparagine and glutamine respectively (Baumann, E. and Gersch, M., 1982).

4 PROCTOLIN

In 1975, Brown, B. and Starratt, A. published a procedure for the isolation from *Periplaneta americana* of a myotropic substance, which they named "proctolin". Proctolin is a pentapeptide, Arg–Tyr–Leu–Pro–Thr, that has been isolated from *Periplaneta* and causes slow graded contractions of the longitudinal muscles of the proctodeum. Such contractions can also be invoked by repetitive nerve stimulation, and it has been suggested that proctolin is either an excitatory neurotransmitter or acts as a neurohormone to modulate muscle excitability. Proctolin acts on preparations other than the hindgut. Its effects on the heart have just been described. At concentrations as low at 10^{-10} to 10^{-9} M it induces myogenic contractions in the locust extensor tibialis. An increased rate of contraction of the hyperneural muscle in *Periplaneta* was elicited in response to proctolin when the innervation of the hyperneural muscle from the ventral nerve cord was intact. When the ventral nerve cord was removed, proctolin was unable to induce muscle contractions. Thus the site of action of proctolin in this preparation is probably distinct from the neuromuscular junctions of the hyperneural muscle (see Miller, Th., 1979). The recent identification of the cell bodies responsible for the synthesis of proctolin and the immunocytochemical mapping of their axon pathways promise to provide much important information (see O'Shea, M. *et al.*, 1982).

4.1 Isolation

Brown, B. and Starratt, A. (1975) report that initially the isolation procedure was developed using gram quantities of excised viscera as the starting material. However it soon became clear that this approach would not yield sufficient quantities for chemical analysis, and so the methods were scaled up to accommodate kilogram quantities of whole cockroaches.

Batches of about 1000 adult cockroaches (approx. 1 kg fresh weight) were processed at a time. The cockroaches were blended in perchloric acid, and the resulting supernatants from 25,000 cockroaches neutralized, concentrated and pooled. This extract was passed through two cation exchange columns, and the extracts from five such preparations combined (125,000 insects extracted). The pooled extract was dissolved in methanol containing the anti-oxidant, 2,6-di-tert.-butyl-*p*-cresol (BHT), which reduced losses of proctolin. The procedure was simplified by Brown, B. (1977) by omitting the final purification steps.

4.2 Determination of primary sequence

Proctolin (180 μg, isolated as described by Brown, B. and Starratt, A., 1975) was used for its structural characterization, the details of which are presented by Starratt, A. and Brown, B. (1975). The purity of the preparation was established by its homogeneity in paper and thin-layer chromatographic systems, and by high-voltage electrophoresis at different pH values. Preliminary investigations showed that acid hydrolysates and leucine aminopeptidase digests of proctolin contained several ninhydrin-positive spots when the products were separated by paper chromatography, indicating that proctolin is a peptide. A positive reaction to the Sakaguchi reagent by proctolin itself and one of the hydrolysis products suggested the presence of arginine, and the UV absorption spectrum of proctolin indicated that tyrosine was also present. The amino acid composition of proctolin was determined after acid hydrolysis by dansylating the products and separating them out by thin-layer chromatography. Spots corresponding to arginine, leucine, proline, threonine and tyrosine were shown to be present in approximately equimolar amounts. Since the peptide had a free NH_2-terminal the dansyl-Edman method was then used to establish the amino acid sequence of the peptide. This was shown to be Arg–Tyr–Leu–Pro–Thr.

4.3 Structure–activity relationships

A comprehensive study of the structure–activity relationships of proctolin has recently been completed by Starratt, A. and Brown, B. (1979). The myotropic activities on cockroach proctodeal muscle of 15 pentapeptide analogues and seven di-, tri- and tetrapeptides with partial amino acid sequences of proctolin were determined. The various peptides were synthesized by methods similar to those used in the synthesis of proctolin itself. The results of the study are summarized in Table 4.

In the first analogue series investigated each amino acid residue was successively replaced by alanine, to see if any single amino acid was essential for activity. Only [Ala^4]-proctolin, in which alanine replaced proline, showed any appreciable activity, being around 15% as active as proctolin. Similarly, successive replacement of each residue with the respective D-amino acid also produced analogues with little activity. [D-Tyr^2]-proctolin, which showed the highest activity, being 11% as active as the natural peptide, was shown by chromatography to contain about 10% proctolin, which therefore accounted for most of the observed activity. Consequently investigation of these two series of analogues suggests that each individual residue in the pentapeptide is essential for activity, and that the natural neuropeptide is made up of L-amino acids as one might expect.

A surprising result was that [Phe(p-OMe)^2]-proctolin, in which the hydroxyl group of the tyrosine residue at position 2 is methylated, was almost three times as active as the natural peptide, suggesting that a free hydroxyl group in this position is not essential for the myotropic activity of the pentapeptide. This conclusion also gains support

Table 4: Myotropic activities of analogs of proctolin on cockroach proctodeal muscle (from Starratt, A. and Brown, B., 1979)

Compound	Residues 1 2 3 4 5	Activity*
1	Arg–Tyr–Leu–Pro–Thr	100.0
2	Ala^1–proctolin	0.05
3	Ala^2–proctolin	0.06
4	Ala^3–proctolin	0.08
5	Ala^4–proctolin	15.70
6	Ala^5–proctolin	0.14
7	(D–Arg)^1–proctolin	3.50
8	(D–Tyr)^2–proctolin	10.97†
9	(D–Leu)^3–proctolin	0.06
10	(D–Pro)^4–proctolin	0.03
11	(D–Thr)^5–proctolin	1.80
12	Phe^2–proctolin	15.40
13	Phe(p–OMe)^2–proctolin	278.80
14	Orn^1–proctolin	1.83
15	Ser^5–proctolin	5.44
16	Proctolin amide	0.04
17	Arg–Tyr–Leu–Pro	0.15
18	Tyr–Leu–Pro–Thr	<0.0001
19	Arg–Tyr–Leu	0.0005
20	Tyr–Leu–Pro	0.0009
21	Leu–Pro–Thr	0.06
22	Arg–Tyr	<0.0001
23	Tyr–Leu	<0.0001

* Relative to the activity of an identical molar quantity of proctolin.
† Activity attributable in large part to proctolin, an impurity in this peptide.
Superior numbers denote position of substitution.

from the fact that [Phe^2]-proctolin, in which the tyrosine hydroxyl is obviously absent, possesses significant myotropic activity (15% of that of proctolin).

Replacement of the basic residue arginine by another basic residue ornithine at position 1 reduced the activity to 3% of that of the neuropeptide and replacement of threonine by the similar residue serine as residue 5 reduced the activity to 5% of the natural peptide. Blocking the free carboxyl group of proctolin with an amide reduced its activity to only 0.04% of that of proctolin. Moreover, shortening the peptide from either end produced compounds with little or no significant activity. When tested at sub-threshold doses, none of the analogues showed antagonistic activity, as measured by inhibition of neurally evoked contractions.

Sullivan, R. and Newcomb, R. (1982) have confirmed some of the structure–function relationships carried out by Starratt and Brown, and further extended this analysis. For example, in their studies replacement of arginine with either neutral or acidic residues resulted in total loss of activity. The importance of a positively charged side-chain for receptor binding was emphasized by the fact that [lys^1]-proctolin is a fully active analogue.

Substitution of [pro^4] to produce either [gly^4]-proctolin or [leu^4]-proctolin produced two compounds with high biological activity. [Ala^4]-proctolin (Starratt, A. and Brown, B., 1979), also showed appreciable activity. Thus it is likely that proline is not a precise requisite for receptor binding when tested on the roach hindgut. It is of interest that Piek, T. *et al.* (1979) have shown that proline is an important residue in eliciting responses in the locust extensor tibia preparations.

Sullivan, R. and Newcomb, R. were also able to elicit responses from the roach hindgut with concentrations of 10^{-6} M RPCH (crustacean red pigment concentrating hormone). The homologous peptide, AKH, was not tested, although it is known to produce contraction in insect heart muscle (Stone, J. and Mordue, W., 1980a), which is also sensitive to proctolin. Peptides with proctolin-like action are present in the corpora cardiaca of a number of insects; however, they are chromatographically distinguishable from proctolin even if pharmacological distinction is not possible. Thus the possibilities of a family of peptides of similar pharmacology, or peptides of quite distinct structure with similar pharmacology, have to be considered. The proposition of Sullivan, R. and Newcomb, R. (1982) for the existence of a glucagon-like prohormone in the corpora cardiaca which contains both AKH and proctolin-like sequences is at present unsubstantiated.

5 HORMONES AND CARBOHYDRATE METABOLISM

(see Steele, vol. 8)

5.1 Hyperglycaemic and glucagon-like hormone(s)

There have been numerous reports confirming the presence of a hyperglycaemic factor in *Periplaneta americana* (Steele, J., 1961; Ralph, C. and McCarthy, C., 1964; Brown, B., 1965; Hanaoka, K. and Takahashi, S., 1976) and other insects. These include *Blaberus* (Bowers, W. and Friedman, S., 1963), *Phormia regina* (Friedman, S., 1967), *Leucophaea màderae* (Wiens, A. and Gilbert, L., 1967), *Calliphora* (Normann, T. and Duve, H., 1969; Vejbjerg, K. and Normann, T., 1974), *Locusta americana* (Goldsworthy, G., 1969; Loughton, B. and Orchard, I., 1981), *Carausius morosus* (Dutrieu, J. and Gourdoux, L., 1965; Gade, G. and Lohr, P., 1982) and *Manduca sexta* (Ziegler, R., 1979). However, in some insects the evidence for a hyperglycaemic factor is equivocal, as it has not been possible to observe a hyperglycaemic response after the injection of a physiological dose of corpus cardiacum extract into normal insects (Friedman, S., 1967; Goldsworthy, G., 1969; Dutrieu, J. and Gourdoux, L., 1965; Normann, T. and Duve, H., 1969; Gade, G., 1979). This lack of response in normal insects may have been due to the release of hypoglycaemic hormone from the neuro-endocrine system (see Normann, T., 1975), after the injection of the hyperglycaemic factor. It has been shown that in insects ligated at the neck, a consistent hyperglycaemic response can be elicited (Loughton, B. and Orchard, I., 1981; Gade, G. and Lohr, P., 1982). However, it should also be noted that an insulin-like (hypoglycaemic) factor has been identified in the corpus cardiacum of *Manduca* (Tager, H. *et al.*, 1976), so that crude aqueous extracts of corpora cardiaca may contain both hyperglycaemic and hypoglycaemic factors. Recent evidence indicates that the source of the hyperglycaemic factor in the locust is the glandular lobe of the corpus cardiacum (Loughton, B. and Orchard, I., 1981), whereas it is speculated that a hypoglycaemic factor is synthesized in the brain neurosecretory cells and stored and released from the storage lobes of the corpus cardiacum (Normann, T., 1975; Duve, H. *et al.*, 1979).

5.2 The chemical identity of hyperglycaemic hormone(s)

The physiological action of the hyperglycaemic hormone is to stimulate the breakdown of glycogen in the fat body, which results in an elevation of haemolymph trehalose. This action is mediated by the activation of the fat body phosphorylase system in a manner that is similar to glucagon activation of liver phosphorylase in vertebrates (Steele, J., 1963), and this has led to the suggestion that glucagon is the insect hyperglycaemic hormone (Tager, H. *et al.*, 1975, 1976). Using radioimmunoassay (RIA) specific for bovine glucagon, it has been shown that extracts of the corpus cardiacum of *Manduca* contain a glucagon-like molecule (Tager, H. *et al.*, 1975, 1976). It has subsequently been shown that a glucagon-like hormone also exists in the honeybee (Maier, V. *et al.*, 1978) and *Calliphora* (Normann, T. and Duve, H., 1978). Further support for a glucagon-like molecule as a hyperglycaemic hormone in insects is provided by the fact that bovine glucagon, when injected into *Calliphora*, has been found to cause a rise in haemolymph trehalose, a fall in glucose, and elevation of haemolymph lipids within 20 min; and hypertrehalosaemia and hyperlipaemia has also been observed when glucagon has been injected into *Schistocerca gregaria* (Normann, T. and Duve, H., 1975). However, although a 40% fall in fat body glycogen was observed, no change in haemolymph trehalose was observed after injection of a supra-maximal dose of bovine glucagon into *Manduca* (Tager, H. *et al.*, 1976). In the same insect Ziegler, R. (1979) reports that no change in the fat body phosphorylase levels can be observed unless a supra-maximal dose of glucagon is injected.

It has also been demonstrated that adipokinetic hormone (AKH I) of the locust, when injected into the cockroach, will induce a hyperglycaemic response (Jones, J. *et al.*, 1977; Holwerda, D. *et al.*, 1977); a hyperglycaemic response can be induced in *Locusta* by AKH II, but not by AKH I (Loughton, B. and Orchard, I., 1981). The amino acid composition of AKH II (Carlsen, J. *et al.*, 1979) is very similar to that of AKH I and suggests some structural similarity (see Fig. 3); yet there is no structural similarity between glucagon and the adipokinetic hormones (Fig. 4). The structural dissimilarity between glucagon and the insect hyperglycaemic hormone is further emphasized by the fact that the insect hyperglycaemic factor is degraded by both chymotrypsin and pepsin, whereas it cannot be deactivated by trypsin; trypsin does deactivate mammalian glucagon (Ziegler, R., 1979).

	AKH I (Stone *et al.*, 1976)	AKH II (Carlsen *et al.*, 1979)	*Periplaneta* hyperglycaemic factor (Jones, 1978)	*Carausius* hyperglycaemic factor (Gade and Mordue, unpublished)	*Periplaneta* heart-accelerating factor (Traina *et al.*, 1976)
Glu	✓	✓	✓	✓	✓
Leu	✓	✓	✓	✓	✓
Asn	✓				
Phe	✓	✓	✓	✓	
Thr	✓	✓		✓	
Pro	✓		✓	?	✓
Asn	✓				
Trp	✓	✓		?	
Gly	✓	✓	✓	✓	✓
Thr	✓				
Asp		✓	✓	✓	✓
Ser		✓	✓	✓	✓
Ala			✓	✓	✓
Val			✓	✓	✓
Lys					✓

FIG. 3. The amino acid compositions of AKH I, AKH II, *Periplaneta americana* and *Carausius morosus* hyperglycaemic factors and *Periplaneta* heart-accelerating factor.

1	2	3	4	5	6	7	8	9	10	11	12	13
His -	Ser -	Gln -	Gly -	Thr -	Phe -	Thr -	Ser -	Asp -	Tyr -	Ser -	Lyr -	Tyr -
14	15	16	17	18	19	20	21	22	23	24	25	26
Leu -	Asp -	Ser -	Arg -	Arg -	Ala -	Gln -	Asp -	Phe -	Val -	Gln -	Trp -	Leu -
27	28	29										
Met -	Asn -	Thr										

Glucagon M.W. 3485 29 amino acid residues

PCA - Leu - Asn - Phe - Thr - Pro - Asn - Trp - Gly - Thr - NH_2

Adipokinetic hormone 10 amino acid residues

FIG. 4. Comparison of the primary structures of Glucagon and AKH I.

Clearly these findings, together with those from studies on the isolation of the hyperglycaemic hormone, indicate that although glucagon-like molecules can be located in insects immunochemically, they are not the only endogenous hyperglycaemic hormones present in insects. The problems and shortcomings of immunochemical identification of hormones are discussed in a subsequent section on the identification of peptides by immunochemistry (see section 5.12 and Kramer, this volume), and they apply equally to glucagon-like hormones. Therefore, demonstrations of the physiological importance of glucagon-like hormones in insects can only be resolved by their structural elucidation, combined with studies to show that the hormones are synthesized and released by endocrine glands.

5.3 Isolation of glucagon-like hormones

Some progress has been made towards the isolation of a glucagon-like factor. Tager, H. *et al.* (1975, 1976) were able to separate a glucagon-like factor, extracted in acid–alcohol from the corpus cardiacum of *Manduca sexta* using a Bio-Gel P-10 gel filtration column with 3 M acetic acid as the eluant. The immunoreactive glucagon eluted at a position corresponding to a molecular weight of about 4500, slightly higher than the molecular weight of bovine glucagon (3500). When subjected to polyacrylamide gel electrophoresis at pH 8.7, the immunoreactive glucagon was found to migrate towards the anode, indicating an acidic molecule of low molecular weight (Tager, H. *et al.*, 1975). Relative to both porcine glucagon and [^{125}I]glucagon, the immunoreactive glucagon was distinctly separated by this electrophoretic method, which must indicate some molecular dissimilarity.

Mammalian glucagon is sensitive to proteolytic digestion, so that the effect of digestion by trypsin of porcine glucagon and glucagon immunoreactivity after gel filtration on a P-10 column was compared. It was demonstrated that the digestion of the latter was either considerably slower or digested to a lesser extent, and it was suggested that this may be the result of the highly acidic nature of the insect peptide (Tager, H. *et al.*, 1975). Alternatively it may be another indication of the molecular dissimilarity between glucagon and the insect immunoreactive material.

However, further support for a glucagon-like hormone in insects is provided by Kramer, K. *et al.* (1980) who demonstrated the presence of immunoreactive glucagon in the haemolymph of *Manduca* by radioimmunoassay. Using a Bio-Gel P-30 column and 3 M acetic acid as the eluant, immunoreactive glucagon was separated from acid–alcohol extracts of larval and pupal haemolymph, and eluted with an apparent molecular weight of about 3000. However, the acceptance of glucagon or a glucagon-like hormone as a true insect hormone must await its structural elucidation.

5.4 Isolation of a hyperglycaemic factor and its chemical nature

There have been numerous studies which have reported methods for isolating hyperglycaemic factors from crude extracts of corpora cardiaca.

Brown, B. (1965) showed that the hyperglycaemic factor from the cockroach *Periplaneta americana* chromatographed into two distinct regions on a paper chromatogram, indicating that the hyperglycaemic factor exists in two forms. Similarly, Mordue, W. and Goldsworthy, G. (1969) demonstrated that the hyperglycaemic activity of the locust *Schistocerca gregaria* could be separated into two regions by paper chromatography. A variety of gel filtration (Sephadex G-10, G-15, G-25, Bio-Gel P-2, P-10, LH-20) and ion exchange (SE-Sephadex C-50 and Cellex CM) columns have been used successfully to isolate the hyperglycaemic hormone from a number of insects (Migliori Natalizi, G. *et al.*, 1970; Traina, M. *et al.*, 1976; Holwerda, D. *et al.*, 1977; Jones, J., 1978; Gade, G., 1981; Loughton, B. and Orchard, I., 1981; Gade, G. and Lohr, P., 1982).

The hyperglycaemic factor from *Periplaneta americana* has been separated as a single peak of hyperglycaemic activity on Sephadex G-15 (eluant 0.1 M acetic acid) and G-25 (eluant pyridine/acetic acid/water) (15:55:30 v/v) columns, and on a Cellex CM ion exchange column eluted in a stepped gradient of 0.01 M ammonium acetate buffer (Migliori Natalizi, G. *et al.*, 1970). Yet the activity was resolved into a second peak on a Sephadex G-10 column eluted in 0.1 M acetic acid. The exclusion limits of the Sephadex gel filtration columns suggest that the cockroach hyperglycaemic factor is a low molecular weight compound; less than 5000 daltons. Attempts to purify the hyperglycaemic factor eluted from a Sephadex G-15 column by a second purification step on an LH-20 column were not successful, as no hyperglycaemic activity could be detected (Traina, M. *et al.*, 1976). However, a second purification step of the hyperglycaemic material eluted from a G-25 column, on a SE-Sephadex C-50 column, resolved the factor into two peaks of activity (Migliori Natalizi, G. and Frontali, N., 1966).

Using a method similar to that used in the isolation and purification of AKH from locusts (Stone, J. *et al.*, 1976), Jones, J. (1978) isolated a hyperglycaemic factor from a methanol extract of 300 corpora cardiaca of *Periplaneta americana*. This extract was separated firstly by gel filtration on a controlled pore glass bead column (G. 75–50), followed by thin-layer chromatography. The purified material isolated on the TLC plate (R_F 0.56) was subjected to acid hydrolysis and the amino acid composition determined. The following residues were found to be present: Asp, Ser, Glu, Pro, Gly, Ala, Val, Leu, Phe, although the non-integral molar proportions suggest that the material was not completely purified. The similarity between the amino acid composition of the hyperglycaemic factor isolated by Jones, J. (1978) and that for the heart-accelerating factor (Traina, M. *et al.*, 1976), which was also of a heterogeneous composition, has been noted (Stone, J. and Mordue, W., 1980b).

The method developed to isolate adipokinetic hormone II (Carlsen, J. *et al.*, 1979) has been used to isolate a hyperglycaemic factor in *Locusta migratoria* (Loughton, B. and Orchard, I., 1981). This involves a first separation step on a Sephadex G-25 column eluted with water, and a second purification step using a Sephadex LH-20 column eluted with aqueous butanol. As it was found that the hyperglycaemic hormone of *Locusta* eluted in the same fraction as AKH II from the LH-20 column, it has been speculated that AKH II may act as the hyperglycaemic factor (Loughton, B. and Orchard, I., 1981). Indeed it would not be surprising if AKH II had both an adipokinetic and glycogenolytic function, similar to the vertebrate hormone glucagon, which stimulates both hepatic glycogenolysis and lipolysis in adipose tissue (Sokal, J., 1973).

Gade, G. and Lohr, P. (1982), using the same two-step purification scheme outlined above (i.e. separation on a G-25 followed by an LH-20 column), have separated the hyperglycaemic factor from *Carausius morosus* into two peaks of activity. These peaks seem to correspond to fractions where AKH I and AKH II isolated from *Locusta* corpora cardiaca would elute (Gade, G., 1981).

The amino acid composition of AKH II elucidated by Carlsen, J. *et al.* (1979) is composed of equimolar proportions of the following amino acids: Asp, Thr, Ser, Glu, Gly, Leu, Phe, Trp. The similarities of the amino acid compositions between the insect peptide AKH I, of known sequence (Stone, J. *et al.*, 1976), AKH II (Carlsen, J. *et al.*, 1979), the hyperglycaemic hormone of *Periplaneta* (Jones, J., 1978), the heart-accelerating factor (Traina, M. *et al.*, 1976) and the *Carausius* hyperglycaemic factor (Gade and Mordue, unpublished) are quite striking (see Fig. 3). This may indicate a closely related family of peptides.

5.5 Hypoglycaemic and insulin-like hormone(s)

In the blowfly, *Calliphora*, cautery of the median neurosecretory cells (MNC) or extirpation of the corpus cardiacum causes an elevation in haemolymph trehalose level (Normann, T., 1975). Normann suggested such hyperglycaemia could be explained by the lack of the normal release of a hypoglycaemic hormone produced in the brain. Similar experiments have demonstrated the presence of a hypoglycaemic hormone in the brain/corpus cardiacum of *Phormia regina* (Chen, A. and Friedman, S., 1977); hyperglycaemia induced by the cautery of the MNC in *Calliphora*, and cardiatectomy in *Phormia*, can be reduced by injection of brain extracts (Chen, A. and Friedman, S., 1977; Duve, H., 1978).

The immunological demonstration of an insulin-like molecule in the neuroendocrine system of a number of insects (Tager, H. *et al.*, 1976; Duve, H. and Thorpe, A., 1979; Duve, H. *et al.*, 1979; Orchard, I. and Loughton, B., 1980; El-Salhy, M. *et al.*, 1980; Yui, R. *et al.*, 1980; see Kramer, this volume), has stimulated speculation as to whether insect hypoglycaemic hormone might be an insulin-like molecule (Normann, T. and Duve, H., 1978). However, the actions of mammalian insulin injected into insects are both contradictory and confusing. In both *Calliphora* and *Vespa* mammalian insulin was found to depress the haemolymph glucose level (Ishay, J. *et al.*, 1976; Duve, H., 1978), and in *Calliphora* a small elevation in the blood trehalose level was observed unless a larger dose of insulin was injected (0.2 IU/fly), which would depress the trehalose level (Normann, T. and Duve, H., 1978). Yet bovine insulin had no effect on either haemolymph trehalose or fat body glycogen in *Manduca sexta* (Tager, H. *et al.*, 1976). In *Hyalophora* insulin was found to inhibit the release of diglycerides and free fatty acids from the pupal fat body (Bhakthan, N. and Gilbert, L., 1968); similarly Orchard, I. and Loughton, B. (1980) found insulin had a hypolipaemic effect in *Locusta*, whereas insulin was reported to stimulate triglyceride release from the fat bodies of *Danaus* and *Agrius* (Chang, F., 1974). The ability of insulin to enhance the growth of embryonic cells of *Drosophila* has also been reported (Seecof, R. and Dewhurst, S., 1974; Mosna, G. and Barigozzi, C., 1976). Therefore, from this evidence it is not clear whether the insect hypoglycaemic hormone and the immunoreactive insulin-like material are one and the same, or as in the case of the hyperglycaemic hormone and the glucagon-like hormone, that they are chemically distinct.

It should be noted that in addition to the neuroendocrine system insulin-like material has been found in both the worker jelly and royal jelly of the honeybee *Apis mellifera* (Dixit, P. and Patel, N., 1964; Kramer, K. *et al.*, 1977, 1982), the haemolymph of *Manduca sexta* and *Drosophila* (Kramer, K. *et al.*, 1977, 1982; Seecof, R. and Dewhurst, S., 1974) and in the intestinal tissues of various *Hymenoptera* (Ishay, J. *et al.*, 1976; Moreau, R. *et al.*, 1981). The source of these insulin-like hormones has not been demonstrated, but it has been suggested that they may be products of the neuroendocrine system (Duve, H. *et al.*, 1979).

5.6 Isolation of an insulin-like hormone(s) from the neuroendocrine system of *Calliphora*

A comprehensive attempt to isolate an insulin-like hormone from the brain of the blowfly of *Calliphora* (involving five separation and purification steps) has been made by Duve, H. *et al.* (1979), who employed many of the procedures used in the isolation and purification of mammalian insulin (Mirsky, I., 1973). Some 120,000 heads of *Calliphora* were subjected to acid ethanol extraction, and the extracted material was applied to a Sephadex G-25 gel filtration column, eluted in 3 M acetic acid (Duve, H. *et al.*, 1979). Those fractions containing insulin-like immunoreactivity were pooled and applied to a second gel-filtration column (Sephadex G-50 superfine), from which three peaks of immunoreactivity eluted. The first peak corresponded to the void volume and the second coincided with the bovine insulin marker; the immunoreactive material in both these peaks showed decreasing immunoreactivity in the bovine immunoassay on dilution. However, the last peak was discounted as an artefact because of its anomalous elution behaviour (Duve, H. *et al.*, 1979).

The next stage in the separation was the use of a cation exchange column; this step is employed to separate the C-peptide fragment from the insulin

(Mirsky, I., 1973). However, Duve, H. *et al.* (1979) found that their immunoreactive material from the G-50 column, when eluted on a Whatman CM-52 cation exchange column, was separated into three peaks of activity, which when subjected to further separation on a Sephadex G-25 column, were resolved into even more peaks of immunological activity. This indicated that the insulin-like material was contaminated with protein (Duve, H. *et al.*, 1979). Using only those peaks which showed decreasing immunological cross-reaction on dilution, the immunoreactive material was resolved (by polyacrylamide gel electrophoresis) into a band with an R_F value of 0.58. This corresponds to the R_F value obtained for bovine insulin. However, one of the criteria that must be fulfilled to demonstrate structural similarity between the "fly-insulin" and bovine insulin is for their displacement behaviours of [^{125}I]-labelled insulin from insulin receptors to be identical; yet their displacement behaviour was found to be dissimilar (Duve, H. *et al.*, 1979). Therefore, these results indicate that although molecules which cross-react with insulin antibodies are present in the fly brain, any structural similarity to bovine insulin is equivocal.

5.7 Purification and amino acid composition of an insulin-like hormone from the haemolymph of *Manduca*

The presence of an insulin-like material in the corpus cardiacum–corpus allatum complex of *Manduca sexta* was observed by Tager, H. *et al.* (1976). Subsequently it was shown that insulin-like material could be detected in the haemolymph of the same insect (Kramer, K., 1980), and it was estimated that the concentration ranges between 0.2 and 1.5 × 10^{-12} g/ml. Considering these two levels, it is perhaps surprising that the haemolymph, rather than the corpus cardiacum–corpus allatum complex, was selected as the source from which to extract the insulin-like material in *Manduca*. However this apart, Kramer, K. *et al.* (1982) have succeeded in purifying and determining an amino acid composition for the insulin-like material from the haemolymph (Table 5; see Kramer, this volume).

The peptide nature of the insulin-like material is implied by its antigenicity, but it has been confirmed by the fact that the proteolytic enzyme, pronase, causes a 90% loss of its specific immunoreactivity when incubated at pH 8 and 37° for 20 min (Kramer, K. *et al.*, 1980). The insulin-like material was extracted from 20 g of freeze-dried haemolymph using 3 M acetic acid; interestingly acid–alcohol extraction was found to be unsuccessful (Kramer, K. *et al.*, 1980). The acidic extract was first separated on a cation-exchange column (Sephadex C-25) at pH 3, and the immunoreactive material was found to elute with 0.5 M NaCl, indicating a cationic property. The second purification step involved separation of the insulin-like material on a Bio-Gel P–10 column eluted in 0.1 M acetic acid, on which the immunoreactive material was resolved into two peaks (Kramer, K. *et al.*, 1982). The lower molecular weight material eluted in a similar position to porcine insulin, indicating a similar molecular weight. The other peak of immunoreactivity material eluted more quickly than the porcine insulin marker, indicating material with a higher molecular weight. However, only the lower molecular weight immunoreactive material was subsequently purified further by affinity chromatography, yielding 10 ng equivalents of immunoreactive insulin (Kramer, K. *et al.*, 1982). This purified material (6 ng) was finally chromatographed on a Bio-Gel P–30 column to remove any contaminating albumin gained during

Table 5: The amino acid composition of "Manduca *insulin*" *and porcine insulin (after Kramer, K.* et al., *1982)*

Amino acid	*Manduca*	Porcine
Asp	3.1	3
Ala	2	2
Val	2.9	3
Ile	2.1	2
His	2.1	2
Lys	0.9	1
Arg	1.1	1
Ser	6.8	3
Glu	6.2	7
Leu	1.8	6
Tyr	2.8	4
Phe	1.0	3
Pro	n.d.	(1)
Met	n.d.	0
Trp	n.d.	0
Cys	n.d.	(6)
Thr + Gly	8	6

The integer values were normalized to 2.0 residues of Ala.
Pro — not detected by HPLC.
Met, Trp and Cys not stable to acid hydrolysis.

the affinity chromatography; a single peak of immunoreactivity was eluted, corresponding to 5 ng of purified material. After hydrolysis for 24 h *in vacuo* in 6 M HCl containing 0.1% phenol, the hydrolysates were analysed as their σ-phthaldiahyde derivatives by reverse-phase HPLC (Waters μ–Bondapak C-18). The amino acid composition obtained for the *Manduca* insulin-like material is shown in Table 5, and it is compared to that for porcine insulin. Kramer, K. *et al.* (1982) suggest that the amino acid compositions of the two materials are remarkably similar. Yet, of the 18 amino acids listed in Table 5, it should be noted that only seven have similar molar ratios, five are different, four amino acids were not detected, and threonine and glycine were not resolved. Therefore, although these data are the best evidence to date for an insulin-like peptide in insects, some discrepancies still remain between the amino acid compositions, which makes any structural similarity between the "*Manduca*-insulin" and porcine insulin uncertain.

5.8 Insulin-like hormones in the royal jelly and midgut of Hymenoptera

Dixit, P. and Patel, N. (1964) were the first to demonstrate the presence of an insulin-like material in insects, in the royal jelly of the honeybee *Apis mellifera*, using a classical insulin bioassay — the epididymal fat pad balance segment method — to detect the insulin-like material. Paper chromatography of their insulin-like material showed that both the latter and mammalian insulin chromatographed to an R_F value of 0.21. More recently Kramer, K. *et al.* (1982) have attempted to purify the insulin-like material from *Apis* royal jelly, using gel filtration followed by affinity chromatography, and found numerous fractions to contain immunoreactivity, for which it is estimated that there is a total yield of 25 pg of porcine insulin equivalent per gram of lyophilized royal jelly (Kramer, K. *et al.*, 1982). This was insufficient to make an amino acid analysis.

Both liquid chromatographic and electrophoretic techniques have been employed to separate the insulin-like material from the midgut of *Apis mellifera*. Each technique revealed that the insulin-like material has an apparent molecular weight of about 18,000 (Moreau, R. *et al.*, 1981), which is about three times the molecular weight of mammalian insulin and twice that of proinsulin. Yet the dilution curve of the *Apis* insulin-like immunoreactive material is superimposable on that of rat insulin. Moreau, R. *et al.* (1981) concluded that either the high molecular weight material has an amino acid sequence similar to that of rat insulin (as part of its structure), giving it common immunological characteristics, or it is a dimer of proinsulin or a trimer of insulin or even a banal molecule.

Despite some extensive studies on the insulin-like material present in various tissues of insects, the structural identity of the insect hypoglycaemic hormone is still not clear. Even if the structural identity of the insulin-like material is determined it will still be necessary to show that it has a physiological function in insects.

5.9 *Calliphora* pancreatic polypeptide

Pancreatic polypeptide (PP) is a peptide molecule of 36 amino acids, which was first isolated from the pancreas of the chicken and chemically characterized by Kimmel, J. *et al.* (1975). Subsequently human, porcine, ovine and bovine pancreatic polypeptides have been purified and sequenced (Floyd, J. *et al.*, 1977), although the sequence of the avian PP differs from bovine PP by more than 50% of the amino acids; the mammalian PPs differ by only one to five amino acids (see Table 6). This is an important consideration where immunochemical studies are undertaken, as the antigenicity of avian and bovine PP are markedly different. PP is produced by a specific type of endocrine cell of the gastrointestinal tract, but more recently immunocytochemical evidence has shown that PP immunochemically similar to avian PP, yet not to bovine or human PP, is located in the central and peripheral nervous system of mammals (Loren, I. *et al*, 1979).

Recently, immunocytochemical studies have demonstrated the presence of pancreatic polypeptide material immunoreactive to bovine PP in the brain and nervous systems of a number of insects, such as *Calliphora erythrocephala* (Duve, H. and Thorpe, A., 1980, 1982), *Bombyx mori* (Yui, R. *et al.*, 1980), *Eristalis aeneus* (El-Salhy, M. *et al.*, 1980; El-Salhy, M., 1981) and *Periplaneta americana* (Endo, Y. *et al.*, 1982). PP-like material has also

Table 6: Amino acid composition of Calliphora *neuropeptide and vertebrate pancreatic polypeptides.*

	Content (residues/mol)							
	Calliphora			Bovine	Porcine	Ovine	Human	Avian
Asx	3.4	3.1	(3)	3	3	3	4	6
Thr	2.4	2.5	(2)	2	2	2	2	2
Ser	1.3	1.2	(1)	0	0	1	0	1
Glx	5.7	6.1	(6)	6	5	6	4	4
Pro	3.8	4.2	(4)	5	5	4	5	4
Gly	1.9	1.9	(2)	1	1	1	1	3
Ala	3.9	3.7	(4)	5	5	5	5	1
Cys	—	—	—	0	0	0	0	0
Val	1.2	1.4	(1)	0	1	0	1	3
Met	1.4	1.6	(1)	2	2	2	2	0
Ile	2.2	1.6	(2)	1	1	1	1	1
Leu	2.5	2.8	(3)	3	3	3	3	3
Tyr	4.0	4.0	(4)	4	4	4	4	4
Phe	0	0	(0)	0	0	0	0	0
His	0	0	(0)	0	0	0	0	1
Lys	0	0	(0)	0	0	0	0	0
Arg	3.1	3.3	(3)	4	4	4	4	3
Trp	—	—	—	0	0	0	0	0

Purified *Calliphora* neuropeptide was analysed by means of an automatic ion-exchange analyser (Joel) with both ninhydrin and *O*-phthaldialdehyde monitoring channels. Two separate analyses were obtained from 24 h acid hydrolysates of approx. 100 pmol each. Values for serine and threonine have been corrected for an estimated 30% loss during hydrolysis. Cysteine and tryptophan were not determined. Results from the two analyses are given (ninhydrin detection) with the nearest integers in parentheses. Integer values were normalized to four residues of tyrosine. Values for vertebrate pancreatic polypeptides are from Floyd, J. *et al.* (1977). Values for *Calliphora* from Duve, H. and Thorpe, A. (1982).

been located in the midgut of two insects, *Periplaneta* (Iwanaga, T. *et al.*, 1981) and *Calliphora* (Duve, H. and Thorpe, A., 1982). However, in both vertebrates and insects the physiological significance of PP is as yet undetermined.

5.10 The isolation and purification of *Calliphora* PP-like material

The amino acid composition of the immunoreactive pancreatic polypeptide from the brain of the blowfly *Calliphora vomitoria* is now known (Duve, H. *et al.*, 1982). This was achieved by extracting the PP-like material from 5 kg of *Calliphora* heads in acid alcohol (65% ethanol acidified to pH 2.8 by conc. phosphoric acid), followed by centrifugation and filtration through a Hyflo bed. The resultant extract was run down a Sephadex G-15 gel filtration column, eluted with 1 M acetic acid (Duve, H. *et al.*, 1981). Those fractions containing PP immunoreactivity were pooled, freeze-dried and then filtered through a Sephadex G-50 superfine column. Again the PP-immunoreactive fractions were pooled; these fractions also contained insulin immunoreactive material. Both the pancreatic polypeptide and the insulin-like materials produced linear dilution curves similar to the bovine standards (Duve, H. *et al.*, 1981).

The total yield of protein from 5 kg of *Calliphora* head after these separation procedures was 960 mg of an off-white powder (Duve, H. *et al.*, 1981). To separate the insulin-like and PP-like materials 462 mg of the powder was reconstituted and eluted on a DEAE-cellulose ion-exchange column. Further purification of those fractions containing PP immunoreactivity was achieved by HPLC, using both Ultrasphere ODS and Cyanopropyl columns, yielding 650 ng of purified PP-like material.

5.11 Amino acid composition of the *Calliphora* PP

Using 100 pmole of purified material, amino acid analysis of the acid hydrolysates was made using an automatic ion-exchange analyser (Joel). The amino acid composition of the pancreatic polypeptide-like material for *Calliphora* from two analyses is shown in Table 6, and compared with those for the mammalian and avian pancreatic polypeptides. The non-

(a)

Arg - Arg - Gly - Glu - Met - Arg - Asp - Ile - Leu - Glu - Trp - Gly -
1 5 10

Ser - Pro - His - Na - Ala - Ala - Pro - Arg -
15 20

(b)

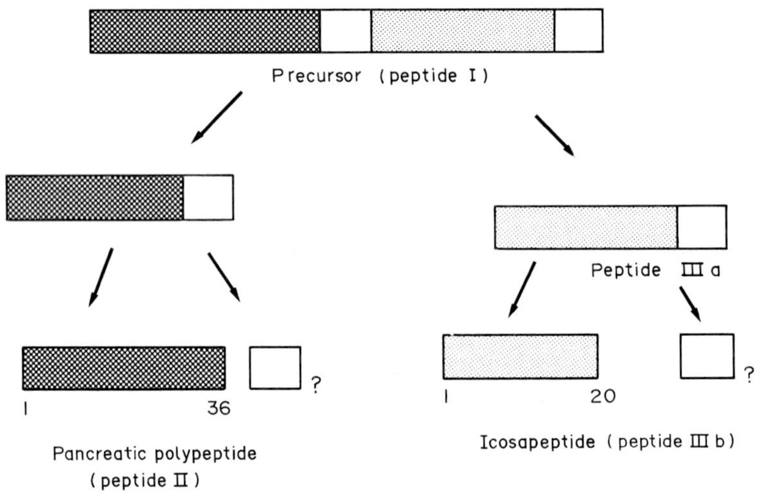

FIG. 5 (**a**) Amino acid sequence of icosapeptide isolated from the duodenal lobes of canine pancreas, determined by sequential Edman degradation. (**b**) Scheme suggested for the conversion of precursor molecule into pancreatic polypeptide and icosapeptide (after Schwartz, T. and Tager, H., 1981).

integral values for a number of the amino acids indicate a heterogeneous composition, implying that the *Calliphora* material is not totally free of impurities. This apart, the *Calliphora* material is similar to its vertebrate counterparts in that it is composed of 36 amino acids. However, when the intergral values of each of the amino acids of the *Calliphora* PP are compared to the mammalian PP there are differences in 6 to 10 of its amino acids; and when compared to the avian PP, 14 amino acid residues are at variance. This shows, therefore, that the structure of PP has been poorly conserved during the course of evolution. However, the physiological importance of PP remains obscure, although according to a recent report by Schwartz, T. and Tager, H. (1981), PP in vertebrates may not be the true hormone. Instead it is suggested that PP may be a linking peptide, derived from a larger precursor molecule, analogous to the C-peptide of proinsulin, and the true hormone may be an identified icosapeptide (see Fig. 5). Obviously, the same arguments can apply to insects (Duve, H. and Thorpe, A., 1982) and if this proves to be the case then it might explain why a physiological role for pancreatic polypeptide has remained elusive.

5.12 Identification of peptides by immunochemistry

The realization that peptide hormones could be visualized and localized by antigenic reaction to specific antibodies resulted in an explosion of immunocytochemical and radioimmunochemical studies. Numerous putative vertebrate peptide and polypeptide hormones have now been shown by these methods to be present in insects. However, these immunochemical studies have outpaced the biochemical physiological investigations, as a result of which the action(s) of most if not all of these vertebrate-like hormones are poorly defined.

One of the major problems, which must be recognized with regard to immunochemical studies, is that they can at best give an indication of the presence of vertebrate-like material (i.e. with a common antigenic determinant); they give no indication as to the chemical structure of the immunoreactive material. This can only be achieved by conventional biochemical purification and characterization procedures. The spurious results that may be obtained from immunochemical studies have recently been highlighted in two studies (Maruo, T. *et al.*, 1979; Juillard, J. *et al.*, 1980). Maruo, T. *et al.* (1979) were able to detect hCG (human chononic gonadotropin)-like factors in extracts of the stomach and hepatopancreas of the lady crab, *Ovalipes ocellatus*, yet after purification procedures it was found that the hCG-like immunoreactivity was due to interference by a trypsin-like protease, giving false, albeit positive, results. Similarly, Julliard, J. *et al.* (1980) were able to immunologically locate β-endorphin-like and ACTH-like material in the human placenta. However, the protein giving the immunological response was subsequently identified as the heavy chain of human immunoglobulin, class IgG_1.

Clearly immunochemical identification of vertebrate hormones in insects cannot be a short cut to their structural elucidation, or a substitute for the well-established biochemical isolation procedures incorporating biological assay of hormonal activity. At best immunochemical studies on uncharacterized hormones can show the presence of molecules of common antigenic determinant(s), but at worst they confuse and muddle the field of insect endocrinology.

6 ADIPOKINETIC HORMONE(S)

Adipokinetic hormone (AKH) from the locust is probably the most extensively investigated insect neurohormone of recent years. Its identity, and much concerning its modes of action, are now known. The existence of an AKH in the corpora cardiaca of locusts which regulated lipid utilization during prolonged flight was first suggested by Mayer, R. and Candy, D. (1969) and Beenakkers, A. (1969) from independent investigations on *Schistocerca gregaria* and *Locusta migratoria*, respectively (see vol. 10). Goldsworthy, G. *et al.* (1972) demonstrated that this lipid-mobilizing activity was confined to the glandular lobes of the corpora cardiaca.

Since these early observations were made much has been learnt of the ways in which the hormone acts (see Steele, vol. 8). The hormone is released in response to specific flight stimuli and acts primarily upon the fat body but also upon the flight muscles to promote mobilization and utilization of diacylglycerides (Chino, vol. 10).

6.1 Isolation

The glandular lobes of *Locusta* and *Schistocerca* corpora cardiaca have been used as the starting material for the isolation procedures outlined below. Goldsworthy and Mordue first showed that the hormone was soluble in methanol. This proved to be a fortuitous finding in developing the isolation procedures, as the preparation of methanolic extracts of glandular lobe tissue was an efficient way of separating the free hormone from many of the proteins and other cell constituents of the glandular lobe cells. Two isolation methods (Stone, J. *et al.*, 1976; Cheeseman, P. *et al.*, 1977) for separating free AKH from glandular lobe tissue have subsequently been developed utilizing these observations.

6.2 Gel filtration chromatography

These methods are given in detail in Stone, J. and Mordue, W. (1980a). Glandular lobes from either *Schistocerca* or *Locusta* were dissected out under simple insect saline and placed in methanol. Routinely, batches of 500 glandular lobes were produced; the pooled tissues disrupted by ultrasonication and the resulting methanolic extract redissolved in water and fractionated by gel filtration on a controlled-pore glass bead column (SIGMA, G-75-50 beads) equilibrated and eluted with distilled water. The hormonal activity eluted near the total column volume. The pooled active column fractions were evaporated to dryness under reduced pressure and subjected to thin-layer chromatography in an isopropanol : acetic acid :

water system. The hormone separated as a single UV absorbing spot with an R_F of 0.67 in this solvent system. Gel permeation using an organic solvent has also been used successfully to isolate large quantities of AKH from glandular lobe tissue and haemolymph (Cheeseman, P. *et al.*, 1976, 1977).

The corpus cardiacum homogenate was ultrasonicated and the resulting supernatant concentrated before application to a Sephadex LH-20 column, previously swollen in 80% methanol. This gel filtration medium is eluted with 80% methanol, and the adipokinetic activity eluted as a single peak. The active fractions were pooled and concentrated under vacuum. The concentrate is subjected to two-dimensional thin-layer chromatography on cellulose.

Both of the procedures described above give similar yields of hormone (approx. 50% of the original activity), and provide a rapid and reliable means of isolating AKH in large quantities from locust corpora cardiaca. It is estimated that one locust contains 0.3–0.8 μg AKH. In order to obtain sufficient

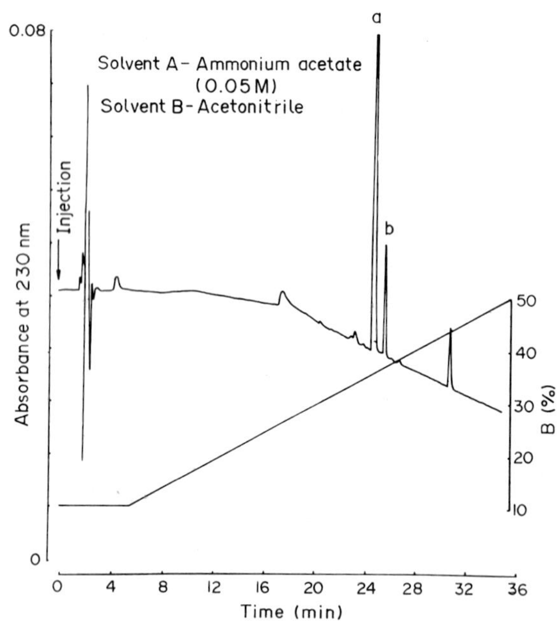

FIG. 6. Separation of five glandular lobes from *Locusta migratoria* by reverse-phase HPLC on a Shandon Hypersil ODS column, using a linear gradient elution. Absorbance was monitored at 230 nm at 0.8 absorbance units full scale. Flow rate — 1.5 ml/min; chart speed — 0.5 cm/min; solvent A — 0.05 M ammonium acetate buffer (pH 5.6); solvent B — acetonitrile. Peak 1 corresponds to AKH and preliminary results indicate that peak 2 may be AKH II.

hormone to determine its structure, the glandular lobes from about 3000 locusts were extracted (see Stone, J. *et al.*, 1976).

More recently, substantial progress has been made with regard to the use of HPLC methods in the isolation of AKH. The methods employed utilize reverse-phase C-18 columns eluted in an aqueous buffer (e.g. 0.05 M ammonium acetate) and an organic modifier (e.g. acetonitrile). A typical separation profile of a methanolic extract of glandular lobes from *Locusta* is shown in Fig. 6. Such methods are sufficiently sensitive to allow the detection of AKH in single pairs of corpora cardiaca.

6.3 Isolation of neurosecretory granules containing AKH

Rademakers, L. and Beenakkers, A. (1977), in an ultrastructural study, showed that axon terminals from the brain penetrate the glandular lobe tissue, and suggest that these axons control the release of AKH from granules in the cells by exocytosis into the haemolymph. To date, isolation of intact granules has proved successful only using the glandular lobes dissected from *Schistocerca gregaria* adults (Stone, J. and Mordue, W., 1979). Glandular lobes were dissected into ice-cold sucrose solution and homogenized using a Teflon pestle in a glass homogenizer. The resulting homogenate subjected to differential centrifugation (1) at 3000 ***g*** for 10 min, (2) at 8000 ***g*** for 10 min, and (3) at 30,000 ***g*** for 1 h, to provide three pellets (Fractions I–III) and a supernatant (Fraction IV). The adipokinetic activity of Fractions I–IV, together with an electron microscope study of the pellets, shows that the 30,000 ***g*** pellet is both rich in neurosecretory granules and contains most adipokinetic activity. This fraction also has the highest specific hormone activity (adipokinetic activity per unit of total protein).

6.4 Determination of primary sequence

AKH is a decapeptide and its amino acid sequence was determined in 1976 by Stone, J. *et al.* It was shown to be an uncharged, terminally blocked decapeptide with the following structure: PCA Leu–Asn–Phe–Thr–Pro–Asn–Trp–Gly–Thr–NH_2 (Fig. 7). Both the availability of sufficient quantities

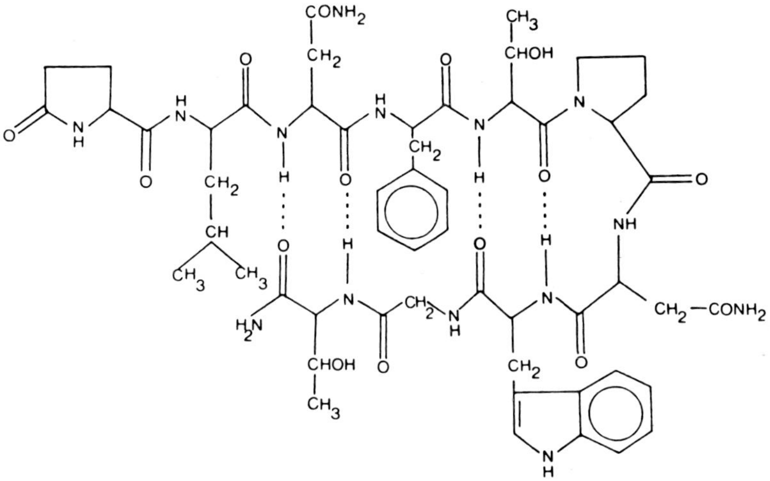

FIG. 7. Adipokinetic hormone (AKH) — secondary structure demonstrating the postulated β-bend involving residues 5–8.

of the natural hormone in pure form (approx. 750 μg or almost 750 nanomol in total was used for the complete characterization), and the use of mass spectrometry as an additional means of determining the sequence of the peptide, together with the more conventional manual techniques (the dansyl-Edman method and enzymatic digestion procedures), enabled the peptide to be characterized completely.

The strategy employed was as follows: because both the NH_2- and COOH-terminals were blocked (demonstrated by lack of mobility in high-voltage electrophoretic studies and lack of derivitization using the dansyl-Edman procedure) the hormone was selectively cleaved into two fragments, using the bacterial protease thermolysin. Its structure could then be determined using manual procedures. The COOH-terminal fragment was partially sequenced using the dansyl-Edman procedure. The structure of the NH_2-terminal fragment was determined in part following digestion with a carboxypeptidase possessing broad specificity. The complete amino acid sequence of the hormone was confirmed by mass spectrometry of derivatives of a mixture of these two fragments and of two derivatives of the intact hormone. The mass spectrometric studies elucidated the identification of the cyclic residue, pyroglutamic acid (PCA), as the blocked NH_2-terminal residue and the presence of amide groups on the aspartic acid residues (i.e. asparagine rather than aspartic acid) and on the threonine COOH-terminal residue. Such fine detail of amino structure is not so easily distinguished using the standard manual techniques.

The primary sequence of AKH was later confirmed by synthesis of the fully active hormone (Broomfield, C. and Hardy, P., 1977; Yamashiro, D. *et al.*, 1981). Synthetic AKH is now available commercially.

6.5 Structure–activity relationships of AKH

Structure–activity studies were able to proceed once the fully active hormone had been synthesized. Attempts have been made to determine which residues and regions of the hormone molecule are necessary to interact with fat body receptors to trigger a response ultimately leading to the release of diglycerides. Peptides structurally related to AKH, as shown in Table 7, were either synthesized or prepared by enzymic cleavage of the synthetic hormone. From the results obtained a number of conclusions could be drawn as to the structural features of AKH that are essential for the hormonal activity of the molecule (Stone, J. *et al.*, 1978; Mordue, W., 1981; Hardy, P. and Sheppard, P., 1983).

6.6 Chain length

Peptides with fewer than eight residues (compounds 1–10), whether blocked or possessing a net charge, were all without activity. The only peptides that exhibited activity were, in fact, uncharged, and they were all octa-, nona-, or decapeptides. All that may be concluded is that larger peptides possess significant adipokinetic activity. (Recent work concerned with immunocytochemical detection of AKH-like peptides has utilized the analogue [Tyr^1]-AKH; however it is understood that this charged decapeptide is without biological activity.) The octapeptides 11, 12, and 13 listed in Table 7 are all active, and it is apparent that these first eight residues (NH_2-terminal is residue 1) are required to produce activity. The similarities in activity between compound 11 (derived from AKH) and compound 13 (RPCH: red-pigment-concentrating-hormone from prawns) indicate that chain length, the presence of eight residues, NH_2-terminal pyroglutamate, and COOH-terminal amide are of prime importance.

The necessity of a minimum of eight residues to trigger an adipokinetic response is demonstrated by octapeptides 11–13, all of which possess some activity. A heptapeptide analogue with the penultimate asparagine residue deleted, compound 10, is inactive, suggesting that the presence of tryptophan amide at the COOH-terminal in a peptide of almost the same size as the octapeptide is insufficient to produce activity.

6.7 Terminal residues

The requirements for the terminal residues are precise. In compound 14 a glycl-amide residue has been added to the partially active octapeptide, compound 11, to produce a blocked nonapeptide; this addition reduced the activity of the nonapeptide to only 35% of that of the octapeptide: addition of glycyl-threonyl-amide to the same octapeptide to produce AKH dramatically increases biological activity. However, if the addition is threonyl-glycyl-amide, compound 16, then the biological activity is markedly reduced, even to below that of the octapeptides. Thus, for the COOH-terminus, it is essential in decapeptides to have the natural sequence of residues to produce maximum activity. The leucine residue is absent in compound 20 and this deletion markedly reduces biological activity. This evidence produces further confirmation for

Table 7: Adipokinetic activities of adipokinetic hormone and structurally related compounds

Compound	Residues (1 2 3 4 5 6 7 8 9 10)	Relative agonist activity*
1	H – Asn – Trp – Thr – NH_2	<0.3
2	Ac – Asn – Trp – Gly – Thr – NH_2	<0.3
3	H – Thr – Pro – Asn – Trp – Gly – Thr – NH_2	<0.3
4	H – Phe – Thr – Pro – Trp – Asn – Gly – Thr – NH_2	<0.3
5	PCA – OH	<0.3
6	PCA – Leu – Asn	<0.3
7	PCA – Leu – Asn – Phe	<0.3
8	PCA – Leu – Asn – Phe – Thr – Pro	<0.3
9	PCA – Leu – Asn – Phe – Thr – Pro – NH_2	<0.3
10	PCA – Leu – Asn – Phe – Thr – Pro – Trp – NH_2	<0.3
11	PCA – Leu – Asn – Phe – Thr – Pro – Asn – Trp – NH_2	20
12	D – PCA – Leu – Asn – Phe – Thr – Pro – Asn – Trp – NH_2	8
13	PCA – Leu – Asn – Phe – Ser – Pro – Gly – Trp – NH_2(RPCH)	20
14	PCA – Leu – Asn – Phe – Thr – Pro – Asn – Trp – Gly – NH_2	7
15	PCA – Leu – Asn – Phe – Thr – Pro – Gly – Trp – Gly – Thr – NH_2	30
16	PCA – Leu – Asn – Phe – Thr – Pro – Asn – Trp – Thr – Gly – NH_2	3
17	D – PCA – Leu – Asn – Phe – Thr – Pro – Asn – Trp – Gly – Thr – NH_2	5
18	PCA – Leu – Asn – Phe – Gly – Pro – Asn – Trp – Gly – Thr – NH_2	33
19	PCA – Leu – Asn – Phe – Ser – Pro – Asn – Trp – Gly – Thr – NH_2	<0.3
20	PCA – Asn – Phe – Thr – Pro – Asn – Trp – Gly – Thr – NH_2	<0.3
21	PCA – Leu – Asn – Phe – Thr – Pro – Asn – Trp – Gly – Thr – NH_2(AKH)	100

*The activity of adipokinetic hormone is defined as 100. Other preparations were compared with the hormone standard on a molar basis. ED_{50} of adipokinetic hormone is 3.0–4.5 pmol per locust. Activities shown as <0.3 indicate that the maximum dose tested (200–250 pmol per locust) did not produce a significant adipokinetic response.

the precise requirements above eight residues: nonapeptides have very low activities.

Evidence that for maximum biological activity the NH_2-terminal residue of the molecule must be the L-enantiomer of pyroglutamic acid (PCA) is provided from comparisons of the activities of compounds 11 and 12 and compounds 17 with 18. For both epimer pairs, replacement of the D-form of pyroglutamic acid with the L-form causes a marked reduction in the activity of the molecule (60% lower in the octapeptides and 95% in the decapeptides).

The decapeptide analogue of AKH, compound 15, in which glycine replaces asparagine as residue 7 (as in RPCH), is only 30% as active as AKH itself. This result is somewhat surprising, as this substitution in position 7 does not reduce the (albeit lower) biological activity in the octapeptide molecule. The result implies that with the decapeptide molecule there are more constraints and a much more precise structural requirement for maximum adipokinetic activity.

Further studies presented by Stone, J. *et al.* (1978) showing that two of the analogues, compounds 11 and 15, exhibit similar time–course relationships to AKH in eliciting the lipid mobilization response in locusts indicated that the analogues and the hormone are transported to their site of action and are degraded over similar time periods. Consequently, the relative agonist activities shown in Table 7 most probably reflect the different abilities of the analogues to activate the receptors. It follows therefore that the structure–activity correlations are indicators of the structural requirements of the locust fat body AKH receptors.

6.8 Conformation

The possible existence of a preferred three-dimensional configuration for AKH has been considered (Stone, J. *et al.*, 1978; Mordue, W. and Stone, J., 1978), and the existence of a β-bend postulated. Application of the secondary structure predictive model of Chou, A. and Fasman, G. (1974) to the hormone indicated a highly favoured conformation in which residues 5–8 form a β-bend, as shown in Fig. 7. Three of the amino acids involved in such a β-turn of the hormone (proline, asparagine and tryptophan) occur at the positions in which they occur in the β-turns of proteins with greatest frequency. The averaged probability of the β-bend occurring here (7.4×10^{-4}) is well above the value below which the conformation is unlikely. The model of Chou and Fasman was developed for globular proteins, but it emphasizes the importance of relatively short-range interactions in the molecule and may therefore be applied to peptides of modest length such as AKH. Indirect evidence in support of the notion that the peptide must adopt this conformation to elicit a response at the receptor is obtained from the fact that a minimum of the first eight residues are necessary for activity, and this octapeptide fragment contains the β-bend. In its β-bend configuration, stability would be conferred on AKH by hydrogen bonding between residues 3 and 10 and between 5 and 8. It is of interest that replacement of threonine as residue 5 by either glycine or serine (compounds 18 and 19) markedly reduces activity and could well weaken any stabilization by hydrogen bonding. However, it must be emphasized that such substitutions in the core residues may well exert their own particular actions other than by disrupting this postulated binding.

6.9 A second peptide from locust corpora cardiaca (AKH II)

Carlsen, J. *et al.* (1979) have separated out a second small peptide from locust corpora cardiaca. Like AKH the peptide is located principally in the glandular lobes of the corpora cardiaca. The authors isolated AKH II from *Schistocerca gregaria*, though they report that it is also present in *Locusta migratoria*. This peptide demonstrated adipokinetic activity in locusts and red pigment concentrating activity in prawns, as does AKH (Mordue, W. and Stone, J., 1976, 1977). Gel filtration on Sephadex G-25 showed that the peptide (AKH II) eluted immediately after AKH. In fact when the column effluent is assayed using the adipokinetic bioassay of Goldsworthy, G. *et al.* (1972) the two peaks of activity cannot be differentiated clearly. However, using the (approx. 100 times) more sensitive prawn red pigment concentration bioassay (devised by Fernlund, P. (1968) during the isolation of the prawn red pigment concentrating hormone) it is possible to distinguish the two separate peaks of hormonal activity.

The peptide was isolated by the following procedure. An aqueous extract of the pooled corpora cardiaca was boiled and lyophilized and then applied to a Sephadex G-25 column equilibrated and eluted with water. Two peaks of activity, one due to AKH and the other due to the second peptide, were distinguished using the prawn erythrophore bioassay. The pooled active fractions of the peptide were rechromatographed on a Sephadex LH-20 column equilibrated and eluted with aqueous butanol. The combined active fractions contain a single peptide, as is judged by its homogeneity in three thin-layer chromatographic systems. Amino acid analyses of this peptide have shown it to contain the following eight amino acid residues in equimolar proportions: Asp, Thr, Ser, Glu, Gly, Leu, Phe and Trp. Recently Orchard, I. and Lange, A. (1983) have described details of other thin-layer systems which are capable of isolating AKH II. However, it is likely that HPLC methods (see section 6.2) will prove more useful and provide more readily reproducible results.

The peptide accounts for about 20% of the total adipokinetic activity of the locust corpora cardiaca. About 30 ng pure peptide was obtained from one adult *Schistocerca gregaria*. On the basis of its amino acid composition and its separation characteristics on the two gel filtration columns it is speculated that its structure probably resembles that of the prawn red pigment concentrating hormone [PCA–Leu–Asn–Phe–Ser–Pro–Gly–Trp–NH_2; Fernlund, P. (1974)] more closely than it resembles AKH [PCA–Leu–Asn–Phe–Thr–Pro–Asn–Trp–Gly–Thr–NH_2; Stone, J. *et al.* (1976)]. At the present time the physiological role of this peptide remains unknown.

REFERENCES

ALTMANN, G. (1956). Die Regulation des Wasserhaltes der Honigbiene. *Insectes Socieaux. 3*, 33–40.

ASTON, R. J. (1979). Studies on the diuretic hormone of *Rhodnius prolixus*. Some observations on the purification and nature of the hormone and the dynamics of its release *in vitro*. *Insect Biochem. 9*, 163–176.

ASTON, R. J. and HUGHES, L. (1980). Diuretic hormone-extraction and chemical properties. In *Neurohormonal Techniques in Insects*. Edited by T. A. Miller. Chap. 4, pages 91–115. Springer-Verlag, New York.

ASTON, R. J. and WHITE, A. F. (1974). Isolation and purification of the diuretic hormone from *Rhodnius prolixus*. *J. Insect Physiol. 20*, 1673–1682.

BAUMANN, E. and GERSCH, M. (1973). Untersuchungen zur Stabilitet des Neurohormons D. *Zool. Jb., Physiol. 77*, 153–160.

BAUMANN, E. and GERSCH, M. (1982). Purification and identification of neurohormone D, a heart accelerating peptide from the corpora cardiaca of the cockroach *Periplaneta americana*. *Insect Biochem. 12*, 7–14.

BEENAKKERS, A. M. TH. (1969). The influence of corpus allatum and corpus cardiacum on lipid metabolism in *Locusta migratoria*. *Gen. Comp. Endocr. 13*, 492.

BERRIDGE, M. J. (1966). The physiology of excretion in the cotton stainer, *Dysdercus fasciatus* Signoret — IV. Hormonal control of excretion. *J. Exp. Biol. 44*, 553–566.

BERRIDGE, M. J. and PATEL, N. G. (1968). Insect salivary glands: Stimulation of fluid secretion by 5-hydroxytryptamine and adenosine-3′,5′-monophosphate. *Science, 162*, 462–463.

BHAKTHAN, N. M. G. and GILBERT, L. I. (1968). Effects of some vertebrate hormones on lipid mobilization in the insect fat body. *Gen. Comp. Endocr. 11*, 186–197.

BOWERS, W. S. and FRIEDMAN, S. (1963). Mobilization of fat body glycogen by an extract of corpus cardiacum. *Nature 198*, 685.

BROOMFIELD, C. E. and HARDY, P. M. (1977). The synthesis of locust adipokinetic hormone. *Tetrahedron. Lett. 25*, 2201–2204.

BROWN, B. E. (1965). Pharmacologically active constituents of the cockroach corpus cardiacum: resolution and some characteristics. *Gen. Comp. Endocr. 5*, 387–401.

BROWN, B. E. (1977). Occurrence of proctolin in six orders of insects. *J. Insect Physiol. 23*, 861–864.

BROWN, B. E. and STARRATT, A. N. (1975). Isolation of proctolin, a myotropic peptide, from *Periplaneta americana*. *J. Insect Physiol. 21*, 1879–1881.

CARLSEN, J., HERMAN, W. S., CHRISTENSEN, M. and JOSEFSSON, L. (1979). Characterization of a second peptide with adipokinetic and red pigment-concentrating activity from the locust corpora cardiaca. *Insect Biochem. 9*, 497–501.

CAZAL, M. and GIRARDIE, A. (1968). Controle humoral de l'equilibre hydrique chez *Locusta migratoria migratorioides*. *J. Insect Physiol. 14*, 655–668.

CHANG, F. (1974). Effect of vertebrate adipokinetic hormones on the rate of *in vitro* lipid release in insects. *Comp. Biochem. Physiol. 49B*, 567–578.

CHEESEMAN, P., JUTSUM, A. R. and GOLDSWORTHY, G. J. (1976). Quantitative studies on the release of locust adipokinetic hormone. *Physiol. Ent. 1*, 115–121.

CHEESEMAN, P., GOLDSWORTHY, G. J. and MORDUE, W. (1977). Studies on the purification of locust adipokinetic hormone. *Life Sci. 21*, 231–236.

CHEN, A. C. and FRIEDMAN, S. (1977). Hormonal regulation of trehalose metabolism in the blowfly, *Phormia regina*: interaction between hypertrahalosemic and hypotrehaloemic hormones. *J. Insect Physiol. 23*, 1223–1232.

CHOU, P. Y. and FASMAN, G. D. (1974). Prediction of protein conformation. *Biochemistry 13*, 222–245.

DIXIT, P. K. and PATEL, N. G. (1964). Insulin-like activity in larval foods of the honeybee. *Nature, 202*, 189–190.

DORES, R. M., DALLMANN, S. H. and HERMAN, W. S. (1979). The regulation of post-eclosion and post-feeding diuresis in the monarch butterfly, *Danaus plexippus*. *J. Insect Physiol. 25*, 895–901.

DUTRIEU, J. and GOURDOUX, L. (1965). Le controle neuroendocrinen de la trehalosemic de *Carausius morosus*. *Gen. Comp. Endocr. 5*, 387–401.

DUVE, H. (1978). The presence of a hypoglucemic and hypotrehalocemic hormone in the neurosecretory system of the blowfly, *Calliphora erythrocephala*. *Gen. Comp. Endocr. 36*, 102–110.

DUVE, H. and THORPE, A. (1979). Immunofluorescent localization of insulin-like material in the median neurosecretory cells of the blowfly, *Calliphora vomitoria* (Diptera). *Cell Tiss. Res. 200*, 187–191.

DUVE, H. and THORPE, A. (1980). Localisation of pancreatic polypeptide (PP)-like immunoreactive material in neurones of the brain of the blowfly, *Calliphora erythrocephala* (Diptera). *Cell Tiss. Res. 210*, 101–109.

DUVE, H. and THORPE, A. (1982). The distribution of pancreatic polypeptide in the nervous system and gut of the blowfly, *Calliphora vomitoria* (Diptera). *Cell Tiss. Res. 227*, 67–77.

DUVE, H., THORPE, A. and LAZARUS, N. R. (1979). Isolation of material displaying insulin-like immunological and biological activity from the brain of the blowfly *Calliphora vomitoria*. *Biochem. J. 184*, 221–227.

DUVE, H., THORPE, A., NEVILLE, R. and LAZARUS, R. (1981). Isolation and partial characterization of pancreatic polypeptide-like material in the brain of the blowfly, *Calliphora vomitoria*. *Biochem. J. 197*, 767–770.

DUVE, H., THORPE, A., LAZARUS, N. R. and LOWRY, P. J. (1982). A neuropeptide of the blowfly *Calliphora vomitoria* with an amino acid composition homologous with vertebrate pancreatic polypeptide. *Biochem. J. 201*, 429–432.

EL-SALHY, M. (1981). Immunohistochemical localization of pancreatic polypeptide (PP) in the brain of the blowfly, *Eristalis aenus* (Diptera). *Experientia 37*, 1009–1010.

EL-SALHY, M., ABOU-EL-ELA, R., FALKMER, S., GRIMELIUS, L. and WILANDER, E. (1980). Immunohistochemical evidence of gastro-entero-pancreatic neuro-hormonal peptides of vertebrate type in the nervous system of the larva of a dipteran insect, the hoverfly, *Eristalis aeneus*. *Regulatory Peptides 1*, 187–204.

ENDO, Y., IWANAGA, T., FUJITA, T. and NISHIITSUTSUJI-UWO, J. (1982). Localization of pancreatic polypeptide (PP)-like immunoreactivity in the central and visceral nervous systems of the cockroach *Periplaneta*. *Cell Tiss. Res. 227*, 1–9.

FERNLUND, P. (1968). Chromactivating hormones of *Pandalus borealis*. Bioassay of the red-pigment concentrating hormone. *Marine Biol. 2*, 13–18.

FERNLUND, P. (1974) Structure of the red pigment concentrating hormone of the shrimp *Pandalus borealis*. *Biochim. Biophys. Acta 371*, 304–311.

FLOYD, J. C., FAJANS, S. S., PEK, S. and CHANCE, R. E. (1977). A newly recognized pancreatic polypeptide: plasma levels in health and disease. *Recent Prog. Horm. Res. 33*, 519–570.

FRIEDMAN, S. (1967). The control of trehalose synthesis in the blowfly, *Phormia regina* Meig. *J. Insect Physiol. 13*, 397–405.

GADE, G. (1979). Adipokinetic and hyperglycaemic factor(s) in the corpora cardiaca/corpora allata complex of the stick insect, *Carausius morosus*. I. Initial characteristics. *Physiol. Ent. 4*, 131–134.

GADE, G. (1981). Studies on a possible hyperglycaemic factor from the corpus cardiaca of stick insects: characterization and preliminary purification. *Zool. Jb., Physiol. 85*, 266–277.

GADE, G. and LOHR, P. (1982). Restricted specificity of a hyperglycaemic factor from the corpus cardiacum of the stick insect *Carausius morosus*. *J. Insect Physiol. 28*, 805–811.

GEE, J. D. (1975). The control of diuresis in the tsetse fly *Glossina austeni:* A preliminary investigation of the diuretic hormone. *J. Exp. Biol. 63*, 391–401.

GEE, J. D. (1977). The hormonal control of excretion. In *Transport of Ions and Water in Animals*. Edited by B. L. Gupta, R, B. Moreton, J. L. Oschman and B. J. Wall. Pages 265–284. Academic Press, Lond.

GERSCH, M. (1967). Experimental examinations on the hormonal control of the water balance and excretion of the larva of *Corethra (Chaoborus)*. *Gen. Comp. Endocr. 9*, 453.

GERSCH, M. and STURZEBECHER, J. (1971). Further studies on chemical characterization of neurohormone D. *Insect Endocrines (Praha)*, 71–80.

GERSCH, M., FISCHER, F., UNGER, H. and KOCH, H. (1960). Die Isolierung neuro-hormonaler Faktoren aus dem Nervensystem der Kuchenschabe *Periplaneta americana*. *Z. Naturf. 15b*, 319–322.

GERSCH, M., RICHTER, K., STÜRZEBECHER, J. and FABIAN, B. (1969). Experimentelle Untersuchungen zun Wirkungsmechanismus des Neurohormons D von *Periplaneta americana* (L.) am "biologischen Modell" der Harnblase von *Bufo bufo* (L.). *Gen. Comp. Endocr. 12*, 40–50.

GOLDBARD, G. A., SAUER, J. R. and MILLS, R. R. (1970). Hormonal control of excretion in the American cockroach. II. Preliminary purification of a diuretic and antidiuretic hormone. *Comp. Gen. Pharmacol. 1*, 82–86.

GOLDSWORTHY, G. J. (1969). Hyperglycemic factors from the corpus cardiacum of *Locusta migratoria*. *J. Insect Physiol. 15*, 2131–2140.

GOLDSWORTHY, G. J. and MORDUE, W. (1974). Neurosecretory hormones in insects. *J. Endocr. 60*, 529–558.

GOLDSWORTHY, G. J., MORDUE, W. and GUTHKELCH, J. (1972). Studies on insect adipokinetic hormones. *Gen. Comp. Endocr. 18*, 545–551.

HANAOKA, K. and TAKAHASHI, S. Y. (1976). Effect of a hyperglycaemic factor on haemolymph trehalose and fat body carbohydrates in the American cockroach. *Insect Biochem. 6*, 621–625.

HARDY, P. M. and SHEPPARD, P. W. (1983). Synthesis of [4,5-dehydrohen^2]- and [3,4 dehydro Pro^6]-locust adipokinetic hormone. *J. Chem. Soc. Perkin Trans. I*, 723–729.

HEARN, M. T. W. (1980). The use of reversed phase high performance liquid chromatography for the structural mapping of polypeptides and proteins. *J. Liquid Chromatogr. 3*, 1255–1276.

HIRIPI, L. and SALANKI-ROZSA, K. (1973). Fluorimetric determination of 5-hydroxytryptamine and catecholamines in the central nervous system and heart of *Locusta migratoria migratorioides*. *J. Insect Physiol. 19*, 1481–1485.

HOLWERDA, D. A., VAN DOORN, J. and BEENAKKERS, A. M. TH. (1977). Characterization of the adipokinetic and hyperglycaemic substances from the locust corpus cardiacum. *Insect Biochem. 7*, 151–157.

HUGHES, L. (1977). High-molecular weight forms of diuretic hormone from *Rhodnius prolixus*. *Biochem. Soc. Trans. 5*, 1060–1063.

HUGHES, L. (1979). Further investigation of the isolation of diuretic hormone from *Rhodnius prolixus*. *Insect Biochem. 9*, 247–255.

ISHAY, J., GRITTEN, S., GALUN, R., DORON, M. and LERON, Z. (1976). The presence of insulin in and some effects of exogenous insulin on Hymenoptera tissues and body fluids. *Comp. Biochem. Physiol. 54A*, 203–206.

ISOBE, M. and GOTO, T. (1980). Diapause hormones. In *Neurohormonal Techniques in Insects*. Edited by T. A. Miller. Pages 216–243. Springer, New York.

IWANAGA, T., FUJITA, T., NISHIITSUTSUJI-UWO, J. and ENDO, Y. (1981). Immunohistochemical demonstration of PP-, somatostatin-, enteroglucagon-immunoreactivities in the cockroach midgut. *Biomed. Res. 2*, 202–207.

JONES, J. (1978). Physiologically active factors in the corpora cardiaca of insects. Ph.D. thesis, University of London.

JONES, J., STONE, J. V. and MORDUE, W. (1977). The hyperglycaemic activity of locust adipokinetic hormone. *Physiol. Ent. 2*, 185–187.

JULLIARD, J. H., SHIBASAKI, T., LING, N. and GUILLEMIN, R. (1980). High-molecular-weight immunoreactive β-endorphin in extracts of human placenta is a fragment of immunoglobulin G. *Science 208*, 183–185.

KIMMEL, J. R., HAYDEN, L. J. and POLLOCK, H. G. (1975). Isolation and characterization of a new pancreatic polypeptide hormone. *J. Biol. Chem. 250* (24), 9369–9376.

KRAMER, K. J. (1980). Insulin-like and glucagon-like hormones in insects. In *Neurohormonal Techniques in Insects*. Edited by T. A. Miller. Pages 116–136. Springer, New York.

KRAMER, K. J., TAGER, H. S., CHILDS, C. N. and SPEIRS, R. D. (1977). Insulin-like hypoglycemic and immunological activities in honeybee royal jelly. *J. Insect Physiol. 23*, 293–295.

KRAMER, K. J., TAGER, H. S. and CHILDS, C. N. (1980). Insulin-like and glucagon-like peptides in insect hemolymph. *Insect Biochem. 10*, 179–182.

KRAMER, K. J., CHILDS, C. N., SPEIRS, R. D. and JACOBS, R. M. (1982). Purification of insulin-like peptides from insect haemolymph and royal jelly. *Insect Biochem. 12*, 91–98.

LOREN, I., ALUMETS, J., HÅKANSON, R. and SUNDLER, F. (1979). Immunoreactive pancreatic polypeptide (PP) occurs in the central and peripheral nervous system: preliminary immunocytochemical observations. *Cell Tiss. Res. 200*, 179–186.

LOUGHTON, B. G. and ORCHARD, I. (1981). The nature of the hyperglycaemic factor from the glandular lobe of the corpus cardiacum of *Locusta migratoria*. *J. Insect Physiol. 27*, 383–385.

MADDRELL, S. H. P. (1963). Excretion in the blood sucking bug, *Rhodnius prolixus* Stål. I. The control of diuresis. *J. Exp. Biol. 40*, 247–256.

MADDRELL, S. H. P. (1971). The mechanisms of insect excretory systems. *Adv. Insect Physiol. 8*, 199–331.

MADDRELL, S. H. P. (1972). The functioning of insect Malpighian tubules. In *Role of Membranes in Secretory Processes*. Edited by L. Bolis, R. D. Keynes and W. Wilbrandt. Pages 338–356. North-Holland, Amsterdam.

MADDRELL, S. H. P. and GEE, J. D. (1974). Potassium-induced release of the diuretic hormones of *Rhodnius prolixus* and *Glossina austeni*: Ca dependence, time course of release and localization of neurohaemal areas. *J. Exp. Biol. 61*, 155–171.

MADDRELL, S. H. P., PILCHER, D. E. M. and GARDINER, B. O. C. (1969). Stimulatory effect of 5-hydroxytryptamine (serotonin) on secretion by Malpighian tubules of insects. *Nature 222*, 784–785.

MADDRELL, S. H. P., PILCHER, D. E. M. and GARDINER, B. O. C. (1971). Pharmacology of the Malpighian tubules of *Rhodnius* and *Carausius*: the structure–activity relationship of tryptamine analogues and the role of cyclic AMP. *J. Exp. Biol. 54*, 779–804.

MAIER, V., WITZNICK, G., KELLER, R. and PFEIFFER, E. F. (1978). Insulin-like and glucagon-like immunoreactivities in the honeybee (*Apis mellifera*). *Acta Endocr. 87*, 69–70.

MARUO, T., SEGAL, S. J. and KOIDE, S. S. (1979). Studies on the apparent human chorionic gonadotropin-like factor in the crab *Ovalipes ocellatus*. *Endocrinology 104*, 932–939.

MAYER, R. J. and CANDY, D. J. (1969). Control of haemolymph lipid concentration during locust flight: an adipokinetic hormone from the corpora cardiaca. *J. Insect Physiol. 15*, 611–620.

MEEK, J. L. and ROSSETTI, Z. L. (1981). Factors affecting retention and resolution of peptides in high-performance liquid chromatography. *J. Chromatogr. 211*, 15–28.

MIGLIORI NATALIZI G. and FRONTALI, N. (1966). Purification of insect hyperglycaemic and heart accelerating hormones. *J. Insect Physiol. 12*, 1279–1287.

MIGLIORI NATALIZI, G., PANSA, M. C., D'AJELLO, V., CASAGLIA, O., BETTINI, S. and FRONTALI, N. (1970). Physiologically active factors from corpora cardiaca of *Periplaneta americana*. *J. Insect Physiol. 16*, 1827–1836.

MILLER, TH. (1979). Nervous versus neurohormonal control of insect heartbeat. *Am. Zool. 19*, 77–86.

MILLS, R. R. (1967). Hormonal control in the American cockroach. I. Release of the diuretic hormone from the terminal abdominal ganglion. *J. Exp. Biol. 46*, 35–41.

MILLS, R. R. and WHITEHEAD, D. L. (1970). Hormonal control of tanning in the American cockroach. Changes in the blood cell permeability during ecdysis. *J. Insect Physiol. 16*, 331–340.

MIRSKY, I. A. (1973). Insulin — purification and biochemical characterization. In *Methods in Investigative and Diagnostic Endocrinology*. Edited by S. A. Berson and R. S. Yalow. Vol. 2B, Part III, Chap. 3, pages 823–827. North Holland, Amsterdam.

MORDUE, W. (1969). Hormonal control of Malpighian tube and rectal function in the desert locust, *Schistocerca gregaria*. *J. Insect Physiol. 15*, 273–285.

MORDUE, W. (1972). Hormones and excretion in locusts. *Gen. Comp. Endocr. Suppl. 3*, 289–298.

MORDUE, W. (1981). Adipokinetic hormone and related peptides. In *Neurosecretion: Molecules, Cells, Systems*. Edited by D. S. Farner and K. Lederis. Pages 391–401. Plenum Press, New York.

MORDUE, W. and GOLDSWORTHY, G. J. (1969). The physiological effects of corpus cardiacum extracts in locusts. *Gen. Comp. Endocr. 12*, 360–369.

MORDUE, W. and STONE, J. V. (1976). Comparison of the biological activities of an insect and a crustacean neurohormone that are structurally similar. *Nature 264*, 287–289.

MORDUE, W. and STONE, J. V. (1977). Relative potencies of locust adipokinetic hormone and prawn red pigment-concentrating hormone in insect and crustacean systems. *Gen. Comp. Endocr. 33*, 103–108.

MORDUE, W. and STONE, J. V. (1978). Structure and metabolism of adipokinetic hormone. In *Comparative Endocrinology*. Edited by P. J. Gaillard and H. H. Boer. Pages 487–490. Elsevier, Amsterdam.

MOREAU, R., RAOELISON, C. and SUTTER, B. C. J. (1981). An intestinal insulin-like molecule in *Apis mellifera* L. (Hymenoptera). *Comp. Biochem. Physiol. 69A*, 79–83.

MORGAN, P. J. and MORDUE, W. (1983). Separation and characteristics of diuretic hormone from the corpus cardiacum of *Locusta*. *Comp. Biochem. Physiol. 75B*, 75–80.

MOSNA, G. and BARIGOZZI, C. (1976). Stimulation of growth by insulin in *Drosophila* embryonic cells in vitro. *Experientia 32*, 855–856.

MOTHES, G. (1960). Weitere Untersuchungen uber den physiologischen Farbwechsel von *Carausius morosus* (BR). *Zool. Jahrb. Physiol. 69*, 133–162.

NICOLSON, S. W. (1976). The hormonal control of diuresis in the cabbage white butterfly, *Pieris brassicae*. *J. Exp. Biol. 65*, 565–575.

NICOLSON, S. W. (1980). Diuresis and its hormonal control in butterflies. *J. Insect Physiol. 26*, 841–846.

NIJHOUT, H. F. and CARROW, G. M. (1978). Diuresis after a bloodmeal in female *Anopheles freeborni*. *J. Insect Physiol. 24*, 293–298.

NORMANN, T. C. (1972). Heart activity and its control in the adult blowfly, *Calliphora erythrocephala*. *J. Insect Physiol. 18*, 1793–1810.

NORMANN, T. C. (1975). Neurosecretory cells in insect brain and production of hypoglycemic hormone. *Nature, 254*, 259–261.

NORMANN, T. C. and DUVE, H. (1969). Experimentally induced release of a neuro-hormone influencing hemolymph trehalose level in *Calliphora erythrocephala* (Diptera). *Gen. Comp. Endocr. 12*, 449–459.

NORMANN, T. C. and DUVE, H. (1978). Insulin-like and glucagon-like neurohormones in the blowfly. In *Comparative Endocrinology*. Edited by P. J. Gaillard and H. H. Boer. Pages 441–444. Elsevier, Amsterdam.

NUNEZ, J. A. (1956). Untersuchungen uber die Reglung des Wasserhaushaltes bei *Anisotarsus cupripennis* Germ. *Z. vergl. Physiol. 38*, 341–354.

ORCHARD, I. and LANGE, A. B. (1983). Release of identified adipokinetic hormones during flight and following neural stimulation in *Locusta migratoria*. *J. Insect Physiol. 29*, 639–642.

ORCHARD, I. and LOUGHTON, B. G. (1980). A hypolipaemic factor from the corpus cardiacum of locusts. *Nature, 286*, 494–496.

O'SHEA, M., ADAMS, M. E. and BISHOP, C. A. (1982). Identification of proctolin-containing neurons. *Fed. Proc. 41*, (13), 2940–2947.

PIEK, T., VISSER, B. J. and MANTEL, P. (1979). Effect of proctolin, BPP_{5a} and related peptides on rhythmic contractions in *Locusta migratoria*. *Comp. Biochem. Physiol. 62C*, 151–154.

PHILLIPS, J. E. (1980). Epithelial transport and control in recta of terrestrial insects. In *Insect Biology in the Future*. Edited by M. Locke and D. Smith. Pages 145–177. Academic Press, New York.

PHILLIPS, J. E., MORDUE, W., MEREDITH, J. and SPRING, J. (1980). Purification and characteristics of the chloride transport stimulating factor from locust corpora cardiaca: a new peptide. *Can. J. Zool. 58*, 1851–1860.

PHILLIPS, J., SPRING, J., HANRAHAN, J., MORDUE, W. and MEREDITH, J. (1981). Hormonal control of salt reabsorption by the excretory system of an insect. Isolation of a new protein. In *Neurosecretion, Molecules, Cells, Systems*. Edited by D. S. Farner and K. Lederis. Chap. 32, pages 371–380. Plenum Press, New York.

PILCHER, D. E. M. (1970). Hormonal control of the Malpighian tubules of the stick insect *Carausius morosus*. *J. Exp. Biol. 52*, 653–665.

PROUX, J. and ROUGON-RAPUZZI, G. (1980). Evidence for vasopressin-like molecule in migratory locust. Radioimmunological measurements in different tissues. Correlation with various states of hydration. *Gen. Comp. Endocr. 42*, 378–383.

PROUX, J., ROUGON-RAPUZZI, G. and REMY, C. (1980). Influence de la *pars inter*-cerebralis et des cellules neurosecretrices sous-ocellaires medianes sur la teneur en substance apparentee a la vasopressine dans la chaine nerveuse ventrale et l'hemolymphe du criquet migrateur. Etude radio-immunologique et immunohistologique. *J. Physiol. (Paris D.) 76*, 227–282.

PROUX, J., ROUGON-RAPUZZI, G. and CUPO, A. (1982). Enhancement of excretion across locust Malpighian tubules by a diuretic vasopressin-like hormone. *Gen. Comp. Endocr. 47*, 449–457.

RADEMAKERS, L. H. P. M. and BEENAKKERS, A. M. TH. (1977). Changes in the secretory activity of the glandular lobe of the corpus cardiacum of *Locusta migratoria* induced by flight. A quantitative electron microscopic study. *Cell Tiss. Res. 180*, 155–171.

RAFAELI, A. and MORDUE, W. (1982). The responses of the Malpighian tubules of *Locusta* to hormones and other stimulants. *Gen. Comp. Endocr. 46*, 130–135.

RALPH, C. L. and MCCARTHY, C. (1964). Effect of brain and corpus cardiacum extracts on haemolymph trehalose of the cockroach *Periplaneta americana*. *Nature, 203*, 1195.

REMY, C. and GIRARDIE, J. (1980). Anatomical organization of two vasopressin-neurophysin-like neurosecretory cells throughout the central nervous system of the migratory locust. *Gen. Comp. Endocr. 40*, 27–35.

REMY, C., GIRARDIE, J. and DUBOIS, M. P. (1977). Exploration immunocytologique des ganglions cérébroïdes et sous-oesophaglen du plasme *Clitumnus extradentatus*. Existence d'une neurosécrétion apparentée à la vasopressine neurophysine. *C.R. Acad. Sci. Paris D. 285*, 1495–1497.

REMY, C., GIRARDIE, J. and DUBOIS, M. P. (1979). Vertebrate neuropeptide-like substances in the suboesophageal ganglion of two insects: *Locusta migratoria* R and F (Orthoptera) and *Bombyx mori* L. (Lepidoptera). Immunocytological investigation. *Gen. Comp. Endocr. 37*, 93–100.

ROUSSEL, J. P. and CAZAL, M. (1969). Action, *in vivo*, d'extraits de corpora cardiaca sur le rythme cardiaque de *Locusta migratoria* L. *C.R. Hebd. Seanc Acad. Sci. Paris D. 268*, 581–583.

SCHWARTZ, L. M. and REYNOLDS, S. E. (1979). Fluid transport in *Calliphora* Malpighian tubules: A diuretic hormone from the thoracic ganglion and abdominal nerves. *J. Insect Physiol. 25*, 847–854.

SCHWARTZ, T. W. and TAGER, H. S. (1981). Isolation and biogenesis of a new peptide from pancreatic islets. *Nature, 294*, 589–591.

SCHWYZER, R. (1982). Peptides and the new endocrinology. *Naturwissenschaften 69*, 15–20.

SEECOF, R. C. and DEWHURST, S. (1974). Insulin is a *Drosophila* hormone and acts to enhance the differentiation of embryonic *Drosophila* cells. *Cell Diff. 3*, 63–70.

SOKAL, J. E. (1973). Glucagon and gut glucagon. In *Methods in investigative and diagnostic endocrinology*. Edited by S. A. Berson and R. S. Yalow. Vol. 2B, Part III, Chap. 4, pages 901–906. North Holland, Amsterdam.

SPRING, J. and PHILLIPS, J. E. (1980a). Studies on locust rectum. I. Stimulants of electrogenic ion transport. *J. Exp. Biol. 86*, 211–223.

SPRING, J. H. and PHILLIPS, J. E. (1980b). Studies on locust rectum. II. Identification of specific ion transport processes regulated by corpora cardiaca and cyclic AMP. *J. Exp. Biol. 86*, 225–236.

STARRATT, A. N. and BROWN, B. E. (1975). Structure of the pentapeptide proctolin, a proposed neurotransmitter in insects. *Life Sci. 17*, 1253–1256.

STARRATT, A. N. and BROWN, B. E. (1979). Analogs of the insect myotropic peptide proctolin: synthesis and structure-activity studies. *Biochem. Biophys. Res. Commun. 90*, 1125–1130.

STEELE, J. E. (1961). Occurrence of a hyperglycaemic factor in the corpus cardiacum of an insect. *Nature 192*, 680–681.

STEELE, J. E. (1963). The site of action of insect hyperglycaemic hormone. *Gen. Comp. Endocr. 3*, 46–52.

STONE, J. V. and MORDUE, W. (1979). Isolation of granules containing adipokinetic hormone from locust corpora cardiaca by differential centrifugation. *Gen. Comp. Endocr. 39*, 543–547.

STONE, J. V. and MORDUE, W. (1980a). Adipokinetic hormone. In *Neurohormonal Techniques in Insects*. Edited by T. A. Miller. Chap. 2, pages 31–80. Springer-Verlag, New York.

STONE, J. V. and MORDUE, W. (1980b). Isolation of insect neuropeptides. *Insect Biochem. 10*, 229–239.

STONE, J. V., MORDUE, W., BATLEY, K. E. and MORRIS, H. R. (1976). Structure of locust adipokinetic hormone, a neurohormone that regulates lipid utilisation during flight. *Nature, 263*, 207–211.

STONE, J. V., MORDUE, W., BROOMFIELD, C. E. and HARDY, P. M. (1978). Structure-activity relationship for the lipid-mobilising action of locust adipokinetic hormone. Synthesis and activity of a series of hormone analogues. *Eur. J. Biochem. 89*, 195–202.

STRAMBI, C., STRAMBI, A., CUPO, A., ROUGON-RAPUZZI, G. and MARTIN, N. (1978). Étude des taux d'une substance apparantee a la vasopressine dans le système nerveux de Grillons soumis à différentes conditions hygrométriques. *C.R. Acad. Sci. Ser. D* 287, 1227–1230.

STRAMBI, C., ROUGON-RAPUZZI, G., CUPO, A., MARTIN, N. and STRAMBI, A. (1979). Mise en évidence immunocytologique d'un composé apparenté à la vasopressine dans le systême nerveux du Grillon *Acheta domesticus. C.R. Acad. Sci. Ser. D 288*, 131–133.

SULLIVAN, R. E. and NEWCOMB, R. W. (1982). Structure function analysis of an arthropod peptide hormone: proctolin and synthetic analogues compared on the cockroach hindgut receptor. *Peptides 3*, 337–344.

TAGER, H. S., MARKESE, J., SPEIRS, R. D. and KRAMER, K. J. (1975). Glucagon-like immunoreactivity in insect corpus cardiacum. *Nature, 254*, 707–708.

TAGER, H. S., MARKESE, J., SPEIRS, R. D. and CHILDS, C. N. (1976). Glucagon-like and insulin-like hormones of the insect neurosecretory system. *Biochem. J. 156*, 515–520.

TRAINA, M. E., BELLINO, M., SERPIETRI, L., MASSA, A. and FRONTALI, N. (1976). Heart-accelerating peptides from cockroach corpora cardiaca. *J. Insect Physiol. 22*, 323–329.

UNGER, H. (1957). Untersuchungen zur neurohormonalen Steuerung der Herztatigkeit bei Schaben. *Biol. Zbl. 76*, 204–227.

UNGER, H. (1965). Der Einfluss der Neurohormone C and D auf die Farbstoffabsorptions-fahigkeit der Malpighischen Gefasse (und des Darmes) der Stabheuschrecke *Carausius morosus* (Br.) *in vitro. Zool. Jb. (Physiol.) 71*, 710–717.

VEJBJERG, K. and NORMANN, T. C. (1974). Secretion of hyperglycaemic hormone from the corpus cardiacum of flying blowflies, *Calliphora erythrocephala. J. Insect Physiol. 20*, 1189–1192.

VIETINGHOFF, U. (1967). Neurohormonal control of renal function in *Carausius morosus. Gen. Comp. Endocr. 9*, 503.

WELSH, J. H. and MOORHEAD, M. (1960). The quantitative distribution of 5-hydroxytryptamine in the invertebrates, especially in their nervous systems. *J. Neurochem. 6*, 146–169.

WIENS, A. W. and GILBERT, L. I. (1967). Regulation of carbohydrate mobilization and utilization in *Leucophaea maderae. J. Insect Physiol. 13*, 779–794.

WILLIAMS, K. W. (1972). Solute-gel interactions in gel filtration. *Lab. Pract. 21*, 667–670.

YAMASHIRO, D., APPLEBAUM, S. W., BIRK, Y. and LI, C. H. (1981). Solid phase synthesis of locust adipokinetic hormone. *Int. J. Peptide Protein Res. 17*, 546–548.

YUI, R., FUJITA, T. and ITO, S. (1980). Insulin-, gastrin-, pancreatic polypeptide-like immunoreactive neurons in the brain of the silkworm, *Bombyx mori. Biomedical Res. 1*, 42–46.

ZIEGLER, R. (1979). Hyperglycaemic factor from the corpora cardiaca of *Manduca sexta* (L.) (Lepidoptera: Sphingidae). *Gen. Comp. Endocr. 39*, 350–357.

6 Chemistry of Ecdysteroids

DENIS H. S. HORN and RINALDO BERGAMASCO†

CSIRO, Melbourne, Victoria, Australia

† Deceased.

1 INTRODUCTION

The literature on the chemistry of ecdysteroids is now so large that in a single chapter it can be covered only in broad outline. Of the earlier reviews, that of Horn, D. (1971) is the most detailed and comprehensive and the present review will form an extension of it. Again all aspects will be reviewed but emphasis will be placed on work published since 1970, and the earlier review should be consulted for a more detailed account of the work before that time. Several other reviews have appeared covering the subject as a whole (Berkoff, C., 1969; Hikino, H. and Hikino, Y., 1970; Rees, H., 1971; and Slama, K. *et al.*, 1974) or covering certain aspects of it. These latter reviews are mentioned later at appropriate places and provide further keys to the literature in special areas.

The process of "moulting" or "ecdysis" by which insects and other arthropods grow takes place uniformly throughout the body and this fact led the Polish biologist Kopeč, S. (1922) to suggest a hormonal controlling mechanism. The classic experiments of Fraenkel, G. (1935) at the University College, London, on the the pupation of the blowfly *Calliphora erythrocephala* and those of Wigglesworth, V. (1934) at Cambridge University with the tropical bug *Rhodnius prolixus* provided convincing support for Kopeč's hypothesis (Wigglesworth, V., this volume). Fraenkel, G. (1935) showed that a mature blowfly larva tightly tied below the ring gland with cotton thread, pupated in the anterior portion only, and that the posterior portion could be induced to pupate by injecting haemolymph from larvae which had just undergone pupation. Later Plagge, E. and Becker, E. (1938), working in Kuhn's laboratory at the Kaiser Wilhelm Institute, Berlin, showed that the unpupated portions of larvae prepared in Fraenkel's experiments could be used to assay for the moulting hormone. With the aid of this "*Calliphora* bioassay" Becker, E. and Plagge, E. (1939) were able to carry out a partial purification of the hormone and show that it was soluble in water but insoluble in lipid solvents.

Following Becker's death during the war, the isolation and determination of the structure of the active principle was continued by Butenandt and Karlson. However, because of the heavy bombing of Berlin they transferred their studies to the Max Planck Institute at Tübingen. As a source of hormone they used pupae of the silkworm *Bombyx mori* which were available in large quantities as a by-product of the silk industry which was principally devoted at that time to the manufacture of parachutes. Finally, after 10 years of painstaking work, Butenandt, A. and Karlson, P. (1954), with the help of large-scale equipment at Hoffmann La Roche, Basel, isolated 25 mg of the pure crystalline hormone named "ecdysone" from ½ ton of silkworm pupae.

In the autumn of 1960 Karlson and collaborators undertook a larger-scale isolation and succeeded in obtaining 250 mg of ecdysone from 1 ton of pupae. With more material available they were able to undertake more extensive structural investigations and came to the conclusion that ecdysone was a steroid containing a conjugated ketone grouping and five hydroxy groups (Karlson, P. *et al.*, 1963). Later studies (Karlson, P. *et al.*, 1965) led to the structure (24)* for ecdysone. This structure was supported by the X-ray studies of Hüber, R. and Hoppe, W. (1965b) which defined the stereochemistry, and the compound was formally designated as 2β,3β,14α,22R,22-pentahydroxy-5β-cholest-7-en-6-one (24). This assignment has since been confirmed by several syntheses (see section 8).

Crustaceans also undergo periodic moulting and Gabe, M. (1953) pointed out the similarity of the Y-organ of certain crustaceans to the prothoracic gland, a source of ecdysone in insects. As

* Numbers in parentheses refer to name and structure of the ecdysteroid in Tables 1 and 2.

crustaceans are more readily available than insects, Karlson, P. and Schmialek, P. (1959) attempted to isolate a moulting hormone from 3 ton of the shrimp *Crangon vulgaris*. The extract obtained, although active in the *Calliphora* bioassay contained very little hormone. In 1963 Horn at the CSIRO, Australia, found that an active extract could be obtained from the seawater crayfish *Jasus lalandei*. However, the concentration of hormone present was exceedingly small and it was found that the crustacean hormone was much more polar and much less stable than ecdysone (24) to chromatography on alumina. Thus it was clear that other milder purification procedures than had been used by Karlson, P. *et al.* (1963) would have to be developed for its isolation.

Fortunately, by chance, it was discovered that the partially purified moulting hormone from crayfish also contained a red chromatophore concentrating hormone that was easy to detect by a simple but extremely sensitive bioassay (Lowe, M. and Horn, D., 1967). With this bioassay it was possible to examine a variety of preliminary extraction procedures and develop an efficient method of isolating the two hormones. Finally, with the aid of large-scale equipment at the Commonwealth Serum Laboratories, Melbourne, it was possible to isolate 2 mg of substantially pure crustacean moulting hormone, named "crustecdysone" from 1 ton of crayfish waste (Hampshire, F. and Horn, D., 1966; Horn, D. *et al.*, 1968). From NMR and mass spectral studies it was identified as 20-hydroxyecdysone (32). This assignment was further confirmed by the synthesis and study of analogous compounds (Horn, D. *et al.*, 1966). The crayfish hormone was found also to be identical with a moulting hormone (MH) isolated from pupae of the silkworm *Antheraea pernyi*, the haemolymph of which had been used by Grace, T. (1962) in the CSIRO at Canberra, Australia, to develop a medium for the culture of insect tissues. It was therefore suggested (Horn, D. *et al.*, 1966) that crustecdysone was identical with β-ecdysone isolated from the silkworm *Bombyx mori*, as was later shown to be the case (Galbraith, M. *et al.*, 1967).

Soon after the isolation of crustecdysone from crayfish, Hocks and Wiechert at Schering AG, Berlin, isolated 20-hydroxyecdysone (32) from the silkworm *Bombyx mori* and showed it to be identical with β-ecdysone (Hocks, P. *et al.*, 1966). In the meanwhile Hoffmeister (1966a,b) reported the isolation of a second silkworm hormone, named "ecdysterone", that was at first considered to be isomeric with ecdysone (24) but was in fact 20-hydroxyecdysone (32) (Hoffmeister, H. and Grützmacher, H., 1966). Also about this time a team at the United States Department of Agriculture at Beltsville reported (Kaplanis, J. *et al.*, 1966b) the isolation of both ecdysone (24) and 20-hydroxyecdysone (32) from the tobacco hornworm *Manduca sexta*. They had ingeniously extracted the pupae at the time the hormone titre was at a maximum and obtained 7.4 mg of 20-hydroxyecdysone (32) (THE I) from a mere 12.4 kg of pupae.

As these isolations of 20-hydroxyecdysone (32) took place about the same time, a variety of different trivial names were given to it (e.g. β-ecdysone, crustecdysone, ecdysterone) and these are found extensively in the early literature. However, it was decided (Goodwin, T. *et al.*, 1978) to use the name "ecdysone" rather than "α-ecdysone" for the hormone isolated by Butenandt, A. and Karlson, P. (1954) and the name "20-hydroxyecdysone" for the more polar analogue.

As 20-hydroxyecdysone (32) was considered to be formed from ecdysone with retention of configuration, the 20R,22R-configuration was proposed (Horn, D. *et al.*, 1966). This assignment was supported by synthesis (Hüppi, G. and Siddall, J., 1968). Confirmation that the synthetic products had the 20R,22R-configuration was obtained by means of NMR studies of derivatives of ponasterone A (58) (Koreeda, M. *et al.*, 1971), sterol analogues (Middleton, E. *et al.*, 1972) and X-ray studies (Dammeier, B. and Hoppe, W., 1971).

Although some twenty other ecdysteroids (zooecdysteroids; see section 2.1) have been isolated from arthropods 20-hydroxyecdysone (32) remains the most commonly occurring ecdysteroid and the most important physiologically at most stages of arthropod development. Numerous derivatives of ecdysone (24) and 20-hydroxyecdysone (32) have subsequently been isolated and identified from both animals and plants. The reader is referred to Table 1 for a listing of phyto- and zooecdysteroids and to Table 2 for the corresponding structural formulae (see Addendum for more recent isolations and structures).

*Table 1: List of phytoecdysteroids (P) and zooecdysteroids (Z)**

Ecdysteroid name and number	Type	References[a-g]
Ajugalactone (1)	P	Koreeda, M. and Nakanishi, K. (1970a)[a,b]
Ajugasterone B (2)	P	Imai, S. *et al.* (1969d)[a,b]
Ajugasterone C (3)	P	Imai, S. *et al.* (1969b)[a,b]
Ajugasterone D (4)	P	Chou W. and Lu, H. (1980)[a,b]
Amarasterone A (5)	P	Takemoto, T. *et al.* (1968a)[a,b]
Amarasterone B (6)	P	Takemoto, T. *et al.* (1968a)[a,b]
Calonysterone (7) (Podecdysone E)	P	Canonica, L. *et al.* (1973a)[a,b]
Capitasterone (8)	P	Takemoto, T. *et al.* (1968b)[a,b]
Carpesterol (9)	P	Tsay, Y. *et al.* (1970)[a,b,c]
Cheilanthone A (10)	P	Faux, A. *et al.* (1970)[a,b]
Cheilanthone B (11)	P	Faux, A. *et al.* (1970)[a,b]
Cyasterone (12)	P	Hikino, H. *et al.* (1968a)[a,b], (1971a)[c]
Cyasterone-2,2-acetate (13)	P	Usmanov, B. *et al.* (1978)[a,b]
Dacrysterone (14)	P	Russell, G. and Fraser, J. (1973)[a,b]; Blunt, J. *et al.* (1979)[d]
3-Dehydroecdysone (15)	Z	Karlson, P. *et al.* (1972)[a,b,f]; Hoffmann, J. *et al.* (1980)[g]
3-Dehydro-20-hydroxyecdysone (16)	Z	Hoffmann, J. and Koolman, J. (1974)[e,f]
24(28)-Dehydromakisterone A (17)	P	Baltaev, U. *et al.* (1978b)[a,b]
2-Deoxyecdysone (18)	P	Chong, Y. *et al.* (1970)[a,b]
	Z	Ohnishi, E. *et al.* (1977b)[a,b]
2-Deoxyecdysone 22-adenosine mono-phosphate (82)a	Z	Tsoupras, G. *et al.* (1982)[a,b]
2-Deoxyecdysone 22-phosphate (82b)	Z	Isaac, R. *et al.* (1982)[a,b]
2-Deoxy-20-hydroxyecdysone (19) (Deoxycrustecdysone)	P	Chong, Y. *et al.* (1970)[a,b]
	Z	Galbraith, M. *et al.* (1968b)[a,b]
Deoxyviperidone (20)	P	Knight, J. and Pettit, G. (1969)[a,b]
2,2-Dideoxyecdysone (21)	Z	Hetru, C. *et al.* (1978)[e]
2,22-Dideoxy-20-hydroxyecdysone (22)	Z	Ikekawa, N. *et al.* (1980)[a,b,g]
20,26-Dihydroxyecdysone (23)	Z	Thompson, M. *et al.* (1967)[a,b]
Ecdysone (24) (α-Ecdysone)	P	Heinrich, G. and Hoffmeister, H. (1967)[a,b]
	Z	Butenandt, A. and Karlson, P. (1954)[a]; Karlson, P. *et al.* (1965)[b]; Huber, R. and Hoppe, W. (1965a,b)[c]
Ecdysone 3-acetate (25)	Z	Isaac, R. *et al.* (1981b)[a,b,g]
Ecdysone 22-phosphate (82c)	Z	Isaac, R. *et al.* (1982)[a,b]
Epicyasterone (26)	P	Hikino, H. *et al.* (1971c)[a,b]
3-Epi-2-deoxyecdysone (27)	P	Russell, G. *et al.* (1981)[a,b]
	Z	Isaac, R. *et al.* (1981a)[a,b,g]
3-Epi-2-deoxyecdysone 3-phosphate (82d)	Z	Tsoupras, G. *et al.* (1982)[a,b]
3-Epi-20,26-dihydroxyecdysone (28)	Z	Kaplanis, J. *et al.* (1979)[a,b]
3-Epi-ecdysone (29)	Z	Kaplanis, J. *et al.* (1979)[a,b]
3-Epi-20-hydroxyecdysone (30)	Z	Thompson, M. *et al.* (1974)[a,b]
14α-Hydroxycarpesterol (31)	P	Kusano, G. *et al.* (1975)[a,b]
20-Hydroxyecdysone (32) (Crustecdysone, Commisterone, β-Ecdysone, Ecdysterone, Isoinokosterone, Polypodine A)	P	Galbraith, M. and Horn, D. (1966)[a,b]
	Z	Hampshire, F. and Horn, D. (1966)[a,b]; Hocks, P. and Weichert, R. (1966)[a,b]; Hoffmeister, H. and Grutzmacher, H. (1966)[a,b]; Dammeier, B. and Hoppe, W. (1971)[c]
26-Hydroxyecdysone (33)	Z	Kaplanis, J. *et al.* (1973)[a,b]
20-Hydroxy-5α-ecdysone (34)	P	Ogawa, S. *et al.* (1971)[a,b]
20-Hydroxyecdysone–2-acetate (35)	P	Chou, W. and Lu, H. (1980)[a,b]
20-Hydroxyecdysone–2-cinnamate (36)	P	Russell, G. *et al.* (1971)[a,b]
20-Hydroxyecdysone–3-*p*-coumarate (37)	P	Russell, G. *et al.* (1972)[a,b]
Inokosterone (38) (Callinecdysone A is probably one of the C25-epimers)	P	Takemoto, T. *et al.* (1967a)[a]; Hikino, H. *et al.* (1976a)[b]
	Z	Faux, A. *et al.* (1969)[a,b]
Integristerone A (39)	P	Baltaev, U. *et al.*(1977)[a,b]
Integristerone B (40)	P	Baltaev, U. *et al.* (1978a)[a,b]
Isocyasterone (41)	P	Hikino, H. *et al.* (1971c)[a,b]
Kaladasterone (42)	P	Canonica, L. *et al.* (1973b)[a], (1977)[b]

*Table 1: List of phytoecdysteroids (P) and zooecdysteroids (Z)**—Continued

Ecdysteroid name and number	Type	References[a-g]
Lobosterol (43)		Tursch, B. *et al.* (1976)[a,b,d]
Makisterone A (44) Callinecdysone B or C24-epimer, Podecdysone D)	P	Imai, S. *et al.* (1968a)[a,b]; Danieli, B. *et al.* (1974)[d]
	Z	Kaplanis, J. *et al.* (1975)[a,b]
Makisterone B (45)	P	Imai, S. *et al.* (1968a)[a]; (1968b)[b]
Makisterone C (46) (Lemmasterone, Podecdysone A)	P	Galbraith, M. *et al.* (1968c)[a,b]; Imai, S. *et al.* (1968a)[a], (1968b)[b]
Makisterone D (47)	P	Imai, S. *et al.* (1968a)[a]; (1968b)[b]
Muristerone A (48)	P	Canonica, L. *et al.* (1972)[a,b]
Osladin (49)	P	Jizba, J. *et al.* (1971b)[a,b]
Paristerone (82e)	P	Singh, P. and Thakur, R. (1982)
Pinnasterol (50)	P	Fukuzawa, A. *et al.* (1981)[a,b]
Pinnasterol–2-acetate (51)	P	Fukuzawa, A. *et al.* (1981)[a,b,c]
Podecdysone B (52)	P	Galbraith, M. *et al.* (1968c)[a], (1969d)[b]
Podecdysone C (53)	P	Galbraith, M. *et al.* (1968c)[a], (1973b)[b]
Polypodine B (54) (Ajugasterone A)	P	Jizba, J. *et al.* (1967b)[a,b]
Polypodine B–2-cinnamate (55)	P	Russell, G. *et al.* (1971)[a,b]
Polypodoaurein (56)	P	Jizba, J. *et al.* (1974b)[a,b]
Polypodosaponin A (57)	P	Jizba, J. *et al.* (1971a)[a,b]
Ponasterone A (58)	P	Nakanishi, K. *et al.* (1966)[a,b]
	Z	McCarthy, J.(1979)[a,b] Lachaise, F. *et al.* (1981)[b]
Ponasterone B (59)	P	Nakanishi, K. *et al.* (1968)[a,b]
Ponasterone C (60)	P	Nakanishi, K. *et al.* (1968)[a,b]; Koreeda, M. and Nakanishi, K. (1970a)[b] Blunt, J. *et al.* (1979)[d]
Ponasterone C–2-cinnamate (61)	P	Russell, G. *et al.* (1972)[a,b]
Ponasteroside A (62) (Warabisterone)	P	Takemoto, T. *et al.* (1968c)[a,b]
Poststerone (63)	P	Hikino, H. *et al.* (1970d)[a,b]
Praemixisterone (64)	P	Saatov, Z. *et al.* (1979b)
Precyasterone (65)	P	Hikino, H *et al.* (1970a)[a,b]
Pterosterone (66)	P	Takemoto, T. *et al.* (1967i)[a], (1968d)[b]; Blunt, J. *et al.* (1979)[d]
Rubrosterone (67)	P	Takemoto, T. *et al.* (1968e)[a,b]
Sengosterone (68)	P	Hikino, H. *et al.* (1969a)[a,b]
Sidasterone A (69)	P	Prakash, A. and Ghosal, S. (1979)[a]
Sidasterone B (70)	P	Prakash, A. and Ghosal, S. (1979)[a]
Silenosterone (71)	P	Saatov, Z. *et al.* (1979a)[a,b]
Sogdysterone (72)	P	Novosel'skaya, I. *et al.* (1975)[a,b]
Stachysterone A (73)	P	Imai, S. *et al.* (1970a)[a,b]
Stachysterone B (74)	P	Imai, S. *et al.* (1970a)[a,b]
Stachysterone C (75)	P	Imai, S. *et al.* (1970b)[a,b]
Stachysterone D (76) (Shidasterone)	P	Imai, S. *et al.* (1970b)[a,b]; Takemoto, T. *et al.* (1969a)[a]; Hikino, H. *et al.* (1975a)[b]
2,14,22,25-Tetradeoxyecdysone (77)	Z	Hetru, C. *et al.* (1978)[e]
2,22,25-Trideoxyecdysone (78)	Z	Hetru, C. *et al.* (1978)[e]
Turkesterone (79)	P	Usmanov, B. *et al.* (1973)[a,b], (1975)[b]
Viperidinone (80)	P	Djerassi, C. *et al.* (1964)[a,b]
Viperidone (81)	P	Djerassi, C. *et al.* (1964)[a,b]
Viticosterone E (82)	P	Rimpler, H. (1969)[a,b]

* See also Addendum.

[a] First isolation; [b] structure identification; [c] X-ray structure determination; [d] determination of stereochemistry; [e] partial identification; [f] labelled metabolite; [g] as conjugate.

*Table 2: Structure of zoo- and phytoecdysteroids**

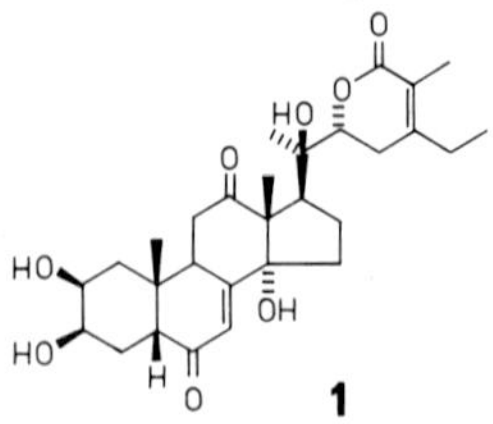
1

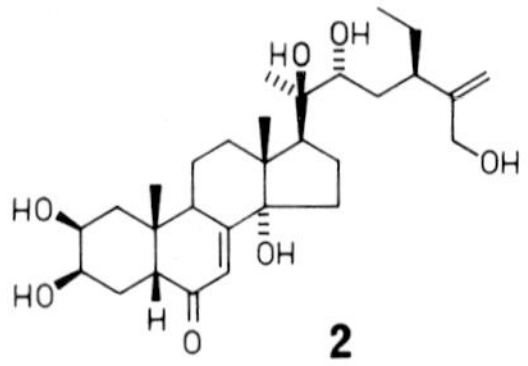
2

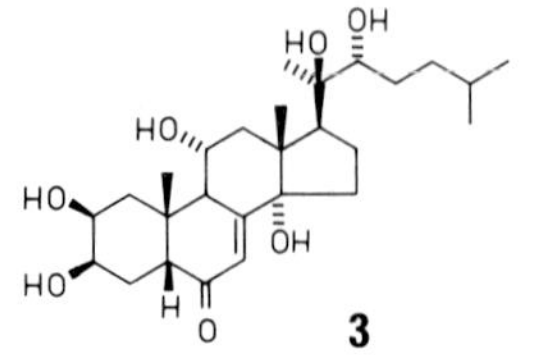
3

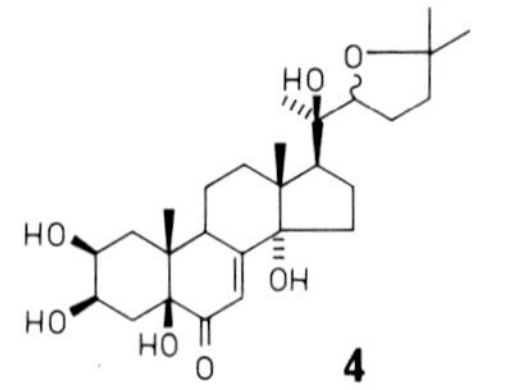
4

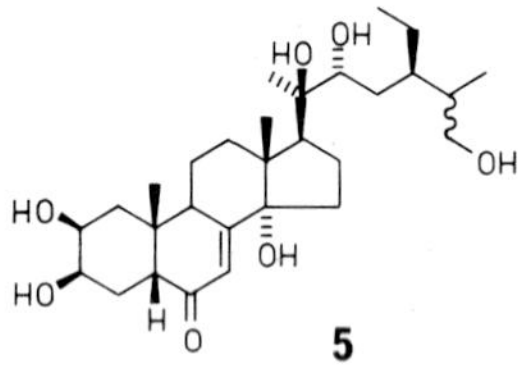
5

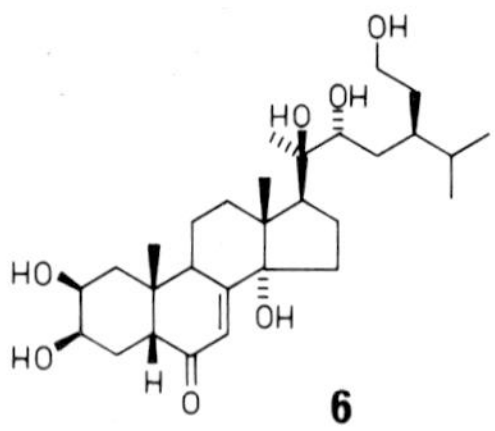
6

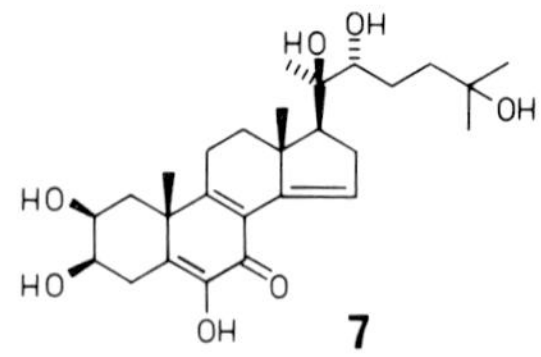
7

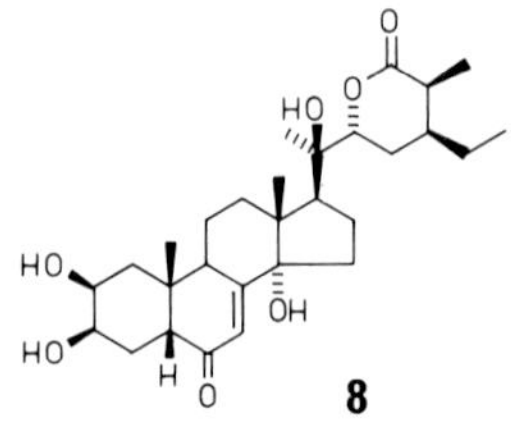
8

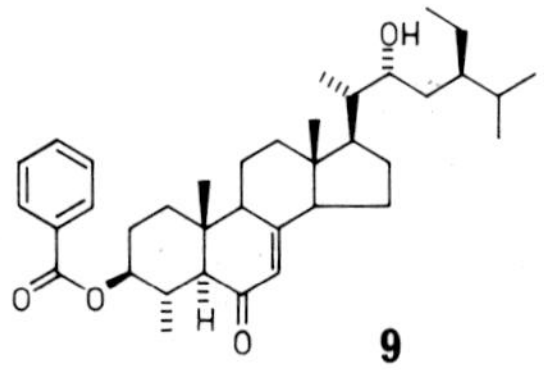
9

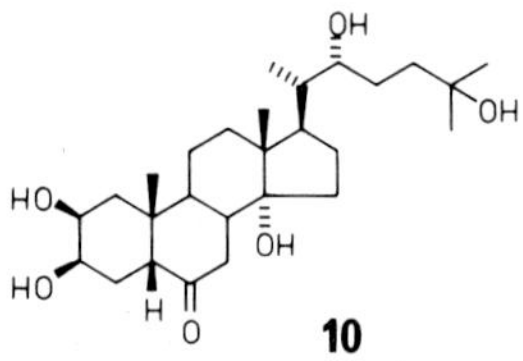
10

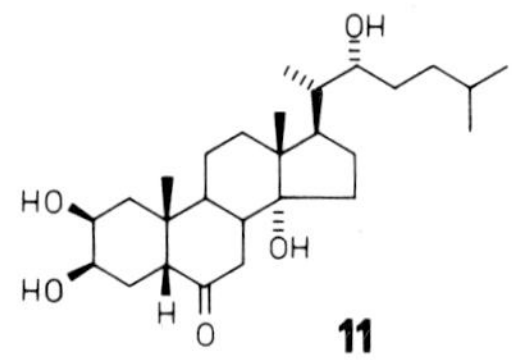
11

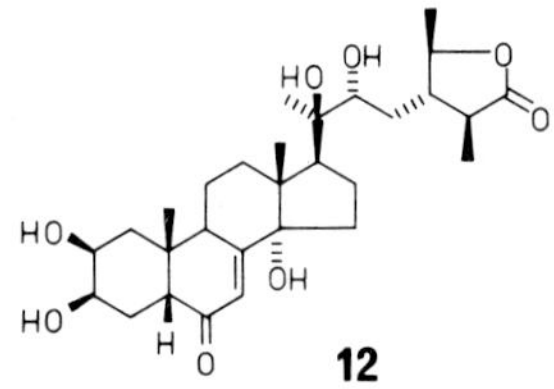
12

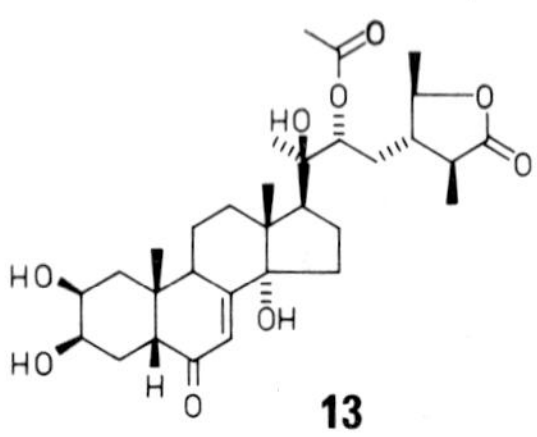
13

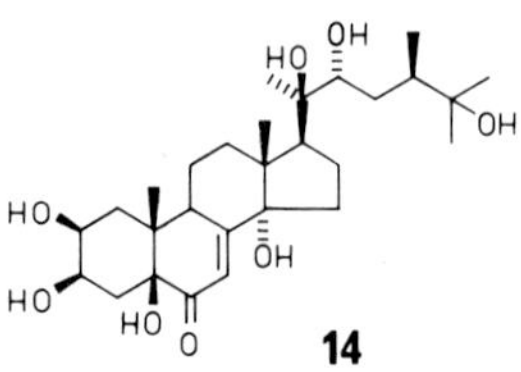
14

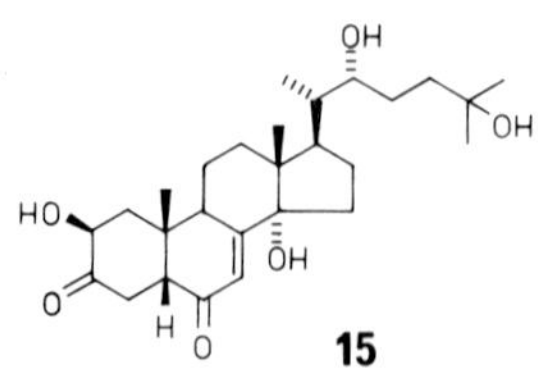
15

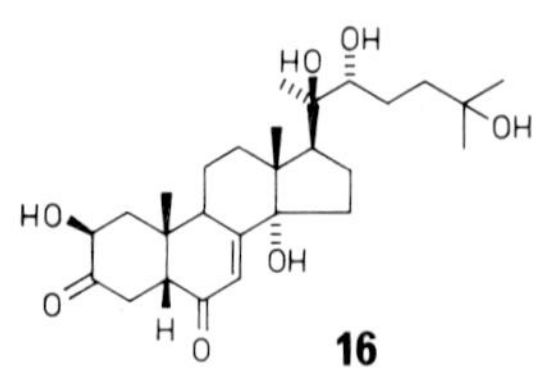
16

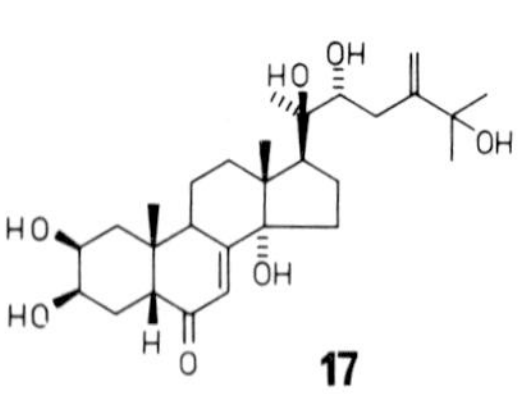
17

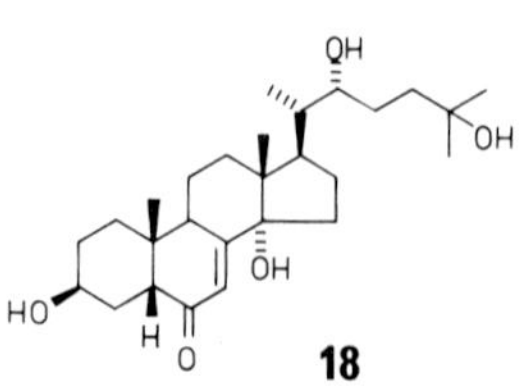
18

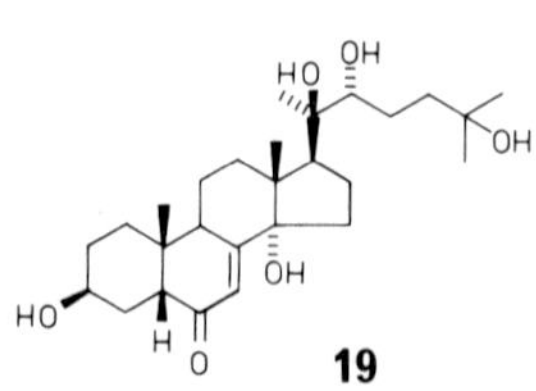
19

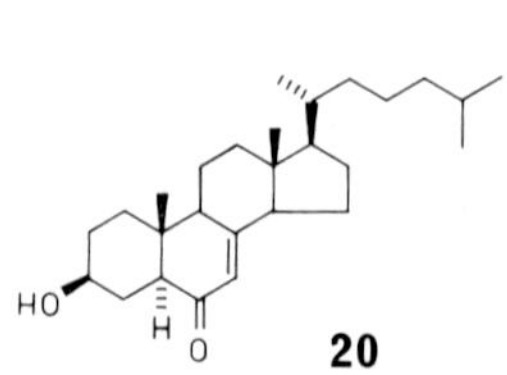
20

* See also Addendum.

Table 2: Structure of zoo- and phytoecdysteroids—Continued

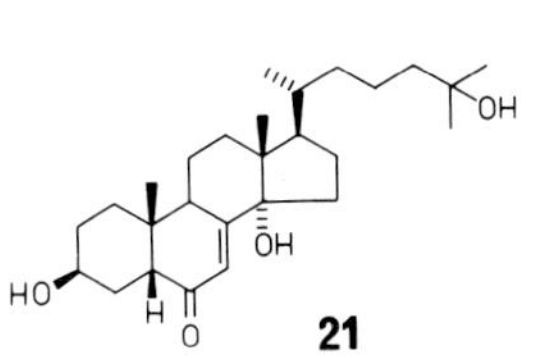
21

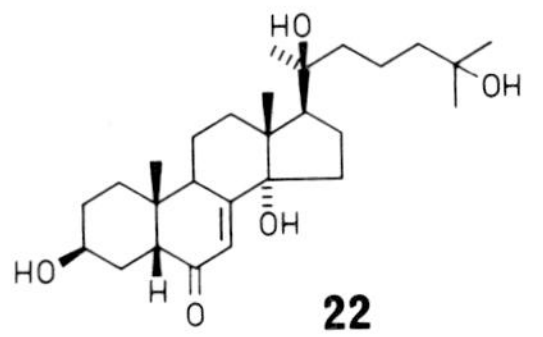
22

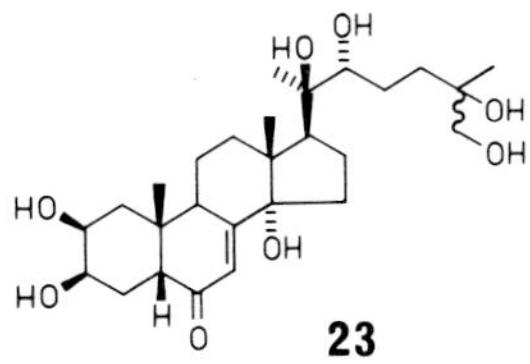
23

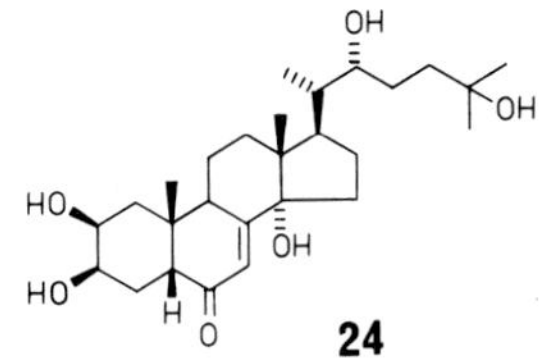
24

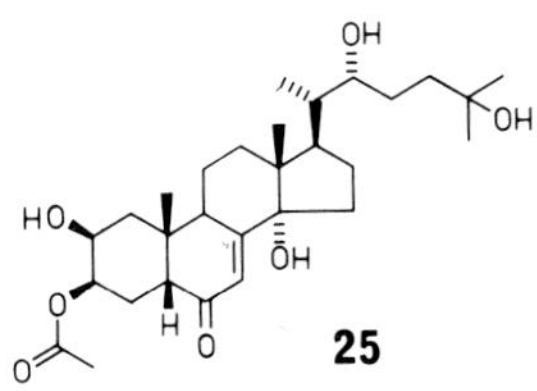
25

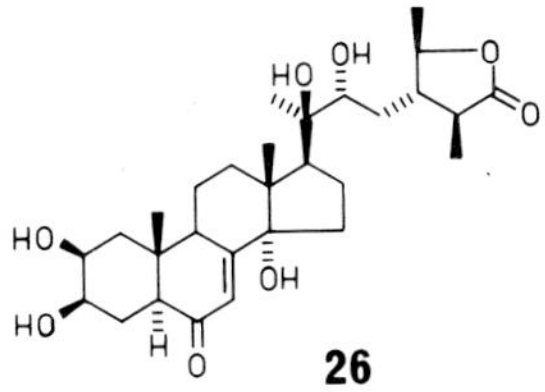
26

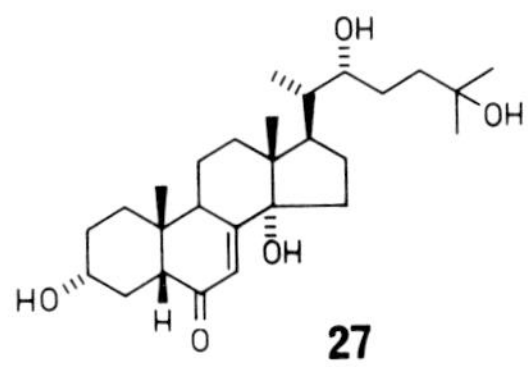
27

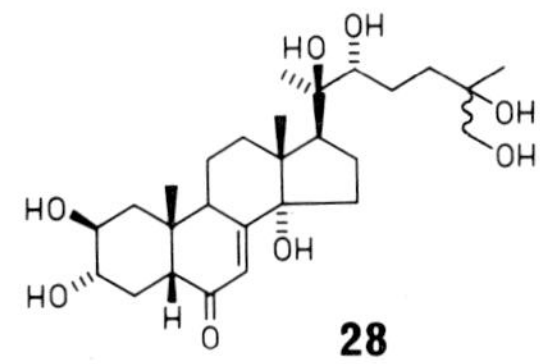
28

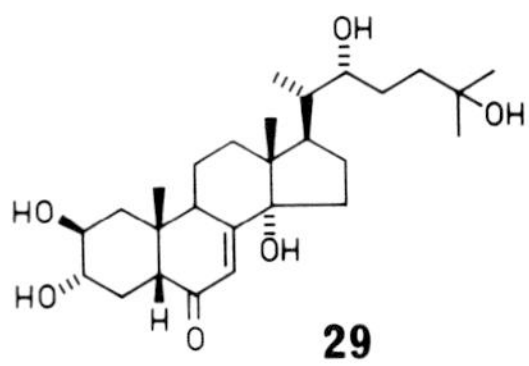
29

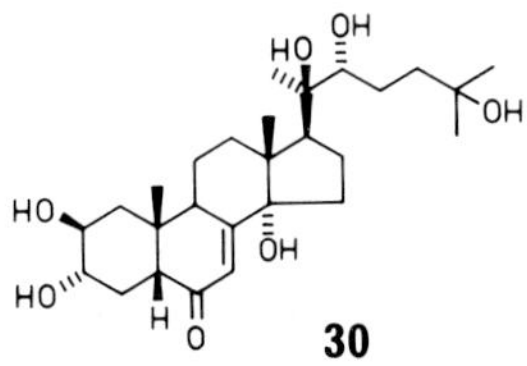
30

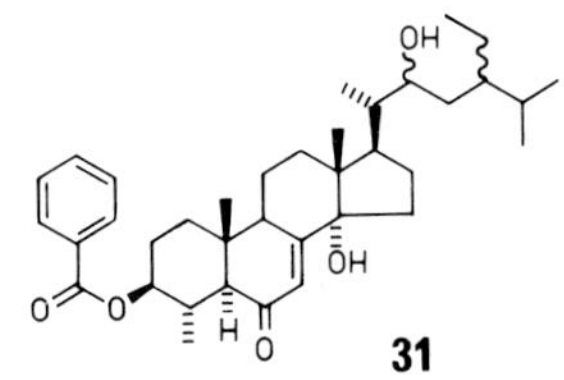
31

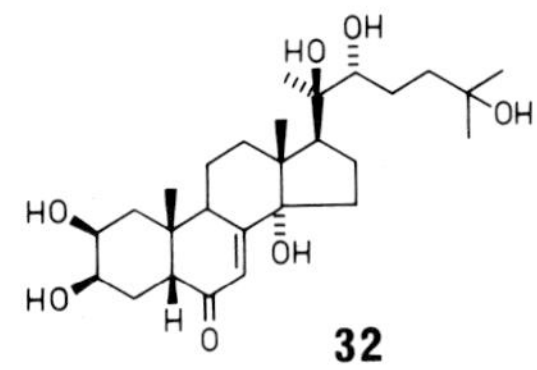
32

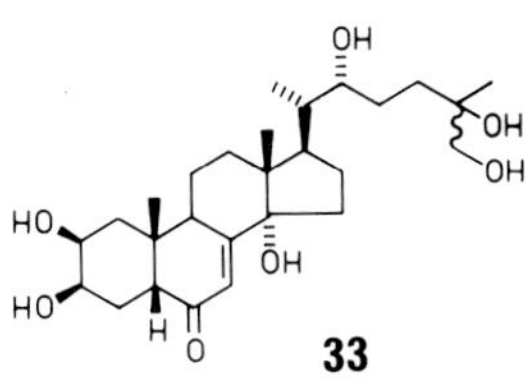
33

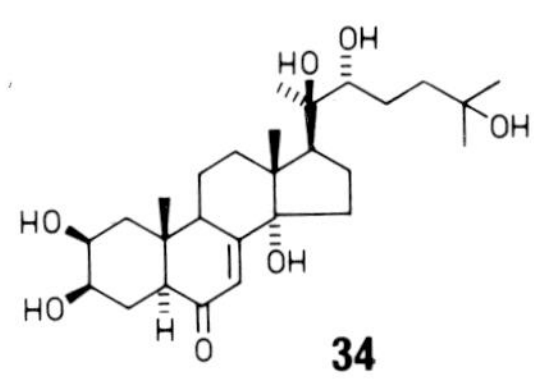
34

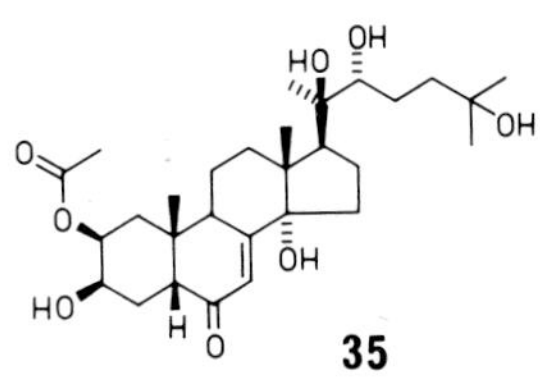
35

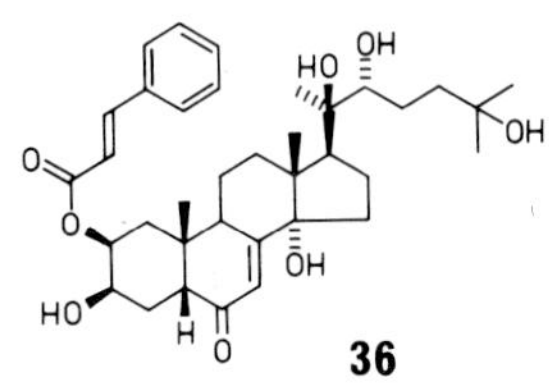
36

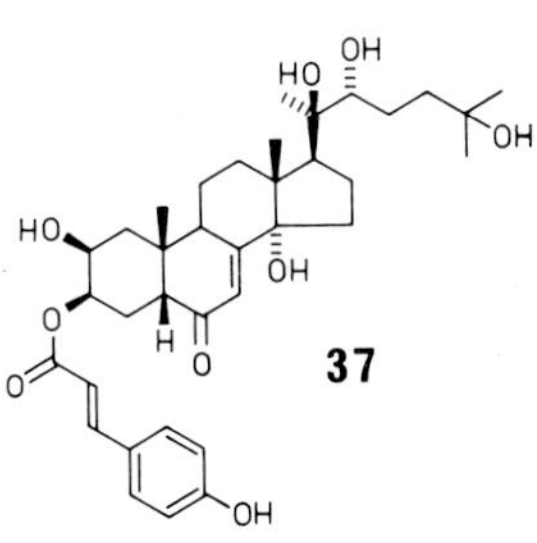
37

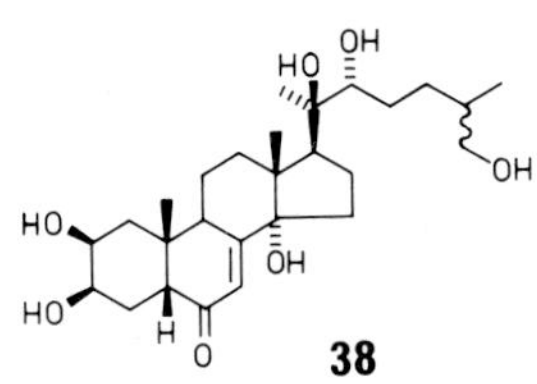
38

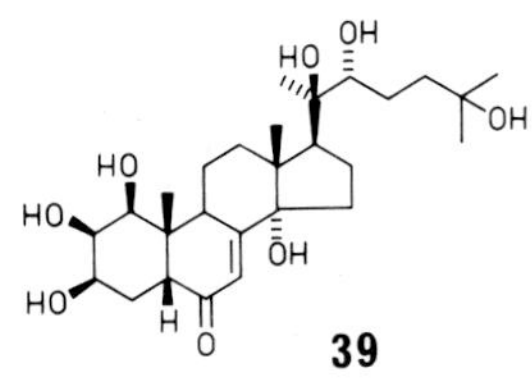
39

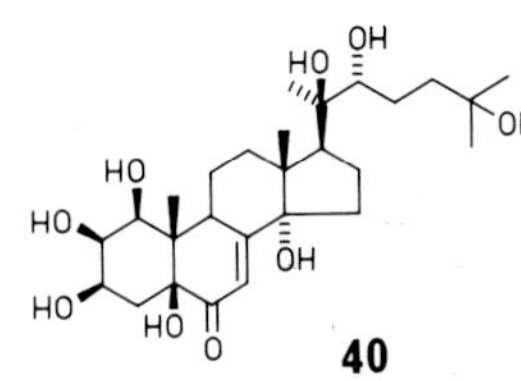
40

Table 2: Structure of zoo- and phytoecdysteroids—Continued

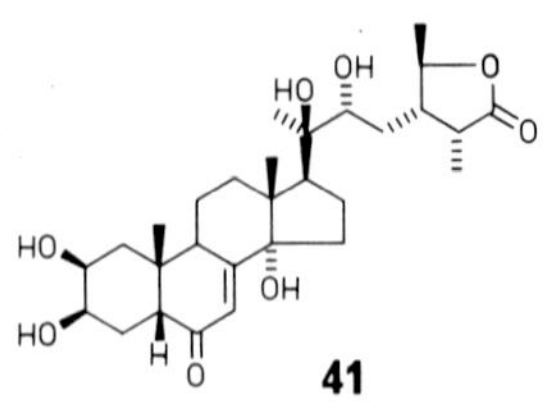
41

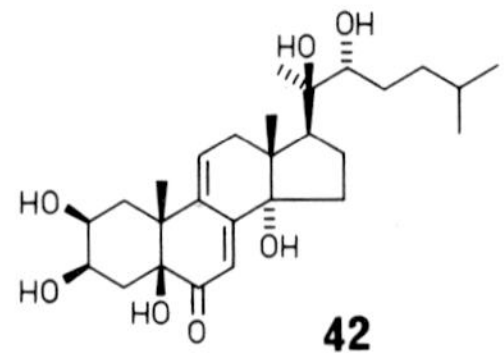
42

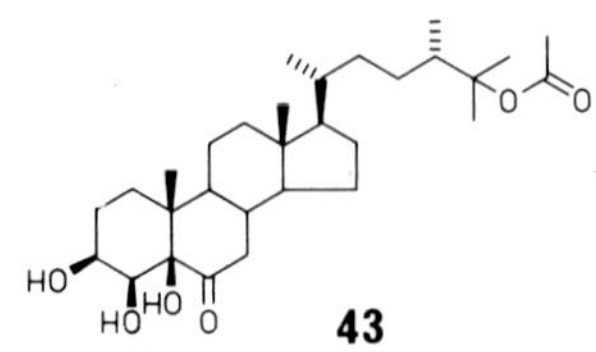
43

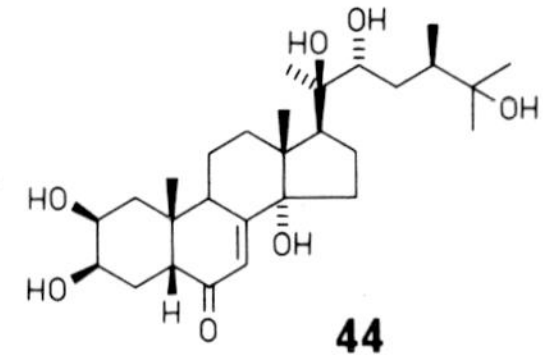
44

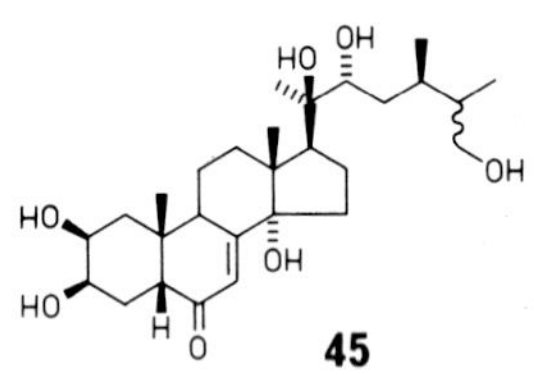
45

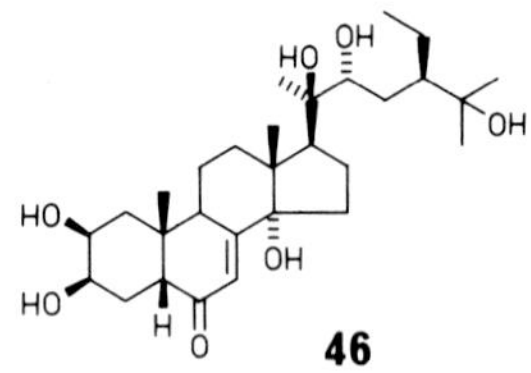
46

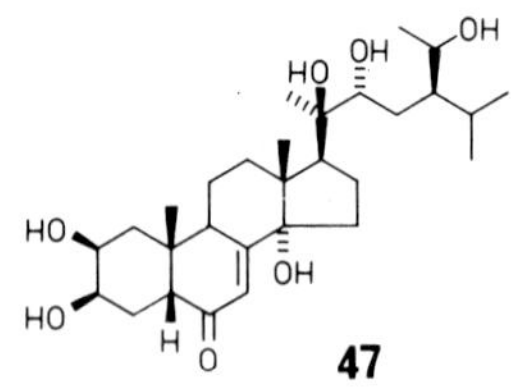
47

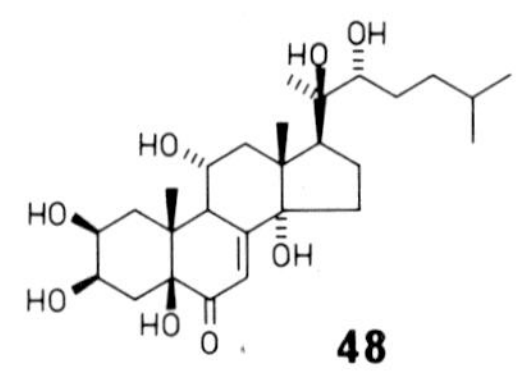
48

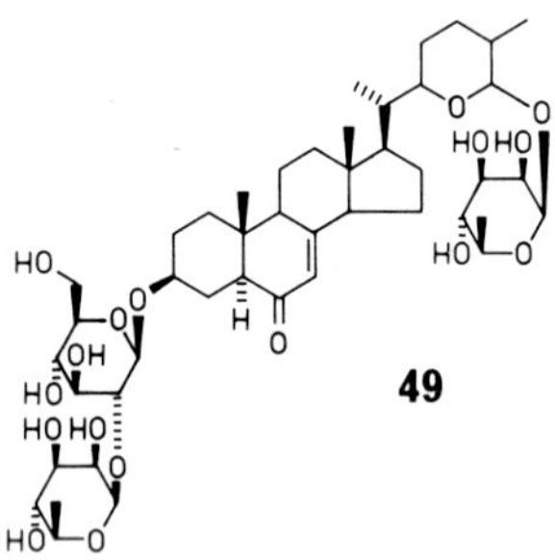
49

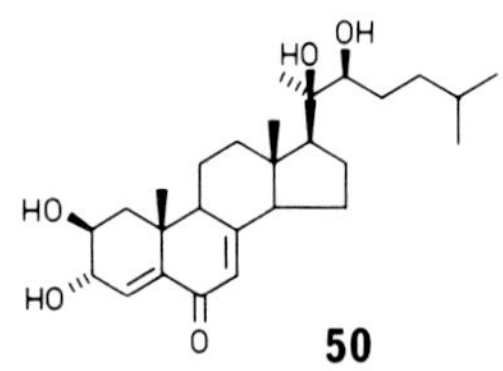
50

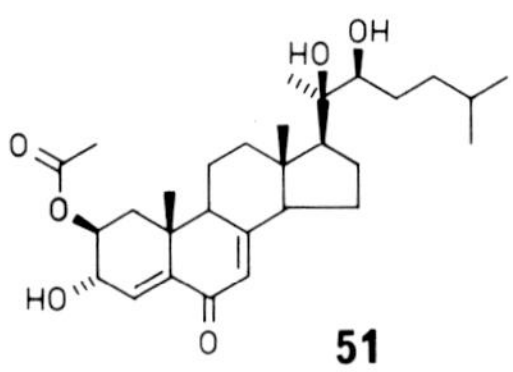
51

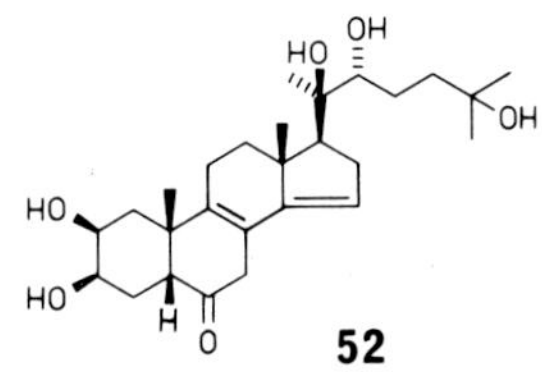
52

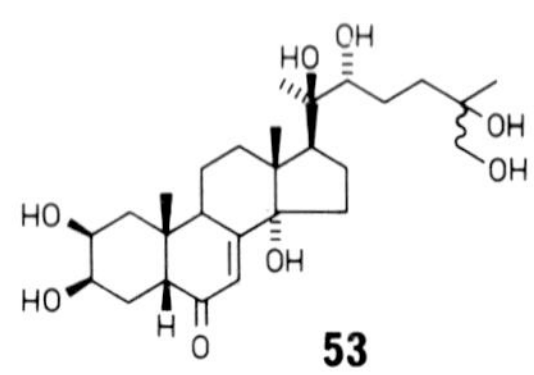
53

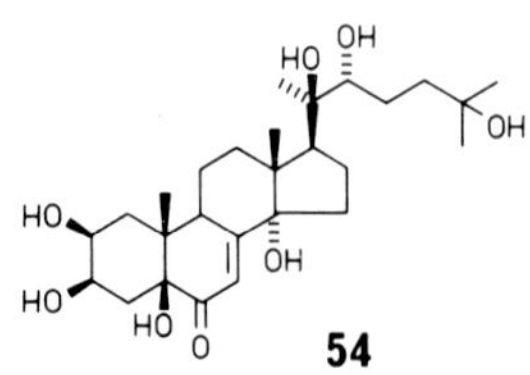
54

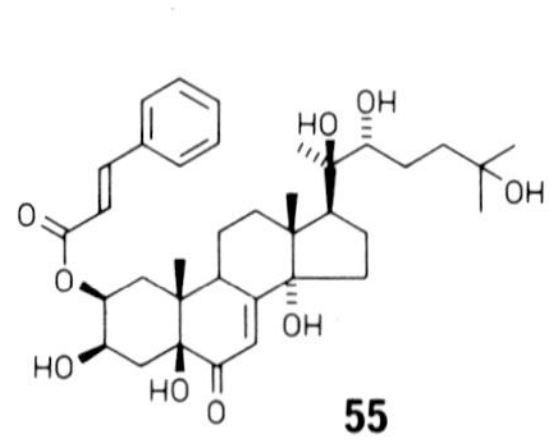
55

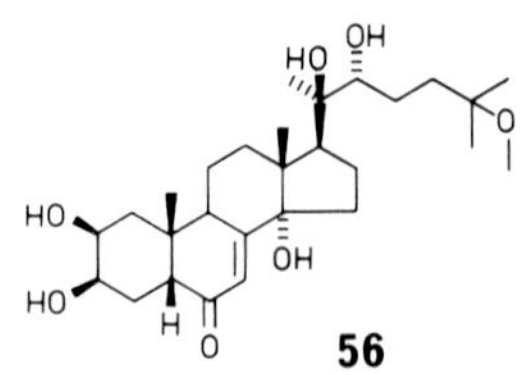
56

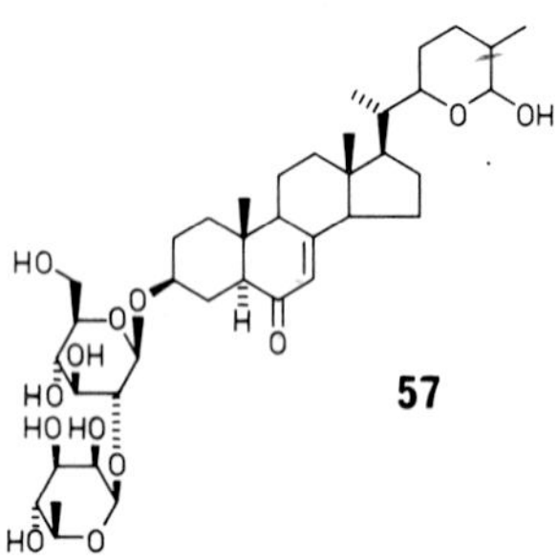
57

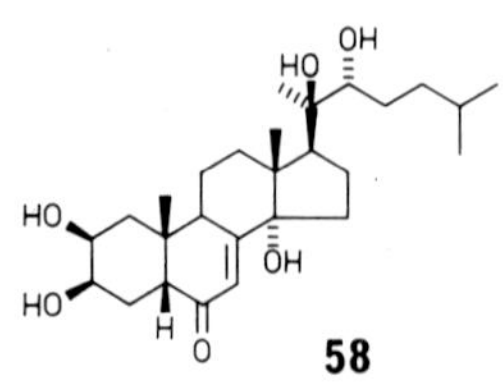
58

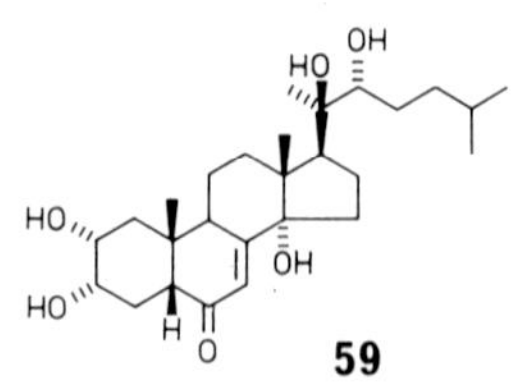
59

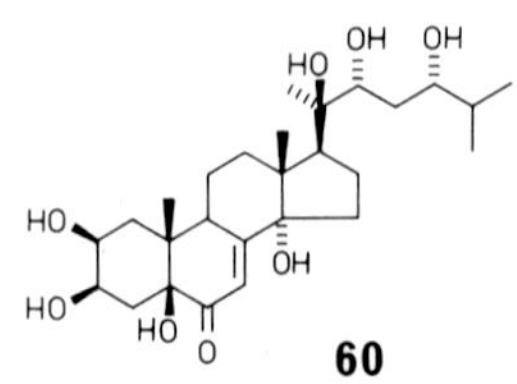
60

Table 2: Structure of zoo- and phytoecdysteroids—Continued

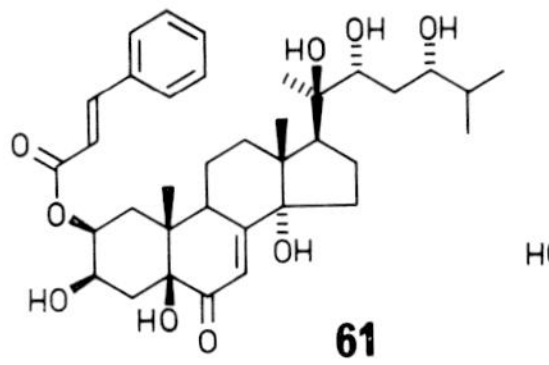
61

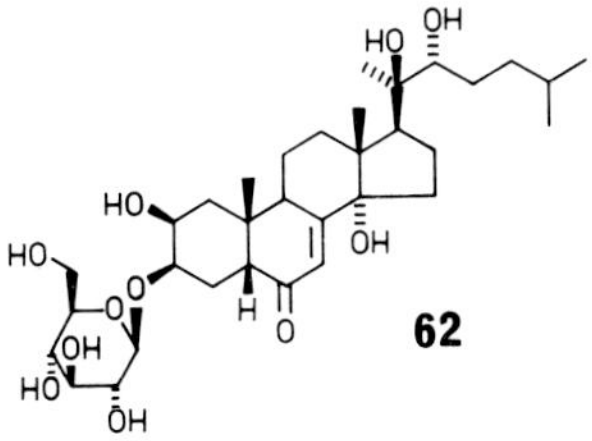
62

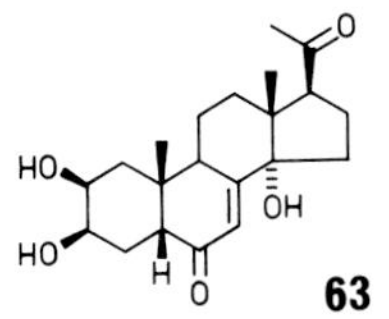
63

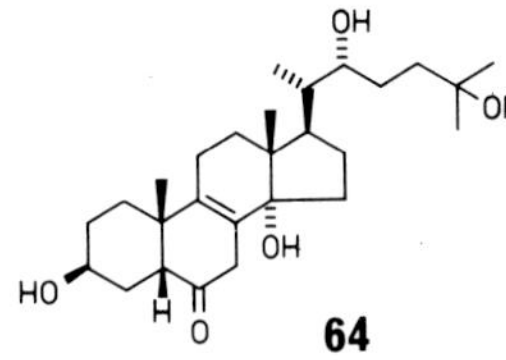
64

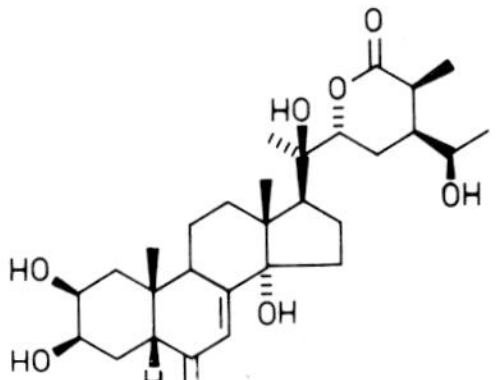
65

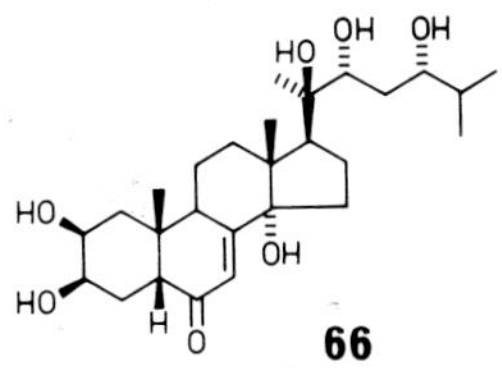
66

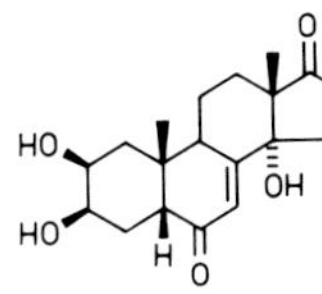
67

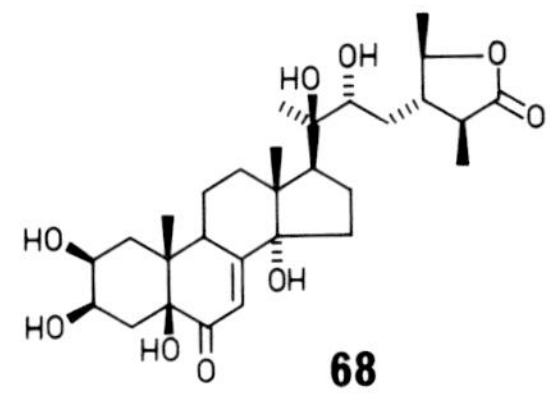
68

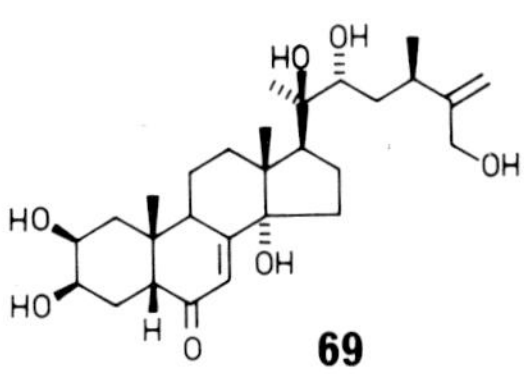
69

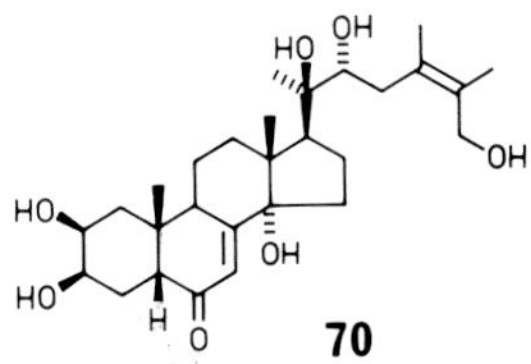
70

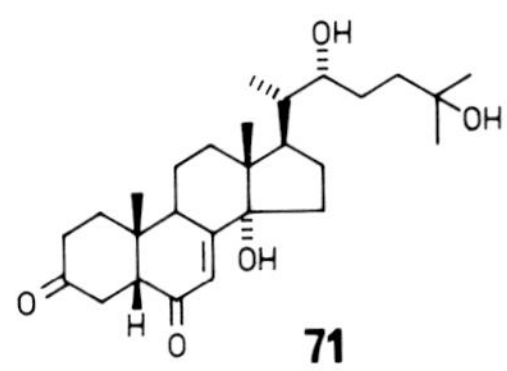
71

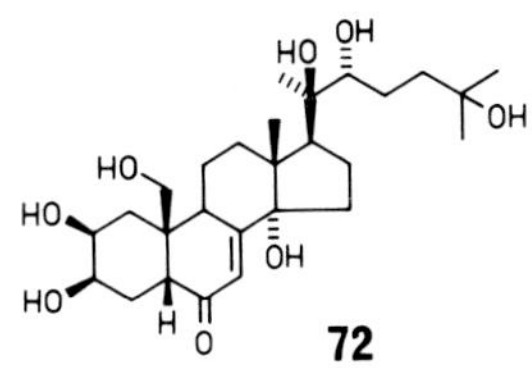
72

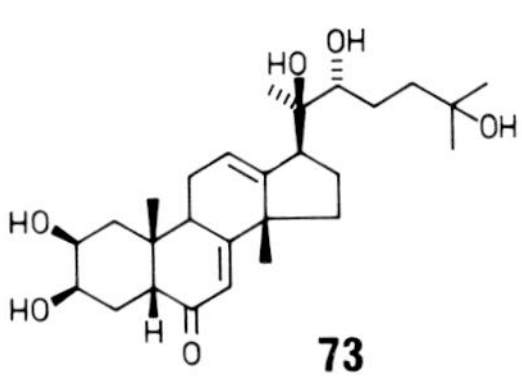
73

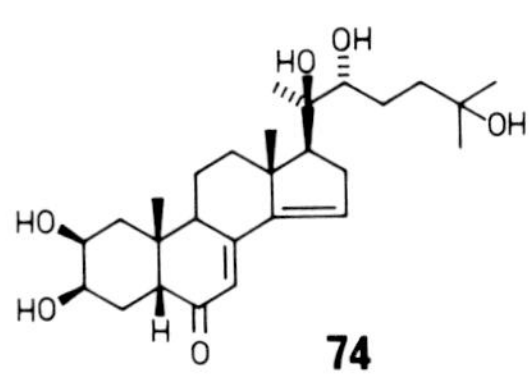
74

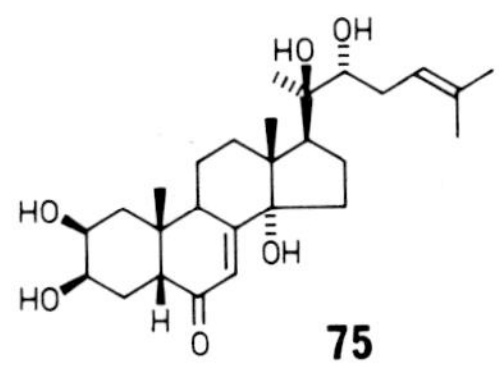
75

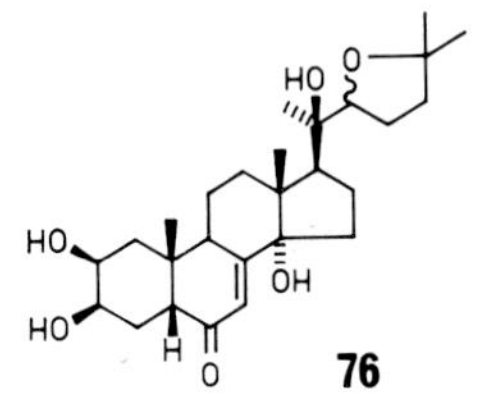
76

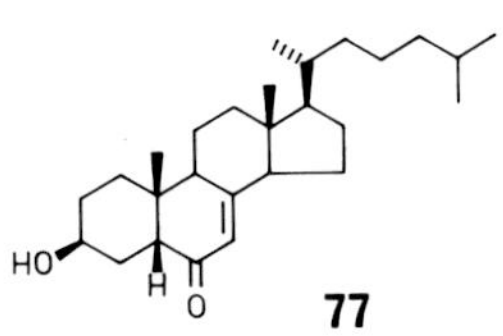
77

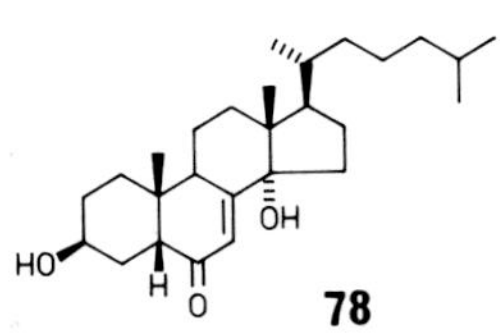
78

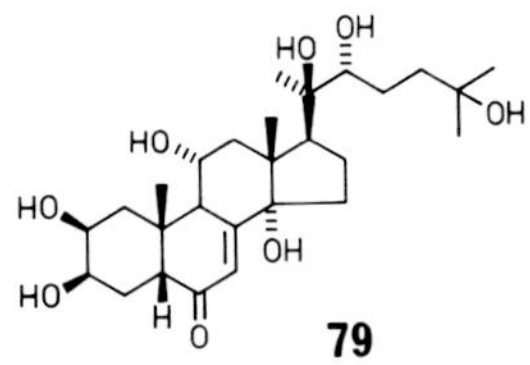
79

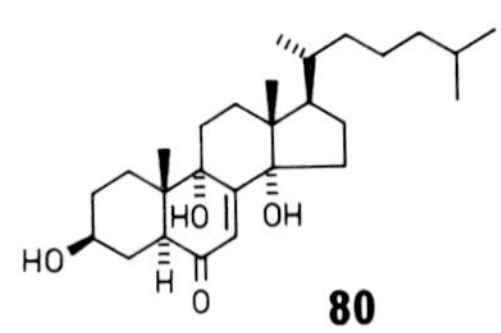
80

Table 2: Structure of zoo- and phytoecdysteroids—Continued

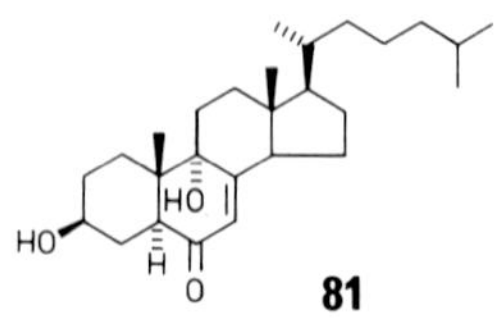

81

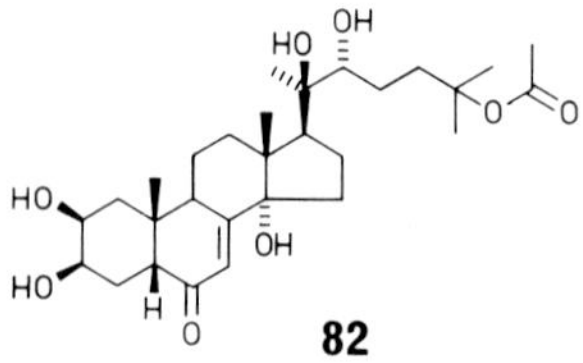

82

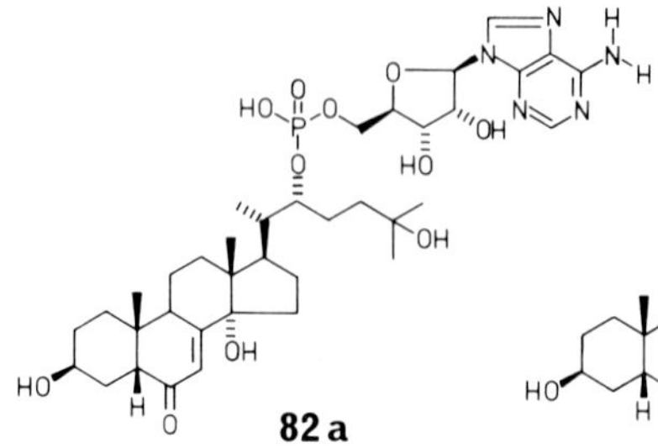

82 a

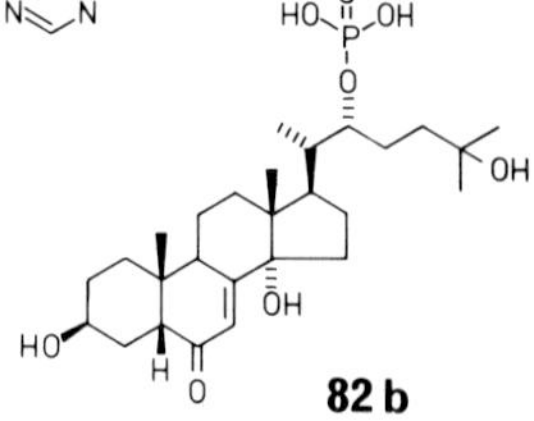

82 b

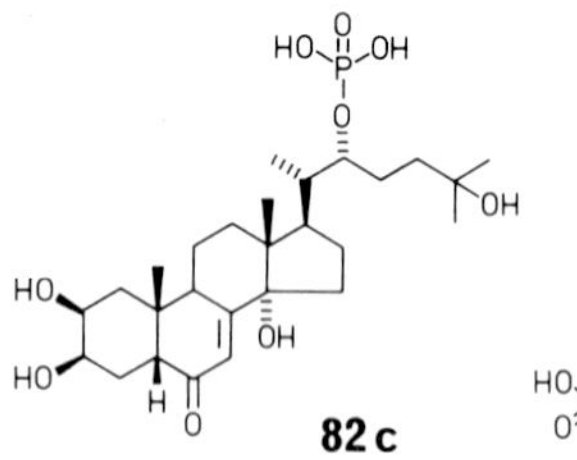

82 c

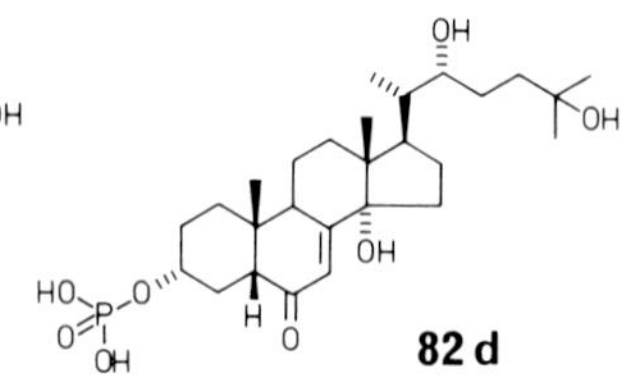

82 d

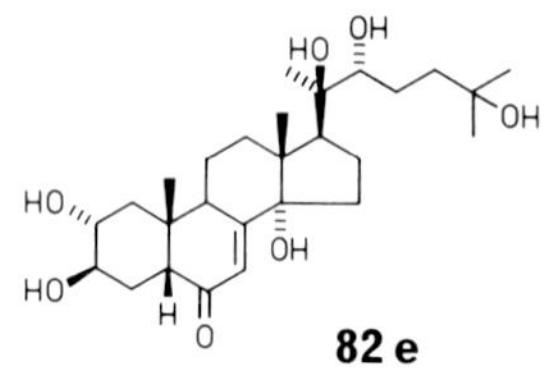

82 e

2 OCCURRENCE

2.1 Arthropods

We have listed in Table 3 the concentrations of 21 ecdysteroids which have been found to occur in a variety of classes of invertebrate animals at different developmental stages. With the exception of the 28-carbon makisterone A (44), and 30-carbon lobosterol (43) all zooecdysteroids are 27-carbon steroids. In some cases the zooecdysteroids have been identified after hydrolysis of conjugating moieties with *Helix pomatia* enzymes which contain mainly glucuronidases and sulphatases. Traces of ecdysteroids in commercial *Helix* enzyme mixtures can usually be ignored (Hoffmann, J. *et al.*, 1980). Most of the zooecdysteroids in Table 3 are biologically active in moulting hormone assays (Bergamasco, R. and Horn, D., 1980). Ecdysone (24) and 20-hydroxyecdysone (32) are the most commonly occurring ecdysteroids and it appears that these two ecdysteroids are the most important moulting hormones of insects and crustaceans (Morgan, E. and Poole, C., 1976a; Highnam, K. and Hill, L., 1977; Spindler, K. *et al.*, 1980; Hoffmann, J. *et al.*, 1983). Ecdysteroids have also been detected in the spider *Pisaura mirabilis* (Arachnida) (Bonaric, J. and De Reggi, M., 1977), the tick *Amblyomma hebraeum* (Arachnida) (Delbecque, J. *et al.*, 1978a) and the crustacean *Limulus polyphemus* (Merostomata) (Winget, R. and Herman, W., 1976). It appears likely, therefore, that ecdysteroids are the moulting hormones of all arthropods.

In recent years it has become apparent that ecdysteroids are not restricted to the phylum Arthropoda but also occur in other invertebrates. Horn, D. *et al.* (1974b) isolated and identified 20-hydroxyecdysone (32) from the nematode *Ascaris lumbricoides*, an intestinal parasite of pigs. Ecdysteroids also occur in the mussel *Mytilus edulis* (Takemoto, T. *et al.*, 1967f) and in the snail *Helix pomatia* (Romer, F., 1979). The polyhydroxylated sterol lobosterol (43), isolated from the soft coral *Lobophytum pauciflorum* (Coelenterata: Octacorallia) (Tursch, B. *et al.*, 1976), has many structural features in common with ecdysteroids. This compound, however, is unlikely to function as a hormone in the coral because it occurs in an uncharacteristically high concentration. 20-hydroxyecdysone also occurs in very large amount in the zoanthid *Gerardia savaglia* (Sturaro, A. *et al.*, 1982) and it is thus likely that in these cases the compounds have a defensive rather than hormonal function.

Before the advent of modern physicochemical techniques the isolation and characterization of ecdysteroids from insects and crustaceans necessitated the laborious extraction of prodigious quantities of material (see section 4). Later, it became possible, using rapid and sensitive techniques such as radioimmunoassay (RIA), gas chromatography–mass spectrometry (GC–MS) and high-pressure liquid chromatography (HPLC), to detect and identify zooecdysteroids in partially purified extracts of small amounts of material. These techniques are now routinely used to monitor ecdysteroid concentrations during arthropod life cycles in different tissues. A particularly elegant example of the powerful use of GC–MS was provided by the identification of apolar precursors of ecdysone (24) in ovaries of adult *Locusta migratoria* (Hetru, C. *et al.*, 1978). These ecdysteroids, namely 2-deoxyecdysone (18), 2,22-dideoxyecdysone (21), 2,22,25-trideoxyecdysone (78) and 2,14,22,25-tetradeoxyecdysone (77), appear to be biosynthetic intermediates of ecdysone (24) and recent experiments with labelled intermediates confirm this assumption (Hetru, C. *et al.*, 1982). More commonly, techniques like RIA are used simply for the detection of ecdysteroids without structure identification. Some examples include the detection of ecdysteroids in eggs of the ant *Pheidole pallidula* (Suzzoni, J. *et al.*, 1980), in ovaries of the adult mosquito *Aedes aegypti* (Hagedorn, H. *et al.*, 1975) and in ovaries and eggs of the termite queens *Macrotermes subhyalinus* and *Macrotermes bellicosus* (Delbecque, J. *et al.*, 1978b).

While zooecdysteroids in whole arthropods occur in the concentration range of 10^{-1}–10^{-5} mg kg^{-1} (10^{-5}–10^{-9}%), the concentrations in insect or crustacean organs can be much higher; thus the eggs of the greater wax moth *Galleria mellonella* contain an ecdysteroid concentration of 74 mg kg^{-1} (Hsiao, T. and Hsiao, C., 1979). Zooecdysteroids are present at all stages of insect and crustacean development (Hoffmann, J. *et al.*, 1983; Spindler, K. *et al.*, 1980). Many ecdysteroids, present during different stages of arthropod development, are as yet unidentified (Lagueux, M. *et al.*, 1979; Hsiao, T. and Hsiao, C., 1979; Hoffmann, J. *et al.*, 1980; Kaplanis, J. *et al.*, 1980). In the orthopteran insect *Locusta migratoria* ecdysteroids appear to control cuticulogenesis in embryonic as well as in post-embryonic development and ecdysteroids are involved in the maturation of ovaries (Hoffmann, J. *et al.*, 1983; Hoffmann, J. and Lagueux, M. vol. 1).

A complex and interesting picture is emerging with regard to the involvement of ecdysteroids in adult insect and crustacean reproduction. In the particular case of *Locusta migratoria* ecdysteroids are biosynthesized by follicular epithelial cells surrounding the terminal oocytes and accumulate in the ovary without secretion into the haemolymph. In an ovulated oocyte ecdysteroids may reach concentrations of 150 μM (Lagueux, M. *et al.*, 1981). The ecdysteroids in the ovary are largely conjugated phosphates (82a, 82b) (Isaac, R. *et al.*, 1982); (82c) (Tsoupras, G. *et al.*, 1982) and (169) (Tsoupras, G. *et al.*, 1983) (see Addendum). During egg-laying the maternal ecdysteroids are almost quantitatively transferred to the newly laid egg (Hoffman, J. *et al.*, 1983). Interestingly, 2-deoxyecdysone (18) accounts for approximately two-thirds of the conjugated ecdysteroids, while ecdysone (24) accounts for the remaining third; 20-hydroxyecdysone (32) conjugates are present only in minor quantities. In most adult insects the presence of ecdysteroids is restricted to the female. In the male blowfly, *Calliphora vicina*, however, the testes contain large quantities (28.6 mg kg^{-1}) of ecdysteroids (Koolman, J. *et al.*, 1979b).

Some significant differences between insect and crustacean ovarian development with regards to ecdysteroids have recently been pointed out by Lachaise, F. *et al.* (1981). Unlike insects, certain crustaceans: (i) accumulate ponasterone A (58) in their ovaries, (ii) do not conjugate ecdysteroids, and (iii) show a decrease in ecdysteroid titre at the end of ovarian development.

The presence of many ecdysteroids during arthropod development raises the question of metabolic interrelationships and the functions of these steroids. During pupal–adult development in the tobacco hornworm *Manduca sexta*, for example, six 27-carbon ecdysteroids have been identified and isolated 5 days after peak-hormone titre (Kaplanis, J. *et al.*, 1979). In order of decreasing concentration these are: 20,26-dihydroxyecdysone (23), 3-epi-20-hydroxyecdysone (30), 20-hydroxyecdysone (32), 3-epi-20,26-dihydroxyecdysone (28), 3-epi-ecdysone (29) and ecdysone (24). The presence of the less biologically active 3α-isomers suggests that epimerization at C3 is a means of inactivation

Table 3: The concentration of ecdysteroids in metazoa[a]

Phylum Class	Species	Source	Method	Ecdysteroids	Concentration (mg kg^{-1})	Comments	References
Arthropoda							
Arachnida	*Amblyomma hebraeum* (tick)	nymphs (day 17) 106 g fr. wt.	RIA, GGCC-MF, TLC	ecdysone (24) + 20-hydroxyecdysone (32)	0.24 (total)		Délbecque, J. *et al.* (1978a)
Meno-stomata	*Limulus polyplemus*	intermoult premoult	RIA, bioassay	20-hydroxyecdysone (32)	10×10^{-3} 30×10^{-3}		Jegla, T. and Costlow, J. (1979b)
Crustaceae	*Balanus balanoides* (acorn barnacle)	1500 kg fr. wt.	GC	ecdysone (24) 20-hydroxyecdysone (32)	6×10^{-6} 1×10^{-3}	stage unspecified	Bebbington, P. and Morgan, E. (1977)
	Callinectes sapidus (shore crab)	early premoult "green" 25 kg	TLC, NMR, UV, MS	callinecdysone A (38)	0.005	callinecdysone A (38) may be one of the C25-epimers of inokosterone (38) callinecdysone B (44) may be the C24-epimer of makisterone A (44).	Faux, A. *et al.* (1969)
		late premoult "peeler" 25 kg	TLC, NMR UV, MS	callinecdysone A (38) 20-hydroxyecdysone (32)	0.02 0.004		
		just after moulting "soft-shell", 25 kg	TLC, NMR UV, MS	20-hydroxyecdysone (32) callinecdysone B (44)	0.28 0.024		
		maturing embryos	TLC, HPLC, bioassay	20-hydroxyecdysone (32) ponasterone A (58)	$6–500 \times 10^{-3}$ $85–475 \times 10^{-3}$	min. and max. concentrations determined by RIA	McCarthy, J. and Skinner, D. (1979); McCarthy, J. (1979)
	Carcinus maenas (shore crab)	ovaries	MS, RIA, TLC	ecdysone (24)	$20-88 \times 10^{-3}$		Lachaise, F. and Hoffmann, J. (1977)
		ovaries (five stages)	TLC, RIA, HPLC, GC-MS	ecdysone (24) 20-hydroxyecdysone (32) ponasterone A (58)	3 – 104 pM g^{-1} 34 – 228 pM g^{-1} 3 – 230 pM g^{-1}	concentrations measured by RIA. No conjugates detected.	Lachaise, F. *et al.* (1981)
		haemolymph		ecdysone (24) 20-hydroxyecdysone (32) ponasterone A (58)	1 – 10 pM ml^{-1} 6 – 25 pM ml^{-1} 1 – 40 pM ml^{-1}		Lachaise, F. *et al.* (1981)
	Gecarcinus lateralis (land crab)	serum of late premoult (stages D_2, D_3)	HPLC, RIA, TLC	ponasterone A (58)	n.d.		McCarthy, J. (1979)
	Homarus americanus (lobster)	postmoult, 5 kg		20-hydroxyecdysone (32)	0.006		Gagosian, R. *et al.* (1974)
	Jasus lalandei (crayfish)	intermoult, 1000 kg		20-hydroxyecdysone (32)	0.002		Hampshire, F. and Horn, D. (1966)
		intermoult, 3000 kg		2-deoxy-20-hydroxy-ecdysone (19)	7×10^{-5}		Galbraith, M. *et al.* (1968b)
	Pachygrapsus crassipes (crab)	haemolymph	RIA, TLC	ecdysone (24) + 20-hydroxyecdysone (32)	$25–120 \times 10^{-3}$ (total)	max. and min. concentrations determined by RIA	Chang, E. *et al.* (1976)
	Sphaeroma serratum		GC	20-hydroxyecdysone (32)	$5–70 \times 10^{-6}$ mg ml^{-1}		Charmantier, G. *et al.* (1976)
	Orconectes limosus (crayfish)	Y-organs	RIA, TLC	ecdysone (24)	125×10^{-9} mg per gland		Keller, R. and Schmid, E. (1979)

Table 3: *Continued*

Phylum Class	Species	Source	Method	Ecdysteroids	Concentration (mg kg^{-1})	Comments	References
		haemolymph	RIA, TLC	20-hydroxyecdysone (32)	$12–120 \times 10^{-6}$ mg ml^{-1}	max. and min. concentrations determined by RIA	Keller, R. and Schmid, E. (1979)
Insecta	*Antheraea pernyi* (oak silkmoth)	pupae, 31 kg		20-hydroxyecdysone (32)	0.006		Horn, D. *et al.* (1966)
	Bombyx mori (silkworm moth)	pupae, 500 kg		ecdysone (24)	0.05		Butenandt, A. and Karlson, P. (1954)
		pupae		20-hydroxyecdysone (32)	n.d.		Karlson, P. (1956a)
		pupae, 1000 kg (dried)		ecdysone (24)	0.25		Karlson, P. *et al.* (1963)
		pupae, 3000 kg		ecdysone (24)	0.068		Hocks, P. and Weichert, R. (1966)
				20-hydroxyecdysone (32)	0.016		Hocks, P. and Weichert, R. (1966)
		pupae, 2200 kg		20-hydroxyecdysone (32)	0.004		Hoffmeister, H. (1966b); Hoffmeister, H. and Grutz-macher, H. (1966)
		developing embryos, 80 g wet wt.	TLC, HPLC, MS; bioassay	ecdysone (24)	0.74		Onishi, E. *et al.* (1977a)
		ovaries (pharate adults) 4 kg		2,22-dideoxy-20-hydroxyecdysone (22)	0.25	total concentration after *Helix pomatia* hydrolysis of conjugates	Ikekawa, N. *et al.* (1980)
		ovaries (pharate adults) 6 days after larval–pupal ecdysis, 270 g	HPLC, MS, TLC bioassay	2-deoxyecdysone (18)	0.31 (free) 0.46 (conj.)	concentrations determined by peak areas in HPLC chromatog; conjugates hydrolysed with *Helix pomatia* enzymes	Onishi, E. *et al.* (1977b)
		ovaries (adults 0.3 h after emergence), 498 g		2-deoxyecdysone (18)	0.22 (free) 0.26 (conj.)		Onishi, E. *et al.* (1977b)
		diapausing embryos, 700 g		2-deoxyecdysone (18)	0.36 (conj.)		Onishi, E. *et al.* (1977b)
	Calliphora erythro-cephala (*C. vicina*) (Blue blowfly)	pupae, 15 kg		ecdysone (24) 20-hydroxyecdysone (32)	0.005 n.d.		Karlson, p. (1956b)
		prepupae, 0.5 kg		20-hydroxyecdysone (32)	0.11		Galbraith, M. *et al.* (1969b)
		adult male testes	RIA, TLC, HPLC	20-hydroxyecdysone (32)	28.6	20-hydroxy-ecdysone (32) tentatively identified as main	Koolman, J. *et al.* (1979b)
		adult male haemolymph		20-hydroxyecdysone (32)	4.4		

Table 3: *Continued*

Phylum Class	Species	Source	Method	Ecdysteroids	Concentration (mg kg^{-1})	Comments	References
		adult female ovaries adult female haemolymph		20-hydroxyecdysone (32) 20-hydroxyecdysone (32)	7.3 30.8	component in mixture	
	Calliphora stygia (brown blowfly)	prepupae, 1.87 kg		20-hydroxyecdysone (32)	0.1		Galbraith, M. *et al.* (1969b)
		pharate adults, 1.795 kg		20-hydroxyecdysone (32)	0.095		Galbraith, M. *et al.* (1969b)
	Dociostaurus maroccanus (Moroccan locust)	adults, 10 kg		ecdysone (24) 20-hydroxyecdysone (32)	1.1 1.3		Stamm, M. (1958)
	Galleria mellonella (greater wax moth)	eggs, 20 g fr. wt.	HPLC, TLC, bioassay	ecdysone (24)* 20-hydroxyecdysone (32)* 3-epi-20-hydroxy-ecdysone (30)* 3-epi-ecdysone (29)	74 (total)	concentrations determined by bioassay; 80-85% of ecdysteroids are conjugated; ecdysone (24) and 20-hydroxy-ecdysone (32) comprise 80% of mixture	Hsaio, T. and Hsiao, C. (1979)
		ovaries (pharate adults) ovaries (1-day-old adults)		ecdysone (24) + 20-hydroxyecdysone (32) ecdysone (24) + 20-hydroxyecdysone (32)	~0.4 μg per female (total) ~0.58μg per female (total)		
	Heliothis zea	pupae (4-day-old)	HPLC	ecdysone (24) 20-hydroxyecdysone (32)	400 ng per pupa 400 ng per pupa	HPLC used for both identification and quantification	Holman, G. and Meola, R. (1978)
	Locusta migratoria (migratory locust)	30 eggs (newly laid)	HPLC, RIA, TLC	2-deoxyecdysone (18) ecdysone (24) 20-hydroxyecdysone (32)	166 nM-eq. 253 nM-eq. 24 nM-eq.	one nM-ecdysone equivalents of highly polar conjugates hydrolysed by *Helix pomatia* and concentrations determined by RIA after separation by TLC, HPLC	Hoffman, J. *et al.* (1980)
		30 eggs (8 days old)	HPLC, RIA, TLC	2-deoxyecdysone (18) ecdysone (24) 20-hydroxyecdysone (32) 20,26-dihydroxy-ecdysone (23) 3-dehydroecdysone (15)	0.41 nM-eq. 2.30 nM-eq. 0.21 nM-eq. 0.04 nM-eq. 0.66 nM-eq.		
		1000 ovaries (at end of first cycle of oocyte maturation)	GC–MS, RIA, TLC	2-deoxyecdysone (18) 2,22-dideoxyecdysone (21)	n.d.	conjugates were hydrolysed with *Helix pomantia*	Hetru, C. *et al.* (1978)

Table 3: Continued

Phylum Class	Species	Source	Method	Ecdysteroids	Concentration (mg kg^{-1})	Comments	References
				2,22,25-trideoxyecdysone (78)		enzymes before identification—	
				2,14,22,25-tetradeoxyecdysone (77)		2,22-dideoxyecdysone (21) only tentatively identified because of lack of reference compound	
	Manduca sexta (tobacco hornworm)	eggs, 5 kg	MS, NMR	26-hydroxyecdysone (33)	5.3		Kaplanis, J. *et al.* (1973)
				20,26-dihydroxyecdysone (23)	0.3		
				ecdysone (24)	0.24		
				20-hydroxyecdysone (33)	0.1		
		eggs (4–18 h old)	TLC, UV, NMR, MS	26-hydroxyecdysone (33)	11.9	total ecdysteroid concentrations were determined after *Helix pomatia* juice hydrolysis of conjugates	Kaplanis, J. *et al.* (1980)
				3-epi-26-hydroxyecdysone	0.56		
				ecdysone (24)	0.47		
				20-hydroxyecdysone (32)	0.017		
				3-epi-20,26-dihydroxyecdysone (28)	0.005		
		meconium fluid, 12.5 l		3-epi-20-hydroxyecdysone (30)	n.d.		Thompson, M. *et al.* (1974)
		pupae (at peak titre of MH) 12.7 kg		ecdysone (24)	0.32		Kaplanis, J. *et al.* (1966b)
				20-hydroxyecdysone (32)	0.29		
		pupae (peak titre of MH) 40.2 kg		ecdysone (24)	0.08		Thompson, M. *et al.* (1967)
				20,26-dihydroxyecdysone (23)	0.08		
		pupae (5 days after peak titre)	bioassay,	20,26-dihydroxyecdysone (23)	0.12		Kaplanis, J. *et al.* (1979)
			TLC, NMR, UV, MS	3-epi-20-hydroxyecdysone (30)	0.09		
				20-hydroxyecdysone (32)	0.056		
				3-epi-20,26-dihydroxyecdysone (28)	0.029		
				3-epi-ecdysone (29)	0.011		
				ecdysone (24)	0.0046		
	Oncopeltus fasciatus (milkweed bug)	embryos (96±4 h. old) 0.5 kg	TLC, MS, bioassay	makisterone A (44)	0.4		Kaplanis, J. *et al.* (1975)
	Schistocerca gregaria (desert locust)	eggs (newly laid) 170 g		ecdysone (24)	2.6 (free) 44.0 (conj.)	26-hydroxyecdysone (33) was tentatively assigned using HPLC-retention times	Dinan, L. and Rees, H. (1981)
				2-deoxyecdysone (18)	0.05 (free) 19.4 (conj.)		
				20-hydroxyecdysone (32)	0.03 (free) 6.1 (conj.)		
				26-hydroxyecdysone (33)	0.08 (conj.)		

Table 3: Continued

Phylum Class	Species	Source	Method	Ecdysteroids	Concentration (mg kg^{-1})	Comments	References
		developing eggs	GC, TLC	ecdysone (24)	21–1900 pM/g (free) 65–4400 pM/g (conj.)	*Helix pomatia* enzymes used for hydrolysis of conjugates—conc. of conj. decreases with increasing development	Gande, A. and Morgan, E. (1979)
				20-hydroxyecdysone (32)	15–710 pM/g (free) 8–450 pM/g (conj.)		
		eggs (17 days old) 160 g		2-deoxy-20-hydroxy-ecdysone (19)	0.63	total ecdysteroids after *Helix pomatia* hydrolysis	Isaac, R. *et al.* (1981a)
				3-epi-2-deoxy-ecdysone (27)	0.78		
		developing eggs		ecdysone-3-acetate (25)	1.88	total, after *Helix pomatia* enzyme hydrolysis	Isaac, R. *et al.* (1981b)
		ovaries (adult females)	GC, TLC	ecdysone (24)	80 ng per insect (free)	determined by *Helix pomatia* enzyme hydrolysis of conjugates	Gande, A. *et al.* (1979)
					2130 ng per insect (conj.)		
				20-hydroxyecdysone (32)	50–345 ng per insect		
		late fifth instars, 32 kg		20-hydroxyecdysone (32)	0.06		Morgan, E. *et al.* (1975a,b); Scalia, S. and Morgan, E. (1982)
Coelenterata Octocorallia							
	Labophytum pauciflorum (soft coral)	dried, whole coral, 7.7 kg		lobosterol (43)	53.3		Tursch, B. *et al.* (1976)
Nematoda	*Ascaris lumbricoides* (parasitic nematode)	adults, 15.5 kg		20-hydroxyecdysone (32)	3×10^{-4}		Horn, D. *et al.* (1974b)
Trematoda	*Schistosoma mansoni*	lung schistosomula		20-hydroxyecdysone equivalent	1.2 ng (mg protein)$^{-1}$		Torpier, G. *et al.* (1982)

[a] Abbreviations used: TLC—thin layer chromatography; HPLC—high performance liquid chromatography; GC—gas chromatography; MS—mass spectrometry; MF—mass fragmentography; NMR—nuclear magnetic resonance spectroscopy; RIA—radioimmunoassay; n.d.—not determined; UV—ultraviolet spectroscopy.

(Kaplanis, J. *et al.*, 1979; see Koolman, J. and Karlson, P., this volume). Presumably, conversion of 3β-isomers to their respective 3α-isomers proceeds *via* 3-keto-intermediates but there is only one report of the detection of a 3-ketoecdysteroid in an insect: Hoffmann, J. *et al.* (1980) have reported that a conjugate of 3-dehydroecdysone (15) occurs in 8-day-old eggs of *Locusta migratoria.* At peak titre of MH activity of *Manduca sexta* during pupal–adult development, however, the ecdysteroid concentrations decrease in the order: 20-hydroxyecdysone (32), ecdysone (24) and 20,26-dihydroxyecdysone (23) (Kaplanis, J. *et al.*, 1966b; Thompson, M. *et al.*, 1967). Thompson, M. *et al.* (1967) have suggested that the changing titre of 20,26-dihydroxyecdysone (23) and the presence of its less biologically active 3α-epimer (28) in *Manduca sexta* implies a specific physiological role in insect development for this ecdysteroid. 26-Hydroxyecdysone (33) is by far the major ecdysteroid during embryonic development in *Manduca sexta* (Kaplanis, J. *et al.*, 1980). However, it is inactive in the housefly bioassay. Kaplanis, J. *et al.* (1980) speculate that 26-hydroxyecdysone (33) may be an inactivation product or that it possesses biological activity related specifically to embryogenesis and emphasize the need for the development of assays to determine ecdysteroid functions other than those of moulting.

The function of ecdysteroid conjugates during ovarian embryonic development is also unclear. In *Locusta migratoria* ecdysteroid conjugates have been implicated as inactive and storage precursor forms of the hormone (Hoffmann, J. *et al.*, 1980). In *Galleria mellonella*, however, the dynamic state of ecdysteroid conjugates led Hsiao, T. and Hsiao, C. (1979) to propose that conjugates themselves play specific roles in development.

2.2 Plants

Nakanishi, K. *et al.* (1966) made the surprising discovery that ponasterone A (58) occurs in the plant *Podocarpus nakaii.* Soon after 20-hydroxyecdysone (32) was found in the conifer *Podocarpus elatus* (Galbraith, M. and Horn, D., 1966), in the fern *Polypodium vulgare* (Jizba, J. *et al.*, 1967a) and in the angiosperm *Achyranthes fauriei* (Takemoto, T. *et al.*, 1967c). 20-Hydroxyecdysone (32) and ecdysone (24) were found to occur in the ferns *Pteridium aquilinum* (Kaplanis, J. *et al.*, 1967) and *Polypodium vulgare* (Heinrich, G. and Hoffmeister, H., 1967). Galbraith, M. *et al.* (1967) showed that 20-hydroxyecdysone (32) derived from plants was identical with the hormone derived from insect sources. Soon it became apparent that many plants contained a variety of ecdysteroids with moulting hormone activity in insect bioassays. We have listed the 69 phytoecdysteroids known to date in Table 1 and their concentration in plants in Table 4 (see also Addendum for ecdysteroids isolated subsequently). The occurrence of ecdysteroids in plants has been reviewed on numerous occasions (Matsuoka, T. *et al.*, 1969; Rimpler, H., 1970; Hikino, H. and Hikino, Y., 1970; Highnam, K., 1970; Heftmann, E., 1970; Herout, V., 1970; Rees, H., 1971; Nakanishi, K., 1971; Horn, D., 1971; Slama, K. *et al.*, 1974; Hikino, H. and Takemoto, T., 1974; Koolman, J., 1977; Prakash, A. and Ghosal, S., 1979; Slama, K., 1979; Hetru, C. and Horn, D., 1980). We have recently reviewed in depth the occurrence and roles of insect hormones in plants (Bergamasco, R. and Horn, D., 1983); our aim here is to briefly summarize the data presented in that review.

It is apparent from Table 4 that phytoecdysteroids occur in relatively large concentrations in many plants. Since phytoecdysteroids commonly comprise between 10^{-3}% and 10^{-1}% of the dry weight of the plant, plants are a far better source of ecdysteroids than arthropods which only contain 10^{-5}–10^{-9}% ecdysteroids. Indeed, the bountiful supply of ecdysteroids from plants greatly stimulated the chemistry and biology of these arthropod hormones. In some plants much higher concentrations of phytoecdysteroids are found: the flowers of *Serratula inermis*, for example, contain 2% of 20-hydroxyecdysone (32) (Yatsyuk, Y. and Segel, G., 1970) and the roots of the herb *Cyanothis arachnoidae* contain 2.9% of 20-hydroxyecdysone (32) (Chou, W. and Lu, H., 1980). It is important to note that different plant parts contain different amounts of ecdysteroids and that ecdysteroid concentrations vary with the season and location of the plant (Hikino, H. *et al.*, 1973; Yen, K. *et al.*, 1974). Further, Imai, S. *et al.* (1969a) noted that phytoecdysteroids are more likely to occur in perenial woody and herbaceous plants than in annual herbaceous plants.

Table 4: The concentration of ecdysteroids in plants

Division Order Family	Species	Ecdysteroid isolated	Plant parts	Concentration (mg kg^{-1})	Reference
PTERIDOPHYTA					
OSMUNDALES					
Osmundaceae	*Plenasium banksiifolium* (Pr.) Pr.	ecdysone (24)	aerial parts	11	Murakami, T. *et al.* (1980)
		20-hydroxyecdysone (32)	aerial parts	270	Murakami, T. *et al.* (1980)
PTERIDALES					
Pteridaceae	*Phymatodes novae-zelandiae*	ecdysone (24)	dry fronds	100	Russell, G. (1972)
		20-hydroxyecdysone (32)	dry fronds	70	Russell, G. (1972)
		polypodine B (54)	dry fronds	180	Russell, G. (1972)
	Pteridium aquilinum Kuhn	ecdysone (24)	leaves	0.5	Kaplanis, J. *et al.* (1967)
		20-hydroxyecdysone (32)	leaves	1	Kaplanis, J. *et al.* (1967)
ASPIDIALES					
Aspidiaceae	*Diplazium donianum* Tardieu-Blot (syn. *Athyrium aphanoneuron* Ohwi)	makisterone A (44)	dry roots and parts	0.2	Hikino, H. *et al.* (1976b)
		makisterone D (47)	dry roots and parts	0.4	Hikino, H. *et al.* (1976b)
BLECHNALES					
Blechnaceae	*Blechnum minus* (R.Br.) Ettingsh.	2-deoxyecdysone (18)		30	Chong, Y. *et al.* (1970)
		2-deoxy-20-hydroxyecdysone (19)		100	Chong, Y. *et al.* (1970)
	Blechnum niponicum Makino	stachysterone D (76)	dry whole plant	0.4	Hikino, H. *et al.* (1975a)
POLYPODIALES					
Polypodiaceae	*Polypodium aureum* L.	20-hydroxyecdysone (32)	rhizomes	7	Jizba, J. *et al.* (1974a)
		polypodoaurein (56)	rhizomes	2	Jizba, J. *et al.* (1974a)
	Polypodium japonicum Makino	20-hydroxyecdysone (32)	rhizomes	500	Imai, S. *et al.* (1969c)
	Polypodium Virginianum	ecdysone (24)	dry whole plant	20	Hikino, H. (1976)
		20-hydroxyecdysone (32)	dry whole plant	200	Hikino, H. (1976)
	Polypodium vulgare Linné	ecdysone (24)	rhizomes	18	Heinrich, G. and Hoffmeister, H. (1967)
		20-hydroxyecdysone (32)	rhizomes	700	Heinrich, G. and Hoffmeister, H. (1967)
		20-hydroxyecdysone (32)	rhizomes	10,000	Jizba, J. *et al.* (1967a)
		polypodine B (54)	rhizomes	400	Jizba, J. *et al.* (1967b)
GYMNOSPERMAE					
CONIFERALES					
Podocarpaceae	*Dacrydium intermedium* Kirk	20-hydroxyecdysone (32)	dry bark	10,000	Russell, G. *et al.* (1972)
		20-hydroxyecdysone 3-p-coumarate (37)	dry bark	100	Russell, G. *et al.* (1972)
		pterosterone (66)	dry bark	100	Russell, G. *et al.* (1972)
		polypodine B (54)	dry bark	5000	Russell, G. *et al.* (1972)
		polypodine B–2-cinnamate (55)	dry bark	600	Russell, G. *et al.* (1972)
		ponasterone C (60)	dry bark	500	Russell, G. *et al.* (1972)
		ponasterone C–2-cinnamate (61)	dry bark	100	Russell, G. *et al.* (1972)
		20-hydroxyecdysone–2-cinnamate (36)	dry bark	600	Russell, G. *et al.* (1971)
		dacrysterone (14)	dry bark	30	Russell, G. and Fraser, J. (1973)

	Podocarpus elatus R. Br.	20-hydroxyecdysone (32)	dry wood	10	Galbraith, M. and Horn, D. (1966, 1969)
		20-hydroxyecdysone (32)	dry bark	4500	Galbraith, M. and Horn, D. (1966, 1969)
	Podocarpus macrophyllus Lamb.	20-hydroxyecdysone (32)	fresh leaves	10	Imai, S. *et al.* (1968a)
		ponasterone A (58)	fresh leaves	500	Imai, S. *et al.* (1968a)
		makisterone A (44)	fresh leaves	10	Imai, S. *et al.* (1968a)
		makisterone B (45)	fresh leaves	1	Imai, S. *et al.* (1968a)
		makisterone C (46)	fresh leaves	1	Imai, S. *et al.* (1968a)
		makisterone D (47)	fresh leaves	1	Imai, S. *et al.* (1968a)
	Podocarpus macrophyllus Lamb. var. *maki* Siebold (= *Podocarpus chinensis* Sweet)	ponasterone A (58)	fresh leaves	300	Imai, S. *et al.* (1967)
	Podocarpus nakai Hay.	ponasterone A (58)	dry leaves	600	Nakanishi, K. *et al.* (1968)
		ponasterone B (59)	dry leaves	70	Nakanishi, K. *et al.* (1968)
		ponasterone C (60)	dry leaves	17	Nakanishi, K. *et al.* (1968)
Taxaceae	*Taxus baccata* Linné	20-hydroxyecdysone (32)	fresh leaves	20	Hoffmeister, H. *et al.* (1967)
			dry leaves	200	Staal, G. (1967)
	Taxus cuspidata Sieb. and Zucc.	20-hydroxyecdysone (32)	fresh leaves	14	Imai, S. *et al.* (1967)
ANGIOSPERMAE					
LILIALES					
Liliaceae	*Trillum smallii* Maximowicz	20-hydroxyecdysone (32)	rhizomes	80	Imai, S. *et al.* (1969c)
	Trillum tschonoskii Maximowicz	20-hydroxyecdysone (32)	rhizomes	100	Imai, S. *et al.* (1969c)
RANUNCULALES					
Ranunculaceae	*Helleborus abchasicus* A. Braun	20-hydroxyecdysone (32)	dry aerial parts and flowers	2400	Hardman, R. and Benjamin, T. (1976)
	Helleborus atrorubens Hort.	20-hydroxyecdysone (32)	dry aerial parts and flowers	2500	Hardman, R. and Benjamin, T. (1976)
		polypodine B (54)	dry aerial parts and flowers	1600	Hardman, R. and Benjamin, T. (1976)
	Helleborus bocconei Ten. ssp. *bocconei*	20-hydroxyecdysone (32)	dry aerial parts and flowers	2600	Hardman, R. and Benjamin, T. (1976)
		polypodine B (54)	dry aerial parts and flowers	500	Hardman, R. and Benjamin, T. (1976)
	Helleborus cyclophyllus Boiss.	20-hydroxyecdysone (32)	dry aerial parts and flowers	2700	Hardman, R. and Benjamin, T. (1976)
		polypodine B (54)	dry aerial parts and flowers	1600	Hardman, R. and Benjamin, T. (1976)
	Helleborus dumentorum Waldst. and Kit. in Willd. ssp. *dumentorum*	20-hydroxyecdysone (32)	dry aerial parts and flowers	2900	Hardman, R. and Benjamin, T. (1976)
		polypodine B (54)	dry aerial parts and flowers	1400	Hardman, R. and Benjamin, T. (1976)
	Helleborus dumentorum ssp. *atrorubens* (Waldst and Kit.) Merzm. and Podl.	20-hydroxyecdysone (32)	dry aerial parts and flowers	2100	Hardman, R. and Benjamin, T. (1976)
		polypodine B (54)	dry aerial parts and flowers	1100	Hardman, R. and Benjamin, T. (1976)
	Heleborus guttatus A. Braun	20-hydroxyecdysone (32)	dry aerial parts and flowers	2700	Hardman, R. and Benjamin, T. (1976)
		polypodine B (54)	dry aerial parts and flowers	700	Hardman, R. and Benjamin, T. (1976)
	Heleborus multifidus ssp. *serbicus* (Adamovic) Merzm. and Podl.	20-hydroxyecdysone (32)	dry aerial parts and flowers	3600	Hardman, R. and Benjamin, T. (1976)
		polypodine B (54)	dry aerial parts and flowers	900	Hardman, R. and Benjamin, T. (1976)
	Heleborus orientalis Hybrids	20-hydroxyecdysone (32)	dry aerial parts and flowers	4500	Hardman, R. and Benjamin, T. (1976)
	Heleborus viridis L. ssp. *occidentalis* (Reuter) Schiffner	20-hydroxyecdysone (32)	dry aerial parts and flowers	2100	Hardman, R. and Benjamin, T. (1976)
URTICALES					
Moraceae	*Morus* ssp.	20-hydroxyecdysone (32)	leaves	1	Takemoto, T. *et al.* (1967e)
		inokosterone (38)	leaves	10	Takemoto, T. *et al.* (1967e)

VIOLALES					
Stachyuraceae	*Stachyurus praecox* (Sieb. and Zucc.)	20-hydroxyecdysone (32)	bark	600	Imai, S. *et al.* (1969c)
		stachysterone A (73)	bark	20–40	Imai, S. *et al.* (1970a)
		stachysterone B (74)	bark		Imai, S. *et al.* (1970a)
		stachysterone C (75)	bark	20	Imai, S. *et al.* (1970b)
		stachysterone D (76)	bark	20	Imai, S. *et al.* (1970b)
COMMELINALES					
Commelinaceae	*Cyanothis arachnoidae* C. B. Clarke	20-hydroxyecdysone (32)	dry whole plant	12,000	Chou, W. and Lu, H. (1980)
		20-hydroxyecdysone (32)	dry roots	29,000	Chou, W. and Lu, H. (1980)
	Cyanothis vaga Schultes	20-hydroxyecdysone (32)	leaves	7000	Santos, A. *et al.* (1970, 1972)
CARYOPHYLLALES					
Aizoaceae	*Sesuvium portulacastrum* Linn.	20-hydroxyecdysone (32)	whole plant	3500	Banerji, A. *et al.* (1971)
	Trianthema portulacastrum Linn.	20-hydroxyecdysone (32)	whole plant	10	Banerji, A. *et al.* (1971)
Amaranthaceae	*Achyranthes aspera* Linn.	20-hydroxyecdysone (32)	seeds	250	Banerji, A. *et al.* (1971)
		20-hydroxyecdysone (32)	stems and leaves	40	Banerji, A. *et al.* (1971)
		20-hydroxyecdysone (32)	roots	90	Banerji, A. *et al.* (1971)
	Achyranthes bidentata Blume	20-hydroxyecdysone (32)	whole plant	90	Ogawa, S. *et al.* (1971)
	Achyranthes fauriei Lev. and Van.	20-hydroxyecdysone (32)	roots	100	Takemoto, T. *et al.* (1967c,d)
		inokosterone (38)	roots	100	Takemoto, T. *et al.* (1967c,d)
	Achyranthes japonica Nakai	20-hydroxyecdysone (32)	dry leaves	2000	Takemoto, T. *et al.* (1968g)
	Achyranthes japonica var. *hachijoensis*	inokosterone (38)	whole plant	70	Ogawa, S. *et al.* (1971)
	Achyranthes mollicola Nakai	20-hydroxyecdysone (32)	whole plant	130	Ogawa, S. *et al.* (1971)
	Achyranthes obtusifolia Lamark	20-hydroxyecdysone (32)	roots	250	Takemoto, T. *et al.* (1967h)
	Achyranthes rubrofusca Wight	20-hydroxyecdysone (32)	roots	490	Takemoto, T. *et al.* (1967g)
		20-hydroxyecdysone (32)	whole plant	19	Ogawa, S. *et al.* (1971)
		inokosterone (38)	whole plant	54	Ogawa, S. *et al.* (1971)
	Aerva tomentosa Forssk.	20-hydroxyecdysone (32)	roots	70	Hardman, R. and Mahar, S. (1978)
	Bosea yervamora Linné	20-hydroxyecdysone (32)	roots	200	Takemoto, T. *et al.* (1967f)
	Cyathula capitata Moquin-Tandon	cyasterone (12)	dry roots	210	Hikino, H. *et al.* (1970c)
		sengosterone (68)	dry roots	3	Hikino, H. *et al.* (1970c)
		poststerone (63)	dry roots	3	Hikino, H. *et al.* (1970b)
		isocyasterone (38)	dry roots	1	Hikino, H. *et al.* (1971b)
	Gomphrena celosiodes Mart.	20-hydroxyecdysone (32)	whole plant	250	Banerji, A. *et al.* (1971)
	Pandiaka involucrata Benth. and Hook.	20-hydroxyecdysone (32)	aerial parts (before flowering)	1200	Hardman, R. and Mahar, S. (1978)
POLEMONIALES					
Convolvulaceae	*Ipomea calonyction* (Choisy) Hallier f. ssp. *nova*	ecdysone (24)	seeds	130	Canonica, L. *et al.* (1975)
		20-hydroxyecdysone (32)	seeds	1500	Canonica, L. *et al.* (1975)
		makisterone A (44)	seeds	250	Canonica, L. *et al.* (1975)
		muristerone A (48)	seeds	250	Canonica, L. *et al.* (1975)
		kaladasterone (42)	seeds	5	Canonica, L. *et al.* (1975)
		calonysterone (7)	seeds	8	Canonica, L. *et al.* (1975)

SCROPHULARIALES					
Solanaceae	*Solanum xanthocarpum* Shrad. and Wendl.	carpesterol (9)	dry fruit	1000	Kusano, G. *et al.* (1975)
		14α-hydroxycarpesterol (31)	dry fruit	1	Kusano, G. *et al.* (1975)
LAMIALES					
Verbenaceae	*Vitex megapotamica* Moldenke	20-hydroxyecdysone (32)	leaves	880	Rimpler, H. and Schulz, G. (1967)
Labiatae (Lamiacae)	*Ajuga chia*	cyasterone (12)	fresh leaves and stems	20	Ikan, R. and Ravid, U. (1971)
	Ajuga decumbens Thunberg	ajugsterone C (3)	dry leaves	15	Imai, S. *et al.* (1969b)
		cyasterone (12)	whole plant	80	Imai, S. *et al.* (1969c)
		20-hydroxyecdysone (32)	whole plant	120	Imai, S. *et al.* (1969c)
		ajugasterone B (2)	whole plant	15	Koreeda, M. and Nakanishi, K. (1970a)
		ajugalactone (1)	whole plant	100	Koreeda, M. and Nakanishi, K. (1970a)
		cyasterone (12)	whole plant	600	Koreeda, M. and Nakanishi, K. (1970a)
		20-hydroxyecdysone (32)	whole plant	1300	Koreeda, M. and Nakanishi, K. (1970a)
		ajugasterone C (3)	whole plant	15	Koreeda, M. and Nakanishi, K. (1970a)
	Ajuga incisa Maximowicz	20-hydroxyecdysone (32)	whole plant	120	Imai, S. *et al.* (1969c)
		cyasterone (12)	whole plant	80	Imai, S. *et al.* (1969c)
	Ajuga iva	20-hydroxyecdysone (32)	fresh whole plant	120	Ikan, R. and Ravid, U. (1970)
	Ajuga japonica Miquel	cyasterone (12)	dry leaves	580	Imai, S. *et al.* (1969b)
		20-hydroxyecdysone (32)	dry leaves	2000	Imai, S. *et al.* (1969b)
		ajugasterone C (3)	dry leaves	100	Imai, S. *et al.* (1969b)
	Ajuga nipponensis Makino	cyasterone (12)	whole plant	80	Imai, S. *et al.* (1969c)
		20-hydroxyecdysone (32)	whole plant	120	Imai, S. *et al.* (1969c)
	Ajuga turkestanica (Rgl.) Briq.	cyasterone (12)	dry roots	100	Usmanov, B. *et al.* (1975)
		20-hydroxyecdysone (32)	dry leaves	480	Usmanov, B. *et al.* (1975)
		turkesterone (79)	dry roots	520	Usmanov, B. *et al.* (1975)
		cyasterone-22-acetate (13)	dry leaves	500	Usmanov, B. *et al.* (1978)
ASTERALES					
Asteraceae (Compositae)	*Rhaponticum integrifolium* C. Winkl.	20-hydroxyecdysone (32)	dry inflorescence	150	Baltaev, U. *et al.* (1978a)
		integristerone A (39)	dry inflorescence	2	Baltaev, U. *et al.* (1978a)
		integristerone B (40)	dry inflorescence	3	Baltaev, U. *et al.* (1978a)
		24(28)-dehydromakisterone A (17)	dry inflorescence	2	Baltaev, U. *et al.* (1978b)
	Serratula sogdiana	sogdysterone (72)	dry inflorescence	3	Novosel'skaya, I. *et al.* (1975)
	Serratula xeranthemoides	20-hydroxyecdysone (32)	dry flower heads at budding time	1600	Kholodova, Y. *et al.* (1979)
		integristerone A (39)	dry flower heads at budding time	1100	Kholodova, Y. *et al.* (1979)
		20-hydroxyecdysone (32)	dry inflorescence	3300	Kholodova, Y. *et al.* (1979)
		integristerone A (39)	dry inflorescence	1800	Kholodova, Y. *et al.* (1979)
		20-hydroxyecdysone (32)	beginning of fruit formation	2800	Kholodova, Y. *et al.* (1979)
		integristerone A (39)	beginning of fruit formation	2100	Kholodova, Y. *et al.* (1979)
		20-hydroxyecdysone (32)	seeds	1100	Kholodova, Y. *et al.* (1979)
		integristerone A (39)	seeds	500	Kholodova, Y. *et al.* (1979)

The photoecdysteroids are widely distributed among the vascular plants: they occur in 27 families of the Pteridophyta, 10 families of the Gymnospermae and 74 families of the Angiospermae. Good sources of phytoecdysteroids are:

(i) the families Osmundaceae, Plagiogyraceae, Cyatheaceae, Aspidiaceae, Blechnaceae and Polypodiaceae of the Pteridophyta (Hikino, H. *et al.*, 1973; Yen, K. *et al.*, 1974);
(ii) the families Podocarpaceae and Taxaceae of the Gymnospermae (Takemoto, T. *et al.*, 1967j; Staal, G., 1967);
(iii) the genera *Achyranthes, Cyathula, Helleborus* and *Ajuga* of the Angiospermae; and
(iv) the family Verbenaceae of the Angiospermae (Imai, S. *et al.*, 1969a; Takemoto, T. *et al.*, 1967j).

Takemoto, T. *et al.* (1967j) claimed that primitive plants like mushrooms and seaweeds do not contain ecdysteroids. Other investigators, however, reported that primitive vascular plants — whisk ferns, horsetails and club mosses — contain MH-active species (Imai, S. *et al.*, 1969a). Recently, Fukuzawa, A. *et al.* (1981) isolated pinnasterol (50) and pinnasterol-2-acetate (51) from the marine red alga *Laurencia pinnata* and claimed that these ecdysteroids were active in the *Sarcophaga* assay.

An analysis of ecdysteroids within the vascular plants may reveal restricted distributions of certain ecdysteroids which may be of taxonomic significance (Bergamasco, R. and Horn, D., 1983). Since only a small number of plant species have been studied for the presence of ecdysteroids, however, the following observations should be considered tentative. Only 20-hydroxyecdysone (32), pterosterone (66) and polypodine B (54) are common to the ferns, gymnosperms and angiosperms. 20-Hydroxyecdysone (32) is often the largest component of phytoecdysteroid mixtures and is also the most commonly occurring phytoecdysteroid. In order of decreasing frequency the next most common ecdysteroids are: ponasterone A (58), polypodine B (54), ecdysone (24) and pterosterone (66). Of the 18 phytoecdysteroids isolated from the Pteridophyta, 8 do not occur in higher plants. Of 17 gymnosperm ecdysteroids 10 are unique to the Gymnospermae. In the flowering plants 36 unique ecdysteroids have been found in addition to 10 ecdysteroids which are common to ferns and gymnosperms. It is interesting that the greatest number of phytoecdysteroids within the flowering plants are found in the order Lamiales and Caryophyllales, which are considered to be among the taxonomically most advanced orders of the Angiospermae (Takhtajan, A., 1980).

The discovery of ecdysteroids in plants led Galbraith, M. and Horn, D. (1966) to suggest that these compounds play a defensive role in plants and since this time it has become widely accepted that secondary plant substances fulfil defensive roles though this is hard to establish rigorously (Whittaker, R. and Feeny, P., 1971; Rhoades, D., 1979). Swain, T. (1978) suggested that the evolution of phytoecdysteroid biosynthesis in plants as a biochemical defence mechanism contributed to the world-wide 60-million-year domination in the Carboniferous period. Phytoecdysteroids are considered to constitute a "qualitative defence directed against non-adapted herbivores" (Bergamasco, R. and Horn, D., 1983). Thus phytoecdysteroids may provide little protection against specialized herbivores, which have evolved efficient detoxification mechanisms, but represent a considerable barrier against non-adapted herbivores. Also phytoecdysteroids may act synergistically with quantitative (e.g. tannins) or other qualitative defences in a complex overall defence strategy. Ample evidence exists to support the notion that phytoecdysteroids represent a qualitative plant defence: ecdysteroids have been shown to act as non-toxic feeding deterrents towards insects (Ma, W., 1972; Jones, C. and Firn, R., 1978) and to affect insect growth and reproduction when incorporated in artificial diets (Staal, G., 1967; Robbins, W. *et al.*, 1968; Matsuoka, T. *et al.*, 1969; Robbins, W. *et al.*, 1970; Riddiford, L., 1970; Singh, P. and Russell, G., 1980; Singh, P. *et al.*, 1982).

A direct physiological and biochemical role of ecdysteroids in plant metabolism has yet to be demonstrated. Heftmann, E. (1975) postulated that the likely gene-level action of ecdysteroids in arthropods may extend to other eukaryotic organisms. Very few reports, however, support this idea. Carlisle, D. *et al.* (1963) claimed that a locust extract containing "ecdysone-λ" significantly stimulated the growth of the internodes of dwarf pea plants in a gibberellin A_3-like fashion. Gibberellin-like effects were also reported by Matsuoka, T. *et al.* (1969) when ponasterone A (58), 20-hydroxyecdysone (32) and inokosterone (38) were applied to second leaf sheaths and roots of rice seedlings. Hendrix, S. and Jones, R. (1972), however, could not induce a gibberellin-like effect by using 20-hydroxyecdysone (32) in four plant assays. Felippe, G. (1980) claimed that 20-hydroxyecdysone (32) stimulates male flower formation in *Cucurbita pepo*. Jacobs, W. and Suthers, H. (1971), however, failed to induce flowering by administering 20-hydroxyecdysone (32) to an *in vitro* culture of apical buds of *Xanthium pennsylvanicum*. However, Geuns, J. (1978) has suggested that the possible roles of ecdysteroids in plant biochemistry need to be re-examined using more reliable and sensitive assays under well-defined conditions.

3 BIOLOGICAL METHODS OF DETECTION

3.1 *In vivo* insect assays

The first bioassay for moulting hormones was developed by Becker, E. and Plagge, E. (1939) and was based on the work of Fraenkel, G. (1935) with larvae of the blowfly *Calliphora erythrocephala* (see Wigglesworth, V., this volume). This bioassay was refined by Karlson, P. and Hanser, G. (1953) into a quantitative assay in the following way. Prepupae of *Calliphora* larvae were ligated below the ring gland and those animals which pupated after 24 h in the anterior portion but not in the posterior portion (about 10% of the total ligated) were ligated a second time below the first ligature, the animals severed between the two ligatures, and the test solution injected into the posterior portion through the severed end. After incubation for a further 24 h the test abdomens were examined for cutical darkening. The dose which produced a 60% sclerotization was taken as the *Calliphora* unit. The bioassay was further improved by Thomson, J. *et al.* (1970) who found that the test solution could be simply injected directly through the cuticle of the abdomen of ligated larvae with a fine glass needle, thus avoiding the necessity of a second ligature. The sensitivity of the *Calliphora* bioassay is 25–50 ng of 20-hydroxyecdysone (32) per abdomen (Thomson, J. *et al.*, 1970). (See Table 5 for detection limits of ecdysteroids by various analytical procedures).

Similar bioassays were developed with larvae of the housefly *Musca domestica* (Kaplanis, J. *et al.*, 1966a; Staal, G., 1967), the fleshfly *Sarcophaga peregrina* (Ohtaki, T. *et al.*, 1967, Takemoto, T.

Table 5: The approximate detection limits for 20-hydroxyecdysone by various methods

Method	Limit	Reference
A. Biological assay		
Musca domestica	5×10^{-9}g per insect or	Kaplanis, J. *et al.* (1966a)
	1.5×10^{-7} g per test	Gilgan, M and Zinck, M. (1972)
Calliphora vicina	7.5×10^{-9} g per insect	Karlson, P. (1956b)
Calliphora Stygia	5×10^{-8} g per insect	Thomson, J. *et al.* (1970)
	3×10^{-5} M	
Aedes atropalpus	3×10^{-6} M	Kelly, T. and Fuchs, M. (1980)
Chromosome puff	1.0×10^{-8} M	Richards, G. (1978)
Imaginal disc	4.2×10^{-9} M	Yund, M. *et al.* (1978), Fristrom, J. and Yund, M. (1980)
Kc Cell lines	1.0×10^{-8} M	Cherbas, L. *et al.* (1980)
Radioimmunoassay	1.0×10^{-9} g	Borst, D and O'Connor, J (1972, 1974)
	7.5×10^{-11}–5.0×10^{-10} g	Horn, D. *et al.* (1976)
	2.0×10^{-12} g	Hirn, M. and Delaage, M. (1980)
B. Instrumental analysis		
TLC	5.0×10^{-7} g	Mayer, R. and Svoboda, J. (1978)
HPLC	5.0×10^{-9} g	Wilson, I. *et al.* (1980)
Fluorescence	2.0×10^{-10} g	Koolman, J. (1980)
GLC–ECD	1.0×10^{-10} g	Wilson I. *et al.* (1980)
GLC–MS	1.0×10^{-15}–1.0×10^{-12} g	Wilson, I. *et al.* (1980)

For a more detailed table of concentrations required for *in vitro* systems see Cherbas, L. *et al.* (1980).

et al., 1967k), the silkworm *Bombyx mori* (Kobayashi, M. *et al.*, 1967) and the desert locust *Schistocerca gregaria* (Morgan, E. *et al.*, 1975a,b); the housefly being the most sensitive (5–6 ng). Sato, Y. *et al.* (1968) developed a simple bioassay in which the ligated larvae of the rice stem borer *Chilo suppressalis* were dipped in a methanolic solution of the test material, and Agui, N. (1973) developed a quantitative bioassay using pieces of integument taken from the unpupated abdomens of ligated larvae of *C. suppressalis*. Kelly, T. and Fuchs, M. (1980) developed a quantitative bioassay using 12 h old adult decapitated female *Aedes atropalpus* mosquitoes. The isolated abdomens were injected with the test solution and the development of the ovaries assessed. This bioassay has been used for structure–activity studies (Fuchs, Kelly and Horn, unpublished) and the results are similar to those obtained with the *Calliphora* bioassay. Williams, C. (1968) devised a bioassay using diapausing pupae of the *Cynthia* silkworm, and Jegla, T. and Costlow, J. (1979a) a bioassay using *Limulus* larvae.

For structure–activity studies (Bergamasco, R. and Horn, D., 1980) it was necessary to bioassay steroid analogues which were poorly soluble in water. This problem was overcome (Galbraith, M. *et al.*, 1981) by adding Tween 80 to the sample to produce an emulsion. The stability of the emulsion was increased by dissolving the material for testing in ethanol saturated with cholic acid. Insect bioassays are expensive and tedious to carry out but offer the best means at present for structure-activity studies, especially with compounds of low water solubility (Bergamasco, R. and Horn, D., 1980).

3.2 Chromosome puff bioassays

Clever, U. and Karlson, P. (1960) showed that ecdysone (24) induced characteristic changes (puff formation) in polytene chromosomes of the salivary glands of dipteran larvae. These changes in the chromosome structure can be used as a means of assaying ecdysteroids (Ashburner, M., 1971, 1973; Richards, G., 1978). The role of ecdysteroids in gene response is reviewed by Richards, G. (1980; chapter 7, volume 2).

3.3 Imaginal disc bioassay

Insect larvae are made up larval cells and imaginal disc cells (see Oberlander, H., vol. 2). The latter differentiate at pupation to form the adult insect. When larvae of *Drosophila melanogaster* are ground up, the imaginal discs can be separated from the larval cells because of their tendency to adhere to glass surfaces (Fristrom, J. and Mitchell, H., 1965). As imaginal discs are sensitive to ecdysteroids (Oberlander, H. and Fulco, L., 1967; Oberlander, H., 1969, 1974; Agui, N. *et al.*, 1969; Chihara, C. *et al.*, 1972; Mandaron, P., 1973; Ohmori, K. and Ohtaki, T., 1973; Fristrom, J. and Yund, M., 1976; see Silvert, D., vol. 2) and the degree of morphological response (evagination) is proportional to the concentration of the hormone in the test solution, the effect can be used as a means of assaying for ecdysteroids and for studying the effect of ecdysteroid structure on imaginal cell development (Bergamasco, R. and Horn, D., 1980). However, the bioassay is limited to compounds which are water-soluble. In general only those compounds which are structurally similar to 20-hydroxyecdysone (32) show a pronounced response.

3.4 Binding assays

Imaginal discs contain ecdysteroid receptors (Yund, M. *et al.*, 1978; Yund, M. and Osterbur, D., this volume) which strongly bind ecdysteroids, particularly ponasterone A (58). Thus discs equilibrated with radioactive ponasterone A (58) of high specific activity can be used to assess the ability of an unlabelled ecdysteroid to displace the radioactive hormone and thus determine its affinity for the receptor (Fristrom, J. and Yund, M., 1976; Yund, M. *et al.*, 1978). A similar assay has been devised using an ecdysteroid-binding protein from *Locusta migratoria* (Feyereisen, R., 1980).

3.5 Kc cell line assays

Kc cell lines derived from primary cultures of *Drosophila melanogaster* embryos exhibit a rapid and reliable morphological response to ecdysteroids (Courgeon, A., 1972; Cherbas, P. *et al.*, 1977; O'Connor, J. vol. 8), which can be used as a quantitative bioassay for these hormones (Cherbas, L. *et al.*, 1980). The bioassay has been used to study structure–activity relationships and it was found that only those compounds active in other tests elicited a response and that the concentrations required were approximately proportional to those required in other systems.

3.6 Radioimmunoassays (RIA)

The detection of ecdysteroids by most bioassays requires relatively large amounts of test solution and is tedious, expensive and time-consuming. By comparison RIA is very simple and is about 10^3 times more sensitive. The use of a RIA to determine ecdysteroids was first introduced by Borst, D. and O'Connor, J. (1972) and without doubt opened up a new era in insect physiology, since it enabled ecdysteroid titres to be determined in tissues of a single insect using simple, inexpensive equipment readily available in most entomological laboratories.

Although ecdysteroids are not immunogenic — that is on injection into mammals they do not induce the formation of antibodies (Ab) — they can be made immunogenic by coupling them covalently to a protein molecule. The antibodies that are raised to such complexes show a strong affinity for free ecdysteroids (E) and in solution an equilibrium is established:

$$^*E + Ab \rightleftharpoons {}^*EAb$$

If the ecdysteroid is radiolabelled (*E) then the percentage-bound can be determined by separating and counting the complex (*EAb).

If unlabelled hormone (E) is also added to the system a new equilibrium is established with the displacement of some of the labelled ecdysteroid from the antibody complex:

$$E + {}^*E + Ab \rightleftharpoons EAb + {}^*EAb$$

If the amount of unlabelled ecdysteroid (E) is systematically varied a series of values for percentage-bound ecdysteroid is obtained which can be used to construct a standard curve of percentage-bound against amount of ecdysteroid added. The quantity of ecdysteroid in a given sample may be found by measuring the percentage-bound value and reading off the amount of ecdysteroid present. In practice it is more convenient to use a computer program to evaluate the experimental results (Ekins, R., 1974).

The first antisera to ecdysteroids were prepared by injecting into a strain of New Zealand rabbit, the antigen formed by coupling the carbonyl group of 20-hydroxyecdysone (32) through, and oximinoacetic acid derivative to, bovine serum albumin (Borst, D. and O'Connor, J., 1972, 1974). The antisera obtained from the immunized rabbits were of relatively low activity and low specificity, showing similar affinities for both ecdysone (24) and 20-hydroxyecdysone (32). A somewhat more active antiserum was obtained by using a mixture of 2-, 3- and 22-hemisuccinates of 20-hydroxyecdysone (32) coupled to human serum albumin (Lauer, R. *et al.*, 1974). This antiserum showed a higher affinity for 20-hydroxyecdysone (32) than ecdysteroids like ecdysone (24) which do not have a 20-hydroxy group. When an antigen prepared by coupling the 22-hemisuccinate of ecdysone (24) to bovine thyroglobulin (Horn, D. *et al.*, 1976) was used, an antiserum of very high activity and a greater affinity for ecdysone (24) than 20-hydroxyecdysone (32) was obtained. In general an antiserum shows greatest affinity for those ecdysteroids which most closely resemble the ecdysteroid used to raise the antibodies. A relatively non-specific antiserum has the advantage of being useful to detect a wide range of ecdysteroids and, if of high sensitivity, can be used to detect unknown moulting hormones and precursor analogues in insect extracts as has been done by Hetru, C. *et al.* (1978). Spindler, K. *et al.* (1978) have introduced a novel method of coupling an ecdysteroid acid to thyroglobulin for RIA. The required inokosterone C26-carboxylic acid (103) was prepared by catalytic oxidation (Spindler, K. *et al.*, 1976) of inokosterone (38). Reum, L. and Koolman, J. (1979) have compared the affinities of several antisera used for RIA of a variety of ecdysteroids.

The above procedures require a tritiated ecdysteroid of high specific activity to obtain good sensitivity. Although such compounds are commercially available they are expensive. Instead it is possible to use [^{125}I]- or[^{131}I]-ecdysteroid derivatives which can be prepared as required. [^{125}I] is preferable to [^{131}I] because of its longer half-life, 60 days as compared with 6 days. Thus ecdysteroid hemisuccinates were reacted (De Reggi, M. *et al.*,

1975; Porcheron, P. *et al.*, 1976; Delaage, M. *et al.*, 1982) with aromatic amino acid derivatives such as tyrosine methyl ester and carrier-free [^{125}I] was introduced into the aromatic ring as required (Greenwood, F. and Hunter, W., 1962). The iodinated derivatives have the advantage over tritiated ones that they are of much higher specific activity. Also their structures more closely parallel the antigen used to raise the antibodies, so that they show a higher affinity for the antibodies and thus a higher sensitivity. They are cheaper than the tritiated ecdysteroids but their preparation is more hazardous. Further information concerning the RIA of ecdysteroids can be found in the detailed reviews of Reum, L. and Koolman, J. (1979) and Hirn, M. and Delaage, M. (1980).

4 METHODS OF ISOLATION

4,1 Arthropods

The initial extraction of arthropod material is preferably carried out with methanol or ethanol containing some water to minimize the extraction of non-polar lipids but not enough to produce a gelatinous extract that is difficult to filter. The filtrate is then extracted with a non-polar solvent such as hexane to remove the non-polar lipids and the aqueous alcohol phase evaporated to an aqueous concentrate, from which the hormones are separated from a large proportion of water-soluble substances by extraction with a higher alcohol.

For the extraction of ecdysone (24) from the aqueous concentrate Karlson, P. *et al.* (1963) used *n*-butanol. However, because of the difficulty of evaporating the higher boiling butanol using vacuum evaporators designed to evaporate water, the lower boiling iso-propanol was initially used to extract the crustacean moulting hormone from aqueous solutions; ammonium sulphate being added to obtain two phases (Horn, D. *et al.*, 1966). Later it was found that *n*-butanol could be removed by vacuum–steam distillation at temperatures below that of the boiling point of water. Accordingly *n*-butanol was used later for the more efficient extraction of the crustacean moulting hormones (Galbraith, M. *et al.*, 1968b; see Chart 1).

In Karlson's procedure (1956b) the *n*-butanol extract obtained above was washed with aqueous acid, base and finally with water. However, these steps are best omitted because of the loss of polar ecdysteroids such as 20-hydroxyecdysone (32) which has a partition coefficient of about 5.0 in such systems. Instead acidic impurities are conveniently removed at a later stage by the addition of

CHART 1: Flow diagram of an improved isolation of 20-hydroxyecdysone (32) (Crustecdysone) from crayfish (*Jasus lalandei*) waste. (CSIRO Procedure.)

Frozen crayfish waste (1000 kg)
↓ Extracted
(a) with 87% ethanol (2270 l)
(b) twice with 67% ethanol (1170 l)
Combined extracts concentrated in vacuum at 15°. Mash discarded.

Aqueous extract (450 l, 35 kg dissolved solids)
↓ Extracted
(a) with hexane (90 l), ethanol (30 l), and water (30 l)
(b) with hexane (45 l)
Combined hexane layers (110 l) backwashed with ethanol (30 l) and water (30 l). Inactive lipids (7 kg) discarded.

Defatted aqueous extract (430 l, 22 kg dissolved solids)
↓ Extracted three time with *n*-butanol (140 l). Combined butanol extracts (520 l) concentrated at 20–30° while bleeding water in (about 500 l). Aqueous layer discarded.

Butanol concentrate (70 l, 13 kg dissolved solids)
↓ Extracted twice with water (18 l) and ethanol (5 l). Aqueous layers backwashed in turn with butanol (23 l) and the combined butanol layers concentrated in vacuum while bleeding in water. Aqueous layers discarded.

Aqueous concentrate (4.5 l, about 500 g dissolved solids)
↓ Countercurrent extraction (three tubes) using chloroform : methanol : water (9 : 8 : 7). The combined aqueous layers (27 l) were concentrated in vacuum while bleeding in a little butanol to prevent foaming. Chloroform layers discarded.

Active extract (2 l, about 80 g dissolved solids)
↓ Countercurrent extraction (three tubes; each phase 600 ml) using chloroform : ethanol : water containing potassium hydrogen carbonate 20 g. Combined chloroform layers (three) concentrated in vacuum. Aqueous layers discarded.

Active syrup (14 g)
↓ Countercurrent extraction (five tubes) using chloroform : methanol : water (2 : 3 : 2). Combined aqueous layers (1.5 l) concentrated in vacuum. Chloroform layers discarded.

Active syrup (4.5 g)
↓ Reversed phase partition chromatography using *n*-butanol : water. Deoxycrustecdysone (2-deoxy-20-hydroxyecdysone, 19) fractions combined.

Crustecdysone fractions (400 mg)
↓ CM-Sephadex chromatography

Crustecdysone fractions (20 mg)
↓ Silica gel chromatography using chloroform : ethanol (85 : 15)

Crustecdysone (20-hydroxyecdysone, 32, 4 mg).

potassium hydrogen carbonate to the first transfer of a counter-current distribution (Horn, D. *et al.*, 1968). Karlson, P. *et al.* (1963) removed polar lipids from the *n*-butanol extract by chromatography on alumina. However, high losses of activity occurred when the hormone was present in the extract at a relatively low concentration (Horn, D. *et al.*, 1966). Instead the polar lipids were removed by a simple countercurrent distribution (CCD) of two to three transfers using chloroform : methanol : water (5 : 4 : 4) without loss of activity (Horn, D. *et al.*, 1968). However, care had to be taken to avoid the formation of emulsions. If the methanol was added to the aqueous extract first, and the extracts were rolled together instead of shaken, emulsions could be largely avoided and completely cleared by additional small proportions of methanol.

Where the yield of recovered hormone was of lesser importance, as in the isolation of ecdysteroids from plants, chromatography under carefully controlled conditions on deactivated alumina (15% water) proved very effective for the preliminary purification of crude extracts of ecdysteroids (see section 4.2). Alumina chromatography could be useful for the separation of moulting hormone precursors from the polar lipid fraction of arthropod extracts.

Karlson, P. (1956b) further purified an insect extract after alumina chromatography by CCD using *n*-butanol : cyclohexane : water (6 : 4 : 10). Subsequently CCD was used extensively by various groups (Kaplanis, J. *et al.*, 1966b, 1972, 1973, 1976; Horn, D. *et al.*, 1968; Galbraith, M. and Horn, D., 1969). Some of the partition coefficients of the hormones in various solvent systems are reported in Table 6. Although the resolution obtained with CCD does not compare with that of partition chromatography, CCD is unsurpassed for handling large amounts of crude natural products containing a wide range of animal or plant constituents. In such cases relatively few transfers can effect a considerable separation of the hormones from other components of the crude extract. To avoid solute interactions it is important that the concentration of

Table 6: Partition coefficients of some ecdysteroids in various solvent systems

Ecdysteroid	K*	Reference
A. cyclohexane–*n*-butanol–water (5:5:10)		
Ecdysone (24)	3.45	Kaplanis, J. *et al.* (1966b)
20-Hydroxyecdysone (32)	0.52	Kaplanis, J. *et al.* (1966b)
3-Epi-20-hydroxyecdysone (30)	0.52	Kaplanis, J. *et al.* (1975)
Makisterone A (44)	1.27	Kaplanis, J. *et al.* (1975)
20,26-Dihydroxyecdysone (23)	0.06	Kaplanis, J. *et al.* (1975)
26-Hydroxyecdysone (33)	0.39	Kaplanis, J. *et al.* (1973)
B. Cyclohexane-*n*-butanol–Water		
Ecdysone (24)	1.28	Karlson, P. (1956a,b)
	1.27	Kaplanis, J. *et al.* (1966b)
20-Hydroxyecdysone (32)	0.13	Karlson, P. (1956a,b)
	0.16	Kaplanis, J. *et al.* (1966b)
C. *n*-Butanol–water (1:1)		
Ecdysone (24)	10 approx.	Karlson, P. *et al.* (1963)
20-Hydroxyecdysone (32)	5.3	Horn, D. (1971)
D. Ethyl acetate–water (1:1)		
Ecdysone (24)	0.32	Kaplanis, J. *et al.* (1966b)
20-Hydroxyecdysone (32)	0.06	Kaplanis, J. *et al.* (1966b)
E. Chloroform–methol–water (2:1:1)		
Ecdysone (24)	0.5	Horn, D. (1971)
20-Hydroxyecdysone (32)	0.2	Horn, D. (1971)
F. Chloroform–Ethanol–Water (1:1:1)		
20-Hydroxyecdysone (32)	1.4	Horn, D. (1971)

*K = Concentration in non-polar phase/concentration in polar phase.

solutes in any phase does not exceed 2–3% (w/v).

An analytical procedure for the isolation of ecdysone (24) and 20-hydroxyecdysone (32) from *Calliphora* pupae using the above procedures was reported by Horn, D. *et al.* (1974a). Borst, D. and O'Connor, J. (1974) described a simplification of this method using TLC as a final step in the preparation of an insect extract for analysis by gas–liquid chromatography (GLC). However, in our experience and that of others (Morgan, E. and Poole, C., 1976a) ecdysteroids are recovered in poor yields from TLC plates, especially when the concentration of the hormone in the extract is low. To overcome this difficulty and avoid the use of chloroform which interferes with the response of the electron capture detector (ECD), Morgan, E. and Poole, C. (1976a) have introduced a further modification, which involves an extraction with ethyl acetate instead of chloroform to remove polar lipids. For high-pressure liquid chromatography (HPLC) Holman, G. and Meola, R. (1978) developed another method using acetonitrile as the extracting solvent, which was reported to extract less lipid constituents and largely avoid the problem of emulsification.

For the determination of ecdysone (24) in insect material Karlson, P. and Shaaya, E. (1964) used a simple isolation procedure which depended on the extraction of the hormone with hot water. In this way soluble proteins were denatured and most of the non-polar lipids remained with the undissolved residue. Lafont, R. *et al.* (1982) have modified this procedure using chloroform also, to assist in the removal of lipids. The aqueous extract was then passed through a short column of C-18 reversed-phase support (Sep-pak C-18 cartridge) to separate the ecdysteroids from water-soluble impurities. Recovery of the ecdysteroids for HPLC analysis was then accomplished by elution with aqueous methanol. This procedure — which is suitable for samples containing more than 30 ng g^{-1} ecdysteroid, though not more efficient than that of Holman, G. and Meola, R. (1978) — is reported to be much faster and provides a simple means of preparing hormone extracts for HPLC analysis.

4.2 Plants

Ecdysteroids can be isolated from plants by procedures similar to those used for arthropods. Generally the concentrations of ecdysteroids in plant extracts are much higher than those in arthropods (see Tables 3 and 4) but most plant extracts contain considerable proportions of phenolic impurities. These can be removed by the judicious extraction of a preliminary *n*-butanol extract with dilute aqueous sodium hydroxide and subsequent neutralization with acetic acid to avoid the formation of artefacts (Galbraith, M. and Horn, D., 1969). When the proportions of phenols in the extract are not excessive and ecdysteroids containing labile lactone functions are present, the phenols are best removed by chromatography on deactivated neutral alumina. Thus alumina chromatography was used to isolate sengosterone (68), which has a side-chain lactone group (Hikino, H. *et al.*, 1970c). Chart 2 shows a general procedure used at the CSIRO, Australia, to isolate and identify ecdysteroids in plant samples. An aliquot of the extract purified in this way was examined by TLC and was pure enough for the ecdysteroids to be clearly distinguished. This procedure was used to screen several hundred Australian and New Guinean plants for ecdysteroids.

Removal of the phenolic materials from the plant extract can also be accomplished by chromatography on polyamide columns (Hänsel, R. *et al.*, 1965; Rimpler, H. and Schulz, G., 1967; Jizba, J. *et al.*, 1967a). Imai, S. *et al.* (1967) have devised a general scheme for the isolation of plant ecdysteroids which depends solely on solvent

CHART 2: Flow diagram of the isolation of ecdysteroids from plants. (CSIRO Procedure.)

Dried and milled sample (10 g)

↓ Extracted with 96% ethanol (200 ml). Extract dried in vacuum and residue partitioned between hexane and 75% ethanol (10 ml of each phase). Hexane extract discarded.

Ethanol phase

↓ Concentrated in vacuum and partitioned between chloroform : ethanol : water (2 ml of each phase). Aqueous phase discarded.

Chloroform phase

↓ Concentrated to dryness and dissolved in ethyl acetate : ethanol (2 : 1) as a 5% solution and filtered through neutral alumina (10% H_2O, 2 g) and eluted with further solvent (25 ml). The total eluate was evaporated to dryness.

Crude ecdysteroids

Dissolved in chloroform : ethanol (2 : 1 to make a 5% solution w/v) and an aliquot (2 μl) transferred to a TLC plate.

partitioning, while Hori, M. (1969) introduced a simple reversed-phase chromatographic procedure using Amberlite resin XAD-2 for the isolation of ecdysteroids from crude plant extracts. However, in some instances it was necessary to use a preliminary purification step before chromatography on the resin. Canonica, L. *et al.* (1975) reported a novel procedure for isolating phytoecdysteroids from Kaladana seeds, while Hardman, R. and Benjamin, T. (1977) have developed an isolation procedure using polyamide chromatography and preparative TLC.

5 SEPARATION AND ANALYSIS

5.1 Partition chromatography (liquid–liquid chromatography, LLC)

Partition chromatography, in which the sample is separated by being partitioned between a non-polar eluting solvent and water retained in a solid support such as cellulose, has not been favoured for the separation of ecdysteroids. However, it should be mentioned that partition chromatography plays a significant role in the case of adsorbents deactivated with large proportions of water (10–15%), for example alumina (Karlson, P. *et al.*, 1963) or silica gel (Galbraith, M. and Horn, D., 1969). Lafont, R. *et al.* (1982) reported the use of the polar bonded HPLC supports Lichrosorb-NH_2 and Lichrosorb-Diol for the separation of ecdysone (24) and 20-hydroxyecdysone (32) by normal partition.

Reversed-phase partition chromatography has proved much more useful for the separation of ecdysteroids and was first introduced by Horn, D. *et al.* (1968) together with a UV detector to monitor the column effluent. Cyclohexane-*n*-butanol (1 : 7) coated onto hydrophobic celite was used as the stationary phase. Water, equilibrated with the stationary phase, was used as the flowing phase. This system was effective in separating 20-hydroxyecdysone (32) and 2-deoxy-20-hydroxyecdysone (19) (Galbraith, M. *et al.*, 1968b), and ecdysone (24) and 20-hydroxyecdysone (32) (Horn, D. *et al.*, 1974a). Although these reversed-phase columns worked well, and were of higher capacity than those used later, gradient elution was difficult to achieve without bleeding of the stationary phase and the columns could not be re-used. These problems were overcome by using an Amberlite XAD-2 resin as stationary phase (Imai, S. *et al.*, 1968a; Hori, M., 1969; Schooley, D. *et al.*, 1972) or Poragel PN resin (Schooley, D. and Nakanishi, K., 1973) as hydrophobic supports. Such column packings have since been superseded by the more efficient HPLC octadecyl supports for analytical separation (Holman, G. and Meola, R., 1978; Gilgan, M., 1976; Ogawa, S. *et al.*, 1977; Lafont, R. *et al.*, 1979, 1980a,b; Dinan, L. *et al.*, 1981; Wilson, I. *et al.*, 1982a,c). A comparison of HPLC and GLC for analysis of ecydsteroids was made by Wilson, I. *et al.* (1980, 1981) and the use of phytoecdysteroids as internal standards evaluated (Wilson, I. *et al.*, 1982b). Details of the use of reversed-phase chromatography for the separation of ecdysteroids can be found in the reviews of Morgan, E. and Poole, C. (1976a), Koreeda, M. and Teicher, B. (1977) and Lafont, R. *et al.* (1982).

5.2 Sephadex chromatography

CM-Sephadex, which shows some reversed-phase and adsorption properties, was found to be invaluable for the separation of both 2-deoxy-20-hydroxyecdysone (19) and 20-hydroxyecdysone (32) from chromatographically similar polar constituents of partially purified crayfish hormones (Horn, D. *et al.*, 1968, 1974a; Galbraith, M. *et al.*, 1969b) and also permitted the separation of 20-hydroxyecdysone (32) and inokosterone (38) (Thomson, J. *et al.*, 1969). Sephadex LH20 chromatography with chloroform–ethanol mixtures as eluting solvent is also invaluable for separating ecdysteroids and for separating them as a group from aromatic compounds of similar polarity (Russell, G. *et al.*, 1972), the aromatic compounds being more strongly retained by this material than ecdysteroids.

5.3 Adsorption chromatography (solid–liquid chromatography, SLC)

Silica gel was found (Horn, D. *et al.*, 1968) to be preferable to alumina for the isolation of

20-hydroxyecdysone (32) and separation of ecdysteroids. Initially gravity flow columns were extensively used (Thompson, M. *et al.*, 1967; Galbraith, M. *et al.*, 1969a; Galbraith, M. and Horn, D., 1969; Kaplanis, J. *et al.*, 1972). In general it was found preferable to use silica gel deactivated with large proportions of water (5–10%) for the chromatography of ecdysteroids because of the strong affinity of these polar compounds for the adsorbent. Also the addition of water to silica gels of larger, less uniform particle size provides an adsorbent with a more linear adsorption isotherm, so that the ecdysteroids are eluted in more symmetrical peaks with less tailing.

Column chromatography was for a time displaced in popularity by TLC (see section 5.4) which is very simple, inexpensive and gives excellent separations. However, it was soon found (Horn, D. *et al.*, 1968) that with column chromatography, especially with small amounts of material, the recovery was higher, there were fewer manipulation losses and the column could be more easily washed free of fine particles and soluble impurities. Resolutions as good as those obtained by TLC can be achieved with preparative gravity columns if the adsorbent is fine, of even particle size and if high ratios of adsorbent to material chromatographed are used. With the advent of inexpensive pumping equipment excellent resolution was obtained with moderate pressures on a preparative scale with silica gel prepared by sedimentation of commercial TLC silica gel (Kinnear, J. *et al.*, 1979).

In recent years solid–liquid HPLC has been extensively used for the separation of ecdysteroids both on a preparative and analytical scale (Nigg, H. *et al.*, 1974a,b; Gilgan, M., 1976; Bollenbacher, W. *et al.*, 1976, 1979; Mayer, R. and Svoboda, J., 1978; Moribayashi, A. and Ohtaki, T., 1978; Lafont, R. *et al.*, 1979, 1980a,b). Detailed reviews of HPLC have been reported by Morgan, E. and Poole, C. (1976a) and Lafont, R. *et al.* (1981).

5.4 Thin-layer chromatography (TLC)

Thin-layer chromatography on silica gel was introduced in the early 1960s and, together with insect bioassays, played an important role in the detection of ecdysteroids in chromatographic fractions. Ecdysteroids on TLC plates give characteristic colour reactions after being sprayed with a reagent (Stahl, E., 1962) consisting of vanillin (1.5 g) in ethanol (50 ml) containing sulphuric acid (1 g) and being heated in an oven at 110° for 5 min (see Table 7). Some ecdysteroid analogues require longer heating or higher temperatures for the colours to develop.

One of the most convenient solvent systems for TLC is 96% ethanol–chloroform (20 : 80). The R_f values of some ecdysteroids are shown in Table 7. TLC makes possible the identification of a wide variety of ecdysteroids with relatively small amounts of material (0.5–10 μg) (Ruh, M. and Black, C., 1976). When fluorescent silica gel plates are used, ecdysteroids are clearly distinguished under UV illumination as dark spots on a fluorescent background. This effect can be used to assay for ecdysteroids (Mayer, R. and Svoboda, J., 1978) and has a similar sensitivity to the method using the vanillin spray reagent.

TLC has also been used extensively for the purification of crude insect and plant extracts and synthetic products (e.g. Hardman, R. and Benjamin, T., 1977). However, it is important to remember that the capacity of the plates is low and effective separation of ecdysteroids can be obtained only if the plates are not overloaded. For this reason the technique is not very useful for the purification of crude extracts which contain relatively small proportions of ecdysteroids. In some instances its use for this purpose has led to confusing results. Column chromatography is certainly superior, as mentioned above.

TLC has proved very useful for the determination of the rate of acetylation of ecdysteroids (Galbraith, M. *et al.*, 1968b). The sample to be studied is reacted with acetic anhydride–pyridine reagent (1 : 2, v/v) and the reaction stopped at selected intervals of time. Aliquots are then run on TLC plates and, after spraying, the intensity of the spots assessed in comparison with standards. This method enables information to be obtained about the configuration or the steric environment of hydroxy groups, even when only microgram quantities of material are available for study as in the case of deoxycrustecdysone (2-deoxy-20-hydroxyecdysone, 19) where the 3-hydroxy group was shown to be axial with the use of only 10 μg of material (Galbraith, M. *et al.*, 1968b). However, TLC has largely been replaced for this purpose by HPLC which gives better resolution and a more accurate measure of the products formed (Bollenbacher, W. *et al.*, 1979).

Comparisons of R_f values and colour reactions of ecdysteroids with the vanillin spray reagent are reported in Table 7 and by Horn, D. (1971), Morgan, E. and Poole, C. (1976a) and Koolman, J. and Karlson, P. (1978), while tables of R_f values of a number of ecdysteroids in various solvent systems are reported by Hikino, H. and Hikino, Y. (1970), Hikino, H. *et al.* (1976b), Sardini, D. and Krepinsky, J. (1974), Kaplanis, J. *et al.* (1976, 1979, 1980), Dinan, L. and Rees, H. (1978), Koolman, J. and Karlson, P. (1978); Koolman, J. *et al.* (1979a); Ikekawa, N. *et al.* (1980), Wilson, I. *et al.* (1981) and Isaac, R. *et al.* (1981a,b).

5.5 Gas–liquid chromatography (GLC)

As ecdysteroids are highly hydroxylated polar steroids they were not at first considered suitable for GLC. However, following the pioneering studies, in

Table 7: TLC colour reactions[a] and R_f values of some ecdysteroids on silica gel[b]

Ecdysteroid	Colour reaction	$R_f \times 100$
A. Developed with Chloroform: 96% ethanol (80:20, v/v)		
20,26-Dihydroxyecdysone (23)	brown-mauve then green	5
20-Hydroxyecdysone (32)	olive-green then yellow-brown	13
Inokosterone (38)	yellow-brown	13
Polypodine B (54)	green then brown	14
20-Hydroxy-22-deoxyecdysone (170)	grey-green	16
Makisterone A (44)	mauve-brown	18
Ecdysone (24)	blue then red-brown	21
Pterosterone (66)	green then yellow-brown	25
Muristerone A (48)	mauve	27
22-Deoxyecdysone	blue	29
2-Deoxy-20-hydroxyecdysone (19)	olive-green then brown	29
Ponasterone C (60)	green then grey	34
Ponasterone A (58)	mauve then green-grey	38
2-Deoxyecdysone (18)	blue then red-brown	39
2,22,25-Trideoxyecdysone (78)	pink then fades	44
Ponasterone B (59)	blue-mauve then brown	52
Acetates of 20-hydroxyecdysone (32)		
22-Acetate	green	23
2- and 3-Acetate	olive-green then brown	36
2,22- and 3,22-Diacetate	green	51
2,3-Diacetate	olive-green then brown	60
2,3,22-Triacetate	green	79
2,3,22,25-Tetracetate	green	90
Acetates of Ecdysone (24)		
22-Acetate	blue	28
2- and 3-Acetate	blue	47
2,22- and 3,22-Diacetates	blue	58
2,3-Diacetate	blue	70
2,3,22-Triacetates	blue	84
2,3,22,25-Tetraacetate	blue	93
B. Developed with Chloroform: 96% ethanol (90:10, v/v)		
2-Deoxy-20-hydroxyecdysone (19)	olive-green then brown	6
2-Deoxyecdysone (18)	blue then red-brown	12
22,25-Dideoxyecdysone	pale orange	19
14,22,25-Trideoxyecdysone	yellow	40
2,22,25-Trideoxyecdysone (78)	pink	41
20-Hydroxyecdysone 2,3,22-triacetate	green	43
Ecdysone 2,3,22-triacetate	blue	48
2,14,22,25-Tetradeoxyecdysone (77)	yellow	60

[a] Plates heated at 100° for 5 min after spraying with vanillin–sulphuric acid reagent (Stahl, E. 1962).
[b] Kieselgel HF_{254} Typ 60, Merck, Darmstadt; 0.2 mm thick.

the United States, of Horning and collaborators (Vandenheuvel, W. and Horning, E., 1967) it was recognized that such compounds could be analysed by GLC if the chromatographic support was carefully deactivated, coated with only a thin film of liquid phase and the compounds were made more volatile and hydrocarbon-like by converting their polar hydroxy groups to trimethylsilyl (TMS) ethers. GLC of ecdysteroids was first introduced by Katz, M. and Lensky, Y. (1970), Morgan, E. and Woodbridge, A. (1971, 1974), and Ikekawa, N. *et al.* (1972). At first a variety of silylating reagents and reaction conditions were used, with confusing results, and it become apparent that carefully controlled silylating conditions and pure reagents were required to obtain a single product and thus a single chromatographic peak. These difficulties led Morgan, E. and Poole, C. (1976b) and Bielby, C. *et al.* (1980) to undertake a study of the products formed by silylation of ecdysteroids under different reaction conditions. They showed that the order of reactivity of the hydroxy groups is: 2, 3, 22, 25 > 20 ≫ 14. At room temperature the reaction of 20-hydroxyecdysone (32) with TMS-imidazole affords the 2,3,22,25-tetrasilyl ether while heating to 100° for 4 h leads to the 2,3,20,22,25-pentasilyl ether. Only by heating to 140° for 18 h (Galbraith, M. *et al.*, 1968a, 1969a) is complete silylation of the 14α-hydroxy group obtained. More reactive reagents can be used to obtain complete silylation but these lead to the formation of unwanted enol–silyl ethers of the unsaturated ketone group (Ikekawa, N. *et al.*, 1972). In view of the poor reactivity of the 14α-hydroxy group it is preferable to use mild silylating conditions and leave the 14α-hydroxy group unchanged (Morgan, E. and Poole, C., 1976b).

The flame ionization detector has been used extensively in GLC. It is reliable but not sensitive enough for the detection of the small amounts of ecdysteroids present in arthropods. To overcome this problem Ikekawa, N. *et al.* (1972) prepared mixed trimethylsilylheptafluorobutyl derivatives of ecdysone (24) so that they could be detected by means of the much more sensitive electron capture detector (ECD). Poole, C. *et al.* (1975) and Poole, C. and Morgan, E. (1975) discovered that TMS ethers of ecdysteroids themselves show a response in the ECD enabling their detection in the range of $5–700 \times 10^{-12}$ g, which is sensitive enough for insect studies (Gande, A. *et al.* 1979).

Steroids as a class do not show electron capturing properties but ecysteroids are exceptional in this regard. The electron capturing properties are due mainly to the presence of the 7-en-6-one grouping, and to some extent the 14α-hydroxy group and the 2- and 3-TMS ethers. Samples monitored by ECD must be carefully prepared and free from electron capturing impurities such as halogenated hydrocarbons. The detectors also require skilled maintenance for optimum sensitivity. GLC with an ECD is more sensitive than HPLC (Wilson, I. *et al.*, 1980).

The modern mass spectrometer can be used as a very sensitive and quantitative gas-chromatographic detector (Ikekawa, N. *et al.*, 1972). This is accomplished by passing the GLC effluent through a semi-permeable membrane or jet separator to remove the carrier gas from the sample and deliver it to the ionization chamber. Spectra of ecdysteroids can then be taken at intervals of time and recorded. This technique has proved invaluable for the identification and characterization of small amounts of ecdysteroids and precursors present in insect tissues (Hetru, C. *et al.*, 1978) and phytoecdysteroids (Hardman, R. and Benjamin, T., 1980).

Alternatively the mass spectrometer can be set to record ions of a specific mass–charge ratio throughout the GLC run. The output from the mass spectrometer is then recorded to provide chromatographic peaks in the usual way. This mass-fragmentographic analysis of ecdysteroid TMS ethers developed by Miyazaki, H. *et al.* (1973) and Chino, H. *et al.* (1974) is useful to monitor ecdysteroid levels in insects (Lafont, R. *et al.*, 1975; Bouthier, A. *et al.*, 1975; Delbecque, J. *et al.*, 1975; Imboden, H. *et al.*, 1978). The abundant ion at m/e 561, corresponding to the fission of the C20–C22 bond, is useful to monitor ecdysteroids with a 20,22-diol function and provides a highly specific means of identifying these compounds in mixtures from insect tissues. Detailed information about the GLC of ecdysteroids can be found in a number of reviews (Morgan, E. and Poole, C., 1976a,b; Koreeda, M. and Teicher, B., 1977; Poole, C. *et al.*, 1980; Morgan, E. and Wilson, I., 1980; Lafont, R. *et al.*, 1980a; Brooks, C. and Gaskell, S., 1980).

6 PHYSICAL METHODS OF CHARACTERIZATION

6.1 General properties

Ecdysteroids are high melting solids which are readily crystallized from ethyl acetate–methanol mixtures or methanol. Often their crystal structures contain 0.5–1 molecule of water per molecule of steroid, and because of this some compounds have double melting points. Melting points of most ecdysteroids are given in Table 8. Many ecdysteroids are soluble in water because of their numerous hydroxy groups. However, those without a terminal 25- or 26-hydroxy group, such as ponasterone A (58), are sparingly soluble in water and, probably because of this fact, show an unusually strong affinity for ecdysteroid receptors *in vitro*.

6.2 Colour reactions

Although ecdysteroids are colourless solids they react with zinc chloride in acetic acid (Chugaev reaction) to give a solution with an absorption maximum at 380 nm, which can be used for quantitative analysis (Kholodova, Y., 1977). Ecdysteroids give characteristic colour reactions with the vanillin–sulphuric acid reagent and this behaviour has proved invaluable in detecting them on TLC plates (see section 5.4).

6.3 Fluorometry

Solutions of ecdysteroids in mixtures of ethanol and sulphuric acid fluoresce strongly when excited with light of a suitable wavelength (Takemoto, T. *et al.*, 1968d) and this effect has been used (Gilgan, M. and Zinck, M., 1972; Mayer, R. and Svoboda, J., 1978; Koolman, J., 1980) to assay for ecdysteroids in the range of 25–2500 ng. Unfortunately the fluorescence is not specific to ecdysteroids and thus cannot be used to determine ecdysteroids in crude insect extracts (Morgan, E. and Wilson, I., 1980). However, Koolman was able, using this method, to determine the concentration of ecdysone (24) and 20-hydroxyecdysone (32) in the blowfly larvae after a careful chromatographic purification of the extract. Inokosterone (38) fluoresces more strongly than 20-hydroxyecdysone (32) in ammonia solution and this behaviour allows the proportions of these compounds to be determined in a mixture of the two (Takemoto, T. *et al.*, 1968f).

6.4 Ultraviolet spectroscopy (UV)

Ecdysteroids are characterized by a strong UV absorption (ε_{max} 10,000–14,000, λ_{max} 236–245 nm in ethanolic solution). With most instruments using cells of about 4 ml capacity, roughly 100 μg of compound is required for an accurate determination. However, it is possible to identify amounts as little as 1 μg. Much less (5–50 ng) can be detected under ideal conditions using HPLC (section 5.3) with a UV detector measuring at 240 nm. In practice most detectors operate at 254 nm and are less sensitive.

6.5 Optical rotatory dispersion (ORD) and circular dichroism (CD)

Ecdysteroids are optically active and show small rotations (about $+60°$) in the visible region but near their UV absorption maximum show larger values (about $+4500°$) with two Cotton effects. The amplitudes of the Cotton effects of ecdysteroids containing a *cis*-fused A/B-ring system are less than those containing the corresponding *trans*-fused system (Djerassi, C., *et al.*, 1964; Nakanishi, K. *et al.*, 1968) and are useful to distinguish them (Scherrer, H., 1969). The ORD spectra of 6-ketosteroids, which do not have a 7-double bond, show negative Cotton effects (Wiechert, R. *et al.*, 1966; Mori, H. *et al.*, 1968a) with larger amplitudes in the 5α-series than the corresponding 5β-series.

Nakanishi, K. (1971) has found that 5β-hydroxy ecdysteroids show characteristic differences in their Cotton effects. He also devised a means of determining the configuration of several hydroxy groups of ecdysteroids by measuring the CD spectra of their benzoate derivatives (Koreeda, M. *et al.*, 1969b). Prakash, A. and Ghosal, S. (1979) have tabulated the amplitudes of the Cotton effects of many ecdysteroids.

6.6 Infrared spectroscopy (IR)

Most ecdysteroids show characteristic strong OH stretching bands at 3330–3400 cm^{-1}, a $C=O$ stretching band at 1640–1667 cm^{-1} with a weaker $C=C$ stretching band at about 1612 cm^{-1}.

Cyasterone (12) (Takemoto, T. *et al.*, 1967b) and sengosterone (68) (Hikino, H. *et al.*, 1969a) are characterized by strong bands at 1752 and 1748 cm^{-1} respectively, ascribed to γ-lactone rings in their side-chains. Capitasterone (8) shows a strong absorption at 1730 cm^{-1} ascribed to the δ-lactone group (Takemoto, T. *et al.*, 1968b) while ajugalactone (1) has an absorption at 1696 cm^{-1} attributed to its unsaturated δ-lactone group (Koreeda, M. and Nakanishi, K., 1970a). Calonysterone (7) with an unusual conjugated unsaturated 1,2-diketone system, shows IR absorptions at 1638, 1619 and 845 cm^{-1} (Canonica, L. *et al.*, 1973a). Nakanishi, K. *et al.* (1968) made use of IR spectroscopy to determine the configuration of the A-ring hydroxyls of ponasterone B (59).

6.7 Mass spectrometry

Mass spectrometry has played an important role in the determination of the structure of ecdysteroids. The spectra of ecdysteroids are characterized by numerous ions 18 mass units apart, corresponding to successive losses of molecules of water from the parent molecule or from fragments formed by side-chain cleavage. The spectra thus provide valuable information about the molecular weight, the number of hydroxy groups present, as well as structural features.

Compounds such as 20-hydroxyecdysone (32), that contain a 20,22-vicinal diol function, undergo facile fission of the C20—C22 bond leading to major ions *m/e* 363 and 345 corresponding to the tetracyclic nucleus and ions at *m/e* 99 and 81 arising from the side-chain fragment (Hampshire, F. and Horn, D., 1966). These side-chain fragments formed from a number of phytoecdysteroids have been tabulated by Nakanishi, K. (1971).

Compounds such as ecdysone (24) that have a hydroxy group at the C20 or C22 position only, undergo side-chain fission less readily and the spectra are more complex with ions due to C17—C20 bond fragmentation becoming more abundant. Rashkes, Y. and Abubakirov, N. (1980) made a study of 15 phytoecdysteroids and noted that with decreasing number of —OR groups the contribution of cleavages of bonds of the steroid skeleton increases. In all spectra the fragmentation

Table 8: Melting points and [^1H] NMR chemical shifts of the methyl signals of some ecdysteroids measured with perdeuteropyridine solutions at about 38°

Ecdysteroid	Melting point (°)	C18	C19	C21	C26	C27	Other
Ajugalactone (1)	225–235 dec.	1.49	1.12	1.65	1.87		C29, 0.74
Ajugasterone B (2)	240 dec.	1.16	1.05	1.54	5.03	4.36	C29, 0.86
					5.53		
Ajugasterone C (3)		1.21	1.27	1.51	0.82	0.82	
Amarasterone A (5)	210–211	1.21	1.06	1.56		1.11	C29, 0.92
Amarasterone B (6)	284–285	1.22	1.07	1.56	0.91	0.91	
Anhydroturkesterone	256–257	1.10		1.45	1.24	1.24	
Calonysterone (7)	234–235	1.40	1.72	1.56	1.42	1.42	
Capitasterone (8)	234–235	1.13	1.07	1.47		1.31	C29, 0.72
Cheilanthone A (10)	235–238	0.83	1.01	1.24	1.36	1.36	
Cheilanthone B (11)	225–228	0.85	1.03	1.24	0.84	0.84	
Cyasterone (12)	164–166	1.19	1.06	1.51		1.33	C29, 1.33
Cyasterone-22-acetate (13)	212–215	0.95	1.06	1.47		1.21	C29, 1.32
Dacrysterone (14)	283–285	1.22	1.15	1.56	1.32	1.32	C28, 107
24(28)-Dehydromakisterone A (17)	244–246	0.95	1.08	1.46	1.40	1.46	C28, 4.97, 5.14
2-Deoxyecdysone (18)	231–232	0.74	1.05	1.28	1.38	1.38	
25-Deoxyecdysone		0.72	1.06	1.25	0.83	0.83	
2-Deoxy-3-epi-20-hydroxyecdysone	234–237	1.23	0.99	1.59	1.38	1.38	
2-Deoxy-20-hydroxyecdysone (19)	250–252	1.21	1.04	1.57	1.35	1.35	
22-Deoxy-20-hydroxyecdysone	131–134	1.14	1.06	1.54	1.35	1.35	
22,25-Dideoxyecdysone	207–209	0.71	1.07	1.00	0.85	0.85	
20,26-Dihydroxyecdysone (23)	149–153	1.22	1.08	1.58		1.48	
Ecdysone (24) (α-Ecdysone)	237–239	0.70	1.05	1.25	1.38	1.38	
Epicyasterone (26)	274–275	1.23	1.41	1.57	1.35	1.35	C29, 1.38
3-Epi-20,26-dihydroxyecdysone (28)		1.22	1.07	1.58	1.48		

Table 8: *Continued*

Ecdysteroid	Melting point (°)	C18	C19	C21	C26	C27	Other
3-Epi-ecdysone (29)		0.75	1.07	1.23	1.40		
3-Epi-20-hydroxyecdysone (30)		1.22	1.07	1.58	1.38	1.38	
20-Hydroxyecdysone (32)	241–242.5	1.20	1.07	1.56	1.38	1.38	
26-Hydroxyecdysone (33)	252–256	0.74	1.08	1.23	1.47		
20-Hydroxyecdysone 2-cinnamate (36)	258–260	1.21	1.11	1.57	1.38	1.38	
20-Hydroxyecdysone 3-p-coumarate (37)	265–267	0.90	1.03	1.16	1.16	1.16	
Inokosterone (38)	255 dec.	1.19	1.07	1.52		1.03	
Integristerone A (39)	244–246	1.07	1.26	1.43	1.24	1.24	
Integristerone B (40)	186–190	1.09	1.43	1.43	1.25	1.25	
Isocyasterone (41)		1.16	1.04	1.52	1.24	1.24	C29, 1.19
22-Iso-ecdysone	251–254	0.79	1.06	1.23	1.41	1.41	
Kaladasterone (42)	242–243						
Makisterone A (44)	263 dec.	1.21	1.09	1.54	1.29	1.32	C28, 1.05
Makisterone B (45)	172–173	1.16	1.04	1.54		1.00	C28, 0.90
Podecdysone A (46) (Makisterone C)	262–265	1.21	1.07	1.55	1.26	1.39	
Makisterone D (47)		1.20	1.06	1.57	0.87	0.97	C29, 1.30
Muristerone A (48)	238–241	1.24	1.38	1.53	0.82	0.82	
Podecdysone B (52)	125–127	1.35	1.08	1.55	1.43	1.43	
Polypodine B (54) (Ajugasterone A)	254–257	1.19	1.10	1.55	1.35	1.35	
Polypodine B 2-cinnamate (55)	268–270	1.19	1.19	1.55	1.38	1.38	
Ponasterone A (58)	259–260 dec.	1.16	1.03	1.51	0.82	0.82	
Ponasterone C (60)	270–272	1.17	1.12	1.54	1.00	1.00	
Ponasterone C 2-cinnamate (61)		0.87	0.92	1.22	0.92	0.92	
Ponasteroside A (62)	278–279.5						
Poststerone (63)	232–235	0.68	1.00	2.12			
Precyasterone (65)		1.11	1.05	1.44	1.45		C29, 1.24
Pterosterone (66)	229–230	1.18	1.05	1.54	1.00	1.00	
Rubrosterone (67)	240–244	0.85	1.03				
Sengosterone (68)	159–161	1.21	1.13	1.56		1.36	C29, 1.34
Sidasterone A (69)	238–240						
Sidasterone B (70)	235–237						
Sogdysterone (72)		1.06		1.46	1.25	1.25	
Stachysterone A (73)			0.92	1.51	1.41	1.41	C14 Me 1.20
Stachysterone B (74)		1.33	1.00	1.49	1.42	1.42	
Stachysterone C (75)	235–240	1.05	1.17	1.55	1.60	1.66	
Stachysterone D (76)	245–250	1.20	1.06		1.37	1.37	
2,14,22,25-Tetradeoxyecdysone (77)	176–177						
Turkesterone (79)		1.23	1.29	1.56	1.35	1.35	
Viperidinone (80)	197–198						
Viperidone (81)	208–210						

For reference to naturally occurring ecdysteroids and synthetic analogues see Tables 1 and 10 respectively.

of the side-chain takes place at the C22—C23, C23—C24 and C24—C25 bonds. In spectra of 20-hydroxyecdysteroids fragmentation of the D-ring at C13—C17 and C14—C15 takes place.

3-Hydroxyecdysteroids in the 5β-series show in their mass spectra a greater tendency to lose one molecule of water as compared with the 5α-series and this behaviour is independent of the configuration of the hydroxy group (Furlenmeier, A. *et al.*, 1966b). Koreeda, M. *et al.* (1969a) have made a detailed study of the fragmentation of ecdysteroid acetates and acetonides. Various derivatives of ecdysteroids have been used for GC–MS and are discussed in the section on GLC (section 5.5).

6.8 Nuclear magnetic resonance (NMR)

Generally most of the proton resonances (CH and CH_2 signals) of ecdysteroids are not well resolved but the signals of the methyl groups usually appear in δ-pyridine solution as separated sharp peaks that provide valuable structural information. The C7 proton appears at about δ6.00 and the protons on carbons bearing hydroxy groups are located at somewhat higher field.

A hydroxy group at C20 produces a marked downfield shift in the signals of the C18 and C20 methyl groups (Hampshire, F. and Horn, D., 1966). Similarly the introduction of a 14α-hydroxy group produces a downfield shift in the chemical shift of the C18 methyl group. However, the loss of the 2-hydroxy group does not affect the chemical shift of the C19 methyl group. The chemical shifts of the C19 methyl signals of ecdysteroids (5β-H) in pyridine solution are 0.3 ppm downfield of the C19 methyl signals of the corresponding 5α-isomers (Furlenmeier, A. *et al.*, 1966b) and this difference provides a useful means of determining the configuration of the A/B-ring junction with relatively small amounts of these compounds. Smith, W. and Newsoroff, G. (1974) have found that the configuration at C5 in ecdysteroid analogues can be determined from the NMR spectra of the corresponding enol acetates since the C7 proton appears as a doublet in the 5α-form but as a singlet in the 5β-form.

Often the hydroxy protons of ecdysteroids appear as sharp signals, sometimes with well-defined splitting. They can usually be identified by exchange with small amounts of D_2O, though the HDO peak often obscures the signals in the middle field. The configuration of the 2- and 3-hydroxy groups can be assigned by examining the shape of the signals of the protons of the same carbons, since the axially oriented protons show greater coupling with protons on the adjacent carbons than the equatorial ones. The hydroxy groups can often be distinguished from one another by selective acetylation, since the protons adjacent to acetyl groups undergo the pronounced acyl shift. The chemical shift of the acetyl protons of the equatorial 2-acetyl group occurs at higher field (δ1.98–2.00) than the axial 3-acetyl or the 22-acetyl group (δ2.10, Furlenmeier, A. *et al.*, 1966b).

The NMR spectra of ecdysteroids are generally taken in deuteropyridine, deuteromethanol or deuterodimethyl sulphoxide because of their poor solubility in deuterochloroform. However, with the newer Fourier transform instruments it is possible to obtain good spectra with nanogram quantities of an ecdysteroid and thus spectra can be obtained in solvents in which the ecdysteroid is only sparingly soluble. The modern high field strength instruments (e.g. operating at 260 or 400 mHz) operate at a lower temperature (20°) than the earlier lower field instruments (38°) and considerable differences in the chemical shifts of the methyl groups are thus often observed. Spectra should be run at 38° with these instruments if a comparison with the older literature values is required.

Koreeda, M. *et al.* (1971) have found that irradiating the 9α-proton of ajugalactone 2,3-diacetate leads to a 9% increase in the integrated intensity of the 2α-proton (Nuclear Overhauser Effect, NOE), indicating that the A-ring adopts a chair conformation as in ecdysone (24).

Chemical shifts of the methyl signals of a large number of ecdysteroids taken in deuteropyridine are reported in Table 8 and in an earlier review (Horn, D., 1971). Other tables of chemical shift values of selected ecdysteroids are also available (Hikino, H. and Hikino, Y., 1970; Hikino, H. *et al.*, 1970a,b, 1971a-c, 1975b, 1976a,b; Imai, S. *et al.*, 1970a; Nakanishi, K. *et al.*, 1968; Canonica, L. *et al.*, 1975; Usmanov, B. *et al.*, 1975; Baltaev, U. *et al.*, 1977, 1978a,b).

The use of [^{13}C]NMR was first explored by Lukacs, G. and Bennett, C. (1972), who assigned the chemical shifts of ecdysone (24). The great advantage of [^{13}C] spectra as compared with [^{1}H] spectra is that almost all the carbon signals are resolved and thus a great deal more structural information can be obtained. Canonica, L. *et al.* (1973a,b) were the first to exploit the technique in the determination of the structures of kaladasterone (42) and calonysterone (7). The solution of the latter would have been difficult without its use. [^{13}C] NMR has also been used in the determination of the structure of shidasterone (stachysterone C, 75) (Hikino, H. *et al.*, 1975a), the configuration at C24 of makisterone A (44) (Danieli, B. *et al.*, 1974) and the configuration at C24 of dacrysterone (14) (Blunt, J. *et al.*, 1979). With the advent of high field strength superconducting magnets it has been possible to reduce the amount of sample necessary to run a spectrum; it is now possible to obtain good spectra with 1 mg or less.

The [^{13}C] chemical shift values of a number of ecdysteroids are reported in Table 9. Tables of [^{13}C] chemical shift values of selected ecdysteroids and analogues have been reported by various authors (Lukacs, G. and Bennett, C., 1972; Hikino, H. *et al.*, 1975a,b 1976a; Smith, W., 1977; Krepinsky, J. *et al.*, 1977; Blunt, J. *et al.*, 1979).

Table 9: ¹³C chemical shifts of some ecdysteroids in perdeutero-pyridine

	ECD[a] (24)	2OHE[a] (32)	POLYB[a] (54)	PNA[b] (58)	MAKA[a] (44)	PTER[c] (66)	MURA[a] (48)	KALA[a] (42)	DAC[c] (14)	PNC[c] (60)	POST[a] (63)	TRIOL[b] (–)[f]	EDOE[d] (27)	DOE[e] (18)
C1	38.2	37.8	38.4	37.9	37.8	37.7	36.2	38.6	34.7	34.6	38.0	37.9	35.6	33.1
C2	68.3	68.0	68.1A	68.0	68.0	68.0	68.2A	67.9A	67.9	67.8	68.0	68.0	31.4	29.7
C3	68.3	68.0	70.0A	68.0	68.0	68.0	69.9A	69.5A	69.7	69.7	68.0	68.0	69.1	64.1
C4	32.5	32.3	36.1	32.4	32.3	32.4	35.9	35.2	35.8	35.9	32.3	32.4	34.3	29.5
C5	51.5	51.1	80.0	51.3	51.1	51.4	80.0	79.7	79.7	79.7	51.5	51.3A	57.2	51.6
C6	203.7	203.2	201.2	203.5	203.1	203.6	201.1	201.1	200.7	200.8	202.8	203.4	201.9	203.2
C7	121.9	121.6	120.0	121.7	121.6	121.6	120.3	116.8	119.8	119.8	122.1	121.5	121.3	121.3
C8	165.8	165.7	167.0	166.0	165.7	166.1	165.0	156.0	166.8	166.8	163.9	166.7	165.8	166.0
C9	34.9	34.5	34.9	34.4	34.5	34.4	46.2	137.4	38.2	38.2	34.5	34.5	34.0	34.4
C10	38.9	38.5	44.9	38.7	38.1	38.6	45.5	45.4	44.7	44.6	38.0	38.6	36.8	37.0
C11	21.4	21.0A	21.0B	21.4	21.4	21.1	69.6	132.9	21.2	21.3	21.2A	20.9	20.8	21.0
C12	31.8A	31.5B	31.8	31.8	31.7A	31.7	44.0	38.6	31.6	31.5	31.9	31.7B	31.6	31.7
C13	47.9	48.1	48.2	48.1	48.1	48.0	48.2	47.4	48.1	48.0	48.1	47.2	47.6	48.0
C14	84.2	84.2	84.2	84.1	84.2	84.1	48.2	83.2	84.0	83.9	84.0	84.2	83.8	84.0
C15	32.2A	32.0B	32.3	31.9	32.3A	31.9	31.7	31.3	32.0	32.0	36.0	31.4B	31.9	31.7
C16	26.9B	21.5A	21.3B	21.4	21.4	21.6	21.4	21.5	22.0	22.0	22.0A	27.3	26.7	26.7
C17	48.6	50.1	50.1	50.0	49.9	50.0	49.4	49.9	49.8	49.8	59.4	51.1A	48.3	48.3
C18	16.1	17.7	17.3	17.9	17.8	17.8	1.74	26.3	17.8	17.8	17.1	15.8	15.8	15.8
C19	24.7	24.3	18.0	24.4	24.3	24.4	18.8	17.9	17.0	17.0	24.3	24.3	23.9	24.3
C20	43.2	76.9	77.0	76.7	76.9	76.7	76.8	76.5	76.9	76.9	208.6	36.0C	43.0	42.9
C21	13.9	21.5	21.5	21.1	21.2	21.4	21.4	21.2B	21.5	21.5	31.1	19.2	13.7	13.6
C22	74.4	77.5	77.7	76.7	74.8	76.7A	76.8	76.8	74.7	76.7B		36.7C	74.0	73.9
C23	25.9B	27.2	27.6	30.2	34.5	35.6	30.1	30.2B	34.5	35.6		24.3D	25.6	25.5
C24	42.5	42.4	42.5	37.1	41.8	77.5A	37.0	37.0	41.7	77.5B		39.7	42.5	42.5
C25	70.2	69.7	69.7	28.1	72.2	33.9	28.2	28.2	72.1	34.0		28.1	69.7	69.6
C26	30.5C	29.9	30.1C	22.3	28.3B	19.5B	22.4B	22.4C	26.4A	19.5		22.6D	30.0	30.0
C27	30.2	29.9	30.3C	23.3	26.2B	17.0B	23.4B	23.2C	28.2A	17.0		22.8D	30.3	30.2
C28									15.3					

Values of: [a] Canonica, L. *et al.* (1977), [b] Hikino, H. *et al.* (1975b), [c] Blunt, J. *et al.* (1979), [d] Russell, G. *et al.* (1981), [e] Hoffmann J. A. (private communication), [f] 22,25-Dideoxyecdysone. Capital letters (A, B, C and D) indicate values that may be reversed.

7 REACTIONS

The reactivity of ecdysteroids is due chiefly to the presence of several secondary, tertiary and, in some cases, primary hydroxy groups, together with a 6-keto-7-ene system.

7.1 Dehydrations

Although ecdysteroids are stable under neutral conditions they are unstable in the presence of strong acids or bases. This instability is associated with the presence of a γ-hydroxy-6-keto-7-ene system, the hydroxy group of which is readily dehydrated under acidic conditions. Ecdysone (24), for example, eliminates the tertiary 14α-hydroxy group in the presence of 1% ethanolic HCl and gives a mixture consisting of the 7,14-dien-7-one (83) (λ_{max} 293 nm) and the thermodynamically more stable 8,14-dien-7-one (84) (λ_{max} 244 nm). Strongly acidic conditions may lead to the additional loss of the tertiary 25-hydroxy group giving two isomeric pairs of trienones (Karlson, P. *et al.*, 1965). When 20-hydroxyecdysone (32), however, was treated with strong acid a complex mixture, consisting of at least eight products, was produced (Galbraith, M. and Horn, D., 1969).

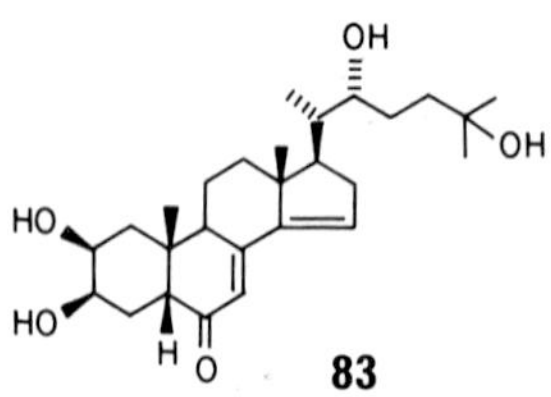

83

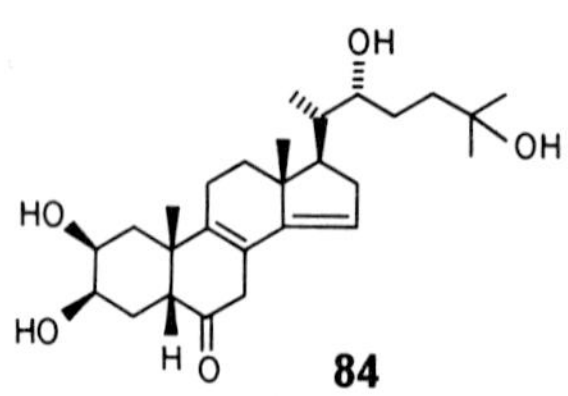

84

The observation of UV absorption maxima at λ 244 and 293 nm following treatment with dilute acid, has been used repeatedly to test the presence of a 14α-hydroxy-6-keto-7-ene system in ecdysteroids: as in the case of cyasterone (12) (Takemoto, T. *et al.*, 1967b), pterosterone (66) (Takemoto, T. *et al.*, 1968d) and rubrosterone (67) (Takemoto, T. *et al.*, 1969b). The resistance of stachysterone A (73) to acid treatment was consistent with the unusual rearranged skeletal structure containing a 14β-methyl group (Imai, S. *et al.*, 1970a). The 7,14-dien-6-one isomer of anhydroponasterone A, synthesized from ponasterone A (58) by brief treatment with methanolic 6 N HCl followed by chromatographic separation from the 8,14-isomer in 11% yield, was used as a model compound in the assignment of the structure of stachysterone B (74) (Imai, S. *et al.*, 1970a). Interestingly, cheilanthone B (11), which lacks the 7-unsaturation, was reported to yield a single 14-dehydration product when refluxed with methanolic HCl (Iyer, R. *et al.*, 1973).

Under carefully controlled conditions, using the aprotic reagents $POCl_3$ and pyridine–SO_3, specific elimination of the 25-hydroxy group may be achieved, the 14α-hydroxy group being too hindered to react (Heinrich, G. and Hoffmeister, H., 1971; Horn and Nearn, unpublished). Although yields are poor, these reactions are useful for the synthesis from 20-hydroxyecdysone (32) of a mixture of stachysterone C (75) and its 25-ene isomer which in turn may be converted by catalytic tritiation to radiolabelled ponasterone A (58) which is invaluable for binding studies (section 3.4).

Ecdysteroids containing 20,22-diol groups may be converted to 17(20)-ene acetate derivatives if acetylation is conducted under vigorous conditions. 20-Hydroxyecdysone (32), for example, gave a 42% yield of the tetraacetate (85) (presumably an E,Z-mixture) when treated with acetic anhydride–anhydrous sodium acetate at 140° (Hikino, H. *et al.*, 1969b).

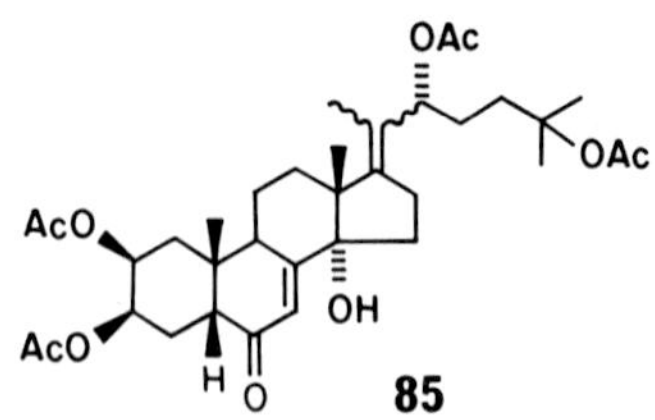

85

If ecdysteroids contain an 11α-hydroxy group then elimination occurs readily, yielding 7,9(11)-dien-6-one derivates. Thus muristerone A (48) lost the equatorial 11α-hydroxy group to give kaladasterone (42) in the presence of 5% NaOH solution in aqueous methanol (Canonica, L. *et al.*, 1973b). Muristerone A 2,3; 20,22-diacetonide-11-tosylate in chloroform gave a quantitative yield of kaladesterone 2,3;20,22-diacetonide when heated in the presence of alumina (Canonica, L. *et al.*, 1973b) and a benzene solution of ajugasterone C 2,3,11,22-tetraacetate was converted to the triacetate by heating in the presence of alumina (Imai, S. *et al.*, 1969b). Usmanov, B. *et al.* (1975) reported that turkesterone 2,3,11,22-tetraacetate was converted to the 7,9(11)-dien-6-one on treatment with potassium bicarbonate. The driving force for this type of elimination, which occurs under unusually mild conditions, must be provided by the resonance energy gained in the formation of the linear, conjugated and planar 7,9(11)-dien-6-ones and elimination of the steric hindrance of the acetyl group.

7.2 Isomerizations

Migration of ester groups attached to the 2,3-vicinal diol system of ecdysteroids may take place in the presence of aqueous acid: Lloyd-Jones, J. *et al.* (1973) found that Jones' oxidation of 20-hydroxyecdysone 2-acetate-20,22-acetonide (86) produced a mixture of the 3β-acetoxy 2-ketone (87) and 2β-acetoxy-3-ketone (88) due to acyl migration (Scheme 1). Similarly, attempted purification of the 2-hemisuccinate of 20-hydroxyecdysone (32) by ion exchange column chromatography led to a 1 : 1 mixture of 2- and 3-hemisuccinates (Horn and Nearn, unpublished).

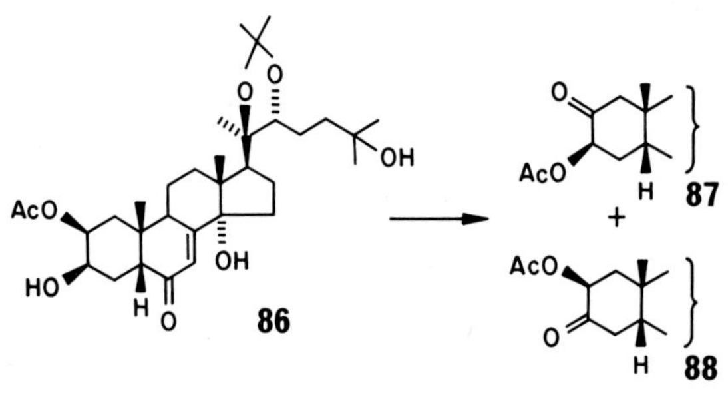

SCHEME 1.

Ecdysteroids are degraded in the presence of strong bases but with mild bases epimerization at C5 occurs, as in the case of 20-hydroxyecdysone (32) (Ogawa, S. *et al.*, 1971). Ecdysone (24) when treated with potassium carbonate in aqueous methanol is equilibrated into a mixture of 5α- and 5β-isomers in a 1 : 4 ratio, respectively (Mori, H. *et al.*, 1968b). The 5β-isomer predominates because a strong axial–axial steric interaction exists between the 2β-hydroxy and 19-methyl groups in the 5α-isomer (see Fig. 1).

In 2-deoxyecdysteroids the above interaction is absent and the 5α-isomer predominates at

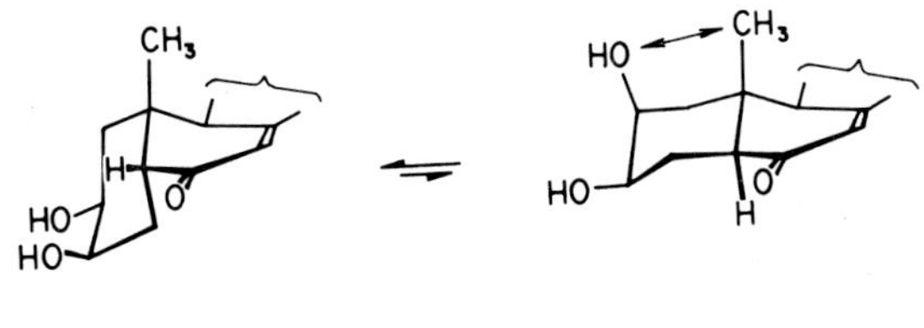

5β 5α

FIG. 1. Equilibrium mixture of 5α- and 5β-ecdysteroids. The 5β-epimer predominates because of the destabilizing axial–axial interaction between the 2β-hydroxy and 19-methyl groups in the 5α-epimer.

equilibrium: 2-deoxy-20-hydroxyecdysone (19), for example, yields an equilibrium mixture of 5α- and 5β-isomers in the ratio of 3 : 2, respectively, when treated with potassium carbonate in aqueous methanol (Chong, Y. *et al.*, 1970). Steroidal 6-ketones which lack the 7-double bond are even more stable as 5α-isomers and the proportion of 5β-isomer under equilibrating conditions is minimal (Mori, H. *et al.*, 1967).

Acid or base equilibration of ecdysteroid analogues lacking a 14α-hydroxy group leads to a complex mixture of products resulting from epimerization at C5 and C14, and 7-double bond isomerization (Furlenmeier, A. *et al.*, 1967; Furutachi, N. *et al.*, 1968; Scherrer, H., 1969).

Precyasterone (65), which contains a δ-lactone ring in the side-chain, can be isomerized to cyasterone (12), which contains a γ-lactone ring, by treatment with aqueous alkali followed by acidification (Hikino, H. *et al.*, 1970d). The mechanism, which involves ring opening and relactonization on the alternative hydroxy group, may parallel the biosynthetic pathway in the plant (Takemoto, T. *et al.*, 1968b).

7.3 Derivatives

On acetylation with acetic anhydride–pyridine at room temperature, 20-hydroxyecdysone (32) forms a 2,3,22-triacetate and a 2,3,22,25-tetraacetate in a ratio of 3:1 respectively (Takemoto, T. *et al.*, 1967c,d). The tertiary and sterically unhindered 25-hydroxy group is partially acetylated but the 14α-hydroxy and the 20-hydroxy groups are too sterically hindered to react. The equatorial 11α-hydroxy groups in ajugasterone C (3) (Imai, S. *et al.*,

1969b), muristerone A (48) (Canonica, L. *et al.*, 1972) and turkesterone (79) (Usmanov, B. *et al.*, 1975) are readily acetylated.

Galbraith, M. and Horn, D. (1969) determined the rate constants at 20° for the acetylation of hydroxy groups in 20-hydroxyecdysone (32) and found that the sterically unhindered equatorial 2β-hydroxy group was acetylated at the fastest rate, followed by the 22-hydroxy group, while the sterically hindered axial 3β-hydroxy group was acetylated at the slowest rate. Using brief reaction times it was possible to isolate the 2-acetate of 20-hydroxyecdysone (32) in 55% yield.

The greater reactivity of the 2β-hydroxy group in 5β-ecdysteroids have been used to advantage in the preparation of several 2-derivatives including:

(i) the 2-tosylate of 20-hydroxyecdysone (32) (Galbraith, M. *et al.*, 1969c);
(ii) the 2-hemisuccinate of 20-hydroxyecdysone (32) (Soumoff, C. *et al.*, 1981);
(iii) 2-hemisuccinyl derivatives of ecdysone (24) (De Reggi, M. *et al.*, 1975);
(iv) 2-benzoate, pivaloate and *N,N*-dimethylsuccinimide derivatives of 20-hydroxyecdysone (32) and ponasterone A (58) (Horn and Nearn, unpublished); and
(v) 2-piperonyl and 2-(3′,4′-methylenedioxy)-cinnamyl derivatives of 20-hydroxyecdysone (32) and polypodine B (54) (Singh, P. *et al.*, 1982).

Since the rates of hydrolysis of ecdysteroid acetate groups vary it is possible to prepare various acetate derivatives of ecdysteroids using controlled acetylation and hydrolysis. Thus Galbraith, M. and Horn, D. (1969) were able to prepare the 3-acetate and the 3,22-diacetate derivates of 20-hydroxyecdysone (32) by hydrolysing the 2,3,22-triacetate with potassium hydrogen carbonate in aqueous methanol. Hydrolysis of 20-hydroxyecdysone-2-cinnamate (36) is possible with potassium hydrogen carbonate in aqueous methanol, but 20-hydroxyecdysone-3-*p*-coumarate (37) is hydrolysed more slowly and requires prolonged treatment with potassium carbonate (Russell, G. *et al.*, 1972). Simple, qualitative studies on rates of acetylation and hydrolysis have become a routine procedure for determining the number, position and configuration of hydroxy groups in ecdysteroids (see section 5.4).

Ecdysteroids with *cis*-2,3-diol functions readily form acetonides as do those with a side-chain 20,22-diol group. Provided mild acid catalysts are used, such as phosphomolybdic acid (Mori, H. *et al.*, 1967), anhydrous copper sulphate (Jizba, J. *et al.*, 1967a) or phosphotungstic acid (Novosel'skaya, I. *et al.*, 1975), epimerization at C5 is avoided. Iyer, R. *et al.* (1973) reported that cheilanthone B (11) gives the 14-en-2,3-acetonide derivative on treatment with acetone in the presence of tosic or sulphuric acid.

Acetonide formation has been used extensively to detect ecdysteroid free diol functions and as protecting groups for subsequent transformations. Baltaev, U. *et al.* (1978a), for example, showed that integristerone A (39) and integristerone B (40) each contained a 1β,2β,3β-triol system because each formed two isomeric diacetonides. Since podecdysone C (53) could be derivatized to a triacetonide two diol functions in the side-chain were indicated (Galbraith, M. *et al.*, 1973b). In order to synthesize the 22-hemisuccinate derivative of ecdysone (24) for the preparation of an anti-ecdysone antiserum, Horn, D. *et al.* (1976) first protected the 2β,3β-diol groups of ecdysone (24) by acetonide formation.

The preparation of mixed derivatives of ecdysteroids allows valuable structural information to be gained. Muristerone A (48), for example, formed an amorphous 2,3;20,22-diacetonide which could be converted into the 20,22-monoacetonide by selective hydrolysis of the 2,3-acetonide group. Side-chain 20,22-diol acetonides are stable to acid hydrolysis because of steric hindrance but 2,3-acetonides are readily hydrolysed. Further benzoylation of the free 2β-,11α- and 3β-hydroxy groups of the muristerone A 20,22-monoacetonide then provided a 20,22-acetonide-2,3,11-tribenzoate for spectroscopic studies (Canonica, L. *et al.*, 1972). Examination of the amplitude of the Cotton effects of this tribenzoate established the α-orientation of the 11-hydroxy group. Koreeda, M. *et al.* (1969b) developed this procedure to establish the α-orientation of the 11-hydroxy group of ajugasterone C (3).

In certain instances, protection of the 20,22-diol group of ecdysteroids with an easily removable group is desirable. During the CSIRO synthesis of the ecdysteroid analogue 2-deoxy-3-epi-20-hydroxyecdysone, the 20,22-cyclic borate function in the intermediate allylic alcohols (89) served to protect the side-chain from oxidative degradation in the next step (Galbraith, M. *et al.*, 1969c). The free diol was readily regenerated by refluxing in methanol.

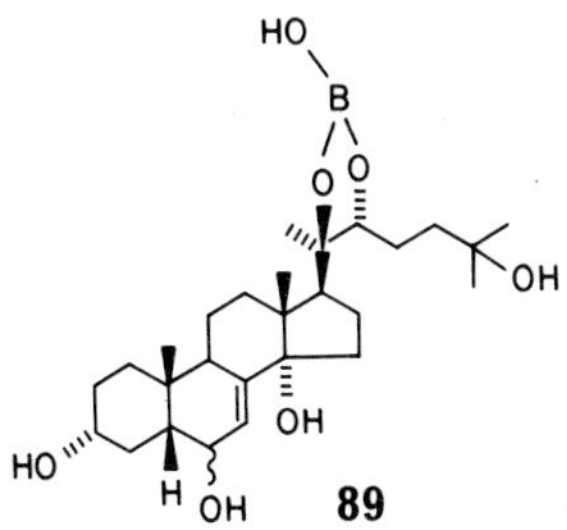

89

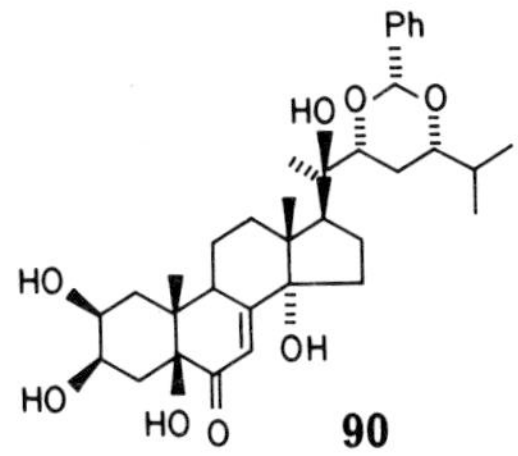

90

Cyclic benzylidene acetal derivatives of ecdysteroids are easily prepared but 2,3-benzylidene acetal derivatives are unstable and very easily hydrolysed (Horn and Wilkie, unpublished). Blunt, J. *et al.* (1979) reacted ponasterone C (60) with benzaldehyde in the presence of zinc chloride for 10 min at room temperature and obtained a 41% yield of the 22,24-benzylidene acetal (90) and a 19% yield of the isomeric 20,22-benzylidene acetal derivatives after silica gel chromatography. Analysis of the [^{13}C] NMR spectrum of the 22,24-benzylidene acetal (90) showed that the side-chain had a 24S-asymmetric centre.

Cyclic carbonate formation by reaction with phosgene in pyridine (Figure 2) has been used to establish *cis*-diaxial relationships between 3β- and 5β-hydroxy groups in a degradation product of polypodine B (54) (Heinrich, G. and Hoffmeister, H., 1968), in a muristerone A (48) derivative (Canonica, L. *et al.*, 1972) and in a sengosterone (68) derivative (Hikino, H. *et al.*, 1970c).

The hydroxy groups of ecdysteroids may be derivatized with silylating reagents in order to improve volatility, thermal stability and sensitivity of detection in GLC (see section 5.5). Derivatization with boron reagents has also been used in the

FIG. 2. Formation of 3,5-cyclic carbonates of ecdysteroids.

fluorometric detection of small quantities of ecdysteroids (see section 6.3).

The 6-keto group in ecdysteroids readily condenses with reagents containing primary amino groups to form Schiff bases for use in GLC (see section 5.5) and in the preparation of suitable ligands for coupling to proteins (see section 3.6). In the case of calonysterone (7) the enolized 6-keto group readily forms the O-methyl ether in the presence of diazomethane (Canonica, L. *et al.*, 1977).

7.4 Oxidations

Free 1,2-diol groups in ecdysteroids are oxidatively cleaved by periodate oxidation. Spectroscopic analysis of the fragments so obtained gives valuable information about the nucleus and side-chains. Ecdysteroids having similar nuclear structures yield identical nuclear fragments: for example pterosterone (66) (Takemoto, T. *et al.*, 1968d), 20-hydroxyecdysone (32) (Hikino, H. *et al.*, 1968a), cyasterone (12) (Takemoto, T. *et al.*, 1967b; Hikino, H. *et al.*, 1968a), ponasterone A (58) (Takemoto, T. *et al.*, 1968h; Moriyama, H. and Nakanishi, K., 1968) and inokosterone (38) (Takemoto, T. *et al.*, 1967a) give the same dialdehyde (91) or its intramolecular condensation product(s) (92) having a rearranged 8-double bond. The dialdehyde (91) is easily converted into the enals (92) by chromatography on silica gel (Takemoto, T. *et al.*, 1967b; Hikino, H. *et al.*, 1968a). Canonica, L. *et al.* (1972) found that muristerone A (48), when reacted with sodium periodate in methanol, consumed two moles of reagent and yielded a pair of isomeric nuclear fragments (93) containing a cyclic acetal moiety. The cyclic acetal may be formed by internal cyclization of an aldehyde–methanol hemiacetal with the favourably orientated 11α-hydroxy group.

The side-chain fragments of ecdysteroids from periodate oxidation are aldehydic and can be converted into crystalline 2,4-dinitrophenylhydrazine (2,4-DNP) derivatives or further oxidized into known compounds. Pterosterone (66) (Takemoto, T. *et al.*, 1968d) and ponasterone C (60) (Nakanishi, K. *et al.*, 1968; Koreeda, M. and Nakanishi, K. 1970b) both of which contain a (24S), 24-hydroxy asymmetric centre in the side-chain, afford the enal (95) (2,4-DNP, m.p. 174–176°), formed by dehydra-

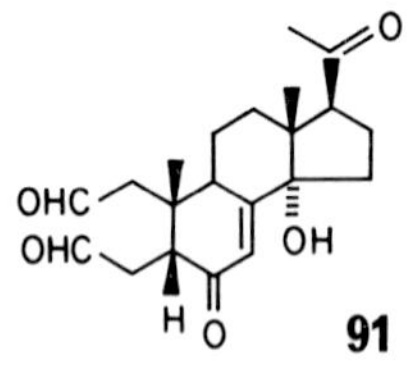

91

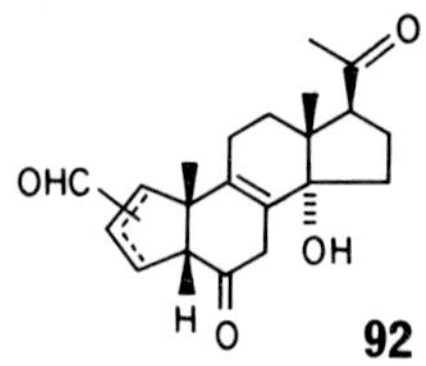

92

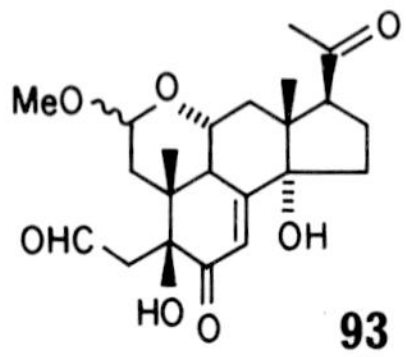

93

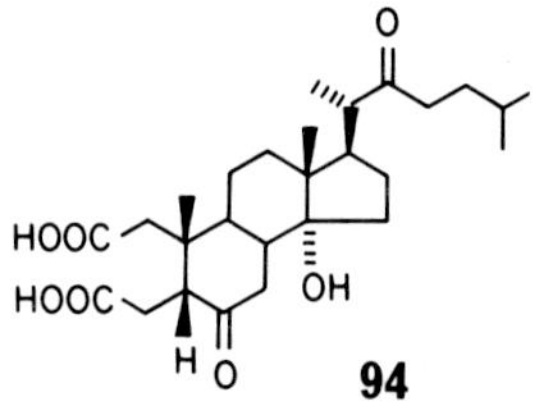

94

tion of the unstable α-hydroxy aldehyde precursor. From similar oxidation of ponasterone A (58) (Nakanishi, K. *et al.*, 1968) the isohexenal (96) is formed, and cyasterone (12) (Hikino, H. *et al.*, 1968a) gives the lactonic aldehyde (97) (2,4-DNP, m.p. 136.5–137.5°). Inokosterone (38) (Takemoto, T. *et al.*, 1967a) furnishes the hydroxyaldehyde (98) and, on further oxidation with potassium permanganate, DL-α-methylglutaric acid (99).

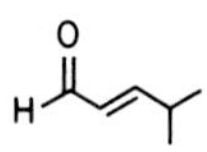

95

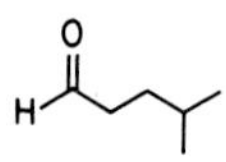

96

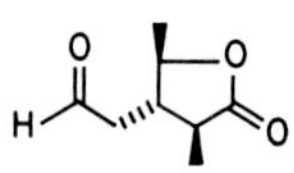

97

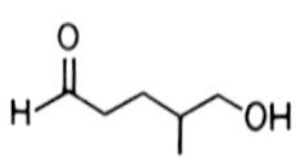

98

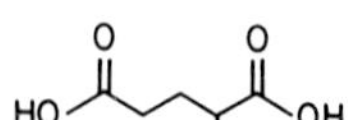

99

Periodate cleavage of the side-chain of the 2-acetate derivates of the following ecdysteroids give the 2-acetate of poststerone (63) as the nuclear fragment: 20-hydroxyecdysone (32) (Galbraith, M. *et al.*, 1968a); makisterone A (44) (Imai, S. *et al.*, 1968a); pterosterone (66), cyasterone (12) and inokosterone (38) (Takemoto, T. *et al.*, 1968h) and ponasterone A (58) (Moriyama, H. and Nakanishi, K., 1968).

By a suitable choice of reagent and careful control of reaction conditions, cleavage of the 20,22-diol group may be achieved without prior protection of the 2,3-diol group. Galbraith, M. *et al.* (1969a) discovered that poststerone (63) could be prepared in 60% yield by directly oxidizing 20-hydroxyecdysone (32) with Jones' reagent; this method was superior to the periodate cleavage-hydrolysis sequence on 20-hydroxyecdysone 2-acetate which gave poststerone (63) in only 35% overall yield. Muristerone A (48) was selectively cleaved at the 20,22-diol group of the side-chain with Collins' reagent at room temperature and afforded the 5β,11α-dihydroxy derivate of poststerone (63) (Canonica, L. *et al.*, 1974). Treatment of cheilanthone B (11) with Jones' reagent afforded the seco-diacid (94) (Iyer, R. *et al.*, 1973).

In some cases useful structural data can be obtained by more extensive oxidative degradation of ecdysteroids. Danieli, B. *et al.* (1974), for example, showed that makisterone A (44) possessed a 24S-asymmetric centre in the side-chain because (+) R-methylsuccinic acid was found among the di- and tri-basic acid mixture resulting from nitric acid–vandium pentoxide oxidation.

Microbial degradation of ecdysteroids has been studied by Canonica, L. *et al.* (1974). Unexpectedly, hydroxylating micro-organisms (*viz. Rhizopus arrhizus* ATCC1145, *Rhizopus nigricans* ATCC6261 and *Curvularia lunata* ATCC12017) known to hydroxylate at positions 11α and 11β in many progesterone derivatives, degraded the ecdysteroid side-chain without hydroxylating the nucleus. To be effective the micro-organism cell walls had to be ruptured by physical methods before reaction. Treatment of 20-hydroxyecdysone (32) with a *R. arrhizus* culture (27°, 110 h) yielded poststerone (63) (70%), rubrosterone (67) (3%) and the side-chain degradation product, 4-hydroxy-4-methyl pentanoic acid lactone. Makisterone A (44) and muristerone A (48) behaved similarly, affording in the latter case the expected 5β,11α-dihydroxy derivative of poststerone (63). Since ecdysone (24) was not transformed, Canonica, L. *et al.* (1974) suggested that the micro-organisms require a (20R,22R)-20,22-diol system in their substrates. Separate experiments showed that rubrosterone (67) was derived from poststerone (63) during the course of the reaction. Wai-Ming, T. *et al.* (1975) later reported that the micro-organism *Fusarium lini*, which is known to hydroxylate pregnanes at the 15α-position and some cardenolides at the 12β-position, transformed ponasterone A (58) into rubrosterone (67) directly.

One of the key reactions (Scheme 2) used by

Koreeda, M. *et al.* (1971) during a study aimed at unravelling the 20,22-diol configuration of ecdysteroids, was the ozonolysis of the triacetate (100), which cleaved the 17(20)-double bond and yielded rubrosterone 2,3-diacetate (101) and the α-acetoxyketone (102).

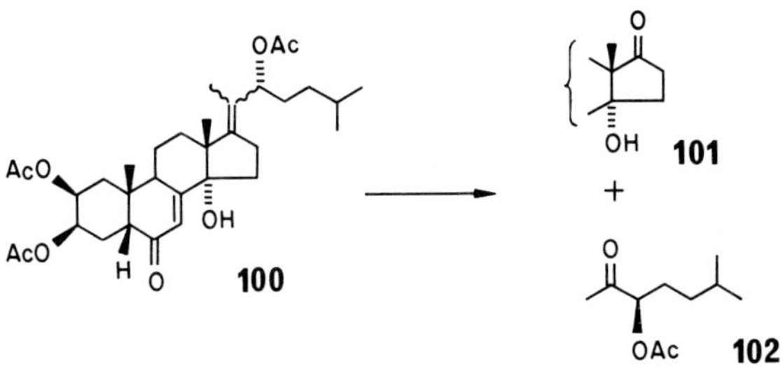

SCHEME 2.

Catalytic oxidation of inokosterone (38) by Spindler, K. *et al.* (1976) afforded a mixture of products including "inokosterone-C26-carboxylic acid" (103), isolated in ≃10% yield after chromatography. Using the same conditions, 3-keto (dehydro) derivatives of ecdysone (24) and 20-hydroxyecdysone (32) were subsequently prepared (Spindler, K. *et al.*, 1977; Dinan, L. and Rees, H., 1978). Dinan, L. and Rees, H. (1978), using TLC to carefully monitor the reactions, prepared 3-dehydro-20-hydroxyecdysone (16) in 67% yield and 3-dehydroecdysone (15) in 49% yield.

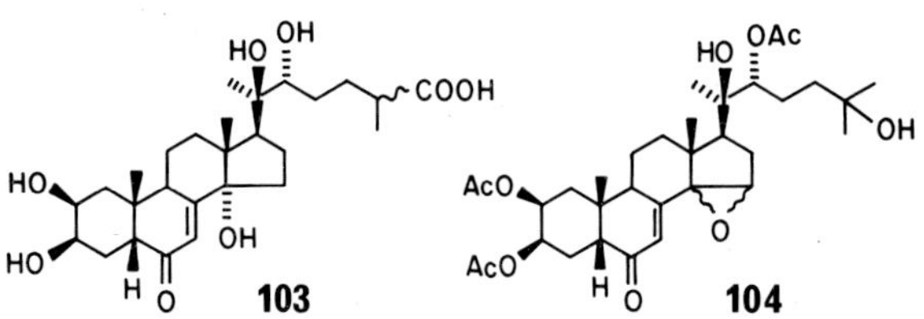

Enzymatic methods have also been used to prepare 3-dehydroecdysone (15) and 3-dehydro-20-hydroxyecdysone (16): Karlson, P. *et al.* (1972) used a partially purified enzyme system from *Calliphora vicina* while Koolman, J. and Spindler, K. (1977) used an ecdysone oxidase (E.C. 1.1.3.16) from the same insect species (see chapter 8, this volume). Davies, T. *et al.* (1980) have reported that both dimethyl sulphoxide–acetic anhydride and manganese dioxide oxidize the 3β-hydroxy group of 20-hydroxyecdysone 2-acetate 20,22-acetonide (86) without acyl migration, which is known to occur under Jones' oxidation conditions (Lloyd-Jones, J. *et al.*, 1973).

The reaction of ecdysteroids with peroxides and peracids has only been marginally investigated. Matsuoka, T. *et al.* (1969) reported that stachysterone B 2,3,22-triacetate is converted to an isomeric pair of 14-epoxides (104) on treatment with perbenzoic acid. Canonica, L. *et al.* (1977) failed to prepare an α-epoxyketone derivative of 20-hydroxyecdysone (32) using standard conditions (H_2O_2, OH^-). However, the allylic alcohols (105), derived from 20-hydroxyecdysone 2,3;20,22-diacetonide by chemical reduction ($NaBH_4$, IPrOH), reacted with *meta*-chloroperbenzoic acid to afford an isomeric mixture of α-hydroxyketones (106) (Scheme 3). The latter were successively oxidized with Collins' reagent and selenium dioxide followed by acid hydrolysis to give a low yield of calonysterone (7) (Canonica, L. *et al.*, 1977).

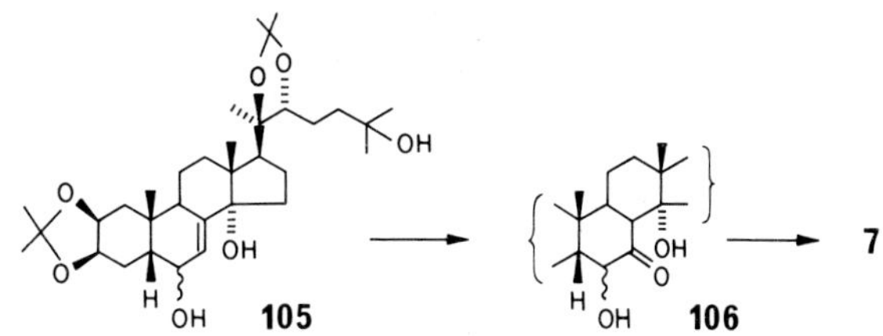

SCHEME 3.

Canonica, L. *et al.* (1977) have shown that 20-hydroxyecdysone (32) can be converted to calonysterone (7) in good yield under base-catalysed autoxidation conditions (Scheme 4). When the

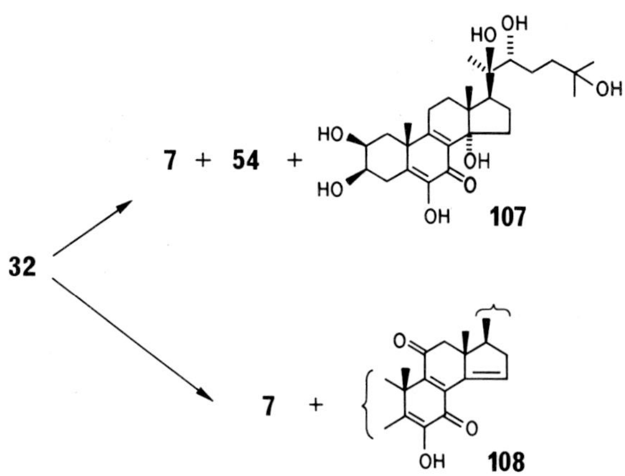

SCHEME 4.

reaction was conducted at 0° a complex mixture was produced and calonysterone (7) (51%), the unstable 14α-hydroxy precursor (107) (< 2%) and polypodine B (54) (2%) were isolated. At room temperature a 71% yield of calonysterone (7) and some 11-ketocalonysterone (108) were produced.

Surprisingly, a quantitative yield of 20-hydroxyecdysone-14α-hydroperoxide (111) was formed during an attempt to reductively cleave the 14α-hydroxy group of 20-hydroxyecdysone (32) (Canonica, L. *et al.*, 1980, 1981). In the proposed mechanism (Scheme 5) the initially formed radical anion (109) undergoes reductive cleavage of the 14α-hydroxy group to give the dienolate anion (110). The latter species then efficiently traps dissolved oxygen and yields the 14α-hydroperoxide (111) on work-up.

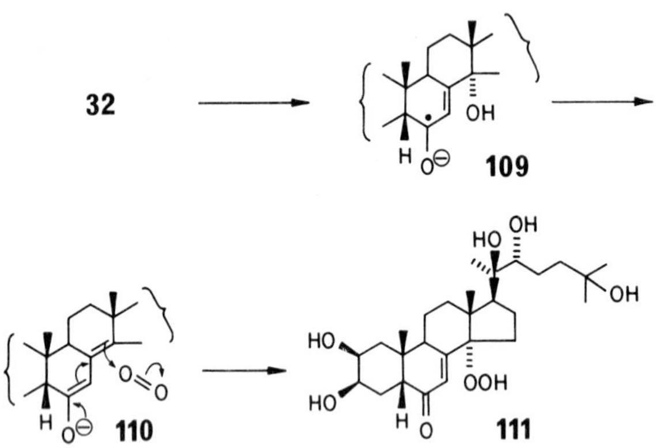

SCHEME 5.

7.5 Reductions

Under catalytic hydrogenations conditions (Pd-C, MeOH) the 14α-hydroxy group of ecdysteroids undergoes hydrogenolysis but the sterically hindered 7-double bond is unaffected (Karlson, P. *et al.*, 1965) and the 6-keto-7-ene moiety of ecdysone (24) is unaffected even at 120° and 100 kPa pressure. Calonysterone (7), containing the unusual 8,14-dien-6,7-diketo system, took up two moles of hydrogen affording the 6,7-diketone (112) (Canonica, L. *et al.*, 1977).

Lithium aluminium hydride (LAH) reduction of the 2,3,22-triacetate of 20-hydroxyecdysone (32) gave a mixture of the 6α- and 6β-allylic alcohols (113) (Matsuoka, T. *et al.*, 1969). Interestingly, the stachysterone B 14-epoxy derivatives (104) were not reduced to the allylic alcohols (113) with LAH (Matsuoka, T. *et al.*, 1969). Dinan, L. and Rees, H. (1978) reported that sodium borohydride in ethanol–tetrahydrofuran (1 : 1, v/v) at room temperature failed to reduce 20-hydroxyecdysone (32) even after 3 days. However, refluxing the solution for 1 h was sufficient to quantitatively produce the allylic alcohols (113).

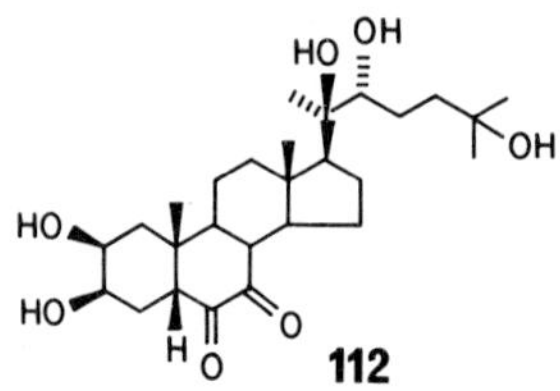
112

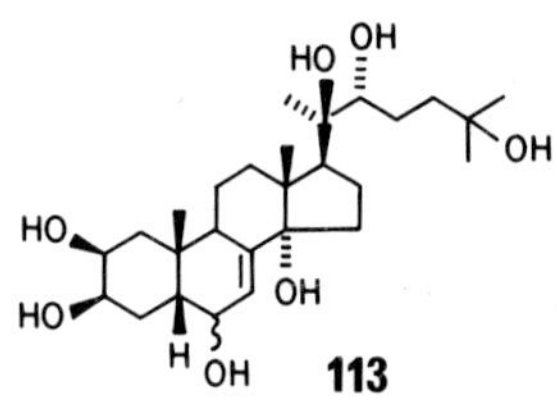
113

Brief treatment (10 min) of 3-dehydroecdysone (15) with sodium borohydride in ethanol–tetrahydrofuran (1 : 1, v/v) at room temperature led to the selective reduction of the 3-keto group to give a mixture of ecdysone (24) and 3-epiecdysone (29) in a ratio of 1 : 2.7, respectively. Similarly, 3-dehydro-20-hydroxyecdysone (16) afforded a mixture of 20-hydroxyecdysone (32) and 3-epi-20-hydroxyecdysone (30) in a ratio of 1 : 2.2 (Dinan, L. and Rees, H., 1978). The yields of 3-epiecdysone (29) and 3-epi-20-hydroxyecdysone (30) were 33% and 52%, respectively.

3-Epiecdysteroids have also been produced enzymatically from their 3β-hydroxy epimers by a specific 3β-ecdysone–dehydrogenase–isomerase isolated from the midgut of the tobacco hornworm *Manduca sexta* (Nigg, H. *et al.*, 1974a). This enzyme was found to readily convert ecdysone (24), 20-hydroxyecdysone (32) and 22-deoxyecdysone into their 3α-epimers, presumably *via* 3-keto intermediates. The fact the 22,25-dideoxyecdysone and 4-[^{14}C]-cholesterol were not epimerized suggested that substrates having an ecdysone-like nucleus and 22, 25-hydroxy groups are required by this enzyme.

An unexpected product was obtained when the 2-tosylate of 20-hydroxyecdysone (32) was used as the starting material in an attempted synthesis of 2-deoxy-20-hydroxyecdysone (19) (Galbraith, M. *et al.*, 1969c). Although this tosylate reacted normally with lithium tri-*t*-butoxyaluminium hydride to give the isomeric allylic alcohols (114), subsequent LAH reduction, cyclic borate formation, allylic oxidation and silica gel chromatography afforded 3-epi-2-

deoxy-20-hydroxyecdysone (117) (Scheme 6). Severe steric hindrance on the α-side of the tosylate (114) was presumed to prevent normal SN_2 displacement by a hydride ion. A new pathway was proposed involving proton abstraction and a concomitant hydride shift to give the 3-keto intermediate (115) followed by reduction from the β-face to yield the 3α-alcohols (116).

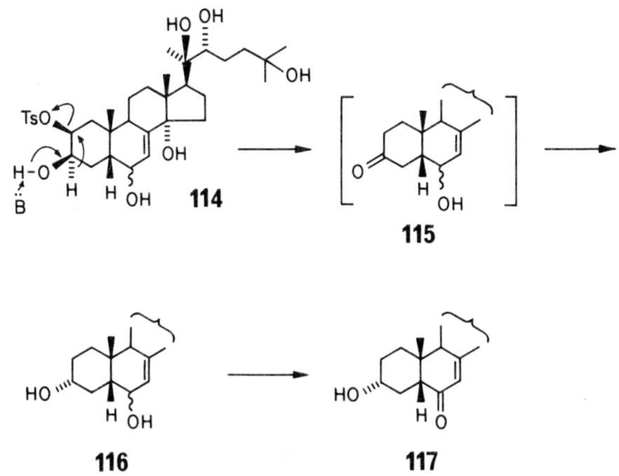

Scheme 6.

20-Hydroxyecdysone (32) underwent reductive cleavage of the 14α-hydroxy group and enol ether formation in the presence of zinc and fluoroboric acid in ethanol (Canonica, L. *et al.*, 1980, 1981) (Scheme 7). The dienol ether (118) was formed in 50% yield while the unconjugated enone (119) and the conjugated pair of enones (120) were formed in 5% and 35% yield, respectively.

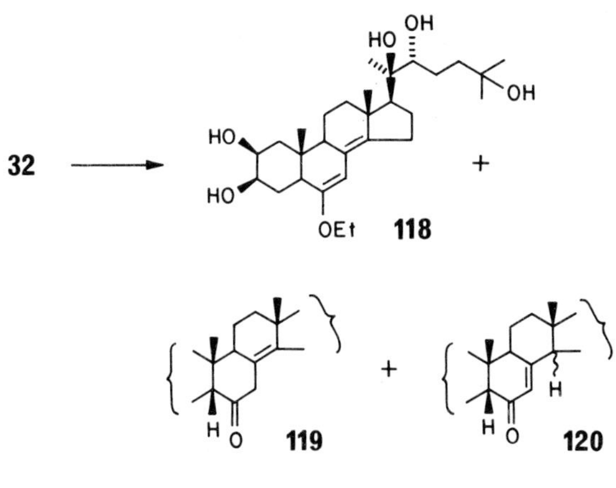

Scheme 7.

7.6 Photochemistry

The effect of UV light on ecdysteroids has been studied by Canonica and his group in Milan (Canonica, L., 1976; Canonica, L. *et al.*, 1980, 1981) in connection with the environmental stability of ecdysteroids. Here we summarize their results obtained with 20-hydroxyecdysone (32) in various solvents. Experiments were carried out under an argon gas atmosphere utilizing a normal, high-pressure UV source equipped with a Pyrex filter.

Photochemical transformations of 20-hydroxyecdysone (32) are markedly solvent-dependent (Scheme 8). In simple, unhindered alcoholic solvents such as isopropanol, a fast reaction takes place giving mainly the isomeric adduct (121) and a small amount of the 14α-hydroperoxide (111). In hindered solvents such as di-isopropylcarbinol, the adducts (122) and the 14α-hydroperoxide (111) are accompanied by the pair of reduction products (113).

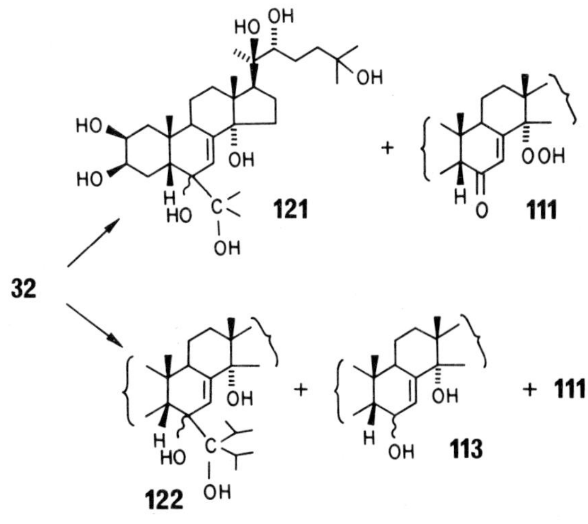

Scheme 8.

In aqueous solutions the reaction is slow and yields a complex six-component mixture: the conjugated enone (120) and unconjugated enone (119) resulting from reductive cleavage of the 14α-hydroxy group; a photochemically isomerized 8-en-6-one (123); the 14α-hydroperoxide (111) and its associated photoproducts (124) and (125) (see below) (Scheme 9).

Additional photolysis experiments demonstrated that the 14α-hydroperoxide (111) was the precursor

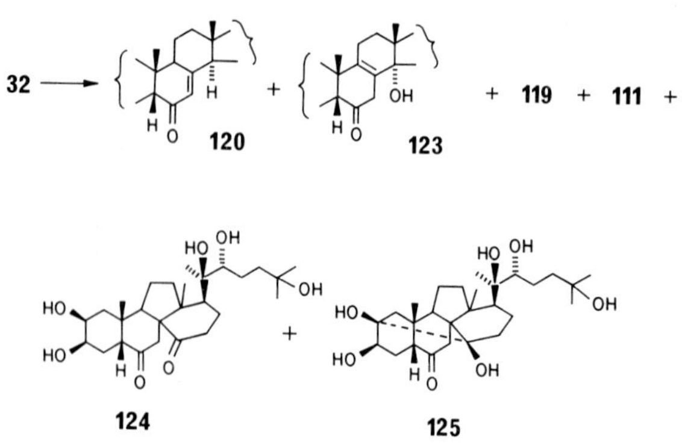

SCHEME 9.

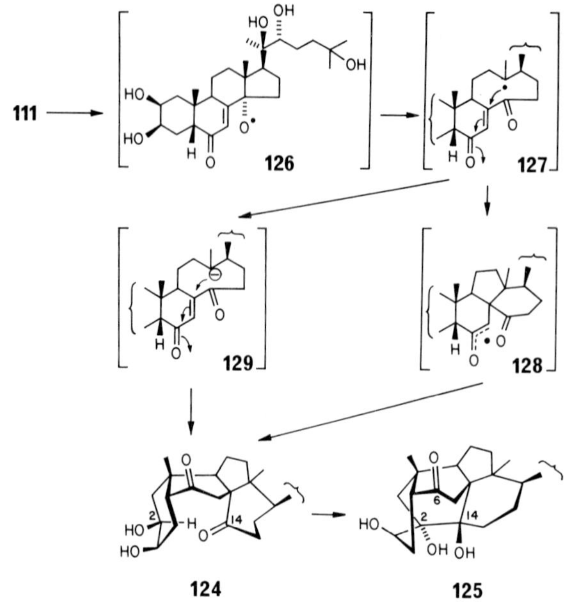

SCHEME 10.

of the spirodiketone (124), which in turn was the precursor of the ketol (125). It was also observed that thermal decomposition of the 14α-hydroperoxide (111) at its melting point or treatment with Fe(II) salts gave a mixture of the spirodiketone (124) and 20-hydroxyecdysone (32).

A probable mechanism for the formation of the spirodiketone (124) and the ketol (125) from the 14α-hydroperoxide (111) is given in Scheme 10. Homolytic cleavage of the O—O bond of peroxide (111) gives the oxygen radical (126) which collapses *via* C—C bond cleavage to the tertiary C13 radical (127). The C13 radical (127) adds intramolecularly to the enedione system to form the delocalized spiro radical (128) which affords the spirodiketone (124) by abstracting H· from the medium. In the case of Fe(II) treatment the C13 radical (127) may be reduced to the anion (129) before conjugate addition and neutralization. A molecular model of the spirodiketone (124) reveals that the 2-proton lies in close proximity to the 14-carbonyl group; the photochemically activated 14-keto group may then abstract the 2-proton and form the 2→14 bond to give the ketone (125).

The composition of the 20-hydroxyecdysone (32) photolysis mixture in aqueous solutions is strongly pH-dependent. At pH 6.85, for example, the 8(14)-en-6-one (119), 6-keto-7-ene (120) and 14α-hydroperoxide (111) comprise 62%, 20% and 10% of the mixture, respectively; at pH 10.4 the 8(14)-en-6-one (119) cannot be detected and the 6-keto-7-ene (120) and 14α-hydroperoxide (111) comprise 10% and 63% of the mixture, respectively.

Addition of a hundredfold excess of the singlet oxygen scavenger DABCO to the irradiated mixture at pH 10.4 completely suppressed the formation of the spirodiketone (124) and spiroketone (125). Under these conditions the 6-keto-7-ene (120) and the 8(14)-en-6-ketone (119) were isolated in 40% and 50% yield, respectively. The 14α-hydroperoxide (111) and its associated photoproducts, (124) and (125), may therefore be derived from activated, singlet oxygen rather than triplet, molecular oxygen.

Of further interest is the reported crosslinking of 20-hydroxyecdysone (32) to polytene chromosomes of *Drosophila melanogaster* salivary glands (Grönemeyer, H. and Pongs, O., 1980; see chapter 13, this volume) by photoactivating the enone chromophore with UV light (320 nm). Immunofluorescence microscopy revealed that 20-hydroxyecdysone (32) was specifically cross-linked at hormone-inducible puff sites. No cross-linking took place if the polytene chromosomes were subjected to heat shock before photoactivation. These results imply that covalent bonding between ecdysteroids and proteins takes place in the presence of UV light at receptor sites only, where the steroid–receptor complex has a relatively long lifetime.

7.7 Metabolism

Biochemical studies have demonstrated that ecdysteroids undergo a variety of oxidations, isomerizations and conjugation reactions both *in vivo* and *in vitro*. Detailed discussions of the chemistry involved are to be found in Chapters 7–9 of this volume.

8 SYNTHESIS

Detailed descriptions of early synthetic schemes leading to ecdysteroids have been published by Hikino, H. and Hikino, Y. (1970), Horn, D. (1971) and Slama, K. *et al.* (1974). Our aim here is to concentrate on important synthetic strategies which have been developed since the above reviews. In Table 10 we present references to syntheses of ecdysteroids and of ecdysteroid analogues prepared for the study of structure–activity relationships (Bergamasco, R. and Horn, D., 1980).

Table 10: *List of synthetic ecdysteroids, analogues*[a] *and intermediates*

Compounds	References
20β-Acetoxy-5 α-pregnane-2,7-dien-6-one	Smith, W. *et al.* (1973)
Calonysterone (7)	Canonica, L. *et al.* (1977)
3-Dehydroecdysone (15)	Spindler, K. *et al.* (1977); Dinan, L. and Rees, H. (1978)
3-Dehydro-20-hydroxyecdysone (16)	Dinan, L. and Rees, H. (1978)
2-Deoxyecdysone (18)	Galbraith, M. *et al.* (1975); Segal, G. *et al.* (1982)
14-Deoxyecdysone	Hocks, P. *et al.* (1966)
22-Deoxyecdysone	Siddall, J. (1969)
22-Deoxy-5α-ecdysone	Lettré, H. *et al.* (1972)
25-Deoxyecdysone	Hafferl, W. *et al.* (1972)
25-Deoxyecdysone-[23,24,25-$^3H_3$]	Hafferl, W. *et al.* (1972)
2-Deoxy-3-epi-20-hydroxyecdysone	Galbraith, M. *et al.* (1969c)
2-Deoxy-20-hydroxyecdysone (19)	Segal, G. *et al.* (1982)
22-Deoxy-20-hydroxyecdysone	Galbraith, M. *et al.* (1968a, 1969a)
14-Deoxymuristerone A	Cherbas, P. *et al.* (1982)
22-Deoxy-24-nor-5α-ecdysone	Lettré, H. *et al.* (1972)
2-Deoxyrubrosterone	Sydykov, Z. *et al.* (1976)
2,25-Dideoxyecdysone	Galbraith, M. *et al.* (1975)
22,25-Dideoxyecdysone	Hocks, P. *et al.* (1966); Thompson, M. *et al.* (1970)
22,25-Dideoxyecdysone 3-methyl ether	Thompson, M. *et al.* (1970)
22,25-Dideoxy-5α-ecdysone	Thompson, M. *et al.* (1970)
22,25-Dideoxy-5α-ecdysone 2,3-diacetate	Mori, H. *et al.* (1970)
22,25-Dideoxy-(24R)-24-ethylecdysone	Thompson, M. *et al.* (1970)

Table 10: *Continued*

Compounds	References
22,25-Dideoxy-5β-hydroxy ecdysone	Thompson, M. *et al.* (1971)
22,25-Dideoxy-20-hydroxyecdysone	Middleton, E. *et al.* (1972)
22,25-Dideoxy-(24R)-24-methylecdysone	Thompson, M. *et al.* (1970)
22,25-Dideoxy-27-nor-ecdysone	Thompson, M. *et al.* (1970)
2,22-Dideoxy-24-nor-ecdysone	Kinnear, J. *et al.* (1976)
22,25-Dideoxy-4,4-dimethyl-5α-ecdysone	Hanna, R. *et al.* (1976)
22,25-Dideoxy, ecdysone 2,3-dinitrate	Gocan, A. and Hodosan, F. (1976)
22,25-Dideoxy-5α-hydroxy-5α-ecdysone	Lettré, H. *et al.* (1972)
22,25-Dideoxy ecdysone 2,3-trimethylsilyl ether	Thompson, M. *et al.* (1970)
Ecdysone (24), 5α-Ecdysone and 22-Isoecdysone	Kerb, U. *et al.* (1966a, 1966b); Furlenmeier, A. *et al.* (1966a,b, 1967); Wiechert, R. *et al.* (1966); Siddall, J. *et al.* (1966a,b); Harrison, I. *et al.* (1966); Edwards, J. *et al.* (1967a,b, 1968); Mori, H. *et al.* (1968b, 1969, 1971); Barton, D. *et al.* (1970); Smith, W. and Newsoroff, G. (1973); Lee, Y. *et al.* (1976, 1980)
Ecdysone (24)-[23,24-$^{14}C_2$]	Smith, W. and Newsoroff, G. (1973)
Ecdysone (24)-[15,16-$^3H_2$]	Lee, Y. *et al.* (1980)
Ecdysone (24)-[23,23,24,24-$^3H_4$]	Hüppi, G. and Siddall, J. (1968); Hafferl, W. *et al.* (1972)
Ecdysone 2-hemisuccinate	De Reggi, M. *et al.* (1975)
Ecdysone 22-hemisuccinate	Horn, D. *et al.* (1976)
3-Epi-ecdysone (29)	Thompson, M. *et al.* (1970); Dinan, L. and Rees, H. (1978)
3-Epi-20-hydroxyecdysone(30)	Dinan, L. and Rees, H. (1978)
20-Hydroxyecdysone (32)	Hüppi, G. and Siddall, J. (1967); Kerb, U. *et al.* (1968); Mori, H. and Shibata, K. (1969)
20-Hydroxyecdysone 2-cinnamate (36)	Russell, G. *et al.* (1972)
20-Hydroxyecdysone 14α-hydroperoxide (109)	Canonica, L. *et al.* (1980, 1981)
Inokosterone-C26-carboxylic acid (103)	Spindler, K. *et al.* (1976)
Kaladasterone (42)	Canonica, L. *et al.* (1973b)
Polypodine B (54)	Canonica, L. *et al.* (1977)
Polypodine B 2-cinnamate (55)	Russell, G. *et al.* (1972)
Ponasterone A (58)	Hüppi, G. and Siddall, J. (1968)
Ponasterone A (58)-[24,25(26)-$^3H_2$]	Heinrich, G. and Hoffmeister, H. (1971), Maroy, P., *et al.* (1978), Sage, B., *et al.* (1982)

Table continued overleaf

Table 10: *Continued*

Compounds	References
Ponasterone A 14α-hydroperoxide	Furatuchi, M. *et al.* (1968)
Poststerone (63)	Siddall, J. *et al.* (1967); Galbraith, M. *et al.* (1969a); Canonica, L. *et al.* (1974)
Rubrosterone (67)	Hikino, H. *et al.* (1968b, 1969b); Hocks, P. *et al.* (1968); Vanbever, W. *et al.* (1970); Shibata, K. and Mori, H. (1971). Caputo, R. *et al.* (1979)
Stachysterone C(75) (and 25-ene isomer)	Heinrich, G. and Hoffmeister, H. (1971); Sage, B. *et al.* (1982)
2,3,22,25-Tetradeoxyecdysone (134)	Kinnear, J. *et al.* (1976); Galbraith, M. *et al.* (1981)
2,3,22,25-Tetradeoxy-5α-ecdysone	Alston, K. *et al.* (1976); Galbraith, M. *et al.* (1981)
2,3,22,25-Tetradeoxy-3,5-cycloecdysone	Galbraith, M. *et al.* (1981)
2,14,22,25-Tetradeoxyecdysone (77)	Galbraith, M. *et al.* (1973a)
2,14,22,25-Tetradeoxyecdysone (77)-[3α-^{3}H]	Galbraith, M. *et al.* (1974)
2,14,22,25-Tetradeoxyecdysone (77)-[11,12-$^3H_2$]	Faux, A. *et al.* (1979)
2,14,22,25-Tetradeoxy-5α-ecdysone-[3α-^{3}H]	Bollenbacher, W. *et al.* (1977)
2,14,22,25-Tetradeoxyecdysone 3-acetate	Rodewald, J. *et al.* (1978, 1979)
2,14,22,25-Tetradeoxy-5α-ecdysone 3-acetate	Rodewald, J. *et al.* (1978, 1979)
2,14,22,25-Tetradeoxy-5α-hydroxy-5α-ecdysone	Kinnear, J. *et al.* (1978)
2,14,22,25-Tetradeoxy-5α-hydroxy-5α-ecdysone-[3α-^{3}H]	Kinnear, J. *et al.* (1978)
2,3,22,25-Tetradeoxy-5α-ecdysone 14-trimethylsilyl ether	Alston, K. *et al.* (1976)
2,22,25-Trideoxyecdysone (78)	Galbraith, M, *et al.* (1973a)
2,22,25-Trideoxyecdysone (78)-[3α-^{3}H]	Galbraith, M. *et al.* (1974); Kinnear, J. *et al.* (1976)
2,22,25-Trideoxyecdysone 3-acetate	Galbraith, M. *et al.* (1981)
2,22,25-Trideoxyecdysone 3-ethyl(methyl)ether	Galbraith, M. *et al.* (1981)
2,22,25-Trideoxyecdysone 14α-hydroperoxide	Furatuchi, N. *et al.* (1968)
2,22,25-Trideoxy-5α-ecdysone	Galbraith, M. *et al.* (1974); Wada, K. (1975)
2,22,25-Trideoxy-5α-ecdysone-[3α-^{3}H]	Galbraith, M. *et al.* (1974)
2,22,25-Trideoxy-5α-ecdysone 3-acetate	Mori, H. *et al.* (1970); Wada, K. (1975); Rodewald J. *et al.* (1978), (1979)
2,22,25-Trideoxy-3-epi-5α-ecdysone	Galbraith, M. *et al.* (1981)
2,22,25-Trideoxy-14-epi-ecdysone 3-acetate	Rodewald, J. *et al.* (1978, 1979)
2,22,25-Trideoxy-14-epi-5α-ecdysone 3-acetate	Rodewald, J. *et al.* (1978, 1979)

Table 10: *Continued*

Compounds	References
2,22,25-Trideoxy-5α-hydroxy-5α-ecdysone 3-acetate	Rodewald, J. *et al.* (1978, 1979)
14,20,25-Trideoxy-5α-hydroxy-5α-ecdysone	Alston, K. *et al.* (1976)
2,22,25-Trideoxy-5α-hydroxy-5α-ecdysone	Kinnear, J. *et al.* (1978)
2,22,25-Trideoxy-5α-hydroxy-5α-ecdysone-[3α-^{3}H]	Kinnear, J. *et al.* (1978)
2,22,25-Trideoxy-5β-hydroxy-5β-ecdysone	Thompson, M. *et al.* (1971)
2,22,25-Trideoxy-4,4-dimethyl-5α-ecdysone	Hanna, R. *et al.* (1976)
2,22,25-Trideoxy-3-epi-3β-methyl(ethyl)ecdysone	Horn, D. *et al.* (1982)
2,22,25-Trideoxy-3α-methyl(ethyl)ecdysone	Horn, D. *et al.* (1982)
2,22,25-Trideoxy-5β-methyl-5β-ecdysone	Horn, D. *et al.* (1982)
Synthesis of intermediates	
20β-Acetoxy-5α-pregna-2,7-dien-6-one	Smith, W. *et al.* (1973)
2β,3β,14α,17β-Tetrahydroxy-5β-androst-7-en-6-one	Cochrane, J. and Hanson, J. (1971)
2β,3β,20-Triacetoxy-5α-pregnan-6-one (157)	Kametani, T. *et al.* (1980a, 1981)[b]

[a] Numerous steroidal and non-steroidal analogues have been synthesized with the aim of producing moulting hormone inhibitors; see Velgova, H. *et al.* (1968, 1969, 1972); Rezabova, B. *et al.* (1968); Robbins, W. *et al.* (1970, 1971, 1975); Thompson, M. *et al.* (1971, 1975); Svoboda, J. and Robbins, W. (1971); Svoboda, J. *et al.* (1972); Machkova, Z. and Cerny, V. (1974); Gocan, A. and Hodosan, F. (1976); Hanna, R. *et al.* (1976); Dias, J. (1977); Kinnear, J. *et al.* (1979); Cambie, R. *et al.* (1981); Galbraith, M. *et al.* (1981); Singh, P. *et al.* (1982).
[b] Total synthesis.

With only one exception (Kametani, T. *et al.*, 1980a, 1981), suitable natural products have been used as synthetic precursors of ecdysteroids and analogues. In the main, tetracyclic androstane, pregnane or 23,24-*bis-nor*cholenic precursors have been employed as starting materials, the nucleus elaborated and the side-chain attached (Siddall, J. *et al.*, 1966a,b; Wiechert, R. *et al.*, 1966; Mori, H. *et al.*, 1968b; Mori, H. and Shibata, K. 1969). In some cases a readily available phytoecdysteroid served as a precursor of another: 20-hydroxyecdysone (32), for example, was converted into poststerone (63) (Siddall, J. *et al.*, 1967), 22-deoxyecdysone (Middleton, E. *et al.*, 1972), rubrosterone (67) (Hikino, H. *et al.*, 1969b), polypodine B (54) and calonysterone (7) (Canonica, L. *et al.*, 1977).

Kinnear, J. *et al.* (1976) developed a general synthesis of 2-deoxyecdysteroid analogues from the

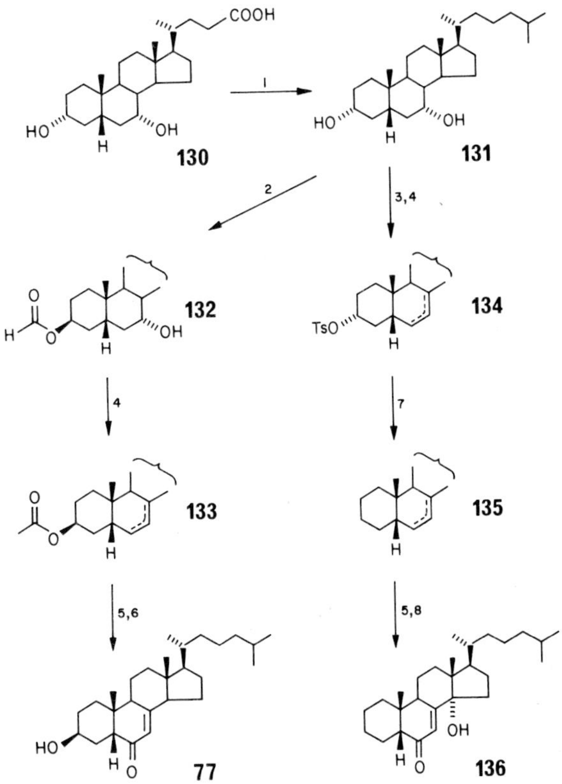

SCHEME 11. *Reagents*: **1**, Anodic coupling, isovaleric acid; **2**, Ph_3P, EtOCO—N═N—COOEt, HCOOH; **3**, TsCl, Pyr; **4**, $POCl_3$, Pyr; **5**, CrO_3 . 2Pyr, CH_2Cl_2; **6**, $KHCO_3$; **7**, LAH; **8**, SeO_2

bile acid, chenodeoxycholic acid (130) (Scheme 11). Reaction sequences were devised so as to retain the natural 5β-configuration of acid (130) throughout the synthesis, thus allowing the easy preparation of 2-deoxy-5β-ecdysteroids which are far more stable in the 5α-series. The synthesis of the ovarian ecdysteroid 2,14,22,25-tetradeoxyecdysone (77) (Hetru, C. *et al.*, 1978) required only five steps: electrolysis of a mixture of chenodeoxycholic acid (130) and isovaleric acid in methanol afforded 5β-cholestan-3α,7α-diol (131) by anodic coupling (Bergström, S. and Krabisch, L., 1957). In this step yields of 50% were obtained if a steady concentration of the more reactive isovaleric acid was maintained by continuous addition during the course of reaction (Faux, A. *et al.*, 1979). Inversion and simultaneous protection of the equatorial 3α-hydroxy group of the diol (131) as the formate ester was achieved using diethyl azodicarboxylate, triphenylphosphine and formic acid (Bose, A. *et al.*, 1973). The hydroxy formate (132) thus produced was then dehydrated to a mixture of 6- and 7-ene isomers (133), resulting from two possible diaxial eliminations. Since the 6-ene isomer did not react with Collin's reagent, the mixture of alkenes (133) was oxidized directly without separation to afford the tetradeoxyecdysone analogue (77) after silica gel chromatography and mild base hydrolysis. In a similar way, 14α-hydroxy-5β-cholest-7-en-6-one (136) was prepared for structure–activity studies. The diol (131) was converted into the 3α-tosyl derivative and dehydrated, without prior purification, into a mixture of 6- and 7-ene 3α-*p*-toluenesulphonates (134). Hydride reduction of the toluenesulphonates (134) afforded an alkene mixture (135) containing about 60% of the 7-ene. Finally, sequential oxidation with Collins' reagent and selenium dioxide provided the ecdysteroid analogue (136) which was found to have a remarkably high activity in the *Calliphora* assay [one-seventh that of ecdysone (24)].

In order to prepare radiolabelled 2,14,22,25-tetradeoxyecdysone-[11,12-$^3H_2$] (77) for biochemical studies the Australian team (Faux, A. *et al.*, 1979) later devised a modified route (Scheme 12) which afford a product of high specific activity (31 Ci mM^{-1}) having tritium at the metabolically inert 11, 12-positions. This material was valuable in biosynthetic studies (Faux, A. *et al.*, 1979; Hetru, C. *et al.*, 1982). Methyl 3α-ethoxycarbonyloxy-7α, 12α-dihydroxycholanate (137) was dehydrated to the 7,11-diene (138) and then hydrolysed to the 7,11-dien-24-oic acid (139) in good yield. Electrolysis of (139) (3.6 A, 110 V) in the presence of isovaleric acid gave the 3α-cholesta-7,11-dienol derivative (140) in 51% yield. Inversion of the 3α-hydroxy group as before gave 5β-cholesta-7,11-dien-3β-yl formate (141) which was then tritiated in the presence of Adams' catalyst affording mainly the [11,12-$^3H_2$]-derivative (142). Oxidation of the 7-ene (142) with Collins' reagent, followed by hydrolysis of the formate group under non-epimerizing conditions, gave 3β-hydroxy-5β-cholest-7-en-6-one-[11,12-$^3H_2$] (77) having a radiochemical purity of 94%.

Two exceptional syntheses of ecdysone (24) from diosgenin (143) have been published by Lee, E. *et al.* (1976, 1980). In each synthesis the spiro-acetal system of rings E and F was skilfully used as a latent, functionalized ecdysteroid side-chain giving the

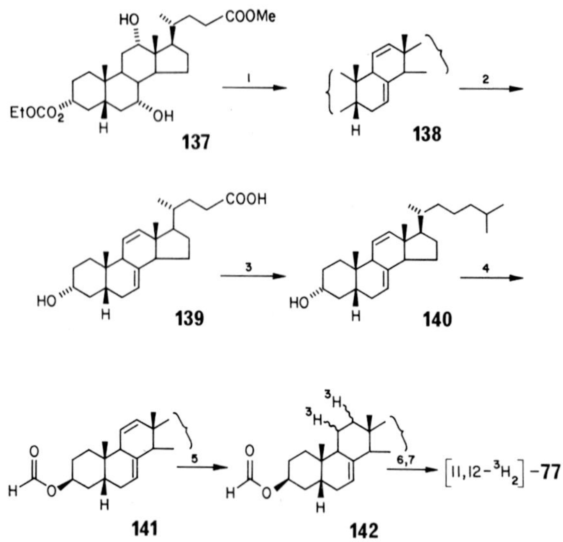

SCHEME 12. *Reagents*: **1**, $POCl_3$, Pyr; **2**, KOH; **3**, Anodic coupling, isovaleric acid; **4**, Ph_3P, EtOCO—N=N—COOEt, HCOOH; **5**, $^3H_2$, PtO_2; **6**, $CrO_3.2Pyr$, CH_2Cl_2; **7**, $KHCO_3$

correct configuration at C20 and C22 in the product. The first synthesis (Scheme 13) began with the stereospecific reduction of diosgenin (143) by a mixed hydride to dihydrodiosgenin (144) in high yield; hydrogenolysis of the 22-oxygen bond proceeded from the less sterically-hindered α-face.

The transformation of dihydrodiosgenin (144) to the 5α-6-keto-7-ene $16 \rightarrow 22$ epoxide (147) was accomplished using methodology already developed for ecdysteroid synthesis. The diol (144) was transformed into the ditosylate, subjected to hydroboration–oxidation to introduce the 6-keto group and heated with lithium bromide in DMF in order to simultaneously eliminate the 3β-tosyloxy group and displace the 26-tosyloxy group with bromide. The 26-bromo-6-keto-2-ene (145) so obtained was subjected to Prévost–Woodward hydroxylation and then acetylation and bromination under equilibrating conditions to provide the 7α,26-dibromoketone (146). Double dehydrobromination of the latter with lithium carbonate in boiling DMF afforded a mixture of 7,25- and 4,25-dien-6-ones which were separated by chromatography. The required 7,25-dien-6-one was hydroxylated at C25 under oxymercuration–demercuration conditions, acetylated by Oppenauer's method and oxidized with selenium dioxide to give the 14α-hydroxyenone (147).

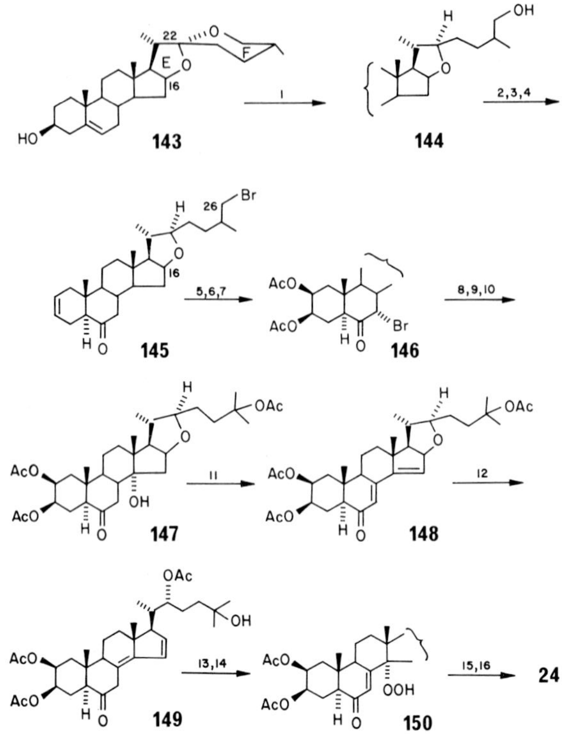

SCHEME 13. *Reagents*: **1**, LAH, $AlCl_3$; **2**, TsCl, Pyr; **3**, B_2H_6, H_2O_2, CrO_3; **4**, LiBr, DMF; **5**, AgOAc, I_2, HOAc; **6**, Ac_2O; **7**, Br_2, HOAc; **8**, $LiCO_3$, DMF; **9**, Hg $(OAc)_2$; $NaBH_4$, OH^-; Ac_2O, CaH_2; **10**, SeO_2; **11**, $(CF_3CO)_2O$, Pyr; **12**, Zn, HOAc; **13**, H_2, Pd(C); **14**, O_2, rose bengal, *hv*; **15**, NaI, HOAc; **16**, K_2CO_3, MeOH

Clean dehydration of the 14α-hydroxy group was effected with trifluoroacetic anhydride to give the 7,14-dien-6-one (148) which served as the key intermediate in the overall strategy. Interestingly, if traces of selenium dioxide from the previous step were removed, the pure 14α-hydroxyenone (147) gave rise to a 14-trifluoroacetate. Treatment of (148) with excess zinc powder in boiling acetic acid led to reductive cleavage of the 16-oxygen bond and concomitant acetyl migration from C25 to C22 to give the (22R)-22-acetoxy-8(14),15-dien-6-one (149). The acetyl migration, although unexpected, was fortunate because the 25-hydroxy group was liberated. Sequential, selective hydrogenation of the 15-double bond and photo-oxygenation in the presence of a sensitizer (Furutachi, N. *et al.*, 1968) afforded the 14α-hydroperoxide (150). Reduction of the hydroperoxide (150) with sodium iodide and acetate hydrolysis-epimerization (Furlenmeier, A. *et al.*, 1967) completed the synthesis to give ecdysone (24) in 1% overall yield. This represented a considerable improvement on earlier syntheses which reported overall yields of ⩽ 0.1%. An added advantage was the possibility of preparing [15,16-$^3H_2$]-ecdysone (24) by tritiating the intermediate 8(14),15-dien-6-one (149). Such a product would be valuable in biochemical studies because the tritiated carbons are located in unreactive nuclear positions.

In their second synthesis Lee, Y. *et al.* (1980) reported an alternative stereospecific conversion of diosgenin (143) to ecdysone (24) and [15,16-$^3H_2$]-ecdysone (24) (Scheme 14). The 26-bromo-6-ketone (151) — obtained in seven steps from diosgenin (143) in the 1976 synthesis — was transformed into the ketal (152) and then into the 7α-bromo-26,27-*bis-nor*-25-methoxycarbonyl intermediate (153) by a sequence involving: (i) ozonolysis, (ii) conversion to an acid *via* the bromoform reaction, (iii) esterification, (iv) acetylation, and (v) bromination under equilibrating conditions. Dehydrobromination of the 7α-bromide (153) gave a mixture of 7- and 4-en-6-ones which were separated by chromatography. The major 7-en-6-one isomer was first hydroxylated at C14 with selenium dioxide and then (without purification) dehydrated with trifluoroacetic anhydride to give the 7,14-dien-6-one (154). In this step it was important to avoid the presence of strong acid which would lead to the formation of the isomeric 8,14-dien-6-one.

Cleavage of the C16—O bond and concomitant lactonization was achieved by treating the 7,14-dien-6-one (154) with activated zinc in degassed, boiling acetic acid, or, simply by passing an acetic acid solution of (154) through an activated zinc column yielding the 8(14),15-dien-6-one 25 → 22 lactone (155). This somewhat unstable lactone (155) was then subjected to catalytic hydrogenation, photo-oxygenation and sodium iodide reduction to give the 14α-hydroxy-7-en-6-one 25 → 22 lactone (156). The synthesis of ecdysone (24) was completed by subjecting the diacetate (156) to basic hydrolysis–epimerization, followed by separation of the desired 5β-epimer and treatment with methylmagnesium bromide. Mori, H. *et al.* (1968b) in their synthesis used a similar strategy to elaborate the side-chain of ecdysone (24).

Although the yield of ecdysone (24) was lower (≈ 0.4% overall) than in their 1976 synthesis

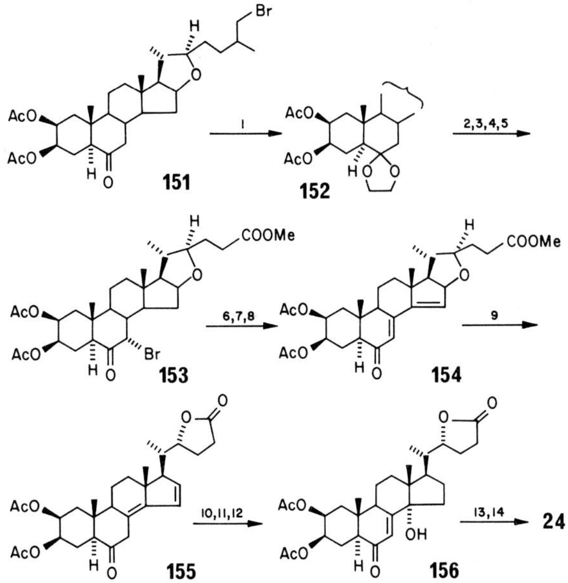

SCHEME 14. *Reagents*: **1**, TsOH, ethylene glycol; **2**, LiBr, Li_2CO_3; **3**, O_3, Zn, HOAc; **4**, NaOBr, NaOH; CH_2N_2; Ac_2O, Pyr; **5**, Br_2, HOAc; **6**, Li_2CO_3, DMF; **7**, SeO_2; **8**, $(CF_3CO)_2O$; **9**, Zn, HOAc; **10**, H_2, Pd(C); **11**, O_2, rose bengal, *hν*; **12**, NaI, MeOH; **13**, K_2CO_3, MeOH; **14**, CH_3MgBr

(Scheme 13), the new synthesis utilized intermediates of improved stability and is amenable to the preparation of doubly labelled [15,16-$^3H_2$]-[26,27-$^{14}C_2$]-ecdysone (24) which would be valuable for insect metabolic and physiological studies. Lee, Y. *et al.* (1980), however, only reported the preparation of [15,16-$^3H_2$]-ecdysone (24) by using a 1 : 1 mixture of tritium and hydrogen to catalytically reduce the 8(14),15-dien-6-one (155) (Scheme 14). The specific activity was 1.0 Ci mM^{-1} after TLC purification.

The stereocontrolled total synthesis of 2β,3β,20β-triacetoxy-5α-pregnan-6-one (168) (Scheme 15) by Kametani, T. *et al.* (1980a, 1981) is an impressive achievement and constitutes a formal total synthesis of 20-hydroxyecdysone (32) since the intermediate (168) had previously been converted to the latter (Mori, H. and Shibata, K., 1969). The starting point for the synthesis was the D-ring aromatic steroid (160) which had been previously prepared (Kametani, T. *et al.*, 1980b) from the vinylcyclobutane (157) by way of the 6α-nitro-8α-cyano-D-ring aromatic steroid (159) using their familiar, and successful, thermolysis strategy involving intramolecular cycloaddition proceeding

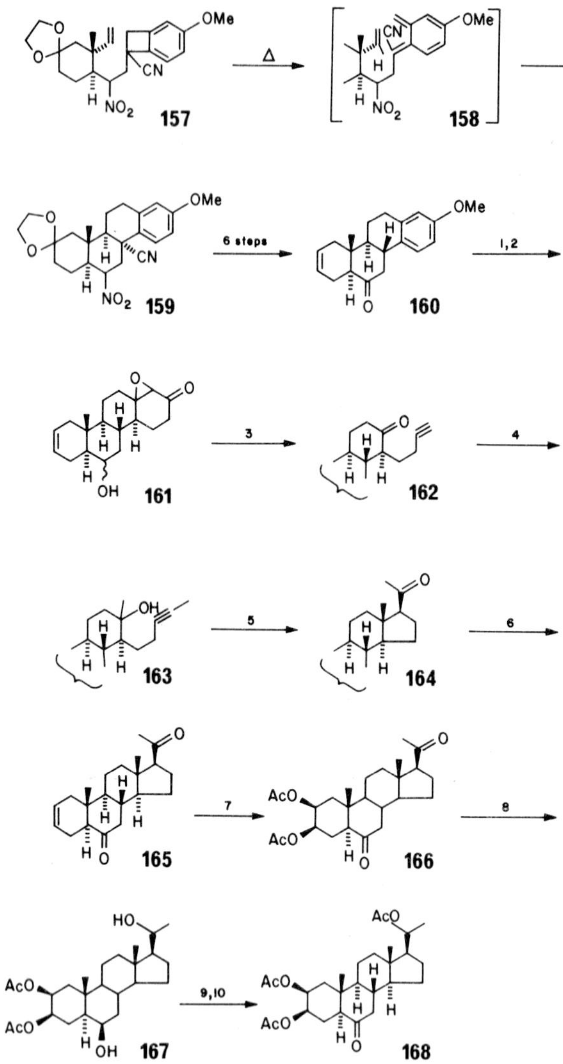

SCHEME 15. *Reagents*: **1**, $NaBH_4$; Li, NH_3; TsOH; **2**, H_2O_2, OH^-; **3**, $TsNHNH_2$; **4**, MeLi; $LiNH_2$, NH_3; MeI; **5**, $(CF_3CO)_2$, CF_3COOH, KOH; **6**, Jones'; **7**, AgOAc, I_2; **8**, $NaBH_4$; **9**, Ac_2O; **10**, Jones'

through the *o*-quinodimethane intermediate (158). The α-epoxyketone (161) was prepared by subjecting the anisole derivative (160) to sodium borohydride and Birch reductions, followed by acid treatment and epoxidation. Eschenmöser ring opening of the α-epoxyketone (161) was achieved using *p*-toluenesulphonyl hydrazine under acidic conditions affording the acetylenic ketone (162). Treatment of the acetylenic ketone (162) sequentially with methyl lithium and methyl iodide in liquid ammonia gave the diol (163). Site-specific and stereospecific cyclization–hydrolysis was achieved by treating the diol (163) with trifluoroacetic anhydride–trifluoroacetic acid followed by base hydrolysis to give 6β-hydroxy-5α-pregn-2-en-20-one (164).

The synthesis was then completed along established lines: Jones' oxidation afforded the 6,20-diketone (165) while Prévost–Woodward oxidation of the double bond and acetylation gave the diacetate (166) which was then reduced to the 6β, 20-diol (167). Selective acetylation of the 20-hydroxy group and Jones' oxidation afforded the target triacetate (168).

9 STRUCTURE AND BIOLOGICAL ACTIVITY

Comprehensive reviews of structure–activity relationships of ecdysteroids using both *in vitro* (Cherbas, L. *et al.*, 1980) and *in vivo* assays (Bergamasco, R. and Horn, D., 1980) have been published. The following structural requirements are essential for both *in vitro* and *in vivo* activity: (i) a *cis*-fused A/B-ring junction, (ii) a 6-keto-7-ene grouping and (iii) a full sterol side-chain.

The presence of a free 14α-hydroxy group in ecdysteroids is essential for high *in vivo* activity and generally increases *in vitro* activity. However, Cherbas, P. *et al.* (personal communication, 1982) have found that 14-deoxymuristerone A is unexpectedly 70–80 times as active in the Kc cell system as 20-hydroxyecdysone (32). 14-Deoxymuristerone A was an unexpected product formed when muristerone A (48) was reacted with sodium iodide and trimethylsilyl chloride in acetonitrile.

Hydroxy groups at the 2β-, 3β- and 20R-positions also enhance biological activity. The 25-hydroxy group diminishes activity in *in vitro* assays. Singh, P. *et al.* (1982) report that 5β-hydroxy ecdysteroids are either less active than, or about as active as, their 5β-hydrogen counterparts in the housefly assay. Thus at a dose-level of 1.0 μg per insect, polypodine B (54), ponasterone C (60), dacrysterone (14) and pterosterone (66) have activities of 42, 42, 51 and 30%, respectively, as compared with 100% for 20-hydroxyecdysone (32). Polypodine B (54) was also found (Horn, D., unpublished) to be less active in the *Calliphora* bioassay than 20-hydroxyecdysone (32). On the other hand Jizba, J.

et al. (1967b) claimed polypodine B (54) to be four times as active as 20-hydroxyecdysone (32) in their assay.

In view of the above requirements for MH activity we are at a loss to explain the activity of pinnasterol (50) and pinnasterol 2-acetate (51) in the *Sarcophaga* assay (Fukuzawa, A. *et al.*, 1981). These ecdysteroids do not possess a *cis* A/B-ring junction, the required configuration of the 22-hydroxy group nor a 14α-hydroxy group.

Galbraith, M. *et al.* (1981) investigated the effect of substituents on the 3-hydroxy group of 3β,14α-dihydroxy-5β-cholest-7-en-6-one on MH activity determined in the *Calliphora* bioassay. The following order of decreasing activities was established relative to 20-hydroxyecdysone (32) as 100%; 3β-hydroxy (33%) > 3β-methoxy (20%) > 3β-acetoxy (8%) > 3β-ethoxy (3%) [The 3β-acetoxy derivative was earlier incorrectly reported as being inactive (Bergamasco, R. and Horn, D., 1980)]. Thus it appears that increasing the size of the 3-substituent increases interference with the receptor, leading to decreased activity. However, the decrease in activity may be due more to a decrease in polarity of the 3-substituent group than to its size, since 20-hydroxyecdysone 3-*p*-coumarate (37) (Singh, P. *et al.*, 1982) and ponasteroside A (62) (Takemoto, T. *et al.*, 1968c) are reported to show high activity in the *Calliphora* assay.

Recent studies at CSIRO indicate that relatively large substituents, such as hemisuccinyl methyl esters and amides, can be accommodated at the 2- and 22-hydroxy groups of 20-hydroxyecdysone (32) without great loss of activity in both the *Calliphora* bioassay and *Drosophila* imaginal disc assay (Yund and Horn, unpublished). The free 2-hemisuccinate of 20-hydroxyecdysone (32) is inactive, presumably as a result of the carboxylate ion hindering access to receptors within target cells. In an earlier study utilizing a chromosome puffing assay Spindler, K. *et al.* (1976) reported weak activities for the 2- and 3-hemisuccinates of 20-hydroxyecdysone (32); the corresponding methyl esters were slightly more active and were of about the same order of activity as the C6-carboxymethyloxime methyl ester. Spindler, K. *et al.* (1976) added that the most active derivative was inokosterone-C26-carboxylic acid (103). Singh, P. *et al.* (1982) have also reported that various 2-cinnamyl derivatives of 20-hydroxyecdysone (32) and polypodine B (54) are active in the *Musca* bioassay. In one case, the 2-(3′,4′-methylenedioxy)-cinnamate derivative of polypodine B (54), the activity exceeded that of the parent compound, perhaps as a result of the synergizing effect of the methylenedioxycinnamate moiety (Singh, P. *et al.*, 1982). The 2- and 22-acetate derivatives of 20-hydroxyecdysone (32) are highly active in both the *Calliphora* bioassay and *Drosophila* imaginal disc assay (Horn and Nearn, unpublished). The 2,3-diacetate derivative, however, was inactive in the *in vitro* test, probably because two ester groups on the A-ring are too bulky to be accommodated in the receptor or because together they greatly reduce the polarity of the A-ring functions.

The above results, though preliminary, help to define essential binding areas with the ecdysteroid receptor. We now consider that an ecdysteroid at the receptor is bound mainly on the β-face of the molecule in the region of the A/B-rings, up to but not over the C2 and C3 oxygen functions, so that non-bulky groups can be accommodated at these positions without greatly interfering with the ecdysteroid–receptor interaction.

The studies of structure–activity relationships have been undertaken partly to gain an insight into the nature of receptor–hormone interactions, and also to find out where bridging groups may be attached to the hormone molecule without interfering with the hormone–receptor interaction. Such derivatives with bridging arms will be required for the preparation of column supports to be used for the isolation of the moulting hormone receptor by affinity chromatography.

Specific ecdysteroid receptors have been discovered in the nuclei of *Drosophila melanogaster* imaginal discs (Fristrom, J. *et al.*, 1978), in cytosol of *Drosophila* cell lines (Maroy, P. *et al.*, 1978) and in the nuclei of hypodermis and midgut glands of the crayfish *Orconectes limosus* (Spindler-Barth, M. *et al.*, 1981).

The isolation of the ecdysteroid receptor from such sources would be an important achievement as it would permit the preparation of receptor antibodies which would be invaluable for further studies of the mechanisms of action of the moulting hormone at the nucleus.

ACKNOWLEDGEMENTS

We are indebted to Professor L. Canonica and Dr. B. Danieli for much unpublished information, to Roland H. Nearn for reading the manuscript, to John S. Wilkie for drawing the formulae and to Cindy L. Sexton for typing the manuscript.

REFERENCES

AGUI, N., YAGI, S. and FUKAYA, M. (1969). Effects of ecdysterone on the *in vitro* development of wing discs of rice stem borer, *Chilo suppressalis*. *Appl. Entomol. Zool. 4*, 158–159.

AGUI, N. (1973). Quantitiative bioassay for moulting hormone *in vitro*. *Appl. Entomol. Zool. 8*, 236–239.

ALSTON, K. T., BEBBINGTON, P. M., GREEN, S. E., MORGAN, E. D. and POOLE, C. F. (1976). The synthesis of some cholest-7-en-6-one derivatives related to ecdysones. *Steroids 27*, 609–616.

ASHBURNER, M. (1971). Induction of puffs in polytene chromosomes of *in vitro* cultured salivary glands of *Drosophila melanogaster* by ecdysone and ecdysone analogues. *Nature New Biol. 230*, 222–224.

ASHBURNER, M. (1973). Sequential gene activation by ecdysone in polytene chromosomes of *Drosophila melanogaster*. I. Dependence upon ecdysone concentration. *Devel. Biol. 35*, 47–61.

BALTAEV, U., GOROVITS, M. B., ABDULLAEV, N. D., YAGUDAEV, M. R. and ABUBAKIROV, N. K. (1977). Phytoecdysones from *Rhaponticum intetgrifolium*. II. Integristerone A. *Khim. Prir. Soedin. 6*, 813–819. *Chem. Abs. 88*, 191233C.

BALTAEV, U., GOROVITS, M. B., ABDULLAEV, N. D., RASHKES, YA. V. and YAGUDAEV, M. R. (1978a). Phytoecdysones of *Rhaponticum integrifolium*. III. Integristerone B. *Khim. Prir. Soedin. 4*, 457–463. *Chem. Abs. 90*, 23392b.

BALTAEV, U., GOROVITS, M. B., RASHKES, YA. V. and ABUBAKIROV, N. K. (1978b). Phytoecdysones from *Rhaponticum integrifolium*. IV. 24(28)-Dehydromakisterone A. *Khim. Prir. Soedin. 4*, 463–465. *Chem. Abs. 90*, 6611e.

BANERJI, A., CHINTALWAR, G. J., JOSHI, N. K. and CHADHA, M. S. (1971). Isolation of ecdysterone from Indian plants. *Phytochemistry 10*, 2225–2226.

BARTON, D. H. R., FEAKINS, P. G., POYSER, J. P. and SAMMES, P. G. (1970). A synthesis of the insect moulting hormone ecdysone and related compounds. *J. Chem. Soc.* (C). 1584–1591.

BEBBINGTON, P. M. and MORGAN, E. D. (1977). Detection and identification of moulting hormone (ecdysones) in the barnacle *Balanus balanoides*. *Comp. Biochem. Physiol. B, 56*, (1B), 77–79.

BECKER, E. and PLAGGE, E. (1939). Uber das pupariumbildung auslosende hormon der fliegen. *Biol. Zbl. 59*, 326–341.

BERGAMASCO, R. and HORN, D. H. S. (1980). The biological activities of ecdysteroids and ecdysteroid analogues. In *Progress in Ecdysone Research*. Edited by J. A. Hoffmann. Pages 299–324. Elsevier/North-Holland Biomedical Press, Amsterdam.

BERGAMASCO, R. and HORN, D. H. S. (1983). Distribution and roles of insect hormones and mimics in plants. In "Insect Endocrinology" vol 1, Edited by R. G. H. Downer and H. Laufer. Alan R. Liss, Inc, N.Y.

BERGSTRÖM, S. and KRABISCH, L. (1957). Bile acids and steroids. XLIX. Preparation of some hydroxycoprostanes; $3\alpha,7\alpha$- and 3α, 12α-dihydroxycoprostane. *Acta Chem. Scand. 11*, 1067.

BERKOFF, C. E. (1969). The chemistry and biochemistry of insect hormones. *Quart. Rev. 23*, 372–391.

BIELBY, C. R., GANDE, A. R., MORGAN, E. D. and WILSON, I. D. (1980). Practical aspects of the preparation and chromatography of the trimethylsilyl ethers of ecdysteroids. *J. Chromatogr. 194*, 43–53.

BLUNT, J. W., LANE, G. A., MUNRO, M. H. G. and RUSSELL, G. B. (1979). The absolute configuration at C24 of the ecdysteroids dacrysterone, pterosterone and ponasterone C. *Aust. J. Chem. 32*, 779–782.

BOLLENBACHER, W. E., GOODMAN, W., VEDECKIS, W. V. and GILBERT, L. I. (1976). The *in vitro* synthesis and secretion of α-ecdysone by the ring glands of the fly *Sarcophaga bullata*. *Steroids 27*, 309–324.

BOLLENBACHER, W. E., GALBRAITH, M. N., GILBERT, L. I. and HORN, D. H. S. (1977). *In vitro* metabolism of 3β-hydroxy-, and $3\beta,14\alpha$-dihydroxy-[3α-^{3}H]-5β-cholest-7-en-6-one by the prothoracic glands of *Manduca sexta*. *Steroids 29*, 47–63.

BOLLENBACHER, W. E., FAUX, A. F., GALBRAITH, M. N., GILBERT, L. I., HORN, D. H. S. and WILKIE, J. S. (1979). *In vitro* metabolism of possible ecdysone precursors by the prothoracic glands of the tobacco hornworm, *Manduca sexta*. *Steroids 34*, 509–526.

BONARIC, J. C. and DEREGGI, M. (1977). Changes in ecdysone levels in the spider *Pisaura mirabilis* nymphs (Araneae, Pisauridae). *Experientia 33*, 1664–1665.

BORST, D. W. and O'CONNOR, J. D. (1972). Arthropod molting hormone: radioimmune assay. *Science 178*, 418–419.

BORST, D. W. and O'CONNOR, J. D. (1974). Trace analysis of ecdysones by gas-liquid chromatography radioimmunoassay and bioassay. *Steroids 24*, 637–656.

BOSE, A. K., LAL, B., HOFFMANN, W. A. (3rd) and MANHAS, M. S. (1973). Steroids. IX. Facile inversion of unhindered sterol configuration. *Tetrahedron Lett. 18*, 1619–1622.

BOUTHIER, A., PENNETIER, J. L., MAUCHAMP, B. and LAFONT, R. (1975). β-Ecdysone level variations in *Locusta migratoria cinerascens* (Orthoptera, Acrididae) haemolymph during the last larval instar. *C.R. Hebd. Seances Acad. Sci., Ser. D 280*, 1837–1840.

BROOKS, C. J. W. and GASKELL, S. J. (1980). *Hormones. C. Hormones of invertebrates.* In *Biochemical applications of mass spectrometry*. Suppl. Vol. 1, pages 639–641. Edited by G. R. Waller and O. C. Dermer. John Wiley & Sons, New York.

BUTENANDT, A. and KARLSON, P. (1954). Über die isolierung eines metamorphose-hormone der insekten in kristallisierten form. *Z. Naturf. 9b*, 389–391.

CAMBIE, R. C., POTTER, G. J., READ, R. W., RUTLEDGE, P. S. and WOODGATE, P. D. (1981). *Solanum* alkaloids. XVIII. Utilization of solasodine for the preparation of nitrogen analogues of ecdysone. *Aust. J. Chem. 34*, 599–622.

CANONICA, L., DANIELI, B., WEISZ-VINCZE, I. and FERRARI, G. (1972). Structure of muristerone A, a new phytoecdysone. *J. Chem. Soc., Chem. Commun.* 1060–1061.

CANONICA, L., DANIELI, B., FERRARI, G., KREPINSKY, J. and RAINOLDI, G. (1973a). Structure of calonysterone, an unusually modified phytoecdysone. *J. Chem. Soc., Chem. Commun.* 737–738.

CANONICA, L., DANIELI, B., FERRARI, G., HAIMOVA, M. A. and KREPINSKY, J. (1973b). The structure of a new phytoecdysone kaladasterone: an application of ^{13}C magnetic resonance spectroscopy to structural problems. *Experientia 29*, 1062–1063.

CANONICA, L., DANIELI, B., PALMISANO, G., RAINOLDI, G. and RANZI, B. M. (1974). The microbial oxidation of insect moulting hormones. *J. Chem. Soc., Chem. Commun.* 656–657.

CANONICA, L., DANIELI, B., FERRARI, G., KREPINSKY, J. and WEISZ-VINCZE, I. (1975). A novel method of isolation of phytoecdysones from kaladana seeds. *Phytochemistry 14*, 525–527.

CANONICA, L. (1976). Phytoecdysones: environmental degradation In *Natural Products and the Protection of Plants*. Edited by G. B. Marini-Bettolo. Ch. II, pages 211–224. Elsevier, Amsterdam.

CANONICA, L., DANIELI, B., PALMISANO, G., FERRARI, G. and KREPINSKY, J. (1977). New phytoecdysones from kaladana. II. Structure of calonysterone. *Gazz. Chim. Ital. 107*, 229–235.

CANONICA, L., DANIELI, B., LESMA, G. and PALMISANO, G. (1980). The photochemistry of 20-hydroxyecdysone. Private communication of results presented at the 12th International Symposium on the Chemistry of Natural Products, Tenerife.

CANONICA, L., DANIELI, B., LESMA, G. and PALMISANO, G. (1981). The photochemistry of 20-hydroxyecdysone. Private communication of results presented at the Symposium: Frontieras de la Investigationes de Productos Naturales, 30/3-2/4, Mexico City.

CAPUTO, R., MANGONI, L., MONACO, P., PALUMBO, G. and PREVITERA, L. (1979). A new synthesis of rubrosterone. *Gazz. Chim. Ital. 109*, 651–654.

CARLISLE, D. B., OSBORNE, D. J., ELLIS, P. E. and MOORHOUSE, J. E. (1963). Reciprocal effects of insect and plant-growth substances. *Nature 200*, 1230.

CHANG, E. S., SAGE, B. A. and O'CONNOR, J. D. (1976). The qualitative and quantitative determinations of ecdysones in tissues of the crab, *Pachygrapsus crassipes*, following molt induction. *Gen. Comp. Endocr. 30*, 21–33.

CHARMANTIER, G., OLLÉ, M. and TRILLES, J. P. (1976). Aspects of ecdysterone determination in *Sphaeroma serratum* (Crustacea: Isopoda: Flabellifera) and first results. *C.R. Acad. Sci. Ser. D. 283*, 1329–1331.

CHERBAS, L., YONGE, C. D., CHERBAS, P. and WILLIAMS, C. M. (1980). The morphological response of Kc-H cells to ecdysteroids: Hormonal specificity. *Wilhelm Roux's Archives 189*, 1–15.

CHERBAS, P., CHERBAS, L. and WILLIAMS, C. M. (1977). Induction of acetylcholinesterase activity by β-ecdysone in *Drosophila* cell line. *Science 197*, 275–277.

CHERBAS, P., TRAINOR, D. A., STONARD, R. J. and NAKANISHI, K. (1982). 14-Deoxymuristerone, a compound exhibiting exceptional moulting hormone activity. *J. Chem. Soc. Chem. Commun.* 1307–1308.

CHIHARA, C. J., PETRI, W. H., FRISTROM, J. W. and KING, D. S. (1972). The assay of ecdysones and juvenile hormones on *Drosophila* imaginal disks *in vitro*. *J. Insect Physiol. 18*, 1115–1123.

CHINO, H., SAKURAI, S., OHTAKI, T., IKEKAWA, N., MIYAZAKI, H., ISHIBASHI, M. and ABUKI, H. (1974). Biosynthesis of α-ecdysone by prothoracic glands *in vitro*. *Science 183*, 529–530.

CHONG, Y. K., GALBRAITH, M. N. and HORN, D. H. S. (1970). Isolation of deoxycrustecdysone, deoxyecdysone and α-ecdysone from the fern *Blechnum minus*. *J. Chem. Soc., Chem. Commun.* 1217–1218.

CHOU, W.-S. and LU, H.-S. (1980). Growth regulation and silk production in *Bombyx mori* L. from phytogenous ecdysteroids. In *Progress in Ecdysone Research*. Edited by J. A. Hoffmann. Pages 281–297. Elsevier/North-Holland Biomedical Press, Amsterdam.

CLEVER, U. and KARLSON, P. (1960). Induktion von puff-veranderungen in den speicheldrusenchromosomen von *Chironomus tentans* durch ecdyson. *Exp. Cell Res. 20*, 623–627.

COCHRANE, J. S. and HANSON, J. R. (1971). An alternative synthesis of 2β,3β,14α,17β-Tetrahydroxy-5β-androst-7-en-6-one. *J. Chem. Soc.* (C). 3730–3732.

COURGEON, A. (1972). Effects of α- and β-ecdysone on *in vitro* diploid cell multiplication in *Drosophila melanogaster*. *Nature New Biol. 238*, 250–251.

DAMMEIER, B. and HOPPE, W. (1971). Die kristall- und molekül strukturanalyse — des-insekten-hautungs-hormons 20-hydroxy-ecdyson (ecdysteron). *Chem. Ber. 104*, 1660–1673.

DANIELI, B., FERRARI, G., KREPINSKY, J., PALMISANO, G. and SARDINI, D. (1974). The determination of absolute configuration at C24 of the phytoecdysone makisterone A. *J. Chem. Soc. Chem. Commun.* 745–746.

DAVIES, T. G., LOCKLEY, J. S., BOID, R., REES, H. H. and GOODWIN, T. W. (1980). Mechanism of formation of the A/B *cis* ring junction of ecdysteroids in *Polypodium vulgare*. *Biochem. J. 190*, 537–544.

DELAAGE, M. A., HIRN, M. H. and DE REGGI, M. L. (1982). Radioimmunoassay of ecdysteroids. *Methods Enzymol. 84* (Immunochem. Tech., Pt. D), 350–358.

DELBECQUE, J. P., PROST, M., MAUME, B. F., DELCHRAMBRE, J., LAFONT, R., MAUCHAMP, B. (1975). Determination of β-ecdysone level by mass fragmentography during the last larval stage of *Tenebrio molitor*. *C.R. Acad. Sci.*, Ser D. *281*, 309–312.

DELBECQUE, J. P., DIEHL, P. A. and O'CONNOR, J. D. (1978a). Presence of ecdysone and ecdysterone in the tick *Amblyomma hebraeum* Koch. *Experientia. 34*, 1379–1381.

DELBECQUE, J. P., LANZREIN, B., BORDEREAU, C., IMBODEN, H., HIRN, M., O'CONNOR, J. D., NOIROT, C. and LUSCHER, M. (1978b). Ecdysone and ecdysterone in physogastric termite queens and eggs of *Macrotermes bellicosus* and *Macrotermes subhyalinus*. *Gen. Comp. Endocr. 36*, 40–47.

DE REGGI, M. L., HIRN, M. H. and DELAAGE, M. A. (1975). Radioimmunossay of ecdysone: an application to *Drosophila* larvae and pupae. *Biochem. Biophys. Research Commun. 66*, 1307–1315.

DIAS, J. R. (1977). Synthesis of 5β-chol-7-en-6-one analogues of ecdysone from cholic acid. *J. Chem. Eng. Data. 22*, 445–448.

DINAN, L. and REES, H. H. (1978). Preparation of 3-epi-ecdysone and 3-epi-20-hydroxyecdysone. *Steroids 32*, 629–638.

DINAN, L. N. and REES, H. H. (1981). The identification and titres of conjugated and free ecdysteroids in developing ovaries and newly-laid eggs of *Schistocerca gregaria*. *J. Insect Physiol. 27*, 51–58.

DINAN, L. N., DONNAHEY, P. L., REES, H. H. and GOODWIN, T. (1981). High-performance liquid chromatography of ecdysteroids and their 3-epi, 3-dehydro and 26-hydroxy derivatives. *J. Chromatogr. 205*, 139–145.

DJERASSI, C., KNIGHT, J. C. and BROCKMANN, H. (1964). Neue Sterine aus dem Kaktus *Wilcoxia viperina*. *Chem. Ber. 97*, 3118–3130.

EDWARDS, J. A., FRIED, J. H. and SIDDALL, J. B. (1967a). 2β,3β-alkylidenedioxy-6-hydroxy-22,23-bisnorchol-7-enoic acid esters. *U.S. Pat.* 3,354,154. *Chem. Abs. 68*, 114873w (1968).

EDWARDS, J. A., FRIED, J. H. and SIDDALL, J. B. (1967b). 2β,3β-alkylidenedioxy-6-oxygenated Δ^7-steroids. *U.S. Pat.* 3,354,152. *Chem. Abs. 69*, 19435s (1968).

EDWARDS, J. A., FRIED, J. H. and SIDDALL, J. B. (1968). 2,3-Acetonides of polyhydroxybisnorchol-7-enoic acid derivatives. *U.S. Pat.* 3,378,549. *Chem. Abs. 69*, 87388d (1968).

EKINS, R. P. (1974). Basic principles and theory. In *Radioimmunoassay and Saturation analysis*. Edited by P. H. Sonksen *British Medical Bulletin 30*, 3–11.

FAUX, A., HORN, D. H. S., MIDDLETON, E. J., FALES, H. M. and LOWE, M. E. (1969). Moulting hormones of a crab during ecdysis. *J. Chem. Soc. Chem. Commun.* 175–176.

FAUX, A., GALBRAITH, M. N., HORN, D. H. S., MIDDLETON, E. J. and THOMSON, J. A. (1970). The structure of two ecdysone analogues, cheilanthones A and B from the fern *Cheilanthes tenuifolia*. *J. Chem. Soc. Chem. Commun.* 243–244.

FAUX, A. F., HORN, D. H. S., KINNEAR, J. F., MARTIN, M.-D., WILLKIE, J. S. and WILLING, R. I. (1979). Evaluation of 3β-hydroxy-5β-cholest-7-en-6-one and related steroids as precursors of ecdysteroids in *Calliphora stygia*. *Insect Biochem. 9*, 101–105.

FELIPPE, G. M. (1980). Insect growth hormones and their effects on some plants. *Cienc. Cult. (Sao Paolo) 32*, 1384–1390. *Chem. Abs. 94*, 78352b.

FEYEREISEN, R. (1980). Ecdysteroid-Protein interactions: specificity of the ecdysteroid binding protein from *Locusta migratoria* haemolymph. In *Progress in Ecdysone Research*. Edited by J. A. Hoffmann. Pages 325–334. Elsevier/North Holland, Amsterdam.

FRAENKEL, G. (1935). A hormone causing pupation in the blowfly *Calliphora erythrocephala*. *Proc. R. Soc. Lond.* B. *118*, 1–12.

FRISTROM, J. W. and MITCHELL, H. K. (1965). The preparative isolation of imaginal discs from larvae of *Drosophila melanogaster*. *J. Cell. Biol. 27*, 445–448.

FRISTROM, J. W. and YUND, M. A. (1976). Characteristics of the action of ecdysones on *Drosophila* imaginal discs, cultured *in vitro*. In *Invertebrate Tissue Culture Research Applications*. Pages 161–178. Edited by Karl Maramorosch. Academic Press Inc., New York.

FRISTROM, J. W., YUND, M. A. and KING, D. S. (1978). Ecdysteroid receptors in imaginal discs of *Drosophila melanogaster*. *Proc. Natl. Acad. Sci. 75*, 6039–6043.

FRISTROM, J. W. and YUND, M. A. (1980). A comparative analysis of ecdysteroid action in larval and imaginal tissues of *Drosophila melanogaster*. In *Progress in Ecdysone Research*. Edited by J. A. Hoffmann. Pages 349–362. Elsevier/North Holland, Amsterdam.

FUKUZAWA, A., KUMAGAI, Y., MASAMUNE, T., FURUSAKI, A., KATAYAMA, C. and MATSUMOTO, T. (1981). Acetyl-pinnasterol and pinnasterol, ecdysone-like metabolites, from the marine red alga *Laurencia pinnata* Yamada. *Tetrahedron Lett. 22*, 4085–4086.

FURUTACHI, N., NAKADAIRA, Y. and NAKANISHI, K. (1968). Photo-oxygenation of βγ-unsaturated ketones. *J. Chem. Soc. Chem. Commun.* 1625–1626.

FURLENMEIER, A., FURST, A., LANGEMANN, A., WALDVOGEL, G., HOCKS, P., KERB, U. and WIECHERT, R. (1966a). The Synthesis of ecdysone. *Experientia 22*, 573.

FURLENMEIER, A., FURST, A., LANGEMANN, A., WALDVOGEL, G., KERB, U., HOCKS, P. and WIECHERT, R. (1966b). Zur synthesen von 2β,3β,14α-trihydroxy-6-keto-Δ^7-A/B-*cis*-steroiden. *Helv. Chim. Acta. 49*, 1591–1601.

FURLENMEIER, A., FURST, A., LANGEMANN, A., WALDVOGEL, G., HOCKS, P., KERB, U. and WIECHERT, R. (1967). Zur Synthese des ecdysons. IX. Mitteilung uber insekten-hormone. *Helv. Chim. Acta. 50*, 2387–2396.

GABE, M. (1953). Sur l'existence, chez quelques crustaces malacostraces d'un organe comparable a la glande de la mue des insectes. *C.R. Acad. Sci. Paris 237*, 1111–1113.

GAGOSIAN, R. B., BOURBONNIERE, R. A., SMITH, W. B., COUCH, E. F., BLANTON, C. and NOVAK, V. (1974). Lobster moulting hormones: isolation and biosynthesis of ecdysones. *Experientia 30*, 723–724.

GALBRAITH, M. N. and HORN, D. H. S. (1966). An insect moulting hormone from a plant. *J. Chem. Soc., Chem. Commun.* 905–906.

GALBRAITH, M. N., HORN, D. H. S., HOCKS, P., SCHULZ, G. and HOFFMEISTER, H. (1967). The identity of the 20-hydroxy-ecdysones from various sources. *Naturwissenschaften 54*, 471–472.

GALBRAITH, M. N., HORN, D. H. S., MIDDLETON, E. J. and HACKNEY, R. J. (1968a). A silyl ether as a protecting group in the synthesis of an isomer of ecdysone. 20S-hydroxy-22-deoxyecdysone. *J. Chem. Soc., Chem. Commun.* 466–467.

GALBRAITH, M. N., HORN, D. H. S., MIDDLETON, E. J. and HACKNEY, R. J. (1968b). Structure of deoxycrustecdysone, a second crustacean moulting hormone. *J. Chem. Soc., Chem. Commun.* 83–85.

GALBRAITH, M. N., HORN, D. H. S., PORTER, Q. N. and HACKNEY, R. J. (1968c). Structure of podecdysone A, a steroid with moulting hormone activity from the bark of *Podocarpus elatus* R. Br. *J. Chem. Soc., Chem. Commun.* 971–972.

GALBRAITH, M. N. and HORN, D. H. S. (1969c). Insect moulting hormones: crustecdysone (20-hydroxyecdysone) from *Podocarpus elatus. Aust. J. Chem. 22*, 1045–1057.

GALBRAITH, M. N., HORN, D. H. S., MIDDLETON, E. J. and HACKNEY, R. J. (1969a). Moulting hormones of insects and crustaceans: the synthesis of 2-deoxy-3-epicrustecdysone. *Aust. J. Chem. 22*, 1059–1067.

GALBRAITH, M. N., HORN, D. H. S., MIDDLETON, E. J. and HACKNEY, R. J. (1969b). Moulting hormones of insects and crustaceans: the synthesis of 22-deoxycrustecdysone. *Aust. J. Chem. 22*, 1517–1524.

GALBRAITH, M. N., HORN, D. H. S., THOMSON, J. A., NEUFELD, G. J. and HACKNEY, R. J. (1969c). Insect moulting hormones: crustecdysone in *Calliphora. J. Insect Physiol. 15*, 1225–1233.

GALBRAITH, M. N., HORN, D. H. S., MIDDLETON, E. J. and HACKNEY, R. J. (1969d). The structure of podecdysone B, a new phytoecdysone. *J. Chem. Soc., Chem. Commun.* 402–403.

GALBRAITH, M. N., HORN, D. H. S., MIDDLETON, E. J. and THOMSON, J. A. (1973a). Biological activity of synthetic moulting hormone analogues in the blowfly *Calliphora stygia. Experientia 29*, 19.

GALBRAITH, M. N., HORN, D. H. S., MIDDLETON, E. J., KAPLANIS, J. N. and THOMPSON, M. J. (1973b). Structure of podecdysone C, a steroid with moulting hormone activity from the bark of *Podocarpus elatus* R. Br. *Experientia 29*, 782.

GALBRAITH, M. N., HORN, D. H. S. and MIDDLETON, E. J. (1974). Insect moulting hormones: the synthesis of possible ecdysone precursors for metabolic studies. *Aust. J. Chem. 27*, 1087–1095.

GALBRAITH, M. N., HORN, D. H. S. and THOMSON, J. A. (1975). Insect moulting hormones: synthesis and biological activity of 2,25-dideoxy-α-ecdysone and deoxyecdysone. *Experienta 31*, 873.

GALBRAITH, M. N., HORN, D. H. S., KELLY, B. A., KINNEAR, J. F., MARTIN, M.-D., MIDDLETON, E. J. and VIRGONA, C. T. F. (1981). Moulting hormones. LIII. The synthesis and biological activity of some ecdysone analogues. *Aust. J. Chem. 34*, 2607–2618.

GANDE, A. R. and MORGAN, E. D. (1979). Ecdysteroids in the developing eggs of the desert locust, *Schistocerca gregaria. J. Insect Physiol. 25*, 289–293.

GANDE, A. R., MORGAN, E. D. and WILSON, I. D. (1979). Ecdysteroid levels throughout the life cycle of the desert locust, *Schistocerca gregaria. J. Insect Physiol. 25*, 669–675.

GEUNS, J. M. C. (1978). Steroid hormones and plant growth and development. *Phytochemistry 17*, 1–14.

GILGAN, M. W. and ZINCK, M. E. (1972). Estimation of ecdysterone from sulphuric acid-induced fluorescence. *Steroids 20*, 95–104.

GILGAN, M. W. (1976). Separation of ecdysterone, inokosterone, makisterone A, α-ecdysone and ponasterone A by a combination of absorptive and reversed-phase liquid chromatography. *J. Chromatogr. 129*, 447–450.

GOCAN, A. and HODOSAN, F. (1976). Exo- and Endohormones. III. New derivaties of synthetic ecdysone analogues and related compounds. *Rev. Roumaine Chem. 21*, 627–629.

GOODWIN, T. W., HORN, D. H. S., KARLSON, P., KOOLMAN, J., NAKANISHI, K., ROBBINS, W. E., SIDDALL, J. B. and TAKEMOTO, T. (1978). Ecdysteroids: a new generic term. *Nature 272*, 122.

GRACE, T. D. C. (1962). Establishment of four strains of cells from insect tissues grown *in vitro. Nature 195*, 788–789.

GREENWOOD, F. C. and HUNTER, W. M. (1962). Preparation of I-131 labelled human growth hormone of high specific activity. *Nature 194*, 495–496.

GRÖNEMEYER, H. and PONGS, O. (1980). Localization of ecdysterone on polytene chromosomes of *Drosophila melanogaster. Proc. Natl. Acad. Sci. 77*, 2108–2112.

HAFFERL, W., WREN, D. L., MARSHALL, J. P., CALZADA, M. C. and SIDDALL, J. B. (1972). Radiochemical synthesis. III. Synthesis of tritiated α-ecdysone and 25-deoxy-α-ecdysone of high specific activity. *J. Label. Compounds 8*, 81–94.

HAGEDORN, H. H., O'CONNOR, J. D., FUCHS, M. S., SAGE, B., SCHLAEGER, D. A. and BOHM, M. K. (1975). The ovary as a source of α-ecdysone in an adult mosquito. *Proc. Natl. Acad. Sci. 72*, 3255–3259.

HAMPSHIRE, F. and HORN, D. H. S. (1966). Structure of crustecdysone, a crustacean moulting hormone. *J. Chem. Soc., Chem. Commun.* 37–38.

HANNA, R., MUCKENSTURM, B. and CHERRY, A. (1976). Analogues diméthylés de l'ecdysone et de la rubrostérone. *Bull. Soc. Chim. France* 1865–1870.

HÄNSEL, R., LEUCHERT, C., RIMPLER, H. and SCHAAF, K. D. (1965). Chemotaxonomische untersuchungen in der gattung *Vitex* L. *Phytochemistry 4*, 19–27.

HARDMAN, R. and BENJAMIN, T. V. (1976). The cooccurrence of ecdysones with bufadienolides and steroidal saponins in the genus *Helleborus. Phytochemistry 15*, 1515–1516.

HARDMAN, R. and BENJAMIN, T. V. (1977). Quantitative determination of phytoecdysones as illustrated by application to species of *Helleborus. J. Chromatogr. 131*, 468–470.

HARDMAN, R. and MAHAR, S. M. A. (1978). Isolation, identification and quantification of ecdysones from *Aerva tomentosa* and *Pandiaka involucrata. Planta Medica 33*, 278–279.

HARDMAN, R. and BENJAMIN, T. V. (1980). The detection of *Helleborus* ecdysteroids by mass spectrometry and by GLC analysis of their TSIM-ethers. *Planta Medica 39*, 148–152.

HARRISON, I. T., SIDDALL, J. B. and FRIED, J. H. (1966). Steroids CCXCVII. Synthetic studies on insect hormones. Part III. An alternative synthesis of ecdysone and 22-isoecdysone. *Tetrahedron Lett., 29*, 3457–3460.

HEFTMANN, E. (1970). Insect moulting hormones in plants. *Rec. Adv. Phytochem. 3*, 211–227.

HEFTMANN, E. (1975). Steroid hormones in plants. *Lloydia 38*, 195–209.

HEINRICH, G. and HOFFMEISTER, H. (1967). Ecdyson als Begleitsubstanz des Ecdysterons in *Polypodium vulgare* L. *Experientia 23*, 995.

HEINRICH, G. and HOFFMEISTER, H. (1968). 5 β-Hydroxyecdysteron, ein Pflanzensteroid mit Häutungshormonaktivität aus *Polypodium vulgare* L. *Tetrhedron Lett. 58*, 6063–6064.

HEINRICH, G. and HOFFMEISTER, H. (1971). Insekten-häutungshormone und ihre Wirkungsweise. Bildung von hormonglycosiden als inaktivierungs-mechanismnus bei *Calliphora erythrocephala. Z. Naturforschung. 25b*, 358–361.

HENDRIX, S. D. and JONES, R. L. (1972). Activity of β-ecdysone in four gibberellin bioassays. *Plant Physiol. 50*, 199–200.

HEROUT, V. (1970). Some relations between plants, insects and their isoprenoids. *Prog. Phytochem. Vol. 2*, 143–202.

HETRU, C., LAGUEUX, M., BANG, L. and HOFFMANN, J. A. (1978). Adult ovaries of *Locusta migratoria* contain the sequence of biosynthetic intermediates for ecdysone. *Life Sci. 22*, 2141–2154.

HETRU, C. and HORN, D. H. S. (1980). Phytoecdysteroids and zooecdysteroids. In *Progress in Ecdysone Research.* Edited by J. A. Hoffmann. Pages 13–28. Elsevier/North Holland Biomedical Press, Amsterdam.

HETRU, C. C., KAPPLER, C., HOFFMANN, J. A., NEARN, R., BANG, L. and HORN, D. H. S. (1982). The biosynthetic pathway of ecdysone: studies with vitellogenic ovaries of *Locusta migratoria* (Orthoptera). *Mol. and Cell. Endocr., 26*, 51–80.

HIGHNAM, K. C. (1970). Mode of action of arthropod steroid and other hormones. In *Advances in Steroid Biochemistry and Pharmacology.* Edited by M. H. Briggs. Vol. 1, pages 1–49. Academic Press, London.

HIGHNAM, K. C. and HILL, L. (1977). *The Comparative Endocrinology of the Invertebrates*. Edward Arnold, London.

HIKINO, H., HIKINO, Y., NOMOTO, K. and TAKEMOTO, T. (1968a). Cyasterone, an insect metamorphosing substance from *Cyathula capitata*: structure. *Tetrahedron 24*, 4895–4906.

HIKINO, H., HIKINO, Y. and TAKEMOTO, T. (1968b). Synthesis of rubrosterone, a metabolite of insect-moulting substances from *Achyranthes rubrofusca*. *Tetrahedron Lett. 40*, 4255–4256.

HIKINO, H., NOMOTO, K. and TAKEMOTO, T. (1969a). Structure of sengosterone, a novel C_{29} insect-moulting substance from *Cyathula capitata*. *Tetrahedron Lett. 18*, 1417–1420.

HIKINO, H., HIKINO, Y. and TAKEMOTO, T. (1969b). Rubrosterone, a metabolite of insect metamorphosing substance from *Achyranthes rubrofusca*: synthesis. *Tetrahedron 25*, 3389–3394.

HIKINO, H. and HIKINO, Y. (1970). Arthropod molting hormones. In *Progress in the Chemistry of Organic Natural Products*. Edited by W. Herz, H. Grisebach and A. I. Scott. Vol. 28, pages 257–312. Springer-Verlag, Vienna.

HIKINO, H., NOMOTO, K., INO, R. and TAKEMOTO, T. (1970a). Structure of precyasterone, novel C_{29} insect-moulting substance from *Cyathula capitata*. *Chem. Pharm. Bull. 18*, 1078–1080.

HIKINO, H., NOMOTO, K. and TAKEMOTO, T. (1970b). Absolute configuration of cyasterone, an insect-moulting substance from *Cyathula capitata*. *Chem. Pharm. Bull. 18*, 2132–2133.

HIKINO, H., NOMOTO, K. and TAKEMOTO, T. (1970c). Sengosterone, an insect metamorphosing substance from *Cyathula capitata*: structure. *Tetrahedron 265*, 887–898.

HIKINO, H., NOMOTO, K. and TAKEMOTO, T. (1970d). Poststerone, a metabolite of insect metamorphosing substances. *Steroids 16*, 393–400.

HIKINO, H., NOMOTO, K. and TAKEMOTO, T. (1971a). Cyasterone, an insect metamorphosing substance from *Cyathula capitata*: absolute configuration. *Tetrahedron 27*, 315–321.

HIKINO, H., NOMOTO, K. and TAKEMOTO, T. (1971b). Isocyasterone, an insect metamorphosing substance from *Cyathula capitata*. *Phytochemistry 10*, 3173–3178.

HIKINO, H., NOMOTO, K. and TAKEMOTO, T. (1971c). Structure of isocyasterone and epicyasterone, novel C_{29} insect-moulting substance from *Cyathula capitata*. *Chem. Pharm. Bull. 19*, 433–435.

HIKINO, H., OKUYAMA, T., JIN, H. and TAKEMOTO, T. (1973). Screenings of Japanese ferns for phytoecdysones. I. *Chem. Pharm. Bull. 21*, 2292–2302.

HIKINO, H. and TAKEMOTO, T. (1974). Ecdysones of plant origin. In *Invertebrate Endocrinology and Hormonal Heterophylly*. Edited by W. J. Burdette. Pages 185–203. Springer-Verlag, New York.

HIKINO, H., OKUYAMA, T., ARIHARA, S., HIKINO, Y., TAKEMOTO, T., MORI, H. and SHIBATA, K. (1975a). Shidasterone, an insect metamorphosing substance from *Blechnum niponicum*: structure *Chem. Pharm. Bull. 23*, 1458–1479.

HIKINO, H., OKUYAMA, T., KONNO, C. and TAKEMOTO, T. (1975b). Carbon-13 nuclear magnetic resonance spectra of phytoecdysones. *Chem. Pharm. Bull. 23*, 125–132.

HIKINO, H. (1976). Steroids XXVIII. Ecdysterone and ecdysone from *Polypodium virginianum*. *Lloydia 39*, 246–247.

HIKINO, H., MOHRI, K., HIKINO, Y., ARIHARA, S., TAKEMOTO, T., MORI, H. and SHIBATA, K. (1976a). Inokosterone, an insect metamorphosing substance from *Achyranthes fauriei*. *Tetrahedron 32*, 3015–3021.

HIKINO, H., MOHRI, K., OKUYAMA, T., TAKEMOTO, T. and YEN, K.-Y. (1976b). Phytoecdysones from *Diplazium donianum*. *Steroids 28*, 649–653.

HIRN, M. H. and DELAAGE, M. A. (1980). Radioimmunological approaches to the quantification of ecdysteroids. In *Progress in Ecdysone Research*. Edited by J. A. Hoffman. Pages 69–82. Elsevier/North Holland, Amsterdam.

HOCKS, P. and WEICHERT, R. (1966). 20-Hydroxy-ecdyson, isoliert aus Insekten. *Tetrahedron Lett. 26*, 2989–2993.

HOCKS, P., JÄGER, A., KERB, U., WIECHERT, R., FURLENMEIER, A., FÜRST, A., LANGEMANN, A. and WALDVOGEL, G. (1966). Synthetic steroids with moulting hormone activity. *Angew Chem. Internat. Edit. 5*, 673.

HOCKS, P., KERB, U. and WIECHERT, R. (1968). Die Synthese des Rubrosterons. *Tetrahedron Lett. 40*, 4281–4284.

HOFFMANN, J. A. and KOOLMAN, J. (1974). Prothoracic glands in the regulation of ecdysone titres and metabolic fate of injected labelled ecdysone in *Locusta migratoria*. *J. Insect Physiol. 20*, 1593–1601.

HOFFMANN, J. A., LAGUEUX, M., HETRU, C., CHARLET, M. and GOLTZENE, F. (1980). Ecdysone in reproductively competent female adults and in embryos of insects. In *Progress in Ecdysone Research*. Edited by J. A. Hoffmann. Pages 431–465. Elsevier/North-Holland Biomedical Press, Amsterdam.

HOFFMANN, J. A., HETRU, C., LAGUEUX, M., CHARLET, M. and HIRN, M. (1983). Steroid hormones in invertebrates. *Nova Acta Leopoldina* (In press.)

HOFFMESITER, H. (1966a). Ecdysterone, a new metamorphosis hormone of insects. *Angew. Chem. Internat. Ed. 5*, 248.

HOFFMEISTER, H. (1966b). Isolierung eines neuen Insektenhormons aus Seidenspinnerpuppen. *Z. Naturf. 21b*, 335–336.

HOFFMEISTER, H. and GRÜTZMACHER, H. F. (1966). Zur Chemie des Ecdysterons. *Tetrahedron Lett. 33*, 4017–4023.

HOFFMEISTER, H., HEINRICH, G., STAAL, G. B. and VAN DER BURG, W. J. (1967). Über das Vorkommen von Ecdysteron in Eiben. *Naturwissenschaften 54*, 471.

HOLMAN, G. M. and MEOLA, R. W. (1978). A high performance liquid chromatography method for the purification and analysis of insect ecdysones: application to measurement of ecdysone titres during pupal-adult development of *Heliothis zea*. *Insect Biochem. 8*, 275–278.

HORI, M. (1969). Automatic column chromatographic method for insect-moulting steroids. *Steroids 14*, 33–46.

HORN, D. H. S., MIDDLETON, E. J., WUNDERLICH, J. A. and HAMPSHIRE, F. (1966). Identity of the moulting hormones of insects and crustaceans. *J. Chem. Soc., Chem. Commun.* 339–340.

HORN, D. H. S., FABBRI, S., HAMPSHIRE, F. and LOWE, M. E. (1968). Isolation of crustecdysone (20R-Hydroxyecdysone) from a crayfish (*Jasus lalandei* H. Milne-Edwards). *Biochem. J. 109*, 399–406.

HORN, D. H. S. (1971). The ecdysones. In *Naturally Occurring Insecticides*. Edited by M. Jacobson and D. G. Crosby. Chap. 9, pages 333–459. Marcel Dekker, New York.

HORN, D. H. S., MIDDLETON, E. J., THOMSON, J. A. and WILKIE, J. S. (1974a). Metabolism of some possible ecdysone precursors in *Calliphora stygia*. *J. Insect Physiol. 20*, 2433–2445.

HORN, D. H. S., WILKIE, J. S. and THOMSON, J. A. (1974b). Isolation of β-ecdysone (20-hydroxyecdysone) from the parasitic nematode *Ascaris lumbricoides*. *Experientia 30*, 1109–1110.

HORN, D. H. S., WILKIE, J. S., SAGE, B. A. and O'CONNOR, J. D. (1976). A high affinity antiserum specific for the ecdysone nucleus. *J. Insect Physiol. 22*, 901–905.

HORN, D. H. S., KELLY, B. A., KINNEAR, J. F., MARTIN, M.-D., MIDDLETON, E. J. and WILLING, R. I. (1982). Insect moulting hormones. LIV. The synthesis and biological activity of some alkylated ecdysone analogues. *Aust. J. Chem. 35*, 629–640.

HSIAO, T. H. and HSIAO, C. (1979). Ecdysteroids in the ovary and the egg of the greater wax moth. *J. Insect Physiol. 25*, 45–52.

HÜBER, R. and HOPPE, W. (1965a). Zur Chemie des Ecdysons. II. Bestimmung des Sterin-Skeletts und Seiner Orientierung mit diffuser Röntgenstreuung in Kristallen von Ecdysone. *Chem. Ber. 98*, 2353–2360.

HÜBER, R. and HOPPE, W. (1965b). Zur Chemie des Ecdysons, VII. Die Kristall-und Molekulstruanalyse des Insektenverpuppungs Hormons Ecdyson mit der automatisierten Faltmolekülmethode. *Chem. Ber. 98*, 2403–2424.

HÜPPI, G. and SIDDALL, J. B. (1967). Steroids CCCXXXIII. Synthetic studies on insect hormones. V. The synthesis of crustecdysone (20-hydroxyecdysone). *J. Amer. Chem. Soc. 89*, 6790–6792.

HÜPPI, G. and SIDDALL, J. B. (1968). Steroids CCCXXXVI. Synthetic studies on insect hormones. Part VI. The synthesis of ponasterone A and its stereochemical identity with crustecdysone. *Tetrahedron Lett. 9*, 1113–1114.

IKAN, R. and RAVID, U. (1970). The isolation and identification of ecdysterone from *Ajuga iva*. *Planta Medica 20*, 33–34.

IKAN, R. and RAVID, U. (1971). The isolation and identification of cyasterone from *Ajuga chia* (Labiatae). *Phytochemistry 10*, 1659–1661.

IKEKAWA, N., HATTORI, F., RUBIO-LIGHTBOURN, J., MIYAZAKI, H., ISHIBASHI, M. and MORI, C. (1972). Gas chromatographic separation of phytoecdysones. *J. Chromatogr. Sci. 10*, 233–242.

IKEKAWA, N., IKEDA, T., MIZUNO, T., OHNISHI, E. and SAKURAI, S. (1980). Isolation of a new ecdysteroid 2,22-dideoxy-20-hydroxyecdysone from the ovaries of the silkworm *Bombyx mori*. *J. Chem. Soc., Chem. Commun.* 448–449.

IMAI, S., FUJIOKA, S., NAKANISHI, K., KOREEDA, M. and KUROKAWA, T. (1967). Extraction of ponasterone A and ecdysterone from Podocarpaceae and related plants. *Steroids 10*, 557–565.

IMAI, S., HORI, M., FUJIOKA, S., MURATA, E., GOTO, M. and NAKANISHI, K. (1968a). Insect hormones. VIII. Isolation of four new phytoecdysones, makisterone A, B, C, D and the structure of makisterone A, a C_{28} steroid. *Tetrahedron Lett. 36*, 3883–3886.

IMAI, S., FUJIOKA, S., MURATA, E., SASAKAWA, Y. and NAKANISHI, K. (1968b). Insect hormones. IX. The structure of three additional phytoecdysones from *Podocarpus macrophyllus*, makisterone B, C, and D. *Tetrahedron Lett. 36*, 3887–3890.

IMAI, S., TOYOSATO, T., SAKAI, M., SATO, Y., FUJIOKA, S., MURATA, E. and GOTO, M. (1969a). Screening results of plants for phytoecdysones. *Chem. Pharm. Bull. 17*, 335–339.

IMAI, S., MURATA, E., FUJIOKA, S., KOREEDA, M. and NAKANISHI, K. (1969b). Structure of ajugasterone C, a phytoecdysone with an 11-hydroxy group. *J. Chem. Soc., Chem. Commun.* 546–547.

IMAI, S., TOYOSATO, T., SAKAI, M., SATO, Y., FUJIOKA, S., MURATA, E. and GOTO, M. (1969c). Isolation of cyasterone and ecdysterone from plant materials. *Chem. Pharm. Bull. 17*, 340–342.

IMAI, S., FUJIOKA, S., MURATA, E., OTSUKA, K. and NAKANISHI, K. (1969d). Structure of the phytoecdysone ajugasterone B. *J. Chem. Soc. Chem. Commun.* 82–83.

IMAI, S., MURATA, E., FUJIOKA, S., MATSUOKA, T., KOREEDA, M. and NAKANISHI, K. (1970a). Structures of stachysterone A, the first natural 27-carbon steroid with a rearranged methyl group, and stachysterone B. *J. Amer. Chem. Soc. 92*, 7510–7512.

IMAI, S., MURATA, E., FUJIOKA, S., MATSUOKA, T., KOREEDA, M. and NAKANISHI, K. (1970b). Structures of stachysterones C and D. *J. Chem. Soc., Chem. Commun.* 352–353.

IMBODEN, H., LANZREIN, B., DELBECQUE, J. P. and LÜSCHER, M. (1978). Ecdysteroids and juvenile hormone during embryogenesis in the ovoviviparus cockroach *Nauphoeta cinerea*. *Gen. Compl. Endocr. 36*, 628–635.

ISAAC, R. E., REES, H. H. and GOODWIN, T. W. (1981a). Isolation of 2-deoxy-20-hydroxyecdysone and 3-epi-2-deoxyecdysone from eggs of the desert locust. *Schistocerca gregaria*, during embryogenesis. *J. Chem. Soc., Chem. Commun.* 418–420.

ISAAC, R. E., REES, H. H. and GOODWIN, T. W. (1981b). Isolation of ecdysone-3-acetate as a major ecdysteroid from the developing eggs of the desert locust *Schistocerca gregaria*. *J. Chem. Soc., Chem. Commun.* 594–595.

ISAAC, R. E., ROSE, M. E., REES, H. H. and GOODWIN, T. W. (1982). Identification of ecdysone-22-phosphate and 2-deoxyecdysone-22-phosphate in eggs of the desert locust *Schistocerca gregaria*, by fast atom bombardment mass spectrometry and n.m.r. spectoscopy. *J. Chem. Soc. Chem. Commun.* 249–251.

IYER, R. T., AYENGAR, K. N. N. and RANGASWAMI, S. (1973). Occurrence of cheilanthone-B in *Cheilanthes mysurensis*. *Indian J. Chem. 11*, 1336–1338.

JACOBS, W. P. and SUTHERS, H. B. (1971). Culture of apical buds of *Xanthium* and their use as a bioassay for flowering activity of ecdysterone. *Amer. J. Bot. 58*, 836–843.

JEGLA, T. C. and COSTLOW, J. D. (1979a). The *Limulus* bioassay for ecdysteroids. *Biol. Bull. 156*, 103–114.

JEGLA, T. C. and COSTLOW, J. D. (1979b). Ecdysteroids in *Limulus* larvae. *Experientia 35*, 554–555.

JIZBA, J., HEROUT, V. and SORM, F. (1967a). Isolation of ecdysterone (crustecdysone) from *Polypodium vulgare* L. rhizomes. *Tetrahedron Lett. 18*, 1689–1691.

JIZBA, J., HEROUT, V. and SORM, F. (1967b). Polypodine B — a novel ecdysone-like substance from plant material. *Tetrahedron Lett. 51*, 5139–5143.

JIZBA, J., DOLEJS, L., HEROUT, V., SORM, F., FEHLHABER, H. W., SNATZKE, G., TSCHESCHE, R. and WULFF, F. (1971a). Polypodosaponin, ein neuer Saponintyp aus *Polypodium vulgare* L. *Chem. Ber. 104*, 837–846.

JIZBA, J., DOLEJS, L., HEROUT, V. and SORM, F. (1971b). The structure of osladin — the sweet principle of the rhizomes of *Polypodium vulgare* L. *Tetrahedron Lett. 18*, 1329–1332.

JIZBA, J., VASICKOVA, S. and HEROUT, V. (1974). Components of the fern *Polypodium aureum* L. *Coll. Czech. Chem. Commun. 39*, 501–505.

JIZBA, J., DOLEJS, L. and HEROUT, V. (1974). Polypodoaurein, a new phytoecdysone from *Polypodium aureum* L. *Phytochemistry 13*, 1915–1916.

JONES, C. G. and FIRN, R. D. (1978). The role of phytoecdysteroids in bracken fern, *Pteridium aquilinum* (L.) Kuhn as a defence against phytophagous insect attack. *J. Chem. Ecol. 4*, 117–138.

KAMETANI, T., TSUBUKI, M. and NEMOTO, H. (1980a). A sterocontrolled total synthesis of 2β,3β,20β-triacetoxy-5α-pregnan-6-one — a total synthesis of 20-hydroxyecdysone. *Tetrahedron Lett. 21*, 4855–4856.

KAMETANI, T., TSUBUKI, M. and NEMOTO, H. (1980b). Total synthesis of steroid hormones. Efficient stereocontrolled synthesis of 17-methoxy-6-oxo-D-homo-18-nor-5β-androsta-2,13,15,17-tetraene. *J. Org. Chem. 45*, 4391–4398.

KAMETANI, T., TSUBUKI, M. and NEMOTO, H. (1981). Total synthesis of steroids by cycloaddition reaction — formal total synthesis of 20-hydroxyecdysone. *Chem. Soc. Jap.* 819–824.

KAPLANIS, J. N., TABOR, L. A., THOMPSON, M. J., ROBBINS, W. E. and SHORTINO, T. J. (1966a). Assay for ecdysone (moulting hormone) activity using the housefly, *Musca domestica* L. *Steroids 8*, 625–631.

KAPLANIS, J. N., THOMPSON, M. J., YAMAMOTO, R. T., ROBBINS, W. E. and LOULOUDES, S. J. (1966b). Ecdysones from the pupa of the tobacco hornworm, *Manduca sexta* (Johannson). *Steroids 8*, 605–623.

KAPLANIS, J. N., THOMPSON, M. J., ROBBINS, W. E. and BRYCE, B. M. (1967). Insect hormones: alpha ecdysone and 20-hydroxyecdysone in bracken fern. *Science 157*, 1436–1438.

KAPLANIS, J. N., THOMPSON, M. J., ROBBINS, W. E. and LINDQUIST, E. L. (1972). α-Ecdysone and 20-hydroxyecdysone from frass of the tobacco hornworm, *Manduca sexta* (L.). *Steroids 20*, 621–625.

KAPLANIS, J. N., ROBBINS, W. E., THOMPSON, M. J. and DUTKY, S. R. (1973). 26-Hydroxyecdysone: new insect moulting hormone from the egg of the tobacco hornworm. *Science 180*, 307–308.

KAPLANIS, J. N., DUTKY, S. R., ROBBINS, W. E., THOMPSON, M. J., LINDQUIST, E. L., HORN, D. H. S. and GALBRAITH, M. N. (1975). Makisterone A: a 28-carbon hexahydroxy moulting hormone from the embryo of the milkweed bug. *Science 190*, 681–682.

KAPLANIS, J. N., ROBBINS, W. E., THOMPSON, M. J. and DUTKY, S. R. (1976). The molting hormones from the embryonated egg of the tobacco hornworm *Manduca sexta* (L.). *Steroids 27*, 675–679.

KAPLANIS, J. N., THOMPSON, M. J., DUTKY, S. R. and ROBBINS, W. E. (1979). The ecdysteroids from the tobacco hornworm during pupal–adult development five days after peak titre of moulting hormone activity. *Steroids 34*, 333–345.

KAPLANIS, J. N., THOMPSON, M. J., DUTKY, S. R. and ROBBINS, W. E. (1980). The ecdysteroids from young embryonated eggs of the tobacco hornworm. *Steroids 36*, 321–335.

KARLSON, P. and HANSER, G. (1953). Bildungsort und Erfolgsorgan des Puparisierungshormons der Fliegen (*Calliphora erythrocephala*). *Z. Naturf. 8b*, 91–96.

KARLSON, P. (1956a). Chemical investigation of the metamorphosis hormone of insects. *Ann. Sci. Nat. Zool. Biol. Animale 18*, 125–137.

KARLSON, P. (1956b). Biochemical studies on insect hormones. *Vit. Horm. 14*, 227–266.

KARLSON, P. and SCHMIALEK, P. (1959). Notiz über die Isolierung von Malonsaurer-amid-[β-phenyl-äthylamid] aus Garneelen. *Hoppe-Seyler's Z. Physiol. Chem. 316*, 83–87.

KARLSON, P., HOFFMEISTER, H., HOPPE, W. and HÜBER, R. (1963). Zur Chemie des Ecdysons. *Justus Liebig's Ann. Chemie 662*, 1–20.

KARLSON, P. and SHAAYA, E. (1964). The ecdysone titre of developing insects. I. A method of determining the ecdysone content. *J. Insect Physiol. 10*, 797–804.

KARLSON, P., HOFFMEISTER, H., HUMMEL, H., HOCKS, P. and SPITELLER, G. (1965). Zur chemie des ecdysons VI. Reaktionen des Ecdysonmoleküls. *Chem. Ber. 98*, 2394–2402.

KARLSON, P., BUGANY, H., DOÖPP, H. and HOYER, G.- A. (1972). 3-Dehydroecdysone, a metabolite of ecdysone in the blowfly *Calliphora erythrocephala*. Meigen. *Hoppe-Seyler's Z. Physiol. Chem. 353*, 1610–1614.

KATZ, M. and LENSKY, Y. (1970). Gas chromatographic analysis of ecdysone. *Experientia 26*, 1043.

KELLER, R. and SCHMID, E. (1979). *In vitro* secretion of ecdysteroids by Y-organs and lack of secretion by mandibular organs of the crayfish following molt induction. *J. Comp. Physiol. 130*, 347–353.

KELLY, T. J. and FUCHS, M. S. (1980). *In vivo* induction of ovarian development in decapitated *Aedes atropalpus* by physiological levels of 20-hydroxyecdysone. *J. Exp. Zool. 213*, 25–32.

KERB, U., HOCKS, P., WIECHERT, R., FURLENMEIER, A., FÜRST, A., LANGEMANN, A. and WALDVOGEL, G. (1966a). Die Synthese des Ecdysons. *Tetrahedron Lett. 13*, 1387–1391.

KERB, U., SCHULZ, G., HOCKS, P., WIECHERT, R., FURLENMEIER, A., FÜRST, A., LANGEMANN, A. and WALDVOGEL, G. (1966b). Zur Synthese des Ecdysons. IV. Mitteilung über Insectenhormone. Die Synthese des natürlichen Häutungshormons. *Helv. Chim. Acta. 49*, 1601–1606.

KERB, U., WIECHERT, R., FURLENMEIER, A. and FÜRST, A. (1968). Über eine Synthese des Crustecdysons (20-Hydroxyecdyson). *Tetrahedron Lett. 40*, 4277–4280.

KHOLODOVA, Y. D. (1977). Use of the Chugaev reaction for the quantitative determination of ecdysones. *Khim. Prir. Soedin.*, 227–230; *Chem. Abs. 87*, 98343h (1977).

KHOLODOVA, Y. D., BALTAEV, U., VOLOVENKO, V. O., GOROVITS, M. B. and ABUBAKIROV, N. K. (1979). Phytoecdysones of *Serratula xeranthemoides. Khim. Prir. Soedin. 2*, 171–174.

KINNEAR, J. F., MARTIN, M.-D., HORN, D. H. S., MIDDLETON, E. J., WILKIE, J. S., GALBRAITH, M. N. and WILLING, R. I. (1976). Insect moulting hormones: The synthesis of ecdysone analogues from a bile acid. *Aust. J. Chem. 29*, 1815–1824.

KINNEAR, J. F., MARTIN, M.-D., CHONG, Y. K., FAUX, A., HORN, D. H. S. and WILKIE, J. S. (1978). Insect moulting hormones: possible intermediates in the metabolism of cholesterol to ecdysteroids in an insect. *Aust. J. Chem. 31*, 2069–2075.

KINNEAR, J. F., MARTIN, M.-D., FAUX, A. F., HORN, D. H. S. and WILKIE, J. S. (1979). Insect moulting hormones: a study of 4-en-3-one steroids as possible ecdysteroid precursors. *Aust. J. Chem. 32*, 2017–2024.

KNIGHT, J. C. and PETTIT, G. R. (1969). Arizona flora: the sterols of *Peniocereus greggii. Phytochemistry 8*, 477–482.

KOBAYASHI, M., NAKANISHI, K. and KOREEDA, M. (1967). The moulting hormone activity of ponasterones on *Musca domestica* (Diptera) and *Bombyx mori* (Lepidoptera). *Steroids 9*, 529–536.

KOOLMAN, J. (1977). Insect hormones. *Natural Methodicum Chemicum. 11*, 208–228.

KOOLMAN, J. and SPINDLER, K. D. (1977). Enzymic and chemical synthesis of 3-dehydroecdysterone, a metabolite of the molting hormone of insects. *Hoppe-Seyler's Z. Physiol. Chem. 358*, 1339–1344.

KOOLMAN, J. and KARLSON, P. (1978). Ecdysone oxidase: Reaction and Specificity. *Eur. J. Biochem. 89*, 453–460.

KOOLMAN, J., REUM, L. and KARLSON, P. (1979a). 26-Hydroxyecdysone, 20,26-dihydroxyecdysone and inokosterone detected as metabolites of ecdysone in the blowfly *Calliphora vicina* by radiotracer experiments. *Hoppe-Seyler's Z. Physiol. Chem. 360*, 1351–1355.

KOOLMAN, J., SCHELLER, K. and BODENSTEIN, D. (1979b). Ecdysteroids in the adult male blowfly *Calliphora vicina. Experientia 35*, 134–135.

KOOLMAN, J. (1980). Analysis of ecdysteroids by fluorometry. *Insect Biochem. 10*, 381–386.

KOPEČ, S. (1922). Studies of the necessity of the brain for the inception of insect metamorphosis. *Biol. Bull. 42*, 323–342.

KOREEDA, M., NAKANISHI, K., IMAI, S., TSUCHIYA, T. and WASADA, N. (1969a). Mass spectrometric studies of ecdysone derivatives. *Mass Spectrosc. 17*, 669–680.

KOREEDA, M., HARADA, N. and NAKANISHI, K. (1969b). Interaction between non-adjacent benzoate groups: an extension of the dibenzoate chirality rule. *J. Chem. Soc., Chem. Commun.* 548–549.

KOREEDA, M. and NAKANISHI, K. (1970a). Ajugalactone, an insect moulting inhibitor as tested by the Chilo dipping method. *J. Amer. Chem. Soc. 92*, 7512–7513.

KOREEDA, M. and NAKANISHI, K. (1970b). 5 β-hydroxy-ecdysones and a revision of the structure of ponasterone C. *J. Chem. Soc. Chem. Commun.* 351–352.

KOREEDA, M., SCHOOLEY, D. A., NAKANISHI, K. and HAGIWARA, H. (1971). The absolute configurations at C-20 and C-22 in ecdysones. *J. Amer. Chem. Soc. 93*, 4084–4085.

KOREEDA, M. and TEICHER, B. A. (1977). Chemical analysis of insect molting hormones. In *Analytical Biochemistry of Insects*. Edited by R. B. Turner. Pages 207–240, Elsevier, Amsterdam.

KREPINSKY, J., FINDLAY, J. A., DANIELI, B., PALMISANO, G., BEYNON, P. and MURAKAMI, S. (1977). Concerning carbon-13 n.m.r. spectroscopy of 5β-hydroxylated phytoecdysones. *Org. Mag. Res. 10*, 255–257.

KUSANO, G., TAKEMOTO, T., BEISLER, J. A. and SATO, Y. (1975). Steroidal constituents of *Solanum xanthocarpum. Phytochemistry 14*, 529–532.

LACHAISE, F. and HOFFMANN, J. A. (1977). Ecdysone and ovarian development in a decapod, *Carcinus maenas. C.R. Acad. Sci. Ser. D. 285*, 701–704.

LACHAISE, F., GOUDEAU, M., HETRU, C., KAPPLER, C. and HOFFMAN, J. A. (1981). Ecdysteroids and ovarian development in the shore crab, *Carcinus maenas. Hoppe-Seyler's Z. Physiol. Chem. 362*, 521–529.

LAFONT, R., MAUCHAMP, B., PENNETIER, J. L., TARROUX, P., DE HYS, L. and DELBECQUE, J. P. (1975). α- and β-Ecdysone levels in insect Haemolymph: correlation with developmental events. *Experientia 31*, 1241–1242.

LAFONT, R., MARTIN-SOMMÉ, G. and CHAMBET, J.-C. (1979). Separation of ecdysteroids by using high-pressure liquid chromatography on microparticulate supports. *J. Chromatogr. 170*, 185–194.

LAFONT, R., SOMMÉ-MARTIN, G., MAUCHAMP, B., MAUME, B. F. and DELBECQUE, J.-P. (1980a). Analysis of ecdysteroids by high-performance liquid chromotography and coupled gas–liquid chromatography–mass spectrometry. In *Progress in Ecdysone Research.* Edited by J. A. Hoffmann. Pages 45–68. Elsevier/North Holland, Amsterdam.

LAFONT, R., BEYDON, P., SOMMÉ-MARTIN, G. and BLAIS, C. (1980b). High-performance liquid chromatography of ecdysone metabolites applied to the cabbage butterfly, *Pieris brassicae* L. *Steroids 36*, 185–207.

LAFONT, R., BEYDON, P., MAUCHAMP, B., SOMMÉ-MARTIN, G., ANDRIANJAFINTRIMO, M. and KRIEN, P. (1981). Recent progress in ecdysteroids analytical methods. In *Regulation of Insect Development and Behaviour. International Conference Part I, Karpaez, Poland, June 1980*. Edited by Kloza, Vol.I, pages 125–144. Wroclaw Technical University Press. Wroclaw, Poland.

LAFONT, R., PENNETIER, J.-L., ANDRIANJAFINTRIMO, M., CLARET, J., MODDE, J.-F. and BLAIS, C. (1982). Sample processing for high-performance liquid chromatography of ecdysteroids. *J. Chromatogr. 236*, 137–149.

LAGUEUX, M., HETRU, C., GOLTZENE, F., KAPPLER, C. and HOFFMANN, J. A. (1979). Ecdysone titre and metabolism in relation to cuticulogenesis in embryos of *Locusta migratoria. J. Insect Physiol. 25*, 709–723.

LAGUEUX, M., HARRY, P. and HOFFMAN, J. A. (1981). Binding of ecdysteroid conjugates to vitelline in *Locusta* oocytes and eggs. *Mol-Cell Endocrinol.* 24, 325–328.

LAUER, R. C., ŞOLOMON, P. H., NAKANISHI, K. and ERLANGER, B. F. (1974). Antibodies to the insect moulting hormone β-ecdysone. *Experientia 30*, 560–562.

LEE, E., LIU, Y.-T., SOLOMON, P. H. and NAKANISHI, K. (1976). Stereospecific conversion of diosgenin to α-edysone. *J. Amer. Chem. Soc. 98*, 1634–1635.

LEE, Y.-W., LEE, E. and NAKANISHI, K. (1980). Stereospecific conversion of diosgenin to labeled α-ecdysone. *Tetrahedron Lett. 21*, 4323–4326.

LETTRÉ, H., GREINER, J., RUTZ, K., HOFMANN, L., EGLE, A. and BIEGER, W. (1972). Steroide mit Strukturmerkmalen des Ecdysons und der Elatericine. *Liebigs Ann. Chem. 758*, 89–110.

LLOYD-JONES, J. G., REES, H. H. and GOODWIN, T. W. (1973). Biosynthesis of ecdysterone from cholesterol in *Taxus baccata. Phytochemistry 12*, 569–572.

LOWE, M. E. and HORN, D. H. S. (1967). Bioassay of the red chromatophore concentrating hormone of the crayfish. *Nature 213*, 408–410.

LUKACS, G. and BENNETT, C. R. (1972). Resonance Magnetique nucleaire du ^{13}C de produits naturels et apparentes VII (18). L'α-ecdysone. *Bull. Soc. Chim. France* 3996–4000.

MA, W. C. (1972). Dynamics of feeding responses in *Pieris brassicae* Linn. as a function of chemosensory input: a behavioural, ultrastructural and electrophysiological study. *Mede-Landbouwhogesch. Wageningen Ned. 72*, 1–62.

McCARTHY, J. F. (1979). Ponasterone A: a new ecdysteroid from the embryos and serum of brachyuran crustaceans. *Steroids 34*, 799–806.

McCARTHY, J. F. and SKINNER, D. M. (1979). Changes in ecdysteroids during embryogenesis of the blue crab, *Callinectes sapidus* Rathbun. *Dev. Biol. 69*, 627–633.

MACHKOVA, Z. and CERNY, V. (1974). Preparation of some amidic 6-ketosteroids as potential inhibitors of postecdysial cuticle sclerotization in *Pyrrhocoris apterus* L. *Collect. Czech. Chem. Commun. 39*, 1091–1097.

MANDARON, P. (1973). Effects of α-ecdysone, β-ecdysone and inokosterone on the *in vitro* evagination of *Drosophila* leg discs and the subsequent differentiation of imaginal integumentary structures. *Develop. Biol. 31*, 101–113.

MAROY, P., DENNIS, R., BECKERS, C., SAGE, B. A. and O'CONNOR, J. D. (1978). Demonstration of an ecdysteroid receptor in a cultured cell line of *Drosophila melanogaster*. *Proc. Natl, Acad. Sci. 75*, 6035–6038.

MATSUOKA, T., IMAI, S., SAKAI, M. and KAMADA, M. (1969). Studies on phytoecdysones – a review of our works. *Ann. Rept. Tekada Res. Lab. 28*, 221–271.

MAYER, R. T. and SVOBODA, J. A. (1978). Thin-layer chromatographic *in situ* analysis of insect ecdysones via fluorescence-quenching. *Steroids 31*, 139–150.

MIDDLETON, E. J., HORN, D. H. S. and GALBRAITH, M. N. (1972). Insect moulting hormones: the synthesis and configuration of some 20-hydroxyecdysones. *Aust. J. Chem. 25*, 1245–1252.

MIYAZAKI, H., ISHIBASHI, M. and MORI, C. (1973). Gas phase microanalysis of zooecdysones. *Anal Chem. 45*, 1164–1168.

MORGAN, E. D. and WOODBRIDGE, A. P. (1971). Insect moulting hormones (Ecdysones). Identification as derivatives by gas chromatography. *J. Chem. Soc., Chem. Commun.* 475–476.

MORGAN, E. D. and WOODBRIDGE, A. P. (1974). Mass spectrometry of insect moulting hormones: trimethyl-silyl-methoxime derivatives of ecdysone and 20-hydroxyecdysone. *Organic Mass Spectrometry 9*, 102–110.

MORGAN, E. D., WOODBRIDGE, A. P. and ELLIS, P. E. (1975a). Studies on the moulting hormones of the desert locust, *Schistocerca gregaria*. *J. Insect Physiol. 21*, 979–993.

MORGAN, E. D., WOODBRIDGE, A. P. and ELLIS, P. E. (1975b). Isolation of moulting hormone from the desert locust, *Schistocerca gregaria* (Forskål). *Acrida 4*, 69–81.

MORGAN, E. D. and POOLE, C. F. (1976a). The extraction and determination of ecdysones in arthropods. In *Advances in Insect Physiology*. Edited by J. E. Treherne, M. J. Berridge and V. B. Wigglesworth. Vol. 12, pages 17–62. Academic Press. London.

MORGAN, E. D. and POOLE, C. F. (1976b). The formation of trimethylsilyl ethers of ecdysones. A reappraisal. *J. Chromatogr. 116*, 333–341.

MORGAN, E. D. and WILSON, I. D. (1980). Progress in the analysis of ecdysteroids. In *Progress in Ecdysone Research*. Edited by J. A. Hoffmann. Pages 29–43. Elsevier/North Holland, Amsterdam.

MORI, H., TSUNEDA, K., SHIBATA, K. and SAWAI, M. (1967). Synthesis of ecdysone. II. Synthesis of and stereochemistry of 2β,3β-dihyroxycholestan-6-ones of their derivatives. *Chem. Pharm. Bull. 15*, 466–473.

MORI, H., SHIBATA, K., TSUNDEDA, K., SAWAI, M. and TSUDA, K. (1968a). The configuration of 22-hydroxycholesterol. *Chem. Pharm. Bull. 16*, 1407–1409.

MORI, H., SHIBATA, K., TSUNEDA, K. and SAWAI, M. (1968b). Synthesis of ecdysone. *Chem. Pharm. Bull. Tokyo 16*, 564–566.

MORI, H. and SHIBATA, K. (1969). Synthesis of ecdysterone. *Chem. Pharm. Bull. 17*, 1970–1973.

MORI, H., SHIBATA, K., TSUNEDA, K. and SAWAI, M. (1969). Synthesis of ecdysone. IV. A novel synthesis of the side chain structure of ecdysone. *Chem. Pharm. Bull. 17*, 690–698.

MORI, H., SHIBATA, K., TSUNDEDA, K., SAWAI, M., SHIODA, M. and MIZUHARA, H. (1970). 14-Hydroxysteroids. Jap. Pat. 7006010. *Chem. Abs. 72*, 121833u (1970).

MORI, H., SHIBATA, K., TSUNEDA, K. and SAWAI, M. (1971). Synthesis of ecdysone- VI. Synthesis of ecdysone from stigmasterol. *Tetrahedron 27*, 1157–1166.

MORIBAYASHI, A. F. and OHTAKI, T. (1978). Inactivation of ecdysone and the possible feed-back control of the titre during pupation of *Sarcophaga peregrina*. *J. Insect. Physiol. 24*, 279–284.

MORIYAMA, H. and NAKANISHI, K. (1968). Insect hormones, VI. Confirmation of the skeletal structure of ponasterone A. *Tetrahedron Lett. 9*, 1111–1112.

MURAKAMI, T., WADA, K., TANAKA, N., YAMAGISHI, T., SAIKI, Y. and CHEN, C.-M. (1980). Chemische und Chemotaxonomische Untersuchungen von Filices. XXXII. Chemische Untersuchungen der Inhaltsstoffe von *Plenasium banksiifolium* (Pr.)Pr. *Chem. Pharm. Bull. 28*, 3137–3139.

NAKANISHI, K., KOREEDA, M., DASAKI, S., CHANG, M. L. and HSU, H. Y. (1966). Insect hormones. I. The structure of ponasterone A, an insect moulting hormone from the leaves of *Podocarpus nakaii* Hay. *J. Chem. Soc., Chem. Commun.* 915–917.

NAKANISHI, K., KOREEDA, M., CHANG, M. L. and HSU, H. Y. (1968). Insect hormones. V. The structures of ponasterone B and C. *Tetrahedron Lett. 9*, 1105–1110.

NAKANISHI, K. (1971). The ecdysones. *Pure Appl. Chem. 25*, 167–195.

NIGG, H. N., SVOBODA, J. A., THOMPSON, M. J., KAPLANIS, J. N., DUTKY, S. R. and ROBBINS, W. E. (1974a). Ecdysone metabolism. Ecdysone dehydrogenase–isomerase. *Lipids 9*, 971–974.

NIGG, H. N., THOMPSON, M. J., KAPLANIS, J. N., SVOBODA, J. A. and ROBBINS, W. E. (1974b). High-pressure liquid–solid chromatography of the ecdysones-insect molting hormones. *Steroids 23*, 507–516.

NOVOSEL'SKAYA, I. L., GOROVITS, M. B. and ABUBAKIROV, N. K. (1975). Phytoecdysones of *Serratula*. IV. Sogdysterone. *Khim. Prir. Soedin. 11*, 429–430. *Chem. Abs. 84*, 74514b (1976).

OBERLANDER, H. and FULCO, L. (1967). Growth and partial metamorphosis of imaginal discs of greater wax moth *Galleria mellonella in vitro*. *Nature 216*, 1140–1141.

OBERLANDER, H. (1969). Effects of ecdysone, ecdysterone and inokosterone on the *in vitro* initiation of metamorphosis of wing disks of *Galleria mellonella*. *J. Insect Physiol. 15*, 297–304.

OBERLANDER, H. (1974). Biological activity of insect ecdysones and analogues *in vitro*. *Experientia 30*, 1409–1410.

OGAWA, S., NISHIMOTO, N., OKAMOTO, N. and TAKEMOTO, T. (1971). Studies on the constituents of Achyranthis radix. VIII. The insect-moulting substances in *Achyranthes* genus. (Supplement 2). *Yakugaku Zasshi 91*, 916–919.

OGAWA, S., YOSHIDA, A. and KATO, R. (1977). Analytical studies on the active constituents in crude drugs. III. High-speed liquid chromatographic determination of ecdysterone and inokosterone in Achyranthis radix. *Chem. Pharm. Bull. 25*, 904–908.

OHMORI, K. and OHTAKI, T. (1973). Effects of ecdysone analogues on development and metabolic activity of wing disks of the fleshfly. *Sorcophaga peregrina in vitro*. *J. Insect Physiol. 19*, 1199–1210.

OHTAKI, T., MILKMAN, R. D. and WILLIAMS, C. M. (1967). Ecdysone and ecdysone analogues: Their assay on the fleshfly *Sarcophaga peregrina*. *Proc. Nat. Acad. Sci. 58*, 981–984.

OHNISHI, E., MIZUNO, T., IKEKAWA, N., AWATA, N. and SAKURAI, S. (1977a). Occurrence of a α-ecdysone in the developing embryos of the silkworm *Bombyx mori*. *J. Insect Physiol. 23*, 317–319.

OHNISHI, E., MIZUNO, T. CHATANI, F., IKEKAWA, N. and SAKURAI, S. (1977b). 2-Deoxy-α-ecdysone from ovaries and eggs of the silkworm, *Bombyx mori*. *Science 197*, 66–67.

PLAGGE, E. and BECKER, E. (1938). Wirkung arteigener und artfremder Verpuppungshormone in Extrakten, *Naturwissenschaften 26*, 430–431.

POOLE, C. F., MORGAN, E. D. and BEBBINGTON, P. M. (1975). Analysis of ecdysone by gas chromatography using electron capture detection. *J. Chromatogr. 104*, 172–175.

POOLE, C. F. and MORGAN, E. D. (1975). Structural requirements for the electron capturing properties of ecdysones. *J. Chromatogr. 115*, 587–590.

POOLE, C. F., ZLATKIS, A., SYE, W.-F., SINGHAWANGCHA, S. and MORGAN, E. D. (1980). The determination of steroids with and without natural electrophores by gas chromatography and electron-capture detection. *Lipids, 15*, 734–744.

PORCHERON, P., FOURCRIER, J., GROS, C., PRADELLES, P., CASSIER, P. and DRAY, F. (1976). Radioimmunoassay of arthropod molting hormone β-ecdysone antibodies production and ^{125}I-iodinated tracer preparation. *FEBS Lett. 61*, 159–162.

PRAKASH, A. and GHOSAL, S. (1979). Phytoecdysones. *J. Scientific and Industrial Res. (India) 38*, 632–647.

RASHKES, YA. V. and ABUBAKIROV, N. K. (1980). Characteristics of the mass spectra of ecdysteroids with a different number of OR groups. *Khim. Prir. Soedin.* 518–28. *Chem. Abs. 94*, 15984p (1981).

REES, H. H., (1971). Ecdysones. In *Aspects of Terpenoid Chemistry and Biochemistry*. Proc. of the phytochemical Symposium Liverpool, 1970. Edited by T. W. Goodwin, pages 181–222. Academic Press, London.

REUM, L. and KOOLMAN, J. (1979). Analysis of ecdysteroids by radioimmunoassay: comparison of three different antisera. *Insect Biochem.* *9*, 135–142.

REZABOVA, B., HORA, J., LANDA, V., CERNY, V. and SORM, F. (1968). On steroids CXIII. Sterilizing effect of some 6-ketosteroids on housefly *musca domestica* L. *Steroids 11*, 475–495.

RHOADES, D. F. (1979). Evolution of plant chemical defence against herbivores. In *Herbivores — Their Interaction with Secondary Plant Metabolites*. Edited by G. A. Rosenthal and D. H. Janzen. Chap. 1, pages 3–54. Academic Press, New York.

RICHARDS, G. (1978). The relative biological activities of α- and β-ecdysone and their 3-dehydro derivatives in the chromosome puffing assay. *J. Insect Physiol. 24*, 329–335.

RICHARDS, G. (1980). Ecdysteroids and puffing in *Drosophila melanogaster*. In *Progress in Ecdysone Research*. Edited by J. A. Hoffmann. Pages 363–378. Elsevier/North Holland, Amsterdam.

RIDDIFORD, L. M. (1970). See Williams, C. M. (1969). *Hormonal interactions between plants and insects*. In *Chemical Ecology*. Edited by E. Sondheimer and J. B. Simeone. Chap. 6, page 114. Academic Press, New York.

RIMPLER, H. and SCHULZ, G. (1967). Vorkommen von 20-Hydroxyecdyson in *Vitex megapotamica*. *Tetrahedron Lett. 22*, 2033–2035.

RIMPLER, H. (1969). Pterosteron, Polypodin B und ein neues ecdysonartiges Steroid (Viticosteron E) aus *Vitex megapotamica* (Verbenaceae). *Tetrahedron Lett. 5*, 329–333.

RIMPLER, H. (1970). Phytoecdysone, Insektenhautungshormone aus Pflanzen. *Präparative Pharmazie 6*, 101–108.

ROBBINS, W. E., KAPLANIS, J. N., THOMPSON, M. J., SHORTINO, T. S., COHEN, C. F. and JOYNER, S. C. (1968). Ecdysones and analogs: effects on development and reproduction of insects. *Science 161*, 1158–1160.

ROBBINS, W. E., KAPLANIS, J. N., THOMPSON, M. J., SHORTINO, T. J. and JOYNER, S. C. (1970). Ecdysones and synthetic analogs: molting hormone activity and inhibitive effects on insect growth, metamorphosis and reproduction. *Steroids 16*, 105–125.

ROBBINS, W. E., KAPLANIS, J. N., THOMPSON, M. J. and SVOBODA, J. A. (1971). Chemistry and biological activity of the ecdysones and ecdysone analogs. In *Chemical Releases in Insects*. Edited by A. S. Tahori, *Proc. 2nd International I.U.P.A.C. Congress of Pesticide Chemistry*, Vol. III, pages 1–31. Gordon & Breach Science Publishers, New York.

ROBBINS, W. E., THOMPSON, M. J., SVOBODA, J. A., SHORTINO, T. J., COHEN, C. F., DUTKY, S. R. and DUNCAN, O. J. (1975). Nonsteroidal secondary and tertiary amines: Inhibitors of insect development and metamorphosis and Δ^{24}-sterol reductase system of tobacco hornworm. *Lipids 10*, 353–359.

RODEWALD, J., SZCZEPEK, J. and GUMULKA, J. (1978). Highly efficient synthesis of 3β-acetoxy-14α-hydroxy-5α-cholest-7-en-6-one. *Bull. Pol. Acad. Sci. Ser. Chim. 26*, 91–94.

RODEWALD, W. J., SZCZEPEK, W. J. and GUMULKA, J. (1979). Chemistry of 7-en-6-oxo steroids. I. New substitution reactions at C-14 in 7-en-6-oxo-cholestanes. *Polish J. Chem. 53*, 139–147.

ROMER, F. (1979). Ecdysteroids in snails. *Naturwissenschaften 66*, 471–472.

RUH, M. F. and BLACK, C. (1976). Separation and detection of α- and β-ecdysone using thin-layer chromatography. *J. Chromatogr. 116*, 480–481.

RUSSELL, G. B., HORN, D. H. S. and MIDDLETON, E. J. (1971). New phytoecdysones from *Dacrydium intermedium*. *J. Chem. Soc., Chem. Commun.* 71.

RUSSELL, G. B., FENEMORE, P. G., HORN, D. H. S. and MIDDLETON, E. J. (1972). Insect moulting hormones: the phytoecdysones of *Dacrydium intermedium*. *Aust. J. Chem. 25*, 1935–1941.

RUSSELL, G. B. (1972). Phytoecdysones from *Phymatodes novae-zelandiae*. *Phytochemistry 11*, 1496.

RUSSELL, G. B. and FRASER, J. G. (1973). Insect moulting hormones: dacrysterone, a new phytoecdysone from *Dacrydium intermedium*. *Aust. J. Chem. 26*, 1805–1807.

RUSSELL, G. B., GREENWOOD, D. R., LANE, G. A., BLUNT, J. W. and MUNRO, M. H. G. (1981). 2-Deoxy-3-epiecdysone from the fern *Blechnum vulcanicum*. *Phytochemistry 20*, 2407–2410.

SAATOV, Z., USMANOV, B. Z. and ABUBAKIROV, N. K. (1979a). Phytoecdysones of *Silene praemixta*. 1. Silenosterone. *Khim. Prir. Soedin. 6*, 793–797.

SAATOV, Z., USMANOV, B. Z. and ABUBAKIROV, N. K. (1979b). Phytoecdysones of *Silene praemixta*. 2. Praemixisterone. *Khim. Prir. Soedin. 6*, 797–799.

SAGE, B. A., TANIS, M. A. and O'CONNOR, J. D. (1982). Characterization of ecdysteroid receptors in cytosol and naive nuclear preparations of *Drosophila* K_C cells. *J. Biol. Chem. 257*, 6373–6379.

SANTOS, A. C., CHUA, M. T., EUFEMIO, N. and ABELA, C. (1970). Isolation of commisterone, a new phytoecdysone from *Cyanotis vaga*. *Experientia 26*, 1053–1054.

SANTOS, A. C., CHUA, M. T., EUFEMIO, N., ABELA, C., HIKINO, H. and TAKEMOTO, T. (1972). Identity of commisterone with ecdysterone. *Planta Medica. 21*, 279–281.

SARDINI, D. and KREPINSKY, J. (1974). Densitometric determination of ecdysones *Farmaco Ed. Prat. 29*, 723–731. *Chem. Abs. 82*, 133870y (1978).

SATO, Y., SAKAI, M., IMAI, S. and FUJIOKA, S. (1968). Ecdysone activity of plant-originated molting hormones applied on the body surface of lepidopterous larvae. *Appl. Entomol. Zool. 3*, 49–51. *Chem. Abs.* 71, 1143b (1969).

SCALIA, S. and MORGAN, E. D. (1982). A re-investigation of the ecdysteroids during embryogenesis in the desert locust, *Schistocerca gregaria*. *J. Insect Physiol. 28*, 647–654.

SCHERRER, H. (1969). Über Einiger Eigenshaften von 6-Keto-Δ^7-steroiden. *Helv. Chim. Acta. 52*, 2428–2436.

SCHOOLEY, D. A., WEISS, G. and NAKANISHI, K. (1972). A simple and general extraction procedure for phytoecdysones based on reversed-phase adsorption chromatgraphy. *Steroids 19*, 377–382.

SCHOOLEY, D. A. and NAKANISHI, K. (1973). Application of high-pressure liquid chromatography to the separation of insect molting hormones. In *Modern Methods of Steroid Analysis*. Edited by E. Heftmann. Pages 37–54. Academic Press, New York.

SEGAL, G. M., SYDYKOV, ZH. S. and TORGOV, I. V. (1982). Partial synthesis of 2-deoxyecdysone and 2-deoxyecdysterone – insect molting hormones. *Izv. Akad. Nauk. SSR, Ser. Khim.* 317–325. *Chem. Abs. 97*, 6644n (1982).

SHIBATA, K. and MORI, H. (1971). Synthetic studies on insect moulting hormones, synthesis of rubrosterone. *Tetrahedron 27*, 1149–1155.

SIDDALL, J. B., MARSHALL, J. P., BOWERS, A., CROSS, A. D., EDWARDS, J. A. and FRIED, J. H. (1966a). Synthetic studies on insect hormones. I. Synthesis of the tetracyclic nucleus of ecdysone. *J. Amer. Chem. Soc. 88*, 379–380.

SIDDALL, J. B., CROSS, A. D. and FRIED, J. H. (1966b). Steroids CCXCII. Synthetic studies on insect hormones II. The synthesis of ecdysone. *J. Amer. Chem. Soc. 88*, 862–863.

SIDDALL, J. B., HORN, D. H. S. and MIDDLETON, E. J. (1967). Synthetic studies on insect hormones. The synthesis of a possible metabolite of crustecdysone (20-hydroxyecdysone). *J. Chem. Soc. Chem. Commun.* 899–900.

SIDDALL, J. B. (1969). Synthetic studies on insect hormones. In *Insect Plant-Interactions*. A report of a work conference at the University of Santa Barbara. *Natl. Acad. Sci*: Washington D.C., pages 70–71.

SINGH, P. and RUSSELL, G. B. (1980). The dietary effects of 20-hydroxyecdysone on the development of housefly. *J. Insect Physiol. 26*, 139–142.

SINGH, P., RUSSELL, G. B. and FREDERICKSEN, S. (1982). The dietary effects of some ecdysteroids on the development of the housefly. *Ent. Exp. and Appl. 32*, 7–12.

SINGH, S. B. and THAKUR, R. S. (1982). Structure and stereochemistry of paristerone, a novel phytoecdysone from the tubers of *Paris polyphylla*. *Tetrahedron 38*, 2189–2194.

SLAMA, K., ROMANUK, M. and SORM, F. (1974). *Insect Hormones and Bioanalogues*. Springer-Verlag, Vienna/New York.

SLAMA, K. (1979). Insect hormones and antihormones in plants. In *Herbivores – Their interaction with Secondary Plant Metabolites*. Pages 683–700. Academic Press, New York.

SMITH, W. B., NEWSOROFF, G. P. and WU, N. Y. (1973). Towards an ecdysone synthesis. 20β-acetoxy-5α-pregna-2,7-dien-6-one. *Steroids 21*, 609–615.

SMITH, W. B. and NEWSOROFF, G. P. (1973). The synthesis of 23,24-^{14}C-ecdysone. *Steroids 22*, 819–828.

SMITH, W. B. and NEWSOROFF, G. P. (1974). An n.m.r. technique for carbon-5 configurations in 6-keto-Δ^7-steroids. *Steroids 23*, 579–583.

SMITH, W. B. (1977). The carbon-13 spectra of steriods on the way to ecdysone. *Org. Mag. Res. 9*, 644–647.

SOUMOFF, C., HORN, D. H. S. and O'CONNOR, J. D. (1981). Production of a new antiserum to arthropod molting hormone and comparison with two other antiserums. *J. Steroid Biochem. 14*, 429–435.

SPINDLER, K.-D., HAMANN, A., SPINDLER-BARTH, M., IHNE, A., BECKERS, C. and EMMERICH, H. (1976). Derivatives of the insect moulting hormone for affinity chromatography, and their biological activities. *Steroids 37*, 553–565.

SPINDLER, K. D., KOOLMAN, J., MOSORA, F. and EMMERICH, H. (1977). Catalytic oxidation of ecdysteroids to 3-dehydro products and their biological activities. *J. Insect Physiol. 23*, 441–444.

SPINDLER, K.-D., BECKERS, C., GRÖSCHEL-STEWART, U. and EMMERICH, H. (1978). A radioimmunoassay for arthropod moulting hormones, introducing a novel method of immunogen coupling. *Hoppe-Seyler's Z. Physiol. Chem. 359*, 1269–1275.

SPINDLER, K.-D., KELLER, R. and O'CONNOR, J. D. (1980). The role of ecdysteroids in the crustacean molting cycle. In *Progress in Ecdysone Research*. Edited by J. A. Hoffman. Pages 247–280. Elsevier/North-Holland Biomedical Press, Amsterdam.

SPINDLER-BARTH, M., BASSEMIR, U., KUPPERT, P. and SPINDLER, K.-D. (1981). Isolation of nuclei from crayfish tissues and demonstration of nuclear ecdysteroid receptors. *Z. Naturf. 36c*, 326–332.

STAAL, G. B. (1967). Plants as a source of insect hormones. *Koninkl. Nederl. Akad. van Wetenschappen Proc. Ser. C. 70(4)*, 409–418.

STAHL, E. (1962). *Dunnschicht-chromatographie*. Page 515. Springer, Berlin.

STAMM, M. D. (1958). Isolement d'hormones de metamorphose dans l'orthoptere *Dociostaurus maroccanus*. *Rev. Exp. Fisiol. 14*, 263–268.

STURARO, A., GUERRIERO, A., DECLAUSER, R. and PIETRA, F. (1982). A new, unexpected marine source of a molting hormone. Isolation of ecdysterone in large amounts from the zoanthia *Gerardia savaglia. Experienta 38*, 1184–1185.

SUZZONI, J. P., PASSERA, L. and STRAMBI, A. (1980). Ecdysteroid titre and caste determination in the ant, *Pheidole pallidula* (Nyl.) (Hymenoptera: Formicidae). *Experientia 36*, 1228–1229.

SVOBODA, J. A. and ROBBINS, W. E. (1971). The inhibitive effects of azasterols on sterol metabolism and growth and development in insects with special reference to the tobacco hornworm. *Lipids 6*, 113–119.

SVOBODA, J. A., THOMPSON, M. J. and ROBBINS, W. E. (1972). Azasteroids: potent inhibitors of insect molting and metamorphosis. *Lipids 7*, 553–556.

SWAIN, T. (1978). Plant-animal coevolution: a synoptic view of the Paleozoic and Mezozoic. In *Biochemical Aspects of Plant and Animal Coevolution*. Edited by J. B. Harborne. Chap. 1, pages 3–19. Academic Press, London.

SYDYKOV, Z. S. and SEGAL, G. M. (1976). Synthesis of 2-deoxyrubrosterone. *Biorg. Khim. 2*, 1634–1641. *Chem. Abs.* 86, 12169g (1977).

TAKEMOTO, T., OGAWA, S. and NISHIMOTO, N. (1967a). Studies on the constituents of Achyranthis radix. III. Structure of inokosterone. *Yakugaku Zasshi 87*, 1474–1477.

TAKEMOTO, T., HIKINO, Y., NOMOTO, K. and HIKINO, H. (1967b). Structure of cyasterone. A novel C29 insect moulting substance from *Cyathula capitata*. *Tetrahedron Lett. 33*, 3191–3194.

TAKEMOTO, T., OGAWA, S. and NISHIMOTO, N. (1967c). Isolation of the moulting hormones of insects from Achyranthis Radix. *Yakugaku Zasshi 87*, 325–327.

TAKEMOTO, T., OGAWA, S. and NISHIMOTO, N. (1967d). Studies on the constituents of Achyranthis Radix. II. Isolation of the insect-moulting hormones. *Yakugaku Zasshi 87*, 1469–1473.

TAKEMOTO, T., OGAWA, S., NISHIMOTO, N., HIRAYAMA, H. and TANIGUCHI, S. (1967e). Isolation of the insect moulting hormones from mulberry leaves. *Yakugaku Zasshi 87*, 748.

TAKEMOTO, T., OGAWA, S., NISHIMOTO, N. and HOFFMEISTER, H. (1967f). Steroide mit Hautungshormon-Aktivitat aus Tieren und Pflanzen. *Z. Naturf. B. 22b*, 681–682.

TAKEMOTO, T., OGAWA, S., NISHIMOTO, N. and TANIGUCHI, S. (1967g). Studies on the constituents of Achyranthis Radix. IV. Isolation of the insect moulting hormones from Formosan *Achyranthes* ssp. *Yakugaku Zasshi 87*, 1478–1480.

TAKEMOTO, T., OGAWA, S., NISHIMOTO, N., YEN, K.-Y., ABE, K., SATO, T., OSAWA, K. and TAKAHASHI, M. (1967h). The isolation of ecdysterone from the Radix of *Achyranthes obtusifolia* Lam. *Yakugaku Zasshi 87*, 1521–1523.

TAKEMOTO, T., HIKINO, Y., ARAI, T., KAWAHARA, M., KONNO, C., ARIHARA, S. and HIKINO, H. (1967i). Isolation of insect moulting substances from *Matteuccia struthiopteris, Lastrea thelypteris* and *Onoclea sensibilis*. *Chem. Pharm. Bull. 15*, 1816.

TAKEMOTO, T., OGAWA, S., NISHIMOTO, N., ARIHARA, S. and BUE, K. (1967j). Insect moulting activity of crude drugs and plants. (1). *Yakugaku Zasshi 87*, 1414–1418.

TAKEMOTO, T., OGAWA, S., NISHIMOTO, N. and MUE, K. (1967k). Studies on the constituents of Achyranthis Radix. V. Insect hormone activity of ecdysterone and inokosterone on the flies. *Yukagaku Zasshi 87*, 1481–1483.

TAKEMOTO, T., NOMOTO, K. and HIKINO, H. (1968a). Insect moulting substances from *Cyathula capitata*. *Tetrahedron Lett. 48*, 4953–4956.

TAKEMOTO, T., NOMOTO, K., HIKINO, Y. and HIKINO, H. (1968b). Structure of capitasterone, a novel C29 insect-moulting substance from *Cyathula capitata*. *Tetrahedron Lett. 47*, 4929–4932.

TAKEMOTO, T., ARIHARA, S. and HIKINO, H. (1968c). Structure of ponasteroside A, a novel glycoside of insect-moulting substance from *Pteridium aquilinum* var. *latiusculum*. *Tetrahedron Lett. 39*, 4199–4202.

TAKEMOTO, T., ARIHARA, S., HIKINO, Y. and HIKINO, H. (1968d). Structure of pterosterone, a novel insect-moulting substance from *Lastrea thelypteris* and *Onoclea sensibilis*. *Tetrahedron Lett. 3*, 375–378.

TAKEMOTO, T., HIKINO, Y., HIKINO, H., OGAWA, S. and NISHIMOTO, N. (1968e). Structure of rubrosterone, a novel C19 metabolite of insect-moulting substances from *Achyranthes rubrofusca*. *Tetrahedron Lett. 26*, 3053–3056.

TAKEMOTO, T., OGAWA, S., MORITA, M., NISHIMOTO, N., DOME, K. and MORISHIMA, K. (1968f). Studies of the constituents of Achyranthis Radix. VI. Determination of insect moulting hormones. *Yakugaku Zasshi 88*, 39–43.

TAKEMOTO, T., OGAWA, S., NISHIMOTO, N., HIRAYAMA, H. and TANIGUCHI, S. (1968g). Studies on the constituents of Achyranthis Radix. VII. The insect-moulting substances of *Achyranthes* and *Cyathula genera*. Supplement. *Yakugaku Zasshi 88*, 1293–1297.

TAKEMOTO, T., HIKINO, Y., ARIHARA, S. and HIKINO, H. (1968h). Absolute configuration of inokosterone, an insect-moulting substance from *Achyranthes fauriei*. *Tetrahedron Lett. 20*, 2475–2478.

TAKEMOTO, T., OKUYAMA, T., ARIHARA, S., HIKINO, Y. and HIKINO, H. (1969a). Isolation of insect moulting substances from *Blechnum amabile* and *Blechnum niponicum*. *Chem. Pharm. Bull. 17*, 1973–1974.

TAKEMOTO, T., HIKINO, Y., HIKINO, H., OGAWA, S. and NISHIMOTO, N. (1969b). Rubrosterone, a metabolite of insect metamorphosing substance from *Achyranthes rubrofusca*: structure and absolute configuration. *Tetrahedron 25*, 1241–1248.

TAKHTAJAN, A. L. (1980). Outline of the classification of flowering plants (Magnoliophyta). *Bot. Rev. 46*, 225–359.

THOMPSON, M. J., KAPLANIS, J. N., ROBBINS, W. E. and YAMAMOTO, R. T. (1967). 20,26-Dihydroxyecdysone a new steroid with moulting hormone activity from the tobacco hornworm, *Manduca sexta* (Johannson). *J. Chem. Soc., Chem. Commun.* 650–653.

THOMPSON, M. J., ROBBINS, W. E., KAPLANIS, J. N., COHEN, C. F. and LANCASTER, S. M. (1970). Synthesis of analogs of α-ecdysone. A simplified synthesis of $2\beta,3\beta,14\alpha$-trihydroxy-7-en-6-one-5β-steroids. *Steroids 16*, 85–104.

THOMPSON, M. J., ROBBINS, W. E., COHEN, C. F., KAPLANIS, J. N., DUTKY, S. R. and HUTCHINS, R. F. N. (1971). Synthesis and biological activity of 5β-hydroxy analogues of α-ecdysone. *Steroids 17*, 399–409.

THOMPSON, M. J., KAPLANIS, J. N., ROBBINS, W. E., DUTKY, S. R. and NIGG, H. N. (1974). 3-Epi-20-hydroxyecdysone from meconium of the tobacco hornworm. *Steroids 24*, 359–366.

THOMPSON, M. J., SERBAN, N. N., ROBBINS, W. E., SVOBODA, J. A., SHORTINO, T. J., DUTKY, S. R. and COHEN, C. F. (1975). Inhibitive effects of structurally modified azasteroids and related nitrogen containing steroids on insect growth and development. *Lipids 10*, 615–622.

THOMSON, J. A., SIDDALL, J. B., GALBRAITH, M. N., HORN, D. H. S. and MIDDLETON, E. J. (1969). The biosynthesis of ecdysones in the blowfly *Calliphora stygia*. *J. Chem. Soc., Chem. Commun.* 669–670.

THOMSON, J. A., IMRAY, F. P. and HORN, D. H. S. (1970). An improved *Calliphora* bioassay for insect moulting hormones. *Aust. J. Exp. Biol. Med. Sci. 48*, 321–328.

TORPIER, G., HIRN, M., NIRDE, P., DE REGGI, M. and CAPRON, A. (1982). Detection of ecdysteroids in the human trematode, *Schistosoma mansoni. Parasitology 84*, 123–130.

TSAY, Y. H., SILVERTON, S. V., BIESLER, J. A. and SATO, Y. (1970). The structure of carpesterol. *J. Amer. Chem. Soc. 92*, 7005–7006.

TSOUPRAS, G., HETRU, C., LUU, B., LAGUEUX, M., CONSTANTIN, E. and HOFFMANN, J. A. (1982). The major conjugates of ecdysteroids in young eggs and in embryos of *Locusta migratoria. Tetrahedron Lett. 23*, 2045–2048.

TURSCH, B., HOOTELE, C., KAISIN, M., LOSMAN, D. and KARLSON, P. (1976). Chemical studies on marine invertebrates. XVI. Structure and absolute configuration of lobosterol, a novel polyoxygenated sterol from the Alcyonacean *Lobophytum pauciflorum* (Coelentorata, Octocorallia). *Steroids 27*, 137–142.

USMANOV, B. Z., GOROVITS, M. B. and ABUBAKIROV, N. K. (1973). Phytoecdysones of *Ajuga turkestanica*. II. *Khim. Prir. Soedin. 9*, 125. *Chem. Abs. 78*, 148125a (1973).

USMANOV, B. Z., GOROVITS, M. B. and ABUBAKIROV, N. K. (1975). Phytoecdysones of *Ajuga turkestanica*. III. Structure of turkesterone. *Khim. Prir. Soedin. 11*, 466–470.

USMANOV, B. Z., RASHKES, YA. V. and ABUBAKIROV, N. K. (1978). Phytoecdysones of *Ajuga turkestanica*. VI. 22-Acetylcyasterone. *Khim. Prir. Soedin. 2*, 215–219. *Chem. Abs.* 89, 163851e (1978).

VANBEVER, W., KOHEN, F., RANADE, V. V. and COUNSELL, R. E. (1970). Synthesis of the steroid rubrosterone. *J. Chem. Soc., Chem. Commun.* 758.

VANDENHEUVEL, W. J. A. and HORNING, E. C. (1967). Conditions for the separation of steroids by gas liquid chromatography. In *The Gas Liquid Chromatography of Steroids*. Edited by J. K. Grant. Memoirs of the Society of Endocrinology. No. 16, pages 39–68. Cambridge University Press.

VELGOVA, H., LABLER, L., CERNY, V., SORM, F. and SLAMA, K. (1968). On steroids. CXI. Some further compounds producing molting deficiencies in an insect. *Collect. Czech. Chem. Commun. 33*, 242–256.

VELGOVA, H., CERNY, V., SORM, F. and SLAMA, K. (1969). On steroids. CXXVI. Further compounds with antisclerotization effect on *Pyrrhocoris apterus* L. larvae; structure and activity correlations. *Collect. Czech. Chem. Commun. 24*, 3354–3376.

VELGOVA, H., CERNY, V. and SORM, F. (1972). On steroids. CXXXVI. Effect of 17-substituent in base-catalysed equilibrium of steroidal $2\beta,3\beta$-disubstituted 6-ketones. *Collect. Czech. Chem. Commun. 37*, 1015–1034.

WADA, K. (1975). Biogenetic-type synthesis of a precursor of ecdysone, $3\beta,14\alpha$-dihydroxycholes-7-en-6-one. *Agr. Biol. Chem. 39*, 1679–1680.

WAI-MING, T., ABDUL-HAJJ, Y. J. and KOREEDA, M. (1975). Microbial oxidation of ecdysones. A convenient preparation of rubrosterone. *J. Chem. Soc., Chem. Commun.* 24–25.

WHITTAKER, R. H. and FEENY, P. P. (1971). Allelochemics: chemical interaction between species. *Science 171*, 757–770.

WIECHERT, R., KERB, U., HOCKS, P., FURLENMEIER, A., FURST, A., LANGEMANN, A. and WALDVOGEL, G. (1966). Zur synthese des ecdysone. II. Mitteilung über Insectenhormone. Synthesen von $2\beta,3\beta$-Dihydroxy-6-keto-A/B-*cis*-steroiden. *Helv. Chim. Acta. 49*, 1581–1590.

WIGGLESWORTH, V. B. (1934). The physiology of ecdysis in *Rhodinius prolixus* (Hemiptera). II. Factors controlling moulting and "metamorphosis". *Quart. J. Mic. Sci. 77*, 191–222.

WILLIAMS, C. M. (1968). Ecdysone and ecdysone-analogues: Their assay and action on diapausing pupae of the *Cynthia* silkworm. *Biol. Bull., 134*, 344–355.

WILSON, I. D., BIELBY, C. R., MORGAN, E. D. and MCLEAN, A. E. M. (1980). Comparison of high-performance liquid chromatography and gas chromatography for the analysis of ecdysteroids. *J. Chromatogr. 194*, 343–352.

WILSON, I. D., SCALIA, S. and MORGAN, E. D. (1981). Reversed-phase thin-layer chromatography for the separation and analysis of ecdysteroids. *J. Chromatogr. 212*, 211–219.

WILSON, I. D., BIELBY, C. R. and MORGAN, E. D. (1982a). Selective effects of mobile and stationary phases in reversed-phase high-performance liquid chromatography of ecdysteroids. *J. Chromatogr. 238*, 97–102.

WILSON, I. D., BIELBY, C. R. and MORGAN, E. D. (1982b). Evaluation of some phytoecdysteroids as internal standards for the chromatographic analysis of ecdysone and 20-hydroxyecdysone from arthropods. *J. Chromatogr. 236*, 224–229.

WILSON, I. D., BIELBY, C. R. and MORGAN, E. D. (1982c). Studies on the reversed-phase thin-layer chromatography of ecdysteroids on C_{12} bonded and paraffin-coated silica. *J. Chromatogr. 242*, 202–206.

WINGET, R. R. and HERMAN, W. S. (1976). Occurrence of ecdysone in the blood of the chelicerate arthropod, *Limulus polyphemus. Experientia 32*, 1345–1346.

YATSYUK, YA. K. and SEGEL, G. M. (1970). On the isolation of ecdysterone. *Khim. Prir. Soedin. 6*, 281.

YEN, K.-Y., YANG, L.-L., OKUYAMA, T., HIKINO, H. and TAKEMOTO, T. (1974). Screening of Formosan ferns for phytoecdysones. I. *Chem. Pharm. Bull. 22*, 805–808.

YUND, M. A., KING, D. S. and FRISTROM, J. W. (1978). Ecdysteroid receptors in imaginal discs of *Drosophila melanogaster. Proc. Natl. Acad. Sci. 75*, 6039–6043.

ADDENDUM

Since the completion of this review a number of important papers on ecdysteroids have appeared. Tsoupras, G. *et al.* (1983) have found that newly laid eggs of *Locusta migratoria* contain a new ecdysteroid conjugate which was shown to be the 22-N^6-(isopentenyl)-adenosine monophosphoric ester of ecdysone (169). Nakano, K. *et al.* (1982) have found in the plant *Taxus cuspidata* 22-deoxy-20-hydroxyecdysone (170) (now named taxisterone) which was synthesised earlier by Galbraith, M. *et al.* (1968a). Two new ecdysteroid glycosides, 22-0-α-D-galactopyranosylintegristerone A (171) (sileneoside C) and 3,22-0-α-D-digalactopyranosyl-20-hydroxyecdysone (172) (sileneoside B) have been isolated (Saatov, Z. *et al.* 1982a, 1982b) from the plant *Silene brahuica*.

Further investigations of ecdysteroids in animals have led to the discovery (Loeb, M. *et al.* 1982) that ecdysteroids are biosynthesised in the testes of the late last-instar larvae of the tobacco budworm. Nirde, P. *et al.* (1983) have detected ecdysone (24) and 20-hydroxy-ecdysone (32) in the human parasite *Schistosoma mansoni* and report that these ecdysteroids stimulate growth and vitellogenesis in this trematode.

A number of compounds have been isolated from plants which inhibit growth and moulting of insects. Plumbagin (173) isolated (Kubo, I. *et al.*, 1983) from the African medicinal plant *Plumbago capensis* is reported to inhibit moulting in four lepidopterous insects when incorporated in an artificial diet. It is possible that compounds of this type may prove to be more ecologically acceptable for the

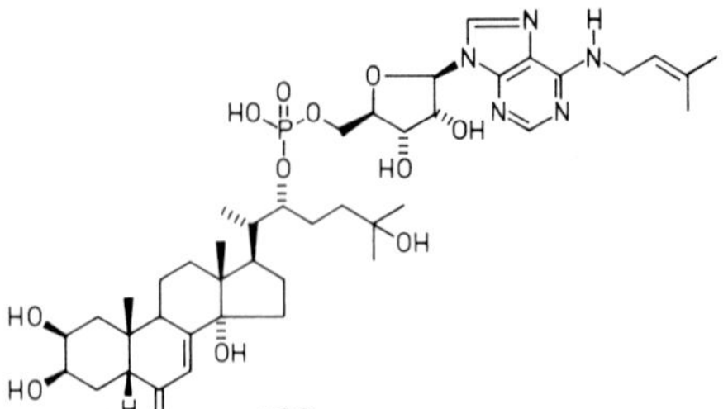

169

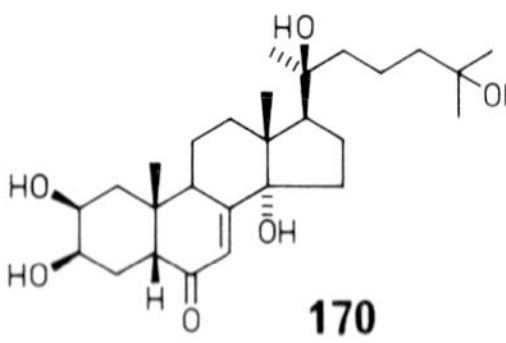

170

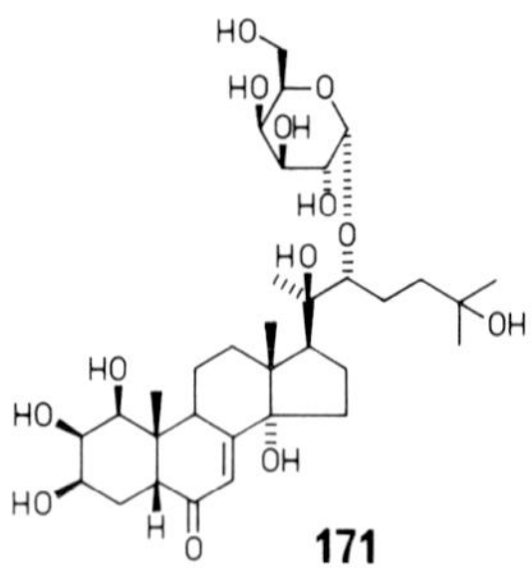

171

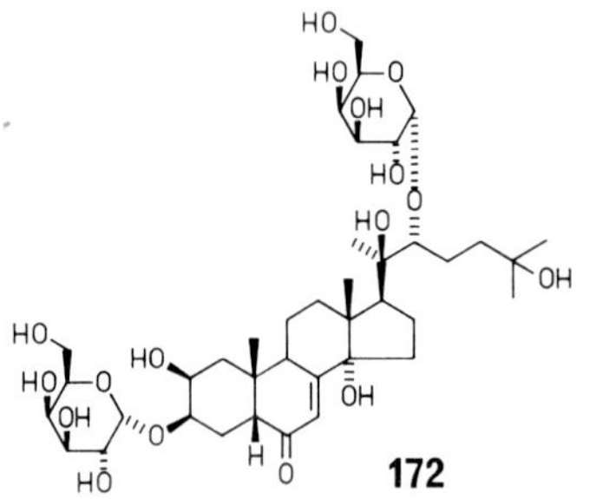

172

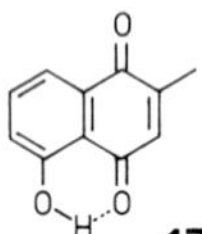

173

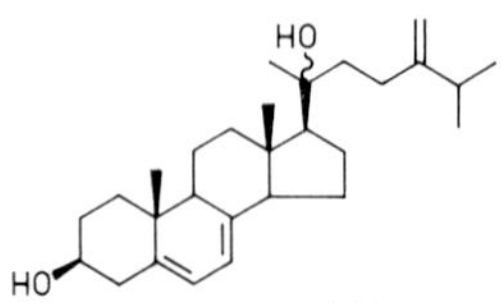

174

control of insects than conventional neurotoxic insecticides. Kubo, I. and Klocke, J. (1983) have found that ajugasterone C (3) acts as a feeding deterrent while ponasterone A (58) inhibits growth and ecdysis of larvae of *Pectinophora gossypiella* and *Bombyx mori.* Dorn, A. and Buhlmann, K. (1982) have found that ecdysteroids administered to females of *Oncopeltus fasciatus* drastically reduced egg hatching while Smith, W. and Nijhout, H. (1983) report that 20-hydroxyecdysone (32) stimulates cell death in the molting glands of *Oncopeltus fasciatus*. Gagalac-Nicolas, M. and Lim-Sylianco, C. (1981) have observed that 20-hydroxyecdysone (32) exhibits antimutagenic behaviour in the Ames test.

Isaac, R. *et al.* (1982) have investigated the separation of ecdysteroids by reversed phase HPLC and ecdysteroid 22-phosphates by ion pair reversed phase and ion exchange chromatography.

Roy, S. *et al.* (1982) have isolated 24-methylenecholest-5,7-dien-3β, 20 ε-diol (174) (amasterol) from the plant *Amaranthus viridis* which also contains ecdysteroids. It is suggested that this sterol may be an intermediate in the biosynthesis of ecdysteroids in this plant. Interestingly amasterol was found to inhibit the growth of lettuce seedlings.

REFERENCES

Dorn, A. and Buhlmann, K. J. (1982). Ovicidal activity of maternally applied ecdysteroids in the large milkweed bug (Heteroptera: Lygaeidae). *J. Econ. Entomol. 75*, 935–938.

Gagalac-Nicolas, M. G. and Lim-Sylianco, C. Y. (1981). Antimutagenic effects of *Cyanotis vaga* Lour. (Schultes) extracts. *Bull. Phillip. Biochem. Soc. 4*, 8–14 (*Chem. Abs.* 1983, *98*, 12367m).

Isaac, R. E., Milner, N. P. and Rees, H. H. (1982). High-performance liquid chromatography of ecdysteroids and ecdysteroid-22-phosphates. *J. Chromatogr. 246*, 317–322.

Kubo, I. and Klocke, J. A. (1983). Isolation of phytoecdysones as insect ecdysis inhibitors and feeding deterrents. *ACS Symp. Ser. 208* (Plant Resist. Insects), 329–346.

Kubo, I., Uchida, M. and Klocke, J. A. (1983). An insect ecdysis inhibitor from the African medicinal plant, *Plumbago capensis* (Plumbaginaceae); a naturally occurring chitin synthetase inhibitor. *Agric. Biol. Chem. 47*, 911–913.

Loeb, M. J., Woods, C. W., Brandt, E. P. and Borkovec, A. B. (1982). Larval testes of the tobacco budworm: a new source of insect ecdysteroids. *Science 218*, 896–897.

Nakano, K., Nohara, T., Tomimatsu, T. and Nishikawa, M. (1982). A phytoecdysteroid, taxisterone, from *Taxus cuspidata. Phytochemistry 21*, 2749–2751.

Nirde, P., Torpier, G., De Reggi, M. L. and Capron, A. (1983). Ecdysone and 20-Hydroxyecdysone: new hormones for the human parasite *Schistosoma mansoni. FEBS Lett. 151*, 223–227.

Roy, S., Dutta, A. K. and Chakraborty, D. P. (1982). Amasterol, an ecdysone precursor and a growth inhibitor from *Amaranthus viridis. Phytochemistry 21*, 2417–2420.

Saatov, Z., Gorovits, M. B., Abdullaev, N. D., Usmanov, B. Z. and Abubakirov, N. K. (1982a). Phytoecdysteroids of plants of the silene genus. IV. Sileneoside C as a galactoside of integristerone A from *Silene brahuica. Khim. Prir. Soedin* (2) 211–214. (*Chem. Abs.* 1982, *97*, 159496r).

Saatov, Z., Gorovits, M. B., Abdullaev, N. D., Usmanov, B. Z. and Abubakirov, N. K. (1982b). Phytoecdysteroids from plants of the genus silene. V sileneoside B-ecdysterone digalactoside from *Silene brahuica. Khim. Prir. Soedin.* (5) 611–615. (*Chem. Abs.* 1982, *98*, 104280f).

Smith, W. A. and Nijhout, H. F. (1983). *In vitro* stimulation of cell death in the molting glands of *Oncopeltus fasciatus* by 20-hydroxyecdysone. *J. Insect Physiol. 29*, 169–176.

Tsoupras, G., Luu, B. and Hoffmann, J. A. (1983). A cytokinin (isopentenyl-Adenosyl-Mononucleotide) linked to ecdysone in newly laid eggs of *Locusta migratoria, Science, 220,* 507–509.

7 Biosynthesis of Ecdysone

H. H. REES

University of Liverpool, Liverpool, UK

1 INTRODUCTION

This chapter will be concerned with the transformation of dietary sterols into ecdysone in both phytophagous and carnivorous insect species. The pathways of phytosterol dealkylation and conversion into cholesterol in phytophagous insects is also considered by Svoboda, J. and Thompson, M., in vol. 10 of this series. Further hydroxylation of ecdysone will not be considered in this chapter except in a cursory manner where necessary for completion. Detailed consideration of the hydroxylation of ecdysone yielding 20-hydroxyecdysone, and of the further metabolism of ecdysteroids, is covered elsewhere in this volume (Smith, S., chapter 8; Koolman, J. and Karlson, P., chapter 9).

Before proceeding to consider the detailed pathways of ecdysone formation, it is pertinent to consider what criteria should be fulfilled before the unequivocal establishment of the intermediacy of a particular compound in a pathway. These criteria are based on analogous ones suggested by Frantz, I. and Schroepfer, G. (1967) for intermediates in cholesterol biosynthesis in vertebrate animals. The compound should be: (a) isolated from tissues and its structure established, (b) formed biologically from the distal precursor, (c) converted biologically into the final product, (d) formed biologically from the postulated immediate precursor, (e) converted biologically into the next postulated compound in the scheme, (f) in addition, each individual step in the pathway should be demonstrated directly.

The intermediary role of any compound in a pathway cannot be regarded as being fully established until all of these criteria are satisfied. Furthermore, establishment of the quantitative importance of a proposed pathway required much additional information which is exceedingly difficult to obtain. It is essential to bear in mind these criteria and considerations during evaluation of the proposed pathways in this chapter. Thus, common observations that particular compounds accumulate in sufficient quantities for isolation and identification from tissues, or that synthetic radioactive precursors are efficiently converted into the final product, do not, of themselves, establish an intermediary role of such substances in a pathway. In fact, potential intermediates may accumulate because they are not further transformed.

In the case of neither the pathways of dealkylation of phytosterols to cholesterol, nor transformation of the latter into ecdysone, have all of the foregoing criteria been satisfied. However, far more information is available for the scheme of phytosterol dealkylation than in the case of ecdysone biosynthesis from cholesterol. In the latter pathway, a particular problem is the extremely low endogenous concentrations of many of the intermediates in the scheme. This stems from the fact that ecdysteroids, being hormones, are only present in tissues in comparatively small amounts. This problem is accentuated by the large endogenous pool of the distal precursor, cholesterol, so that administered exogenous radioactively labelled cholesterol is diluted to such an extent that it is impossible to detect the minute amounts of transient intermediates. The situation is further complicated by the ready production of polar steroids by autoxidation of cholesterol.

When considering such a diverse class of animals as the Insecta, there is always a danger of attempting to unify their metabolic processes into common schemes. In the pathways considered in this chapter, it is apparent that there is considerable species variation, which results in further difficulties in our current understanding of ecdysone biosynthesis.

2 DEALKYLATION OF PLANT STEROLS

In contrast to vertebrate animals and plants, insects (in common with many other classes of invertebrates) cannot synthesize sterols *de novo* from small molecules such as acetate or mevalonate. Consequently, for normal growth, development, and reproduction, insects require a dietary source of sterols, the only exception being those species where associated symbionts or intestinal micro-organisms may provide a source of these components (for review see Clayton, R. 1964). Sterols function in insects primarily as structural components of membranes and as precursors of ecdysteroids, the overwhelming majority of the sterol being required for the former function in view of the comparatively low levels of these developmental hormones. In most species of insects investigated to date, cholesterol (5)* fulfils the dietary sterol requirement for normal growth, development and reproduction. However, two species, *Drosophila pachea* and *Xyleborus ferrugineus*, are apparently incapable of introducing the Δ^7 bond into the sterol nucleus and require a dietary sterol containing either a Δ^7 bond or a $\Delta^{5,7}$-diene system (Heed, W. and Kircher, H. 1965; Chu, H. -M. *et al.*, 1970). This is pertinent in view of the occurrence of a Δ^7 bond in ecdysteroids. It has also been reported that sitosterol (1) in the

* Denotes structures in scheme.

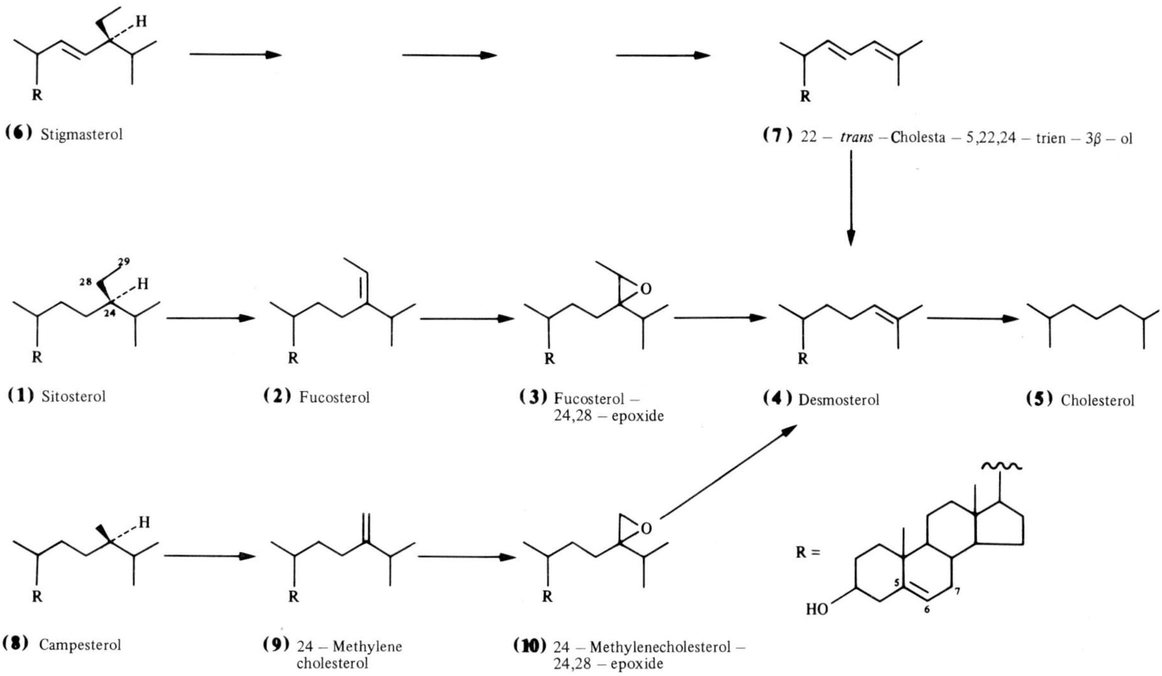

SCHEME 1. Conversion of sitosterol, stigmasterol, and campesterol into cholesterol in insects.

diet supported normal growth and survival of the lepidopteran, *Crambus trisectus*, whereas cholesterol was ineffective (Dupnik, T. and Kamm, J. 1970). The reason for this finding has not been elucidated.

Whereas carnivorous insects obtain their cholesterol directly from their food, the natural dietary sterols of phytophagous species are mostly ones bearing an alkyl group at C-24, e.g. sitosterol (1), stigmasterol (6), and campesterol (8). Most phytophagous and also omnivorous species investigated to date, can convert C_{29} and C_{28} sterols into C_{27} sterols, mainly cholesterol (5). However, three phytophagous species have been reported to be incapable of effecting such dealkylation reactions, *viz.* the large milkweed bug, *Oncopeltus fasciatus* (Svoboda, J. *et al.*, 1977), the khapra beetle, *Trogoderma granarium* (Svoboda, J. *et al.*, 1980a), and the honeybee, *Apis mellifera* (Svoboda, J. *et al.*, 1981). In the case of the latter species, a previous report (Allais, J. -P. *et al.*, 1971) suggesting that [^{3}H]sitosterol (1) was converted into cholesterol (5) is at variance with this; the reason for this discrepancy is not clear. Presumably, these phytophagous species which cannot dealkylate plant sterols have no major quantitative requirement for C_{27} sterols. The ability to dealkylate C_{28} and C_{29} sterols is not confined to insects, and has been demonstrated in members of other invertebrate phyla, including a protozoan, a coelenterate, an insect-parasitic and a free-living nematode, several crustaceans, together with some molluscs. If, as in the case of insects, there is a specific requirement by the dealkylating invertebrates for a C_{27} sterol for normal development and reproduction, the dealkylation pathway provides a means of satisfying such a requirement from dietary phytosterols. In fact, there is an apparent correlation between the ability of invertebrate animals to dealkylate phytosterols and absence of *de novo* sterol synthesis (Rees, H. 1984).

The capacity of different insect species to utilize (24*R*)- and (24*S*)-alkyl sterols differs appreciably. For example, the tobacco hornworm, *Manduca sexta*, can utilize both stereoisomers of 24-methylcholesterol (8 and 11), but on the basis of different growth rates, the 24*S*-methyl group of dihydrobrassicasterol (11) may be less efficiently dealkylated than the 24*R*-methyl group of campesterol (8) (Svoboda, J. and Robbins, W. 1968). This

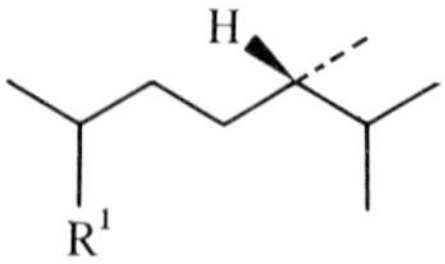

(11) Dihydrobrassicasterol

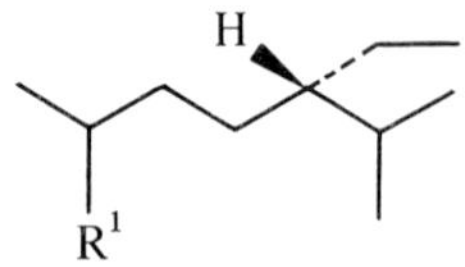

(12) Clionasterol

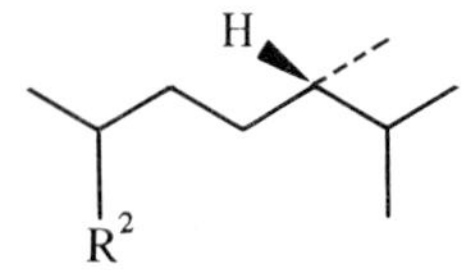

(13) (24*S*) – 24 – Methylcholesta - 5,7 –dien – 3β – ol

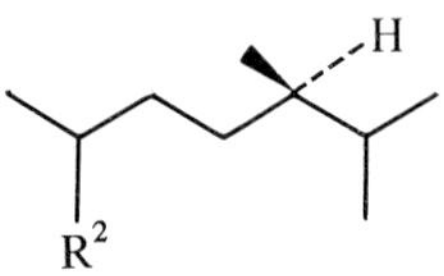

(14) (24*R*) – 24 – Metnylcholesta – 5,7 – dien – 3β – ol

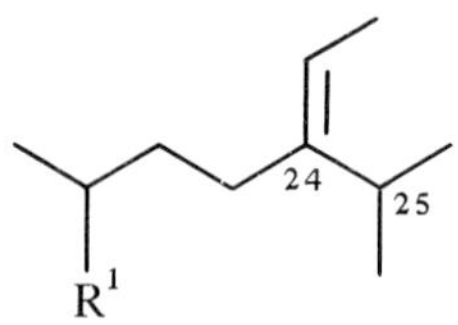

(15) 28 – Isofucosterol

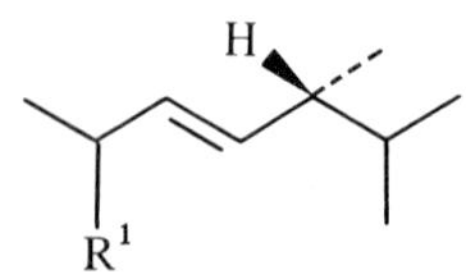

(16) Brassicasterol

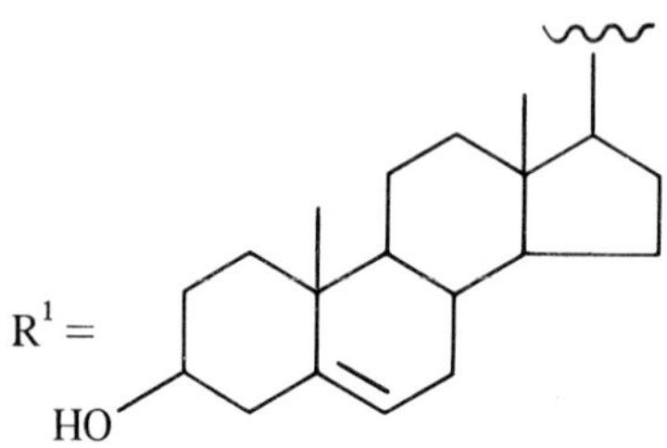

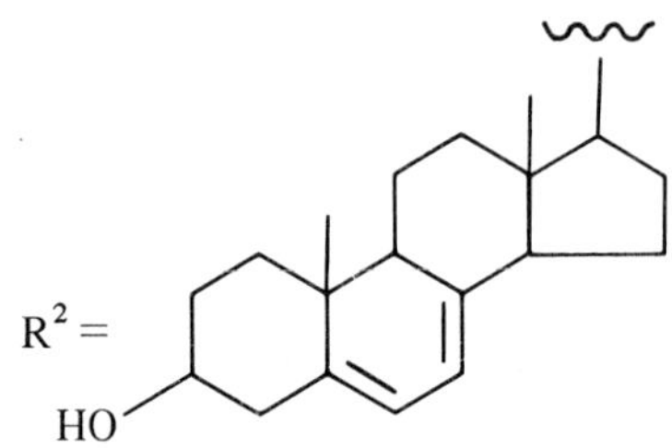

may be an adaptation of the insect to its major host plant, tobacco, which contains predominantly stigmasterol (24*S*; 6) and sitosterol (24*R*; 1). (The absolute stereochemistry at C-24 in sitosterol is the same as that in stigmasterol, although the presence of the Δ^{22} bond in the latter reverses the nomenclature assignment from 24*R* to 24*S*.) In the silkworm, *Bombyx mori*, both stereoisomers of 24-ethylcholesterol (sitosterol (1) and clionasterol (12)) completely satisfied the sterol requirment (cited in Fujimoto, Y. *et al.*, 1980a), whereas there was a preference for (24*R*)- alkylsterols over (24*S*)-isomers in *Dermestes maculatus* (Ikan, R. *et al.*, 1970). *Drosophila* species apparently differ in their capacity to utilize stereoisomers of 24-methyl-cholesta-5,7-dien-3β-ol to support growth, development and reproduction. Whereas the (24*S*)-isomer (13) satisfies the sterol requirement of *Drosophila melanogaster* and *Drosophila mojavensis*, but not *Drosophila pachea*, the (24*R*)-isomer (14) suffices for *D. melanogaster* and *D. pachea*, but not for *D. mojavensis* (Kircher, H. and Rosenstein, F., 1974). The reason for these differing stereochemical requirements at C-24 in alkylated sterols for support of normal development and reproduction in different species is not clear.

Since cholesterol is known to serve as a distal precursor of C_{27} ecdysteroids, it is pertinent to consider the pathways of formation of this sterol from dietary plant sterols in phytophagous and omnivorous insect species. This subject has been covered in several reviews (Ritter, F. and Wientjens, W., 1967; Thompson, M. *et al.*, 1973; Svoboda, J. *et al.*, 1975a, 1978). There is evidence for the operation of the pathways shown in scheme 1 as major routes in the dealkylation of plant sterols in several species of insects. Much of this scheme is based on work on the tobacco hornworm, *Manduca sexta* (Thompson, M. *et al.*, 1973). We shall now consider the components of this scheme, in turn. Modifications of these routes in certain species will be considered in section 2.4.

2.1 Sitosterol

Since sitosterol (1) has been the most widely used substrate for investigation of dealkylation in insects, the species in which dealkylation of this sterol

Table 1: Dealkylation of sitosterol

Order	Species	Reference
Diplura	*Thermobia domestica*	Svoboda, J. and Robbins, W. (1971)
Dictyoptera	*Blattella germanica*	Robbins, W. *et al.* (1962)
	Eurycotis floridana	Ritter, F. and Wientjens, W. (1967)
	Periplaneta americana	Svoboda, J. and Robbins, W. (1971)
Orthoptera	*Locusta migratoria*	Allais, J.-P. and Barbier, M. (1971)
	Schistocerca gregaria	Majumder, M. (1978)
Coleoptera	*Anthonomus grandis*	Earle, N. *et al.* (1967)
	Tribolium confusum	Svoboda, J. *et al.* (1972a)
	Tenebrio molitor	Nicotra, N. *et al.* (1978)
Lepidoptera	*Bombyx mori*	Ikekawa, N. *et al.* (1966)
	Manduca sexta	Svoboda, J. *et al.* (1967)
	Heliothis zea	Svoboda, J. and Robbins, W. (1971)
	Spodoptera frugiperda	Svoboda, J. and Robbins, W. (1971)
	Spodoptera littoralis	Clarke, G. (1981)
Hymenoptera	*Neodiprion pratti*	Schaefer, C. *et al.* (1965)

have been demonstrated using either radioactive or non-radioactive substrate are summarized in Table 1.

2.1.1 Fucosterol

That fucosterol (2) might be an intermediate in the dealkylation of sitosterol in the cockroaches *Blattella germanica* and *Eurycotis floridana*, was originally suggested by its tentative identification as a metabolite during [3-^{3}H]sitosterol (1) dealkylation on the basis of gas–liquid chromatographic characteristics. Furthermore, [3-^{3}H]fucosterol (2) was further metabolized to cholesterol (5), probably via a compound which had the same gas–liquid chromatographic properties as desmosterol (4) (Ritter, F. and Wientjens, W. 1967). However, the intermediacy of fucosterol (2) in the dealkylation of sitosterol (1) was first firmly established in the tobacco hornworm, *Manduca sexta*. In this case, [^{3}H]fucosterol (2) was isolated and identified as a metabolite of [^{3}H]sitosterol (1), and was also converted efficiently into [^{3}H]cholesterol (5) via [^{3}H]desmosterol (4) (Svoboda, J. *et al.*, 1971). The ready utilization and conversion of fucosterol (2) into cholesterol (5) via desmosterol (4) in *Manduca sexta* had been demonstrated earlier (Svoboda, J. and Robbins, W. 1968). Fucosterol (2) has also been implicated as an intermediate in dealkylation of sitosterol (1) in several other species (Table 2).

2.1.2 28-Isofucosterol

The question now arises as to whether introduction of the $\Delta^{24(28)}$ bond into sitosterol (1) is completely stereospecific or is the geometric isomer of fucosterol (2), 28-isofucosterol (15), also involved in the dealkylation. In this respect it is significant that 28-isofucosterol (15) satisfies the sterol requirement in *Bombyx* and that this sterol is dealkylated to cholesterol, albeit not as efficiently as fucosterol (2), in that species (Morisaki, M. *et al.,* 1972, 1974). Furthermore, both fucosterol (2) and smaller quantities of 28-isofucosterol (15) have been identified in *Bombyx mori*, whereas mulberry leaves, on which the insects were reared, contained no appreciable amount of these sterols (Morisaki, M. *et al.*, 1981). 28-Isofucosterol (15) is also converted into cholesterol in the yellow mealworm, *Tenebrio molitor* (Randall, P. *et al.*, 1972), and formation of both fucosterol (2) and smaller amounts of 28-isofucosterol (15) from [^{3}H]sitosterol (1) has been demonstrated in that species (Nicotra, F. *et al.*, 1978). However, comparison of the relative rates of dealkylation of fucosterol (2) and 28-isofucosterol (15) in *Tenebrio* awaits elucidation. Although 28-isofucosterol (15) but not fucosterol (2) has been identified in prepupal honeybees, *Apis mellifera*, from brood reared by workers fed chemically

Table 2: Evidence for intermediacy of fucosterol (2) in sitosterol (1) dealkylation

Species	Substrate	Products	Other observations	References
Blattella germanica	[³H]Sitosterol	fucosterol?, desmosterol?, cholesterol	GLC characterisation	Ritter, F. and Wientjens, W. (1967)
Eurycotis floridana	[³H]Fucosterol	desmosterol?, cholesterol		
Manduca sexta	[³H]Sitosterol	fucosterol, desmosterol, cholesterol	fucosterol identified by mass;	Svoboda, J. *et al.* (1971)
	[³H]Fucosterol	desmosterol, cholesterol		
			fucosterol supports growth and development	Svoboda, J. and Robbins, W. (1968)
Locusta migratoria	[³H]Sitosterol	fucosterol, desmosterol, cholesterol		Allais, J.-P. and Barbier, M. (1971)
Tribolium confusum	[³H]Sitosterol	fucosterol, desmosterol, cholesta-5,7, 24-trien-3β-ol, 7-dehydrocholesterol, cholesterol		Svoboda, J. *et al.* (1972a)
Bombyx mori	[³H]Fucosterol	fucosterol-24, 28-epoxide, cholesterol		Morisaki, M. *et al.* (1972)
			fucosterol identified by mass; fucosterol supports growth and development	Morisaki, M. *et al.* (1981), Morisaki, M. *et al.* (1974)
Tenebrio molitor	[³H]Sitosterol	fucosterol, 28-isofucosterol		Nicotra, F. *et al.* (1978)
Schistocerca gregaria	[¹⁴C]Sitosterol	fucosterol, desmosterol, cholesterol, 7-dehydrocholesterol		Majumder, M. (1978)
	[³H]Fucosterol	fucosterol-24, 28-epoxide, desmosterol, cholesterol	in midgut and Malpighian tubule homogenate	
Spodoptera littoralis	[¹⁴C]Sitosterol	fucosterol, fucosterol-24, 28-epoxide, desmosterol, cholesterol, 7-dehydro-cholesterol		Clarke, G. (1981)
	[³H]Fucosterol	fucosterol-24, 28-epoxide, desmosterol, cholesterol, 7-dehydrocholesterol	fucosterol identified by mass	

defined synthetic diets, the former sterol (15) does not apparently arise from dealkylation of sitosterol (1) (Svoboda, J. *et al.*, 1980b) and could ultimately be of plant origin. In the cotton leafworm, *Spodoptera littoralis*, fucosterol (2), but not 28-isofucosterol (15), could be detected in mature larvae reared on either non-radioactive or [¹⁴C]-labelled sitosterol-supplemented semi-defined diet (Clarke, G., 1981).

In summary, the foregoing considerations indicate that in both *Bombyx* and *Tenebrio*, some dealkylation of sitosterol (1) via 28-isofucosterol (15) does occur. It should be appreciated that inability to detect 28-isofucosterol (15) in *Spodoptera* or the detection of smaller quantities of 28-isofucosterol than of fucosterol (2) in *Tenebrio* do not, in themselves, demonstrate that the route via fucosterol predominates over that via 28-isofucosterol, since the latter sterol could conceivably be rapidly metabolized. Current evidence is at least consistent with the operation of a major route via fucosterol (2) in all investigated species where sitosterol dealkylation does occur. The results in *Spodoptera* do suggest that the dealkylation of sitosterol in this species may occur essentially, exclusively via fucosterol (2). However, it is premature on the basis of data currently available, to speculate whether there are species differences with regard to the nature of the $\Delta^{24(28)}$ geometrical isomer(s) involved in sitosterol dealkylation.

2.1.3 Mechanism of formation of 24(28)-ethylidene group

Two conceivable mechanisms of formation of the 24(28)-ethylidene grouping from 24-ethylsterols have been considered: (a) direct elimination of two hydrogen atoms, or (b) hydroxylation at either C-24 or C-28 followed by dehydration. However, experimental evidence does not support operation of the latter route. For example, stigmast-5-en-3*β*,24-diol is not utilised in *Bombyx* (Morisaki, M. *et al.*, 1974) or *Tenebrio* (Nicotra, F. *et al.*, 1979). Although 24(28)-ethylidene sterols and cholesterol are formed from stigmast-5-en 3*β*, 24-diol in *Tenebrio*, it has not been possible to detect formation of the latter sterol from sitosterol (1). These results suggest that

Table 3: Evidence for intermediacy of desmosterol (4) in sitosterol (1) dealkylation

Species	Substrate	Products	Other observations	References
Manduca sexta	[^{3}H]Sitosterol	desmosterol, cholesterol	dealkylation also occurred after by-passing gut by injection	Svoboda, J. *et al.* (1967)
	[^{14}C]Desmosterol	cholesterol		Also see Table 2
	Sitosterol-supplemented diet	desmosterol identified by mass		
	[^{14}C]Desmosterol	cholesterol	in gut homogenate	Svoboda, J. *et al.* (1969b)
Blatella germanica, *Eurycotis floridana*				See Table 2
Thermobia domestica, *Blattella germanica*, *Periplaneta americana*, *Spodoptera frugiperda*, *Manduca sexta*, *Heliothis zea*	Sitosterol-supplemented diet in combination with 20,25-diazacholesterol	desmosterol, cholesterol		Svoboda, J. and Robbins, W. (1971)
Locusta migratoria				See Table 2
Tribolium confusum	[^{14}C]Desmosterol	7-dehydrocholesterol, cholesterol		Svoboda, J. *et al.* (1972a) Also see Table 2
Tenebrio molitor	[^{3}H]Clionasterol + triparanol	desmosterol, cholesterol		Pettler, P. *et al.* (1974)
Bombyx mori	[^{3}H]Sitosterol	desmosterol, cholesterol		Fujimoto, Y. *et al.* (1974a)
Schistocerca gregaria	[^{3}H]Fucosterol-24,28-epoxide	desmosterol, cholesterol	in midgut and Malpighian tubule microsomes	Majumder, M. (1978)
	[^{14}C]Desmosterol	cholesterol		
Spodoptera littoralis	[^{3}H]Fucosterol-24,28-epoxide	desmosterol, cholesterol	midgut microsomes	Clarke, G. (1981)
	[^{14}C]Desmosterol	cholesterol		

the 24(28)-ethylidene group probably arises by direct dehydrogenation of the 24-ethyl group (Nicotra, F. *et al.*, 1979).

2.1.4 Desmosterol

Evidence for the intermediacy of desmosterol (4) in the conversion of sitosterol (1) into cholesterol (5) was facilitated by the finding that certain vertebrate hypocholesterolaemic agents (in particular triparanol (17) and some azasteroids, e.g. 20,25-diazacholesterol (18)) which block reduction of the Δ^{24} bond during cholesterol biosynthesis in vertebrate animals have a similar action in insects leading to accumulation of desmosterol (Svoboda, J. and Robbins, W., 1967, 1971).

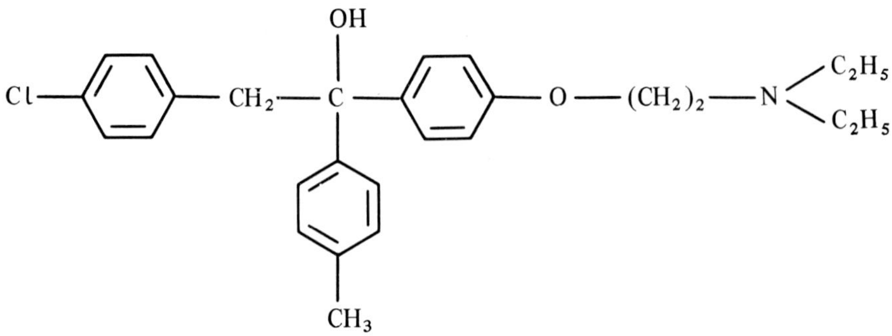

(17) Triparanol

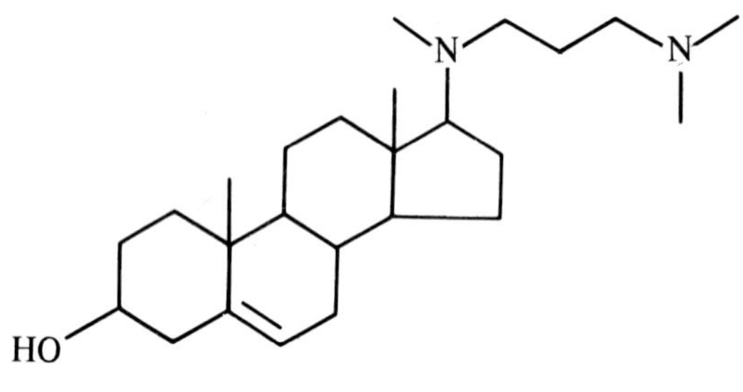

(18) 20,25 – Diazacholesterol

was originally furnished in *Manduca sexta* (Svoboda, J. *et al.*, 1967). This has been corroborated in several other species (Table 3). Desmosterol (4) was also demonstrated to be the terminal intermediate in the conversion of campesterol (8), stigmasterol (6), fucosterol (2), 24-methylenecholesterol (9), brassicasterol (16) and dihydrobrassicasterol (11) into cholesterol (5) (Svoboda, J. and Robbins, W., 1968). In many of the foregoing studies the detection of demosterol (4)

Demosterol reductase activity, catalysing the conversion of desmosterol (4) into cholesterol (5), has been demonstrated *in vitro* in a preparation from *Manduca* midgut (Svoboda, J. *et al.*, 1969b) and in microsomal fractions of midgut tissue from *Schistocerca gregaria* (Majumder, S., 1978) and *Spodoptera littoralis* (Clarke, G., 1981). Inhibition of desmosterol reductase from microsomes of midgut and Malpighian tubules of the locust, *Schistocerca gregaria*, by *p*-hydroxy-mercuribenzoate and extensive reversal of this inhibition in the presence of dithiothreitol indicates the involvement of a sulphydryl group in the Δ^{24}-sterol reductase activity (Majumder, S. 1978). This is analogous to the rat liver desmosterol reductase (Avigan, J. and Steinberg, D. 1961).

2.1.5 Fucosterol-24,28-epoxide

The demonstration that desmosterol (4) was amongst the products formed upon treatment of fucosterol-24,28-epoxide (3) with boron trifluoride etherate, prompted the suggestion that this epoxide might be an intermediate in the transformation of fucosterol (2) to desmosterol (4) during dealkylation in insects (Ikekawa, N. *et al.*, 1971; Ohtaka, H. *et al.*, 1973). This suggestion was substantiated by the demonstration in *Bombyx mori* that fucosterol-24,28-epoxide (3) was formed from fucosterol (2) and was converted into cholesterol (5) *in vivo* as well as into desmosterol (4) by a cell-free system (1500***g*** supernatant) from gut tissue (Morisaki, M. *et al.*, 1972; Awata, N. *et al.*, 1975). Similarly, in *Locusta migratoria*, the conversion of fucosterol (2) into the epoxide (3), and of the latter into desmosterol (4) and cholesterol (5), has been demonstrated (Allais, J.-P. *et al.*, 1973; Allais, J.-P. and Barbier, M., 1977). The transformation *in vivo* of fucosterol-24,28-epoxide (3) into cholesterol in *Tenebrio molitor* has also been reported recently (Nicotra, F. *et al.*, 1981). Additional evidence for the intermediacy of the epoxide (3) in fucosterol (2) dealkylation is furnished by the direct demonstration of each step in this conversion ((2) → (3) → (4)) in cell-free systems from gut tissues (*Bombyx mori* (Fujimoto, Y. *et al.*, 1980a); microsomal fractions from *Schistocerca gregaria* (Majumder, S., 1978) and *Spodoptera littoralis* (Clarke, G., 1981)). Further substantiation of the involvement of the epoxide (3) is provided by the recent mass isolation of fucosterol-24,18-epoxide (3) from larvae of *Bombyx mori* (Ikekawa, N. *et al.*, 1980a) and by the trapping of [4-^{14}C]fucosterol-24,28-epoxide formed from [4-^{14}C]sitosterol *in vivo* in *Spodoptera littoralis* (Clarke, G., 1981). The foregoing evidence leaves little doubt that a major route of sitosterol dealkylation is that shown in scheme 1. The extensive inhibition of sitosterol (1) dealkylation in *Bombyx in vivo* by a structural analogue of fucosterol-24,28-epoxide (3), 24,28-iminofucosterol (19), which is a potent inhibitor of the conversion of this epoxide into desmosterol (4) (Fujimoto, Y. *et al.*, 1974a; Awata, N. *et al.*, 1975), is consistent with the involvement of fucosterol-24,28-epoxide (3) in sitosterol dealkylation.

During investigation of [6-^{3}H]fucosterol-24,28-epoxide (3) metabolism by midgut microsomes from *Spodoptera littoralis*, it became apparent that this substrate was not only metabolized to desmosterol (4) and cholesterol (5), but also by a branch pathway via stigmast-5-en-3β, 24,28-triol (20) to 3β,24-dihydroxy-stigmast-5-en-24-one (21) (Clarke, G., 1981). Since the latter sterol was not further metabolized in this system, and does not support growth and development in *Bombyx mori* (Morisaki, M. *et al.*, 1974), this branch pathway does not apparently form part of the dealkylation sequence. Formation of stigmast-5-en-3β,24,28-triol (20) from fucosterol-24,28-epoxide (3) can be inhibited by 3,3,3-trichloro-1,2-epoxypropane (Clarke, G., 1981) and probably results from the action of an epoxide hydrolase enzyme of low substrate specificity. It could conceivably be the same epoxide hydrolase reported recently in midgut microsomes of the related species, *Spodoptera eridania*, although in this case the trisubstituted epoxide substrates investigated were refractory to enzymic attack (Mullin, C. and Wilkinson, C., 1980a,b).

The few reported attempts to demonstrate metabolism of sitosterol (1) by *in vitro* systems from insects, either in tissue slices or in cell-free preparations, have so far been unsuccessful; the reason for this is obscure (Awata, N. *et al.*, 1975; Majumder, M., 1978; Clarke, G., 1981). However, transformation of [6-^{3}H]fucosterol (2) to desmosterol (4) and cholestrol (5) has been demonstrated in tissues from *Spodoptera littoralis in vitro* (Clarke, G. 1981). Although this conversion occurred to a small extent in several tissues, it was primarily located in the midgut. In *Bombyx mori* (Awata, N. *et al.*, 1975) and *Schistocerca gregaria* (Majumder, M. 1978), although enzymic activity catalysing the transformation of fucosterol-24,28-epoxide (3) to cholesterol (5) occurred primarily in the gut, appreciable activity was also detected in other tissues. Fucosterol-24,28-epoxide dealkylating activity is located primarily in the microsomal

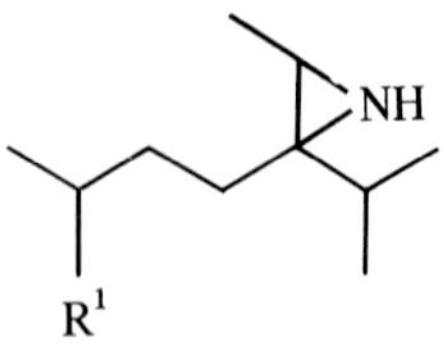

(19) 24,28 – Iminofucosterol

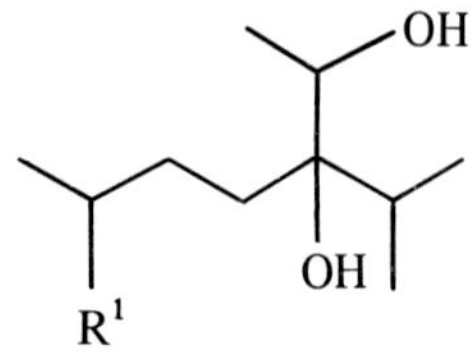

(20) Stigmast – 5 – en – 3β,24,28 – triol

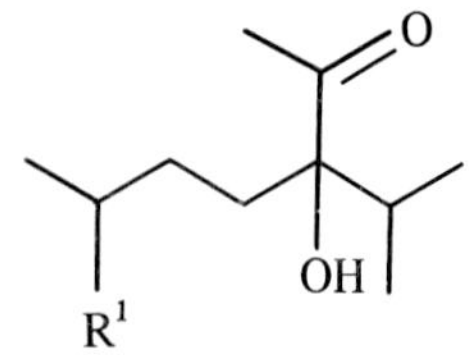

(21) 3β,24 – Dihydroxystigmast – 5 – en – 24 – one

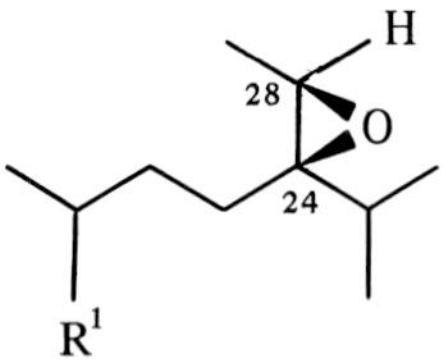

(22) (24*R*, 28*R*) – Fucosterol – 24,28 – epoxide

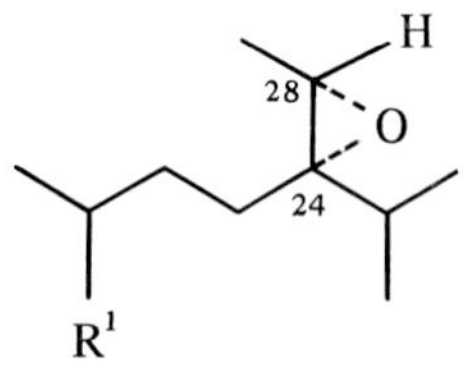

(23) (24*S*, 28*S*) – Fucosterol– 24,28 – epoxide

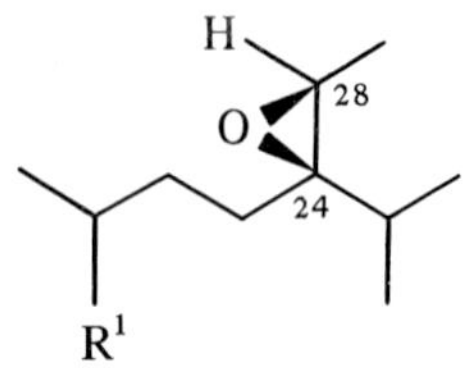

(24) (24*R*, 28*S*) – 28 – Isofucosterol – 24,28 – epoxide

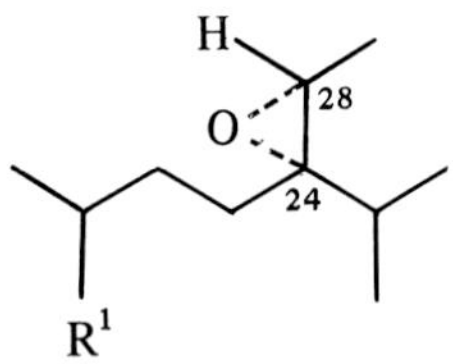

(25) (24*S*, 28*R*) – 28 – Isofucosterol – 24,28 – epoxide

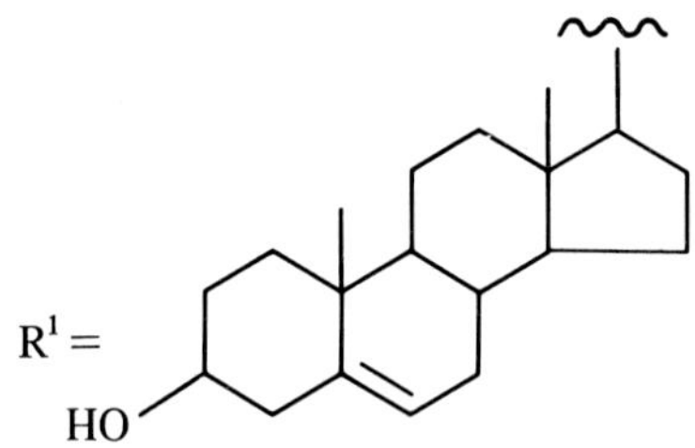

fraction of *S. littoralis* midgut and *S. gregaria* midgut and Malpighian tubules (Clarke, G., 1981; Majumder, M., 1978). Some properties of the fucosterol-24,28-epoxide cleavage enzyme catalysing transformation of the epoxide (3) to desmosterol (4) have been investigated in a microsomal preparation from *Spodoptera littoralis* midgut, utilizing 3,3,3-trichloro-1,2-epoxypropane to inhibit formation of stigmast-5-en-3β,24,28-triol (20) (Clarke, G. 1981). The cleavage enzyme activity [(3) → (4)] in this *in vitro* preparation had V_{max} 45 nmol h^{-1} mg^{-1} protein and an apparent K_m for fucosterol-24,28-epoxide (3) of 111 nM. However, the latter value does not take into account the concentration of endogenous substrate in the microsomes. Lack of extensive inhibition by 5,5[1]-dithiobis(2-nitrobenzoicacid) (DTNB) suggests that a sulphydryl group is not critically involved in the activity of the enzyme. Similarly, lack of inhibition with carbon monoxide, or with low concentrations of cyanide or metopirone, indicate that cytochrome P-450 is not involved in the reaction. However, it is difficult to rationalize the reduction (88%) in the enzymic activity observed anaerobically compared with aerobic conditions in terms of the mechanism shown in scheme 2 (section 2.1.7).

Stereochemistry. Both the (24*R*,28*R*)- and (24*S*, 28*S*)-fucosterol-24,28-epoxide stereoisomers (22, and 23, respectively) have been synthesized and the absolute configurations at C-24 and C-28 established by chemical correlation with sitosterol (1) and clionasterol (12) (Fujimoto, Y. *et al.*, 1980b). This work revealed that previous stereochemical assignments for (22) and (23) should be reversed (Chen, S.-M. *et al.*, 1975). When [3α-^{3}H]-labelled (24*R*, 28*R*)- and (24*S*, 28*S*)-fucosterol-24,28-epoxides were incubated with a cell-free preparation from *Bombyx mori* gut tissue, both isomers were effectively converted into desmosterol (4) and cholesterol (5), the (24*R*, 28*R*)-isomer (22) being a slightly better substrate. The latter isomer (22) was also consumed at a faster rate than the (24*S*, 28*S*)-isomer (23). Incubation of [3α-^{3}H]fucosterol (2) with this cell-free preparation in the presence of either 24,28-iminofucosterol (19) or a fucosterol-24,28-epoxide "trap" yielded both fucosterol epoxide isomers (22) and (23) in approximately equal amounts; under these conditions, appreciable desmosterol (4) and cholesterol (5) were also formed. *In vivo*, slightly preferential formation of the (24*R*, 28*R*)-isomer (22) from [3α^{3}H]fucosterol was observed in the presence of cold "traps" of either epoxide isomer (Fujimoto, Y. *et al.*, 1980a). Since in these experiments on the formation of the fucosterol-24,28-epoxide isomers from fucosterol, the epoxides are also further metabolized to desmosterol and cholesterol, exact quantitative conclusions are difficult. However, taken together with the slightly preferential utilization of the (24*R*, 28*R*)-isomer (22), these experiments suggest that metabolism via the latter isomer might predominate to a slight extent. It is apparent at least in *Bombyx* that both the formation of the epoxide from fucosterol (2) and its transformation into desmosterol (4) proceed with a low degree of stereospecificity. In agreement with this is the isolation of endogenous mass of both fucosterol-24,28-epoxide isomers (22 and 23) from *Bombyx mori* (Ikekawa, N. *et al.*, 1980a), and the demonstration that both isomers support normal growth and development in this species (Fujimoto, Y. *et al.*, 1980a). The situation is apparently similar in the yellow mealworm, *Tenebrio molitor*, where both fucosterol-24,28-epoxide isomers are transformed into cholesterol to about the same extent, suggesting low substrate stereospecificity in this species also (Nicotra, F. *et al.*, 1981). As in *Bombyx*, there was a slightly greater conversion of the (24*R*, 28*R*)-isomer (22).

2.1.6 28-Isofucosterol-24,28-epoxide

In *Bombyx mori*, as alluded to earlier (section 2.1.2), 28-isofucosterol (15) satisfies the sterol requirement (Morisaki, M. *et al.*, 1974) and is also transformed into cholesterol (Morisaki, M. *et al.*, 1972). Therefore, the demonstration that both stereoisomers of 28-isofucosterol-24,28-epoxide (24 and 25) were unable to support growth and development in that species is somewhat of an enigma (Fujimoto, Y. *et al.*, 1980a). These workers suggested that the transformation of 28-isofucosterol (15) into cholesterol (5) in *Bombyx* occurs via a pathway not involving the corresponding epoxide; other explanations are also tenable. In apparent agreement with this is the isolation of both fucosterol (2) and smaller amounts of 28-isofucosterol (15) from *Bombyx*, whereas only endogenous fucosterol-24,28-epoxide isomers (22 and 23) but not the 28-isofucosterol-24,28-epoxides (24 and 25) could be detected (Ikekawa, N. *et al.*, 1980a).

The efficient conversion of [7-$^3H_2$](24*R*, 28*S*)-28-isofucosterol-24,28- epoxide (24), and to a much smaller extent of the corresponding (24*S*, 28*R*)-isomer (25), into cholesterol in *Tenebrio molitor* (Nicotra, F. *et al.*, 1981) suggests that the situation is different in this species and that dealkylation of 28-isofucosterol (15) occurs via the corresponding epoxide. However, in *T. molitor*, the transformation of 28-isofucosterol (15) into the corresponding epoxides (24 and 25) has not been reported as yet. It is interesting that the substrate stereospecificity shown by this species is marked for the 28-isofucosterol-24,28-epoxides (24 and 25), but is quite low for the fucosterol-24,28-epoxides (22 and 23).

2.1.7 Mechanism of fucosterol-24,28-epoxide dealkylation

Simultaneously with the demonstration in *Bombyx mori* of the probable intermediacy of fucosterol-24,28-epoxide (3) in the dealkylation of fucosterol (2) (Morisaki, M. *et al.*, 1972), it was shown that the C-25 hydrogen of 28-isofucosterol (15) is retained during its transformation to cholesterol (5) in *Tenebrio molitor* (Randall, P. *et al.*, 1972). This result suggested that the C-25 hydrogen atom must migrate (presumably to C-24) during dealkylation, since the Δ^{24} sterol, desmosterol (4), is involved in the process. The C-25 hydrogen atoms of clionasterol (12) and sitosterol (1) have been located at C-24 in the desmosterol (4) formed during dealkylation in *Tenebrio molitor* (Pettler, P. *et al.*, 1974, 1978) and *Bombyx mori* (Fujimoto, Y. *et al.*, 1974a),

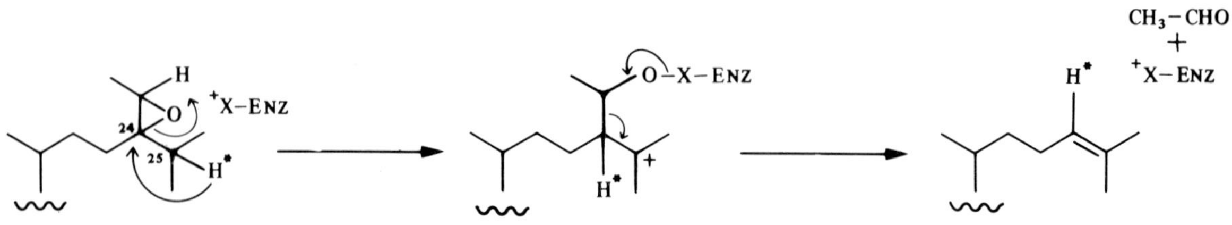

(3) Fucosterol – 24,28 – epoxide **(4)** Desmosterol

SCHEME 2. Possible mechanism of dealkylation of fucosterol-24,28-epoxide (3) (Pettler, P. *et al.*, 1978).

respectively. This hydrogen migration may be rationalized in terms of a possible mechanism (Scheme 2) analogous to that proposed previously for the chemical transformation of fucosterol 24,28-epoxide (3) into desmosterol (4) by boron trifluoride etherate (Ohtaka, H. *et al.*, 1973).

2.1.8 OVERALL PATHWAY OF DEALKYLATION OF SITOSTEROL

Demonstration that the Δ^{24} sterol, desmosterol (4), is a terminal intermediate in sitosterol dealkylation prompted the suggestion that the pathway of dealkylation may, at least in part, be the reverse of C-24 alkylation of sterols in plants (Svoboda, J. *et al.*, 1967). In the latter process, the extra carbon atoms are added sequentially, a 24-methylene sterol being an intermediate. However, there is no evidence that the C-29 and C-28 carbon atoms are removed sequentially during dealkylation (Allais, J.-P. and Barbier, M., 1966), and 24-methylenecholesterol (9) could not be detected as an intermediate during the conversion of sitosterol (1) into cholesterol (5) in *Locusta migratoria* (Allais, J.-P. and Barbier, M. 1971).

The possibility has also been considered that the pathway of sitosterol (1) dealkylation might conceivably mimic the side-chain cleavage of cholesterol or pregnenolone in vertebrates (Randall, P. *et al.*, 1972; Morisaki, M. *et al.*, 1974). However, the observations that none of the sterols on such conceivable routes supported growth and development of *Bombyx* (Morisaki, M. *et al.*, 1974) and that 24-hydroxycholesterol, a common intermediate in these pathways, was negligibly transformed to cholesterol, indicate that such routes may be excluded.

2.2 Campesterol

There is evidence that the pathway of dealkylation of campesterol (8) is analogous to that for sitosterol (1) (Scheme 1). In *Manduca sexta*, the transformation of [2,4-^{3}H]campesterol (8) to 24-methylene cholesterol (9), desmosterol (4), and cholesterol (5) has been demonstrated, the 24-methylene sterol also being identified by mass (Svoboda, J. *et al.*, 1972b). The conversion of [2,4-^{3}H]24-methylene-cholesterol (9) into desmosterol and cholesterol has also been demonstrated in that species. Similarly, 24-methylenecholesterol (9) and cholesterol were identified as metabolites of the 24*S* isomer of campesterol, dihydrobrassicasterol (11), in *Manduca* (Svoboda, J. *et al.*, 1972b), a conversion tentatively demonstrated earlier in the German cockroach, *Blattella germanica* (Ritter, F. and Wientjens, W., 1967). In the latter species, a 24-methylene sterol, ergosta-7,22,24(28)-trien-3β-ol (26), was also tentatively identified as a metabolite of ergosta-7,22-dien-3β-ol (27) (Ritter, F. and Wientjens, W., 1967).

By analogy with the pathway of sitosterol dealkylation, it would be expected that the metabolism of 24-methylenecholesterol (9) would proceed via the corresponding 24,28-epoxide (10) (Scheme 1). However, 24-methylenecholesterol-24,28-epoxide (10), in contrast to fucosterol epoxide (3), when treated with boron trifluoride etherate did not yield desmosterol as a fragmentation product (Ohtaka, H. *et al.*, 1973). Evidence for the intermediacy of 24-methylenecholesterol-24,28-epoxide (10) in the dealkylation of 24-methylenecholesterol (9) is furnished by the demonstration of the conversion of the latter sterol (9) into the corresponding 24,28-epoxide (10), desmosterol (4) and cholesterol (5), and by the transformation of the epoxide (10) into desmosterol (4) and cholesterol (5) by a microsomal fraction from midgut and Malpighian tubules from *Schistocerca gregaria* (Majumder, M. 1978). In this cell-free system, fucosterol-24,28-epoxide (3) and 24-methylenecholesterol-24,28-epoxide (10) were

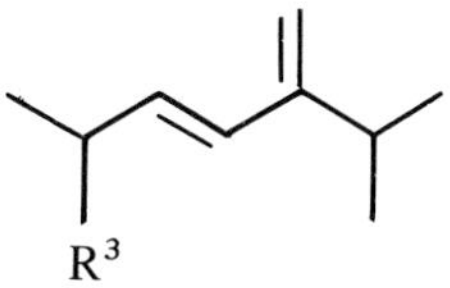

(26) Ergosta – 7,22,24 (28) – trien – 3β – ol

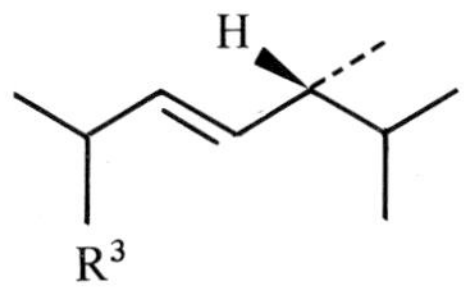

(27) Ergosta – 7,22 – dien – 3β – ol

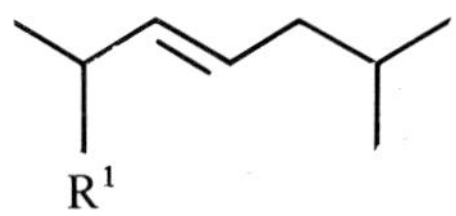

(28) *Trans* – 22 – dehydro cholesterol

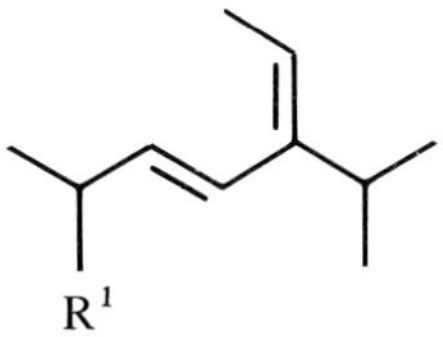

(29) 22 – *Trans* – stigmasta – 5,22,24 (28) – trien – 3β – ol

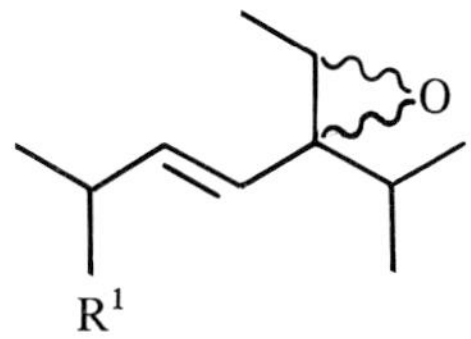

(30) 24,28 – Epoxystigmasta – 5,22 – dien – 3β – ol

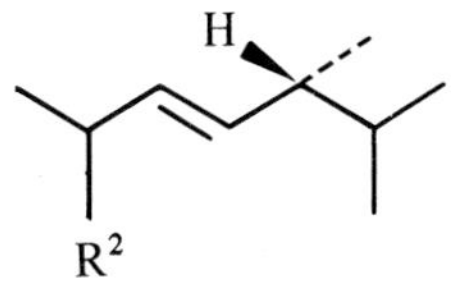

(31) Ergosterol

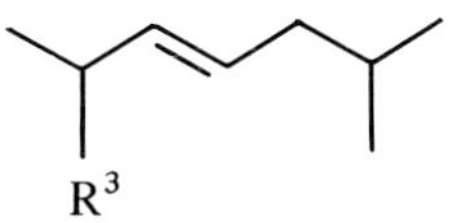

(32) 5a – Cholesta – 7,22 – dien – 3β – ol

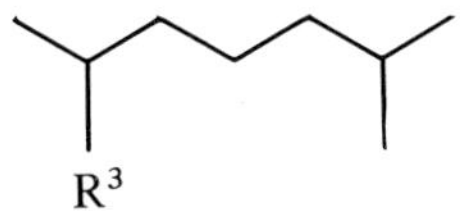

(33) 5a – Cholest – 7 – en 3β – ol

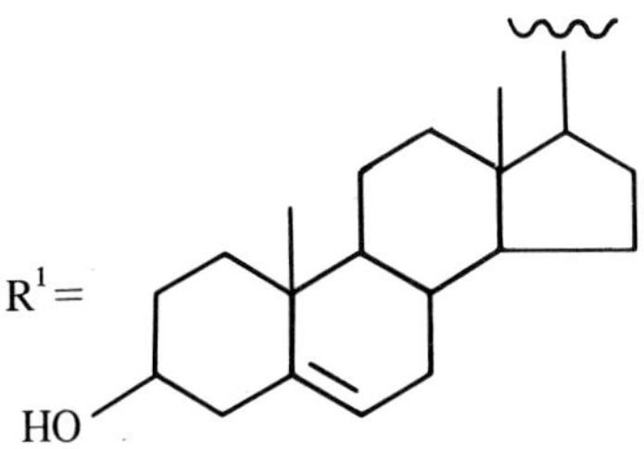

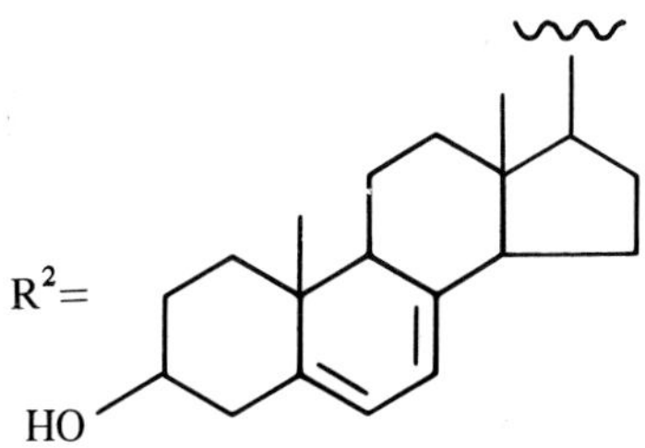

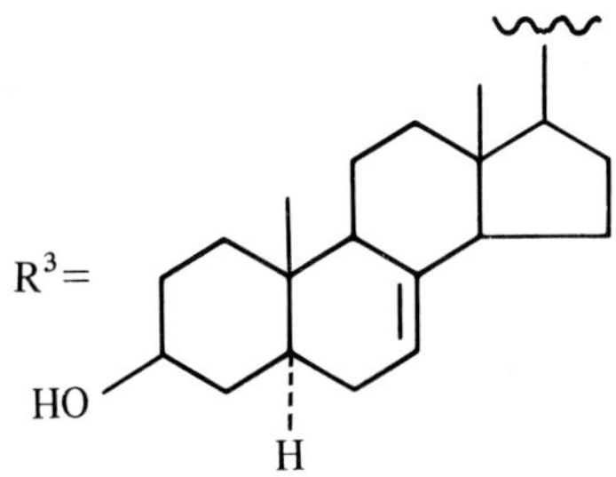

metabolized to desmosterol (4) and cholesterol (5) to approximately the same extent.

2.3 Stigmasterol

Administration of the inhibitor, 20,25-diazacholesterol (18), together with [^{3}H]stigmasterol (6) in the diet of *Manduca sexta* caused marked inhibition of growth and development and accumulation not only of desmosterol (4), but also of a compound identified as 22-*trans*-cholesta-5,22,24-trien-3β-ol (7) in the insect sterols (Svoboda, J. *et al.*, 1969a; Hutchins, R. *et al.*, 1970). The latter compound also occurred to a much lesser extent in the sterols of insects fed [^{3}H]stigmasterol in the absence of inhibitor. That 22-*trans*-cholesta-5,22,24-trien-3β-ol (7) is a precursor of desmosterol (4) and cholesterol (5) was shown by feeding this sterol alone or in combination with the diazasterol to *M. sexta*. The sterols of the insects reared on the 22-*trans*-cholesta-5,22,24-trien-3β-ol (7) alone contained 77% cholesterol (5) and 15% desmosterol (4), whereas in the case of insects reared on the inhibitor diet, cholesterol and desmosterol accounted for 3% and 9%, respectively, the remaining sterol being unchanged triene (7). This indicates inhibition of $\Delta^{22,24}$-sterol reductase activity by the diazasterol. In neither case was any 22-dehydrocholesterol (28) detected, indicating that during the conversion of 22-*trans*-cholesta-5,22,24-trien-3β-ol (7) into cholesterol, the Δ^{22} bond is reduced before the Δ^{24} bond. When *trans*-22-dehydrocholesterol (28) was used as the dietary sterol, there was little, if any, conversion of this sterol into cholesterol (5), suggesting that the presence of a Δ^{24} bond is necessary for reduction of the Δ^{22} bond of 22-*trans*-cholesta-5,22,24-trien-3β-ol (7) by *Manduca* (Thompson, M. *et al.*, 1972). 22-*trans*-cholesta-5,22,24-trien-3β-ol (7) has also been isolated as an intermediate in stigmasterol (6) dealkylation in *Blattella germanica* (German cockroach), *Heliothis zea* (corn earworm), and *Tribolium confusum* (Svoboda, J. *et al.*, 1969a, 1972a).

By analogy with the established intermediacy of fucosterol (2) in the dealkylation of sitosterol (1), it would be expected that 22-*trans*-stigmasta-5,22,24(28)-trien-3β-ol (29) is a probable intermediate preceding 22-*trans*-cholesta-5,22,24-trien-3β-ol (7) during dealkylation of stigmasterol (6) to cholesterol (5; Thompson, M. *et al.*, 1972); this has not yet been established. However, the corresponding epoxide, 24,28-epoxy-stigmasta-5,22-dien-3β-ol (30), together with 22-*trans*-stigmasta-5,22,24-trien-3β-ol (7) have been isolated from *Spodoptera littoralis* reared on a semi-defined diet containing stigmasterol (6) as the sterol supplement, in the presence of the inhibitor, 20,25-diazacholesterol (18) (Wilson, T. and Rees, H., unpublished results). Thus, the pathway of stigmasterol (6) dealkylation would appear to be analogous to that of sitosterol dealkylation (Scheme 1).

Although many insect species can saturate the Δ^{22} bond of stigmasterol (6) during its transformation to cholesterol (5), the German cockroach, *Blattella germanica*, is unable to effect reduction of such a bond. This explains why *B. germanica* reared aseptically on a sterol-free diet supplemented with the Δ^{22} sterols, ergosterol (31) ergosta-7,22-dien-3β-ol

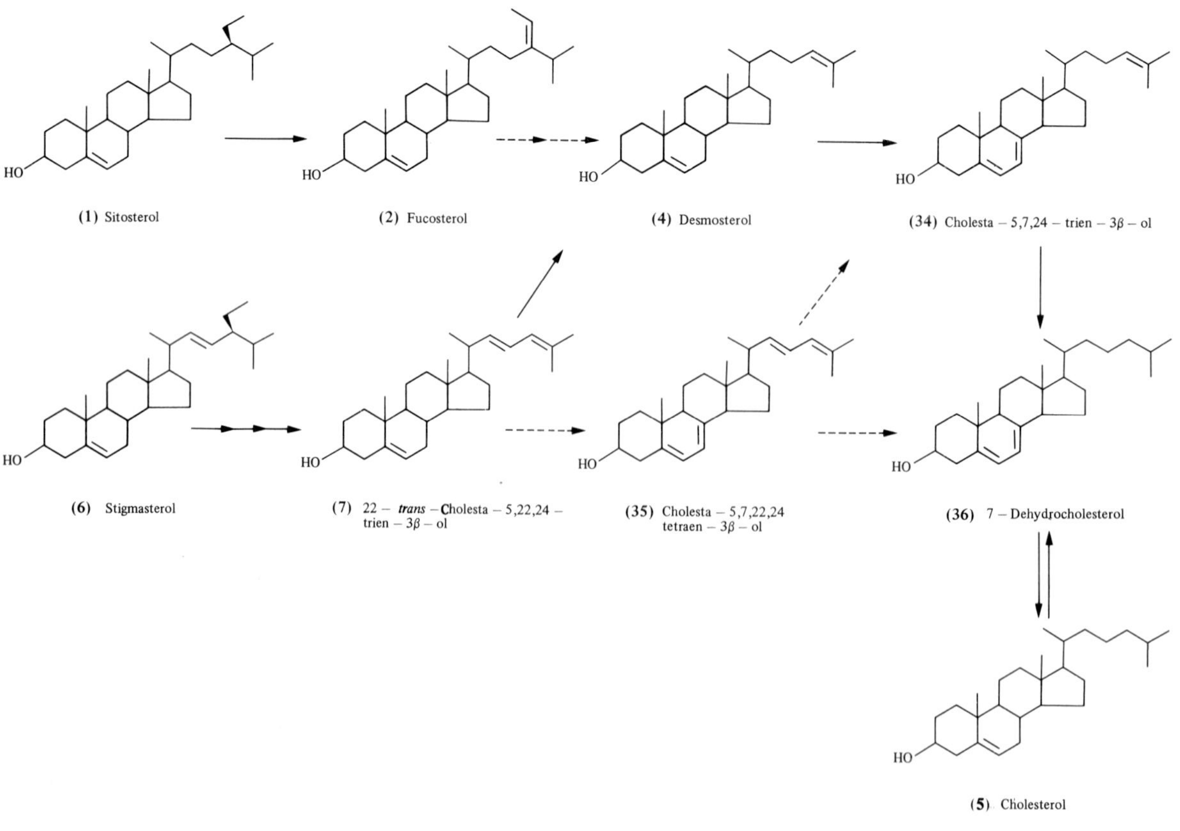

SCHEME 3. Conversion of sitosterol and stigmasterol into cholesterol in *Tribolium confusum* (Svoboda, J. *et al.*, 1972a).

(27), or stigmasterol (6) grew badly, whereas normal growth ensued when the sterol supplements contained no Δ^{22} bond (Ritter, F. and Wientjens, W. 1967). 22-Dehydrocholesterol (28) was a major dealkylation product of stigmasterol (6) and ergosterol (31), the latter transformation also indicating the ability of the cockroach to saturate the Δ^7 bond of a $\Delta^{5,7}$ diene. Ergosta-7,22-dien-3β-ol (27) was dealkylated to both 5α-cholesta-7,22-dien-3β-ol (32) and 22-dehydrocholesterol (28), indicating that rearrangement of a Δ^7 bond to a Δ^5 bond occurs.

2.4 Species variation

Although the pathways of dealkylation shown in Scheme 1 are applicable to many phytophagous insect species investigated, some species show interesting fundamental differences in the utilization and metabolism of phytosterols (Svoboda, J. *et al.*, 1978, 1982). For example, in the confused flour beetle, *Tribolium confusum*, in which more than 50% of the tissue sterols consists of 7-dehydrocholesterol (36) (Beck, S. and Kapadia, G., 1957), metabolism of sitosterol (1) and stigmasterol (6) follow the routes shown in Scheme 3. It is apparent that this species can introduce the Δ^7 bond into several sterol substrates. Despite this, there are also a number of similarities between these pathways and those demonstrated in other species. The reason for such high proportions of 7-dehydrocholesterol (36) in this species is unclear. As we shall consider in a later section (4.2), small amounts of 7-dehydrocholesterol (36) have been detected in numerous insect species. It is always possible that metabolic pathways such as those shown in Scheme 3 could also operate to a limited extent in such phytophagous species (Svoboda, J. *et al.*, 1972a).

The Mexican bean beetle, *Epilachna varivestis*, shows another interesting variation of phytosterol metabolism, where the formation of saturated sterols is emphasized. Comparison of the sterols in the insects with those in the dietary soybean leaves showed the presence of little cholesterol in both cases, with appreciable lathosterol (5α-cholest-7-en-3β-ol; 33) in the insects but not in the foodplant (Svoboda, J. *et al.*, 1975b). Large amounts of saturated C_{27}, C_{28} and C_{29} sterols (cholestanol (37), campestanol (38) and stigmastanol (39)) were present in the insects, whereas the soybean leaves contained only very small quantities of stanols. Studies with [^{3}H]-labelled Δ^5-phytosterols showed that they are first reduced to stanols, which are then dealkylated yielding cholestanol, into which a Δ^7 bond is introduced to form lathosterol (33). The exact significance of this emphasis on metabolism to saturated sterols in this species is unclear.

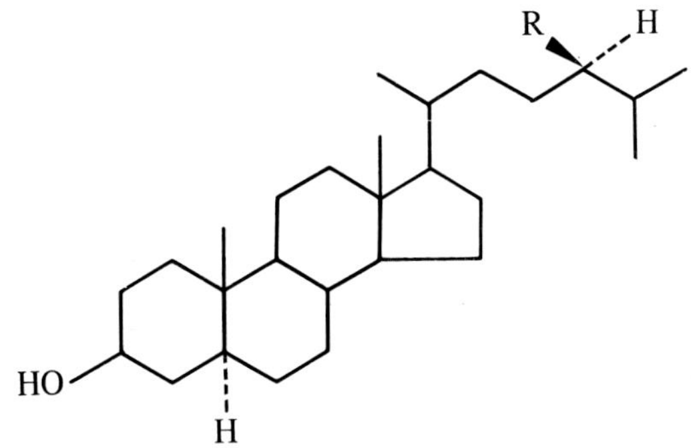

(37) R=H ; 5a – Cholestanol

(38) R=CH_3 ; 5a – Campestanol

(39) R=C_2H_5; 5a – Stigmastanol

Of course, another variation in sterol metabolism in phytophagous insects occurs in the species alluded to earlier (p. 251), which cannot dealkylate phytosterols. In at least one of these species, the milkweed bug, *Oncopeltus fasciatus*, this metabolic feature is reflected in the fact that the major ecdysteroid occurring in the eggs is the C_{28} compound, makisterone A (40) (Kaplanis, J. *et al.*, 1975; Hoffmann, J. and Lagueux, M., vol. 1).

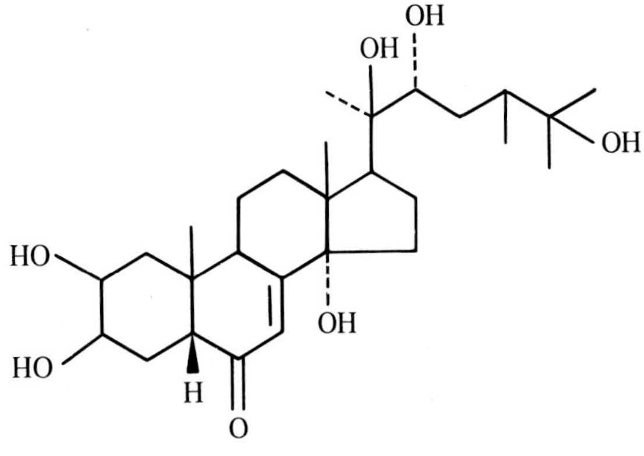

(40) Makisterone A

It is thus clear that there is considerable diversification in the ways insects satisfy their mandatory sterol requirement.

3 SITES OF ECDYSONE SYNTHESIS

The conversion of [^{3}H]cholesterol into ecdysone was first demonstrated by Karlson, P. and Hoffmeister, H. (1963) *in vivo* in *Calliphora erythrocephala*. Subsequently, the transformation of radioactive cholesterol into ecdysteroids has been substantiated (e.g. Galbraith, M. *et al.*, 1970; Willig, A. *et al.*, 1971). However, as alluded to in the Introduction to this chapter, the percentage incorporations are generally low, primarily because ecdysteroids as a rule occur only in small amounts, coupled with the occurrence of a large cholesterol pool in insects.

3.1 Prothoracic glands

According to the classical scheme of hormonal control of insect post-embryonic development, the prothoracic glands (or the ring glands in *Cyclorrhapha*) are the source of ecdysteroids in larval and pupal stages (see chapters 1–3, this volume). Although some moulting hormone activity as detected in bioassays has been reported in larval prothoracic glands (Ellis, P. *et al.*, 1972; Romer, F. 1973; Ruhland, U. and Romer, F. 1977), these organs certainly do not store significant amounts of ecdysteroid (Agui, N. *et al.*, 1972; King, D. *et al.*, 1974; Chino, H. *et al.*, 1974; Bollenbacher, W. *et al.*, 1976; Hirn, M. *et al.*, 1979). The notion that the prothoracic glands were the source of moulting hormone rested for many years primarily on the results of classical extirpation and implantation experiments, without direct evidence that these organs actually produced ecdysteroid (see Gilbert, L. 1964). That prothoracic glands (or ring glands) do secrete ecdysone without any detectable 20-hydroxyecdysone has now been directly established by culture *in vitro* in appropriate media of such organs from larvae/prepupae of several insect species (*Manduca sexta*, King, D. *et al.*, 1974; *Bombyx mori*, Chino, H. *et al.*, 1974; *Leucophaea maderae*, Borst, D. and Engelmann, F. 1974; King, D. and Marks, E. 1975; *Sarcophaga bullata*, Bollenbacher, W. *et al.*, 1976; *Locusta migratoria*, Hirn, M. *et al.*, 1979). This technique allowed the accumulation of sufficient ecdysteroid product of the prothoracic glands for characterization, in many cases by physical techniques, thus obviating the difficulties caused by the lack of accumulation of the hormone in the glands. In many of these studies it was apparent that the ecdysone was released into the culture medium immediately after being synthesized, thus explaining the low endogenous ecdysteroid pool generally observed in prothoracic glands. Since ecdysone was not detectable in culture media in which other tissues (including oenocytes) of *Manduca* prepupae were cultured, the prothoracic glands are apparently the primary, if not the sole, source of ecdysteroid at that stage in this species (King, D. *et al.*, 1974). There is no direct evidence reported as yet that the ecdysteroid product of the prothoracic glands from the pupal stage of development is also ecdysone.

Despite the foregoing evidence for the production of mass of ecdysone by prothoracic glands cultured *in vitro*, several groups of workers have experienced considerable difficulty in demonstrating conversion of radioactive cholesterol into ecdysone by such cultured glands (Willig, A. *et al.*, 1971; King, D. 1972a; see Gilbert, L. and King, D. 1973, and Gilbert, L. *et al.*, 1977). No doubt this may be explained, at least partly, by the apparently large cholesterol pool in prothoracic glands (Sakurai, S. *et al.*, 1977; Hirn, M. *et al.*, 1979). When haemolymph, haemolymph lipoprotein fractions, or cholesterol emulsified in Tween-80 were added to synthetic culture media for *Bombyx* prothoracic glands, there was a marked enhancement (4–7-fold) in the amount of ecdysone produced (Chino, H. *et al.*, 1974; Sakurai, S. *et al.*, 1977). The fact that a similar enhancement was also observed when Tween-80 alone was the supplement, led to the suggestion that the major precursors for ecdysone synthesis are contained within the glands. This was corroborated by the fact that only 0.4% of the total ecdysone produced by such glands was derived from [4^{14}-C]cholesterol added to a synthetic medium as a Tween-80 emulsion (Sakurai, S. *et al.*, 1977). The production of comparatively large amounts of ecdysone by *Locusta migratoria* prothoracic glands cultured in a defined medium devoid of sterols, suggests that in this species also, the endogenous sterol can serve as a major source of precursor for ecdysone synthesis (Hirn, M. *et al.*, 1979).

Despite the difficulties considered in the preceding paragraphs, the incorporation, albeit low, of [4-^{14}C]cholesterol into ecdysone by prothoracic glands cultured in vitro has been reported for *Bombyx mori* (Sakurai, S. *et al.*, 1977) and *Tenebrio molitor* (Romer, F. *et al.*, 1974). However, in the latter case the labelled product was characterized solely on the basis of the TLC properties of the free and acetylated compound. Additionally, in the case of *Locusta migratoria*, efficient incorporation of label into a compound presumed to be ecdysone on the basis of co-chromatography on TLC, was observed when prothoracic glands were cultured in a medium containing haemolymph derived from locusts which had been injected previously with [^{3}H]cholesterol (Hoffmann, J. *et al.*, 1975). Since significant metabolism of the cholesterol would have occurred in the intact insect, the nature of the true labelled precursor utilized by the glands *in vitro* for hormone synthesis cannot be ascertained.

In conclusion, the foregoing considerations leave little doubt that larval prothoracic glands can effect ecdysone, but not 20-hydroxyecdysone, synthesis. It is now apparent that after being released into the haemolymph, the ecdysone may undergo C-20 hydroxylation to the generally more active hormone, 20-hydroxyecdysone, in several peripheral tissues (see Smith, S., this volume).

3.2 Ovaries

Since adult insects do not generally moult, and their prothoracic glands degenerate shortly after emergence, there was no reason to suspect that adults would synthesize ecdysteroids. As far back as 1956, Karlson, P. and Stamm-Menéndez, M. observed that female adults of *Bombyx mori* contained moulting hormone activity. However, it was nearly 20 years later before it became apparent that ecdysteroids were associated with ovaries, and that ecdysteroid synthesis occurred during ovarian development either in adults or pharate adults. A full discussion of ovarian ecdysteroids is outside the scope of this chapter and only salient features will be considered (for reviews see Hoffmann, J. *et al.*, 1980; Hoffmann, J. and Lagueux, M., vol. 1; Hagedorn, H., vol. 8). That the ovaries are a site of ecdysteroid synthesis has been demonstrated in several insect species on the basis of the abolishment of hormone production *in vivo* as a result of ovariectomy and of an increase in ecdysteroids (detected by radioimmunoassay) during incubation of ovaries *in vitro*. At least in *Locusta migratoria*, the follicle cells are the site of hormone synthesis within the ovary, as indicated by the secretion of ecdysteroid into the medium when such cells from maturing adult females were cultured *in vitro* (Goltzené, F. *et al.*, 1978). No evidence for significant ecdysteroid synthesis in any tissue besides ovaries has been obtained in adult females of this species. When various tissues, including oenocytes and prothoracic glands from reproductively competent female adults of *Locusta*, were incubated *in vitro* in the presence of radioactive cholesterol which had been bound to haemolymph lipoprotein *in vivo*, incorporation of label into ecdysteroids was only observed in ovaries (Hoffmann, J. *et al.*, 1980). Of course, the true nature of the labelled precursor incorporated into hormone cannot be ascertained in this case, since metabolism of the cholesterol would have occurred during the *in vivo* lipoprotein binding procedure. Ecdysteroid synthesis in ovaries of *Locusta migratoria* (Charlet, M., *et al.*, 1979) and the mosquito, *Aedes aegypti* (Hagedorn, H. *et al.*, 1979; Hagedorn, H., vol. 8), is stimulated by a neurohormone produced in the median neurosecretory cells of the pars intercerebralis and secreted into the haemolymph via the corpora cardiaca. However, it is too early to speculate whether these tropic hormones are identical to the prothoracicotropic hormones in these species (Bollenbacher, W. and Granger, N., this volume).

In numerous species the ovarian ecdysteroids are almost entirely passed into the oöplasm of the eggs, where they may function during embryogenesis (see Hoffmann, J. *et al.*, 1980; Hoffman, J. and Lagueux, M., vol. 1). Ovarian and egg ecdysteroids have been shown to occur in several species to a large extent as polar conjugates hydrolysable with a crude enzyme from *Helix pomatia* (e.g. Ohnishi, E. *et al.*, 1977; Hsiao, T. and Hsiao, C. 1979; Gande, A. *et al.*, 1979; Dinan, L. and Rees, H. 1981a). However, in many studies on ovaries and eggs involving radioimmunoassay of ecdysteroids, such conjugates would have gone largely undetected since several antisera raised against free ecdysteroids do not cross-react significantly with ecdysteroid conjugates before hydrolysis. The major conjugate constituents of eggs of the desert locust, *Schistocerca gregaria*, have been fully characterized by physical techniques as ecdysone-22-phosphate and 2-deoxyecdysone-22-phosphate (Isaac, R. *et al.*, 1982). These are the first ecdysteroid conjugates to be fully identified. During embryogenesis, enzymic hydrolysis of the conjugates may occur, releasing free, active ecdysteroid (Rees, H. *et al.*, 1981).

In mosquitoes the ovarian ecdysone appears to be largely released into the haemolymph and, after conversion into 20-hydroxyecdysone in the fat body, may initiate vitellogenin synthesis (Hanaoka, K. and Hagedorn, H. 1980).

Whereas 20-hydroxyecdysone is generally the predominant ecdysteroid during larval and early pupal stages, this compound is usually a minor constituent of ovaries and eggs, where ecdysone is often the major hormone.

In the author's laboratory, advantage has been taken of the comparatively large synthesis of ecdysteroids in ovaries of the desert locust, *Schistocerca gregaria*, for relatively efficient incorporation of radioactive cholesterol into the hormones. In this procedure, substrate is injected into maturing adult females before the period of ecdysteroid synthesis, and after extraction of the deposited eggs, the phosphate conjugates (which account for 95% of the ecdysteroids) are hydrolysed with a so-called "aryl sulphatase preparation" from *Helix pomatia* to release the ecdysteroids (primarily ecdysone and 2-deoxyecdysone with smaller amounts of others) (Dinan, L. and Rees, H. 1981a,b). The presence of acid phosphatase activity in this *Helix* enzyme preparation accounts for hydrolysis of the ecdysteroid phosphates. This labelling procedure generally gives 0.4–1% incorporation of radioactive cholesterol into ecdysteroids, which is far greater than the yield generally obtained in larval systems (generally in the region of 0.01–0.12%).

3.3 Other sites

Although, as discussed earlier, larval prothoracic glands from several insect species have been shown to secrete ecdysone, there are reports in the literature suggesting that abdominal tissue(s) may also be involved in moulting and/or ecdysteroid synthesis in some species. For example, when final-instar larvae of the lepidopteran, *Calpodes ethlius*, are ligated between the thorax and abdomen, the anterior portion does not pupate unless part of the abdomen is also present (Weir, S. 1970). Posterior halves are also apparently required for pupation in larvae of the flesh-fly, *Sarcophaga peregrina* (Ohtaki, T., 1972). Other studies have shown that metamorphosis continues in isolated larval *Tenebrio molitor* abdomens (Stellwaag-Kittler, F., 1954) and that growth, development, regeneration, and reproduction are unimpeded in several cockroach species deprived of their prothoracic glands (Nutting, W. 1955; Chadwick, L. 1955, 1956; Bullière, D. and Bullière, F. 1974; Gersch, M. 1977), suggesting that there are sources of ecdysteroid besides the prothoracic glands in these species. Similarly, extirpation of the prothoracic glands in *Galleria mellonella* does not prevent subsequent ecdyses up to the adult stages (Piepho, H. 1948).

Reports of ecdysteroid synthesis from radioactive cholesterol in *Periplaneta americana* larvae devoid of prothoracic glands (Gersch, M. and Eibisch, J., 1977) and in isolated larval abdomens of *Mamestra brassicae* (Gersch, M. and Stürzebecher, J. 1971), *Bombyx mori* (Nakanishi, K. *et al.*, 1972) and *Musca domestica* (Studinger, G. and Willig, A. 1975) indicate that tissues besides the prothoracic glands are capable of moulting hormone formation. The incorporation was generally low, and in some cases the ecdysteroids were not extensively purified. In *Bombyx*, incorporation of cholesterol into ecdysteroid could be demonstrated in isolated abdomens from late, but not from early, fifth-instar larvae. The appearance of small but significant amounts of moulting hormone activity in isolated larval abdomens of the Colorado beetle, *Leptinotarsa decemlineata*, has also been demonstrated (Hsiao, T. *et al.*, 1976). Since ovarian synthesis of ecdysteroids generally does not occur until ovarian maturation in adult/pharate adult stages, a contribution from this source in these studies can be discounted.

That the oenocytes, which are abdominal cells possessing an ultrastructure characteristic of steroid-producing cells, are a possible site of ecdysteroid synthesis was suggested by Locke, M. (1969). The occurrence of moulting hormone activity in oenocytes from *Tenebrio molitor* and *Bombyx mori* larvae (Romer, F., 1971, 1973; Ruhland, U. and Romer, F., 1977) does not of itself indicate synthesis *in situ* of ecdysteroid in that tissue, especially as the prothoracic glands would be functional in such larvae. However, the apparent incorporation of [4-^{14}C]cholesterol primarily into a compound co-chromatographing on TLC with 20-hydroxyecdysone by isolated abdominal oenocytes of *T. molitor* larvae suggests that such cells can effect ecdysteroid formation *de novo* (Romer, F. *et al.*, 1974). This is unlikely to be a general phenomenon, since incorporation of [^{3}H]cholesterol into ecdysteroid could not be demonstrated in oenocytes from *Locusta migratoria in vitro* under conditions where labelling of hormone in isolated prothoracic glands could be achieved (Hoffmann, J. *et al.*, 1975).

It appears that abdominal synthesis of moulting hormone has been most extensively investigated in the yellow mealworm, *Tenebrio molitor*. In this species a particularly large peak of ecdysteroid occurs just before the middle of the pupal instar (Delbecque, J.-P. *et al.*, 1978b). However, a ligation between thorax and abdomen does not prevent the pupal–adult moult when applied long before the ecdysteroid peak, either to newly-moulted *Tenebrio* pupae (Janda, V. 1933) or to late last-instar larvae (i.e. after the last larval critical period) (Stellwaag-Kittler, F., 1954). Thus, these observations, demonstrating the occurrence of the adult moult without prothoracic glands raise the question as to the source of ecdysteroid for this moult. In fact, isolated abdomens of *Tenebrio* of both sexes ligated at pupal ecdysis show an impressive peak of moulting hormone (primarily 20-hydroxyecdysone with less ecdysone) 6–10 days after ligation (Delbecque, J.-P. *et al.*, 1978a). This hormone is released into the haemolymph, the concentrations at peak titre being similar to those in intact pupae and far higher than those detected in isolated abdomens of other species. Furthermore, the abdominal hormone titre peak is correlated with initiation of adult development (i.e. apolysis and deposition of adult cuticle). These observations, taken in conjunction with the fact that involution of the prothoracic glands in *Tenebrio* is completed before the pupal moulting hormone peak (Glitho, I. *et al.*, 1979), suggest that abdominal ecdysteroid production could be responsible for triggering pupal–adult development of normal animals. Although the exact site of synthesis of this hormone in isolated pupal abdomens is not established, it could be the oenocytes which can, at least in larvae, effect hormone synthesis. It remains to be demonstrated that oenocytes from pupae can account for this ecdysteroid synthesis responsible for pupal–adult development (Delbecque, J.-P. *et al.*, 1978a). An ovarian origin for this moulting hormone can be eliminated, since isolated abdomens of both males and females show similar hormone production (Delbecque, J.-P. *et al.*, 1978a) and since ovarian ecdysteroid production does not occur until much later in the adults, where the hormone is largely passed into the eggs rather than being released into the haemolymph (Laverdure, A.-M. *et al.*, 1977).

That the prothoracic glands are necessary for induction of every larval moulting cycle and for the induction of larval–pupal development in *Tenebrio* is supported by the results of ligation experiments (Stellwaag-Kittler, F., 1954; Delbecque, J.-P. *et al.*, 1978a). After the pupal moult the prothoracic glands are replaced by an abdominal organ, possibly the oenocytes, for ecdysteroid production (Delbecque, J.-P. *et al.*, 1978a). This explains why abdomens of late last-instar larvae ligated after the last critical period of prothoracic gland action, can proceed not only through the pupal moult but also the adult moult (Stellwaag-Kittler, F. 1954). Although this scheme suggested for *Tenebrio* might be applicable to other species (e.g. the Colorado beetle, *Leptinotarsa decelineata* (Hsiao, T. *et al.*, 1975) and the hide beetle, *Dermestes vulpinus* (Slama, K. 1975)) which undergo two successive moults after thoracic–abdominal ligation, it is unlikely to be widely occurring (see Delbecque, J.-P. *et al.*, 1978a).

In conclusion, there is little doubt that the prothoracic glands are the prime, if not the only, physiologically important source of moulting hormone during larval–larval and larval–pupal transformation. In those species where abdominal larval moulting hormone production has been demonstrated, the levels are generally very low. This must be the case, otherwise moulting hormone bioassays in isolated abdomens would be impossible. It had been suggested that the prothoracic glands may be the normal site of larval ecdysone synthesis, but that other tissues (e.g. oenocytes) may have the genetic ability to effect ecdysone synthesis and may only do so under experimental stress (Gilbert, L. 1974). However, there is now a compelling case for physiological ecdysteroid production in abdomens of *Tenebrio molitor* pupae, but it remains to be seen whether such a system operates in other species. Clearly, numerous facets of ecdysteroid synthesis outside the prothoracic glands and ovaries remain to be clarified.

4 PATHWAYS OF ECDYSONE BIOSYNTHESIS

The pathways of ecdysone biosynthesis have been considered in several reviews (Rees, H. 1971; Thompson, M. *et al.*, 1972, 1973; Gilbert, L. and King, D. 1973; Gilbert, L. *et al.*, 1977; Svoboda, J. *et al.*, 1978; Rees, H. *et al.*, 1980, 1981; Rees, H., 1983). Since the percentage incorporations *in vivo* of radioactive cholesterol into ecdysteroids in larvae are generally very low (often in the range 0.01–0.12%), particular care is required in the purification of the hormones, especially as some potential labelled metabolites (e.g. steryl glycosides) have similar chromatographic properties to ecdysteroids in some systems. The role of cholesterol as a precursor of ecdysteroids in insects is consistent with the observation that only few investigated species cannot be maintained on a diet containing cholesterol as the sole sterol (see section 2). Two species, *Drosophila pachea* and *Xyleborus ferrugineus*, apparently cannot introduce a Δ^7 bond into the sterol nucleus and require either a Δ^7 or $\Delta^{5,7}$ dietary sterol (Heed, W. and Kircher, H. 1965; Chu, H.-M. *et al.*, 1970). Presumably these species utilize such sterols as distal precursors of their ecdysteroids. Of course, in the case of the C_{28} ecdysteroid, makisterone A (40), isolated from eggs of *Oncopeltus fasciatus* (Kaplanis, J. *et al.*, 1975), the precursor would be expected to be a C_{28} sterol, such as campesterol (8).

4.1 Nuclear double-bond transformations of sterols

It is now clear that during the conversion of cholesterol into ecdysone, modification of the nucleus occurs before side-chain hydroxylation (Galbraith, M. *et al.*, 1972; Thompson, M. *et al.*, 1973; Gilbert, L. *et al.*, 1977). Therefore, it is germane to consider the sterol nuclear double-bond transformations which have been demonstrated in insects (Table 4). Of course, it is quite possible in many of these studies where aseptic organisms were not used that associated microorganisms might make a significant contribution to this metabolism (see Clayton, R. 1964; Ritter, F. and Wientjens, W., 1967). These combined data, which have been obtained mainly utilizing sterols of the cholestane series in various insect species (but not in a single species) may be summarized as follows:

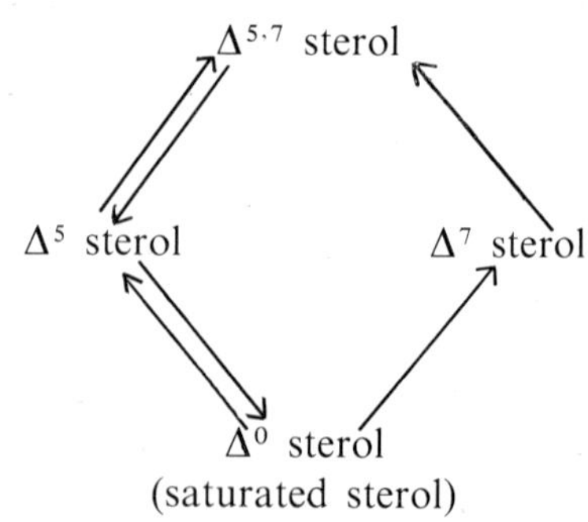

However, it is impossible on the basis of the data available to established whether some of the transformations included in Table 4 occur via a direct or indirect route. It is not clear whether the transformations, $\Delta^{5,7} \rightarrow \Delta^7$ and $\Delta^7 \rightarrow \Delta^0$ occur. Consequently, the diagram does not cover all possible reactions, although it can explain all possible transformations as it stands.

On the basis of sterol analysis of the cricket, *Acheta domesticus*, it was suggested, without direct evidence, that the formation of $\Delta^{5,7}$ sterols from

Table 4: Sterol nuclear double-bond transformations

Substrate	Product(s)	Trans-formation	Species	Comments	Reference
$[^{14}\text{C}]$Cholesterol (5)	7-dehydrocholesterol (36)	$\Delta^5 \rightarrow \Delta^{5,7}$	*Periplaneta americana*		Ishii, S. *et al.* (1963)
			Blattella germanica	aseptic	Robbins, W. *et al.* (1964)
			Musca domestica		Kaplanis, J. *et al.* (1960)
			Musca domestica	aseptic	Monroe, R. *et al.* (1967)
			Calliphora erythrocephala	aseptic	Johnson, P. *et al.* (1975)
Cholesterol (5)	7-dehydrocholesterol (36)	$\Delta^5 \rightarrow \Delta^{5,7}$	*Tribolium confusum*		Svoboda, J. *et al.* (1972a)
Sitosterol (1)	7-dehydrositosterol	$\Delta^5 \rightarrow \Delta^{5,7}$	*Musca domestica*		Kaplanis, J. *et al.* (1963)
$[^{14}\text{C}]5\alpha$-cholestan-3β-ol (37)	5α-cholest-7-en-3β-ol (33)	$\Delta^0 \rightarrow \Delta^7$	*Blattella germanica*		Louloudes, S. *et al.* (1962)
			Calliphora erythrocephala	aseptic	Johnson, P. *et al.* (1975)
$[^{3}\text{H}]5\alpha$-Cholestan-3β-ol (37)	5α-cholest-7-en-3β-ol (33)	$\Delta^0 \rightarrow \Delta^7$	*Eurycotis floridana*	aseptic	Clayton, R. and Edwards, A. (1963)
			Blattella germanica	aseptic	
5α-Cholest-7-en-3β-ol (33)	7-dehydrocholesterol (36)	$\Delta^7 \rightarrow \Delta^{5,7}$	*Drosophila pachea*	aseptic	Goodnight, K. and Kircher, H. (1971)
$[^{14}\text{C}]5\alpha$-Cholestan-3β-ol (37)	cholesterol (5)	$\Delta^0 \rightarrow \Delta^5$	*Thermobia domestica*		Thompson, M. *et al.* (1973)
$[^{14}\text{C}]$Cholesterol (5)	5α-cholestan-3β-ol (37)	$\Delta^5 \rightarrow \Delta^0$	*Epilachna varivestis*		Svoboda, J. *et al.* (1975b)
$[^{3}\text{H}]$Sitosterol (1)	stigmastanol (39)				
$[^{14}\text{C}]$Ergosterol (31)	22-dehydrocholesterol (28)	$\Delta^{5,7} \rightarrow \Delta^5$	*Blattella germanica*		Clarke, G. and Bloch, K. (1959)
$[^{3}\text{H}]$7-Dehydrocholesterol (36)	cholesterol (5)	$\Delta^{5,7} \rightarrow \Delta^5$	*Tribolium confusum*		Svoboda, J. *et al.* (1972a)
Ergosta-7,22-dien-3β-ol (27)	22-dehydrocholesterol (28)	$\Delta^7 \rightarrow \Delta^5$?	*Blattella germanica*	aseptic	Ritter, F. and Wientjens, W. (1967)
$[^{14}\text{C}]$Cholesterol (5) or $[^{3}\text{H}]$7-Dehydrocholesterol (36) or $[^{3}\text{H}]5\alpha$-Cholest-7-en-3β-ol (33)	cholesterol (5) 7-dehydrocholesterol (36) 5α-cholest-7-en-3β-ol (33) 5α-cholestan-3β-ol (37)	Δ^0, Δ^5 Δ^7 & $\Delta^{5,7}$ completely interconvertible	*Schistocerca gregaria*		Johnson, P. and Rees, H. (1977)

Δ^5 compounds might occur via saturated and Δ^7 sterols, i.e. $\Delta^5 \rightarrow \Delta^0 \rightarrow \Delta^7 \rightarrow \Delta^{5,7}$ (Martin, M. and Carls, G. 1968). However, in aseptic *Calliphora erythrocephala*, significant 5α-cholestan-3β-ol (Δ^0; 37) and 5α-cholest-7-en-3β-ol (Δ^7; 33) were not detected during the transformation of $[^{14}\text{C}]$cholesterol to 7-dehydrocholesterol (36), and $[^{14}\text{C}]5\alpha$-cholestan-3β-ol was only metabolized to 5α-cholest-7-en-3β-ol (33) but not to 7-dehydrocholesterol, suggesting that the transformation of cholesterol (Δ^5) to 7-dehydrocholesterol ($\Delta^{5,7}$) is probably direct (Johnson, P. *et al.*, 1975). In *Schistocerca gregaria* larvae, cholesterol, 7-dehydrocholesterol, 5α-cholestan-3β-ol and 5α-cholest-7-en-3β-ol are apparently completely interconvertible, but in this case the contribution of associated micro-organisms cannot be ascertained (Johnson, P. and Rees, H. 1977).

When $[^{3}\text{H}]5\alpha$-cholest-7-en-3β-ol (33) was examined as a possible precursor of 20-hydroxyecdysone in fifth-instar *Schistocerca gregaria*, the incorporation was low (0.01%) compared with that of cholesterol (0.12%), suggesting that this Δ^7 sterol is not a normal intermediate in ecdysteroid biosynthesis (Johnson, P. and Rees, H. 1977). This low incorporation of 5α-cholest-7-en-3β-ol (33) into ecdysteroid probably occurs via its initial transformation into cholesterol, a conversion already demonstrated in this species. The observation that non-radioactive 5α-cholest-7-en-3β-ol (33) injected together with [1α-^{3}H]cholesterol into *Calliphora stygia* did not significantly reduce the incorporation of [1α-^{3}H]cholesterol into 20-hydroxyecdysone might also suggest that this Δ^7 sterol is not an intermediate in the transformation of cholesterol to the hormone. However, it is always possible that the non-radioactive sterol does not penetrate to the site of ecdysteroid synthesis in sufficient quantity to produce a significant effect (Horn, D. *et al.*, 1974).

4.2 7-Dehydrocholesterol

A possible early intermediate in the transformation of cholesterol into ecdysone is 7-dehydrocholesterol (Thompson, M. *et al.*, 1973). This sterol, which can be detected and be formed from cholesterol in numerous species (Table 4), also satisfies the essential sterol requirements of many, but apparently not all, insects investigated (Clayton, R. 1964). It may be significant that of a number of individual tissues from *Manduca sexta* and *Periplaneta americana*, only the prothoracic glands possess a comparatively large percentage of 7-dehydrocholesterol (60% and 25% of the total sterols, respectively). This implicates the prothoracic glands as a site of synthesis or storage of 7-dehydrocholesterol (Thompson, M. *et al.*, 1973). However, incubation of prothoracic glands with 7-dehydrocholesterol *in vitro* does not increase the rate of ecdysone synthesis (Gilbert, L. *et al.*, 1977).

Horn, D. and co-workers (1974) demonstrated that the incorporation of [^{3}H]7-dehydrocholesterol into 20-hydroxyecdysone in *Calliphora stygia* was slightly greater than that of [^{3}H]- or [^{14}C]cholesterol. Although this might be interpreted as suggesting that 7-dehydrocholesterol is an intermediate in ecdysone formation from cholesterol, the authors indicate that a much higher incorporation of the $\Delta^{5,7}$ sterol might have been expected since the endogenous pool of cholesterol is much higher than that of 7-dehydrocholesterol in *C. stygia*. Since the incorporation of [^{3}H]cholesterol was only slightly decreased upon simultaneous administration of non-radioactive 7-dehydrocholesterol, it was suggested that the biosynthesis of 20-hydroxyecdysone may not proceed exclusively via 7-dehydrocholesterol, without excluding the possibility that the latter sterol does not penetrate to the site of synthesis in sufficient quantity to produce a significant effect.

To obtain accurate comparisons of the incorporation of different sterols into ecdysteroids, a double-label technique has been utilized in studies on the desert locust, *Schistocerca gregaria*. This approach allows comparison of two sterols as precursors in a single experiment, and thus obviates any variation owing to differences in the exact developmental stages of different batches of insects. When [^{3}H]7-dehydrocholesterol and [^{14}C]cholesterol were administered simultaneously to fifth-instar *S. gregaria*, the incorporation of both sterols into 20-hydroxyecdysone was similar (Johnson, P. and Rees, H., 1977). Whereas simultaneous administration of non-radioactive 7-dehydrocholesterol reduced the incorporation of both labelled substrates into 20-hydroxyecdysone, the conversion of [^{14}C]cholesterol was reduced more than that of [^{3}H]7-dehydrocholesterol. This result was interpreted as being consistent with the intermediacy of 7-dehydrocholesterol in ecdysone biosynthesis from cholesterol, but was not by any means conclusive.

Owing to the difficulty of interpretation of the foregoing data on the status of 7-dehydrocholesterol in ecdysteroid formation, the relative incorporations of [^{3}H]7-dehydrocholesterol and [^{14}C]cholesterol into ecdysteroid were then compared in an entirely different system, viz. in maturing ovaries of adult female *S. gregaria*. In this case the incorporation of cholesterol into the egg ecdysone conjugates approached twice that of 7-dehydrocholesterol (Gibson, J. and Rees, H., unpublished results). The latter compound has been identified as a constituent sterol in ovaries of *S. gregaria* (Dinan, L., 1980; Dinan, L. and Rees, H., 1980b). Interpretation of these experiments in both larval and adult *S. gregaria* is complicated by the very extensive conversion of 7-dehydrocholesterol into cholesterol. In an attempt to suppress as far as possible this transformation of 7-dehydrocholesterol to cholesterol, AY 9944 (*trans*-1,4-bis(2-chlorobenzylaminomethyl) cyclohexane dihydrochloride), which is known to inhibit this reduction of the

Δ^7 bond in vertebrates (Humber, L. *et al.*, 1966), was administered simultaneously with the substrates to another group of maturing adult female *S. gregaria*. Compared to the foregoing experiment, the inhibitor caused a similar (approx. 38%) decrease in the amount of [^{3}H]cholesterol (i.e. derived from [^{3}H]7-dehydrocholesterol) isolated from the eggs as in the incorporation of [^{3}H]7-dehydrocholesterol into ecdysone (Gibson, J. and Rees, H., unpublished results). Although these data might be interpreted as suggesting that the incorporation of [^{3}H]7-dehydrocholesterol into ecdysone occurs via the intermediacy of cholesterol (7-dehydrocholesterol ⇄ cholesterol ---→ ecdysone) rather than directly (cholesterol ⇄ 7-dehydrocholesterol ---→ ecdysone), it is by no means conclusive.

Interpretation of all the foregoing relative incorporation data on 7-dehydrocholesterol and cholesterol is complicated by factors such as the possible differing membrane permeabilities of these sterols, as well as the possible existence of several pools and quite different endogenous concentrations of both compounds. In conclusion, the intermediacy of 7-dehydrocholesterol in ecdysone biosynthesis from cholesterol must remain as an open question in view of conflicting data which are difficult to interpret.

4.3 Stereochemistry of Δ^7 bond formation

The stereochemistry of Δ^7 bond formation during ecdysteroid biosynthesis has been investigated by incorporation of [7α-^{3}H]- and [7β-^{3}H]cholesterol in *Calliphora erythrocephala*. The results showed that the 7β-hydrogen is stereospecifically eliminated, whereas the 7α-hydrogen is retained (Cook, I. *et al.*, 1973). This involves the same stereochemistry of hydrogen elimination as the formation of 7-dehydrocholesterol in this species (Johnson, P. *et al.*, 1975) and as the conversion of 5α-cholestan-3β-ol (38) into 5α-cholest-7-en-3β-ol (33) in a cockroach (Clayton, R. and Edwards, A., 1963). The same stereochemistry was also observed during ecdysteroid formation in the plants, *Taxus baccata* and *Polypodium vulgare* (Cook, I. *et al.*, 1973). However, in the latter cases, a significant amount (40–60%) of the 7β-tritium of cholesterol was reincorporated, primarily at C-3 in the ecdysteroid. This suggests that in these plants a substantial amount of the tritium from C-7 of cholesterol presumably enters a small compartmentalized pool of NADPH and is reincorporated at C-3 during a subsequent reductive step (see Davies, T. *et al.*, 1980). However, the results demonstrate clearly that introduction of the Δ^7 bond during ecdysteroid formation in the insect and plants occur by removal of the 7β- and, of necessity, the 8β-hydrogens i.e. *cis* hydrogens.

Although formation of the Δ^7 bond of ecdysteroids involves removal of the *cis*-orientated 7β,8β-hydrogens, the process does not necessarily have to be a *cis* dehydrogenation. Plausible mechanisms involving two successive *trans* processes or molecular oxygen have been suggested (Cook, I. *et al.*, 1973).

It is apparent that double-bond formation is influenced by electron-withdrawing groups at positions α to the potential bond; in the absence of such groups removal of *cis* hydrogens is observed, whereas in their presence removal of *trans* hydrogens occurs (Wilton, D. and Akhtar, M., 1970). If this generalization holds, the removal of the 7β,8β-hydrogens observed during ecdysteroid biosynthesis would be consistent with the introduction of the Δ^7 bond before formation of the 6-oxo group.

4.4 Formation of the A/B *cis* ring junction

It is apparent that formation of the A/B *cis* ring junction occurs fairly early in the ecdysone biosynthetic pathway (Thompson, M. *et al.*, 1973; Galbraith, M. *et al.*, 1975). Although several plausible mechanisms could account for this process (Rees, H., 1971), it has been frequently assumed that a 5β-ecdysteroid precursor is derived from the corresponding 5α-compound by enolization of a 6-oxo group. However, this does not explain adequately why numerous 6-oxo steroids with a 5α-hydrogen are hormonally inactive, whereas the corresponding 5β-compounds are active (Galbraith, M. *et al.*, 1973; Bergamasco, B. and Horn, D., 1980; Horn, D. and Bergamasco, R., this volume). In the presence of a hydroxyl group at C-2, an A/B *cis* ring junction (i.e. 5β-H) is favoured, owing to the proximity of the C-2 hydroxl group to the C-19 methyl in the A/B *trans* ring configuration, whereas in the absence of such a hydroxyl, an A/B *trans* ring junction (i.e. 5α-H) is favoured.

During the formation of certain steroid hormones and bile acids in animals, the analogous transformations of a Δ^5 bond to give a 5β-H occurs via a 3-oxo-Δ^4-steroid intermediate (Samuels, L. and Eik-Nes, K., 1968). The possibility that a 3-oxo-Δ^4-steroid could be an intermediate in ecdysone biosynthesis has been investigated by examining the fates of the C-3 and C-4 hydrogens of cholesterol during its incorporation into the hormone. For this, [4α-$^3H_1$]-, [4β-$^3H_1$]- and [3α-$^3H_1$] cholesterol species were prepared (Lockley, W. *et al.*, 1978) and incorporated in turn in admixture with [4-^{14}C]cholesterol into ecdysteroids in larvae of both *Calliphora erythrocephala* and *Schistocerca gregaria* (Lockley, W. *et al.*, 1975; Davies, T. *et al.*, 1981). There was essentially complete retention of the [4α-^{3}H] and essentially complete removal of the [4β-^{3}H] of cholesterol during incorporation into 20-hydroxyecdysone in both species. However, during transformation of the [3α-^{3}H] cholesterol into the ecdysteroid in both species, there was extensive elimination of the tritium with consistent retention of a variable amount up to approximately 20–30% in the hormone. Owing to this complication, the same three stereospecifically labelled [^{3}H]cholesterol substrates in admixture with [4-^{14}C]cholesterol were incorporated into the conjugates of 2-deoxyecdysone (52) and ecdysone (47) during oögenesis in adult female *S. gregaria* and the ecdysteroids isolated from the eggs. In this case, there was essentially complete removal of the tritium from [3α-^{3}H]- and [4β-^{3}H]cholesterol during incorporation into the ecdysteroids (Davies, T. *et al.*, 1981). This elimination of the 3α- and 4β-hydrogens of cholesterol has been explained by postulating the involvement of a 3-oxo-Δ^4 steroid intermediate (41; scheme 4) in formation of the A/B *cis* ring junction of ecdysteroids in insects, without excluding other interpretations. The stereochemistry of hydrogen elimination from C-4 of cholesterol during formation of steroid hormones and bile acids in vertebrates is the same as that observed during ecdysone synthesis, *viz.* 4β (Murota, S. *et al.*, 1971; Akhtar, M. *et al.*, 1980). It is interesting that the mechanism of formation of the A/B *cis* ring junction during ecdysteroid formation in plants is apparently different from that in insects (Davies, T. *et al.*, 1980).

The retention of up to 30% of the tritium observed during synthesis of ecdysteroids from [3α-^{3}H]cholesterol in larvae of *Schistocerca* and *Calliphora* has been explained by invoking transfer of the 3α-tritium to a small compartmentalized pool of NAD(P)$^+$ by 3β-hydroxysteroid dehydrogenase action during formation of the 3-oxo group, with reintroduction of some of this tritium (located at a position other than C-3) during a subsequent step requiring NAD(P)H (Davies, T. *et al.*, 1981; also see Rees, H. *et al.*, 1980, 1981, for a more comprehensive coverage). Several precedents exist in the case of steroid metabolism in rats and microbes for analogous efficient incorporation of ^{3}H (or ^{2}H) at certain positions (especially C-3) in steroids from labelled, compartmentalized NAD(P)$^+$ pools, which of necessity must be small (Cronholm, T., 1972; Björkhem, J. and Gustafsson, J. -Å., 1971). Thus, the dangers of using label at the 3α-position in putative ecdysteroid precursors are clear. Furthermore, as alluded to in section 4.3, during 20-hydroxyecdysone formation in the fern, *Polypodium vulgare*, there was significant transfer of label, presumably via NAD(P)$^+$, from the 7β-position of cholesterol to the 3α-position of the ecdysteroid (Cook, I. *et al.*, 1973).

4.5 Hydroxylation sequences

4.5.1 Putative intermediates

Except in the case of certain ovarian systems (see section 4.5.2), as alluded to in the Introduction, it has proved generally difficult to detect intermediates in the conversion of cholesterol into ecdysone in larval or prepupal systems, where such compounds do not accumulate to a significant extent. Even during the incorporation of radioactive cholesterol into ecdysteroids in the latter systems, there is generally no indication of label in intermediates. Consequently, studies on ecdysteroid biosynthesis have been approached largely by investigating the incorporation of radioactively labelled putative intermediates either *in vivo* or *in vitro*. Although this approach has furnished much useful information, there is considerable scope for drawing unjustified or erroneous conclusions from the results. It is relevant at this juncture to remind ourselves of the criteria in the Introduction for establishment of the intermediacy of compounds in a biosynthetic pathway. In particular, we should

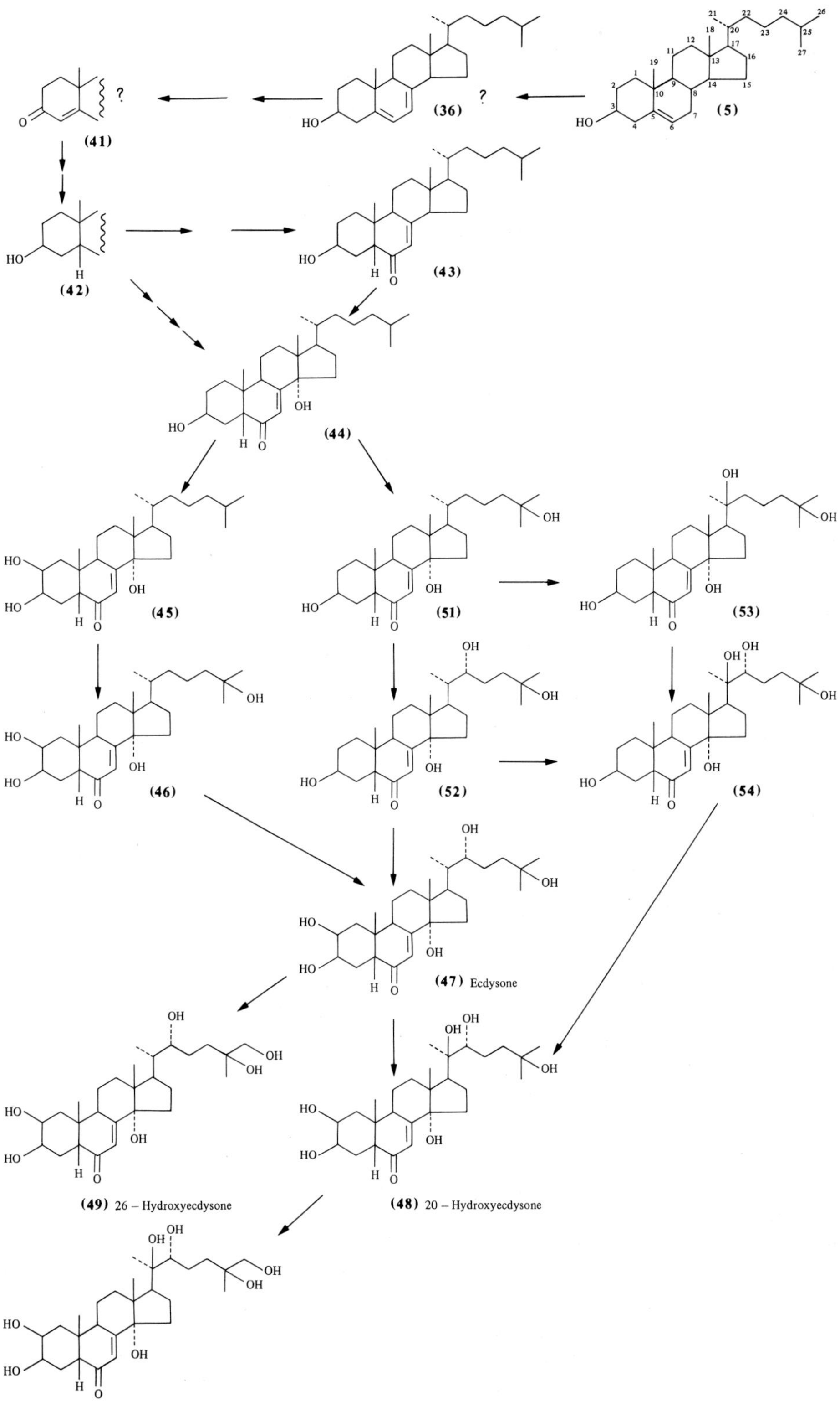

SCHEME 4. Possible composite scheme of diverging pathways of ecdysteroid biosynthesis.

bear in mind that the isolation of a compound from tissues or the metabolism of a radioactive putative precursor to ecdysteroids, do not, of themselves, establish the intermediacy of such compounds in the pathway. Whether an exogenous compound is metabolized depends upon whether or not it can reach the site of synthesis, and on the degree of substrate specificity of the relevant enzyme(s). It is conceivable that a true intermediate supplied exogenously might not be metabolized due to permeability barriers, and that unnatural compounds may well be accepted if the specificity of the enzymes is low.

As discussed earlier (section 3.1), prothoracic glands from several species of insects, when cultured *in vitro*, produce ecdysone. Such *in vitro* systems have been used successfully in the case of both *Bombyx mori* (Sakurai, S. *et al.*, 1977) and *Manduca sexta* (Gilbert, L. *et al.*, 1980) for investigation of the biosynthetic pathway. This approach obviates certain complications inherent in the *in vivo* studies, such as variability in catabolism/excretion, non-specific metabolism by peripheral tissues, and relative permeabilities to different tissues of putative labelled intermediates. In the case of *M. sexta*, as in the case of several species, it has not been possible to demonstrate the incorporation of [^{3}H]cholesterol into ecdysone in prothoracic glands cultured *in vitro*. Although the exact reason for this is somewhat of an enigma, it has been suggested that *in vivo* the cholesterol could be metabolized by other tissues (e.g. the fat body) to an intermediate appropriate for transformation to ecdysone by the glands (Gilbert, L. *et al.*, 1977).

The results of studies on the metabolism of putative ecdysteroid precursors other than 2-deoxyecdysteroids in insect systems are summarized in Table 5. Many of these findings are incorporated into the possible composite scheme of diverging pathways of ecdysteroid biosynthesis shown in Scheme 4.

Early precursors The negligible incorporation of [22-^{3}H]-(22*R*)-22-hydroxycholesterol and [26-^{14}C]-25-hydroxycholesterol into 20-hydroxyecdysone (48) in *Calliphora stygia* indicates that the structure of the sterol side-chain is also important for the enzymes catalyzing modification of the nucleus, and that at least some modification of the cholesterol nucleus occurs before side-chain hydroxylation (Galbraith, M. *et al.*, 1972).

The demonstration that the incorporation of cholesterol into 20-hydroxyecdysone in *Calliphora stygia* is at least ten times higher than that of 5β-cholest-7-en-3β-ol and 5β-cholest-8(14)-en-3β-ol suggests that the latter compounds are not inter-

Table 5: Metabolism of putative ecdysteroid precursors in insect systems

Substrate	Products	Species	Developmental stage	System	Reference
[1α-^{3}H]7-Dehydrocholesterol (36)	20-hydroxyecdysone (0.025%)	*Calliphora stygia*	puparium formation	*in vivo*	Horn, D. *et al.* (1974)
[1α-^{3}H]Cholesterol (5)	20-hydroxyecdysone (0.02%)				
[4-^{14}C]Cholesterol (5)	20-hydroxycedysone (0.12%)	*Schistocerca gregaria*	fifth instar	*in vivo*	Johnson, P. and Rees, H. (1977)
[1α, 2α(n)-^{3}H]7-Dehydrocholesterol (36)	20-hydroxycedysone (0.11%)				
[24-^{3}H]5α-Cholest-7-en-3β-ol (33)	20-hydroxycedysone (0.14%)				
[4-^{14}C]Cholesterol (5)	ecdysone (0.25%)	*Schistocerca gregaria*	maturing adult females	*in vivo*	Gibson, J. M. and Rees, H. H. (unpublished)
[1α, 2α(n)-^{3}H]7-Dehydrocholesterol (36)	ecdysone (0.13%)				
*[1α-^{3}H]Cholesterol (5)	20-hydroxyecdysone (0.02%)	*Calliphora stygia*	puparium formation	*in vivo*	Galbraith, M. *et al.* (1972)
[26-^{14}C]25-Hydroxycholesterol	20-hydroxyecdysone (0.0008%)				
[22-^{3}H]-(22*R*)-22-Hydroxycholesterol	20-hydroxyecdysone (< 0.0001%)				

Table 5: Continued

Substrate	Products	Species	Developmental stage	System	Reference
*[1α-^{3}H]Cholesterol (5) [11,12-^{3} H_2]5β-Cholest-7-en-3β-ol [11,12-$^3H_2$]5β-Cholest-8(14)-en-3β-ol 3β-Hydroxy-[11,12-$^3H_2$]5β-Cholest-7-en-6-one (43)	20-hydroxycedsone (0.02%) 20-hydroxycedsone (< 0.0023%) 20-hydroxycedsyone (< 0.0017%) 20-hydroxyecdysone (0.003%)	*Calliphora stygia*	puparium formation	*in vivo*	Faux, A. *et al.* (1979
[4-^{14}C]Cholesterol	3β-Hydroxy-5α-cholestan-6-one (55)	*Bombyx mori*	fifth instar	*in vivo*	Sakurai, S. *et al.* (1977)
[4-^{14}C]Cholesterol 3β-hydroxy-[4-^{14}C]-5α-cholestan-6-one (55)	ecdysone (0.0082%) ecdysone (0.014%)			prothoracic glands *in vitro*	
[1α, 2α-$^3H_2$]Cholesterol [1α, 2α-$^3H^2$]Cholest-4,7-dien-3-one	20-hydroxyecdysone (0.07%) 20-hydroxycedysone (0.004%)	*Calliphora stygia*	puparium formation	*in vivo*	Kinnear, J. *et al.* (1979)
[1α, 2α-$^3H_2$]Cholesterol [4-^{14}C]Cholest-4-en-3-one [4-^{14}C]Cholesta-4,6-dien-3-one	ecdysone (0.40%) ecdysone (< 0.003%) ecdysone (< 0.002%)	*Schistocerca gregaria*	adult females and fifth instar	*in vivo*	Nali, M., Gibson, J. M., Rees, H. H. and Goodwin, T. W. (unpublished)
3β-Hydroxy-[4-^{14}C]-5α-cholestan-6-one (55)	ecdysone (< 0.002%)		adult females		
3β-Hydroxy-[3α-^{3}H]-5β-cholest-7-en-6-one (43) 3β, 14α-Dihydroxy-[3α-^{3}H]-5β-cholest-7-en-6-one (44) (2,22,25-Trideoxycedysone)	14-deoxyecdysone (7%) 2,14-dideoxyecdysone 2,14,22-trideoxy-ecdysone? 14,22-dideoxyecdysone? 14,22,25-trideoxyecdysone ecdysone (38%) 22-deoxyecdysone? (46)	*Manduca sexta*	fifth instar	prothoracic glands *in vitro*	Bollenbacher, W. *et al.* (1977)
3β, 14α-Dihydroxy-[3α-^{3}H]-5β-cholest-7-en-6-one (44) 2,22,25-Trideoxyecdtsone)	20-hydroxyecdysone(0.5%) ecdysone 22,25-dideoxyecdysone? (45) 22-deoxyecdysone? (46)	*Calliphora stygia*	puparium formation	*in vivo*	Galbraith, M. *et al.* (1975)
	20-hydroxyecdysone (0.2%)	*Calliphora stygia*	puparium formation	isolated abdomens	
3β, 14α-Dihydroxy-[3α-^{3}H]-5α-cholest-7-en-6-one	2,22-dideoxy-5α-ecdysone 2-deoxy-5α-ecdysone	*Manduca sexta*	fifth instar	prothoracic glands *in vitro*	Bollenbacher, W. *et al.* (1979)

Table 5: Continued

Substrate	Products	Species	Developmental stage	System	Reference
3β, 5α, 14α-Trihydroxy-[3α-^{3}H]-cholest-7-en-6-one	3β, 5α, 14, 22, 25-pentahydroxycholest-7-en-6-one 3β, 5α, 14, 25-tetrahydroxycholest-7-en-6-one?	*Manduca sexta*	fifth instar	prothoracic glands *in vitro*	Bollenbacher, W. *et al.* (1979)
3β, 14α,-Dihydroxy-[3α-^{3}H]-cholesta-4,7-dien-6-one	not metabolized				
3β-Hydroxy-5α-cholestan-6-one (55) 3β-Hydroxy-5α-cholest-7-en-6-one	not metabolized to ecdysone	*Manduca sexta*	fifth instar	prothoracic glands *in vitro*	Gilbert, L. *et al.* (1980)
2β, 3β, 14α-Trihydroxy-[1α-^{3}H]-5β-cholest-7-en-6-one (22,25-Dideoxyecdysone; 45)	22-deoxyecdysone (trace) ecdysone 20-hydroxyecdysone 20,26-dihydroxyecdysone	*Manduca sexta*	diapausing pupae	*in vivo*	Kaplanis, J. *et al.* (1969, 1972a)
	no detectable labelled ecdysteroids	*Musca domestica*	adult females	*in vivo*	Thompson, M. *et al.* (1972)
2β, 3β, 14α-Trihydroxy-[22,23,24,25-^{3}H]-5β-cholest-7-en-6-one (22,25-Dideoxyecdysone; 45)	22-deoxyecdysone (46) ecdysone 20-hydroxyecdysone 20,26-dihydroxyecdysone	*Manduca sexta*	prepupae	*in vivo* and *in vitro*; fat body, Malpighian tubules	King, D. (1972a)
	ecdysone (50%)	*Manduca sexta*	prepupae	prothoracic glands *in vitro*	King, D. (1972a)
	22-deoxyecdysone (25%; 46) ecdysone (2%) 20-hydroxyecdysone (4%)	*Calliphora stygia*	puparium formation	*in vivo*	Thomson, J. *et al.* (1971)
	negligible incorporation into ecdysteroids	*Sarcophaga bullata* *Gastrimargus africanus* *Dermestes maculatus*	prepupae fourth instar	*in vivo* *in vivo* *in vivo*	King, D. (1972b)
2β, 3β, 14α-Trihydroxy-[4-^{14}C]-5β-cholest-7-en-6-one (22,25-Dideoxyecdysone; 45)	22-deoxyecdysone (46) 22,25-dideoxy-20,26-dihydroxyecdysone (22-deoxyinokosterone) 22-deoxy-26-hydroxyecdysone 22-deoxy-20-hydroxyecdysone 20-hydroxyecdysone (trace)	*Manduca sexta*	larval development	*in vivo*	Kaplanis, J. *et al.* (1972a)
	22-deoxyecdysone (46) 22-deoxy-26-hydroxyecdysone 22,25-dideoxy-26-hydroxyecdysone 22-deoxy-20-hydroxyecdysone	*Leucophaea maderae*	late-instar larvae	fatbody *in vitro*	Thompson, M. *et al.* (1978)

Table 5: Continued

Substrate	Products	Species	Developmental stage	System	Reference
2β, 3β, 14α, 25-Tetrahydroxy-[22,23,24-^{3}H]-5β-cholest-7-en-6-one (22-Deoxyecdysone; 46)	20-hydroxyecdysone (25%) ecdysone (5%)	*Calliphora stygia*	puparium formation	*in vivo*	Thomson, J. *et al.* (1971)
	ecdysone (30%) 20-hydroxyecdysone (14%) 20,26-dihydroxyecdysone? (4%)	*Manduca sexta*	prepupae	*in vivo* and *in vitro*; fat body, Malpighian tubules	King, D. (1972a)
	negligible incorporation into ecdysteroids	*Sarcophaga bullata* *Gastrimargus africanus*	prepupae fourth instar	*in vivo* *in vivo*	King, D. (1972a)
2β, 3β, 14α, 22-Tetrahydroxy-[23,24,25-$^3H_5$]-5β-cholest-7-en-6-one (25-Deoxyecdysone)	20-hydroxyecdysone ponasterone A (56) (25-Deoxy-20-hydroxyecdysone) inokosterone (63) (25-deoxy-20,26-dihydroxyecdysone	*Calliphora stygia*	puparium formation	*in vivo*	Thomson, J. *et al.* (1969)
	20-hydroxyecdysone ponasterone A (56)	*Manduca sexta*	prepupae	*in vivo*	King, D. (1972a)

* Incorporation of cholesterol into ecdysteroid in the same system included for comparison.

mediates in the biosynthesis of the hormone from cholesterol (Faux, A. *et al.*, 1979).

When fifth-instar *Bombyx mori*, were injected with [4-^{14}C]cholesterol, labelled 3β-hydroxy-5α-cholestan-6-one (55) could be detected in the prothoracic glands. Incubation of such glands *in vitro* resulted in the disappearance of this labelled steroid from the tissue and the concomitant appearance of [^{14}C]ecdysone in the culture medium, suggesting the intermediacy of 3β-hydroxy-5α-cholestan-6-one (55) in ecdysteroid formation. The incorporation of 3β-hydroxy-[4-^{14}C]-5α-cholestan-6-one (55) into ecdysone (0.014%) was slightly greater than that of [4-^{14}C]cholesterol (0.0082%), when these substrates were incubated *in vitro* with prothoracic glands. These results suggest that 3β-hydroxy-5α-cholestan-6-one (55) might be a precursor of ecdysone in prothoracic glands of *Bombyx mori* (Sakurai, S. *et al.*, 1977). However, labelled 3β-hydroxy-5α-cholestan-6-one (55) was not incorporated significantly into ecdysone either by *Manduca sexta* prothoracic glands cultured *in vitro* (Gilbert, L. *et al.*, 1980) or *in vivo* by ovaries of maturing adult female desert locusts, *Schistocerca gregaria* (Nali, M., Gibson, J., Rees, H. and Goodwin, T., unpublished). In the latter *Schistocerca* work, it may be significant that the 3β-hydroxy-[^{14}C]-5α-cholestan-6-one was poorly incorporated into the eggs as compared to [^{3}H]cholesterol. It is difficult to reconcile these conflicting data, especially for the analogous prothoracic gland systems from the Lepidoptera, *B. mori* and *M. sexta*. Thus, the intermediacy of 3β-hydroxy-5α-cholestan-6-one (55) in ecdysone biosynthesis awaits further substantiation.

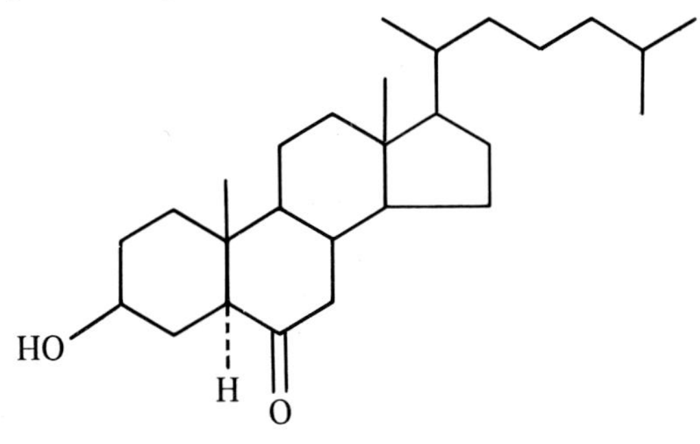

(55) 3β – Hydroxy – 5a – cholestan – 6 – one

As discussed in section 4.4, the observed elimination of the 3α- and 4β-hydrogens of cholesterol during ecdysteroid formation in insects has been interpreted by invoking the intermediacy of a 3-oxo-Δ^4 steroid intermediate (Davies, T. *et al.*, 1981). By analogy with vertebrate systems, such a 3-oxo-Δ^4 intermediate would most probably be formed by first oxidation at C-3 followed by isomerization of the Δ^5 bond to the Δ^4 position. This would not be possible with 3β-hydroxy-5β-cholestan-6-one (55), which lacks the Δ^5 bond. Therefore, it is difficult to envisage the mutual intermediacy of a 3-oxo-Δ^4 steroid and 3β-hydroxy-5α-cholestan-6-one (55) in ecdysone biosynthesis from cholesterol.

It was reported originally that 3β-hydroxy-[3α-^{3}H]-5β-cholest-7-en-6-one (2,14,22,25-tetradeoxyecdysone; often referred to as "ketol"; 43) was incorporated into 20-hydroxyecdysone (48) in pupariating *Calliphora stygia in vivo* to approximately the same extent (0.02%) as [1α-^{3}H]cholesterol, with 91% of the tritium in the 20-hydroxyecdysone being located at C-3 (Horn, D. *et al.*, 1976). However, this labelled substrate (43) was metabolized to 14-deoxyecdysone, but not to ecdysone, in *Manduca sexta* prothoracic glands cultured *in vitro* (Bollenbacher, W. *et al.*, 1977). When the 3β-hydroxy-5β-cholest-7-en-6-one (43) was labelled in more stable positions (C-11 and C-12), the incorporation (0.003%) into 20-hydroxyecdysone in pupariating *Calliphora stygia* was much lower than that of cholesterol (0.02%) (Faux, A. *et al.*, 1979). This suggests that the higher apparent incorporation observed with the [3α-^{3}H] compound could be explained by transfer of some of the 3α-^{3}H to a small compartmentalized pool of NAD(P)$^+$ by 3β-hydroxysteroid dehydrogenase action on the substrate, resulting in formation of a 3-oxo group. Some of this ^{3}H could be subsequently reintroduced at C-3 during reduction of a 3-oxo steroid, which might conceivably be an endogenous postulated 3-oxo-Δ^4 steroid precursor (41; Davies, T. *et al.*, 1981) or a 3-dehydroecdysteroid produced after formation of the ecdysteroid by ecdysone oxidase reaction (Koolman, J., 1980; also see Koolman, J. and Karlson, P., this volume). This illustrates the danger of using putative ecdysteroid precursors radioactively labelled at C-3 in *in vivo* studies; this problem does not arise in the case of *Manduca* prothoracic glands cultured *in vitro*, since no apparent incorporation of ^{3}H into ecdysone from the 3β-hydroxy-[3α-^{3}H]-5β-cholest-7-en-6-one (43) was observed in that system (Bollenbacher, W. *et al.*, 1977). The somewhat analogous transfer of ^{3}H, presumably via NAD(P)$^+$, from the 7β-position of cholesterol to the 3α-position of 20-hydroxyecdysone during ecdysteroid formation in the fern, *Polypodium vulgare*, has been alluded to earlier (section 4.3). Thus, on the basis of the foregoing evidence, 3β-hydroxy-5β-cholest-7-en-6-one (43) is apparently not a precursor of ecdysone (47) in prepupae.

During the metabolism of 3β-hydroxy-[3α-^{3}H]-5β-cholest-7-en-6-one (43) to 14-deoxyecdysone by *Manduca sexta* prothoracic glands *in vitro*, all the other metabolites identified also lacked a C-14 hydroxyl group, *viz.* 14,22,25-trideoxyecdysone, 14,22-dideoxyecdysone (tentative), 2,14,22-trideoxyecdysone (tentative), and 2,14-dideoxyecdysone (Bollenbacher, W. *et al.*, 1977). The occurrence of these metabolites suggests that the sequence of hydroxylation can be initiated on either the steroid nucleus or the side-chain: C-2 → C-25 → C-22 and C-25 → C-22 → C-2. However, it cannot be concluded on the basis of these data that this apparently diverging hydroxylation sequence occurs normally during ecdysone formation. The lack of hydroxylation of 3β-hydroxy-5β-cholest-7-en-6-one (43) at C-14 by the prothoracic glands has been interpreted in more than one way (Bollenbacher, W. *et al.*, 1977). It is possible that the glands lack the enzyme system catalyzing introduction of the 14α-hydroxyl group and that this hydroxylation occurs elsewhere in the insect. A 14α-hydroxylated steroid would then be the appropriate precursor for ecdysone synthesis by the prothoracic glands. Secondly, it is possible that the A/B *cis* ring fusion hindered the interaction of the 14-hydroxylase enzyme system with the precursor. Alternatively, 14α-hydroxylation could occur before introduction of the 6-oxo group and/or Δ^7 bond. Consequently, the 3β-hydroxy-5β-cholest-7-en-6-one (43) would not be a suitable substrate for 14α-hydroxylation, but would be acceptable to the enzyme complexes which introduce the other hydroxyl groups (C-2, C-22 and C-25). If the latter alternative is valid, synthesis of ecdysone in the prothoracic glands starts from a relatively simple sterol, such as cholesterol or 7-dehydrocholesterol.

It has been reported that 3β,14α-dihydroxy-[3α-^{3}H]-5β-cholest-7-en-6-one (2,22,25-trideoxyecdysone; often referred to as "ketodiol"; 44) is much more efficiently incorporated (0.5%) than [1α-^{3}H]cholesterol (0.02%) into 20-hydroxyecdysone (48) (and ecdysone) in pupariating *Calliphora stygia* (Galbraith, M. *et al.*, 1975). Despite the shortcomings of utilizing label at the 3α-position, this result should be valid since the incorporation is much higher than that (0.02%) ascribed to anomalous ^{3}H transfer from 3β-hydroxy-[3α-^{3}H]-5β-cholest-7-en-6-one (43), and since efficient (38%) transformation of 3β,14α-dihydroxy-[3α-^{3}H]-5β-cholest-7-en-6-one (44) into ecdysone by *Manduca* prothoracic glands *in vitro* has been demonstrated (Bollenbacher, W. *et al.*, 1977). Thus, these results suggest an intermediary role for 3β,14α-dihydroxy-5β-cholest-7-en-6-one (44) in the transformation of cholesterol into ecdysteroid. In the *Calliphora* work, compounds which were probably 22,25-dideoxyecdysone (45) and 22-deoxyecdysone (46) were also isolated as metabolites of 3β,14α-dihydroxy-5β-cholest-7-en-6-one (44). Similarly, a compound tentatively identified as 22-deoxyecdysone (46) was isolated as a metabolite of 3β,14α-dihydroxy-5β-cholest-7-en-6-one (44) by *Manduca* prothoracic glands.

Following the initial report that the 4β-hydrogen of cholesterol is eliminated during ecdysteroid biosynthesis and the suggestion that the A/B *cis* ring junction might arise by saturation of a 3-oxo-Δ^{4} intermediate (Lockley, W. *et al.*, 1975), some putative precursors which contain a Δ^{4} bond or could conceivable give rise to such a feature have been investigated. Although 3β,5α,14α-trihydroxy-[3α-^{3}H]-5α-cholest-7-en-6-one and 3β,14α-dihydroxy-[3α-^{3}H]cholesta-4,7-dien-6-one were apparently more efficiently incorporated than [1α-^{3}H]cholesterol into 20-hydroxyecdysone in *Calliphora stygia in vivo* (Kinnear, J. *et al.*, 1978), these substrates were not incorporated into ecdysone in *in vitro* cultures of prothoracic glands from *Manduca sexta* (Bollenbacher, W. *et al.*, 1979). In fact, the 3β,14α-dihydroxy-[3α-^{3}H]cholesta-4,7-dien-6-one was not detectably metabolized by the prothoracic glands. These results again suggested that the apparent incorporation of these [3α-^{3}H]-labelled substrates into 20-hydroxyecdysone in *Calliphora in vivo* was probably due to anomalous reincorporation of released tritium into the hormone, a phenomenon considered earlier during metabolism of 3β-hydroxy-[3α-^{3}H]-5β-cholest-7-en-6-one (43). This complication apparently does not arise in the *in vitro* prothoracic gland system. None of the 3-oxo-Δ^{4} steroids which have been investigated as putative ecdysteroid precursors was incorporated, *viz.* [1α,2α-$^{3}H_{2}$]cholesta-4,7-dien-3-one (pupariating *Calliphora stygia;* Kinnear, J. *et al.*, 1979), [4-^{14}C]cholest-4-en-3-one and [4-^{14}C]cholesta-4,6-dien-3-one (*Schistocerca gregaria*, fifth-instar and adult females; Nali, M., Gibson, J., Rees, H. and Goodwin, T., unpublished). However, this does not eliminate the possibility that a steroid possessing a 3-oxo-Δ^{4} grouping in addition to other obligate structural features could be an ecdysteroid precursor.

Certain 5α-steroids have been explored as possible ecdysteroid precursors in *Manduca sexta* prothoracic glands cultured *in vitro*. The metabolism of 3β,14α-dihydroxy-[3α-^{3}H]-5α-cholest-7-en-6-one and of 3β,5α,14α-trihydroxy-[3α-^{3}H]-cholest-7-en-6-one was inefficient in comparison with the 5β-steroid, 3β,14α-dihydroxy-[3α-^{3}H]-5β-cholest-7-en-6-one (44) which, as indicated earlier, was metabolized to ecdysone (38% incorporation) probably via 22-deoxyecdysone (46), in this *in vitro* system (Bollenbacher, W. *et al.*, 1979). Neither of these 5α-steroids was transformed into ecdysone, only hydroxylation on the side-chain (C-25 and C-22) occurring, with retention of the 5α-configuration. Particularly relevant to subsequent considerations was the observation that hydroxylation at C-25 preceded that at C-22. The lack of hydroxylation at C-2 of these 5α-steroids suggests that the enzyme systems catalysing C-2 hydroxylation recognize only the 5β-configuration, whereas the enzyme complexes that introduce the side-chain hydroxyls do not discriminate significantly between the 5β- and 5α-epimers (i.e. A/B *cis* and *trans* fusions). Even in the case of the side-chain hydroxylases some specificity for the nuclear structure is suggested by the inability of the prothoracic glands to metabolize 3β,14α-dihydroxy[3α-^{3}H]-cholesta-4,7-dien-6-one, which has a fairly planar A/B ring conformation. Neither 3β-hydroxy-5α-cholest-7-en-6-one, nor as indicated earlier, its 5β-epimer, is metabolized to ecdysone by *Manduca* prothoracic glands *in vitro* (Gilbert, L. *et al.*, 1980).

(b) *Side-chain hydroxylations.* Information on the sequence of side-chain hydroxylations during ecdysteroid biosynthesis has been almost entirely derived from investigation of the metabolism of appropriate radioactive putative intermediates. This is already evident from the foregoing section, particularly in the case of 3β,14α-dihydroxy-5β-cholest-7-en-6-one (44) metabolism, where 22,25-dideoxyecdysone (45; *Calliphora*) and 22-deoxyecdysone (46; *Calliphora* and *Manduca*) have been tentatively identified as metabolites during ecdysteroid formation, suggesting that hydroxylation at C-25 precedes that at C-22. Further evidence for this has been furnished by investigation of the metabolism of the latter two deoxyecdysteroids.

22,25-Dideoxyecdysone (45) in addition to exhibiting moulting hormone activity in bioassays, also shows a variety of biological activities (Robbins, W. *et al.*, 1970) including the termination of pupal diapause in *Manduca sexta* (Kaplanis, J. *et al.*, 1969). [1α-^{3}H]22,25-Dideoxyecdysone (45) when injected into male diapausing hornworm, pupae terminated diapause and was efficiently converted into ecdysone (47), 20-hydroxyecdysone (48) and 20,26-dihydroxyecdysone (50) with detection of traces ($< 1\%$) of 22-deoxyecdysone (46) (Kaplanis, J. *et al.*, 1969, 1972a). Similarly, [22,23,23,25-^{3}H]-22,25-dideoxyecdysone (45) injected into *Manduca* prepupae was metabolized to ecdysone (47), 20-hydroxyecdysone (48) and 20,26-dihydroxyecdysone (50; King, D., 1972a). The observation that appreciable radioactivity (20%) was present in 22-deoxyecdysone (46), in this case with no detectable label in ponasterone A (25-deoxy-20-hydroxyecdysone; 56), 25-deoxyecdysone or 22-deoxy-20-hydroxyecdysone (59), suggests that side-chain hydroxylation occurs in an ordered sequence with the C-25 group being introduced before that at C-22 (King, D. 1972b). Further studies with *Manduca* prepupal tissues *in vitro* demonstrated that whereas the fat body and Malpighian tubules converted the [^{3}H]22,25-dideoxyecdysone (45) into 22-deoxyecdysone (46) and the three ecdysteroids (47, 48 and 50), the prothoracic glands did not metabolize the substrate beyond ecdysone (King, D. 1972a). This was the earliest indication that ecdysone was the ecdysteroid product of the prothoracic glands, a notion that has since been substantiated in several species (see section 3.1).

Calliphora stygia prepupae also efficiently transform [22,23,24,25-^{3}H]22,25-dideoxyecdysone to 22-deoxyecdysone (25%; 46), ecdysone (2%) and 20-hydroxyecdysone (4%) (Thomson, J. *et al.*, 1971). These authors suggested that — because ecdysone was formed in significant amounts from [^{3}H]22,25-dideoxyecdysone (and from [^{3}H]22-deoxyecdysone), whereas mass of ecdysone was normally undetectable at the time of puparium formation — these two substrates may not be normal precursors of 20-hydroxyecdysone in this species. However, labelled ecdysone may accumulate under these experimental conditions merely due to its rapid formation from exogenous substrate and the consequent inability of ecdysone-metabolizing enzymes to remove the compound as rapidly as it is produced. Whereas the foregoing evidence suggests that 22,25-dideoxyecdysone (45) may be an intermediate in ecdysone biosynthesis in *Manduca* and *Calliphora*, this substrate is not transformed significantly to ecdysone and 20-hydroxyecdysone in the blowfly, *Sarcophaga bullata*, the grasshopper, *Gastrimargus africanus*, and the hide beetle, *Dermestes maculatus* (King, D. 1972b).

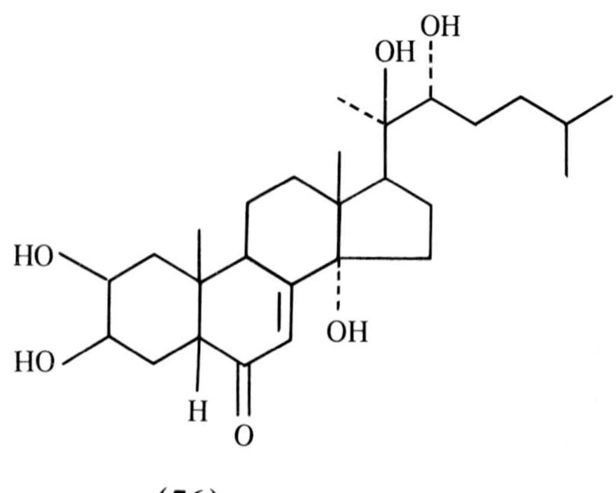

(56) Ponasterone A

In contrast to the transformations of 22,25-dideoxyecdysone (45) demonstrated in prepupae of *Calliphora stygia* and *Manduca sexta*, and during pupal–adult development of *M. sexta*, the metabolism of this substrate differs appreciably during larval development of the latter species. Thus, when newly hatched *Manduca* larvae were reared to the early prepupal stage on a [4-^{14}C]22,25-dideoxyecdysone-supplemented diet, the major labelled metabolites identified in larval frass and early prepupae consisted of a complex mixture of steroid metabolites, the majority of which lacked a C-22 hydroxyl group (Scheme 5; Kaplanis, J. *et al.*, 1972a). Whereas 22-deoxyecdysone (46) was a major metabolite, only minute amounts of ecdysone and 20-hydroxyecdysone were produced. However, the only ecdysteroids which could be isolated from unlabelled frass from fifth-instar *Manduca* larvae were ecdysone and 20-hydroxyecdysone (Kaplanis, J. *et al.*, 1972b). This difference in the ecdysteroid metabolic pattern between animals treated with the labelled 22,25-dideoxyecdysone and untreated ones probably results from exposure of the treated animals to concentrations of that compound above endogenous levels. The results strongly suggest that the 22-hydroxylase system has low activity and may be a rate-limiting step in ecdysone biosynthesis during larval development (Thompson, M. *et al.*, 1981). Since endogenous ecdysteroid titres during larval stages are generally very much lower than in prepupae and pupae, a low rate of ecdysone/20-hydroxyecdysone biosynthesis would be expected in larvae.

The metabolism of [4-^{14}C]22,25-dideoxyecdysone (45) by *in vitro* cultures of cockroach (*Leucophaea maderae*) fat body from late-instar larvae appears to be analogous to that in *Manduca sexta* larvae. Again, in the case of the fat body cultures, 22-deoxyecdysteroids were produced with 22-deoxyecdysone (46) predominating (scheme 6; Thompson, M. *et al.*, 1978). In this case, no evidence for even limited occurrence of 22-hydroxylated metabolites was obtained, suggesting lack of the C-22 hydroxylation system. This is not surprising, since the fat body is not a normal source of ecdysone. However, fat body from *M. sexta* prepupae can effect efficient C-25 and C-22 hydroxylation of 22,25-dideoxyecdysone (King, D. 1972a).

It is significant that during 22,25-dideoxyecdysone (45) metabolism by *M. sexta* larvae and by *Leucophaea maderae* larval fat body cultures, hydroxylation at C-20 can occur in the absence of a C-22 hydroxyl group (compounds 57, 59 and 62). This can also occur in *Bombyx mori* ovaries, which contain 2,22-dideoxy-20-hydroxyecdysone (53; Ikekawa, N. *et al.*, 1980b). However, as alluded to earlier, in *M. sexta* prepupae, no detectable 22-deoxy-20-hydroxyecdysone (59) was formed from [^{3}H]22-deoxyecdysone (46) (King, D. 1972b), suggesting that an analogous process does not occur in this species at a stage of development when marked ecdysone (47) and 20-hydroxyecdysone (48) synthesis occurs. A major sequence of side-chain hydroxylation in the formation of 20-hydroxyecdysone in insects undoubtedly involves hydroxylation at C-22 followed by C-20 (Smith, S., this volume). This is also the preferred sequence in formation of 20,22-dihydroxycholesterol during side-chain cleavage of cholesterol during steroid hormone formation in vertebrates, although the reverse sequence may also operate to a limited extent (Burstein, S. and Gut, M. 1971).

As already indicated, the studies on the metabolism of 2,22,25-trideoxyedysone (44) and 22,25-

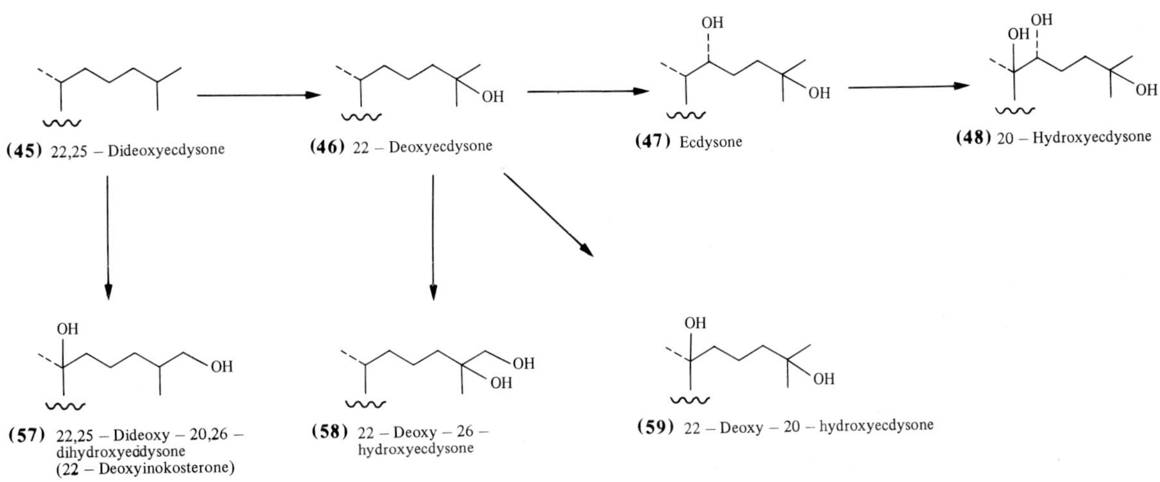

SCHEME 5. Metabolites of 22,25-dideoxyecdysone during larval development of *Manduca sexta* (Kaplanis, J. *et al.*, 1972a).

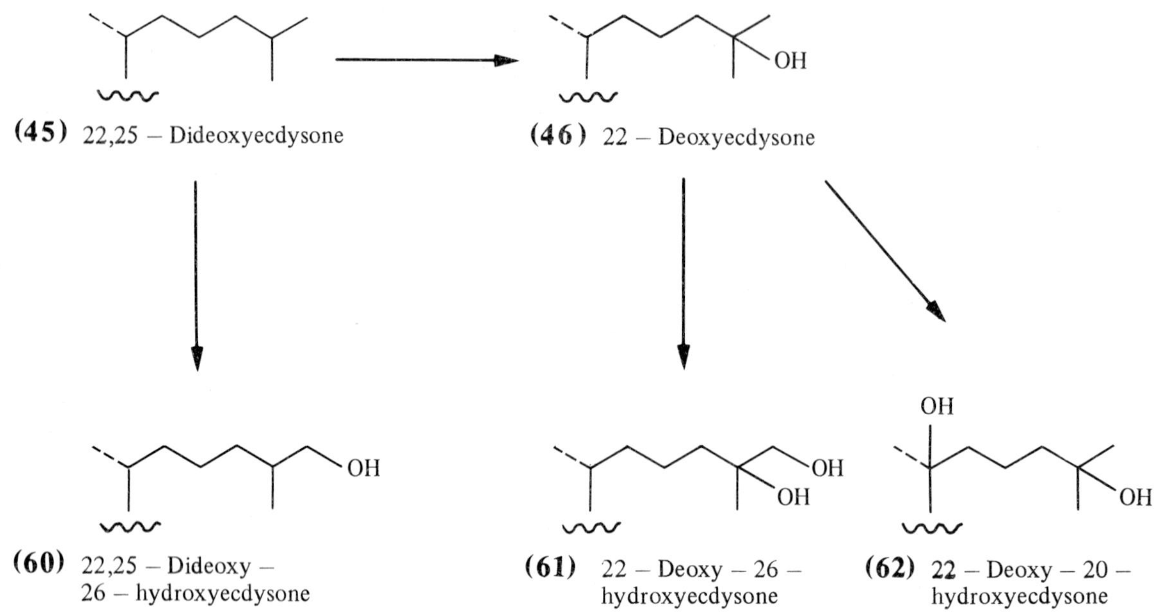

SCHEME 6. Metabolism of 22,25-dideoxyecdysone *in vitro* by fat body from late-instar larvae of *Leucophaea maderae* (Thompson, M. *et al.*, 1978).

dideoxyecdysone (45), suggest that 22-deoxyecdysone (46) is an intermediate in ecdysteroid formation in *Calliphora stygia* and *Manduca sexta*. Further evidence for this is furnished by the efficient metabolism of [22,23,24-^{3}H]22-deoxyecdysone (46) to ecdysone (47) and 20-hydroxyecdysone (48) in *Calliphora stygia* during puparium formation (Thomson, J. *et al.*, 1971) and into these ecdysteroids as well as into a compound tentatively identified as 20,26-dihydroxyecdysone (50) in prepupae of *Manduca sexta* (King, D. 1972a). In the latter species, formation of [^{3}H]22-deoxy-20-hydroxyecdysone (59) from this substrate could not be detected, again suggesting that hydroxylation at C-22 is a prerequisite for hydroxylation at C-20. However, [22,23,24-^{3}H]22-deoxyecdysone (46), when injected into larvae of the grasshopper, *Gastrimargus africanus*, and third-instar larvae of the blowfly, *Sarcophaga bullata*, was metabolized to a complex mixture of polar steroids. Ecdysone was not detectable, and only small amounts ($< 5\%$) of a product tentatively identified as 20-hydroxyecdysone (48) were observed in *S. bullata* only (King, D. 1972a). This was interpreted as indicating that 22-deoxyecdysone (46) is not a normal precursor of ecdysone (47) in *Gastrimargus* and *Sarcophaga*.

Evidence has already been cited indicating that hydroxylation at C-25 precedes that at C-22 in ecdysone formation, but the possibility that the reverse process may occur (i.e. C-22 hydroxylation before C-25 hydroxylaton) has also been checked by investigation of the metabolism of [23,24,25-$^3H_5$] 25-deoxyecdysone. This compound was metabolized to 20-hydroxyecdysone (48) as well as to inokosterone (63) and ponasterone A (56) during puparium formation in *Calliphora stygia* (Thomson, J. *et al.*, 1969), and into 20-hydroxyecdysone

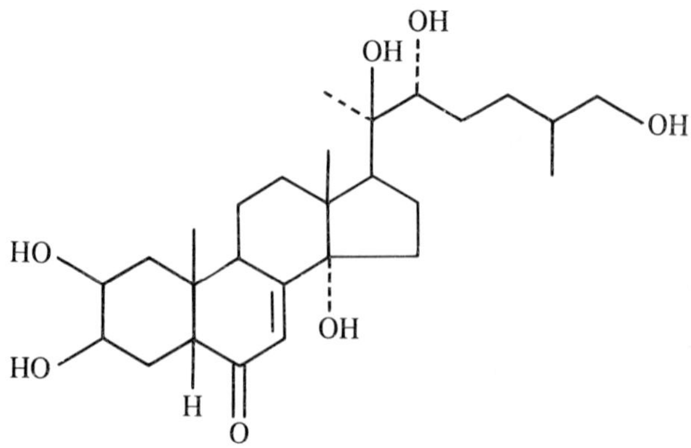

(63) Inokosterone

(48) and ponasterone A (56) in *Manduca sexta* prepupae (King, D. 1972a). Since ponasterone A (56) and inokosterone (63) are not normally detectable in these two species, it was concluded that 25-deoxyecdysone is not a normal precursor of ecdysteroids.

In summary, the foregoing combined data suggest that a sequence of terminal hydroxylations during ecdysone formation in prepupae, at least in *Calliphora* and *Manduca*, involves hydroxylation of 2,22,25-trideoxyecdysone (44) at C-2 followed by C-25 and then C-22 ((44) → (45) → (46) → (47); Scheme 4).

4.5.2 Ovarian ecdysteroid precursors

The occurrence of 2-deoxyecdysteroids such as 2-deoxyecdysone (52) and 2-deoxy-20-hydroxyecdysone (54) in Nature (Table 6) suggests that hydroxylation at C-2 might be the terminal step in ecdysteroid biosynthesis. In fact, 2-deoxyecdysteroids, especially 2-deoxyecdysone (52), are apparently prominent in the metabolism of insect ovaries, where they occur largely as polar conjugates in many species investigated (Table 6). In the case of ovaries of the migratory locust, *Locusta migratoria*, compounds which probably represent a series of biosynthetic precursors of ecdysone (47) have been identified, *viz* 2,14,22,25-tetradeoxyecdysone (43), 2,22,25-trideoxyecdysone (44), 2,22-dideoxyecdysone (51), and 2-deoxyecdysone (52) (Hetru, C. *et al.*, 1978, 1982). Thus, in this system, these compounds may form a biosynthetic pathway involving the hydroxylation sequence C-14, C-25, C-22 followed by C-2. Evidence supporting the operation of such a pathway in *L. migratoria* has been obtained by incubating vitellogenic ovaries *in vitro*, in turn, with [1α,2α-^{3}H]cholesterol (5), [11,12-^{3}H]2,14,22,25-tetradeoxyecdysone (43) and [11,12-^{3}H]2,22,25-trideoxyecdysone (44) (Hetru, C. *et al.*, 1982). The conversions (in the range 0.01–1%) demonstrated directly are summarized in Scheme 7 and lend support for the possible operation in *L. migratoria* ovaries of the pathway:

Table 6: Occurrence of 2-deoxyecdysteroids and accompanying ecdysteroids in insects

Compound	Form detected	Species	Tissue(s)	References
2-Deoxy-20-hydroxyecdysone (54) 20-Hydroxyecdysone (48)	free	*Antheraea pernyi*	pupae	Horn, D. *et al.* (1966) Galbraith, M. *et al.* (1968)
2-Deoxyecdysone (52)			ovaries, diapausing embryos	Ohnishi, E. *et al.* (1977)
2,22-Dideoxy-20-hydroxyecdysone (53)	free and conjugated	*Bombyx mori*	ovaries	Ikekawa, N. *et al.* (1980b)
2-Deoxy-20-hydroxyecdysone (54) Ecdysone (47) 20-Hydroxyecdysone (48)			ovaries	Ohnishi, E. *et al.* (1981)
2-Deoxyecdysone (52) 2-Deoxy-20-hydroxyecdysone (54)	primarily conjugated (phosphates)	*Schistocerca gregaria*	ovaries and eggs	Dinan, L. and Rees, H. (1981a)
Ecdysone (47) 20-Hydroxyecdysone (48) 26-Hydroxyecdysone (49) 20,26-Dihydroxyecdysone (50)				Isaac, R. *et al.* (1981, 1982, 1984)
Ecdysone (47) 20-Hydroxyecdysone (trace; 48) 2-Deoxyecdysone (52)	primarily conjugated	*Locusta migratoria*	ovaries	Hetru, C. *et al.* (1978, 1982) Lagueux, M. *et al.* (1981)
2,22-Dideoxyecdysone (51) 2,22,25-Trideoxyecdysone (44) 2,14,22,25-Tetradeoxyecdysone (43)	form not established			
2-Deoxyecdysone? (52) Ecdysone (47) 20-Hydroxyecdysone (48)	primarily conjugated	*Galleria mellonella*	ovaries, eggs	Bollenbacher, W. *et al.* (1978) Hsiao, T. and Hsiao, C. (1979)

(5) → (43) → (44) →(51) → (52) → (47) (Scheme 4). Although 20-hydroxyecdysone (48) is barely detectable in vitellogenic ovaries of *Locusta in situ*, this was the only ecdysteroid which was labelled from [^{3}H]cholesterol in these experiments, with no radioactivity being reported in ecdysone (47) and 2-deoxyecdysone (52), the major ecdysteroids. The reason for this discrepancy is not clear. Furthermore, the relative proportion of labelled free/conjugated ecdysone in the metabolites formed from [^{3}H]2,14,22,25-tetradeoxyecdysone (43) and 2,22,25-trideoxyecdysone (44) *in vitro* was very much greater than normally observed for mass of these compounds in the ovaries *in situ* (Lagueux, M. *et al.*, 1981) suggesting that the conjugating system is not functioning efficiently under the *in vitro* conditions. It cannot be ascertained whether compounds (43), (44) and (51) normally occur in both free and conjugated forms, since the procedure employed for isolation of mass of these steroids involved enzymic hydrolysis (*Helix pomatia* enzyme) of the extracts (Hetru, C. *et al.*, 1982). Thus, the stage of the biosynthetic pathway at which conjugation occurs remains to be established.

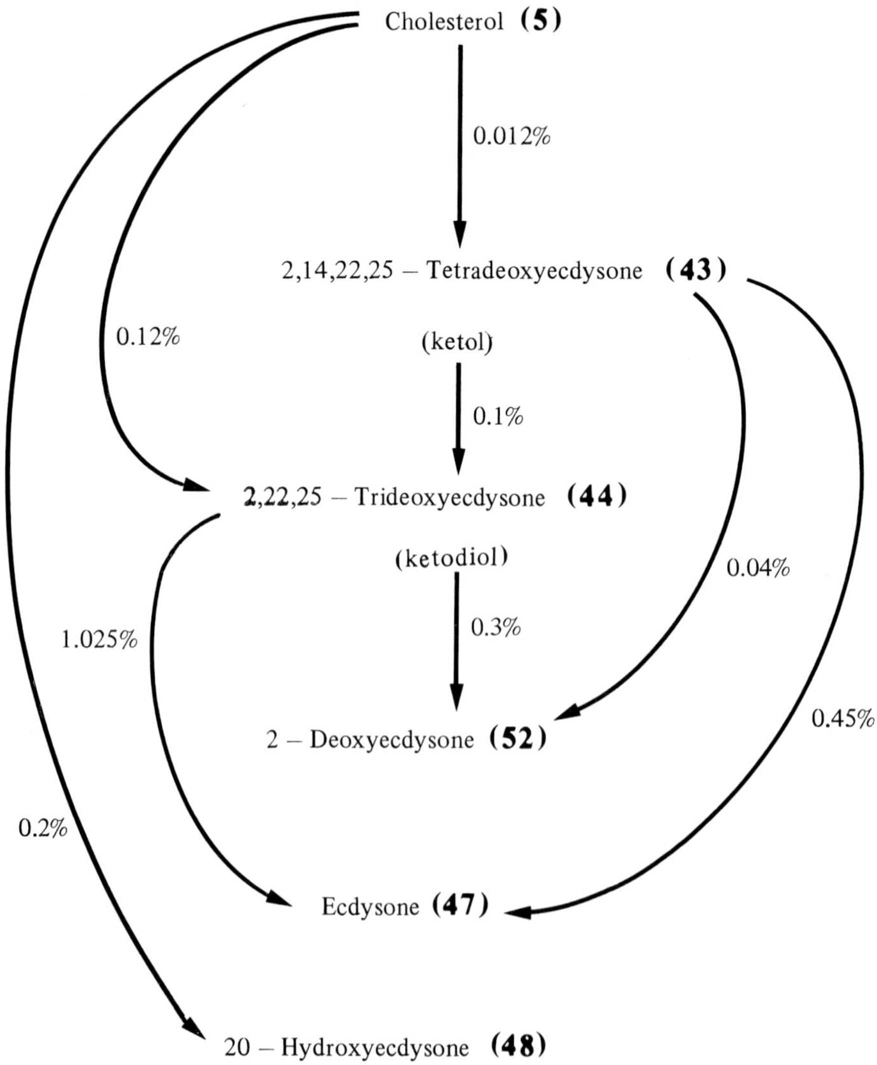

SCHEME 7. Metabolism *in vitro* of [1α,2α-^{3}H]cholesterol (5), [11,12-^{3}H]2,14,22,25-tetradeoxyecdysone (43), [11,12-^{3}H]2,22,25-trideoxyecdysone (44) by vitellogenic ovaries from *Locusta migratoria* (Hetru, C. *et al.*, 1982). The percentage conversions refer to both free and conjugated products.

That ecdysone (47) biosynthesis in ovaries of locusts occurs via 2-deoxyecdysone (52) is further supported by the incorporation of [^{3}H]cholesterol (5) into 2-deoxyecdysone (52) and ecdysone (47) in maturing ovaries (*Schistocerca gregaria*, Dinan, L. and Rees, H. 1981b; *Locusta migratoria*, Lagueux, M. *et al.*, 1981).

Although a linear pathway of biosynthesis of ecdysone in *Locusta* ovaries has been suggested (Hetru, C. *et al.*, 1982), a branched pathway of 20-hydroxyecdysone (48) formation may exist in ovaries of *Bombyx mori* and *Schistocerca gregaria*, based solely on the structures of ecdysteroids isolated. The co-occurrence of 2-deoxyecdysone (52), ecdysone (47), 2,22-dideoxy-20-hydroxyecdysone (53) and 2-deoxy-20-hydroxyecdysone (54) in *Bombyx* ovaries suggests that a precursor such as 2,22-dideoxyecdysone (51) may not only undergo hydroxylation at C-22 and C-2 to yield ecdysone (47), which could undergo C-20 hydroxylation ((51) → (52) → (47) → (48); scheme 4), but may also be hydroxylated sequentially at C-20, C-22 and C-2 to form 20-hydroxyecdysone (48) ((51) → (53) → (54) → (48)) (Ohnishi, E. *et al.*, 1981). Similarly, the co-occurrence of 2-deoxyecdysone (52), ecdysone (47), 20-hydroxyecdysone (48) and 2-deoxy-20-hydroxyecdysone (54) in *Schistocerca gregaria* ovaries/eggs suggests that 20-hydroxyecdysone (54) may arise not only by C-20 hydroxylation of ecdysone (47) [(52) → (47) → (48)], but also by hydroxylation of 2-deoxy-20-hydroxyecdysone (54) at C-2 [(52) → (54) → (48)] (Isaac, R. *et al.*, 1981). However, it is always possible that certain compounds may be detectable because they are not metabolized further.

4.5.3 Overall Sequences

Having considered the detailed facets of the hydroxylation sequences during ecdysone biosynthesis, it is now necessary to evaluate the available evidence for the different overall routes. It is not implied that all the reactions constituting the diverging pathways of ecdysteroid biosynthesis shown in composite Scheme 4 occur simultaneously in one species, or even in a single species at different stages of development.

Examination of the metabolism of administered radioactive putative precursors has been a prominent approach in investigation of the later stages of ecdysone biosynthesis. However, the exogenous radioactive substrates may be subject to different degrees of permeability barriers in reaching the biosynthetic sites. This complication could yield anomalous results.

The possibility existed that the comparatively high moulting hormone activity in the *Calliphora* bioassay of certain deoxyecdysone analogues was due, at least in part, to their metabolism to 20-hydroxyecdysone (48) in the test organism (see Gilbert, L. and King, D., 1973). However, it is now apparent that there is no correlation necessarily between the biological activity and transformation of such compounds to 20-hydroxyecdysone (48). For example, in the series, 22-deoxyecdysone (46), 22,25-dideoxyedysone (45) and 2,22,25-trideoxyecdysone (44), the moulting hormone activity increases with decreasing number of hydroxyl groups, whereas the percentage conversion into 20-hydroxyecdysone (48) and ecdysone (47) decreases with decreasing number of hydroxyl groups. Thus, formation of ecdysone and 20-hydroxyecdysone cannot account for the observed activity of these deoxyecdysteroids. Rather, the greater activity of the less-hydroxylated analogues probably results from their lower water solubility and consequent greater intrinsic binding constant to the receptor because of an increase in non-specific hydrophobic binding (Bergamasco, B. and Horn, D. 1980; Horn and Bergamasco, this volume).

Let us now assume that 2,22,25-trideoxyecdysone (44) is a universal intermediate in the biosynthesis of ecdysone (47) and 20-hydroxyecdysone (48) and consider the possible metabolic routes emanating from this precursor (Scheme 4). In considering a metabolic sequence, it is the overall rate, itself determined by the rate-limiting reaction(s), that is critical in determining the quantitative importance of a pathway. Both the rates of formation and further transformation of a reputed intermediate are important in evaluating the overall significance of a pathway. This is illustrated in the simple example shown below, where the distal precursor may be readily transformed into intermediate X, which is not efficiently transformed into the final product. Conversely, intermediate A may be readily transformed into the final product but not be readily formed from the distal precursor. This underlines that a misleading picture may be obtained if only either the efficient formation or transformation of a suspected intermediate is demonstrated.

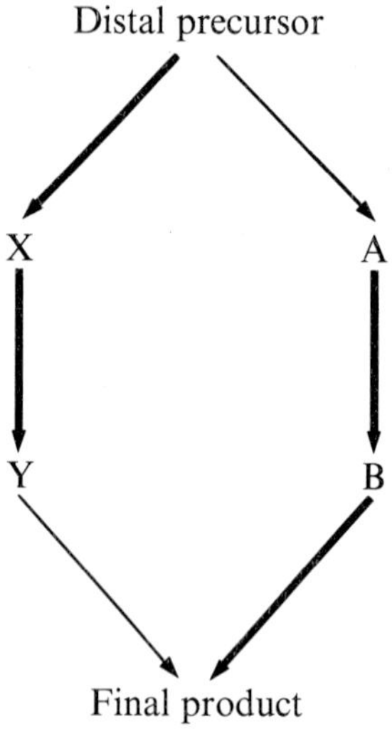

It is well known that during steroid hydroxylations in vertebrates, cytochrome P-450 (a haemoprotein) serves as the terminal oxygenase for a system of electron carriers which channel electrons from NADPH (or NADH) to oxygen. The involvement of cytochrome P-450 in C-20 hydroxylation of ecdysone (47) in various insect species has also been established (see Smith, S., this volume). Particularly relevant to the present discussion of hydroxylation sequences in ecdysone formation is the demonstration that a pure cytochrome P-450 species (P-450 11β) isolated from vertebrate adrenal cortex mitochondria not only catalyses 11β-hydroxylation but also, albeit less efficiently, 18-hydroxylation of various C_{21} steroids as well as 11β- and 19-hydroxylation of certain C19 steroids (Sato, H. *et al.*, 1978). Thus, the specificity associated with the various cytochrome P-450 species is not absolute with regard to substrate or position hydroxylated within the steroid. Consequently, if an analogous situation exists with regard to the various hydroxylases involved in ecdysone formation, a single cytochrome P-450 species could be involved in hydroxylations at more than one position and also act on more than one substrate. Both these factors would, by themselves, lead to diverging pathways. However, the rates of these reactions would not be equivalent, leading to pathways of differing quantitative importance. Furthermore, formation of some compounds which are unsuitable as substrates for further hydroxylation might occur.

We have already considered evidence consistent with the transformation of 2,22,25-trideoxyecdysone (44) via 22,25-dideoxyecdysone (45) and 22-deoxyecdysone (46) into ecdysone (47) and 20-hydroxyecdysone (48) in *Calliphora stygia* and *Manduca sexta* prepupae. The fact that detectable 2-deoxysteroids did not apparently accumulate as metabolites of 2,22,25-trideoxyecdysone (46) in these studies does not necessarily preclude an alternative route via 2-deoxy-intermediates; such compounds could be rapidly metabolized. Despite this, the evidence is suggestive that at least in prepupae of these species, the complete ecdysone nucleus is formed before side-chain hydroxylation ((44) → (45) → (46) → (47); Scheme 4). However, the transformation of cholesterol into compounds (44)–(46) has not been reported. The lack of incorporation of 22,25-dideoxyecdysone (45) and 22-deoxyecdysone (46) into ecdysone (47) and 20-hydroxyecdysone (48) in *Sarcophaga bullata* prepupae and *Gastrimargus africanus* fourth-instar larvae suggests that the foregoing route may not be significant in these species. However, it is premature to speculate on present evidence whether or not this reflects a true species difference.

In contrast to the prepupal systems, *Locusta* ovaries metabolize 2,22,25-trideoxyecdysone (44) to 2-deoxyecdysone (52) which, although not demonstrated directly, is presumably transformed to ecdysone (47). In this study, formation of compounds (45) and (46) possessing a complete ecdysone nucleus was not detected. This evidence, taken together with the occurrence of 2-deoxyecdysteroids in ovaries/eggs (Table 6), the incorporation of [^{3}H]cholesterol into 2-deoxyecdysone (52) and ecdysone (47) in developing ovaries, and the results summarized in Scheme 7, supports the operation of a route via 2-deoxyecdysteroids in ovaries, e.g. ((44) → (51) → (52) → (47)). The additional possibility of operation of branched pathways invoking compounds (53) and (54) in the formation of 20-hydroxyecdysone (48) also exists (Scheme 4).

Thus, on the evidence available, it does appear that ecdysone formation in certain ovarian systems may involve C-2 hydroxylation as the last step, whereas at least in *Calliphora* and *Manduca* prepupae the complete nucleus may be formed before side-chain hydroxylation. Of course, the criteria outlined in the Introduction for establishment of the intermediacy of a compound in a pathway are far from being fulfilled for intermediates of both routes. It is too early to speculate whether or not different biosynthetic routes operate in different species at equivalent stages of development or in a single species at different developmental stages.

4.6 Stereochemistry of hydroxylations

Several enzymic hydroxylations of steroids (Bergstrom, S. *et al.*, 1958; Hayano, M. *et al.*, 1958) and fatty acids (Morris, L. 1970) have been shown to occur by direct hydroxylation and result in retention of configuration at the respective carbon atoms. In the case of ecdysone biosynthesis, the stereochemistry of hydroxylation at C-22 and C-2 has been investigated.

For investigation of the hydroxylation mechanism at C-22, advantage was taken of the fact that cholesterol formed from [2−^{3}H]mevalonic acid (MVA) in rat liver is labelled at C-22. Thus, 3*R*-[2-^{14}C,(2*R*)-2-$^3H_1$]MVA (64) and 3*R*-[2-^{14}C, (2*S*)-2-$^3H_1$]MVA (65) were incubated with a rat liver preparation yielding cholesterol species known to be labelled at the 22*R* (66) and 22*S* (67) positions, respectively (Scheme 8). These cholesterol samples were then incorporated into ecdysone (phosphate) in maturing adult female *Schistocerca gregaria*. The 22*S*-tritium of the cholesterol species (67) was retained during formation of ecdysone (69), but was eliminated upon chemical transformation into the 22-oxo derivative, whereas the 22*R*-tritium in cholesterol (66) was eliminated during biosynthesis of ecdysone (68). This demonstrates that introduction of the 22*R* hydroxyl group in ecdysone involves direct hydroxylation with retention of configuration (Greenwood, D. and Rees, H. 1982). Similarly, introduction of a 22*R* hydroxyl into cholesterol during formation of (20*R*,22*R*)-20,22-dihydroxycholesterol in the biosynthesis of pregnenolone by bovine adrenocortical mitochondria involves elimination of the 22-*pro*-R hydrogen with retention of configuration (Duque, C. *et al.*, 1979).

The stereochemistry of hydroxylation at C-2 in ecdysone has been investigated by incorporation of [1α,2α-^{3}H]cholesterol in admixture with [4-^{14}C]cholesterol into 2-deoxyecdysone (52) and ecdysone (47) phosphates in maturing adult female *S. gregaria*. Tritium in the substrate was localized in the 1α (46%), 1β (17%) and 2α (37%) positions, as revealed by ^{3}H NMR studies. The [^{3}H,^{14}C]cholesterol was incorporated into 2-deoxyecdysone (52) and ecdysone (47) with only minimal decrease in ^{3}H/^{14}C ratio, suggesting that the 2α-hydrogen was retained during ecdysone formation. Oxidation of the 2β-hydroxyl of the ecdysone to an oxo group under non-equilibrating conditions resulted in removal of 38% of the total tritium in the molecule from C-2. This confirmed that the ^{3}H at the 2α-position in the cholesterol (37% of the total) was completely retained during ecdysone biosynthesis and indicated that hydroxylation at C-2 in the hormone, of necessity, involved elimination of the 2β-hydrogen of cholesterol. This result again suggests that hydroxylation at C-2 during ecdysone formation occurs with retention of configuration at that position (Greenwood, D. *et al.*, 1982).

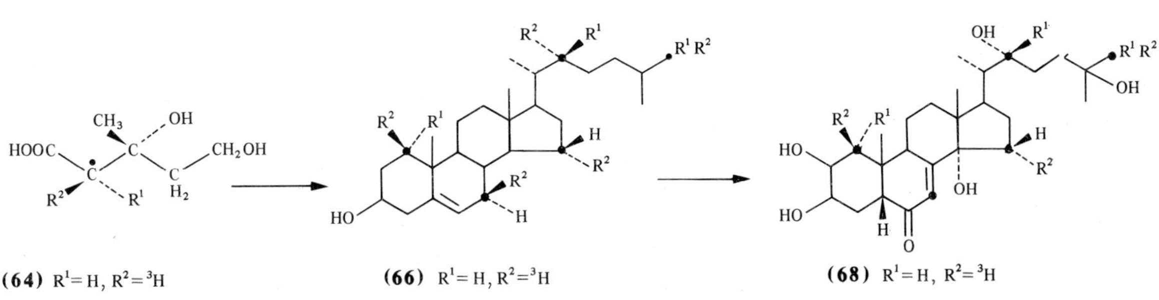

(64) R^1=H, R^2=^{3}H (66) R^1=H, R^2=^{3}H (68) R^1=H, R^2=^{3}H

(65) R^1=^{3}H, R^2=H (67) R^1=^{3}H, R^2=H (69) R^1=^{3}H, R^2=H

SCHEME 8. Incorporation of [2-^{14}C, (2*R*)-2-^{3}H]MVA (64) and [2-^{14}C, (2*S*)-2-^{3}H]MVA (65) into cholesterol and ecdysone.

5 SUMMARY AND CONCLUSIONS

The criteria which should be satisfied for unequivocal establishment of the intermediary role of any postulated intermediate in biosynthetic pathways are discussed.

Evidence is presented supporting the operation of the following pathway of sitosterol dealkylation in phytophagous insects: sitosterol → fucosterol → fucosterol-24,28-epoxide → desmosterol → cholesterol. The possible involvement of 28-isofucosterol and 28-isofucosterol-24,28-epoxide in sitosterol dealkylation is also considered. Evidence for the existence of analogous pathways in the dealkylation of campesterol and stigmasterol is presented. However, some phytophagous insects lack the capacity to dealkylate phytosterols, whereas others, although capable of dealkylation emphasize the formation of $\Delta^{5,7}$ or saturated sterols.

Although the occurrence of ecdysone synthesis has been established in prothoracic glands in immature stages of insects and in maturing ovaries in pupal (pharate adult) or adult stages, there is evidence for ecdysone synthesis in other unidentified tissues in larvae and pupae of some species.

During the biosynthesis of ecdysone from cholesterol, at least some modification of the nucleus occurs before side-chain hydroxylation. The interconversion of Δ^5, Δ^7, $\Delta^{5,7}$ and Δ^0 sterols is discussed. Although 7-dehydrocholesterol is incorporated into ecdysteroids, it is not clear whether or not this transformation occurs directly or via cholesterol. The *cis* 7β- and 8β-hydrogens are removed during introduction of the Δ^7 bond in ecdysteroids. Elimination of the 3α- and 4β-hydrogens of cholesterol during ecdysteroid formation in insects is probably best explained by postulating the involvement of a 3-oxo-Δ^4-steroid in formation of the A/B *cis* ring junction.

Whereas the transformation, albeit low, of cholesterol into ecdysone by prothoracic glands *in vitro* has apparently been demonstrated in the case of some species, this has not been possible in others. Thus it is not clear, at least in some species, whether cholesterol is the starting substrate utilized in prothoracic glands for ecdysone synthesis, or whether some other derived steroid produced in another tissue is taken up by these glands for conversion into hormone.

The sequence of the hydroxylation steps during ecdysone biosynthesis has been largely investigated by examination of the metabolism of synthetic putative radioactive precursors either *in vivo* or in prothoracic glands cultured *in vitro*. These studies support the involvement of 2,22,25-trideoxyecdysone in ecdysone synthesis. Evidence is presented that the later hydroxylation steps in prepupae of *Calliphora* and *Manduca* occur in the sequence, C-2, C-25, followed by C-22, without definitely precluding the possible involvement of a pathway via 2-deoxysteroids. However, the isolation of a series of 2-deoxyecdysteroids of ovarian origin in some species, together with the demonstration of several metabolic transformations in ovaries of *Locusta migratoria in vitro* support the operation in ovaries of the hydroxylation sequence, C-14, C-25, C-22, followed by C-2. In ovaries of some species, 20-hydroxyecdysone may arise not only from ecdysone but also possibly via a branched pathway involving 2,22-dideoxy-20-hydroxyecdysone and/or 2-deoxy-20-hydroxyecdysone, as suggested by the occurrence of these ecdysteroids.

Introduction of the C-2 and C-22 hydroxyl groups during ecdysone formation occurs by direct hydroxylation and results in retention of configuration at the respective carbon atoms.

ACKNOWLEDGEMENTS

Work from author's laboratory was supported by the Science and Engineering Research Council and the Agricultural Research Council. I am grateful to Dr J. F. Pennock for valuable discussions.

REFERENCES

AGUI, N., KIMURA, Y. and FUKAYA, M. (1972). Action of the prothoracic gland on the insect integument *in vitro*. *Appl. Ent. Zool. 7*, 71–78.

AKHTAR, M., CALDER, M., SMITH, T. and WRIGHT, J. N. (1980). Mechanistic and stereochemical studies on 3-oxo steroid Δ^4-Δ^5-isomerase from human placenta. *Biochem. J. 185*, 411–421.

ALLAIS, J.-P. and BARBIER, M. (1966). Observations sur l'utilisation d'un β-sitostérol-28,29^{14}C par le criquet. *Locusta migratoria. C.R. Acad. Sci. Paris (D) 263*, 1252–1254.

ALLAIS, J.-P. and BARBIER, M. (1971). Sur les intermédiaires de la désalkylation en C_{24} du β-sitostérol par le criquet *Locusta migratoria* L. *Experientia 27*, 506.

ALLAIS, J.-P. and BARBIER, M. (1977). Le fucosterol-24,28-epoxyde, intermediaire dans la biodegradation oxydative du fucosterol en cholesterol par le criquet *Locusta migratoria* (R. et F.). *FEBS Lett. 82*, 333–336.

ALLAIS, J.-P., ALCAIDE, A. and BARBIER, M. (1973). Fucosterol-24, 28 epoxide and 28-oxo-β-sitosterol as possible intermediates in the conversion of β-sitosterol into cholesterol in the locust *Locusta migratoria* L. *Experientia 29*, 944–945.

ALLAIS, J.-P., PAIN, J. and BARBIER, M. (1971). La désalkylation en C_{24} du β-sitostérol par l'Abeille *Apis mellifica* L. *C.R. Acad. Sci. Paris (D), 272*, 877–879.

AVIGAN, J. and STEINBERG, D. (1961). Studies on cholesterol biosynthesis. III. The desmosterol reductase system in liver. *J. Biol. Chem. 236*, 2898–2900.

AWATA, N., MORISAKI, M. and IKEKAWA (1975). Carbon-carbon bond cleavage of fucosterol-24,28-oxide by cell-free extracts of silkworm, *Bombyx mori. Biochem. Biophys. Res. Commun. 64*, 157–161.

BECK, S. D. and KAPADIA, G. G. (1957). Insect nutrition and metabolism of sterols. *Science 126*, 258–259.

BERGAMASCO, B. and HORN, D. H. S. (1980). The biological activities of ecdysteroids and ecdysteroid analogues. In *Progress in Ecdysone Research*. Edited by J. A. Hoffmann. Pages 299–324. Elsevier/North Holland Biomedical Press, Amsterdam.

BERGSTROM, S., LINDSTREDT, S., SAMUELSON, B., COREY, E. J. and GREGORIOU, G. A. (1958). The stereochemistry of 7α-hydroxylation in the biosynthesis of cholic acid from cholesterol. *J. Amer. Chem. Soc. 80*, 2337–2338.

BJÖRKHEM, J. and GUSTAFSSON, J.-Å. (1971). Mechanism of microbial transformation of cholesterol into coprostanol. *Eur. J. Biochem. 21*, 428–432.

BOLLENBACHER, W. E., FAUX, A. F., GALBRAITH, M. N., GILBERT, L. I., HORN, D. H. S. and WILKIE, J. S. (1979). *In vitro* metabolism of possible ecdysone precursors by the prothoracic glands of the tobacco hornworm, *Manduca sexta. Steroids 34*, 509–526.

BOLLENBACHER, W. E., GALBRAITH, M. N., GILBERT, L. I. and HORN, D. H. S. (1977). *In vitro* metabolism of 3β-hydroxy-, and 3β,14α-dihydroxy-[3α-^{3}H]-5β-cholest-7-en-6-one by the prothoracic glands of *Manduca sexta. Steroids 29*, 47–63.

BOLLENBACHER, W. E., GOODMAN, W., VEDECKIS, W. V. and GILBERT, L. I. (1976). The *in vitro* synthesis and secretion of α-ecdysone by the ring glands of the fly, *Sarcophaga bullata. Steroids 27*, 309–324.

BOLLENBACHER, W. E., ZVENKO, H., KUMARAN, A. K. and GILBERT, L. I. (1978). Changes in ecdysone content during postembryonic development of the wax moth, *Galleria mellonella*: the role of the ovary. *Gen. Comp. Endocr. 34*, 169–179.

BORST, D. W. and ENGELMANN, F. (1974). *In vitro* secretion of α-ecdysone by prothoracic glands of a hemimetabolous insect. *Leucophaea maderae* (Blattaria). *J. Exp. Zool. 189*, 413–419.

BULLIÈRE, D. and BULLIÈRE, F. (1974). Recherch des sites de synthèse de l'hormone de mue chez les embryons de *Blabera craniifer* Burm., Insecte Dictyoptère. *C.R. Acad. Sci. Paris (D), 274*, 377–380.

BURSTEIN, S. and GUT, M. (1971). Biosynthesis of pregnenolone. *Recent Prog. Hormone Res. 27*, 303–349.

CHADWICK, L. E. (1955). Molting of roaches without prothoracic glands. *Science 121*, 435.

CHADWICK, L. E. (1956). Removal of prothoracic glands from the nymphal cockroach. *J. Exp. Zool. 131*, 291–305.

CHARLET, M., GOLTZENÉ, F. and HOFFMANN, J. A. (1979). Experimental evidence for a neuroendocrine control of ecdysone biosynthesis in adult females of *Locusta migratoria. J. Insect Physiol.* 25, 463–466.

CHEN, S.-M. L., NAKANISHI, K., AWATA, N., MORISAKI, M., IKEKAWA, N. and SHIMIZU, Y. (1975). Stereospecificity in the conversion of fucosterol 24,28-epoxide to desmosterol in the silkworm, *Bombyx mori. J. Amer. Chem. Soc. 97*, 5297–5299.

CHINO, H., SAKURAI, S., OHTAKI, T., IKEKAWA, N., MIYAZAKI, H., ISHIBASHI, M. and ABUKI, H. (1974). Biosynthesis of α-ecdysone by prothoracic glands *in vitro. Science 183*, 529–530.

CHU, H.-M., NORRIS, D. M. and KOK, L. T. (1970). Pupation requirement of the beetle, *Xyleborus ferrugineus*: sterols other than cholesterol. *J. Insect Physiol. 16*, 1379–1387.

CLARK, A. J. and BLOCH, K. (1959). Conversion of ergosterol to 22-dehydrocholesterol in *Blattella germanica. J. Biol. Chem. 234*, 2589–2594.

CLARKE, G. S. (1981). Ph.D thesis, University of Liverpool.

CLAYTON, R. B. (1964). The utilization of sterols by insects. *J. Lipid Res. 5*, 3–19.

CLAYTON, R. B. and EDWARDS, A. M. (1963). Conversion of 5α-cholestan 3β-ol to Δ^7-5α-cholesten-3β-ol in cockroaches. *J. Biol. Chem. 238*, 1966–1972.

COOK, I. F., LLOYD-JONES, J. G., REES, H. H. and GOODWIN, T. W. (1973). The stereochemistry of hydrogen elimination from C-7 during biosynthesis of ecdysones in insects and plants. *Biochem. J. 136*, 135–145.

CRONHOLM, T. (1972). Steroid metabolism in rats given [1-$^2H_2$]ethanol Biliary metabolites of corticosterone and administered 4-androstene-3,17-dione. *Eur. J. Biochem. 27*, 10–22.

DAVIES, T. G., DINAN, L. N., LOCKLEY, W. J. S., REES, H. H. and GOODWIN, T. W. (1981). Formation of the A/B *cis* ring junction of ecdysteroids in the locust, *Schistocerca gregaria. Biochem. J. 194*, 53–62.

DAVIES, T. G., LOCKLEY, W. J. S., BOID, R., REES, H. H. and GOODWIN, T. W. (1980). Mechanism of formation of the A/B *cis* ring junction of ecdysteroids in *Polypodium vulgare. Biochem. J. 190*, 537–544.

DELBECQUE, J.-P. A., DELACHAMBRE, J., HIRN, M. and DE REGGI, M. (1978a). Abdominal production of ecdysterone and pupal-adult development in *Tenebrio molitor* (Insecta Coleoptera). *Gen. Comp. Endocr. 35*, 436–444.

DELBECQUE, J.-P., HIRN, M., DELACHAMBRE, J. and DE REGGI, M. (1978b). Cuticular cycle and molting hormone levels during the metamorphosis of *Tenebrio molitor* (Insecta Coleoptera). *Devel. Biol. 64*, 11–30.

DINAN, L. N. (1980). Ph.D. thesis, University of Liverpool.

DINAN, L. N. and REES, H. H. (1981a). The identification and titres of conjugated and free ecdysteroids in developing ovaries and newly laid eggs of *Schistocerca gregaria. J. Insect Physiol. 27*, 51–58.

DINAN, L. N. and REES, H. H. (1981b). Incorporation *in vivo* of [4-^{14}C]cholesterol into the conjugated ecdysteroids in ovaries and eggs of *Schistocerca americana gregaria. Insect Biochem. 11*, 255–265.

DUPNIK, T. D. and KAMM, J. A. (1970). Development of an artificial diet for *Crambus trisectus. J. Econ. Ent. 63*, 1578–1581.

DUQUE, C., MORISAKI, M., IKEKAWA, N. and SHIKITA, M. (1979). Stereochemistry of cholesterol hydroxylation at C-22 during pregnenolone biosynthesis. *Tet. Lett.* 4479–4482.

EARLE, N. W., LAMBREMONT, E. N., BURKS, M. L., SLATTEN, B. H. and BENNETT, A. F. (1967). Conversion of β-sitosterol to cholesterol in the boll weevil and the inhibition of larval development by two azasterols. *J. Econ. Ent. 60*, 291–293.

ELLIS, P. E, MORGAN, E. D. and WOODBRIDGE, A. P. (1972). Moult-inducing hormones of the prothoracic gland of insects. *Nature 238*, 274–276.

FAUX, A. F., HORN, D. H. S., KINNEAR, J. F., MARTIN, M.-D., WILKIE, J. S. and WILLING, R. I. (1979). Evaluation of 3β-hydroxy-5β-cholest-7-en-6-one and related steroids as precursors of ecdysteroids in *Calliphora stygia. Insect Biochem. 9*, 101–105.

FRANTZ, I. D. and SCHROEPFER, G. J. (1967). Sterol biosynthesis *Ann. Rev. Biochem. 36*, 691–726.

FUJIMOTO, Y., AWATA, N., MORISAKI, M. and IKEKAWA, N. (1974a). Migration of C-25 hydrogen of sitosterol to C-24 during the conversion into desmosterol in the silkworm, *Bombyx mori. Tet. Lett.* 4335–4338.

FUJIMOTO, Y., MORISAKI, M. and IKEKAWA, N. (1974b). Synthesis of 24,28-iminofucosterol and its inhibitory effects on growth and steroid metabolism in the silkworm, *Bombyx mori. Steroids, 24*, 367–375.

FUJIMOTO, Y., MORISAKI, M. and IKEKAWA, N. (1980a). Stereochemical importance of fucosterol epoxide in the conversion of sitosterol into cholesterol in the silkworm, *Bombyx mori. Biochemistry 19*, 1065–1069.

FUJIMOTO, Y., MURAKAMI, K. and IKEKAWA, N. (1980b). Synthesis of (24*R*, 28*R*)- and (24*S*, 28*S*)-fucosterol epoxides. Revision of C-24,28 configurations. *J. Org. Chem. 45*, 566–569.

GALBRAITH, M. N., HORN, D. H. S. and MIDDLETON, E. J. (1970). The biosynthesis of crustecdysone in the blowfly, *Calliphora stygia. J. Chem. Soc. Chem. Comm.* 179–180.

GALBRAITH, M. N., HORN, D. H. S., MIDDLETON, E. J. and HACKNEY, R. J. (1968). Structure of deoxycrustecdysone a second crustacean moulting hormone. *J. Chem. Soc. Chem. Comm.* 83–85.

GALBRAITH, M. N., HORN, D. H. S., MIDDLETON, E. J. and THOMSON, J. A. (1972). Side-chain hydroxylation in the biosynthesis of β-ecdysone (20-hydroxyecdysone) in the blowfly, *Calliphora stygia. Experientia 28*, 264.

GALBRAITH, M. N., HORN, D. H. S., MIDDLETON, E. J. and THOMSON, J. A. (1973). Biological activity of synthetic moulting hormone analogues in the blowfly *Calliphora stygia. Experientia 29*, 19.

GALBRAITH, M. N., HORN, D. H. S., MIDDLETON, E. J., THOMSON, J. A. and WILKIE, J. S. (1975). Metabolism of 3β,14α-dihydroxy-5β-[3α-^{3}H]-cholest-7-en-6-one in *Calliphora stygia*. *J. Insect Physiol. 21*, 23–32.

GANDE, A. R., MORGAN, E. D. and WILSON, I. D. (1979). Ecdysteroid levels throughout the life cycle of the desert locust, *Schistocerca gregaria*. *J. Insect Physiol. 25*, 669–675.

GERSCH, M. (1977). Haütungen von insekten ohne haütungsdrüse: befunde mit larven von *Periplaneta americana*. *Experientia 33*, 228–230.

GERSCH, M. and EIBISCH, J. (1977). Synthesis of ecdysone-^{14}C and ecdysterone-^{14}C from cholesterol-^{14}C in cockroaches (*Periplaneta americana*) without moulting glands. *Experientia 33*, 468.

GERSCH, M. and STÜRZEBECHER, J. (1971). Über eine synthese von ecdysone-^{3}H und ecdysteron-^{3}H aus cholesterin-^{3}H in geschnürten abdomina von *Mamestra brassicae* — Raupen. *Experientia 27*, 1475–1476.

GILBERT, L. I. (1964). Physiology of growth and development: endocrine aspects. In *The Physiology of Insecta*. Edited by M. Rockstein. 1st edn., Vol. 1, pages 149–225. Academic Press, New York.

GILBERT, L. I. (1974). Endocrine action during insect growth. *Recent Prog. Hormone Res. 30*, 347–390.

GILBERT, L. I and KING, D. S. (1973). Physiology of growth and development: endocrine aspects. In *The Physiology of Insecta*. Edited by M. Rockstein. 2nd edn., Vol. 1, pages 249–370. Academic Press, New York.

GILBERT, L. I., BOLLENBACHER, W. E., GOODMAN, W., SMITH, S. L., AGUI, N., GRANGER, N. and SEDLAK, B. J. (1980). Hormones controlling insect metamorphosis. *Recent Prog. Horm. Res. 36*, 401–449.

GILBERT, L. I., GOODMAN, W. and BOLLENBACHER, W. E. (1977). Biochemistry of regulatory lipids and sterols in insects. In *International Review of Biochemistry. Biochemistry of Lipids*. II. Edited by T. W. Goodwin. Vol. 14, pages 1–50. University Park Press, Baltimore.

GLITHO, I., DELBECQUE, J. P. and DELACHAMBRE, J. (1979). Prothoracic gland involution related to moulting hormone levels during the metamorphosis of *Tenebrio molitor L. J. Insect Physiol. 25*, 187–191.

GOLTZENÉ, F., LAGUEUX, M., CHARLET, M. and HOFFMANN, J. A. (1978). The follicle cell epithelium of maturing ovaries of *Locusta migratoria*: a new biosynthetic tissue for ecdysone. *Hoppe-Seyler's Z. Physiol. Chem. 359*, 1427–1434.

GOODNIGHT, K. C. and KIRCHER, H. W. (1971). Metabolism of lathosterol by *Drosophila pachea*. *Lipids 6*, 166–169.

GREENWOOD, D. R., DINAN, L. N. and REES, H. H. (1982). Mechanism of hydroxylation at C-2 during the biosynthesis of ecdysone in the ovaries of the locust, *Schistocera gregaria*. *217*, 783–789.

GREENWOOD, D. R. and REES, H. H. (1982). Mechanism of hydroxylation at C-22 during the biosynthesis of ecdysteroids in the locust, *Schistocerca gregaria*. *Biochem. J. 208*, 857–864.

HAGEDORN, H. H., SHAPIRO, J. P. and HANAOKA, K. (1979). Ovarian ecdysone secretion is controlled by a brain hormone in an adult mosquito. *Nature 282*, 92–94.

HANAOKA, K. and HAGEDORN, H. H. (1980). Brain hormone control of ecdysone secretion by the ovary in a mosquito. In *Progress in Ecdysone Research*. Edited by J. A. Hoffmann. Pages 467–480. Elsevier/North-Holland Biomedical Press, Amsterdam.

HAYANO, M., GUT, M., DORFMAN, R. I., SEBEK, O. K. and PETERSON, D. H. (1958). Steric considerations in the enzymatic course of the hydroxylation of steroids. *J. Amer. Chem. Soc. 80*, 2336–2337,

HEED, W. B. and KIRCHER, H. W. (1965). Unique sterol in the ecology and nutrition of *Drosophila pachea*. *Science 149*, 758–761.

HETRU, C., KAPPLER, C., HOFFMANN, J. A., NEARN, R., BANG, L. and HORN, D. H. S. (1982). The biosynthetic pathway of ecdysone: studies with vitellogenic ovaries of *Locusta migratoria* (Orthoptera). *Mol. Cell. Endocr. 26*, 51–81.

HETRU, C., LAGUEUX, M., LUU, B. and HOFFMANN, J. A. (1978). Adult ovaries of *Locusta migratoria* contain the sequence of biosynthetic intermediates for ecdysone. *Life Sci. 22*, 2141–2154.

HIRN, M., HETRU, C., LAGUEUX, M. and HOFFMANN, J. A. (1979). Prothoracic gland activity and blood titres of ecdysone and ecdysterone during the last larval instar of *Locusta migratoria L. J. Insect Physiol., 25*, 255–261.

HOFFMANN, J. A., KOOLMAN, J. and BEYLER, CH. (1975). Rôle des glandes prothoraciques dans la production d'ecdysone au cours du dernier stade larvaire de *Locusta migratoria*. *C.R. Acad. Sci. Paris (D) 280*, 733–736.

HOFFMANN, J. A., LAGUEUX, M., HETRU, C., CHARLET, M. and GOLTZENÉ, F. (1980). Ecdysone in reproductively competent female adults and in embryos of insects. In *Progress in Ecdysone Research*. Edited by J. A. Hoffmann. Pages 431–465. Elsevier/North-Holland Biomedical Press, Amsterdam.

HORN, D. H. S., MIDDLETON, E. J., THOMSON, J. A. and WILKIE, J. S. (1974). Metabolism of some possible ecdysone precursors in *Calliphora stygia*. *J. Insect Physiol. 20*, 2433–2445.

HORN, D. H. S., MIDDLETON, E. J., THOMSON, J. A. and WILKIE, J. S. (1976). Metabolism of 3β-hydroxy-5β-(3α-^{3}H)-cholest-7-en-6-one in *Calliphora stygia*. *Insect Biochem. 6*, 601–604.

HORN, D. H. S., MIDDLETON, E. J., WUNDERLICH, J. A. and HAMPSHIRE, F. (1966). Identity of the moulting hormones of insects and crustaceans. *J. Chem. Soc. Chem. Commun.* 339–341.

HSIAO, T. H. and HSIAO, C. (1979). Ecdysteroids in the ovary and the eggs of the greater wax moth. *J. Insect Physiol. 25*, 45–52.

HSIAO, T., HSIAO, C. and DE WILDE, J. (1975). Moulting hormone production in the isolated larval abdomen of the Colorado beetle. *Nature, 255*, 727–728.

HSIAO, T. H., HSIAO, C. and DE WILDE, J. (1976). Molting hormone titer changes and their significance during development of the Colorado beetle, *Leptinotarsa decemlineata*. *J. Insect Physiol. 22*, 1257–1261.

HUMBER, L. G., CHAPPEL, C. I., MARTON, A. V., KRAML, M. and DUBUC, J. (1966). The relationship between structure and activity in a novel series of serum cholesterol-lowering agents. *J. Med. Chem. Soc. 9*, 329–337.

HUTCHINS, R. F. N., THOMPSON, M. J. and SVOBODA, J. A. (1970). The synthesis and the mass and nuclear magnetic resonance spectra of side chain isomers of cholesta-5,22-dien-3β-ol and cholesta-5,22,24-trien-3β-ol. *Steroids 15*, 113–130.

IKAN, R., MARKUS, A. and BERGMANN, E. D. (1970). Synthesis of campesteryl acetate ((24*R*)-24-methyl-3β-acetoxycholesten-5-ene) and its 24*S*-epimer. *Steroids 16*, 517–522.

IKEKAWA, N., FUJIMOTO, Y., TAKASU, A. and MORISAKI, M. (1980a). Isolation of fucosterol epoxide from larvae of the silkworm, *Bombyx mori*. *J. Chem. Soc. Chem. Commun.* 709–711.

IKEKAWA, N., IKEDA, T., MIZUNO, T., OHNISHI, E. and SAKURAI, S. (1980b). Isolation of a new ecdysteroid, 2,22-dideoxy-20-hydroxyecdysone, from the ovaries of the silkworm, *Bombyx mori*. *J. Chem. Soc. Chem. Commun.* 448–449.

IKEKAWA, N., MORISAKI, M., OHTAKA, H. and CHIYODA, Y. (1971). Reaction of fucosterol 24,28-epoxide with boron trifluoride etherate. *J. Chem. Soc. Chem. Commun.* 1498.

IKEKAWA, N., SUZUKI, M., KOBAYASHI, M. and TSUDA, K. (1966). Studies on the sterol of *Bombyx mori* L. IV. Conversion of the sterol in the silkworm. *Chem. Pharm. Bull. 14*, 834–836.

ISAAC, R. E., DINAN, L. N., REES, H. H. and GOODWIN, T. W. (1984). Biosynthesis and metabolism of ovarian ecdysteroids in *Schistocerca gregaria*. In *Comparative endocrinology*. Edited by B. Lofts. Hong Kong University Press. (In press.)

ISAAC, R. E., REES, H. H. and GOODWIN, T. W. (1981). Isolation of 2-deoxy-20-hydroxyecdysone and 3-epi-2-deoxyecdysone from eggs of the desert locust, *Schistocerca gregaria*, during embryogenesis. *J. Chem. Soc. Chem. Commun.* 418–420.

ISAAC, R. E., ROSE, M. E., REES, H. H. and GOODWIN, T. W. (1982). Identification of ecdysone-22-phosphate and 2-deoxyecdysone-22-phosphate in eggs of the desert locust, *Schistocera gregaria*, by fast atom bombardment mass spectrometry and n.m.r. spectroscopy. *J. Chem. Soc., Chem. Commun.* 249–251.

ISHII, S., KAPLANIS, J. N. and ROBBINS, W. E. (1963). Distribution and fate of 4-^{14}C-cholesterol in the adult male American cockroach. *Ann. Ent. Soc. Amer. 56*, 115–119.

JANDA, V. (1933). Über die Entwicklung und den Sauerstoffverbrauch isolierter Fragmente der Puppen von *Tenebrio molitor* L. *Acta Soc. Entomol. Cechoslov. 30*, 95–96.

JOHNSON, P., COOK, I. F., REES, H. H. and GOODWIN, T. W. (1975). Mode of formation of cholesta-5,7-dien-3β-ol from cholest-5-en-3β-ol by larvae of *Calliphora erythrocephala*. *Biochem. J. 152*, 303–311.

JOHNSON, P. and REES, H. H. (1977). Biosynthesis of ecdysones: metabolism of 7-dehydrocholesterol in *Schistocerca gregaria*. *J. Insect Physiol. 23*, 1387–1392.

KAPLANIS, J. N., DUTKY, S. R., ROBBINS, W. E., THOMPSON, M. J., LINDQUIST, E. L., HORN, D. H. S. and GALBRAITH, M. N. (1975). Makisterone A: a 28-carbon hexahydroxy molting hormone from the embryo of the milkweed bug. *Science, 190*, 681–682.

KAPLANIS, J. N., MONROE, R. E., ROBBINS, W. E. and LOULOUDES, S. J. (1963). The fate of dietary H^3-β-sitosterol in the adult house fly. *Ann. Ent. Soc. Amer. 56*, 198–201.

KAPLANIS, J. N., ROBBINS, W. E. and TABOR, L. A. (1960). The utilization and metabolism of 4-^{14}C-cholesterol by the adult house fly. *Ann. Ent. Soc. Amer. 53*, 260–264.

KAPLANIS, J. N., ROBBINS, W. E., THOMPSON, M. J. and BAUMHOVER, A. H. (1969). Ecdysone analog: conversion to alpha ecdysone and 20-hydroxyecdysone by an insect. *Science 166*, 1540–1541.

KAPLANIS, J. N., THOMPSON, M. J., DUTKY, S. R., ROBBINS, W. E. and LINDQUIST, E. L. (1972a). The metabolism of 4-^{14}C-22,25-bisdeoxyecdysone during larval development in the tobacco hornworm, *Manduca sexta* (L.). *Steroids 20*, 105–120.

KAPLANIS, J. N., THOMPSON, M. J., ROBBINS, W. E. and LINDQUIST, E. L. (1972b). α-Ecdysone and 20-hydroxyecdysone from frass of the tobacco hornworm, *Manduca sexta*. *Steroids 20*, 621–625.

KARLSON, P. and HOFFMEISTER, H. (1963). Zur biogenese des ecdysons, I. Umwandlung von cholesterin in ecdyson. *Hoppe–Seyler's Z. Physiol. Chem. 331*, 298–300.

KARLSON, P. and STAMM-MENÉNDEZ, M. D. (1956). Notiz über den nachweis von metamorphose-hormon in den imagines von *Bombyx mori*. *Hoppe–Seyler's Z. Physiol. Chem. 305*, 109–111.

KING, D. S. (1972a). Metabolism of α-ecdysone and possible immediate precursors by insects *in vivo* and *in vitro*. *Gen Comp. Endocr. Suppl. 3*, 221–227.

KING, D. S. (1972b). Ecdysone metabolism in insects. *Amer. Zool. 12*, 343–345.

KING, D. S. and MARKS, E. P. (1975). The secretion and metabolism of α-ecdysone by cockroach *(Leucophaea maderae)* tissues *in vitro*. *Life Sci. 15*, 147–154.

KING, D. S., BOLLENBACHER, W. E., BORST, D. W., VEDECKIS, W. V., O'CONNOR, J. D., ITTYCHERIAH, P. I. and GILBERT, L. I. (1974). The secretion of α-ecdysone by the prothoracic glands of *Manduca sexta in vitro*. *Proc. Nat. Acad. Sci. USA 71*, 793–796.

KINNEAR, J. F., MARTIN, M.-D., CHONG, Y. K., FAUX, A., HORN, D. H. S. and WILKIE, J. S. (1978). Insect moulting hormones: possible intermediates in the metabolism of cholesterol to ecdysteroids in an insect. *Aust. J. Chem. 31*, 2069–2075.

KINNEAR, J. F., MARTIN, M.-D., FAUX, A. F., HORN, D. H. S. and WILKIE, J. S. (1979). Insect moulting hormones: a study of 4-en-3-one steroids as possible ecdysteroid precursors. *Aust. J. Chem. 32*, 2017–2024.

KIRCHER, H. W. and ROSENSTEIN, F. U. (1974). Preparation of 5, 7-ergostadien 3β-ol and its utilization by three *Drosophila* species. *J. Amer. Oil Chem. Soc. 51*, 525A.

KOOLMAN, J. (1980). Ecdysteroids in the blowfly, *Calliphora vicina*. In *Progress in Ecdysone Research*. Edited by J. A. Hoffmann. Pages 187–209. Elsevier/North-Holland Biomedical Press, Amsterdam.

LAGUEUX, M., SALL, C. and HOFFMANN, J. A. (1981). Ecdysteroids during embryogenesis in *Locusta migratoria*. *Amer. Zool. 21*, 715–726.

LAVERDURE, A.-M., LAGUEUX, M. and HOFFMANN, J. A. (1977). Ecdysone et développment ovarien chez la femelle adulte de *Tenebrio molitor*. *Bull. Soc. Zool. France, 102*, 311–312.

LOCKE, M. (1969). The ultrastructure of the oenocytes in the molt/intermolt cycle of an insect. *Tissue Cell, 1*, 103–154.

LOCKLEY, W. J. S., BOID, R., LLOYD-JONES, J. G., REES, H. H. and GOODWIN, T. W. (1975). Fate of the C-4 hydrogen atoms of cholesterol during its transformation into ecdysones in insects and plants. *J. Chem. Soc. Chem. Commun.*, 346–348.

LOCKLEY, W. J. S., REES, H. H. and GOODWIN, T. W. (1978). Synthesis of stereospecifically labelled [4-^{3}H] cholesterol. *J. Labelled Compounds and Radiopharmaceuticals, 15*, 413–423.

LOULOUDES, S. J., THOMPSON, M. J., MONROE, R. E. and ROBBINS, W. E. (1962). Conversion of cholestanol to Δ^7-cholestenol by the German cockroach. *Biochem. Biophys. Res. Commun. 8*, 104–106.

MAJUMDER, M. S. I. (1978). Ph.D. thesis, University of Liverpool.

MARTIN, M. M. and CARLS, G. A. (1968). The lipids of the common house cricket, *Acheta domesticus* L. III. Sterols. *Lipids 3*, 256–259.

MONROE, R. E., HOPKINS, T. L. and VALDER, S. A. (1967). Metabolism and utilization of cholesterol-4-^{14}C for growth and reproduction of aseptically reared houseflies, *Musca domestica* L. *J. Insect Physiol. 13*, 219–233.

MORISAKI, M., OHTAKA, H., AWATA, N. and IKEKAWA, N. (1974). Nutritional effect of possible intermediates of phytosterol dealkylation in the silkworm, *Bombyx mori*. *Steroids 24*, 165–176.

MORISAKI, M., OHTAKA, H., OKUBAYASHI, M. and IKEKAWA, N. (1972). Fucosterol-24,28-epoxide, as a probable intermediate in the conversion of β-sitosterol to cholesterol in the silkworm. *J. Chem. Soc. Chem. Commun.* 1275–1276.

MORISAKI, M., YING, B. and IKEKAWA, N. (1981). Identification of both fucosterol and isofucosterol in the silkworm, *Bombyx mori*. *Experientia 37*, 336–337.

MORRIS, L. J. (1970). Mechanisms and stereochemistry in fatty acid metabolism. *Biochem. J. 118*, 681–693.

MULLIN, C. A. and WILKINSON, C. F. (1980a). Purification of an epoxide hydrolase from the midgut of the southern armyworm *(Spodoptera eridania)*. *Insect Biochem. 10*, 681–691.

MULLIN, C. A. and WILKINSON, C. F. (1980b). Insect epoxide hydrolase: properties of a purified enzyme from the southern armyworm *(Spodoptera eridania)*. *Pestic. Biochem. Physiol. 14*, 192–207.

MUROTA, S., FENSELAU, C. C. and TALALAY, P. (1971). Partial purification of a beef adrenal Δ^5-3-ketosteroid isomerase and studies of its mechanism of action. *Steroids, 17*, 25–37.

NAKANISHI, K., MORIYAMA, H., OKAUCHI, T., FUJIOKA, S. and KOREEDA, M. (1972). Biosynthesis of α- and β-ecdysones from cholesterol outside the prothoracic gland in *Bombyx mori*. *Science, 176*, 51–52.

NICOTRA, F., PIZZI, P., RONCHETTI, F., RUSSO, G. and TOMA, L. (1981). Substrate stereospecificity in the metabolism of fucosterol and isofucosterol 24,28- epoxides in *Tenebrio molitor*. *J. Chem. Soc. Perkin I*, 480–483.

NICOTRA, F., RONCHETTI, F. and RUSSO, G. (1978). Conversion of β-sitosterol into both fucosterol and isofucosterol in *Tenebrio molitor*. *Experientia 34*, 699–700.

NICOTRA, F., RONCHETTI, F., RUSSO, G. and TOMA, L. (1979). Role of 24- and 28- hydroxylated intermediates in the metabolism of β-sitosterol in the insect, *Tenebrio molitor*. *Biochem. J. 183*, 495–499.

NUTTING, W. L. (1955). Extirpation of roach prothoracic glands. *Science 122*, 30–31.

OHNISHI, E., MIZUNO, T., CHATANI, F., IKEKAWA, N. and SAKURAI, S. (1977). 2-Deoxy-α-ecdysone from ovaries and eggs of the silkworm, *Bombyx mori*. *Science 197*, 66–67.

OHNISHI, E., MIZUNO, T., IKEKAWA, N. and IKEDA, T. (1981). Accumulation of 2-deoxy-ecdysteroids in ovaries of the silkworm *Bombyx mori*. *Insect Biochem. 11*, 155–159.

OHTAKA, H., MORISAKI, M. and IKEKAWA, N. (1973). Reaction of 24,28-epoxides of sterol side chain with boron trifluoride etherate. *J. Org. Chem. 38*, 1688–1691.

OHTAKI, T. (1972). A possible role of posterior half of larval body on the pupation of the fleshfly, *Sarcophaga peregrina*. *Jpn. J. Med. Sci. Biol. 25*, 33–41.

PETTLER, P. J., LOCKLEY, W. J. S., REES, H. H. and GOODWIN, T. W. (1974). Migration of the C-25 hydrogen of clionasterol to the C-24 position during dealkylation by the insect, *Tenebrio molitor*. *J. Chem. Soc. Chem. Commun.* 844–846.

PETTLER, P. J., LOCKLEY, W. J. S., REES, H. H. and GOODWIN, T. W. (1978). Mechanism of dealkylation of clionasterol in the insect *Tenebrio molitor*. *Biochem. J. 174*, 397–404.

PIEPHO, H. (1948). Zur Frage der Bildungsorgane des Häutungswirkstoffs bei Schmetterlingen. *Naturwissenschaften, 35*, 94–95.

RANDALL, P. J., LLOYD-JONES, J. G., COOK, I. F., REES, H. H. and GOODWIN, T. W. (1972). The fate of the C-25 hydrogen of 28-isofucosterol during conversion into cholesterol in the insect *Tenebrio molitor*. *J. Chem. Soc. Chem. Comm.* 1296–1298.

REES, H. H. (1971). Ecdysones. In *Aspects of Terpenoid Chemistry and Biochemistry*. Edited by T. W. Goodwin. Pages 181–22. Academic Press, London.

REES, H. H. (1984). Biosynthesis of steroids — comparative aspects. *Nova Acta Leopoldina, 56*, 267–295.

REES, H. H. (1983). Ecdysteroids. In *Biosynthesis of Isoprenoid Compounds*. Vol. 2. Edited by J. W. Porter and S. L. Spurgeon. Pages 463–505. Wiley, New York.

REES, H. H., DAVIES, T. G., DINAN, L. N., LOCKLEY, W. J. S. and GOODWIN, T. W. (1980). The biosynthesis of ecdysteroids. In *Progress in*

Ecdysone Research. Edited by J. A. Hoffmann. Pages 125–137. Elsevier/North-Holland Biomedical Press, Amsterdam.

REES, H. H., GREENWOOD, D. R., DINAN, L. N., ISAAC, R. E. and GOODWIN, T. W. (1981). Biosynthesis and metabolism of ecdysteroids. In *Regulation of Insect Development and Behaviour.* Edited by F. Sehnal, A. Zabza, J. J. Menn and B. Cymborowski. Pages 71–92. Wroclaw Technical University Press, Wroclaw, Poland.

RITTER, F. J. and WIENTJENS, W. H. J. M. (1967). Sterol metabolism in insects. *T.N.O. Nieuws 22*, 381–392.

ROBBINS, W. E., DUTKY, R. C., MONROE, R. E. and KAPLANIS, J. N. (1962). The metabolism of ^{3}H-β-sitosterol by the German cockroach. *Ann. Ent. Soc. Amer. 55*, 102–104.

ROBBINS, W. E., KAPLANIS, J. N., THOMPSON, M. J., SHORTINO, T. J. and JOYNER, S. C. (1970). Ecdysones and synthetic analogs: molting hormone activity and inhibitive effects on insect growth, metamorphosis and reproduction. *Steroids 16*, 105–125.

ROBBINS, W. E., THOMPSON, M. J., KAPLANIS, J. N. and SHORTINO, T. J. (1964). Conversion of cholesterol to 7-dehydrocholesterol in aseptically reared German cockroaches, *Steroids 4*, 635–644.

ROMER, F. (1971). Häutungshormone in den oenocyten des mehlkäfers. *Naturwissenschaften 58*, 324–325.

ROMER, F. (1973). Moulting hormone content in prothoracic glands and oenocytes of *Tenebrio molitor* within a moulting cycle. *Experientia 29*, 1417–1418.

ROMER, F., EMMERICH, H. and NOWOCK, J. (1974). Biosynthesis of ecdysones in isolated prothoracic glands and oenocytes of *Tenebrio molitor in vitro. J. Insect Physiol. 20*, 1975–1987.

RUHLAND, U. and ROMER, F. (1977). Nachweis von häutungsaktiven stoffen in isolierten prothorakaldrüsen und oenocyten bei *Bombyx mori* während des 5. larvenstadiums. *Wilhelm Roux Arch. Devel. Biol. 181*, 123–134.

SAKURAI, S., IKEKAWA, N., OHTAKI, T. and CHINO, H. (1977). 3β-Hydroxy-5α-cholestan-6-one: a possible precursor of α ecdysone biosynthesis. *Science 198*, 627–629.

SAMUELS, L. T. and EIK-NES, K. B. (1968). Metabolism of steroid hormones. In *Metabolic Pathways.* Edited by D. M. Greenberg. Vol. 2. Pages 169–220. Academic Press, New York.

SATO, H., ASHIDA, N., SUHARA, K., ITAGAKI, E., TAKEMORI, S. and KATAGIRI, M. (1978). Properties of an adrenal cytochrome P-450 (P-$450_{11\beta}$) for the hydroxylations of corticosteroids. *Arch. Biochem. Biophys. 190*, 307–314.

SCHAEFER, C. H., KAPLANIS, J. N., ROBBINS, W. E. (1965). The relationship of the sterols of the Virginia pine sawfly, *Neodiprion pratti.* Dyar, to those of two host plants *Pinus virginiana*, Mill and *Pinus rigida*, Mill. *J. Insect Physiol. 11*, 1013–1021.

SLAMA, K. (1975). Old concepts and new findings on hormonal control of morphogenesis in *Dermestes.* In *Coll. Int. CNRS No. 251. (Lille): Actualités sur les hormones d'invertébrés.* Pages 499–506. CNRS, Paris.

STELLWAAG-KITTLER, F. (1954). Zur physiologie der käferhäutung untersuchungen am mehlkäfer. *Tenebrio molitor* L. *Biol. Zentralbl. 73*, 12–49.

STUDINGER, G. and WILLIG, A. (1975). Biosynthesis of α and β-ecdysone in isolated abdomens of larvae of *Musca domestica. J. Insect Physiol. 21*, 1793–1798.

SVOBODA, J. A. and ROBBINS, W. E. (1967). Conversion of beta sitosterol to cholesterol blocked in an insect by hypocholesterolemic agents. *Science 156*, 1637–1638.

SVOBODA, J. A. and ROBBINS, W. E. (1968). Desmosterol as a common intermediate in the conversion of a number of C_{28} and C_{29} plant sterols to cholesterol by the tobacco hornworm. *Experientia 24*, 1131–1132.

SVOBODA, J. A. and ROBBINS, W. E. (1971). The inhibitive effects of azasterols on sterol metabolism and growth and development in insects with special reference to the tobacco hornworm. *Lipids 6*, 113–119.

SVOBODA, J. A., THOMPSON, M. J. and ROBBINS, W. E. (1967). Desmosterol, an intermediate in dealkylation of β-sitosterol. *Life Sci. 6*, 395–404.

SVOBODA, J. A., HUTCHINS, R. F. N., THOMPSON, M. J. and ROBBINS, W. E. (1969a). 22-*Trans*-cholesta-5,22,24-trien-3β-ol-an intermediate in the conversion of stigmasterol to cholesterol in the tobacco hornworm, *Manduca sexta*, (Johannson). *Steroids 14*, 469–476.

SVOBODA, J. A., WOMACK, M., THOMPSON, M. J. and ROBBINS, W. E. (1969b). Comparative studies on the activity of 3β-hydroxy-Δ^5-norcholenic acid on the Δ^{24}-sterol reductase enzyme(s) in an insect and the rat. *Comp. Biochem. Physiol. 30*, 541–549.

SVOBODA, J. A., THOMPSON, M. J. and ROBBINS, W. E. (1971). Identification of fucosterol as a metabolite and probable intermediate in conversion of sitosterol to cholesterol in the tobacco hornworm. *Nature New Biol. 230*, 57–58.

SVOBODA, J. A., ROBBINS, W. E., COHEN, C. F. and SHORTINO, T. J. (1972a). Phytosterol utilization and metabolism in insects: recent studies with *Tribolium confusum.* In *Insect and Mite Nutrition.* Edited by J. G. Rodriguez. Pages 505–516. North-Holland Publishing Co., Amsterdam.

SVOBODA, J. A., THOMPSON, M. J. and ROBBINS, W. E. (1972b). 24-Methylenecholesterol: isolation and identification as an intermediate in the conversion of campesterol to cholesterol in the tobacco hornworm. *Lipids 7*, 156–158.

SVOBODA, J. A., KAPLANIS, J. N., ROBBINS, W. E. and THOMPSON, M. J. (1975a). Recent developments in insect steroid metabolism. *Ann. Rev. Ent. 20*, 205–220.

SVOBODA, J. A., THOMPSON, M. J., ROBBINS, W. E. and ELDEN, T. C. (1975b). Unique pathways of sterol metabolism in the Mexican bean beetle, a plant-feeding insect. *Lipids 10*, 524–527.

SVOBODA, J. A., DUTKY, S. R., ROBBINS, W. E. and KAPLANIS, J. N. (1977). Sterol composition and phytosterol utilization and metabolism in the milkweed bug. *Lipids 12*, 318–321.

SVOBODA, J. A., THOMPSON, M. J., ROBBINS, W. E. and KAPLANIS, J. N. (1978). Insect steroid metabolism. *Lipids 13*, 742–753.

SVOBODA, J. A., NAIR, A. M. G., AGARWAL, N. and ROBBINS, W. E. (1980a). Lack of conversion of C_{29}-phytosterols to cholesterol in the khapra beetle, *Trogoderma granarium.* Everts. *Experientia 36*, 1029–1030.

SVOBODA, J. A., THOMPSON, M. J., HERBERT, E. W. (JR.) and SHIMANUKI, H. (1980b). Sterol utilization in honey bees fed a synthetic diet: analysis of prepupal sterols. *J. Insect Physiol. 26*, 291–294.

SVOBODA, J. A., HERBERT, E. W. (JR.), THOMPSON, M. J. and SHIMANUKI, H. (1981). The fate of radiolabelled C_{28} and C_{29} phytosterols in the honey bee. *J. Insect Physiol. 27*, 183–188.

SVOBODA, J. A., THOMPSON, M. J., HERBERT, E. W., SHORTINO, T. J. and SZCZEPANIK-VANLEEUWEN, P. A. (1982). Utilization and metabolism of dietary sterols in the honey bee and yellow fever mosquito. *Lipids 17*, 220–225.

THOMPSON, M. J., KAPLANIS, J. N., ROBBINS, W. E. and SVOBODA, J. A. (1973). Metabolism of steroids in insects. *Adv. Lipid Res. 11*, 219–265.

THOMPSON, M. J., KAPLANIS, J. N., WEIRICH, G. F., SVOBODA, J. A. and ROBBINS, W. E. (1981). Moulting hormones of the tobacco hornworm. In *Regulation of Insect Development and Behaviour.* Edited by F. Sehnal, A. Zabza, J. J. Menn and B. Cymborowski. Pages 107–124. Wroclaw Technical University Press, Wroclaw, Poland.

THOMPSON, M. J., ROBBINS, W. E., DUTKY, S. R., MARKS, E. P., FILIPI, P. A. and FINEGOLD, H. (1978). Isolation and identification of the metabolites of 22,25-dideoxyecdysone from cockroach fat body cultures. *Lipids 13*, 783–790.

THOMPSON, M. J., SVOBODA, J. A., KAPLANIS, J. N. and ROBBINS, W. E. (1972). Metabolic pathways of steroids in insects. *Proc. R. Soc. Lond. B., 180*, 203–221.

THOMSON, J. A., HAFFERL, W., GALBRAITH, M. N., HORN, D. H. S. and MIDDLETON, E. J. (1971). Possible biosynthetic precursors of β-ecdysone in *Calliphora stygia. J. Chem. Soc. Chem. Commun.* 1023–1024.

THOMSON, J. A., SIDDALL, J. B., GALBRAITH, M. N., HORN, D. H. S. and MIDDLETON, E. J. (1969). The biosynthesis of ecdysones in the blowfly *Calliphora stygia. J. Chem. Soc. Chem. Commun.* 669–670.

WEIR, S. B. (1970). Control of moulting in insects. *Nature 228*, 580–581.

WILLIG, A., REES, H. H. and GOODWIN, T. W. (1971). Biosynthesis of insect moulting hormones in isolated ring glands and whole larvae of *Calliphora. J. Insect Physiol. 17*, 2317–2326.

WILTON, D. C. and AKHTAR, M. (1970). The stereochemistry of hydrogen elimination during 7,8-double bond formation by *Tetrahymena pyriformis. Biochem. J. 116*, 337–339.

ADDENDUM

Further evidence has been furnished that the honeybee, *Apis mellifera*, is unable to dealkylate phytosterols at the C–24 position to produce cholesterol (Svoboda, J. *et al.*, 1983). Absence of phytosterol dealkylation has been established in *Dysdercus fasciatus* and the major ecdysteroid in fifth instar larvae has been identified as makisterone A (C_{28}), with no detectable C_{27} and C_{29} ecdysteroids (Gibson, J. *et al.*, 1983). Makisterone A has also been reported as a major ecdysteroid with little, if any, ecdysone and 20-hydroxyecdysone in the haemolymph of fifth-stage larvae of the heteropterans, *Nezara viridula*, *Podisus maculiventris*, *Dysdercus cingulatus*, and *Oncopeltus fasciatus* (Aldrich, J. *et al.*, 1982; Kelly, T. *et al.*, 1981).

In *Drosophila melanogaster*, it is likely that the ring gland is not a major source of ecdysteroids during pupal development at a time when the endogenous ecdysteroid titre is increasing (Redfern, C., 1983). However, in *Manduca sexta* pupae, the prothoracic glands are a major source of ecdysone (Bowen, M. *et al.*, 1983). The role of prothoracic glands in ecdysteroid production during the metamorphosis of holometabolous insects has been reviewed (Redfern, C. and Bowes, M., 1984).

The four major ecdysteroid conjugates present in the newly laid eggs of the desert locust, *Schistocerca gregaria* have been identified as 22-phosphate esters (Isaac, R. *et al.*, 1983).

REFERENCES

Aldrich, J. R., Kelly, T. J. and Woods, C. W. (1982). Larval moulting hormone of Trichophoran Hemiptera-Heteroptera: Makisterone A, not 20-hydroxyecdysone. *J. Insect Physiol. 28*, 857–861.

Bowen, M. F., Bollenbacher, W. E. and Gilbert, L. I. (1983). *In vitro* studies on the role of the brain and prothoracic glands in the pupal diapause of *Manduca sexta*. *J. Exp. Biol. 108*, 9–24.

Gibson, J. M., Majumder, M. S. I., Mendis, A. H. W. and Rees, H. H. (1983). Absence of phytosterol dealkylation and identification of the major ecdysteroid as makisterone A in *Dysdercus fasciatus* (Heteroptera, Pyrrhocoridae). *Arch. Insect Biochem. Physiol. 1*, 105–120.

Isaac, R. E., Rose, M. E., Rees, H. H. and Goodwin, T. W. (1983). Identification of the 22-phosphate esters of ecdysone, 2-deoxyecdysone, 20-hydroxyecdysone and 2-deoxy-20-hydroxyecdysone from newly laid eggs of the desert locust, *Schistocerca gregaria*. *Biochem. J. 213*, 533–541.

Kelly, T. J., Woods, C. W., Redfern, R. E., and Borkovec, A. B., (1981). Makisterone A: the moulting hormone of larval *Oncopeltus*? J. Exp. Zool. 218, 127–132.

Redfern, C. P. F. (1983). Ecdysteroid synthesis by the ring gland of *Drosophila melanogaster* during late-larval, prepupal and pupal development. *J. Insect Physiol. 29*, 65–71.

Redfern, C. P. F. and Bownes, M. (1984). Prothoracic glands and the production of steroid hormones during the metamorphosis of holometabolous insects. (in press).

Svoboda, J. A., Herbert, E. W. (Jr.) and Thompson, M. J. (1983). Definitive evidence for lack of phytosterol dealkylation in honey bees. *Experientia 39*, 1120–1121.

8 Regulation of Ecdysteroid Titer: Synthesis

STAN L. SMITH

Bowling Green State University, Bowling Green, Ohio, USA

1 INTRODUCTION

Hormones are chemical regulatory agents which elicit specific physiological responses in specific target tissues. As such, hormones may function in the homeostatic regulation of physiological processes and/or may serve as signals for the initiation or regulation of developmental events. In insects, the postembryonic developmental events — molting and metamorphosis — are known to be controlled by at least three major types of hormones; namely, the prothoracicotropic hormones (PTTH), the ecdysteroids or molting hormones (MH), and the juvenile hormones (JH). More recent studies have also indicated that ecdysteroids and JH may be implicated in reproductive and embryogenic events. It follows that for such events to occur normally the titers of each of the hormones involved, as well as the interplay amongst these titers, must be choreographically precise. To understand the physiological and developmental significance of these insect hormone titers it is essential that we also understand the biochemical events and processes which contribute to, and are reflected by, the hormone titer; e.g. hormone synthesis, release, transport, storage, degradation, and excretion, as well as the factors which regulate each of these contributing processes. This chapter will be concerned specifically with an overview of ecdysteroid titers and ecdysteroid titer regulation; particular emphasis will be placed on 20-hydroxyecdysone biosynthesis during insect postembryonic development.

2 AN OVERVIEW OF ECDYSTEROIDS

Ecdysteroids were traditionally viewed as being molting hormones (MH) in insects based on their biological activity in several MH bioassays (see Karlson, P., 1980). Utilizing one such MH bioassay, the *Calliphora* test (Fraenkel, G., 1935), studies on the chemistry of MH were initiated in the 1940s (see V. Wigglesworth, this volume). However, it was not until the mid-1950s that the first two MH were isolated in crystalline form from *Bombyx mori* pupae (Butenandt, A. and Karlson, P., 1954; Karlson, P., 1956). These two MH were provisionally termed α- and β-ecdysone but are now designated as ecdysone and 20-hydroxyecdysone, respectively (Goodwin, T. *et al.*, 1978). Based on several derivatizations, nuclear magnetic resonance spectroscopy, mass spectrometry, X-ray structural analysis and chemical synthesis, the structures of these MH were eventually and unequivocally determined by the mid-1960s (Huber, R. and Hoppe, W., 1965; Karlson, P. *et al.*, 1965; Hoffmeister, H. and Grützmacher, H., 1966; Kerb, U. *et al.*, 1966; Siddall, J. *et al.*, 1966; Dammeier, B. and Hoppe, W., 1971). Both MH proved to be steroids which exhibit a number of striking peculiarities in comparison with vertebrate steroid hormones. Both ecdysone and 20-hydroxyecdysone possess the full C_{27} carbon skeleton (cholestane) of cholesterol, exhibit a *cis* fusion (5β) of the A and B rings, have two functional moieties in the B ring (a 6-keto-7-ene system), and possess hydroxyl groups at several positions rendering these molecules appreciably water-soluble (see D. Horn and R. Bergamasco, this volume). Accordingly, ecdysone is 2β,3β,14α,22R,25-pentahydroxy-5β-cholest-7-en-6-one and 20-hydroxyecdysone is 2β,3β,14α,20R,22R,25-hexahydroxy-5β-cholest-7-en-6-one (Fig. 1).

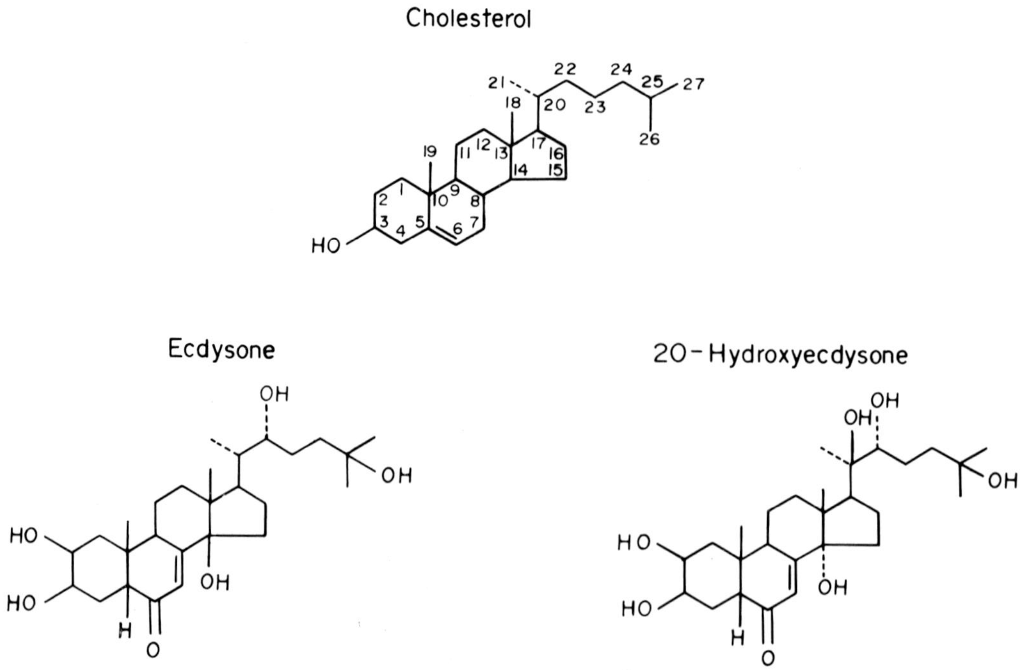

FIG. 1. Structures of cholesterol, ecdysone and 20-hydroxyecdysone.

Following the structural characterization of the first two ecdysteroids, progress in the chemistry of these steroids has been very rapid. We now know that ecdysone and 20-hydroxyecdysone are simply the more common members of a larger group of naturally occurring or structurally related polyhydroxylated ketosteroids generically referred to as ecdysteroids (Goodwin, T. *et al.*, 1978). These steroids are probably ubiquitous in arthropods and have also been isolated from species in other animal phyla such as annelids, flat-worms, molluscs and nematodes (see Gilbert, L. *et al.*, 1977; Nirde, Ph. *et al.*, 1983). To date 20 or so ecdysteroids have been tentatively or unequivocally identified as natural products in insects or crustaceans (Table 1). In addition to these zooecdysteroids, at least 60 structurally distinct ecdysteroids (phytoecdysteroids) have been isolated from numerous and phylogenetically diverse families of plants, in most cases at much higher concentrations than in arthropods (Horn, D., 1971; Nakanishi, K., 1971; Rees, H., 1971; Hetru, C. and Horn, D., 1980).

Along with the progress in ecdysteroid chemistry, significant advances have also occurred recently in our understanding of the biochemistry and biology of ecdysteroids. However, despite this progress, and

Table 1: *Insect ecdysteroids*

Ecdysteroid	Organism	References
Ecdysone (α-ecdysone)	*Bombyx mori* pupae	Butenandt, A. and Karlson, P., 1954; Huber, R. and Hoppe, W., 1965; Karlson, P. *et al.*, 1965
20-Hydroxyecdysone (β-ecdysone, ecdysterone)	*Bombyx mori* pupae	Karlson, P., 1956; Hoffmeister, H. and Grützmacher, H., 1966; Dammeier, B. and Hoppe, W., 1971
20,26-Dihydroxyecdysone	*Manduca sexta* pupae	Thompson, M. *et al.*, 1967
3-Dehydroecdysone*	*Calliphora erythrocephala* pupae	Karlson, P., *et al.*, 1972
26-Hydroxyecdysone	*Manduca sexta* eggs, *Calliphora vicina* larvae	Kaplanis, J., *et al.*, 1973; Koolman, J. *et al.*, 1979
3-Dehydro-20-hydroxyecdysone*	*Locusta migratoria* larvae	Hoffmann, J. *et al.*, 1974
3-Epiecdysone	*Manduca sexta* prepupae and pupae	Nigg, H. *et al.*, 1974; Kaplanis, J. *et al.*, 1979
3-Epi-20-hydroxyecdysone	*Manduca sexta* meconium	Thompson, M. *et al.*, 1974
24-Methyl-20-hydroxyecdysone (makisterone A)	*Oncopeltus fasciatus* eggs and larvae	Kaplanis, J. *et al.*, 1975; Kelly, T. *et al.*, 1981
2-Deoxyecdysone	*Bombyx mori* ovaries and eggs, *Galleria mellonella* pharate adults	Ohnishi, E. *et al.*, 1977; Bollenbacher, W. *et al.*, 1978
2,22-Dideoxyecdysone	*Locusta migratoria* ovaries	Hetru, C. *et al.*, 1978
2,22,25-Trideoxyecdysone	*Locusta migratoria* ovaries	Hetru, C. *et al.*, 1978
2,14,22,25-Tetradeoxyecdysone	*Locusta migratoria* ovaries	Hetru, C. *et al.*, 1978
3-Epi-20,26-dihydroxyecdysone	*Manduca sexta* pupae	Kaplanis, J. *et al.*, 1979
25-Deoxy-20,26-dihydroxyecdysone* (inokosterone)	*Calliphora vicina* larvae	Koolman, J. *et al.*, 1979
2,22-Dideoxy-20-hydroxyecdysone	*Bombyx mori* ovaries	Ikekawa, N. *et al.*, 1980
3-Epi-26-hydroxyecdysone	*Manduca sexta* eggs	Kaplanis, J. *et al.*, 1980a
2-Deoxy-20-hydroxyecdysone	*Schistocerca gregaria* eggs	Isaac, R. *et al.*, 1981a
3-Epi-2-deoxyecdysone	*Schistocerca gregaria* eggs	Isaac, R. *et al.*, 1981a
Ecdysone-3-acetate	*Schistocerca gregaria* eggs	Isaac, R. *et al.*, 1981b
Ecdysone-22-phosphate	*Schistocerca gregaria* eggs	Isaac, R. *et al.*, 1982
2-Deoxyecdysone-22-phosphate	*Schistocerca gregaria* eggs	Isaac, R. *et al.*, 1982
3-Epi-2-deoxyecdysone-3-phosphate	*Locusta migratoria* eggs	Tsoupras, G. *et al.*, 1982
2-Deoxyecdysone-22-adenosine monophosphate	*Locusta migratoria* eggs	Tsoupras, G. *et al.*, 1982
Ecdysone-22-N^6-(isopentenyl)-adenosine monophosphate	*Locusta migratoria* eggs	Tsoupras, G. *et al.*, 1983
25-Deoxy-20-hydroxyecdysone (ponasterone A)	isolated from the crustaceans, *Callinectes sapidus* and *Gecarcinus lateralis*	McCarthy, J., 1979

The zooecdysteroids listed in this table have been isolated either as natural products or as products formed by insect enzyme preparations (*) incubated with naturally occurring ecdysteroid substrates (i.e. ecdysone or 20-hydroxyecdysone). The only crustacean ecdysteroid listed, ponasterone A, has not been isolated to date from insects. References are listed on the basis of primogenitorship of reporting and/or physicochemical characterization of the ecdysteroid.

in some cases because of it, many longstanding and recent questions concerning ecdysteroids remain to be answered.

The increasing number of ecdysteroids which have been identified as natural products in insects, as well as the recent findings that these compounds are present in some insect adults and embryonated eggs (Table 1), suggest that the traditional view of ecdysteroids as MH may have to be expanded in the near future as the functions of these bioreactive steroids are more clearly delineated. The recent discoveries of ecdysteroid cell receptors which do not completely follow the vertebrate paradigm (Yund, M. *et al.*, 1978; Maróy, P. *et al.*, 1978; Bonner, J., 1982; see Yund, M. and Osterbur, D., this volume) suggest that much more needs to be known about the interactions between ecdysteroids and their receptors, the properties of these receptors and, *in extenso*, the factors that influence receptor synthesis and degradation. Lastly, but by no means exhaustively, another area which needs to be studied more intensively at a concerted level is the nature and interactions of the environmental, neural and endocrine factors which contribute to the regulation of ecdysteroid metabolism and the ecdysteroid titer. It is to this last area that this chapter is addressed.

3 ECDYSTEROID TITERS AND INSECT DEVELOPMENT–REPRODUCTION

While the postembryonic development of insects has been observed with awe and curiosity since antiquity, only within the last few decades have we come to an understanding of the endocrine nature of the processes and factors responsible for the regulation of molting and metamorphosis. Based on the pioneering studies by Kopeć, S. (1917, 1922), Wigglesworth, V. (1934, 1936, 1940), Fukuda, S. (1940a,b, 1944), Williams, C. (1947, 1948, 1952), and others (see Wigglesworth, V., this volume), there emerged by the early 1950s a basic model of the endocrine control of insect postembryonic development. According to this basic model, the brain in response to appropriate environmental stimuli synthesizes and releases via neurohemal organs (corpora cardiaca or corpora allata) a brain hormone(s). This brain hormone, now referred to as prothoracicotropic hormone (PTTH), stimulates the prothoracic glands (PG) or analogous structures (ring gland, ventral glands) to synthesize and secrete MH, which initiates the events generically referred to as a molt. However, the character of a molt, i.e. whether it is larval–larval, larval–pupal or pupal–adult, is dictated by the relative titer of juvenile hormone (JH) which is synthesized by the corpora allata (CA) and released at or before the molt. The basic model hypothesized that molting in the presence of a high titer of JH would result in a larval–larval molt, whereas molting in the presence of a reduced or absent titer of JH would result in a larval–pupal or pupal–adult molt, respectively. It is a tribute to the scientists who contributed to this basic model, using such classical endocrinological techniques as extirpation, implantation, parabiosis and ligation, that their basic tenets have not only withstood the test of time but have been reaffirmed consistently by more recent studies which utilized sophisticated microchemical and instrumental techniques.

Although early studies demonstrated that exogenous ecdysone administered to insects could initiate molting and metamorphosis (see Karlson, P., 1980), one of the implicit assumptions of the basic model, i.e. that MH titers fluctuated during insect postembryonic development, was not supported by direct experimental evidence until the early to mid-1960s. At this time Burdette, W. (1962) and Shaaya, E. and Karlson, P. (1965a,b), using the *Calliphora* bioassay, succeeded in demonstrating that MH titers varied during larval–pupal–adult development of *Calliphora erythrocephala* and *Bombyx mori*. In accord with the basic model, their data suggested that an increase or peak in ecdysteroid activity preceded or occurred concomitantly with the molt.

Despite the potential significance of these initial titer studies real progress in the determination of insect ecdysteroid titers was not forthcoming for several years. In part this was due to various limitations imposed by the insect materials themselves and to the lack of sufficiently sensitive methods for the accurate detection and quantification of ecdysteroids. With the advances and development in the early 1970s of newer and more sensitive methods for the microchemical analysis of ecdysteroids, investigations on the titers of these compounds proliferated exponentially.

Although these more recent titer studies have contributed to our understanding of the biology of ecdysteroids, it is important to recognize that the titer data themselves are intrinsically complex, difficult to interpret with certainty, and may be as much a reflection of several methodological considerations as of the hormonal status of the animal. Accordingly, before discussing the general characteristics of ecdysteroid titers during the insect life cycle, it is appropriate to present a brief review of ecdysteroid titer methodology.

3.1 Ecdysteroid titer methodology

Ecdysteroids as hormones are titered for the same reasons as vertebrate hormones. The titer data give us an approximation of the physiological concentrations of the hormone bathing or in the tissues and the changes in these concentrations during development or adult life. Hormone titer data are important among other things, therefore, both for the design of *in vitro* experiments and for clarifying suspect times during the life cycle when the hormone may be eliciting its physiological effects.

The value of hormone titer data is predicated on the thoroughness of the titer determination. There are four essential aspects to all hormone titers: (1) the tissue (usually hemolymph or blood) or system being assayed; (2) the qualitative composition of the titer, i.e. the chemical form(s) of the hormone present; (3) the quantitative composition of the titer, i.e. the total amounts of each of the hormones present; and (4) the temporal profile of the titer, i.e. the developmental or physiological times and durations of the fluctuations in the titer.

In contrast to vertebrates, insects pose several unique disadvantages with respect to determining hormone titers and interpreting titer data. Firstly, both the phylogeny and diversity of insects makes extrapolation of titer data from one species to another more tenuous than in vertebrates. Secondly, the small size of insects imposes severe restrictions with respect to determining all the aspects of a hormone titer. Although ecdysteroids are present in insects usually in concentrations ranging from 10^{-5} to 10^{-8} M or g/g fresh weight, approximately one to four orders of magnitude greater than vertebrate steroid hormone titers, the small size of many insects makes it extremely difficult to collect sufficient hemolymph or other specific tissues for accurate titer determinations. In many situations, therefore, ecdysteroid titers have been determined using whole-body extracts. The obvious disadvantage with this is that the resultant titer data may not reflect the hemolymph or specific tissue titer, as the different amounts and types of ecdysteroids in the various tissues, or excreted into the gut, may contribute to this whole-body titer (Kaplanis, J. *et al.*, 1980b; see J. Koolman and P. Karlson, this volume).

A third major disadvantage with respect to determining the ecdysteroid titer is the relatively short life cycle of insects compared to vertebrates. Given this abbreviated life span, interrupted by one to several molts depending on the species, any fluctuations in the ecdysteroid titer may and do occur somewhat rapidly in time. Accordingly, for accurate determinations of ecdysteroid titers it is necessary that precisely staged animals be used, that the titer determinations be made sufficiently close in time so as not to miss any fluctuations, and that sufficiently large numbers of animals be assayed at each stage so that an accurate estimate of the population variability can also be determined. Unfortunately, as with determinations of ecdysteroid titers in whole-body extracts, in many insect studies these aspects of the titer determination are either not feasible in practice or have not been taken into account.

In addition to the aforementioned factors, the methods used to determine the ecdysteroid titer are extremely important. Inasmuch as several excellent reviews on the microdetermination of ecdysteroids have recently been published, only salient features of these methods will be presented here (see Gilbert, L. *et al.*, 1977; Chang, E. and O'Connor, J., 1979; Morgan, E. and Wilson, I., 1980; Delaage, M. *et al.*, 1982).

The various methods most commonly used to determine ecdysteroid titers include: (1) bioassays; (2) radioimmunoassays; (3) simple chromatographic techniques such as thin-layer chromatography; and (4) instrumental methods such as high-pressure liquid chromatography, gas–liquid chromatography, and mass spectrometry/fragmentography. Ideally, since no one category of these methods meets all the criteria of a perfect assay (i.e. accurate, precise, sensitive, specific, rapid, simple and inexpensive), determinations of ecdysteroid titers should be based on experimental regimens which incorporate two or more complementary types of methods. In addition, the methods used to extract, concentrate and quantify ecdysteroids should incorporate procedures for determining an index of recovery as well as assessing the degree to which contaminants may interfere with the determination of the titer.

3.1.1 Bioassay

Bioassay, as a means of quantifying ecdysteroids, is the oldest method. Most ecdysteroid bioassays are based on the *Calliphora* test developed by Fraenkel, G. (1935) and refined by Karlson, P. and Shaaya, E. (1964). The most commonly used organisms in these bioassays are *Calliphora erythrocephala* and *Calliphora stygia* (Karlson, P. and Shaaya, E., 1964; Thomson, J. *et al.*, 1970), *Musca domestica* (Kaplanis, J. *et al.*, 1966b), *Sarcophaga peregrina* (Ohtaki, T. *et al.*, 1967) and *Chilo suppressalis* (Sato, Y. *et al.*, 1968; Agui, N., 1973). With the exception of the *Chilo in vitro* test (Agui, N., 1973), these bioassays are systemic, assessing and quantifying ecdysteroids on the basis of their ability to elicit pupariation, sclerotization and tanning when injected into ligated last larval instar dipteran abdomens. The advantage of such bioassays is their sensitivity; the *Musca* test is sensitive down to approximately 5–6 ng of 20-hydroxyecdysone per abdomen. In addition to their sensitivity, these bioassays have proved essential in determining the structural properties of ecdysteroids which are necessary for high MH activity, viz., the full C_{27} skeleton, *cis* fusion of the A-B rings, the 6-keto-7-ene system, and the 3β, 14α and 22R hydroxyl groups (see Bergamasco, R. and Horn, D., 1980). Nevertheless, as a means of quantifying ecdysteroids, bioassays have many disadvantages. They are expensive, time-consuming, quantitatively imprecise and qualitatively non-specific. Since these bioassays measure biological activity rather than the intrinsic chemical properties of the hormone administered they fail to distinguish different structural forms of ecdysteroids with the same MH activity; in addition, one can never be sure whether the bioassay result is due to the injected test compound, a subsequent metabolite(s) or the sum of these. Given these inherent limitations, bioassays have been replaced for the most part by more modern methods of quantification.

3.1.2 Radioimmunoassay

The development and proliferation of ecdysteroid radioimmunoassays (RIA) has proved invaluable in the quantification of these steroids (Borst, D. and O'Connor, J., 1972; Chang, E. and O'Connor, J., 1979; Hirn, M. and Delaage, M., 1980; Delaage, M. *et al.*, 1982). Presupposing that one has antibodies prepared against an ecdysteroid immunogen, a radiolabelled ecdysteroid of high specific activity as the labelled ligand, an unlabelled ecdysteroid as the standard, and access to a liquid scintillation counter, RIA is a fairly simple, precise, rapid and sensitive method for quantifying ecdysteroids.

The RIA method is essentially a competitive protein-binding assay in which unlabelled ecdysteroids (both standards and biological samples) compete with labelled ecdysteroids for a finite number of binding sites on the antibodies. As little as 2–20 pg of hormone can be detected with some RIA. Despite its advantages, however, there are limitations and factors which must be considered when using this method. In the first place, the specificity, affinity and sensitivity of ecdysteroid antibodies to the various ecdysteroids and related steroids may vary considerably depending on both the immunogen used to generate the antibody and the particular antiserum bleed for any given immunogen. Accordingly, all of these parameters should be ascertained with every batch of antiserum. Secondly, even with a "well-defined antibody" the RIA method cannot be used by itself to determine which ecdysteroids are responsible for the immunoreactivity. RIA activity can only be expressed in standard ligand equivalents. It is essential, therefore, that the RIA method be used in conjunction with a chromatographic technique so as to identify both the types and amounts of each ecdysteroid present.

3.1.3 Chromatographic and instrumental methods

Several chromatographic and instrumental methods have been used extensively both for the separation and characterization of ecdysteroids and, usually in conjunction with RIA, for determining ecdysteroid titers. The simplest and least expensive of these methods is thin-layer chromatography (TLC), with the ecdysteroids being identified on the basis of R_f values. Numerous solvent systems for TLC have been reported, the most common being a solution of chloroform: 95% ethanol (4 : 1, v/v); the stationary phases for TLC are usually silica or alumina (Morgan, E. and Poole, C., 1976b; Morgan, E. and Wilson, I., 1980). Although TLC does provide a useful means of separating ecdysteroids on the basis of polarity, it is the least sensitive method for their quantification with a limiting value of approximately 500 ng. In addition to this limitation, TLC usually requires extensive pre-clean-up of biological samples before running, and caution must be exercised in identifying ecdysteroids on the basis of R_f values alone.

In contrast to TLC, high-pressure liquid chromatography (HPLC) offers a number of advantages for the detection, quantification and characterization of ecdysteroids (Morgan, E. and Poole, C., 1976b; Gilbert, L. *et al.*, 1977; Lafont, R. *et al.*, 1980a; Morgan, E. and Wilson, I., 1980). Despite the initial set-up expense, HPLC is a fairly rapid, precise and sensitive method. Since the detection and quantification of ecdysteroids by HPLC is usually based on ultraviolet (UV) absorption by their non-specific 6-keto-7-ene-system ($\lambda_{max} \sim$ 242 nm; $\varepsilon \sim$ 12,400), ecdysteroid derivative formation is not required and these compounds can be detected in amounts as small as 10 ng with sensitive UV detectors. Although ecdysteroids are non-specifically identified on the basis of retention times with HPLC, the judicious use of both normal and reversed-phased separations will usually lead to an accurate identification of most free ecdysteroids.

Gas–liquid chromatography (GLC) has also been used for several years as a method of separating and quantifying ecdysteroids (Morgan, E. and Poole, C., 1976b; Gilbert, L. *et al.*, 1977; Morgan, E. and Wilson, I., 1980; Lafont, R. *et al.*, 1980a). A distinct disadvantage of this method, in addition to the initial set-up expense, is that the ecdysteroids must be derivatized to less polar compounds, usually trimethylsilyl (TMS) ethers, which precludes recovery of the native steroid for further analysis or study. However, GLC does offer several advantages over other chromatographic techniques. When interfaced with flame ionization detection (FID), GLC is sensitive down to about 50 ng of derivatized 20-hydroxyecdysone and this sensitivity can be enhanced down to approximately 5 pg of hormone (e.g. hexakis-TMS-20-hydroxyecdysone) when GLC is interfaced with electron capture detection (ECD). An added advantage to GLC is that it can also be interfaced with a mass spectrometer (MS) as the chromatographic detector. Mass spectrometry, unlike FID, ECD and UV absorption, is fairly specific, gives an accurate determination of molecular mass, and can be sensitive down to the picogram or femtogram level with single or multiple ion monitoring, i.e. mass fragmentography (MF).

3.2 Ecdysteroid titers during postembryonic development

3.2.1 Temporal and quantitative profiles of ecdysteroid titers

Utilizing one or more of the previously mentioned methods, ecdysteroid titers have been measured in numerous species of insects during various stages of postembryonic development. Some of the more major species for which such ecdysteroid titers have been determined are listed in Table 2 (see L. Riddiford, vol. 8).

Although the data from these titer determinations vary somewhat in detail, several generalizations and patterns have emerged from these studies. For one thing, it appears that ecdysteroids are always present during insect postembryonic development, usually in concentrations ranging from 10^{-5} to 10^{-8} M or g/g fresh weight, depending on the species and stage of development. Within any given species the ecdysteroid titer fluctuates considerably during postembryonic development, varying from basal levels ($\sim 10^{-8}$ M) by a factor of from 20 to 30 (e.g. in *Drosophila melanogaster*) up to approximately 10^2–10^3 (e.g. in *Manduca sexta* and *Schistocerca gregaria*) (see Dean, R. *et al.*, 1980; Richards, G., 1981).

The dramatic variations in the ecdysteroid titer during insect postembryonic development do not appear to occur in a random fashion. At least one major peak in the ecdysteroid titer has been found during every insect larval and pupal stadium examined (Figs 2–4). For the most part, these major

Table 2: Major species of insects for which ecdysteroid titers have been determined during postembryonic development

Species	Stages/components assayed	Assay methods	References
Aeshna cyanea	penultimate and last instar larvae; hemolymph	RIA; HPLC	Schaller, F. and Charlet, M., 1980
Gryllus bimaculatus	8th-instar larvae; whole body	*Calliphora* bioassay;	Maas, H., 1974
Schistocerca gregaria	all five instar larval stages; whole body	GLC-ECD	Morgan, E. and Poole, C., 1976a; Wilson, I. and Morgan, E., 1978; Gande, A. *et al.*, 1979
Locusta migratoria	4th- and 5th-instar larvae; whole body and hemolymph	*Calliphora* bioassay; TLC; RIA	Hoffmann, J. *et al.*, 1974; Baehr, J.-C. *et al.*, 1979; Hirn, M. *et al.*, 1979
Blattela germanica	5th- and 6th-instar larvae; whole body	*Musca* bioassay	Masner, P. *et al.*, 1975
Periplaneta americana	6th- and last-instar larvae; whole body and hemolymph	*Calliphora* bioassay; RIA; TLC	Shaaya, E., 1978; Gersch, M., 1980
Oncopeltus fasciatus	last-instar larvae; whole body and hemolymph	*Calliphora* bioassay; RIA; HPLC	Feir, D. and Winkler, G., 1969; Kelly, T. *et al.*, 1981; Redfern, R. *et al.*, 1982
Rhodnius prolixus	5th-instar larvae; hemolymph	RIA; TLC; HPLC; MS	Steel, C. *et al.*, 1982
Leptinotarsa decemlineata	4th-instar larvae and pupae; whole body	*Musca* bioassay; RIA	Hsiao, T. *et al.*, 1976; DeWilde, J. *et al.*, 1980
Tenebrio molitor	last-instar larvae, pupae and pharate adults; hemolymph	RIA; GLC-MF/MS	Delbecque, J.-P. *et al.*, 1975, 1978a,b; Delachambre, J. *et al.*, 1979, 1980; Glitho, I. *et al.*, 1979
Xyleborus ferrugineus	pupae and pharate adults; whole body	RIA; HPLC	Rao, K. *et al.*, 1981
Calliphora vicina (erythrocephala)	all three instar larval stages, pupae and pharate adults; whole body and hemolymph	*Calliphora* bioassay; RIA; TLC	Shaaya, E. and Karlson, P., 1965a; Berreur, P. *et al.*, 1979a; Koolman, J. 1980
Drosophila melanogaster	all three instar larval stages, pupae and pharate adults; whole body and hemolymph	RIA; HPLC; TLC; GLC-ECD	Borst, D. *et al.*, 1974; De Reggi, M. *et al.*, 1975; Hodgetts, R. *et al.*, 1977; Garen, A. *et al.*, 1977; Berreur, P. *et al.*, 1979b; Klose, W. *et al.*, 1980; Kraminsky G. *et al.*, 1980; Maróy, P. *et al.*, 1980; Handler, A. 1982
Chironomus thummi	3rd- and 4th-instar larvae, pupae; whole body	RIA; TLC; HPLC	Valentin, M. *et al.*, 1978
Drosophila hydei	3rd-instar larvae; whole body	RIA; GLC	Beckers, C. and Emmerich, H., 1976
Sarcophaga crassipalpis	puparia and pupae; whole body	*Musca* bioassay	Walker, G. and Denlinger, D., 1980
Sarcophaga bullata	larviposition to adult eclosion; whole body	RIA	Wentworth, S. *et al.*, 1981
Stomoxys calcitrans	puparia; whole body	HPLC	O'Neill, M. *et al.*, 1977
Bombyx mori	all five instar larval stages, pupae and pharate adults; whole body and hemolymph	*Calliphora* bioassay; *Sarcophaga* bioassay; RIA; TLC; HPLC	Burdette, W., 1962; Shaaya, E. and Karlson, P., 1965b; Hanaoka, K. and Ohnishi, E., 1974; Calvez, B. *et al.*, 1976; Coulon, M., 1977; Coulon, M. and Feyereisen, R., 1977; Coulon, M. *et al.*, 1979; Calvez, B., 1981; Kiguchi, K. and Agui, N., 1981
Calpodes ethlius	5th-instar larvae, pupae and pharate adults; hemolymph	RIA; TLC; HPLC; MS	Dean, R. *et al.*, 1980
Cerura vinula	last-instar larvae and pupae; whole body	*Calliphora* bioassay	Shaaya, E. and Karlson, P., 1965b
Choristoneura fumiferana	last-instar larvae; whole body	*Calliphora* bioassay	Lagueux, M. *et al.*, 1976

Table 2: Continued

Species	Stages/components assayed	Assay methods	References
Galleria mellonella	penultimate- and last-instar larvae, pupae and pharate adults; whole-body and hemolymph	*Musca* bioassay; RIA; TLC; GLC-ECD	Hsiao, T. and Hsiao, C., 1977; Bollenbacher, W. *et al.*, 1978; Maróy, P. and Tarnóy, K., 1978; Rohner, R. and Meyer, D., 1980; Sehnal, F. *et al.*, 1981
Heliothis zea	pupae and pharate adults; whole body	HPLC	Holman, G. and Meola, R., 1978
Heliothis virescens	5th-instar larvae, pupae and pharate adults; hemolymph	RIA	Loeb, M., 1982
Manduca sexta	4th- and 5th-instar larvae, pupae and pharate adults; whole body and hemolymph	*Musca* bioassay; RIA; TLC; HPLC; MS	Kaplanis, J. *et al.*, 1966a; Bollenbacher, W. *et al.*, 1975, 1981; Wielgus, J. *et al.*, 1979; Smith, S. *et al.*, 1983
Philosamia cynthia	4th- and 5th-instar larvae, and pupae; hemolymph	RIA	Calvez, B. *et al.*, 1976
Pieris brassicae	last-instar larvae, pupae, and pharate adults; hemolymph	RIA; GLC; MS	Lafont, R. *et al.*, 1975, 1977; Claret, J. *et al.*, 1977
Mamestra brassicae	last-instar larvae; hemolymph	RIA	Agui, N. and Hiruma, K., 1982

ecdysteroid peaks are detected in both hemolymph and whole-body extracts irrespective of the method used for quantification.

During larval–larval and larval–pupal development (larval–adult in hemimetabolous insects) the major peak in the ecdysteroid titer is usually present during the latter half of each stadium (Figs 2 and 3). Temporally and quantitatively, this titer peak is characterized by a rapid rise from basal levels before apolysis; maximal levels either slightly before, at, or slightly after apolysis (pupariation in dipterans); and an equally rapid drop back to low or basal levels by the time of ecdysis. Although both the magnitude and duration of these major titer peaks during larval stadia are highly variable depending on the species, it appears that peaks preceding larval–pupal and larval–adult metamorphic molts are of slightly greater magnitude and/or duration than peaks preceding the non-metamorphic larval–larval molts. As might be expected, there are exceptions to this general pattern. For example, in *Tenebrio molitor* the major peak in the ecdysteroid titer during larval–pupal development seems to occur after apolysis (Delbecque, J.-P. *et al.*, 1975, 1978b; Delachambre, J. *et al.*, 1979, 1980) and in *Calliphora vicina* the major peak in the ecdysteroid titer which occurs with the onset of pupariation does not drop back to basal levels until pharate adult development is almost complete (Koolman, J., 1980).

In contrast to larval–larval and larval–pupal development, during pupal–adult development the major ecdysteroid titer peak occurs in the first half or middle of the pupal stadium (see Dean, R. *et al.*, 1980). This titer peak is characterized by a slow rise from basal levels with the onset of pupal–adult apolysis; this rise continues over an extended period during pharate–adult development, followed usually by a slow return to basal levels before pupal–adult ecdysis (Fig. 4). The major ecdysteroid titer peak in pupal stadia not only differs in temporal aspects from those which occur in larval stadia but is also much greater in magnitude and duration (both by a factor of about 1.5 to 20). Again, there are also exceptions to this generalization. For example, in *Sarcophaga bullata* there are two major peaks reported to occur in the pupal stadium (Wentworth, S. *et al.*, 1981) and in female *Galleria mellonella* the major ecdysteroid titer peak early in the pupal stadium is followed by a second major increase in the titer just before pupal–adult ecdysis (Bollenbacher, W. *et al.*, 1978).

The exceptions notwithstanding, what physiological significance can be attributed to these major ecdysteroid titer peaks? Given the temporal and quantitative profiles of these peaks, as well as their apparent ubiquity in larval and pupal stadia, it is thought that these are the classical peaks predicted by the basic model. As such these major peaks in the ecdysteroid titer are responsible for initiating

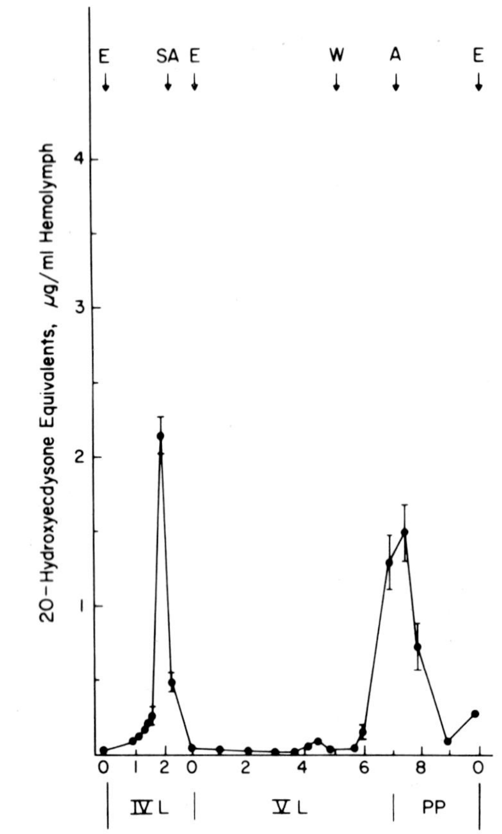

FIG. 2. The hemolymph ecdysteroid titer during the fourth larval (IVL) and fifth larval (VL)–pharate pupal (PP) stadia of *Manduca sexta*. The developmentai events during these stages are ecdysis (E), spiracle apolysis (SA), wandering behavior (W) and apolysis (A). Each datum point is the mean ± SEM of four to eight separate determinations. (Figure reproduced with permission from Bollenbacher, W. *et al.*, 1981.)

and/or maintaining the events associated with molting and the pharate development which occurs during the molt, i.e. between apolysis and ecdysis. Presumably, the greater magnitude and duration of the ecdysteroid titer peak during pharate–adult development may be a reflection of the greater need for these compounds during this time of extensive remodelling and differentiation as compared to earlier molts. Based on the available evidence, therefore, it is appropriate to suggest that the single major ecdysteroid titer peak which is detected during each larval and pupal stadium be considered a "molting peak".

In addition to the major molting peaks, it is becoming increasingly apparent that there may be one or more additional peaks in the ecdysteroid titer during certain stadia, particularly the stadium leading to the first metamorphic molt (see Dean, R. *et al.*, 1980). Such multiple peaks have been found in the last larval stadia of *Pieris brassicae* (Lafont, R. *et al.*, 1975, 1977), *Tenebrio molitor* (Delbecque, J.-P. *et al.*, 1975, 1978b; Delachambre, J. *et al.*, 1979, 1980), *Manduca sexta* (Bollenbacher, W. *et al.*, 1975, 1981; Wielgus, J. *et al.*, 1979; Smith, S. *et al.*, 1983), *Locusta migratoria* (Baehr, J.-C. *et al.*, 1979; Hirn, M. *et al.*, 1979), *Galleria mellonella* (Bollenbacher, W. *et al.*, 1978; Maróy, P. and Tarnóy, K., 1978; Sehnal, F. *et al.*, 1981), *Calliphora vicina* (Koolman, J., 1980), *Calpodes ethlius*

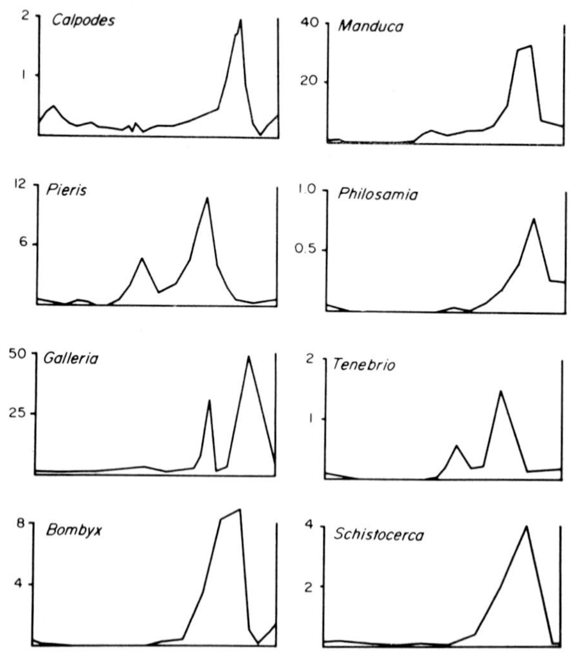

FIG. 3. The ecdysteroid titer profiles during the last larval stadium of several representative insect species. Data for *Calpodes ethlius* from Dean, R. *et al.* (1980), *Pieris brassicae* from Lafont, R. *et al.* (1977), *Galleria mellonella* from Bollenbacher, W. *et al.* (1978), *Bombyx mori* from Calvez, B. *et al.* (1976), *Manduca sexta* from Bollenbacher, W. *et al.* (1975), *Philosamia cynthia* from Calvez, B. *et al.* (1976), *Tenebrio molitor* from Delbecque, J.-P. *et al.* (1975) and *Schistocerca gregaria* from Morgan, E. and Poole, C. (1976a). Abscissae: relative duration of stadium. Ordinates: *Calpodes*, *Bombyx* and *Philosamia* — RIA activity in ng 20-hydroxyecdysone equivalents ml^{-1} hemolymph $\times 10^{-2}$; *Manduca* and *Galleria* — RIA activity in ng 20-hydroxyecdysone equivalents per animal $\times 10^{-2}$; *Tenebrio* and *Schistocerca* — ng 20-hydroxyecdysone per larva $\times 10^{-2}$; *Pieris* — ng 20-hydroxyecdysone ml^{-1} hemolymph $\times 10^{-2}$. (Figure reproduced with permission from Dean, R. *et al.*, 1980.)

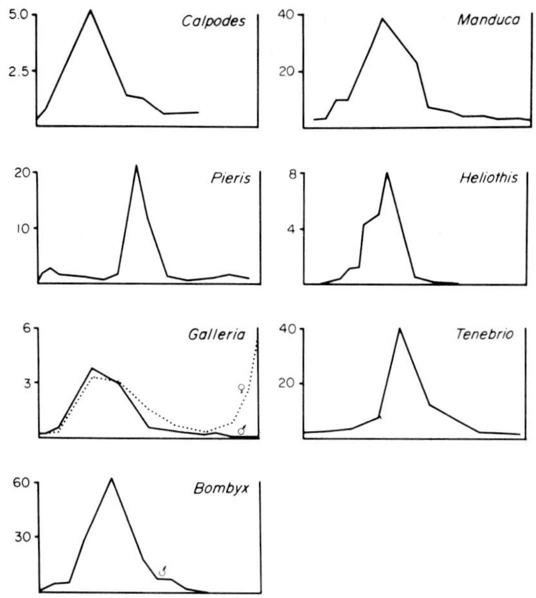

FIG. 4. The ecdysteroid titer profiles during the pupal stadium of several representative insect species. Data for *Calpodes ethlius* from Dean, R. *et al.* (1980), *Pieris brassicae* from Lafont, R. *et al.* (1975), *Galleria mellonella* from Bollenbacher, W. *et al.* (1978), *Bombyx mori* from Calvez, B. *et al.* (1976), *Manduca sexta* from Bollenbacher, W. *et al.* (1981), *Heliothis zea* from Holman, G. and Meola, R. (1978) and *Tenebrio molitor* from Delbecque, J.-P. *et al.* (1978b). Abscissae: relative duration of stadium. Ordinates: *Calpodes*, *Bombyx* and *Tenebrio* — RIA activity in ng 20-hydroxyecdysone equivalents ml^{-1} hemolymph $\times 10^{-2}$; *Manduca* — RIA activity in ng 20-hydroxyecdysone equivalents g^{-1} fresh weight $\times 10^{-2}$; *Pieris* — ng 20-hydroxyecdysone ml^{-1} hemolymph $\times 10^{-2}$; *Heliothis* — ng 20-hydroxyecdysone per animal $\times 10^{-2}$; *Galleria* — RIA activity in ng 20-hydroxyecdysone equivalents per animal $\times 10^{-2}$. (Figure reproduced with permission from Dean, R. *et al.*, 1980.)

(Dean, R. *et al.*, 1980), *Aeshna cyanea* (Schaller, F. and Charlet, M., 1980) and *Drosophila melanogaster* (see Richards, G., 1981).

In contrast to the major molting peaks, these "non-molting peaks" are usually present during the intermolt period and are much more subtle in magnitude and duration (both by a factor of about 0.05 to 0.5) than the larval molting peaks (Figs 2 and 3). There are exceptions, however, to this pattern. During the last larval stadium of *Tenebrio molitor* both the subtle ecdysteroid titer peak and the major molting peak appear to occur after apolysis (Delbecque, J.-P. *et al.*, 1975, 1978b; Delachambre, J. *et al.*, 1979, 1980) and in the last larval stadium of *Galleria mellonella* an intermolt peak which is fairly comparable in magnitude to the molting peak has been reported (Bollenbacher, W. *et al.*, 1978).

In some insects, such as *Bombyx mori* and *Rhodnius prolixus*, such intermolt peaks have not been detected during the last larval stadium (Calvez, B. *et al.*, 1976; Steel, C. *et al.*, 1982), whereas in other species such as *Drosophila melanogaster* and *Locusta migratoria* these subtle peaks both have and have not been detected (see Richards, G., 1981; Hoffmann, J. *et al.*, 1974; Baehr, J.-C. *et al.*, 1979; Hirn, M. *et al.*, 1979). While clearly methodological considerations are responsible for some of the above differences, it remains to be established if subtle intermolt peaks in the ecdysteroid titer are a widespread phenomenon in insects. However, given their confirmed presence in some insects what physiological or developmental significance can be attributed to these subtle peaks?

While some of the intermolt peaks have been found to correlate temporally with such activities as DNA synthesis, RNA synthesis, lipid synthesis, and organelle formation, it is still too early to draw any firm conclusions at the cellular level (see Dean, R. *et al.*, 1980). Tentatively, however, it does appear that one type of intermolt peak may be significant developmentally. In *Manduca sexta* a subtle intermolt ecdysteroid titer peak has been detected on about day 4 (Fig. 2) of the last larval stadium (Bollenbacher, W. *et al.*, 1975, 1981; Wielgus, J. *et al.*, 1979; Smith, S. *et al.*, 1983). This ecdysteroid peak and comparable intermolt peaks reported for other holometabolous insects (e.g. *Calpodes*) are thought to be responsible for reprogramming the epidermis from a larval to a pupal commitment (see Dean, R. *et al.*, 1980; L. Riddiford, vol. 8).

In *Manduca* the intermolt ecdysteroid titer peak is known to be temporally coincident with the termination of the first head critical period, i.e. this "reprogramming" peak occurs shortly after, and in response to, the first release of PTTH during the last larval stadium (Truman, J., 1972; Truman, J. and Riddiford, L., 1974; Gilbert, L. *et al.*, 1981). Further studies have also shown that the amounts of ecdysteroids in this subtle peak ($\sim 10^{-7}$ M) are both capable of eliciting reprogramming in the larval epidermis in the absence of JH (Riddiford, L., 1976, 1978, 1981) and commensurate with an ecdysteroid receptor ($K_d \sim 10^{-7}$ M) mediated event (Bonner, J., 1982). Finally, and consistent with a reprogramming function for the intermolt peak, a recent determination of the ecdysteroid titer in *Manduca* (Fig. 2) has revealed that an intermolt peak is not present during the non-metamorphic fourth larval stadium (Bollenbacher, W. *et al.*, 1981).

3.2.2 Composition of ecdysteroid titers

In addition to the temporal and quantitative characteristics, many of the recent titer studies have also contributed much to our understanding of the qualitative composition of ecdysteroid titers. Several ecdysteroids have been found to be present in various insects during postembryonic development (Table 1). These include, in addition to ecdysone and 20-hydroxyecdysone, 20,26-dihydroxy-ecdysone, 3-epiecdysone, 3-epi-20-hydroxyecdysone, makisterone A, 2-deoxyecdysone, 3-epi-20,26-dihydroxyecdysone, and several incompletely characterized ecdysteroid conjugates (see Koolman and Karlson, this volume). Although not isolated as natural products yet, it is also possible that 3-dehydroecdysone, 3-dehydro-20-hydroxyecdysone, and inokosterone may be present in some insects based on their formation by insect enzyme preparations.

With the exception of ecdysone and 20-hydroxyecdysone, it appears that the other ecdysteroids are present only as minor or incompletely identified constituents of the hemolymph ecdysteroid titer. However, with respect to whole-body ecdysteroid titers there may be times during postembryonic development when these other ecdysteroids are present in concentrations greater than ecdysone and 20-hydroxyecdysone. For instance, during late pupal–adult development in *Manduca sexta* six ecdysteroids have been detected and quantified in whole-body extracts. The most prevalent ecdysteroid was 20,26-dihydroxyecdysone, followed in order of decreasing quantities by 3-epi-20-hydroxyecdysone, 20-hydroxyecdysone, 3-epi-20,26-dihydroxyecdysone, 3-epiecdysone and ecdysone (Kaplanis, J. *et al.*, 1979). Similarly, during late pharate–adult development in *Galleria mellonella* 2-deoxyecdysone has been found to be present in ovariectomized females in concentrations greater than ecdysone or 20-hydroxyecdysone (S. Smith and W. Bollenbacher, unpublished observations).

Functionally, based on their MH activity in bioassay, it is thought that the 26-hydroxy-, 3-epi-, and 3-dehydro-ecdysteroids are degradation or inactivation products of ecdysone and 20-hydroxyecdysone, whereas 2-deoxyecdysone is presumed to be an ecdysone precursor. The picture is less clear, however, with respect to the ecdysteroid conjugates. Although these conjugates have been postulated to be inactivation products or storage forms (Hoffmann, J. and Lagueux, M., vol. 1), only tentative evidence exists to show that free ecdysteroids may be liberated from conjugates during insect postembryonic development (Moribayashi, A. and Ohtaki, T., 1978, 1980). Since hydrolytic digestions of conjugates with extracts of the snail, *Helix pomatia*, have revealed that most ecdysteroids are present in such conjugates, e.g. ecdysone, 20-hydroxyecdysone, 3-dehydro-ecdysteroids and the 3α-epimers of ecdysteroids, the possibility exists that some conjugates may serve as storage forms of ecdysteroids while others are inactivation products (see Koolman, J. and Karlson, P., this volume).

In contrast to the other ecdysteroids, a substantial body of information exists to support the tenet that ecdysone and 20-hydroxyecdysone are the major MH during postembryonic development. In addition to the numerous studies which have shown that exogenous ecdysone and 20-hydroxyecdysone elicit molting and metamorphosis when administered to insects (see Karlson, P., 1980), the body of ecdysteroid titer studies have shown that ecdysone and 20-hydroxyecdysone are the major hemolymph ecdysteroids present during postembryonic development. One exception to this tenet, however, is *Oncopeltus fasciatus* which appears to have makisterone A as its major hemolymph ecdysteroid (Kelly, T. *et al.*, 1981).

The ratio of ecdysone to 20-hydroxyecdysone is not constant during postembryonic development but varies dramatically. Such variations in the ratio of ecdysone to 20-hydroxyecdysone have been reported for example in *Manduca sexta* (Bollenbacher, W. *et al.*, 1975, 1981; Smith, S. *et al.*, 1983), *Pieris brassicae* (Lafont, R. *et al.*, 1975, 1977), *Schistocerca gregaria* (Morgan, E. and Poole, C., 1976a; Gande, A. *et al.*, 1979), *Bombyx mori* (Calvez, B. *et al.*, 1976), *Heliothis zea* (Holman, G. and Meola, R., 1978), *Locusta migratoria* (Hirn, M. *et al.*, 1979) and *Xyleborus ferrugineus* (Rao, K. *et al.*,

1981). In *Pieris brassicae*, there are even distinct ecdysone and 20-hydroxyecdysone peaks (Lafont, R. *et al.*, 1975, 1977).

Based on the reported ratios of ecdysone to 20-hydroxyecdysone, it appears as if 20-hydroxyecdysone is the predominant constituent of the ecdysteroid titer molting peaks during larval stadia. For example, ecdysone to 20-hydroxyecdysone ratios of 1:14.6 and 1:5.4 have been reported for the molting peaks in the penultimate and last larval stadia, respectively, of *Manduca sexta* (Bollenbacher, W. *et al.*, 1981); a ratio of 1:6.6 for the molting peak in the last larval stadium of *Calpodes ethlius* (Dean, R. *et al.*, 1980); and a ratio of 1:3.4 for the molting peak in the last larval stadium of *Rhodnius prolixus* (Steel, C. *et al.*, 1982). In some larval stadia, e.g. in *Sarcophaga bullata*, *Drosophila melanogaster*, and *Schistocerca gregaria*, the ecdysone to 20-hydroxyecdysone ratio may be as low as 1:27 or ecdysone may not be present in detectable levels (Bollenbacher, W. *et al.*, 1976; Gande, A. *et al.*, 1979; see Richards, G., 1981).

In contrast to these ratios in larval stadia, however, there are times during the pupal stadium when ecdysone may be in greater concentration in the hemolymph than 20-hydroxyecdysone. For example, on day 7 of pharate adult development in *Manduca sexta*, the molting peak is near maximum levels and the hemolymph ratio of ecdysone to 20-hydroxyecdysone is 3.3:1 (Bollenbacher, W. *et al.*, 1981). Similar "reverse ratios" of ecdysone to 20-hydroxyecdysone, i.e. ecdysone in greater concentration than 20-hydroxyecdysone, have also been noted in the pupal stadia of *Bombyx mori*, *Heliothis zea* and *Pieris brassicae* (Calvez, B. *et al.*, 1976; Holman, G. and Meola, R., 1978; Lafont, R. *et al.*, 1975, 1977).

Although it remains to be established, there are reasons for suspecting that the changing ratios of ecdysone to 20-hydroxyecdysone may be of some developmental import. Even though ecdysone and 20-hydroxyecdysone elicit numerous and comparable effects *in vivo*, it is probable that many of the effects attributed to ecdysone may be due to its *in vivo* conversion to 20-hydroxyecdysone. This conversion is known to be one of the major reactions in the metabolism of ecdysone in insects, and the changing ratios of ecdysone to 20-hydroxyecdysone may be a reflection of changes in the rate of this conversion (see section 4.3).

In *in vitro* systems not capable of converting ecdysone to 20-hydroxyecdysone (King, D., 1972a,b), 20-hydroxyecdysone is one to several hundred times more active than ecdysone in eliciting a response, e.g. puffing in *Drosophila melanogaster* polytene chromosomes (Ashburner, M., 1971), interganglionic ventral nerve cord shortening in *Galleria mellonella* (Robertson, J. and Pipa, R., 1973), and evagination and development of the imaginal disks of *Drosophila melanogaster* and *Sarcophaga bullata* (Chihara, C. *et al.*, 1972; Chen, T. and Hodgetts, R., 1974; Milner, M. and Sang, J., 1974). Since 20-hydroxyecdysone can elicit these normal developmental responses *in vitro* at physiological concentrations ($\sim 10^{-7}$ M), while ecdysone cannot elicit comparable responses even at unphysiological concentrations of 10^{-5} M, it has been assumed that 20-hydroxyecdysone is the active form of MH and that ecdysone serves as a prohormone (King, D., 1972b; Gilbert, L. and King, D., 1973; King, D. *et al.*, 1974; Gilbert, L. *et al.*, 1977, 1980). More recent evidence in support of this "prohormone hypothesis" is the finding that the ecdysteroid cellular receptors show a ~1000-fold greater affinity for 20-hydroxyecdysone than for ecdysone (Fristrom, J., 1981; O'Connor, J. and Chang, E., 1981; see Yund, M. and Osterbur, D., this volume; O'Connor, J., vol. 8).

Although the preponderance of data support the contention that 20-hydroxyecdysone is the active MH during postembryonic development, does this mean that ecdysone functions solely as a prohormone? In addition to the developmentally significant times when ecdysone exceeds 20-hydroxyecdysone concentration in some insects, there is suggestive evidence that ecdysone has specific developmental effects on, e.g., puff induction in *Chironomus tentans* salivary gland polytene chromosomes (Clever, U. *et al.*, 1973); DNA and RNA synthesis, tracheole migration and evagination in wing and leg imaginal disks of *Galleria mellonella*, *Plodia interpunctella*, *Cadra cautella*, *Paramyelois transitella* and *Drosophila melanogaster* (Oberlander, H., 1969, 1972a,b; Mandaron, P., 1973; Oberlander, H. *et al.*, 1973; Bergstrom, G. and Oberlander, H., 1975); and cell division in Kc cells from *Drosophila melanogaster* embryos

(Courgeon, A., 1972). Thus, it remains a distinct possibility that ecdysone may function in some capacity other than as a prohormone during postembryonic development (see Koolman, J. and Karlson, P., this volume). Other studies have implied that ecdysteroids may elicit effects independent of cellular receptors, i.e. there may be alternative or additional mechanisms of action for some of these compounds (see Fristrom, J., 1981). Judged in this light, the varying ratios of ecdysone to 20-hydroxyecdysone may be extremely important.

3.3 Ecdysteroid titers during adult life and embryogenesis

Although it had been observed as early as the 1950s that MH activity is present in adult *Bombyx mori* (Karlson, P. and Stamm-Menéndez, M., 1956), the rediscovery in the mid-1970s of ecdysteroids and ecdysteroid activity in adult female insects and insect eggs came as something of a surprise (see Hoffmann, J. *et al.*, 1980; Hoffmann, J. and Lagueux, M., vol. 1). Since these stages of the insect life cycle were thought to lack the biosynthetic capacity and/or requirement for MH, the most obvious questions raised by these early studies were what types of ecdysteroids were present, where did they come from, and what function did they serve? The story quickly became much more complicated but, since it is the major topic of two reviews in this series (see Hoffmann, J. and Lagueux, M., vol. 1; Hagedorn, H., vol. 8), only salient features of it will be presented here.

3.3.1 Chemistry and occurrence

Following the initial discoveries of ecdysteroids in adult female insects and insect eggs (Kaplanis, J. *et al.*, 1973, 1975; Hagedorn, H. *et al.*, 1975), it soon became apparent that this was not a limited phenomenon. At present, for example, ecdysteroids and ecdysteroid-like materials have been found in adult female insects and/or eggs from such representative species as the lepidopterans, *Manduca sexta* (Kaplanis, J. *et al.*, 1973; Bollenbacher, W. *et al.*, 1981), *Galleria mellonella* (Bollenbacher, W. *et al.*, 1978; Hsiao, T. and Hsiao, C., 1979), and *Bombyx mori* (Hirn, M. *et al.*, 1977; Ohnishi, E. *et al.*, 1977; Coulon, M., 1977); the orthopterans, *Locusta migratoria* (Hoffmann, J. *et al.*, 1980; Lagueux, M. *et al.*, 1977, 1979, 1981) and *Schistocerca gregaria* (Gande, A. and Morgan, E., 1979); the isopterans, *Macrotermes bellicosus* and *Macrotermes subhyalinus* (Delbecque, J.-P. *et al.*, 1978c); the hemipteran, *Oncopeltus fasciatus* (Kaplanis, J. *et al.*, 1975; Dorn, A. and Romer, F., 1976); the dipterans, *Aedes aegypti* (Hagedorn, H. *et al.*, 1975), *Aedes atropalpus* (Masler, E. *et al.*, 1980), *Anopheles stephensi* (Redfern, C., 1982) and *Drosophila melanogaster* (see Richards, G., 1981; Garen, A. *et al.*, 1977; Rubenstein, E. *et al.*, 1982); the collembolan, *Folsomia candida* (Palevody, C. *et al.*, 1977); and the thysanuran, *Thermobia domestica* (DeLaPaz, A. *et al.*, 1983).

As research continued in this area a number of the ecdysteroids extracted from insect adults and eggs were subjected to physiochemical characterization. Intriguingly, it was soon realized that, in addition to ecdysone and 20-hydroxyecdysone, there were structurally unique forms of ecdysteroids present in some insect adults and eggs. To date, at least 16 ecdysteroids have been isolated from insect adults and eggs, including such novel ecdysteroid conjugates as ecdysone-3-acetate, ecdysone-22-phosphate, 2-deoxyecdysone-22-phosphate, 3-epi-2-deoxyecdysone-3-phosphate, 2-deoxyecdysone-22-adenosine monophosphate and ecdysone-22-N^6-(isopentenyl)-adenosine monophosphate (see Table 1).

3.3.2 Functions of adult and embryonic ecdysteroids

In contrast to the role of ecdysteroids as MH during insect postembryonic development, no unified role for these steroids during adult life and embryogenesis has emerged. In part this is due to the qualitative complexity of the ecdysteroids present in insect adults and eggs and in part because these ecdysteroids may in fact serve diverse physiological roles.

The ovaries are the apparent source of ecdysteroids during the adult life of female insects (see Hoffmann, J. and Lagueux, M., vol. 1, and Hagedorn, H., vol. 8, this series). However, both the types of ecdysteroids synthesized and the fate of these ecdysteroids are highly variable depending on the species. It appears, however, as if most insects may fall within one of three general patterns.

The first pattern is that exemplified by such species as *Aedes aegypti* (Hagedorn, H. *et al.*, 1975; Hanaoka, K. and Hagedorn, H., 1980) and *Aedes atropalpus* (Masler, E. *et al.*, 1980). The ovaries of these species are characterized by the presence of an ecdysteroid RIA immunoreactive material which, for the most part, is not sequestered in the ovary but released (at least under *in vitro* conditions). This RIA-active material has been identified as ecdysone in *Aedes aegypti* and it is assumed to be ecdysone in *Aedes atropalpus*. However, 20-hydroxyecdysone appears to be the predominant if not sole ecdysteroid present in adult female whole-body extracts of these species. It has been inferred, therefore, that the ovarian ecdysone in these species is released and quantitatively hydroxylated to 20-hydroxyecdysone.

The whole-body ecdysteroid titer in both these species is characterized by a major peak (2–4-fold change in magnitude) which occurs about 20 h post blood-feeding in the anautogenous mosquito *Aedes aegypti* and about 30 h post-eclosion in the autogenous mosquito *Aedes atropalpus*. This major ecdysteroid titer peak in both species is thought to initiate and/or regulate various events of the gonadotropic cycle, e.g. fat body vitellogenesis, dopa-decarboxylase activity (Sekeris, C. and Fragoulis, E., vol. 8), and separation of secondary follicles in the ovary. However, based on discrepancies in the literature the exact physiological function(s) of ecdysone and 20-hydroxyecdysone in adult female *Aedes* remains to be established (see Fuchs, M. and Kang, S.-H., 1981; Hagedorn, H., vol. 8, this series).

The second major pattern of ecdysteroid formation and fate in adult female insects is that exemplified for example by *Locusta migratoria, Schistocerca gregaria, Bombyx mori* and *Macrotermes bellicosus* (see Hoffmann, J. *et al.*, 1980). In these species the ovaries are a source of several different types of ecdysteroids. For the most part these ecdysteroids are not released to the hemolymph or body but are sequestered by the ovaries and subsequently packaged, predominantly as conjugates of 2-deoxyecdysone and ecdysone, as maternal factors in the eggs.

Current data suggest that these maternal ecdysteroids are metabolized during embryogenesis (Lagueux, M. *et al.*, 1981). Concomitant with their metabolism, free ecdysteroids and other different ecdysteroid conjugates appear. The free ecdysteroids isolated to date from *Locusta* (e.g. 2,14,22,25-tetradeoxyecdysone, 2,22,25-trideoxyecdysone, 2,22-dideoxyecdysone, 2-deoxyecdysone, ecdysone) are compatible with the hypothesis that the maternal conjugates function as a storage form of ecdysone or intermediates for the biosynthesis of ecdysone/20-hydroxyecdysone. Accepting the plausibility of this hypothesis, what are the functions of ecdysone/20-hydroxyecdysone during embryogenesis?

During the embryogenesis of *Locusta*, four distinct peaks in the titer of free ecdysteroids have been noted to occur. These ecdysteroid titer peaks are composed predominantly of ecdysone and to a lesser extent 20-hydroxyecdysone and 20,26-dihydroxyecdysone. These peaks temporally correlate with various events of embryonic cuticulogenesis. The first ecdysteroid titer peak occurs concomitantly with the onset of the deposition of the first cuticle, i.e. the serosal cuticle. The latter three ecdysteroid titer peaks each occur concomitantly with apolysis of the cuticle from the previous cycle of cuticulogenesis, with the last embryonic cuticle functioning as the first-instar larval cuticle. On the basis of these data, therefore, it would appear that the major function of ecdysone/20-hydroxyecdysone during *Locusta* embryogenesis may be similar to the role of these compounds as MH during postembryonic development. Similar temporal correlations between ecdysteroid titer peaks and cuticle depositions have been noted in the eggs of *Oncopeltus fasciatus* (Dorn, A. and Romer, F., 1976), *Blaberus craniifer* (Bullière, D. *et al.*, 1979), *Carausius morosus* (Louvet, J. *et al.*, 1979) and *Leucophaea maderae* (Matz, G., 1980).

The third major pattern of ecdysteroid formation and fate in adult female insects is that exemplified by *Galleria mellonella* (Bollenbacher, W. *et al.*, 1978; Smith, S. and Bollenbacher, W., unpublished observations). The *Galleria* pattern shows characteristics of both the *Aedes* and *Locusta* patterns. Like *Locusta*, the ovaries of *Galleria* are a source of several ecdysteroids, including 2-deoxyecdysone, ecdysone, 20-hydroxyecdysone and polar conjugates. Furthermore, it appears as if a significant portion of these maternal ecdysteroids is sequestered in the ovaries and packaged in the eggs

(Bollenbacher, W. *et al.*, 1978; Hsiao, T. and Hsiao, C., 1979). However, in contrast to *Locusta* where only minor amounts of ovarian ecdysteroids are released to the body or hemolymph, in *Galleria* a significant portion (30–35%) of the ovarian ecdysteroids can be detected in ovariectomized females. In this sense the *Galleria* pattern shows a similarity to the *Aedes* pattern. However, in *Aedes* only a single ovarian ecdysteroid (ecdysone) is released, whereas in *Galleria* all of the ecdysteroids isolated from the ovaries have also been isolated from ovariectomized bodies. At present the physiological importance of this ovarian/body distribution of ecdysteroids in *Galleria* remains to be established.

4 ECDYSTEROID METABOLISM AND ECDYSTEROID TITERS

4.1 General processes

Ecdysteroid titers are inherently complex when examined with respect to metabolic processes. The hemolymph ecdysteroid titer at any given moment is probably a reflection of the rates of several metabolic processes including: ecdysteroid biosynthesis and secretion, transport to target tissues, tissue uptake, sequestration, degradation and excretion. When viewed over time, ecdysteroid titers are even more complex, encompassing varying changes in the rates of these various metabolic processes. While the remainder of this chapter will be primarily concerned with the significance of ecdysone and 20-hydroxyecdysone biosynthesis and secretion during postembryonic development, it is necessary to briefly discuss the other metabolic processes which may contribute to the regulation of the ecdysteroid titer (see Koolman, J. and Karlson, P., this volume).

4.1.1 Ecdysteroid transport and tissue uptake

Although ecdysone and 20-hydroxyecdysone are appreciably water-soluble, there is suggestive evidence that these steroids may be transported to target tissues by hemolymph binding proteins in some insects (see Yund, M. and Osterbur, D., this volume). A high-affinity ecdysteroid binding protein has been demonstrated to be present in the hemolymph of *Locusta migratoria* (Feyereisen, R., 1977a, 1980) and it is possible that similar binding proteins may also be present in *Drosophila hydei* (Emmerich, H., 1970a; Butterworth, F. and Berendes, H., 1974), *Pyrrhocoris apterus* (Emmerich, H., 1970b) and *Calliphora vicina* (Thamer, G. and Karlson, P., 1972; Koolman, J., 1980). It is not readily apparent at present how widespread such hemolymph ecdysteroid binding proteins may be; nor is it also known at present whether such hemolymph binding proteins might contribute to the regulation of the ecdysteroid titer by facilitating target tissue uptake of the hormone or perhaps by protecting ecdysteroids from degradation, excretion, or non-specific tissue uptake.

In a similar sense it at present also remains obscure to what degree the tissue uptake of ecdysteroids may influence the hemolymph ecdysteroid titer. In large part this is because very little is known about the mechanism(s) of ecdysteroid translocation across cell membranes. Preliminary studies with *Sarcophaga peregrina* fat body have suggested that the translocation of 20-hydroxyecdysone across cell membranes may be by active transport (Natori, S. and Ohtaki, T., 1976). However, other studies with the *Kc* cell line from *Drosophila melanogaster* have suggested that the tissue uptake of 20-hydroxyecdysone is not carrier-mediated (O'Connor, J. *et al.*, 1980; vol. 8). Considering that the transport and mechanism(s) of tissue uptake of hemolymph ecdysteroids might have profound influences both on the regulation of the ecdysteroid titer and on the actions of ecdysteroids, clearly, these two processes need to be studied more thoroughly.

4.1.2 Ecdysteroid conjugation, degradation and excretion

Ecdysteroid conjugates are known to be present in varying concentrations during all stages of insect development. Although the qualitative composition of conjugate digests suggests that some conjugates may be storage forms for the release or synthesis of ecdysone or 20-hydroxyecdysone (particularly in eggs and possibly in pupae), no direct evidence exists to show the release of free ecdysteroids from conjugates during postembryonic development. Numerous studies, however, have shown that injected radiolabelled ecdysone or 20-hydroxyecdysone may eventually end up as conjugated forms (see Koolman, J. and Karlson, P., this volume). Based on these data, therefore, it would appear that at least during postembryonic development conjugation contributes to the regulation of the ecdysteroid titer by irreversibly withdrawing these steroids from the free ecdysteroid pool. However, until more is known about the structural composition of ecdysteroid conjugates, as well as the putative enzymes responsible for forming (transferases) and breaking (hydrolases) conjugates, the exact role(s) of these substances in ecdysteroid titer regulation remains speculative (Koolman, J., 1980).

Similarly, the exact role of several inactivation products of ecdysone and 20-hydroxyecdysone (i.e. 3-dehydro-, 3-epi-, and 26-hydroxylated ecdysteroids) in ecdysteroid titer regulation remains speculative. These ecdysteroids not only are formed as metabolites of injected radiolabelled ecdysone/20-hydroxyecdysone in many insects, several of them have been isolated as natural products (Table 1) and the enzymes for the formation of 3-dehydroecdysteroids (ecdysone oxidase) and 3-epi-ecdysteroids (ecdysone 3-epimerase) have also been partially characterized (see Nigg, H. *et al.*, 1974; Koolman, J. and Karlson, P., 1975, 1978; Mayer, R. *et al.*, 1979).

Suggestive evidence that the formation of these inactivation products may contribute to the regulation of the ecdysteroid titer has come from several studies, primarily with *Locusta migratoria*, *Calliphora erythrocephala* and *Sarcophaga peregrina* (Ohtaki, T. *et al.*, 1968; Hoffmann, J. *et al.*, 1974; Hoffmann, J. and Koolman, J., 1974; Feyereisen, R. *et al.*, 1976; Young, N., 1976a,b; Young, N. and Young, P., 1976; Koolman, J., 1980; Ohtaki, T., 1981). These studies and others have demonstrated that there are dramatic variations during postembryonic development in the catabolism of radiolabelled ecdysone and 20-hydroxyecdysone (see Koolman, J. and Karlson, P., this volume).

Excretion, either of free ecdysteroids or of ecdysteroid inactivation products, has also been suggested to contribute to the regulation of the ecdysteroid titer. It is known that free ecdysone and 20-hydroxyecdysone, as well as conjugates and inactivation products of ecdysteroids, are excreted at varying rates in insects. Ecdysteroids have been detected in the feces of, for example, *Manduca sexta* (Moriyama, H. *et al.*, 1970; Kaplanis, J. *et al.*, 1972), *Locusta migratoria* (Hoffmann, J. and Koolman, J., 1974; Hoffmann, J. *et al.*, 1974; Koolman, J. *et al.*, 1975; Feyereisen, R. *et al.*, 1976), *Gryllus bimaculatus* (Wildmann, H. and Romer, F., 1977), *Tenebrio molitor* (Weinheimer, G. and Romer, F., 1977), *Schistocerca gregaria* (Wilson, I. and Morgan, E., 1978) and *Rhodnius prolixus* (Steel, C. *et al.*, 1982).

While insufficient data exist at present to quantify the degree to which ecdysteroid conjugation, catabolism and excretion contribute to the regulation of the ecdysteroid titer, there can be little doubt that at certain times during development these processes may have a major influence on the ecdysteroid titer. Perhaps the best evidence, albeit indirect, in favor of this comes from investigations on the half-lives of injected radiolabelled ecdysone. These studies have revealed that the half-life can vary dramatically during postembryonic development. Half-lives as short as 4 min and as long as 940 min have been reported for ecdysone in *Calliphora vicina* third-instar larvae and pupae, respectively (Koolman, J., 1980). Similarly, 20-hydroxyecdysone has been found to have a much longer half-life in *Calliphora* pupae (2100 min) than in third-instar larvae (200 min) (Koolman, J., 1980).

4.2 Ecdysone biosynthesis

Although increases and decreases in ecdysteroid catabolism and excretion could effect changes in the ecdysteroid titer, the biosynthesis and secretion of ecdysone must be viewed as the primary process contributing to the ecdysteroid titer during postembryonic development. While it cannot be ruled out entirely, the possibility that some phytophagous insects might obtain their required ecdysteroids from host plants has not been shown to date. Indeed, the preponderance of data suggests that such ingested ecdysteroids are either inactivated/conjugated in the gut or elicit abnormalities in growth and development (Feyereisen, R. *et al.*, 1976; Kubo, I. *et al.*, 1981, 1983). In at least one case, however, dietary ecdysteroids have been found to restore normal postembryonic development in an ecdysteroid-deficient mutant of *Drosophila melanogaster* (Garen, A. *et al.*, 1977).

Since insects do not obtain their ecdysteroids from the diet and since, like most invertebrates, they are incapable of synthesizing steroids from simple precursors such as acetate or propionate (Svoboda, J. and Thompson, M., vol. 10), insects have an obligatory requirement for a sterol nucleus in order for ecdysteroid biosynthesis to occur (Hobson, R., 1935a,b; Clayton, R., 1964; Rees, H., 1971). This sterol requirement is usually fulfilled in insects by cholesterol or various phytosterols obtained either dietetically or via endosymbiosis with steroidogenic organisms (Thompson, M. *et al.*, 1973; Svoboda, J. *et al.*, 1975). The various transformations required to convert phytosterols to cholesterol, and cholesterol to ecdysone, are the topics of two other reviews in this series and will not be discussed further in this chapter (see Rees, H., this volume; Svoboda, J. and Thompson, M., vol. 10).

4.2.1 Sites and kinetics of ecdysone synthesis

Based on the data from several classical studies in the 1940s and early 1950s, it generally became accepted that the PG or their analogs (ring or ventral glands) were the source of MH during insect postembryonic development (see Wigglesworth, V., this volume). This tenet persisted after the structural elucidation of ecdysone and 20-hydroxyecdysone even in the absence of any direct evidence that the PG actually synthesized ecdysteroids. By the early 1970s, however, a body of evidence which suggested that the PG might not be the source of MH was beginning to accumulate (Moriyama, H. *et al.*, 1970; Gilbert, L. and King, D., 1973).

To resolve this situation, during the mid to late 1970s the PG or their analogs from several species of insect larvae were cultured *in vitro* and the gland product formed and secreted into the culture medium was characterized and quantified using RIA, bioassay and physicochemical techniques. The results of these experiments established unequivocally that the larval PG of *Manduca sexta* (King, D. *et al.*, 1974), *Bombyx mori* (Chino, H. *et al.*, 1974), *Leucophaea maderae* (Borst, D. and Engelmann, F., 1974; King, D. and Marks, E., 1974), *Tenebrio molitor* (Romer, F. *et al.*, 1974), *Locusta migratoria* (Hoffmann, J. *et al.*, 1975; Hirn, M. *et al.*, 1979), and *Sarcophaga bullata* (Bollenbacher, W. *et al.*, 1976) synthesize and secrete ecdysone. Since these studies demonstrated that the PG do not contain significant levels of ecdysteroid activity themselves, as others had previously reported (Agui, N. *et al.*, 1972; Ellis, P. *et al.*, 1972; Romer, F., 1973), there could be little doubt that the ecdysone was being synthesized by the glands *in vitro* and released almost immediately following its synthesis.

Although it was also found in the case of *Manduca* that no other tissues (including brain, salivary glands, fat body, Malpighian tubules, oenocytes, body wall and midgut) other than the PG were capable of effecting the synthesis of ecdysone (King, D. *et al.*, 1974), several lines of evidence continue to suggest that non-PG tissues may also possess the capability of synthesizing ecdysone during postembryonic development and may effect this capability under certain situations. For example, isolated or ligated larval abdomens of *Tenebrio molitor* (Stellwaag-Kittler, F., 1954), *Leptinotarsa decemlineata* (Hsiao, T. *et al.*, 1975), and *Dermestes vulpinus* (Slama, K., 1976) undergo molting and/or metamorphosis in the absence of the PG, and ecdysteroid titers have been detected in isolated abdomens of *Leptinotarsa* (Hsiao, T. *et al.*, 1975). Similarly, extirpation of the PG has been found to have no effect on the subsequent development of *Galleria mellonella* or several species of cockroaches (Piepho, H., 1948; Nutting, W., 1955; Chadwick, L., 1956; Gersch, M., 1977). It also has been reported that ecdysteroid biosynthesis from radiolabelled cholesterol can occur in isolated larval abdomens or prothoracectomized larvae of *Mamestra brassicae* (Gersch, M. and Stürzebecher, J., 1971), *Bombyx mori* (Nakanishi, K. *et al.*, 1972), *Musca domestica* (Studinger, G. and Willig, A., 1975) and *Periplaneta americana* (Gersch, M. and Eibisch, J., 1977).

Clearly, the strongest evidence in support of the view that non-PG tissues may be a physiological source of ecdysone during postembryonic development comes from studies with *Tenebrio molitor* pupae. In this species PG involution is normally completed before the major ecdysteroid titer peak during the pupal stadium (Glitho, I. *et al.*, 1979) and even in isolated pupal abdomens substantial and physiological levels of ecdysteroids can be detected several days after ligation (Delbecque, J.-P. *et al.*, 1978a). The ovaries cannot be the source of these ecdysteroids in the pupal stage as there were no detectable titer differences noted between males and females.

Given this apparent ability of non-PG organs to synthesize ecdysteroids, surprisingly little is known about the identity of such tissues, at least during postembryonic development. It has been suggested that oenocytes may be capable of ecdysone synthesis (Locke, M., 1969) and, indeed, abdominal oenocytes from *Tenebrio molitor* larvae have been shown to convert radiolabelled cholesterol to 20-hydroxyecdysone (Romer, F. *et al.*, 1974). However, similar experiments with oenocytes from *Locusta migratoria* have failed to show any ecdysteroid synthesis (Hoffmann, J. *et al.*, 1975) and, as mentioned before, *Manduca sexta* oenocytes were not found to synthesize any ecdysone under *in vitro* conditions (King, D. *et al.*, 1974).

A more recent *in vitro* study with larval testes of *Heliothis virescens* has revealed that the testicular sheath may be a potential source of several ecdysteroids, including ecdysone (Loeb, M. *et al.*, 1982). Based on RIA activity, the larval testes of this species were found to release five times more ecdysteroid into the culture medium than was retained in the testes. However, since the pattern of ecdysteroids released *in vitro* was not the same as the ecdysteroid pattern in the hemolymph or testes, it remains to be established what affect such testicular ecdysteroid synthesis and secretion may have on the circulating ecdysteroid titer.

Another tissue which may synthesize and secrete ecdysone during postembryonic development, particularly during late pharate adult development, is the ovary. Although it is certain that the ovaries are a source of ecdysone and other ecdysteroids in several adult and pharate adult female insects (see Hoffmann, J. and Lagueux, M., vol. 1, and Hagedorn, H., vol. 8, this series) it remains to be determined to what extent such ovarian ecdysteroids are released to the hemolymph during postembryonic development.

The exceptions notwithstanding, there is no compelling reason at present for doubting that the PG are the primary, if not sole, physiological source of ecdysone during larval–larval and larval–pupal development. Indeed, the preponderance of data suggests that variations in the hemolymph titer of ecdysteroids result essentially from variations in the rates of ecdysone synthesis and secretion by the PG

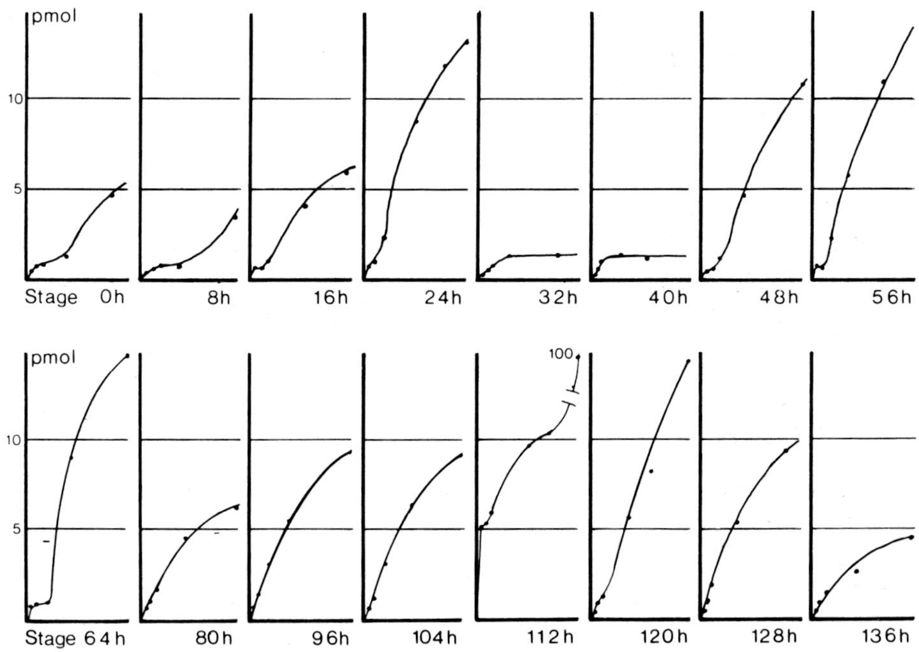

Fig. 5. The kinetics of *in vitro* ecdysone secretion by PG from last-instar larvae of *Locusta migratoria*. Abscissae: hours of fifth-instar post-ecdysis. Ordinates: ecdysone secreted (pmol equivalents) into the medium during 0.5, 2, 4 and 8h of incubation. (Figure reproduced with permission from Hirn, M. *et al.*, 1979.)

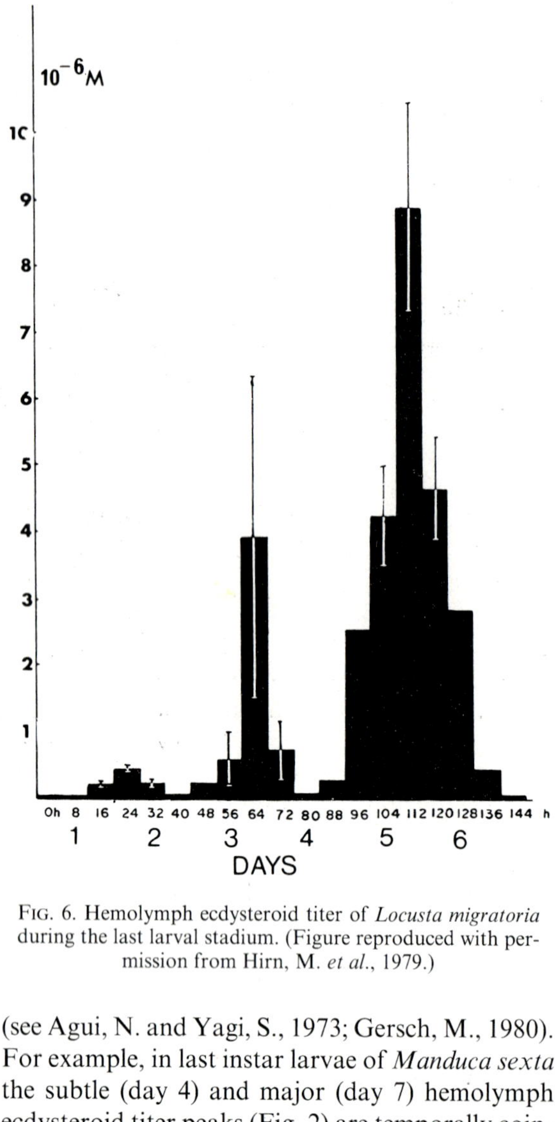

FIG. 6. Hemolymph ecdysteroid titer of *Locusta migratoria* during the last larval stadium. (Figure reproduced with permission from Hirn, M. *et al.*, 1979.)

(see Agui, N. and Yagi, S., 1973; Gersch, M., 1980). For example, in last instar larvae of *Manduca sexta* the subtle (day 4) and major (day 7) hemolymph ecdysteroid titer peaks (Fig. 2) are temporally coincident with a subtle increased rate and the maximum rate, respectively, of ecdysone secretion by PG extirpated on various days of the last instar (Bollenbacher, W. *et al.*, 1975). Similarly, in last instar larvae of *Locusta migratoria*, the kinetics of ecdysone secretion by PG *in vitro* showed three periods of increased activity at 24, 64 and 112 h into the stadium (Fig. 5). These three periods of intense ecdysone production by the PG were found to be coincident temporally with the three peaks in the hemolymph ecdysteroid titer during the last larval instar (Fig. 6) (Hirn, M. *et al.*, 1979). Even in the less advanced paleopterous insects it appears that variations in ecdysone production by the PG may be the major source of variation in the hemolymph ecdysteroid titer. In last instar larvae of the dragonfly, *Aeshna cyanea*, the kinetics of ecdysone secretion by their ventral glands (PG analogs) *in vitro* was found to vary. Two periods of increased ecdysone secretion by these ventral glands were found during the instar, and these periods of increased secretion were temporally and, in a relative sense, quantitatively coincident with the two peaks in the hemolymph ecdysteroid titer (Schaller, F. and Charlet, M., 1980).

4.2.2 Regulation of Ecdysone Synthesis

Much remains to be determined regarding the regulation of ecdysone biosynthesis during insect postembryonic development. Since the PG are the primary, if not sole, physiological source of ecdysone during these stages of development for most insects, it follows that the factors which regulate the production and secretion of ecdysone by the PG are themselves important regulators of the hemolymph ecdysteroid titer. Furthermore, inasmuch as the PG have been shown to release ecdysone immediately following its synthesis (see previous section), it also follows that the factors which regulate PG activity probably do so at the level of ecdysone synthesis and not at the level of its release.

PTTH. There is little doubt that the principal factor regulating PG activity is the tropic hormone, PTTH. However, even here much of what we know unequivocally is based upon studies with a single insect species, *Manduca sexta* (see Bollenbacher, W. and Granger, N., this volume).

Despite the long history of its putative existence (Kopeć, S., 1917, 1922) and tropic nature (Fukuda, S., 1940a,b, 1944; Williams, C., 1947, 1948, 1952) real progress in elucidating the chemistry, titer and tropic mechanisms of PTTH was hampered by the lack of a sufficiently sensitive and specific PTTH assay. Indeed, much of what was known about PTTH and its "regulation" of ecdysone biosynthesis was obtained indirectly using classical endocrinological techniques and non-specific dauer-pupae PTTH bioassays.

With the advances in insect organ culture and the development of ecdysteroid RIA (see section 3.1) in the early to mid-1970s, the requisite methods existed for the development of a defined and specific *in vitro* assay for directly assessing PTTH activity on the PG. Just such an *in vitro* assay was developed with *Manduca sexta* and the specifics of this assay — as well as its applicability for studying the chemistry, site of synthesis, titer and actions of PTTH, are discussed in another chapter of this series (see Bollenbacher, W. and Granger, N., this volume).

Utilizing this *in vitro* assay, it has recently been demonstrated unequivocally that PTTH from the brain directly activates the PG of *Manduca sexta* by eliciting an increase in the basal rate of ecdysone biosynthesis (Bollenbacher, W. *et al.*, 1979, 1980). This activation was found to be both specific and dose-dependent.

While this activation confirmed a basic tenet of insect endocrinology it could have been argued that such an *in vitro* activation might not be physiological. A more recent study, however, strongly suggests that PTTH is acting similarly under *in vivo* conditions and that it is probably one of the primary physiological regulators of PG activity.

A preliminary titer of PTTH in the hemolymph of fourth and fifth (last) instar larvae of *Manduca sexta* has been determined recently using the *in vitro* PTTH assay (Gilbert, L. *et al.*, 1981). The data from this study demonstrated that the hemolymph titer of PTTH is dramatically elevated just prior to, or concomitant with, the increases in the hemolymph titer of ecdysteroids (Fig. 7). Furthermore, these periods of elevated PTTH titer in the hemolymph were found to correspond temporally to the times of putative PTTH release predicted by classical head ligation studies, i.e. the so-called head critical periods (HCPs).

Previous determinations of the HCPs in fourth and fifth instar larvae of *Manduca* have revealed that there is one HCP in the fourth larval instar and two HCPs in the fifth instar (Truman, J., 1972; Truman, J. and Riddiford, L., 1974). The single

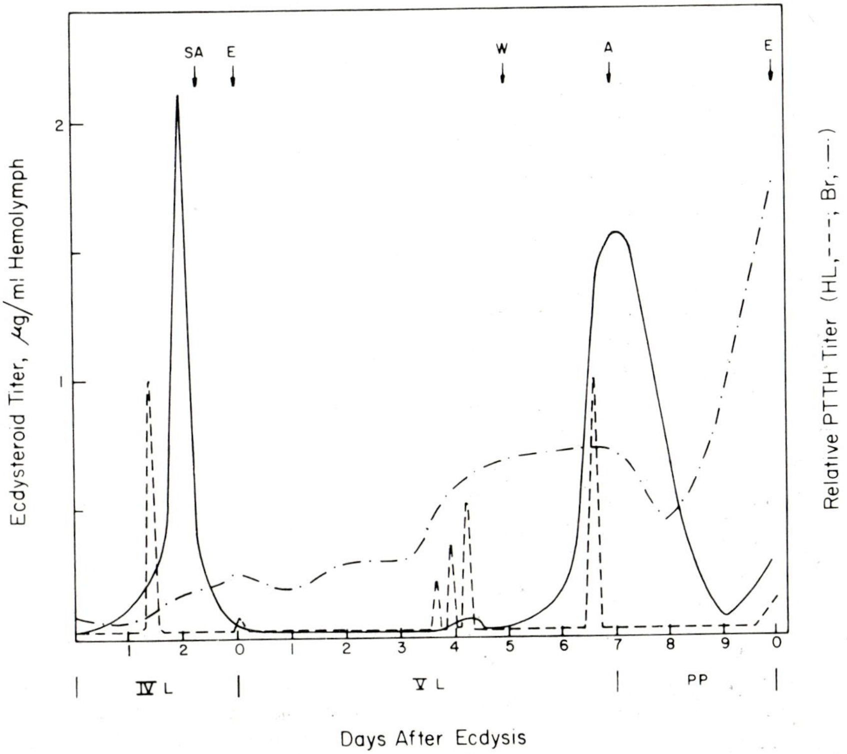

FIG. 7. The ecdysteroid and PTTH titers during the fourth larval (IVL) and fifth larval (VL)–pharate pupal (PP) stadia of *Manduca sexta*. The PTTH titers were determined for hemolymph (HL) and brain (Br) and are expressed relative to highest level. Developmental events include spiracle apolysis (SA), ecdysis (E), wandering behavior (W) and apolysis (A). (Figure reproduced with permission from Bollenbacher, W. and Gilbert, L., 1981.)

HCP in the fourth instar, and the second HCP in the last larval instar, are followed respectively by a single major increase in the hemolymph ecdysteroid titer which constitutes the "molting peak" for each instar (see section 2.1); whereas the first HCP of the fifth larval instar occurs just prior to the subtle peak in the hemolymph ecdysteroid titer which is presumably responsible for reprogramming the animal from larval to pupal synthesis (Riddiford, L., 1976, 1978, 1981).

On the basis of the hemolymph titer data in *Manduca*, it appears that PTTH is released as a single burst over a 4–6 h period during both the single HCP of the fourth instar and the second HCP of the fifth instar. By contrast, during the first HCP of the fifth instar the data suggest that PTTH may be released in a pulsatile or episodic fashion as three smaller bursts over an 18–20 h period (see Gilbert, L. *et al.*, 1981).

Although these studies with *Manduca* have demonstrated that PTTH directly activates PG *in vitro* and probably does so *in situ*, the exact mechanism(s) by which PTTH effects an increase in ecdysone biosynthesis is far from being fully understood. Since PTTH is probably a peptide, it has been assumed that this hormone may affect the PG in fashions similar to the more thoroughly studied actions of ACTH on the adrenal cortex. Preliminary studies with *Manduca* suggest, however, that the ACTH–adrenal cortex model may be only partially valid.

ACTH is known to stimulate steroidogenesis in the adrenal cortex by enhancing the rate-limiting step, i.e. the conversion of cholesterol to pregnenolone as catalyzed by the cytochrome P450-dependent cholesterol side chain cleavage system (see Simpson, E., 1979). The relationship between ACTH and cholesterol side chain cleavage activity is complex, involving both a cyclic AMP (cAMP)-dependent activation of cholesterol ester hydrolase activity and a labile protein factor. This latter factor may be "activated" via a cAMP-dependent protein kinase or perhaps by cAMP-independent factors such as calcium. In any case, both cAMP-dependent and cAMP-independent processes have been implicated in the actions of ACTH (Sharma, R., 1973).

Evidence that PTTH may be acting similarly to ACTH comes from several studies with *Manduca* PG. These studies have revealed that the PG from last-instar larvae of *Manduca* exhibit both an active adenylate cyclase and phosphodiesterase activity, enzyme activities associated with the synthesis and breakdown, respectively, of cAMP (Vedeckis, W. and Gilbert, L., 1973; Vedeckis, W. *et al.*, 1974). The adenylate cyclase activity has been found to be higher in active PG than in inactive glands, suggesting that cAMP may be a second messenger for PTTH (Smith, W. and Combest, W., vol. 8). It has also been demonstrated that PG incubated with cAMP and/or the phosphodiesterase inhibitors, aminophylline and 1-methyl-3-isobutyl xanthine, exhibit a 2–3-fold increase in ecdysone secretion (Fig. 8) (Vedeckis, W. *et al.*, 1974, 1976).

When the levels of endogenous cAMP were measured in the PG during the last larval instar of *Manduca*, however, it was found that the cAMP titer rose from basal levels some 15-fold during the first HCP but then fell back to basal levels and remained there for the rest of the instar (Fig. 9) (Vedeckis, W. *et al.*, 1976). These data suggested, therefore, that cAMP may mediate the actions of PTTH during the first HCP of the last larval instar of *Manduca* but not its actions during the second HCP.

Although the view that PTTH effects ecdysone biosynthesis by both cAMP-dependent and -independent processes is compatible with the actions of ACTH on the adrenal cortex, more recent studies suggest an alternative hypothesis. With the development of the *in vitro* PTTH assay, it has been possible

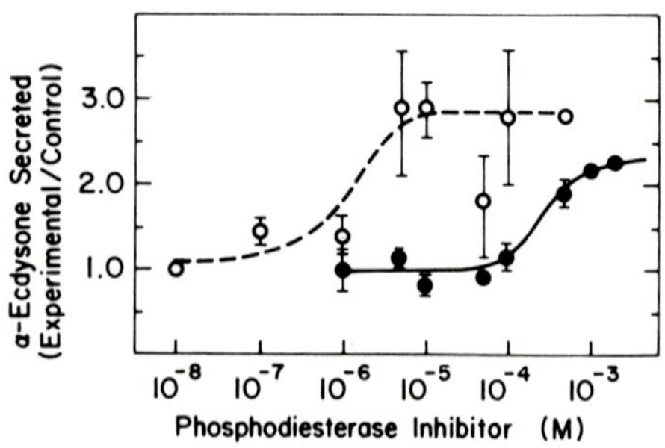

FIG. 8. The effects of the phosphodiesterase inhibitors, aminophylline (●——●) and 1-methyl-3-isobutyl-xanthine (○——○), on *in vitro* ecdysone secretion by day 3 PG from last-instar larvae of *Manduca sexta*. Contralateral glands served as controls. Each point is the mean ± SEM of replicate values from 2 to 3 separate experiments. (Figure reproduced with permission from Vedeckis, W. *et al.*, 1976.)

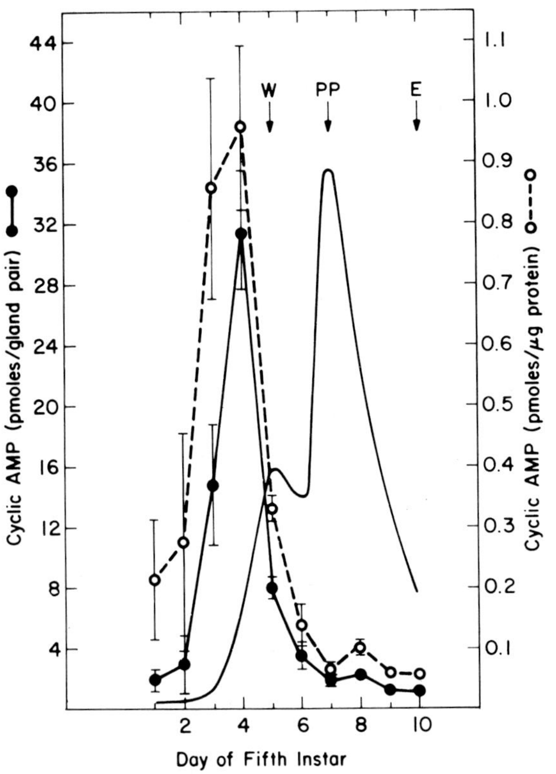

FIG. 9. The endogenous titer of cyclic AMP in the PG during the fifth larval stadium of *Manduca sexta*. Each point is the mean ±SEM of three determinations. The solid line represents the ecdysone secretory activity of PG during the same developmental period (from Bollenbacher, W. *et al.*, 1975). Developmental events include: wandering behaviour (W), pharate pupal development (PP) and ecdysis (E). (Figure reproduced with permission from Vedeckis, W. *et al.*, 1976.)

to partially purify PTTH from day 0–1 pupal brains of *Manduca* (Gilbert, L. *et al.*, 1981; see Bollenbacher, W. and Granger, N., this volume). The data from these studies suggest that PTTH is not a single moiety. Preliminary analyses suggest that there is a big and small form of PTTH (Gilbert, L. *et al.*, 1981). These two putative forms of PTTH not only differ in size but also in their chemical properties and ability to activate the PG. The big form of PTTH has been found to activate ecdysone biosynthesis in the PG to twice the level achieved by the small form of PTTH. Intriguingly, although to date only the big form of PTTH has been recovered from the hemolymph of *Manduca* during the HCP of the fourth larval instar, the kinetics of activation of the PG during the fifth larval instar suggest the possibilities that the small form of PTTH may be released during the first HCP and the large form during the second HCP (Fig. 7).

Although this hypothesis remains to be confirmed, if true it could indicate that the small form of PTTH may effect small amounts of ecdysone biosynthesis in the PG via cAMP-dependent processes, whereas the big form of PTTH may more fully activate the PG via cAMP-independent processes. From a purely speculative point of view, such a two-hormone–two-mechanisms hypothesis seems logical if one accepts that the minor and major peaks in the hemolymph ecdysteroid titer during the fifth larval instar of *Manduca* serve different functions, viz., reprogramming and molting, respectively (see section 3.2). Such a two-hormone–two-mechanisms system could serve as a failsafe device to insure that sufficient ecdysone is synthesized during the first HCP to elicit reprogramming but not so much as to be "misread" by the animal as a "molting" peak.

Other factors. In addition to PTTH, it is becoming increasingly apparent that others factors may also influence the rate of ecdysone biosynthesis, either by indirectly affecting the synthesis and release of PTTH or by directly acting on the PG. Such factors include environmental and neural stimuli, JH, hemolymph proteins and ecdysteroids. Although many of these factors are thought to be modulatory rather than regulatory in function, they may nevertheless serve roles as critical as PTTH in the control of ecdysone synthesis and secretion.

Clearly, the exact mechanism by which environmental factors such as light and temperature affect ecdysone synthesis remains to be determined. Such factors undoubtedly influence PTTH synthesis and release, and it is known that a light-sensitive circadian clock is involved in the release of PTTH (Truman, J., 1972; Truman, J. and Riddiford, L., 1974; Gelman, D. and Hayes, D., 1982; see Page, T., vol. 6, and Denlinger, D., vol. 8, this series). Intriguingly, the possibility exists that some environmental factors may also act directly on the PG. A recent study with *Samia cynthia ricini* larvae has suggested, based on transplantation experiments, that in addition to a light-sensitive cephalic circadian clock for the release of PTTH these organisms also possess a light-sensitive PG clock for the release of ecdysone (Mizoguchi, A. and Ishizaki, H., 1982).

Neural regulation of ecdysone biosynthesis, either indirectly via regulation of PTTH or directly on the PG, has been suggested by several studies. Studies with *Rhodnius prolixus* and *Locusta migratoria* have implicated that PTTH synthesis and release may be under direct neural regulation (Steel, C., 1978; Girardie, A. and DeReggi, M., 1978; see Steel, C. and Davey, K., vol. 8, this series), and more recently, it has also been demonstrated that PTTH release in *Manduca sexta* can be elicited by a calcium-dependent K^+-induced depolarization of the PTTH neurohemal organs (Carrow, G. *et al.*, 1981).

There is also an accumulating body of cytological evidence which suggests possible direct neural regulation or modulation of ecdysone biosynthesis and release at the PG. Although proof of such neural regulation has not been demonstrated (Blazsek, I. *et al.*, 1975; Herman, W. and Gilbert, L., 1966; Srivastava, K. *et al.*, 1977), it is known that the PG are innervated by various neural processes, including neurosecretory axons, from several ventral ganglia (Benedeczky, I. *et al.*, 1980; see Granger, N. and Bollenbacher, W., 1981). In the focal organism for many of these studies, *viz. Galleria mellonella* larvae, the PG are known to be innervated by the prothoracic, mesothoracic and subesophageal ganglia (Singh, H. and Sehnal, F., 1979) and all three of these ganglia have been reported to modify PG secretory activity either when implanted into decapitated larvae with PG *in situ* (Malá, J. *et al.*, 1977) or when incubated *in vitro* with PG (Gersch, M., 1978).

In addition to environmental and neural factors, ecdysone biosynthesis may be affected by JH and hemolymph proteins. JH and JH analogs have been reported to inhibit the synthesis and/or release of PTTH in *Chilo suppressalis* (Fukaya, M. and Mitsuhashi, J., 1961; Yagi, S. and Fukaya, M., 1974), *Manduca sexta* (Nijhout, H. and Williams, C., 1974a,b; Nijhout, H., 1975) and *Mamestra brassicae* (Hiruma, K. *et al.*, 1978). It has furthermore been suggested that such JH inhibition of PTTH synthesis and release may be a major factor contributing to larval or prepupal diapause in *Diatraea grandiosella* (Yin, C. and Chippendale, G., 1973) and *Monema flavescens* (Takeda, N., 1978; see D. Denlinger, vol. 8).

In contrast to the reported inhibitory effects of JH on PTTH synthesis and release, numerous studies have suggested that JH may directly activate the PG. Most of the earlier studies which suggested this JH stimulation, however, were based on the observations that CA implanted into brainless lepidopteran pupae would elicit adult development (see Gilbert, L. and King, D., 1973). Since it recently has been demonstrated in *Manduca sexta* that the CA are the neurohemal organs for the release of PTTH (Agui, N. *et al.*, 1980), the possibility exists that the tropic action of the CA on the PG may have actually been due to PTTH rather than JH. Nevertheless, several studies utilizing JH or JH analogs do support the premise that JH does affect PG activity. As early as the late 1950s it was demonstrated that ether extracts of adult male *Hyalophora cecropia* abdomens (a rich source of JH) would elicit adult development when injected into debrained lepidopteran pupae (Gilbert, L. and Schneiderman, H., 1959). More recently, it has also been reported that JH analogs topically applied to isolated abdomens of *Mamestra brassicae* larvae and pupae will activate PG implants (Hiruma, K. *et al.*, 1978; Hiruma, K., 1980).

This putative JH stimulation of the PG appears also to be affected by other factors. Recent experiments with *Mamestra brassicae* suggest that prior exposure of the PG to PTTH or ecdysone will enhance PG sensitivity to JH stimulation (Hiruma, K., 1982). Similar mechanisms may explain why JH analogs have been found to inhibit PG activity in early last-instar larvae but stimulate PG activity in late-instar larvae of *Spodoptera littoralis* (Cymborowski, B. and Stolarz, G., 1979) and *Manduca sexta* (Safranek, L. *et al.*, 1980).

Although all of these *in vivo* studies certainly suggest that JH affects PG activity, they do not demonstrate a direct affect of JH or JH analogs on the PG. Only recently have the effects of JH and JH analogs on PG activity, both *in vivo* and *in vitro*, been examined (Gruetzmacher, M., 1982). When JH-I or the JH analog ZR512 were topically applied to neck-ligated last-instar larvae of *Manduca sexta*, the results indicated that both compounds stimulated PG activity. However, when PG were extirpated from last-instar larvae or early pupae of *Manduca* and incubated with various concentrations of JH homologs, JH homolog acids, or ZR512, it was found that none of these compounds could significantly affect *in vitro* PG activity, either in an inhibitory or stimulatory fashion. Therefore, although this study confirmed the existence of an *in vivo* JH–PG axis, the data suggested that the JH effect on the PG is not a direct one.

Further experiments with *Manduca* have indicated that the JH activation of the PG may be mediated by a putative PG stimulatory factor which is induced in the fat body by JH. A preliminary analysis of the properties of this stimulatory factor has demonstrated that it is not PTTH. However, based on several lines of evidence, including its activation kinetics of the PG, it has been suggested that this PG stimulatory factor may be a lipoprotein providing sterol substrate to the PG (Gruetzmacher, M., 1982).

It remains to be demonstrated whether or not the PG stimulatory factor from the fat body is indeed a sterol carrier protein (SCP) for the PG in *Manduca* and whether such a factor may function as an intermediate in the JH–PG axis of other insects. However, it is known that lipoprotein SCPs are involved in the regulation of vertebrate steroidogenesis (see Gwynne, J. and Strauss, J., 1982) and preliminary studies have also suggested that hemolymph lipoproteins may be involved in the regulation of ecdysone biosynthesis. Lipoproteins, which are present in insect hemolymph, have been suggested to function as SCPs by carrying the substrate (i.e. cholesterol or some more oxygenated intermediate) for the synthesis of ecdysone to the PG (Chino, H. *et al.*, 1974; Gilbert, L. and Chino, H., 1974; H. Chino, vol. 10). In support of this, hemolymph and hemolymph lipoproteins have been demonstrated to markedly stimulate ecdysone biosynthesis in *Bombyx mori* PG *in vitro* (Chino, H. *et al.*, 1974; Sakurai, S. *et al.*, 1977).

More direct evidence for the existence of an SCP involved in ecdysone biosynthesis has come from recent and preliminary studies with *Manduca sexta* (Gunnar, W., 1979; W. Bollenbacher, W. Gunnar, M. Fulco and L. Gilbert, unpublished observations). In these studies a protein was isolated from the hemolymph of fifth-instar larvae of *Manduca* which markedly stimulates ecdysone biosynthesis in PG *in vitro*. The kinetics of PG activation by this factor were found to be several-fold greater than that which could be achieved by PTTH, suggesting that this protein might be a SCP. Based on several lines of evidence, including buoyant density centrifugation, gel filtration and thermal lability studies, it is fairly certain that this PG stimulatory hemolymph protein is not PTTH but a lipoprotein.

Although much remains to be determined concerning the properties of this hemolymph protein and its role in the regulation of ecdysone biosynthesis, a preliminary titer of this putative SCP during the fifth larval instar of *Manduca* has shown that it is present in maximal amounts at the time (day 7) of the hemolymph ecdysteroid titer molting peak (Fig. 2). Given the ability of this protein to activate the PG to a higher level than that achieved by PTTH, the SCP titer data suggest that this protein may be a major regulator of the hemolymph ecdysteroid titer, particularly at the times of the molting peaks.

In addition to these factors, ecdysteroids themselves may contribute to the regulation of ecdysone biosynthesis by feedback loops to the brain or directly to the PG. Ecdysone and 20-hydroxyecdysone have been reported to stimulate the synthesis and release of PTTH in *Leucophaea maderae* (Marks, E. *et al.*, 1972) and *Mamestra brassicae* (Agui, N. and Hiruma, K., 1977, 1978); and in *Rhodnius prolixus* evidence exists to suggest that these steroids may stimulate PTTH synthesis but inhibit its release (Steel, C., 1973, 1975, 1978; C. Steel and K. Davey, vol. 8).

In contrast to these cephalic feedback loops there is suggestive evidence that ecdysteroids directly inhibit ecdysone biosynthesis in the PG. It was demonstrated in the early 1970s that injections of 20-hydroxyecdysone would cause a decrease in RNA synthesis in saturniid pupal PG (Siew, Y. and Gilbert, L., 1971). More recently, evidence that 20-hydroxyecdysone acts as a negative feedback regulator of ecdysone biosynthesis was reported in a study with *Pieris brassicae* pupae (Beydon, P. and Lafont, R., 1983). Injections of ecdysone or 20-hydroxyecdysone into *Pieris* pupae were found to inhibit ecdysone biosynthesis, with 20-hydroxyecdysone being the more effective inhibitor. Since 20-hydroxyecdysone was also found to elicit such inhibition in decapitated pupae, it was suggested that this steroid acts directly on the PG. Direct evidence in support of this hypothesis has been recently reported for *Manduca sexta* (Scott, D., 1982). In this study 20-hydroxyecdysone, but not ecdysone, was found to inhibit ecdysone biosynthesis in fifth-instar larval PG *in vitro*.

In conclusion, it is fairly clear that the regulation of ecdysone biosynthesis in the PG is an extremely complex process involving not only PTTH but several other factors as well. At present we do not know the exact mechanisms for any of these factors, the quantitative extent to which each of these factors contribute to the regulation of ecdysone biosynthesis, nor the integrative events occurring between and among these factors at the PG. Despite this apparent complexity of PG regulation, it is imperative that much more research be focused on this problem in the near future. Particular emphasis needs to be placed on both elucidating the biosynthesis pathway(s) for ecdysone in the PG and characterizing the enzymes catalyzing the various biosynthetic steps (see Rees, H., this volume). Only then will we be in a position to properly address and eventually understand how ecdysone biosynthesis is regulated in the PG.

4.3 20-Hydroxyecdysone biosynthesis

Following the initial demonstrations that ecdysone can be hydroxylated to 20-hydroxyecdysone in specific tissues peripheral to the PG (King, D. and Siddall, J. 1969; King, D., 1969, 1972a,b), this conversion has become the most thoroughly studied reaction (both *in vivo* and at a subcellular level) in the biosynthetic pathway(s) of ecdysteroids. Indeed, to date this reaction is the only step in the biosynthesis of 20-hydroxyecdysone from cholesterol for which we have achieved partial characteristics of the enzyme systems involved.

There are several reasons why this particular reaction has received so much attention. From a practical standpoint the conversion of ecdysone to 20-hydroxyecdysone has been studied extensively because radiolabelled ecdysone substrate of high specific activity became available, thus facilitating trace detection and quantification of this reaction (King, D. and Siddall, J., 1969; Hafferl, W. *et al.*, 1972). Another cogent reason this conversion has received so much attention is that the reaction is a single hydroxylation. As such, information obtained about this conversion can serve as a paradigm for investigations on the several other hydroxylase reactions which must be involved in the biosynthesis of ecdysone from cholesterol (see Rees, H., this volume). Most importantly, however, this reaction has been studied intensively because of its potential biological significance.

Although it has been established that the larval PG of insect synthesize and secrete only ecdysone (see section 4.2.1), an accumulating body of evidence indicates that 20-hydroxyecdysone, and not ecdysone, is both the physiologically active form of MH and the predominant hemolymph ecdysteroid at least during larval–larval and larval–pupal development (see section 3.2). Accordingly, the post-PG conversion of ecdysone to 20-hydroxyecdysone must be viewed as a reaction critical to insect postembryonic development. Indeed, it has been suggested that this reaction may be a rate-limiting or regulated step which contributes to the overall control of the hemolymph ecdysteroid titer (Feyereisen, R. and Hoffmann, J., 1977; Smith, S. *et al.*, 1979, 1980; Feyereisen, R. and Durst, F., 1980a,b; Gilbert, L. *et al.*, 1980).

4.3.1 Ecdysone 20-mono-oxygenase systems

Although numerous earlier studies suggested that the conversion of ecdysone to 20-hydroxyecdysone is a major reaction in ecdysteroid metabolism (Table 3), they provided little insight into the properties of the enzyme(s) catalyzing this hydroxylation. By the mid-1970s, however, several laboratories began work on the characterization of this enzyme activity using an *in vitro* approach. Since nothing was known at that time concerning the properties of insect steroid hydroxylases, most of these studies were initiated using vertebrate steroid hydroxylase systems as models.

In vertebrates, it was known at that time that steroid hormone hydroxylations are catalyzed by complex enzyme systems referred to as mixed function oxidases or mono-oxygenases. These steroid hydroxylases are located in the mitochondria or endoplasmic reticulum of steroidogenic tissues in the adrenal cortex, testes, ovaries, and placenta. Such vertebrate steroid hydroxylases were also known to be actually short electron transport chains consisting of a flavoprotein reductase, a non-heme iron–sulfur protein, i.e. a ferrodoxin (mitochondrial steroid hydroxylases only), and a cytochrome P-450 as the terminal oxidase. In

Table 3: Ecdysone 20-mono-oxygenase activity in insects

Species	Tissue(s)	References
Calliphora vicina (erythrocephala)	whole body; fat body, hemocytes, Malpighian tubules	King, D., 1969; King, D. and Siddall, J., 1969; Young, N., 1976a,b; Young, N. and Young, P., 1976; Koolman, J., 1976; Koolman, J. *et al.*, 1979
Calliphora stygia	whole body	Galbraith, M. *et al.*, 1970
Antheraea polyphemus	whole body	Cherbas, L. and Cherbas, P., 1970
Bombyx mori	whole body	Moriyama, H. *et al.*, 1970
Hyalophora cecropia	whole body	Gorell, T. *et al.*, 1972
Manduca sexta	whole body; fat body, Malpighian tubules, midgut, integument, salivary glands(?)	King, D., 1972a,b; Nigg, H. *et al.*, 1976; Bollenbacher, W. *et al.*, 1977; Smith, S. *et al.*, 1977, 1979, 1980, 1983; Mayer, R. *et al.*, 1978; Kaplanis, J. *et al.*, 1980b
Chironomus tentans	whole body	Clever, U. *et al.*, 1973
Locusta migratoria	whole body; fat body, Malpighian tubules, hemocytes, gonads, midgut plus hindgut	Hoffmann, J. and Koolman, J., 1974; Hoffmann, J. *et al.*, 1974; Koolman, J. *et al.*, 1975; Feyereisen, R. *et al.*, 1976; Feyereisen, R. and Hoffmann, J., 1977; Feyereisen, R., 1977b, 1978; Feyereisen, R. and Durst, F., 1978, 1980a,b
Leucophaea maderae	leg regenerates, fat body	King, D. and Marks, E., 1974; Thompson, M. *et al.*, 1978
Aeshna cyanea	whole body	Schaller, F. *et al.*, 1975; Schaller, F. and Hoffmann, J., 1976
Sarcophaga bullata	fat body, carcass	Bollenbacher, W. *et al.*, 1976
Choristoneura fumiferana	whole body	Lagueux, M. *et al.*, 1976
Drosophila hydei	fat body	Beckers, C. and Emmerich, H., 1976
Gryllus bimaculatus	whole body	Wildmann, H. and Romer, F., 1977
Tenebrio molitor	whole body	Weinheimer, G. and Romer, F., 1977
Schistocerca gregaria	Malpighian tubules, fat body	Johnson, P. and Rees, H., 1977
Galleria mellonella	whole body	Rohner, R. and Meyer, D., 1980
Pieris brassicae	whole body	Lafont, R. *et al.*, 1980b
Drosophila melanogaster	whole body	Klose, W. *et al.*, 1980; Mitchell, M., 1983; Mitchell, M. and Smith, S., 1983
Aedes aegypti	whole body; body wall and fat body	Smith, S., 1983a

addition, it was also known by the mid-1970s that vertebrate steroid hydroxylases, and indeed all cytochrome P-450-dependent enzymes, require NADPH and molecular oxygen for activity (Cooper, D. *et al.*, 1969).

Given the stereochemical nature of the conversion of ecdysone to 20-hydroxyecdysone, viz., a C_{20}-hydroxylation or ecdysone 20-mono-oxygenation, coupled with the fact that by the mid-1970s cytochrome P-450-dependent enzyme activities were known to occur in insects (Ray, J., 1965, 1967; Agosin, M. and Perry, A., 1974), it seemed plausible that ecdysone 20-mono-oxygenase systems would be cytochrome P-450-dependent and, by extension, would require NADPH and molecular oxygen for activity (Fig. 10).

Tissue and subcellular distribution. Before ecdysone 20-mono-oxygenase activity could be characterized *in vitro*, however, it was necessary to find tissue sources of high enzyme activity and to develop sensitive and specific *in vitro* assays for its detection and quantification. Although two laboratories opted to develop *in vitro* ecdysone 20-mono-oxygenase assays which utilized unlabelled ecdysone as substrate (Nigg, H. *et al.*, 1976; Mayer, R. *et al.*, 1978), several laboratories developed more sensitive *in vitro* radioassays for assessing ecdysone 20-mono-oxygenase activity (Feyereisen, R., 1977b; Feyereisen, R. and Durst, F., 1978; Bollenbacher, W. *et al.*, 1977; Johnson, P. and Rees, H., 1977; Smith, S. *et al.*, 1977, 1979, 1980; Kaplanis, J. *et al.*, 1980b). In these latter cases, radiolabelled ecdysone of varying specific activity (0.0014 to 60 Ci mmol^{-1}) was incubated with NADPH and suspect tissues/subcellular fractions, and the ecdysone 20-mono-oxygenase activity was quantified by the amount of

$$RH + NADPH + H^+ + O_2 \longrightarrow ROH + NADP^+ + H_2O$$

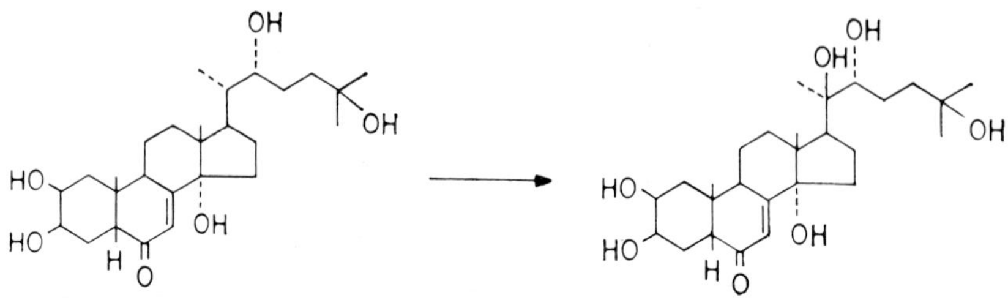

FIG. 10. The hydroxylation of ecdysone (left) to 20-hydroxyecdysone (right) as catalyzed by a cytochrome P-450-dependent steroid hydroxylase.

radiolabelled 20-hydroxyecdysone formed, as judged by TLC separations and liquid scintillation spectrometry. That both of these types of *in vitro* assays were actually measuring ecdysone 20-mono-oxygenase activity was confirmed by characterizing the *in vitro* assay product (or derivatized product) as 20-hydroxyecdysone (or a derivative) using HPLC, GLC-ECD and MS.

Utilizing these *in vitro* assays, it has been demonstrated that ecdysone 20-mono-oxygenase activity is localized in specific tissues peripheral to the PG, as earlier studies had suggested (King, D., 1969, 1972a,b). Although the data vary somewhat in detail, based on studies primarily with last-instar larvae of *Calliphora vicina, Manduca sexta, Locusta migratoria* and *Schistocerca gregaria*, it appears as if the major tissue sources of ecdysone 20-mono-oxygenase activity are the fat body, Malpighian tubules, midgut, integument and possibly gonads (see Table 3). It is also fairly clear that other tissues do not possess ecdysone 20-mono-oxygenase activity, including the PG, muscle, ventral nerve cord and imaginal disks. Because of unconfirmed or conflicting data, however, it remains to be established whether other tissues such as hemolymph/hemocytes, salivary glands, fore- and hindgut, and oenocytes possess ecdysone 20-mono-oxygenase activity, and if so, to what extent.

In any case, based on their relatively high enzyme activity, characterizations of ecdysone 20-mono-oxygenase systems were undertaken in the fat body (Bollenbacher, W. *et al.*, 1977; Smith, S. *et al.*, 1977, 1979, 1980) and midgut (Nigg, H. *et al.*, 1976; Mayer, R. *et al.*, 1978; Kaplanis, J. *et al.*, 1980b) of last-instar larvae of *Manduca sexta*, and in Malpighian tubules of last-instar larvae of *Locusta migratoria* (Feyereisen, R. *et al.*, 1977b; Feyereisen, R. and Durst, F., 1978) and *Schistocerca gregaria* (Johnson, P. and Rees, H., 1977). These studies initially focused on the subcellular distribution of ecdysone 20-mono-oxygenase activity.

In *Manduca sexta* fifth-instar larval fat body, the subcellular localization of ecdysone 20-mono-oxygenase activity was accomplished using an *in vitro* mono-oxygenase radioassay in conjunction with differential and sucrose gradient centrifugation, marker enzyme analysis and electron microscopy (EM) (Bollenbacher, W. *et al.*, 1977; Smith, S. *et al.*, 1977, 1979, 1980). Ecdysone 20-mono-oxygenase activity and subcellular marker enzyme activities were measured in fat body homogenates, subcellular fractions obtained by differential centrifugation and partially or totally reconstituted homogenates. The data from these experiments revealed that the mitochondrial fraction contained 67% of the total fat body ecdysone 20-mono-oxygenase activity with the remainder being recovered in the nuclear (22%) and microsomal fractions (11%). Since the mono-oxygenase activity exhibited by partially or totally reconstituted homogenates was found to be additive, the data implied that ecdysone 20-mono-oxygenase is mitochondrial in origin, with the activity in the other fractions being due to mitochondrial contamination. In support of this supposition, marker enzyme analysis of the subcellular fractions revealed that ecdysone 20-mono-oxygenase activity had the same subcellular distribution as cytochrome *c* oxidase activity, a mitochondrial marker; whereas no correlations were found between the subcellular distribution of ecdysone 20-mono-oxygenase activity and the microsomal marker enzyme activity, glucose-6-phosphatase. Further confirmation of the mitochondrial location of fat body ecdysone 20-mono-oxygenase was provided by EM and sucrose gradient centrifugation. EM of the subcellular fractions obtained by differential centrifugation revealed a distribution of mitochondria in the various fractions which was approximately proportional to the percentage distribution of mono-oxygenase activity present in each fraction. Similarly, *in vitro* radioassay of fat body subcellular fractions obtained by sucrose-step gradient centrifugation revealed that 84% of the total postnuclear mono-oxygenase activity was present in mitochondrial fractions and 16% in the microsomal fractions. More recently, based on EPR spectroscopy, it has been unequivocally demonstrated that an iron–sulfur protein of the ferrodoxin type is present in fat body mitochondria of *Manduca* (Smith, S. *et al.*, 1980). This would be expected if a steroid hydroxylase system is present in these mitochondria (see Sato, R. and Omura, T., 1978).

Although these studies with *Manduca* fat body reported the first unequivocal evidence for the existence of a mitochondrial steroid hydroxylase in invertebrates, the results were not entirely surprising. An earlier preliminary study with midgut from fifth-instar larvae of *Manduca sexta* had suggested that the ecdysone 20-mono-oxygenase activity in this tissue might be mitochondrial in location based on its detectable presence only in the 10,000 ***g*** pellet obtained by differential centrifugation (Nigg, H. *et al.*, 1976). An earlier study had also demonstrated that microsomal fractions of the house fly, *Musca domestica*, were incapable of converting ecdysone to 20-hydroxyecdysone (Terriere, L. and Yu, S., 1976).

Shortly following these initial studies with *Manduca sexta* it was reported that the ecdysone 20-mono-oxygenase activity in the Malpighian tubules of last-instar larvae of *Schistocerca gregaria* was also mitochondrial in location, based on differential centrifugation and marker enzyme analysis (Johnson, P. and Rees, H., 1977). However, similar to *Manduca* fat body, at least some mono-oxygenase activity in this tissue was also found to be associated with the microsomal fraction.

Collectively, the *Manduca* and *Schistocerca* studies seemed to suggest that insect ecdysone 20-mono-oxygenase systems were localized within mitochondria. Certainly, this was an attractive hypothesis from an evolutionary standpoint, since C_{20} steroid hydroxylation in vertebrates was also known to occur (as part of cholesterol side-chain cleavage) within the mitochondria of their steroidogenic tissues (Cooper, D. *et al.*, 1969; see Mitani, F., 1979). However, it soon became apparent that in insects ecdysone 20-mono-oxygenase systems could also be microsomal in location.

Preliminary studies with *Locusta migratoria* fat body, epidermis and Malpighian tubules suggested that the ecdysone 20-mono-oxygenase activity in these tissues might be microsomal in location (see Lagueux, M. *et al.*, 1976; Feyereisen, R., 1977b). Later, based on differential and sucrose gradient centrifugation, EM and marker enzyme analysis, it was demonstrated that the majority of ecdysone 20-mono-oxygenase activity in the Malpighian tubules and fat body from last-instar larvae of *Locusta* is localized in the microsomal fraction with a smaller amount associated with the mitochondrial fraction (Feyereisen, R. and Durst, F., 1978). Since in *Locusta* Malpighian tubules more mono-oxygenase activity was found associated with microsomal fractions obtained by sucrose gradient centrifugation than obtained by differential centrifugation (with mitochondrial fractions exhibiting just the opposite effect), it was felt that the previous localizations of ecdysone 20-mono-oxygenase systems in the mitochondrial fractions of *Manduca* and *Schistocerca* might be artifactual and due to contamination of these mitochondrial fractions by microsomes (Feyereisen, R. and Durst, F., 1978). Clearly, in retrospect this was not the case.

More recent studies of ecdysone 20-mono-oxygenase systems have revealed that these enzyme systems can be both mitochondrial and/or microsomal in location depending on the insect species and tissue in question. A recent evaluation of the subcellular distribution of ecdysone 20-mono-oxygenase activity in *Manduca* midgut revealed that 79% of the total mono-activity was associated with the mitochondrial fraction, with

11% of the activity associated with the microsomal fraction. The mitochondrial fraction was also found to have a higher specific activity (49 pmol 20-hydroxyecdysone $min^{-1} mg^{-1}$ protein) than the microsomal fraction (34 pmol 20-hydroxyecdysone $min^{-1} mg^{-1}$ protein) (Mayer, R. *et al.*, 1978). However, a more recent re-evaluation of *Manduca* midgut ecdysone 20-mono-oxygenase activity has revealed that although the mitochondrial fraction has a higher total activity than the microsomal fraction (70 versus 30%), the microsomal fraction has a higher specific activity than the mitochondrial fraction (42 versus 78 pmol 20-hydroxyecdysone $min^{-1} mg^{-1}$ protein) (Kaplanis, J. *et al.*, 1980b). Thus, in a single species and tissue, *Manduca* midgut, ecdysone 20-mono-oxygenase activity has been shown to be localized in both the microsomal and mitochondrial fractions. Even more recently, a preliminary study with *Drosophila melanogaster* third-instar larvae has also suggested that ecdysone 20-mono-oxygenase activity in a single species may be both mitochondrial (25% total activity) and microsomal (75% total activity) in location (Mitchell, M., 1983; Mitchell, M. and Smith, S., 1983).

Although more insect steroid hydroxylase systems need to be examined, the dual subcellular localization of ecdysone 20-mono-oxygenase systems in insects should not be viewed as controversial in itself, as precedents for this exist with vertebrates. Both C_{17}-steroid hydroxylase and C_{18}-steroid hydroxylase activities in vertebrates have been shown to be present in both mitochondrial and microsomal fractions (see Gower, D., 1974).

Cofactor requirements of ecdysone 20-mono-oxygenase systems. Studies on the cofactor requirements of ecdysone 20-mono-oxygenase systems have revealed that these insect systems exhibit similarities to vertebrate steroid hydroxylases. Vertebrate steroid hydroxylase systems require NADPH as a source of reducing equivalents for hydroxylation. This requirement is more or less obligatory for microsomal steroid hydroxylases, whereas mitochondrial hydroxylases can use other secondary sources of reducing equivalents such as NADH or Krebs cycle intermediates to support hydroxylation (see Sato, R. and Omura, T., 1978; Péron, F. *et al.*, 1975; Mitani, F., 1979). Presumably these secondary sources contribute to the intramitochondrial generation of NADPH via generation of NADH coupled to transhydrogenation of $NADP^+$.

Most insect ecdysone 20-mono-oxygenase systems examined appear to be supported by exogenous NADPH or NADPH generating systems (Bollenbacher, W. *et al.*, 1977; Smith, S. *et al.*, 1977, 1979, 1980; Johnson, P. and Rees, H., 1977; Feyereisen, R. and Durst, F., 1978; Mayer, R. *et al.*, 1978; Mitchell, M., 1983; Mitchell, M. and Smith, S., 1983). This requirement for NADPH is obligatory for the microsomal ecdysone 20-mono-oxygenase systems in *Locusta migratoria* Malpighian tubules (Feyereisen, R. and Durst, F., 1978) and *Drosophila melanogaster* larvae (Mitchell, M., 1983; Mitchell, M. and Smith, S., 1983). However, in the mitochondrial ecdysone 20-mono-oxygenase systems of *Manduca* fat body (Bollenbacher, W. *et al.*, 1977; Smith, S. *et al.*, 1977, 1979, 1980) and *Drosophila* larvae (Mitchell, M., 1983; Mitchell, M. and Smith, S., 1983) NADH has been noted to be about 25% and 50%, respectively, as effective as NADPH on a molar basis in supporting ecdysone 20-mono-oxygenation.

Although oxidized pyridine dinucleotides, ATP, ADP and Krebs cycle intermediates were found to be relatively ineffective in supporting mitochondrial hydroxylation in *Manduca* fat body and *Drosophila* larvae, several of these compounds have been demonstrated to support the mitochondrial ecdysone 20-mono-oxygenase activity of *Manduca* midgut (Mayer, R. *et al.*, 1978). Indeed, isocitrate, malate, succinate, ADP, NADH and NADPH have all been found to support this midgut mono-oxygenase activity.

Whether these cofactor differences between *Manduca* midgut and the other insect mitochondrial steroid hydroxylases are a reflection of genuine biochemical differences between the various mitochondrial mono-oxygenase systems, or perhaps a reflection of procedural differences, remains to be determined. Certainly, vertebrate mitochondrial steroid hydroxylases are known to be supported by NADPH or factors such as Krebs cycle intermediates. However, such vertebrate hydroxylases are usually supported better by NADPH when incubated under hypotonic conditions which disrupt the outer mitochondrial membrane; under isotonic conditions with intact mitochondria, Krebs cycle intermediates are the more effective hydrogen donors (Harding, B. *et al.*, 1968). Since *Manduca* midgut mitochondrial mono-oxygenase activity was supported by both types of factors under hyptonic conditions, this may suggest that midgut mitochondria have effective secondary pathways for providing reducing equivalents to the mono-oxygenase even under hypotonic and outer membrane disrupting conditions (Mayer, R. *et al.*, 1978).

Clearly much remains to be determined regarding how reducing equivalents are passed to mitochondrial ecdysone 20-mono-oxygenase systems. It also remains to be determined exactly where the insect mitochondrial steroid hydroxylases are situated. In vertebrates such steroid hydroxylases are associated with the inner mitochondrial membrane and, at present, there is no compelling reason for suspecting that insect mitochondrial hydroxylase systems are located anywhere other than this site (see Sato, R. and Omura, T., 1978; Mitani, F., 1979). If true, however, this poses a whole realm of questions as to how ecdysone and cofactors penetrate the outer mitochondrial membrane, and whether other factors may regulate this process.

Some data regarding these possibilities have been obtained by a recent preliminary study with *Manduca* midgut mitochondria (Kaplanis, J. *et al.*, 1980b). When such mitochondria were isolated and incubated under isotonic conditions, they could not effect ecdysone hydroxylation in the absence of added hydrogen donor(s). However, when incubated under hypotonic conditions it was also found that these mitochondria did not require added hydrogen donor(s) to effect significant levels of 20-hydroxyecdysone synthesis. These data suggested that *Manduca* midgut mitochondria contain endogenous hydrogen donor for mono-oxygenase activity but that the outer membrane may serve as a permeability barrier to the passage of ecdysone. Intriguingly, succinate was found to overcome this mitochondrial permeability barrier but was effective only in the presence of phosphate ions. Since malonate was found to inhibit the succinate effect, it was suggested that succinate functions by energizing the mitochondria resulting in an increased permeability or transport of extra-mitochondrial ecdysone to the intramitochondrial site of hydroxylation.

Enzyme kinetics Although most studies of insect ecdysone 20-mono-oxygenase systems are in agreement that these enzymes exhibit the hyperbolic saturation substrate kinetics characteristic of Michaelis–Menten systems, there clearly are reported differences in the kinetic properties of these various insect steroid hydroxylases. On the basis of reported apparent K_m and apparent V_{max} values it appears as if ecdysone 20-mono-oxygenase systems may fall into one of two general categories: (1) those characterized by relatively lower apparent K_m values for ecdysone and lower apparent V_{max} values and (2) those characterized by relatively higher apparent K_m values for ecdysone and higher apparent V_{max} values.

In the former category are the ecdysone 20-mono-oxygenase systems in: *Calliphora erythrocephala* whole body ($K_m = 4.5 \times 10^{-7}$ M; $V_{max} \cong 10$–40 pg min^{-1} mg^{-1} tissue) (Young, N., 1976a); *Manduca sexta* fat body mitochondria ($K_m = 1.6 \times 10^{-7}$ M; $V_{max} = 2.5$ ng min^{-1} mg^{-1} protein) (Bollenbacher, W. *et al.*, 1977; Smith, S. *et al.*, 1977, 1979, 1980); *Locusta migratoria* Malpighian tubule microsomes ($K_m = 2.7 \times 10^{-7}$ M; $V_{max} \cong 2.2$ pmol min^{-1} mg^{-1} protein) (Feyereisen, R. and Durst, F., 1978); and *Drosophila melanogaster* whole-body mitochondria ($K_m = 6.4 \times 10^{-8}$ M; $V_{max} = 4.1$ pg min^{-1} mg^{-1} tissue) and microsomes ($K_m = 9.9 \times 10^{-8}$ M; $V_{max} = 12.8$ pg min^{-1} mg^{-1} tissue) (Mitchell, M., 1983; Mitchell, M. and Smith, S., 1983). In the latter category are the ecdysone 20-mono-oxygenases in the mitochondria of *Schistocerca gregaria* Malpighian tubules ($K_m = 1.25 \times 10^{-6}$ M; $V_{max} = 2.5$ nmol min^{-1} mg^{-1} protein) (Johnson, P. and Rees, H., 1977) and *Manduca sexta* midgut ($K_m = 1.8 \times 10^{-5}$ M; $V_{max} = 46.6$ pmol min^{-1} mg^{-1} protein) (Mayer, R. *et al.*, 1978).

Accepting the premise that the apparent K_m values represent a relative index of the enzyme's affinity for ecdysone and that the apparent V_{max} values are a relative index of the enzyme concentration or titer (Segel, I., 1976), it would appear that the mono-oxygenase systems in the former category are higher-affinity–lower-capacity systems, whereas those hydroxylases in the latter category are lower-affinity–higher-capacity systems. On the basis of this it has been suggested that the former category of hydroxylases may function at times of relatively low ecdysone concentrations with a limited capacity for conversion, whereas the latter type of hydroxylases may function at times of high ecdysone titers and rapid turnover (Kaplanis, J. *et al.*, 1980b).

Clearly, however, caution needs to be exercised in interpreting such kinetics data. Although the low K_m values of several of the mono-oxygenase systems may help to explain why so little ecdysone is present in the hemolymph of many insects during larval and pupal development, the K_m values of these enzymes should nevertheless approximate the endogenous concentration of their ecdysone substrate (Segel, I., 1976). This is certainly not the case for most of the mono-oxygenase systems. With the exceptions of the *Manduca* fat body and *Locusta* Malpighian tubule mono-oxygenase systems whose K_m values do approximate the highest endogenous ecdysone titer in the last-instar larvae of these organisms, i.e. 2 to 4×10^{-7} M (Hirn, M. *et al.*, 1979; Smith, S. *et al.*, 1983), the other mono-oxygenase systems have apparent K_m values which either slightly or greatly exceed the endogenous ecdysone titer. These elevated K_m values may indicate that there are "activators" which function *in vivo* to effectively lower the K_m, that there are higher intracellular concentrations of ecdysone near the mono-oxygenase systems, or that the procedures used to determine the various apparent K_m values led to an overestimation (Segel, I., 1976).

There is suggestive evidence to support the phenomenon of *in vivo* activation of insect mono-oxygenases by various factors but it remains to be demonstrated whether such "activators" function as regulators of ecdysone 20-mono-oxygenase activity (see section 4.3.2). Similarly, it is a very real possibility that some tissues or subcellular fractions may have *in situ* concentrations of ecdysone which exceed the hemolymph titer and approximate the mono-oxygenase apparent K_m (Karlson, P. *et al.*, 1964; Gorell, T. *et al.*, 1972; Natori, S. and Ohtaki, T., 1976). However, it cannot be dismissed that some of the apparent K_m values, particularly the higher ones, may have been overestimated. For example, the apparent K_m of the *Manduca* midgut mitochondrial mono-oxygenase system was determined from kinetics data generated by a relatively insensitive non-radioassay and analyzed only by Lineweaver–Burk plots (Mayer, R. *et al.*, 1978). Such assay methods would not be sensitive enough to detect lower levels of conversion or slight changes in conversion rates, and any errors in the determination of velocity would be magnified when reciprocals of the data are calculated (Segel, I., 1976).

Another potential source of error with the K_m determinations has been the failure in several studies to determine if the mono-oxygenase apparent K_m values are a constant of the enzyme system and independent of tissue concentration (Segel, I., 1976). Only with *Manduca* fat body and *Drosophila* larvae have the mono-oxygenase apparent K_m values been demonstrated to be independent of tissue concentration (Bollenbacher, W. *et al.*, 1977; Smith, S. *et al.*, 1977, 1979, 1980, Mitchell, M., 1983).

As an example of how such considerations may affect the apparent K_m, when this parameter was redetermined recently for *Manduca* midgut mono-oxygenase using an *in vitro* radioassay coupled with analysis of the kinetics data by both Michaelis–Menten and Lineweaver–Burk plots, the results revealed that the midgut mono-oxygenase activity exhibited an apparent K_m for ecdysone of 4.7×10^{-7} M (Smith, S. *et al.*, 1983). This apparent K_m approximated the highest physiological titer of ecdysone in the animal, and was independent of tissue concentration. The fact that this apparent K_m is two orders of magnitude lower than the previously reported value serves to illustrate that the apparent K_m may be as much a reflection of methodology as of the intrinsic properties of the mono-oxygenase system.

Cytochrome P-450 properties of ecdysone 20-mono-oxygenase systems. In addition to their subcellular localizations and cofactor requirements, the various insect ecdysone 20-mono-oxygenase systems exhibit many of the properties of vertebrate cytochrome P450-dependent systems. These properties include: inhibition by both specific and non-specific P450-dependent steroid hydroxylase inhibitors; demonstrable presence of cytochrome P-450 in mono-oxygenase active subcellular fractions; and, in one case, demonstrable action of cytochrome P-450 as the biocatalyst in ecdysone 20-mono-oxygenation.

To help establish the mixed function oxidase and cytochrome P-450 nature of ecdysone 20-mono-oxygenase systems, the effects of various vertebrate steroid hydroxylase inhibitors on the hydroxylation of ecdysone have been investigated (Cooper, D. *et al.*, 1969; Gower, D., 1974; Sato, R. and Omura, T., 1978). The specific steroid hydroxylase inhibitors, metyrapone and *p*-aminoglutethimide, were both found to inhibit in a dose–response fashion the ecdysone 20-mono-oxygenase systems in *Manduca* fat body mitochondria (Bollenbacher, W. *et al.*, 1977; Smith, S. *et al.*, 1977, 1979, 1980) and in *Drosophila* whole-body mitochondria and microsomes (Mitchell, M., 1983). The I_{50} values for these inhibitions (range from 5×10^{-5} M to 5×10^{-6} M) were in essential agreement with those observed for vertebrate steroid hydroxylases. Similarly, metyrapone was found to inhibit the ecdysone 20-mono-oxygenase activity in *Schistocerca* Malpighian tubule mitochondria (Johnson, P. and Rees, H., 1977) and *Locusta* Malpighian tubule microsomes (Feyereisen, R. and Durst, F., 1978).

In addition to these specific inhibitors of vertebrate steroid hydroxylases, various non-specific inhibitors of general cytochrome P-450-dependent systems have also been investigated. The non-specific inhibitor, *p*-chloromercuribenzoate, was found to inhibit in a dose–response fashion the ecdysone 20-mono-oxygenase activities in *Manduca* fat body mitochondria and *Locusta* Malpighian tubule microsomes (Bollenbacher, W. *et al.*, 1977; Smith, S. *et al.*, 1977, 1979, 1980; Feyereisen, R. and Durst, F., 1978). Several other general inhibitors of P-450 systems, including 7,8-benzoflavone, SKF525-A, piperonyl butoxide and *N*-ethylmaleimide were also found to inhibit the microsomal mono-oxygenase activity in *Locusta* Malpighian tubules (Feyereisen, R. and Durst, F., 1978). Collectively, these inhibition data provide strong evidence, albeit indirect, that ecdysone 20-mono-oxygenase systems are P-450-dependent.

A general characteristic, however, of all cytochrome P-450-dependent mono-oxygenases is their strong and reversible inhibition by carbon monoxide (CO) which competes with O_2 for the reduced heme protein (Cooper, D. *et al.*, 1969; Sato, R. and Omura, T., 1978). The degree of such inhibition is approximately proportional to the ratio of CO/O_2, indicating the competitive nature of the inhibitory effect.

The effects of CO on ecdysone 20-mono-oxygenase activity have been examined in several insects, and in all cases CO was found to inhibit these enzymes. In *Manduca* fat body mitochondria, *Locusta* Malpighian tubule microsomes and *Drosophila* whole-body homogenates, the CO effect was dose-dependent in regard to the CO/O_2 (Smith, S. *et al.*, 1979; Feyereisen, R. and Durst, F., 1978; Mitchell, M., 1983). However, the dose–response of inhibition was not the same in all cases. *Manduca* fat body and *Drosophila* homogenate ecdysone 20-mono-oxygenase activities were inhibited approximately 50% by a CO/O_2 of 8:1(v/v), whereas comparable inhibition of *Locusta* Malpighian tubule mono-oxygenase activity was achieved with a CO/O_2 of approximately 1.5:1 (v/v). Similar to the *Manduca* fat body system, the mono-oxygenase activity in *Schistocerca* Malpighian tubule mitochondria was found to be inhibited approximately 50% by a CO/O_2 of 9:1 (Johnson, P. and Rees, H., 1977). However, in *Manduca* midgut the mitochondrial mono-oxygenase activity was reported to be inhibited approximately 80% by a CO/O_2 of 4:1 (Mayer, R. *et al.*, 1978).

It is not readily apparent whether these differences in sensitivity to CO inhibition are a reflection of dramatic biochemical differences among the various ecdysone 20-mono-oxygenases or perhaps due to procedural differences. However, it is known that the CO inhibition of *Manduca* fat body, *Drosophila* larvae and *Locusta* Malpighian tubule mono-oxygenase systems is a specific effect of CO on the mono-oxygenase systems and not on energy metabolism. Cyanide (up to 10^{-3} M) had no effect on these mono-oxygenase systems. Interestingly, however, even low concentrations of cyanide have been found to inhibit the ecdysone 20-mono-oxygenase activity in *Schistocerca* Malpighian tubule mitochondria (Johnson, P. and Rees, H., 1977).

Although these inhibition experiments indicate similarities between insect ecdysone 20-mono-oxygenase systems and vertebrate steroid hydroxylases, and suggest a role for cytochrome P-450 in such insect systems, they cannot establish the presence of cytochrome P-450 in the mono-oxygenase active tissues or subcellular fractions. To demonstrate the presence of heme P-450 in any biological sample it is necessary to utilize the technique of CO-difference spectroscopy (Cooper, D. *et al.*, 1969; Sato, R. and Omura, T., 1978). A nephelometric technique, CO difference spectroscopy can establish the presence of heme P-450 in a sample based on the fact that reduced heme P-450–CO adducts will absorb maximally at 450 nm.

When various insect subcellular fractions were subjected to CO difference spectroscopy, the spectra revealed the presence of heme P-450 in the mitochondrial fractions from *Manduca* fat body (Bollenbacher, W. *et al.*, 1977; Smith, S. *et al.*, 1977, 1979, 1980) and midgut (Mayer, R. *et al.*, 1978); the mitochondrial fraction from *Schistocerca* Malpighian tubules (Johnson, P. and Rees, H., 1977) and the microsomal fraction from *Locusta* Malpighian tubules (Feyereisen, R. and Durst, F., 1978). In the cases of *Manduca* fat body and midgut mitochondria the peak of absorbance at 450 nm was also accompanied by a much larger Soret band at about 430 nm (Fig. 11). This latter peak is probably due to the presence of large amounts of cytochrome a-a_3-CO in these mitochondria.

Based on the millimolar extinction coefficient of vertebrate P-450 systems, i.e. 91 $cm^{-1} mM^{-1}$ (see Sato, R. and Omura, T., 1978), it was calculated that the amount of heme P-450 in the various insect fractions was ~10 to 15 pmol mg^{-1} protein for *Manduca* fat body and midgut mitochondria, ~910 pmol mg^{-1} protein for *Schistocerca* Malpighian tubule mitochondria and ~100–150 pmol mg^{-1} protein for *Locusta* Malpighian tubule mitochondria. With the exception of the *Schistocerca* system, therefore, the heme P-450 concentrations in the various insect systems are low in comparison to the heme P-450 content in vertebrate adrenal cortex systems (~1.5 nmol mg^{-1} protein). However,

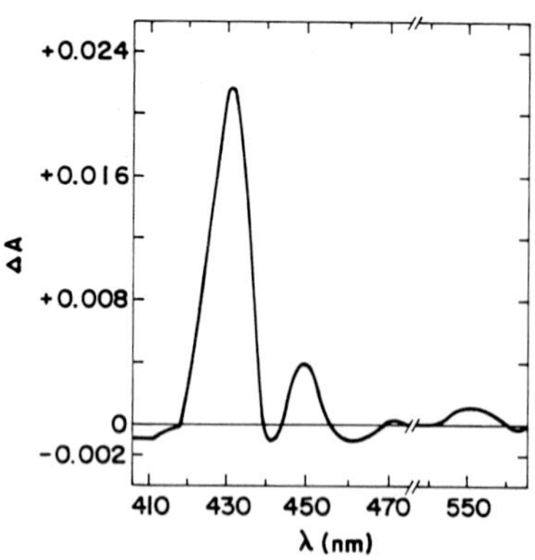

FIG. 11. The CO difference spectrum of fat body mitochondria from day 1 last-instar larvae of *Manduca sexta*. (Figure reproduced with permission from Smith, S. *et al.*, 1979.)

this may be a reflection of the generally lower qualitative and/or quantitative specialization of insect fat body, midgut and Malpighian tubules for steroid hydroxylation in comparison to vertebrate adrenal cortex.

Although CO difference spectra establish the presence of P-450 in a suspect tissue or subcellular fraction, they do not establish that the P-450 is the biocatalyst for the mono-oxygenation in question. To demonstrate unequivocally that a given steroid hydroxylase is cytochrome P-450-dependent it is necessary to determine the photochemical action spectrum of that hydroxylase activity (or to purify and reconstitute the system).

The photochemical action spectrum determination measures the ability of various wavelengths of monochromatic light of equal intensities to reverse the CO inhibition of an enzyme system. Since CO inhibited P-450 systems absorb maximally at 450 nm, this wavelength of light should be the most effective in reversing CO inhibition (see Cooper, D. *et al.*, 1969; Smith, S. *et al.*, 1980). It has been demonstrated with *Schistocerca* Malpighian tubule mitochondria and *Locusta* Malpighian tubule microsomes that white light will reverse CO inhibition of ecdysone 20-mono-oxygenase activity (Johnson, P. and Rees, H., 1977; Feyereisen, R. and Durst, F., 1978). However, in the case of *Manduca* fat body mitochondria, a complete photochemical action spectrum of ecdysone 20-mono-oxygenase activity has been determined (Smith, S. *et al.*, 1979, 1980). The results of this determination revealed unequivocally that cytochrome P-450 was the biocatalyst in the ecdysone 20-mono-oxygenase system (Fig. 12). The data further revealed, however, that a putative second pigment absorbing around 420 nm may also be involved in the conversion of ecdysone to 20-hydroxyecdysone.

Clearly, in retrospect, insect ecdysone 20-mono-oxygenases are cytochrome P-450-dependent systems which exhibit many properties similar to vertebrate P-450 systems. It is also becoming increasingly apparent that cytochrome P-450 systems play numerous roles in the endocrinology and physiology of insects. P-450 systems have been suggested to be involved in the epoxidation of JH precursors, the metabolism of precocenes and other allelochemics, and are well known for their roles in xenobiotic metabolism (Feyereisen, R. *et al.*, 1981;

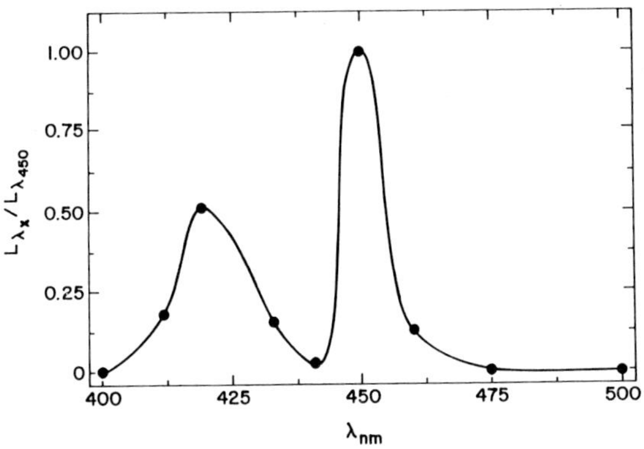

FIG. 12. The photochemical action spectrum of ecdysone 20-mono-oxygenase activity in mitochondria from the fat body of day 1 last-instar larvae of *Manduca sexta*. Each point is the average of 5 to 20 determinations. (Figure reproduced with permission from Smith, S. *et al.*, 1979.)

Hodgson, E., 1983; see Hodgson, E., vol. 11, this series). In addition, P-450 systems will undoubtedly prove to be the biocatalysts for the various other hydroxylations required to convert cholesterol to ecdysone.

Substrate specificity. Based on product characterization by various analytical techniques, there can be no doubt that the insect ecdysone 20-mono-oxygenase systems do in fact hydroxylate ecdysone to 20-hydroxyecdysone. However, little is known about the substrate specificity of such mono-oxygenase systems.

There are reasons for suspecting that ecdysone 20-mono-oxygenase systems might accept several substrates. For one thing, although vertebrate steroid hydroxylases exhibit substrate specificity it is not absolute. It has recently been demonstrated with a purified adrenal cortex mitochondrial 11β-hydroxylase system that such a system will not only catalyze 11β-hydroxylation, but also, somewhat less efficiently, 18-hydroxylation and 19-hydroxylation of various C_{21} and C_{19} steroids (e.g. testosterone, deoxycorticosterone, 4-androstene-3,17-dione, deoxycortisol) (Sato, H. *et al.*, 1978). Thus, the specificity associated with steroid hydroxylases may be an operational property determined by both the hydroxylase and substrate. Secondly, it has been known for several years that the fat body of *Manduca sexta* is capable of hydroxylating 22,25-dideoxyecdysone to ecdysone or 20-hydroxyecdysone (King, D., 1972a).

In addition to these factors, the specificity of ecdysone 20-mono-oxygenase systems takes on added significance when judged in the light of the various reports that ecdysone can be synthesized in isolated abdomens and prothoracectomized insects (see section 4.2.1) or the various suggestions that the PG substrate for conversion to ecdysone may be some compound more oxygenated than cholesterol (Gilbert, L. *et al.*, 1977). However, despite the potential significance of the specificity of ecdysone 20-mono-oxygenase systems only preliminary studies have been carried out.

In *Locusta migratoria* the Malpighian tubule microsomal ecdysone 20-mono-oxygenase system has been demonstrated not to hydroxylate 5α-ecdysone, 2β-acetoxy-ecdysone, 3-dehydroecdysone and 3-dehydro-20-hydroxyecdysone (Feyereisen, R. and Durst, F., 1978). Furthermore, this

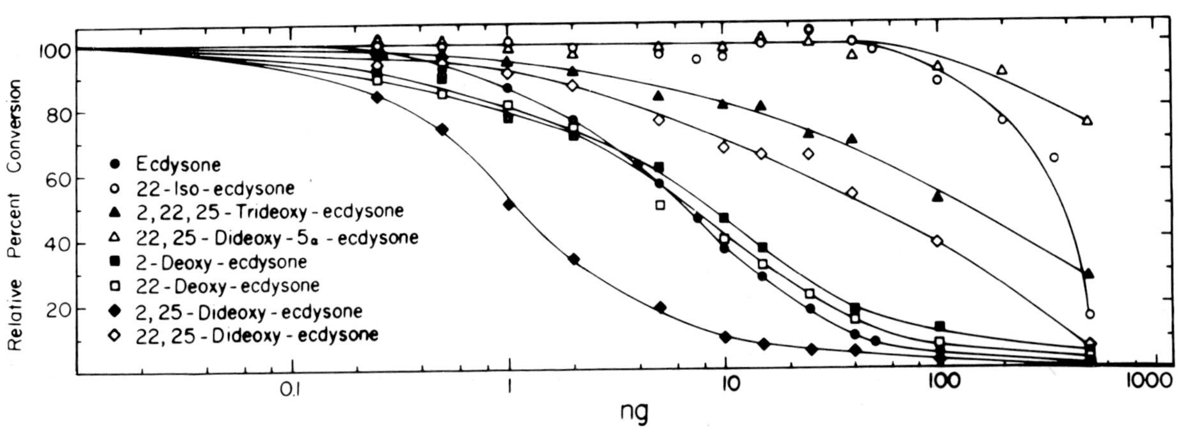

FIG. 13. The apparent substrate specificity of the ecdysone 20-mono-oxygenase system in mitochondria from the fat body of day 1 last-instar larvae of *Manduca sexta*. (Figure reproduced with permission from Smith, S. *et al.*, 1980.)

mono-oxygenase system did not further hydroxylate ecdysone beyond 20-hydroxyecdysone, nor did the system exhibit any side chain cleavage activity.

By contrast, however, a preliminary study with the *Manduca* fat body mitochondrial mono-oxygenase has suggested that this system may accept various substrates depending on their polarity and structure (Smith, S. *et al.*, 1980). Based solely upon competition curves, i.e. the presumed ability of a test substance to compete with [^{3}H]ecdysone for the active site on the mono-oxygenase system, it appears as if several ecdysteroids may serve as suitable substrates for this ecdysone 20-mono-oxygenase system (Fig. 13). Surprisingly, the most effective competitior was not ecdysone but 2,25-dideoxyecdysone, followed in order of decreasing competability by ecdysone, 22-deoxyecdysone, 2-deoxyecdysone, 22,25-dideoxyecdysone, 2,22,25-trideoxyecdysone, 22-isoecdysone and 22,25-dideoxy-5α-ecdysone.

If indeed these various ecdysteroids are actually serving as substrates for the ecdysone 20-mono-oxygenase system, which cannot be distinguished until the products are chemically identified, this would indicate that the mono-oxygenase may hydroxylate a variety of substrates and, in some cases, efficiently enough to be biologically significant. As a logical extension, this would also raise the possibility of multiple functions for the ecdysone 20-mono-oxygenase system.

4.3.2 Regulation of ecdysone 20-mono-oxygenase activity

Developmental variations in ecdysone 20-mono-oxygenase activity. Based on several *in vivo* studies on the metabolic fate of injected radiolabelled ecdysone, it has been suggested that the apparent rate of conversion of ecdysone to 20-hydroxyecdysone varies considerably during postembryonic development (see Table 3). Furthermore, data from most of these studies seem to suggest that the conversion of ecdysone to 20-hydroxyecdysone is maximal at the times when the MH titer is expected to be high. On the basis of these data, therefore, it has been inferred that ecdysone 20-mono-oxygenase activity may vary dramatically during development, reaching maximal activity at times of high MH titer. While there is little doubt that the apparent conversion rate of ecdysone to 20-hydroxyecdysone varies during development, such relative conversion rates determined *in vivo* cannot be taken *a priori* as a relative index of ecdysone 20-mono-oxygenase activity. At best such *in vivo* data are highly equivocal.

For one thing, it is known that exogenous ecdysone may not be metabolized along the same lines as endogenous ecdysone (Clever, U. *et al.*, 1973). Secondly, the apparent conversion rate of ecdysone to 20-hydroxyecdysone *in vivo* is a reflection of a multitude of reactions in addition to ecdysone 20-mono-oxygenase systems. Changes in the rate of excretion or catabolism of ecdysone, changes in the rate of ecdysone biosynthesis, changes in the rate of tissue uptake can all alter the pool of ecdysone available to the ecdysone 20-mono-oxygenase systems and consequently be reflected as apparent changes in ecdysone 20-mono-oxygenase activity. Given these inherent limitations with *in vivo* systems, therefore, it is necessary that *in vitro* systems be utilized to determine what changes, if any, are taking place with respect to ecdysone 20-mono-oxygenase activity during postembryonic development.

Unfortunately, very few studies have been directed toward analyzing ecdysone 20-mono-oxygenase activity under *in vitro* conditions during postembryonic development. A preliminary study with *Manduca sexta* midgut mitochondria suggested that the ecdysone 20-mono-oxygenase activity in this tissue was maximal during the wandering stage (prepupal) of last-instar larvae but was absent during most of the pupal stage and on the first days of each of the five larval instars (Nigg, H. *et al.*, 1976). Similarly, a preliminary study of mitochondrial ecdysone 20-mono-oxygenase activity in *Schistocerca gregaria* Malpighian tubules during the fifth larval instar revealed that activity was low during the early stages of the instar but rose approximately 3-fold by the fifth day and then slowly decreased by approximately 40% by day 7 of the instar (Johnson, P. and Rees, H., 1977). The peak in mono-oxygenase activity was found to precede the maximum titer of MH by about 2 days (Gande, A. *et al.*, 1979).

More recently, several enzyme activities, including ecdysone 20-mono-oxygenase, were measured in the microsomes from the fat body and Malpighian tubules of *Locusta migratoria* during the fifth larval instar (Feyereisen, R. and Durst, F., 1980a). These activities were compared to the tissue microsomal P-450 levels and the hemolymph titer of 20-hydroxyecdysone. The data revealed that ecdysone 20-mono-oxygenase activity was low in both tissues during the first 4 days of the instar but then underwent a rapid and approximately 7-fold increase in both tissues by day 5 of the instar. This increase was followed by an equally rapid decrease back to basal levels by day 7 of the instar. Interestingly, the peaks in both tissue mono-oxygenase activities coincided with the peak of 20-hydroxyecdysone in the hemolymph. Accompanying the peaks of both tissue mono-oxygenase activities were also peaks in the microsomal cytochrome P-450 levels, NADPH cytochrome c-reductase activities and P-450 linked lauric acid ω-hydroxylase activity (fat body only). The data not only suggested that the increases in ecdysone 20-mono-oxygenase activities were due to increases in the titers of the mono-oxygenase systems, but were also in essential accord with previous *in vivo* studies which suggested that ecdysone 20-mono-oxygenase activity is maximal at the times of the MH titer peak. Furthermore, the data from this *Locusta* study were also consistent with the view that the ratio of ecdysone to 20-hydroxyecdysone may be determined by the relative mono-oxygenase activity, as the dramatic increase in ecdysone 20-mono-oxygenase activity on day 5 was concomitant with the dramatic drop noted in the hemolymph ecdysone to 20-hydroxyecdysone ratio (Hirn, M. *et al.*, 1979).

More recently, the ecdysone 20-mono-oxygenase activities in the fat body and midgut of *Manduca sexta* were determined during larval–pupal development (Smith, S. *et al.*, 1983). Similar to *Locusta*, the mono-oxygenase activities for both *Manduca* tissues were found to exhibit a single major fluctuation during the 10-day stadium of the last larval instar. Fat body mono-oxygenase exhibited a 10-fold fluctuation in activity, whereas midgut exhibited a 60-fold fluctuation in activity. Substrate kinetics of the mono-oxygenases revealed that the apparent K_m values for fat body and midgut systems were fairly constant during the instar at 2.4×10^{-7} M and 4.7×10^{-7} M, respectively. By contrast, the apparent V_{max} values of the mono-oxygenase systems in each tissue fluctuated in a manner quantitatively and temporally coincident with the fluctuations in mono-oxygenase activity. Consistent with the *Locusta* study, therefore, and more direct, these findings suggested that the changes in the mono-oxygenase activities were a function of changes in the titers of the enzymes.

However, in contrast to the *Locusta* study, the fluctuations in *Manduca* fat body and midgut mono-oxygenase activity were not temporally coincident with one another; nor were either of them temporally coincident with the major hemolymph ecdysteroid titer peak. Fat body mono-oxygenase exhibited peak activity on day 4 (Fig. 14), whereas midgut mono-oxygenase exhibited peak activity on day 5 of the instar (Fig. 15). The peak activities for fat body and midgut mono-oxygenases were temporally coincident and succedent, respectively, with the minor hemolymph ecdysteroid titer peak on day

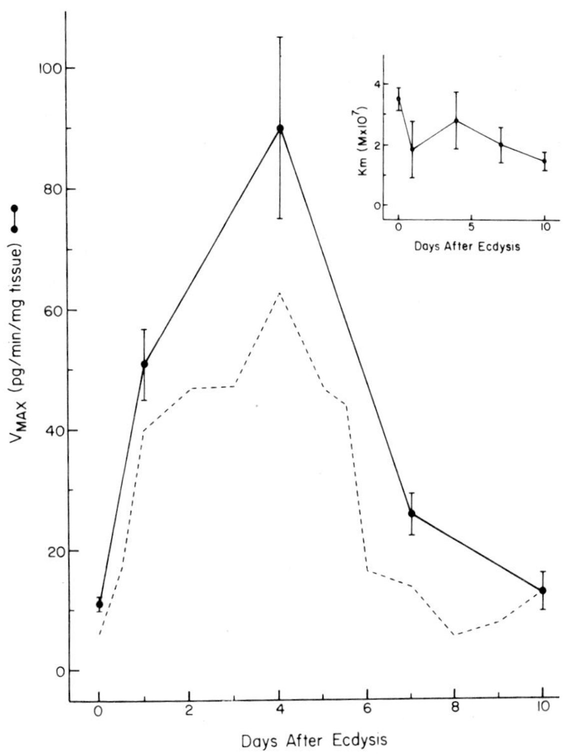

FIG. 14. The profiles of the apparent K_m (insert) and V_{max} of the fat body ecdysone 20-mono-oxygenase activity during the last larval stadium of *Manduca sexta*. The dotted line depicts the ecdysone 20-mono-oxygenase activity of fat body when incubated with substrate levels (3.4×10^{-7} M) which approximate K_m. (Figure reproduced with permission from Smith, S. *et al.*, 1983.)

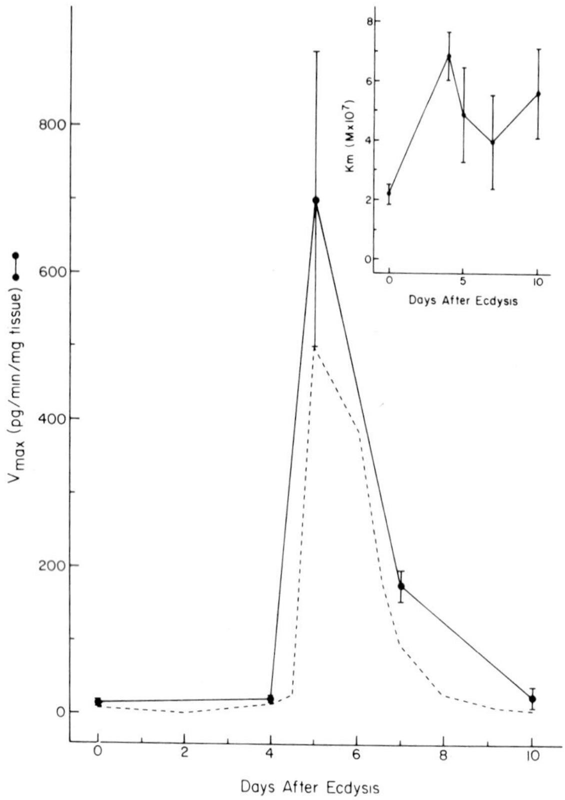

FIG. 15. The profiles of the apparent K_m (insert) and V_{max} of the midgut ecdysone 20-mono-oxygenase activity during the last larval stadium of *Manduca sexta*. The dotted line depicts the ecdysone 20-mono-oxygenase activity of midgut when incubated with substrate levels (3.4×10^{-7} M) which approximate K_m. (Figure reproduced with permission from Smith, S. *et al.*, 1983.)

4 (the ecdysteroid peak presumably responsible for larval–pupal reprogramming). However, both tissue mono-oxygenase activities were at basal levels at the time of the major hemolymph ecdysteroid titer peak on day 7–8 (see Fig. 2).

Clearly, these data indicated that the relationship between ecdysone 20-mono-oxygenase activity and the hemolymph ecdysteroid titer is much more complex in holometabolous insects than previously anticipated. While the significance of the fluctuations in *Manduca* mono-oxygenase activities is not fully understood at present, it has been speculated that the increase in fat body mono-oxygenase activity may be important to ensure sufficient 20-hydroxyecdysone for larval–pupal reprogramming, whereas the increase in midgut mono-oxygenase activity may occur as part of an excretory pathway to rid the hemolymph of ecdysteroids so that the animal can "read" the reprogramming peak as distinct from the molting peak. It has also been speculated that the basal activities of both mono-oxygenases during the molting peak may serve as a mechanism for conserving ecdysone (Smith, S. *et al.*, 1983).

Factors affecting ecdysone 20-mono-oxygenase activity. If little is known about the exact role, or even the developmental times, of the fluctuations in ecdysone 20-mono-oxygenase activity in insects, even less is known about the factors which may be responsible for these fluctuations. Studies have suggested that the product of ecdysone 20-mono-oxygenase systems, i.e. 20-hydroxyecdysone, may via feedback inhibit the enzymes (Cherbas, L. and Cherbas, P., 1970; Feyereisen, R. and Hoffmann, J., 1977), and it has been demonstrated *in vitro* and *in vivo* that 20-hydroxyecdysone can competitively inhibit the ecdysone 20-mono-oxygenase activity in *Manduca* fat body mitochondria, $K_I = 2.7 \times 10^{-5}$ M (Smith, S. *et al.*, 1979), *Locusta* Malpighian tubule microsomes, $K_I = 7.5 \times 10^{-7}$ M (Feyereisen, R. and Durst, F., 1978) and *Calliphora erythrocephala* whole body, $K_I = 2.1 \times 10^{-7}$ M (Young, N., 1976a). Based on their K_I values, it is possible that 20-hydroxyecdysone may inhibit *Locusta* and *Calliphora* mono-oxygenase activity at the times of peak ecdysteroid titer; however, it is doubtful whether 20-hydroxyecdysone would affect *Manduca* mono-oxygenase activity under physiological conditions.

In contrast to regulation by inhibition, other studies primarily with *Locusta migratoria* larvae, larval abdomens, and permanent larvae (prothoracectomized or X-rayed) have suggested that ecdysone or 20-hydroxyecdysone may be required for the activation or induction of ecdysone 20-mono-oxygenase systems (Feyereisen, R. and Hoffmann, J., 1977; Feyereisen, R. and Durst, F., 1978, 1980a,b). It has been demonstrated with this organism that injections of ecdysone or 20-hydroxyecdysone will selectively induce cytochrome P-450 systems, including ecdysone 20-mono-oxygenase activity. This induction appears to be a specific effect of active molting hormones as 22-isoecdysone and 22,25-dideoxyecdysone were found to be ineffective. It also appears as if a head factor is not required for this induction as it can be

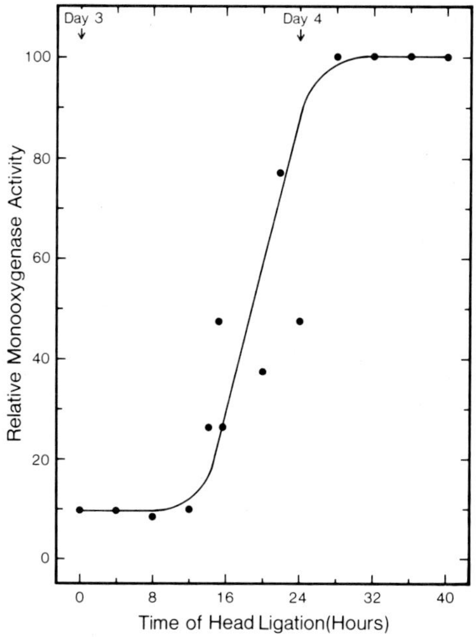

FIG. 16. The head critical period for the induction of midgut ecdysone 20-mono-oxygenase activity in last instar larvae of *Manduca sexta*.

achieved with ecdysone injected into isolated abdomens. Lastly, it also appears as if such induction requires protein synthesis as animals simultaneously injected with ecdysone and actinomycin D, puromycin or cycloheximide failed to show any induction.

In addition to the *Locusta* studies, other preliminary studies with *Manduca* fat body and midgut mono-oxygenase systems have suggested the possibility that these steroid hydroxylases may be regulated by dietary or endogenous factors. The gradual 10-fold increase in *Manduca* fat body mono-oxygenase activity which occurs during the feeding period, i.e. the first 4 days, of the last larval instar has suggested that dietary factors may contribute to the increase in mono-oxygenase activity. Indeed, the precedent for dietary factors, such as allelochemics or xenobiotics, to induce or activate vertebrate and insect P-450 systems is well established (Brattsten, L. *et al.*, 1977; Lasker, J. *et al.*, 1982; see Hodgson, E., vol. 11, this series). Not surprisingly, preliminary studies have demonstrated that some fairly common plant allelochemics (e.g. flavonoids) do have the ability to activate fat body ecdysone 20-mono-oxygenase activity (Smith, S., 1983b). Similarly, there is suggestive evidence that the major increase in *Manduca* midgut mono-oxygenase activity may be regulated by endogenous factors. Preliminary studies with *Manduca* have demonstrated that there is a head critical period (HCP) for the induction of midgut mono-oxygenase activity (Fig. 16). Interestingly, this mono-oxygenase HCP is temporally coincident with the first HCP for PTTH release in the last larval instar (S. Smith and W. Bollenbacher, unpublished observations).

Certainly, it remains to be determined whether the midgut mono-oxygenase induction is mediated by PTTH, a head factor released concomitantly with PTTH or perhaps by the small burst of ecdysone resulting from PTTH release. However, the precedent for ecdysone induction of mono-oxygenase activity has been demonstrated with *Locusta* and vertebrate P-450 steroid hydroxylases are known to be induced by neuropeptides, e.g. the induction of cholesterol side chain cleavage cytochrome P-450 by ACTH (DuBois, R. *et al.*, 1981).

5 GENERAL SUMMARY AND PROSPECTS

Although many facets of the regulation of ecdysteroid titers and ecdysteroid biosynthesis remain to be elucidated, clearly much has already been accomplished within the last decade. We now know that ecdysone and 20-hydroxyecdysone are the major hemolymph ecdysteroids. We know that the PG are the primary physiological sources of ecdysone during most of postembryonic development and that 20-hydroxyecdysone is formed from ecdysone by cytochrome P-450-dependent mono-oxygenase systems. In large measure, we also understand that the ecdysteroid titer is a reflection of ecdysone biosynthesis on the one hand, and its catabolism and excretion on the other hand.

Much of insect endocrinology, like vertebrate endocrinology, is now on a threshold where we cease examining gross regulatory phenomena and move to a finer level of investigating modulatory phenomena. It is anticipated that the next decade will bring us much closer to understanding how ecdysone biosynthesis is effected by PTTH and other factors, how ecdysteroid catabolism and excretion are regulated, and how the various endocrine axes interact with one another to effect homeostasis and development. On a more detailed level, it is probable that some of the enzymes catalyzing the biosynthesis of ecdysone may be isolated and partially characterized in the near future. Significant advances will no doubt also be accomplished in isolating and identifying *in situ* factors which act on ecdysone 20-mono-oxygenase and other ecdysteroid-metabolizing enzyme systems. As progress continues in these areas the potential clearly exists for the development of new pest control strategies and agents which will be based on and exploit the various "windows of vulnerability" in ecdysteroid metabolism.

REFERENCES

AGOSIN, M. and PERRY, A. S. (1974). Microsomal mixed-function oxidases. In *The Physiology of Insecta*, 2nd edn. Edited by M. Rockstein. Vol. 5, pages 537–596. Academic Press, New York.

AGUI, N. (1973). Quantitative bioassay of moulting hormone *in vitro*. *Appl. Ent. Zool. 9*, 236–238.

AGUI, N. and HIRUMA, K. (1977). Ecdysone as a feedback regulator for the neurosecretory brain cells in *Mamestra brassicae*. *J. Insect Physiol. 23*, 1393–1396.

AGUI, N. and HIRUMA, K. (1978). *In vitro* activation of neurosecretory cells in *Mamestra brassicae* by β-ecdysone. *Gen. Comp. Endocr. 33*, 467–472.

AGUI, N. and HIRUMA, K. (1982). Ecdysteroid titer and its critical period during larval pupal ecdysis in the cabbage armyworm, *Mamestra brassicae* L. (Lepidoptera: Noctuidae). *Appl. Ent. Zool. 17*, 144–146.

AGUI, N. and YAGI, S. (1973). Relationship between endogenous ecdysteroid titer and *in vitro* activity of prothoracic glands of *Mamestra brassicae* (Lepidoptera: Noctuidae). *Appl. Ent. Zool. 8*, 239–240.

AGUI, N., BOLLENBACHER, W. E., GRANGER, N. A. and GILBERT, L. I. (1980). Corpus allatum is release site for insect prothoracicotropic hormone. *Nature 285*, 669–670.

AGUI, N., KIMURA, Y. and FUKAYA, M. (1972). Action of the prothoracic gland on the insect integument *in vitro*. *Appl. Ent. Zool. 7*, 71–78.

ASHBURNER, M. (1971). Induction of puffs in polytene chromosomes of *in vitro* cultured salivary glands of *Drosophila melanogaster* by ecdysone and ecdysone analogues. *Nature New Biol. 230*, 222–224.

BAEHR, J.-C., PORCHERON, P., PAPILLON, M. and DRAY, F. (1979). Haemolymph levels of juvenile hormone, ecdysteroids and protein during the last two larval instars of *Locusta migratoria*. *J. Insect Physiol. 25*, 415–421.

BECKERS, C. and EMMERICH, H. (1976). The ecdysone titer in *Drosophila hydei* and the metabolism of moulting hormones in the salivary glands. In *Actualités sur les Hormones D'Invertébrés*. Edited by M. M. Durchon. Pages 351–357. *Colloques Internationaux C.N.R.S.* No. 251, Paris.

BENEDECZKY, I., MALÁ, J. and SEHNAL, F. (1980). Ultrastructural study on the innervation of prothoracic glands in *Galleria mellonella*. *Gen. Comp. Endocr. 41*, 400–407.

BERGAMASCO, R. and HORN, D. H. S. (1980). The biological activities of ecdysteroids and ecdysteroid analogues. In *Progress in Ecdysone Research*. Edited by J. A. Hoffmann. Pages 299–324. Elsevier, Amsterdam.

BERGTROM, G. and OBERLANDER, H. (1975). RNA synthesis in imaginal disks of *Galleria mellonella*: effects of alpha- and beta-ecdysone and fat body *in vitro*. *J. Insect Physiol. 21*, 39–52.

BERREUR, P., PORCHERON, P., BERREUR-BONNENFANT, J. and DRAY, F. (1979a). External factors and ecdysone release in *Calliphora erythrocephala*. *Experientia 35*, 1031.

BERREUR, P., PORCHERON, P., BERREUR-BONNENFANT, J. and SIMPSON, P. (1979b). Ecdysteroid levels and pupariation in *Drosophila melanogaster*. *J. Exp. Zool. 210*, 347–352.

BEYDON, P. and LAFONT, R. (1983). Feedback inhibition of ecdysone production by 20-hydroxyecdysone in *Pieris brassicae* pupae. *J. Insect Physiol. 29*, 529–533.

BLAZSEK, I., BALAZS, A., NOVAK, V. J. A. and MALÁ, J. (1975). Ultrastructural study of the prothoracic glands of *Galleria mellonella* L. in the penultimate, last larval and pupal stages. *Cell Tissue Res. 158*, 269–280.

BOLLENBACHER, W. E. and GILBERT, L. I. (1981). Neuroendocrine control of postembryonic development in insects: the prothoracicotropic hormone. In *Neurosecretion: Molecules, Cells and Systems*. Edited by D. S. Farner and K. Lederis. Pages 361–370. Plenum, New York.

BOLLENBACHER, W. E., VEDECKIS, W. V., GILBERT, L. I. and O'CONNOR, J. D. (1975). Ecdysone titers and prothoracic gland activity during larval-pupal development of *Manduca sexta*. *Devel. Biol. 44*, 46–53.

BOLLENBACHER, W. E., GOODMAN, W., VEDECKIS, W. V. and GILBERT, L. I. (1976). The *in vitro* synthesis and secretion of α-ecdysone by the ring glands of the fly, *Sarcophaga bullata*. *Steroids 27*, 309–324.

BOLLENBACHER, W. E., SMITH, S. L., WIELGUS, J. J. and GILBERT, L. I. (1977). Evidence for an α-ecdysone cytochrome P-450 mixed function oxidase in insect fat body mitochondria. *Nature 268*, 660–663.

BOLLENBACHER, W. E., ZVENKO, H., KUMARAN, A. K. and GILBERT, L. I. (1978). Changes in ecdysone content during postembryonic development of the wax moth, *Galleria mellonella*: the role of the ovary. *Gen. Comp. Endocr. 34*, 169–179.

BOLLENBACHER, W. E., AGUI, N., GRANGER, N. A. and GILBERT, L. I. (1979). *In vitro* activation of insect prothoracic glands by the prothoracicotropic hormone. *Proc. Nat. Acad. Sci. 76*, 5148–5152.

BOLLENBACHER, W. E., AGUI, N., GRANGER, N. A. and GILBERT, L. I. (1980). Insect prothoracic glands *in vitro*: a system for studying the prothoracicotropic hormone. In *Invertebrate Systems In Vitro*. Edited by E. Kurstak, K. Maramorosch and A. Dubendorfer. Pages 253–271. Elsevier, Amsterdam.

BOLLENBACHER, W. E., SMITH, S. L., GOODMAN, W. and GILBERT, L. I. (1981). Ecdysteroid titer during larval-pupal-adult development of the tobacco hornworm, *Manduca sexta*. *Gen. Comp. Endocr. 44*, 302–306.

BONNER, J. J. (1982). An assessment of the ecdysteroid receptor of *Drosophila*. *Cell 30*, 7–8.

BORST, D. W. and ENGELMANN, F. (1974). *In vitro* secretion of α-ecdysone by prothoracic glands of a hemimetabolous insect, *Leucophaea maderae* (Blattaria). *J. Exp. Zool. 189*, 413–419.

BORST, D. W. and O'CONNOR, J. D. (1972). Arthropod molting hormone: radioimmune assay. *Science 178*, 418–419.

BORST, D. W., BOLLENBACHER, W. E., O'CONNOR, J. D., KING, D. S. and FRISTROM, J. W. (1974). Ecdysone levels during metamorphosis of *Drosophila melanogaster*. *Devel. Biol. 39*, 308–316.

BRATTSTEN, L. B., WILKINSON, C. F. and EISNER, T. (1977). Herbivore-plant interactions: mixed function oxidases and secondary plant substances. *Science 196*, 1349–1352.

BULLIÈRE, D., BULLIÈRE, F. and DEREGGI, M. (1979). Ecdysteroid titres during ovarian and embryonic development in *Blaberus craniifer*. *Wilhelm Roux's Arch. Devel. Biol. 186*, 103–114.

BURDETTE, W. J. (1962). Changes in titer of ecdysone in *Bombyx mori* during metamorphosis. *Science 135*, 432.

BUTENANDT, A. and KARLSON, P. (1954). Über die isolierung eines metamorphose-hormons der insekten in Kristellisierter form. *Z. Naturforsch. 9b*, 389–391.

BUTTERWORTH, F. M. and BERENDES, H. D. (1974). Ecdysone binding proteins in haemolymph and tissues of *Drosophila hydei* larvae. *J. Insect Physiol. 20*, 2195–2204.

CALVEZ, B. (1981). Progress of developmental programme during the last larval instar of *Bombyx mori*: relationships with food intake, ecdysteroids and juvenile hormone. *J. Insect Physiol. 27*, 233–239.

CALVEZ, B., HIRN, M. and DEREGGI, M. (1976). Ecdysone changes in the haemolymph of two silkworms (*Bombyx mori* and *Philosamia cynthia*) during larval and pupal development. *FEBS Lett. 71*, 57–61.

CALVEZ, M. B. (1976). Taux d'ecdysone circulante aux derniers âges larvaires et induction de la diapause nymphale chez *Pieris brassicae*. *C.R. Acad. Sci. Paris (D) 282*, 1367–1370.

Carrow, G. M., Calabrese, R. L. and Williams, C. M. (1981). Spontaneous and evoked release of prothoracicotropin from multiple neurohemal organs of the tobacco hornworm. *Proc. Nat. Acad. Sci. 78*, 5866–5870.

Chadwick, L. E. (1956). Removal of prothoracic glands from the nymphal cockroach. *J. Exp. Zool. 131*, 291–305.

Chang, E. S. and O'Connor, J. D. (1979). Arthropod molting hormones. In *Methods of Hormone Radioimmunoassay* 2nd edn. Edited by B. M. Jaffe and H. R. Behrman. Pages 797–814. Academic Press, New York.

Charlet, M., Goltzene, F. and Hoffmann, J. A. (1979). Experimental evidence for the neuroendocrine control of ecdysone biosynthesis in adult females of *Locusta migratoria. J. Insect Physiol. 25*, 463–466.

Chen, T. T. and Hodgetts, R. B. (1974). The appearance of dopa decarboxylase activity in imaginal discs of *Sarcophaga bullata* undergoing development *in vitro. Devel. Biol. 38*, 271–284.

Cherbas, L. and Cherbas, P. (1970). Distribution and metabolism of α-ecdysone in pupae of the silkworm *Antheraea polyphemus. Biol. Bull. 138*, 115–128.

Chihara, C. J., Petri, W. H., Fristrom, J. W. and King, D. S. (1972). The assay of ecdysones and juvenile hormones on *Drosophila melanogaster* imaginal disks *in vitro. J. Insect Physiol. 18*, 1115–1123.

Chino, H., Sakurai, S., Ohtaki, T., Ikekawa, N., Miyazaki, H., Ishibashi, M. and Abuki, H. (1974). Biosynthesis of α-ecdysone by prothoracic glands *in vitro. Science 183*, 529–530.

Claret, J., Porcheron, P. and Dray, F. (1977). Critical periods and ecdysone titers in the pupae of *Pieris brassicae. Experientia 33*, 1389–1390.

Clayton, R. B. (1964). The utilization of sterols by insects. *J. Lipid Res. 5*, 3–19.

Clever, U., Clever, I., Storbeck, I. and Young, N. L. (1973). The apparent requirement of two hormones, α- and β-ecdysone, for molting induction in insects. *Devel. Biol. 31*, 47–60.

Cooper, D. Y., Schleyer, H., Estabrook, R. W. and Rosenthal, O. (1969). The role of cytochrome P-450 in the biosynthesis and metabolism of steroids. In *Progress in Endocrinology*. Edited by C. Gual and F. J. G. Ebling. Pages 784–802. Excerpta Medica, Amsterdam.

Coulon, M. (1977). Relations between feeding and radiosensivity cycles and the ecdysone cycle in *Bombyx mori*, at the end of embryonic development and during the first four larval instars. *Devel. Growth Diff. 19*, 181–185.

Coulon, M. and Feyereisen, R. (1977). Effets de l'irradiation de larves de *Bombyx mori* sur l'evolution du titre des ecdysteroides de l'hemolymphe et le metabolisme de l'ecdysone. *C.R. Acad. Sci. Paris (D) 284*, 1689–1692.

Coulon, M., Calvez, B., deReggi, M., Legay, J. M. and Hirn, M. (1979). Variations of ecdysteroid content during *Bombyx mori* development: relation to growth and morphogenetic phases. *Experientia 35*, 1120–1122.

Courgeon, A. M. (1972). Effects of α- and β-ecdysone on *in vitro* diploid cell multiplication in *Drosophila melanogaster. Nature New Biol. 238*, 250–251.

Cymborowski, B. and Stolarz, G. (1979). The role of juvenile hormone during larval-pupal transformation of *Spodoptera littoralis*: switch over in the sensitivity of the prothoracic glands to juvenile hormone. *J. Insect Physiol. 25*, 939–942.

Dammeier, B. and Hoppe, W. (1971). Die Kristall- und molekulstrukturanalyse des insektenhautungshormons 20-hyhdroxy-ecdyson (ecdysteron). *Chem. Ber. 104*, 1660–1673.

Dean, R. L., Bollenbacher, W. E., Locke, M., Smith, S. L. and Gilbert, L. I. (1980). Haemolymph ecdysteroid levels and cellular events in the intermoult/moult sequence of *Calpodes ethlius. J. Insect Physiol. 26*, 267–280.

Delaage, M. A., Hirn, M. H. and deReggi, M. L. (1982). Radioimmunoassay of ecdysteroids. In *Methods in Enzymology*, vol. 84: *Immunochemical Techniques*, Pt. D. Edited by J. J. Langone and H. VanVunakis. Pages 350–358. Academic Press, New York.

Delachambre, J., Delbecque, J.-P., Provansal, A., Grillot, J. P., deReggi, M. L. and Cailla, H. L. (1979). Total and epidermal cyclic AMP levels related to the variations of ecdysteroids and bursicon during the metamorphosis of the mealworm *Tenebrio molitor* L. *Insect Biochem. 9*, 95–99.

Delachambre, J., Besson, M. T., Connat, J.-L. and Delbecque, J.-P., (1980). Ecdysteroid titers and integumental events during the metamorphosis of *Tenebrio molitor*. In *Progress in Ecdysone Research*. Edited by J. A. Hoffmann. Pages 211–234. Elsevier, Amsterdam.

DeLaPaz, A. R., Delbecque, J.-P., Bitsch, J. and Delachambre, J. (1983). Ecdysteroids in the haemolymph and the ovaries of the firebrat *Thermobia domestica* (Packard) (Insecta, Thysanura): Correlations with integumental and ovarian cycles. *J. Insect Physiol. 29*, 323–329.

Delbecque, J.-P., Prost, M., Maume, B. F., Delachambre, J., Lafont, R. and Mauchamp, B. (1975). Dosage de la β-ecdysone par fragmentographie de masse au cours du denier stade larvaire de *Tenebrio molitor* L. (Insecte, Coléoptere). *C.R. Acad. Sci. Paris (D) 281*, 309–312.

Delbecque, J.-P., Delachambre, J., Hirn, M. and deReggi, M. (1978a). Abdominal production of ecdysterone and pupal-adult development in *Tenebrio molitor* (Insecta, Coleoptera). *Gen. Comp. Endocr. 35*, 436–444.

Delbecque, J.-P., Hirn, M., Delachambre, J. and deReggi, M. L. (1978b). Cuticular cycle and molting hormone levels during the metamorphosis of *Tenebrio molitor* (Insecta Coleoptera). *Devel. Biol. 64*, 11–30.

Delbecque, J.-P., Lanzrein, B., Bodereau, C., Imboden, C., Hirn, M., O'Connor, J. D., Noirot, C. and Lüscher, M. (1978c). Ecdysone and ecdysterone in physogastric termite queens and eggs of *Macrotermes bellicosus* and *Macrotermes subhyalinus. Gen. Comp. Endocr. 36*, 40–47.

DeReggi, M. L., Hirn, M. H. and Delaage, M. A. (1975). Radioimmunoassay of ecdysone. An application to *Drosophila* larvae and pupae. *Biochem. Biophys. Res. Comm. 66*, 1307–1315.

DeWilde, J., Hsiao, T. H. and Hsiao, C. (1980). The regulation of the metamorphic moults in the Colorado potato beetle *Leptinotarsa decemlineata* Say. In *Progress in Ecdysone Research*. Edited by J. A. Hoffmann. Pages 83–98. Elsevier, Amsterdam.

Dorn, A. and Romer, F. (1976). Structure and function of prothoracic glands and oenocytes in embryos and last larval instars in *Oncopeltus fasciatus* Dallas. (Insecta, heteroptera) *Cell Tiss. Res. 171*, 331–350.

DuBois, R. N., Simpson, E. R., Kramer, R. E. and Waterman, M. R. (1981). Induction of synthesis of cholesterol side chain cleavage cytochrome P-450 by adrenocorticotropin in cultured bovine adrenocortical cells. *J. Biol. Chem. 256*, 7000–7005.

Ellis, P. E., Morgan, E. D. and Woodbridge, A. P. (1972). Moult-inducing hormones of the prothoracic gland of insects. *Nature 238*, 274–276.

Emmerich, H. (1970a). Ecdysonebindende proteinfraktionen in den speicheldrüsen von *Drosophila hydei. Z. vergl. Physiol. 68*, 385–402.

Emmerich, H. (1970b). Über die hämolymphproteine von *Pyrrhocoris apterus* und über die bindung von ecdyson durch hämolymphproteine. *J. Insect Physiol. 16*, 725–747.

Feir, D. and Winkler, G. (1969). Ecdysone titres in the last larva and adult stages of the milkweed bug. *J. Insect Physiol. 15*, 899–904.

Feyereisen, R. (1977a). A specific binding protein for the moulting hormone ecdysterone in locust haemolymph. *Experientia 33*, 1111–1112.

Feyereisen, R. (1977b) Cytochrome P-450 et hydroxylation de l'ecdysone en ecdystérone chez *Locusta migratoria in vitro. C.R. Acad. Sci. Paris (D) 284*, 1831–1834.

Feyereisen, R. (1978). Role of cytochrome P-450 in ecdysone hydroxylation in the locust (*Locusta migratoria* L., Orthoptera). *Gen. Comp. Endocr. 34*, 105–106.

Feyereisen, R. (1980). Ecdysteroid-protein interactions: specificity of the ecdysteroid binding protein from *Locusta migratoria* haemolymph. In *Progress in Ecdysone Research*. Edited by J. A. Hoffmann. Pages 325–334. Elsevier, Amsterdam.

Feyereisen, R. and Durst, F. (1978). Ecdysterone biosynthesis: a microsomal cytochrome P-450 linked ecdysone 20-monooxygenase from tissues of the African migratory locust. *Eur. J. Biochem. 88*, 37–47.

Feyereisen, R. and Durst, F. (1980a). Development of microsomal cytochrome P-450 monooxygenases during the last larval instar of the locust, *Locusta migratoria*: correlation with the hemolymph 20-hydroxyecdysone titer. *Mol. Cell. Endocr. 20*, 157–169.

FEYEREISEN, R. and DURST, F. (1980b). Control of cytochrome P-450 monooxygenases in an insect by the steroid moulting hormone. In *Microsomes, Drug Oxidations and Chemical Carcinogenesis*. Edited by M. J. Coon. Pages 595–598. Academic Press, New York.

FEYEREISEN, R. and HOFFMANN, J. A. (1977). Regulation of ecdysone hydroxylation in *Locusta migratoria*: role of the moulting hormone level. *J. Insect Physiol. 23*, 1175–1181.

FEYEREISEN, R., LAGUEUX, M. and HOFFMANN, J. A. (1976). Dynamics of ecdysone metabolism after ingestion and injection in *Locusta migratoria*. *Gen. Comp. Endocr. 29*, 319–327.

FEYEREISEN, R., PRATT, G. E. and HAMNETT, A. F. (1981). Enzymatic synthesis of juvenile hormone in locust corpora allata: evidence for a microsomal cytochrome P-450 linked methyl farnesoate epoxidase. *Eur. J. Biochem. 118*, 231–238.

FRAENKEL, G. (1935). A hormone causing pupation in the blowfly, *Calliphora erythrocephala*. *Proc. R. Soc., Lond. B. 118*, 1–12.

FRISTROM, J. W. (1981). *Drosophila* imaginal discs as a model system for the study of metamorphosis. In *Metamorphosis: A Problem in Developmental Biology* 2nd ed. Edited by L. I. Gilbert and E. Frieden. Pages 217–240. Plenum, New York.

FUCHS, M. S. and KANG, S.-H. (1981). Ecdysone and mosquito vitellogenesis: a critical appraisal. *Insect Biochem. 11*, 627–633.

FUKAYA, M. and MITSUHASHI, J. (1961). Larval diapause in the rice stem borer with special reference to its hormonal mechanism. *Bull. Nat. Inst. Agric. Res. Japan 13*, 1–30.

FUKUDA, S. (1940a). Induction of pupation in silkworm by transplanting the prothoracic gland. *Proc. Imp. Acad. (Tokyo) 16*, 414–416.

FUKUDA, S. (1940b). Hormonal control of molting and pupation in the silkworm. *Proc. Imp. Acad. (Tokyo) 16*, 417–420.

FUKUDA, S. (1944). The hormonal mechanism of larval molting and metamorphosis in the silkworm. *J. Fac. Sci. Tokyo Univ. 6*, 477–532.

GALBRAITH, M. N., HORN, D. H. S., MIDDLETON, E. J. and THOMSON, J. A. (1970). The biosynthesis of crustecdysone in the blowfly *Calliphora stygia*. *Chem. Comm.* 179–180.

GANDE, A. R. and MORGAN, E. D. (1979). Ecdysteroids in the developing eggs of the desert locust, *Schistocerca gregaria*. *J. Insect Physiol. 25*, 289–293.

GANDE, A. R., MORGAN, E. D. and WILSON, I. D. (1979). Ecdysteroid levels throughout the life cycle of the desert locust, *Schistocerca gregaria*. *J. Insect Physiol. 25*, 669–675.

GAREN, A., KAUVAR, L. and LEPESANT, J. A. (1977). Roles of ecdysone in *Drosophila* development. *Proc. Nat. Acad. Sci. 74*, 5099–5103.

GELMAN, D. and HAYES, D. (1982). Critical periods for the brain and prothoracic glands of 5th instars of the European corn borer, *Ostrinia nubilalis* Hubner. *Comp. Biochem. Physiol. 73A*, 81–87.

GERSCH, M. (1977). Haütungen von insekten ohne haütungsdrüse: Befunde mit larven von *Periplaneta americana*. *Experientia 33*, 228–230.

GERSCH, M. (1978). Hormonale und humorale steverung durch neurosekretorische peptide. In *Probleme der Korrelation Neuronaler und Endokriner Regulation bei Evertebraten*. Edited by H. Penzlin. Pages 8–31. Friedrich-Schiller Universität, Jena.

GERSCH, M. (1980). Control of activity of the prothoracic glands in insects at the cellular level. In *Progress in Ecdysone Research*. Edited by J. A. Hoffmann. Pages 111–123. Elsevier, Amsterdam.

GERSCH, M. and EIBISCH, J. (1977). Synthesis of ecdysone-^{14}C and ecdysterone-^{14}C from cholesterol-^{14}C in cockroaches (*Periplaneta americana*) without moulting glands. *Experientia 33*, 468.

GERSCH, M. and STÜRZEBECHER, J. (1971). Über eine synthese von ecdysone-^{3}H und ecdysteron-^{3}H aus cholesterin-^{3}H in geschürten abdomina von *Mamestra brassicae* Raupen. *Experientia 27*, 1475–1476.

GILBERT, L. I. and CHINO, H. (1974). Transport of lipids in insects. *J. Lipid. Res. 15*, 439–456.

GILBERT, L. I. and KING, D. S. (1973). Physiology of growth and development: endocrine aspects. In *Physiology of Insects*, 2nd edn. Edited by M. Rockstein. Vol. 1, pages 249–370. Academic Press, New York.

GILBERT, L. I. and SCHNEIDERMAN, H. A. (1959). Prothoracic gland stimulation by juvenile hormone extracts of insects. *Nature 184*, 171–173.

GILBERT, L. I., GOODMAN, W. and BOLLENBACHER, W. E. (1977). Biochemistry of regulatory lipids and sterols in insects. In *International Review of Biochemistry: Biochemistry of Lipids*. II. Edited by T. W. Goodwin. Vol. 14, pages 1–50. University Park Press, Baltimore.

GILBERT, L. I., BOLLENBACHER, W. E., GOODMAN, W., SMITH, S. L., AGUI, N., GRANGER, N. and SEDLAK, B. J. (1980). Hormones controlling insect metamorphosis. *Rec. Prog. Hormone Res. 36*, 401–449.

GILBERT, L. I., BOLLENBACHER, W. E., AGUI, N., GRANGER, N. A., SEDLAK, B. J., GIBBS, D. and BUYS, C. M. (1981). The prothoracicotropes: source of the prothoracicotropic hormone. *Amer. Zool. 21*, 641–653.

GIRARDIE, A. and DEREGGI, M. (1978). Moulting and ecdysone release in response to electrical stimulation of protocerebral neurosecretory cells in *Locusta migratoria*. *J. Insect Physiol. 24*, 797–802.

GLITHO, I., DELBECQUE, J. P. and DELACHAMBRE, J. (1979). Prothoracic gland involution related to moulting hormone levels during the metamorphosis of *Tenebrio molitor* L. *J. Insect Physiol. 25*, 187–191.

GOODWIN, T. W., HORN, D. H. S., KARLSON, P., KOOLMAN, J., NAKANISHI, K., ROBBINS, W. E., SIDDALL, J. B. and TAKEMOTO, T. (1978). Ecdysteroids: a new generic term. *Nature 272*, 122.

GORELL, T. A., GILBERT, L. I. and TASH, J. (1972). The uptake and conversion of α-ecdysone by the pupal tissues of *Hyalophora cecropia*. *Insect Biochem. 2*, 94–106.

GOWER, D. B. (1974). Modifiers of steroid-hormone metabolism: a review of their chemistry, biochemistry and clinical applications. *J. Steroid Biochem. 5*, 501–523.

GRANGER, N. A. and BOLLENBACHER, W. E. (1981). Hormonal control of insect metamorphosis. In *Metamorphosis: A Problem in Developmental Biology* 2nd ed. Edited by L. I. Gilbert and E. Frieden. Pages 105–137. Plenum, New York.

GRUETZMACHER, M. C. (1982). The role of juvenile hormone in regulating prothoracic gland activity in *Manduca sexta* (L.) (Lepidoptera). Ph.D. dissertation, Northwestern University, Evanston, Illinois.

GUNNAR, W. P. (1979). The role of a hemolymph protein in the regulation of ecdysone synthesis by the prothoracic glands of *Manduca sexta*. B.S. Honors thesis, Northwestern University, Evanston, Illinois.

GWYNNE, J. T. and STRAUSS, J. F. (1982). The role of lipoproteins in steroidogenesis and cholesterol metabolism in steroidogenic glands. *Endocr. Rev. 3*, 299–328.

HAFFERL, W., WREN, W. L., MARSHALL, J. P., CALZADA, M. C. and SIDDALL, J. B. (1972). Radiochemical synthesis. III. Synthesis of tritiated α-ecdysone and 25-deoxy-α-ecdysone of high specific activity. *J. Label. Compounds 8*, 81–94.

HAGEDORN, H. H., O'CONNOR, J. D., FUCHS, M. S., SAGE, B., SCHLAEGER, D. A. and BOHM, M. K. (1975). The ovary as a source of α-ecdysone in adult mosquitoes. *Proc. Nat. Acad. Sci. 72*, 3255–3259.

HANAOKA, K. and HAGEDORN, H. H. (1980). Brain hormone control of ecdysone secretion by the ovary in a mosquito. In *Progress in Ecdysone Research*. Edited by J. A. Hoffmann. Pages 467–480. Elsevier, Amsterdam.

HANAOKA, K. and OHNISHI, E. (1974). Changes in ecdysone titre during pupal-adult development in the silkworm, *Bombyx mori*. *J. Insect Physiol. 20*, 2375–2384.

HANDLER, A. M. (1982). Ecdysteroid titers during pupal and adult development in *Drosophila melanogaster*. *Devel. Biol. 93*, 73–82.

HARDING, B. W., BELL, J. J., OLDHAM, S. B. and WILSON, L. D. (1968). Cortico-steroid biosynthesis in adrenal cortical mitochondria. In *Functions of the Adrenal Cortex* II. Edited by K. W. McKerns. Pages 831–896. Appleton-Century-Crofts, New York.

HERMAN, W. S. and GILBERT, L. I. (1966). The neuroendocrine system of *Hyalophora cecropia* (L.) (Lepidoptera: Saturniidae). I. The anatomy and histology of the ecdysial glands. *Gen. Comp. Endocr. 7*, 275–291.

HETRU, C. and HORN, D. H. S. (1980). Phytoecdysteroids and zooecdysteroids. In *Progress in Ecdysone Research*. Edited by J. A. Hoffmann. Pages 13–28. Elsevier, Amsterdam.

HETRU, C., LAGUEUX, M., LUU, B. and HOFFMANN, J. A. (1978). Adult ovaries of *Locusta migratoria* contain the sequence of biosynthetic intermediates for ecdysone. *Life Sci. 22*, 2141–2154.

HIRN, M. H. and DELAAGE, M. A. (1980). Radioimmunological approaches to the quantification of ecdysteroids. In *Progress in Ecdysone Research*. Edited by J. A. Hoffmann. Pages 69–82. Elsevier, Amsterdam.

HIRN, M., COULON, M. and DEREGGI, M. (1977). Taux d'ecdystéroides dans l'oeuf de *Bombyx mori* et variations au cours de l'embryogenèse. *C.R. Acad. Sci. Paris (D) 284*, 2147–2150.

HIRN, M., HETRU, C., LAGUEUX, M. and HOFFMANN, J. A. (1979). Prothoracic gland activity and blood titres of ecdysone and ecdysterone during the last larval instar of *Locusta migratoria* L. *J. Insect Physiol. 25*, 255–261.

HIRUMA, K. (1980). Possible roles of juvenile hormone in the prepupal stage of *Mamestra brassicae. Gen. Comp. Endocr. 41*, 392–399.

HIRUMA, K. (1982). Factors affecting change in sensitivity of prothoracic glands to juvenile hormone in *Mamestra brassicae. J. Insect Physiol. 28*, 193–199.

HIRUMA, K., SHIMADA, H. and YAGI, S. (1978). Activation of the prothoracic gland by juvenile hormone and prothoracicotropic hormone in *Mamestra brassicae. J. Insect Physiol. 24*, 215–220.

HOBSON, R. P. (1935a). On a fat soluble growth factor required by blowfly larvae. I. Distribution and properties. *Biochem. J. 29*, 1292–1296.

HOBSON, R. P. (1935b). On a fat soluble growth factor required by blowfly larvae. II. Identity of the growth factor with cholesterol. *Biochem. J. 29*, 2023–2026.

HODGETTS, R. B., SAGE, B. and O'CONNOR, J. D. (1977). Ecdysone titers during postembryonic development of *Drosophila melanogaster. Devel. Biol. 60*, 310–317.

HODGSON, E. (1983). The significance of cytochrome P-450 in insects. *Insect Biochem. 13*, 237–246.

HOFFMANN, J. A. and KOOLMAN, J. (1974). Prothoracic glands in the regulation of ecdysteroid titres and metabolic fate of injected labelled ecdysone in *Locusta migratoria. J. Insect Physiol. 20*, 1593–1601.

HOFFMANN, J. A., KOOLMAN, J., KARLSON, P. and JOLY, P. (1974). Molting hormone titer and metabolic fate of injected ecdysone during the fifth larval instar and in adults of *Locusta migratoria* (Orthoptera). *Gen. Comp. Endocr. 22*, 90–97.

HOFFMANN, J. A., KOOLMAN, J. and BEYLER, C. (1975). Rôle des glandes prothoraciques dans la production d'ecdysone au cours du dernier stade larvaire de *Locusta migratoria. C.R. Acad. Sci. Paris (D). 280*, 733–736.

HOFFMANN, J. A., LAGUEUX, M., HETRU, C., CHARLET, M. and GOLTZENE, F. (1980). Ecdysone in reproductively competent female adults and in embryos of insects. In *Progress in Ecdysone Research*. Edited by J. A. Hoffmann. Pages 431–465. Elsevier, Amsterdam.

HOFFMEISTER, H. and GRÜTZMACHER, H. F. (1966). Zur chemie des ecdysterons. *Tetrahedron Lett. 33*, 4017–4023.

HOLMAN, G. M. and MEOLA, R. W. (1978). A high-performance liquid chromatography method for the purification and analysis of insect ecdysones: application to measurement of ecdysone titers during pupal-adult development of *Heliothis zea. Insect Biochem. 8*, 275–278.

HORN, D. H. S. (1971). The ecdysones. In *Naturally Occurring Insecticides*. Edited by M. Jacobson and D. G. Crosby. Pages 333–459. Dekker, New York.

HSIAO, T. H. and HSIAO, C. (1977). Simultaneous determinations of molting and juvenile hormone titers of the greater wax moth. *J. Insect Physiol. 23*, 89–93.

HSIAO, T. H. and HSIAO, C. (1979). Ecdysteroids in the ovary and the egg of the greater wax moth. *J. Insect Physiol. 25*, 45–52.

HSIAO, T., HSIAO, C. and DEWILDE, J. (1975). Moulting hormone production in the isolated larval abdomen of the Colorado beetle. *Nature 255*, 727–728.

HSIAO, T. H., HSIAO, C. and DEWILDE, J. (1976). Moulting hormone titre changes and their significance during development of the Colorado beetle, *Leptinotarsa decemlineata. J. Insect Physiol. 22*, 1257–1261.

HUBER, R. and HOPPE, W. (1965). Zur chemie des ecdysterons. VII. Die Kristall-und molekül-struktur-analyse des insektenuerpuppungshormons ecdyson mit der automatisierten faltmolekülmethode. *Chem. Ber. 98*, 2403–2424.

IKEKAWA, N., IKEDA, T., MIZUNO, T., OHNISHI, E. and SAKURAI, S. (1980). Isolation of a new ecdysteroid, 2,22-dideoxy-20-hydroxyecdysone, from the ovaries of the silkworm, *Bombyx mori. J. Chem. Soc., Chem. Commun.* 448–449.

ISAAC, R. E., REES, H. H. and GOODWIN, T. W. (1981a). Isolation of 2-deoxy-20-hydroxyecdysone and 3-epi-2-deoxyecdysone from eggs of the desert locust, *Schistocerca gregaria*, during embryogenesis. *J. Chem. Soc., Chem. Commun.* 418–420.

ISAAC, R. E., REES, H. H. and GOODWIN, T. W. (1981b). Isolation of ecdysone-3-acetate as a major ecdysteroid from the developing eggs of the desert locust, *Schistocerca gregaria. J. Chem. Soc., Chem. Commun.* 594–595.

ISAAC, R. E., ROSE, M. E., REES, H. H. and GOODWIN, T. W. (1982). Identification of ecdysone-22-phosphate and 2-deoxyecdysone-22-phosphate in eggs of the desert locust, *Schistocera gregaria*, by fast atom bombardment mass spectroscopy and NMR spectroscopy. *J. Chem. Soc., Chem. Commun.* 249–251.

JOHNSON, P. and REES, H. H. (1977). The mechanism of C-20 hydroxylation of α-ecdysone in the desert locust *Schistocerca gregaria. Biochem. J. 168*, 513–520.

KAPLANIS, J. N. THOMPSON, M. J., YAMAMOTO, R. T. and LOULOUDES, S. J. (1966a). Ecdysones from the pupa of the tobacco hornworm, *Manduca sexta* (Johannson). *Steroids 8*, 605–623.

KAPLANIS, J. N., TABOR, L. A., THOMPSON, M. J., ROBBINS, W. E. and SHORTINO, T. J. (1966b). Assay for ecdysone (moulting hormone) activity using the housefly, *Musca domestica* L. *Steroids 8*, 625–631.

KAPLANIS, J. N., THOMPSON, M. J., ROBBINS, W. E. and LINDQUIST, E. L. (1972). α-Ecdysone and 20-hydroxyecdysone from frass of the tobacco hornworm, *Manduca sexta* (L.) *Steroids 20*, 621–625.

KAPLANIS, J. N., ROBBINS, W. E., THOMPSON, M. J. and DUTKY, S. R. (1973). 26-Hydroxyecdysone: new insect molting hormone from the egg of the tobacco hornworm. *Science 180*, 307–308.

KAPLANIS, J. N., DUTKY, S. R., ROBBINS, W. E., THOMPSON, M. J., LINDQUIST, E. L., HORN, D. H. S. and GALBRAITH, M. N. (1975). Makisterone A: a 28-carbon hexahydroxy molting hormone from the embryo of the milkweed bug. *Science 190*, 681–682.

KAPLANIS, J. N., THOMPSON, M. J., DUTKY, S. R. and ROBBINS, W. E. (1979). The ecdysteroids from the tobacco hornworm during pupal-adult development five days after peak titer of molting hormone activity. *Steroids 34*, 333–345.

KAPLANIS, J. N., THOMPSON, M. J., DUTKY, S. R. and ROBBINS, W. E. (1980a). The ecdysteroids from young embryonated eggs of the tobacco hornworm. *Steroids 36*, 321–336.

KAPLANIS, J. N., WEIRICH, G. F., SVOBODA, J. A., THOMPSON, M. J. and ROBBINS, W. E. (1980b). Moulting hormones of the tobacco hornworm and milkweed bug: their chemistry and biochemistry. In *Progress in Ecdysone Research*. Edited by J. A. Hoffmann. Pages 163–186. Elsevier, Amsterdam.

KARLSON, P. (1956). Biochemical studies on insect hormones. *Vit. Horm. 14*, 227–266.

KARLSON, P. (1980). Ecdysone in retrospect and prospect. In *Progress in Ecdysone Research*. Edited by J. A. Hoffmann. Pages 1–11. Elsevier, Amsterdam.

KARLSON, P. and SHAAYA, E. (1964). Der ecdysontiter während der insektenentwicklung I. Eine methode zur bestimmung des ecdysongehalts. *J. Insect Physiol. 10*, 797–804.

KARLSON, P. and STAMM-MENÉNDEZ, M. D. (1956). Notiz über den nachweis von metamorphose-hormon in den imagines von *Bombyx mori. Hoppe-Seyler's Z. Physiol. Chem. 306*, 109–111.

KARLSON, P., SEKERIS, C. E. and MAURER, R. (1964). Zum wirkungsmechanismus der hormone. I. Verteilung von tritium-markierten ecdyson in larven von *Calliphora erythrocephala. Hoppe-Seyler's Z. Physiol. Chem. 336*, 100–106.

KARLSON, P., HOFFMEISTER, H., HUMMEL, H., HOCKS, P. and SPITELLER, G. (1965). Reaktionen des ecdysonmolekuls. *Chem. Ber. 98*, 2394–2402.

KARLSON, P., BUGANY, H., DOÖPP, H. and HOYER, G.-A. (1972). 3-Dehydroecdyson, ein stoffwechselprodukt der ecdysons bei der schmeissfliege *Calliphora erythrocephala* Meigen. *Hoppe-Seyler's Z. Physiol. Chem. 353*, 1610–1614.

KELLY, T. J., WOODS, C. W., REDFERN, R. E. and BOŘKOVEC, A. B. (1981). Makisterone A: the molting hormone of larval *Oncopeltus? J. Exp. Zool. 218*, 127–132.

KERB, U., SCHULZ, G., HOCKS, P., WIECHERT, R., FURLENMEIER, A., FÜRST, A., LANGEMANN, A. and WALDVOGEL, G. (1966). Zur synthese des ecdysons. IV. Mitleilung über insektenhormone die synthese des naturlichen hautungshormons. *Helv. Chem. Acta 49*, 1601–1606.

KIGUCHI, K. and AGUI, N. (1981). Ecdysteroid levels and developmental events during larval moulting in the silkworm, *Bombyx mori. J. Insect Physiol. 27*, 805–812.

KING, D. S. (1969). Evidence for peripheral conversion of α-ecdysone to β-ecdysone in crustaceans and insects. *Gen. Comp. Endocr. 13*, 512.

KING, D. S. (1972a). Metabolism of α-ecdysone and possible immediate precursors by insects *in vivo* and *in vitro*. *Gen. Comp. Endocr.* (Suppl. 3), 221–227.

KING, D. S. (1972b). Ecdysone metabolism in insects. *Amer. Zool. 12*, 343–345.

KING, D. S. and MARKS, E. P. (1974). The secretion and metabolism of α-ecdysone by cockroach (*Leucophaea maderae*) tissues *in vitro*. *Life Sci. 15*, 147–154.

KING, D. S. and SIDDALL, J. B. (1969). Conversion of α-ecdysone to β-ecdysone by crustaceans and insects. *Nature 221*, 955–956.

KING, D. S., BOLLENBACHER, W. E., BORST, D. W., VEDECKIS, W. V., O'CONNOR, J. D., ITTYCHERIAH, P. I. and GILBERT, L. I. (1974). The secretion of α-ecdysone by the prothoracic glands of *Manduca sexta in vitro*. *Proc. Nat. Acad. Sci. 71*, 793–796.

KLOSE, W., GATEFF, E., EMMERICH, H. and BEIKIRCH, H. (1980). Developmental studies on two ecdysone deficient mutants of *Drosophila melanogaster*. *Wilhelm Roux's Archiv. 189*, 57–67.

KOOLMAN, J. (1976). Ecdysone oxidase, an enzyme of the ecdysone metabolism in insects. In *Actualités sur les Hormones D'Invertébrés*. Edited by M. M. Durchon. Pages 403–412. *Colloques Internationaux C.N.R.S.* No. 251, Paris.

KOOLMAN, J. (1980). Ecdysteroids in the blowfly, *Calliphora vicina*. In *Progress in Ecdysone Research*. Edited by J. A. Hoffmann. Pages 187–209. Elsevier, Amsterdam.

KOOLMAN, J. and KARLSON, P. (1975). Ecdysone oxidase, an enzyme from the blowfly *Calliphora erythrocephala* (Meigen). *Hoppe-Seyler's Z. Physiol. Chem. 356*, 1131–1138.

KOOLMAN, J. and KARLSON, P. (1978). Ecdysone oxidase: Reaction and specificity. *Eur. J. Biochem. 89*, 453–460.

KOOLMAN, J., HOFFMANN, J. A. and DREYER, M. (1975). Moulting hormone in *Locusta migratoria*: Rate of ecdysone 20-hydroxylation and excretion during the last larval instar. *Experientia 31*, 247–248.

KOOLMAN, J., REUM, L. and KARLSON, P. (1979). 26-Hydroxyecdysone, 20,26-dihydroxyecdysone and inokosterone detected as metabolites of ecdysone in the blowfly *Calliphora vicina* by radiotracer experiments. *Hoppe-Seyler's Z. Physiol. Chem. 360*, 1351–1355.

KOPEĆ, S. (1917). Experiments on metamorphosis of insects. *Bull. Int. Acad. Sci., Cracovie, Classe Sci. Math. Nat. Ser. B.*, 57–60.

KOPEĆ, S. (1922). Studies on the necessity of the brain for the inception of insect metamorphosis. *Biol. Bull. 42*, 323–342.

KRAMINSKY, G. P., CLARK, W. C., ESTELLE, M. A., GIETZ, R. D., SAGE, B. A., O'CONNOR, J. D. and HODGETTS, R. B. (1980). Induction of translatable mRNA for dopa decarboxylase in *Drosophila*: an early response to ecdysterone. *Proc. Nat. Acad. Sci. 77*, 4175–4179.

KUBO, I., KLOCKE, J. A. and ASANO, S. (1981). Insect ecdysis inhibitors from the East African medicinal plant, *Ajuga remota* (Labiatae). *Agric. Biol. Chem. 45*, 1925–1927.

KUBO, I., KLOCKE, J. A. and ASANO, S. (1983). Effects of ingested phytoecdysteroids on the growth and development of two lepidopterous larvae. *J. Insect Physiol. 29*, 307–316.

LAFONT, R., MAUCHAMP, B., PENNETIER, J. L., TARROUX, P., DEHYS, L. and DELBECQUE, J. P. (1975). α- and β-ecdysone levels in insect hemolymph: correlation with developmental events. *Experientia 31*, 1241–1242.

LAFONT, R., MAUCHAMP, B., BLAIS, C. and PENNETIER, J. L. (1977). Ecdysones and imaginal disc development during the last larval instar of *Pieris brassicae*. *J. Insect Physiol. 23*, 277–283.

LAFONT, R., SOMMÉ-MARTIN, G., MAUCHAMP, B., MAUME, B. F. and DELBECQUE, J.-P. (1980a). Analysis of ecdysteroids by high-performance liquid chromatography and coupled gas-liquid chromatography-mass spectrometry. In *Progress in Ecdysone Research*. Edited by J. A. Hoffmann. Pages 45–68. Elsevier, Amsterdam.

LAFONT, R., BEYDON, P., SOMMÉ-MARTIN, G. and BLAIS, C. (1980b). High performance liquid chromatography of ecdysone metabolites applied to the cabbage butterfly, *Pieris brassicae* L. *Steroids 36*, 185–207.

LAGUEUX, M., PERRON, J.-M. and HOFFMANN, J. A. (1976). Ecdysone metabolism and endogenous moulting hormone titer during larval-pupal development in *Choristoneura fumiferana*. *J. Insect Physiol. 22*, 57–62.

LAGUEUX, M., HIRN, M. and HOFFMANN, J. A. (1977). Ecdysone during ovarian development in *Locusta migratoria*. *J. Insect Physiol. 23*, 109–119.

LAGUEUX, M., HETRU, C., GOLTZENE, F., KAPPLER, C. and HOFFMANN, J. A. (1979). Ecdysone titer and metabolism in relation to cuticulogenesis in embryos of *Locusta migratoria*. *J. Insect Physiol. 25*, 709–723.

LAGUEUX, M., SALL, C. and HOFFMANN, J. A. (1981). Ecdysteroids during embryogenesis in *Locusta migratoria*. *Amer. Zool. 21*, 715–726.

LASKER, J. M., HUANG, M.-T. and CONNEY, A. H. (1982). *In vivo* activation of zoxazolamine metabolism by flavone. *Science 216*, 1419–1421.

LOCKE, M. (1969). The ultrastructure of the oenocytes in the molt/intermolt cycle of an insect. *Tissue Cell 1*, 103–154.

LOEB, M. J. (1982). Diapause and development in the tobacco budworm, *Heliothis virescens*: a comparison of haemolymph ecdysteroid titres. *J. Insect Physiol. 28*, 667–673.

LOEB, M. J., WOODS, C. W., BRANDT, E. P. and BOŘKOVEC, A. B. (1982). Larval testes of the tobacco budworm: a new source of insect ecdysteroids. *Science 218*, 896–898.

LOUVET, J. P., FOURNIER, B., CAVALLIN, M., DELACHAMBRE, J., DELBECQUE, J. P. and GAUTHIER, J. (1979). Variations of total ecdysteroids response to RIA during embryonic development and cuticular cycles in the phasmid *Carausius*. *Fourth Europ. Symp. Ecdysone*, Strasbourg, France.

MAAS, H. (1974). Ecdysontiterbestimmung wahrend des vorletzten larvenstadiums bei *Gryllus bimaculatus*. Unveröffentlichte Diplomarbeit, Mainz.

MALÁ, J., GRANGER, N. A. and SEHNAL, F. (1977). Control of prothoracic gland activity in larvae of *Galleria mellonella*. *J. Insect Physiol. 23*, 343–347.

MANDARON, P. (1973). Effects of α-ecdysone, β-ecdysone and inokosterone on the *in vitro* evagination of *Drosophila* leg discs and the subsequent differentiation of imaginal integumentary structures. *Devel. Biol. 31*, 101–113.

MARKS, E. P., ITTYCHERIAH, P. I. and LELOUP, A. M. (1972). The effect of β-ecdysone on insect neurosecretion *in vivo*. *J. Insect Physiol. 18*, 847–850.

MARÓY, P. and TARNÓY, K. (1978). Moulting hormone level during the last instar of *Galleria mellonella* larva. *J. Insect Physiol. 24*, 325–327.

MARÓY, P., DENNIS, R., BECKERS, C., SAGE, B. A. and O'CONNOR, J. D. (1978). Demonstration of an ecdysteroid receptor in a cultured cell line of *Drosophila melanogaster*. *Proc. Nat. Acad. Sci. 75*, 6035–6038.

MARÓY, P., KOCZKA, K., FEKETE, E. and VARGHA, J. (1980). Molting hormone titer of *D. melanogaster* larvae. *Dros. Inform. Serv. 55*, 98–99.

MASLER, E. P., FUCHS, M. S., SAGE, B. and O'CONNOR, J. D. (1980). Endocrine regulation of ovarian development in the autogenous mosquito, *Aedes atropalpus*. *Gen. Comp. Endocr. 41*, 250–259.

MASNER, P., HANGARTNER, W. and SUCHY, M. (1975). Reduced titers of ecdysone following juvenile hormone treatment in the german cockroach, *Blattella germanica*. *J. Insect Physiol. 21*, 1755–1762.

MATZ, G. (1980). Variations du taux des ecdystéroides au cours de l'embryogénèse de *Leucophaea maderae* Fabr. (Insecte Dictyoptère). *C.R. Acad. Sci. Paris (D) 291*, 501–504.

MAYER, R. T., SVOBODA, J. A. and WEIRICH, G. F. (1978). Ecdysone 20-hydroxylase in midgut mitochondria of *Manduca sexta*. (L.) *Hoppe-Seyler's Z. Physiol. Chem. 359*, 1247–1257.

MAYER, R. T., DURRANT, J. L., HOLMAN, G. M., WEIRICH, G. F. and SVOBODA, J. A. (1979). Ecdysone 3-epimerase from the midgut of *Manduca sexta* (L.). *Steroids 34*, 555–562.

MCCARTHY, J. P. (1979). Ponasterone A: a new ecdysteroid from the embryos and serum of *Brachyuran* crustaceans. *Steroids 34*, 799–806.

MCDANIEL, C. N. (1979). Haemolymph ecdysone concentrations in *Hyalophora cecropia* pupae, dauer pupae and adults. *J. Insect Physiol. 25*, 143–145.

MILNER, M. J. and SANG, J. H. (1974). Relative activities of α-ecdysone and β-ecdysone for the differentiation *in vitro* of *Drosophila melanogaster* imaginal discs. *Cell 3*, 141–143.

MITANI, F. (1979). Cytochrome P450 in adrenocortical mitochondria. *Mol. Cell. Biochem. 24*, 21–43.

MITCHELL, M. J. (1983). Characterization and quantification of ecdysone 20-monooxygenase activity in wandering stage larvae of *Drosophila melanogaster*. M.S. thesis, Bowling Green State University, Bowling Green, Ohio.

MITCHELL, M. J. and SMITH, S. L. (1983). Characterization of ecdysone 20-monooxygenase activity in third instar larvae of *Drosophila melanogaster*. *Ohio J. Sci. 83*, 68.

MIZOGUCHI, A. and ISHIZAKI, H. (1982). Prothoracic glands of the saturniid moth *Samia cynthia ricini* possess a circadian clock controlling gut purge timing. *Proc. Nat. Acad. Sci. 79*, 2726–2730.

MORGAN, E. D. and POOLE, C. F. (1976a). The pattern of ecdysone levels during development in the desert locust, *Schistocerca gregaria*. *J. Insect Physiol.* *22*, 885–889.

MORGAN, E. D. and POOLE, C. F. (1976b). The extraction and determination of ecdysones in arthropods. In *Advances in Insect Physiology*. Edited by J. E. Treherne, M. J. Berridge and V. B. Wigglesworth. Vol. 14, pages 17–62. Academic Press, New York.

MORGAN, E. D. and POOLE, C. F. (1977). Chemical control of insect moulting. *Comp. Biochem. Physiol.* *57B*, 99–109.

MORGAN, E. D. and WILSON, I. D. (1980). Progress in the analysis of ecdysteroids. In *Progress in Ecdysone Research*. Edited by J. A. Hoffmann. Pages 29–43. Elsevier, Amsterdam.

MORIBAYASHI, A. F. and OHTAKI, T. (1978). Inactivation of ecdysone and the possible feedback control of the titre during pupation of *Sarcophaga peregrina*. *J. Insect Physiol.* *24*, 279–284.

MORIBAYASHI, A. and OHTAKI, T. (1980). Inactivation and reactivation of 20-hydroxyecdysone during pupal-adult development of the fleshfly *Sarcophaga peregrina*. *Japan J. Med. Sci. Biol.* *33*, 189–201.

MORIYAMA, H., NAKANISHI, K., KING, D. S., OKAUCHI, T., SIDDALL, J. B. and HAFFERL, W. (1970). On the origin and metabolic fate of α-ecdysone in insects. *Gen. Comp. Endocr.* *15*, 80–87.

NAKANISHI, K. (1971). The ecdysones. *Pure Appl. Chem.* *25*, 167–195.

NAKANISHI, K., MORIYAMA, H., OKAUCHI, T., FUJIOKA, S. and KOREEDA, M. (1972). Biosynthesis of α- and β-ecdysones from cholesterol outside the prothoracic gland in *Bombyx mori*. *Science* *176*, 51–52.

NATORI, S. and OHTAKI, T. (1976). Uptake of β-ecdysone by the fat body and membrane vesicles of fat body cells of *Sarcophaga peregrina* larvae. *J. Insect Physiol.* *22*, 227–234.

NIGG, H. N., SVOBODA, J. A., THOMPSON, M. J., KAPLANIS, J. N., DUTKY, S. R. and ROBBINS, W. E. (1974). Ecdysone metabolism: ecdysone dehydrogenase-isomerase. *Lipids* *9*, 971–974.

NIGG, H. N., SVOBODA, J. A., THOMPSON, M. J., DUTKY, S. R., KAPLANIS, J. N. and ROBBINS, W. E. (1976). Ecdysone 20-hydroxylase from the midgut of the tobacco hornworm (*Manduca sexta* L.). *Experientia* *32*, 438–439.

NIJHOUT, H. F. (1975). A threshold size for metamorphosis in the tobacco hornworm, *Manduca sexta* (L.) *Biol. Bull.* *149*, 214–225.

NIJHOUT, H. F. and WILLIAMS, C. M. (1974a). Control of moulting and metamorphosis in the tobacco hornworm, *Manduca sexta* (L.): growth of the last instar larva and the decision to pupate. *J. Exp. Biol.* *61*, 481–491.

NIJHOUT, H. F. and WILLIAMS, C. M. (1974b). Control of moulting and metamorphosis in the tobacco hornworm, *Manduca sexta* (L.): cessation of juvenile hormone secretion as a trigger for pupation. *J. Exp. Biol.* *61*, 493–501.

NIRDE, PH., TORPIER, G., DEREGGI, M. L. and CAPRON, A. (1983). Ecdysone and 20-hydroxyecdysone: new hormones for the human parasite *Schistosoma mansoni*. *FEBS Lett.* *151*, 223–227.

NUTTING, W. L. (1955). Extirpation of roach prothoracic glands. *Science* *122*, 30–31.

OBERLANDER, H. (1969). Effects of ecdysone, ecdysterone and inokosterone on the *in vitro* initiation of metamorphosis of wing disks of *Galleria mellonella*. *J. Insect Physiol.* *15*, 297–304.

OBERLANDER, H. (1972a). α-Ecdysone induced DNA synthesis in cultured wings of *Galleria mellonella*: inhibition by 20-hydroxyecdysone and 22-isoecdysone. *J. Insect Physiol.* *18*, 223–228.

OBERLANDER, H. (1972b). The hormonal control of development of imaginal disks. In *The Biology of Imaginal Disks*. Edited by H. Ursprung and R. Nöthiger. Pages 155–172. Springer-Verlag, New York.

OBERLANDER, H., LEACH, C. E. and TOMBLIN, C. (1973). Cuticle deposition in imaginal disks of three species of Lepidoptera: effects of ecdysones *in vitro*. *J. Insect Physiol.* *19*, 993–998.

O'CONNOR, J. D. and CHANG, E. S. (1981). Cell lines as a model for the study of metamorphosis. In *Metamorphosis: A Problem in Developmental Biology* 2nd edn. Edited by L. I. Gilbert and E. Frieden. Pages 241–261. Plenum, New York.

O'CONNOR, J. D., MARÓY, P., BECKERS, C., DENNIS, R., ALVAREZ, C. M. and SAGE, B. A. (1980). Ecdysteroid receptors in cultured *Drosophila* cells. In *Gene Regulation by Steroid Hormones*. Edited by A. K. Roy and J. H. Clark. Pages 263–277. Springer-Verlag, New York.

OHNISHI, E., MIZUNO, T., CHATANI, F., IKEKAWA, N. and SAKURAI, S. (1977). 2-Deoxy-α-ecdysone from ovaries and eggs of the silkworm, *Bombyx mori*. *Science* *197*, 66–67.

OHTAKI, T. (1981). Ecdysteroid metabolism in insects. *Amer. Zool.* *21*, 727–731.

OHTAKI, T., MILKMAN, R. D. and WILLIAMS, C. M. (1967). Ecdysone and ecdysone analogues: their assay on the fleshfly *Sarcophaga peregrina*. *Proc. Nat. Acad. Sci.* *58*, 981–984.

OHTAKI, T., MILKMAN, R. D. and WILLIAMS, C. M. (1968). Dynamics of ecdysone secretion and action in the fleshfly *Sarcophaga peregrina*. *Biol. Bull.* *135*, 322–334.

O'NEILL, M. P., HOLMAN, G. M. and WRIGHT, J. E. (1977). β-Ecdysone levels in pharate pupae of the stable fly, *Stomoxys calcitrans* and interaction with the chitin inhibitor diflubenzuron. *J. Insect Physiol.* *23*, 1243–1244.

PALEVODY, C., DELBECQUE, J.-P. and DELACHAMBRE, J. (1977). Variations du taux des ecdystéroides en relation avec les cycles de nue et les cycles ovariens chez *Folsomia candida* (Insecta. Collembola). *C.R. Acad. Sci. Paris (D)* *285*, 1323–1326.

PÉRON, F. G., HAKSAR, A. and LIN, M.-T. (1975). Sources of reducing equivalents for cytochrome P-450 mitochondrial steroid hydroxylations in rat adrenal cortex cells. *J. Steroid Biochem.* *6*, 411–417.

PIEPHO, H. (1948). Zur frage der bildungsorgane des häutungswirkstoffs bei schmetterlingen. *Naturwissenschaften* *35*, 94–95.

RAO, K. D. P., NORRIS, D. M. and CHU, H. M. (1981). Ecdysteroids during pupal development of female *Xyleborus ferrugineus*. *Ent. Exp. Appl.* *30*, 151–156.

RAY, J. W. (1965). In *Pest Infestation Research*, p. 59. Her Majesty's Stationery Office, London.

RAY, J. W. (1967). The epoxidation of aldrin by housefly microsomes and its inhibition by carbon monoxide. *Biochem. Pharmac.* *16*, 99–107.

REDFERN, C. P. F. (1982). 20-Hydroxy-ecdysone and ovarian development in *Anopheles stephensi*. *J. Insect Physiol.* *28*, 97–109.

REDFERN, R. E., KELLY, T. J., BOŘKOVEC, A. B. and HAYES, D. K. (1982). Ecdysteroid titers and molting aberrations in last-stage *Oncopeltus* nymphs treated with insect growth regulators. *Pest. Biochem. Physiol.* *18*, 351–356.

REES, H. H. (1971). Ecdysones. In *Aspects of Terpenoid Chemistry and Biochemistry*. Edited by T. W. Goodwin. Pages 181–222. Academic Press, New York.

RICHARDS, G. (1981). The radioimmune assay of ecdysteroid titres in *Drosophila melanogaster*. *Mol. Cell. Endocr.* *21*, 181–197.

RIDDIFORD, L. M. (1976). Hormonal control of insect epidermal cell commitment *in vitro*. *Nature* *259*, 115–117.

RIDDIFORD, L. M. (1978). Ecdysone-induced change in cellular commitment of the epidermis of the tobacco hornworm, *Manduca sexta*, at the initiation of metamorphosis. *Gen. Comp. Endocr.* *34*, 438–446.

RIDDIFORD, L. M. (1981). Hormonal control of epidermal cell development. *Amer. Zool.* *21*, 751–762.

ROBERTSON, J. and PIPA, R. (1973). Metamorphic shortening of interganglionic connections of *Galleria mellonella* (Lepidoptera) *in vitro*: stimulation by ecdysone analogues. *J. Insect Physiol.* *19*, 673–679.

ROHNER, R. M. and MEYER, D. R. (1980). The titers and the clearance of ecdysteroids from the blood of last instar larvae of *Galleria mellonella* L. *Experientia* *36*, 1327–1329.

ROMER, F. (1973). Moulting hormone content in prothoracic glands and oenocytes of *Tenebrio molitor* within a moulting cycle. *Experientia* *29*, 1417–1418.

ROMER, F., EMMERICH, H. and NOWOCK, J. (1974). Biosynthesis of ecdysones in isolated prothoracic glands and oenocytes of *Tenebrio molitor in vitro*. *J. Insect Physiol.* *20*, 1975–1987.

RUBENSTEIN, E. C., KELLY, T. J., SCHWARTZ, M. B. and WOODS, C. W. (1982). *In vitro* synthesis and secretion of ecdysteroids by *Drosophila melanogaster* ovaries. *J. Exp. Zool.* *223*, 305–308.

SAFRANEK, L., CYMBOROWSKI, B. and WILLIAMS, C. M. (1980). Effects of juvenile hormone on ecdysone-dependent development in the tobacco hornworm, *Manduca sexta*. *Biol. Bull.* *158*, 248–256.

SAKURAI, S., IKEKAWA, N., OHTAKI, T. and CHINO, H. (1977). 3β-Hydroxy-5α-cholestan-6-one: a possible precursor of ecdysone biosynthesis. *Science* *198*, 627–629.

SATO H., ASHIDA, N., SUHARA, K., ITAGAKI, E., TAKEMORI, S. and KATAGIRI, M. (1978). Properties of an adrenal cytochrome P-450 (P-$450_{11}\beta$) for the hydroxylations of corticosteroids. *Arch. Biochem. Biophys.* *190*, 307–314.

SATO, R. and OMURA, T. (1978). *Cytochrome P-450*. Academic Press, New York.

SATO, Y., SAKAI, M., IMAI, S. and FUJIOKA, S. (1968). Ecdysone activity of plant originated moulting hormones applied to the body surface of lepidopterous larvae. *Appl. Ent. Zool. 3*, 49–51.

SCHALLER, F. and CHARLET, M. (1980). Neuroendocrine control and rate of ecdysone biosynthesis in larvae of a paleopteran insect: *Aeshna cyanea* Müller. In *Progress in Ecdysone Research*. Edited by J. A. Hoffmann. Pages 99–110. Elsevier, Amsterdam.

SCHALLER, F., HOFFMANN, J. A. and KOOLMAN, J. (1975). Métabolisme de l'ecdysone chez *Aeshna cyanea* (Insecte, Odonata). Résultats d'injections d' x-ecdysone radioactive á des animaux et á des larves au cycle de mue bloqué. *C.R. Acad. Sci. Paris D. 280*, 1813–1816.

SCHALLER, F. E. and HOFFMANN, J. A. (1976). Metabolisme de L'α-ecdysone *in vivo* et *in vitro* chez *Aeshna cyanea* (Insecte, Odonate). In *Actualités sur les Hormones D'Invertébrés*. Edited by M. M. Durchon. Pages 393–401. *Colloques Internationaux CNRS* No. 251, Paris.

SCOTT, D. (1982). Communication to Insect Hormone Discussion Group, workshop 1982, Royal Entomological Society, London.

SEGEL, I. H. (1976). *Biochemical Calculations*, 2nd edn. John Wiley and Sons, New York.

SEHNAL, F., MARÓY, P. and MALÁ, J. (1981). Regulation and significance of ecdysteroid titre fluctuations in lepidopterous larvae and pupae. *J. Insect Physiol. 27*, 535–544.

SHAAYA, E. (1978). Ecdysone and juvenile hormone activity in the larvae of the cockroach *Periplaneta americana*. *Insect Biochem. 8*, 193–195.

SHAAYA, E. and KARLSON, P. (1965a). Der ecdysontiter während der insektenentwicklung II. Die postembryonale entwicklung der schmeissfliege *Calliphora erythrocephala* Meig. *J. Insect Physiol. 11*, 65–69.

SHAAYA, E. and KARLSON, P. (1965b). Der ecdysontiter während der insektenentwicklung IV. Die entwicklung der lepidopteren *Bombyx mori* L. and *Cerura vinula* L. *Devel. Biol. 11*, 424–432.

SHARMA, R. K. (1973). Regulation of steroidogenesis by adrenocorticotropic hormone in isolated adrenal cells of rat. *J. Biol. Chem. 248*, 5473–5476.

SIDDALL, J. B., CROSS, A. D. and FRIED, J. H. (1966). Steroids. CCXCII. Synthetic studies on insect hormones. II. The synthesis of ecdysone. *J. Amer. Chem. Soc. 88*, 862–877.

SIEW, Y. C. and GILBERT, L. I. (1971). Effects of moulting hormone and juvenile hormone on insect endocrine gland activity. *J. Insect Physiol. 17*, 2095–2104.

SIMPSON, E. R. (1979). Cholesterol side-chain cleavage, cytochrome P-450, and the control of steroidogenesis. *Mol. Cell. Endocr. 13*, 213–227.

SINGH, H. H. and SEHNAL, F. (1979). Lack of specific neurons in the ventral nerve cord for the control of prothoracic glands. *Experientia 35*, 1117–1119.

SLAMA, K. (1976). Old concepts and new findings on hormonal control of morphogenesis in *Dermestes*. In *Actualités sur les Hormones D' Invertébrés*. Edited by M. M. Durchon. Pages 499–506. *Colloques Internationaux CNRS*, No. 251, Paris.

SMITH, S. L. (1983a). Ecdysone 20-monooxygenase activity in adult female *Aedes aegypti* mosquitoes. *Ohio J. Sci. 83*, 68–69.

SMITH, S. L. (1983b). The effects of plant allelochemics on *Manduca sexta* ecdysone 20-monooxygenase activity. *Ohio J. Sci. 83*, 69.

SMITH, S. L., BOLLENBACHER, W. E., COOPER, D. Y. SCHLEYER, H., WIELGUS, J. J. and GILBERT, L. I. (1977). Characterization of ecdysone 20-monooxygenase activity in *Manduca sexta*. *Amer. Zool. 17*, 928. (Abs., 387).

SMITH, S. L., BOLLENBACHER, W. E., COOPER, D. Y., SCHLEYER, H., WIELGUS, J. J. and GILBERT, L. I. (1979). Ecdysone 20-monooxygenase: characterization of an insect cytochrome P-450 dependent steroid hydroxylase. *Mol. Cell. Endocr. 15*, 111–133.

SMITH, S. L., BOLLENBACHER, W. E. and GILBERT, L. I. (1980). Studies on the biosynthesis of ecdysone and 20-hydroxy-ecdysone in the tobacco hornworm, *Manduca sexta*. In *Progress in Ecdysone Research*. Edited by J. A. Hoffmann. Pages 139–162. Elsevier, Amsterdam.

SMITH, S. L., BOLLENBACHER, W. E. and GILBERT, L. I. (1983). Ecdysone 20-monooxygenase activity during larval-pupal development of *Manduca sexta*. *Mol. Cell. Endocr. 31*, 227–251.

SRIVASTAVA, K. P., TIWARI, R. K. and KUMAR, P. (1977). Effect of sectioning the prothoracic gland nerves in the larva of the lemon butterfly, *Papilio demoleus* L. *Experientia 33*, 98–99.

STEEL, C. G. H. (1973). Humoral regulation of the cerebral neurosecretory system of *Rhodnius prolixus* (Stal) during growth and moulting. *J. Exp. Biol. 58*, 177–187.

STEEL, C. G. H. (1975). A neuroendocrine feedback mechanism in the insect moulting cycle. *Nature 253*, 267–269.

STEEL, C. G. H. (1978). Nervous and hormonal regulation of neurosecretory cells in the insect brain. In *Comparative Endocrinology*. Edited by P. J. Gaillard and H. H. Boer. Pages 327–330. Elsevier, Amsterdam.

STEEL, C. G. H., BOLLENBACHER, W. E., SMITH, S. L. and GILBERT, L. I. (1982). Haemolymph ecdysteroid titres during larval–adult development in *Rhodnius prolixus*: correlations with moulting hormone action and brain neurosecretory cell activity. *J. Insect Physiol. 28*, 519–525.

STELLWAAG-KITTLER, F. (1954). Zur physiologie der käferhäutung untersuchungen am mehlkäfer *Tenebrio molitor* L. *Biol. Zentralbl. 73*, 12–49.

STUDINGER, G. and WILLIG, A. (1975). Biosynthesis of α- and β-ecdysone in isolated abdomens of larvae of *Musca domestica*. *J. Insect Physiol. 21*, 1793–1798.

SVOBODA, J. A., KAPLANIS, J. N., ROBBINS, W. E. and THOMPSON, M. J. (1975). Recent developments in insect steroid metabolism. *Ann. Rev. Ent. 20*, 205–220.

TAKEDA, N. (1978). Hormonal control of prepupal diapause in *Monema flavescens* (Lepidoptera). *Gen. Comp. Endocr. 34*, 123–131.

TERRIERE, L. C. and YU, S. J. (1976). Interaction between microsomal enzymes of the house fly and the moulting hormones and some of their analogs. *Insect Biochem. 6*, 109–114.

THAMER, G. and KARLSON, P. (1972). Nachweis der proteinbindung von ecdyson bei der schmeissfliege *Calliphora erythrocephala* Meigen. *Z. Naturf. 27b*, 1191–1195.

THOMPSON, M. J., KAPLANIS, J. N., ROBBINS, W. E. and YAMAMOTO, R. T. (1967). 20,26-Dihydroxyecdysone, a new steroid with moulting hormone activity from the tobacco hornworm, *Manduca sexta* (Johannson). *J. Chem. Soc., Chem. Commun.* 650–653.

THOMPSON, M. J., KAPLANIS, J. N., ROBBINS, W. E. and SVOBODA, J. (1973). Metabolism of steroids in insects. *Adv. Lipid Res. 11*, 219–265.

THOMPSON, M. J., KAPLANIS, J. N., ROBBINS, W. E., DUTKY, S. R. and NIGG, H. N. (1974). 3-Epi-20-hydroxyecdysone from meconium of the tobacco hornworm. *Steroids 24*, 359–366.

THOMPSON, M. J., ROBBINS, W. E., DUTKY, S. R., MARKS, E. P., FILIPI, P. A. and FINEGOLD, H. (1978). Isolation and identification of the metabolites of 22,25-dideoxyecdysone from cockroach fat body cultures. *Lipids 13*, 783–790.

THOMSON, J. A., IMRAY, F. P. and HORN, D. H. S. (1970). An improved *Calliphora* bioassay for insect moulting hormones. *Aust. J. Exp. Biol. Med. Sci. 48*, 321–328.

TRUMAN, J. W. (1972). Physiology of insect rhythms I. Circadian organization of the endocrine events underlying the moulting cycle of larval tobacco hornworms. *J. Exp. Biol. 57*, 805–820.

TRUMAN, J. W. and RIDDIFORD, L. M. (1974). Physiology of insect rhythms III. The temporal organization of the endocrine events underlying pupation of the tobacco hornworm. *J. Exp. Biol. 60*, 371–382.

TSOUPRAS, G., HETRU, C., LUU, B., LAGUEUX, M., CONSTANTIN, E. and HOFFMANN, J. A. (1982). The major conjugates of ecdysteroids in young eggs and in embryos of *Locusta migratoria*. *Tetrahedron Lett. 23*, 2045–2048.

TSOUPRAS, G., LUU, B. and HOFFMANN, J. A. (1983). A cytokinin (isopentenyl-adenosyl-mononucleotide) linked to ecdysone in newly laid eggs of *Locusta migratoria*. *Science 220*, 507–509.

VALENTIN, M., BOLLENBACHER, W. E., GILBERT, L. I. and KROEGER, H. (1978). Alterations in ecdysone content during the post-embryonic development of *Chironomus thummi*: correlations with chromosomal puffing. *Z. Naturf. 33c*, 557–560.

VEDECKIS, W. and GILBERT, L. I. (1973). Production of cyclic AMP and adenosine by the brain and prothoracic glands of *Manduca sexta*. *J. Insect Physiol. 19*, 2445–2457.

VEDECKIS, W. V., BOLLENBACHER, W. E. and GILBERT, L. I. (1974). Cyclic AMP as a possible mediator of prothoracic gland activation. *Zool. Jahrb. Physiol. 78*, 440–448.

VEDECKIS, W. V., BOLLENBACHER, W. E. and GILBERT, L. I. (1976). Insect prothoracic glands: a role for cyclic AMP in the stimulation of α-ecdysone secretion. *Mol. Cell. Endocr. 5*, 81–88.

WALKER, G. P. and DENLINGER, D. L. (1980). Juvenile hormone and moulting hormone titers in diapause- and non-diapause destined flesh flies. *J. Insect Physiol. 26*, 661–664.

WEINHEIMER, G. and ROMER, F. (1977). Der metabolismus von ^{3}H-α-ecdyson bei larven von *Tenebrio molitor. J. Insect Physiol. 23*, 619–624.

WENTWORTH, S. L., ROBERTS, B. and O'CONNOR, J. D. (1981). Ecdysteroid titres during postembryonic development of *Sarcophaga bullata* (Sarcophagidae: Diptera). *J. Insect Physiol. 27*, 435–440.

WIELGUS, J. J., BOLLENBACHER, W. E. and GILBERT, L. I. (1979). Correlations between epidermal DNA synthesis and haemolymph ecdysteroid titre during the last larval instar of the tobacco hornworm *Manduca sexta. J. Insect Physiol. 25*, 9–16.

WIGGLESWORTH, V. B. (1934). The physiology of ecdysis in *Rhodnius prolixus* (Hemiptera). II. Factors controlling moulting and metamorphosis. *Quart. J. Mic. Sci. 77*, 191–222.

WIGGLESWORTH, V. B. (1936). The function of the corpus allatum in the growth and reproduction of *Rhodnius prolixus* (Hemiptera). *Quart. J. Mic. Sci. 79*, 91–121.

WIGGLESWORTH, V. B. (1940). The determination of characters at metamorphosis in *Rhodnius prolixus* (Hemiptera). *J. Exp. Biol. 17*, 201–222.

WILDMANN, H. and ROMER, F. (1977). Metabolism of ^{3}H-α-ecdysone in *Gryllus bimaculatus* during the eighth larval instar. *J. Insect Physiol. 23*, 343–349.

WILLIAMS, C. M. (1947). Physiology of insect diapause. II. Interaction between the pupal brain and prothoracic glands in the metamorphosis of the giant silkmoth, *Platysamia cecropia. Biol. Bull. 93*, 89–98.

WILLIAMS, C. M. (1948). Physiology of insect diapause. III. The prothoracic glands in the cecropia silkworm, with special reference to their significance in embryonic and postembryonic development. *Biol. Bull. 94*, 60–65.

WILLIAMS, C. M. (1952). Physiology of insect diapause. IV. The brain and prothoracic glands as an endocrine system in the cecropia silkworm. *Biol. Bull. 103*, 120–138.

WILSON, I. D. and MORGAN, E. D. (1978). Variations in ecdysteroid levels in 5th instar larvae of *Schistocerca gregaria* in gregarious and solitary phases. *J. Insect Physiol. 24*, 751–756.

YAGI, S. and FUKAYA, M. (1974). Juvenile hormone as a key factor regulating larval diapause of the rice stem borer, *Chilo suppressalis* (Lepidoptera: Pyralidae). *Appl. Ent. Zool. 9*, 247–255.

YIN, C. M. and CHIPPENDALE, G. M. (1973). Juvenile hormone regulation of the larval diapause of the southwestern cornborer, *Diatraea grandiosella. J. Insect Physiol. 19*, 2403–2420.

YOUNG, N. L. (1976a). The metabolism of ^{3}H-molting hormone in *Calliphora erythrocephala* at the mature larval and white puparial stages. *Insect Biochem. 6*, 1–12.

YOUNG, N. L. (1976b). The metabolism of ^{3}H-molting hormone in *Calliphora erythrocephala* during larval development. *J. Insect Physiol. 22*, 153–155.

YOUNG, N. L. and YOUNG, P. R. (1976). Biochemical control of moulting hormone titer in *Calliphora erythrocephala* during puparium formation. *Insect Biochem. 6*, 169–177.

YUND, M. A., KING, D. S. and FRISTROM, J. W. (1978). Ecdysteroid receptors in imaginal discs of *Drosophila melanogaster. Proc. Nat. Acad. Sci. 75*, 6039–6043.

9 Regulation of Ecdysteroid Titer: Degradation

JAN KOOLMAN and PETER KARLSON

Philipps-Universität Marburg, FRG

1 INTRODUCTION

1.1 Hormonal regulatory systems

One of the most important concepts in general physiology is that of homeostasis: It comprises a type of regulation in which one feature, e.g. the concentration of some ions, or an important metabolite or energy source is kept at as constant a level as possible. There are many types of biochemical regulations involved in homeostasis (see Volumes 4 and 10).

As far as animal hormones are concerned, homeostatic control of their concentration is the exception rather than the rule, since hormones generally act as signal substances. It is by means of the hormones that homeostasis of other physiological functions, including control of metabolite concentration, is achieved. Other hormones, sometimes called morphogenetic hormones, are involved in the control of developmental processes. The hormone is secreted at the onset of the developmental process, its titer rises to a certain peak and then falls again when the process is either completed or at least completely triggered. This type of titer change is characteristic of ecdysteroids (chapter 8, this volume).

From the physiological point of view, it is useful to look not only at the hormone titer but at the whole system. As shown in Fig. 1, a hormonal regulatory system involves several biochemical and/or biological processes. The first is the biosynthesis and secretion from the hormone-producing cell into the circulation. That part of the hormonal system of ecdysteroids is dealt with in chapters 7 and 8 in this volume. Through the circulation, the hormone reaches the target tissue, where it exerts its action. Certain aspects of the hormonal system are also dealt with in volume 8, and to some extent in volumes 1 and 2. In the present chapter, emphasis is laid on the fate of ecdysone, 20-hydroxyecdysone, and other ecdysteroids in the circulation and in the tissues.

As can be seen in Fig. 1, some of the tissues, notably the fat body can metabolize the ecdysone to 20-hydroxyecdysone and to inactive products which are either excreted or stored. Storage is necessary in those insect stages that comprise closed systems, i.e. the egg (chapter 13, volume 1) and the pupa.

1.2 Regulation of hormone titers

1.2.1 Hydroxylation

The most important reaction in this respect is the hydroxylation of ecdysone in the 20-position to

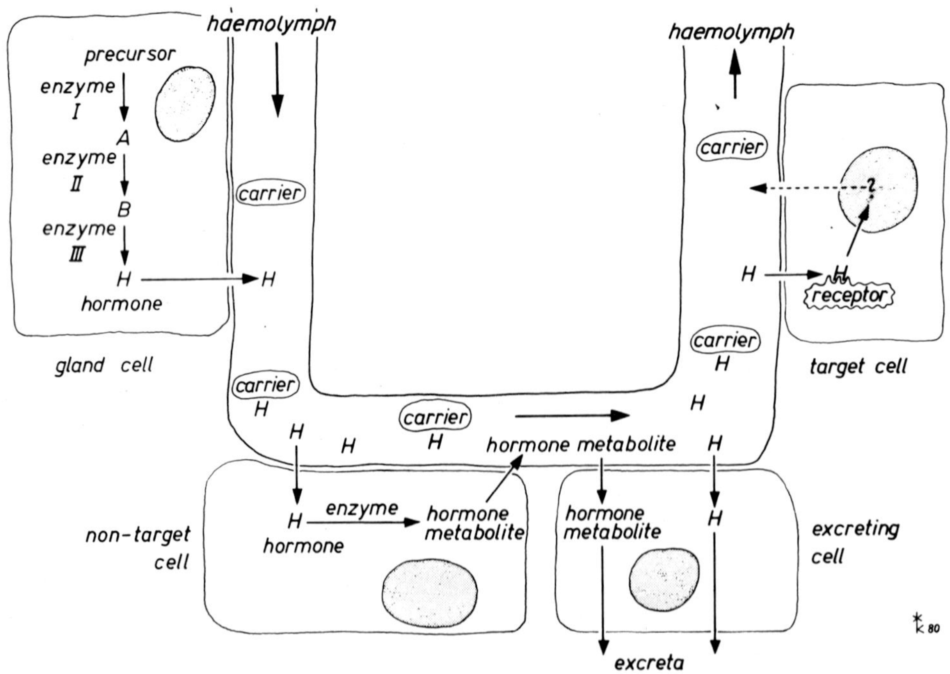

FIG. 1. General scheme of a steroid hormonal system. This scheme applies to all animals which use a steroid for hormonal control and regulation, including the chordates and arthropods. For insects, a few modifications have to be introduced: (i) instead of blood circulating in vessels haemolymph is circulating in an open system; (ii) the hormone secreted by the gland cells (ecdysone) prior to its action has to be converted to its active form (20-hydroxyecdysone) in other cell types. This is valid at least for dipteran larvae in which ecdysteroids act as molting hormone.

yield 20-hydroxyecdysone (ecdysterone). This reaction has been demonstrated in all species investigated (see Table 1 for References) and seems to be a general event. It is known that in most bioassays, 20-hydroxyecdysone is more active than ecdysone (chapter 2, volume 8). Hence the hydroxylation at the 20-position has been described as "activation" of the hormone. Some authors even regard ecdysone as a pro-hormone with 20-hydroxyecdysone being the "true" hormone (King, D. *et al.*, 1974). However, hormones are defined as regulatory substances *secreted* by endocrine glands or specialized tissues. The product of the prothoracic and homologous glands is ecdysone; thus, according to the original definition of Starling, E. (1907), ecdysone is the hormone, and 20-hydroxyecdysone would be an active metabolite. In most insect species studied, the conversion of ecdysone to 20-hydroxyecdysone occurs rather rapidly (chapter 8, this volume), and the circulating hormone is mainly 20-hydroxyecdysone which is mainly responsible for eliciting the molting process. However, it is quite possible that ecdysone, and perhaps also 2-deoxyecdysone, have definite functions in some developmental processes (Koolman, J. 1982).

Though hydroxylation of ecdysone at the 20-position leads to a compound with high biological activity, further hydroxylation at C-26 yields a product with much reduced hormonal activity (Bergamasco. R. and Horn, D. 1980). This hydroxylation can therefore be regarded as the first step of inactivation.

1.2.2 Dehydroxylation

Hydroxylation at specific positions of the steroid nucleus is a well-known reaction in steroid biochemistry. On the other hand, removal of hydroxyl groups is a rather rare event which so far has only been observed in micro-organisms (Hayakawa, S. 1973). In this context, it is remarkable that inokosterone (25-deoxy-20,26-dihydroxyecdysone) has been identified as a metabolite of labeled ecdysone in *Calliphora* (Koolman, J. *et al.*, 1979). It is conceivable that dehydroxylation occurs through the elimination of sulfuric acid from a 25-sulfate of either 20-hydroxyecdysone or 20,26-dihydroxyecdysone, and that the product was then converted into inokosterone.

1.2.3 Oxidation and epimerization

Ecdysone and other ecdysteroids are readily oxidized at C-3 to yield 3-dehydroecdysone. Similar oxidations are quite common in steroid biochemistry; the hydrogen is normally transferred to NAD^+. However, this type of reduction has not yet been demonstrated in insects. Instead ecdysone is oxidized by an oxidase, at least in some species (see below). This oxidation may be followed by a reduction yielding the 3α-stereoisomers of ecdysteroids (= 3-epiecdysteroids).

1.2.4 Carboxylic acids

In Fig. 3 a further oxidation of 26-hydroxyecdysteroids to the corresponding carboxylic acids is shown. In a recent review (Koolman, J. 1982) the name "ecdysonoic acid" for the carboxylic acid derived from ecdysone was proposed, since the systematic name, $2\beta,3\beta,14\alpha$,22R,25-pentahydroxy-6-oxo-cholest-7-en-26-oic acid, is rather cumbersome. Corresponding oxidation products may be expected from 20-hydroxyecdysone and possibly also from 3-epiecdysone and 3-epi-20-hydroxyecdysone (Lafont, R. *et al.*, 1980).

The evidence for the occurrence of these metabolites stems from investigations of the fractionation of polar metabolites, mainly conjugates (see below). Among these polar compounds are some acidic ecdysteroids with a p*K* of 4 to 5, i.e. in the range of carboxylic acids. They cannot be hydrolyzed to neutral ecdysteroids. It is therefore likely that they represent acids in which one of the terminal methyl groups has been oxidized to a carboxyl group. However, none of the ecdysonoic acids have yet been isolated in crystalline form.

1.2.5 Isolation of unlabelled metabolites

The results presented above and summarized in Fig. 3 and Table 1 have all been obtained with radiolabelled ecdysteroids. It is, however, also desirable to isolate ecdysone metabolites from insect sources without interfering with hormone metabolism by the use of exogenous hormone. This would corroborate the identification of the radioactive metabolite, and it could further illuminate the

Table 1: Ecdysone metabolites identified in radiotracer experiments

Species	Stage; Tissue	Steroid applied by injection	Ecdysteroid formed*	Method of separation and of analysis	References
Aeshna cyanea	larvae	ecdysone	2, 16, 17, 25, 27	TLC	Schaller, F. *et al.* (1975); Schaller F. and Hoffman, J. (1976)
Locusta migratoria	embryos, larvae, female adults	ecdysone (injected as well as ingested)	2, 16, 17, 25, 27	TLC, HPLC	Koolman, J. *et al.* (1973); Hoffmann, J. *et al.* (1974); Feyereisen, R. *et al.* (1976); Lagueux, M. *et al.* (1979); Hoffmann, J. *et al.* (1980)
Schistocerca americana	ovaries and eggs	cholesterol	1, 2, 25, 26, 27	TLC, HPLC, GC, derivative formation	Dinan, L. and Rees, H. (1981)
Gryllus bimaculatus	larvae	ecdysone	2, 25. 26, 27	TLC	Wildmann, H. and Romer, F. (1977)
Leucophaea maderae	fat body	22,25-dideoxy-ecdysone	25, 26, 27	TLC, column chromatography, NMR, MS, derivative formation	Thompson, M. *et al.* (1978)
	leg regenerates	ecdysone	2, 27	TLC, derivative formation	King, D. and Marks, E. (1974)
Pyrrhocoris apterus	larvae	ecdysone	27	TLC, electrophoresis	Emmerich, H. (1970a)
Manduca sexta	prepupae and several tissues	22,25-dideoxy-ecdysone	1, 2, 3, 25, 27	TLC, CCD, derivative formation	King, D. (1972); King, D. and Siddall, J. (1974); Kaplanis, J. *et al.* (1969, 1972a, 1974)
Antheraea polyphemus	pupae	ecdysone	2, 27	TLC	Cherbas, L. and Cherbas, P. (1970)
Hyalophora cecropia	pupae and several tissues	ecdysone	2, 27	TLC	Gorell, T. *et al.* (1972)
Bombyx mori	larvae	ecdysone (injected as well as ingested)	2, 25, 26, 27	HPLC, liquid chromatography	Moriyama, H. *et al.* (1970)
	larvae and pupae	ponasterone A and 20-hydroxyecdysone (injected as well as ingested)	2, 25, 26, 27	liquid chromatography, TLC	Hikino, H. *et al.* (1975a,b)
Galleria mellonella	larvae	ecdysone	2, 27	TLC	Rohner, R. and Meyer, D. (1980)
Pieris brassicae	pharate pupae, wing discs	ecdysone	2, 3, 7, 11, 8, 12, 26, 27	TLC, HPLC, derivative formation	Lafont, R. *et al.* (1980); Blais, C. and Lafont, R. (1980)
Tenebrio molitor	larvae	25-deoxy-ecdysone	2, 27	TLC	Weinheimer, G. and Romer, F. (1977)
Choristoneura fumiferana	larvae	ecdysone	2, 16, 27	TLC	Lagueux, M. *et al.* (1976)
Chironomus tentans	larvae	ecdysone, 20-hydroxyecdysone	2, 27	TLC	Clever, U. *et al.* (1973)
Calliphora vicina	larvae and pupae	ecdysone, 20-hydroxyecdysone	2, 3, 7, 25, 26, 27	TLC, HPLC, derivative formation, cocrystallization	King, D. and Siddall, J. (1969); Moriyama, H. *et al.* (1970); Karlson, P. and Koolman, J. (1973); Sannasi, A., and Karlson, P. (1974);

Table 1: Continued

Species	Stage; Tissue	Steroid applied by injection	Ecdysteroid formed*	Method of separation and of analysis	References
Calliphora vicina (continued)					Young, N. (1976b); Russel, G. and Price, G. (1977); Greenwood, D. and Russel, G. (1978); Koolman, J. *et al.* (1979)
	larvae	ponasterone A	25	TLC	Heinrich, G. and Hoffmeister, H. (1970)
Calliphora stygia	puparia	25-deoxy-ecdysone	27	TLC, column chromatography, cocrystallization	Thomson, J. *et al.* (1969)
		20-hydroxy-ecdysone	27	GC-cocrystallization	Galbraith, M. *et al.* (1969)
Sarcophaga peregrina	larvae	22,25-dideoxy-ecdysone	2, 27	TLC, derivative formation	King, D. (1972)
	pupae	20-hydroxy-ecdysone	8, 25, 26, 27	TLC, HPLC	Moribayashi, A. and Ohtaki, T. (1978, 1980)
Glossina morsitans	female adults	ecdysone (ingested)	27	TLC	Whitehead, D. (1981)
Drosophila melanogaster	larvae and larval mutants	ecdysone, 20-hydroxyecdysone	2, 27	TLC	Klose, W. *et al.* (1980)
Drosophila hydei	salivary glands	ecdysone, 20-hydroxyecdysone	25, 27	TLC	Beckers, C. and Emmerich, H. (1976)
Musca domestica	larvae	22,25-dideoxy-ecdysone	25, 27	TLC, CCD,MS column chromatography	Kaplanis, J. *et al.* (1976)

* The numbers refer to the following ecdysteroids:
1 = ecdysone; 2 = 20-hydroxyecdysone; 3 = 20,26-dihydroxyecdysone; 7 = 26-hydroxyecdysone; 8 = 3-epi-20-hydroxyecdysone; 11 = 3-epi-ecdysone; 12 = 3-epi-20,26-dihydroxyecdysone; 16 = 3-dehydroecdysone; 17 = 3-dehydro-20-hydroxyecdysone; 25 = hydrolyzable highly polar ecdysteroids; 26 = highly polar ecdysteroids resistant to hydrolysis (possibly ecdysonoic acid (18) and -oic acids of other ecdysteroids (19–21); 27 = other or unidentified.

major routes of ecdysone metabolism in the course of normal insect development.

Experiments of this kind are hampered by the small amounts of metabolites present, which were for a long time difficult to detect and identify. Only recently methods (e.g. radioimmunoassay and fluorometry) have become available that measure not only active ecdysteroids but also inactive metabolites.

In spite of these difficulties, studies on ecdysone metabolites have been published, and a few metabolites have been unequivocally identified. Among them are 26-hydroxyecdysone, 20,26-dihydroxyecdysone and several 3-epiecdysteroids (see Table 2).

It can be easily visualized that the titer of any hormone is the result of two main processes: biosynthesis and secretion on the one hand, and metabolic inactivation and excretion on the other. In most hormonal systems the amount of hormone consumed by the target tissue is rather small compared with the amount that is metabolically inactivated. This is borne out by the rather short half-lives of many animal hormones which are in the order of a few minutes to one hour.

Steroid hormones circulating in the blood are often protected from metabolic degradation by binding to blood protein(s). This phenomenon may be important in the regulation of ecdysteroid titers (see chapter 13, this volume). It suffices here to mention that an ecdysteroid-binding protein has been demonstrated in the haemolymph of *Locusta migratoria* (Feyereisen, R. 1977a, 1980). Suggestive evidence for the existence of binding proteins in the haemolymph of insects was also obtained in experiments with other insect species including *Pyrrhocoris apterus* (Emmerich, H. 1970a), *Drosophila hydei* (Emmerich, H. 1970b; Butterworth, F.

Table 2: Ecdysteroids isolated from arthropods

Ecdysteroid	Formula	Source	References	Biological activity
Ecdysone	(1)	*Bombyx mori* pupae	Butenandt, A. and Karlson, P. (1954) Karlson, P. *et al.* (1965) Huber, R. and Hoppe, W. (1965)	100%
20-Hydroxyecdysone (ecdysterone, β-ecdysone)	(2)	*Bombyx mori* pupae	Karlson, P. (1956) Hoffmeister, A. and Grützmacher, H. (1966)	100%
		Bombyx mori pupae	Hocks, P. and Wiechert, R. (1966)	
		*Jasus lalandei** total	Hampshire, F. and Horn, D. (1966)	
20,26-Dihydroxyecdysone	(3)	*Manduca sexta* pupae	Thompson, M. *et al.* (1967)	10%
2-Deoxy-20-hydroxyecdysone	(4)	*Jasus lalandei** total	Galbraith, M. *et al.* (1968)	100%
25-Deoxy-20,26-dihydroxyecdysone (callinecdysone A = inokosterone or 25-epimer)	(5)	*Callinectes sapidus**, total	Faux, A. *et al.* (1969)	100%
24-Methyl-20-hydroxyecdysone (callinecdysone B = possibly 24-epi-makisterone A)	(6)	*Callinectes sapidus**, total	Faux, A. *et al.* (1969)	active
26-Hydroxyecdysone (two isomers: 25R and 25S?)	(7)	*Manduca sexta* eggs	Kaplanis, J. *et al.* (1973)	inactive
3-Epi-20-hydroxyecdysone	(8)	*Manduca sexta* meconium	Thompson, M. *et al.* (1974)	10–20%
24-Methyl-20-hydroxyecdysone (= makisterone A)	(9)	*Oncopeltus fasciatus* embryonated eggs	Kaplanis, J. *et al.* (1975)	100%
2-Deoxyecdysone	(10)	*Bombyx mori* ovaries	Ohnishi, E. *et al.* (1977)	100%
3-Epiecdysone	(11)	*Manduca sexta* pupae	Kaplanis, J. *et al.* (1979)	10–20%
3-Epi-20,26-dihydroxyecdysone	(12)	*Manduca sexta* pupae	Kaplanis, J. *et al.* (1979)	1%
3-Epi-26-hydroxyecdysone	(13)	*Manduca sexta* embryonated eggs	Kaplanis, J. *et al.* (1980b)	inactive
2,22-Dideoxy-20-hydroxyecdysone	(14)	*Bombyx mori* ovaries	Ikekawa, N. *et al.* (1980)	10%
25-Deoxy-20-hydroxyecdysone (= ponasterone A)	(15)	*Callinectes sapidus**, embryos *Gecarcinus lateralis**, haemolymph	McCarthy, J. (1979)	100%
Ecdysone-3-acetate	(—)	*Schistocerca gregaria* eggs	Isaac, R. *et al.* (1981a)	?
3-Epi-2-deoxyecdysone	(—)	*Schistocerca gregaria* eggs	Isaac, R. *et al.* (1981b)	?
Ecdysone-22-phosphate	(—)	*Schistocerca gregaria* eggs	Isaac, R. *et al.* (1982a)	?
2-Deoxyecdysone-22-phosphate	(—)	*Schistocerca gregaria* eggs	Isaac, R. *et al.* (1982b)	?
3-Epi-2-deoxyecdysone-3-phosphate	(—)	*Locusta migratoria* embryos	Tsoupras, G. *et al.* (1982)	?
2-Deoxyecdysone-22-adenosinemono-phosphate	(—)	*Locusta migratoria* eggs	Tsoupras, G. *et al.* (1982)	?

Only the first isolation which included a physicochemical characterization is mentioned. The biological activity measured *in vivo* in a dipteran bioassay is expressed as percentage of the hormonal activity of 20-hydroxyecdysone (partly taken from Bergamasco and Horn, 1980). Crustacean species are labelled with an asterisk.

and Berendes, H. 1974) and *Calliphora vicina* (Thamer, G. and Karlson, P. 1972). In *Calliphora,* the protein calliphorin (chapter 9, volume 10) seems to have the ability to bind ecdysteroids (Reum, L. *et al.*, 1982).

In vertebrates, several steroid hormones with quite different physiological functions are known. In insects, only one steroid hormone class — the ecdysteroids — has been discovered up to now. However, the juvenile hormones bear some biochemical relations to the ecdysteroids since they also belong to the family of isoprenoids (chapter 10, this volume). All other insect hormones are peptides (chapter 5, this volume). Hormonal systems using peptides as signal substances have some features which are quite different from those of steroid-mediated regulatory systems (see Karlson, P. 1982; Gersch, M. and Karlson, P. 1982).

2 PATHWAYS OF ECDYSTEROID METABOLISM

2.1 General considerations and overview

In discussing steroid hormones in insects, one is always tempted to draw analogies to vertebrate steroid hormones. It is well known that many vertebrate steroid hormones bear an oxo group at C-3 and an adjacent (conjugated) double bond between C-4 and C-5 as shown in partial formula (I). Ecdysteroids, the insect steroid hormones, have a similar α,β-unsaturated oxo group in ring B (partial formula II; chapter 6, this volume).

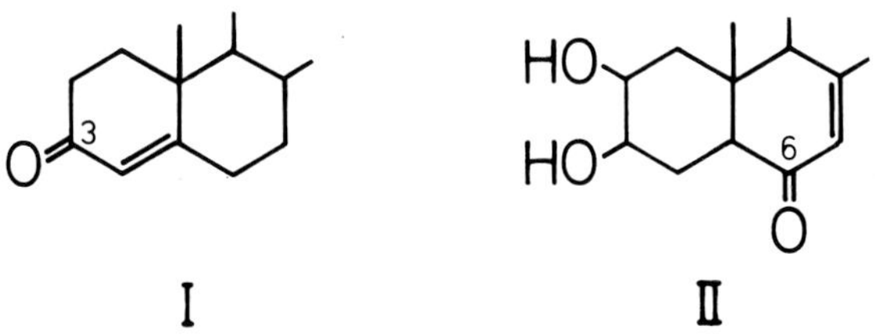

FORMULAE

Vertebrate hormones showing this structural feature are inactivated by reduction in ring A. The oxo group is reduced mainly to a 3α-hydroxyl group; reduction of the double bond yields either 5α- or 5β-steroids. This type of reduction apparently does not occur in ecdysteroid metabolism. Instead, the main reactions are (Fig. 2):

(a) hydroxylation of ecdysone to 20-hydroxyecdysone,
(b) epimerization at C-3 to yield a stereoisomer,
(c) hydroxylation at C-26, eventually followed by oxidation to a carboxylic acid, and
(d) conjugation.

An overview of the main routes is given in Fig. 3. Details and experimental evidence will be discussed below.

2.2 *In vivo* studies on ecdysone metabolism

In vivo studies of the fate of injected ecdysone or other hormonally active ecdysteroids have only become possible with the availability of a labeled ecdysteroid. In the pioneering studies, ecdysone tritiated by the Wilzbach technique (Karlson, P. *et al.*, 1963) has been used. Later, a synthesis of ecdysone was developed in which the key intermediate contains a triple bond in the side chain. This synthetic precursor could be catalytically tritiated and then converted into tritiated ecdysone of high specific activity (King, D. and Siddall, J. 1969; Hafferl, W. *et al.*, 1972).

Using the tracer technique, it was shown that ecdysone is metabolized in various ways, as outlined in Fig. 3. The metabolites shown in this chart have been identified in various insect species (Table 1).

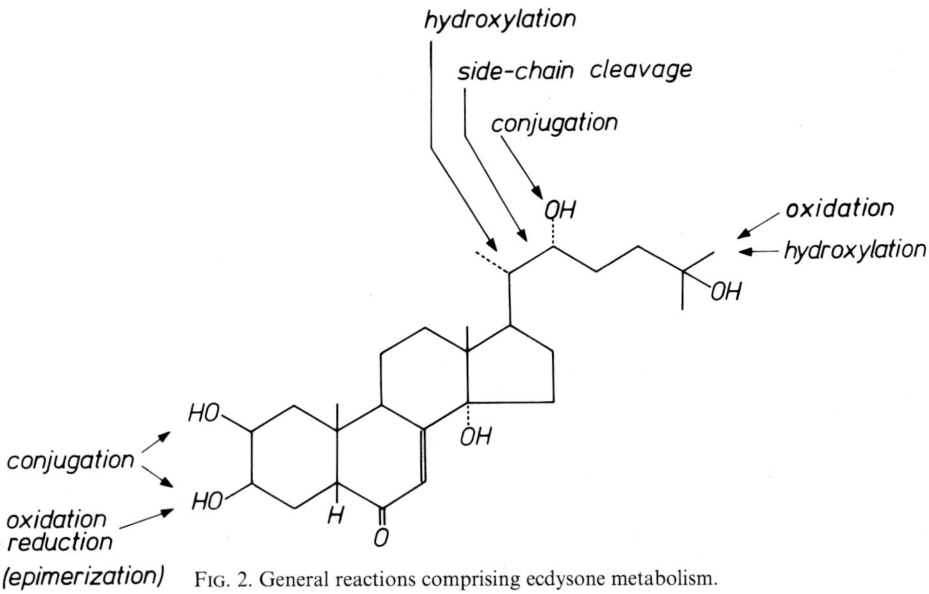

FIG. 2. General reactions comprising ecdysone metabolism.

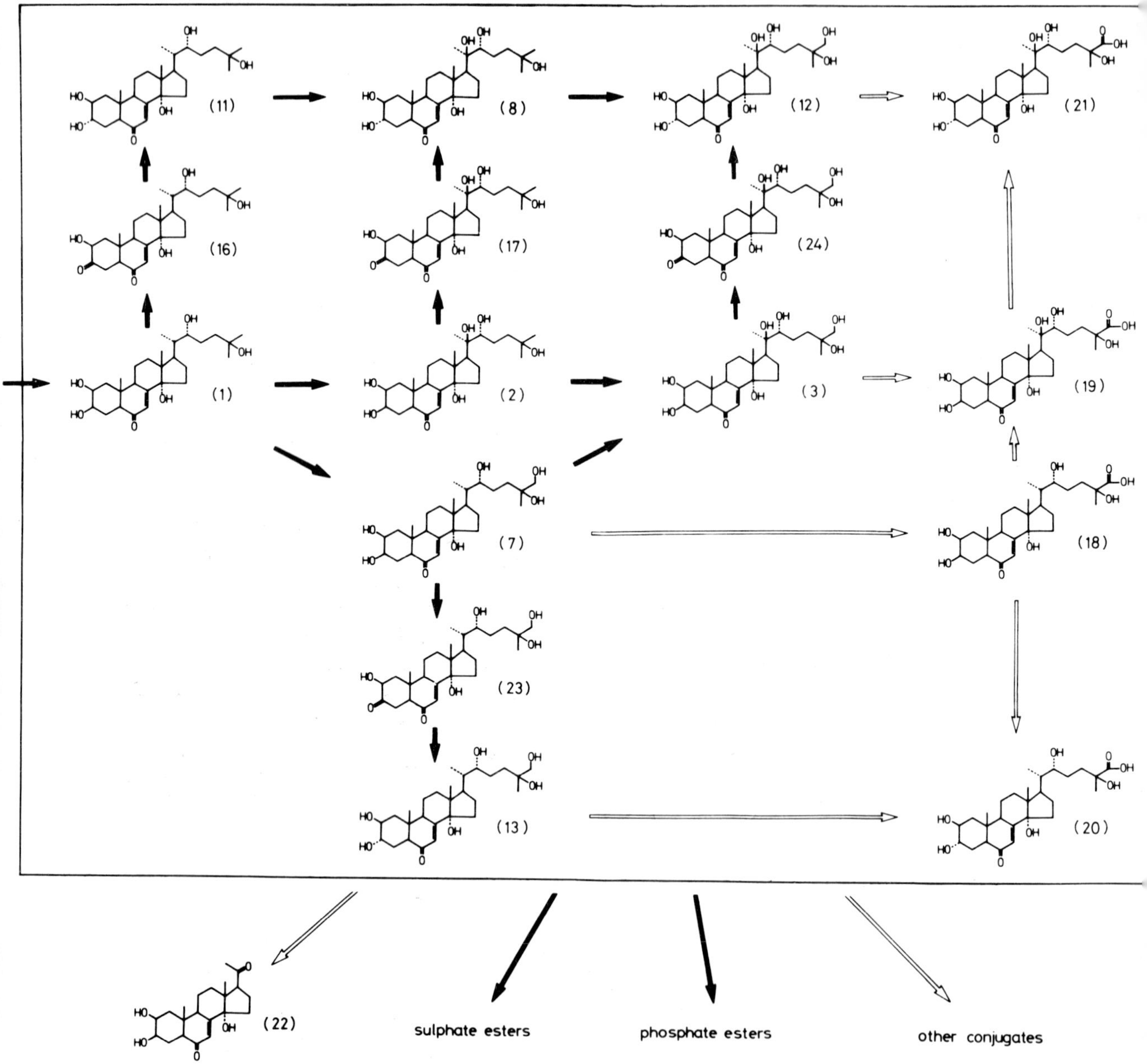

FIG. 3. Main routes of the metabolism of ecdysone in insects.
The ecdysteroids are:

(1) = ecdysone;
(2) = 20-hydroxyecdysone;
(3) = 20, 26-dihydroxyecdysone;
(7) = 26-hydroxyecdysone;
(8) = 3-epi-20-hydroxyecdysone;
(11) = 3-epiecdysone;
(12) = 3-epi-20, 26-dihydroxyecdysone;
(13) = 3-epi-26-hydroxyecdysone;
(16) = 3-dehydroecdysone;
(17) = 3-dehydro-20-hydroxyecdysone;
(18) = ecdysonoic acid;
(19) = 20-hydroxyecdysonoic acid;
(20) = 3-epiecdysonoic acid;
(21) = 3-epi-20-hydroxyecdysonoic acid;
(22) = poststerone;
(23) = 3-dehydro-26-hydroxyecdysone;
(24) = 3-dehydro-20, 26-dihydroxyecdysone.

Heavy arrows indicate reactions leading to ecdysteroids isolated or identified by radiotracer experiments (see also Tables 1 and 2).

2.3 *In vitro* studies on ecdysone metabolism

The metabolic capacity of insect tissues can best be investigated using tissue preparations and labelled ecdysteroids as precursors. By means of such experiments the tissue distribution of various degradative reactions can be investigated. This approach has its limitations but is a useful additional source of information on ecdysone metabolism.

The tissues most active in metabolizing ecdysone and 20-hydroxyecdysone are the fat body, the Malpighian tubules and the midgut. The main reactions are again hydroxylation of ecdysone in the 20-position, hydroxylation at C-26 and conjugate formation (Table 3).

There is one "tissue", the K cell line of *Drosophila*, that has receptors for ecdysteroids and is able to respond to the hormone but apparently does not metabolize ecdysteroids (chapter 3, volume 8). It can thus be concluded that metabolism is not a prerequisite for biological action, and that the hormone molecule is not altered in the course of its action.

2.4 Conjugates of ecdysone and its metabolites

After injection of labelled ecdysone into insects, a large portion of the radioactivity can be traced to highly polar compounds. They are in part carboxylic acids (see above) and in part conjugates. After incubation of the conjugate fraction with hydrolytic enzymes, e.g. with an extract from *Helix pomatia*, a mixture of ecdysteroids is recovered.

The fact that the conjugates are hydrolyzed by enzyme preparations has led to the tentative identification of the conjugates as sulfate or phosphate esters, glucuronidates, and glucosides. However, caution is necessary. In most of these experiments commercial enzyme preparations were used, and they were not checked for contaminating enzymes with different specificity. We found that the commercial sulfatase from snails (*Helix pomatia*) contains a large number of different hydrolases (Sannasi, A. and Karlson, P. 1974).

On the other hand, these digestion experiments allow valid conclusions concerning the nature of the ecdysteroids. Nearly all metabolites of ecdysone listed in Table 2 are found in these digests, notably

Table 3: Localization of reactions of the ecdysone metabolism as detected by incubation of tissues in vitro.

Reaction	Tissue	Species	References
C-20-hydroxylation	Malpighian tubules, fat body, haemocytes, gonads, gut	*Locusta migratoria*	Feyereisen, R. *et al.* (1976)
	Malpighian tubules, fat body	*Schistocerca gregaria*	Johnson, P. and Rees, H. (1977)
	leg regenerates	*Leucophaea maderae*	King, D. and Marks, E. (1974)
	fat body	*Leucophaea maderae*	Thompson, M. *et al.* (1978)
	fat body, Malpighian tubules, midgut, integument	*Manduca sexta*	King, D. (1972) Nigg, H. *et al.* (1976)
	fat body, haemocytes, Malpighian tubules	*Calliphora vicina*	Koolman, J. (1976); Koolman, J. *et al.* (1979)
C-26-hydroxylation	fat body	*Leucophaea maderae*	Thompson, M. *et al.* (1978)
	imaginal wing discs	*Pieris brassicae*	Blais, C. and Lafont, R. (1980)
	fat body, Malpighian tubules	*Manduca sexta*	King, D. (1972)
	fat body	*Calliphora vicina*	Koolman, J. *et al.* (1979)
C-3-oxidation	gut, fat body, haemocytes	*Locusta migratoria*	Feyereisen, R. *et al.* (1976)
	fat body, Malpighian tubules, gut, salivery glands, haemocytes	*Calliphora vicina*	Koolman, J. (1976, 1978, 1980); Russel, G. and Price, G. (1977)
Conjugation	gut, Malpighian tubules, fat body, ovary	*Locusta migratoria*	Feyereisen, R. *et al.* (1976); Lagueux, M. *et al.* (1977)
	fat body	*Leucophaea maderae*	Thompson, M. *et al.* (1978)
	fat body, Malpighian tubules	*Manduca sexta*	King, D. (1972)
	fat body, Malpighian tubules, gut, haemocytes	*Calliphora vicina*	Heinrich, G. and Hoffmeister, H. (1970); Russel, G. and Price, G. (1977); Koolman, J. *et al.* (1979)

ecdysone, 20-hydroxyecdysone, 20,26-dihydroxy-ecdysone and several 3-dehydro- and 3-epiecdy-steroids. Thus the conjugating enzymes are apparently very efficient and not specific for a particular metabolite.

In an attempt to characterize the conjugates they were subjected to high voltage electrophoresis (Koolman, J. *et al.*, 1973). It was shown that (at least in larvae of *Locusta migratoria* and *Calliphora vicina*) all highly polar compounds are negatively charged, thus excluding simple glucosides of ecdysteroids as components of the mixture, within the limits of sensitivity. It is very likely that sulfates are present; phosphates are also present. We will now consider the various types of conjugates.

2.4.1 Sulfates

The presence of sulfate esters of ecdysteroids in the highly polar fraction is well established. The evidence is based not only on enzymatic hydrolysis, the limitations of which have been discussed, but also on the incorporation of [^{35}S] *in vivo* after administration of labelled sulfate, and on the migration of the conjugates in high voltage electrophoresis (Koolman, J. *et al.*, 1973).

Attempts to synthesize sulfates of ecdysteroids have been in part successful. Though the structure of the synthetic product derived from ecdysone is not yet fully established, a comparison of the synthetic sulfate and a fraction of the polar metabolites of ecdysone shows great similarity. It is likely that the natural conjugate is further modified, possibly by acetylation.

2.4.2 Phosphates

The presence of ecdysteroid phosphate esters has been inferred from electrophoretic studies and digestion experiments (Sannasi, A. and Karlson, P. 1974). Only recently have the compounds been identified: ecdysone 22-phosphate and 2-deoxyecdysone 22-phosphate in extracts of eggs of *Schistocerca gregaria* (Isaac, R. *et al.*, 1982a) and 3-epi-2- deoxyecdysone 3-phosphate in embryos from *Locusta migratoria* (Tsoupras, G. *et al.*, 1982).

Apparently more complex phosphate esters are also formed. From newly laid eggs of *Locusta migratoria* the 22-ester of 2-deoxyecdysone with adenosine monophosphoric acid has been isolated (Tsoupras, G. *et al.*, 1982; see chapter 13, volume 1).

2.4.3 Glucosides and Glucuronides

Heinrich, G. and Hoffmeister, H. (1970) reported the isolation and identification of an α-glucoside of 20-hydroxyecdysone as one of the compounds in the highly polar metabolite fraction. It is well known that insects synthesize glucosides; e.g. *N*-acetyl-dopamin-4-O-β-glycoside has been found in the fruitfly (Okubo, S. 1958), and the glucoside of protocatechuic acid in the cockroach (Brunet, P. and Kent, P. 1955). However, the existence of ecdysteroid glucosides in insects has not been confirmed convincingly by other authors.

Similarly, the presence of glucuronides among the highly polar compounds has often been proposed or at least discussed, but no proof has yet been given to the occurrence of glucuronic acid as one of the conjugating components.

2.4.4 Hydrolysis of conjugates

The hydrolysis of ecdysteroid conjugates *in vivo* has often been postulated. So far only indirect or tentative evidence has been obtained from the embryonic (Gande, A. *et al.*, 1979; Lagueux, M. *et al.*, 1979; Hoffmann, J. *et al.*, 1980); (chapter 13, volume 1) and pupal stage (Moribayashi, A. and Ohtaki, T. 1978, 1980) of a few insect species, i.e. ecdysteroids can be released from a certain fraction of the conjugates. This would infer that there are two types of ecdysteroid conjugates which differ in structure and function: inactivation products and storage forms.

It would make sense that developing embryos lacking differentiated prothoracic glands need a special pathway for the synthesis of ecdysteroids to induce deposition of cuticle. On the other hand the role of ecdysteroid conjugates in pupae of holometabolous insects as storage form of the hormone, if confirmed, remains obscure at present.

2.5 Kinetics of the catabolism of ecdysteroids

Biologically, the most important aspect of variations in ecdysone metabolism are the changes in ecdysteroid titer. They are discussed in detail in chapter 8 of this volume. It may suffice here to point out that there are marked peaks of ecdysteroid titers during embryonic and post-embryonic development of all insects studied. The question is, then: to what extent do metabolic reactions contribute to these titer changes?

Since only few detailed studies on the time-course of ecdysone metabolism have been published, a complete answer to this question is not yet possible. The half-life of active hormone has been measured by two different methods (bioassay and radioassay) and found to be in the order of few minutes to many hours, depending on the species and stage under study (Table 4). Whether these values reflect correctly the half-life of endogenous ecdysteroids, is still an open question.

A more detailed study of the time course of appearance and disappearance of some ecdysone metabolites has been made with *Calliphora* larvae (Fig. 4; Karlson, P. and Koolman, J. 1973; Young, N. 1976a; Koolman, 1981).

Similar studies on the metabolism and excretion of ecdysone during the last larval instar of *Locusta migratoria* (Hoffmann, J. *et al.*, 1974) are also of relevance here.

Thus, the rate of ecdysone metabolism shows large variations during insect development and is presumably physiologically regulated (cf. Koolman, J. 1982).

3 ENZYMES OF ECDYSTEROID METABOLISM

As can be inferred from the metabolites noted above, several enzymes must be involved in the formation of these compounds. They can be classi-

Table 4: Half-life of injected ecdysteroids

Ecdysteroid	Species	Stage	Half-life (min)	References
Ecdysone	*Pyrrhocoris apterus*	larvae	80–100	Emmerich, H. (1970a)
	Bombyx mori	larvae	~ 15	Moriyama, H. *et al.*, (1970)
	Antheraea polyphemus	pupae	~ 240	Cherbas, L. and Cherbas, P. (1970)
	Calliphora vicina	larvae	4	Koolman, J. (1980)
	Calliphora vicina	prepupae	106	Koolman, J. (1980)
	Calliphora vicina	pupae	940	Koolman, J. (1980)
	Calliphora vicina	larvae	5/53	Young, N. (1976a)
Ecdysone (+ 20-hydroxyecdysone)				
	Samia cynthia	pupae	360	Ohtaki, T. and Williams, C. (1970)
	Sarcophaga peregrina	larvae	60	Ohtaki, T. and Williams, C. (1970)
	Calliphora vicina	larvae	~ 180	Shaaya, E. (1969)
20-Hydroxyecdysone				
	Gryllus bimaculatus	larvae	40/1440	Wildmann, H. and Romer, F. (1977)
	Bombyx mori	larvae	~ 60	Hikino, H. *et al.*, (1975a)
	Bombyx mori	larvae	240	Moriyama, H. *et al.*, (1970)
	Manduca sexta	larval abdomens	190–240	Nijhout, H. (1976)
	Galleria mellonella	larvae	15	Rohner, R. and Meyer, D. (1980)
	Calliphora vicina	larvae	200	Koolman, J. (1980)
	Calliphora vicina	prepupae	710	Koolman, J. (1980)
	Calliphora vicina	pupae	2100	Koolman, J. (1980)
	Calliphora vicina	larvae	90	Young, N. (1976a)
	Calliphora vicina	prepupae	1200	Young, N. (1976a)
Cyasterone				
	Samia cynthia	pupae	1900	Ohtaki, T. and Williams, C. (1970)
	Sarcophaga peregrina	larvae	180–240	Ohtaki, T. and Williams, C. (1970)

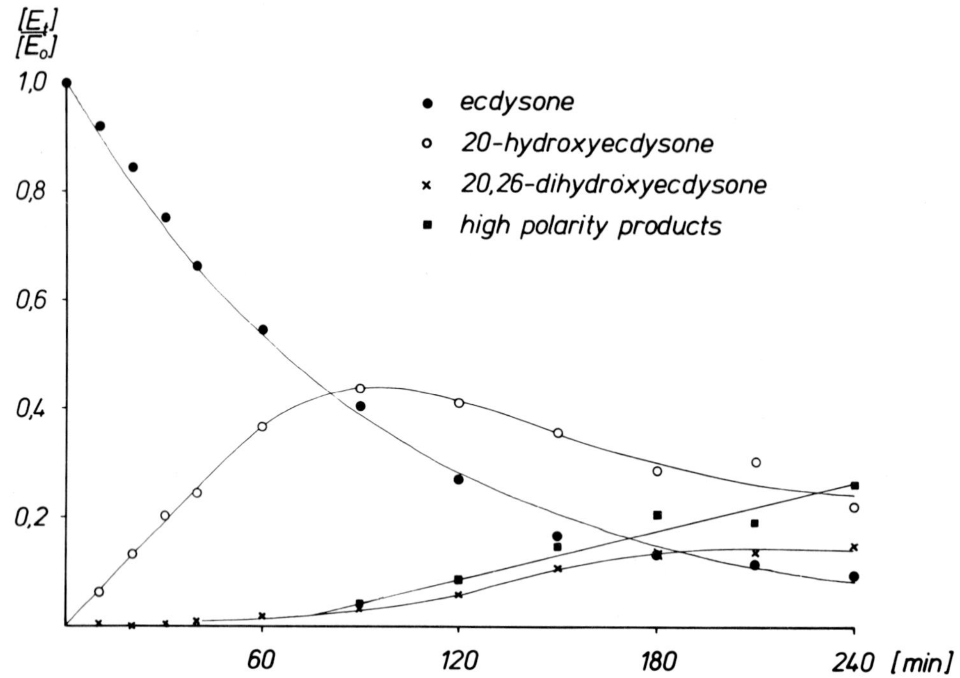

FIG. 4. Short-term kinetic of the metabolism of ecdysone by fat body of *Calliphora vicina* larvae *in vitro*. Six isolated fat bodies from day-6 larvae were incubated in *Calliphora* Ringer. Following the addition of radiolabelled ecdysone (final concentration around 0.5 μM) small aliquots were withdrawn from the incubation fluid and analyzed by TLC and HPLC in order to obtain a time course of the relative concentrations of ecdysone and some of its metabolites (from Koolman, J. 1981).

fied as hydroxylases, oxidases, and transferases yielding conjugates, and are listed in Table 5.

3.1 Mono-oxygenases (hydroxylases)

3.1.1 Ecdysone 20-mono-oxygenase (see chapter 8, this volume)

Steroid hydroxylases are well known as a special group of cytochrome P-450 containing enzymes. This also holds for ecdysteroid hydroxylases from insects. One of these enzymes is sufficiently characterized to be incorporated into the enzyme list, namely ecdysone 20-mono-oxygenase (EC 1.14.99.22). The reaction can be written as follows:

$$\text{ecdysone} + O_2 + \text{NADPH} + H^+ = \text{20-hydroxyecdysone} + H_2O + \text{NADP}^+$$

The reaction was first demonstrated *in vivo* in *Calliphora vicina* and two crustacean species (King, D. and Siddall, J. 1969), but later also in many other species from various orders. First detection of the enzyme was reported from *Manduca sexta* (Nigg, H. *et al.*, 1976).

Isolation of the enzyme in pure form has not yet been described; this may be due to the fact that mono-oxygenases are complex enzyme systems. They need an electron transport chain for their function and cannot be isolated as easily as cytosolic enzymes. However, the partially purified preparation allows one to obtain some data on the kinetics which are given in Table 5.

3.1.2 Other hydroxylases

According to the metabolite pattern, at least one other hydroxylase, i.e. ecdysone 26-mono-oxygenase (or 20-hydroxyecdysone 26-mono-oxygenase) should be present in insect tissues. While the enzyme activity has not yet been demonstrated *in vitro* it may be assumed from the ecdysteroids observed *in vivo* that the ecdysone 26-hydroxylase has a broad substrate specificity and reacts with many ecdysteroids. Moreover, a number of hydroxylases involved in the biosynthesis of ecdysone should be mentioned here; they are discussed in chapter 7 of this volume.

Table 5: Enzymes of ecdysteroid metabolism

Name and EC Number	Source	Preparation	K_m values[a] (μM)	References
Ecdysone 20-monooxygenase (1.14.99.22) (= ecdysone 20-hydroxylase)	*Calliphora vicina*	analyzed *in vivo*	I: 0.45	Young, N. (1976a)
	fat body mitochondria of *Manduca sexta*	enriched	I: 0.16	Bollenbacher, W. *et al.*, (1977); Smith, S. *et al.*, (1977, 1979, 1980)
	midgut mitochondria of *Manduca sexta*	enriched	I: 18.3	Nigg, H. *et al.*, (1976); Mayer, R. *et al.*, (1978); Kaplanis, J. *et al.*, (1980)
	Malpighian tubule mitochondria of *Schistocerca gregaria*	enriched	I: 0.71	Johnson, P. and Rees, H. (1977); Rees, H. *et al.*, (1981)
	Malpighian tubule microsomes of *Locusta migratoria*	enriched	I: 0.27	Feyereisen, R. (1977b); Feyereisen, R. and Durst, F. (1978)
Ecdysone oxidase (1.1.3.16)	*Calliphora vicina*	purified (2000 ×)	I: 98 II: 31	Koolman, J. and Karlson, P. (1975, 1978), Koolman, J. (1978)
Ecdysone 3-epimerase	midgut of *Manduca sexta*	enriched (5–8 ×)	I: 17	Nigg, H. *et al.*, (1974); Mayer, R. *et al.*, (1979); Kaplanis, J. *et al.*, (1980a)

a. In this column the substrates are: I = ecdysone; II = 20-hydroxyecdysone

3.2 Ecdysone oxidase and ecdysone 3-epimerase

3.2.1 Ecdysone oxidase

This enzyme (EC 1.1.3.16) catalyzes the oxidation of the hydroxyl group at C-3 to an oxo group. The hydrogen is transferred to molecular oxygen, and the reaction can be written as:

$$\text{ecdysone} + O_2 = \text{3-dehydro-ecdysone} + H_2O_2$$

The enzyme has been highly purified from homogenates of *Calliphora* prepupae (Koolman, J. and Karlson, R. 1975, 1978). From the reaction, it can be inferred that it has a prosthetic group. It seems not to be a flavoprotein or a metallo-enzyme, but this negative evidence may not be completely convincing. The molecular mass has been estimated at 240,000 Daltons.

Instead of oxygen, artificial substrates (dyes) can serve as hydrogen acceptors e.g. methylene blue or dichloroindophenol. The K_m values for ecdysone and 20-hydroxyecdysone are listed in Table 5. Ecdysone oxidase does not oxidize cholesterol, 20-hydroxycholesterol or 22,25-dideoxyecdysone, but other ecdysteroids besides ecdysone, notably 20-hydroxyecdysone, can serve as substrates.

3.2.2 Ecdysone 3-epimerase

The enzyme (or enzyme system) converts ecdysteroids to their 3-epimers. A tentative equation for this reaction may be given:

$$\text{ecdysone} + NAD(P)H + H^+ + O_2$$
$$= \text{3-epiecdysone} + NAD(P)^+ + H_2O \text{ (or } H_2O_2\text{)}$$

The similarity of this reaction with that of ecdysone oxidase indicates that the reaction of ecdysone 3-epimerase may actually be a combination of two reactions: an oxidation catalyzed by ecdysone oxidase, and a reduction due to a presumed 3-dehydroecdysone reductase (Koolman, J. 1976). However, attempts to demonstrate the formation of 3-dehydroecdysteroids during the reaction of ecdysone 3-epimerase were not successful (Mayer, R. *et al.*, 1979).

Ecdysone 3-epimerase was discovered in homogenate of *Manduca sexta* midgut and was purified 5–8-fold from the post-microsomal supernatant (Nigg, H. *et al.*, 1974; Mayer, R. *et al.*, 1979). The enzyme reacts irreversibly with the 3β-hydroxyl group of ecdysteroids to yield 3α-hydroxyecdysteroids. The reaction was found to be dependent on either NADH (K_m = 85 μM) or NADPH (K_m =

51 μM) and oxygen (Mayer *et al.*, 1979). The Michaelis constants for ecdysone and 20-hydroxyecdysone are given in Table 5.

3.3 Transferases involved in conjugation

The structure of the various ecdysteroid conjugates has been discussed above. Corresponding enzymes must be postulated to form these conjugates, but the experimental evidence is meager.

A sulfate transferase yielding a sulfate ester of ecdysteroids was first detected in *Prodenia erydania* (Yang, R. and Wilkinson, C. 1972, 1973) and was recently studied *in vitro* in more detail (Enderle, U. and Koolman, J. 1981). The donor of the sulfate group was identified as 3′-phospho-adenylylsulfate (PAPS). This is in line with the coenzyme requirement of all known sulfate transferase. This enzyme has not yet been purified but the fact that insects are able to synthesize PAPS was recently demonstrated with homogenates of tissues from *Spodoptera littoralis* (Isaac, R. *et al.*, 1982b).

The activity of a glycosyl transferase was described in *Calliphora vicina* (Heinrich, G. and Hoffmeister, H. 1970). The reaction was reported to occur with fat body tissue kept *in vitro*; ponasterone A and 20-hydroxyecdysone were used as substrates. The product of the reaction was a very polar metabolite to which was assigned the structure of an α-glucoside of the hydroxyl group at C-3. The evidence for this structure is insufficient. Moreover, one would expect a β-glucoside, similar to other carbohydrate conjugates in insects. A reinvestigation of this question appears necessary.

In conclusion, it is obvious that the enzymology of ecdysone degradation is still in its infancy. Much more work is needed for a detailed knowledge of the ecdysone-metabolizing enzymes.

4 EXCRETION OF ECDYSONE METABOLITES

Metabolic end-products are normally excreted. There is no indication that ecdysteroids are fully degraded to CO_2 and H_2O; the rather rapid conversion to conjugates indicates that these conjugates are earmarked for excretion via the Malpighian tubules.

Free ecdysteroids as well as conjugates have been found in the faeces of various insect species. The main ecdysteroids identified are listed in Table 6, together with relevant references.

In *Locusta*, the excretion of labelled ecdysone injected into larvae has been followed in some detail (Hoffmann, J. *et al.*, 1974; Hoffmann, J. and Koolman, J. 1974; Koolman, J. *et al.*, 1975; Feyereisen, R. *et al.*, 1976). There are great differences in the fate in early, mid and late fifth-instar insects. In early fifth instar larvae, when the hormone titer is low, ecdysone is excreted virtually unchanged. At the peak of the ecdysteroid titer the rate of excretion is lower and the products are mainly conjugates. Thus, the rate of excretion may play a role in the regulation of ecdysteroid titer in the insect. However, more data, especially with other species, are necessary before any generalizations can be drawn.

As mentioned above, rather high activities of ecdysteroid-metabolizing enzymes can be found in the intestine. It is not yet known if this high activity is related to the excretion of the metabolite formed. It may also serve a very important function for phytophageous insects feeding on ecdysteroid-containing plants for the inactivation of ingested ecdysteroids. This may be concluded from experiments in which the metabolic fate of ecdysone injected into locusts was compared to that of ingested ecdysone (Feyereisen, R. *et al.*, 1976). Ecdysteroids were rapidly taken up from the food by gut cells and inactivated *via* 3-dehydroecdysteroids to conjugates. On the other hand, feeding experiments showed that in other insect species (e.g. termites, flies) ecdysteroids at least partially retain their biological activity when applied with the food (e.g. Lüscher, M. and Karlson, P. 1958; Garen, A. *et al.*, 1977). The conflicting observations are possibly explained by species differences. Neither termites nor fly larvae are confronted with exogenous ecdysteroids.

4.1 Role of Malpighian tubules and gut

Both the Malpighian tubules and the gut may be involved in the excretion of free and conjugated steroids from the circulatory system. Published evidence on which part of the excretory system secretes ecdysteroids is still lacking. However, during *in vitro* experiments with Malpighian tubules from adult male tsetse flies (*Glossina morsitans*)

Table 6: Ecdysteroids detected in insect faecal material

Ecdysteroid	Species	Stage	Concentration in the frass	References
Ecdysone	*Schistocerca gregaria*	larvae	1.0–1.5 $\mu g\, g^{-1}$	Hoffmeister, H. *et al.*, (1965)
Hormonally active ecdysteroids	*Locusta migratoria*	larvae	0.4 $\mu g\, g^{-1}$	Hoffmeister, H. *et al.*, (1965)
		adults	0.0 $\mu g\, g^{-1}$	Hoffmeister, H. *et al.*, (1965)
Hormonally active ecdysteroids	*Periplaneta americana*	larvae	0.8 $\mu g\, g^{-1}$	Hoffmeister, H. *et al.*, (1965)
Hormonally active ecdysteroids	*Bombyx mori*	larvae	0.1 $\mu g\, g^{-1}$	Hoffmeister, H. *et al.*, (1965)
Hormonally active ecdysteroids	*Calliphora vicina*	larvae	0.2 $\mu g\, g^{-1}$	Hoffmeister, H.*et al.*, (1965)
20-Hydroxyecdysone	*Schistocerca gregaria*	larvae	$\sim$0.3 $\mu g\, g^{-1}$	Wilson, I. and Morgan, E. (1978)
Conjugated 20-hydroxyecdysone	*Schistocerca gregaria*	larvae	$\sim$0.1 $\mu g\, g^{-1}$	Wilson, I. and Morgan, E. (1978)
Ecdysone and metabolites	*Locusta migratoria*	larvae and adults	not determined	Hoffmann, J. and Koolman, J. (1974); Hoffmann, J. *et al* (1974); Koolman, J. *et al.*, (1975); Feyereisen, R. *et al.*, (1976)
Ecdysone and metabolites	*Gryllus bimaculatus*	larvae	not determined	Wildmann, H. and Romer, F. (1977)
Ecdysone and metabolites	*Tenebrio molitor*	larvae	not determined	Weinheimer, G. and Romer, F. (1977)
20-Hydroxyecdysone and metabolites	*Bombyx mori*	larvae	not determined	Hikino, H. *et al.*, (1975a)
Ecdysone and metabolites	*Manduca sexta*	larvae	not determined	Moriyama, H. *et al.*, (1970)
Ecdysone and 20-hydroxyecdysone	*Manduca sexta*	larvae	0.1–0.2 $\mu g\, g^{-1}$	Kaplanis, J. *et al.*, (1972a)
20-Hydroxyecdysone and metabolites	*Manduca sexta*	larvae	not determined	Kaplanis, J. *et al.*, (1972b)
Ecdysone and metabolites	*Glossina morsitans*	female adult	not determined	Whitehead, D. (1981)

(Gee, J. *et al.*, 1977) it was observed that the tubules were able to excrete ecdysone and 20-hydroxyecdysone (unpublished).

4.2 Fate of metabolites in pupae and eggs

During the stages of insect development which have no active excretory systems (i.e. in eggs, embryos and pupae), hormonally active ecdysteroids have to be removed from the circulatory system by means other than excretion. This can be achieved either by enzymatic inactivation or local separation. Apparently both means are used in insects. Developing embryos of *Locusta migratoria* reveal an active ecdysone metabolism (Hoffmann, J. *et al.*, 1980). Considerable variations in the content of ecdysone, 20-hydroxyecdysone and 20,26-dihydroxyecdysone were observed during embryonic development, indicating both an efficient biosynthesis and catabolism of ecdysteroids (chapter 13, volume 1). The activities of hormone-inactivating enzymes, notably ecdysone oxidase, ecdysone 26-hydroxylase and conjugate-forming enzymes, were detected *in vivo*. During the pupal stage of several holometabolous insect species an efficient inactivation of ecdysteroids by ecdysone oxidase, ecdysone 3-epimerase and conjugating enzymes was also detected (*Calliphora vicina* Koolman, J. 1978; *Manduca sexta* Kaplanis, J. *et al.*, 1980a; *Bombyx mori* Moribayashi, A. and Ohtaki, T. 1978, 1980).

The removal of ecdysone and its metabolites from the circulatory system was observed in pupae of *Manduca sexta* (Kaplanis, J. *et al.*, 1980a) with high ecdysteroid concentrations being detected in the pupal meconium. From this material several ecdysteroids could be isolated in crystalline form, including 3-epi-20-hydroxyecdysone, 20-hydroxyecdysone and 3-epiecdysone (Thompson, M. *et al.*, 1974). Obviously the meconium serves as a depository for metabolic waste products.

5 GENERAL CONCLUSIONS AND FUTURE PROBLEMS

The ecdysteroid titer is regulated in large measure by the secretory activity of the hormone-producing gland. The steady-state level of hormone in the haemolymph is a balance between synthesis on the one hand and catabolism, excretion and local storage on the other hand. In this context, several functions can be attributed to the ecdysone metabolism.

(1) It activates ecdysone by hydroxylation at C-20. This reaction is of special importance for those stages in which 20-hydroxyecdysone is the active form of the molting hormone.
(2) It inactivates the molting hormone (ecdysone, 20-hydroxyecdysone, 2-deoxyecdysone ?) using three different pathways: (i) 26-hydroxylation often followed by C-26-oxidation, (ii) 3-oxidation and/or 3-epimerization, and (iii) conjugate formation.
(3) It converts the hormone into a molecular form (anionic ecdysteroid) which is presumed to have a higher elimination rate than the hormone. Thus it improves the process of ecdysteroid excretion.
(4) It is likely to convert ecdysteroids into a storage form of the hormone (special type of conjugate). Active hormone is released from this storage form by hydrolysis at a certain stage of development.

It is well known that the activity of the enzymes involved in ecdysone metabolism is physiologically regulated (review: Koolman, J. 1982). However, the details of the regulation (kinetics, mechanism) and its relevance still have to be analyzed. In particular the structure of some ecdysone metabolites awaits elucidation (e.g. highly polar ecdysteroids). The metabolites have to be synthesized chemically or enzymatically to test their biological properties and their metabolic fate. Assays of higher specificity are needed to follow developmental profiles of titers of certain ecdysone metabolites in insects. As a final result of these efforts it should be possible to search for natural substances and to design synthetic substances which interfere with the metabolism of ecdysone. Such compounds have the potential to become the "fourth-generation pesticides" of the future.

REFERENCES

Beckers, G. and Emmerich, H. (1976). The ecdysone titer in *Drosophila hydei* and the metabolism of the moulting hormones in the salivary gland. In *Actualités sur les hormones d'invertébrés*. Edited by M. M. Durchon, *Coll. Internat. C.N.R.S.* No. 251, pp.351–357. Paris.

Bergamasco, H. and Horn, D. H. S. (1980). The biological activities of ecdysteroids and ecdysteroid analogues. In *Progress in Ecdysone Research*. Edited by J. A. Hoffmann. pages 299–324. Elsevier/North Holland, Amsterdam.

Blais, C. and Lafont, R. (1980). *In vitro* differentiation of *Pieris brassicae* imaginal wing discs: effects and metabolism of ecdysone and ecdysterone. *Wilhelm Roux's Archiv. 188*, 27–36.

Bollenbacher, W. E., Smith, S. L., Wielgus, J. J. and Gilbert, L. I (1977). Evidence for an α-ecdysone cytochrome P-450 mixed function oxidase in insect fat body mitochondria. *Nature 268*, 660–663.

Brunet, P. C. J. and Kent, P. W. (1955). Observations on the mechanism of a tanning reaction in *Periplaneta* and *Blatta*. *Proc. R. Soc. Lond. B. 144*, 259–274.

Butenandt, A. and Karlson, P. (1954). Über die Isolierung eines Metamorphosehormons der Insekten in kristallisierter Form. *Z. Naturf. 9b*, 389–391.

Butterworth, F. M. and Berendes, H. D. (1974). Ecdysone-binding proteins in haemolymph and tissues of *Drosophila hydei* larvae *J. Insect Physiol. 20*, 2195–2204.

Cherbas, L. and Cherbas, P. (1970). Distribution and metabolism of α-ecdysone in pupae of the silkworm *Antheraea polyphemus*. *Biol. Bull. 138*, 115–128.

Clever, U., Clever, I., Storbeck, I. and Young, N. L. (1973). The apparent requirement of two hormones, α- and β-ecdysone, for molting induction of insects. *Devel. Biol. 31*, 47–60.

Dinan, L. N. and Rees, H. A. (1981). The identification and titers of conjugated and free ecdysteroids in developing ovaries and newly-laid eggs of *Schistocerca gregaria*. *J. Insect Physiol. 27*,51–58.

Emmerich, H. (1970a). Über die Hämolymphproteine von *Pyrrhocoris apterus* und über die Bindung von Ecdyson durch Hämolymphproteine. *J. Insect Physiol. 16*, 725–747.

Emmerich, H. (1970b). Ecdysonbindende Proteinfraktionen in den Speicheldrüsen von *Drosophila hydei*. *Z. Vergl. Physiol. 68*, 385–402.

Enderle, U. and Koolman, J. (1981). Ecdysteroid sulphotransferase: analysis of the enzyme *in vitro*. *Biochem. Soc. Trans. 9*, 156p.

Faux, A., Horn, D. H. S. and Middleton, E. J. (1969). Moulting hormones of a crab during ecdysis. *Chem. Commun. 4*, 175–176.

Feyereisen, R. (1977a). A specific binding protein for the moulting hormone ecdysterone in locust haemolymph. *Experientia 33*, 1111–1112.

Feyereisen, R. (1977b). Cytochrome P-450 et hydroxylation de l'ecdysone en ecdystérone chez *Locusta migratoria in vitro*. *C.R. Acad. Sci. Paris D, 284*, 1831–1834.

Feyereisen, R. (1980). Ecdysteroid-protein interactions: specificity of the ecdysteroid binding protein from *Locusta migratoria* haemolymph. In *Progress in Ecdysone Research*. Edited by J. A. Hoffmann. pages 325–334. Elsevier/North Holland, Amsterdam.

Feyereisen, R. and Durst, F. (1978). Ecdysterone biosynthesis: a microsomal cytochrome-P-450-linked ecdysone 20-mono-oxygenase from tissues of the African migratory locust. *Eur. J. Biochem. 88*, 37–47.

FEYEREISEN, R., LAGUEUX, M. and HOFFMANN, J. A. (1976). Dynamics of ecdysone metabolism after ingestion and injection in *Locusta migratoria*. *Gen. Comp. Endocr. 29*, 319–327.

GALBRAITH, M. N., HORN, D. H. S., MIDDLETON, E. J. and HACKNEY, R. J. (1968). Structure of deoxycrustecdysone, a second crustacean moulting hormone. *Chem. Commun.* 83–85.

GALBRAITH, M. N., HORN, D. H. S., MIDDLETON, E. J., THOMSON, J. A., SIDDALL, J. B. and HAFFERL, W. (1969). Catabolism of crustecdysone in the blowfly *Calliphora stygia*. *Chem. Commun.* 1134–1135.

GANDE, A. R., MORGAN, E. D. and WILSON, I. D. (1979). Ecdysteroid levels throughout the life cycle of the desert locust, *Schistocerca gregaria*. *J. Insect Physiol. 25*, 669–675.

GAREN, A., KAUVAR, L. and LEPESANT, J.-A. (1977). Roles of ecdysone in *Drosophila* development. *Proc. Nat. Acad. Sci. 74*, 5099–5103.

GEE, J. D., WHITEHEAD, D. L. and KOOLMAN, J. (1977). Steroids stimulate secretion by insect malpighian tubules. *Nature 269*, 238–239.

GERSCH, M. and KARLSON, P. (1982). The evolution of hormonal systems. *Nova Acta Leopoldina*. Halle, DDR (in press).

GORELL, T., GILBERT, L. I. and TASH, J. (1972). The uptake and conversion of α-ecdysone by the pupal tissues of *Hyalophora cecropia*. *Insect Biochem. 2*, 94–106.

GREENWOOD, D. R. and RUSSEL, G. B. (1978). 26-Hydroxy-β-ecdysone, a metabolite of β-ecdysone in the blowfly, *Calliphora erythrocephala*. *Experientia 34*, 687–688.

HAFFERL, W., WREN, W. L., MARSHALL, J. P., CALZADA, M. C. and SIDDALL, J. B. (1972). Radiochemical synthesis. III. Synthesis of tritiated α-ecdysone and 25-deoxy-α-ecdysone of high specific activity. *J. Label. Compounds 8*, 81–94.

HAMPSHIRE, F. and HORN, D. H. S. (1966). Structure of crustecdysone, a crustacean moulting hormone. *Chem. Commun.* 37–38.

HAYAKAWA, S. (1973). Microbiological transformation of bile acids. In *Adv. Lipid Res. 2*, pp.143–192.

HEINRICH, G. and HOFFMEISTER, H. (1970). Insect-moulting hormones and their mode of action. Formation of hormone glycosides as inactivation mechanism in *Calliphora erythrocephala*. *Z. Naturf. B. 25*, 358–361.

HIKINO, H., OHIZUMI, Y. and TAKEMOTO, T. (1975a). Detoxication mechanism of *Bombyx mori* against exogenous phytoecdysone ecdysterone. *J. Insect Physiol. 21*, 1953–1963.

HIKINO, H., OHIZUMI, Y. and TAKEMOTO, T. (1975b). Steroid metabolism in *Bombyx mori*. I. Catabolism of ponasterone A and ecdysterone in *Bombyx mori*. *Hoppe-Seyler's Z. Physiol. Chem. 356*, 309–314.

HOCKS, P. and WIECHERT, R. (1966). 20-Hydroxy-Ecdyson, isoliert aus Insekten. *Tetrahedron Lett. 33*, 2989–2993.

HOFFMANN, J. A. and KOOLMAN, J. (1974). Role of the prothoracic glands in the regulation of ecdysone titer and the metabolic fate of injected labelled ecdysone in *Locusta migratoria*. *J. Insect Physiol. 20*, 1593–1601.

HOFFMANN, J. A., KOOLMAN, J., KARLSON, P. and JOLY, P. (1974). Molting hormone titer and metabolic fate of injected ecdysone during the fifth larval instar and in adults of *Locusta migratoria* (Orthoptera). *Gen. Comp. Endocr. 22*, 90–97.

HOFFMANN, J. A., LAGUEUX, M., HETRU, C., CHARLET, M. and GOLTZENE, F. (1980). Ecdysone in reproductively competent female adults and in embryos of insects. In *Progress in Ecdysone Research*. Edited by J. A. Hoffmann. pp.431–466. Elsevier/North Holland, Amsterdam.

HOFFMEISTER, H. and GRÜTZMACHER, H. F. (1966). Zur Chemie des Ecdysterons. *Tetrahedron Lett. 33*, 4017–4023.

HOFFMEISTER, H., RUFER, C. and AMMON, H. (1965). Ausscheidung von Ecdyson bei Insekten. *Z. Naturf. 20b*, 130–133.

HUBER, R. and HOPPE, W. (1965). Zur Chemie des Ecdysons VII, Die Kristall- und Molekülstrukturanalyse des Insektenhormons Ecdyson mit der automatisierten Faltmolekülmethode. *Chem. Ber. 98*, 2403–2424.

IKEKAWA, N., IKEDA, T., MIZUNO, T., OHNISHI, E. and SAKURAI, S. (1980). Isolation of a new ecdysteroid, 2,22-dideoxy-20-hydroxy-ecdysone, from the ovaries of the silkworm, *Bombyx mori*. *J. Chem. Soc. Chem. Commun.* 448–449.

ISAAC, R. E., REES, H. H. and GOODWIN, T. W. (1981a). Isolation of ecdysone-3-acetate as a major ecdysteroid from the developing eggs of the desert locust, *Schistocerca gregaria*. *J. Chem. Soc. Chem. Commun.* 594–595.

ISAAC, R. E., REES, H. H. and GOODWIN, T. W. (1981b). Isolation of 2-deoxy-20-hydroxyecdysone and 3-epi-2-deoxyecdysone from eggs of the desert locust, *Schistocerca gregaria*, during embryogenesis. *J. Chem. Soc. Chem. Commun.* 418–420.

ISAAC, R. E., ROSE, M. E., REES, H. H. and GOODWIN, T. W. (1982a). Identification of ecdysone-22-phosphate and 2-deoxyecdysone-22-phosphate in eggs of the desert locust, *Schistocerca gregaria*, by fast atom bombardment mass spectroscopy and n.m.r. spectroscopy. *J. Chem. Soc. Chem. Commun.* 249–251.

ISAAC, R. E., PHUA, K. K. and REES, H. H. (1982b). 3′-Phosphoadenosine-5-phosphosulphate synthesis and involvement in sulphotransferase reactions in the insect, *Spodoptera littoralis*. *Biochem. J. 204*. 127–133.

JOHNSON, P. and REES, H. H. (1977). The mechanism of C-20-hydroxylation of α-ecdysone in the desert locust *Schistocerca gregaria*. *Biochem. J. 168*, 513–520.

KAPLANIS, J. N., ROBBINS, W. E., THOMPSON, M. J. and BAUMHOVER, A. H. (1969). Ecdysone analog: conversion to α-ecdysone and 20-hydroxyecdysone by an insect. *Science 166*, 1540–1541.

KAPLANIS, J. N., THOMPSON, M. J., DUTKY, S. R., ROBBINS, W. E. and LINDQUIST, E. L. (1972a). The metabolism of 4-^{14}C-22, 25-bisdeoxyecdysone during larval development in the tobacco hornworm, *Manduca sexta* (L.). *Steroids 20*, 105–120.

KAPLANIS, J. M., THOMPSON, M. J., ROBBINS, W. E. and LINDQUIST, E. L. (1972b). α-Ecdysone and 20-hydroxyecdysone from frass of the tobacco hornworm, *Manduca sexta* (L.) *Steroids 20*, 621–625.

KAPLANIS, J. N., ROBBINS, W. E., THOMPSON, M. J. and DUTKY, S. R. (1973). 26-Hydroxyecdysone: new insect molting hormone from the egg of the tobacco hornworm. *Science 180*, 307–308.

KAPLANIS, J. N., DUTKY, S. R., ROBBINS, W. E. and THOMPSON, M. J. (1974). The metabolism of the synthetic ecdysone analog, 1α-^{3}H-22, 25-bisdeoxyecdysone in relation to certain of its biological effects in insects. In *Invertebrate Endocrinology and Hormonal Heterophylly* pp.161–175. Edited by W. J. Burdette. Springer-Verlag, New York, Heidelberg, Berlin.

KAPLANIS, J. N., DUTKY, S. R., ROBBINS, W. E., THOMPSON, M. J., LINDQUIST, E. L., HORN, D. H. S. and GALBRAITH, M. N. (1975). Makisterone A: a 28-Carbon hexahydroxy molting hormone from the embryo of the milkweed bug. *Science 190*, 681–682.

KAPLANIS, J. N., THOMPSON, M. J., DUTKY, S. R. and ROBBINS, W. E. (1979). The ecdysteroids from the tobacco hornworms during pupal-adult development five days after peak titer of molting hormone activity. *Steroids 34*, 333–345.

KAPLANIS, J. N., WEIRICH, G. F., SVOBODA, J. A., THOMPSON, M. J. and ROBBINS, W. E. (1980a). Moulting hormones of the tobacco hornworm and milkweed bug: their chemistry and biochemistry. In *Progress in Ecdysone Research*. Edited by J. A. Hoffmann. pp.163–186. Elsevier/North Holland, Amsterdam.

KAPLANIS, J. N., THOMPSON, M. J., DUTKY, S. R. and ROBBINS, W. E. (1980b). The ecdysteroids from young embryonated eggs of the tobacco hornworm. *Steroids 36*, 321–336.

KARLSON, P. (1956). Biochemical studies on insect hormones. *Vit. Hormones 14*, 227–266.

KARLSON, P. (1982).The concept of hormonal systems in retrospect and prospect. In *The Evolution of Hormonal Systems* (ed. M. Gersch and P. Karlson). *Nova Acta Leopoldina*, Halle, DDR (in press).

KARLSON, P. and KOOLMAN, J. (1973). On the metabolic fate of ecdysone and 3-dehydroecdysone in *Calliphora vicina*. *Insect Biochem. 3*, 409–417.

KARLSON, P., MAURER, R. and WENZEL, M. (1963). Eine Mikromethode zur Markierung von Steroiden und von Ecdyson mit Tritium. *Z. Naturf. 18b*, 219–224.

KARLSON, P., HOFFMEISTER, H., HUMMEL, H., HOCKS, P. and SPITELLER, G. (1965). Zur Chemie des Ecdysons, VI. Reaktionen des Ecdysonmoleküls. *Chem. Ber. 98*, 2394–2402.

KING, D.S. (1972). Metabolism of α-ecdysone and possible immediate precursors by insects *in vivo* and *in vitro*. *Gen. Comp. Endocr.*, Suppl. *3*, 221–227.

KING, D. S. and MARKS, E. P. (1974). The secretion and metabolism of α-ecdysone by cockroach (*Leucophaea maderae*) tissues *in vitro*. *Life Sci. 15*, 147–154.

KING, D. S. and SIDDALL, J. B. (1969). Conversion of α-ecdysone to β-ecdysone by crustaceans and insects. *Nature 221*, 955–956.

King, D. S. and Siddall, J. B. (1974). Biosynthesis and inactivation of ecdysone. In *Invertebrate Endocrinology and Hormonal Heterophylly*, pp.147–152. Edited by W. J. Burdette. Springer-Verlag, New York, Heidelberg, Berlin.

King, D. S., Bollenbacher, W. E., Borst, D. W., Vedeckis, W. V., O'Connor, J. D., Ittycheriah, P. I. and Gilbert, L. I. (1974). The secretion of α-ecdysone by the prothoracic glands of *Manduca sexta in vitro*. *Proc. Nat. Acad. Sci. 71*, 793–796.

Klose, W., Gateff, E., Emmerich, H. and Beikirch, H. (1980). Developmental studies on two ecdysone deficient mutants of *Drosophila melanogaster*. *Wilhelm Roux's Archiv. 189*, 57–67.

Koolman, J. (1976). Ecdysone oxidase, an enzyme of the ecdysone metabolism in insects. In *Actualités sur les Hormones d'Invertébrés*, pp.403–412. Edited by M. M. Durchon, *Coll. Internat. C.N.R.S.* No. *251*, Paris.

Koolman, J. (1978). Ecdysone oxidase in insects. *Hoppe-Seyler's Z. Physiol. Chem. 359*, 1315–1321.

Koolman, J. (1980). Ecdysteroids in the blowfly, *Calliphora vicina*. In *Progress in Ecdysone Research*. Edited by J. A. Hoffmann. pp.187–209. Elsevier/North Holland, Amsterdam.

Koolman, J. (1981). Kinetics of the enzymatic degradation of moulting hormones in the blowfly *Calliphora vicina*. In *Regulation of Insect Development and Behaviour*. Edited by F. Sehnal, J. J. Menn and B. Cymborowsky. Technical University of Wroclaw, No. 22, *Conference 7*, pp.93–106.

Koolman, J. (1982). Ecdysone metabolism. *Insect Biochem 12*, 225–250.

Koolman, J. and Karlson, P. (1975). Ecdysone oxidase, an enzyme from the blowfly *Calliphora erythrocephala* (Meigen). *Hoppe-Seyler's Z. Physiol. Chem. 356*, 1131–1138.

Koolman, J. and Karlson, P. (1978). Ecdysone oxidase: reaction and specificity. *Eur. J. Biochem. 89*, 453–460.

Koolman, J., Hoffmann, J. A. and Karlson, P. (1973). Sulfate esters as inactivation products of ecdysone in *Locusta migratoria*. *Hoppe-Seyler's Z. Physiol. Chem. 354*, 1043–1048.

Koolman, J., Hoffmann, J. A. and Dreyer, M. (1975). Moulting hormone in *Locusta migratoria*: rate of ecdysone 20-hydroxylation and excretion during the last larval instar. *Experientia 31*, 247–248.

Koolman, J., Reum, L. and Karlson, P. (1979). 26-Hydroxyecdysone, 20,26-dihydroxyecdysone and inokosterone, metabolites of ecdysone in the blowfly, *Calliphora vicina*. *Hoppe-Seyler's Z. Physiol. Chem. 360*, 1351–1355.

Lafont, R., Beydon, P., Somme-Martin, G. and Blais, C. (1980). High performance liquid chromatography of ecdysone metabolites applied to the cabbage butterfly, *Pieris brassicae L. Steroids 36*, 185–207.

Lagueux, M., Perron, J.-M. and Hoffmann, J. A. (1976). Ecdysone metabolism in relation to endogenous moulting hormone titre during larval-pupal development in *Choristoneura fumiferana* (Clem.). *J. Insect Physiol. 22*, 57–62.

Lagueux, M., Hirn, M. and Hoffmann, J. A. (1977). Ecdysone during ovarian development in *Locusta migratoria* (Orthoptera). *J. Insect Physiol. 23*, 109–119.

Legueux, M., Hetru, C., Goltzene, F., Kappler, C. and Hoffmann, J. A. (1979). Ecdysone titre and metabolism in relation to cuticulogenesis in embryos of *Locusta migratoria*. *J. Insect Physiol. 25*, 709–723.

Lüscher, M. and Karlson, P. (1958). Experimentelle Auslösung von Häutungen bei der Termite *Kalotermes flavicollis* (Fabr.). *J. Insect Physiol. 2*, 341–345.

Mayer, R. T., Svoboda, J. A. and Weirich, G. F. (1978). Ecdysone 20-hydroxylase in midgut mitochondria of *Manduca sexta*. (L.)*Hoppe-Seyler's Z. Physiol. Chem. 359*, 1247–1257.

Mayer, R. T., Durrant, J. L., Holman, G. M., Weirich, G. F. and Svoboda, J. A. (1979). Ecdysone 3-epimerase from the midgut of *Manduca sexta* (L.). *Steroids 34*, 555–562.

McCarthy, J. P. (1979). Ponasterone A: A new ecdysteroid from the embryos and serum of *Brachyuran* crustaceans. *Steroids 34*, 799–806.

Moribayashi, A. F. and Ohtaki, T. (1978). Inactivation of ecdysone and the possible feedback control of the titre during pupation of *Sarcophaga peregrina*. *J. Insect Physiol. 24*, 279–284.

Moribayashi, A. and Ohtaki, T. (1980). Inactivation and reactivation of 20-hydroxyecdysone during pupal-adult development of the fleshfly *Sarcophaga peregrina*. *Japan J. Med. Sci. Biol. 33*, 189–201.

Moriyama, H., Nakanishi, K., King, D. S., Okauchi, T., Siddall, J. B., and Hafferl, W. (1970). Origin and metabolic fate of α-ecdysone in insects. *Gen. Comp. Endocr. 15*, 80–87.

Nigg, H. N., Svoboda, J. A., Thompson, M. J., Kaplanis, J. N., Dutky, S. R. and Robbins, W. E. (1974). Ecdysone metabolism: Ecdysone dehydrogenase-isomerase. *Lipids 9*, 971–974.

Nigg, H. N., Svoboda, J. A., Thompson, M. J., Dutky, S. R., Kaplanis, J. N. and Robbins, W. E. (1976). Ecdysone 20-hydroxylase from the midgut of the tobacco hornworm (*Manduca sexta* L.). *Experientia 32*, 438–439.

Nijhout, H. F. (1976). The role of ecdysone in pupation of *Manduca sexta*. *J. Insect Physiol. 22*, 453–463.

Ohnishi, E., Mizuno, T., Chatani, F., Ikekawa, N. and Sakurai, S. (1977). 2-Deoxy-α-ecdysone from ovaries and eggs of the silkworm, *Bombyx mori*. *Science 197*, 66–67.

Ohtaki, T. and Williams, C. M. (1970). Inactivation of α-ecdysone and cyasterone by larvae of the fleshfly, *Sarcophaga peregrina*, and pupae of the silkworm, *Samia cynthia*. *Biol. Bull. 138*, 326–333.

Okubo, S. (1958). Occurrence of N-acetylhydroxytyramine glucoside in *Drosophila melanogaster*. *Med. J. Osaka Univ. 9*, 327–337.

Rees, H. H., Greenwood, D. R., Dinan, L. N., Isaac, R. E. and Goodwin, T. W. (1981). Biosynthesis and metabolism of ecdysteroids. In *Regulation of Insect Development and Behaviour*. Edited by F. Sehnal, A. Zabza, J. J. Menn and B. Cymborowsky. Technical University of Wroclaw, No. 22, *Conferences 7*, pp.71–92.

Reum, L., Käuser, G., Enderle, U. and Koolman, J. (1982). A steroid-binding protein from insect haemolymph isolated by photoaffinity labelling and immunoadsorption. *Z. Naturf. 37c*, 967–976.

Rohner, R. M. and Meyer, D. R. (1980). The titers and clearance of ecdysteroids from the blood of last instar larvae of *Galleria mellonella*. *Experientia 36*, 1327–1328.

Russel, G. B. and Price, G. M. (1977). Metabolism of β-ecdysone during the larval and white puparial stage of the blowfly, *Calliphora erythrocephala*. *Insect Biochem. 7*, 197–202.

Sannasi, A. and Karlson, P. (1974). Metabolism of ecdysone: Phosphate and sulphate esters as conjugates of ecdysone in *Calliphora vicina*. *Zool. Jahrb. Physiol. 78*, 378–386.

Schaller, F. E. and Hoffmann, J. A. (1976). Métabolisme de l'α-ecdysone *in vivo* et *in vitro* chez *Aeshna cyanea* (Insecte, Odonate). In *Actualités sur les Hormones d' Invertébrés*. Edited by M. M. Durchon. *Coll. Internat. C.N.R.S.* No. *251*, pp.393–402.

Schaller, F., Hoffmann, J. A. and Koolman, J. (1975). Métabolisme de l'ecdysone chez *Aeshna cyanea* (Insecte Odonate). Résultats d'injections d'α-ecdysone radioactive á des animaux et á des larves au cycle de mue bloqué *C.R. Acad. Sci. Paris D 280*, 1813–1816.

Shaaya, E. (1969). Untersuchungen über die Verteilung des Ecdysons in verschiedenen Geweben von *Calliphora erythrocephala* und über seine biologische Halbwertszeit. *Z. Naturf. 24b*, 718–721.

Smith, S. L., Bollenbacher, W. E., Cooper, D. Y., Schleyer, H., Wielgus, J. J. and Gilbert, L. I. (1977). Characterization of ecdysone 20-mono-oxygenase activity in *Manduca sexta*. *Amer. Zool. 17*, 928. (Abstract 387)

Smith, S. L., Bollenbacher, W. E., Cooper, D. Y., Schleyer, H., Weilgus, J. J. and Gilbert, L. I. (1979). Ecdysone 20-mono-oxygenase: Characterisation of an insect cytochrome P-450 dependent steroid hydroxylase. *Mol. Cell. Endocr. 15*, 111–133.

Smith, S. L., Bollenbacher, W. E. and Gilbert, L. I. (1980). Studies on the biosynthesis of ecdysone and 20-hydroxyecdysone in the tobacco hornworm, *Manduca sexta*. In *Progress in Ecdysone Research*. Edited by J. Hoffmann. pp.139–162. Elsevier/North Holland, Amsterdam.

Starling, E. H. (1907). Die chemische Koordination der Körpertätigkeiten. *Verh. Ges. Dtsch. Naturforsch. u. Ärzte 78*. Vers., pp.246–260. Leipzig, F. C. W. Vogel.

Thamer, G. and Karlson, P. (1972). Nachweis der Proteinbindung von Ecdyson bei der Schmeißfliege *Caliphora erythrocephala* Meigen. *Z. Naturf. 27b*, 1191–1195.

Thompson, M. J., Kaplanis, J. N., Robbins, W. E. and Yamamoto, R. T. (1967). 20, 26-Dihydroxyecdysone, a new steroid with moulting hormone activity from the tobacco hornworm, *Manduca sexta* (Johannson). *J. Chem. Soc. Chem. Commun. 13*, 650–653.

THOMPSON, M. J., KAPLANIS, J. N., ROBBINS, W. E., DUTKY, S. R. and NIGG, H. N. (1974). 3-Epi-20-hydroxyecdysone from meconium of the tobacco hornworm. *Steroids 24*, 359–366.

THOMPSON, M. J., ROBBINS, W. E., DUTKY, S. R., MARKS, E. P., FILIPI, P. A. and FINEGOLD, H. (1978). Isolation and identification of the metabolites of 22, 25-dideoxyecdysone from cockroach fat body cultures. *Lipids 13*, 783–790.

THOMSON, J. A., SIDDALL, J. B., GALBRAITH, M. N., HORN, D. H. S. and MIDDLETON, E. J. (1969). Biosynthesis of ecdysones in the blowfly *Calliphora stygia*. *J. Chem. Soc. Chem. Commun.*, 669–670.

TSOUPRAS, G., HETRU, C., LUU, B. LAGUEUX, M., CONSTANTIN, E. and HOFFMANN, J. A. (1982). The major conjugates of ecdysteroids in young eggs and in embryos of *Locusta migratoria*. *Tetrahedron Lett. 23*, 2045–2048.

WEINHEIMER, G. and ROMER, F. (1977). Der Metabolismus von ^{3}H-α-Ecdyson bei Larven von *Tenebrio molitor*. *J. Insect Physiol. 22*, 619–624.

WHITEHEAD, D. L. (1981). The effect of phytosterols on tsetse reproduction. *Insect Sci. Application 1*, 281–288.

WILDMANN, H. and ROMER, F. (1977). Metabolism of ^{3}H-α-ecdysone in *Gryllus bimaculatus* during the eighth larval instar. *J. Insect Physiol. 23*, 343–349.

WILSON, I. D. and MORGAN, E. D. (1978). Variations in ecdysteroid levels in 5th instar larvae of *Schistocerca gregaria* in gregarious and solitary phases. *J. Insect Physiol. 24*, 751–756.

YANG, R. S. H. and WILKINSON, C. F. (1972). Enzymatic sulphation of *p*-nitrophenol and steroids by larval gut tissues of the southern army worm (*Prodenia eridania* Cramer). *Biochem J. 130*, 487–493.

YANG, R. S. H. and WILKINSON, C. F. (1973). Sulphotransferases and phosphotransferases in insects. *Comp. Biochem. Physiol. 46B*, 717–726.

YOUNG, N. L. (1976a). The metabolism of ^{3}H-molting hormone in *Calliphora erythrocephala* at the mature larval and white puparial stages. *Insect Biochem. 6*, 1–12.

YOUNG, N. L. (1976b) The metabolism of ^{3}H-molting hormone in *Calliphora erythrocephala* during larval development. *J. Insect Physiol. 22*, 153–155.

10 Juvenile Hormone Biosynthesis

DAVID A. SCHOOLEY and FRED. C. BAKER

Zoecon Corporation, Palo Alto, California, USA

1 INTRODUCTION

The characterization of the first insect juvenile hormone, JH-I, from *Hyalophora cecropia* by Röller, H. *et al.* (1967) intrigued chemists and biochemists, as it constituted the first known natural product with a dihomosesquiterpenoid skeleton. This biosynthetic "curiosity" would have been notable even had JH-I been an obscure secondary metabolite from a plant, rather than a hormone with practical significance. The rapid subsequent identification by Meyer, A. *et al.* (1968) of a congeneric hormone, JH-II, as a similar structure but with one homologous ethyl side-chain, further piqued interest in JH biosynthesis.

At this time the biosynthesis of the common sesquiterpenoid farnesol, whose pyrophosphate (FPP) is a general precursor of sesquiterpenoids, was intimately understood thanks to many scrupulously detailed studies, especially those of Cornforth, Popják, and their collaborators (reviewed by Cordell, G., 1976). The genesis of the extra methylene units in JH-I and JH-II could be postulated to occur in two ways: first, subsequent to formation of FPP, an unknown intermediate (from FPP) could be alkylated by methionine. Such a pathway is analogous to the known biosynthetic pathway generating 24-alkyl sterols in plants. Alternatively, propionate could substitute for acetate in formation of homologs of acetoacetate and mevalonate. While no evidence for such a homomevalonate intermediate existed, propionate was known to substitute for acetate in the biosynthesis of methyl-branched fatty acids in many organisms, and to be responsible for biogenesis of methyl branches in macrolides isolated from micro-organisms.

Aside from this most unprecedented biosynthetic question, the pathway leading to formation of a farnesoic acid* moiety as its methyl ester was previously unreported, although farnesoic acid is known and biological formation of methyl esters is uncommon but documented. A terminal epoxide functionality in terpenoids such as occurs in JH, has a notable biochemical precedent in squalene-2,3-oxide, the reactive intermediative which is enzymatically cyclized to sterols. (The epoxide moiety, which is sensitive to heat, attack by nucleophilic species, and tends to cyclize in the presence of many Lewis acids, imparts most of the chemical instability of juvenile hormones.) Unraveling the biosynthetic details of the genesis of these functionalities is still not complete, despite the apparent simplicity of these acyclic modified sesquiterpenoids.

The family of JHs has continued to expand. The discovery of the sesquiterpenoid, JH-III, by Judy, K. *et al.* (1973) seemed to complete a series of homologs. However, more recently we discovered the trihomosesquiterpenoid JH-0 in eggs of *Manduca sexta* (Bergot, B. *et al.*, 1980) together with an isomer, 4-methyl JH-I, whose gross skeletal structure (Fig. 1) is known (Bergot, B. *et al.*, 1981a), but with an as yet unknown stereochemical configuration of the 4-methyl and 10,11-epoxy moieties. The structures of these trihomosesquiterpenoid JH-like materials from *M. sexta* eggs raise even more questions concerning biogenesis of the homologous sesquiterpenoids. In this context it is necessary to mention that the Dufour's glands of several *Myrmica* ant species produce farnesene and homo-, dihomo-, and trihomo-derivatives (Morgan, E. and Wadhams, L., 1972; Morgan, E. *et al.*, 1979; Attygalle, A. and Morgan, E., 1982), which are suspected of being trail pheromone constituents. Moreover, the dihomosesquiterpenoid faranal (Fig. 1) (Ritter, F. *et al.*, 1977; Kobayashi, M. *et al.*, 1980) is known to be a trail pheromone of the Pharaoh ant, *Monomorium pharaonis*, and is the only known isoprenoid, in addition to 4-methyl JH-I, to have the peculiar methyl branch at C-4 in a position alpha to another methyl (C-3). It is curious that the only known natural homosesquiterpenoids appear to function as chemical messengers in the Insecta!

In concluding the introduction, the question of identity of JHs in various insect orders should be addressed. Subsequent to the discovery of JH-III, the identity of the JHs of a number of insects was investigated by several groups, especially that of K. Trautmann *et al.* (1974; reviewed by Schooley, D., 1977). The status *ca.* 1975 was that only JH-III was isolated from most orders of insects analyzed. Subsequently, a number of reports claimed the presence of JH-I and/or JH-II in a cockroach (*Nauphoeta*), in *Locusta*, in *Leptinotarsa*, and certain other insects from a number of orders. Recently, reinvestigation of many of these reports has shown the *absence* of any hormone *other than JH-III* (Bergot, B. *et al.*, 1981c; de Kort, C. *et al.*, 1982; Baker *et al.*, unpublished data). Thus, we appear to be returning to the *status quo* of 1975: only in the Lepidoptera have JHs other than JH-III been isolated. Finally, for other recent reviews on JH biosynthesis, see Law, J. (1983), King, D. (1983), and Tobe, S. and Feyereisen, R. (1983).

2 METHODS EMPLOYED FOR JH BIOSYNTHETIC STUDIES

2.1 *In vivo* studies

The earliest research on JH biosynthesis was conducted in the laboratory of Röller and Dahm (Metzler, M. *et al.*, 1971, 1972) using *in vivo* incubation of radiolabeled precursors injected into adult male *Hyalophora cecropia*. These studies provided valuable information on the biogenesis of the methyl group of the ester moiety from [*methyl*-^{14}C]methionine, and on the incorporation of the epoxy acid into the JH-I molecule. However, these *in vivo* studies were also characterized by the inability to detect incorporation of labeled precursors such as mevalonate and propionate into JH-I. Adult *H. cecropia* would seem to be a good model for such *in vivo* experiments, since the males are known to contain extraordinary amounts of JH-I and JH-II and might be expected to show high biosynthetic rates. However, it is now known that adult males of this animal are quite atypical in that

* Preferred (Christophe, J. and Popják, G., 1961) over the alternative trivial name farnesenic acid, because: farnesol (generally accepted) → farnesal (generally accepted) → farnesoic acid; cf octanol → octanal → octanoic acid. Other abbreviations: homofarnesol = 11-homofarnesol, dihomofarnesol = 7,11-dihomofarnesol, homomevalonate = 3′-homomevalonate. *E* and *Z* are used rather than *trans* and *cis*, respectively.

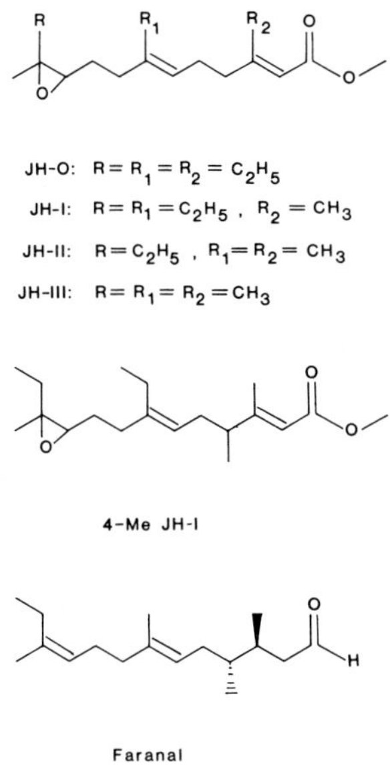

FIG. 1. Chemical structures of the juvenile hormones JH-0, JH-I, JH-II, JH-III, 4-methyl JH-I, and the trail pheromone, faranal.

the corpora allata (CA) secrete only JH acid, which is then methylated and stored in the accessory sex gland (Shirk, P. *et al.*, 1976; Weirich, G. and Culver, M., 1979; Peter, M. *et al.*, 1981).

The difficulty of *in vivo* biosynthetic experiments is apparent from considering the relative size of the CA: in adult female *M. sexta*, the CA are ~ 20 μg in a 5 g animal or 1/250,000 of the whole. If an injected precursor distributes uniformly and passively throughout the animal, obviously very little reaches the CA. Moreover, competing metabolic pathways which occur outside the CA may divert precursor to formation of unrelated labeled intermediates, greatly confusing isolation and analysis of JH. As an example, Metzler, M. *et al.* (1971) found that [5-^{3}H]mevalonate incorporated readily into [^{3}H]farnesol in whole *H. cecropia*, but not into JH-I or JH-II. In retrospect it should be pointed out that JH-I biosynthesized from the intermediate [1-^{3}H]7,11-dihomofarnesol expected from incorporation of [5-^{3}H]mevalonate would not retain any tritium, as oxidation at C-1 would remove all label, whereas farnesol would retain radiolabel from all three equivalents of mevalonate incorporated. Similarly, JH-II would be expected to retain radiolabel from one of two mevalonate moieties; however, JH-II is produced to a much lesser extent by *H. cecropia* than JH-I and furthermore, it was not clear if JH-II was analyzed for [5-^{3}H]mevalonate incorporation (Metzler, M. *et al.*, 1971). JH-III (if formed) would only be expected to retain radiolabel from two of three [5-^{3}H] mevalonate units (Fig. 2).

The incorporation of radioactivity from methionine observed *in vivo* is noteworthy: the extent of incorporation was a full 0.1% of the amount injected, while the specific activity of JH-I recovered was about 1000 times lower than that of the precursor. This methylation occurs in the accessory sex glands, which are larger than the CA, but are still a

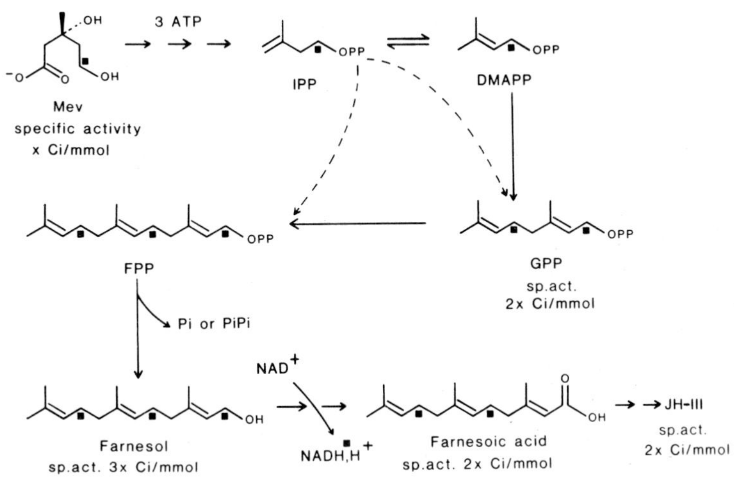

FIG. 2. Mode of incorporation of [5-^{3}H]Mev into JH-III showing the relative specific activities of expected intermediates. The location of each tritium on carbon atoms is denoted with a solid square.

small fraction of the animal. The importance of the discovery of methionine incorporation is inestimable for the many subsequent *in vitro* studies with CA for the purpose of JH identification, as well as studies on JH biosynthetic pathways and their regulation. It is fortunate for these purposes that methionine is involved in very few pathways of lipid biosynthesis, minimizing unrelated radiolabeled metabolites. The lack of incorporation of [*methyl*-^{14}C]methionine into the ethyl branches of the JH-I skeleton from these *in vivo* studies tended to diminish enthusiasm for an alkylation mechanism generating these groups, although biosynthetic data based on lack of incorporation are never conclusive.

Subsequent to the discovery that [^{14}C]propionate was incorporated efficiently into JH-II by *Manduca sexta* CA cultured *in vitro* (Schooley, D. *et al.*, 1973), Rodé-Gowal, H. *et al.* (1975) repeated the *in vivo* incorporation studies with [1-^{14}C]propionate in adult male *H. cecropia*, but with 20 times as much labeled substrate. Use of these larger quantities now allowed the detection of incorporation of [1-^{14}C] propionate into JH-I and JH-II amounting to about 0.001% of the injected dose. Each purified hormone was also hydrogenated to a derivative. In the intact hormones and derivatives radiolabel was chromatographically inseparable from the mass of substance, supporting authentic incorporation into the hormones. Simultaneously, farnesol was isolated and found to be devoid of radiocarbon (see Fig. 3 for pathway). In contrast, parallel incorporation studies with [1-^{14}C]butyrate resulted in isolation of both radiolabeled JH and farnesol. Since [2-^{14}C]acetate had been shown previously by Metzler, M. *et al.* (1971) to incorporate into JH-I in *H. cecropia*, the conclusion was reached by H. Rodé-Gowal *et al.* that [1-^{14}C]butyrate was degraded to the level of acetate or acetoacetate, which then incorporated into farnesol and JH. The incorporation of [1-^{14}C]propionate into JH-I and JH-II, but not into farnesol, was taken as good evidence for specific incorporation not involving the intermediacy of [^{14}C]acetate from propionate.

The latter studies (Rodé-Gowal, H. *et al.*, 1975) again reveal disadvantages of *in vivo* methodology. When [5-^{14}C]mevalonate was injected into fifth-instar *H. cecropia* larvae, which were allowed to metamorphose and then extracted as adults, no incorporation of radioactivity into JH could be detected. In contrast, incubation of [2-^{14}C]mevalonate with CA of *M. sexta in vitro* resulted in incorporation into JH-II and JH-III (Schooley, D. *et al.*, 1973). Quantitative double-label studies *in vitro* using labeled methionine as a "mass-marker" for total JH produced have shown in fact substantial permeability problems for homomevalonate in *M. sexta* CA (Schooley, D. *et al.*, 1976) and mevalonate in *Diploptera punctata* CA (Feyereisen, R. *et al.*, 1981a). These difficulties *in vitro* are presumably magnified with *in vivo* systems.

Rodé-Gowal, H. *et al.* (1975) also conducted studies on *in vivo* incorporation of precursors containing the preformed JH-I skeleton. They injected several JH-I precursors into adult male *H. cecropia*;

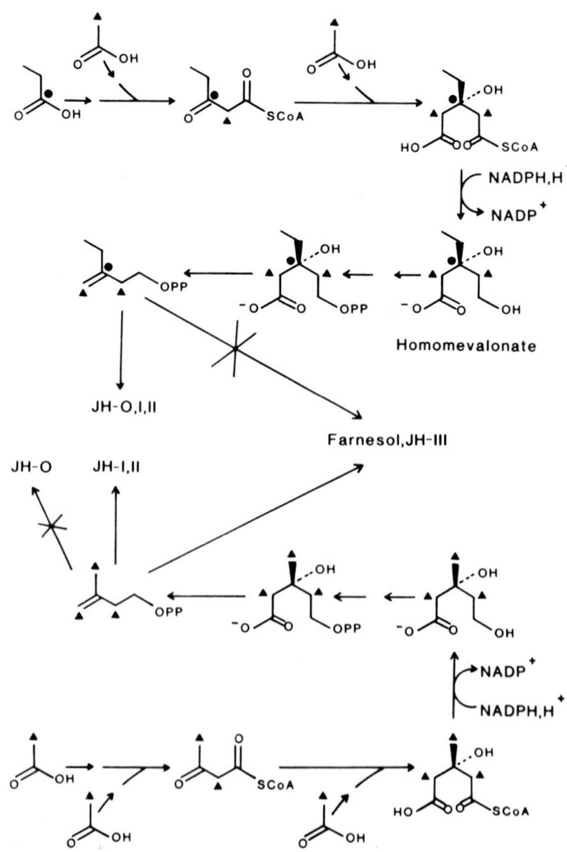

FIG. 3. Fate of [1-^{14}C]propionate and [2-^{14}C]acetate during JH biosynthesis; emphasizing the absence of incorporation of intact propionate (or homomevalonate) into farnesol or JH-III. Solid circles indicate position of label from [1-^{14}C]propionate; solid triangles mark position of label from [2-^{14}C]acetate.

all precursors were labeled with ^{3}H in the C-11 methyl group: methyl 7,11-dihomofarnesoate, 7,11-dihomofarnesol, and 7,11-dihomofarnesoic acid. None of these precursors showed appreciable incorporation into JH-I under conditions nominally identical to those where 10,11-epoxy-7,11-dihomofarnesoic acid incorporated efficiently. In retrospect this result may be less surprising, as we would now surmise that the unepoxidized acid and the dihomofarnesol would have to be converted to the corresponding epoxy acid by the idiosyncratic *H. cecropia* CA, and thence secreted for methylation in the accessory sex gland.

In contrast, numerous studies from Pratt, Tobe, and their co-workers have shown that CA of all insect species examined *in vitro* readily assimilate farnesol and farnesoic acid into JH-III, and dihomofarnesoic acid into JH-I. In fact, farnesoic acid generally stimulates JH-III biosynthesis *in vitro* (Pratt, G. and Tobe, S., 1974; see sections 4 and 5). Tobe, S. and Pratt, G. (1976) have attributed physiological significance to measurements of the ratio of the spontaneous rate of biosynthesis of JH by CA *in vitro* to that of farnesoic acid stimulated JH-III biosynthesis, a ratio that they quaintly (Judy, K., 1976) term "FEAR" (for fractional endocrine activity ratio).

In the context of the above results with dihomofarnesoic and farnesoic acids, we have explored alternative methods to study effects of precursors on JH titers in *M. sexta in vivo*. Baker, F. *et al.* (1981, unpublished) treated third-instar larvae of *M. sexta* topically with mevalonolactone in acetone, or alternatively, injected them with mevalonolactone or farnesoic acid in DMSO. In the succeeding instar the larvae were sacrificed for JH titer determination by the method of Bergot, B. *et al.*, (1981b). No detectable amount of JH-III was observed with farnesoic acid treatments as high as 200 μg or mevalonolactone as high as 500 μg; from which data it could be calculated that less than 1 part in 10^{7} of the exogenous farnesoic acid or mevalonolactone was utilized for conversion to JH-III. Inasmuch as farnesoic acid is readily assimilated and converted to JH-III by *M. sexta* larval CA *in vitro* (Kramer, S. and Staal, G., 1981), these data, together with those of Rodé-Gowal, H. *et al.* (1975) suggest that exogenous farnesoic acid, or its homolog, fail to reach the CA for unknown reasons.

2.2 *In vitro* studies

The first demonstration of *in vitro* JH biosynthesis was that of Röller, H. and Dahm, K. (1970), who incubated brain–corpora cardiaca–CA complexes from *Hyalophora cecropia* male pupae in Grace's medium together with sterile, heat-treated hemolymph (6% v/v) from pharate adult *H. cecropia*. After relatively long-term incubation (7 days), organic solvent extracts were found to possess biological activity (*Tenebrio* assay), and a purified extract was found to contain several micrograms of non-labeled JH-I as determined by GC/MS. After 7 days in culture the glands still retained physiological activity on reimplantation into *Galleria mellonella* larvae. A curious aspect of this study, recently pointed out by Law, J. (1983), applies also to a study by Ajami, A. (1974) reporting *in vitro* incorporation of [*methyl*-^{14}C]methionine into JH-I in cultured CA of *H. cecropia*: more recent investigations from Röller and Dahm's laboratory (Peter, M. *et al.*, 1981; Dahm, K. *et al.*, 1981) have shown that CA lack the requisite *O*-methyl transferase to make JH-I, and that they secrete only acids of JH-I and JH-II. The possibility of an artifactual chemical esterification in the early studies must be considered.

Judy, K. *et al.* (1973) adapted the *in vitro* procedure of Röller and Dahm, substituting 1% BSA for the treated hemolymph and adding [*methyl*-^{14}C]methionine to Grace's medium, prepared devoid of methionine. Long-term cultures of CA from adult female *M. sexta* produced organic solvent extractable radiolabeled materials. Separation by TLC and normal-phase LC gave two pure materials which contributed virtually all of the organosoluble radioactivity and bioactivity (*Galleria* wax test). Identification of these hormones by GC/MS and microchemical techniques showed one of them to be JH-II, and the other to be JH-III. (While JH-III was previously unknown in nature, Bowers, W. *et al.* (1965) had performed biological tests of synthetic methyl 10,11-epoxyfarnesoate (racemic JH-III) and predicted that *H. cecropia* JH would have a similar structure when it was identified!) The novel combination of organ culture in medium with labeled substrate, together with LC isolation of the resultant hormones, provided a simple system that was widely adapted and modified for studies on JH identification in other species, as well as for JH biosynthesis research.

Pratt, G. and Tobe, S. (1974) developed a short-term *in vitro* system for culturing CA of *Schistocerca gregaria*. An incubation time of 3 h was found to be quite adequate, rather than as many days, obviating the need for scrupulously sterile conditions (subsequently they used incubations as short as 10 min). These workers substituted TC-199 medium for Grace's and, perhaps more significantly, used vigorous aeration of the medium through agitation. This system has now been utilized for a variety of insect species, including *Manduca sexta* (Kramer, S. and Law, J., 1980). CA of the latter insect in fact produce JH more slowly than those from the hemimetabolous insects investigated by Pratt and Tobe and others (reviewed by Tobe, S. and Feyereisen, R., 1983; Feyereisen, R., this volume).

The unique advantages of the system described by Judy, K. *et al.* (1973) for JH biosynthesis research were quickly demonstrated. Substitution of [2-^{14}C]acetate, [2-^{14}C]mevalonate, or [1-^{14}C]propionate for labeled methionine revealed that all three precursors incorporated into JH-II, but only acetate and mevalonate incorporated into JH-III (Schooley, D. *et al.*, 1973). Partial degradations (see section 2.4) were still essential to provide the first experimental evidence that the ethyl side branch of JH-II has its genesis from propionate, presumably through the intermediacy of homomevalonate (Fig. 3).

Additional successful applications of the *in vitro* approach were quick to materialize. Pratt and Tobe and their co-workers (reviewed by Tobe, S. and Feyereisen, R., 1983; Feyereisen, R., this volume) studied many aspects of the regulation of JH-III biosynthesis in several species of Orthoptera and Dictyoptera. Jennings, R. *et al.* (1975a) demonstrated the incorporation of [5-^{3}H]homomevalonate into JH-II in cultured CA of *M. sexta*. However, additional studies revealed that the dilution of specific radioactivity on incorporation of homomevalonate into JH-II was 96:1 (Schooley, D. *et al.*, 1976), whereas with propionate dilution was only about 1.2:1, and dilution was unmeasurable with methionine as precursor. Such permeability barriers mandate biosynthesis studies with cell-free enzyme preparations.

Other characteristics of *in vitro* CA culture systems deserve mention in closing. A typical short-term culture utilizes four CA pairs in 100 μl medium. For *M. sexta*, such a culture contains from 4 μg of wet tissue (fifth instar) to 80 μg of wet tissue (adult female). On a dry weight basis, the corresponding weights are 0.8 μg and 16 μg, respectively (Kramer and Schooley, unpublished). If the 100 μl of culture medium contains a relatively modest precursor concentration (0.1 mM of homomevalonate, for example), the weight of precursor in solution (1.6 μg) is quite massive compared to the total dry weight of the tissue! Accordingly, endogenous pools of precursor are subject to overwhelming dilution if there are not permeability barriers to the added test substrate. Another aspect realized only recently is that CA extirpated from *M. sexta* larvae when the JH specific esterase titer is high give aberrantly low results when assayed for JH biosynthesis by [*methyl*-^{14}C]methionine incorporation, as the product of the glands is hydrolyzed in the assay system (Kramer, S. and Kalish, F., 1984).

2.3 Studies with cell-free homogenates of corpora allata

Reibstein, D. and Law, J. (1973) discovered that *M. sexta* CA homogenates supplemented with [*methyl*-^{3}H]-*S*-adenosyl-L-methionine would synthesize JH labeled in the ester moiety; the first successful enzyme-level JH biosynthesis study. Other applications of cell-free enzyme preparations are adequately discussed in subsequent sections reviewing specific enzymatic steps. However, certain general methodological difficulties of such studies deserve emphasis. In particular, the minute amounts of tissue available (see above) render such enzymatic studies extremely difficult. Accordingly, assays must monitor conversion of high specific radioactivity substrate-to-product transformations, rather than relying on simple spectrophotometric methods. With such very limited quantities, enzymes from CA have been purified only crudely by ultracentrifugal fractionation into subcellular components. The only report of a higher degree of purification of CA enzymes is the use of a sucrose density gradient centrifugation to analyze the *O*-methyl transferase and epoxidase enzymes from *Locusta* CA (Feyereisen, R. *et al.*, 1981b). Finally, of necessity, the rates of substrate conversion are expressed as pmol- or nmol/gland pair/h, rather than per mg of protein, because of difficulties in determining routinely the minuscule amounts of protein in CA homogenates.

2.4 Analytical and degradative techniques

A requisite of any definitive study on JH biosynthesis is an isolation procedure which will provide the juvenile hormone(s) in a pure state. Additionally, in studies of identity of intermediates in the biosynthetic pathway, microchemical or spectrometric methods are required which will provide reliable data on the localization of those atoms in the product(s) which are derived from the known position of labeled atom in the precursor molecule.

High-pressure, high-resolution liquid chromatography (LC) has been used for final separation of the individual JHs in all but the earliest biosynthetic studies. Use of gas–liquid chromatography for such micro-preparative isolations is hampered by lower recoveries and the poor thermostability of JH. Isolation of JH from *in vivo* sources requires extensive prepurifications, whereas with *in vitro* studies prepurification is minimal; extraction of culture medium followed by TLC analysis is usually sufficient. However, recent indications suggest that the iso-octane partitioning procedure of Hammock, B. and Sparks, T. (1977), as modified by Feyereisen, R. and Tobe, S. (1981), alone is adequate. Methods for isolation of JH from tissue have been reviewed extensively (Schooley, D., 1977). A recent advance which deserves mention is a simple procedure for separating JH from lipid extracts of tissues. In most adsorption chromatography systems, triglycerides and JH are separable only with difficulty. However, lipids have poor solubility in acetonitrile, so that several acetonitrile extractions of a neat, crude lipid extract dissolves JH leaving much lipid behind. Passage of this acetonitrile extract through disposable cartridges containing a hydrophobic, reversed-phase type support (e.g., Waters Associates C_{18} Sep-Pak) quantitatively removes the remaining glycerides from JH (Bergot, B. *et al.*, 1981b). The selectivity of this process is similar to the efficient but extremely time-consuming molecular distillation procedure used for delipidation by Meyer, A. *et al.* (1968).

Biosynthetic studies on formation of intermediates in the JH biosynthetic pathway require substantially different isolation conditions from those used for JH. Because the analytical requirements are usually specific for the particular intermediate and/or enzyme system under study, it is not of general interest to review these procedures here. As most intermediates in the JH biosynthetic pathway are polar and/or unstable, reversed-phase LC has become the usual method of choice for their analysis.

Numerous types of microchemical degradative procedures have been used to localize those atoms which become labeled in biochemical conversion of a specifically labeled precursor to JH or an intermediate from this precursor. Metzler, M. *et al.* (1971) rigorously proved that [*methyl*-^{3}H] methionine incorporated solely into the methyl of the ester moiety as shown in Fig. 4. Hydrogenation of JH-I and JH-II derived from whole animals injected with radiolabeled methionine yielded derivatives with different chromatographic behavior, but still containing radioactivity, thus providing evidence that the radioactivity was not incorporated into an inseparable, but unrelated, metabolite possibly generated in the complex *in vivo* system. Subsequently they reduced this derivative with lithium aluminum hydride; the resultant

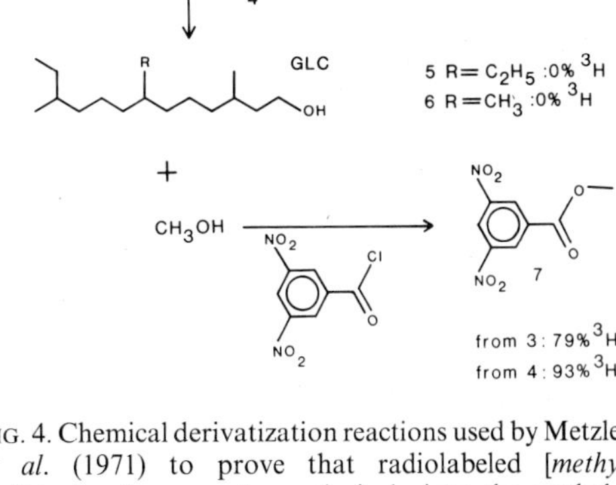

FIG. 4. Chemical derivatization reactions used by Metzler, M. *et al.* (1971) to prove that radiolabeled [*methyl*-^{3}H] methionine incorporates exclusively into the methyl ester moiety of JH-I and JH-II in *Hyalophora cecropia in vivo.*

C_{17}-alcohol from the chain was devoid of label. The methanol fragment (isolated as its 3,5-dinitrobenzoate, *7* in Fig. 4) contained all of the radioactivity. Subsequent workers have resorted to base-catalyzed transesterification of JH biosynthesized *in vitro* from labeled methionine to prove that all methionine-derived radioactivity is in the methyl of the ester moiety (e.g., Judy, K. *et al.*, 1973; Reibstein, D. and Law, J., 1973). Loss of all radioactivity from the methyl ester on transesterification provides a simple reconfirmation of the results of Metzler, M. *et al.* (1971).

For studies on the biosynthesis of the carbon skeletons of JH-II and JH-III, Schooley, D. *et al.* (1973) developed a procedure for oxidative degradation of the two hormones at the unsaturated bonds and the epoxy ring. As shown in Fig. 5, this procedure affords three fragments of each structure (*12a,b, 13a,* and *14*), each isolated as a crystalline derivative whose specific radioactivity can be readily determined. Carbon atoms C-1 and C-2 are lost as an oxidatively unstable oxalic acid half-ester *11*. This procedure allowed the determination of the number of labeled atoms in each fragment as compared to the parent JH, but did not include the far more laborious task of degradation of the three fragments to single carbon atom derivatives to prove *specific* incorporation. Proof of *specific* incorporation of three precursors for the two different hormones would have required a truly monumental effort. However, comparison of data from degradation fragments of JH-II and JH-III biosynthesized from [2-^{14}C]mevalonate, [2-^{14}C]acetate, and [1-^{14}C]propionate allowed determination of the specific radioactivity *ratio* of *each fragment as compared to the parent JH*. This body of data demonstrated the probable biosynthesis of JH-II from two equivalents of mevalonate, with the non-mevalonate-derived portion arising from one propionate and two acetates (presumably *via* homomevalonate). JH-III was biosynthesized from three equivalents of mevalonate (or nine acetates) as expected, but did not incorporate significant radiolabel from [1-^{14}C]propionate. Subsequently, Peter, M. and Dahm, K. (1975) used a modification of this degradation scheme in a study on specific incorporation of [1-^{14}C]propionate into JH-I in *Hyalophora cecropia in vivo*. They reacted the two ketonic fragments from JH-I (*8b, 13b* in Fig. 5) with

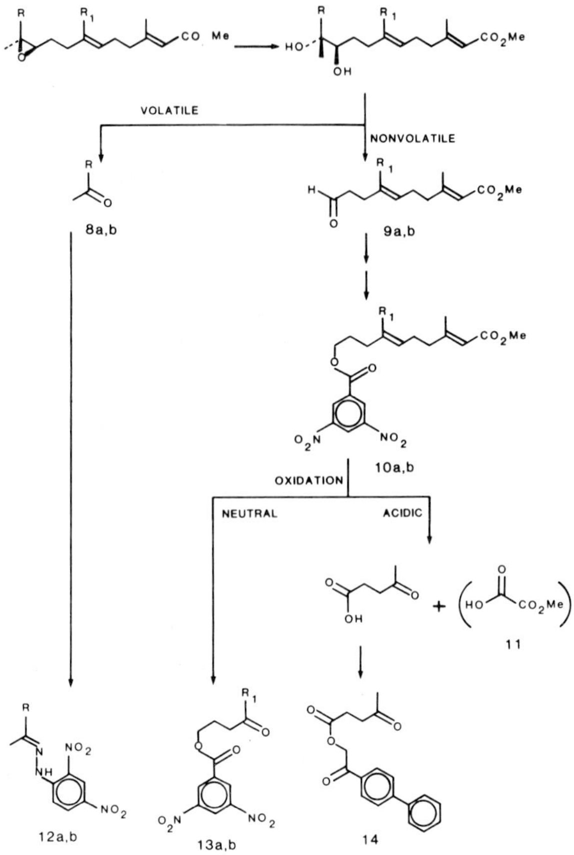

FIG. 5. Degradation scheme used to localize radiolabeled atoms in JH-III, JH-II, or JH-I biosynthesized from [2-^{14}C]mevalonate, [2-^{14}C]acetate and [1-^{14}C]- and [2-^{14}C]propionate. Certain derivatized JH fragments are identified with a number followed by the letter *a* or *b*; *a* signifies R or R′ = Me; *b* is for R or R′ = Et.

a phenyl Grignard reagent. The substituted, radiolabeled benzyl alcohol derivatives so obtained were oxidatively cleaved to benzoic acid, whose carboxyl carbon corresponded to C-11 or C-7 of JH-I. All radioactivity from [1-^{14}C]propionate was in fact contained at these two positions, rigorously proving the specificity of incorporation of this precursor.

Another modification of the procedure of Schooley, D. *et al.* (1973) was used by Jennings, R. *et al.* (1975a) to study incorporation of [5-^{3}H] homomevalonate into JH-II in *Manduca sexta* CA maintained *in vitro*. After cleavage of the 10,11-diol, all tritium remained in the fragment *9a* (Fig. 5) containing C-1 through C-10. Oxidation of *9a* to a C-10 carboxylate followed by methylation removed the remaining hydrogen on C-10. All tritium remained in this diester, allowing Jennings, R. *et al.* (1975a) to conclude that the tritium could only be at C-9, since hydrogen atoms on all other carbons are known to be mevalonate-derived; thus proving the specificity of incorporation of [5-^{3}H]homomevalonate.

Those molecules, like the JHs, which are inherently asymmetric and can exist in two optical antipodes, are usually racemic when chemically synthesized but are nearly always a single antipode when biosynthesized. Judy, K. *et al.* (1973) developed a novel technique for determining the absolute configuration of JH-III in the course of its identification, which is also useful for analysis of enantiomeric composition of JH. Thus, conversion of JH-III to its diol and then to the diol monoester with (+)-α-methoxy-α-trifluoromethylphenylacetic (MTP) acid gave derivatives from racemic JH-III which (as diastereomers) were separated on LC. Localization of radiolabel from a minute amount of carrier-diluted biosynthetic JH-III with the faster-eluting diastereomer allowed determination of its optical purity and absolute configuration. This method has been used widely to study the chirality of labeled JH from *in vitro* studies, either to prove that the material is not derived from contamination with synthetic JH, or to study the stereospecificity of the epoxidase, *O*-methyl transferase or esterase (e.g. Jennings, R. *et al.*, 1975a; Hammock, B. and Mumby, S., 1978; Peter, M. *et al.*, 1979, 1981; *inter alia*). A limitation of the method is a 4% epimerization at C-10 on hydrolysis of the epoxide. Schooley, D. *et al.* (1978a) attempted to improve the method by using a methanolysis of the epoxide and conversion of the resultant 11-methoxy-10-hydroxy derivatives to their MTP esters. This method gave no racemization at C-10 but the diastereomeric esters were resolvable only in the case of JH-III; Feyereisen, R. *et al.* (1981a) used this technique to study the chirality of JH-III produced by epoxidase enzyme preparations. Most recently, Hamnett, A. *et al.* (1981) have further refined this approach by making (−) menthoxyacetate esters of the JH methoxyhydrin derivatives; they used this method in a study of the substrate specificity of the *O*-methyl transferase enzyme. The latter diastereomeric esters separate dramatically and appear to be the derivative of choice for this analytical procedure.

3 FORMATION OF THE METHYL ESTER AND EPOXIDE MOIETIES

Although the *O*-methyl transferase enzyme catalyzes one of the last steps in JH biosynthesis, it was one of the first reactions to be characterized biochemically and proved instrumental in the detailed study of JH biosynthesis. Accordingly, we will review the methylation reaction before biogenesis of the JH carbon skeleton. Furthermore, because there appears to be a dichotomy in the order of the last two biosynthetic steps leading to JH (Fig. 6), depending on the insect species studied, it is appropriate to consider the 10,11-epoxidase enzyme in conjunction with the *O*-methyl transferase.

The methyl ester moiety common to all of the known JHs is rare in natural products as a whole. Fatty acid methyl esters have been reported from several sources but some of these can be explained as artifacts of the work-up procedure (Zatz, M. *et al.*, 1981, and references therein). Akamatsu, Y. and Law, J. (1970) demonstrated the methylation of fatty acids (e.g. oleic acid) by a soluble enzyme preparation from *Mycobacterium phlei* using *S*-adenosyl-[*methyl*-^{14}C]methionine as the methyl donor and were the first to intimate a similar alkylation reaction for generation of the JH-I methyl ester moiety (Fig. 7). *In vivo* JH biosynthesis studies by Metzler, M. *et al.* (1971) showed that when radiolabeled [*methyl*-^{14}C or ^{3}H]methionine was injected into adult male *Hyalophora cecropia*, the resulting JH-I and JH-II were radiolabeled only in

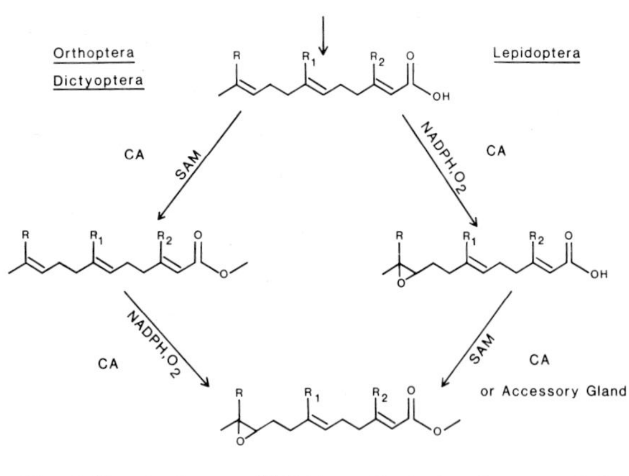

FIG. 6. Terminal steps of JH biosynthesis showing suspected differences in sequence of enzymatic steps according to insect order studied.

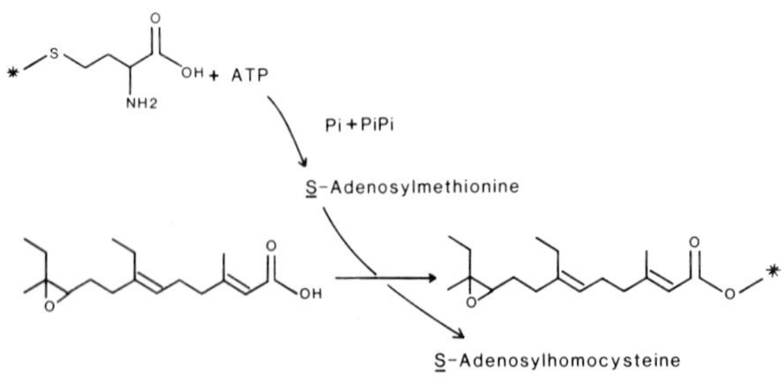

FIG. 7. *O*-Methyl transferase catalyzed formation of JH-I from the corresponding epoxy acid plus *S*-adenosyl methionine (SAM). The asterisk indicates radiolabel (^3H or ^{14}C) in the methyl group.

the methyl ester moiety (and not in the sesquiterpenoid backbone). In a subsequent study, Metzler, M. *et al.* (1972) demonstrated that radiolabeled JH-I acid could be methylated to JH-I by male *H. cecropia in vivo*. Later, Judy, K. *et al.* (1973) found that *Manduca sexta* CC–CA maintained *in vitro* also were able to incorporate [*methyl*-^{14}C]methionine solely into the ester moiety of JH-II, with no incorporation into the homosesquiterpenoid skeleton.

Judy, K. *et al.* (1973) found that homogenized CA from *M. sexta* lost the ability to incorporate [*methyl*-^{14}C]methionine into JH. Reibstein, D. and Law, J. (1973) concluded that this was probably due to the lability of the methionine activating enzyme, as they demonstrated incorporation of radioactivity from *S*-adenosyl-[*methyl*-^3H]methionine (SAM) into JHs using a cell-free preparation from adult female CC–CA of the same species. Conversely, SAM was not incorporated into JHs by intact CC–CA, presumably due to permeability barriers (Reibstein, D. and Law, J., 1973). JH biosynthesis by CC–CA homogenates plus SAM could be greatly stimulated by inclusion of JH acids in the incubation (Reibstein, D. *et al.*, 1976); furthermore, methyl farnesoate was formed from farnesoic acid in the absence of NADPH (co-factor for the 10,11-epoxidase) suggesting a lack of specificity of the *O*-methyl transferase (see section 5). However, in the presence of NADPH , JH-III was produced from farnesoic acid with little methyl farnesoate being detected. Since Reibstein, D. *et al.* (1976) were unable to observe conversion of methyl farnesoate to JH-III by *M. sexta* homogenates in the presence of NADPH, they concluded that the JH acid ("epoxy acid") is the natural (endogenous) substrate for the *O*-methylation step in this animal. These studies showed that both JH-III and JH-I acids were converted to their corresponding esters in the presence of SAM. The rate of JH-III production by gland homogenates was nearly saturated at 50 μM SAM, but even at 1 mM the epoxyfarnesoic acid substrate (perhaps not in true solution at this concentration?) was not yet saturating. The rate of the methylation reaction increased ~3-fold on increasing the pH from 5 to 7.

In contrast, in species of Orthoptera (e.g. *Locusta, Schistocerca*) and Dictyoptera (e.g. *Periplaneta, Blaberus*) there is evidence which suggests that the last two enzymatic steps in JH biosynthesis are in the opposite order (see e.g. Pratt, G. and Tobe, S., 1974; Tobe, S. and Pratt, G., 1974). These workers found, for example, that locust CA maintained *in vitro* with [*methyl*-^{14}C]methionine and [12-^3H]farnesoic acid accumulate substantial amounts of labeled methyl farnesoate within the gland, whereas epoxyfarnesoic acid does not accumulate. Furthermore, the intact glands utilize labeled epoxyfarnesoic acid at a lower rate than farnesoic acid. Certainly, taking into account that microsomal preparations from CA of *Blaberus giganteus* (Hammock, B., 1975) and *Locusta migratoria* (Feyereisen, R. *et al.*, 1981b) are able to convert methyl farnesoate to JH-III, the evidence is very strong for epoxidation preceding methyl transfer in these orders.

At this point it is worth elaborating on the unusual JH biosynthetic events in adult *H. cecropia*. CA of males of this species generate and secrete JH acid but lack the *O*-methyl transferase enzyme (Peter, M. *et al.*, 1981); instead the JH acids generated in the CA are secreted into the hemolymph and finally methylated in the accessory sex glands (Fig. 7, Weirich, G. and Culver, M., 1979; Peter, M. *et al.*, 1981; Dahm, K. *et al.*, 1981). The *O*-methyl transferase from the accessory glands is soluble and can be found in the gland epithelium as well as in the glandular secretion, but not in any other part of the genital tract of the unmated male. Adult female *H. cecropia* CA do contain methyl transferase activity (Peter, M. *et al.*, 1981). Accessory sex glands from male *Antheraea pernyi* and *M. sexta* (species which do not sequester high levels of JH) were found to lack methyl transferase activity (Weirich, G. and Culver, M., 1979).

Based on the levels of *in vitro* JH biosynthesis by CA from several orthopteran and dictyopteran species, and the degree of stimulation of biosynthesis by exogenous farnesoic acid, one would suspect high levels of *O*-methyl transferase and 10,11-epoxidase enzymes in these animals. Hammock, B. (1975) studied inhibition of the NADPH and O_2-dependent microsomal epoxidase from *B. giganteus*. The epoxidation was inhibited by carbon monoxide, methylene blue, SKF 525A, and piperonyl butoxide, substances known to inhibit cytochrome P-450-linked enzymes. No formation of the 6,7-10,11-diepoxide was observed. Pratt and co-workers have conducted extensive studies with cell-free preparations of *Locusta migratoria* CA and have found the *O*-methyl transferase to be a soluble enzyme which uses SAM as co-factor (Feyereisen, R. *et al.*, 1981b). In the same study, the 10,11-epoxidase was found to be an exclusively microsomal enzyme which used NADPH as co-factor (K_m values for substrates methyl farnesoate and NADPH were ~ 8μM and 3 μM respectively). The epoxidase banded along with NADPH–cytochrome-*c* reductase on a sucrose density gradient and was inhibited by oxidized cytochrome-*c*, $NADP^+$, and carbon monoxide; from this and reversal of CO inhibition by white light the authors concluded that the epoxidase is probably a microsomal cytochrome P-450 linked enzyme. The properties of the *Locusta* epoxidase with respect to co-factors and inhibitors are similar to those of the *Blaberus* enzyme (Hammock, B., 1975).

4 GENESIS OF THE CARBON SKELETON OF THE FIVE KNOWN JHs

The carbon skeleton of JH-III can be envisaged as arising from the acyclic sesquiterpenoid farnesyl pyrophosphate by a series of simple enzymatic modifications, namely hydrolysis of the pyrophosphate ester, oxidation of the resultant alcohol, and epoxidation of the 10,11 double bond (Fig. 8). The biosynthesis of FPP has been extensively investigated in a variety of organisms, as it is the precursor of sesquiterpenes (see Cordell, G., 1976, for review), steroids, and triterpenes. However, the ethyl branches found in JH-II, JH-I, JH-0, and 4-methyl JH-I (iso JH-0), as well as the 4-methyl group in the latter hormone are extraordinarily uncommon and their origin is of particular interest from a biosynthetic standpoint. Considerable efforts were devoted initially to determining the origin of the ethyl side branches of JH-I and JH-II. As these earlier studies were concerned with biosynthesis of intact JHs from simple precursors (e.g., propionate) *in vivo* or *in vitro*, we review these

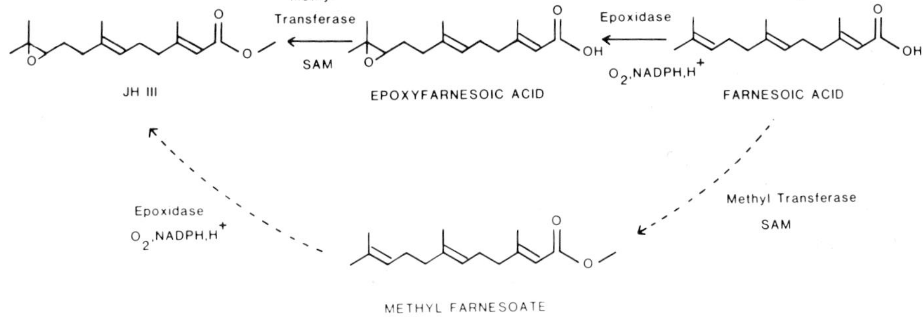

FIG. 8. Expected route for JH-III biosynthesis from farnesyl pyrophosphate. There are two pathways possible for conversion of farnesoic acid to JH-III (see Fig. 6).

results prior to subsequent sections detailing separately the various enzymatic steps.

As mentioned in section 1, two main explanations for generation of the ethyl side branches could be postulated, i.e. methylation of "modified" methyl groups by a C-donor (e.g. SAM) or condensation of homoisoprenoid units presumably formed from propionyl-CoA plus two acetyl-CoA units *via* the unknown (in nature) intermediate, homomevalonate (Fig. 9). The results from *in vitro* studies with intact CC–CAs of *Manduca sexta* (Schooley, D. *et al.*, 1973) provided strong evidence for the latter theory. The degradative procedures used in these studies (outlined in section 2) allowed the localization of labeled atoms from precursors into one of several fragments of the JH molecule (Fig. 5), but not the precise assignment of particular labeled atoms. However, a synthesis of all degradative data from three precursors (acetate, propionate, and mevalonate) incorporating into JH-II and JH-III provides very convincing support for the homomevalonate hypothesis. Subsequently, Jennings, R. *et al.* (1975b) degraded radiolabeled JH-I and JH-II biosynthesized from [1-^{14}C]propionate by CA of *Heliothis virescens* maintained *in vitro*; they found incorporation of two equivalents of propionate into JH-I, and one equivalent into JH-II; those fragments of the molecule found to contain the label had been expected to do so in accord with the homomevalonate hypothesis. Peter, M. and Dahm, K. (1975) performed degradation studies on JH-I biosynthesized *in vivo* by *Hyalophora cecropia* from [1-^{14}C]propionate; they were able to demonstrate specific incorporation of the C-1 of propionate only into the C-7 and C-11 atoms of JH-I. Also in this year, Jennings, R. *et al.* (1975a) reported degradation studies on JH-II biosynthesized by CA of *M. sexta in vitro* from [5-^{3}H]homomevalonate and [*methyl*-^{14}C]methionine. The tritium at C-5 of homomevalonate was localized at C-9 of JH-II as anticipated from analogy with the homomevalonate/mevalonate pathways. Based on this biosynthetic pathway three homomevalonate units would be expected to form the carbon skeleton of JH-0 (Fig. 9).

Following these early studies the trend has been toward experiments with homogenates of CAs in an effort to overcome difficulties associated with what appear to be permeability barriers resulting in poor incorporation of some precursors, or rather, substantial dilution of specific radioactivity on incorporation. The cell-free approach is also necessary to study individual enzymatic steps, and particularly to identify the reputed homologous intermediates.

4.1 Biosynthesis of HMG-CoA and HEG-CoA

Based on known biosynthetic pathways one would anticipate that homomevalonate should arise by reduction of 3-hydroxy-3-ethylglutaryl-CoA (HEG-CoA, a homolog of HMG-CoA), generated in turn from two units of acetyl-CoA plus one

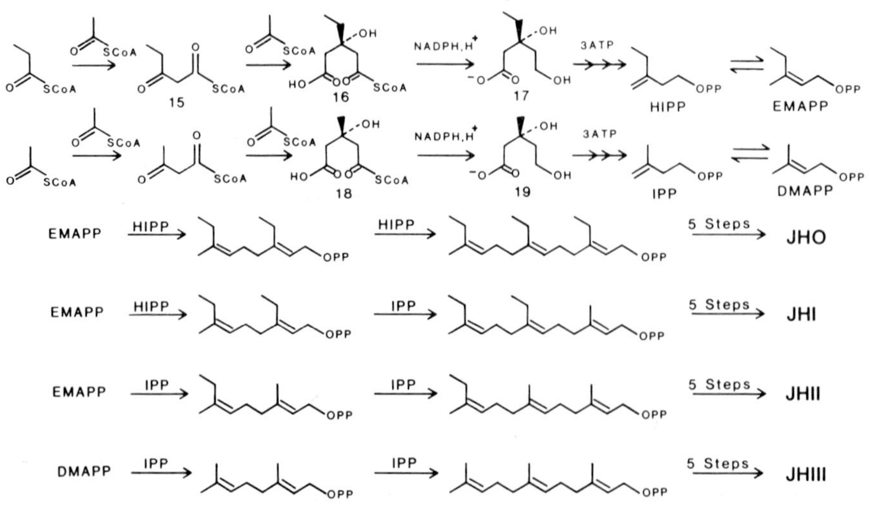

FIG. 9. Overall scheme for biosynthesis of JH-0, JH-I, JH-II and JH-III from acetyl-and propionyl-CoA.

propionyl-CoA unit *via* 3-ketovaleryl-CoA (homologous with acetoacetyl-CoA, Fig. 9). Studies with cytosolic preparations of *M. sexta* CC–CAs incubated with [1-^{14}C]acetyl-CoA plus unlabeled propionyl-CoA, followed by hydrolysis and derivatization of the products, generated a mixture of the radiolabeled diacids HMG and HEG (identified as their diethyl esters by GC/MS). Under similar conditions but with [1-^{14}C]propionyl-CoA plus unlabeled acetyl-CoA as precursors, only the radiolabeled HEG diethyl ester could be detected (Baker, F. and Schooley, D., 1978; Schooley, D. *et al.*, 1978b). Further research verified that the natural products were indeed thioesters, HMG-CoA (*18*) and HEG-CoA (*16*), and that the absolute configuration of both compounds was 3*S* as anticipated (Bergot, B. *et al.*, 1979). Limited attempts to isolate 3-ketovaleryl-CoA (*15*, precursor of HEG-CoA) were unsuccessful (E. Lee, unpublished data).

4.2 Formation of mevalonate and homomevalonate

There has been considerable interest in the biosynthesis and metabolism of mevalonate (Mev) by insect tissue because of its key role in biosynthesis of steroids and other isoprenoids, and thus its important link with the observed absolute requirement for dietary sterols by all insect species investigated. Goodfellow's group conducted a detailed investigation of Mev biosynthesis in the fly *Sarcophaga bullata*. All tissues studied were able to synthesize Mev; however, the ring gland (equivalent to the CA) was not investigated (Goodfellow, R. *et al.*, 1973). In the presence of NADPH, cell-free preparations from *M. sexta* CA convert [1-^{14}C]acetyl-CoA plus non-labeled propionyl-CoA into both labeled (3*R*)-Mev (*19*) and low levels of chiral homomevalonate (HMev, Lee, E. *et al.*, 1978). With [1-^{14}C]propionyl-CoA plus unlabeled acetyl-CoA, homomevalonate (HMev) but not Mev was radiolabeled. Later studies showed that the absolute configuration of HMev (*17*), like Mev, was 3*R* (Bergot, B. *et al.*, 1979) although initial indications based on LC elution order of diastereomeric derivatives had raised the possibility that HMev could have the opposite 3*S* stereochemistry (Lee, E. *et al.*, 1978). Much higher rates of formation of Mev and HMev were observed when the immediately preceding substrates [*methyl*-^{3}H]HMG-CoA or [3-^{14}C]HEG-CoA were utilized rather than acetyl- and propionyl-CoA (Baker, F. and Schooley, D., 1981). Presumably this is because only one enzyme (HMG-/HEG-CoA reductase) in the pathway must act instead of a sequence of three enzymes (namely thiolase(s), synthase(s), and reductase(s)) acting in concert when acetyl- and propionyl-CoA are incubated under similar conditions. The HMG/HEG-CoA reductase was confined to the microsomal fraction and NADPH appeared to be the preferred electron donor (Baker, F. and Schooley, D., 1981). Similar levels of HMG-CoA reductase activity ($\sim$ 20 pmol/CC–CA pr/h) were demonstrated by Kramer, S. and Law, J. (1980) for animals of the same age; the activity was shown to vary according to the age/stage of animals used. Much higher levels of HMG-CoA reductase activity were found in homogenates of CA from *Diploptera puncata*, in keeping with the observed high rates of *in vitro* JH-III biosynthesis by this animal (Feyereisen, R. *et al.*, 1981a) and very high *in vivo* titers (Tobe, S. *et al.*, in preparation). The *Diploptera* enzyme had a K_m of 2.7 μM for (3*S*)-HMG-CoA as measured in CA homogenates (Feyereisen, R. *et al.*, 1981a).

It is well known that the enzyme HMG-CoA reductase plays an important role in mammalian steroid biosynthesis and is subject to multivalent feedback control by its immediate or distant products (see e.g., Brown, M. and Goldstein, J., 1980). This inactivation process appears to be *via* a phosphorylation of the HMG-CoA reductase enzyme, a process which can be reversed by a specific phosphoprotein phosphatase (for a review see Gibson, D. and Ingebritsen, T., 1978). Studies with cell-free preparations from *Manduca sexta* and *Diploptera punctata* CA suggest that insect HMG-CoA reductase is under similar phosphorylation/dephosphorylation control (Feyereisen, R. *et al.*, 1981a; Monger, D. and Law, J., 1982).

4.3 Generation of the isoprenol/homoisoprenol pyrophosphates

The series of reactions encompassing mevalonate kinase, 5-phosphomevalonate kinase and finally, ATP dependent-mevalonate 5-pyrophosphate decarboxylase have not been studied in depth in insects in general and rarely in CA preparations.

Several metabolism studies have been conducted on whole insects treated with labeled Mev. Kaplanis, J. *et al.* (1961), and Goodfellow, R. *et al.* (1974), found that flies (*Musca domestica* and *Sarcophaga bullata*, respectively) metabolize Mev to complex mixtures of lipids, and Nes, W. *et al.* (1982) found that the aphid *Schizaphis graminum* converts Mev into long-chain fatty alcohols. Incorporation of Mev into acetate-derived lipids appears to be quantitatively important in these insects; this metabolic pathway was unknown until Edmond, J. and Popják, G. (1974) demonstrated that in rat tissue Mev can be diverted through a shunt to 3-methylglutaconyl-CoA, 3-hydroxy-3-methylglutaryl-CoA, free acetoacetate and acetyl-CoA, and thence lipids. The proportion of Mev diverted into non-isoprenoid pathways in rats is strongly dependent on the tissue studied, being a minor pathway in liver but significant in kidney (Righetti, M. *et al.*, 1976).

Mevalonate kinase has been studied in fly tissue; it was apparently absent from fat body and gut of 6-day-old third-instar *Sarcophaga* larvae but was distributed in the muscle and brain complex cytosolic fractions (Barnes, F. and Goodfellow, R., 1971). The enzyme activity (specific activity and enzyme units per organism) was shown to fluctuate during development. Extensive purification of the kinase from axenic third-instar larvae (Goodfellow, R. and Barnes, F., 1971) established that the enzyme has a MW of ~113,000 with K_m for Mev and ATP of 0.62 mM and 4.7 mM respectively; the enzyme was activated by either Mn^{2+} or Mg^{2+}. As expected, only one isomer (presumably 3*R*) of a (3*RS*)-Mev mixture was phosphorylated. (Since radiolabeled Mev was converted to C_5, C_{10}, C_{15}, and C_{20} isoprenoid products by *Sarcophaga in vivo* (Goodfellow, R. *et al.*, 1972) the enzymatic machinery must be available for overall sesquiterpenoid biosynthesis in this species.)

Incubation of *M. sexta* CC–CA cell-free preparations (supernatant following 10,000 ***g*** for 10 min) with either [2-^{14}C]Mev or [5-^{3}H]HMev plus ATP resulted in production of radiolabeled polar products tentatively identified as mevalonate phosphate and homomevalonate phosphate plus pyrophosphate (E. Lee, unpublished results). Furthermore, following hydrolysis and derivatization (with diphenylcarbamoyl chloride) homoisopentenyl diphenylcarbamate (DC) could be demonstrated to be a product from [5-^{3}H]HMev metabolism (Baker, F. *et al.*, 1981) while radiolabeled isopentenyl and dimethylallyl DC were also formed from [2-^{14}C]Mev. Crude homogenates of adult female *Schistocerca nitens* CA plus ATP were also able to convert [2-^{14}C]Mev to isopentenyl-, dimethylallyl-, and farnesyl intermediates (Baker *et al.*, unpublished results). These results all support the anticipated conversion of Mev and HMev to isoprenoid products. Analysis of the isoprenoid products indicated that some were present as free alcohols while others were polar materials which migrated little on silica gel TLC. Incubation of the polar radioactive metabolites with alkaline phosphatase liberated organic-extractable materials; these were derivatized with diphenylcarbamoyl chloride to form diphenylcarbamates, which co-chromatographed with the diphenylcarbamates of authentic isoprenol markers. These data are strongly supportive of the existence of phosphate or pyrophosphate intermediates.

4.4 Isopentenyl/homoisopentenyl pyrophosphate isomerase

This system is of particular interest in light of what is known of the substrate specificity of the enzyme from other sources (see section 5 for pertinent discussion). The studies described in the prior section on metabolism of radiolabeled Mev by *M. sexta* and *S. nitens* CA enzymes had detected the radiolabeled dimethylallyl- and isopentenyl pyrophosphates and thus confirmed the presence of an active isomerase; however, results from incubations of *M. sexta* and *S. nitens* CA enzymes with [5-^{3}H]HMev provided no evidence for isomerization of the homoisopentenyl product to the suspected ethylmethylallyl intermediate required for chain initiation of the JHs with an ethyl substituent at C-11 (all JHs but JH-III). The absence of this intermediate was presumably due to the successively lower levels of products generated metabolically from HMev *via* three consecutive enzymatic steps.

Accordingly, the isomerase enzyme has also been studied with the immediate substrates, rather than relying on enzymatic conversion of Mev to IPP (or HMev to HIPP). [1-^{14}C]Isopentenyl pyrophosphate was converted readily to the dimethylallyl pyrophosphate product by *M. sexta* CC–CA

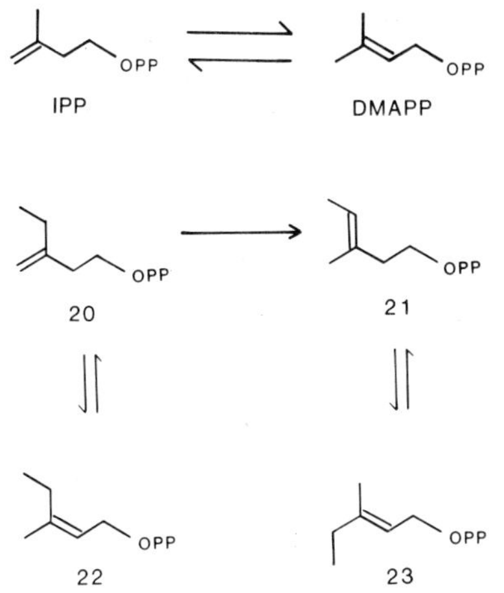

FIG. 10. Above: Conversion of isopentenyl pyrophosphate (IPP) to dimethylallyl pyrophosphate by IPP isomerase. Below: Reaction of HIPP (*20*) with IPP isomerase yields various products. The scheme depicted (from Koyama, T. *et al.*, 1973a) shows the reaction catalyzed by pig liver IPP isomerase, in which case the expected product *22* is minor, with *21* the major product at equilibrium. In contrast, IPP isomerase from *Manduca sexta* CA isomerizes *20* to *22* without formation of isomers *21*, *23*, or *24* (Fig. 11; Baker, F. *et al.*, 1981).

homogenates (Fig. 10) while [1-^{3}H]homoisopentenyl pyrophosphate (HIPP, *20*, Fig. 10) was converted to the expected product, ethylmethylallyl pyrophosphate (EMAPP, *22*, Baker, F. *et al.*, 1981). The rate of isomerization of HIPP was much lower than that of IPP, which may in part explain the inability to detect this product when the multi-step conversion from HMev was monitored.

4.5 Prenyl transferase

This important enzyme, which catalyzes formation of farnesyl pyrophosphate from one equivalent of DMAPP and two equivalents of IPP (Figs 2 and 9), has been studied in great detail from mammalian and plant sources (reviewed by Rilling, H., 1979; see Nishino, T. *et al.*, 1973; Popják, G. *et al.*, 1969). Prenyl transferase has been found to possess a low order of substrate specificity, particularly with respect to the allylic substrate (DMAPP or its analogs). Studies with insect tissue have been sparse; we have demonstrated low levels of farnesyl pyrophosphate production from IPP (plus or minus DMAPP) by *M. sexta* CC–CA homogenates; with *S. nitens* CC–CA homogenates levels of farnesyl products generated were much higher and low levels of the geranyl intermediate were also detected (Baker, F. *et al.*, 1981). Ogura's group have demonstrated farnesyl pyrophosphate synthetase in cell-free extracts from whole *Bombyx mori* (Koyama, T. *et al.*, 1981). In *Manduca* CC–CA enzyme preparations the level of prenyl transferase activity for FPP synthesis appears very low compared to that of the corresponding isomerase. Moreover, since the rate of homoisopentenyl pyrophosphate isomerization is very low compared to that of isopentenyl pyrophosphate, it is not surprising that the homologs of geranyl or farnesyl pyrophosphate have not been detected at all as products from [^{3}H]HIPP incubated with *Manduca sexta* CA preparations. The most plausible explanation for this low prenyl transferase activity is that optimum incubation conditions have not been realized, or that unstable enzyme preparations have been utilized. It is curious that the *S. nitens* enzyme is very active in (nominally) identically prepared homogenates. Because *M. sexta* adult female CA secrete JH-II and JH-III *in vitro* (Judy, K. *et al.*, 1973), and are able to metabolize [^{3}H]HMev into JH-II in intact CA (Jennings, R. *et al.*, 1975a), it is difficult to believe that *M. sexta* CA could not contain an active prenyl transferase capable of accepting homologous substrates.

Ogura's group was able to demonstrate the biogenesis of the di- and trihomofarnesyl pyrophosphate skeletons of JH-I and JH-0 using mammalian prenyl transferase and appropriate homologous isoprenoid substrates (see section 5). Accordingly, attempts to study biosynthesis of homo-, dihomo- or trihomofarnesyl pyrophosphate by insect CC–CA may hardly be expected to produce any novel discoveries. However, a particular anomaly in the JH biosynthetic pathway is the generation of the 4-methyl JH-I skeleton in *M. sexta* eggs. Based on the known substrate specificity of mammalian prenyl transferase (Koyama, T. *et al.*, 1980), the formation of the 4-methyl group would mandate the generation of an homoisopentenyl pyrophosphate isomer corresponding to *21* or *24* (Fig. 11); formation of isomers *21* and *24* could not be demonstrated in adult female *M. sexta* CC–CA (Baker, F. *et al.*, 1981), but has not been investigated in eggs. Concomitant formation of 4-methyl JH-I with JH-0 would suggest very lax prenyl transferase substrate specificity or mandate

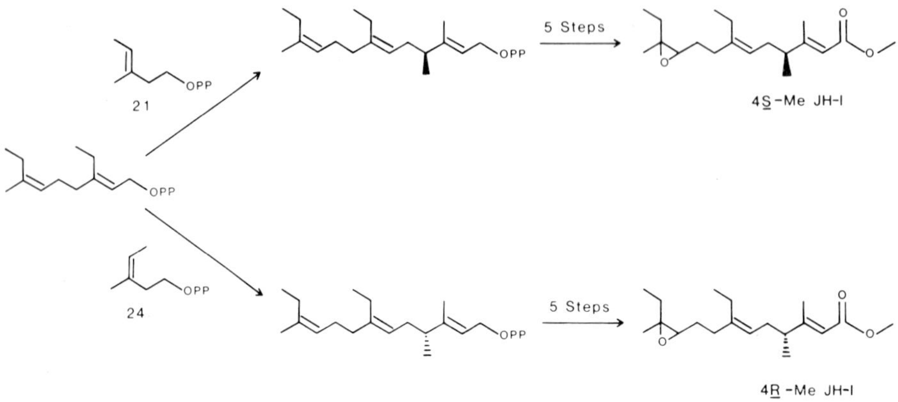

FIG. 11. Scheme showing the theoretical influence of geometry of homoisoprenoid precursors *21* and *24* on the anticipated stereochemistry of product 4-methyl JH-I.

the existence of two discrete prenyl transferase enzymes at this stage; either of these alternatives would be difficult to investigate with the insect enzyme systems presently available.

4.6 Conversion of the sesquiterpenoid pyrophosphates to the corresponding acids

It is known that in rat liver the first step in the conversion of allylic pyrophosphates to their acids is their irreversible dephosphorylation into free prenols *via* a phosphatase or pyrophosphatase enzyme (Christophe, J. and Popják, G., 1961). Reibstein, D. *et al.* (1976) showed the conversion of [^{3}H]farnesyl pyrophosphate by *M. sexta* CC–CA homogenates to farnesoic and epoxyfarnesoic acids, thus providing circumstantial evidence for the existence of a phosphatase (Fig. 8). In the previous section we discussed studies with CC–CA homogenates and labeled IPP and HIPP, during which we observed considerable generation of the free alcohols (Baker, F. *et al.*, 1981). We presumed this to be due to endogenous phosphatases, particularly since the effect was inhibited in phosphate buffer, a known inhibitor of phosphatases (Christophe, J. and Popják, G., 1961). The properties of the phosphatase have not been investigated; it can certainly be said that its existence greatly complicates studies with the pyrophosphate intermediates!

Prior literature evidence supports the existence of a farnesol/farnesal dehydrogenase enzyme from sources other than insect CAs. A mammalian liver cytosolic preparation could convert farnesol or farnesal to (2*E*,6*E*)-farnesoate; the reaction rate was enhanced in the presence of NAD^+ (Christophe, J. and Popják, G., 1961). Farnesol and farnesal oxidation (*via* "octanol dehydrogenase") has been demonstrated in several organs including the ring gland of *Drosophila melanogaster* (Madhavan, K. *et al.*, 1973). However, the latter workers assigned farnesol dehydrogenase and farnesal dehydrogenase a degradative function rather than a synthetic role in JH biosynthesis. Farnesol dehydrogenase activity has been indirectly demonstrated using *Diploptera punctata* CA *in vitro*, as farnesol can stimulate JH-III biosynthesis 2–3-fold (Feyereisen, R. *et al.*, 1981a). Farnesol stimulation of JH-III biosynthesis by intact *M. sexta* CA *in vitro* has also been observed (S. J. Kramer, unpublished results). We have conducted preliminary studies on farnesol and farnesal dehydrogenase in cell-free preparations of CA from *M. sexta* adult females (Baker, F. *et al.*, 1983) and have been able to show for the first time the generation of farnesal from (2*E*,6*E*)-farnesol; farnesoic acid was also produced and levels of the latter could be increased by addition of NAD^+ to the incubations (Fig. 8). Limited studies (Baker, F. *et al.*, 1983) have indicated that glands from wandering stage *M. sexta* larvae (at which stage the *in vivo* JH titer is known to be undetectable) still show considerable farnesol dehydrogenase activity. This may suggest either involvement in metabolism in addition to JH biosynthesis, or the lack of a regulatory function of the alcohol dehydrogenase on the rate of overall JH biosynthesis.

5 SUBSTRATE SPECIFICITY OF ENZYMES CATALYZING JH BIOSYNTHESIS

The co-occurrence of the homologous JHs in Lepidoptera is an enigma which has prompted numerous studies on the substrate specificity of several enzymes which are likely important in this pathway. All current evidence is consistent with biogenesis of the JH skeletons occurring *via* condensation of the appropriate isoprenoid and/or homoisoprenoid intermediates. In any species producing the higher JH homologs each enzyme in the biosynthetic pathway must be able to transform both the homologous and normal intermediates or, alternatively, separate enzymes must exist for these transformations. Attempts to unravel the differences between JH-III- *vs.* mixed JH-producing organisms, as well as to learn how the relative ratios of JHs are altered during the life span of mixed JH-producing insects (e.g. *M. sexta*), have been rather unsuccessful to date. (The situation might be less complicated if the ethyl branches of the higher homologs were generated *in situ* by methylation of pre-existing JH-III or one of its precursors!) We present here many examples of loose substrate specificity in enzymes of JH biosynthesis; it will become apparent to the reader that these are not restricted to mixed JH-producing organisms. In fact, some highly relevant data come from non-insect systems.

We have mentioned earlier that CC–CA homogenates from adult female *M. sexta* (a stage known to produce JH-II as well as JH-III) catalyzed the formation of both HMG- and HEG-CoA from acetyl-CoA plus propionyl-CoA (Baker, F. *et al.*, 1978; Schooley, D. *et al.*, 1978) and that the product HEG-CoA had the same 3*S* configuration as HMG-CoA (Bergot, B. *et al.*, 1979; Fig. 9). However, in other experiments it was shown that CA homogenates from *Tenebrio molitor* and *Schistocerca nitens* (species which contain only JH-III) could not only generate the higher homolog HEG, but moreover that the amount of the higher homolog HEG produced by *Tenebrio* (as measured by the ratio of HEG:HMG) was greater than that for *Manduca* (Baker, F. and Schooley, D., 1981). Furthermore, in these studies it was shown that a rat liver preparation was similarly able to generate both diacids HEG and HMG (presumably as their CoA esters), from acetyl-CoA plus propionyl-CoA. Although homoisoprenoids are not normally found in mammalian systems, previous studies of acetoacetyl-CoA thiolase in pig heart preparations (Stern, J., 1955) did not rule out generation of the intermediate β-ketovaleryl-CoA (*15*) from appropriate precursors, and β-ketovalerate accumulation has been shown in human patients suffering from disorders in which normal propionate metabolism is arrested (e.g. Sweetman, L. *et al.*, 1978). It is also interesting that HEG acid has been detected in human urine (Spiteller, M. and Spiteller, G., 1979).

Cell-free preparations from *Manduca sexta* adult female CC–CA were also able to generate Mev and HMev from HMG-CoA and HEG-CoA respectively, in the presence of NADPH (Baker, F. and Schooley, D., 1981). Although the individual substrates were reduced to a similar degree, a mixed incubation containing equimolar levels of each substrate resulted in higher ratios of Mev to HMev, possibly indicating a preference of the enzyme for HMG-CoA. However, *S. nitens* CA homogenates reduced HEG-CoA at a greater rate than HMG-CoA while no HMG-/HEG-CoA reductase activity could be demonstrated in *T. molitor* CA (Baker, F. and Schooley, D., 1981). In this study rat liver microsomal preparations were found to catalyze the reduction of HEG-CoA to HMev, although at a reduced rate as compared to the reduction of HMG-CoA to Mev.

Only scant information is available on the specificity of enzymes responsible for conversion of Mev to isopentenyl pyrophosphate in insect systems. [5-^{3}H]HMev was converted to homoisopentenyl pyrophosphate (HIPP, *20*) by *M. sexta* cell-free preparations and the HMev monosphosphate and pyrophosphate intermediates were tentatively identified (E. Lee, unpublished results). However, *Schistocerca nitens* CA homogenates were also able to convert HMev to HIPP (Baker, F. *et al.*, 1981). Beedle, A. *et al.* (1975) have shown that HMev was phosphorylated by a mammalian liver preparation but that the final phosphorylation/decarboxylation step to HIPP did not occur. We have also monitored the phosphorylation of (3*RS*)-[^{3}H]HMev by a preparation similar to that of Beedle, A. *et al.* (1975); following alkaline phosphatase hydrolysis of the products and subsequent derivatization with chiral reagents we found that only (3*R*)-HMev had been phosphorylated (F. C. Baker, unpublished results).

The series of reactions catalyzed by IPP isomerase and prenyl transferase are particularly interesting as the substrate specificity of those enzymes from plant and mammalian sources has been studied in some detail. Much of the information thus obtained with artificial substrates has relevance with respect to the processing of homoisoprenoid intermediates of insect JH biosynthesis. Koyama, T. *et al.* (1973a) found that pig liver IPP isomerase converted 3-ethylbut-3-enyl (homoisopentenyl) pyrophosphate (*20*) mainly into (3*E*)-3-methylpent-3-enyl pyrophosphate (*21*, Fig. 10). The product anticipated by analogy with isomerization of IPP is (2*Z*)-3-methylpent-2-enyl pyrophosphate (ethylmethylallyl pyrophosphate, EMAPP, *22*, Fig. 10), but it in fact was produced to a lesser extent. Investigation of the IPP isomerase in CC–CA homogenates from adult female *M. sexta* revealed that [1-^{3}H]HIPP was converted to the expected product, EMAPP, in agreement with its projected role in JH-II biosynthesis at this stage and therefore pinpointing an apparent difference between the enzyme specificity of the mammalian and insect systems (Baker, F. *et al.*, 1981). The mono- and triphosphate esters corresponding to HIPP were not utilized by the enzyme preparations. Rigorous identification of products indicated that no isomerization of HIPP or EMAPP to other possible isomers (e.g. *21, 23,* or *24*) had occurred, thus confirming the expected isomerization and highlighting an obvious difference from the enzymatic specificity of pig liver isopentenyl pyrophosphate isomerase incubated with HIPP (Koyama, T. *et al.*, 1973a). However, the biogenesis of 4-methyl JH-I (recently discovered in eggs of *M. sexta*; Bergot, B. *et al.*, 1981a) would mandate the intermediacy of either (3*E*)- or (3*Z*)-3-methylpent-3-enyl pyrophosphate (*21* or *24*), depending upon the absolute configuration at the 4-position of the 4-Me JH-I product (see below and Fig. 11).

No studies have been conducted with respect to substrate specificity of prenyl transferase from insects. In *M. sexta* such studies may be difficult considering the low enzyme activity obtained to date (Baker, F. *et al.*, 1981) although similar experimentation with species known to produce high levels of JH-III may be enlightening. It is known that prenyl transferase (farnesyl pyrophosphate synthetase) from mammalian and plant sources shows rather broad substrate specificity. Ogura's group were able to show biogenesis of the homo-, dihomo-, and trihomofarnesyl skeletons by incubating the appropriate pyrophosphate precursors with pig liver farnesyl pyrophosphate synthetase (Koyama, T. *et al.*, 1973b). Further studies allowed the asymmetric biosynthesis of intermediates used in the determination of the absolute configuration of faranal (Koyama, T. *et al.*, 1980; Kobayashi, M. *et al.*, 1980). Faranal, the trail pheromone of the Pharaoh ant, is the only previously known natural product with alkyl branching like that of 4-methyl JH-I; it has the 4*R* configuration (Fig. 1). The carbon skeleton with the proper 4*R* configuration can be obtained enzymatically *via* condensation of (3*Z*)-3-methylpent-3-enyl pyrophosphate (*24*) with homogeranyl pyrophosphate (Kobayashi, M. *et al.*, 1980).

Hydrolysis of dimethylallyl-, geranyl-, and (2*E*,6*E*)-farnesyl pyrophosphates to their corresponding alcohols has been demonstrated in a rat liver microsomal preparation (Christophe, J. and Popják, G., 1961); whether this was due to the effect of a single or several enzyme(s) was undetermined. We also found that IPP and HIPP, and their allylic pyrophosphate products resulting from isomerase treatment, were partially hydrolyzed to the corresponding alcohols by *M. sexta* CC–CA homogenates in experiments intended to focus on the isomerase enzyme (Baker, F. *et al.*, 1981). Rat liver dehydrogenase(s) can convert (C_5, C_{10}, C_{15}) prenols into the corresponding acids and appear to be specific for the 2E isomers. These prenols were also better substrates for rat liver alcohol dehydrogenase than was ethanol, but they could not act as substrates for yeast alcohol dehydrogenase. Preliminary results from our laboratory also suggest that the farnesol dehydrogenase(s) from *M. sexta* CC–CA utilize(s) (2*E*,6*E*)-farnesol and geraniol rather than the corresponding 2Z,6E isomers (Baker, F. *et al.*, 1983). Octanol dehydrogenase from the ring gland of *Drosophila melanogaster* oxidizes primary alcohols, including farnesol; *in situ* detection of enzymatic activity on agar gels, following electrophoresis of wholefly homogenates, showed that the alcohol (ethanol) dehydrogenase zone, which separated from octanol dehydrogenase, also possessed some farnesol dehydrogenase activity (Madhavan, K. *et al.*, 1973). Intact CC–CA from adult female *M. sexta* were also able to convert (4*R*)- or (4*S*)-3,4,11-trimethyl-7-ethyl-2,6,10-tridecatrien-1-ol *in vitro* to the corresponding JH (4-methyl JH-I) and thus further demonstrated lack of enzyme specificity throughout the terminal biosynthetic steps (F. C. Baker *et al.*, unpublished results).

The last two steps of JH biosynthesis have been shown to be particularly lacking in substrate specificity in many species studied. This was initially demonstrated when dihomofarnesoic acid incubated with intact CA from animals which normally generate only JH-III was converted to JH-I as the sole product (see e.g. Pratt, G. and Tobe, S., 1974; Pratt, G. *et al.*, 1978). Pratt's group have conducted extensive substrate specificity studies of the *O*-methyl transferase from *Locusta* CA; compared to the farnesoic acid control a large number of analogs showed moderate activity and in fact some, including 10,11-epoxyfarnesoic acid, were methylated at a greater rate (Pratt, G. *et al.*, 1981). Only the 2E isomers of analogs were effectively methylated while the 2Z isomers were not substrates (Pratt, G. *et al.*, 1981).

In the presence of SAM, adult *Manduca sexta* CA homogenates could methylate added JH-I acid to the corresponding ester (little or no JH-I is normally present at this stage of development) but were unable to convert several long-chain fatty acids (including 9,10-epoxyoctadecanoic acid) to their corresponding methyl esters (Reibstein, D. *et al.*, 1976). Surprisingly, the *O*-methyl transferase from adult male *Hyalophora cecropia* accessory glands possesses uncommonly high substrate specificity; the preferred substrates were JH-I acid > JH-II acid ≫ JH-III acid, which correlates with the predominance of JH-I at this stage (Peter, M. *et al.*, 1981). Furthermore, the latter enzyme system would only methylate the (10*R*,11*S*)-JH-I acid isomer of a racemic mixture (Peter, M. *et al.*, 1979). On the contrary, methylation of racemic epoxyfarnesoic acid by *Locusta migratoria* CA cell-free preparations affords racemic JH-III (Hamnett, A. *et al.*, 1981).

We are aware only of limited studies on the substrate specificity of the epoxidase. Hammock, B. and Mumby, S. (1978) obtained the surprising result that *Blaberus* CA homogenates were able to epoxidize (2*Z*,6*E*)-methyl farnesoate faster than the natural 2E isomer. Precocene appears to serve as an alternative substrate for the epoxidase (Hamnett, A. and Pratt, G., 1983; see also section 7); the resultant epoxide is extraordinarily reactive and either hydrolyzes or binds to nucleophiles in biopolymers. The dihydrodiols from precocene epoxide hydrolysis have the 3*R* configuration, suggesting that the epoxide has the same spatial configuration as JH-III produced by this enzyme (Hamnett, A. and Pratt, G., 1983).

Thus, overwhelming evidence suggests rather loose substrate specificity at most steps in sesquiterpenoid biosynthesis in enzyme preparations even from species which produce only JH-III or from diverse organisms which do not produce JH at all. While it is true that forcing conditions have often been employed to explore these reactions, and more detailed studies taking K_m values and related parameters into account might significantly alter the scenario, it seems more likely that one or two, as yet undetermined, specific steps account for generation of the higher JH homologs. If this were true, then investigation of the very early steps responsible for generation or metabolism of propionate or propionyl-CoA may be rewarding.

Rather little is known concerning propionate metabolism in insects. Degradation studies on JH biosynthesized from propionate labeled with carbon-14 at C-1 have shown clear-cut incorporation data without any indication of scrambling of label both *in vitro* (JH-II from *Manduca sexta*, Schooley, D. *et al.*, 1973; JH-I and JH-II from *Heliothis virescens*, Jennings, R. *et al.*, 1975b) and *in vivo* (JH-I from *H. cecropia*, Peter, M. and Dahm, K., 1975). However, degradation studies on the products from incubations with [2-^{14}C]propionate *in vitro* (Schooley, D. *et al.*, 1973) and [2-^{14}C]- and [3-^{14}C]propionate *in vivo* (Peter, M. and Dahm, K., 1975) revealed substantial scrambling of label into portions of the JH molecule metabolically derived from acetate, in addition to incorporation into expected positions. Recently Blomquist's group has studied the incorporation of [1-^{13}C]-, [2-^{13}C]-, and [3-^{13}C]propionate into methyl-branched hydrocarbons in the housefly. From ^{13}C NMR spectra of isolated products it was clear that propionate incorporates as an intact unit of which the C-3 moiety gives rise to the hydrocarbon methyl branch. However, from the same studies it was found that propionate also undergoes metabolism *via* a pathway in which C-1 is lost by decarboxylation, while C-2 is converted to the methyl carbon of acetate, and C-3 to the carboxyl carbon of acetate (Dillwith, J. *et al.*, 1982). This distribution of ^{13}C label can be explained by metabolism of precursor *via* the propionate/malonate semialdehyde pathway. The latter has been demonstrated in plants and micro-organisms but not in animals, where propionate metabolism is normally *via* carboxylation to methylmalonyl-CoA followed by conversion to succinate. Isomerization of methylmalonyl-CoA to succinyl-CoA requires vitamin B_{12}, a co-factor which is absent from plants and, interestingly, from *Musca* (Dillwith, J. *et al.*, 1982). Additional research on the fate in insects of propionate, and of intermediates metabolized thereto, could be of great value in assessing why only certain insect species produce homologous JHs.

6 EFFECT OF SUBSTRATE CONCENTRATION ON THE RATE OF JH BIOSYNTHESIS AND THE JH PROFILE

Several studies have been directed toward 'forcing' increased rates of JH biosynthesis by intact CA *in vitro* as well as attempting to alter the profile of JHs secreted by providing various substrates in the medium. It is well known that the terminal enzymatic steps are particularly amenable to such manipulations in intact CA; thus inclusion of farnesoic acid or farnesol in incubation medium containing an optimal concentration of methionine leads to increased rates of JH-III biosynthesis, data which suggest that the terminal enzymatic steps are not rate-limiting (Tobe, S. and Pratt, G., 1976; Feyereisen, R. *et al.*, 1981a). Additionally, the substrate specificity of the terminal biosynthetic steps appears low, as discussed above. Rates of *in vitro* JH biosynthesis can also be stimulated by addition of appropriate early precursors; mevalonate (1 mg ml^{-1} of culture medium = 7.7 mM) caused up to a 10-fold increase in JH-III production by CA of several species of JH-III-producing insects (Dahm, K. *et al.*, 1976). Inclusion of similar concentrations of mevalonate plus homomevalonate (i.e. 1 mg of each per ml of culture medium) also enhanced JH production from *M. sexta* and *Heliothis virescens* (Dahm, K. *et al.*, 1976) and JH acid production from male *Hyalophora cecropia* CA (Peter, M. *et al.*, 1981). Using minimal medium we noticed an apparent inhibition of JH biosynthesis rather than enhancement by 1 mM Mev or HMev (Baker *et al.*, unpublished results). However, CA from insects which normally secrete only JH-III could not be induced to produce any of the homologous JHs when incubated in medium containing 1 mg ml^{-1} each of Mev and HMev (Dahm, K. *et al.*, 1976). In contrast, when CA from adult female *M. sexta* or *H. virescens* were incubated in medium containing similar levels of Mev plus HMev the levels of the higher JH homologs (I, II) were somewhat enhanced relative to JH-III as compared to ratios of product obtained in medium lacking Mev plus HMev (Dahm, K. *et al.*, 1976). JH-I was not normally produced by adult female *M. sexta* CA incubated in medium lacking HMev. The ratio of JH-II : JH-I acids produced from *H. cecropia* CA was >1 when 183 μM [2-^{14}C]Mev was used as a precursor while the ratio was <1 when 38 μM [1-^{14}C]propionate was the radiolabeled precursor (Peter, M. *et al.*, 1981); thus demonstrating a modest effect of even fairly low substrate concentration on the final JH profile. In these studies with *H. cecropia*, CA maintained in medium either with or without Mev and HMev were unable to produce any JH-III acid. In the context of these *in vitro* experiments with lepidopteran CAs it should be recalled that *in vivo M. sexta* shift their JH production from exclusively homoisoprenoid hormones in embryos (JH-0 and 4-Me JH-I, Bergot, B. *et al.*, 1980, 1981a), to predominantly JH-II and JH-I in larvae (Peter, M. *et al.*, 1976; Bergot, B. *et al.*, unpublished results), and lastly to JH-II and JH-III in adults (Peter, M. *et al.*, 1976; Bergot, B. *et al.*, unpublished results). These dramatic changes *in vivo* contrast sharply with relatively minor differences in JH profile observed in the *in vitro* studies, and highlight how little we really understand about the intraglandular milieu *in vivo*.

The inability of intact CA from JH-III-producing animals to biosynthesize the higher homologs when supplied with homologous plus regular precursors, and the stimulation of JH-I and JH-II rather than JH-III production by lepidopteran CA under similar conditions, has led to speculation that biosynthesis of the homologous JHs is controlled by highly specific enzymes rather than by supply of precursors. However, it is possible that inability of CA from JH-III-producing insects to biosynthesize higher JH homologs when supplied with HMev is due to permeability barriers. Indeed, radiolabeled [5-^{3}H]HMev underwent ~100-fold dilution as compared to propionate, on incorporation into JH-II in *M. sexta* CA (Schooley, D. *et al.*, 1976). Similar permeability barriers may also exist for Mev, but there do not appear to be comparative quantitative measurements such as exist for HMev. While propionate has been shown to function as a homoisoprenoid precursor in both *in vitro* and *in vivo* studies, it is most likely utilized by glands as propionyl-CoA. It is conceivable that the energy-requiring conversion of propionate to its thioester is a process that does not normally occur in CA of most species, especially those producing only JH-III. It is also possible that within the CA cells of lepidopteran species a specialized pathway generating propionyl-CoA from an undetermined source is operational; CA of solely JH-III-producing species may lack this pathway.

Failure to demonstrate homologous JH production by cultured CA from a JH-III-producer may also be due to inadvertent selection of experimental conditions favoring JH-III biosynthesis. Most published studies have utilized tissue culture media which contain high levels of glucose, most or all amino acids, and other intermediary metabolites, several of which may penetrate the CA rapidly and give rise to high concentrations of acetate or its activated equivalent. For this reason we conducted some preliminary studies on JH biosynthesis by *Manduca sexta* CA in minimal medium (Hanks' salts). Although levels of JH produced (monitored by [*methyl*-^{14}C]methionine incorporation) were low and somewhat variable, we obtained evidence that inclusion of 1 mM propionate or HMev increased the ratio of JH-II : JH-III produced. A similar effect was caused by 1 mM valine or isoleucine (both generate propionyl-CoA upon metabolism) while the opposite effect was observed using 1 mM leucine, which is metabolized to HMG-CoA (and ultimately acetyl-CoA). (Parenthetically, the [*methyl*-^{14}C]methionine used to monitor JH production should eventually be metabolized *via* 2-oxobutyrate to non-labeled propionyl-CoA). In related experiments where 20 μM farnesoic acid was used as substrate the rate of JH (JH-III) production by CA from adult female *M. sexta* was about 30 pmol/ gland pair/hour, a value comparable to that obtained using 20 μM farnesoic acid and complete TC-199 medium. Similar levels of JH-III production were obtained with 20 μM farnesol in minimal medium. In minimal medium lacking terminal intermediates (farnesoic acid, farnesol), the rate of JH output was 5–10 times lower than in a more complete tissue culture medium (TC-199) but the major JH was again JH-III (ratio JH-III : JH-II = 10 : 1); in TC-199 medium the ratio of JH-III : JH-II produced is about 1 : 1 in glands of this stage (S. Kramer, unpublished results). Studies in minimal medium may be useful for measuring effects of appropriate precursors on the JH profile from CA of species which normally secrete only JH-III.

The change in the relative levels of the homoisoprenoid JHs during the life cycle of *M. sexta* is very intriguing from the standpoint of understanding enzymatic mechanisms and their control. The physiological significance of the change in relative ratios of the various JHs at each stage in development is another enigma. JH-I and JH-II are more potent than JH-III in morphogenetic assays on species of most insect orders (Henrick, C., 1982); however, the level of biological activity has little relationship to the identity of the natural JH of a species, and in most cases studied, species other than lepidopterans contain only JH-III. The data of Staal, G. (1977a) suggest that allatectomized *M. sexta* third-instar larvae can develop normally up to the fifth instar by feeding with diet containing either JH-I, JH-II or JH-III individually, although more JH-III must be supplied as compared to JH-I and JH-II.

A synthesis of all data on substrate specificity of individual enzymic steps would suggest that there are no clear-cut enzymic differences between insects which produce homologous JHs and those that produce only JH-III. From such data it is tempting to conclude that mere availability of homologous substrates (propionyl-CoA→→HEG-CoA→→HMev, etc., Fig. 9) determines the identity of the JHs secreted. It is, however, difficult to reconcile such a hypothesis (attractive but perhaps simplistic) with the data available from experiments in which cultured CA are "challenged" to biosynthesize homologous JHs from homologous precursors; experiments which are certainly open to criticism but which show at best minor effects of precursors on the JH identity. Further, experiments on the substrate specificity of two or three sequentially-acting enzymes in the JH biosynthetic pathway have shown a much higher apparent specificity (Lee, E. *et al.*, 1978) than is revealed on examining single enzymatic steps. We can only conclude that much additional research is required before we can understand how the homologous JHs are made, and what their role may be.

7 DISRUPTION OF JH BIOSYNTHESIS

Enzyme inhibitors are invaluable tools for studying properties of enzymes and in simplifying studies in preparations containing several sequentially acting enzymes. Moreover, enzyme inhibition is frequently accompanied by dramatic changes in physiological processes. There has been considerable interest in the discovery of chemical inhibitors of JH biosynthesis, as a sufficiently potent agent with this property should cause a "chemical allatectomy" of treated larvae or adults. From a practical standpoint, such a chemical should cause premature metamorphosis of larvae to non-feeding stages (in the case of holometabolous insects), thus halting damage to host plants treated with the inhibitor, and causing sterilization of adult insects exposed to the inhibitor.

In 1976, W. Bowers *et al.* reported the isolation of two substances from the plant *Ageratum houstonianum* which caused premature metamorphosis in *Oncopeltus fasciatus*. These previously known natural products (Kasturi, T. and Manithomas, T., 1967; Alertson, A., 1955; Fig. 12) were rechristened precocene I and II for their morphogenetic effects on a number of Hemiptera and Orthoptera (Bowers, W. *et al.*, 1976; vol. 8).

When incubated *in vitro* with CA, precocene II was shown to inactivate glands of *Periplaneta americana* at 100 μM (3 h incubation; Pratt, G. and Bowers, W., 1977) and *Oncopeltus fasciatus* at 50 μM (24 h incubation; Müller, P. *et al.*, 1979). These results do not appear to correlate readily with *in vivo* morphological effects, as the former species is not sensitive to treatment with precocene II whereas the latter does respond. However, when *Locusta migratoria* CA are maintained *in vitro* with [4-^{3}H]precocene I, the 3,4-dihydrodiols are rapidly formed and macromolecules suffer extensive labelling (Pratt, G. *et al.*, 1980). These results are consistent with a hypothesis that reactive epoxides of the precocenes are formed by JH epoxidase and then either hydrolyze to the diol or alkylate this (and perhaps other) enzyme(s). This cytotoxicity to locust CA (which are very sensitive to precocenes) may then be attributable to the precocenes serving as alternative substrates for an enzyme of JH biosynthesis and yielding lethal epoxide metabolities. Hammock, B. and Mumby, S. (1978) had earlier studied the effect of precocene II on epoxidation of methyl farnesoate in CA homogenates from the cockroach *Blaberus giganteus*; they found no detectable inhibition of the epoxidase enzyme at 100 μM precocene II. While this would appear to be at odds with an epoxidase–"suicide substrate" hypothesis, it is possible that precocene epoxide is formed but alkylates other cell components rather than the epoxidase *per se*. It should also be noted that this cockroach (unlike locusts) does not show morphogenetic effects on treatment with precocene.

Despite synthesis of a great number of analogs of precocene by many investigators, there has been no

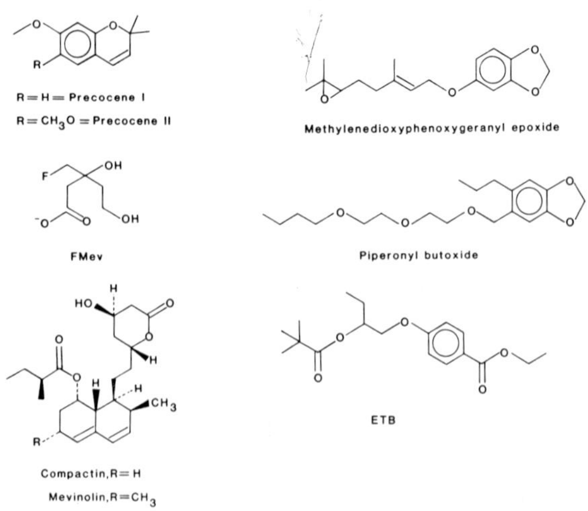

FIG. 12. Chemical structures of several inhibitors of JH biosynthesis and the JH agonist/antagonist, ETB (see text for discussion of effects of these agents *in vivo* and/or *in vitro*).

breakthrough in expanding the scope of the narrow species specificity of this family of chemicals. Furthermore, Hsia, M. *et al.* (1981) have shown that treatment of rats *in vivo* with precocene II causes hepatotoxicity, and that incubation of [4-^{3}H]precocene II with rat liver microsomes *in vitro* results in formation of the dihydrodiols and covalent binding of radioactive molecules to macromolecules, just as when *Locusta* CA are incubated with [4-^{3}H]precocene I.

More recently fluoromevalonate (FMev, Fig. 12) has been shown to act as a JH antagonist in lepidopterans *in vivo* (Quistad, G. *et al.*, 1981). Larvae of six moth species responded in various degrees to topical treatment with the lactone form of FMev, showing coloration changes, patches of pupal cuticle, prepupal behavior, or even precocious pupae after the molt subsequent to treatment. It is presumed that fluoromevalonolactone hydrolyzes rapidly at physiological pH to fluoromevalonate. As a close structural analog of Mev and HMev, FMev is probably a competitive inhibitor of enzymes utilizing these substrates, and preliminary results (Quistad, G. *et al.*, 1981) indeed suggested that it did block metabolism of both natural substrates in CA homogenates. Moreover, application of FMev to *Manduca sexta* larvae results in dramatic lowering of hemolymph JH titer (Edwards, J. *et al.*, 1983). In an *in vitro* study using cultured intact CA of *M. sexta*, Kramer, S. and Staal, G. (1981) found that FMev inhibited JH biosynthesis from [*methyl*-^{14}C]methionine, with an ID_{50} of 50 μM. Fluoromevalonate has not yet been shown to have JH antagonist properties in insects other than lepidopterans, and its activity on Lepidoptera which are important pests is too low for practical utility. Synthesis of analogs of FMev has resulted only in decreased biological potency (Quistad, G. *et al.*, 1981). This discovery was made based on knowledge of JH biosynthesis pathways and the known hypocholesterolemic properties of FMev (see references in Quistad, G. *et al.*, 1981), highlighting the potential for the discovery of additional rationally designed inhibitors of JH biosynthesis.

In addition to these dramatic *in vivo* results, a number of chemicals show evidence interpreted as inhibition of JH biosynthesis in CA cultured *in vitro*. The most potent and interesting of these substances is compactin (ML236B, Fig. 12), a fungal metabolite which inhibits HMG-CoA reductase and which has been used in clinical trials in humans as an hypocholesterolemic agent (see Endo, A., 1981, for a review on compactin and its homolog mevinolin = monacolin k). Monger, D. *et al.* (1982) have shown that compactin inhibits JH biosynthesis in intact *M. sexta* CA at 10 nM; the assay utilized adult female CA maintained and monitored according to Kramer, S. and Law, J. (1980). Similar experiments with intact CA of *Periplaneta americana* have shown an even greater potency for compactin (ID_{50} ~ 1 nM, Edwards, J. and Price, N., 1983). In homogenates of *M. sexta* CA, the k_I for inhibition of HMG-CoA reductase was 0.9 nM (Monger, D. *et al.*, 1982).

In these studies it was found that compactin has no discernible anti-JH effects *in vivo* on either *M. sexta* or *P. americana* when applied as a single dose, topically or by injection. However, repeated injection of compactin into *M. sexta* larvae caused black pigmentation characteristic of JH deficiency in this species (Monger, D. *et al.*, 1982). The lack of *in vivo* effects is surprising in view of the extreme potency of the inhibitor *in vitro* (more than 5000 times as potent as FMev!); but examination of the structure of compactin reveals several moieties that are likely to be very subject to oxidative metabolism. More metabolically stable analogs of compactin could have useful *in vivo* activity as antagonists of JH. Industrial enthusiasm for this approach has been tempered by reports from the Merck group (Robertson, R. *et al.*, 1981; Minsker, D. *et al.*, 1983) that mevinolin (a compactin homolog, Fig. 12) has teratogenic activity in rats *in vivo*, and that the teratogenic activity is relieved by co-administration of Mev, the product of HMG-CoA reductase. Based on the latter result, it can be speculated that any inhibitor of HMG-CoA reductase will likely be a teratogen in rats. The relevance of this finding for ultimately assessing potential safety to humans of HMG-CoA reductase inhibitors used on food crops is debatable.

Other substances have been studied as potential *in vitro* inhibitors of JH biosynthesis (Pratt, G. and Finney, J., 1977; Hammock, B. and Mumby, S., 1978; Kramer, S. and Staal, G., 1981). Many of these chemicals have been tested at very high concentrations, as high as 0.1–10 mM. It is not clear that many of these candidate inhibitors are in fact soluble in culture medium at the stated concentrations. However, interesting results have been obtained even though potencies do not approach that of compactin. Pratt, G. and Finney, J. (1977) found that methylenedioxyphenoxygeranyl epoxide (Fig. 12) inhibits JH-III biosynthesis from [*methyl*-^{14}C]methionine by *P. americana* glands at 1 μM. Kramer, S. and Staal, G. (1981) found inhibition of JH biosynthesis by several substances in a similar system using CA of fifth-instar larval *M. sexta*. In addition to FMev (see above), piperonyl butoxide (Fig. 12) and a compound originally described as a JH analog, ETB (Fig. 12), appeared to block JH biosynthesis. Piperonyl butoxide causes black pigmentation characteristic of JH deficiency in *Manduca* treated *in vivo* with high doses (Staal, G., 1977b). Kramer, S. and Staal, G. (1981) reported that co-administration of farnesoic acid with piperonyl butoxide *in vitro* does not result in a restoration of biosynthesis. From this they concluded that piperonyl butoxide blocks esterification and/or epoxidation, in accord with its known ability to block epoxidative-type enzymes. ETB shows mixed agonist/antagonist activity when applied to *M. sexta* larvae *in vivo* (Staal, G., 1977b) ; in virtually all other species it shows only JH agonist activity. Kramer, S. and Staal, G. (1981) incubated *M. sexta* CA with solutions of ETB. They reported an inhibition of JH biosynthesis; an extremely steep dose–response curve was found with an abrupt transition from 0 to 100% inhibition over less than a 5-fold increase in concentration. Whether this observation represents an inhibition of biosynthesis, or a membrane effect reached abruptly at the water solubility limit of ETB, remains to be seen. Hammock, B. and Mumby, S. (1978) studied the ability of various chemicals to inhibit the epoxidation of methyl farnesoate to JH-III in homogenates of *Blaberus giganteus* CA. The 70 chemicals that they screened included precocene II, methylenedioxyphenyl-type mixed function oxidase inhibitors, electron transport inhibitors, and a variety of other substances. The best inhibitors were two substituted imidazole derivatives (I_{50} values of 1.5 and 0.4 μM) which were previously known as inhibitors of epoxidation and hydroxylation in rat liver microsomal preparations. However, these inhibitors did not evince morphogenetic effects when assayed *in vivo* on two unrelated insect species, *Tenebrio molitor* and *Oncopeltus fasciatus*. Several methodological facets of this study deserve consideration. First, Hammock and Mumby chose a concentration of methyl farnesoate of 0.5 μM because it is well below the critical micelle concentration of ~ 10 μM. Secondly, all *in vitro* incubations were conducted in Carbowax 50M-treated glassware to prevent binding of lipophilic precursor, product, and inhibitors to the vessels; and lastly, no inhibitors were tested at concentrations above 0.1 mM. Similarly, in studies of *Locusta* CA epoxidase, Feyereisen, R. *et al.* (1981b) conducted incubations in Carbowax 20M-treated glassware, and discovered that it was not possible to obtain homogeneous solutions of the hydrophobic substrate methyl farnesoate without addition of 1% bovine serum albumin. Unfortunately, many studies using intact CA, or CA enzyme preparations, have utilized high concentrations of lipophilic enzyme inhibitors and/or appropriate sesquiterpenoid substrates without the precautions necessary to ensure homogeneous solution of these materials.

From the many studies discussed above we may conclude that exploitation of inhibitors of JH biosynthesis for pest control agents is of importance but fraught with difficulties. The precocenes, FMev, and compactin appear to attack enzymatic processes that may cause lesions in mammals as well. In theory, the most insect-specific enzymatic processes in the biosynthetic pathway may be the conversion of the farnesyl pyrophosphate moiety to a methyl ester of a modified farnesoic acid, or the generation and/or utilization of HMev as a JH-II/JH-I/JH-0 precursor in the Lepidoptera. *A priori* design of agents capable of disrupting these processes is not simple, but will be aided by further investigation of the biosynthetic pathways outlined in this chapter.

ACKNOWLEDGEMENTS

We thank the NSF (Grant No. PCM82–08665) for partial financial support, Dr René Feyereisen for his comments on the manuscript, and Mrs Noreen Collins for diligent assistance in preparing the manuscript.

REFERENCES

AJAMI, A. M. (1974). Approaches to the biosynthesis of the *Cecropia* C_{18}-juvenile hormone. *J. Insect Physiol. 20*, 2497–2511.

AKAMATSU, Y. and LAW, J. H. (1970). The enzymatic synthesis of fatty acid methyl esters by carboxyl group alkylation. *J. Biol. Chem. 245*, 709–713.

ALERTSON, A. R. (1955). Ageratochromene, a heterocyclic compound from the essential oils of some *Ageratum* species. *Acta Chem. Scand. 9*, 1725–1726.

ATTYGALLE, A. B. and MORGAN, E. D. (1982). Structures of homofarnesene and bishomofarnesene isomers from *Myrmica* ants. *J. Chem. Soc. Perkin Trans. I 4*, 949–951.

BAKER, F. C. and SCHOOLEY, D. A. (1981). Biosynthesis of 3-hydroxy-3-methylglutaryl-CoA, 3-hydroxy-3-ethylglutaryl-CoA, mevalonate and homomevalonate by insect corpus allatum and mammalian hepatic tissues. *Biochim. Biophys. Acta 664*, 356–372.

BAKER, F. C. and SCHOOLEY, D. A. (1978). Juvenile hormone biosynthesis: identification of 3-hydroxy-3-ethylglutarate and 3-hydroxy-3-methylglutarate in cell-free extracts from *Manduca sexta* incubated with propionyl- and acetyl-CoA. *J. Chem. Soc. Chem. Commun.* 292–293.

BAKER, F. C., LEE, E., BERGOT, B. J. and SCHOOLEY, D. A. (1981). Isomerization of isopentenyl pyrophosphate and homoisopentenyl pyrophosphate by *Manduca sexta* corpora cardiaca–corpora allata homogenates. In *Juvenile Hormone Biochemistry*. Edited by G. E. Pratt and G. T. Brooks. Pages 67–80. Elsevier, Amsterdam.

BAKER, F. C., MAUCHAMP, B., TSAI, L. W. and SCHOOLEY, D. A. (1983). Farnesol and farnesal dehydrogenase(s) in corpora allata of the Tobacco Hornworm moth *Manduca sexta*. *J. Lipid Res. 24*, 1586–1594.

BARNES, F. J. and GOODFELLOW, R. D. (1971). Mevalonate kinase: localization and variation in activity during the development of *Sarcophaga bullata*. *J. Insect Physiol. 17*, 1415–1427.

BEEDLE, A. S., REES, H. H. and GOODWIN, T. W. (1975). Phosphorylation of homomevalonic acid by an enzyme system from rat liver. In *Biochemical Society Transactions*, 557th Meeting, Liverpool, Vol. 3, pages 738–739.

BERGOT, B. J., BAKER, F. C., CERF, D. C., JAMIESON, G. and SCHOOLEY, D. A. (1981a). Qualitative and quantitative aspects of juvenile hormone titers in developing embryos of several insect species: discovery of a new JH-like substance extracted from eggs of *Manduca sexta*. In *Juvenile Hormone Biochemistry*. Edited by G. E. Pratt and G. T. Brooks. Pages 33–45. Elsevier, Amsterdam.

BERGOT, B. J., BAKER, F. C., LEE, E. and SCHOOLEY, D. A. (1979). Absolute configuration of homomevalonate and 3-hydroxy-3-ethylglutaryl and 3-hydroxy-3-methylglutaryl coenzyme A., produced by cell-free extracts of insect corpora allata. A cautionary note on prediction of absolute stereochemistry based on liquid chromatographic elution order of diastereomeric derivatives. *J. Amer. Chem. Soc. 101*, 7432–7434.

BERGOT, B. J., JAMIESON, G. C., RATCLIFF, M. A. and SCHOOLEY, D. A. (1980). JH zero: new naturally occurring insect juvenile hormone from developing embryos of the tobacco hornworm. *Science 210*, 336–338.

BERGOT, B. J., RATCLIFF, M. A. and SCHOOLEY, D. A. (1981b). Method for quantitative determination of the four known juvenile hormones in insect tissue using gas chromatography-mass spectroscopy. *J. Chromatog. 204*, 231–244.

BERGOT, B. J., SCHOOLEY, D. A. and DE KORT, C. A. D. (1981c). Identification of JH III as the principal juvenile hormone in *Locusta migratoria*. *Experientia 37*, 909–910.

BOWERS, W. S., OHTA, T., CLEERE, J. S. and MARSELLA, P. A. (1976). Discovery of insect anti-juvenile hormones in plants. *Science 193*, 542–547.

BOWERS, W. S., THOMPSON, M. J. and UEBEL, E. C. (1965). Juvenile and gonadotropic hormone activity of 10,11-epoxyfarnesenic acid methyl ester. *Life Sci. 4*, 2323–2331.

BROWN, M. S. and GOLDSTEIN, J. L. (1980). Multivalent feedback regulation of HMG-CoA reductase, a control mechanism coordinating isoprenoid synthesis and cell growth. *J. Lipid Res. 21*, 505–517.

CHRISTOPHE, J. and POPJÁK, G. (1961). Studies on the biosynthesis of cholesterol. XIV. the origin of prenoic acids from allyl pyrophosphates in liver enzyme systems. *J. Lipid Res. 2*, 244–257.

CORDELL, G. A. (1976). Biosynthesis of sesquiterpenes. *Chem. Rev. 76*, 425–460.

DAHM, K. H., BHASKARAN, G., PETER, M. G., SHIRK, P. D., SESHAN, K. R. and RÖLLER, H. (1976). On the identity of the juvenile hormone in insects. In *The Juvenile Hormones*. Edited by L. I. Gilbert. Pages 19–47. Plenum Press, New York.

DAHM, K. H., BHASKARAN, G., PETER, M. G., SHIRK, P. D., SESHAN, K. R. and RÖLLER, H. (1981). The juvenile hormones of Cecropia. In *Regulation of Insect Development and Behaviour: International Conference*. Edited by M. Kloza, F. Sehnal, A. Zabza, J. J. Menn, and B. Cymborowski. Pages 183–198. Wroclaw Technical University Press, Wroclaw, Poland.

DE KORT, C. A. D., BERGOT, B. J. and SCHOOLEY, D. A. (1982). The nature and titer of juvenile hormone in the Colorado potato beetle, *Leptinotarsa decemlineata*. *J. Insect Physiol. 28*, 471–474.

DILLWITH, J. W., NELSON, J. H., POMONIS, J. G., NELSON, D. R. and BLOMQUIST, G. J. (1982). A ^{13}C NMR study of methyl-branched hydrocarbon biosynthesis in the housefly. *J. Biol. Chem. 257*, 11305–11314.

EDMOND, J. and POPJÁK, G. (1974). Transfer of carbon atoms from mevalonate to *n*-fatty acids. *J. Biol. Chem. 249*, 66–71.

EDWARDS, J. P., BERGOT, B. J. and STAAL, G. B. (1983). Effects of three compounds with anti-juvenile hormone activity and a juvenile hormone analogue on endogenous juvenile hormone levels in the tobacco hornworm, *Manduca sexta*. *J. Insect Physiol. 29*, 83–89.

EDWARDS, J. P. and PRICE, N. R. (1983). Inhibition of juvenile hormone III biosynthesis in *Periplaneta americana* with the fungal metabolite compactin (ML-236B). *Insect Biochem. 13*, 185–189.

ENDO, A. (1981). Biological and pharmacological activity of inhibitors of 3-hydroxy-3-methylglutaryl coenzyme A reductase. *Trends Biochem. Sci. 6*, 10–12.

FEYEREISEN, R. and TOBE, S. S. (1981). A rapid partition assay for routine analysis of juvenile hormone release by insect corpora allata. *Analyt. Biochem. 111*, 372–375.

FEYEREISEN, R., KOENER, J. and TOBE, S. S. (1981a). *In vitro* studies with C_2, C_6, and C_{15} precursors of C_{16} JH biosynthesis in the corpora allata of adult female *Diploptera punctata*. In *Juvenile Hormone Biochemistry*. Edited by G. E. Pratt and G. T. Brooks. Pages 81–92. Elsevier, Amsterdam.

FEYEREISEN, R., PRATT, G. E. and HAMNETT, A. F. (1981b). Enzymic synthesis of juvenile hormone in locust corpora allata: evidence for a microsomal cytochrome P-450 linked methyl farnesoate epoxidase. *Eur. J. Biochem. 118*, 231–238.

GIBSON, D. M. and INGEBRITSEN, T. S. (1978). Reversible modulation of liver hydroxymethylglutaryl CoA reductase. *Life Sci. 23*, 2649–2664.

GOODFELLOW, R. D. and BARNES, F. J. (1971). Mevalonate kinase from the larva of the fleshfly, *Sarcophaga bullata*. *Insect Biochem. 1*, 271–282.

GOODFELLOW, R. D., HUANG, Y.-S. and RADTKE, JR., H. E. (1972). Isoprenol biosynthesis in the fly, *Sarcophaga bullata*. *Insect Biochem. 2*, 467–475.

GOODFELLOW, R. D., LIU, G. C. K., STEIN, J. P. and HARKER, K. (1973). Mevalonate biosynthesis in the fly, *Sarcophaga bullata*. *Insect Biochem. 3*, 113–122.

GOODFELLOW, R. D., HUANG, Y.-S. and WU, J.-C. (1974). *In vivo* incorporation of mevalonic acid into saponifiable lipids of the fleshfly. *Amer. Zool. 14*, 1291.

HAMMOCK, B. D. (1975). NADPH dependent epoxidation of methyl farnesoate to juvenile hormone in the cockroach *Blaberus giganteus* L. *Life Sci. 17*, 323–328.

HAMMOCK, B. D. and MUMBY, S. M. (1978). Inhibition of epoxidation of methyl farnesoate to juvenile hormone III by cockroach corpus allatum homogenates. *Pestic. Biochem. Physiol. 9*, 39–47.

HAMMOCK, B. D. and SPARKS, T. C. (1977). A rapid assay for insect juvenile hormone esterase activity. *Analyt. Biochem. 82*, 573–579.

HAMNETT, A. F., PRATT, G. E., STOTT, K. M. and JENNINGS, R. C. (1981). The use of radio HRLC in the identification of the natural substrate of the *O*-methyl transferase and substrate utilization by the enzyme. In *Juvenile Hormone Biochemistry*. Edited by G. E. Pratt and G. T. Brooks. Pages 93–105. Elsevier, Amsterdam.

HAMNETT, A. F. and PRATT, G. E. (1983). The absolute configuration of precocene I dihydrodiols produced by metabolism of precocene I by corpora allata of *Locusta migratoria, in vitro*. *Life Sci. 32*, 2747–2753.

HENRICK, C. A. (1982). Juvenile hormone analogs: structure–activity relationships. In *Insecticide Mode of Action*. Edited by J. R. Coats. Pages 315–402. Academic Press, New York.

HSIA, M. T. S., GROSSMAN, S. and SCHRANKEL, K. R. (1981). Hepatotoxicity of the anti-juvenile hormone precocene II and the generation of dihydrodiol metabolites. *Chem.-Biol. Interactions 37*, 265–277.

JENNINGS, R. C., JUDY, K. J. and SCHOOLEY, D. A. (1975a). Biosynthesis of the homosesquiterpenoid juvenile hormone JH II [methyl (2*E*,6*E*,10*Z*)-10,11-epoxy-3,7,11-trimethyltridecadienoate] from [5-^{3}H]homomevalonate in *Manduca sexta*. *J. Chem. Soc. Chem. Commun.* 21–22.

JENNINGS, R. C., JUDY, K. J., SCHOOLEY, D. A., HALL, M. S. and SIDDALL, J. B. (1975b). The identification and biosynthesis of two juvenile hormones from the tobacco budworm moth *(Heliothis virescens)*. *Life Sci. 16*, 1033–1040.

JUDY, K. J. (1976). Summary of Session II; Biosynthesis and Metabolism of Juvenile Hormone. In *The Juvenile Hormones*. Edited by L. I. Gilbert. Pages 96–100. Plenum, New York.

JUDY, K. J., SCHOOLEY, D. A., DUNHAM, L. L., HALL, M. S., BERGOT, B. J. and SIDDALL, J. B. (1973). Isolation, structure, and absolute configuration of a new natural insect juvenile hormone from *Manduca sexta*. *Proc. Nat. Acad. Sci. USA 70*, 1509–1513.

KAPLANIS, J. N., DUTKY, R. C. and ROBBINS, W. E. (1961). The incorporation of 2-^{14}C-mevalonate into house fly lipids. *Ann. Ent. Soc. Amer. 54*, 114–116.

KASTURI, T. R. and MANITHOMAS, T. (1967). Essential oil of *Ageratum conyzoides* — isolation and structure of two new constituents. *Tetrahedron Lett.*, 2573–2575.

KING, D. S. (1983). Chemistry and metabolism of the juvenile hormones. In *Invertebrate Endocrinology; Vol. 1*. Edited by H. Laufer and R. G. H. Downer. Pages 57–64. A. R. Liss, New York.

KOBAYASHI, M., KOYAMA, T., OGURA, K., SETO, S., RITTER, F. J. and BRÜGGEMANN-ROTGANS, I. E. M. (1980). Bioorganic synthesis and absolute configuration of Faranal. *J. Amer. Chem. Soc. 102*, 6602–6604.

KOYAMA, T., OGURA, K. and SETO, S. (1981). Enzymatic synthesis of insect juvenile hormone skeleton. *Nippon Kagaku Kaishi (J. Chem. Soc. Japan, Chem. Ind. Chem.) 5*, 689–696.

KOYAMA, T., OGURA, K. and SETO, S. (1973a). Studies on isopentenyl pyrophosphate isomerase with artificial substrates. *J. Biol. Chem. 248*, 8043–8051.

KOYAMA, T., OGURA, K. and SETO, S. (1973b). Enzymatic synthesis of bishomofarnesyl pyrophosphate with the insect juvenile hormone skeleton. *Chem. Lett.* 401–404.

KOYAMA, T., SAITO, A., OGURA, K. and SETO, S. (1980). Substrate specificity of farnesyl pyrophosphate synthetase. Application to asymmetric synthesis. *J. Amer. Chem. Soc. 102*, 3614–3618.

KRAMER, S. J. and LAW, J. H. (1980). Control of juvenile hormone production: the relationship between precursor supply and hormone synthesis in the tobacco hornworm, *Manduca sexta*. *Insect Biochem. 10*, 569–575.

KRAMER, S. J. and STAAL, G. B. (1981). *In vitro* studies on the mechanism of action of anti-juvenile hormone agents in larvae of *Manduca sexta*. In *Juvenile Hormone Biochemistry*. Edited by G. E. Pratt and G. T. Brooks. Pages 425–437. Elsevier, Amsterdam.

KRAMER, S. J. and KALISH, F. (1984). Regulation of the corpora allata in the black mutant of *Manduca sexta*. *J. Insect Physiol. 30*, 311–316.

LAW, J. H. (1983). Biosynthesis of juvenile hormones in insects. In *Biosynthesis of Isoprenoid Compounds*. Vol. 2. Edited by J. W. Porter and S. L. Spurgeon. Pages 507–534. John Wiley & Sons, New York.

LEE, E., SCHOOLEY, D. A., HALL, M. S. and JUDY, K. J. (1978). Juvenile hormone biosynthesis: homomevalonate and mevalonate synthesis by insect corpus allatum enzymes. *J. Chem. Soc. Chem. Commun.* 290–292.

MADHAVAN, K., CONSCIENCE-EGLI, M., SIEBER, F. and URSPRUNG, H. (1973). Farnesol metabolism in *Drosophila melanogaster*: ontogeny and tissue distribution of octanol dehydrogenase and aldehyde oxidase. *J. Insect Physiol. 19*, 235–241.

METZLER, M., DAHM, K. H., MEYER, D. and RÖLLER, H. (1971). On the biosynthesis of juvenile hormone in the adult *Cecropia* moth. *Z. Naturf. 26b*, 1270–1276.

METZLER, M., MEYER, D., DAHM, K. H., RÖLLER, H. and SIDDALL, J. B. (1972). Biosynthesis of juvenile hormone from 10-epoxy-7-ethyl-3,11-dimethyl-2,6-tridecadienoic acid in the adult *Cecropia* moth. *Z. Naturf. 27b*, 321–322.

MEYER, A. S., SCHNEIDERMAN, H. A., HANZMANN, E. and KO, J. H. (1968). The two juvenile hormones from the *Cecropia* silk moth. *Proc. Nat. Acad. Sci. USA 60*, 853–860.

MINSKER, D. H., MACDONALD, J. S., ROBERTSON, R. T. and BOKELMAN, D. L. (1983). Mevalonate supplementation in pregnant rats suppresses the teratogenicity of mevinolinic acid, an inhibitor of 3-hydroxy-3-methylglutaryl-Coenzyme A reductase. *Teratology 28*, 449–456.

MONGER, D. J. and LAW, J. H. (1982). Control of juvenile hormone biosynthesis. Evidence for phosphorylation of the 3-hydroxy-3-methylglutaryl coenzyme A reductase of insect corpus allatum. *J. Biol. Chem. 257*, 1921–1923.

MONGER, D. J., LIM, W. A., KEZDY, F. J. and LAW, J. H. (1982). Compactin inhibits insect HMG-CoA reductase and juvenile hormone biosynthesis. *Biochem. Biophys. Res. Commun. 105*, 1374–1380.

MORGAN, E. D. and WADHAMS, L. J. (1972). Chemical constituents of Dufour's gland in the ant, *Myrmica rubra*. *J. Insect Physiol. 18*, 1125–1135.

MORGAN, E. D., PARRY, K. and TYLER, R. C. (1979). The chemical composition of the Dufour gland secretion of the ant *Myrmica scrabrinodis*. *Insect Biochem. 9*, 117–121.

MÜLLER, P. J., MASNER, P., KALIN, M. and BOWERS, W. S. (1979). *In vitro* inactivation of corpora allata of the bug *Oncopeltus fasciatus* by precocene II. *Experientia 35*, 704–705.

MÜLLER, P. J., MASNER, P., TRAUTMANN, K. H., SUCHÝ, M. and WIPF, H.-K. (1974). The isolation and identification of juvenile hormone from cockroach corpora allata *in vitro*. *Life Sci. 15*, 915–921.

NES, W. D., CAMPBELL, B. C., STAFFORD, A. E., HADDON, W. F. and BENSON, M. (1982). Metabolism of mevalonic acid to long chain fatty alcohols in an insect. *Biochem. Biophys. Res. Commun. 108*, 1258–1263.

NISHINO, T., OGURA, K. and SETO, S. (1973). Comparative specificity of prenyltransferase of pig liver and pumpkin with respect to artificial substrates. *Biochim. Biophys. Acta 302*, 33–37.

PETER, M. G. and DAHM, K. H. (1975). Biosynthesis of juvenile hormone in the *Cecropia* moth. Labelling pattern from [1-^{14}C]-propionate through degradation to single carbon atom derivatives. *Helv. Chim. Acta 58*, 1037–1048.

PETER, M. G., DAHM, K. H. and RÖLLER, H. (1976). The juvenile hormones in blood of larvae and adults of *Manduca sexta* (Joh.). *Z. Naturf. 31c*, 129–131.

PETER, M. G., GUNAWAN, S. and EMMERICH, H. (1979). Preparation of optically pure juvenile hormone I labelled in the ester methyl group with tritium at very high specific activity. *Experientia 35*, 1141–1142.

PETER, M. G., SHIRK, P. D., DAHM, K. H. and RÖLLER, H. (1981). On the specificity of juvenile hormone biosynthesis in the male *Cecropia* moth. *Z. Naturf. 36c*, 579–585.

POPJÁK, G., RABINOWITZ, J. L. and BARON, J. M. (1969). Artificial substrates for prenyltransferase. *Biochem. J. 113*, 861–868.

PRATT, G. E. and BOWERS, W. S. (1977). Precocene II inhibits juvenile hormone biosynthesis by cockroach corpora allata *in vitro*. *Nature 265*, 548–550.

PRATT, G. E. and FINNEY, J. R. (1977). Chemical inhibitors of juvenile hormone biosynthesis *in vitro*. In *Crop Protection Agents — their Biological Evaluation*. Edited by N. R. McFarlane. Pages 113–132. Academic Press, New York.

PRATT, G. E. and TOBE, S. S. (1974). Juvenile hormones radiobiosynthesised by corpora allata of adult female locusts *in vitro*. *Life Sci. 14*, 575–586.

PRATT, G. E., HAMNETT, A. F., FINNEY, J. R. and WEAVER, R. J. (1978). The biosynthesis of radiolabeled *Cecropia* (C18) juvenile hormone and several analogs by isolated corpora allata of the American cockroach (*Periplaneta americana*) during incubation with methyl-labeled methionine and analogs of farnesenic acid. *Gen. Comp. Endocrinol. 34*, 113.

PRATT, G. E., JENNINGS, R. C., HAMNETT, A. F. and BROOKS, G. T. (1980). Lethal metabolism of precocene-I to a reactive epoxide by locust corpora allata. *Nature 284*, 320–323.

PRATT, G. E., STOTT, K. M., BROOKS, G. T., JENNINGS, R. C., HAMNETT, A. F. and ALEXANDER, B. A. J. (1981). Utilisation of farnesoic acid analogues by the *O*-methyl transferase from corpora allata of adult female *Locusta migratoria*. In *Juvenile Hormone Biochemistry*. Edited by G. E. Pratt and G. T. Brooks. Pages 107–121. Elsevier, Amsterdam.

QUISTAD, G. B., CERF, D. C., SCHOOLEY, D. A. and STAAL, G. B. (1981). Fluoromevalonate acts as an inhibitor of insect juvenile hormone biosynthesis. *Nature 289*, 176–177.

REIBSTEIN, D. and LAW, J. H. (1973). Enzymatic synthesis of insect juvenile hormones. *Biochem. Biophys. Res. Commun. 55*, 266–272.

REIBSTEIN, D., LAW, J. H., BOWLUS, S. B. and KATZENELLENBOGEN, J. A. (1976). Enzymatic synthesis of juvenile hormone in *Manduca sexta*. In *The Juvenile Hormones*. Edited by L. I. Gilbert. Pages 131–146. Plenum Press, New York.

RIGHETTI, M., WILEY, M. H., MURRILL, P. A. and SIPERSTEIN, M. D. (1976). The *in vitro* metabolism of mevalonate by sterol and non-sterol pathways. *J. Biol. Chem. 251*, 2716–2721.

RILLING, H. C. (1979). Prenyltransferase. *Pure Appl. Chem. 51*, 597–608.

RITTER, F. J., BRÜGGEMANN-ROTGANS, I. E. M., VERWIEL, P. E. J., PERSOONS, C. J. and TALMAN, E. (1977). Trail pheromone of the Pharaoh's ant, *Monomorium pharaonis*: isolation and identification of faranal, a terpenoid related to juvenile hormone II. *Tetrahedron Lett.*, 2617–2618.

ROBERTSON, R. T., MINSKER, D. H., MACDONALD, J. S., BOKELMAN, D. L. and CHRISTIAN, M. S. (1981). Mevalonic acid antagonism of the teratogenic effects of mevinolinic acid, a potent inhibitor of hydroxymethylglutaryl-coenzyme-A reductase. *Teratology 23*, 58.

RODÉ-GOWAL, H., ABBOTT, S., MEYER, D., RÖLLER, H. and DAHM, K. H. (1975). Propionate as a precursor of juvenile hormone in the *Cecropia* moth. *Z. Naturf. 30c*, 392–397.

RÖLLER, H. and DAHM, K. H. (1970). The identity of juvenile hormone produced by corpora allata *in vitro*. *Naturwissenschaften 57*, 454–455.

RÖLLER, H., DAHM, K. H., SWEELEY, C. C. and TROST, B. M. (1967). The structure of the juvenile hormone. *Angew. Chem. Int. Edit. 6*, 179–180.

SCHOOLEY, D. A. (1977). Analysis of the naturally occurring juvenile hormones — their isolation, identification, and titer determination at physiological levels. In *Analytical Biochemistry of Insects*. Edited by R. B. Turner. Pages 241–287. Elsevier, New York.

SCHOOLEY, D. A., BERGOT, B. J., GOODMAN, W. and GILBERT, L. I. (1978a). Synthesis of both optical isomers of insect juvenile hormone III and their affinity for the juvenile hormone-specific binding protein of *Manduca sexta*. *Biochem. Biophys. Res. Commun. 81*, 743–749.

SCHOOLEY, D. A., JUDY, K. J., BAKER, F. C., LEE, E., BERGOT, B. J. and HALL, M. S. (1978b). Biosynthesis of the juvenile hormones: the role of the homoisoprenoid intermediates. In *Comparative Endocrinology*. Edited by P. J. Gaillard and H. H. Boer. Pages 499–502. Elsevier, Amsterdam.

SCHOOLEY, D. A., JUDY, K. J., BERGOT, B. J., HALL, M. S. and JENNINGS, R. C. (1976). Determination of the physiological levels of juvenile hormones in several insects and biosynthesis of the carbon skeletons of the juvenile hormones. In *The Juvenile Hormones*. Edited by L. I. Gilbert. Pages 101–117. Plenum Press, New York.

SCHOOLEY, D. A., JUDY, K. J., BERGOT, B. J., HALL, M. S. and SIDDALL, J. B. (1973). Biosynthesis of the juvenile hormones of *Manduca sexta*: labeling pattern from mevalonate, propionate, and acetate. *Proc. Nat. Acad. Sci. USA 70*, 2921–2925.

SHIRK, P. D., DAHM, K. H. and RÖLLER, H. (1976). The accessory sex glands as the repository for juvenile hormone in male *cecropia* moths. *Z. Naturf. 31c*, 199–200.

SPITELLER, M. and SPITELLER, G. (1979). Trennung und charakterisierung saurer harnbestand-teile. *J. Chromatogr. 164*, 253–317.

STAAL, G. B. (1977a). Essential differences between natural juvenile hormones and juvenile hormone analogs elucidates by use of a substitution assay. *Scr. Varia Pont. Acad. Sci. 41*, 333–348.

STAAL, G. B. (1977b). Insect control with insect growth regulators based on insect hormones. *Scr. Varia Pont. Acad. Sci. 41*, 353–377.

STERN, J. R. (1955). Enzymes of acetoacetate formation and breakdown. In *Methods in Enzymology*. Vol. I. Edited by S. P. Colowick and N. O. Kaplan. Pages 573–585. Academic Press, New York.

SWEETMAN, L., WEYLER, W., NYHAN, W. L., DE CESPEDES, C., LORIA, A. R. and ESTRADA, Y. (1978). Abnormal metabolites of isoleucine in a patient with proplionyl-CoA carboxylase deficiency. *Biomed. Mass Spectrom 5*, 198–207.

TOBE, S. S. and FEYEREISEN, R. (1983). Juvenile hormone biosynthesis: regulation and assay. In *Invertebrate Endocrinology; Vol. 1*. Edited by H. Laufer and R. G. H. Downer. Pages 161–178. A. R. Liss, New York.

TOBE, S. S. and PRATT, G. E. (1974). The influence of substrate concentrations on the rate of insect juvenile hormone biosynthesis by corpora allata of the desert locust *in vitro*. *Biochem. J. 144*, 107–113.

TOBE, S. S. and PRATT, G. E. (1976). Farnesenic acid stimulation of juvenile hormone biosynthesis as an experimental probe in corpus allatum physiology. In *The Juvenile Hormones*. Edited by L. I. Gilbert. Pages 147–163. Plenum, New York.

TRAUTMANN, K. H., MASNER, P., SCHULER, A., SUCHÝ, M. and WIPF, H. -K. (1974). Evidence of the juvenile hormone methyl (2*E*,6*E*)-10,11-epoxy-3,7,11-trimethyl-2,6-dodecadienoate (JH-3) in insects of four orders. *Z. Naturf. 29c*, 757–759.

WEIRICH, G. F. and CULVER, M. G. (1979). *S*-Adenosylmethionine: juvenile hormone acid methyltransferase in male accessory reproductive glands of *Hyalophora cecropia* (L.). *Archiv. Biochem. Biophys. 198*, 175–181.

ZATZ, M., DUDLEY, P. A., KLOOG, Y. and MARKEY, S. P. (1981). Nonpolar lipid methylation. Biosynthesis of fatty acid methyl esters by rat lung membranes using *S*-adenosylmethionine. *J. Biol. Chem. 256*, 10028–10032.

11 Regulation of Juvenile Hormone Titer: Synthesis

RENÉ FEYEREISEN

Oregon State University, Corvallis, Oregon, USA

INTRODUCTION

Determination of juvenile hormone (JH) titer gives only a static image of a dynamic reality. Maintenance of a particular level of JH in the hemolymph, or change of JH titer as a signal for physiological changes in the insect, both require that the insect must be able to control the overall balance of JH synthesis, JH uptake and release by tissues, and JH clearance by degradation and excretion. In addition the insect must be able to control the hemolymph volume. Control of JH synthesis is achieved by regulating the endocrine activity of the corpora allata (CA), the identified source of juvenile hormone(s). Control of JH degradation is achieved by regulating enzymes of JH catabolism (see Hammock, B. this volume), but little is known about JH distribution in tissues (Hammock, B. *et al.*, 1975; Nowock, J. *et al.*, 1976) or about JH excretion (Erley, D. *et al.*, 1975; Röseler, P. F. and Röseler, I., 1978). Possible control of hemolymph volume is implicit in the definition of a titer (i.e. a concentration of hormone, which is the ratio of a certain quantity of hormone over a certain volume). Obviously the hormone titer can change simply by changing hemolymph volume. Is the hemolymph titer of JH an important determinant of the hormonal effects of JH? De Wilde, J. *et al.* (1971) have defined an effective titer as "the hormone concentration observed in the direct vicinity of the target organ during induction of the response". If the mode of action of JH involves binding to JH receptors (Goodman, W. and Chang, E.; this volume), a change in "effective titer" would directly cause a change in JH receptor occupancy. The "effective titer" of JH is thus a theoretical value related to the hemolymph titer but limited by such factors as hemolymph protein binding, hemolymph mixing, etc. In fact it is the concentration of JH in a pharmacokinetic "compartment" (Cutler, D., 1978; Segre, G., 1982) within the hemolymph.

With these limitations on the definition of JH titer in mind, one must say at the outset that very little is known about the regulation of JH titer. Akamatsu, Y. *et al.* (1975) have tried to integrate data on JH synthesis, degradation and hemolymph volume obtained in larvae of *Manduca sexta* into a simple equation for JH titer. They felt "the control of JH levels in the hemolymph at the time of metamorphosis could well be determined primarily by changes in JH esterase levels". In contrast, Kramer, S. (1978) concluded, on the basis of JH biosynthetic rates *in vitro*, half-life of injected JH *in vivo* and JH titer, that "control of the activity of the CA is probably of prime importance in the regulation of the JH titer in the Colorado potato beetle". The use of models describing the regulation of JH titer will eventually increase as data on quantitative JH biochemistry begin to accumulate. This chapter focuses on the patterns of JH synthesis and on the regulation of the CA by allatotropic (stimulatory) and allatostatic (inhibitory) influences. No attempt will be made to review the considerable body of information gained by classical and indirect methods (Reviews by Gilbert, L., 1964; Wigglesworth, V., 1964, 1970; Engelmann, F., 1970; De Wilde, J. and De Loof, A., 1973; Gilbert, L. and King, D., 1973; Doane, W., 1973; Sláma, K. *et al.*, 1974, etc.). Instead, studies using direct methods (i.e. *in vitro* assays for JH biosynthesis) will be reviewed extensively and reference to studies which relied solely on indirect methods will be made if they bring cogent points to the subject matter.

1 ASSAYS OF CA ACTIVITY

Classically, the activity or inactivity of the CA has been deduced from experiments involving ligation, decapitation or allatectomy, all aimed at establishing "critical periods", when JH is or is not required during insect development. A great deal of our present understanding of the functions of JH during the life of an insect is based on experiments which tell us what happens when the source of JH is removed. The concepts derived from JH deficiency syndromes were confirmed by replacement therapy experiments, i.e. implantation of CA, parabiosis or the administration of JH or its analogs. Replacement therapy experiments also introduced the idea that JH titer needs to be very low or nil at a specific period of development to allow metamorphosis to occur. It is only recently that *in vitro* assays for JH biosynthesis have allowed a rational study of CA activity and regulation.

1.1 Criteria of CA activity *in situ*

1.1.1 Morphometric changes

One indirect method, which is perhaps the most widely used, is the measure of CA volume. Wigglesworth, V. (1936) showed that the volume of the CA of adult female *Rhodnius prolixus* was considerably increased during periods of egg maturation triggered by a blood-meal. A correlation between CA volume and secretory activity was often assumed, perhaps because gland volume was the easiest parameter which could be quantified during growth, metamorphosis or reproduction. CA volume estimations have been performed by a number of techniques, either on fixed tissue or on fresh tissue just after dissection (Joly, L., 1960; Staal, G., 1961). Sectioning of fixed glands gave estimates of the largest CA area, or an integration of each section was used to calculate volume. Mass of the glands (Gilbert, L. and Schneiderman, H., 1961) can also be considered as an estimate of gland volume. Nuclear size and its variations have been followed in great detail, e.g. during post-embryonic development in honeybees (Pflugfelder, O., 1948). Scharrer, B. and Von Harnack, M. (1958) measured changes in cell number during a reproductive cycle in *Leucophaea maderae*, but gland volume can also increase without an increase in cell number, i.e. by cell enlargement (Legay, J., 1950). A nucleocytoplasmic ratio first used by Engelmann, F. (1957) gave the volume of CA tissue per million nuclei and was "considered to be the decisive measurement" (Engelmann, F., 1970). For instance, the variations of nucleocytoplasmic ratios and rates of JH-III synthesis occur synchronously during a gonotrophic cycle in *Diploptera punctata* (Engelmann, F., 1959a; Tobe, S. and Stay, B., 1977; Szibbo, C. and Tobe, S., 1981a). It should be emphasized, however, that this apparent relation between nucleocytoplasmic ratio and CA activity is an exception rather than a rule, and that estimates of CA activity based on volume determinations are generally misleading, as we shall see later in this chapter. The plethora of studies examining changes in CA volume will not be reviewed here but extensive discussions or references to earlier work can be found in other reviews (Pflugfelder, O., 1958; Engelmann, F., 1970; De Wilde, J. and De Loof, A., 1973; Sláma, K. *et al.*, 1974).

Changes in CA morphometry have been directly related to changes in JH titer in relatively few studies. Johnson, R. and Hill, L. (1973a) showed that the changes in volume and number of cells of the CA in the fourth and fifth (last) larval stages of *Locusta migratoria* reflect somatic growth of the whole insect but are not related to levels of circulating JH as measured by the *Galleria* bioassay. However, during the first gonotrophic cycle of female *L. migratoria*, as well as in adult male *L. migratoria* and *Schistocerca gregaria*, the levels of JH in the hemolymph could be related to volumes of the CA (Johnson, R. and Hill, L., 1973b, 1975). Röseler, P.-F. (1977) showed that in queenless workers of *Bombus terrestris* an increase in CA volume coincides with an increasing JH titer, but this appears to be a trend rather than an exact correlation. Lanzrein, B. *et al.* (1978) indicated a positive correlation between CA volume and JH titer during the oocyte maturation phase of *Nauphoeta cinerea*. In *Monema flavescens*, prepupal diapause is controlled by a high JH titer. Takeda, N. (1978) showed that changes in CA volume during diapause occur in parallel with a rise and fall in hemolymph levels of JH as measured by the *Galleria* bioassay. Baehr, J.-C. *et al.* (1982) have used a radioimmunoassay for JH-III to show that successive cycles of vitellogenesis and egg care in *Labidura riparia* are characterized by parallel changes in CA volume and JH-III titers.

1.1.2 Gland ultrastructure

Criteria for the secretory activity of the CA have been sought by histological, microscopical, and especially ultrastructural comparisons of glands of presumed high and low activity. Such criteria (reviewed by Cassier, P., 1979 and Sedlak, B. this volume) obviously cannot give a quantitative evaluation of CA activity.

Chu, H. and Norris, D. (1979) used both scanning and transmission electron microscopy to compare morphological and ultrastructural features of presumably high and low CA activity of *Xyleborus ferrugineus*. Joly, L. *et al.* (1968) pointed out that comparing the ultrastructure of CA taken at different periods of their ontogenic development, periods in which the CA are considered active and periods in which the CA are considered inactive, is not a valid method because a given structure can be related to a developmental stage or age of the glands, rather than to their activity. Instead, these authors suggested that comparing the ultrastructure of CA of the same age, but of different activity, should yield better criteria of activity. These criteria should then be used to interpret the ultrastructure of CA. In *L. migratoria*, the destruction of the C cells of the pars intercerebralis can be used to obtain inactive CA at any time during ontogenic development. This inactivity is judged by the symptoms of allatectomy observed in locusts after destruction of the C cells (Girardie, A., 1966, 1967). Joly, L. *et al.* (1968) therefore proceeded to compare the ultrastructure of CA from 8- to 15-day-old adult females (terminal oocyte length of 1.15–1.85 mm; "active" CA) and from females of the same age but deprived of their C cells just after adult emergence ("inactive" CA; terminal oocyte length of 0.5–1.0 mm). Whorls of smooth endoplasmic reticulum were the most conspicuous features of "inactive" glands. Using a combination of morphometric criteria, biological tests as well as the ultrastructural criteria that they so carefully defined, Joly, L. *et al.* (1968) inferred the ontogenic cycle of activity of the CA in *L. migratoria*. The influence of the milieu of the host on transplanted CA was also studied using ultrastructural criteria (Joly, L. *et al.*, 1969). It is now possible to study the ultrastructure of CA after *in vitro* assay of JH biosynthesis, and thus make relevant comparisons of ultrastructure and activity of single glands; the effects of precocene on CA ultrastructure and activity were studied in *D. punctata* (Johnson, G. *et al.*, 1980; Feyereisen, R. *et al.*, 1981a).

1.1.3 Macromolecular synthesis in the CA

Before the advent of *in vitro* assays for JH synthesis by the CA few assays were developed to assess the synthetic abilities of the glands, because the biochemistry of JH synthesis was unknown. However, it was believed that macromolecular synthesis in the CA would give an indirect and crude measure of JH synthetic rates. This measure would be more accurate than increases in CA volume which were interpreted by some authors as reflecting a net storage of JH. Abd-el-Wahab, A. and Sirlin, J. (1959) used [^{14}C]adenine to label nucleic acids in the ring glands of *Drosophila melanogaster*, and Nair, K. (1963) stained CA sections from *Chrysocoris purpurens* at various stages of ovarian development to locate the presence of nucleic acids. Szibbo, C. and Tobe, S. (1981a) measured the percentage of CA cell nuclei labelled after [^{3}H]thymidine injection in adult female *Diploptera punctata*. Only few cells (~1%) located primarily at the periphery of the glands were labelled just before the peak in cell number and JH biosynthetic activity observed in this insect (see section 2.4). Ulrich, G. and Rembold, H. (1983) have measured DNA by photometry in CA nuclei stained with the Feulgen reaction. These results were compared with levels of [^{3}H]thymidine incorporation to suggest several phases of endomitosis in the CA of maturing larvae of the honeybee *Apis mellifera*, but no possible relationship with JH synthesis can be inferred from such studies. Siew, Y. and Gilbert, L. (1971) used [^{3}H]uridine to study RNA synthesis by autoradiography in the CA of *Philosamia cynthia* pupae. Transient increases in labelling of the CA cells (interpreted as activation of the glands) were observed after injection of 20-hydroxyecdysone or of a mixture of JH-I isomers. Gillott, C. and Dogra, G. (1972) followed peaks of [^{3}H]uridine incorporation in the CA of adult female *Melanoplus sanguinipes* and considered this technique more accurate than volume changes to estimate CA activity. The same technique was used to study changes in synthetic activity of the CA during caste differentiation in a termite, *Zootermopsis angusticollis* (Yin, C. and Gillott, C., 1975). Protein synthesis has also been used as a probe. Vogel, H. *et al.* (1977) proposed that the rate of incorporation of amino acids by the CA should reflect periods of rest and activity, and followed [^{3}H]histidine incorporation by autoradiography in *A. mellifera* as a function of the daily foraging schedule. In contrast to morphometric changes in the CA which have been compared to changes in JH biosynthesis *in vitro* in a few studies, there is no indication of the validity of assays which measure macromolecular synthesis in the CA. Weber, V. and Emmerich, H. (1976) have shown that peaks of *in vitro* RNA synthesis and ecdysone synthesis by *Tenebrio molitor* larval prothoracic glands occur synchronously. It would thus be interesting to follow RNA synthesis in CA of a species where it could be directly compared with JH biosynthesis.

1.2 Bioassays of CA activity

In contrast to the assays which try to estimate CA activity *in situ* by changes in morphometry, ultrastructure or macromolecular synthesis, a number of assays have been devised to monitor the effects of transplanted CA on physiological processes in the host. In most cases these experiments were only extensions of the classical endocrinological concept of replacement therapy, designed to better understand the functions of the corpus allatum hormone (Pflugfelder, O., 1940; Wigglesworth, V., 1940). In some cases, however, transplantation experiments were turned into bioassays of CA activity and selected examples are presented below. Williams, C. (1961) utilized a "pupal assay" for CA activity in *Hyalophora cecropia*. Four CA were implanted into abdomens of previously chilled female pupae which initiate adult development when returned to 25° (Williams, C., 1946). About 1 month after implantation of the CA into the host pupae, the organisms were scored using both internal and external anatomical characters. High activity of the implanted CA was indicated by the formation of "second pupal stages". Thus, an overall index of CA activity from the third larval stage to the adult stage was deduced, suggesting a sharp drop in CA activity during the final (fifth) larval stage and no activity during the pupal stage.

Ozeki, K. (1962, 1963) assayed the activity of CA from the earwig *Anisolabis maritima* using a similar approach: two pairs of CA + CC were implanted into larvae of the last stage, 2–4 days after the penultimate ecdysis. After the last molt, the activity of the implanted CA was quantified by the percentage of animals (from 20 to 30 replicates) with incomplete or abnormal metamorphosis. The presumed cycle of CA activity was thus obtained for the penultimate and the last larval stages of *A. maritima*. Activity declined at the end of the penultimate larval stage, and no activity was observed until the last third of the last larval stage, suggesting total inactivity of the CA during metamorphosis. Odhiambo, T. (1966) designed an assay for adult male *Schistocerca gregaria* in which two CA of known age were implanted into each newly emerged locust soon after allatectomy. Three characters — sexual behavior, body coloration and the condition of the accessory glands — were noted 12 days after implantation. Odhiambo reported that this assay gave results consistent with the morphometric changes of the CA observed in adult male desert locusts. Joly, P. (1967) took advantage of color polymorphism in *Locusta migratoria* for a bioassay of CA activity. The glands to be tested were implanted in larvae, 24 h before ecdysis to the last larval stage. Glands presumed to be active induced a green coloration which was scored 48 h after ecdysis. Thus, the implanted CA were influenced by the hormonal milieu of the host for only 72 h in this bioassay. In a modification of the *Antheraea polyphemus* test of Gilbert, L. and Schneiderman, H. (1960), Watson, J. (1967) implanted CA from *Thermobia domestica* into the midrib of the antenna of *Antheraea polyphemus* pupae just before the end of the critical period for JH action, thus presumably measuring the secretory activity of the CA over a short term. Lea, A. and Thomsen, E. (1969) designed a bioassay for CA of adult female *Calliphora erythrocephala*. They followed the effect of an implanted CA on the volume of the nuclei of the medial neurosecretory cells of an allatectomized host. Enlarged nuclei found 1 or 2 days after implantation testified to the activity of the implanted CA. On the basis of this elaborate bioassay, Lea and Thomsen concluded that CA activity in *C. erythrocephala* was not related to CA volume as Strangways-Dixon, J. (1961) had suggested. Although quite different in its principle, a bioassay in which the growth of two implanted oocytes was measured over a 27 h period in adult female *Nauphoeta cinerea* was reported by Wilhelm, R. and Lüscher, M. (1970) to give a useful measure of CA activity of the hosts. The observation that the functional activity of the left colleterial gland in female *Periplaneta americana* is controlled by the CA (Bodenstein, D. and Sprague, I., 1959) suggested the development of a bioassay (Willis, J. and Brunet, P., 1966). Dixon, S. and Moser, E. (1972) used that bioassay, measuring the amount of glucoside secretion of the left colleterial gland of decapitated roaches 8 days after implantation with three pairs of CA from honeybee larvae (heterologous donors). In this assay the implantation of CA into intact virgin cockroaches, instead of decapitated cockroaches, gave erroneous results because host and implanted CA interacted (De Wilde, J., 1976). Bioassays were also used to estimate the decline in CA activity after incubation of the glands with precocenes *in vitro* (Müller, P. *et al.*, 1979; Bowers, W. and Feldlaufer, M., 1982).

Limitations to such bioassays of CA activity were already recognized by Pflugfelder, O. (1939), who showed that implantation of CA from an early larval stage of *Carausius morosus* into a recipient of a later larval stage caused "unmistakable signs of degeneration" of the recipient's glands. Wigglesworth, V. (1948) stated that "from the standpoint of experimental method it is obvious that the transplantation of the corpus allatum alone may give a misleading impression of its function", and Gilbert, L. (1964) expressed similar concerns. Denervation of the donor CA and its transfer into a different hormonal environment is likely to affect both host and donor CA. Influence of the host's milieu on implanted CA is now used in conjunction with *in vitro* assays of JH biosynthesis to provide direct evidence for the existence of hormonal modulators of CA activity (see section 4). All the bioassays of CA activity are dependent upon a biological response to JH, and thus upon a determined JH titer. They involve far-reaching assumptions concerning the pharmacokinetics of JH (e.g. that JH titer is directly correlated to JH synthesis) and about the pharmacodynamics of JH (e.g. that sensitivity to JH of the test organism is directly proportional to JH titer). These assumptions are of course untenable, and the multiple assays for JH titer can thus not qualify in this discussion as assays for CA activity. Assays involving the transplantation of CA are obviously suspect in those species where the CA serve as a neurohemal organ for PTTH (Agui, N. *et al.*, 1980; Carrow, G. *et al.*, 1981) and this limitation has long been recognized (Ichikawa, M. and Nishiitsutsuji-Uwo, J., 1959; Kobayashi, M. and Yamashita, Y., 1959). The bioassays described above cannot, however, be brushed aside without mention in a chapter like this one, because they illustrate both the ingenuity of insect endocrinologists, and the need for a reliable assay of CA activity.

1.3 *In vitro* assays for JH biosynthesis

The most direct way to measure CA activity is of course to measure, in a non-invasive manner, the amount of JH synthesized by the CA and released into the hemolymph over a defined period of time (i.e. the rate of JH production). This tautology is simply intended to point out the inherent extreme difficulties of this direct *in vivo* approach. It implies that the newly released JH be somehow differentiated from the circulating JH present before the start of the observation. It also implies that the pharmacokinetics of the JH be precisely known in order to account for all the JH produced, because distribution and degradation of JH does not stop during the observation period. One possible way to measure JH biosynthesis *in vivo* would be to use radiolabelled precursors, but this introduces another series of difficulties, because the pharmacokinetics of the precursor used must now be defined with precision to ascertain a constant incorporation ratio. Elaborate determinations of specific activities may then yield rates of JH production. If labelled methionine is chosen as the precursor (Metzler, M. *et al.*, 1971), then changes in the hemolymph pool size of this amino acid must be followed during development (De Kort, C. and Kramer, S., 1976). Another possible way would be to measure the rate of dilution of injected tracer amounts of radiolabelled JH by newly biosynthesized JH. Hemolymph mixing, JH half-life and determination of initial JH titer may cause most of the difficulties in this approach. Cogent use of tracer kinetic models (Carson, E. *et al.*, 1982) may in the future facilitate the measurement of CA activity *in vivo*.

The *in vitro* approach, in which the CA are transferred from their natural environment into a test tube and the amount of JH produced over a period of time is measured, presently offers the best assay of CA activity. Two different methods have been used to date to measure the amount of JH produced by CA *in vitro*, a radiochemical assay, and radioimmunoassays (RIAs). Table 1 gives a list of insect species in which these methods have been used for studies of CA physiology.

Röller, H. and Dahm, K. (1970) provided the first evidence that JH is produced and released by the CA. They showed that brain–CC–CA complexes of male pupae of *Hyalophora cecropia* (taken 2–3 days before adult emergence) could be cultured *in vitro* for 7 days without losing their potential to release

Table 1: Insect species in which JH synthesis has been studied in vitro

Order		Method	References
Orthoptera	*Schistocerca gregaria*	RCA	Tobe, S. and Pratt, G. (1975); Pratt, G. *et al.* (1975a)
	Locusta migratoria	RCA	Ferenz, H.-J. and Kaufner, I. (1981); Girardie, J. *et al.* (1981)
		RIA	Baehr, J.-C. *et al.* (1981)
	Acheta domesticus	RIA	Strambi, C. (1981)
	Melanoplus sanguinipes	RCA	McCaffery, A. and McCaffery, V. (1983)
Dictyoptera	*Periplaneta americana*	RCA	Weaver, R. *et al.* (1975); Pratt, G. *et al.* (1975a)
	Diploptera punctata	RCA	Tobe, S. and Stay, B. (1977); Stay, B. and Tobe, S. (1981)
	Nauphoeta cinerea	RCA	Lanzrein, B. *et al.* (1978)
Isoptera	*Macrotermes subhyalinus*	RCA	Lanzrein, B. *et al.* (1977)
Coleoptera	*Leptinotarsa decemlineata*	RCA	Kramer, S. (1978); Khan, M. *et al.* (1982a)
	Tenebrio molitor	RCA	Weaver, R. *et al.* (1980)
	Pterostichus nigrita	RCA	Ferenz, H.-J. (1981)
Hymenoptera	*Bombus terrestris*	RCA	Röseler, P.-F. and Röseler, I. (1978)
	Polistes gallicus	RCA	Röseler, P.-F. *et al.* (1980)
Lepidoptera	*Manduca sexta*	RIA	Granger, N. *et al.* (1979, 1982b)
		RCA	Kramer, S. and Law, J. (1980)

RCA: radiochemical assay; RIA: radioimmunoassay.
Only representative references of physiological interest are listed. JH identification studies are not reported in this table.

JH. Significant quantities of JH-I, active in the *Tenebrio* bioassay and identified by gas chromatography and mass spectrometry, were isolated from the culture medium. Transplantation bioassay of the CA and brain–CC parts of the cultured brain–CC–CA complex showed that the CA was the source of JH. However, Röller and Dahm's original experiments could not be reproduced in subsequent studies (Peter, M. *et al.*, 1981) in so far as JH-I acid, not JH-I, is a product of the adult male CA in *H. cecropia* (see below).

Metzler, M. *et al.* (1971) showed that the S-methyl group of methionine is incorporated *in vivo* into both JH-I and JH-II by adult male *H. cecropia.* Up to 0.1% of the dose injected could be recovered in JH. These authors also showed that when both [2-^{14}C]JH-I acid and [*methyl*-^{3}H]methionine are injected into adult male silkmoths, double-labelled JH-I can be isolated (Metzler, M. *et al.*, 1972). These experiments suggested that in the JH biosynthetic pathway, methionine could serve as methyl donor. But since the experiments were done *in vivo*, they did not establish whether this (or these) reaction(s) occurred within the CA. It turns out that the adult male *H. cecropia* silkmoth is a rather unique case where methyl transfer from methionine (through S-adenosylmethionine) to JH-I acid and JH-II acid does not occur in the CA, but in the accessory sex glands (Weirich, G. and Culver, M., 1979; Dahm, K. *et al.*, 1981; Peter, M. *et al.*, 1981). Judy, K. *et al.* (1973a,b), however, discovered JH-III by culturing CA from *Manduca sexta* and *Schistocerca vaga in vitro* with labelled methionine, thus establishing the role of the *S*-methyl group of methionine in the biosynthesis of the methyl ester of JH by the CA.

Although these studies were extremely important for all subsequent work on JH biochemistry (see Schooley, D. and Baker, F., this volume), their importance to the development of a physiologically relevant assay of CA activity was not directly apparent. Indeed, a large number of glands had to be cultured over a long term *in vitro*, thus making any possible relationship with putative *in vivo* rates of synthesis impossible to establish.

Pratt, G. and Tobe, S. (1974) reported that CA of adult female *Schistocerca gregaria* utilize both [^{3}H]farnesoic acid and [*methyl*-^{14}C]methionine in short-term *in vitro* incubations to produce very large amounts of double-labelled JH-III. It thus became possible to show that the rates of JH-III synthesis from these two exogenous, labelled precursors was linear over periods of up to 4 h after removal from the insect (Tobe, S. and Pratt, G., 1974a). These experiments also made it possible to measure the stoichiometry of [^{3}H]farnesoic acid and [*methyl*-^{14}C]methionine utilization: a molar incorporation ratio close to unity was rapidly obtained *in vitro*, indicating that under those conditions radiolabelled methionine was virtually the exclusive source of the methyl ester group of JH-III (Pratt, G. and Tobe, S., 1974; Tobe, S. and Pratt, G., 1974a). The same conclusion was reached by Judy, K. *et al.*

(1973a) who analyzed by mass fragmentography the ratio of [^{14}C] to [^{12}C] in JH-III biosynthesized by *M. sexta* CA.

1.3.1 The radiochemical assay

The conclusion that radioactivity from [*methyl*-^{14}C]methionine is incorporated into the methyl ester moiety of JH without significant dilution has led to the development of a short-term radiochemical assay for the quantitative measurement of the endocrine activity of CA (Pratt, G. *et al.*, 1975a; Tobe, S. and Pratt, G., 1975a).

The radiochemical assay for CA activity is truly unique in terms of assays for the activity of an endocrine gland. Indeed, its simplicity is that hormone production *in vitro* can be readily measured because *all* the hormone produced is radiolabelled at the methyl ester group. Insect hemolymph usually contains large amounts of free methionine (Florkin, M. and Jeuniaux, C., 1974). It is believed, although not specifically established, that the CA take up methionine and convert it to S-adenosylmethionine, a universal methyl donor. As part of the normal biosynthetic pathway, a methyl group is then transferred enzymatically from S-adenosylmethionine to a carboxylic acid precursor of JH (Reibstein, D. and Law, J., 1973; Feyereisen, R. *et al.*, 1981b). Fortunately for insect endocrinologists, the pool of methionine and S-adenosylmethionine in the CA is extremely small, so that transfer of the glands to an *in vitro* culture medium readily dilutes the non-radioactive methionine from the CA with radioactive methionine from the medium. Consequently, after a short lag period (Pratt, G. *et al.*, 1975a) which represents the time necessary to achieve isotopic equilibrium within the CA (i.e. finish dilution) all biosynthesized JH is labelled. (Accuracy of the assay may be improved by replacing the incubation medium after a short time, e.g. 45 min, to discard any isotopically diluted JH synthesized during this lag period). Extraction, (purification) and quantification of JH produced is then simple, but careful procedures should be followed (Pratt, G. and Tobe, S., 1974; Hamnett, A. and Pratt, G., 1978; Feyereisen, R. and Tobe, S., 1981). The specific activity of the biosynthesized JH should be equal to the specific activity of the precursor radiolabelled methionine, thus determining the sensitivity of the assay. It should be stressed that the radiochemical assay does not in any way involve biochemical reactions not performed by the glands *in vivo*, nor does it modify or measure the rate of just one biochemical reaction in the CA. It simply replaces a non-radioactive precursor by a radioactive precursor. Routine use of the radiochemical assay is subordinate to the establishment of several parameters (Pratt, G. and Weaver, J., 1978; Tobe, S. and Feyereisen, R., 1983); some are of general validity for *in vitro* assays of CA activity, another is specific for this assay (Table 2).

(1) The JH synthesized *in vitro* must be identified by non-equivocal techniques. Because the JH biosynthesized in the radiochemical assay is radiolabelled, identification is relatively easy and has been achieved by radio-GLC or HPLC with or without derivatization of the hormone. When high specific activity [*methyl*-^{3}H]methionine is used, it is possible to achieve a very low detection limit (Hamnett, A. and Pratt, G., 1978). Once the *in vitro* product of the CA has been identified, it is justified to use simple methods for the routine quantification of synthesized or released JH such as thin-layer chromatography (Pratt, G. and Tobe, S., 1974; Hamnett, A. and Pratt, G., 1978) or rapid solvent partition (Feyereisen, R. and Tobe, S., 1981). It is, of course, desirable to confirm that the JH biosynthesized *in vitro* is identical to the hormone found *in vivo* (i.e. extractable from the hemolymph). In Lepidoptera which produce a mixture of JH homologs, this may be quite difficult, because

Table 2: Validation of in vitro *assays for JH biosynthesis*

(RCA, RIA)	The JH(s) produced *in vitro* should be identified.
(RCA, RIA)	Optimal incubation conditions should be found and the CA should not be damaged during dissection or incubation.
(RCA, RIA)	Synthesis and/or release of JH during the incubation period should be linear for glands of high and low activity.
(RCA only)	A fixed molar incorporation ratio of label from methionine should be demonstrated.
(RIA only)	Antibodies used to quantify JH production should be fully characterized in terms of their cross-reactivity with JH homologs, JH precursors and metabolites as well as non-JH molecules.

RCA: radiochemical assay: RIA: radioimmunoassay.

various homologs synthesized in a particular ratio may not be metabolized at the same rate (see Hammock, B., this volume), and thus the contribution of JH synthesis to JH titer may not be easily understood.

(2) An optimal culture medium should be found (Weaver, R. *et al.*, 1980). This is of course a prerequisite for any type of *in vitro* work, but parameters such as pH and Na^+/K^+ ratio should certainly not be overlooked. It is desirable to use chemically defined media, rather than media supplemented with vertebrate or insect serum fractions etc., because the latter may introduce unknown modulators of CA function. Optimal incubation conditions include the establishment of a dose–response for methionine concentration (Tobe, S. and Pratt, G., 1974a; Weaver, R. *et al.*, 1980; Feyereisen, R. *et al.*, 1981c; Khan, M. *et al.*, 1982a). Typically JH synthesis is optimal over a wide range of methionine concentrations, which is in good agreement with reported hemolymph levels of this amino acid. Integrity of the CA during dissection and incubation should be maintained to prevent leakage of intracellular components such as precursors (Pratt, G. and Weaver, J., 1978) and to prevent loss of biosynthesized JH. Because the methyl ester group of JH is labelled, any JH esterase activity present or released in the medium during the incubation might attenuate apparent rates of JH synthesis. Hemolymph JH esterase can easily be removed by careful dissection and preincubation rinses of the glands (Hammock, B. and Mumby, S., 1978), but the continued presence of JH esterase in (or around) the CA in species such as *Manduca sexta* (Reibstein, D. *et al.*, 1976; Granger, N. *et al.*, 1979) may invalidate results obtained with the radiochemical assay.

(3) Linearity of JH synthesis or release should be established. The standard *in vitro* incubation time for the glands must be long enough to ensure that sufficient amounts of JH are produced to permit precise quantification, and yet short enough to exclude any time at which linearity is not maintained. Thus linearity of JH synthesis or release must be shown for low-activity glands as well as for high-activity glands because limits of linearity may differ with levels of activity (Pratt, G. *et al.*, 1975a). If linearity of hormone production is not documented, the results cannot be expressed as rates of hormone production. In most studies an incubation time of 3 h was found satisfactory. In Lepidoptera the CA may produce more than one JH homolog, and thus linearity must be shown for each homolog. Dahm, K. *et al.* (1976) suggested that the ratio of JH homologs may vary in long-term incubations of *M. sexta* CA.

(4) A crucial, but more difficult, parameter is that radioactivity from the *S*-methyl group of methionine must be incorporated at a fixed molar ratio, ideally a ratio of 1 : 1 with the carbon skeleton of JH. In other words, the specific activity of radiobiosynthesized JH must be equal to the specific activity of methionine in the incubation medium. As pointed out above, this feature is the basis of the radiochemical assay. It nonetheless needs to be confirmed for each species investigated, because it justifies the calculation of molar quantities of biosynthesized JH. Various techniques have been used to establish the molar incorporation ratio. Judy, K. *et al.* (1973a) have used mass spectral studies to show an incorporation rate of at least 91% in JH-II and JH-III from *M. sexta* CA, but unfortunately a coupled GC–MS is not available in every insect endocrinology laboratory. Molar incorporation ratios from [^{3}H]farnesoic acid and [*methyl*-^{14}C]methionine indicate stoichiometry (1.02 ± 0.17, Pratt, G. and Tobe, S., 1974) for JH-III biosynthesized from farnesoic acid by *Schistocerca gregaria* CA. This method has also been used with the product of *Periplaneta americana* and *Tenebrio molitor* CA (Pratt, G. and Bowers, W., 1977; Weaver, R. *et al.*, 1980). Addition of exogenous farnesoic acid results in a considerable stimulation of JH-III biosynthesis in these species and thus double-labelling of the JH-III produced under these conditions provides good evidence for the origin of the methyl ester moiety. However, labelled farnesoic acid and its higher homologs are not commercially available. Specific activity of JH-III radiobiosynthesized by bumblebees was studied by Röseler, P.-F. and Röseler, I. (1978) using the *Galleria* bioassay, but it is not known how accurately this technique can detect a small but significant difference in specific activity. Careful use of RIA (Baehr, J.-C., *et al.*, 1976; Strambi, C. *et al.*, 1981) might be appropriate to determine specific activity. Another method to ascertain stoichiometric incorporation of radioactivity from methionine into JH

biosynthesized *in vitro* was used by Feyereisen, R. *et al.* (1981c). Samples of radiobiosynthesized JH-III from *Diploptera punctata* CA were analyzed by reversed-phase HPLC. The slope of a regression line between peak area and radioactivity in the JH-III peak was established and compared to the slope of a standard calibration line established using synthetic JH of known specific activity. This method allowed a precise determination of specific activity. It confirmed that the JH produced by *D. punctata* CA *in vitro* was not diluted to a significant extent by potential methyl-donors other than labelled methionine. This method is relatively easy and GC can be used instead of HPLC, but in both cases a clean chromatography profile should be obtained to allow precise computation of peak areas. Hamnett, A. and Pratt, G. (1978) have described the use of capillary column radio-GC for specific activity determinations of biosynthesized JH and JH precursors; calibration was done by co-injection of internal standard [^{3}H]methyl palmitate.

Isotopic dilution of the radiolabel used as mass-marker for biosynthesized JH has been shown to be negligible in all species studied to date, thus fulfilling an essential criterion for the validity of the radiochemical assay (see Table 2). In long-term incubations of *Tenebrio molitor* CA, a significant pool of non-labelled methyl donor contributed (30%) to the biosynthesized JH-III (Judy, K. *et al.*, 1975). However, the methionine concentration was about 10 times below the optimal concentration measured by Weaver, R. *et al.* (1980) for this species, and the incubation medium used (Grace's) was one of the worst tested by the latter authors. Thus the interdependence of the criteria listed in Table 2 is obvious.

In addition to isotopic dilution, a possible but very remote possibility is that isotopic discrimination might invalidate the radiochemical assay. However, no isotopeic effect was observed in *D. punctata* (Feyereisen, R. *et al.*, 1981c).

Advantages of the radiochemical assay are its extreme simplicity and its high sensitivity, allowing the activity of single pairs of CA or single glands to be monitored. Disadvantages are that it cannot be used to determine the activity of glands which do not esterify their secretory products, such as the CA from adult male *Hyalophora cecropia* (Peter, M. *et al.*, 1981). Also, it cannot be used in those species where dissection is so difficult that the CA cannot be obtained free of tissue (including hemolymph) contaminating the medium with JH esterase activity.

1.3.2 Radioimmunoassays (RIAs)

Röller, H. and Dahm, K. (1970) used the *Tenebrio* bioassay to quantify JH-I produced by CA *in vitro*. Other assays for JH can be used to quantify the amounts of JH released *in vitro*, for example radioimmunoassays (RIAs) which are theoretically sensitive and straightforward. Two different types of RIAs for JH have been developed. The RIA of Baehr, J.-C. *et al.* (1976, 1979), in which antigens and tracers are prepared via the carboxylic position of JH acids, is designed to measure JH-I, II and III with three different antisera. Iodinated histamine conjugates of those three JHs are used as tracers. The RIA of Strambi, C. *et al.* (1981), in which antigen and tracer are prepared via the diol derivative of JH-I, is designed to measure the diols of JH-I, II and III with a single antiserum and an iodinated succinyl–(JH-I)–glycyltyrosine tracer. Both types of RIAs have been used to study CA activity *in vitro* (Baehr, J.-C. *et al.*, 1981; Strambi, C., 1981). In addition, the Baehr *et al.* antisera have been used in conjunction with tritiated JH-I (Granger, N. *et al.*, 1979) and JH-III (Granger, N. *et al.*, 1982a) to study JH biosynthesis by larval *M. sexta* CA *in vitro*. Several parameters must be defined before RIAs can be used as routine tools for *in vitro* assays of JH biosynthesis (Table 2). Identification of the JH synthesized *in vitro*, linearity of JH release and optimization of incubation conditions are essential for any *in vitro* assay and have been discussed above.

RIA procedures can be straightforward when a single JH homolog is released *in vitro*, and possible interference by intermediates of JH biosynthesis or by JH acid can be minimized by appropriate dissection and incubation procedures. When more than one JH homolog is produced by the CA *in vitro*, as in Lepidoptera, the ratio of the homologs produced cannot be assumed to be constant in normal or experimental conditions. The various homologs may be assayed on aliquots of the same sample with antisera to each of the homologs, but this procedure is valid only if the antisera show a high degree of specificity (typically less than 1% cross-reactivity with other JHs). Thus the various JH homologs should be separated before assay in order to prevent cross-reaction in the RIA. Alternatively, the sum of all JH homologs may be assayed together, but this procedure is valid only if the antiserum displays an equal cross-reactivity towards all homologs (as in the Strambi, C. *et al.* (1981) RIA where the diols of JH-I, II and III are equally recognized). If individual levels of different JH homologs are to be measured by RIA procedures, separation of the JH diols by HPLC is mandatory in the Strambi, C. *et al.* (1981) RIA, whereas the Baehr, J.-C. *et al.* (1979) RIA calls for the use of as many specific antisera as JH homologs are produced. Granger, N. *et al.* (1979, 1981, 1982b) have measured developmental changes in rates of JH-I release *in vitro* by larval *Manduca sexta* CA. These authors used an antiserum to JH-I which cross-reacts (12%) with JH-II. On day 5 of the fifth larval stage, the ratio of JH-II to JH-I was determined to be 4:1 (Granger, N. *et al.*, 1979). Determination of this ratio throughout development would legitimize the quantitative expression of the rates of JH release in these studies.

1.3.3 Validation of *in vitro* assays for JH biosynthesis

How do *in vitro* assays for JH biosynthesis relate to CA activity *in vivo*? The assumption is made that the rate of release or synthesis measured *in vitro* is the same as the rate *in vivo*. This can only be determined *a posteriori* when both JH titer and rates of JH degradation are known. Also of concern is "that the normal secretory function [may be] ensured only when the integrity of the central nervous system-corpus allatum complex is unbroken" (Wigglesworth, V., 1948). The available evidence (De Kort, C. and Granger, N., 1981; Tobe, S. and Feyereisen, R., 1983) indicates that the rate of JH synthesis of CA *in vitro* shortly after extirpation of the glands is probably a good indicator of the rate of synthesis of those glands *in situ* (Pratt, G. *et al.*, 1976). In particular, glands which are dependent on intact nervous connections for sustained activity *in vivo* (Tobe, S. *et al.*, 1977) maintain a constant rate of activity for 3 h *in vitro*. This is a sufficient time to allow precise quantification of the JH produced (Pratt, G. *et al.*, 1975a). On the other hand, glands which are controlled by "nervous inhibition" do not activate spontaneously *in vitro* (Stay, B. and Tobe, S., 1977; Tobe, S. *et al.*, 1981), suggesting that low activity measured *in vitro* reflects low activity *in vivo*.

2 PATTERNS OF CA ACTIVITY

2.1 Synthesis and release of JH

The possibility that JH may be stored in the CA, and that hormone release from the glands may be an important facet of CA regulation, was advanced by various authors trying to interpret morphometric or ultrastructural changes (e.g. Thomsen, E. and Thomsen, M., 1970). *In vitro* assays of CA activity have conclusively shown that the rate of JH release is proportional to the rate of JH synthesis over a wide range of CA activity in the following species: *Schistocerca gregaria* (Tobe, S. and Pratt, G., 1974b), *Periplaneta americana* (Pratt, G. *et al.*, 1975a) and *Diploptera punctata* (Tobe, S. and Stay, B., 1977). This correlation between synthesis (i.e. the sum of JH found in the glands and in the incubation medium) and release (i.e. JH found in the incubation medium) is also observed for glands stimulated by exogenous farnesoic acid (Tobe, S. and Pratt, G., 1974b; Feyereisen, R. *et al.*, 1981c). In *S. gregaria*, labelled JH-III is detected in the incubation medium within 10 min of the addition of labelled precursors (Tobe, S. and Pratt, G., 1974a). The species studied to date do not have any mechanism for the storage of biosynthesized JH-III, and intraglandular hormone levels are proportional to synthetic rates. Thus, high intraglandular levels of JH are related to high rates of synthesis and release of JH. Intraglandular JH was not detected by the *Tenebrio* bioassay (Röller, H. and Dahm, K., 1970) or by RIA (Granger, N. *et al.*, 1979) in some species but was detected by *Galleria* bioassay and by GC–ECD (Johnson, R. and Hill, L., 1973b; Huibregtse-Minderhoud, L. *et al.*, 1980) in the CA of adult locusts. Absence of intraglandular JH in single experiments does not by itself prove that storage of the hormone does not occur; this can only be shown by establishing the relationship between synthesis and release using CA of widely different activities.

Available data thus indicate that JH release is not regulated but "suggests that the export of JH is governed by laws of simple physical diffusion" (Tobe, S. and Pratt, G., 1947b). Fick's law expresses passive diffusion as a function of the concentration gradient, the surface area of the membrane, the partition coefficient of the molecule between the membrane and the outside phase, and the diffusion coefficient in the membrane. Nothing is known about those factors. The concentration gradient is possibly never limiting, at least *in vitro*, and one can only speculate about the importance of surface area which can increase with increased gland/cell volume and with the proliferation of extracellular spaces, especially at the periphery of the glands (Cassier, P., 1979). Although JH is secreted *in vitro* as a simple solute (Pratt, G. and Tobe, S., 1974), the presence of JH binding proteins may facilitate diffusion of JH *in vivo*. Such a mechanism was postulated for steroid binding proteins in vertebrates which were reported to facilitate hormone passage from the site of synthesis into the bloodstream (Westphal, U., 1970; Ewing, L. *et al.*, 1976; Westphal, U. *et al.*, 1978). Because JH-III is released from the CA, whereas its precursor methyl farnesoate is not released (or at least 100 times less efficiently, Pratt, G. and Weaver, J., 1978) one may question whether this selectivity simply reflects a large difference in diffusion coefficients. Other physical parameters, e.g. aqueous solubility, are less than 100-fold different, and methyl farnesoate concentrations in the glands can be considerable (Pratt, G. *et al.*, 1975b). Thus passive diffusion of biosynthesized JH as a mechanism of release from the glands may not be sufficient to explain the rapid release of JH. Morphological evidence of JH release by CA cells is lacking, a situation which is similar to steroid hormone secretion.

2.2 Symmetry of right and left CA

Sensitive assays are required to study the effect of endogenous or synthetic chemicals which would potentially stimulate or inhibit the CA. Because CA are usually paired organs (except in some Hemiptera, Diptera, etc.), it would be tempting to use one member of a pair as a control and to subject the other member to experimental conditions (Tobe, S., 1977a). Such a protocol would be particularly interesting in those insects where linearity of JH synthesis is too short to allow both a control and an experimental incubation to be performed on an individual pair of CA. Symmetry of right and left CA function therefore needs to be established. Tobe, S. (1977a) investigated the symmetry of CA function in adult female *S. gregaria*. In only 25% of the CA assayed did the rate of JH synthesis by individual members of the pair differ by less than 2-fold; in 35% of the CA assayed the difference was more than 10-fold. This considerable asymmetry (up to 1000 : 1) was observed whether synthetic rates by the pair were low or high, and it was randomly distributed between the right and left glands. In addition, asymmetry was not related to the age of the insects. Weaver, R. (1979) obtained very different results with CA of adult female *P. americana*. The activity of one gland differed from that of the contralateral gland by less than 2-fold in 72% of the pairs assayed. This very low degree of asymmetry was randomly distributed and had no relation with either age or total synthetic activity of the CA pair. Symmetry of the glands was even more pronounced in adult female *Diploptera punctata*, with less than 10% of the glands exhibiting a functional asymmetry greater than 2-fold (Szibbo, C. and Tobe, S., 1981a). Cell number and gland volume were also similar between right and left CA and thus in *D. punctata* symmetry is both structural and functional. Ferenz, H.-J. and Kaufner, I. (1981) observed a considerable degree of asymmetry in volume and activity of adult female *Locusta migratoria* CA. One member of the pair often showed little or no activity, an observation also made by Weaver, R. (1979) on young adult *Periplaneta americana*. What is the significance of these widely varying degrees of asymmetry in different species? Strong, L. (1965) had suggested that the lateral neurosecretory cells of the brain (LNSC) could activate the ipsilateral CA in *Schistocerca paranensis* and thus that the CA could be regulated independently of one another. Schooneveld, H. (1970) showed that right and left CA of adult *Leptinotarsa decemlineata* sometimes differed considerably in size, up to a factor of 10, but this difference in size was dependent on the number of CA cells, and not dependent on changes in LNSC. Weaver, R. (1979) suggested that symmetry is consistent with a predominant humoral control of CA activity (e.g. *P. americana*) whereas asymmetry reflects predominant control by nervous connections with the brain (e.g. *Schistocerca* sp.).

Obviously, the degree of symmetry between the activity of left and right CA cannot be assumed for any species, and has to be measured on individual members of a pair. Grouping six left and six right CA (Granger, N. *et al.*, 1979) obliterates any randomly distributed asymmetry.

2.3 CA activity during larval development and metamorphosis

A basic tenet of the "classical scheme" for the hormonal control of molting and metamorphosis is that a decline in JH titer/CA activity controls metamorphosis. Until very recently, "turning off the CA" and "cessation of JH secretion" were hardly disputed concepts. There is however, remarkably little direct information on JH synthesis during metamorphosis.

In the Holometabola, only *Manduca sexta* has been studied to date. The rates of JH release were studied in the penultimate (fourth) and last (fifth) larval stages as well as in the early pupal stage. The release rate of JH-I/II as well as of JH-III was almost constant during the penultimate larval stage, and almost negligible in the early pupal stage. During the last larval stage, rates of JH release were apparently much higher (5–6 times) than in the penultimate larval stage with a sharp decline after pupal apolysis. A trough in the release rate of JH-I/II *in vitro* was seen at the time of pupal "commitment", 1 day before the onset of the "wandering" stage (Granger, N. *et al.*, 1979, 1981, 1982a,b). The precise molar ratio of JH-I, II, and III was not obtained in these studies, or in that of Kramer, S. and Law, J. (1980) and remains to be elucidated throughout metamorphosis. The relative importance of JH-III synthesis appeared to increase steadily during the two last larval stages (Granger, N. *et al.*, 1982b). JH-II is the most prominent homolog in larval *M. sexta* hemolymph (Bergot, B. *et al.*, in preparation) and JH-II is also an important (Granger, N. *et al.*, 1979) if not major (Kramer, S. and Law, J., 1980) product of larval CA *in vitro*. Interestingly, a relation between rates of JH release *in vitro* and JH titers (bioassay, Fain, M. and Riddiford, L., 1975; GC–MS, Bergot, B. *et al.*, in preparation) is obtained only when changing hemolymph volumes are taken into account (Granger, N. *et al.*, 1979, 1982b). It thus appears that an increased hemolymph volume, rather than a decreased CA activity, determines JH titer in the penultimate and possibly last larval stage of *M. sexta*.

In the Hemimetabola, JH synthesis by larval CA *in vitro* has been studied in a locust and in a cockroach. Injeyan, H. and Tobe, S. (1981) studied JH-III synthesis in *Schistocerca gregaria* during the penultimate and last larval stages of solitarious and gregarious females. The rates of JH synthesis were highest in the beginning (second day) of each larval stage. CA activity was not turned off during the last larval stage, and was not related to the changes in CA volume which appear to follow somatic growth. Synthetic rates were higher in solitarious larvae, suggesting that the higher JH titer measured by bioassay in isolated *L. migratoria* (Joly, L. and Joly, P., 1974) is caused by higher CA activity. The rates of JH-III synthesis by larval female *D. punctata* CA were determined by Szibbo, C. *et al.* (1982) (Fig. 1). JH synthesis remained between 1 and 3 pmol h^{-1} per pair in the penultimate larval stage. In the last larval stage the rate of JH synthesis reached 2.5 pmol h^{-1} per pair on the first day after ecdysis, but rapidly declined thereafter, and fell below the limit of detection (0.1 pmol) during the last half of the stage. An assay of higher sensitivity (i.e. using labelled methionine of higher specific activity) would be necessary to measure CA activity in the second half of the final larval stage when JH-III titer in hemolymph is approximately 130 nM (Tobe, S. *et al.*, in preparation). The changes in CA volume and

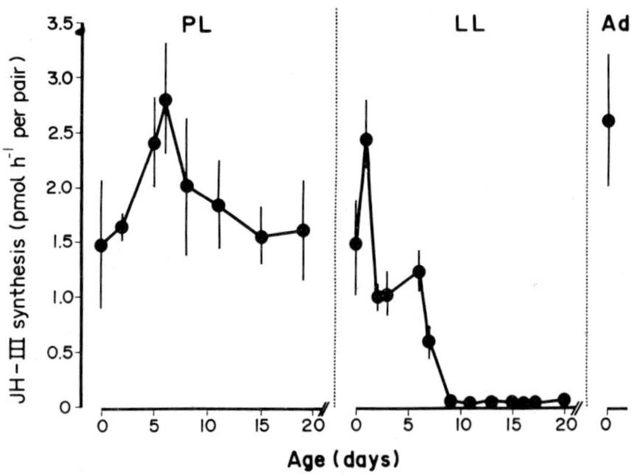

FIG. 1. Rate of JH synthesis by CA from female larvae of *Diploptera punctata*. PL: Penultimate larval stage; LL: last larval stage; Ad: adult stage. Dotted lines indicate the time of ecdysis. (From Szibbo, C. *et al.*, 1982.)

CA cell number, as well as somatic growth (body wet weight), were also recorded in the final two stages of *Diploptera punctata* (Szibbo, C. *et al.*, 1982). CA activity is not related to either CA volume or CA cell number, because the latter two parameters are greatest in the second half of both stages, when CA activity is lowest. Hemolymph JH esterase levels follow almost a mirror image of JH synthetic rates, but the JH titer has not been determined throughout the last two larval stages of *D. punctata* and therefore the relative importance of synthesis and degradation cannot be assessed.

The paucity of studies on JH synthesis during larval development hardly allows general conclusions. It is, however, established that: (1) CA volume has no relation whatsoever to CA activity during metamorphosis; (2) CA activity is not necessarily turned *off* for metamorphosis to occur.

2.4 CA activity and reproduction

The radiochemical assay for JH biosynthesis has considerably improved our understanding of the role played by the CA in the initiation and regulation of ovarian development. The study of CA regulation in adult insects, heretofore conducted only with indirect assays (Highnam, K., 1964; Engelmann, F., 1970; De Wilde, J. and De Loof, A., 1973), has also found an invaluable tool in the radiochemical assay (Tobe, S., 1980). Various patterns of CA activity (JH-III synthesis) have been described in relation to reproduction, and can be related to the degree of synchrony of oocyte growth as well as to the continuity of oocyte production (i.e. timing of successive gonotrophic cycles) (Tobe, S., 1980).

Examples of species where oocyte growth is synchronous, but where successive gonotrophic cycles are well separated, are *Diploptera punctata* (a viviparous cockroach) and *Nauphoeta cinerea* (an ovoviviparous cockroach). In *D. punctata* the CA of mated females shows a cycle of activity which is precisely correlated with growth of the basal oocytes (Fig. 2). At ovulation, when the oocytes are fertilized and retracted into the brood sac, JH synthesis is low. JH synthesis remains low during gestation when the developing embryos are incubated in the brood sac and no additional oocytes are matured (Tobe, S. and Stay, B., 1977). The peak

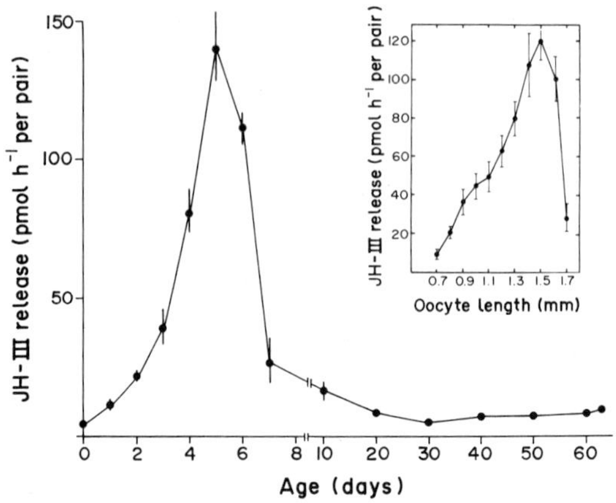

FIG. 2. Rate of JH release by CA from adult female *Diploptera punctata*. Mating occurs on the day of adult emergence (day 0). Eggs are ovulated on day 8 and retracted in the brood sac. Gestation lasts until around day 63, when first stage larvae are released (parturition). Inset shows the relation between oocyte length and rate of JH release during the first seven days of the adult stage. (Adapted from Feyereisen, R. *et al.*, 1981c; see also Tobe, S. and Stay, B., 1977 and Szibbo, C. and Tobe, S., 1982a.)

in nucleocytoplasmic ratio in the CA, already described by Engelmann, F. (1959a), coincides with the peaks in CA volume, CA cell number and JH synthesis (Szibbo, C. and Tobe, S., 1981a). However, changes in JH synthesis during this cycle are over 10-fold, whereas the morphometric parameters change less dramatically. Thus, under these normal conditions, CA morphometry has correctly predicted the timing but not the level of CA activity. The cycle of JH synthesis by CA of mated female *D. punctata* has been correlated with the hemolymph titer of JH determined by GC–MS (Tobe, S. *et al.*, in preparation). Remarkably, the very high rates of JH synthesis observed during the period of rapid vitellogenic growth of the oocytes are associated with very high (6 μM) hemolymph titers of JH. The results of this study suggest that CA activity is the essential modulator of JH titer during a gonotrophic cycle in this insect, and that the radiochemical assay gives an accurate insight of CA activity *in vivo*. The pattern of JH synthesis during the gonotrophic cycle of *N. cinerea* is very similar to that described for *D. punctata* and has been related to CA volume and hemolymph titers of JH, measured by the *Galleria* bioassay (Lanzrein, B. *et al.*, 1978). The results of an *in vivo* assay of CA activity, the stimulation of ovarian growth measured 7 days after implantation of CA into decapitated females, do not correlate well with the results of *in vitro* assays of CA activity, and this again demonstrates the inadequacy of *in vivo* bioassays (see section 1.2). When JH titer and *in vitro* CA activity were compared in individual insects no obvious correlation was seen and Lanzrein, B. *et al.* suggested that JH may be synthesized in pulses rather than continuously, possibly in response to a negative feedback mechanism.

Examples of species where oocyte growth is synchronous, but where successive gonotrophic cycles are contiguous, are *Schistocerca gregaria* and *Locusta migratoria*. In these locusts the second and subsequent gonotrophic cycles are shorter than the first, with vitellogenesis of the penultimate oocyte starting just after chorionation and ovulation of the basal oocytes (Tobe, S. and Pratt, G., 1975a; Tobe, S., 1977b; Ferenz, H.-J. and Kaufner, I., 1981). During the first gonotrophic cycle, a peak of JH synthesis is observed at the onset of the previtellogenic period of the basal oocytes (Tobe, S. and Pratt, G., 1975a; Ferenz, H.-J. and Kaufner, I., 1981; Girardie, J. *et al.*, 1981). The locust CA then show signs of cyclical activity, but the relation to ovarian development is imprecise. Tobe, S. and Pratt, G. (1975) suggest that for *S. gregaria* the later periods of high JH synthesis correspond with the onset of previtellogenic growth in each batch of penultimate oocytes. However, they also correspond with periods of vitellogenesis in the basal oocytes. No relationship between length of the basal oocyte and JH synthetic rate is apparent in *L. migratoria* (Girardie, J. *et al.*, 1981; Ferenz, H.-J. and Kaufner, I., 1981). JH titers in whole-body extracts of *L. migratoria* also show a high degree of scatter during the period of vitellogenic growth (Rembold, H., 1981).

Examples of species where oocyte growth is synchronous, but where successive gonotrophic cycles are overlapping, are *Periplaneta americana* and *Melanoplus sanguinipes*. In mated *P. americana* the penultimate oocyte becomes vitellogenic before the basal oocyte of the same ovariole is ovulated, allowing rapid cycles of ootheca formation. JH synthesis by CA *in vitro* shows a small peak prior to the onset of vitellogenesis in the first wave of oocytes, followed 4–5 days later by the first of a succession of larger peaks (Fig. 3). Each of these larger peaks (3–5-fold increase in rates of JH synthesis) correspond with late vitellogenesis in basal oocytes and early vitellogenesis in penultimate oocytes. Thus, in mated *P. americana* there is a regular succession of peaks of CA activity and ootheca formation (Weaver, R. *et al.*, 1975; Weaver, R. and Pratt, G., 1977). In *M. sanguinipes*, overlap of gonotrophic cycles is even more pronounced, with the antepenultimate oocyte often entering vitellogenesis before ovulation of the basal oocyte. This makes it difficult to demonstrate a functional relationship between CA activity and maturation of any given wave of oocytes. There is however a temporal correspondence between peaks of JH release by CA *in vitro* and the beginning of rapid yolk deposition in each cycle of oocyte development (McCaffery, A. and McCaffery, V., 1983).

In contrast to the above examples of species where oocyte growth is synchronous, complete intra-ovarian asynchrony and continuous egg production is observed in the telotrophic ovaries of *Tenebrio molitor*. The CA become active just before the largest oocytes become vitellogenic. Thereafter wide variations of JH synthesis in individual females and of basal oocyte lengths in each ovary make any correlation impossible (Weaver, R. *et al.*, 1978, 1980). Although not commented upon by Kramer, S. (1978) and Khan, M. *et al.* (1982a) (see also section 3.4), a similar situation may be occurring in adult *Leptinotarsa decemlineata* reared under long-day conditions.

The patterns of CA activity during reproduction suggest that a precisely regulated reproductive mode is associated with precisely regulated CA activity. It has been suggested that the degree of regulation of CA activity may be related to the sensitivity of the physiological processes associated with oocyte growth (Tobe, S. and Feyereisen, R., 1983). Loose regulation of CA activity would occur when JH titer simply needs to exceed a given threshold. Precise regulation of the CA would occur when the rate of oocyte growth shows a dose–response to JH.

2.5 Sexual dimorphism

Differences in size or ultrastructure of male and female CA have been reported frequently, especially during the adult stage (Pflugfelder, O., 1958; Cassier, P., 1979). In some studies a functional dimorphism was shown by transplantation bioassays (Fukuda, S., 1963; Pener, M., 1967; Brousse-Gaury, P. and Cassier, P., 1975). Dimorphism is often dramatic in social insects (e.g. Lüscher, M., 1976; Lanzrein, B. *et al.*, 1977). In adult male *Hyalophora cecropia* the CA are about 5 times larger than in females (Gilbert, L. and Schneiderman, H., 1961), but when activity was tested by transplantation into brainless pupae, both male and female glands were reported to have the same effect, i.e. a small percentage of pupae transformed into a mixture of pupa and adult (Williams, C., 1959). In contrast, Bhaskaran, G. *et al.* (1980a) observed that female CA are much less active than male CA, and reported that JH-I acid is quite active in the host organism for this CA bioassay, brainless *H. cecropia* dauerpupae. The accessory sex gland is the repository for the large amounts of JH found in male *H. cecropia* moths (Shirk, P. *et al.*, 1976). Peter, M. *et al.* (1981) reported an intriguing functional dimorphism in the CA of this species, with male CA apparently lacking the *O*-methyl transferase activity and thus releasing JH-I acid and JH-II acid *in vitro* instead of the esterified hormones. The male accessory sex glands store JH after methylation of the acids by an *O*-methyl transferase with the same specificity as the enzyme from female CA (Peter, M. *et al.*, 1981; Weirich, G. and Culver, M., 1979; Dahm, K. *et al.*, 1981).

Another type of functional sexual dimorphism of the CA was studied in adult *Diploptera punctata* by Szibbo, C. and Tobe, S. (1982). Male CA are smaller and contain fewer cells than female CA, and this may be related to differences in cumulative somatic growth (one less larval stadium in males). Male CA produce JH-III at continuous low levels when compared to females (Tobe, S. *et al.*, 1979) and are less effective than female CA in supporting complete oocyte growth in allatectomized females (Szibbo, C. and Tobe, S., 1982). Male CA implanted into male or female hosts synthesize much less JH than female CA implanted in insects of either sex, as measured by the radiochemical assay. This dimorphism is not only quantitative, but it appears that male CA are unable to respond to the stimulatory signals to which female CA can respond (see section 4.2.2.). Thus in males, the rate of JH synthesis expressed per CA cell is much lower than in females. It does not increase in a female host as opposed to the "compensatory response" exhibited by female CA (Szibbo, C. and Tobe, S., 1981b, 1982). A lack of functional dimorphism between male and female CA was reported by Khan, M. *et al.* (1982a) in adult *L. decemlineata*.

3 MODULATION OF CA ACTIVITY BY EXTERNAL FACTORS

Making the correct physiological adjustments in response to environmental stimuli is a well recognized *raison d'être* of neuroendocrine integration. An insect's perception of social, chemical or physical factors may ultimately result in altered CA activity. The effects of external factors on CA activity have provided experimental frameworks for the study of CA control. This section analyzes the few studies in which modulation of CA activity by such factors has been studied by following JH synthesis *in vitro*.

3.1 Mating

Mating in female *Periplaneta americana* usually occurs 4–5 days after adult emergence, and the first ootheca is produced 4 days later. Only 50% of virgin females produce an ootheca in the first 28 days following emergence. This is caused by a dramatic reduction in the rate of vitellogenic growth of the first group of basal oocytes (Weaver, R. and Pratt, G., 1977). Whereas rates of JH synthesis are similar in virgin and mated females during the first 6 days after adult emergence, the cyclic activity of the CA is then lost in virgins and the rate of JH synthesis gradually declines to a plateau of low activity (Fig. 3) (Weaver, R. and Pratt, G., 1977). Significantly, mating after 3 weeks of enforced virginity restores (after a brief delay) the high and cyclic activity of the CA, as well as normal ovarian cycles (Weaver, R. and Pratt, G., 1977). Mating in female *D. punctata* usually occurs immediately following adult emergence and is essential for a normal rate of oocyte growth (Engelmann, F., 1959a; Roth, L. and Stay, B., 1961). CA activity of virgin *D. punctata* is very low compared to the activity of mated female CA, and vitellogenic oocytes are observed in only very few insects (Stay, B. and Tobe, S., 1977). Mating in

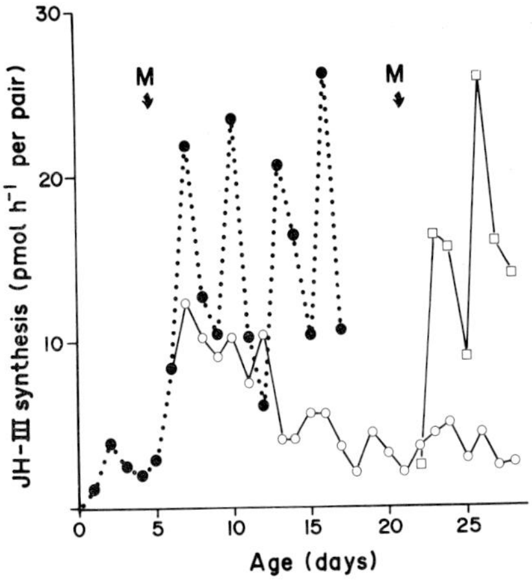

FIG. 3. Rate of JH synthesis by CA from adult female *Periplaneta americana*. Mating normally occurs on days 4–5 (arrow) and is followed by a rapid succession of CA activity peaks (●···●), which do not occur in virgin females (○—○). Mating of females deprived of males for 21 days of adult life (arrow) restores the cyclic activity of the CA (□—□). (Adapted from Weaver, R. and Pratt, G., 1977.)

female *Nauphoeta cinerea* occurs 3–4 days after adult emergence. Whereas ovulation first occurs on day 12 in mated females, it is delayed until day 24 in virgin females, and only fragile egg cases are expelled (Lanzrein, B. *et al.*, 1981a). JH synthesis by CA of virgin females shows a peak on day 9 as in mated females, but it is much smaller (5-fold), and JH titers measured by the *Galleria* bioassay are at least 10 times lower in virgins than in mated females (Lanzrein, B. *et al.*, 1981a,b). In these three species of cockroaches, mating stimuli are necessary for the expression of a normal pattern of JH synthesis by the CA of adult females. In *D. punctata* mating can be simulated by denervation of the CA of virgin females (Stay, B. and Tobe, S., 1977; Tobe, S. *et al.*, 1981). Mating stimuli affect reproductive parameters in varying degrees, depending on the species (reviews by Cassier, P., 1967; Engelmann, F., 1970). However, in some species mating has no effect on JH synthesis. In adult female *Leptinotarsa decemlineata* reared in long-day conditions, mating increases the rate of oviposition but the rate of JH synthesis by CA of mated and virgin females is identical (Dortland, D., 1979).

3.2 Social hierarchy

Röseler, P.-F. and Röseler, I. (1978) have measured rates of JH-III synthesis by CA of bumblebee workers. Complete egg maturation occurs only in the absence of a queen in *Bombus terrestris* workers. In the presence of a queen (queenright colonies), the workers do not produce eggs. Within 1 day after adult emergence the rate of JH synthesis in queenless workers is higher than in queenright workers, and this difference is maintained throughout the first gonotrophic cycle (5 days). Excretion of radioactivity from injected [^{3}H]JH-I occurs at the same rate in queenright and queenless workers, but JH titers obtained by the *Galleria* bioassay are much higher in queenless workers. Röseler, P.-F. and Röseler, I. (1978) conclude that the JH titer in *B. terrestris* workers is regulated mainly by variations in the rate of JH synthesis. The lower activity of CA from queenright workers is the result of inhibition by a pheromone from the queen's mandibular glands (Röseler, P.-F. *et al.*, 1981). At the end of hibernation in adult *Polistes gallicus* wasps one female becomes the dominant egg-layer while the others become subordinated. Little oocyte growth occurs in subordinated females, but these insects help raise the larvae. Röseler, P.-F. *et al.* (1980) have shown that CA activity of dominant females is generally higher than CA activity of subordinated females. This difference in JH synthetic rates *in vitro* is related to differences in CA volumes. Causal relationships between JH levels and aggressive behavior in the determination of social dominance in wasps are not yet fully understood (Barth, R. *et al.*, 1975). A number of social signals have been reported to directly or indirectly affect CA activity (Cassier, P., 1967; Doane, W., 1973), but most studies rely essentially on measures of CA volume and thus need to be verified by *in vitro* techniques.

3.3 Nutrition

The commitment to metamorphosis can be delayed in some insects by starvation. In some Lepidoptera a threshold size must be reached for metamorphosis to occur (Allegret, P., 1964; Nijhout, H., 1975a). Allatectomy prevents the supernumerary larval molt induced by starvation in fifth-stage larvae of *Manduca sexta* (Bhaskaran, G. and Jones, G., 1980). Consequently, the CA of larvae may be involved in the neuroendocrine response to starvation which is usually attributed to PTTH (Doane, W.,

1973). However, Cymborowski, B. *et al.* (1982) observed a decrease in JH esterase levels and in CA activity in starved *M. sexta* larvae, as well as an elevated JH titer. Thus, delayed metamorphosis may not necessarily result from the release of an allatotropin.

The influence of nutrition on CA activity is best documented in the adult stage in relation to reproduction. Whereas starvation would seem to prevent a decline of CA activity in larvae of some insects, the opposite is often seen in adults (Johansson, A., 1964; Cassier, P., 1967; Engelmann, F., 1970). Effects of starvation on JH biosynthesis were studied by Tobe, S. and Chapman, C. (1979) in adult female *Schistocerca gregaria*. Starvation results in a decline in JH synthesis within a few days. When starved locusts are allowed to feed again, JH synthesis increases somewhat, but an irreversible decrease of the reproductive potential may have taken place. Weaver, R. and Pratt, G. (1981) studied JH synthesis by CA from starved and re-fed *Periplaneta americana* adults (Fig. 4). Starvation, when initiated at the time of ootheca formation, delays deposition of the next ootheca by 2–3 days and rapidly inhibits ovarian activity. CA activity rapidly loses its cyclicity and after 11 days of starvation JH synthesis falls below the limit of detection. However, feeding of starved females rapidly leads to a resumption of CA activity and oocyte growth. Changes in CA morphometry and rates of oocyte growth after starvation and subsequent feeding were previously studied by Brousse-Gaury, P. (1977). In adult female *Leptinotarsa decemlineata*, starvation also causes a rapid decline in JH synthesis (Khan, M. *et al.*, 1982b) and prevents oviposition. Five days of starvation also considerably lowers the hemolymph JH-III titer. The CA of newly emerged beetles increase in activity in the first 4–16 h, whether fed or not. Starvation-induced CA inhibition is, however, apparent within 24 h of emergence. Inhibition of JH biosynthesis in starved *L. decemlineata* is dependent upon intact innervation of the CA (Khan, M. *et al.*, 1982b). The cockroach *Diploptera punctata* is one of the few examples of an insect in which a gonotrophic cycle can occur without delay even when starved (Roth, L. and Stay, B., 1961). Starvation does not influence the cycle of JH synthesis observed in adult females of this species (Rankin and Stay, personal communication).

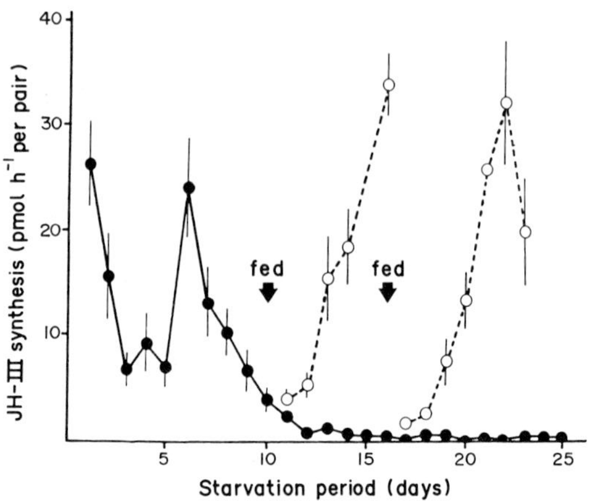

FIG. 4. The effects of starvation and subsequent feeding on rates of JH synthesis by CA from adult female *Periplaneta americana*. The cockroaches were withdrawn from food at the time of ootheca formation. Feeding was resumed after 10 or 16 days of starvation (arrows). (From Weaver, R. and Pratt, G., 1981.)

Engelmann, F. (1965, 1970) has advocated the idea that CA activity can be affected directly by the internal nutritional milieu, in addition to those stimuli which control the CA via the central nervous system. A decrease in hemolymph protein levels, for instance, was thought to inactivate the CA directly during starvation or at the end of an egg maturation period. Although numerous control mechanisms for CA activity have now been confirmed or discovered (section 4), Engelmann's hypothesis has not been conclusively invalidated. In adult female *D. punctata*, hemolymph protein levels are already minimal on day 3, i.e. 2 days before the peak of JH synthesis, and no change in hemolymph protein level is observed when JH synthesis decreases at the end of this cycle (Mundall, E. *et al.*, 1981). However, such a direct comparison should be done in an insect species where starvation does result in a decrease in CA activity. Except for the absolute requirement for methionine, rather little is known about the metabolic requirements of the CA. Thus, the "internal nutritional milieu" hypothesis still needs to be tested by biochemical techniques.

3.4 Photoperiod

Adult diapause is under photoperiodic control in many insects. In *L. decemlineata*, photoperiod, temperature and food availability all contribute to the "diapause syndrome" (De Wilde, J. and De Loof, A., 1973) which has also been described as a case of "pseudo allatectomy" (De Wilde, J. and De Boer, J., 1961). Rearing adult beetles under short-day conditions leads to diapause, whereas rearing under long-day conditions leads to reproductive activity. De Kort, C. *et al.* (1981, 1982) have measured JH titers in the Colorado potato beetle by the GC–MS method of Bergot, B. *et al.* (1981). JH-III was the only JH homolog detected, in high levels in long-day beetles and in low levels in short-day prediapause and diapause beetles. These results are in agreement with the bioassay data of De Wilde, J. *et al.* (1968). Rates of JH synthesis are higher in long-day beetles than in short-day beetles (Kramer, S., 1978; Khan, M. *et al.*, 1982a). Consequently, CA activity is the most important factor in JH titer regulation in *L. decemlineata* (Kramer, S., 1978). Interestingly enough, CA volume (or the surface of its largest cross-section) has been considered to be a good indicator of CA activity in this species, because changes in CA volume are in general agreement with bioassay data for JH titer (De Wilde, J. *et al.*, 1968; Schooneveld, H. *et al.*, 1977, 1979). However, radiochemical assay of JH biosynthesis by CA from long-day females did not confirm this relationship, especially in larger glands which exhibited a great variability (Fig. 5).

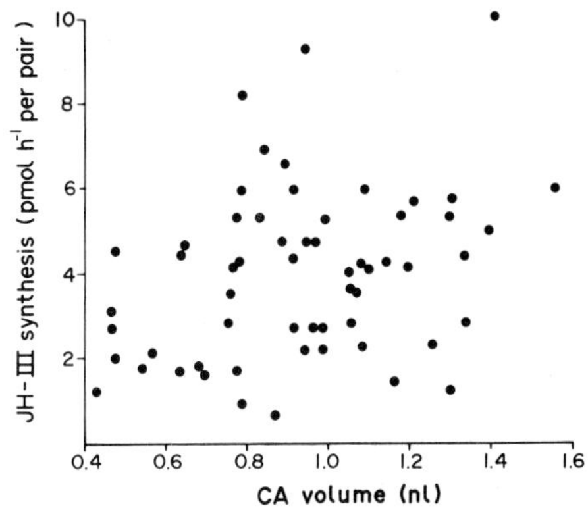

Fig. 5. Rate of JH synthesis by CA from adult female *Leptinotarsa decemlineata* reared under long-day conditions, in relation to the volume of the paired glands. (From Khan, M. *et al.*, 1982a.)

Ferenz, H.-J. (1981) has reported another example of photoperiod influence on CA activity. In the carabid beetle *Pterostichus nigrita* short-day conditions after adult emergence induce the previtellogenic growth of the oocytes which is correlated with low CA activity. When short-day beetles are transferred to long-day conditions a considerable increase in JH synthesis is observed. This increase in CA activity, which is concomitant with the onset of vitellogenesis, occurs between 5 and 10 days after the switch in photoperiod and is maintained for at least 30 days. The activation of CA activity following transfer from short-day to long-day photoperiod is also observed in males although to a lesser extent. The physiological function of this effect is unclear in males, but it can be concluded that the neuroendocrine integration of the photoperiodic signals is not sex-specific in *P. nigrita*. Interestingly, when newly emerged beetles are reared under long-day conditions, JH synthesis is barely detectable and no previtellogenic growth of the oocytes is observed (Ferenz, H.-J., 1981). Accordingly, previtellogenic growth of the oocytes is correlated in *P. nigrita* with a low CA activity, and vitellogenesis with a high CA activity, suggesting a different threshold or a different control mechanism for these two events (see section 2.4). Signals other than photoperiod can help differentiate between previtellogenic and vitellogenic processes and thus CA involvement. Measuring CA activity in an insect requiring a blood-meal to complete a gonotrophic cycle (Gwadz, R. and Spielman, A., 1973) would thus be of great interest.

4 CONTROL OF CA ACTIVITY

The various patterns of CA activity during development and their modulation by external factors, imply the existence of control mechanisms for CA activity. Paradoxically, there is so much indirect information on so diverse control mechanisms, that a consensus has not emerged. Whereas most "classical schemes" for the hormonal control of post-embryonic development clearly indicate that molting hormone production by the prothoracic glands is under the control of PTTH, there is usually no depiction of a control mechanism for JH production (see e.g. Gilbert, L., 1964; Doane, W., 1973). The anatomical connection between the CA and the brain is probably sufficient to imply the possibility of control. Fraenkel, G. (1980) has observed that "the role of allatotropic hormones, which mobilize the juvenile hormone, has been postulated or claimed ever since the parallel activity of the prothoracicotropic hormone was recognized. Evidence for such hormones seems overwhelming ..., but no attempts of isolation or characterization seem to have ever been forthcoming." The development of radiochemical and radioimmunological procedures for the assay of JH synthesis *in vitro* has provided a new impetus to studies of the control of CA activity. This section analyzes the contributions of *in vitro* assays for JH synthesis to our understanding of CA regulation. "A common principle has not been found and probably should not be expected to exist" (Engelmann, F., 1970). Therefore, individual regulating factors and mechanisms will be reviewed, with the understanding that their relative importance may vary tremendously from one species to another, and indeed also from one developmental stage to another.

4.1 Innervation of the CA

The innervation of the CA is complex and varies from species to species, making any generalization impossible. However, connections of the CA to the corpora cardiaca (CC), to the brain and to the suboesophageal ganglion are of course suggestive of functional relationships. Figure 6 shows a "composite model" of CA innervation. The anatomical mapping of both conventional and neurosecretory axonal pathways is a necessary complement to the physiological studies on CA regulation but shall not be reviewed here (see e.g. Mason, C., 1973; Nijhout, H., 1975b; Fraser, J. and Pipa, R., 1977; Pipa, R., 1978). Axons from the CC to the CA constitute the nervi corporis allati I (NCA-I) and may originate in the contralateral medial neurosecretory cells (MNSC) or lateral neurosecretory cells (LNSC), as well as in the ipsilateral LNSC or tritocerebral neurosecretory cells (TNSC) of the brain. Depending on their anatomical origin, these axons reach the CC via one of the nervi corporis cardiaci (NCC), namely the NCC-I for axons originating in the MNSC, NCC-II for LNSC and NCC-III for TNSC respectively. In addition to the brain–NCC–NCA-I pathways, the CA are also innervated via the NCA-II originating in the suboesophageal ganglion. For the sake of clarity, we shall consider here only those experiments involving nerve section, and discuss destruction of brain NSC later. It may seem difficult to assess correctly the role of CA innervation by measuring CA activity *in vitro*, which implies severance of all anatomical connections of the CA. Two different approaches have been taken to solve this apparent dilemma.

The first approach is to section one or all connections of the CA, leave the CA in place and measure CA activity after a certain time. This is the classical approach, where CA activity is measured by a specific biological response, such as supernumerary larval molt or growth of oocytes (review in Raabe, M., 1982). Of course, CA activity *in vitro* can also be measured after leaving the denervated glands *in vivo* for various lengths of time, and thus the effect of the hormonal milieu on innervated or denervated glands can be compared.

In the second approach, CA are incubated *in vitro* either alone or in the presence of attached brains and/or CC. A comparison of rates of JH release in both types of incubations gives an idea of the importance of anatomical connections of the CA with the brain and/or CC but does not exclude humoral factors

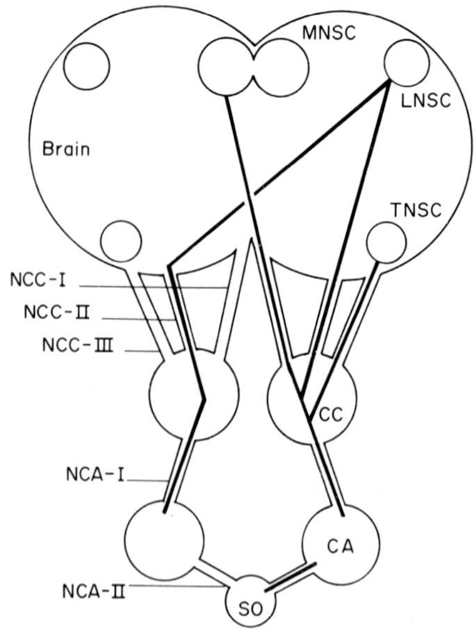

FIG. 6. Scheme of CA innervation in insects, showing the nervi corporis cardiaci (NCC) and the nervi corporis allati (NCA). Axonal pathways demonstrated by cobalt chloride fills are also schematized, but do not necessarily occur simultaneously in every species. Anatomical accuracy is not intended. CC: Corpus cardiacum; CA: corpus allatum; SO: suboesophageal ganglion.

released *in vitro* and reaching the CA via the incubation medium.

Severance of the NCA-I at any time during the first gonotrophic cycle of *Schistocerca gregaria* results in a rapid decline of CA activity (Tobe, S. *et al.*, 1977). JH synthesis was measured 1–4 days after the operation and compared to the CA activity of sham-operated insects; it appeared that the CA had virtually lost the ability to synthesize JH within 1 day of NCA section. JH synthesis *in vitro* (during a 3-h incubation period) was also measured 3–24 h after section of the NCA-I to determine how rapidly the effect of nerve section manifests itself *in vivo*. The effect was already detectable after 3 h, and by 6 h after NCA-I section, *in vitro* incubation of the CA revealed an almost negligible level of JH synthesis (Tobe, S. *et al.*, 1977). The results can be related to *in vitro* incubations which showed that the rate of JH release declines rapidly after approximately 3 h (Pratt, G. *et al.*, 1975a). These results suggest that the CA of adult female *S. gregaria* receive a stimulatory or sustaining "factor" via the NCA-I and not via the hemolymph. Because the NCA-I carries both conventional and neurosecretory axons, the physiological nature of this "factor" is unknown. Neurosecretory axon endings have been observed in the CA of a large number of species from various orders (review by Raabe, M., 1982).

Whereas the CA of adult female *S. gregaria* provide an example of stimulatory influence of intact nerve connections, a few examples of inhibitory influences have been described. In adult female *Diploptera punctata*, total denervation of the CA in virgins mimics the effect of mating (Engelmann, F., 1959a), i.e. the CA allowed to remain *in vivo* undergo a normal cycle of JH synthesis (Stay, B. and Tobe, S., 1977). Stimulation of CA activity in virgin *D. punctata* is observed within 4 h of NCA-I section, but no stimulation of JH synthesis is observed *in vitro* when CA from virgin females are incubated for up to 24 h (Tobe, S. *et al.*, 1981). Thus, mating releases the CA from strong inhibitory signals travelling via the NCA-I, but denervated CA can increase their activity only in a suitable hormonal milieu. After unilateral NCA-I section in virgin females, the fully innervated CA remains inactive, whereas the denervated CA is activated *in vivo* (Tobe, S. *et al.*, 1981). This shows that CA activation after NCA-I section does not result from the decreased level of a humorally acting inhibitor. Rapid activation of the ipsilateral CA is observed in virgin females after unilateral section of the NCC-I or of both NCC-I and NCC-II. Thus, inhibitory signals from the brain are transmitted to the CA via the NCC-I/NCA-I pathway (Tobe, S. *et al.*, 1981). Again, whether these signals are of neural or neurosecretory nature is unknown. "Nervous inhibition" is observed not only in virgin females, but occurs with a lower intensity in mated females. Total denervation of the CA in mated females accelerates the cycle of JH synthesis to a slight degree, and the operation may even result in a small stimulation of CA activity (Stay, B. and Tobe, S., 1977; Tobe, S. and Stay, B., 1980). During the last larval stage of *D. punctata*, the very low levels of JH synthesis (Fig. 1) are caused, in part, by "nervous inhibition" since denervation causes both an increase in JH synthesis and a supernumerary larval molt (Szibbo, C. and Tobe, S., 1983). Intact nervous connections of the CA are also needed to implement negative feedback responses to an elevated JH titer (see section 4.2.1.).

Inhibition of JH biosynthesis in starved adult *Leptinotarsa decemlineata* (see section 3.3) occurs only if the nervous connections of the CA are intact (Khan, M. *et al.*, 1982b). Denervated CA of starved animals or CA implanted in starved hosts all show higher rates of JH synthesis *in vitro* than innervated glands. Starvation therefore does not seem to influence CA activity directly because of a deficient nutritional milieu, but rather through inhibitory stimuli reaching the CA via nervous connections, as also suggested for starved *Periplaneta americana* by Pipa, R. (1982). In normally fed *L. decemlineata*, severance of nervous connections does not modify either the rate of JH synthesis or the JH titer (Schooneveld, H. *et al.*, 1979).

In conclusion, signals arriving at the CA by way of nervous tracts (NCA-I) can either maintain or suppress JH biosynthesis, depending on the species. Restraining signals can be lifted by appropriate external factors integrated by the brain, such as feeding and mating, leaving the CA free to respond to humoral factors. Control of CA activity by signals from the suboesophageal ganglion via the NCA-II has not been conclusively demonstrated. The existence of humoral factors directly controlling CA activity has been postulated for a long time, but none has been characterized to date. The combination of *in vivo* experiments and *in vitro* determination of CA activity has, however, confirmed their existence.

4.2 Role of the JH titer: feedback loops

Mechanisms for the control of CA activity by experimentally modified titers of JH have been described, and demonstrate the existence of feedback loops. The structure of these loops may, however, be complex. The simplest feedback loop would involve a direct effect of JH on CA activity. Such an effect should be amenable to direct *in vitro* testing, but no clear evidence has been found. CA from *P. americana* are not affected by nominal JH-III concentrations of up to 1 mM *in vitro* (Pratt, G. and Finney, J., 1977), and CA from *Manduca sexta* were reported to be inhibited only by unphysiological (0.1 mM) concentrations of either JH-I, II or III (Kramer, S. and Staal, G., 1981). Khan, M. *et al.* (1982c) also reported that *L. decemlineata* CA are not affected by *in vitro* JH-III concentrations up to 20 μM. In addition, large amounts of JH biosynthesized *in vitro* do not affect the linearity of JH release rates in *D. punctata* over a 48 h period (Friedel, T. *et al.*, 1980). When more than five pairs of *Locusta migratoria* CA are incubated together in 1 ml of medium, the rate of JH release measured by RIA drops considerably after 24 h *in vitro*, but is restored by transfer of the glands to a fresh medium (Baehr, J.-C. *et al.*, 1981); either feedback inhibition of the CA by JH or exhaustion of the incubation medium could explain these results.

4.2.1 Control of CA activity by elevated JH titer

Elevated JH titer can be obtained either by implantation of additional CA or by treatment of insects with JH or JH analogs. Such experiments have revealed a suppression of CA activity, which had already been postulated on the basis of indirect evidence (reviews by Cassier, P., 1967; Sláma, K. *et al.*, 1974). Daily applications of 30 μg of JH-I isomers to adult female *Leptinotarsa decemlineata* reared under long-day conditions result in a sharp reduction of JH synthesis (Schooneveld, H. *et al.*, 1979). This JH analog treatment leads to (in this species) an 80% inhibition of CA activity within 5 days. A single injection of JH-III (50 μg) can also inhibit CA activity, an effect which is apparent within 12 h after injection and is maintained for at least 72 h (Khan, M. *et al.*, 1982c). Because implantation of three pairs of CA also causes a significant suppression of JH synthesis by the host glands, it can be concluded that the extra CA caused at least a temporary elevation in JH titer. Khan, M. *et al.* (1982c) showed that this suppression of JH synthesis is mediated by signals which reach the CA both through the hemolymph and through nervous tracts. Indeed, although injection of JH-III into beetles with denervated CA caused an inhibition of CA activity, the same treatment to beetles implanted with a pair of extra CA caused a greater inhibition of the host innervated CA than of the implanted CA.

The effects of elevated JH titer on CA activity were also studied in adult female *Diploptera punctata.* Topical application of the JH analog hydroprene (ZR 512) to newly emerged mated females caused a suppression of the normal cycle of JH synthesis which is more pronounced at 100 μg than at 25 μg, suggesting a dose-dependent effect (Tobe, S. and Stay, B., 1979). Treatment with the JH analog hydroprene does in fact amount to an elevation in JH titer, because oocyte growth is stimulated in treated females or restored in allatectomized insects. (These experiments can also be interpreted otherwise, i.e. suppression of CA activity may result from the early completion of vitellogenesis, which is known to inhibit the CA, see section 4.3.) Elevated JH titer can also be obtained by implantation of supernumerary CA of the same age, which results in accelerated oocyte growth and advanced spermatophore release and oviposition (Stay, B. and Tobe, S., 1978; Tobe, S. and Stay, B., 1980). This elevated JH titer can have different effects on host and donor CA as shown by Tobe, S. and Stay, B. (1980). When CA from newly emerged females are implanted into insects of the same age, the implanted CA undergo a normal cycle of JH synthesis, but the innervated host glands are considerably suppressed (Fig. 7). Denervation of the host glands lifts this inhibition and results in similar rates of JH synthesis by host and implanted CA. As a result, the gonotrophic cycle is accelerated and vitellogenesis is completed ahead of the normal schedule. Thus, CA innervation is necessary in *D. punctata* for the implementation of a negative feedback suppressing JH synthesis in the presence of an elevated JH titer. Treatment of *M. sexta* larvae with the JH agonist hydroprene causes a reduction of JH-I and JH-II levels measured on whole-body extracts and a reduction of JH synthesis, indicating the effectiveness of the negative feedback loop in this species (Staal, G. *et al.*, 1981; Kramer, S. and Staal, G., 1981; Edwards, J. *et al.*, 1983). The negative feedback response to elevated JH titer is not always fully effective, as evidenced by the numerous reports of deranged metamorphosis etc. caused by CA implantations, but such indirect evidence cannot exclude the possibility of "nervous inhibition" which is ineffective on implanted CA.

The existence of a positive feedback loop, which would stimulate JH synthesis in response to an elevated JH titer, was suggested (Tobe, S. and Stay, B., 1979). Topical application of 2.5 μg hydroprene to newly emerged female *D. punctata* reportedly stimulated JH synthesis within a narrow temporal gate on day 4. Positive feedback loops are, however, highly unstable and their operation in the control of JH synthesis remains conjectural.

4.2.2 Compensating stimulation of JH synthesis following unilateral allatectomy

Unilateral allatectomy of mated females of *D. punctata* up to 3 days after emergence does not result in a significant delay of either spermatophore release, oocyte growth or oviposition (Stay, B. and Tobe, S., 1978; Tobe, S. and Stay, B., 1980; Szibbo, C. and Tobe, S., 1981b). Are these JH-dependent physiological events fully satisfied with a JH production apparently cut in half? Quite the contrary, unilateral allatectomy stimulates JH synthesis of the remaining CA, resulting in a perfect cycle of JH synthesis, where the single gland has exactly doubled its rate of JH synthesis. This remarkable functional compensation phenomenon is observed whether the single remaining CA is innervated or not (Tobe, S. and Stay, B., 1980), indicating that stimulation of the CA is the response to a change in the hormonal milieu. The normal cycle of JH synthesis during the gonotrophic cycle of *D. punctata* is accompanied by changes in CA volume and CA cell number (see section 2.4; Szibbo, C. and Tobe, S., 1981a). After unilateral allatectomy the remaining CA undergoes changes in volume and cell number which are exactly parallel to those of control glands, and no increase in the percentage of cells labelled with [^{3}H]thymidine is observed when compared to control glands during the gonotrophic cycle (Szibbo, C. and Tobe, S., 1981b). Thus, compensation is not the result of gland hypertrophy, but the rate of JH release per cell or per unit CA volume may simply be doubled during the gonotrophic cycle. Alternatively, the number of CA cells actually engaged in JH synthesis may be increased, but in any case, CA volume or cell number are obviously not good indicators of CA activity. As discussed above (see section 2.5), CA of adult male *D. punctata* do not respond to the stimulatory signals, i.e. they do not compensate (Szibbo, C. and Tobe, S., 1982).

Schooneveld, H. *et al.* (1979) studied the effects of unilateral allatectomy in adult female *Leptinotarsa decemlineata* reared under long-day conditions. This operation leads to a doubling of the remaining CA activity as measured *in vitro* after 7 days, and as also evidenced by a normal JH titer. In *L. decemlineata* this functional compensation is followed by a significant increase in CA volume. Thus, a functional compensation following unilateral allatectomy has been demonstrated in two species, but is suggested in other species as well (e.g. Scharrer, B., 1946; Staal, G., 1961; Cassier, P., 1966, 1967; Sehnal, F. and Granger, N., 1975).

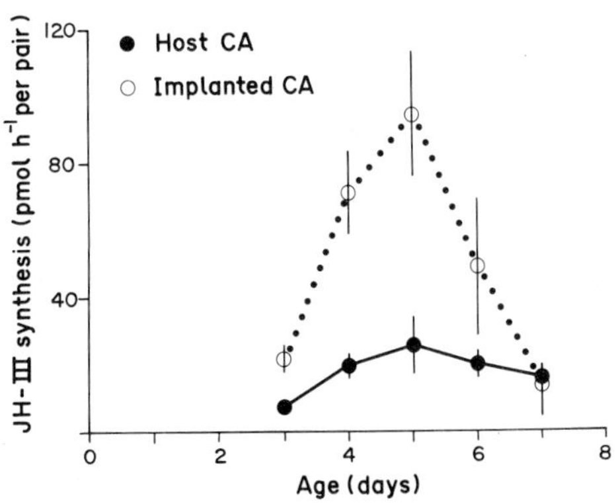

FIG. 7. Rate of JH synthesis by implanted CA (○) and host, innervated CA (●) in adult female *Diploptera punctata*. A pair of CA from day 0 cockroaches was implanted into mated female hosts on day 0. Oocyte growth was accelerated in the experimental animals. (From Tobe, S. and Stay, B., 1980.)

In conclusion, experiments involving an experimental elevation or lowering of the JH titer have clearly demonstrated the existence of negative feedback control of JH synthesis. Such a mechanism may not be equally effective, or sensitive to changes in JH titer, in different species (Huibregtse-Minderhoud, L. *et al.*, 1980; Lanzrein, B. *et al.*, 1981a). Brain neurosecretory cells are possibly involved in feedback loops controlling JH titer (McCaffery, A. and Highnam, K., 1975a,b) but direct evidence is still lacking (see Steele and Davey, vol. 8).

4.3 Control of CA activity by the ovary

4.3.1 Ovarian stimulation of CA activity

Ovariectomy of mated adults leads to a rapid decline of JH synthesis in *D. punctata*, and the operation performed on last-stage larvae also appears to suppress CA activity in adults (Stay, B. and Tobe, S., 1978). Lanzrein, B. *et al.* (1981a,b) showed that ovariectomy of newly emerged *Nauphoeta cinerea* adults abolishes the normal cycle of JH synthesis observed in this species. The rates of JH synthesis remain extremely low for at least 20 days, in fact even lower than in virgin females, resulting in very low hemolymph JH titers. Reimplantation of an ovary restores CA activity, suggesting that a stimulatory factor may be transmitted from the ovary by humoral means (Lanzrein, B. *et al.*, 1981a). Stay, B. *et al.* (1983) confirmed those results in *D. punctata* and showed that reimplantation of three ovarioles is sufficient to restore high CA activity. Low activity of CA in ovariectomized insects cannot be modified very much by denervation, suggesting that the putative humoral factor from the ovary does not activate the CA by lifting a "nervous inhibition". The nature of this ovarian stimulating factor is presently unknown; it is also unknown whether its effect on the CA is direct or indirect.

Ovariectomy considerably suppresses JH synthesis by CA of adult *Polistes gallicus*, even though some glands were considerably enlarged following the operation (Röseler, P.-F. *et al.*, 1980). Ovariectomy of adult *Acheta domesticus* also increases CA volume, but determinations of JH titer and CA activity *in vitro* revealed that JH synthesis is reduced in those conditions (Strambi, C., 1981). Ovariectomy in the last larval instar of *Periplaneta americana* leads to an overall decrease in CA activity of adults, and apparently to a loss of the precise, cyclic activity of the CA (Weaver, R., 1981). Thus, ovariectomy suppresses CA activity in a number of species, and CA volume increase after the operation does not necessarily indicate hyperactivity. The ultrastructural features of the CA in ovariectomized *Leucophaea maderae* (Scharrer, B., 1978) as well as their morphometric features (Scharrer, B. and Von Harnack, M., 1961) suggest hyperactivity, but this needs to be verified by *in vitro* procedures. Continued vitellogenin synthesis occurs after long-term ovariectomy in *L. maderae* (Engelmann, F., 1978), but a high JH titer or a high CA activity are not necessarily implied, as shown in ovariectomized *A. domesticus* (Strambi, C., 1981; Bradley, J. and Edwards, J., 1978) and *Diploptera punctata* (Stay, B. *et al.*, 1983). Interestingly, ovariectomy in *Labidura riparia* results in an increase in CA volume and an increase in hemolymph JH III titer (Baehr, J.-C. *et al.*, 1982). Although stimulation of the CA by the ovary has been inferred by indirect methods in some species (e.g. Wilhelm, R. and Lüscher, M., 1974; Srivastava, K. and Tiwari, R., 1978), ovariectomy has more often been thought to cause CA hyperactivity (reviews by Highnam, K., 1964; Cassier, P., 1967; Engelmann, F., 1970).

4.3.2 Ovarian inhibition of CA activity

At the completion of vitellogenesis in *D. punctata* CA activity drastically declines (Fig. 2). This phenomenon is independent of the age of the insect or of the CA, but seems to be linked to an appropriate physiological event in the ovaries. This is shown by CA implantation experiments between females of different ages (Tobe, S. and Stay, B., 1980). When CA from 2-day females are implanted into 0-day hosts, or when CA from 0-day females are implanted into 3-day hosts, the rate of JH synthesis of both pairs of CA declines synchronously at the completion of vitellogenesis. The youngest CA, even denervated, never reaches the high rates of JH synthesis that would be observed in a normal cycle (Tobe, S. and Stay, B., 1980). This inhibition of the CA is therefore not related to the negative feedback mechanism which in *D. punctata* operates *via* intact nervous connections (see section 4.2.1.)

but is dependent on the presence of an ovary. Humoral inhibition of CA activity by the ovary was shown by Stay, B. *et al.* (1980). CA from newly emerged females undergo a normal cycle of JH synthesis when implanted into newly emerged males, as would be expected from denervated CA maintained in females. The male milieu is, however, different in at least two respects: (1) female CA are activated more than 10-fold in the male milieu in the absence of an ovary; (2) at the end of this apparently normal cycle of activity, 7 days after implantation into males, the activity of female CA does not decline to the very low level (less than 20 pmol h^{-1} per pair) observed in normal females. Instead, their activity declines (for an unknown reason) only to about 50–60 pmol h^{-1} per pair (Stay, B. *et al.*, 1980; Friedel, T. *et al.*, 1980). However, implantation of both an ovary and a pair of CA from newly emerged females into males perfectly mimics the normal cycle of JH synthesis that would be observed in females (Stay, B. *et al.*, 1980). Towards completion of vitellogenesis the ovary therefore releases an inhibitory signal which, by humoral means only, achieves a dramatic decline in JH biosynthesis. Lanzrein, B. *et al.* (1981a,b) reached the same conclusion for *N. cinerea* but used a much simpler experimental protocol: if ovariectomy is performed just before the normal peak of JH synthesis (observed on day 9 in *N. cinerea*), this peak of CA activity is not observed (see above). However, if ovariectomy is performed on day 9 or 10, CA activity decreases only very slowly, showing that the rapid inactivation after day 9 is dependent on the presence of a maturing ovary (Lanzrein, B. *et al.*, 1981a,b). The nature of the ovarian factor(s) which inhibits the CA in *D. punctata* and *N. cinerea* is unknown, but indirect evidence suggests that ecdysteroids produced by the ovary may be involved. Injection of 20-hydroxyecdysone into males implanted with female CA mimics the implantation of an ovary, resulting in a dose-dependent inhibition of CA activity which becomes apparent *in vivo* within 24 h and is maintained for at least 4 days (Stay, B. *et al.*, 1980; Friedel, T. *et al.*, 1980). Incubation of *D. punctata* CA *in vitro* for up to 42 h in a medium containing up to 10 μM 20-hydroxyecdysone did not modify the rate of JH release, suggesting that the ecdysteroid effect *in vivo* is indirect or requires the presence of another factor absent in the *in vitro* conditions (Friedel, T. *et al.*, 1980). A decrease in CA activity can also be provoked by 20-hydroxyecdysone injections in normal females or by larval prothoracic gland implantation in ovariectomized females (Friedel, T. *et al.*, 1980; Lanzrein, B. *et al.*, 1981a,b; see also Engelmann, F., 1959b). Finally, hemolymph ecdysteroid titers rise before ovulation in cockroaches (Bullière, D. *et al.*, 1979; Lanzrein, B. *et al.*, 1981b; Zhu, X. *et al.*, 1983), a temporal relation which again suggests that ovarian ecdysteroids may be natural regulators of JH synthesis in cockroaches. It is tempting to speculate about the allatostatic effect of ecdysteroids in other developmental stages or even in insect species from other orders. The virtual inactivity of CA taken at the end of the last larval stage of *D. punctata*, is caused in part by "nervous inhibition", but is also caused by the larval milieu which is either inhibitory or devoid of any stimulatory factors (Szibbo, C. and Tobe, S., 1983). Indeed, larval CA are capable of synthesizing JH at high rates when implanted into allatectomized adult females and conversely, CA from adult females are inhibited when implanted into larvae, late during the last stage. A high ecdysteroid titer may contribute to the unfavorable milieu of larvae (Szibbo, C. and Tobe, S., 1983). Khan, M. *et al.* (1982c) reported that injection of 20-hydroxyecdysone had no effect on CA activity in adult *Leptinotarsa decemlineata*, and this is consistent with the ovarian asynchrony and continuous egg production in this species.

In conclusion, the ovary may control CA activity in adults by two mechanisms: a stimulatory mechanism, and an inhibitory mechanism. Sequential operation of these mechanisms may allow a normal cycle of CA activity in those species where cyclicity in ovarian function and CA activity are paired. Stimulation and inhibition have been demonstrated conclusively in only two cockroach species. It is not yet known if the two processes can occur independently of each other, i.e. if species can be found where only one of the two control mechanisms operates. It is not yet known how widely the ovarian regulation of CA activity is distributed among Insecta (see e.g. Aldrich, J., 1980; Rossignol, P. *et al.*, 1981). Its presence in the phylogenetically ancient cockroaches may indicate that this mode of regulation is primitive as is, indeed, the CA control of reproduction. Previously, complex interactions between the JH and ecdysteroid hormonal systems were thought to occur only during development and metamorphosis. Work during the last decade, particularly in Diptera, has provided evidence for such complex interactions during reproduction as well. A primitive (unifying?) model may be found in Thysanura which show a cyclicity of reproduction and molting (Watson, J., 1964; Hagedorn, H. and Kunkel, J., 1979; Rojo de la Paz, A. *et al.*, 1983).

4.4 Role of the brain

Because denervation of the CA can have such dramatic effects on gland activity, it is logical to study the source of the axonally transmitted signals, i.e. the brain. Integration of external and internal stimuli is likely to occur at the level of brain neurosecretory cells. The brain has been implicated as a source of neurosecretory factors which control CA activity. These factors may be released in neurohemal organs such as the CC and act via the hemolymph, but may also be released inside the CA themselves (non-humoral neurosecretion; antisera to substance P, β-endorphin and LH–RH react with peptidergic neurons in the CA of *Leucophaea maderae*, Hansen, B. *et al.*, 1982). In addition, the brain may be the origin of neural signals controlling the release of neurosecretion at the periphery, inside or outside the CA. Neural signals (i.e. neurotransmitter release) may also have direct effects on the activity of CA cells. All these signals may be inhibitory (allatostatic) or stimulatory (allatotropic), and identification of their chemical nature may not be easy. Evidence for the control of CA activity by brain NSC has been reviewed by Engelmann, F. (1970); De Wilde, J. and De Loof, A. (1973); Doane, W. (1973), etc. More recent papers (e.g. Baehr, J.-C., 1973, 1976; Bhaskaran, G. and Jones, G., 1980; Bhaskaran, G. *et al.*, 1980b; Friedel, T., 1974; Girardie, A. *et al.*, 1974; Granger, N. and Sehnal, F., 1974; Hodková, M., 1977, 1979a; Sehnal, F. and Granger, N., 1975, etc.) were reviewed by De Kort, C. and Granger, N. (1981) and Raabe, M. (1982) (see also Loughton and Orchard, this volume). Studies using *in vitro* assays for JH biosynthesis will be reviewed below; they include decapitation experiments, brain NSC cautery and/or electrostimulation, as well as *in vitro* incubations of brain–CC–CA complexes.

Lanzrein, B. *et al.* (1981a) implanted decapitated female *Nauphoeta cinerea* with their own CA on day 4 after adult emergence. The CA were retrieved and assayed for activity *in vitro* 6–16 days later. A normal cycle of JH synthesis was reported, resulting in ovulation with a 3–4-day delay. Barth, R. and Sroka, P. (1975) had, however, shown that the brain is needed only very early (a few hours) after emergence to allow oocyte growth in *N. cinerea*. The activity of CA transplanted into decapitated insects is determined by such factors as: the half-life of humoral effects to which the CA were subjected in the donor; the replacement of neuroendocrine factors of the brain by other neuroendocrine centers; not to mention the traumatic effects of decapitation, usually benign by the standards of insect endocrinologists (!). Thus, the experiments of Lanzrein, B. *et al.* (1981a) and Stay, B. *et al.* (1983) suggesting that CA can respond to stimulatory or inhibitory ovarian signals in the absence of the brain do not rule out the possibility that the brain is normally involved.

In vitro assays of JH synthesis after cautery/electrostimulation experiments brought new evidence for the existence of an allatotropin in adult female *Locusta migratoria*. Electrocoagulation of the NCC-I tracts at the point of decussation (under the median ocellar nerve) inhibited ovarian development and also CA activity (Girardie, J. *et al.*, 1981). When the operation, which amounts to an intracerebral axotomy, was performed the day after adult emergence and CA activity measured 20 days later, a 6-fold lower rate of JH synthesis was observed. Pratt, G. and Pener, M. (1983) measured JH synthesis up to 10 days after cautery of the MNSC (pars intercerebralis). After 6 days the difference in CA activity between sham- and MNSC-cauterized locusts was over 100-fold, and both CA volume increase and oocyte growth were inhibited. Tobe, S. *et al.* (1982) showed that electrostimulation of the MNSC resulted in an increase in JH synthesis, most noticeable 2–3 days after stimulation. No stimulation of JH synthesis was observed in locusts in which the intracerebral NCC-I tracts had been previously cauterized, confirming that release of an allatotropin had been blocked by cautery. Stimulation of the LNSC suggested that these cells may also be the source of an allatotropic factor, but only when the MNSC are inactivated. These results are in agreement with previous, indirect, studies of the allatotropic activity of brain NSC in *L. migratoria* (e.g. Girardie, A., 1966, 1967; Girardie, J. and Girardie, A., 1977; Lazarovici, P. and Pener, M., 1978; McCaffery, A., 1976; Moulins, M. *et al.*, 1974; Pener, M. *et al.*, 1972, etc.). The young adult female milieu in *Diploptera punctata* has a stimulatory influence on CA, because denervated CA of mated or virgin females and implanted CA from last-stage larvae or from pregnant females are all activated (Stay, B. and Tobe, S., 1977; Szibbo, C. and Tobe, S., 1983). The presence of an allatotropin is suggested by unilateral allatectomy experiments (section 4.2.2). In the absence of the ovaries, the young adult female milieu is not stimulatory any more. It is not known whether the ovary has a direct allatotropic influence, or whether it stimulates the release of an allatotropin by NSC from the brain. It is also difficult to differentiate the effects of presence of allatotropin and absence of allatostatin. However, implantation of female CA into males allowed a direct test of the allatotropic function of brain NSC, because activation of the CA occurs in the absence of an ovary (Stay, B. *et al.*, 1980; Friedel, T. *et al.*, 1980). Cautery of the brain MNSC of males had a dramatic effect on the activity of implanted female CA (Tobe, S. *et al.*, 1981). Whether the cautery was performed before, at the same time or after implantation of female CA, a significantly lower rate of JH release was observed when compared to sham-operated controls. Thus, destruction of the host's MNSC at emergence prevented the activity increase of CA implanted on day 0 or on day 4; cautery of the MNSC of the host 7 days after implantation of female CA provoked a rapid decline of gland activity. Replacement therapy, i.e. implantation of a day 0 female brain, partially reversed the effect of MNSC cautery of the host (Tobe, S. *et al.*, 1981). These experiments strongly suggest that male MNSC are the source of a humorally acting allatotropin.

Granger, N. *et al.* (1981, 1982b) have used *in vitro* procedures to investigate the neuroendocrine regulation of larval *Manduca sexta* CA. The rate of JH release by isolated CA was compared to the rate of JH release by brain–CC–CA complexes during the fourth and fifth larval stages. A ratio different from unity would indicate either an overall stimulatory or inhibitory effect of the brain–CC complex on the CA. This ratio fluctuated rather dramatically, with JH-I/II production apparently inhibited at the time of pupal commitment, and JH-III production stimulated after the onset of the wandering stage (Granger, N. *et al.*, 1981, 1982b). Thus, both the timing and the amplitude of brain–CC effects could be assessed by this method. Most remarkably from the biochemical standpoint, those studies suggested that the synthesis of JH-III and of the higher homologs JH-I/II are regulated independently of each other. Because the nervous tracts are essentially intact in incubations of brain–CC–CA complexes, it is difficult to distinguish between humoral and axonal regulators of CA activity, and it is also assumed that the brain functions "normally" *in vitro*. The rates of JH release by "intact" brain–CC–CA complexes were compared to rates of JH release in co-incubations of CA and brain–CC complexes of larvae of the same age (Granger, N. *et al.*, 1981). In other words, the effects of NCA section were assessed in short-term *in vitro* assays. CA activities in co-incubations were never higher than activities reported for intact complexes, an indication that most effects of the brain–CC complex were operating via the incubation medium rather than via the NCA. Brain–CC–CA complexes from the fourth and fifth larval stages of *M. sexta* were incubated in a constant volume *in vitro*, whereas hemolymph volume changes about 20-fold during this period. Thus, the effect of humoral regulators of CA activity may have been underestimated in the early fourth larval stage and overestimated in the late fifth larval stage. Feedback interactions *in vitro* might also interfere with the detection of CA regulators. For instance, Granger, N. and Borg, T. (1976) have suggested that long-term co-incubation of *Galleria mellonella* brains with CC–CA complexes may reduce the allatotropic activity of the brain. Nonetheless, *in vitro* incubations of isolated CA in medium preconditioned by brains, brain–CC complexes or containing active extracts, are most likely to allow progress in the characterization and isolation of the elusive allatotropin(s) and other regulators of CA activity.

Antibodies to medial parts of the brains from mature adult *L. migratoria* are reportedly able to delay or inhibit oocyte growth and vitellogenesis when injected daily into locusts (Rembold, H. *et al.*, 1980). Necrosis *or* hypertrophy of the CA are also observed after multiple injections of antibodies. In the absence of any determination of either JH titer or CA activity in these insects, the conclusion that an immunogenic allatotropic factor is present in the brain of mature female locusts seems premature. However, if the effects of the antibodies obtained by Rembold *et al.* can be better characterized and indeed attributed to the inhibition of CA activity, purification of an allatotropin from *Locusta migratoria* may be greatly facilitated.

4.4.1 On "Programming of the CA"

Recent evidence obtained by measuring CA activity *in vitro* has confirmed that the glands are controlled by internal and external signals, and do not follow a rigid, autonomous pattern of activity. To quote Wigglesworth, V. (1948): "it is clear that the functional activity of the corpus allatum is not determined wholly by humoral means; nor is this determination wholly autonomous; it does not, for instance, produce juvenile hormone for definite number of moulting stages. In other words, it is not the corpus allatum itself which 'counts the instars'". On the basis of transplantation bioassays, Hodková, M. (1979b) has concluded that the "counting the instars" mechanism in *Pyrrhocoris apterus* is located in the brain.

Application of certain JH analogs to late embryonic stages was reported to inhibit metamorphosis in *Hyalophora cecropia*, *P. apterus* and *Oncopeltus fasciatus* (Riddiford, L. and Williams, C., 1967; Riddiford, L., 1970). CA transplantation experiments in *P. apterus* led Riddiford, L. and Truman, J. (1972) to explain this delayed effect of JH as evidence "that JH can somehow interfere with the programming of the embryonic corpus allatum", in such a way that larvae of the fifth stage fail to cease JH synthesis. These experiments (Riddiford, L., 1971) were most challenging, because they implied that activity of the CA in the fifth larval stage was determined by exogenous JH application before hatching of the eggs. Willis, J. and Lawrence, P. (1970) suggested that these effects, in *O. fasciatus*, could be due to the persistence of the JH analog until metamorphosis. Several attempts to reproduce the presumed delayed effect of JH agonists on metamorphosis and Riddiford and Truman's experiments on *P. apterus* failed (Sláma, K. *et al.*, 1974; Sláma, K. and Socha, R., 1979; Socha, R. and Hodková, M., 1981). The original observations were attributed to persistence of, and contamination by, the JH analogs.

5 RATE-LIMITATION AND BIOCHEMICAL CONTROL OF JH SYNTHESIS

Whatever the origin or chemical nature of allatotropic and allatostatic factors, the problem of how such stimulatory or restraining signals are translated into higher or lower rates of JH biosynthesis remains a challenge for biochemists. The mode of action of allatotropins and allatostatins could include the regulation of JH precursor availability, the regulation of one (rate-limiting) or more enzymes in the JH biosynthetic pathway or the regulation of total CA biosynthetic capacity, i.e. cell number/volume; and of course any combinations of the above possibilities. Storage of JH in the glands and regulation of JH release does not appear to occur, but this has been demonstrated for only three species! (section 2.1). Two antagonistic factors may be present at the same time, so their respective modes of action must allow for either interaction or exclusion. For instance, an allatostatic effect conveyed by the NCC-I–NCA-I pathway in *Diploptera punctata* virgins takes precedence over an allatotropin present in the hemolymph. Mating lifts this block to the action of allatotropin, but a residual allatostatic influence is apparently still present in mated females and can be removed by denervation.

Non-storing endocrine glands, such as steroid-producing glands of vertebrates, are known to regulate precisely the availability of the precursor cholesterol (Brown, M. *et al.*, 1979; Gwynne, J. and Strauss, J., 1982) and gonadotropin-induced steroidogenesis in the rat luteal cells for instance, is acutely dependent on exogenous lipoprotein-bound cholesterol (Azhar, S. *et al.*, 1981). In the CA, exogenous methionine (C_1), acetate (C_2) and propionate (C_3), as well as later intermediates, can serve as precursors (Schooley, D. and Baker, F. this volume). Precursor pool size in the CA, or origin of the C_2 and C_3 units *in vivo*, are essentially unknown. Schooley, D. *et al.* (1976) have shown that *Manduca sexta* CA are strongly dependent on exogenous propionate for JH-II synthesis *in vitro*, and this suggests that propionate availability may play a role in determining the ratio between the homoisoprenoids JH-I, II and the "normal" isoprenoid JH-III which is not dependent on propionate for biosynthesis. Metabolic regulation and rate-limitation can be detected by various methods (Rollerston, F., 1972): (1) a teleological approach, where inspection of the pathway indicates those early steps which, if controlled, may prevent metabolic waste; (2) an enzymological approach, where analysis of the properties of enzymes *in vitro* can pinpoint a low and modifiable activity as being consistent with control and rate-limitation; (3) a physiological approach, where intact CA are subject to various treatments to give different rates of JH synthesis and where steady-state concentrations of intermediates are used to deduce sites of rate-limitation and control. The three approaches have been taken simultaneously, but the CA may be particularly suited for the third approach (Feyereisen, R. *et al.*, 1981d). The concept of rate-limitation is more than just finding the slowest enzyme in the biosynthetic pathway. If meaningful regulation of the pathway is to occur at the level of the rate-limiting enzyme, this enzyme has to be considerably slower than the second slowest enzyme. Indeed, the greater the difference between the two slowest enzymes, the more influence a small change in the rate-limiting enzyme will have on the overall rate of synthesis (Yagil, G. and Hoberman, H., 1969). The smaller the difference, the more likely it is that rate-limitation will shift from one step to another.

The enzyme HMG-CoA reductase is potentially a site of rate-limitation and regulation of JH biosynthesis, as the teleological and enzymological approaches would suggest. Indeed, HMG-CoA reductase plays an essential role in the regulation of cholesterol biosynthesis in vertebrates and has been recognized as a rate-limiting enzyme in isoprenoid biosynthesis. The enzyme is regulated by feedback mechanisms mediated by a variety of factors including cholesterol bound to low-density lipoproteins (LDL), oxygenated sterols, or non-sterol isoprenoids derived from mevalonate (Rodwell, V. *et al.*, 1976; Kandutsch, A. *et al.*, 1978; Brown, M. and Goldstein, J., 1980). These feedback mechanisms operate via short-term phosphorylation of the enzyme, long-term changes in enzyme levels, as well as by yet unknown mechanisms (Beg, Z. and Brewer, H., 1981, 1982; Arebalo, R. *et al.*, 1982). Indirect evidence for the reversible phosphorylation of HMG-CoA reductase of insect CA has been obtained in *D. punctata* (Feyereisen, R. *et al.*, 1981d) and *M. sexta* (Monger, D. and Law, J., 1982).

Reversible phosphorylation would be a means of hormonal or neural control of the enzyme, and could be caused by initial changes in cAMP or Ca^{2+}-calmodulin levels (Cohen, P., 1982), in response to allatotropic or allatostatic stimuli, or indeed by the level of phosphorylation of the hormonal receptor itself (Roth, R. and Cassell, D., 1983). Although reversible phosphorylation of HMG-CoA reductase is a property consistent with metabolic control of the enzyme, there is as yet no conclusive evidence that this enzyme actually does control JH biosynthesis *in vivo*.

Exogenous precursors such as farnesoic acid and farnesol have been used to study rate-limitation and control (Tobe, S. and Feyereisen, R., 1983). They are also useful in the study of total CA biosynthetic capacity (Tobe, S. and Pratt, G., 1976; Feyereisen, R. *et al.*, 1981c,d). Farnesoic acid is almost certainly an obligatory intermediate in *de novo* synthesis of JH-III (Feyereisen, R. *et al.*, 1981d; Hamnett, A. *et al.*, 1981). As an exogenous precursor of JH-III it has been shown to stimulate biosynthesis *in vitro* at optimal concentrations usually around 0.04 mM and the physiological rate-limiting step in JH synthesis is thus prior to the entry of exogenous farnesoic acid into the biosynthetic pathway. The degree of stimulation of JH synthesis is, however, not constant; in fact it varies considerably under different physiological conditions. The rate of JH biosynthesis (*de novo*, or "spontaneous") divided by the rate of farnesoic acid-stimulated JH biosynthesis was called the "fractional endocrine activity ratio" (FEAR) by Tobe, S. and Pratt, G. (1976). A small FEAR value indicates that CA activity can be considerably stimulated by the addition of farnesoic acid, whereas a FEAR value close to 1.0 indicates that the flux through the rate-limiting step(s) is close to the maximal capacity of the last two enzymes (*O*-methyl transferase and epoxidase), i.e. esterification and/or epoxidation become themselves rate-limiting. FEAR values undergo rapid and dramatic changes during reproduction in *Schistocerca gregaria* and range from 0.003

to 0.750; this indicates that there is a rapid and efficient control of the rate-limiting step(s) (Fig. 8) (Tobe, S. and Pratt, G., 1976). Whereas JH synthesis expressed per unit CA volume ("volume specific activity") changes considerably (up to 200-fold) during ovarian development of *S. gregaria*, under conditions of farnesoic acid stimulation the "volume specific activity" does not change significantly. Thus, there is no relation between CA volume and activity (Tobe, S. and Pratt, G., 1975b). Moreover, it was concluded that a causal relationship might exist between volume of the CA and its ability to utilize exogenous farnesoic acid (Tobe, S. and Pratt, G., 1976). The farnesoic acid-stimulated rate may be an expression of the total synthetic capacity of the glands, i.e. it may dictate the maximum possible rate of *de novo* JH synthesis.

In adult female *Diploptera punctata*, only very small fluctuations of FEAR values were observed during a reproductive cycle (Feyereisen, R. *et al.*, 1981c), and only during the 3 days of maximal CA activity were FEAR values elevated. Remarkably, the changes in JH synthesis per cell or per unit CA volume exactly parallel changes in FEAR (Fig. 9). During a gonotrophic cycle in *D. punctata* JH synthesis is "following" CA volume or CA cell number, however, on day 4 a drastic change in the rate-limiting step modifies this relationship and allows very high rates of JH synthesis during 3 days.

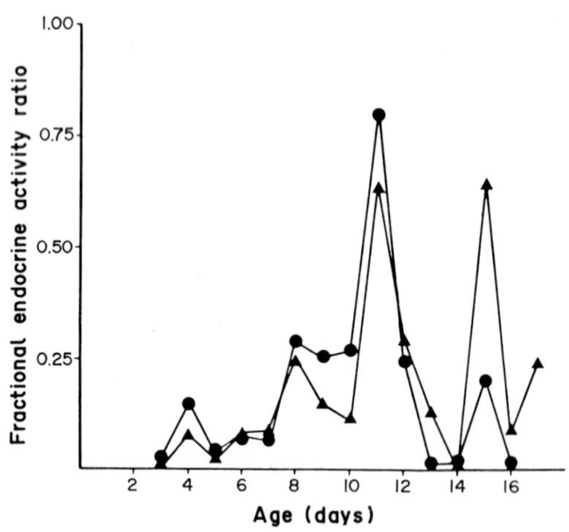

FIG. 8. Changes in the fractional endocrine activity ratio (FEAR) during sexual maturation and early reproductive life of adult female *Schistocerca gregaria*. FEAR represents the extent to which the rate of *de novo* JH synthesis approaches the rate of JH synthesis by CA incubated in the presence of exogenous farnesoic acid (stimulated rate). Circles and triangles represent two different experimental series. (From Tobe, S. and Pratt, G., 1976.)

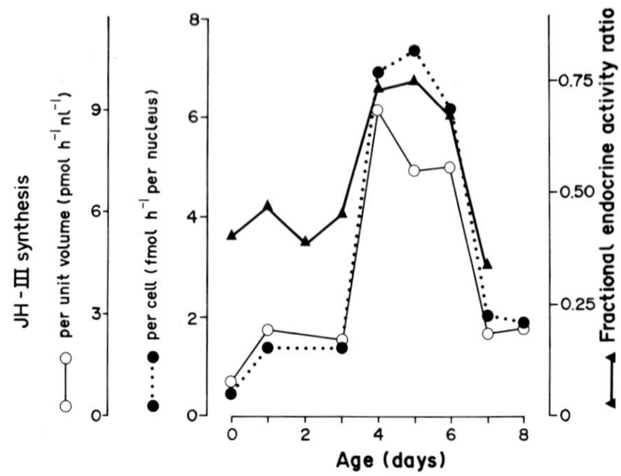

FIG. 9. Fractional endocrine activity ratio (FEAR) and rates of JH synthesis expressed per unit CA volume and per cell, in adult female *Diploptera punctata*. (Adapted from Feyereisen, R. *et al.*, 1981c and from Szibbo, C. and Tobe, S., 1982a.)

In adult female *Periplaneta americana*, farnesoic acid stimulates JH synthesis only 1.5–5-fold during an oviposition cycle (Pratt, G. *et al.*, 1975b; Weaver, R. and Pratt, G., 1981). This lower capacity to stimulate JH synthesis is not due to a restricted availability of farnesoic acid at its intracellular site of utilization, because large quantities of methyl farnesoate accumulate in the CA during incubation. Thus, esterification (*O*-methyl transferase activity) is much higher than epoxidation in *P. americana* CA. Because methyl farnesoate does not accumulate to any significant extent in CA not stimulated by exogenous farnesoic acid ("spontaneous" or *de novo* synthesis), an intracellular feedback mechanism may limit the rate of farnesoic acid synthesis (Pratt, G. *et al.*, 1975b). In farnesoic acid-stimulated *D. punctata* CA, methyl farnesoate does not accumulate to the extent seen in *P. americana* and the rates of epoxidation are never less than 91% of the rates of esterification. Farnesol-stimulated CA cannot be stimulated further by farnesoic acid, thus suggesting that farnesol and farnesal dehydrogenase activities are both higher than *O*-methyl transferase and epoxidase activities (Feyereisen, R. *et al.*, 1981c,d). There is as yet no evidence for physiological saturation of the epoxidase.

During gonotrophic cycles of *P. americana*, the farnesoic acid-stimulated rate of JH synthesis is relatively constant, indicating that under normal physiological conditions, changes in an early, rate-limiting step(s) control the activity of the CA, whereas the total synthetic capacity of the CA remains constant (Weaver, R. and Pratt, G., 1981). However, after starvation, even though esterification and epoxidation do not become rate-limiting, the farnesoic acid-stimulated rate of JH synthesis drops considerably. This indicates that starvation causes both a decrease in the activity of the rate-limiting step(s) *and* a drop in the total synthetic capacity of the CA. In re-fed animals, both parameters are increased again (Weaver, R. and Pratt, G., 1981). A similar situation was observed in adult *L. decemlineata* (Khan, M. *et al.*, 1982b,c) where starvation or JH-III injection causes a rapid decrease in JH synthesis, followed later by a decrease of the farnesoic acid-stimulated rate of JH synthesis. Thus, both in *P. americana* and *L. decemlineata*, a prolonged "nervous inhibition" decreases not only the rate of JH synthesis, but also the total synthetic capacity of the CA as measured by the non-rate-limiting esterification and epoxidation reactions.

Rate-limitation and control of CA activity is outlined in Fig. 10. Although theoretical and derived only from data on a few insects, it provides a framework for discussion and future research. Regulation of CA activity by allatotropic and allatostatic factors can be separated into two classes of physiological/biochemical responses (Tobe, S. and Pratt, G., 1976; Feyereisen, R. *et al.*, 1981c,d). Both rapid and slow responses can change the rate-limiting step which, by definition, determines the rate of JH synthesis. However, slow responses also affect the total biosynthetic capacity of the CA, such as non-rate-limiting enzymes, cell volume or number, and cell ultrastructure. Slow responses will thus modify the rate of JH synthesis as stimulated by late intermediates (it is implicit that late intermediates enter the biosynthetic pathway beyond the step(s) which respond "fast"). Because the rate-limiting step and thus the rate of JH synthesis is affected by both slow and rapid responses, and because the stimulated rate of JH synthesis can be affected only by slow responses, the relative importance of rapid responses can only be deduced from the analysis of FEAR changes (Fig. 10). Rapid responses are prevalent in *S. gregaria* (Tobe, S. and Pratt, G., 1976; Fig. 8) and in this species, section of the NCA-I removes a rapid allatotropic response (Tobe, S. *et al.*, 1977). Rapid responses are compatible with "neural" control and may lead to a pulse-like, on/off type of CA activity which would also explain the high asymmetry found for *S. gregaria* CA. Reversible phosphorylation of a key enzyme (HMG-CoA reductase?) may turn out to be such a rapid response. Slow responses are probably responsible for the cyclic changes in CA morphometry and/or ultrastructure, and Fig. 10 clearly shows that *there is no relation between CA activity* (rate of JH synthesis) *and CA morphometry*. Johansson, A. (1958) had already proposed that "growth of the corpus allatum and its production of hormone seem to be controlled in two different ways" in *Oncopeltus fasciatus*, and Staal, G. (1961) suggested that the CC might be involved in changing CA volume, and that "the growth of the corpora allata in normal individuals and the gradual increase in secretory activity during development are two different processes". Pratt, G. and Pener, M. (1983) have measured the conversion of precocene I to precocene I dihydrodiol by intact *Locusta migratoria* CA; this activity admittedly represents a measure of methyl farnesoate epoxidase. Electrocoagulation of the pars intercerebralis prevented the normal increase in both CA volume and epoxidase activity, and had a drastic effect on rates of JH synthesis, but had no effect on the epoxidase activity per unit CA volume. Thus, *L. migratoria* allatotropin(s) from the brain MNSC "regulates independently the volume of the CA, including the levels of non-rate limiting enzymes like the epoxidase, and the rate of actual hormone synthesis" (Pratt, G. and Pener, M., 1983). Factors which cause slow responses may be more difficult to test (Pratt, G. *et al.*, 1976) and identify because they may affect cell growth and/or division. If we assume (or safely predict?) that allatotropins are peptides, it would be reasonable to envision a large "proallatotropin" providing both a peptide responsible for a rapid allatotropic response and another peptide with mitogenic activity. Such a mechanism operates in vertebrates where proopiomelanocortin, for instance, provides a number of bioactive peptides including ACTH and a stimulator of adrenal mitogenesis (Estivariz, F. *et al.*, 1982). Whether all CA cells are controlled by both rapid and slow responses is unknown. All CA

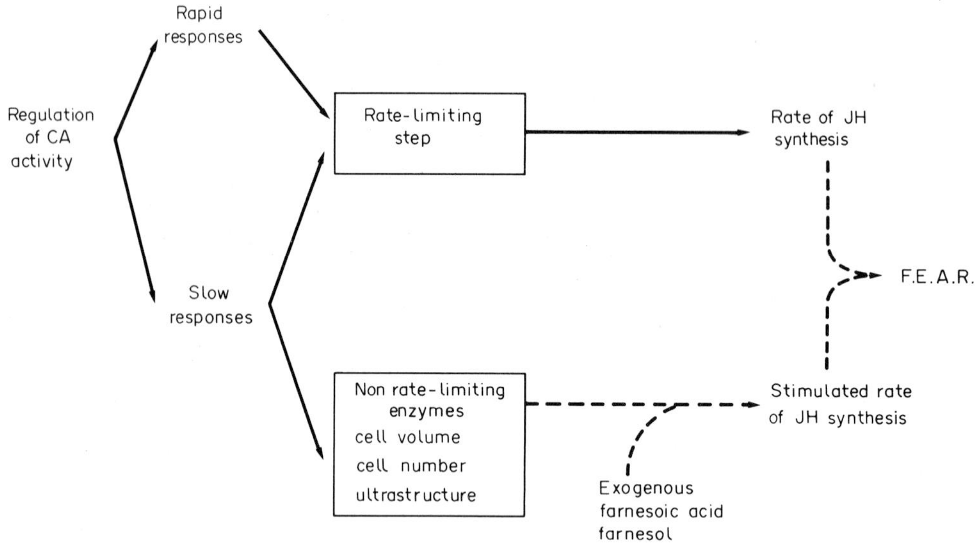

FIG. 10. Theoretical scheme of CA regulation, showing the effects of rapid responses (for example, changes in enzyme activity following phosphorylation/dephosphorylation, etc.) and slow responses (for example, enzyme synthesis or degradation, change in cell number, etc.). See text for details.

cells in *Schistocerca gregaria* carry out JH synthesis from exogenous farnesoic acid (Tobe, S. and Pratt, G., 1976) and are thus presumably capable of slow responses, but the possibility that *de novo* JH synthesis is not equally distributed among CA cells has not yet been tested.

6 CONCLUSION AND PROSPECTS

This review of CA endocrinology is deliberately selective, in that emphasis is given to the discussion of data obtained by radiochemical assay and by RIA. These techniques are now available to all insect endocrinologists and should replace indirect assays of CA activity. It is hoped that this review will soon be outdated by the chemical characterization and identification of regulators of CA activity. Allatotropins and allatostatins may serve as models for agonists or antagonists which are likely to be added to the arsenal of biorational pest control agents of tomorrow. Progress in peptide chemistry and plant molecular genetics is so rapid that apparently far-fetched ideas may soon become respectable industrial goals in plant protection. *In vitro* assays for JH biosynthesis will also be useful to solve other problems, such as the identification of the JH(s) in various insect orders; the regulation of JH homolog ratio in Lepidoptera; the genetic control of CA activity, e.g. in the black larval mutant of *Manduca sexta* (Safranek, L. and Riddiford, L., 1975); the changes in CA activity during the aging process (Injeyan, H. *et al.*, 1981), etc. Most important, control of CA activity should be studied in more than the handful of insect species which have provided most of the material for this chapter. This should be done not so much in an effort to find common principles, but rather to document the evolutionary diversity of endocrine control mechanisms.

ACKNOWLEDGEMENTS

I thank Drs D. W. Bean, V. J. Brookes, G. E. Pratt and P. D. Shirk for critically reading the manuscript, and Ms. Karen Lundholm for her excellent secretarial assistance.

REFERENCES

ABD-EL-WAHAB, A. and SIRLIN, J. L. (1959). Nuclear RNA and hormone production in the ring gland of *Drosophila*. *Exp. Cell Res.* *18*, 301–312.

AGUI, N., BOLLENBACHER, W. E., GRANGER, N. A. and GILBERT, L. I. (1980). Corpus allatum is release site for insect prothoracicotropic hormone. *Nature 285*, 669–670.

AKAMATSU, Y., DUNN, P. E., KEZDY, F. J., KRAMER, K. J., LAW, J. H., REIBSTEIN, D. and SANBURG, L. L. (1975). Biochemical aspects of juvenile hormone action in insects. In *Control Mechanisms in Development*. Edited by R. H. Meints and E. Davies. Pages 123–149. Plenum Press, New York.

ALDRICH, J. R. (1980). Inhibition of juvenile hormone synthesis by β-ecdysone in adult female milkweed bugs. *Amer. Zool. 20*, 900.

ALLEGRET, P. (1964). Interrelationship of larval development, metamorphosis and age in a pyralid Lepidopteran, *Galleria mellonella* (L.), under the influence of dietetic factors. *Exp. Gerontol. 1*, 49–66.

AREBALO, R. E., TORMANEN, C. D., HARDGRAVE, J. E., NOLAND, B. J. and SCALLEN, T. J. (1982). *In vivo* regulation of rat liver 3-hydroxy-3-methylglutaryl-coenzyme A reductase: immunotitration of the enzyme after short-term mevalonate or cholesterol feeding. *Proc. Nat. Acad. Sci. 79*, 51–55.

AZHAR, S., MENON, M. and MENON, K. M. J. (1981). Receptor-mediated gonadotropin action in the ovary: demonstration of acute dependence of rat luteal cells on exogenously supplied steroid precursor (sterols) for gonadotropin-induced steroidogenesis. *Biochim. Biophys. Acta 665*, 362–375.

BAEHR, J.-C. (1973). Contrôle neuroendocrine du fonctionnement du corpus allatum chez *Rhodnius prolixus*. *J. Insect Physiol. 19*, 1041–1055.

BAEHR, J.-C. (1976). Etude du contrôle neuro-endocrine du fonctionnement du corpus allatum chez les Larves du quatrième stade de *Rhodnius prolixus*. *J. Insect Physiol. 22*, 73–82.

BAEHR, J.-C., PRADELLES, P., LEBREUX, C., CASSIER, P. and DRAY, F. (1976). A simple and sensitive radioimmunoassay of insect juvenile hormone using an iodinated tracer. *FEBS Lett. 69*, 123–128.

BAEHR, J.-C., PRADELLES, P., DRAY, F. (1979). A radioimmunological assay for naturally occurring insect juvenile hormones using iodinated tracers: Its use in the analysis of biological samples. *Ann. Biol. Anim. Biochem. Biophys. 19*, 1827–1836.

BAEHR, J.-C., CARUELLE, J. P., PORCHERON, P. and CASSIER, P. (1981). Quantification of juvenile hormones using a radioimmunological assay in the analysis of biological samples. In *Juvenile Hormone Biochemistry*. Edited by G. E. Pratt and G. T. Brooks. Pages 47–57. Elsevier/North Holland, Amsterdam.

BAEHR, J.-C., CASSIER, P., CAUSSANEL, C. and PORCHERON, P. (1982). Activity of corpora allata, endocrine balance and reproduction in female *Labidura riparia* (Dermaptera). *Cell Tiss. Res. 225*, 267–282.

BARTH, R. H. and SROKA, P. (1975). Initiation and regulation of oocyte growth by the brain and corpora allata of the cockroach, *Nauphoeta cinerea*. *J. Insect Physiol. 21*, 321–330.

BARTH, R. H., LESTER, L. J., SROKA, P., KESSLER, T. and HEARN, R. (1975). Juvenile hormone promotes dominance behavior and ovarian development in social wasps (*Polistes annularis*). *Experientia 31*, 691–692.

BEG, Z. H. and BREWER, H. B. JR. (1981). Regulation of liver 3-hydroxy-3-methylglutaryl coenzyme A reductase. *Curr. Top. Cell. Reg. 20*, 139–184.

BEG, Z. H. and BREWER, H. B. JR. (1982). Modulation of rat liver 3-hydroxy-3-methylglutaryl-CoA reductase activity by reversible phosphorylation. *Fed. Proc. 41*, 2634–2638.

BERGOT, B. J., RATCLIFF, M. and SCHOOLEY, D. A. (1981). Method for quantitative determination of the four known juvenile hormones in insect tissue using gas chromatography/mass-spectrometry. *J. Chromatogr. 204*, 231–244.

BHASKARAN, G. and JONES, G. (1980). Neuroendocrine regulation of corpus allatum activity in *Manduca sexta*: The endocrine basis for starvation-induced supernumerary larval moult. *J. Insect Physiol. 26*, 431–440.

BHASKARAN, G., DELEON, G., LOOMAN, B., SHIRK, P. D. and RÖLLER, H. (1980a). Activity of juvenile hormone acid in brainless, allatectomized diapausing *Cecropia* pupae. *Gen. Comp. Endocr. 42*, 129–133.

BHASKARAN, G., JONES, G. and JONES, D. (1980b). Neuroendocrine regulation of corpus allatum activity in *Manduca sexta*: Sequential neurohormonal and nervous inhibition in the last-instar larva. *Proc. Nat. Acad. Sci. 77*, 4407–4411.

BODENSTEIN, D. and SPRAGUE, I. B. (1959). The developmental capacities of the accessory sex glands in *Periplaneta americana*. *J. Exp. Zool. 142*, 177–202.

BOWERS, W. S. and FELDLAUFER, M. F. (1982). *In vitro* inactivation of *Tenebrio molitor* corpora allata by a synthetic precocene analog. *Gen. Comp. Endocr. 47*, 120–124.

BRADLEY, J. T. and EDWARDS, J. S. (1978). Yolk proteins in the house cricket, *Acheta domesticus*: Identification, characterization, and effect of ovariectomy upon their synthesis. *J. Exp. Zool. 204*, 239–248.

BROUSSE-GAURY, P. (1977). Starvation and reproduction in *Periplaneta americana* L.: control of mating behaviour in the female. *Adv. Invertebr. Reprod. 1*, 328–343.

BROUSSE-GAURY, P. and CASSIER, P. (1975). Contribution à l'étude de la dynamique des corpora allata chez *Blabera fusca*. Influence du jeune, du groupement et du régime photopériodique. *Bull. Biol. Fr. Belg. 109*, 253–277.

BROWN, M. S. and GOLDSTEIN, J. L. (1980). Multivalent feedback regulation of HMG-CoA reductase, a control mechanism coordinating isoprenoid synthesis and cell growth. *J. Lipid Res. 21*, 505–517.

BROWN, M. S., KOVANEN, P. T. and GOLDSTEIN, J. L. (1979). Receptor-mediated uptake of lipoprotein-cholesterol and its utilization for steroid synthesis in the adrenal cortex. *Rec. Prog. Hormone Res. 35*, 215–257.

BULLIÈRE, D., BULLIÈRE, F. and DE REGGI, M. (1979). Ecdysteroid titres during ovarian and embryonic development in *Blaberus craniifer*. *Wilhelm Roux's Archiv. 186*, 103–114.

CARROW, G. M., CALABRESE, R. L. and WILLIAMS, C. M. (1981). Spontaneous and evoked release of prothoracicotropin from multiple neurohemal organs of the tobacco hornworm. *Proc. Nat. Acad. Sci. 78*, 5866–5870.

CARSON, E. R., GODFREY, K. R. and REEVE, J. (1982). A review of modelling and the role of dynamic tracer studies in metabolism. In *Quantitative Approaches to metabolism*. Edited by D. G. Cramp. Pages 1–72. J. Wiley & Sons, New York.

CASSIER, P. (1966). Effets de l'ablation d'un corps allate sur la fécondité et la descendance des femelles isolées du Criquet migrateur (*Locusta migratoria migratorioides* R. et F.). *Insectes Sociaux 12*, 17–28.

CASSIER, P. (1967). La reproduction des insectes et la régulation de l'activité des corps allates. *Annee Biol. 6*, 595–670.

CASSIER, P. (1979). The corpora allata of insects. *Int. Rev. Cytol. 57*, 1–73.

CHU, H. M. and NORRIS, D. M. (1979). Comparative morphology and ultrastructure of the corpora allata in newly emerged and sexually mature female *Xyleborus ferrugineus* (Fabr.) (Coleoptera: Scolytidae). *Int. J. Insect Morph. Embryol. 8*, 359–374.

COHEN, P. (1982). The role of protein phosphorylation in neural and hormonal control of cellular activity. *Nature 296*, 613–620.

CUTLER, D. J. (1978). On the definition of the compartment concept in pharmacokinetics. *J. Theor. Biol. 73*, 329–345.

CYMBOROWSKI, B., BOGUS, M., BECKAGE, N. E., WILLIAMS, C. M. and RIDDIFORD, L. M. (1982). Juvenile hormone titres and metabolism during starvation-induced supernumerary larval moulting of the tobacco hornworm, *Manduca sexta* L. *J. Insect Physiol. 28*, 129–135.

DAHM, K. H., BHASKARAN, G., PETER, M. G., SHIRK, P. D., SESHAN, K. R. and RÖLLER, H. (1976). On the Identity of the Juvenile Hormone in Insects. In *The Juvenile Hormones*. Edited by L. I. Gilbert. Pages 19–47. Plenum Press, New York.

DAHM, K. H., BHASKARAN, G., PETER, M. G., SHIRK, P. D., SESHAN, K. R. and RÖLLER, H. (1981). The juvenile hormones of *Cecropia*. In *Regulation of Insect Development and Behaviour*. Edited by F. Sehnal, A. Zabza, J. J. Menn and B. Cymborowski. Part I, pages 183–198. Wroclaw Technical University Press, Wroclaw.

DE KORT, C. A. D. and GRANGER, N. A. (1981). Regulation of the juvenile hormone titer. *Ann. Rev. Ent. 26*, 1–28.

DE KORT, C. A. D. and KRAMER, S. J. (1976). Age-related changes in amino acid composition and turnover in the Colorado potato beetle, *Leptinotarsa decemlineata*. *Insect Biochem. 6*, 339–343.

DE KORT, C. A. D., KHAN, M. A., BERGOT, B. J. and SCHOOLEY, D. A. (1981). The JH titre in the Colorado beetle in relation to reproduction and diapause. In *Juvenile Hormone Biochemistry*. Edited by G. E. Pratt and G. T. Brooks. Pages 125–134. Elsevier/North Holland, Amsterdam.

DE KORT, C. A. D., BERGOT, B. J. and SCHOOLEY, D. A. (1982). The nature and titre of juvenile hormone in the Colorado Potato Beetle, *Leptinotarsa decemlineata*. *J. Insect Physiol. 28*, 471–474.

DE WILDE, J. (1976). Juvenile hormone and caste differentiation in the honey bee (*Apis mellifera* L.). In *Phase and Caste Determination in Insects*. Edited by M. Lüscher. Pages 5–20. Pergamon Press, Oxford and New York.

DE WILDE, J. and DE BOER, J. A. (1961). Physiology of diapause in the adult Colorado beetle. II. Diapause as a case of pseudo-allatectomy. *J. Insect Physiol. 6*, 152–161.

DE WILDE, J. and DE LOOF, A. (1973). Reproduction. Endocrine Control. In *The Physiology of Insecta*. 2nd edition. Edited by M. Rockstein. Vol. 1, pages 97–157. Academic Press, New York.

DE WILDE, J., STAAL, G. B., DE KORT, C. A. D., DE LOOF, A. and BAARD, G. (1968). Juvenile hormone titer in the haemolymph as a function of photoperiodic treatment in the adult Colorado potato beetle (*Leptinotarsa decemlineata* Say). *Proc. K. Ned. Akad. Wet. Ser. C. 71*, 321–326.

DE WILDE, J., DE KORT, C. A. D. and DE LOOF, A. (1971). The significance of juvenile hormone titers. *Mitt. Schw. Ent. Ges. 44*, 79–86.

DIXON, S. E. and MOSER, E. (1972). Duality in function in corpora allata of honeybee larvae. *Canad. J. Zool. 50*, 593–595.

DOANE, W. W. (1973). Role of hormones in insect development. In *Developmental systems: Insects*. Edited by S. J. Counce and C. H. Waddington. Vol. 2, pages 291–497. Academic Press, London and New York.

DORTLAND, D. F. (1979). The hormonal control of vitellogenin synthesis in the fat body of the female Colorado potato beetle. *Gen. Comp. Endocr. 38*, 332–344.

EDWARDS, J. P., BERGOT, B. J. and STAAL, G. B. (1983). Effects of three compounds with anti-juvenile hormone activity and a juvenile hormone analogue on endogenous juvenile hormone levels in the tobacco hornworm, *Manduca sexta*. *J. Insect Physiol. 29*, 83–89.

ENGELMANN, F. (1957). Die Steuerung der Ovarfunktion bei der ovoviviparen Schabe *Leucophaea maderae* (Fabr.). *J. Insect Physiol. 1*, 257–278.

ENGELMANN, F. (1959a). The control of reproduction in *Diploptera punctata* (Blattaria). *Biol. Bull. 116*, 406–419.

ENGELMANN, F. (1959b). Uber die Wirkung Implantierter Prothoraxdrüsen im Adulten Weibchen von *Leucophaea maderae* (Blattaria). *Z. Vergl. Physiol. 41*, 456–470.

ENGELMANN, F. (1965). The mode of regulation of the corpus allatum in adult insects. *Arch. Anat. Micr. 54*, 387–404.

ENGELMANN, F. (1970). *The Physiology of Insect Reproduction*. Pergamon Press, Oxford.

ENGELMANN, F. (1978). Synthesis of vitellogenin after long-term ovariectomy in a cockroach. *Insect Biochem. 8*, 149–154.

ERLEY, D., SOUTHARD, S. and EMMERICH, H. (1975). Excretion of juvenile hormone and its metabolites in the locust, *Locusta migratoria*. *J. Insect Physiol. 21*, 61–70.

ESTIVARIZ, F. E., ITURRIZIA, F., MCLEAN, C., HOPE, J. and LOWRY, P. J. (1982). Stimulation of adrenal mitogenesis by N-terminal proopiocortin peptides. *Nature 297*, 419–422.

EWING, L. L., CHUBB, C. E. and ROBAIRE, B. (1976). Macromolecules, steroid binding and testosterone secretion by rabbit testes. *Nature 264*, 84–86.

FAIN, M. J. and RIDDIFORD, L. M. (1975). Juvenile hormone titers in the hemolymph during late larval development of the tobacco hornworm, *Manduca sexta* (L.). *Biol. Bull. 149*, 506–521.

FERENZ, H.-J. (1981). Effect of the photoperiod on corpus allatum activity *in vitro* in the beetle, *Pterostichus nigrita* F. *Experientia 37*, 1211–1212.

FERENZ, H.-J. and KAUFNER, I. (1981). Juvenile hormone synthesis in relation to oogenesis in *Locusta migratoria*. In *Juvenile Hormone Biochemistry*. Edited by G. E. Pratt and G. T. Brooks. Pages 135–145. Elsevier/North Holland, Amsterdam.

FEYEREISEN, R. and TOBE, S. S. (1981). A rapid partition assay for routine analysis of juvenile hormone release by insect corpora allata. *Analyt. Biochem. 111*, 372–375.

FEYEREISEN, R., JOHNSON, G., KOENER, J., STAY, B. and TOBE, S. S. (1981a). Precocenes as pro-allatocidins in adult female *Diploptera punctata*: A functional and ultrastructural study. *J. Insect Physiol. 27*, 855–868.

FEYEREISEN, R., PRATT, G. E. and HAMNETT, A. F. (1981b). Enzymic synthesis of juvenile hormone in locust corpora allata: evidence for a microsomal cytochrome P-450 linked methyl farnesoate epoxidase. *Eur. J. Biochem. 118*, 231–238.

FEYEREISEN, R., FRIEDEL, T. and TOBE, S. S. (1981c). Farnesoic acid stimulation of C_{16} juvenile hormone biosynthesis by corpora allata of adult female *Diploptera punctata*. *Insect Biochem. 11*, 401–409.

FEYEREISEN, R., KOENER, J. and TOBE, S. S. (1981d). *In vitro* studies with C_2, C_6 and C_{15} precursors of C_{16}JH biosynthesis in the corpora allata of adult female *Diploptera punctata*. In *Juvenile Hormone Biochemistry*. Edited by G. E. Pratt and G. T. Brooks. Pages 81–92. Elsevier/North Holland, Amsterdam.

FLORKIN, M. and JEUNIAUX, C. (1974). Hemolymph: Composition. In *The Physiology of Insecta*. Edited by M. Rockstein. Vol. V, pages 256–307. Academic Press, New York.

FRAENKEL, G. (1980). Foreword and overview. In *Neurohormonal techniques in insects*. Edited by T. A. Miller, pages IX–XV. Springer, New York.

FRASER, J. and PIPA, R. (1977). Corpus allatum regulation during the metamorphosis of *Periplaneta americana*: axon pathways. *J. Insect Physiol. 23*, 975–984.

FRIEDEL, T. (1974). Endocrine control of vitellogenesis in the harlequin bug, *Dindymus versicolor*. *J. Insect Physiol. 20*, 717–733.

FRIEDEL, T., FEYEREISEN, R., MUNDALL, E. C. and TOBE, S. (1980). The allatostatic effect of 20-hydroxyecdysone on the adult viviparous cockroach, *Diploptera punctata*. *J. Insect Physiol. 26*, 665–670.

FUKUDA, S. (1963). Induction of supernumerary molting in silkworm larvae by implantation of corpora allata from female moths. *Annot. Zool. Jap. 36*, 14–20.

GILBERT, L. I. (1964). Physiology of growth and development: Endocrine aspects. In *Physiology of Insecta* Vol. I. Edited by M. Rockstein. Pages 149–225. Academic Press, New York.

GILBERT, L. I. and KING, D. S. (1973). Physiology of growth and development: Endocrine aspects. In *The Physiology of Insecta*, 2nd edition. Edited by M. Rockstein. Vol. 1. Pages 249–370. Academic Press, New York.

GILBERT, L. I. and SCHNEIDERMAN, H. A. (1960). The development of a bioassay for the juvenile hormone of insects. *Trans. Amer. Micr. Soc. 79*, 38–67.

GILBERT, L. I. and SCHNEIDERMAN, H. A. (1961). The content of juvenile hormone and lipid in Lepidoptera: sexual differences and developmental changes. *Gen. Comp. endocr. 1*, 453–472.

GILLOTT, C. and DOGRA, G. S. (1972). Neurosecretory cell and corpus allatum activity during production of successive egg batches in virgin *Melanoplus sanguinipes* (Fab). *Gen. Comp. Endocr. 18*, 126–132.

GIRARDIE, A. (1966). Contrôle de l'activité génitale chez *Locusta migratoria*. Mise en évidence d'un facteur allatotrope dans la *Pars intercerebralis*. *Bull. Soc. Zool. Fr. 91*, 423–439.

GIRARDIE, A. (1967). Contrôle neuro-hormonal de la métamorphose et de la pigmentation chez *Locusta migratoria cinerascens*. *Bull. Biol. Fr. Belg. 101*, 79–114.

GIRARDIE, J. and GIRARDIE, A. (1977). Libération provoquée *in vitro* du produit de neurosécrétion des cellules protocérébrales médianes chez le criquet migrateur. *J. Physiol. (Paris) 73*, 707–721.

GIRARDIE, A., MOULINS, M. and GIRARDIE, J. (1974). Rupture de la diapause ovarienne *d'Anacridium aegyptium* par stimulation électrique des cellules neurosécrétrices médianes de la pars intercerebralis. *J. Insect Physiol. 20*, 2261–2275.

GIRARDIE, J., TOBE, S. and GIRARDIE, A. (1981). Biosynthèse de l'hormone juvénile C_{16} (JH-III) et maturation ovarienne chez le criquet migrateur. *C.R. Acad. Sci. 293 III*, 443–446.

GRANGER, N. A. and BORG, T. K. (1976). The allatotropic activity of the larval brain of *Galleria mellonella* cultured *in vitro*. *Gen. Comp. Endocr. 29*, 349–359.

GRANGER, N. A. and SEHNAL, F. (1974). Regulation of larval corpora allata in *Galleria mellonella*. *Nature 251*, 415–417.

GRANGER, N. A., BOLLENBACHER, W. E., VINCE, R., GILBERT, L. I., BAEHR, J.-C. and DRAY, F. (1979). *In vitro* biosynthesis of juvenile hormone by the larval corpora allata of *Manduca sexta*: quantification by radioimmunoassay. *Mol. Cell. Endocr. 16*, 1–17.

GRANGER, N. A., BOLLENBACHER, W. E. and GILBERT, L. I. (1981). An *in vitro* approach for investigating the regulation of the corpora allata during larval-pupal metamorphosis. In *Current Topics in Insect Endocrinology and Nutrition*. Edited by G. Bhaskaran, S. Friedman and J. G. Rodriguez. Pages 83–105. Plenum Press, New York.

GRANGER, N. A., NIEMIEC, S. M., GILBERT, L. I. and BOLLENBACHER, W. E. (1982a). Juvenile hormone III biosynthesis by the larval corpora allata of *Manduca sexta*. *J. Insect Physiol.* *28*, 385–391.

GRANGER, N. A., NIEMIEC, S. M., GILBERT, L. I. and BOLLENBACHER, W. E. (1982b). Juvenile hormone synthesis *in vitro* by larval and pupal corpora allata of *Manduca sexta*. *Mol. Cell. Endocr.* *28*, 587–604.

GWADZ, R. W. and SPIELMAN, A. (1973). Corpus allatum control of ovarian development in *Aedes aegyptii*. *J. Insect Physiol.* *19*, 1441–1448.

GWYNNE, J. T. and STRAUSS, J. F. III (1982). The role of lipoproteins in steroidogenesis and cholesterol metabolism in steroidogenic glands. *Endocrine Rev.* *3*, 299–329.

HAGEDORN, H. H. and KUNKEL, J. G. (1979). Vitellogenin and vitellin in insects. *Ann. Rev. Ent.* *24*, 475–505.

HAMMOCK, B. D. and MUMBY, S. M. (1978). Inhibition of epoxidation of methyl farnesoate to juvenile hormone III by cockroach corpus allatum homogenates. *Pestic. Biochem. Physiol.* *9*, 39–47.

HAMMOCK, B., NOWOCK, J., GOODMAN, W., STAMOUDIS, V. and GILBERT, L. I. (1975). The influence of hemolymph-binding protein on juvenile hormone stability and distribution in *Manduca sexta* fat body and imaginal discs *in vitro*. *Mol. Cell. Endocr.* *3*, 167–184.

HAMNETT, A. F. and PRATT, G. E. (1978). Use of automated capillary column radio gas chromatography in the identification of insect juvenile hormones. *J. Chromatogr.* *158*, 387–399.

HAMNETT, A. F., PRATT, G. E., STOTT, K. M. and JENNINGS, R. C. (1981). The use of radio HRLC in the identification of the natural substrate of the O-methyl transferase and substrate utilization by the enzyme. In *Juvenile Hormone Biochemistry*. Edited by G. E. Pratt and G. T. Brooks. Pages 93–105. Elsevier/North Holland, Amsterdam.

HANSEN, B. L., HANSEN, G. N. and SCHARRER, B. (1982). Immunoreactive material resembling vertebrate neuropeptides in the corpus cardiacum and corpus allatum of the insect *Leucophaea maderae*. *Cell Tiss. Res.* *225*, 319–329.

HIGHNAM, K. C. (1964). Endocrine relationships in insect reproduction. *Roy. Ent. Soc. London Symp.* *2*, 26–42.

HODKOVÁ, M. (1977). Nervous pathways in photoperiodic regulation of reproduction in females of *Pyrrhocoris apterus* (Hemiptera). *Acta Ent. Bohemoslov.* *74*, 353–361.

HODKOVÁ, M. (1979a). Hormonal and nervous inhibition of reproduction by brain in diapausing females of *Pyrrhocoris apterus* L. (Hemiptera). *Zool. Jb. Physiol.* *83*, 126–136.

HODKOVÁ, M. (1979b). Does the insect brain count larval instars. *Experientia* *35*, 135–136.

HUIBREGTSE-MINDERHOUD, L., VAN DEN HONDEL-FRANKEN, M. A. M., VAN DER KERK-VAN HOOF, A. C., BIESSELS, H. W. A., SALEMINK, C. A., VAN DER HORST, D. J. and BEENAKKERS, A. M. Th. (1980). Quantitative determination of the juvenile hormones in the haemolymph of *Locusta migratoria* during normal development and after implantation of corpora allata. *J. Insect Physiol.* *26*, 627–631.

ICHIKAWA, M. and NISHIITSUTSUJI-UWO, J. (1959). Studies on the role of the corpus allatum in the Eri-silkworm, *Philosamia cynthia ricini*. *Biol. Bull.* *116*, 88–94.

INJEYAN, H. S. and TOBE, S. S. (1981). Phase polymorphism in *Schistocerca gregaria*: Assessment of juvenile hormone synthesis in relation to vitellogenesis. *J. Insect Physiol.* *27*, 203–210.

INJEYAN, H. S., DALE, J. F. and TOBE, S. S. (1981). Increased juvenile hormone synthesis and decreased fertility in aging *Schistocerca gregaria*. *Canad. J. Zool.* *59*, 1744–1748.

JOHANNSON, A. S. (1958). Relation of nutrition to endocrine-reproductive functions in the milkweed bug *Oncopeltus fasciatus* (Dallas) (Heteroptera: Lygaeidae). *Nytt. Mag. Zool.* Vol. *7*, Oslo University Press.

JOHANNSON, A. S. (1964). Feeding and nutrition in reproductive processes in insects. *Roy. Ent. Soc. London Symp.* *2*, 43–55.

JOHNSON, G., FEYEREISEN, R., STAY, B. and TOBE, S. (1980). Effects of precocene on corpora allata activity and ultrastructure in adult female *Diploptera punctata*. *Amer. Zool.* *20*, 859.

JOHNSON, R. A. and HILL, L. (1973a). The activity of the corpora allata in the fourth and fifth larval instars of the migratory locust. *J. Insect Physiol.* *19*, 1921–1932.

JOHNSON, R. A. and HILL, L. (1973b). Quantitative studies on the activity of the corpora allata in adult male *Locusta* and *Schistocerca*. *J. Insect Physiol.* *19*, 2459–2467.

JOHNSON, R. A. and HILL, L. (1975). Activity of the corpora allata in the adult female migratory locust. *J. Insect Physiol.* *21*, 1517–1519.

JOLY, L. (1960). Fonctions des corpora allata chez *Locusta migratoria* (L.) Thèse, Faculté des Sciences de l'Université de Strasbourg.

JOLY, L. and JOLY, P. (1974). Comparaison de la phase grégaire et de la phase solitaire de *Locusta migratoria migratorioides* (Orthoptère) du point de vue de la teneur de leur hemolymphe en Hormone Juvénile. *C.R. Acad. Sci.* *279D*, 1007–1009.

JOLY, L., JOLY, P., PORTE, A. and GIRARDIE, A. (1968). Etude physiologique et ultrastructurale des corpora allata de *Locusta 78migratoria* L. (Orthoptère) en phase grégaire. *Archiv. Zool. Exp. Gen.* *109*, 703–728.

JOLY, L., JOLY, P., PORTE, A. and GIRARDIE, A. (1969). Analyse ultrastructurale de l'activité des *corpora allata* de *Locusta migratoria* L. et ses conséquences sur la structure que l'on doit attribuer au mécanisme humoral controlant la métamorphose. *Archiv. Zool. Exp. Gen.* *110*, 617–628.

JOLY, P. (1967). Comparaison du volume et de l'activité physiologique des corpora allata de *Locusta migratoria* L. *Ann. Soc. Ent. Fr.* *3*, 601–608.

JUDY, K. J., SCHOOLEY, D. A., DUNHAM, L. L., HALL, M. S., BERGOT, B. J. and SIDDALL, J. B. (1973a). Isolation, structure, and absolute configuration of a new natural insect juvenile hormone from *Manduca sexta*. *Proc. Nat. Acad. Sci.* *70*, 1509–1513.

JUDY, K. J., SCHOOLEY, D. A., HALL, M. S., BERGOT, B. J. and SIDDALL, J. B. (1973b). Chemical structure and absolute configuration of a juvenile hormone from grasshopper corpora allata *in vitro*. *Life Sci.* *13*, 1511–1516.

JUDY, K. J., SCHOOLEY, D. A., TROETSCHLER, R. G., JENNINGS, R. C., BERGOT, B. J. and HALL, M. S. (1975). Juvenile hormone production by corpora allata of *Tenebrio Molitor in vitro*. *Life Sci.* *16*, 1059–1066.

KANDUTSCH, A. A., CHEN, H. W. and HEINIGER, H.-J. (1978). Biological activity of some oxygenated sterols. *Science* *201*, 498–501.

KHAN, M. A., DODERER, A., KOOPMANSCHAP, A. B. and DE KORT, C. A. D. (1982a). Improved assay conditions for measurement of corpus allatum activity *in vitro* in the adult Colorado potato beetle, *Leptinotarsa decemlineata*. *J. Insect Physiol.* *28*, 279–284.

KHAN, M. A., KOOPMANSCHAP, A. B., PRIVEE, H. and DE KORT, C. A. D. (1982b). The mode of regulation of the corpus allatum activity during starvation in adult females of the Colorado potato beetle, *Leptinotarsa decemlineata* (Say). *J. Insect Physiol.* *28*, 791–796.

KHAN, M. A., KOOPMANSCHAP, A. B. and DE KORT, C. A. D. (1982c). The effects of juvenile hormone 20-hydroxyecdysone and precocene II on activity of corpora allata and the mode of negative-feedback regulation of these glands in the adult Colorado potato beetle. *J. Insect Physiol.* *28*, 995–1001.

KOBAYASHI, M. and YAMASHITA, Y. (1959). A function of corpus allatum in neurosecretory system in the silkworm, *Bombyx mori*. *J. Sericult. Sci. Jpn.* *28*, 335–339.

KRAMER, S. J. (1978). Age-dependent changes in corpus allatum activity *in vitro* in the adult Colorado potato beetle, *Leptinotarsa decemlineata*. *J. Insect Physiol.* *24*, 461–464.

KRAMER, S. J. and LAW, J. H. (1980). Control of Juvenile Hormone production: The relationship between precursor supply and hormone synthesis in the tobacco hornworm, *Manduca sexta*. *Insect Biochem.* *10*, 569–575.

KRAMER, S. J. and STAAL, G. B. (1981). *In vitro* studies on the mechanism of action of anti-juvenile hormone agents in larvae of *Manduca sexta*. In *Juvenile Hormone Biochemistry*. Edited by G. E. Pratt and G. T. Brooks. Pages 425–437. Elsevier/North Holland, Amsterdam.

LANZREIN, B., GENTINETTA, V., FEHR, R. and LÜSCHER, M. (1978). Correlation between haemolymph juvenile hormone titre, corpus allatum volume, and corpus allatum *in vivo* and *in vitro* activity during oocyte maturation in a cockroach (*Nauphoeta cinerea*). *Gen. Comp. Endocr.* *36*, 339–345.

LANZREIN, B., GENTINETTA, V. and LÜSCHER, M. (1977). *In vivo* and *in vitro* studies on the endocrinology of the reproductives of the termite *Macrotermes subhyalinus*. In *Proceedings of the Eighth International Congress of the International Union for the Study of Social Insects*. Pages 265–268. Wageningen, The Netherlands.

LANZREIN, B., WILHELM, R. and BÜSCHOR, J. (1981a). On the regulation of the *corpora allata* activity in adult females of the ovoviparous cockroach *Nauphoeta cinerea*. In *Juvenile Hormone Biochemistry*. Edited by G. E. Pratt and G. T. Brooks. Pages 147–160. Elsevier/North Holland, Amsterdam.

LANZREIN, B., WILHELM, R. and GENTINETTA, V. (1981b). On relations between corpus allatum activity and oocyte maturation in the cockroach *Nauphoeta cinerea*. In *Regulation of Insect Development and Behaviour*. Edited by F. Sehnal, A. Zabza, J. J. Menn and B. Cymborowski, Part II, pages 523–534. Wroclaw Technical University Press, Wroclaw.

LAZAROVICI, P. and PENER, M. P. (1978). The relations of the pars intercerebralis, corpora allata, and juvenile hormone to oocyte development and oviposition in the African migratory locust. *Gen. Comp. Endocr. 35*, 375–386.

LEA, A. O. and THOMSEN, E. (1969). Size independent secretion by the corpus allatum of *Calliphora erythrocephala*. *J. Insect Physiol. 15*, 477–482.

LEGAY, J. M. (1950). Note sur l'évolution des corpora allata au cours de la vie larvaire de *Bombyx mori*. *C.R. Soc. Biol. 144*, 512–513.

LÜSCHER, M. (1976). Evidence for an endocrine control of caste determination in higher termites. In *Phase and Caste Determinations in Insects*. Edited by M. Lüscher. Pages 91–103. Pergamon Press, New York.

MASON, C. A. (1973). New features of the brain-retrocerebral neuroendocrine complex of the locust *Schistocerca vaga* (Scudder). *Z. Zellforsch. 41*, 19–32.

MCCAFFERY, A. R. (1976). Effects of electrocoagulation of cerebral neurosecretory cells and implantation of corpora allata on oocyte development in *Locusta migratoria*. *J. Insect Physiol. 22*, 1081–1092.

MCCAFFERY, A. R. and HIGHNAM, K. C. (1975a). Effects of corpora allata on the activity of the cerebral neurosecretory system of *Locusta migratoria migratorioides* R. & F. *Gen. Comp. Endocr. 25*, 358–372.

MCCAFFERY, A. R. and HIGHNAM, K. C. (1975b). Effects of corpus allatum hormone and its mimics on the activity of the cerebral neurosecretory system of *Locusta migratoria migratorioides* R. & F. *Gen. Comp. Endocr. 25*, 373–386.

MCCAFFERY, A. R. and MCCAFFERY, V. A. (1983). Corpus allatum activity during overlapping cycles of oocyte growth in adult female *Melanoplus sanguinipes*. *J. Insect Physiol. 29*, 259–266.

METZLER, M., DAHM, K. H., MEYER, D. and RÖLLER, H. (1971). On the biosynthesis of juvenile hormone in the adult *Cecropia* moth. *Z. Naturf. 26b*, 1270–1276.

METZLER, M., MEYER, D., DAHM, K. H. and RÖLLER, H. (1972). Biosynthesis of juvenile hormone from 10-epoxy-7-ethyl-3, 11-dimethyl-2, 6-tridecadienoic acid in the adult *Cecropia* moth. *Z. Naturf. 27b*, 321–322.

MONGER, D. J. and LAW, J. H. (1982). Control of juvenile hormone biosynthesis. Evidence for phosphorylation of the 3-hydroxy-3-methyl-glutaryl coenzyme A reductase of insect corpus allatum. *J. Biol. Chem. 257*, 1921–1923.

MOULINS, M., GIRARDIE, A. and GIRARDIE, J. (1974). Manipulation of sexual physiology by brain stimulation in insects. *Nature 250*, 339–340.

MÜLLER, P. J., MASNER, P., KÄLIN, M. and BOWERS, W. S. (1979). *In vitro* inactivation of corpora allata of the bug *Oncopeltus fasciatus* by precocene II. *Experientia 35*, 704–705.

MUNDALL, E. C., TOBE, S. S. and STAY, B. (1981). Vitellogenin fluctuations in haemolymph and fat body and dynamics of uptake into oocytes during the reproductive cycle of *Diploptera punctata*. *J. Insect Physiol. 27*, 821–827.

NAIR, K. M. (1963). The localization of ribonucleic acid and hormone production in the corpus allatum of *Chrysocoris purpurens* (Westw.) *J. Histochem. Cytochem. 11*, 495–499.

NIJHOUT, H. F. (1975a). A threshold size for metamorphosis in the tobacco hornworm, *Manduca sexta* (L.). *Biol. Bull. 149*, 214–225.

NIJHOUT, H. F. (1975b). Axonal pathways in the brain-retrocerebral neuroendocrine complex of *Manduca sexta* (L.) (Lepidoptera: Sphingidae) *Int. J. Insect Morph. Embryol. 4*, 529–538.

NOWOCK, J., HAMMOCK, B. and GILBERT, L. I. (1976). The binding protein as a modulator of juvenile hormone stability and uptake. In *The Juvenile Hormones*. Edited by L. I. Gilbert. Pages 354–373. Plenum Press, New York and London.

ODHIAMBO, T. R. (1966). Morphometric changes and the hormonal activity of the corpus allatum in the adult male of the desert locust. *J. Insect Physiol. 12*, 655–664.

OZEKI, K. (1962). Studies on the secretion of the juvenile hormone in the earwig, *Anisolabis maritima*. *Sci. Pap. Coll. Gen. Educ. Univ. Tokyo 12*, 65–72.

OZEKI, K. (1963). Further studies on the secretion of juvenile hormone by the corpus allatum in the earwig, *Anisolabis maritima*. *Sci. Pap. Coll. Gen. Educ. Univ. Tokyo 13*, 213–221.

PENER, M. P. (1967). Effects of allatectomy and sectioning of the nerves of the corpora allata on oocyte growth, male sexual behaviour, and colour change in adults of *Schistocerca gregaria*. *J. Insect Physiol. 13*, 665–684.

PENER, M. P., GIRARDIE, A. and JOLY, P. (1972). Neurosecretory and corpus allatum controlled effects on mating behavior and color change in adult *Locusta migratoria migratorioides* males. *Gen. Comp. Endocr. 19*, 494–508.

PETER, M. G., SHIRK, P. D., DAHM, K. H. and RÖLLER, H. (1981). On the specificity of juvenile hormone biosynthesis in the male *Cecropia* moth. *Z. Naturf. 36c*, 579–585.

PFLUGFELDER, O. (1939). Wechselwirkungen von Drüsen innerer Sekretion bei *Dixippus morosus Br.* *Z. Wiss. Zool. 152*, 384–408.

PFLUGFELDER, O. (1940). Austausch verschiedenalter Corpora Allata bei *Dixippus morosus*. *Z. Wiss. Zool. 153*, 384–408.

PFLUGFELDER, O. (1948). Volumetrische Untersuchungen an den Corpora Allata der Honigbiene *Apis mellifera*. *Biol. Zbl. 67*, 223–241.

PFLUGFELDER, O. (1958). *Entwicklungsphysiologie der Insekten*. 2nd Edition. Akademische Verlagsgesellschaft, Leipzig.

PIPA, R. L. (1978). Locations and central projections of neurons associated with the retrocerebral neuroendocrine complex of the cockroach *Periplaneta americana* (L.). *Cell Tiss. Res. 193*, 443–455.

PIPA, R. L. (1982). Neural influence on corpus allatum activity and egg maturation in starved virgin *Periplaneta americana*. *Physiol. Ent. 7*, 449–455.

PRATT, G. E. and BOWERS, W. S. (1977). Precocene II inhibits juvenile hormone biosynthesis by cockroach corpora allata *in vitro*. *Nature 265*, 548–550.

PRATT, G. E. and FINNEY, J. R. (1977). Chemical inhibitors of juvenile hormone biosynthesis *in vitro*. In *Crop Protection Agents — their Biological Evaluation*. Edited by N. R. McFarlane. Pages 113–132. Academic Press, London, New York and San Francisco.

PRATT, G. E. and PENER, M. P. (1983). Precocene sensitivity of corpora allata in adult female *Locusta migratoria* after electrocoagulation of the pars intercerebralis neurosecretory cells. *J. Insect Physiol. 29*, 33–39.

PRATT, G. E. and TOBE, S. S. (1974). Juvenile hormones radiobiosynthesized by corpora allata of adult female locusts *in vitro*. *Life Sci. 14*, 575–586.

PRATT, G. E. and WEAVER, J. R. (1978). The rate of juvenile hormone biosynthesis by insect corpora allata *in vitro*. In *Comparative Endocrinology*. Edited by P. J. Gaillard and H. H. Boer. Pages 503–506. Elsevier/North Holland, Amsterdam.

PRATT, G. E., TOBE, S. S., WEAVER, R. J. and FINNEY, J. R. (1975a). Spontaneous synthesis and release of C_{16} juvenile hormone by isolated corpora allata of female locust *Schistocerca gregaria* and female cockroach *Periplaneta americana*. *Gen. Comp. Endocr. 26*, 478–484.

PRATT, G. E., TOBE, S. S. and WEAVER, R. J. (1975b). Relative oxygenase activities in juvenile hormone biosynthesis of corpora allata of an african locust (*Schistocerca gregaria*) and american cockroach (*Periplaneta americana*). *Experientia 31*, 120–122.

PRATT, G. E., WEAVER, R. J. and HAMNETT, A. F. (1976). Continuous monitoring of juvenile hormone release by superfused corpora allata of *Periplaneta americana*. In *The Juvenile Hormones*. Edited by L. I. Gilbert. Pages 164–178. Plenum Press, New York.

RAABE, M. (1982). *Insect Neurohormones*. Plenum Press, New York and London.

REIBSTEIN, D. and LAW, J. H. (1973). Enzymatic synthesis of insect juvenile hormones. *Biochem. Biophys. Res. Commun. 55*, 266–272.

REIBSTEIN, D., LAW, J. H., BOWLUS, S. B. and KATZENELLENBOGEN, J. A. (1976). Enzymatic synthesis of juvenile hormone in *Manduca sexta*. In *The Juvenile Hormones*. Edited by L. I. Gilbert. Pages 131–146. Plenum Press, New York.

REMBOLD, H. (1981). Modulation of JH III-titer during the gonotrophic cycle of *Locusta migratoria*, measured by gas chromatography-selected ion monitoring mass spectrometry. In *Juvenile Hormone Biochemistry*. Edited by G. E. Pratt and G. T. Brooks. Pages 11–20. Elsevier/North Holland, Amsterdam.

REMBOLD, H., EDER, J. and ULRICH, G. M. (1980). Inhibition of allatotropic activity and ovary development in *Locusta migratoria* by anti-brain-antibodies. *Z. Naturf. 35c*, 1117–1119.
RIDDIFORD, L. M. (1970). Prevention of metamorphosis by exposure of insect eggs to juvenile hormone analogs. *Science 167*, 287–288.
RIDDIFORD, L. M. (1971). Juvenile hormone and insect embryogenesis. *Mitt. Schw. Ent. Ges. 44*, 177–186.
RIDDIFORD, L. M. and TRUMAN, J. W. (1972). Delayed effects of juvenile hormone on insect metamorphosis are mediated by the corpus allatum. *Nature 237*, 458.
RIDDIFORD, L. M. and WILLIAMS, C. M. (1967). The effects of juvenile hormone on the embryonic development of silkworms. *Proc. Nat. Acad. Sci. 57*, 595–601.
RODWELL, V. W., NORDSTROM, J. L. and MITSCHELEN, J. J. (1976). Regulation of HMG-CoA reductase. *Adv. Lipid. Res. 14*, 1–74.
ROJO DE LA PAZ, A., DELBECQUE, J. P., BITSCH, J. and DELACHAMBRE, J. (1983). Ecdysteroids in the haemolymph and the ovaries of the firebrat *Thermobia domestica* (Packard) (Insecta, Thysanura): Correlations with integumental and ovarian cycles. *J. Insect Physiol. 29*, 323–329.
RÖLLER, H. and DAHM, K. H. (1970). The identity of juvenile hormone produced by corpora allata *in vitro*. *Naturwissenschaften 57*, 454–455.
ROLLERSTON, F. S. (1972). A theoretical background to the use of measured concentrations of intermediates in study of the control of intermediary metabolism. *Curr. Top. Cell Reg. 5*, 47–75.
RÖSELER, P.-F. (1977). Juvenile hormone control of oogenesis in bumblebee workers, *Bombus terrestris*. *J. Insect Physiol. 23*, 985–992.
RÖSELER, P.-F. and RÖSELER, I. (1978). Studies on the regulation of the juvenile hormone titre in bumblebee workers, *Bombus terrestris*. *J. Insect Physiol. 24*, 707–713.
RÖSELER, P.-F., RÖSELER, I. and STRAMBI, A. (1980). The activity of corpora allata in dominant and subordinated females of the wasp *Polistes gallicus*. *Insectes Sociaux 27*, 97–107.
RÖSELER, P.-F., RÖSELER, I. and VAN HONK, C. G. J. (1981). Evidence for inhibition of corpora allata activity in workers of *Bombus terrestris* by a pheromone from the queen's mandibular glands. *Experientia 37*, 348–351.
ROSSIGNOL, P. A., FEINSOD, F. M. and SPIELMAN, A. (1981). Inhibitory regulation of corpus allatum activity in mosquitoes. *J. Insect Physiol. 27*, 651–654.
ROTH, L. M. and STAY, B. (1961). Oocyte development in *Diploptera punctata* (Eschscholtz) (Blattaria). *J. Insect Physiol. 7*, 186–202.
ROTH, R. A. and CASSELL, D. J. (1983). Insulin receptor: Evidence that it is a protein kinase. *Science 219*, 299–301.
SAFRANEK, L. and RIDDIFORD, L. M. (1975). The biology of the black larval mutant of the tobacco hornworm, *Manduca sexta*. *J. Insect Physiol. 21*, 1931–1938.
SCHARRER, B. (1946). The relationship between corpora allata and reproductive organs in adult *Leucophaea maderae* (Orthoptera). *Endocrinology 38*, 46–55.
SCHARRER, B. (1978). Histophysiological studies on the corpus allatum of *Leucophaea maderae*. VI. Ultrastructural characteristics in gonadectomized females. *Cell. Tiss. Res. 194*, 533–545.
SCHARRER, B. and VON HARNACK, M. (1958). Histophysiological studies on the corpus allatum of *Leucophaea maderae*. I. Normal life cycle in male and female adults. *Biol. Bull. 115*, 508–520.
SCHARRER, B. and VON HARNACK, M. (1961). Histophysiological studies on the corpus allatum of *Leucophaea maderae*. III. The effect of castration. *Biol. Bull. 121*, 193–208.
SCHOOLEY, D. A., JUDY, K. J., BERGOT, B. J., HALL, M. S. and JENNINGS, R. C. (1976). Determination of the physiological levels of juvenile hormones in several insects and biosynthesis of the carbon skeletons of the juvenile hormones. In *The Juvenile Hormones*. Edited by L. I. Gilbert. Pages 101–117. Plenum, New York.
SCHOONEVELD, H. (1970). Structural aspects of neurosecretory and corpus allatum activity in the adult Colorado beetle, *Leptinotarsa decemlineata* Say, as a function of daylength. *Neth. J. Zool. 20*, 151–237.
SCHOONEVELD, H., SANCHEZ, A. O. and DE WILDE, J. (1977). Juvenile hormone-induced break and termination of diapause in the Colorado potato beetle. *J. Insect Physiol. 23*, 689–696.
SCHOONEVELD, H., KRAMER, S. J., PRIVEE, H. and VAN HUIS, A. (1979). Evidence of controlled corpus allatum activity in the adult Colorado potato beetle. *J. Insect Physiol. 25*, 449–453.
SEGRE, G. (1982). Pharmacokinetics-compartmental representation. *Pharmac. Ther. 17*, 111–127.
SEHNAL, F. and GRANGER, N. A. (1975). Control of corpora allata function in larvae of *Galleria mellonella*. *Biol. Bull. 148*, 106–116.
SHIRK, P. D., DAHM, K. H. and RÖLLER, H. (1976). The accessory sex glands as the repository for juvenile hormone in male *Cecropia* moths. *Z. Naturf. 31c*, 199–200.
SIEW, Y. C. and GILBERT, L. I. (1971). Effects of moulting hormone and juvenile hormone on insect endocrine gland activity. *J. Insect Physiol. 17*, 2095–2104.
SLÁMA, K. and SOCHA, R. (1979). Delay of juvenoid action from embryogenesis until metamorphosis unfounded. *Acta Ent. Bohemslov. 76*, 289–299.
SLÁMA, K., ROMANŇUK, M. and SŎRM, F. (1974). *Insect Hormones and Bioanalogues*. Springer, Wien and New York.
SOCHA, R. and HODKOVÁ, M. (1981). Delayed effects of juvenile hormone analogues on metamorphosis of *Pyrrhocoris apterus* are not mediated by the corpus allatum. *Experientia 37*, 1029–1030.
SRIVASTAVA, K. P. and TIWARI, R. K. (1978). Effect of ovariectomy on the corpus allatum area, gut contents and fat body in the red cotton bug *Dysdercus koenigii*. *Indian J. Exp. Biol. 16*, 109–110.
STAAL, G. B. (1961). *Studies on the physiology of phase induction in Locusta migratoria migratorioides* R. & F. H. Veeman & Zonen N.V., Wageningen.
STAAL, G. B., HENRICK, C. A., BERGOT, B. J., CERF, D. C., EDWARDS, J. P. and KRAMER, S. J. (1981). Relationships and interactions between JH and anti-JH analogues in Lepidoptera. In *Regulation of Insect Development and Behaviour*. Edited by F. Sehnal, A. Zabza, J. J. Menn and B. Cymborowski. Part I, pages 323–340. Wroclaw Technical University Press, Wroclaw.
STAY, B. and TOBE, S. S. (1977). Control of juvenile hormone biosynthesis during the reproductive cycle of a viviparous cockroach. I. Activation and inhibition of corpora allata. *Gen. Comp. Endocr. 33*, 531–540.
STAY, B. and TOBE, S. S. (1978). Control of juvenile hormone biosynthesis during the reproductive cycle of a viviparous cockroach. II. Effects of unilateral allatectomy, implantation of supernumerary corpora allata, and ovariectomy. *Gen. Comp. Endocr. 34*, 276–286.
STAY, B. and TOBE, S. S. (1981). Control of the corpora allata during a reproductive cycle in a viviparous cockroach. *Amer. Zool. 21*, 663–674.
STAY, B., FRIEDEL, T., TOBE, S. S. and MUNDALL, E. C. (1980). Feedback control of juvenile hormone synthesis in cockroaches: Possible role for ecdysterone. *Science 207*, 898–900.
STAY, B., TOBE, S. S., MUNDALL, E. C. and RANKIN, S. (1983). Ovarian stimulation of juvenile hormone biosynthesis in the viviparous cockroach, *Diploptera punctata*. *Gen. Comp. Endocr.* (In press.)
STRAMBI, C. (1981). Some data obtained by radioimmunoassay of juvenile hormones. In *Juvenile Hormone Biochemistry*. Edited by G. E. Pratt and G. T. Brooks. Pages 59–63. Elsevier/North Holland, Amsterdam.
STRAMBI, C., STRAMBI, A., DE REGGI, M. L., HIRN, M. H. and DELAAGE, M. A. (1981). Radioimmunoassay of insect juvenile hormones and of their diol derivatives. *Eur. J. Biochem. 118*, 401–406.
STRANGWAYS-DIXON, J. (1961). The relationships between nutrition, hormones and reproduction in the blowfly *Calliphora erythrocephala* (Meig). II. The effect of removing the ovaries, the corpus allatum and the median neurosecretory cells upon selective feeding, and the demonstration of the corpus allatum cycle. *J. Exp. Biol. 38*, 637–646.
STRONG, L. (1965). The relationships between the brain, corpora allata, and oocyte growth in the Central American locust, *Schistocerca* sp. II. The innervation of the corpora allata, the lateral neurosecretory complex, and oocyte growth. *J. Insect Physiol. 11*, 271–280.
SZIBBO, C. M. and TOBE, S. S. (1981a). Cellular and volumetric changes in relation to the activity cycle in the corpora allata of *Diploptera punctata*. *J. Insect Physiol. 27*, 655–665.
SZIBBO, C. M. and TOBE, S. S. (1981b). The mechanism of compensation in juvenile hormone synthesis following unilateral allatectomy in *Diploptera punctata*. *J. Insect Physiol. 27*, 609–613.
SZIBBO, C. M. and TOBE, S. S. (1982). Intrinsic differences in juvenile hormone synthetic ability between corpora allata of males and females of the cockroach *Diploptera punctata*. *Gen. Comp. Endocr. 46*, 533–540.

Szibbo, C. M. and Tobe, S. S. (1983). Nervous and humoral inhibition of C_{16} juvenile hormone synthesis in last instar females of the viviparous cockroach, *Diploptera punctata*. *Gen. Comp. Endocr. 49*, 437–445.

Szibbo, C. M., Rotin, D., Feyereisen, R. and Tobe, S. S. (1982). Synthesis and degradation of C_{16} juvenile hormone (JH III) during the final two stadia of the cockroach, *Diploptera punctata*. *Gen. Comp. Endocr. 48*, 25–32.

Takeda, N. (1978). Hormonal control of prepupal diapause in *Monema flavescens* (Lepidoptera). *Gen. Comp. Endocr. 34*, 123–131.

Thomsen, E. and Thomsen, M. (1970). Fine structure of the corpus allatum of the female blow-fly *Calliphora erythrocephala*. *Z. Zellforsch. 110*, 40–60.

Tobe, S. S. (1977a). Asymmetry in hormone biosynthesis by insect endocrine glands. *Canad. J. Zool. 55*, 1509–1514.

Tobe, S. S. (1977b). Inhibition of growth in developing oocytes of the desert locust. *Experientia 33*, 343–345.

Tobe, S. S. (1980). Regulation of the corpora allata in adult female insects. In *Insect Biology in the Future*. Edited by M. Locke and D. S. Smith. Pages 345–367. Academic Press, New York.

Tobe, S. S. and Chapman, C. S. (1979). The effects of starvation and subsequent feeding on juvenile hormone synthesis and oocyte growth in *Schistocerca americana gregaria*. *J. Insect Physiol. 25*, 701–708.

Tobe, S. S. and Feyereisen, R. (1983). Juvenile hormone biosynthesis: regulation and assay. In *Invertebrate Endocrinology. Endocrinology of Insects*. Edited by R. G. H. Downer and H. Laufer. Vol 1, pages 161–178. Alan R. Liss, New York.

Tobe, S. S. and Pratt, G. E. (1974a). The influence of substrate concentration on the rate of insect juvenile hormone biosynthesis by corpora allata of the desert locust *in vitro*. *Biochem. J. 144*, 107–113.

Tobe, S. S. and Pratt, G. E. (1974b). Dependence of juvenile hormone release from corpus allatum on intraglandular content. *Nature 252*, 474–476.

Tobe, S. S. and Pratt, G. E. (1975a). Corpus allatum activity *in vitro* during ovarian maturation in the desert locust, *Schistocerca gregaria*. *J. Exp. Biol. 62*, 611–627.

Tobe, S. S. and Pratt, G. E. (1975b). The synthetic activity and glandular volume of the corpus allatum during ovarian maturation in the desert locust *Schistocerca gregaria*. *Life Sci. 17*, 417–422.

Tobe, S. S. and Pratt, G. E. (1976). Farnesenic acid stimulation of juvenile hormone synthesis as an experimental probe in corpus allatum physiology. In *The Juvenile Hormones*. Edited by L. I. Gilbert. Pages 147–163. Plenum Press, New York.

Tobe, S. S. and Stay, B. (1977). Corpus allatum activity *in vitro* during the reproductive cycle of the viviparous cockroach, *Diploptera punctata* (Eschscholtz). *Gen. Comp. Endocr. 31*, 138–147.

Tobe, S. S. and Stay, B. (1979). Modulation of juvenile hormone synthesis by an analogue in the cockroach. *Nature 281*, 481–482.

Tobe, S. S. and Stay, B. (1980). Control of juvenile hormone biosynthesis during the reproductive cycle of a viviparous cockroach. III. Effects of denervation and age on compensation with unilateral allatectomy and supernumerary corpora allata. *Gen. Comp. Endocr. 40*, 89–98.

Tobe, S. S., Chapman, C. S. and Pratt, G. E. (1977). Decay in juvenile hormone biosynthesis by insect corpus allatum after nerve transection. *Nature 286*, 728–730.

Tobe, S. S., Musters, A. and Stay, B. (1979). Corpus allatum function during sexual maturation of male *Diploptera punctata*. *Physiol. Ent. 4*, 79–86.

Tobe, S. S., Stay, B., Friedel, T., Feyereisen, R. and Paulson, C. (1981). The role of the brain in regulation of the corpora allata in female *Diploptera punctata*. In *Juvenile Hormone Biochemistry*. Edited by G. E. Pratt, and G. T. Brooks. Pages 161–174. Elsevier/North Holland, Amsterdam.

Tobe, S. S., Girardie, J. and Girardie, A. (1982). Enhancement of juvenile hormone biosynthesis in locusts following electrostimulation of cerebral neurosecretory cells. *J. Insect Physiol. 28*, 867–871.

Ulrich, G. M. and Rembold, H. (1983). Caste-specific maturation of the endocrine system in the female honey bee larva. *Cell Tiss. Res. 230*, 49–55.

Vogel, H., Heinzeller, T. and Renner, M. (1977). Daily protein synthesis in the pars intercerebralis and corpora allata of *Apis mellifera* L. with and without training to a foraging schedule. *J. Comp. Physiol. 118*, 51–60.

Watson, J. A. L. (1964). Moulting and reproduction in the adult firebrat, *Thermobia domestica* (Packard) (Thysanura, Lepismatidae) — II. The reproductive cycles. *J. Insect Physiol. 10*, 399–408.

Watson, J. A. L. (1967). The growth and activity of the corpora allata in the larval firebrat, *Thermobia domestica* (Packard) (Thysanura, Lepismatidae). *Biol. Bull. 132*, 277–291.

Weaver, R. J. (1979). Comparable activities of left and right corpora allata consistent with humoral control of juvenile hormone biosynthesis in the cockroach. *Canad. J. Zool. 57*, 343–345.

Weaver, R. J. (1981). Radiochemical assays of corpus allatum activity in adult female cockroaches following ovariectomy in the last nymphal instar. *Experientia 37*, 435–436.

Weaver, R. J. and Pratt, G. E. (1977). The effect of enforced virginity and subsequent mating on the activity of the corpus allatum of *Periplaneta americana* measured *in vitro*, as related to changes in the rate of ovarian maturation. *Physiol. Ent. 2*, 59–76.

Weaver, R. J. and Pratt, G. E. (1981). Effects of starvation and feeding upon corpus allatum activity and oocyte growth in adult female *Periplaneta americana*. *J. Insect Physiol. 27*, 75–83.

Weaver, R. J., Pratt, G. E. and Finney, J. R. (1975). Cyclic activity of the corpus allatum related to gonotrophic cycles in adult female *Periplaneta americana*. *Experientia 31*, 597–598.

Weaver, R. J., Pratt, G. E. and Hamnett, A. F. (1978). Radiochemical assays of juvenile hormone biosynthesis by cultured corpora allata in relation to oogenesis in the beetle, *Tenebrio molitor*, and the cockroach *Periplaneta americana*. *Gen. Comp. Endocr. 34*, 112–113.

Weaver, R. J., Pratt, G. E., Hamnett, A. F. and Jennings, R. C. (1980). The influence of incubation conditions on the rates of juvenile hormone biosynthesis by corpora allata isolated from adult females of the beetle *Tenebrio molitor*. *Insect Biochem. 10*, 245–254.

Weber, V. and Emmerich, H. (1976). RNA biosynthesis in isolated prothoracic glands of *Tenebrio molitor in vitro*. *Experientia 32*, 1609–1610.

Weirich, G. F. and Culver, M. G. (1979). S-Adenosylmethionine: Juvenile hormone acid methyltransferase in male accessory reproductive glands of *Hyalophora cecropia* (L.). *Archiv. Biochem. Biophys. 198*, 175–181.

Westphal, U. (1970). Corticosteroid-binding globulin and other steroid hormone carriers in the blood stream. *J. Reprod. Fertil. Suppl. 10*, 15–38.

Westphal, U., Stroupe, S. D., Cheng, S. and Harding, G. B. (1978). Mechanism of steroid binding to serum proteins. *J. Toxicol. Env. Health 4*, 229–247.

Wigglesworth, V. B. (1936). The function of the corpus allatum in the growth and reproduction of *Rhodnius prolixus* (Hemiptera). *Quart. J. Mic. Sci. 79*, 91–121.

Wigglesworth, V. B. (1940). The determination of characters at metamorphosis in *Rhodnius prolixus*. *J. Exp. Biol. 17*, 201–222.

Wigglesworth, V. B. (1948). The functions of the corpus allatum in *Rhodnius prolixus* (Hemiptera). *J. Exp. Biol. 25*, 1–14.

Wigglesworth, V. B. (1964). The hormonal regulation of growth and reproduction in insects. *Adv. Insect Physiol. 2*, 247–336.

Wigglesworth, V. B. (1970). *Insect Hormones*. Oliver & Boyd, Edinburgh.

Wilhelm, R. and Lüscher, M. (1970). Uber die Reifung transplantierter Oocyten unter verschiedenen Bedingungen bei der Schabe *Nauphoeta cinerea*. *Rev. Suisse Zool. 77*, 621–624.

Wilhelm, R. and Lüscher, M. (1974). On the relative importance of juvenile hormone and vitellogenin for oocyte growth in the cockroach *Nauphoeta cinerea*. *J. Insect Physiol. 20*, 1887–1894.

Williams, C. M. (1946). Physiology of insect diapause. I. The role of the brain in the production and termination of pupal dormancy in the giant silkworm *Platysamia cecropia*. *Biol. Bull. 90*, 234–243.

Williams, C. M. (1959). The juvenile hormone. I. Endocrine Activity of the corpora allata of the adult *Cecropia* silkworm. *Biol. Bull. 116*, 323–338.

Williams, C. M. (1961). The juvenile hormone. II. Its role in the endocrine control of molting, pupation, and adult development in the *Cecropia* silkworm. *Biol. Bull. 121*, 572–585.

Willis, J. H. and Brunet, P. C. J. (1966). The hormonal control of collaterial gland secretion. *J. Exp. Biol. 44*, 363–378.

Willis, J. H. and Lawrence, P. A. (1970). Deferred action of juvenile hormone. *Nature 225*, 81–83.

YAGIL, G. and HOBERMAN, H. D. (1969). Rate of isotope exchange in enzyme catalyzed reactions. *Biochemistry 14*, 352–360.

YIN, C. and GILLOTT, C. (1975). Endocrine activity during caste differentiation in *Zootermopsis angusticollis* Hagen (Isoptera): a morphometric and autoradiographic study. *Canad. J. Zool. 53*, 1690–1700.

ZHU, X. X., GFELLER, H. and LANZREIN, B. (1983). Ecdysteroids during oogenesis in the ovoviviparous cockroach *Nauphoeta cinerea*. *J. Insect Physiol. 29*, 225–235.

12 Regulation of Juvenile Hormone Titer: Degradation

BRUCE D. HAMMOCK

University of California, Davis, California, USA

1 JUVENILE HORMONE METABOLIC PATHWAYS

Before the metabolism of the juvenile hormones (JHs) was ever investigated, numerous scientists speculated that the JHs were metabolized by epoxide hydration and ester cleavage. Several research groups even initiated synthetic programs aimed at replacing these supposedly labile functionalities to develop more stable JH mimics (juvenoids). Had people realized that both the epoxide and ester moieties of JH are much more chemically stable than most of the model epoxides and esters studied, they possibly would not have jumped to the conclusion that hydrolysis of these two moieties represented the major routes of JH metabolism. Ignorance was on this occasion very useful because these hydrolytic pathways do account for the majority of JH metabolism in the insects so far studied. However, it is becoming more obvious that very specialized enzymes are responsible for this metabolism. The evidence to date strongly supports the hypothesis that all pathways of JH degradation represent loss of biological activity.

The basic scheme of JH metabolism proposed by Slade, M. and Zibitt, C. (1971, 1972) has been only slightly modified in the subsequent decade (Fig. 1). Several reviews have now appeared which cover in reasonable depth the metabolism of the JHs (Hammock, B. and Quistad, G. 1976, 1981; de Kort, C. and Granger, N. 1981; Richards, G. 1981). For those readers desiring such information, these reviews provide bibliographies and keys to much of the literature on JH metabolism, and still other reviews (referenced in Hammock, B. and Quistad, G. 1976, 1981) and other chapters in this series have considered JH metabolism in less detail. This chapter covers the pathways of JH metabolism in insects from a more biochemical and physiological perspective. Then the regulation of JH degradation in the Coleoptera and Lepidoptera is considered in greater detail. In most endocrine systems so far studied, hormone action is regulated largely by biosynthesis and receptor presence, while hormone degradation remains rather constant. An argument is made in this chapter that in at least some insects hormone degradation plays a key role in JH regulation. Also discussed are some practical aspects of this work, first in relation to insecticide resistance and secondly as an avenue leading to new concepts in insect control.

In interpreting results in this field it is especially important to appreciate some of the difficulties in studying xenobiotic metabolism. *In vivo* studies are tedious, especially when using physiological levels of substrates. One can never be sure that the radiolabeled hormone to be used is entering the correct *in vivo* pool. *In vitro* studies are at best suspect and, in general, the more clearly defined the system, the further it is from biological reality. The controls necessary to integrate *in vivo* and *in vitro* data are not impossible, but they are numerous.

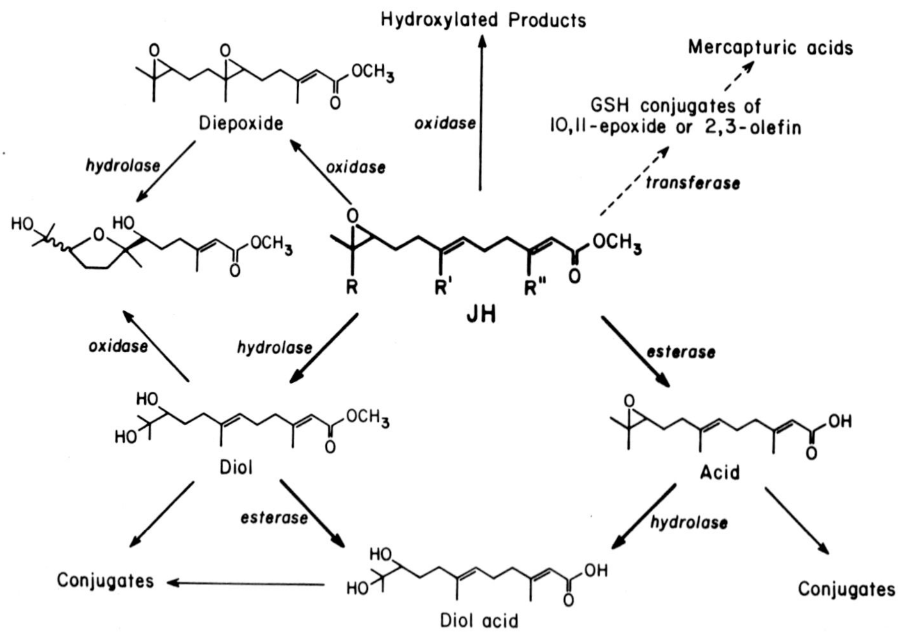

FIG. 1. Metabolic pathways and major metabolites of JH discussed in the text. JH 0, $R = R' = R'' = C_2H_5$; JH I, $R = R' = C_2H_5$, $R'' = CH_3$; JH II, $R = C_2H_5$, $R' = R'' = CH_3$; JH III, $R = R' = R'' = CH_3$. The major routes of metabolism are by epoxide hydration and/or ester cleavage (⟶ bold) followed by conjugation. A variety of other oxidative pathways may be significant in some insects (⟶). The role of glutathione conjugation has not been established clearly in any species (- - ->).

Synthesis and analysis of JH metabolite standards are not nearly so trivial as one may glean from the literature. To further complicate matters, JH and its metabolites may stick to glass, dissolve in lipid micelles or plastics, be oxidized by solvent peroxides, or may hydrolyze or rearrange in the presence of acids, bases or certain ions. In light of these difficulties, it is surprising that such a large amount of excellent work has been accomplished.

1.1 Epoxide hydration

Ethers comprise a rather stable set of organic molecules with only a few exceptions. The three-membered ring of the epoxide moiety is, however, highly strained and possesses properties which are quite distinct from many of the usual properties attributed to ethers. Because of their highly strained nature many epoxides are very reactive, but it is important to remember that even here there is a continuum of reactivity from the very unstable to the environmentally persistent. For example, the aflatoxins are thought to be mutagenic and cytotoxic due to the presence of a metabolically formed epoxide ring. In fact, many of the most potent mutagens and carcinogens known act by alkylation through a very reactive epoxide. Slight changes in the structure surrounding the epoxide change the target for alkylation, and the allatotoxin, precocene (see Bowers, W., vol. 8), is also thought to act via a reactive epoxide. Somewhat less reactive epoxides have found industrial application, as evidenced by the epoxy plastics and cements. Still other epoxides, as represented by some of the sterically hindered cyclodiene insecticides, do not fit the unstable stereotypes and are, in fact, environmentally persistent. The JHs and related epoxy-containing juvenoids fall between the latter two groups on the reactivity scale. The epoxide moiety is certainly not environmentally persistent (outside of a hydrophobic microcosm), but neither is it likely to polymerize or be a potent alkylating agent. Even if the reactivity of the epoxide moiety were much greater, the JHs lack the planar structure needed to intercalate with DNA; and Ivie, G. *et al.* (1980) have shown that even reactive, trisubstituted epoxides are exceptionally poor mutagens when compared to their monosubstituted homologs. The structurally related squalene oxide does cyclize to yield the steroid nucleus, and the carbonium ion formed when JH epoxide is exposed to mineral acids does invite cyclization and rearrangement. However, with the exception of some minor metabolites discussed in section 1.3, cyclization and rearrangement of the JH epoxide appear biologically unimportant.

Thus, the JHs of insects are unlikely to represent a mutagenic risk (should western societies shift to an insect diet) (Hsia, M. *et al.,* 1979), alkylation of an active site is an unlikely explanation of JH biological activity, and the insect will not be able to rely upon spontaneous hydrolysis or cyclization for rapid clearance of the JHs at critical times in its life. Two direct mechanisms are known which lead to metabolism of the epoxide moiety of JH: glutathione (GSH) conjugation (discussed in section 1.4) and epoxide hydration. Epoxide hydration involves addition of water to the epoxide moiety to yield 1,2-diols or glycols. The enzymes responsible for this reaction are known as epoxide ether hydrolases (E.C. 3.3.2.3, formerly known as hydrases or hydratases, E.C. 4.2.1.63) (Fig. 1). Interest first in the cyclodiene insecticides, and subsequently in mutagenicity, has led to much work on the epoxide hydrolases of mammals (Oesch, F. 1973; Hammock, B. *et al.,* 1980; Lu. A. and Miwa, G. 1980). Unfortunately, knowledge of insect epoxide hydrolases, related either to xenobiotic or JH metabolism, lags far behind.

The early observations on JH metabolism by Slade, M. and Zibitt, C. (1971, 1972) were extended to other insects by White, A. (1972); Ajami, A. and Riddiford, L. (1973) and numerous other workers. It is clear from these studies that the contribution of epoxide hydration to JH metabolism varies dramatically in different insects. Although detailed data are lacking, epoxide hydration appears to vary from the major pathway of JH metabolism in *Sarcophaga bullata*, to a role of equal importance to JH esterase in *Musca domestica*, to a minor role in *Trichoplusia ni*, and *Manduca sexta* (Table 1). There also can be a tremendous variation within a single order as in the Diptera or Coleoptera where *Tribolium castaneum* produces largely the diol but *Sitophilus granarius* produces largely the acid and diol acid (Edwards, J. and Rowlands, D. 1977, 1978). Since it seems very likely that epoxide hydrolases play a major role in JH degradation in some insects, and

Table 1: Involvement of epoxide hydration in juvenile hormone metabolism

Order; Genus, species	References
Thysanura	
Thermobia domestica (firebrat)	Ajami, A. and Riddiford, L. (1973)
Orthoptera	
Blattella germanica (German cockroach)	Ajami, A. and Riddiford, L. (1973)
Gromphadorhina portentosa (Madagascar cockroach)	Slade, M. and Wilkinson, C. (1974)
Locusta migratoria (migratory locust)	Emmerich, H. and Hartmann, R. (1973); Erley, D. *et al.* (1978)
Periplaneta americana (American cockroach)	Ajami, A. and Riddiford, L. (1973); Sams, G. *et al.* (1978)
Schistocerca americana (American grasshopper)	Gill, S. *et al.* (1972)
Schistocerca gregaria (gregarious grasshopper)	White, A. (1972)
Schistocerca vaga (vagrant grasshopper)	Slade, M. and Zibitt, C. (1972)
Hemiptera	
Oncopeltus fasciatus (larger milkweed bug)	Ajami, A. and Riddiford, L. (1973)
Pyrrhocoris apterus (cotton stainer)	Ajami, A. and Riddiford, L. (1973)
Rhodnius prolixus (Assassin bug)	White, A. (1972)
Homoptera	
Dactynotus ambrosiae	Ajami, A. and Riddiford, L. (1973)
Lepidoptera	
Antheraea pernyi (antheraea moth)	Ajami, A. and Riddiford, L. (1973)
Hyalophora cecropia (cecropia moth)	Slade, M. and Zibitt, C. (1971, 1972); Ajami, A. and Riddiford, L. (1973)
Manduca sexta (tobacco hornworm)	Slade, M. and Zibitt, C. (1971, 1972); Hammock, B. *et al.* (1975b); Mitsui, T. *et al.* (1979)
Samia cynthia (cynthia moth)	Ajami, A. and Riddiford, L. (1973)
Spodoptera eridania (southern armyworm, formerly *Prodenia*)	Slade, M. and Wilkinson, C. (1974); Slade, M. *et al.* (1976); Mullin, C. and Wilkinson, C. (1980a,b)
Trichoplusia ni (cabbage looper)	Wing, K. *et al.* (1981)
Coleoptera	
Hippodamia convergens (convergent lady beetle)	Ajami, A. and Riddiford, L. (1973)
Leptinotarsa decemlineata (Colorado potato beetle)	Ajami, A. and Riddiford, L. (1973)
Sitophilus granarius (grain weevil)	Edwards, J. and Rowlands, D. (1978)
Tenebrio molitor (yellow mealworm)	Ajami, A. and Riddiford, L. (1973); Hammock, B. *et al.* (1975a); Mumby, S. and Hammock, B. (1979)
Tribolium castaneum (red flour beetle)	Edwards, J. and Rowlands, D. (1977)
Diptera	
Chaoborus americanus (phantom midge)	Ajami, A. and Riddiford, L. (1973)
Calliphora erythrocephala (blowfly)	Slade, M. and Wilkinson, C. (1974)
Drosophila melanogaster (vinegar fly)	Ajami, A. and Riddiford, L. (1973); Wilson, T. and Gilbert, L. (1978)
Musca domestica (house fly)	Ajami, A. and Riddiford, L. (1973); Hammock, B. *et al.* (1974a,b, 1975a, 1977a); Yu, S. and Terriere, L. (1975, 1978a,b); Mumby, S. and Hammock, B. (1979)
Phormia regina (black blowfly)	Yu, S. and Terriere, L. (1978a,b)
Sarcophaga bullata (flesh fly)	Slade, M. and Zibitt, C. (1972); Yu, S. and Terriere, L. (1978a,b)
Stomoxys calcitrans (stable fly)	Hammock, B. *et al.* (1975a)
Hymenoptera	
Apis mellifera (honey bee)	Ajami, A. and Riddiford, L. (1973); de Kort, F. *et al.* (1977)
Atta texana (Texas leaf cutting ant)	Ajami, A. and Riddiford, L. (1973)
Solenopsis invicta (red imported fire ant)	Ajami, A. and Riddiford, L. (1973)

probably in xenobiotic metabolism in many insects, it is unfortunate that so little is known about these enzymes. For more comprehensive discussions of epoxide hydration the reader should see Brooks, G. (1977); Mullin, C. and Wilkinson, C. (1980a,b), and Hammock, B. and Quistad, G. (1981).

A discussion of epoxide hydration in insects is somewhat complicated by intraspecific and interspecific variation, but of greater significance is the variation in the substrates used to measure epoxide hydration. Since potent, selective inhibitors of ester hydrolysis have only recently become available, JH and JH acid have only occasionally been used as substrates. Commonly used substrates include the cyclodiene analog HEOM, styrene and octene 1,2-oxides, stilbene oxides, and epoxy-containing juvenoids. Reasonable circumstantial evidence supports the hypothesis that JH and epoxy juvenoids are degraded by the same enzyme(s); however, even in this case they may be metabolised by different enzymes than those which degrade JH acid. Evidence is building that the enzymes which metabolize monsubstituted epoxides, and possibly the cyclodienes, are quite different from those which hydrate JH.

With the substrates and insects so far studied, epoxide hydrolases of insects appear to be largely microsomal or at least membrane-bound. Such subcellular distribution was demonstrated using JH I and/or the juvenoid R 20458 in *M. domestica, P. regina, S. bullata, M. sexta,* and *T. ni.* As other substrates become routinely available, hydrolase activity may be found in other subcellular compartments. Epoxide hydrolase activity on JH and epoxy-containing juvenoids is maximal at a neutral pH, while the insect epoxide hydrolases acting on several other xenobiotics tend to have a more basic pH optimum in the several insects studied. This situation is similar in mammals where both the microsomal and cytosolic epoxide hydrolases which metabolize JH and juvenoids have a neutral pH optimum, while those metabolizing epoxides such as benzopyrene oxide have a basic pH optimum. These data suggest the presence of different forms of the microsomal hydrolase acting on JH and xenobiotics such as HEOM and octene 1,2-oxide. It has been shown in mammals that the pH optimum of a single hydrolase can vary depending upon the substrate and the hydrophobic environment of the enzyme, but such large differences are unlikely (Brooks, G. *et al.*, 1970; Hammock, B. *et al.*, 1974a; Slade, M. and Wilkinson, C. 1974; Slade, M. *et al.*, 1975; Hammock, B. *et al.*, 1975a; Yu. S. and Terriere, L. 1978a,b; Wing, K. *et al.*, 1981; Mullin, C. and Wilkinson, C. 1980a,b).

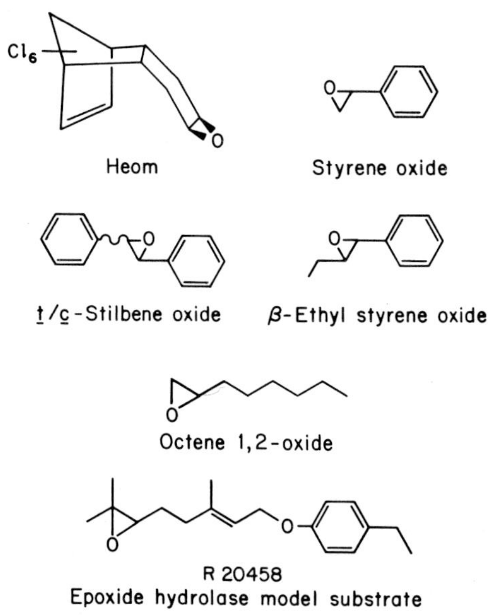

SCHEME A

A variety of compounds have been screened as potential inhibitors of epoxide hydration in insects, but as yet no potent inhibitors have been found. From the numerous compounds screened in *M. domestica* as inhibitors of the hydration of the epoxy-containing juvenoid R 20458, only the lack of inhibition caused by trichloropropene and cyclohexene oxides was noteworthy. These compounds are the best-known inhibitors of the microsomal epoxide hydrolase in mammals, but interestingly they fail to inhibit the hydration of R 20458 or JH hydration in mammalian liver cytosol or microsomes as well. Using the dieldrin analog HEOM as a substrate, a variety of compounds including JH, juvenoids, organophosphates, methylene dioxyphenyl and benzothiozole mixed function oxidase (MFO) inhibitors, trichloropropene oxide, and naphthalene, 1,2-epoxide all caused some inhibition of epoxide hydration. Interestingly, some glycidyl ethers proved to be quite potent inhibitors of HEOM hydration in *Calliphora erythrocephala* and *Spodoptera eridania* (previously *prodenia*). Unfortunately, only a few of these compounds have been screened (without success) as inhibitors of JH epoxide hydration. The action of several inhibitors of octene oxide hydration in *S. eridania* supports the presence of an essential lysine or histidine. However, no information is yet available on the mechanism of JH hydration. Preliminary studies, including the inhibition data discussed above, provide further evidence that JH is degraded in insects by a rather unique hydrolase(s).

It should, however, be noted that such differential inhibition is likewise not definitive. In mammalian systems trichloropropene oxide has been found to be a moderate inhibitor of the hydration of some substrates which have a rapid turnover but low affinity for the enzyme. However, other substrates which have a higher affinity for the same enzyme suffer no inhibition with this material.

Just as selective inhibitors have proven useful in elucidating the role of esterases in JH regulation, selective, potent inhibitors of the epoxide hydrolases acting on JH could be very important in elucidating the role of these enzymes in insect development. Unfortunately such inhibitors are not yet known. Possibly the discovery of chalcone oxides as potent inhibitors of the mammalian cytosolic and mitochondrial epoxide hydrolases will lead to useful inhibitors of JH epoxide hydration in insect systems (Brooks, G. 1973, 1974; Hammock, B. *et al.,* 1974a; Slade, M. *et al.,* 1975; Craven, A. *et al.,* 1976; Yu, S. and Terriere, L. 1978a,b; Mullin, C. and Wilkinson, C. 1980b; Mullin, C. and Hammock, B. 1982). Phenobarbital has been reported to induce JH I epoxide hydrolase activity in *Musca domestica*, while hydrolase activity on octene oxide was induced in *S. eridania* with pentamethylbenzene. Potent inducers of JH epoxide hydrolase may place a strain on the mechanism regulating endocrine homeostasis (Yu, S. and Terriere, L. 1978b; Mullin, C. and Wilkinson, C. 1980a).

Several workers have reported the absence of epoxide hydrolase activity in the hemolymph from a variety of insects including *Periplaneta americana*, *Gromphadorhina portentosa*, *Trichoplusia ni*, *Spodoptera eridania*, *Manduca sexta*, *Hyalophora cecropia*, and *Tenebrio molitor*. Epoxide hydrolase activity is fairly widely distributed in other tissues, and in various insects it has been reported high in midguts, fat bodies, and imaginal disks (Hammock, B. *et al.,* 1974a; Slade, M. and Wilkinson, C. 1974; Hammock, B. *et al.,* 1975a; Slade, M. *et al.,* 1976; Wing, K. *et al.,* 1981).

Epoxide hydrolase activity has been titered during the development of several insects. In *M. domestica* epoxide hydrolase activity, expressed in terms of metabolism per whole insect equivalent, increased to high levels towards the end of the last larval stadium, dropped in the pupa and increased again during the first 2 weeks of adult life. Epoxide hydrolase activity on R 20458 was very high throughout the pupal stage of male and female *T. molitor*. Two separate studies using HEOM, JH I, and subsequently octene oxide led to the finding that the midgut epoxide hydrolase was most active during the middle or wandering-burrowing phase of the last larval instar. Similar peak activities of epoxide hydrolase acting on R 20458 were found in the fat body and midgut of last larval instar *T. ni*. However, midgut activity peaked 1 day earlier and fell off more rapidly than that in the fat body. Although detailed kinetic studies are lacking on the enzymes involved in JH metabolism in larval tissues, epoxide hydration appears relegated to a secondary role in the midguts, imaginal disks and fat bodies of *M. sexta* and *T. ni* larvae. At low substrate concentrations JH acid and JH acid diol are found as exclusive metabolites in homogenates or with organs held *in vitro*. Only at very high substrate concentrations is the diol found to be a significant metabolite. Several interpretations are possible from this observation, but one interpretation is consistent with other circumstantial evidence including *in vivo* data. JH ester cleavage appears to be the major route of initial metabolism and the hydrolase(s) involved appear to prefer JH acid over JH itself as a substrate. The situation appears different in *H. cecropia* and *S. eridania* where the diol is detected as a metabolite (Slade, M. and Zibitt, C. 1971, 1972; Slade, M. *et al.,* 1976; Yu, S. and Terriere, L. 1978a,b; Mullin, C. and Wilkinson, C. 1980a,b; Wing, K. *et al.,* 1981).

Only two comprehensive studies have been published on the biochemistry of any insect epoxide hydrolases. Mullin, C. and Wilkinson, C. (1980a,b) report the most catalytically active epoxide hydrolase ever purified from a eukaryotic source. By monitoring the relative activity on different substrates during purification, these workers clearly demonstrated the presence of multiple forms of microsomal epoxide hydrolases in midguts from *S. eridania*. Unfortunately for endocrinologists, the form purified had no hydrolytic activity on JH. Clearly, this excellent work needs to be duplicated on the hydrolase(s) metabolizing JH. The information needed to assess the role of epoxide hydrolases in insect endocrinology is illustrated by the more comprehensive information on JH esterases in sections 1.2, 2, and 3. With improved assay methods such as the use of cellulose prelayer thin-layer plates, and with the availability of model substrates, hopefully this information will be available soon. Such expanded research on epoxide hydrolases could be justified, based upon the role of these enzymes in the degradation of both natural and man-made xenobiotics, intermediary metabolism such as prostaglandin or leukotriene metabolism, or the conversion of sitosterol into cholesterol (Morisaki, M. *et al.*, 1981), as well as insect endocrinology.

1.2 Ester hydrolysis

Ester hydrolysis appears to be the major route of JH metabolism in most insects so far studied, and by far the greatest amount of work has centered on this pathway (Table 2). It would be nice to think that this emphasis on esterase research stemmed from a knowledge of relative rates of metabolism by different pathways. Such is not the case. Most of the studies listed in Tables 1 and 2 focused immediately upon the biochemistry or physiology of the enzyme investigated, with total disregard of competing pathways and of *in vivo* studies. This lack of experimental information on *in vivo* pathways stems from two causes. First, although experiments on *in vivo* metabolism are conceptually simple, they are difficult to carry out. Second, JH esterase activity is much easier to measure than epoxide hydrolase activity. Several surveys of JH metabolism have been made, but these studies largely ignored the biology of the test organisms. Other studies have simply examined hemolymph metabolism, which would not detect epoxide hydration. Thus, the blanket statements in this chapter — as well as others — that ester hydrolysis is the major route of metabolism are based on a disappointingly small database.

There are numerous studies involving the monitoring of insect esterases, probably resulting from the ease by which esterase activity can be measured using esters of various phenols such as α and β-naphthol, nitrophenols, and β-methylumbelliferone or thiols such as thiocholine. It is tempting to extrapolate this information on esterases to JH esterases, but one should be wary of such extrapolation. These stains are likely to detect a variety of esterases, proteases, phosphatases and other hydrolytic enzymes. They may not, however, detect JHE for three rather obvious reasons. First, the JHs are α,β-unsaturated and thus conjugated esters. Not only does this conjugation restrict rotation and lend planarity to this portion of the molecule; it also stabilizes the JH ester to chemical hydrolysis. Acid and base hydrolysis, as well as enzymatic hydrolysis, are thought to proceed through a tetrahedral intermediate at the carbonyl carbon as shown by Scheme C. As the trigonal carbon takes on tetrahedral character, it must move out of conjugation with the 2,3-olefin and thus is resistant to attack. Second, JH is an ester of a very small alcohol and a very large acid, while most of the model substrates are esters of very large alcohols such as naphthol and very small acids such as acetic acid. Several workers have found that even ethyl esters of JH and related juvenoids are very slow to hydrolyze in the presence of insect esterases, while isopropyl esters are still more stable. The concept of resonance stabilization of the ester and steric hindrance of JH esterases found practical utilization in development of the hydrolytically stable dodecadienoate esters hydroprene (ethyl ester) and methoprene (isopropyl ester). Ester hydrolysis must proceed through an intermediate in which the carbonyl group moves out of conjugation with two olefinic functionalities. It is unlikely that an enzyme which distinguishes between a methyl and ethyl ester will hydrolyze a naphthyl ester rapidly. In some species the $2E$ geometry is required for rapid enzymatic hydrolysis, and evidence is mounting in several species that the esterase's recognition of JH may be as specific as, or more specific than, that of the lepidopteran hemolymph carriers — or possibly some JH receptors. Finally, JH is a highly lipophilic molecule normally present at very low physiological concentrations. Although JH esterases may be very active, it is unlikely that they are present in high concentrations. Therefore, it is unlikely that JH esterases will be detected in the presence of more active enzymes using millimolar concentrations of spectrophotometric substrates. For any real chance of detecting JH esterases with classical substrates the enzyme(s) will probably have to be purified or at least enriched (Weirich, G. *et al.*, 1973; Weirich, G. and Wren, J., 1973, 1976; Yu, S. and Terriere, L., 1975; Hammock, B. and Quistad, G., 1976, 1981; Hammock, B. *et al.*, 1977a,b; Sparks, T. and Hammock, B., 1979a).

Table 2: Involvement of ester cleavage in juvenile hormone metabolism

Order; Genus, species	References
Thysanura	
Thermobia domestica (firebrat)	Ajami, A. and Riddiford, L. (1973)
Orthoptera	
Blaberus giganteus	Hammock, B. (1975); Hammock, B. *et al.* (1977b); Hammock, B. and Mumby, S. (1978); Mumby, S. *et al.* (1979)
Blattella germanica (German cockroach)	Ajami, A. and Riddiford, L. (1973)
Diploptera punctata (Pacific beetle cockroach)	Rotin, D. *et al.,* (1982); Szibbo, C. *et al.* (1982)
Locusta migratoria (migratory locust)	Emmerich, H. and Hartmann, R. (1973); Erley, D. *et al.* (1975); de Kort, C. *et. al.* (1979); Peter, M. *et al.* (1979)
Periplaneta americana (American cockroach)	Ajami, A. and Riddiford, L. (1973); Sams, G. *et al.* (1978); de Kort, C. *et al.* (1979)
Periplaneta brunnea (brown cockroach)	Weirich, G. and Wren, J. (1976)
Schistocerca gregaria (gregarious grasshopper)	White, A. (1972); Pratt, G. (1975)
Schistocerca vaga (vagrant grasshopper)	Slade, M. and Zibitt, G. (1972); Weirich, G. and Wren, J. (1976)
Hemiptera	
Dysdercus cingulatus	Sláma K. and Jarolím, V. (1980)
Oncopeltus fasciatus (large milkweed bug)	Ajami, A. and Riddiford, L. (1973)
Pyrrhocoris apterus (cotton stainer)	Ajami, A. and Riddiford, L. (1973); Sláma, K. and Jarolím, V. (1980)
Rhodnius prolixus (assassin bug)	White, A. (1972)
Homoptera	
Dactynotus ambrosiae	Ajami, A. and Riddiford, L. (1973)
Lepidoptera	
Anagasta kuehniella (Mediterranean flour moth)	Kramer, K. and Childs, C. (1977)
Antheraea pernyi (antheraea moth)	Ajami, A. and Riddiford, L. (1973)
Colias eurytheme (alfalfa butterfly)	Wing, K. *et al.* (1982)
Danus plexippus (monarch butterfly)	Lessman, C. and Herman, W. (1981); Jones, D. *et al.* (1982); Wing, K. *et. al.* (1982)
Diatraea grandiosella (Southwestern corn borer)	Brown, J. *et al.* (1977)
Diatraea saccharalis (sugarcane border)	Roe, R. *et al.* (1983)
Ephestia elutella (tobacco moth)	Hammock, B. *et al.* (1981); Jones, D. *et al.* (1982); Wing, K. *et al.* (1982)
Estigmene acraea (salt marsh caterpillar)	Jones, D. *et al.* (1982); Hammock, B. *et al.* (1981)
Galleria mellonella (greater wax moth)	McCaleb, D. and Kumaran, A. (1980); McCaleb, D. *et al.* (1980); Reddy, G. *et al.* (1979, 1980); Rudnicka, M. *et al.* (1979) Sláma, K. and Jarolím, V. (1980); Hammock, B. *et al.* (1981); Kumaran, A. *et al.* (1981); Wing, K. *et al.* (1982); Jones, D. *et al.* (1982)
Heliothis virescens (tobacco budworm)	Hammock, B. *et al.* (1981); Wing, K. *et al.* (1982)
Hemileuca nevadensis (Nevada buck-moth)	Hammock, B. *et al.* (1981); Jones, D. *et al.* (1982); Wing, K. *et al.* (1982)

Table 2: Involvement of ester cleavage in juvenile hormone metabolism (Continued)

Order; Genus, species	References
Hyalophora cecropia (cecropia moth)	Slade, M. and Zibitt, C. (1971, 1972); Ajami, A. and Riddiford, L. (1973); Whitmore, D. *et al.* (1974)
Hyalophora gloveri (saturniid moth)	Whitmore, D. *et al.* (1972, 1974)
Junonia coenia (buckeye butterfly)	Hammock, B. *et al.* (1981); Jones, D. *et al.* (1982); Wing, K. *et al.* (1982)
Manduca quinquemaculata (tomato hornworm)	Kramer, K. and Childs, C. (1977)
Manduca sexta (tobacco hornworm)	Slade, M. and Zibitt, C. (1971, 1972); Ajami, A. and Riddiford, L. (1973); Weirich, G. and Wren, J. (1973); Weirich, G. *et al.* (1973); Ajami, A. (1975); Akamatsu, Y. *et al.* (1975); Hammock, B. *et al.* (1975b); Sanburg, L. *et al.* (1975a, b); Nowock, J. *et al.* (1976); Nowock, J. and Gilbert, L. (1976); Kramer, K. and Childs, C. (1977); Vince, R. and Gilbert, L. (1977); de Kort, C. *et al.* (1979); Mitsui, T. *et al.* (1979); Kramer, S. and Law, J. (1980); Coudron, T. *et al.* (1981); Hammock, B. *et al.* (1981); Wing, K. *et al.* (1982); Jones, D. *et al.* (1982); Sparks, T. *et al.* (1983)
Orgia vetusta (western tussock moth)	Hammock, B. *et al.* (1981); Jones, D. *et al.* (1982); Wing, K. *et al.* (1982)
Pectinophora gossypiella (pink bollworm)	Hammock, B. *et al.* (1981); Jones, D. *et al.* (1982); Wing, K. *et al.* (1982)
Pieris rapae (cabbage butterfly)	Jones, D. *et al.* (1982); Wing, K. *et al.* (1982)
Plodia interpunctella (Indian meal moth)	Oberlander, H. and Silhacek, D. (1976)
Pseudoplusia includens (soybean looper)	Sparks, T. *et al.* (1982)
Samia cynthia (cynthia moth)	Ajami, A. and Riddiford, L. (1973); Weirich, G. and Wren, J. (1976)
Sphinx cheris (great ash sphinx)	Kramer, K. and Childs, C. (1977)
Spodoptera eridania (southern armyworm, formerly *Prodenia*)	Slade, M. and Wilkinson, C. (1974)
Spodoptera exigua (beet armyworm)	Hammock, B. *et al.* (1981); Jones, D. *et al.* (1982); Wing, K. *et al.* (1982)
Spodoptera littoralis (army worm)	Sláma, K. and Jarolím, V. (1980)
Trichoplusia ni (cabbage looper)	Hammock, B. *et al.* (1977a,b); Hammock, B. and Sparks, T. (1977); Mumby, S. *et al.* (1979); Sparks, T. *et al.* (1979a,b); Sparks, T. and Hammock, B. (1979a,b); Hammock, B. *et al.* (1981); Sparks, T. and Hammock, B. (1980); Jones, G. *et al* (1981); Jones, D. *et al.* (1982); Wing, K. *et al.* (1981, 1982) Rudnicka, M. and Hammock, B. (1981a,b); Yuhas, D. *et al.* (1983)
Coleoptera	
Hippodamia convergens (convergent lady beetle)	Ajami, A. and Riddiford, L. (1973)
Leptinotarsa decemlineata (Colorado potato beetle)	Ajami, A. and Riddiford, L. (1973); Kramer, S. and de Kort, C. (1976a,b); Kramer, S. (1978a,b); de Kort, C. *et al.* (1978, 1979); Kramer, S. and Law, J. (1980); de Kort, C. *et al.* (1981); de Kort, C. and Granger, N. (1981)
Leptinotarsa haldemani	de Kort, C. *et al.* (1979)
Tenebrio molitor (yellow mealworm)	Ajami, A. and Riddiford, L. (1973); Weirich, G. and Wren, J. (1976); McCaleb, D. *et al.* (1980); Sláma, A. and Jarolím, V. (1980); Sparks, T. and Hammock, B. (1980)
Tribolium castaneum (red flour beetle)	Edwards, J. and Rolands, D. (1977)

Table 2: Involvement of ester cleavage in juvenile hormone metabolism (Continued)

Order; Genus, species	References
Diptera	
Culex pipiens pipiens (northern house mosquito)	Hooper, G. (1976);Brown, T. and Hooper, G. (1979)
Chaoborus americanus (phantom midge)	Ajami, A. and Riddiford, L. (1973)
Drosophila hydei (fruit fly)	Klages, G. and Emmerich, H,. (1979a,b); Bisser, A. and Emmerich, H. (1981)
Drosophila melanogaster (vinegar fly)	Chihara, A. *et al.* (1972); Ajami, A . and Riddiford, L. (1973); Wilson, T. and Gilbert, L. (1978)
Musca domestica (house fly)	Ajami, A. and Riddiford, L. (1973); Hammock, B. *et al.* (1977a); Yu, S. and Terriere, L. (1978a); de Kort, C. *et al.* (1979); Mumby, S. *et al.* (1979)
Phormia regina (black blowfly)	Yu, S. and Terriere, L. (1978a)
Sarcophaga bullata (flesh fly)	Slade, M. and Zibitt, C. (1972); Weirich, G. and Wren, J. (1976); Yu, S. and Terriere, L. (1978a)
Hymenoptera	
Apis mellifera (honey bee)	Ajami, A. and Riddiford, L. (1973); de Kort, C. *et al.* (1977, 1979); Mane, S. and Rembold, H. (1977)
Atta texana (Texas leaf cutting ant)	Ajami, A. and Riddiford, L. (1973)
Bombus terrestris (Bumble bee)	Röseler, P. and Röseler, I. (1978)
Solenopsis invicta (Red imported fire ant)	Ajami, A. and Riddiford, L. (1973)

Any definitive detection of JH esterase activity must employ radioactive JH as a substrate, and a number of analytical methods, as well as radiosynthesis of the JHs, have been reviewed by Hammock, B. and Quistad, G. (1981). If JH acid is not clearly the sole metabolite, chromatographic methods are required such as thin-layer, high-performance liquid, or gas–liquid chromatography. The most rapid chromatographic technique now employed involves direct spotting of biological extracts on the cellulose pre-layer of silica gel thin-layer plates, visualization of suspected metabolites by co-chromatography with ultraviolet dense standards, and radiotracer analysis. When JH acid is the only metabolite, two methods have proven very useful. The first method involves release of radioactive methanol from methoxy-labeled JH. Unfortunately the methyl-labeled JHs are not commercially available, although several published procedures describe their preparation. The second method involves partition of chain-radiolabeled JH and JH acid between a basic aqueous phase and a hydrocarbon organic phase. Iso-octane or dodecane are often used to retard volatilization, but thorough mixing is necessary (Sanburg, L. *et al.*, 1975a; Nowock, J. *et al.*, 1976; Hammock, B. and Quistad, G., 1976, 1981; Hammock, B. *et al.*, 1977b; Hammock, B. and Sparks, T., 1977; Vince, R. and Gilbert, L., 1977; Hwang-Hsu, K. *et al.*, 1979; Klages, G. and Emmerich, H., 1979a,b; Sparks, T. *et al.*, 1979b; Peter, M. 1981).

1.3 Oxidation

A student of xenobiotic metabolism could look at the JH structure and envision numerous possible oxidative metabolites involving epoxidation, hydroxylation alpha to the olefins or epoxide, hydroxylation on methyl or ethyl side-chains, and other such pathways. Surprisingly, oxidative pathways of metabolism appear unimportant in most of the insects examined. However, in only a few cases, such as the investigation of JH metabolism in *M. sexta* fat body and wing disks, have techniques been rigorous enough to exclude moderate levels of oxidative metabolism. Ajami, A. and Riddiford, L. (1973) examined JH metabolism in an impressive number of species and report some evidence for oxidative degradation. This work cannot be taken too seriously because the analytical method used

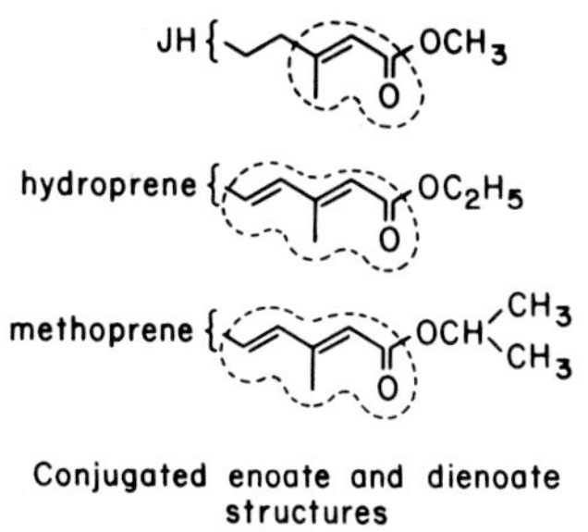

SCHEME B

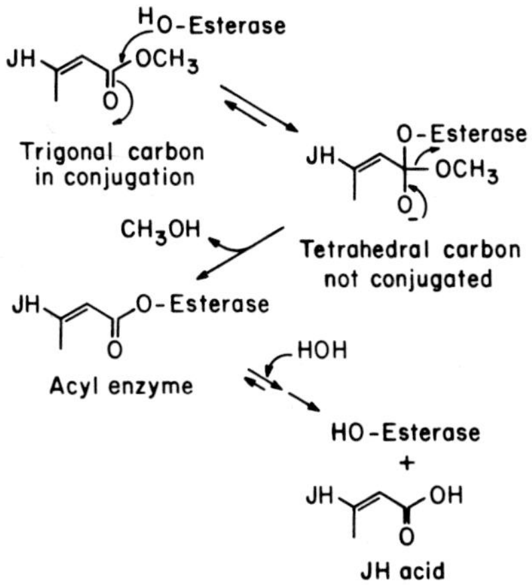

SCHEME C

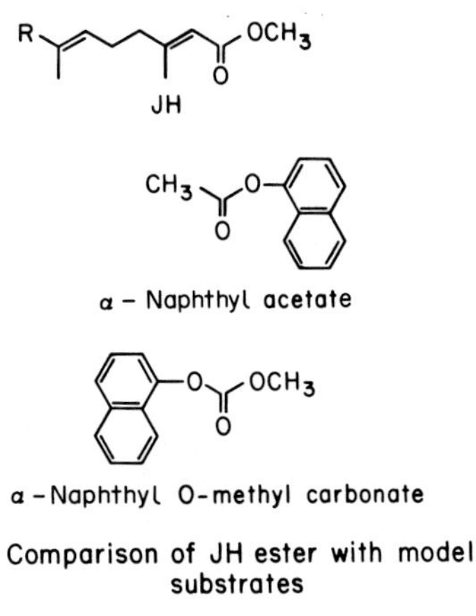

SCHEME D

yields poor resolution. The solvent system employed did not resolve JH diepoxide from JH, and the authenticity of their standards is equivocal. However, there is some evidence from this study, as well as from earlier work by Slade, M. and Zibitt, C. (1972), that oxidase activity may contribute to metabolism.

The best demonstration of oxidative metabolism of JH comes from studies involving the flies *Phormia regina*, *Sarcophega bullata* and especially *Musca domestica*. The initial studies by Hammock, B. *et al.* (1977a) and the more comprehensive work by Yu, S. and Terriere, L. (1978a) are surprisingly consistent considering the disparate strains and analytical techniques. The autoradiographic method of Hammock, B. *et al.* (1977a) facilitated the detection of minor discrete metabolites as distinct spots, while the work of Yu, S. and Terriere, L. (1978a) was able to discern lower levels of oxidase-involved metabolism.

Hammock, B. *et al.* (1977a) found that roughly equal amounts of JH acid and diol, with traces of JH diol acid, were produced from JH when microsomes from both susceptible (NAIDM) and resistant (methoprene-R and dimethoate-R-unpublished) *M. domestica* were incubated without NADPH. When NADPH was added, little change was noted with the susceptible strains, but extensive metabolism was noted with the resistant strains. This observation indicates that some insecticide-resistant *M.domestica* have an enhanced ability to metabolize JH. Most oxidative metabolites were unidentified but discrete radioactive spots co-chromatographed with the *cis*- and *trans*-tetrahydrofuran diol acids (Fig. 1 and below). The tetrahydrofuran diol esters were also present. However, in the solvent system used they chromatographed with JH acid and were only detected after the metabolite band was reacted with diazomethane. Although oxidative metabolism of JH I was very minor in microsomes from susceptible flies, metabolism of the juvenoid R 20458 by oxidative pathways was extensive and rapid. Since these compounds are of similar structure and polarity, this discrepancy is interesting.

Yu, S. and Terriere, L. (1978a) reported metabolic pathways similar to those above involving 6, 7-epoxidation. Using a variety of techniques they provided good evidence that an NADPH-dependent, CO-sensitive microsomal oxidase was involved in the metabolism of JH I in *M. domestica*. They also found a large increase in oxidase activity in two resistant strains of *M. domestica*, and this increase could even be underestimated due to

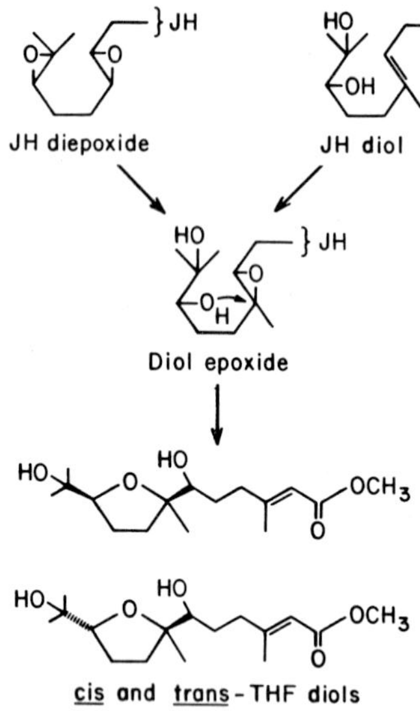

SCHEME E

co-chromatography of esterase and oxidase bands. In one of the two susceptible strains examined, oxidative metabolism of JH I was also found to be a major route of metabolism. Oxidative as well as hydrolytic metabolism was much slower in *P. regina* and *S. bullata* than in *M. domestica*. In an excellent subsequent experiment, oxidation as well as conjugation were demonstrated *in vivo* (Yu, S. and Terriere, L., 1978b).

Yu, S. and Terriere, L. (1978a) also followed the relative contribution of oxidation, epoxide hydration, and ester cleavage throughout the development of *M. domestica*. Hydrolase and esterase levels are high in the early larval instars and drop as pupation approaches, while oxidase levels increase before pupation. All three activities are low during the early pupal stage, and later increase. Epoxide hydrolase activity peaks shortly after eclosion, while esterase and oxidase activites continue to increase for 5 or more days into the adult stage. Possibly all three pathways are involved in JH regulation.

The origin of the tetrahydrofuran diols may seem complex, but it is actually rather straightforward as illustrated by some earlier work with 1,5-diepoxyhexane and the juvenoid R 20458. When one of the two epoxides of the corresponding JH diepoxide is opened to the diol by acid-catalyzed hydrolysis, one of the alcohols can attack the remaining epoxide to give a 5-, 6- or 7-membered ring or a bicyclic ether. By far the major products are the two tetrahydrofuran diols which are *cis* or *trans* across the ring. When the JH diol is oxidized under neutral conditions, the JH diol epoxide is an intermediate that can be isolated. This compound will slowly cyclize to yield almost exclusively the tetrahydrofuran diols. Similar reactions can occur enzymatically (Gill, S. *et al.*, 1974; Hammock, B. *et al.*, 1974a,b; 1975b; 1977a; Hammock, B. and Quistad, G. 1976, 1981; Yu, S. and Terriere, L., 1978a; Hammock, B., 1981).

With the juvenoid R 20458 reduction of the epoxide to the corresponding olefin was found to occur rapidly in the gut contents of rats and steers (Hoffman, L. *et al.*, 1973; Ivie, G. *et al.*, 1976; Ivie, G., 1976). This reaction has been demonstrated to occur in the gut contents of man and several other mammals with several compounds including JH (Hammock, unpublished). The pathway has not been detected in insects, but the search has not been exhaustive, and it is complicated by the facile reoxidation of the 10,11 olefin by solvent peroxides. However, this pathway may be of use to entomologists for providing radiolabeled precursors including methyl farnesoate to study JH biosynthesis.

1.4 Conjugation

Compared to research on the microsomal oxidases, very little research on conjugating reactions has been reported. This disparity is disappointing because conjugating reactions are likely to play a major role in the metabolism of a variety of xenobiotics. It is thus not surprising that there is still less information on conjugating reactions in insects than in vertebrates (Yang, R., 1976) and almost none on the conjugation reactions involving JH.

Slade, M. and Zibitt, C. (1972) found that apparent conjugates were produced from JH I *in vivo* in *Manduca sexta*, *Hyalophora cecropia*, *Schistocerca vaga* and *Sacrophaga bullata*. For example, in 2 h, 45% of the hormone injected into *S. vaga* appeared as conjugated material. Cleavage with Glusulase

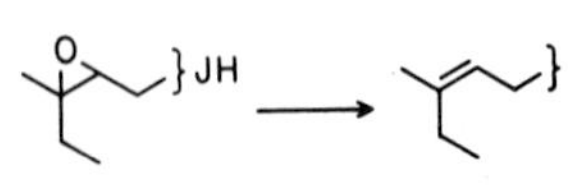

SCHEME F

and rechromatography resulted in firm evidence for conjugation of JH diol and possibly JH acid and diol acid. Reasonable evidence was also presented for sulfate conjugation of JH diol in *S. bullata*. In his study of JH III metabolism in *Schistocerca gregaria* and *Rhodnius prolixus*, White, A. (1972) demonstrated the presence of JH diol, JH acid, and JH diol acid as sulfate and/or glucose conjugates. The most detailed study of the conjugation of JH to date was by Slade, M. and Wilkinson, C. (1974) in *Spodoptera eridania*. Using sulfatase cleavage and subsequent enzymatic synthesis of a sulfate conjugate of the diol acid, they presented good evidence that sulfate conjugation of the diol acid was the major, if not the only, route of conjugation.

If the presence of highly polar metabolites is taken as evidence of conjugation, then essentially every *in vivo* metabolism study involving JH has resulted in detection of conjugates. It is possible that some of the highly polar *in vivo* metabolites of JH result from extensive metabolism of the JH backbone rather than from conjugation of a primary metabolite. In most studies the conjugates have not been hydrolyzed to the aglycones. Even when hydrolytic enzymes are used, one must be very careful about their integrity. In no study has the structure of a JH conjugate been established or the mechanism of conjugation studied in detail.

In their 1973 survey Ajami, A. and Riddiford, L. found the appearance of water-soluble metabolites to be very common. Three water-soluble products were obtained from extracts of *H. cecropia* and *M. sexta*, and when hydrolyzed about 60% of the radioactivity co-chromatographed with JH acid and JH acid diol. Apparent conjugation has also been found in *Drosophila melanogaster*, *Leptinotarsa decemlineata*, *Locusta migratoria*, *Musca domestica*, *Tribolium castaneum*, and *Trichoplusia ni* (Edwards, J. and Rowlands, D. 1977; Erley, D. *et al.*, 1975; Hammock, B. *et al.*, 1981; Kramer, S. *et al.*, 1977; Sparks, T. and Hammock, B. 1980; Wilson, T. and Gilbert, L. 1978; Yu, S. and Terriere, L., 1978a).

Thus, there is excellent evidence for sulfate conjugation of JH diol and/or diol acid, and less evidence for other possible routes of conjugation.

Glutathione conjugation may superficially appear to be a likely route of metabolism, but conclusive evidence for this route is so far lacking in insects. Glutathione is known to attack epoxides both spontaneously and enzymatically. However, using the juvenoid epoxide R 20458, Gill, S. *et al.* (1974) were unable to find evidence for glutathione conjugation catalyzed by mammalian liver homogenates. Analogous studies failed to show R 20458 conjugation with insect systems, which is not surprising since conjugation is usually very slow with polysubstituted epoxides (Chasseaud, L., 1979). However, Morello, A. and Agosin, M. (1979) report that the glutathione conjugate leading to the subsequent formation of the mercapturate at the 10 position of JH is the major metabolite in rat hepatocyte suspensions. Another possible route of metabolism is a Michael-like attack of GSH at the conjugated 2,3-olefin. Even a reversible attack at this position could result in partial loss of biological activity due to *E*/*Z* isomerization. This reaction between glutathione and JH was only demonstrated to occur with JH I at 37° and higher. Addition of *M. domestica* 100,000 ***g*** supernatant to JH caused a negligible increase in the production of water-soluble radioactivity or 2*E*/*Z* isomerization as monitored by liquid chromatography (Hammock, B. *et al.*, 1977a; Hammock, B. and Mumby, S. unpublished). Possibly the degradation of 2*Z* juvenoids in deproteinated insect hemolymph observed by Sláma, K. and Jarolím, V. (1981) was due to glutathione. Thus, there are no firm data supporting glutathione conjugation of JH as an important route of metabolism in insects (Fig. 1). However, reaction between JH and glutathione or another sulfhydryl group could lead to primary metabolism, especially in the presence of an effective catalyst. Also, artifacts could arise if high levels of glutathione are used to inhibit tyrosinase action.

Thus, the few *in vivo* studies performed indicate that conjugation (probably sulfate conjugation) is important in the ultimate conversion of JH into water-soluble metabolites. Although this question has not been addressed, it is unlikely that conjugation is the rate-limiting step in JH inactivation. However, further research is needed on the mechanisms of JH conjugation if we wish to glean an appreciation of its ultimate disposition.

1.5 JH dynamics

An appreciation of the dynamics of JH biosynthesis,

distribution, action, metabolism and elimination is ultimately very important to our understanding of insect endocrinology. Yet there is no study even on the kinetics of JH degradation. Hammock, B. *et al.* (1974b, 1975a) employed simplistic kinetic arguments to explain the action of some juvenoids. The best study of the *in vivo* kinetics of JH degradation was reported by Erley, D. *et al.* (1975) in *L. migratoria.* These workers found very rapid elimination of JH from the hemolymph and appearance of radioactivity in the feces. There was little difference in the rates of elimination of JH in males or females of either sex. It was interesting that the Malpighian tubules appeared to be the site of most JH excretion. A similar approach by Edwards, J. and Rowlands, D. (1977) correlated JH metabolism with development in *T. castaneum*, while Kramer, S. (1978b) attempted to summarize his extensive investigation of JH metabolism in *L. decemlineata* with a balance sheet of biosynthesis and degradation.

Several approaches need to be taken in the investigation of the kinetics of JH metabolism and excretion. One approach involves careful *in vivo* studies such as those performed by Erley, D. *et al.* (1975) or Edwards, J. and Rowlands, D. (1978). Such studies could indicate the relative importance of metabolism in the regulation of JH, yet one must consider the effects of *in vivo* pools. Appropriate models could involve a comparison of species where JH metabolism appears slow, such as in *D. melanogaster* or *Apis mellifera*, with species where JH metabolism is rapid and appears important in endocrine regulation such as *M. sexta* or *T. ni*. Such studies become even more crucial in insects falling in an intermediate class. A second approach is to correlate *in vivo* metabolism with *in vitro* metabolism by organ culture and tissue homogenates. For instance, Wing, K. *et al.* (1981) have found in *T. ni* that ester hydrolysis by a single enzyme in the fat body and hemolymph accounts for most of the *in vivo* degradation of JH. From this information simplistic models of the kinetics of JH degradation can be developed. Such models can then be coupled with binding data and partition characteristics of JH as first approached by Nowock, J. *et al.* (1976). As additional information becomes available, such as JH metabolism by the small but very active imaginal disks, a model of increasing complexity can be devised.

2 JH ESTERASES

As indicated in section 1.2, much more is known about JH ester hydrolysis than about other pathways of metabolism. Thus, in this section the biochemical properties of JH esterase are discussed in more detail.

2.1 JH specific esterases, possible definitions

Sanburg, L. *et al.* (1975) coined the term JH specific esterase to refer to those esterases in *M. sexta* hemolymph which are difficult to inhibit with the organophosphate di-isopropylphosphofluoridate (DFP), show marginal hydrolytic activity on α-naphthyl acetate, and hydrolyze JH on or off the carrier protein. At the onset, it must be remembered that in this context specific means at most highly selective. As discussed in more detail below, one of the key experiments which Sanburg, L. *et al.* (1975a) used to distinguish general and specific JH esterases cannot be duplicated in the same or several other laboratories (Coudron, T. *et al.*, 1981). More alarming has been the widespread and indiscriminate application of the term JH specific esterase to a variety of insects. In general the workers who have applied this terminology have failed to experimentally verify Sanburg's criteria for a JH specific esterase. In spite of these problems, the concept that JH specific esterases exist is extremely useful, but the term should be defined clearly in the publication using it. To this end several possible definitions of varying biological and biochemical rigor are discussed below and in Table 3.

Sanburg, L. *et al.* (1975a) indicated the presence of two predominant peaks of JH esterase activity. The major peak of JH hydrolytic activity also hydrolyzed α-naphthyl acetate. This peak of activity failed to hydrolyze JH bound to the carrier protein and was DFP-sensitive. The smaller peak of JH hydrolytic activity had very low activity on α-naphthyl acetate, was DFP-resistant, and metabolized JH in the presence of carrier protein. One could thus have the impression that insect hemolymph is full of very active general esterases which will rapidly hydrolyze any ester, including JH. Only the carrier protein then saves JH from rapid catabolism. At critical stages a DFP-resistant JH specific esterase degrades JH regardless of the presence of carrier protein. The attractiveness of this hypothesis has led to its being applied to numerous insect systems, and in some it may be true. At least in *Diatraea saccharalis*, *Manduca sexta* and *Trichoplusia ni* the hypothesis is quantitatively misleading. The predominant high molecular weight activity peak or JH general esterase of Sanburg, L. *et al.* (1975a) has not been found by subsequent workers. JH is undoubtedly degraded by general esterases, proteases, lipases, phosphatases and other enzymes in insect tissue, but in *D. saccharalis, M. sexta* and *T. ni* these enzymes are orders of magnitude less active, on a tissue wet weight basis, than the JH specific esterase. That is not to say these enzymes are unimportant, and the carrier protein may be important in protecting JH from these enzymes. After all, JH hydrolysis is usually assayed in hemolymph diluted 10–1000-fold with incubation times of only a few minutes, while JH must survive in the presence of general esterases for long enough to exert its action. An example of the potential role of general esterases in degrading JH was obtained using hemolymph from fourth-instar *T. ni*. The JH specific esterase from the fifth instar rapidly hydrolyzes both JH I and III, but the carrier protein has a K_d of 4.2×10^{-7} M for JH I but only 3.7×10^{-6} for JH III (Hammock, B. *et al.*, 1977b). Thus, in concentrated fourth-instar hemolymph JH I was relatively stable (largely protected by binding to the carrier protein) while JH III was hydrolyzed very quickly to the free acid, presumably by general or DFP-sensitive esterases. Thus, even if the corpora allata produced relatively high ratios of JH III to JH I and II, the circulating titer of III would be much lower than that expected from the rate of biosynthesis. This observation could explain some of the differences observed between the JH homolog ratios determined *in vivo* and *in vitro*.

In a variety of last-stadium lepidopterous larvae, a single enzyme or a closely related group of enzymes appear responsible for most JH degradation. In general, these enzymes appear DFP-resistant and have low activity on α-naphthyl acetate. However, these two characteristics of the JH esterase are not essential to the definition of JH specific esterase. It should also be recalled that the ester bond of α-naphthyl acetate is not stabilized by conjugation. Thus, if it can fit into the active site of a JH esterase, it would not be surprising if it were hydrolyzed more quickly than JH. One would not exclude an enzyme from biological relevance simply because it metabolized both JH and α-naphthyl acetate.

Table 3: Possible definitions (criteria) for the term "JH specific esterase"

Biological
(1) A JH hydrolyzing esterase whose role has been shown to be essential for the clearance of JH from an insect's body.
(2) A JH hydrolyzing esterase whose titer correlates with suspected declines in JH titer.
Biochemical
(3) An esterase capable of rapidly hydrolyzing JH in the presence or absence of carrier protein.
(4) An enzyme which hydrolyzes JH with a high k_{cat}/K_m ratio.
(5) An enzyme which hydrolyzes JH with a low apparent K_m.

The arguments put forward in section 1.2 support the concept that in many insects specific enzymes are involved in JH metabolism. The definitive proof that an esterase is important in endocrine regulation rests upon the biological effects observed when that enzyme is chemically or biochemically eliminated. Such information is likely to be slow in coming (definition 1, Table 3). Circumstantial evidence, as first presented by Weirich, G. *et al.* (1973), that JH esterase levels correlated with declines in JH titers is also reasonable (definition 2, Table 3). One can also make biochemical arguments that an enzyme capable of hydrolyzing a lipophilic, stable, planar methyl ester must be somewhat special. For the Lepidoptera the definition that a JH specific esterase will degrade JH in the presence of carrier protein seems very reasonable (Fig. 2) (definition 3, Sanburg, L. *et al.*, 1975a). However, many insects lack a carrier protein. It is possible to define JH specific esterase more biochemically. The term specificity is greatly overused in biochemistry, but conservatively it is based on a ratio of k_{cat}/K_m (Fersht, A., 1977). Thus, one could rigorously define a JH specific esterase as an enzyme which shows a higher k_{cat}/K_m ratio for JH than for an alternative substrate such as α-naphthyl acetate. Such a definition requires separation of the JH esterase from other esterases if not in fact purification to homogeneity (definition 4).

Based upon knowledge of JH titers a less rigorous definition could also be useful. Since circulating JH

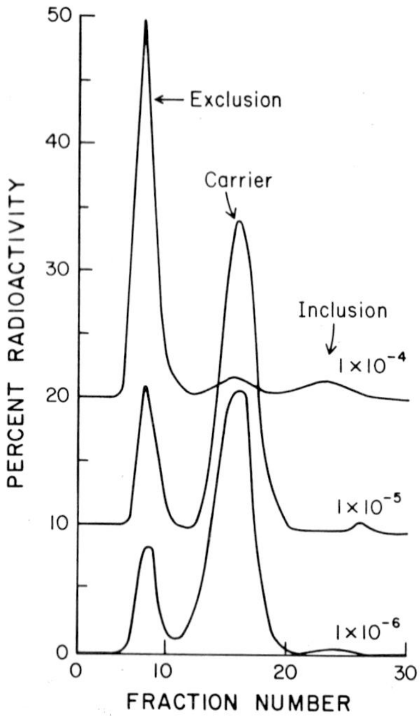

FIG. 2. Binding of varying concentrations of [^{3}H]JH I to hemolymph proteins from *Manduca sexta*. Hemolymph from day 4 last-instar larvae was collected and treated with 2×10^{-3} M DFP under N_2 to inhibit esterase activity. Varying concentrations of JH I were incubated with 1 ml of undiluted hemolymph (1×10^{-4} [top] to 1×10^{-6} M [bottom]) at 30° and the hemolymph passed rapidly through a Sephadex G-100 column held at 1°. Equilibrium was apparently reached almost immediately. The radioactivity in each fraction was determined and expressed as a percentage of the amount applied to the column. This figure shows qualitatively how the JH carrier can be overlooked at unphysiologically high JH concentrations. It also indicates that the equilibrium between the specific carrier and lipid depots (represented by the lipoprotein in the exclusion volume) may be significant.

titers are orders of magnitude lower than the K_m values of JH esterases, it is unlikely that a JH molecule will encounter a JH esterase which is saturated with hormone or in the acyl form. Thus, k_{cat} has less relevance than in the classical definition of enzyme specificity. Arbitrarily one could define a JH specific esterase as an esterase hydrolyzing JH with a low apparent K_m (for instance, 1×10^{-6} M or below) (definition 4). A JH specific esterase is not, however, an esterase that happens to be resistant to DFP inhibition. In spite of the widespread usage of the term JH specific esterase, very few workers have defined the term in their publications, and the data presented do not justify using JH specific esterase by any of the four definitions discussed. Any enzyme which degrades JH is of biological significance, and thus should be examined experimentally regardless of its ability to hydrolyze other substrates, or of its susceptibility to DFP. In some insects it is appropriate to further consider the influence of a carrier protein, and the original concept of Sanburg, L. *et al.* (1975a) may, in fact, be applicable to some, but certainly not to all, insects.

2.2 JH esterase tissue and subcellular distribution

Much of the work on JH ester hydrolysis has been concentrated on the hemolymph. However, work which defines the tissue(s) responsible for most JH metabolism in the whole animal, as well as experiments which define the JH metabolism in target tissues, is wanting. A nice approach was used with *Leptinotarsa decemlineata* where it was first found that the majority of the metabolism occurred in hemolymph and that contributions from the fat body were minor by comparison. Subsequent work concentrated on the hemolymph (Kramer, S. and de Kort, C., 1976a,b; Kramer, S. *et al.*, 1977; de Kort, C. *et al.*, 1978). In *Tenebrio molitor* larvae and pupae over 95% of the JH esterase activity was found to be in the cytosolic fraction (Sparks, T. and Hammock, B., 1980).

Although the thrust of the work concentrated on other topics, the subcellular distribution of JH esterase activity in *Musca domestica* larvae, pupae and adults was monitored in two laboratories. The highest JH esterase activity was found in mature larvae where over 80% of the activity appeared to be microsomal. The relative amounts in the microsomal and cytosolic fractions appeared rather constant at several developmental stages. Although the JH esterase activities in the two fractions appear similar, they can be differentially inhibited. The microsomal JH esterase activity was even less stable than the cytosolic activity. Possibly this instability has led workers to overlook esterase activity in the Diptera and Hymenoptera, especially in whole-body homogenates (Hammock, B. *et al.*, 1977a; Yu, S. and Terriere, L., 1978a,b; Mumby, S. *et al.*, 1979; Sparks, T. and Hammock, B., 1980). In an early study using *Drosophila melanogaster*, Chihara, C. *et al.* (1972) first reported the ability of imaginal disks to inactivate JH very quickly. The metabolism of JH in target tissues has still not been well studied although Chihara's work illustrates the importance of such an approach. Klages, G. and Emmerich, H. (1979b) performed the more aesthetically pleasing study of monitoring subcellular distribution in isolated tissues of *Drosophila hydei*. The fat body contains high levels of both esterases and epoxide hydrolases, while the body wall metabolizes JH, largely by ester hydrolysis. Slightly higher activity was reported in the cytosolic fraction. It is not known if the JH esterase activity in the two fractions is identical. The subsequent work of Bisser, A. and Emmerich, H. (1981) raises the possibility that large aggregates of a cytosolic enzyme are pelleted with the microsomal fraction, or that a lipophilic microsomal enzyme is dislodged during tissue preparation. The hemolymph of *D. hydei* (along with a number of other dipterous and hymenopterous species) appeared remarkably inactive in metabolizing JH esterase. An exception was the hemolymph and body fluid of the prepupa and pupa. High levels of JH esterase activity have also been found in the molting fluid of *Manduca sexta*. Possibly an examination of molting fluids could provide a clue to the evolution of JH esterase.

Day 4 last-instar larvae of *Galleria mellonella* have very high JH esterase activity in the midgut and fat body when expressed as activity per unit protein. In fact, both the gonads and wing disks had higher specific activity than the hemolymph. The JH esterase activity in the wing disks was shown to be dependent upon developmental stage and peaked before the hemolymph esterase activity, then rapidly fell to negligible levels. Thus, JH metabolism could play a role in reducing the sensitivity of the disks to both endogenous and exogenous JH (Reddy, G. *et al.*, 1980).

High esterase activity occurs in the epidermis, fat body, hemolymph, and imaginal disks of *M. sexta*, and in the fat body the activity is largely cytosolic. The JH esterase activity in the epidermis, fat body and hemolymph all increased at similar times for the pre-wandering peak (Fig. 3 e,g) (Hammock, B. *et al.*, 1975b; Nowock, J. and Gilbert, L. 1976; Vince, R. and Gilbert, L. 1977; Mitsui, T. *et al.*, 1979), possibly indicating similar regulatory mechanisms.

In terms of total *in vivo* activity, the fat body and hemolymph account for the majority of the JH esterase activity. For instance, during the pre-wandering peak of JH esterase activity (Fig. 3 e,g) 37% of the total body activity is in the fat body and 50% is in the hemolymph; 90% of activity in the fat body is cytosolic and due to the same enzyme which is present in the hemolymph. This work illustrates the importance of expressing metabolism on a larval equivalent basis if the data are to be useful in constructing pharmacokinetic models of JH (Wing, K. *et al.*, 1981).

2.3 JH esterase substrate selectivity

Very little unequivocal information is available on JH esterase substrate specificity since the pure enzymes have not been well studied. It is possible that some of the enzymatic activities attributed to JH esterases are due to other enzymes which co-purify

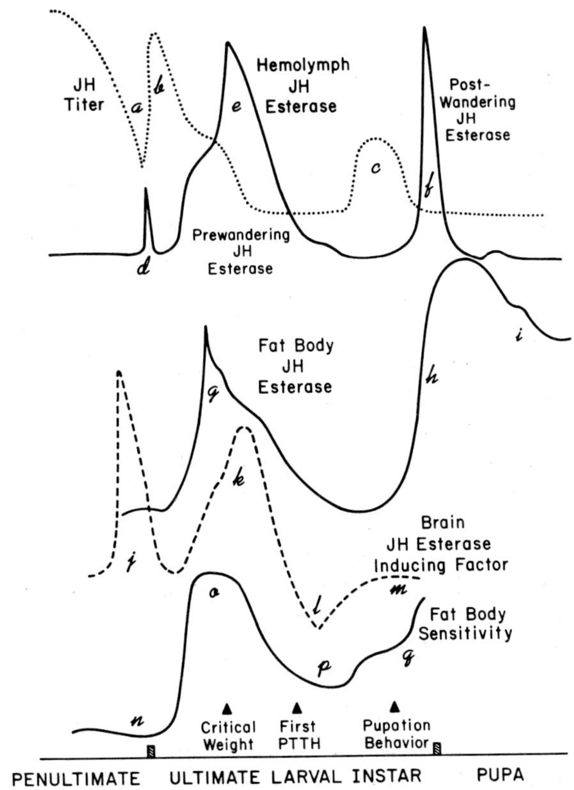

FIG. 3. Temporal variations in JH and factors influencing its metabolism during the larval-pupal transformation in a generalized lepidopterous larva. Data are largely from *Manduca sexta* and *Trichoplusia ni*. Relative JH titer (......), hemolymph JH esterase activity (————), fat body JH esterase activity (————), brain JH esterase-inducing factor (– – – – – – –), and sensitivity of the fat body to the brain factor (————). The lower-case letters refer to events discussed in more detail in the text.

with them. In some cases selective inhibitors have been used to eliminate the activity of other enzymes. More commonly, the substrate specificity of crude hemolymph was examined. The resulting information is only valuable in saying certain structures cannot be hydrolyzed by the crude enzyme mixtures. It must again be noted that true substrate specificity can only be obtained from the ratio k_{cat}/K_m or by mixing two potential substrates and monitoring enzymatic hydrolysis when combined.

Hemolymph from the cockroach *Blaberus giganteus* hydrolyzed both JH I and III, but as observed with numerous other species, was unable to hydrolyze ethyl or isopropyl esters of several juvenoids and model substrates such as esters of 3,3-dimethyl acrylic and cinnamic acid (Hammock, B. *et al.*, 1977b). With crude hemolymph from *Locusta migratoria*, JH I is hydrolyzed faster than III and the unnatural enantiomer of III is hydrolyzed faster than the natural enantiomer. These results may reflect a combined interaction of carrier protein(s) (see Goodman, W. and Chang, E., this volume) and esterases, but also reflect true substrate specificity of the crude hemolymph since JH I and III were combined in a single incubation (Peter, M. *et al.*, 1979). K_m values of 10.6×10^{-6} M and similar rates of hydrolysis were found for JH I and III in *Leptinotarsa decemlineata* hemolymph. The ethyl ester of hydroprene and isopropyl ester of methoprene were not hydrolyzed under these conditions (Kramer, S. and de Kort, C., 1976a,b). Weirich, G. and Wren, J. (1976) found that *Tenebrio molitor* hemolymph rapidly hydrolyzed natural 2*E* but not 2*Z* JH I or its isopropyl ester. In this and most reported experiments, high protein concentrations preclude experimental detection of carrier interactions. Larval homogenates of *Culex pipiens pipiens* were similarly unable to degrade hydroprene and methoprene (Hooper, G., 1976; Brown, T. and Hooper, G., 1979). *Musca domestica* are also unable to hydrolyze methoprene, but can slowly hydrolyze hydroprene (Hammock, B. *et al.*, 1977a; Terriere, L. and Yu, S., 1977). Thus, the results of Bisser, A. and Emmerich, H. (1981) that ethyl 10,11-epoxyfarnesoate is an excellent competitive inhibitor of JH hydrolysis by *Drosophila hydei* epidermis are not surprising.

Weirich, G. and Wren, J. (1973, 1976) reported that last-instar *Manduca sexta* hemolymph could hydrolyze the 2*E* but not the 2*Z* isomers of JH or the isopropyl ester of JH. The hydrolysis of the ethyl ester of JH by *M. sexta* has been questioned by subsequent workers. These observations were extended to *Samia cynthia* which is similarly capable of hydrolyzing the ethyl ester but not the 2*Z* isomer of JH I. Coudron, T. *et al.* (1981) reported rapid hydrolysis of methyl farnesoate by homogeneous JH esterase from *M. sexta* hemolymph, indicating that the epoxide moiety is not crucial for enzyme binding. Similar conclusions can be drawn from M. Peter's (1981) study showing the C10 enantiomers of both JH I and III are hydrolyzed at equal rates in *M sexta* hemolymph. The homogeneous *M. sexta* JH esterase appears to have no hydrolytic activity on α-naphthyl acetate, while the *Trichoplusia ni* enzymes show slight activity (Sparks, T. and Hammock, B., 1979a; Yuhas, D. *et al.*, 1982).

The interest in such unnatural substrates stems from several sources. The resistance of the dodecadienoate (2,4-di-unsaturated) juvenoids to hydrolysis probably stems from increased stability from additional conjugation, poorer fit due to decreased rotation of the C3–4 and C4–5 bonds, and steric hindrance due to ethyl and isopropyl esters. This resistance to hydrolysis may increase the biological activity of these compounds. If JH esterases hydrolyzed ethyl, naphthyl, umbelliferyl or other esters, inexpensive, continuous spectrophotometric assays could be devised. With the present state of knowledge, the current tedious assays employing JH itself must be retained. However, additional experiments are needed to determine if the esterases from *Galleria mellonella*, *Pyrrhocoris apterus*, and *T. molitor* hydrolyzing 2*E* and *Z* juvenoid umbelliferyl esters are the same as those hydrolyzing JH (Sláma, K. and Jarolìm, V., 1980, 1981). If the JH esterases of *G. mellonella* and *P. apterus* do hydrolyze these substrates, their specificity for the alcohol moiety is much less restrictive than the other esterases studied.

The K_m values for JH esterase activities in the hemolymph of several lepidopterous species are shown in Table 4. Since K_m can be taken as a reasonable estimate of enzyme-substrate affinity with serine esterases and proteases, there is an obvious high affinity for the presumed natural substrates, although there is some variation in the data from different laboratories. Even within the Lepidoptera, there are huge differences in the JH esterase activity in the hemolymph. It will be interesting to see if this effect is due to quantitative changes in enzyme titer or actual enzyme specific activity. Coudron, T. *et al.* (1981) and Peter, M. (1981) both report that JH I is hydrolyzed slightly faster than JH III with *M. sexta* hemolymph, although JH III apparently binds more tightly than I. Similar conclusions can be made from the data of Sparks, T. *et al.* (1982) showing that TFT is a much more potent inhibitor of JH III than JH I hydrolysis in *M. sexta*. Observations of Sparks, T. and Hammock, B. (1979a) that JH I is hydrolyzed faster than JH III in *T. ni* hemolymph have been confirmed in recent work. However, Wing, K. *et al.* (1982) found that JH III apparently binds more tightly to the esterase than JH I. The rate-limiting reaction for most serine esterases and proteases is deacylation, and other workers have shown that deacylation of acylchymotrypsin depends upon the size of the acyl group. Thus, more rapid deacylation may explain these superficially anomalous observations. Since natural JH titers are far below the K_m of the enzyme, some future work should concentrate on the acylation step of hydrolysis. This step is the rate-limiting step of JH inactivation under physiologically relevant conditions, while the deacylation step is the rate-limiting step of the overall reaction. Thus, the V_m values, and possibly even the K_m values, are kinetically valid at artificially high JH concentrations where deacylation is the rate-limiting step. However, different experimental designs are needed to evaluate the kinetic interactions of JH and JH esterase at physiological concentrations of the hormone.

Thus, not a great deal is known about substrate specificity of JH esterases. The limited data available certainly suggest enzymes which are highly selective, if not specific, for their natural substrates. Although the receptor sites, carrier molecules and apparently some hydrolytic enzymes all show high specificity for the JH structure, the structural features for recognition seem somewhat different in each case.

2.4 JH esterase purification

Only in the Lepidoptera has much progress been made on purification of JH esterases. Sanburg, L. *et al.* (1975a) estimated a molecular weight of *Manduca sexta* JH esterase at 67,000 based upon a sedimentation coefficient of 4.98S and a diffusion coefficient of 6.4×10^{-7} cm^2 s^{-1}. A similar molecular weight was indicated by gel filtration. Several JH esterase bands of different apparent pI values were found with column isoelectric focusing, but this technique often leads to artifacts. For instance, Wing, K. *et al.* (1982) found JH esterase activity at only a single pI. Coudron, T. *et al.* (1981) purified *M. sexta* JHE to homogeneity (1700×) by gel filtration, cellulose DE-32 anion exchange, hydroxyapatite, and Sephadex C-50 cation exchange chromatography. The resulting single band on SDS gel electrophoresis had a specific activity of 84 nmol JH hydrolyzed min^{-1} mg^{-1} protein and no activity on α-naphthyl acetate. JH I was hydrolyzed at about twice the rate of JH III. The fact that molecular weight estimates based upon gel filtration, ultracentrifugation and SDS electrophoresis all agree indicates the presence of a single polypeptide chain.

Two approaches have been reported for the purification of JH esterase from *T. ni* Although polyethylene glycol precipitation will denature some enzymes, it proved very useful with JHE, probably because most of the unwanted proteins are of a much higher or much lower molecular weight than JHE (and also JH carrier protein). A combination of polyethylene glycol precipitation and DEAE chromatography yielded 134-fold purification. Electrophoresis and subsequent isoelectric focusing yielded a single protein band. Although *Trichoplusia ni* JH esterase is an exceptionally stable enzyme in crude hemolymph or even whole body homogenates, the enzyme proved quite unstable in the absence of other protein (Rudnicka, M. and Hammock, B. 1981a,b). Yuhas, D. *et al.* (1982) employed a combination of ammonium sulfate precipitation anion exchange chromatography, gel filtration chromatography and ultracentrifugation to achieve a specific activity of 56 nmol JH hydrolyzed min^{-1} mg^{-1} following a 417-fold purification. The single polypeptide chain was smaller than that from *M. sexta* with an estimated molecular weight of 48,000. The enzyme appeared stable in the sucrose from ultracentrifugation but rapidly lost activity when exposed to charged environments. Unlike *M. sexta* JH esterase, *T. ni* JH esterase seems to maintain some hydrolytic activity on α-naphthyl acetate (Sparks, T. and Hammock, B., 1979a).

Although initial studies indicated that a variety of esterases were hydrolyzing JH in *Galleria mellonella* hemolymph (Rudnicka, M. *et al.*, 1979), more recent work indicates the involvement of a single enzyme. McCaleb, D. *et al.* (1980) estimate the molecular weight of the *G. mellonella* hemolymph JH esterase to be 60,000, and Wing, K. *et al.* (1982) report a single, discrete pI for hemolymph JH esterase activity. Rudnicka, M. and Kochman, M. (unpublished) report a 2360-fold purification of JH esterase of *G. mellonella* which hydrolyzed JH I at a rate of 920 nmol/min-mg. The analytical conditions used may emphasize the acylation rather than the slower deacylation step.

Some progress has been made on the purification of JH esterases from several insects. When hemolymph from *Blaberus giganteus* was chromatographed on Sephadex G-150, three major peaks of JH esterase activity were found using JH I or III as substrate, while hydrolysis of ethyl or isopropyl esters was minimal (Hammock, B. *et al.*, 1977b). Chromatography of the hemolymph of *Locusta migratoria* on Sepharose 6-B led to a major peak of activity ($> 1.5 \times 10^5$ daltons) and a minor peak (6.7×10^4 daltons) (Peter, M. *et al.*, 1979). Klages, G. and Emmerich, H. (1979b) reported a subunit molecular weight of 35,000 from *Drosophila hydei*, while subsequent work indicates that huge enzyme aggregates are formed (Bisser, A. and Emmerich, H., 1981).

Although a great deal of information can be obtained from physiological studies with JH esterase, homogeneous enzymes are ultimately needed for many studies. The methods described which lead to homogeneous enzymes from *M. sexta* and *T. ni* would be considered adequate for most vertebrate work. However, considering the very limited insect material available, these preparations must be heroic if one hopes to have reasonable quantities of enzyme for subsequent biochemical work. Hopefully, more sensitive and high-yield biochemical techniques for protein purification and characterization will facilitate research in this area.

2.5 JH esterase inhibition

The first groups to investigate JH ester hydrolysis assumed that the reaction was mediated by typical carboxylesterases. In fact, Whitmore, D. *et al.* (1972) reported that the α-naphthyl acetate esterase activity, which probably co-migrated with JH esterase activity when the hemolymph of *Hyalophora gloveri* was electrophoresed, could be inhibited by *O,O*-di-isopropyl phosphorofluoridate (DFP). Kramer, S. *et al.* (1974) first demonstrated that JH esterases from *Manduca sexta* are not typical esterases, by his failure to inhibit them with DFP. Only by using massive concentrations of DFP and extended incubation times can the JH esterase of *M. sexta* be inhibited (Hammock, B. *et al.*, 1975b; Sanburg, L. *et al.*, 1975a). Pratt, G. (1975) found that paraoxon (diethyl-4-nitrophenyl phosphate, $I_{50} = 1.7 \times 10^{-7}$ M) and dichlorvos (O-2,2-dichlorovinyl-O,O-dimethyl phosphate, $I_{50} = 6 \times 10^{-6}$ M) were good inhibitors of the hemolymph JH esterase from *Schistocerca gregaria*, while DEF (*S,S,S*-tributyl phosphorotrithioate) was a notably poor inhibitor. Di-isopropyl paraoxon was used by Hammock, B. (1975) to inhibit JH esterase activity in corpus allatum homogenates from *B. giganteus*. Hammock, B. *et al* (1977b) screened 42 potential inhibitors on three partially purified JH esterase fractions from *B. giganteus* hemolymph. In some, but not all, cases the JH esterase and α-naphthyl acetate esterases could be differentially inhibited. As shown below, several very potent inhibitors of JH esterase activity were found, including the phosphoramidothiolates and 5-phenylphosphates with I_{50} values ranging from 10^{-7} to 10^{-10} molar. Once again DEF, and all carbamates tested, proved to be very poor inhibitors. Phenylmethylsulfonylfluoride (PMSF) was a marginal inhibitor ($I_{50} \cong 10^{-6}$ M).

Peter, M. *et al.* (1979) report a very interesting situation in the hemolymph of *L. migratoria*. Although the JH esterases are in want of further purification, they do seem to hydrolyze α-naphthyl acetate while showing a high specificity for both JH homologs and enantiomers about the epoxide moiety. This high specificity could qualify them as JH specific esterases, although they are susceptible to inhibition with DFP.

The dipterous (and possibly hymenopterous) insects examined appear to be quite different from other insects in regard to their JH esterases. In an extensive study of esterase action on malathion, α-

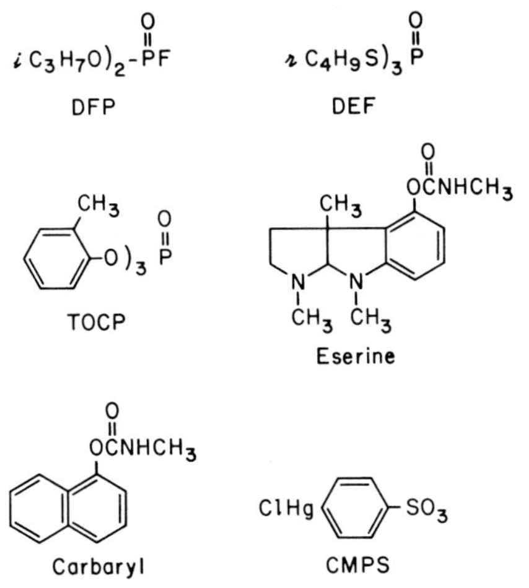

Compounds generally shown to be poor JH esterase inhibitors

SCHEME G

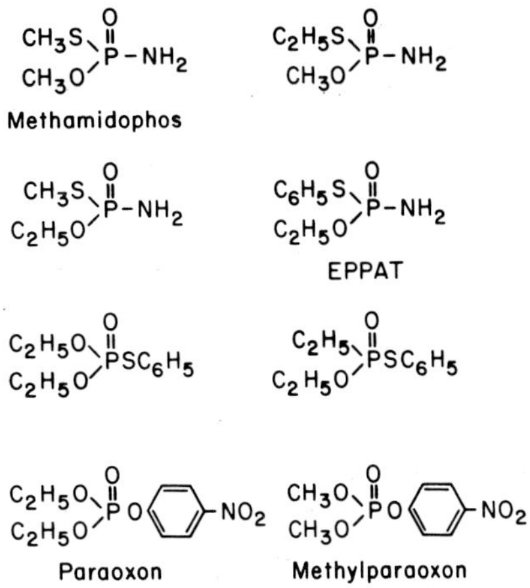

Organophosphorous compounds shown to be good JH esterase inhibitors

SCHEME H

naphthyl acetate, JH, methoprene, and hydroprene in *Culex pipiens pipiens*, Hooper, G. (1976) concluded that JH was hydrolyzed by different enzymes than those hydrolyzing malathion and α-naphthyl acetate. JH was rapidly metabolized by these mosquito homogenates with JH acid as the major metabolite, and the esterase was again surprisingly resistant to tri-orthocresol phosphate (TOCP), *p*-chloromercuriphenyl sulfonic acid (CMPS) and several other classical esterase inhibitors. High concentrations of triphenyl phosphate and paraoxon were needed for significant inhibition of the activity. In *Musca domestica* JH hydrolyzing enzymes were found both in the cytosolic and microsomal fractions of whole-body homogenates. Yu, S. and Terriere, L. (1978a) successfully used paraoxon to inhibit microsomal esterase activity, while Mumby, S. *et al.* (1979) found that an *N*-ethyl carbamate of a geraniol-based juvenoid was a potent inhibitor of both the microsomal and cytosolic forms of the esterase. However, the *N*-ethyl, propyl and butyl carbamates of 1-naphthol were also excellent inhibitors and are simple to synthesize and purify. The *N*-propyl carbamate appears to be most useful in inhibiting both forms of the enzyme, while the *N*-ethyl was most potent in the microsomal form. Sparks, T. and Hammock, B. (1980) were further

Table 4: Apparent K_m values of hemolymph JH esterases from several species

Species	Substrate	$K_m(\times 10^{-7}$ M)	References
Diatraea saccharalis	JH I	11	Roe, R. *et al.* (1982)
	JH III	40	Roe, R. *et al.* (1982)
Galleria mellonella	JH I	18	McCaleb, D. *et al.* (1980)
	JH III	1.7	Wing, K. *et al.* (1982)
Heliothis virescens	JH III	3.1	Wing, K. *et al.* (1982)
Manduca sexta	JH I	8.7 (5.0)*	Peter, M. (1981)
	JH III	14	Vince, R. and Gilbert, L. (1977)
		9.3 (1.4)*	Peter, M. (1981)
Spodoptera exigua	JH III	1.0	Wing, K. *et al.* (1982)
Trichoplusia ni	JH I	3.2	Wing, K. *et al.* (1982)
		11	Yuhas, D. *et al.* (1982)
	JH III	2.0	Hammock, B. *et al.* (1982)
		1.2	Wing, K. *et al.* (1982)
		4.0	Yuhas, D. *et al.* (1983)

*Pre-wandering JH esterase peak (post-wandering).

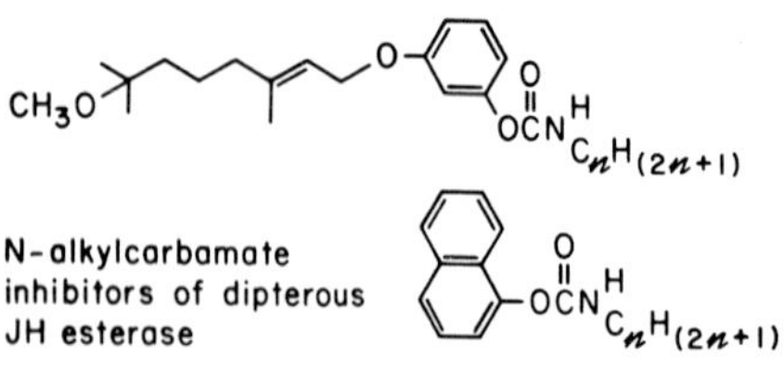

SCHEME I

able to distinguish the two forms by inhibiting the microsomal form with CMPS and EPPAT (O-ethyl S-phenyl phosphoramidothiolate).

As in all other insects so far examined, the JH esterase activity in the hemolymph of the beetle *Leptinotarsa decemlineata* is resistant to DFP, but it is surprisingly susceptible to inhibition by the 4-*t*-octylphenyl ethoxylate, Triton X-100 (Kramer, S. and de Kort, C., 1976b; Kramer, S. *et al.*, 1977). In *Tenebrio molitor* Triton X-100 also selectively inhibits hemolymph JH esterase, while DFP and paraoxon are poor inhibitors (McCaleb, D. *et al.*, 1980). EPPAT proved to be a useful inhibitor of both microsomal and cytosolic JH esterase from *T. molitor* (Sparks, T. and Hammock, B., 1980). Mane, S. and Rembold, H. (1977) report that high levels of Triton X-100 will also inhibit JH esterase activity in *Apis mellifera*.

Most research has centered on the Lepidoptera. Whitmore, D. *et al.* (1972) used DFP, CMPS, and eserine in their study of JH esterases induced with JH in *Hyalophora gloveri* pupae. DFP caused major, but not total, inhibition of the enzyme stained in the gel with α-naphthyl acetate. This observation raises the possibility that the JH esterase was partially co-electrophoresing with other enzymes. Sixteen potential esterase inhibitors were screened by Ajami, A. (1975) on hemolymph from pupal *M. sexta*. If the hemolymph is free of fat body, JH esterase activity is very low in the pupa. Thus, it is possible that these data refer to inhibition of general esterase activity on JH. PMSF, Abate and Dursban were shown to stabilize radioactive JH in the hemolymph *in vitro*, as well as possibly synergizing hormonal action *in vivo*.

As discussed earlier, the discovery by Kramer, S. *et al.* (1974) that JH esterases from *M. sexta* were surprisingly resistant to inhibition by DFP was a major contribution to the study of JH metabolism. DFP has subsequently been used, and occasionally misused, in many studies. The information from Mitsui, T. *et al.* (1979) that JH is metabolized by both DFP-susceptible and -resistant esterases in *M. sexta* epidermis is possibly the clearest indication of general *vs.* specific JH esterase action. Sláma, K. and Jarolím, V. (1980) found that DFP was a potent inhibitor of the hydrolysis of 4-methylumbelliferyl esters of JH III from *Galleria mellonella*, and several other insects. However, the lipase-like enzymes which hydrolyzed oleic acid esters were rather insensitive to this inhibitor. The JH esterases of some insects, although not necessarily the ones studied, will not hydrolyze esters of large alcohols such as methylumbelliferol. However, the resistance of some lipases to DFP inhibition should lead to caution in the use of DFP to distinguish JH esterases. Slade, M. and Wilkinson, C. (1973 and unpublished) took a similar approach and found that some inhibitors of epoxide hydration also inhibited esterase activity in *Spodoptera eridania*. Sparks, T. and Hammock, B. (1979a, 1980 and unpublished) surveyed a large number of esterase inhibitors. Some useful trends emerge from the 30 compounds reported. A variety of "typical" esterase inhibitors including TOCP, DEF, DFP, eserine and several other carbamates, PMSF, CMPS, and EPTA failed to inhibit JH esterase. EPPAT and several phosphoramidothilates appeared to be good inhibitors, while paraoxon and dichlorvos are more commonly available and useful inhibitors. EPPAT and several paraoxon homologs proved to be excellent inhibitors in *Manduca sexta* (Sparks, T. unpublished). It is somewhat difficult to interpret the data of Brown, J. *et al.* (1977) who studied the propensity of larval α-naphthyl acetate esterases to bind JH and several mimics. Possibly they were observing co-chromatography of the JH carrier with an esterase band on gel elctrophoresis, or possibly the acyl enzyme of a conjugated ester is stable enough to isolate.

For *in vivo* studies selective inhibition of JH esterase is important for determining the role of the esterase and for experimental stabilization of JH. At most stages of the insect's life it is important that the inhibitor not be a potent inhibitor of acetylcholine esterase and thus highly toxic. EPPAT and 1,1,1-trifluorotetradecane-2-one (TFT) meet this criterion in *Trichoplusia ni*. When comparing the ratio of inhibition (I_{50}) of hemolymph JH esterase and α-naphthyl acetate esterase for various inhibitors, it was found that DFP had a ratio of >25,000, carbaryl >14, EPPAT 10 and TFT 0.001, indicating that selective inhibition was certainly possible. The ideal JH esterase inhibitor for physiological studies should have minimal toxicity to insect and experimenter, show high selectivity for JH esterase inhibition over other esterases, penetrate rapidly, have a useful persistence *in vivo,* and allow almost total inhibition in an essentially irreversible manner. EPPAT approaches this ideal for *G. mellonella, M. sexta, T. ni,* and some other insects in that it irreversibly phosphorylates JH esterase (below). Hopefully better probes can be found. It should be noted that the inhibition patterns and potencies of EPPAT, TFT and related compounds vary with the species examined, and they are by no means specific JH esterase inhibitors.

Towards this end a variety of polarized ketones were synthesized by Hammock, B. *et al.* (1982). The most active inhibitor of *T. ni* JH esterase in this series was TFT, whose carbon chain length is identical to that of JH I and II. In contrast to irreversible inhibitors such as EPPAT, DFP and paraoxon, the polarized ketones are reversible inhibitors showing pure competitive inhibition kinetics. These polarized ketones are excellent examples of a transition state analog inhibitor. Note how the polarized ketone (below) mimics the tetrahedral intermediate on the reaction coordinate leading to the JH-acyl enzyme (Scheme C, section 1.2). Pauling, L. (1946, 1948) calculated that the affinity of an enzyme for a transition state of a reaction must be much higher than for either the substrate or the product of that reaction. Transition state theory has become a central philosophy in explaining enzyme

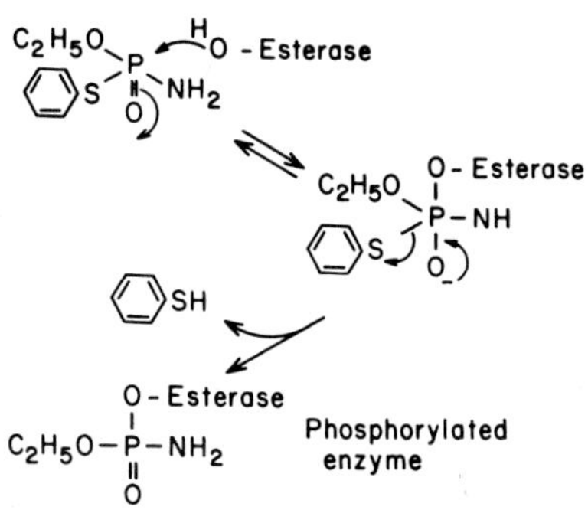

SCHEME J

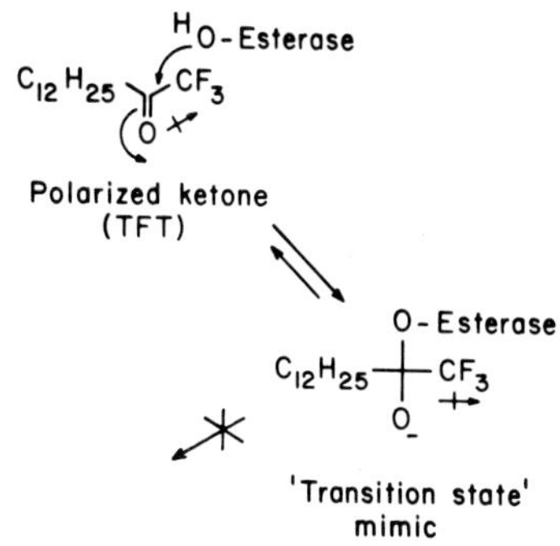

SCHEME K

action, and this theory may be very useful in the design of new enzyme inhibitors. Many inhibitors have been designed to mimic substrates, but even classical insecticides such as paraoxon act because they mimic the transition state (before phosphorylation) rather than the substrate. This effect can be seen above in the diagram showing inhibition of JH esterase by EPPAT.

Unfortunately, TFT has some limitations as a tool for studying JH esterase. In spite of its high selectivity and potency in *T. ni,* it is less active in *M. sexta* and some other insects. The slope of its inhibition curve is shallow, thus obtaining total inhibition of the esterase is difficult with TFT. Also, reversible inhibitors must be much more potent than irreversible inhibitors in order to obtain equivalent *in vivo* effects. However, the class of compounds promises to be very useful as NMR probes for the esterase active site, affinity ligands and as direct inhibitors if the potency can be improved.

Several research groups have used esterase inhibitors to evaluate the similarity of esterases from subcellular fractions, from developmental stages, from different tissues, from different species as well as esterases acting on different substrates. For instance, Mumby, S. *et al.* (1979) used a variety of carbamates to show that the esterase activity of the cytosolic and microsomal fractions of *Musca domestica* could be differentially inhibited. Although purification is ultimately needed to characterize an enzyme, a battery of esterase inhibitors were used in *T. ni* to test the similarity of JH esterase activities (Hammock, B. *et al.,* 1977b; Sparks, T. and Hammock, B. 1979b; Jones, G. *et al.,* 1981; Wing, K. *et al.,* 1981). Of course, similar patterns of inhibition do not prove that two enzymes are identical, but different patterns of inhibition serve to distinguish them without intense effort. G. Hooper's (1976) study in mosquitoes is an excellent example of the use of inhibitors to distinguish esterases acting on different substrates, as was the use of DFP in *M. sexta* to distinguish esterases acting on JH and α-naphthyl acetate (Kramer, S. *et al.*, 1974).

Inhibitors traditionally have been used to classify mammalian esterases into four major groups (Aldridge, W., 1953; Holmes, R. and Masters, C., 1967). Some general trends are evident from current knowledge of JH esterases. However, the application of mammalian classification systems should be cautious because insects are exceedingly diverse and all classification shemes used to date are clearly artificial. JH esterase activity does not depend upon the presence of divalent cations, as evidenced by the failure of the enzymes to be inhibited by EDTA. The sulfhydryl reagents tested are also remarkably poor inhibitors of JH esterases with the possible exception of some *M. domestica* esterases. However, the instability of partially purified *T. ni* esterase may indicate the presence of sulfhydryl groups necessary for activity. Although JH esterases are not inhibited by a variety of organophosphates widely used as carboxylesterase inhibitors, they are very susceptible to inhibition by some organophosphates. Eserine is a unversally poor inhibitor of enzymes hydrolyzing JH ester, indicating that JH esterase and acetylcholine esterases can be differentially inhibited. Other carbamates tested were similarly poor inhibitors of JH esterase activities in all species tested except for the Diptera where some *N*-alkyl-carbamates proved to be good inhibitors. Thus, JH esterases can best be classified as carboxylesterases (E.C. 3.1.1.1). This classification is certainly realistic since their primary substrate is, in fact, a carboxylic acid ester, and hog liver carboxylesterase does hydrolyze JH. However, this classification must not be over-interpreted since numerous lines of evidence distinguish JH esterases from many other more typical carboxylesterases.

JH esterase inhibitors have found utility in a variety of biochemical and biological studies. For example, Yu, S. and Terriere, L. (1978b) successfully blocked ester hydrolysis of JH in order to study epoxide hydration, and such inhibitors have also been used to investigate JH biosynthesis. Inhibition of JH metabolism may in fact prove crucial in JH receptor work and in other investigations of JH action (Chang, E. and Emmerich, H., personal communication). In all of these studies it is, of course, necessary to run controls to ensure that the inhibitors used do not influence the biochemical parameter under investigation. To this end it may be useful to test a variety of inhibitors.

A critical physiological application of JH esterase inhibitors is to explore the *in vivo* role of JH ester hydrolysis. For instance, Kramer, S. *et al.* (1977) was able to stabilize radioactive JH *in vivo* with Triton X-100 in *Leptinotarsa decemlineata*, and thus demonstrate a role for the esterase. Sparks, T. and Hammock, B. (1980) found that topical application of EPPAT to pre-wandering *Trichoplusia ni* led to a decrease in hemolymph JH esterase activity, an increase in the *in vivo* stability of radioactive JH and a delay in pupation (discussed further in section 3.3). Inhibition of the second last-instar JH esterase peak blocked pupation (Fig. 4). Although the data on *L. decemlineata, T. ni,* and recently *Manduca sexta* (Safranek, L. and Williams, C. personal communication; Sparks, T., unpublished) support the concept that in some insects JH esterase has an *in vivo* role, the inhibitors used are still not as specific as one might hope.

JH esterase inhibitors are exceedingly valuable probes, but for practical insect control their utility is likely to be very limited, in spite of early hopes as illustrated by the work of Solomon, K. and Metcalf, R. (1974) and Ajami, A. (1975). The inhibitors would share the limitations of juvenoids in prolonging the destructive instars of many pests and only acting at specific periods in development. As discussed below (section 3) degradative metabolism may not be crucial in JH regulation in some insects, and in others regulation may involve both biosynthesis and degradation of JH. In *T. ni* JH esterase appears to be in such large excess that the majority of the enzyme must be inhibited before *in vivo* changes are seen. Thus, even potent inhibitors could delay without totally blocking metamorphosis. However, the JH esterase still may be a useful target for biochemical and molecular approaches to insect control. If the enzyme can be caused to appear precociously in the hemolymph, metamorphosis may be accelerated or reproduction inhibited. Antagonistic effects could be obtained by blocking the production or action of the enzyme.

2.6 JH esterase interaction with carrier molecules

Following the early work on JH metabolism, it seemed reasonable that JH would need to have some protection if it were to circulate effectively in those insects which rapidly metabolize the hormone (Slade, M. and Zibitt, C. 1971, 1972; Weirich, G. *et al.*, 1973; see Goodman and Chang, this volume). Although the JHs are water-soluble at physiological concentrations (Kramer, S. *et al.*, 1974), JH will behave as a surfactant. For instance, several independent methods indicate a critical micelle concentration for JH of approximately 10^{-5} M (Kramer, S. *et al.*, 1974, 1976a,b; Mayer, R. and Burke, M. 1976; Hammock, B. *et al.*, 1977b). It is known that such surfactants will accumulate at membrane cytosol interfaces and, in fact, may travel through the cell on membrane surfaces. Akamatsu, Y. *et al.* (1975) thus pointed out that JH carriers may be important in preventing such a surface excess. While the moderate water-solubility of JH is surprising, its lipophilicity is not. The octanol water partition coefficient for JH I of 3.7 certainly explains its tendency to be rapidly absorbed into any lipophilic sink (Hammock, B. *et al.*, 1975b; Nowock, J. *et al.*, 1976; Mumby, S. and Hammock, B. 1979). Thus, for rapid circulation in an aqueous compartment, a JH carrier molecule offers several advantages, as further discussed below. The biochemistry and physiology of JH carriers has been recently reviewed by several authors (Gilbert, L. *et al.*, 1978; Kramer, S. and Law, J. 1980; Hammock, B. and Quistad, G. 1981), most notably in this volume (see Goodman, W. and Chang, E. this vol.). Thus, this section will largely be restricted to the interaction of carrier molecules with JH esterases.

Outside the order Lepidoptera, there are few reports of high-affinity hemolymph carrier proteins. In many of the insects so far examined, high molecular weight lipoproteins, or proteins with hydrophobic sites, may assume the additional role as JH carrier. Although these proteins have a high capacity, the limited number of studies to date indicate that they have low affinity. No data are yet available, but one might also predict that mass transfer will be slower from some of these proteins than the lepidopteran JH carrier molecules. Thus, JH may not reach equilibrium as quickly with the aqueous environment. It is still questionable if these proteins make a major contribution to JH dynamics due to the low affinity of the proteins for the hormone and the low molarity of JH in circulation. In work with *Leptinotarsa decemlineata* Kramer, S. and de Kort, C. (1978) ultimately demonstrated a minimal role of hemolymph lipoproteins in protecting JH from ester hydrolysis. However, the high levels of lipoproteins required led these workers to question the biological significance of this protection. Sams, G. *et al.* (1978) report that lipoproteins from *Periplaneta americana* hemolymph retard JH metabolism in short-term organ culture. Clearly more work is needed on JH transport in insects lacking high-affinity carriers and on the relative roles of high- and low-affinity carriers in insects with both forms (Trautmann, K. 1972; Emmerich, H. and Hartmann, R. 1973; Emmerich, H. 1976; Kramer, S. *et al.*, 1976a,b; Kramer, S. and de Kort, C. 1976b, 1978; Bassi, S. *et al.*, 1977; Hammock, B. *et al.*, 1977b; de Kort, C. *et al.*, 1977).

R. Hartmann, in 1978, reported an exciting discovery that the hemolymph of the grasshopper *Gomphocerus rufus* contains a glycolipoprotein (molecular weight 220,000) which shows a very high affinity for JH ($K_d = 5 \times 10^{-8}$ M). The protein is apparently produced, at least in part, by the fat body throughout the life of male and female, normal and allatectomized insects. Hartmann, R. (1978) argued that a major role of the carrier was to distribute apparently large amounts of JH from the corpora allata through the insect. Excellent evidence for a physiological role for the protein was provided when the injection of anti-carrier antibodies blocked JH-dependent reproduction in adult females. Peter, M. *et al.* were probably examining a similar protein from the hemolymph of *Locusta migratoria* in their 1979 study. The high-affinity carrier in this case binds JH III more tightly than JH I and the natural enantiomer of JH III more tightly than the unnatural. By first examining the specificity of the hemolymph esterase and then the enantiomeric selectivity of JH metabolism in crude hemolymph, a clear protective role for the carrier was demonstrated. The esterase activity in the hemolymph of *L. migratoria* was much lower than that described in the hemolymph of many Lepidoptera, but it was sufficient for metabolizing unbound hormone. As discussed for *Trichoplusia ni*, the carrier protein and esterase may function in concert in *L. migratoria* to increase the proportion of circulating JH III should other homologs be produced. Both groups also found high-capacity, low-affinity JH binding proteins of questionable significance in JH circulation.

Carrier proteins in the Lepidoptera have been the subject of a much more intensive investigation. Whitmore, E. and Gilbert, L. (1972, 1974) found that in six lipoproteins from the hemolymph of the saturniids *Hyalophora cecropia*, *Hyalophora gloveri* and *Antheraea polyphemus* will bind JH *in vitro* but that only two will bind JH *in vivo*. The tremendous capacity of these lipoproteins tends to overshadow the presence of a high-affinity carrier similar to that found in other Lepidoptera. The distribution of 10^{-8} M radioactive JH I in undiluted *H. cecropia* pupal hemolymph run on a Sephadex G-100 column is shown below. These data indicate quite clearly the binding of JH to high molecular weight lipoproteins. Rechromatography of the intermediate molecular weight fractions yields a pattern somewhat similar to that seen for *M. sexta* and numerous other Lepidoptera in which a protein of low molecular weight is critical in JH binding (Hammock, B. and Gilbert, L., unpublished).

Certainly in the majority of lepidopterous species examined, the low molecular weight carrier plays the major role in JH distribution in spite of the tremendous capacity of the lipoproteins (Kramer, S. *et al.*, 1974). As shown in Fig. 2, when last-stadium day 4 hemolymph esterase activity is inhibited and the undiluted *Manduca sexta* hemolymph then incubated with varying concentrations of JH I, quantitative binding of JH to very high molecular weight proteins only occurs at unphysiologically high concentrations of JH. In general, such gel permeation columns have been run following equilibration of JH with dilute hemolymph. Such experiments emphasize the equilibrium between the JH carrier and free hormone. The data shown in Fig. 2 emphasize that *in vivo* an important equilibrium is between the JH carrier and lipophilic depots including hemolymph lipoproteins. Even at very low JH concentrations (not shown), a surprisingly large proportion of JH is bound to very high molecular weight proteins. A variety of juvenoids including R 20458, methoprene, hydroprene and Ro 8-4374 bind largely to the lipoprotein fraction under these conditions, but some binding to the JH specific carrier is observed. Such juvenoid binding to the carrier molecule is deemphasized by competitive binding experiments (Hammock, B. and Gilbert, L., unpublished). Thus, these lipoproteins probably have a less significant role than the JH carrier protein in JH distribution and turnover in the Lepidoptera. However, their role must not be totally discounted with either JH or juvenoids.

The hemolymph JH carrier has been most extensively studied in *M. sexta*. The carrier has an acidic pI, a molecular weight of 28,000, and a single JH binding site with a K_d of 3×10^{-7} M for JH I. The carrier molecule shows a greater affinity for JH I and 0 than for II or III, as well as ability to recognize the appropriate geometrical and optical isomers of the hormone. Its specificity is further indicated by its very low affinity for JH metabolites and for a variety of biologically active juvenoids. The fat body appears to be a major site of synthesis of this protein which is then released into the hemolymph (Kramer, S. *et al.*, 1974, 1976a,b; Akamatsu, Y. *et al.*, 1975; Sanburg, L. *et al.*, 1975b; Nowock, J. *et al.*, 1975, 1976; Nowock, J. and Gilbert, L. 1976; Goodman, W. *et al.*, 1976, 1978a,b; Goodman, W. and Gilbert, L. 1978; Peterson, R. *et al.*, 1977; Law, J. 1978; Gilbert, L. *et al.*, 1978; Schooley, D. *et al.*, 1978). A similar situation has now been observed in a variety of lepidopterous species. Although the concentrations and apparent equilibrium dissociation constants of the JH carriers from different species vary, several workers have calculated that the majority of JH I or II at physiological concentrations will be bound.

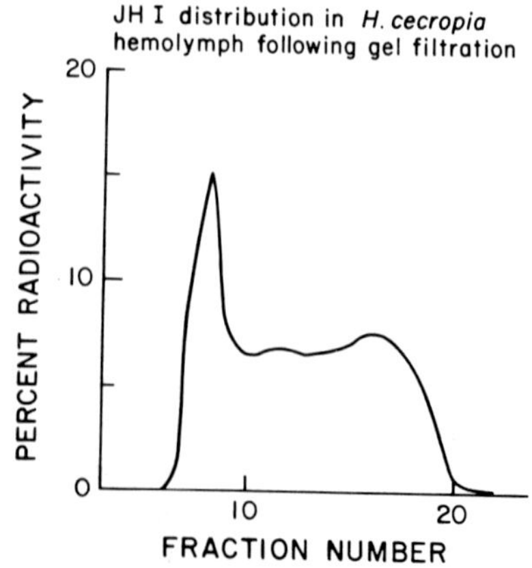

SCHEME L

Goodman, W. and Gilbert, L. (1978) reported that the carrier protein undergoes only slight fluctuations correlated with hemolymph protein concentration during the penultimate larval stadium of *M. sexta*. In *T. ni,* as in *M. sexta,* a single protein appears responsible for high-affinity JH binding in the hemolymph. The same number of binding sites occur whether JH I or III is employed as ligand. However, JH I is bound about 10-fold more tightly. JH I binding activity fluctuates only slightly during the last larval stadium of *T. ni,* in contrast to the dramatic changes in JH esterase activity. Since the binding activity is highest in mid-instar, the hypothesis that a disappearance of binding protein leads to a release of free JH into the hemolymph is not supported (Hammock, B. *et al.*, 1977a; Sparks, T. *et al.*, 1979a; Sparks, T. and Hammock, B. 1979b; Wing, K. *et al.*, 1981, 1982). No information is yet available on JH carrier regulation. In spite of the minor fluctuations in titer so far observed, such a study of regulation could be very informative.

The concept that hemolymph JH carriers play a multitude of roles in influencing JH metabolism and distribution has evolved from several laboratories (Table 5) (Kramer, S. *et al.,* 1974, 1976a,b; Ferkovich, S. *et al.*, 1975, 1977; Hammock, B. *et al.*, 1975b; Sanburg, L. *et al.,* 1975a,b; Ferkovich, S. and Rudder, R. 1976; Kramer, S. and Childs, C. 1977; Goodman, W. and Gilbert, L. 1978; Goodman, W. *et al.*, 1976, 1978a; Gilbert, L. *et al.*, 1976, 1978; Sparks, T. and Hammock, B. 1979a; Kramer, S. and Law, J. 1980; Hammock, B. and Quistad, G. 1981; Wing, K. *et al.,* 1981, 1982). The qualitative significance of the protection of JH from general esterases has not been established. However, such a protective role clearly exists. For instance, in *T. ni* JH I is metabolized slightly faster than JH III by the last-instar JH esterase, but in concentrated penultimate larval instar hemolymph JH III (which is weakly bound by the carrier molecule) is hydrolyzed much more rapidly than JH I. The approach of Peter, M. (1981) promises also to yield useful information along this line.

It is likely that the JH carrier plays a major role in effecting rapid distribution of the hormone. The apparent K_d measured for numerous species appears sufficient to account for binding most of the JH present in the hemolymph. Measurements of k_a and k_d (association and dissociation rates) have been less successful but appear very rapid (Goodman, W., personal communication; Peterson, R. and Law, J., unpublished; Peterson, R. 1977; Wing, K. *et al.,* 1982). The k_a is sufficiently high to ensure almost immediate binding when JH is released from the corpora allata. Thus, changes in rates of JH synthesis are rapidly reflected in changes in circulating hormone titers. In contrast to cells and even vesicles which very slowly release lipophilic molecules to an aqueous environment, the k_d of the JH carrier is also very fast. For serine esterases K_m is often proportional to substrate affinity for the enzyme. Under assay conditions where V_{max} is approached, the rate-limiting step in ester hydrolysis is deacylation of the enzyme. However, under physiological conditions with very low JH concentrations, the acylation of JH esterases will be the rate-limiting step in the removal of JH from circulation. The V_{max} of this step is higher, and the K_m lower, than that of the overall reaction as estimated by classical Lineweaver–Burk methods. Many studies only employed substrate concentrations above the K_m. Thus, by mass action JH can be transferred rapidly and irreversibly from hemolymph carrier molecules to JH esterases without the need to envision direct interaction of the esterase and the carrier. Thus, the JH carrier serves to protect and distribute JH in the absence of JH esterases and serves to affect rapid hydrolysis in the presence of these esterases. Without the carrier either maintaining JH in solution or extracting JH from lipophilic depots by mass action, JH catabolism would be much slower. As more is learned of the distribution of JH carriers, it may be possible to correlate their presence with lower JH synthetic rates and with insects or insect stages in which rapid, precise changes in circulating JH titers are critical for development. As with JH esterases, it will be important to somehow block JH binding if

Table 5: Possible roles of JH carrier proteins

(1) Protection from general esterases (and possibly other enzymes).
(2) Maintenance of JH in solution rather than in lipophilic depots or at membrane aqueous interfaces to effect rapid and equal distribution of JH throughout the larva.
(3) Extraction of JH from lipophilic depots by mass action.
(4) Interaction with JH esterases to effect rapid degradation.
(5) Vary ratio of JH homologs by preferential binding and protection from metabolism.
(6) Modulation of JH action on target issues.

the physiological roles of the carrier molecules are to be established.

3 REGULATION OF JH DEGRADATION

Most models for endocrine regulation in the vertebrates are based upon the assumption that degradation and elimination of hormones remain rather constant and that circulating hormone titer is determined largely by biosynthesis. Similar arguments have been made for JH regulation in insects. However, evidence is building in at least three orders that changes in the rates of JH degradation, as well as changes in biosynthetic rates, are crucial in regulating circulating as well as tissue hormone titers. In some Lepidoptera, for instance, the insect seems to turn down but not turn off JH biosynthesis. Thus, an increase in JH metabolism is necessary for reducing JH titers to a level which allows prothoracicotropic hormone (PTTH) release and the initiation of metamorphosis (see Steel and Davey, vol. 8). Thus, the factor(s) which causes the production of JH esterases may initiate the cascade of endocrine events leading to pupation. For proper maturation and reproduction in several insects it also appears that JH metabolism must be precisely turned off. Thus, it is likely that regulation of JH degradation will be an important area of investigation in the future.

3.1 Regulation in the Orthoptera

Retnakaran, A. and Joly, P. (1976) found that cautery of the A and B neurosecretory cells of *Locusta migratoria* led to a reduction in hemolyph JH esterase activity. Further work on regulation of JH metabolism in the Orthoptera was not pursued until very recently. Sparks, T. and Hammock, B. (unpublished) found increases in JH esterase activity during the last stadium of *Blaberus giganteus*. This work was halted due to difficulties encountered in timing experimental animals. Using well-timed animals, Szibbo, C. *et al.* (1982) found that the hydrolysis of JH by hemolymph enzymes increased dramatically during the last half of the last stadium of *Diploptera punctata*. This increase in JH esterase activity correlated well with a reduction in *in vitro* JH biosynthesis. JH esterase activity was high at adult emergence and then declined in mated females during the time that JH appears to stimulate vitellogenesis. A subsequent sharp rise in JH esterase activity was then observed and maintained during pregnancy. The low titers of JH esterase observed in allatectomized females were restored by juvenoid application. Whether this induction occurs directly or through another hormone is not yet known (Rotin, D. *et al.*, 1982). Thus, the hemolymph JH esterase activities correlate well with suspected declines in JH titers, as well as with declines in JH biosynthesis in *D. punctata*.

3.2 Regulation in the Coleoptera

The regulation of JH degradation in larval Coleoptera has not been well studied, possibly because the insects are hard to stage. However, as discussed earlier, high JH esterase and epoxide hydrolase activities occur in the late larvae and pupae of *Leptinotarsa decemlineata* and *Tenebrio molitor* (Kramer, S. and de Kort, C. 1976a; Sparks, T. and Hammock, B. 1980) which suggests that JH catabolism in these larvae may play an analogous role to that observed in the Lepidoptera.

Regulation of JH metabolism has been studied in more detail in adult *L. decemlineata*. Although polar conjugates were the major *in vivo* metabolites, tissue levels of JH esterase and epoxide hydrolase were very low, allowing the workers to concentrate on the hemolymph enzyme. Hemolyph JH esterase activity increases in both long and short photophase adults, but the levels rise much higher in the short-day animals which will enter diapause soon after adult emergence. The increases in JH esterase activity correlate with drops in *in vivo* JH titer in both long- and short-day animals. Since low JH titers are needed for short-day *L. decemlineata* to enter diapause, the appearance of very high hemolymph JH esterase activity suggests a biological role. However, JH titers are dramatically reduced before peak esterase activity is detected. This information led de Kort, C. and Granger, N. (1981) to conclude that JH esterase is not a primary factor in reducing JH titer. However, studies of the rates of *in vivo* biosynthesis indicate that a decrease in biosynthesis rather than a cessation of biosynthesis correlates with decreases in *in vivo* JH titers. Thus, the data of Kramer, S. (1978b) suggest that a combination of degradation and decreased synthesis account for the observed reduction in JH titer. Possibly the use of selective esterase inhibitors or pharmacokinetic calculations will allow the role of hemolymph esterases to be accurately assessed (de Wilde, J. *et al.*, 1968; Kramer, S. and de Kort, C. 1976a,b, 1978; Kramer, S. *et al.,* 1977; Kramer, S. 1978a,b).

Kramer, S. (1978a) further reports that JH treatment rapidly increases JH esterase activity in diapausing beetles. Several experiments indicate that this effect is not direct. For instance, when short-day beetles were transferred to a long-day photoperiod, the expected rise in JH esterase did not occur, and its rise could also be blocked by ligation even with JH application. The rise in hemolymph JH esterase in short-day beetles also could be blocked by JH application shortly after emergence, possibly suggesting that the hemolymph JH titers had to be reduced before the beetle could reach the stage necessary for the high JH esterase titers needed for complete clearance of JH from the hemolymph. Such a cascade effect where one endocrine change signals another could be an effective mechanism for timing such events. Based on his studies Kramer, S. (1978a) concluded that a complex of neuroendocrine events control hemolymph JH esterase activity in the adult. A model in which JH esterase activity is induced by a head factor (other than JH) in response to a circulating titer of JH seems to fit the data.

3.3 Regulation in the Lepidoptera

The regulation of JH metabolism is best understood in the Lepidoptera, and this regulation deals largely with fat body and hemolymph JH esterases. As indicated earlier in *Manduca sexta* and *Trichoplusia ni*, epoxide hydration and conjugation seem unlikely to be rate-limiting steps of JH degradation. The midgut JH esterase activity and the fat body and midgut epoxide hydrolase activities are approximately an order of magnitude lower than the peak fat body and hemolymph JH esterase activities. As discussed earlier, the changes in hemolymph JH binding activities are not dramatic. Models of JH dynamics may ultimately show these changes to be significant, but the crude titers in *T. ni* suggest the JH carrier has as much a role in clearance as in preservation of the hormone. Although many tissues are capable of rapid JH metabolism, the high specific activity of the JH esterases in hemolymph and fat body (and to a much lesser extent epidermis) undoubtedly account for the majority of JH hydrolysis. This is not to say that metabolism in a target tissue is unimportant in limiting JH action, but rather that this metabolism does little to alter total *in vivo* JH titers. The high rates of JH degradation in epidermis and imaginal disks in fact support the concept that metabolism may have a crucial microscopic role in the tissue concerned. At peak levels of JH hydrolytic activity a single enzyme appears responsible for the vast majority of JH hydrolysis. Using a number of biochemical techniques, no differences have been found among the JH esterases collected at several times during development or from the major tissues of *M. sexta* and *T. ni*. Only the JH esterase activity in the midgut appears to be due to different enzymes (Sparks, T. and Hammock, B. 1979a,b; Hammock, B. *et al.*, 1981; Wing, K. *et al.*, 1981).

In most of the lepidopterous larvae examined, two major bursts of hemolymph JH esterase activity are observed in the last larval instar (Fig. 3). One burst occurs early in the instar (e) supposedly correlated with a decline in circulating JH titers (b). Near this time the insect reaches a critical weight which can be used as an indication of commitment to pupation. When the JH titer reaches a sufficiently low level, PTTH is released which initiates a cascade of events leading to pupation. These subjects are discussed in greater detail elsewhere (Steel and Davey, vol. 8). This burst is referred to as the pre-wandering JH esterase peak. The second (prepupation) burst occurs late in the last stadium (f), presumably in response to a prepupal pulse of JH (c). In various organisms the relative activities of these two peaks of JH esterase activity vary from *Pectinophora gossypiella* and *Ephestia elutella* where the prepupal activity is larger than the pre-wandering activity, to *Spodoptera exigua* and *Trichoplusia ni* where peaks e and f are of similar specific activity, to *Manduca sexta* and *Pieris rapae* where the second peak (f) is relatively minor. In a few insects a second peak has not been detected (Weirich, G. *et al.*, 1973; Weirich, G. and Wren, J. 1976; Vince, R. and Gilbert, L. 1977; Reddy, G. *et al.*, 1979; Sparks, T. *et al.*, 1979a; Hammock, B. *et al.*, 1981; Jones, D. *et al.*, 1982; Oźyhar, A. *et al.*, 1982).

The fat body seems to be the major source of hemolymph JH esterase activity in both the pre-wandering (e *vs.* g) and the prepupation peak (f *vs.* h). Although the work of Whitmore, D. *et al.* (1972, 1974) has been criticized because α-naphthyl acetate was used to follow JH esterase activity; it is possible that the JH esterases of the saturniids studied hydrolyze both substrates. It is more likely that the JH esterase activity studied co-chromatographed with other esterases since DFP inhibition was incomplete. Regardless of these criticisms, the conclusion of Whitmore, D. *et al.* (1974) that JH esterase arises largely from the fat body has been supported by subsequent experiments. Hammock, B. *et al.* (1975b) noted that *M. sexta* fat bodies released JH esterase activity in culture, and Nowock, J. and Gilbert, L. (1976) found that the JH esterase activity in fat body homogenates or released by fat body *in vitro* correlated with hemolymph JH titers in pre-wandering *M. sexta* and similar results were obtained in *Galleria mellonella* by Reddy, G. *et al.* (1979). Wing, K., *et al.* (1981) monitored fat body midgut and hemolymph JH esterase activity during *T. ni* development. The activity in the fat body (g) correlated well with the pre-wandering burst of hemolymph JH esterase, and the rise in fat body JH esterase also correlated with the rise in hemolymph JH esterase leading to the post-wandering peak (h). However, JH esterase activity in the fat body remained high into the pupal stage, indicating that synthesis and release are under separate control if JH esterase is in fact produced by the fat body (i). Subsequent studies with *T. ni* indicate that release rates from fat body, and to a lesser extent the epidermis *in vitro,* correlate well with *G. mellonella* and *T. ni* hemolymph JH esterase levels (Reddy, G. *et al.,* 1979; Wing, K., 1981). Thus, the definitive experiments to demonstrate unequivocally the fat body as the source of hemolymph JH esterase have yet to be run, but all evidence to date points to the fat body as the major source of JH esterase.

Although the two peaks of hemolymph JH esterase activity are due to the same enzyme which seems to be produced by the fat body at two separate times during development, the regulation of each peak appears to be quite distinct. A head factor other than juvenile hormone appears necessary for the pre-wandering peak (e) (Fig. 4a). The pre-wandering level of JH esterase drops dramatically following head ligation in *G. mellonella, M. sexta* or *T. ni.* Reimplantation of the brain or subesophageal ganglion (or extracts thereof) will partially restore the JH esterase activity while JH and juvenoids have little effect in isolated abdomens of pre-wandering larvae. JH seems to speed the elimination of JH esterase from the hemolymph of debrained *G. mellonella* larvae. Application of JH to very early last-instar *G. mellonella* decreases hemolymph JH esterase activity. Since appearance of the first JH esterase peak (e) can be blocked by prior treatment with JH or juvenoids, the esterase is probably not released until the insect determines that it is in the last larval instar. In normal and black mutant pre-wandering *M. sexta* and *T. ni* larvae, application of JH causes a slight increase in JH esterase activity. Allatectomy at this stage of development leads to a decrease, but not an elimination, of JH esterase activity. However, these minor effects of JH are either mediated through the head or JH acts in concert with a head factor. Possibly at this stage JH acts in a positive feedback role through the head to fine tune its own metabolism (Fig. 4) (Reddy, G. *et al.,* 1979; Sparks, T. and Hammock, B. 1979a,b; Hammock, B. *et al.,* 1981; Jones, G. *et al.,* 1981; Wing, K. *et al.,* 1981; Hammock, B. and Riddiford, L., unpublished; Jones, G. unpublished).

Jones, G. *et al.* (1981) monitored this head factor in *T. ni* by implanting brains and/or subesophageal ganglia or injecting their homogenates into isolated abdomens and then monitoring hemolymph JH esterase activity. Two peaks of apparent JH esterase inducing factor activity were observed. One peak (j) correlates with a very minor burst of JH esterase activity that occurs at the larval'pupal molt (d); the second peak (k) correlates with the major pre-wandering burst of JH activity (e). Homogenates from still later times actually reduce the JH esterase activity in isolated abdomens (l), while the brain, subesophageal ganglion or homogenates thereof from late last-instar larvae (m) have no effect in inducing the prepupal peak of JH esterase activity (f). Thus, the JH esterase-inducing factor from the head is present during the major pre-wandering JH esterase peak (e), but it also occurs where there is a very minor JH esterase peak (d) and fails to occur where there is a post-wandering peak (f).

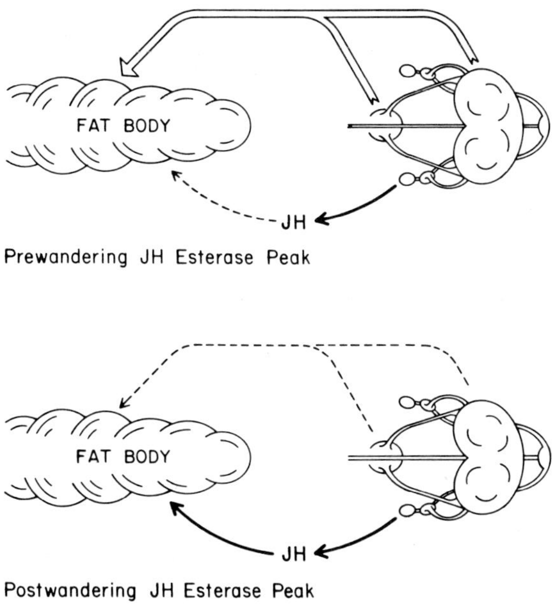

FIG. 4. Possible mechanism of regulation of the two major JH esterase peaks in *Manduca sexta* and *Trichoplusia ni*. Production of JH esterase leading to the pre-wandering peak appears to be stimulated by a water-soluble factor from the brain and/or subesophageal ganglion (⟹). JH will inhibit JH esterase production if applied early and slightly stimulate it if applied later, but it has no effect on isolated abdomens (– – →). Production of the JH esterase comprising the second peak is directly stimulated by JH itself, even in isolated abdomens (⟶). The process which controls this change in tissue response to JH is as yet not understood. Thus, in some insects JH titers may be controlled at times by a decrease in biosynthesis coupled with an increase in degradation.

This apparent anomaly was explained by monitoring the sensitivity of the abdomen to inducing factor. The sensitivity of isolated abdomens is low in the penultimate and early last instar (n). Thus, the first burst of inducing factor (j) may act on an unresponsive abdomen to produce a minor amount of JH esterase (d) at the larval–larval molt (a). Possibly this event prepares the larva to respond with a major peak of hemolymph JH esterase (e) activity when the second burst of inducing factor occurs in the pre-wandering stage (k). During the mid instar, the fat body again becomes unresponsive (p). By the time of the second major peak of hemolymph JH esterase, the fat body is again competent to respond (q).

A great deal remains to be done with the inducing factor. It is not known if it is one or several factors, its chemical nature is unknown, and its (or their) apparent presence in both the brain and the subesophageal ganglion (but not other nervous tissue) is disconcerting. This factor may have a number of roles in inducing adult development, and its role in changing the response of the fat body to JH is also unclear (Hammock, B. *et al.*, 1981; Jones, G. *et al.*, 1981).

Ożyhar, A. *et al.* (1981) state that they were unable to cause an increase in hemolymph JH esterase levels in *Galleria mellonella* isolated abdomens by brain implantation, although they did find increases in other esterases apparently involved in lipid metabolism. They were able to obtain increased JH esterase activity by dipping the larvae in 20-hydroxyecdysone. Similar results were obtained by Sparks, T. and Hammock, B. (1979b) who concluded that the esterase induction observed was due, in part, to fat body lysis. However, much lower ecdysteroid levels were active in *G. mellonella*. The role of ecdysone, either alone or in concert with other endocrine factors in JH esterase induction, needs re-examination.

The pre-wandering JH esterase activity is sensitive to the external environment of the insect. For instance McCaleb, D. and Kumaran, A. (1978, 1980) demonstrated in *G. mellonella* that chilling and injury affected JH esterase activity. Reddy, G. *et al.* (1979) demonstrated in the same insect that starvation reduced hemolymph JH esterase activity, and this work was then repeated in *Trichoplusia ni* and *Manduca sexta* (Fig. 5) (Sparks, T. and Hammock, B., 1979b; Hammock, B. *et al.*, 1981; Cymborowski, B., *et al.*, 1982; Sparks, T., Hammock, B. and Riddiford, L., unpublished). Further studies in *T. ni* indicate that the JH esterase titers are not influenced by the mere bulk of the food consumed, but rather the quality of the food. Protein appears to be the single most important nutritional factor in influencing JH catabolism in that the hemolymph JH esterase levels were directly proportional to the amount of casein added to a minimal diet. A variety of plant natural products which lengthened the feeding stage of the insect also decreased the JH esterase titer, indicating a relationship between plant antibiosis and insect endocrinology (Levinson, and Hammock, B., unpublished; Hammock, B. *et al.*, 1981).

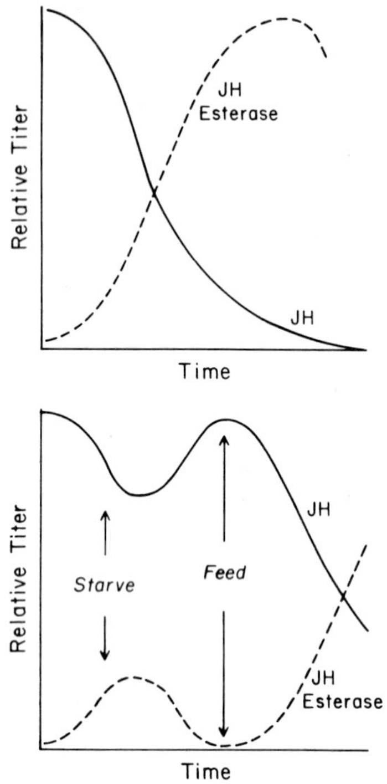

FIG. 5. Schematic representation of the effect of starvation and refeeding on hemolymph JH and JH esterase titers in the early last-larval instar of *Manduca sexta* and *Trichoplusia ni*. In normal larvae (top) JH titers drop as the insect approaches the wandering stage while JH esterase titers increase. If the larvae are starved or placed on an inadequate food supply, JH titers soon increase while JH esterase titers drop. This treatment may result in supernumerary molting or a lengthened prewandering stage. Subsequent refeeding may lead to normal development. Drawn from Cymborowski, B. *et al.*, 1982; Hammock, B. and Quistad, G. 1981; and Hammock, B. *et al.*, 1981.

The above information seems to explain the pre-wandering, but not the post-wandering, JH esterase peak. The clue to its regulation was found by Sparks, T. and Hammock, B. (1979b) who demonstrated that JH and some JH mimics could induce the appearance of JH esterases in post-wandering isolated abdomens of *T. ni* (Fig. 4b). In a series of subsequent studies this induction was shown to depend upon juvenoids or JH homologs having an ethyl branch at the epoxide carbon. The induction phenomenon occurred at JH doses which yielded physiologically reasonable levels of circulating JH. Of crucial significance were classical corpora allata extirpation–replacement experiments. Larvae allatectomized before the prepupation burst of JH occurred produced little JH esterase in both *M. sexta* and *T. ni,* while allatectomy later in life had a smaller effect on reducing JH esterase titers. The allatectomized larvae retained their capacity to respond to topical JH application by producing JH esterase. Corpus allatum reimplantation or JH treatment led to normal JH esterase levels. The ability of JH mimics to induce JH esterase could be blocked by ETB (as could be predicted from its hormone antagonist role; see Bowers, vol. 8) as well as by the biosynthesis inhibitors *o*-bromophenoxymethylimidazole and fluoromevalonate. Induction of JH esterase activity in prepupal fat body slightly preceded, then paralleled hemolymph induction *in vivo* and the post-wandering (but not the pre-wandering) fat body could be induced *in vitro*. The ability of the prepupation burst of JH (c) to induce its own hydrolysis has now been demonstrated in several moth and butterfly species (Sparks, T. and Hammock, B. 1979b; Sparks, T. *et al.,* 1979b; McCaleb, D. and Kumaran, A. 1980; Hammock, B. *et al.,* 1981; Hammock, B. and Quistad, G. 1981; Jones, G. *et al.,* 1981; Jones, D. *et al.,* 1982; Kumaran, A. *et al.,* 1981; Wing, K. *et al.,* 1981; Wing, K. 1981; Sparks, T., Hammock, B. and Riddiford, L., unpublished).

Whitmore, D. *et al.* (1972, 1974) and Reddy, G. *et al.* (1979) found that JH could induce JH esterase in saturniid and *G. mellonella* pupae. At this time fat body JH esterase levels are very high, but in saturniids the fact that transcription and translation appear necessary for release from the fat body supports the hypothesis that JH leads to synthesis of JH esterase. In *T. ni* pupae Wing, K. *et al.* (1981) demonstrated that more JH esterase is detectable in the pupa following JH treatment than can be accounted for by that previously present in the fat body. These workers also demonstrated that fat body lysis does not account for the apparent induction. Whether JH treatment and JH esterase-inducing factors lead to *de novo* synthesis, or simply activation of the enzyme, awaits further study. The roles of both the pre-wandering and the prepupal bursts of JH esterase are explained below, and possibly the JH esterase activity in the fat body remains high (i) to keep JH titers low during the early pupal stage. However, why hemolymph JH esterase levels fall dramatically in the early pupal stage, yet remain responsive to JH treatment, is unknown. The mechanism by which the fat body's production of

JH esterase shifts from being unresponsive to JH in the pre-wandering stage (e) to being responsive to JH in the post-wandering stage (f) similarly is unknown. This shift in sensitivity occurs simultaneously with wandering in *T. ni*. However, both normal and allatectomized *M. sexta* are not sensitive to JH induction of JH esterases until shortly before pupation.

Before one can invoke a biological rationale for the mechanisms by which insects regulate JH esterase, the role for these enzymes in insect development must be established. Nijhout, H. (1975) was justified in questioning the role of JH metabolism in hormone regulation. Simply correlating the appearance of an esterase which metabolizes JH with the theoretical disappearance of JH does not prove its role in regulation. The so-called JH esterase could be, for instance, an amidase which just happens to degrade JH, or the enzyme could be serving as a scavenger which is only important in case of a failure of corpora allata to halt biosynthesis. In many insects, as apparently with most vertebrate hormones, degradation may play a secondary role to biosynthesis and release in endocrine regulation.

In the Lepidoptera the properties of the JH esterases (discussed in section 2) provide strong circumstantial evidence for a role in JH regulation (Fig. 3). The best direct evidence for the role of JH esterase in JH titer regulation came from the application of the JH esterase inhibitor EPPAT to *T. ni* larvae. When JH esterase was inhibited in prewandering larvae (e), the feeding stage was extended and thus pupation was delayed. This effect could be duplicated by the application of JH or juvenoids. EPPAT was shown to inhibit JH esterases and thus slow JH metabolism *in vivo*. One can thus envision the JH titer not being sufficiently reduced to allow PTTH release and thus blocking the cascade of events leading to pupation. EPPAT and paraoxon could similarly inhibit the prepupation burst (f) of JH esterase activity in *T. ni*. Application of these compounds led to leakage of molting fluid, an inability to shed the larval cuticle, and some traits suggesting a larval–pupal intermediate. These results are difficult to rationalize with our current understanding of the pupation process. Like other lepidopterous larvae, *T. ni* were shown by ligation studies to be committed to pupation long before the application of the JH esterase inhibitors. However, application of JH and juvenoids to the prepupa gave similar symptoms to those following inhibitor application, while treatment with the potent acetylcholine and α-naphthyl acetate esterase inhibitor DFP failed to block pupation. These studies with EPPAT do not prove unequivocally that JH esterase is crucial for the regulation of JH titer. However, they provide strong evidence that JH esterase is important in JH regulation and that the release of JH esterase may be the critical factor in initiating the decline in JH titer allowing PTTH release and ultimately pupation (Sparks, T. and Hammock, B. 1980).

With this role in mind it becomes easier to hypothesize reasons for regulatory mechanisms involved in JH degradation. Since high JH titers are needed to maintain normal development in the larval stage, it would be very uneconomical for JH to directly induce its own degradation until the insect is well into its last larval instar. In addition, there are definite advantages to having the esterase under neurosecretory control. As discussed earlier, JH esterase titers drop very quickly when *Galleria mellonella*, *Manduca sexta* or *Trichoplusia ni* are either starved or placed on an inadequate diet, while Cymborowski, B. *et al.* (1982) found that JH titers increase in starved insects (Fig. 5). Since a circulating titer of JH blocks PTTH release, the insect can control the initiation of pupation by regulating JH esterase and JH biosynthesis. This ability is likely to be critical to the insect's survival. Due to high mortality in the larval stage, there is a selective advantage for insects to reach the reproductive stage quickly. However, the insect must have sufficient dietary reserves to survive the pupal stage and efficiently reproduce. Thus, the ability to delay pupation by increasing the length of the feeding stage or inducing a supernumerary molt in response to inadequate food supply offers a distinct advantage in some insects such as *M. sexta* and *T. ni*. Neuroendocrine regulation of JH degradation, as well as biosynthesis, provides a mechanism to accomplish this delay in response to dietary and other factors. Such regulation in the future may be shown to be involved in a variety of developmental and behavioral responses of insects to environmental cues. However, a diversity of insect responses can be anticipated. For instance, when early-instar *G. mellonella* larvae are exposed to an inadequate diet, they may undergo precocious pupation.

In contrast to the regulation of events discussed above, the release of PTTH initiates a cascade of events leading inexorably to pupation. Once PTTH is released, the timing of the numerous events involved in the pupation sequence becomes too complex for any one of the events to be tied to environmental cues. Such is the case with JH esterase. No longer is the esterase titer controlled by neuroendocrine factors which can respond to environmental cues. Rather, in the post-wandering larvae the JH esterase is induced by the presence of JH itself.

Numerous workers have noted the presence of elevated JH titers in lepidopterous prepupae (c) (see Jones, D. *et al.*, 1982, for review), but what this JH does has remained a mystery. Kiguchi, K. and Riddiford, L. (1978) report that when *M. sexta* are allatectomized so that this prepupal burst of JH is eliminated, adultoid structures are observed in the pupa. Thus in *M. sexta* the JH may serve to synchronize developmental processes by retarding the rate of maturation of tissues which are especially sensitive to ecdysone in the absence of JH. Allatectomy in *T. ni* actually stops development at a specific prepupal marker. These prepupae can be induced to pupate by the application of JH. Therefore, the prepupal pulse of JH serves to synergize the effects or stimulate the production of ecdysone in facilitating pupation of *T. ni* (Jones, G. and Hammock, B., unpublished). Since in both *M. sexta* and in *T. ni* this prepupal pulse of JH (c) is critical for normal development, it is essential that the first peak of JH esterase activity is eliminated after JH reaches low enough levels for PTTH release. Thus, the appearance of a factor which reduces JH esterase levels (l) may be of physiological significance (Jones, G. *et al.*, 1981). Although complete elucidation of the role of the prepupal pulse of JH will require additional research, the disastrous effects of later application of JH to the late prepupa and early pupa have been long known. Thus, it is essential for normal development that the hemolymph titer of JH be dramatically reduced before pupation. The Lepidoptera seem to have a very effective means of carrying this reduction out if JH in fact induces its own catabolism at this developmental stage. The model for JH esterase regulation shown in Fig. 4 is consistent with most of the data currently available.

Little is known about the regulation of JH metabolism in adult Lepidoptera. The possibility that JH acid may be remethylated by extra-allatal tissues raises the possibility that epoxide hydrolases may be important in the irreversible inactivation of the hormone (Weirich, G. and Culver, M. 1979). Lessman, C. and Herman, W. (1981) raise the possibility that JH metabolism may increase during the migration of *Danus plexippus* and thus prevent reproduction during migration. As more developmental and behavioral events are found to be controlled by JH, it seems reasonable to examine JH degradation as well as biosynthesis in endocrine regulation. In fact, it may be prudent to reexamine the rates of degradation of some vertebrate hormones.

4 RESISTANCE AND JH METABOLISM

In his 1967 article on third-generation pesticides, C. Williams enumerated the many potential advantages which could be obtained by designing pesticides based upon an insect's own hormones. Many of these potential attributes have proven to be true. However, the statement that "insect hormones promise to provide insecticides that are ... proof against the evolution of resistance" was immediately questioned by Ellis, P. (1968) and Schneiderman, H. (1971, 1972). Shortly thereafter Dyte, C. (1972), and Cerf, D. and Georghiou, G. (1972) showed that insect strains resistant to pesticides could also be resistant to JHs and to their mimics or juvenoids. Vinson, S. and Plapp, F. (1974) examined the potential of insects to become resistant to juvenoids, and in a review of insect resistance to juvenoids and benzoylphenyl ureas, Sparks, T. and Hammock, B. (1982) cataloged cross-resistance to juvenoids in 13 insect species, and induced resistance in still other species. Such resistance is not surprising since any exogenous molecule, even if it is the natural hormone, could be expected to be treated as a xenobiotic or foreign compound. Exogenously applied hormones are thus likely to be subjected to the same penetration and distribution barriers and inactivation processes associated with other forms of pesticide resistance. Also, insects have the ability to regulate their own hormones, and thus have some ability to adjust their physiology towards normal development even in the face of exogenous hormone application. Based upon the structures of most of the juvenoids under serious consideration for development, insects previously selected for high mixed-function oxidase levels could be expected to rapidly become resistant when subsequently pressured with juvenoids (Hammock, B. 1981; Sparks, T. and Hammock, B. 1982).

However, Williams, C. (1967) predictions regarding resistance have some validity. In the cases of resistance and cross-resistance to juvenoids so far studied, the resistance seems to result largely from reduced penetration and increased metabolism. There is no indication of resistance at the sites of juvenoid action. As illustrated by recent problems with the pyrethroid insecticides, such resistance mechanisms have the potential to devastate insect control programs. Since potent juvenoids probably act by being imperfect hormone mimics (mimicking JH at some but not all sites and possibly acting as antihormones at still others), a site of action resistance is likely to develop if insect populations are intensely pressured.

There is, however, cause for optimism about juvenoids. It is unlikely that their use will be heavy enough to induce direct resistance, and cross-resistance problems due to metabolism can be overcome by using different structures. If an insect population can become resistant to a new class of compounds using the same mechanisms of increased metabolism and decreased penetration that were effective for other classes of insecticides, little additional pressure is put on a population. However, if the insect population must overcome a new mechanism of action in order to become resistant, additional pressure is put on that population. Thus, the juvenoids, benzoylphenyl ureas and possibly new insect growth regulators which hit still other novel sites may be very valuable, although possibly limited, control agents for many years.

Hammock, B. *et al.* (1977a) and Yu, S. and Terriere, L. (1978a) demonstrated that *Musca domestica* strains which were either resistant or cross-resistant to juvenoids due to oxidative metabolism also metabolized JH at a greatly increased rate. Neither the esterases nor the epoxide hydrolases which normally are involved in JH metabolism appeared induced, further indicating that these enzymes are involved in hormone regulation. The fact that JH was rapidly metabolized by oxidases bodes both ill and well for insect control. First, it probably indicates that insects have mechanisms to regulate hormone titer in spite of perturbing forces. Second, it may indicate that high oxidase levels due to selection or induction may be biologically expensive, since high levels of resistance commonly decline when selection pressure is removed. Edwards, J. and Rowlands, D. (1978) found that JH was metabolized somewhat faster in a resistant strain of *Sitophilus granarius* by hydrolytic pathways.

In addition to the influence of insecticide resistance upon JH titer, any insect hormone or hormone mimic has profound changes on the biology of the insect. As the insect's biology changes under the influence of a hormone, it is likely that its response to the environment, biological control agents, behavioral control agents as well as insecticides also will change. This possibility has recently been illustrated by the work of Gayen, A. and Knowles, C. (1982) who demonstrated alterations in chlordimeform metabolism were correlated with JH-mediated physiological changes in *Diatraea grandiosella*.

5 CONCLUSION: THE PRACTICAL EXPLOITATION OF INSECT REGULATORY BIOLOGY

5.1 Research in insect neuroendocrinology — why bother?

The investigation of the biosphere, and especially its six-legged inhabitants, has provided entertainment for generations of scientists. Considering the diversity and complexity of insects, there should be ample entertainment for many generations to come. The reasons for such investigations into insect biochemistry and physiology, as with any basic science, are clear to any who have studied the evolution of science or technology. The greatest applications for the research are likely to be the ones that are unforeseen. The elucidation of the concept of neuroendocrinology by a handful of entomologists is one of many examples (Scharrer, B. 1976, 1978).

Yet, we are at a time when all science is being questioned by numerous governments around the world. The concept that all aspects (good and ill) of life in all advanced societies rests upon the basic research of previous generations appears too complex or too radical for most politicians to grasp. The general public has only the vaguest appreciation that cheap food, television, and even canned beer are available only thanks to basic science. The long-term fate of research rests upon a worldwide need for education. In the short term each discipline must also tell society not what we will do for man in the long term, but how basic research will benefit him immediately. This goal has been difficult for insect endocrinologists, especially neuroendocrinologists. Since vertebrate researchers have embraced the concept of neuroendocrinology, the "practical" justification for insect research has been tenuous. Rapid strides in a variety of biological fields have put us on the threshold of a new concept of insect control. The basic investigation of insect regulatory systems is the cornerstone of this concept which could revolutionize the procedures which are used to control insect pests in the future.

5.2 Classical exploitation of discoveries in neuroendocrinology

It is still widely held that peptide hormones and neurotransmitters cannot be exploited in terms of practical insect control. In fact, there are numerous rather conventional ways in which these materials could be exploited. The active portion of small peptides could be mimicked by a classical chemical, as illustrated by the opiate drugs. Alternatively, such peptides could be derivatized to alter their stability or lipophilicity. Many chemical procedures applicable to such derivatization have been known for years and used in peptide synthesis and analysis. Recently, such techniques have been used to profoundly alter the biological and chemical properties of carbamate insecticides, herbicides and other chemicals (Fukuto, T. and Fahmy, M. 1981). Alternatively, one could disrupt processing enzymes involved in the activation of a prohormone or enzyme. In mammalian systems this approach has been successfully exploited with the development of the drug Captopril which blocks the conversion of angiotensin I to II. Possibly the benzoylphenyl ureas such as Dimilin® act as "transition state" analog inhibitors of proteases which act as processing enzymes. Many ramifications of the above approach could be employed, yet these are not radically different from some current approaches in pesticide chemistry. It will soon be possible to employ fundamentally new approaches to exploit insect peptide chemistry for insect control.

5.3 Third-generation concept of pesticide development

Sparks, T. and Hammock, B. (1982) argued that one should define first-, second- and third-generation pesticides in terms of the approach used to develop the compound rather than the final product. This concept only slightly alters the arguments put forward by Ordish, G. (1967) and Williams, C. (1967), yet makes the concept more useful in describing present and future trends in pesticide chemistry. Thus, the first generation of pesticides are those which arose from decades or centuries of a trial-and-error approach to insect control. The arsenicals, oils and botanicals clearly fit into this category. DDT, carbamates and organophosphates characterize the second-generation approach in which effective structures were elucidated using a vigorous synthesis-screening program (Ordish, G. 1967). Those pesticides based upon insect hormones and pheromones appear to have been classified as third-generation compounds by Williams, C. (1967). Extension of this procedure could lead to fourth-, fifth-, and sixth-generation pesticides every time a new lead structure is elucidated (Bowers, W. 1976). This problem is avoided if we consider the approach rather than the end-point to classify pesticides. Thus, the second-generation approach to pesticide development involves the process of screening and bioactivity-directed synthesis. A third-generation approach involves an appreciation of the physiology and ecology of target and non-target species which directs a synthetic program to produce environmentally acceptable chemicals.

In most cases modern pesticides will be developed using a combination of second- and third-generation approaches. J. Siddall's (1976) review of the development of some juvenoids illustrates this concept. The natural juvenile hormones clearly represented the lead structure, but much of the structure optimization resulted from a second-generation approach. The biodegradable DDT mimics and the organophosphates and carbamates which have been derivatized to increase safety and selectivity are excellent examples of the third-generation approach. The parent compounds obviously resulted from a second generation approach, but the subsequent modifications were based upon knowledge of target and non-target biology, as well as environmental chemistry.

Other insect growth regulators such as the benzoylphenyl ureas and anti-juvenile hormones were discovered by a second-generation approach using innovative screening methods. Even the microbial insecticides such as *Bacillus thuringiensis* are classic examples of the second-generation approach. Thus, both second- and third-generation approaches alone and in combination are very useful in pesticide development, and will likely continue to lead to valuable compounds (Sparks, T. and Hammock, B. 1982). Yet, we are starting an era of a radically new concept in pesticide development in which the biologist will play a central role.

5.4 The evolving role of the biologist

The role of the biologist has changed as new approaches to pesticide development have evolved. In the classical second-generation approach the biologist's role is largely that of a technician, screening compounds which emerge from a chemistry program or isolates of micro-organisms. There is certainly room for innovation in terms of improved bioassays, suggestions of possible natural extracts to screen, and other ways. However critical the role of the biologist to the success of the development program, his role is clearly limited. A more exciting role exists in the third-generation approach. The biologist may provide lead structures, as in the case of JHs, he may elucidate differences in mode of action between target and non-target organisms, and he may elucidate a mechanism of action for a new class of compounds. However innovative his approach, the biologist soon finds himself again involved in a screening program. This screen may be at a physiological or biochemical level, yet it is nevertheless a screen. At best the biologist can hope to provide leads which can be exploited chemically. Regardless of these leads, an empirical approach is invariably demanded, requiring luck, as well as a tremendous investment in exploratory chemistry. As a new or fourth-generation approach to pesticide development is initiated, the biologist faces the exciting and demanding new role of directing pesticide development at each step from the observation of natural history through commerical exploitation. This route may be exceedingly challenging, yet the goal is clear and the approach is no longer empirical.

5.5 The fourth-generation approach to pesticide development

The way in which one can exploit basic entomological research can be illustrated by the hypothetical fate of a cotton bollworm that feeds on a very special, and only slightly futuristic, cotton plant in California's Imperial Valley. The insect's finely tuned physiology is allowing it to survive the intense, desiccating conditions of its environment until it consumes plant material containing its own diuretic hormone. If the exogenous diuretic hormone reaches the insect's hemolymph in effective levels, the life expectancy of such an organism is very short.

At a time when plant genetic engineers are seriously considering changing the entire biology of crops to enhance salt tolerance, fix nitrogen, or even alter the photosynthetic process, the insertion and expression of a single gene will be technically foreseeable in the near future.

The research on vectors, expression, and regulation in plants is progressing rapidly, and the technology to carry out the above manipulations is likely to be available in a few years, but there are also courses of action which can be taken immediately. The technology to synthesize small genes and to insert and express synthetic or natural genes in yeast or bacteria now exists. The micro-organism acts not only as the factory to produce the critical peptide; it also serves as an encapsulation agent to help protect and distribute the biologically active ingredient. The potential of *Bacillus thuringiensis* and *Bacillus israeliensis* as major insecticides illustrates the possible utility of this approach. Although the genetics are less well understood, commercial organisms of both plants and insects also offer possibilities for such manipulation. If we had the messages for critical regulatory peptides from pest insects, they could be exploited now for insect control. Considering the relative support for plant molecular biology and insect regulatory biology, it is likely that we will be able to manipulate the plant genome long before we know how to manipulate it to affect insect control. Numerous gaps in our knowledge exist, and many problems can be foreseen with this approach, but knowledge can be expanded and the problems will be solved. For example, although there is evidence that some insects absorb even large proteins through the gut wall intact, our knowledge of absorption and digestion in insects is woefully lacking. As we exploit such new concepts in insect control, new problems — as well as new ramifications of old problems — will arise. How does one effect patent protection; how does one prevent the escape of a gene from a crop cucurbit to a closely related weed; will new forms of resistance develop; will such research ultimately be used to justify further reduction in the natural plant gene pool?

Almost every chapter in this volume describes insect systems which can be exploited as described here to affect practical insect control. This chapter alone describes a variety of leads. The time to move on this fourth-generation approach to pest control is now. In a social context, such a move is critical. Metcalf, R. (1980) has charged that our war on insect vectors of human and animal disease and pests of crop plants is faltering. Even new leads from natural products chemistry will provide at best stop-gap measures. Here we simply summon the "experience" of plants in screening xenobiotics against insect pests. The chances of a plant successfully breaking the sequence code of even a short regulatory peptide by random evolutionary pressure are frustratingly small. Yet, the production of certain insect neurohormones by a crop plant should provide it with almost absolute protection against phytophagous insects. When entomologists instruct the plant as to the nature of these critical codes, science will have enhanced greatly the powers of man's already powerful tool in the war on insects — the plant itself. With a fourth-generation approach to pesticide development, we have the ability to take basic knowledge directly to commercial application using the tools of genetic engineering. This new horizon in pesticide chemistry certainly supports the concept that the most applied research that our society can support is, in fact, basic research.

ACKNOWLEDGEMENTS

Original research in this chapter was supported in part by NIEHS grant ES 02710–02, while the author was supported by NIEHS research career development award ES 00107–04. The constructive comments of a variety of colleagues, notably T. Hanzlik, G. Jones, P. Kaplan, J. Roe and M. Roe are gratefully acknowledged.

The study of juvenile hormone metabolism would have been impossible without the radiolabeled hormones provided in part by Walter Hafferl, John Siddall and Edward White. May we remember them.

REFERENCES

AJAMI, A. M. (1975). Inhibitors of ester hydrolysis as synergists for biological activity of cecropia juvenile hormone. *J. Insect Physiol. 21,* 1017–1025.

AJAMI, A. M. and RIDDIFORD, L. M. (1973). Comparative metabolism of the cecropia juvenile hormone. *J. Insect Physiol. 19,* 635–645.

AKAMATSU, Y., DUNN, P. E., KÉZDY, F. J., KRAMER, K. J., LAW, J. H., REIBSTEIN, D. and SANBURG, L. L. (1975). Biochemical Aspects of Juvenile Hormone Action in Insects. In *Control Mechanisms in Development*. Edited by R. H. Meints and E. Davies. Pages 123–149. Plenum Press, New York.

ALDRIDGE, W. N. (1953). Serum esterases: two types of esterase (A and B) hydrolysing *p*-nitrophenyl acetate, propionate and butyrate, and a method for their determination. *Biochem. J. 53,* 110–117.

BASSI, S. D., GOODMAN, W., ALTENHOFEN, C. and GILBERT, L. I. (1977). The binding of exogenous juvenile hormone by the haemolymph of *Oncopeltus fasciatus*. *Insect Biochem. 7,* 309–312.

BENDER, M. L., KÉZDY, F. J. and WEDLER, F. C. (1967). α-Chymotrypsin enzyme concentration and kinetics. *J. Chem. Ed. 44,* 84–88.

BISSER, A. and EMMERICH, H. (1981). Hydrolysis of juvenile hormone and juvenile hormone analogues in the larval integument of *Drosophila hydei*. In *Juvenile Hormone Biochemistry*. Edited by G. E. Pratt and G. T. Brooks. Pages 199–212. Elsevier, New York.

BOWERS, W. S. (1976). Discovery of insect antiallatotropins. In *The Juvenile Hormones*. Edited by L. I. Gilbert. Pages 394–408. Plenum Press, New York.

BROOKS, G. T. (1973). Insect epoxide hydrase inhibition by juvenile hormone analogues and metabolic inhibitors. *Nature, New Biol. 245,* 382–384.

BROOKS, G. T. (1974). Inhibitors of cyclodiene epoxide ring hydrating enzymes of the blowfly, *Calliphora erythrocephala*. *Pestic. Sci. 5,* 177–183.

BROOKS, G. T. (1977). Epoxide hydratase as a modifier of biotransformation and biological activity. *Gen. Pharmac. 8,* 221–226.

BROOKS, G. T., HARRISON, A. and LEWIS, S. E. (1970). Cyclodiene epoxide ring hydration by microsomes from mammalian liver and houseflies. *Biochem. Pharmac. 19,* 255–273.

BROWN, T. M. and HOOPER, G. H. S. (1979). Metabolic detoxication as a mechanism of methoprene resistance in *Culex pipiens pipiens*. *Pestic. Biochem. Physiol. 12,* 79–86.

Brown, J. J., Chippendale, G. M. and Turunen, S. (1977). Larval esterases of the southwestern corn borer, *Diatraea grandiosella:* temporal changes and specificity. *J. Insect Physiol. 23,* 1255–1260.

Cerf, D. C. and Georghiou, G. P. (1972). Evidence of cross-resistance to a juvenile hormone analogue in some insecticide-resistant house flies. *Nature 239,* 401–402.

Chasseaud, L. F. (1979). The role of glutathione and glutathione-S-transferases in the metabolism of chemical carcinogens and other electrophilic agents. *Adv. Cancer Res. 29,* 175–274.

Chihara, C. J., Petri, W. H., Fristrom, J. W. and King, D. S. (1972). The assay of ecdysones and juvenile hormones on *Drosophila* imaginal disks *in vitro. J. Insect Physiol. 18,* 1115–1123.

Coudron, T. A., Dunn, P. E., Seballos, H. L., Wharen, R. E., Sanburg, L. L. and Law, J. H. (1981). Preparation of homogeneous juvenile hormone specific esterase from the haemolymph of the tobacco hornworm, *Manduca sexta. Insect Biochem. 11,* 453–461.

Craven, A. C. C., Brooks, G. T. and Walker, C. H. (1976). The inhibition of HEOM epoxide hydrase in mammalian liver microsomes and insect pupal homogenates. *Pestic. Biochem. Physiol. 6,* 132–141.

Cymborowski, B., Bogus, M., Beckage, N. E., Williams, C. M. and Riddiford, L. M. (1982). Juvenile hormone titres and metabolism during starvation-induced supernumerary larval moulting of the tobacco hornworm *Manduca sexta* L. *J. Insect Physiol. 28,* 129–135.

Dyte, C. E. (1972). Resistance to synthetic juvenile hormone in a strain of flour beetle, *Tribolium castaneum. Nature 238,* 48–49.

Edwards, J. P. and Rowlands, D. G. (1977). Metabolism of a synthetic insect juvenile hormone (JH-I) during the development of *Tribolium castaneum* (Herbst) (Coleoptera, Tenebrionidae). *Pestic. Biochem. Physiol. 7,* 194–201.

Edwards, J. P. and Rowlands, D. G. (1978). Metabolism of a synthetic juvenile hormone (JH-I) in two strains of the grain weevil, *Sitophilus granarius. Insect Biochem. 8,* 23–28.

Ellis, P. (1968). Can insect hormones and their mimics be used to control pests? *Pest Articles and News Summaries 14,* 329–342.

Emmerich, H. (1976). Summary of Session IV: Juvenile hormone effects at the molecular level (Binding and transport). In *The Juvenile Hormones.* Edited by L. I. Gilbert. Pages 323–326. Plenum Press, New York.

Emmerich, H. and Hartmann, R. (1973). A carrier lipoprotein for juvenile hormone in the haemolymph of *Locusta migratoria. J. Insect Physiol. 19,* 1663–1675.

Erley, D., Southard, S. and Emmerich, H. (1975). Excretion of juvenile hormone and its metabolism in the locust, *Locusta migratoria. J. Insect Physiol. 21,* 61–70.

Ferkovich, S. M. and Rutter, R. R. (1976). Influence of a haemolymph protein fraction on the binding of juvenile hormone in homogenates of insect epidermis (*Plodia interpunctella* (Hübner)). *Wilhelm Roux's Arch. Dev. Biol. 179,* 243–248.

Ferkovich, S. M., Silhacek, D. L. and Rutter, R. R. (1975). Juvenile hormone binding proteins in the haemolymph of the Indian meal moth. *Insect Biochem. 5,* 141–150.

Ferkovich, S. M., Oberlander, H. and Rutter, R. R. (1977). Release of a juvenile hormone binding protein by fat body of the Indian meal moth *Plodia interpunctella, in vitro. J. Insect Physiol. 23,* 297–302.

Fersht, A. (1977). *Enzyme structure and mechanism.* W. H. Freeman, San Francisco.

Fukuto, T. R. and Fahmy, M. A. H. (1981). Sulfur and Pesticide Metabolism in action. In *Sulfur and Propesticide Action.* Edited by J. D. Rosen, P. S. McGee and J. E. Casida. *American Chemical Society Symposium Series 158,* pages 35–49. American Chemical Society Publications, Washington, D.C.

Gayen, A. K. and Knowles, C. O. (1982). Chlordimeform penetration, metabolism, and excretion in diapausing and nondiapausing larvae of the southwestern corn borer. *J. Agric. Food Chem. 30,* 122–126.

Gilbert, L. I., Goodman, W. and Nowock, J. (1976). The possible roles of binding proteins in juvenile hormone metabolism and action. In *Actualitiés sur les Hormones D'Invertebrés.* Pages 413–434. Colloques Internationaux CNRS No. 251. Paris.

Gilbert, L. I., Goodman, W. and Granger, N. (1978). Regulation of juvenile hormone titer in the Lepidoptera. In *Comparative Endocrinology.* Edited by P. J. Gaillard and H. H. Boer. Pages 471–486. Elsevier, Amsterdam.

Gill, S. S., Hammock, B. D., Yamamoto, I. and Casida, J. E. (1972). Preliminary chromatographic studies on the metabolites and photodecomposition products of the juvenoid 1-(4′-ethylphenoxy)-6,7-epoxy-3,7-dimethyl octene. In *Insect Juvenile Hormones: Chemistry and Action.* Edited by J. J. Menn and M. Beroza. Pages 177–189. Academic Press, New York.

Gill, S. S., Hammock, B. D. and Casida, J. E. (1974). Mammalian metabolism and environmental degradation of the juvenoid 1-(4′-ethylphenoxy)-3,7-dimethyl-6,7-epoxy-*trans*-2-octene and related compounds. *J. Agric. Food Chem. 22,* 386–395.

Goodman, W. and Gilbert, L. I. (1978). The hemolymph titer of juvenile hormone binding protein and binding sites during the fourth larval instar of *Manduca sexta. Gen. Comp. Endocr. 35,* 27–34.

Goodman, W., Bollenbacher, W. E., Zvenko, H. and Gilbert, L. I. (1976). A competitive protein binding assay for juvenile hormone. In *The Juvenile Hormones.* Edited by L. I. Gilbert. Pages 75–95. Plenum Press, New York.

Goodman, W., Schooley, D. A. and Gilbert, L. I. (1978a). Specificity of the juvenile hormone binding protein: The geometrical isomers of juvenile hormone I. *Proc. Natl. Acad. Sci. U.S.A. 75,* 185–189.

Goodman, W., O'Hern, P. A., Zaugg, R. H. and Gilbert, L. I. (1978b). Purification and characterization of a juvenile hormone binding protein from the hemolymph of the fourth instar tobacco hornworm, *Manduca sexta. Mol. Cel.. Endocr. 11,* 225–242.

Hammock, B. D. (1975). NADPH dependent epoxidation of methyl farnesoate to juvenile hormone in the cockroach *Blaberus giganteus* L. *Life Sci. 17* (3), 323–328.

Hammock, B. D. (1981). Metabolism and environmental fate of juvenoids: A decade's research in perspective and an assessment of the future. In *Regulation of Insect Development and Behavior.* Edited by F. Sehnal, A. Zabża, J. J. Menn, and B. Cymborowski. Pt. 2, pages 841–852. Wrocław Technical University Press, Wrocław, Poland.

Hammock, B. D. and Quistad, G. B. (1976). The degradative metabolism of juvenoids by insects. In *The Juvenile Hormones.* Edited by L. I. Gilbert. Pages 374–393. Plenum Press, New York.

Hammock, B. D. and Quistad, G. B. (1981). Metabolism and mode of action of juvenile hormone, juvenoids, and other insect growth regulators. In *Progress in Pesticide Biochemistry.* Edited by D. H. Hutson and T. R. Roberts. Vol. 1, pages 1–83. John Wiley & Sons, New York.

Hammock, B. D. and Mumby, S. M. (1978). Inhibition of epoxidation of methyl farnesoate to juvenile hormone III by cockroach corpus allatum homogenates. *Pestic. Biochem. Physiol. 9,* 39–47.

Hammock, B. D. and Sparks, T. C. (1977). A rapid assay for insect juvenile hormone esterase activity. *Anal. Biochem. 82,* 573–579.

Hammock, B. D., Gill, S. S. and Casida, J. E. (1974a). Insect metabolism of a phenyl epoxygeranyl ether juvenoid and related compounds. *Pestic. Biochem. Physiol. 4,* 393–406.

Hammock, B. D., Gill, S. S. and Casida, J. E. (1974b). Synthesis and morphogenetic activity of derivatives and analogs of aryl geranyl ether juvenoids. *J. Agric. Food Chem. 22,* 379–385.

Hammock, B. D., Gill, S. S., Hammock, L. and Casida, J. E. (1975a). Metabolic O-dealkylation of 1-(4′-ethylphenoxy)-3,7-dimethyl-7-methoxy or ethoxy-*trans*-2-octene, potent juvenoids. *Pestic. Biochem. Physiol. 5,* 12–18.

Hammock, B. D., Nowock, J., Goodman, W., Stamoudis, V. and Gilbert, L. I. (1975b). The influence of hemolymph binding protein on juvenile hormone stability and distribution in *Manduca sexta* fat body and imaginal discs *in vitro. Mol. Cell. Endoc. 3,* 167–184.

Hammock, B. D., Mumby, S. M. and Lee, P. W. (1977a). Mechanisms of resistance to the juvenoid methoprene in the house fly *Musca domestica L. Pestic. Biochem. Physiol. 7,* 261–272.

Hammock, B. D., Sparks, T. C. and Mumby, S. M. (1977b). Selective inhibition of JH esterases from cockroach hemolymph. *Pestic. Biochem. Physiol. 7,* 517–530.

Hammock, B. D., Gill, S. S., Mumby, S. M. and Ota, K. (1980). Current status and possible implication of research on a soluble mammalian epoxide hydrase. In *Molecular Basis of Environmental Toxicity.* Edited by R. S. Bhatnager. Chapter 10, pages 229–272. Ann Arbor Science Publishers, Ann Arbor, Michigan.

HAMMOCK, B. D., JONES, D., JONES, G., RUDNICKA, M., SPARKS, T. C. and WING, K. D. (1981). Regulation of juvenile hormone esterase in the cabbage looper, *Trichoplusia ni*. In *Regulation of Insect Development and Behavior*. Edited by F. Sehnal, A. Zabża, J. J. Menn and B. Cymborowski. Pages 219–235. Wrocław Technical University Press, Wrocław, Poland.

HAMMOCK, B. D., WING, K. D., MCLAUGHLIN, J., LOVELL, V. M. and SPARKS, T. C. (1982). Trifluoromethylketones as possible transition state analog inhibitors of juvenile hormone esterase. *Pestic. Biochem. Physiol. 17*, in press.

HARTMANN, R. (1978). The juvenile hormone carrier in the haemolymph of the acridine grasshopper *Gomphocerus rubus* L.: Blocking of the juvenile hormone's action by means of repeated injections of an antibody to the carrier. *Wilhelm Roux's Arch. Dev. Biol. 184*, 301–324.

HOFFMAN, L. J., ROSS, J. H. and MENN, J. J. (1973). Metabolism of 1-(4′-ethylphenoxy)-6,7-epoxy-3,7-dimethyl-2-octene (R 20458) in the rat. *J. Agric. Food Chem. 21*, 156–163.

HOLMES, R. S. and MASTERS, C. J. (1967). The developmental multiplicity and isoenzyme status of cavian esterases. *Biochim. Biophys. Acta 132*, 379–399.

HOOPER, G. H. S. (1976). Esterase mediated hydrolysis of naphthyl esters, malathion, methoprene, and cecropia juvenile hormone in *Culex pipiens pipiens*. *Insect Biochem. 6*, 255–266.

HSIA, M. T. S., ADAMOVICS, J. A. and KREAMER, B. L. (1979). Microbial mutagenicity studies of insect growth regulators and other potential insecticidal compounds in *Salmonella typhimurium*. *Chemosphere 8*, 521–529.

HWANG-HSU, K., REDDY, G., KUMARAN, A. K., BOLLENBACHER, W. E. and GILBERT, L. I. (1979). Correlations between juvenile hormone esterase activity, ecdysone titre and cellular reprogramming in *Galleria mellonella*. *J. Insect Physiol. 25*, 105–111.

IVIE, G. W. (1976). Epoxide to olefin: a novel biotransformation in the rumen. *Science 191*, 959–961.

IVIE, G. W., WRIGHT, J. E. and SMALLEY, H. E. (1976). Fate of the juvenile hormone mimic 1-(4′-ethylphenoxy)-3,7-dimethyl-6,7-epoxy-*trans*-2-octene (Stauffer R-20458) following oral and dermal exposure to steers. *J. Agric. Food Chem. 24*, 222–227.

IVIE, G. W., MACGREGOR, J. T. and HAMMOCK, B. D. (1980). Mutagenicity of psoralen epoxides. *Mutation Res. 79*, 73–77.

JONES, D., JONES, G., WING, K. D., Rudnicka, M. and HAMMOCK, B. D. (1982). Juvenile hormone esterases of Lepidoptera. I. Activity in the hemolymph during the last larval instar of 11 species. *J. Comp. Physiol. 148*, 1–10.

JONES, G., WING, K. D., JONES, D. and HAMMOCK, B. D. (1981). The source and action of head factors regulating juvenile hormone esterase in larvae of the cabbage looper, *Trichoplusia ni*. *J. Insect Physiol. 27*, 85–91.

KIGUCHI, K. and RIDDIFORD, L. M. (1978). A role of juvenile hormone in pupal development of the tobacco hornworm, *Manduca sexta*. *J. Insect Physiol. 24*, 673–680.

KLAGES, G. and EMMERICH, H. (1979a). Juvenile hormone binding proteins in the haemolymph of third instar larvae of *Drosophila hydei*. *Insect Biochem. 9*, 23–30.

KLAGES, G. and EMMERICH, H. (1979b). Juvenile hormone metabolism and juvenile hormone esterase titer in hemolymph and peripheral tissues of *Drosophila hydei*. *J. Comp. Physiol. 132*, 319–325.

DE KORT, C. A. D. and GRANGER, N. A. (1981). Regulation of the juvenile hormone titer. *Ann. Rev. Ent. 26*, 1–28.

DE KORT, C. A. D., WIETEN, M., KRAMER, S. J. and GOEWIE, E. (1977). Juvenile hormone degradation and carrier proteins in honey bee larvae. *Proc. K. Ned. Akad. Wet. Amsterdam, Ser. C, 80*, 297–301.

DE KORT, C. A. D., KRAMER, S. J. and WIETEN, M. (1978). Regulation of juvenile hormone titres in the adult Colorado beetle: Interaction with carboxylesterases and carrier proteins. In *Comparative Endocrinology*. Edited by P. J. Gaillard and H. H. Boer. Pages 507–512. Elsevier/North-Holland, Amsterdam.

DE KORT, C. A. D., WIETEN, M. and KRAMER, S. J. (1979). The occurrence of juvenile hormone specific esterases in insects: a comparative study. *Prok. K. Ned. Akad. Wet. Ser. C. Biol. Med. Sci. 82*, 325–332.

KRAMER, K. J. and CHILDS, C. N. (1977). Interaction of juvenile hormone with carrier proteins and hydrolases from insect haemolymph. *Insect Biochem. 7*, 397–403.

KRAMER, K. J., SANBURG, L. L., KÉZDY, F. J. and LAW, J. H. (1974). The juvenile hormone binding protein in the hemolymph of *Manduca sexta* Johannson (Lepidoptera: Sphingidae). *Proc. Natl. Acad. Sci. USA 71*, 493–497.

KRAMER, K. J., DUNN, P. E., PETERSON, R. C. and LAW, J. H. (1976a). Interaction of juvenile hormone with binding proteins in insect hemolymph. In *The Juvenile Hormones*. Edited by L. I. Gilbert. Pages 327–341. Plenum Press, New York.

KRAMER, K. J., DUNN, P. E., PETERSON, R. C., SEBALLOS, H. L., SANBURG, L. L. and LAW, J. H. (1976b). Purification and characterization of the carrier protein for juvenile hormone from hemolymph of the tobacco hornworm, *Manduca sexta* (Lepidoptera: Sphingidae). *J. Biol. Chem. 251*, 4979–4985.

KRAMER, S. J. (1978a). Age-dependent changes in corpus allatum activity *in vitro* in the adult Colorado potato beetle, *Leptinotarsa decemlineata*. *J. Insect Physiol. 24*, 461–464.

KRAMER, S. J. (1978b). Regulation of the activity of JH specific esterases in the Colorado potato beetle, *Leptinotarsa decemlineata*. *J. Insect Physiol. 24*, 743–747

KRAMER, S. J. and DE KORT, C. A. D. (1976a). Age-dependent changes in juvenile hormone esterase and general carboxyesterase activity in the hemolymph of the Colorado potato beetle, *Leptinotarsa decemlineata*. *Mol. Cell. Endocr. 4*, 43–53.

KRAMER, S. J. and DE KORT, C. A. D. (1976b). Some properties of hemolymph esterases from *Leptinotarsa decemlineata* Say. *Life Sciences 19*, 211–218.

KRAMER, S. J. and DE KORT, C. A. D. (1978). Juvenile hormone carrier lipoproteins in the haemolymph of the Colorado potato beetle, *Leptinotarsa decemlineata*. *Insect Biochem. 8*, 87–92.

KRAMER, S. J. and LAW, J. H. (1980). Synthesis and transport of juvenile hormones in insects. *Accounts of Chem. Research, 13*, 297–303.

KRAMER, S. J., WIETEN, M. and DE KORT, C. A. D. (1977). Metabolism of juvenile hormone in the Colorado potato beetle, *Leptinotarsa decemlineata*. *Insect Biochem. 7*, 231–236.

KUMARAN, A. K., BROOK, M. M. and KHIPPLE, S. (1981). Juvenile hormone esterase activity in Galleria: Effect of JH, ligation and brain implantation. In *Juvenile Hormone Biochemistry*. Edited by G. E. Pratt and G. T. Brooks. Pages 177–184. Elsevier/North-Holland, Amsterdam.

LAW, J. H. (1978). Interaction of juvenile hormone with the hemolymph carrier protein. In *Comparative Endocrinology*. Edited by P. J. Gaillard and H. H. Boer. Pages 511–514. Elsevier/North-Holland, Amsterdam.

LESSMAN, C. A. and HERMAN, W. S. (1981). Flight enhances juvenile hormone inactivation in *Danaus plexippus plexippus* L. (Lepidoptera: Danaidae). *Experientia 37*, 599–601.

LU, A. Y. H. and MIWA, G. T. (1980). Molecular properties and biological functions of microsomal epoxide hydrase. *Ann. Rev. Pharmacol. Toxicol. 20*, 513–531.

MCCALEB, D. C. and KUMARAN, A. K. (1978). Effect of factors that influence metamorphosis on JH esterase activity in *Galleria mellonella*. *Amer. Zool. 18*, 626.

MCCALEB, D. C. and KUMARAN, A. K. (1980). Control of juvenile hormone esterase activity in *Galleria mellonella* larvae. *J. Insect Physiol. 26*, 171–177.

MCCALEB, D. C., REDDY, G. and KUMARIN, A. K. (1980). Some properties of the haemolymph juvenile hormone esterases in *Galleria mellonella* larvae and *Tenebrio molitor* pupae. *Insect Biochem. 10*, 273–277.

MANE, S. D. and REMBOLD, H. (1977). Developmental kinetics of juvenile hormone inactivation in queen and worker castes of the honey bee, *Apis mellifera*. *Insect Biochem. 7*, 463–467.

MAYER, R. T. and BURKE, M. D. (1976). Albumin and cytochrome P450 binding characteristics of juvenile hormone and its analogs. *Pestic. Biochem. Physiol. 6*, 377–385.

METCALF, R. L. (1980). Changing role of insecticides in crop protection. *Ann. Rev. Ent. 25*, 219–256.

MITSUI, T., RIDDIFORD, L. M. and BELLAMY, G. (1979). Metabolism of juvenile hormone by the epidermis of the tobacco hornworm *Manduca sexta*. *Insect Biochem. 9*, 637–643.

MORELLO, A. and AGOSIN, M. (1979). Metabolism of juvenile hormone with isolated rat hepatocytes. *Biochem. Pharmacol. 28*, 1533–1539.

MORISAKI, M., YING, B. and IKEKAWA, N. (1981). Identification of both fucosterol and isofucosterol in the silkworm, *Bombyx mori*. *Experientia 37*, 336–337.

MULLIN, C. and WILKINSON, C. F. (1980a). Insect epoxide hydrolase: Properties of a purified enzyme from the southern armyworm (*Spodoptera eridania*). *Pestic. Biochem. Physiol. 14*, 192–207.

MULLIN, C. and WILKINSON, C. F. (1980b). Purification of an epoxide hydrolase from the midgut of the southern army worm (*Spodoptera eridania*). *Insect Biochem. 10*, 681–691.

MULLIN, C. and HAMMOCK, B. D. (1982). Chalcone oxides — potent selective inhibitors of cytosolic epoxide hydrolase. *Arch. Biochem. Biophys. 216*, 423–439.

MUMBY, S. M. and HAMMOCK, B. D. (1979). Stability of epoxide containing juvenoids to dilute aqueous acid. *J. Agric. Food Chem. 27*, 1223–1228.

MUMBY, S. M., HAMMOCK, B. D., SPARKS, T. C. and OTA, K. (1979). Synthesis and bioassay of carbamate inhibitors of the juvenile hormone hydrolyzing esterases from the housefly, *Musca domestica. J. Agric. Food Chem. 27*, 763–765.

NIJHOUT, H. F. (1975). Dynamics of juvenile hormone action in larvae of the tobacco hornworm, *Manduca sexta* (L.). *Biol. Bull. 149*, 568–579.

NOWOCK, J. and GILBERT, L. I. (1976). In vitro analysis of factors regulating the juvenile hormone titer of insects. In *Invertebrate Tissue Culture*. Edited by E. Kurstak and K. Maramorosch. Chapter 16, Pages 203–212. Academic Press. New York.

NOWOCK, J., GOODMAN, W., BOLLENBACHER, W. E. and GILBERT, L. I. (1975). Synthesis of juvenile hormone binding proteins by the fat body of *Manduca sexta. Gen. Comp. Endocrinol. 27*, 230–239.

NOWOCK, J., HAMMOCK, B. and GILBERT, L. I. (1976). The binding protein as a modulator of juvenile hormone stability and uptake. In *The Juvenile Hormones*. Edited by L. I. Gilbert. Pages 354–373. Plenum Press. New York.

OBERLANDER, H. and SILHACEK, D. L. (1976). Action of juvenile hormone on imaginal discs of the Indian meal moth. In *The Juvenile Hormones*. Edited by L. I. Gilbert. Pages 220–233. Plenum Press, New York.

OESCH, F. (1973). Mammalian epoxide hydrases: inducible enzymes catalyzing the inactivation of carcinogenic and cytotoxic metabolites derived from aromatic and olefinic compounds. *Xenobiotica 3*, 305–340.

ORDISH, G. (1967). Biological Methods in Crop Pest Control. Pages 200–242. Constable, London.

OŻYHAR, A., SEHNAL, F., WIŚNIEWSKI, J. R. and KOCHMAN, M. (1981). Regulation of juvenile hormone hydrolysis in the haemolymph. In *Juvenile Hormone Biochemistry*. Edited by G. E. Pratt and G. T. Brooks. Pages 193–198. Elsevier, New York.

OŻYHAR, A., WIŚNIEWSKI, J. R., SEHNAL, F. and KOCHMAN, M. (1982). Age dependent changes in the binding and hydrolysis of juvenile hormone in the haemolymph of last instar larvae of *Galleria mellonella*. (Submitted).

PAULING, L. (1946). Molecular architecture and biological reactions. *Chem. Eng. News 24*, 1375–1377.

PAULING, L. (1948). Chemical achievement and hope for the future. *Amer. Sci. 36*, 51–58.

PETER, M. (1981). Stereochemistry of juvenile hormone hydrolysis in insect haemolymph. In *Regulation of Insect Development and Behavior*. Edited by F. Sehnal, A. Zabża, J. J. Menn and B. Cymborowski. Pages 237–244. Wrocław Technical University Press, Wrocław, Poland.

PETER, M. G., GUNAWAN, S., GELLISSEN, G. and EMMERICH, H. (1979). Differences in hydrolysis and binding of juvenile hormones in *Locusta migratoria* hemolymph. *Z. Naturf. 34C*, 588–598.

PETERSON, R. (1977). Juvenile hormone carrier protein from *Manduca sexta* (L.): An improved purification, determination of binding specificity and selective modification. Ph.D. dissertation, University of Chicago.

PETERSON, R. C., REICH, M. F., DUNN, P. E., LAW, J. H. and KATZENELLENBOGEN, J. A. (1977). Binding specificity of the juvenile hormone carrier protein from the hemolymph of the tobacco hornworm, *Manduca sexta* Johannson (Lepidoptera: Sphingidae). *Biochemistry 16*, 2305–2311.

PRATT, G. E. (1975). Inhibition of juvenile hormone carboxyesterase of locust haemolymph by organophosphates *in vitro*. *Insect Biochem. 5*, 595–607.

REDDY, G., HWANG-HSU, K. and KUMARAN, A. K. (1979). Factors influencing juvenile hormone esterase activity in the wax moth, *Galleria mellonella. J. Insect Physiol. 25*, 65–71.

REDDY, G., MCCALEB, D. C. and KUMARAN, A. K. (1980). Tissue distribution of juvenile hormone hydrolytic activity in *Galleria mellonella* larvae. *Experientia 36*, 461–462.

RETNAKARAN, A. and JOLY, P. (1976). Neurosecretory control of juvenile hormone inactivation in *Locusta migratoria* (L.). In *Actualitiés sur Les Hormones D'Invertebrés*. Pages 317–323. Colloques Internationaux CNRS No. 251, Paris.

RICHARDS, G. (1981). Insect hormones in development. *Biol. Rev. 56*, 501–549.

ROE, R. M., HAMMOND, A. M. Jr. and SPARKS, T. C. (1982). Characterization of the plasma juvenile hormone esterase in synchronous last stadium female larvae of the sugar cane borer, *Diatraea saccharalis. J. Insect Physiol. 13*, 163–170.

RÖSELER, P. F. and RÖSELER, I. (1978). Studies on the regulation of the juvenile hormone titre in bumblebee workers, *Bombus terrestris. J. Insect Physiol. 24*, 707–713.

ROTIN, D., FEYEREISEN, R., KOENER, J. and TOBE, S. S. (1982). Haemolymph juvenile hormone esterase activity during the reproductive cycle of the viviparous cockroach, *Diploptera punctata Insect Biochem. 12*, 263–268.

RUDNICKA, M. and HAMMOCK, B. D. (1981a). Approaches to the purification of the juvenile hormone esterase from the cabbage looper, *Trichoplusia ni. Insect Biochem. 11*, 437–444.

RUDNICKA, M. and HAMMOCK, B. D. (1981b). Purification of juvenile hormone esterase from *Trichoplusia ni*. In *Regulation of Insect Development and Behavior*. Edited by F. Sehnal, A. Zabża, J. J. Menn and B. Cymborowski. Page 217. Wrocław Technical University Press, Wrocław, Poland.

RUDNICKA, M., SEHNAL, F., JAROLÍM, V. and KOCHMAN, M. (1979). Hydrolysis and binding of the juvenile hormone in the haemolymph of *Galleria mellonella. Insect Biochem. 9*, 569–575.

SAMS, G. R., COCCHIARO, G. F. and BELL, W. J. (1978). Metabolism of juvenile hormone in cultures of ovaries and fat body in the cockroach *Periplaneta americana. In Vitro 14*, 956–960.

SANBURG, L., KRAMER, K., KÉZDY, F. and LAW, J. H. (1975a). Juvenile hormone specific esterases in the haemolymph of the tobacco hornworm, *Manduca sexta. J. Insect Phsyiol. 21*, 873–888.

SANBURG, L. L., KRAMER, K. J., KÉZDY, F. J., LAW, J. H. and OBERLANDER, H. (1975b). Role of juvenile hormone esterases and carrier proteins in insect development. *Nature, London 253*, 266–267.

SCHARRER, B. (1976). Neurosecretion — Comparative and Evolutionary Aspects. In *Perspectives in Brain Research*. Edited by M. A. Corner and D. F. Swaab. *Progress in Brain Research*. Vol. 45, pages 125–137. Elsevier, Amsterdam.

SCHARRER, B. (1978). Peptidergic neurons: Facts and trends. *Gen. Comp. Endocr. 34*, 50–62.

SCHNEIDERMAN, H. A. (1971). The strategy of controlling insect pests with growth regulators. *Mitt. Schweiz Ent. Ges. 44*, 141–150.

SCHNEIDERMAN, H. A. (1972). Insect Hormone and Insect Control. In *Insect Juvenile Hormones: Chemistry and Action*. Edited by J. J. Menn and M. Beroza. Pages 3–27. Academic Press, New York.

SCHOOLEY, D. A., BERGOT, B. J., GOODMAN, W. and GILBERT, L. I. (1978). Synthesis of both optical isomers of insect juvenile hormone III and their affinity for the juvenile hormone-specific binding protein of *Manduca sexta. Biochem. Biophys. Res. Commun. 81*, 743–749.

SIDDALL, J. B. (1976). Insect growth regulators and insect control: A critical appraisal. *Environ. Health Perspect. 14*, 119–126.

SLADE, M. and WILKINSON, C. F. (1973). Juvenile hormone analogs: a possible case of mistaken identity? *Science 181*, 672–674.

SLADE, M. and WILKINSON, C. F. (1974). Degradation and conjugation of cecropia juvenile hormone by the southern armyworm (*Prodenia eridania*). *Comp. Biochem. Physiol. 49B*, 99–103.

SLADE, M. and ZIBITT, C. H. (1971). Metabolism of cecropia juvenile hormone in lepidopterans. In *Proceedings of the Second International IUPAC Congress of Pesticide Chemistry*. Edited by A. S. Tahori. *Chemical Releases in Insects*. Vol. 3. Gordon & Breach.

SLADE, M. and ZIBITT, C. H. (1972). Metabolism of cecropia juvenile hormone in insects and mammals. In *Insect Juvenile Hormones: Chemistry and Action*. Edited by J. J. Menn and M. Beroza. Pages 155–176. Academic Press, New York.

SLADE, M., BROOKS, G. T., HETNARSKI, H. K. and WILKINSON, C. F. (1975). Inhibition of the enzymatic hydration of the epoxide HEOM in insects. *Pestic. Biochem. Physiol. 5*, 35–46.

SLADE, M., HETNARSKI, H. K. and WILKINSON, C. F. (1976). Epoxide hydrase activity and its relationship to development in the southern armyworm, *Prodenia eridania*. *J. Insect Physiol.* *22*, 619–622.

SLÁMA, K. and JAROLÍM, V. (1980). Fluorimetric method for the determination of juvenoid esterase activity in insects. *Insect Biochem.* *10*, 73–80.

SLÁMA, K. and JAROLÍM, V. (1981). Hydrolysis of 2(*E*) and 2(*Z*) isomers of juvenoid esters related to their juvenile hormone activity in insects. In *Juvenile Hormone Biochemisry*. Edited by G. E. Pratt and G. T. Brooks. Pages 185–192. Elsevier, New York.

SOLOMON, K. R. and METCALF, R. L. (1974). The effect of piperonyl butoxide and triorthocresyl phosphate on the activity and metabolism of altosid (isopropyl 11-methoxy-3,7,11-trimethyldodeca-2,4-dienoate) in *Tenebrio molitor* L. and *Oncopeltus fasciatus* (Dallas). *Pestic. Biochem. Physiol.* *4*, 127–134.

SPARKS, T. C. and HAMMOCK, B. D. (1979a). A comparison of the induced and naturally occurring juvenile hormone esterases from last instar larvae of *Trichoplusia ni*. *Insect Biochem.* *9*, 411–421.

SPARKS, T. C. and HAMMOCK, B. D. (1979b). Induction and regulation of juvenile hormone esterases during the last larval instar of the cabbage looper, *Trichoplusia ni*. *J. Insect Physiol.* *25*, 551–560.

SPARKS, T. C. and HAMMOCK, B. D. (1980). Comparative inhibition of the juvenile hormone esterases from *Trichoplusia ni*. *Tenebrio molitor* and *Musca domestica*. *Pestic. Biochem. Physiol.* *14*, 290–302.

SPARKS, T. C. and HAMMOCK, B. D. (1982). Insect Growth Regulators: Resistance and the Future. In *Pest Resistance to Pesticides: Challenges and Prospects*. Edited by G. P. Georghiou and T. Saito. Plenum Press, New York. (In press.)

SPARKS, T. C., WING, K. D. and HAMMOCK, B. D. (1979a). Effects of the anti hormone-hormone mimic ETB on the induction of insect juvenile hormone esterase in *Trichoplusia ni*. *Life Sci.* *25*, 445–450.

SPARKS, T. C., WILLIS, W. S., SHOREY, H. H. and HAMMOCK, B. D. (1979b). Hemolymph juvenile hormone esterase activity in synchronous last instar larvae of the cabbage looper, *Trichoplusia ni*. *J. Insect Physiol.* *25*, 125–132.

SPARKS, T. C., HAMMOCK, B. D. and RIDDIFORD, L. M. (1983). The haemolymph juvenile hormone esterase of *Manduca sexta* (L.): inhibition and regulation. *Insect Biochem.* *13*, 529–541.

SZIBBO, C. M., ROTIN, D., FEYEREISEN, R. and TOBE, S. S. (1982). Synthesis and degradation of C_{16} juvenile hormone (JH III) during the final two stadia of the cockroach, *Diploptera punctata*. *Gen. Comp. Endocr.* *48*, 25–32.

TERRIERE, L. C. and YU, S. J. (1977). Juvenile hormone analogs: *in vitro* metabolism in relation to biological activity in blow flies and flesh flies. *Pestic. Biochem. Physiol.* *7*, 161–168.

TRAUTMANN, K. H. (1972). In vitro study of the carrier proteins of ^{3}H-labelled juvenile hormone active compounds in the haemolymph of *Tenebrio molitor* L. larvae. *Z. Naturf.* *27B*, 263–273.

VINCE, R. K. and GILBERT, L. I. (1977). Juvenile hormone esterase activity in precisely timed last instar larvae and pharate pupae of *Manduca sexta*. *Insect Biochem.* *7*, 115–120.

VINSON, S. B. and PLAPP, F. W. (1974). Third generation pesticides: The potential for the development of resistance by insects. *J. Agric. Food Chem.* *22*, 356–360.

WEIRICH, G. and WREN, J. (1973). The substrate specificity of juvenile hormone esterase from *Manduca sexta* haemolymph. *Life Sci.* *13*, 213–226.

WEIRICH, G. and WREN, J. (1976). Juvenile-hormone esterase in insect development: a comparative study. *Physiol. Zool.* *49*, 341–350.

WEIRICH, G. F. and CULVER, M. G. (1979). S-adenosylmethionine: juvenile hormone acid methyltransferase in male accessory reproductive glands of *Hyalophora cecropia* (L.). *Arch. Biochem. Biophys.* *198*, 175–181.

WEIRICH, G., WREN, J. and SIDDALL, J. B. (1973). Developmental changes of the juvenile hormone esterase activity in haemolymph of the tobacco hornworm, *Manduca sexta*. *Insect Biochem.* *3*, 397–407.

WHITE, A. F. (1972). Metabolism of the juvenile hormone analogue methyl farnesoate-10,11-epoxide in two insect species. *Life Sci.* *11*, 201–211.

WHITMORE, E. and GILBERT, L. I. (1972). Haemolymph lipoprotein transport of juvenile hormone. *J. Insect Physiol.* *18*, 1153–1167.

WHITMORE, E. and GILBERT, L. I. (1974). Haemolymph proteins and lipoproteins in Lepidoptera — a comparative electrophoretic study. *Comp. Biochem. Physiol.* *47B*, 63–78.

WHITMORE, D., Jr., WHITMORE, E. and GILBERT, L. I. (1972). Juvenile hormone induction of esterases: A mechanism for the regulation of juvenile hormone titer. *Proc. Natl. Acad. Sci. USA* *69*, 1592–1595.

WHITMORE, D., Jr., GILBERT, L. and ITTYCHERIAH, P. I. (1974). The origin of hemolymph carboxylesterases 'induced' by the insect juvenile hormone. *Mol. Cell. Endocr.* *1*, 37–54.

DE WILDE, J., STAAL, G. B., DE KORT, C. A. D., DE LOOF, A. and BAARD, G. (1968). Juvenile hormone titer in the haemolymph as a function of photoperiodic treatment in the adult Colorado beetle (*Leptinotarsa decemlineata* Say). *Proc. K. Ned. Akad. Wet. Ser. C.* *71*, 321–326.

WILLIAMS, C. M. (1967). Third-generation pesticides. *Sci. Am.* *217*, 13–17.

WILSON, T. G. and GILBERT, L. I. (1978). Metabolism of juvenile hormone I in *Drosophila melanogaster*. *Comp. Biochem. Phsyiol.* *60A*, 85–89.

WING, K. D. (1981). Juvenile hormone esterases of *Trichoplusia ni* and other Lepidoptera: characteristics, site of synthesis, interaction with other proteins and inhibition. Ph.D. dissertation. University of California at Riverside.

WING, K. D., SPARKS, T. C., LOVELL, V. M., LEVINSON, S. O. and HAMMOCK, B. D. (1981). The distribution of juvenile hormone esterase and its interrelationship with other proteins influencing juvenile hormone metabolism in the cabbage looper, *Trichoplusia ni*. *Insect Biochem.* *11*, 473–485.

WING, K. D., RUDNICKA, M., JONES, G., JONES, D. and HAMMOCK, B. D. (1982). Juvenile hormone esterases of Lepidoptera II. Isoelectric points and binding affinities of hemolymph juvenile hormone esterase and binding protein activities. *J. Comp. Physiol.* (Accepted.)

YANG, R. S. H. (1976). Enzymatic conjugation and insecticide metabolism. In *Insecticide Biochemistry and Physiology*. Edited by C. F. Wilkinson. Pages 177–228. Plenum Press, New York and London.

YU, S. J. and TERRIERE, L. C. (1975). Microsomal metabolism of juvenile hormone analogs in the house fly, *Musca domestica* L. *Pestic. Biochem. Physiol.* *5*, 418–430.

YU, S. J. and TERRIERE, L. C. (1978a). Metabolism of juvenile hormone I by microsomal oxidases, esterase, and epoxide hydrase of *Musca domestica* and some comparisons with *Phormia regina* and *Sarcophaga bullata*. *Pestic. Biochem. Physiol.* *9*, 237–246.

YU, S. J. and TERRIERE, L. C. (1978b). Juvenile hormone epoxide hydrase in house flies, flesh flies and blow flies. *Insect Biochem.* *8*, 349–352.

YUHAS, D. A., ROE, R. M., SPARKS, T. C. and HAMMOND, A. M., Jr. (1983). Purification and kinetics of juvenile hormone esterase from the cabbage looper, *Trichoplusia ni* (Hübner). *Insect Biochem.* *13*, 129–136.

13 Ecdysteroid Receptors and Binding Proteins

MARY ALICE YUND and DAVID L. OSTERBUR

University of California, Berkeley, California, USA

1 INTRODUCTION

Insect steroid hormones, the ecdysteroids, mediate or coordinate many processes in the insect life cycle including molting and cuticle production, metamorphosis and vitellogenesis. The responding tissues, termed target tissues, detect the presence of physiological concentrations of hormone and modulate internal cellular activity in response. The specialized molecules which have evolved for detection of hormone and transduction of the signal in the target tissue are called receptors. The key to the study of hormone action and response at the cellular and molecular level is the receptor. The receptor can be identified experimentally by its ability to bind hormone with properties appropriate for the observed hormone response. An investigation of the properties and function of the receptor should lead to identification of other components of the hormone response mechanism in the target tissue.

In this chapter we will be examining the properties of receptors for ecdysteroids and the role of receptors in hormone action. Much of the work in the molecular endocrinology of ecdysteroid receptors has utilized the most popular insect for genetics and molecular biology, *Drosophila melanogaster*. For this reason, and because our own research utilizes *Drosophila*, the examples used and the results presented are derived from *Drosophila*. Consideration of ecdysteroid receptors will be divided into three sections:

(1) The role of hormone receptors in hormonal regulation and coordination of cell activities during development and in homeostasis.
(2) The experimental techniques, criteria and problems in identifying and characterizing hormone receptors and receptor function.
(3) The experimental investigation of ecdysteroid receptors and comparison to a generalized scheme of vertebrate steroid hormone action.

2 THE ROLE OF RECEPTOR IN HORMONE RESPONSE IN TARGET TISSUES

2.1 The hormone–receptor complex

The first step in hormone detection and response by a target tissue is the interaction, or binding, of the hormone to the receptor. In all cases examined the receptor has been found to be a protein. The receptor recognizes and binds the hormone at a specific active site analogous to the catalytic site on an enzyme. Binding of the hormone to the receptor results in changes in the receptor and its associations in the cell. The modified receptor or the hormone–receptor complex interacts with other cellular components to initiate the response preprogrammed, or determined, for that cell. In the best-described mechanism of ecdysteroid and vertebrate steroid hormone action, binding of hormone–receptor complex in the nucleus results in changes in transcription of specific genes. However, the complexity of hormone response must involve more than one molecular mechanism. For example, there are hormone-dependent events, such as permeability changes, which occur too soon after hormone exposure to be mediated via changes in transcription (e.g. Raikow, R. and Fristrom, J., 1971).

2.2 Tissue sensitivity and sequential hormone action

Only those cells which have receptors are competent to respond to hormone. The presence and number of receptors in a cell is crucial to hormone response. Little is known about the onset of receptor synthesis during development or the rate of synthesis and turnover of receptors in the cell. The fate of receptors in hormone response is also unknown. Receptors may be recycled and reused after binding hormone and initiating the response. Alternatively, hormone action may in some cases require a period of reduced sensitivity to hormone. Inactivating receptor or removing receptors from potential hormone-binding locations in the cell, termed down-regulation of receptors, would be one way of isolating the cell from additional hormonal stimulation. Subsequent synthesis of receptor would be necessary to restore hormone responsiveness to the cell. In some cases the continuous presence of the hormone, and presumably the hormone–receptor complex, is required to sustain the hormone response. In other cases the hormone initiates a sequence of events which continues after the hormone is withdrawn or proceeds only in the absence of hormone (Fristrom, J. and Yund, M., 1980; Fristrom, J. *et al.*, 1982; Doctor, J. and Fristrom, J., volume 2).

2.3 Tissue specificity of the hormone response

Because the response to the hormone is different in different tissues and at different developmental stages, the specificity of the response must lie in the target tissue, not the hormone. How much of the tissue-specific response is inherent in the receptor remains to be determined. There are two alternative hypotheses. There may be only one receptor protein — one gene and one gene product — which is the same in all target tissues. Differences in the response would lie in the other intercellular components or in the chromatin-binding sites which are accessible to the receptor or hormone–receptor complex in that tissue. These sites for hormone–receptor complex interaction would be different in different cell types. The extreme alternative hypothesis is that a specific receptor — a separate gene and gene product — exists for each cell type or class of hormone response. Each receptor would have both a hormone-binding site and a binding site for a particular chromatin or DNA conformation or other intracellular molecular structure. The hormone–binding domains would be similar or identical for all receptors, in agreement with experimental observations. The other binding domains would be gene-specific and would specify a particular set of hormone responses. A given cell type, e.g. fat body, might express the fat body receptor or, alternatively, some subset of the total family of receptor genes corresponding to the spectrum of ecdysteroid responses in fat body cells.

Understanding the mechanism of ecdysteroid action will require a detailed molecular description of the structure and function of ecdysteroid receptors from several tissues. Formidable as this task may appear, the interaction of hormone with the receptor is only the first step in hormone response. A great deal remains to be learned about hormone action.

3 EXPERIMENTAL INVESTIGATION OF HORMONE RECEPTORS

3.1 Techniques for detecting the receptor and the hormone–receptor complex

Three general methods have been used for the identification of hormone receptors:

(1) binding of radiolabeled hormone or hormone analog to the receptor;
(2) photoaffinity labeling of the receptor; covalently linking a fluorescent or radiolabeled ligand to the receptor; and
(3) binding of anti-receptor antibody to receptor.

Of these three the most widely used method is binding of radiolabeled ligand to receptor in the target tissue or a target tissue extract. This experimental approach requires a radiolabeled hormone and samples of target tissue. Most of the basic information about hormone receptors and hormone–receptor interaction has been obtained using radiolabeled ligands. This assay for the receptor depends on the ability of the receptor to bind and retain or to rebind hormone. Any condition or procedure not conducive to hormone–receptor binding invalidates the assay and makes identification of the receptor impossible.

Because the radiolabeled hormone binding is too labile for some manipulations, photoaffinity labeling techniques have been developed for several hormone–receptor systems. A target tissue or tissue extract is incubated with a radiolabeled hormone analog having a photoactive group and then exposed to light of the photoactive wavelength. The activated ligand forms a covalent bond with the receptor, permanently attaching the radiolabeled ligand to the receptor. A photoaffinity procedure applied to *Drosophila* target tissues uses underivatized ecdysteroids. Ecdysteroids contain an α, β-unsaturated ketone which can be photoactivated at 330 nm to form a carbene radical. This radical will form covalent bonds to any molecule it contacts. Presumably, under the appropriate conditions, some portion of these would be the receptor. Complexed hormone is identified by radiolabel on the hormone molecule or by antibody to the hormone. Because the efficiency of photoaffinity labeling is generally of the order of 5% or less, and because the hormone is covalently linked (a nonphysiological condition), photoaffinity labeling is not a general replacement for radiolabeled ligand binding assay in determining receptor properties.

Production of antibody to receptor requires extensive information about the properties of the receptor, and both a purified or partially purified receptor and a means of assaying the antibody for ability to bind receptor. The major advantage of an antibody assay is that it allows identification of the receptor by a criterion other than a functional hormone-binding site. An antibody is useful in situations in which the receptor protein might no longer be in its native active state, or for situations in which a radioligand binding assay is not feasible or not sensitive enough, e.g. cytological localization of receptor. Antibody affinity columns have been used to purify receptor from crude extract (see e.g. Fraser, C. and Venter, J., 1980). In insects, because of their small size, the sensitive antibody technique would be very useful in study of the ontogeny of the receptor and the distribution of receptor in the tissues of the insect.

3.1.1 HORMONE-BINDING ASSAYS

Most binding assays use a radiolabeled ligand and rely on identification of the hormone–receptor complex by detecting ligand radioactivity. The most common fractions analyzed for binding activity are a high-speed supernatant (100,000 ***g***, 1 h) of the whole cell homogenate (often referred to as cytosol), a moderate salt (0.3–0.4 M KCl) extract of isolated nuclei, or, less commonly, isolated nuclei. In a typical binding assay a radiolabeled ligand–extract mixture is incubated under conditions which allow the hormone–receptor complex to form and reach its equilibrium concentration. The hormone bound to macromolecules is separated from free hormone, and the amount of radioactivity in the bound fraction is determined. Alternatively, radiolabeled ligand may be injected into the organism or incubated with explanted target tissue *in vitro*. In this instance the target tissue extract when prepared would contain the radiolabeled hormone–receptor complex. The presence of endogenous hormone or other molecules capable of interacting with the receptor may prevent accurate quantitation and characterization of the receptor.

Separation of bound and free hormone to allow identification and quantitation of receptor has been accomplished by several methods. A standard method is sucrose gradient analysis (see Fig. 1). The assay mixture is loaded onto the top of a sucrose gradient. In the ultracentrifuge the hormone–receptor complex sediments more rapidly, moving ahead of the free hormone. A radioactivity profile across the gradient locates the hormone–receptor complex. In addition to quantifying the hormone–receptor complex in the assay mixture, the gradient peak indicates that the hormone is binding a discrete molecular species, not a spectrum of macromolecules. With appropriate standards the sedimentation coefficient (S) of the hormone–receptor complex can be determined. Other separation strategies include precipitating the free hormone (e.g. with dextran-coated charcoal), precipitating the macromolecules including the hormone–receptor complex (e.g. with protamine sulfate or hydroxylapatite) or separating macromolecules from small molecules by size exclusion chromatography.

3.1.2 Choosing a ligand for hormone-binding assays

A successful binding assay based on the affinity of

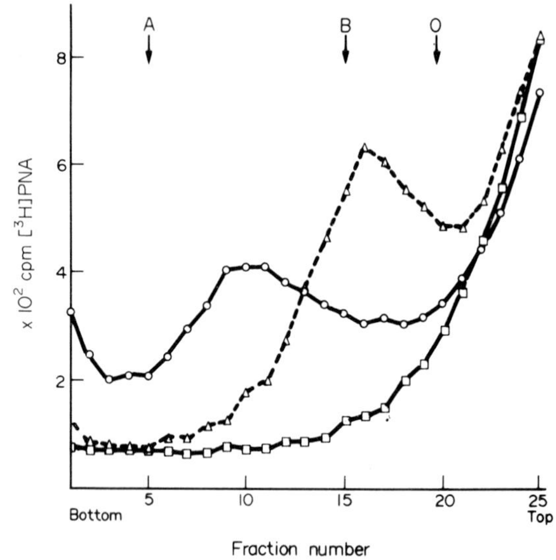

FIG. 1. Sucrose density gradient centrifugation of ecdysteroid receptors from imaginal discs. A 0.3 M KCl extract of imaginal disc nuclei prepared as described in Yund, M. *et al.* (1978) was dialyzed against 0.01 M Tris, 0.015 M EDTA, pH 7.4, preincubated with 3.2×10^{-9} M [³H]ponasterone A, and run in a 5–20% linear sucrose gradient containing either 0.1 M KCl (○–○) or 0.3 M KCl (△–△). A nuclear extract preincubated with 3.2×10^{-9} M [³H]ponasterone A plus 2×10^{-7} M 20-hydroxyecdysone as unlabeled competitor was run on a 0.3 M KCl 5–20% linear sucrose gradient (□–□). Sedimentation standards were aldolase (A, 7.35S), bovine serum albumin (B, 4.35S) and ovalbumin (O, 3.6S) (Osterbur, D. and Yund, M. 1982).

hormone for receptor depends upon a radioligand with the appropriate properties.

(1) The ligand must have high enough affinity for the receptor to permit separation of bound and free ligand in a binding assay. Characterization of the receptor requires a ligand–receptor complex which is stable during experimental manipulations.

(2) The ligand must have high enough specific activity to bind detectable amounts of labeled ligand to the receptor in an aliquot of tissue extract of convenient assay size (0.1–1.0 ml). A simplified example may illustrate this point. Suppose that the aliquot of target tissue extract contains 10^{12} receptors. The maximum number of molecules which can be bound by this extract is 10^{12}, no matter how much ligand might be added. If the specific activity of the ligand is such that 1 molecule in 10^{12} decays per minute, the maximum radioactivity bound to receptor would be 1 cpm. If the specific activity of

the ligand corresponds to a decay of 1 mole ule in 10^9 per minute, the radioactivity bound at saturation would be approximately 10^3 cpm, a level reliably and conveniently detectable in routine assays. Because receptors are a very small percentage of the total protein of a cell, the limitation of protein solubility makes it impractical to concentrate the extract significantly to increase receptor concentration. Synthesizing a high specific activity ligand is often easier than manipulating the receptor, particularly in the early stages of an investigation when little is known about the properties of the receptor.

(3) The ligand must be stable during the assay. It is necessary to verify that the radioactivity bound to the putative receptor is the original ligand, i.e. that no metabolism has occurred. This is particularly important if radioligand has been injected into whole organisms or incubated with tissues or tissue extracts at metabolic temperatures. Both radiolabeled material from the entire extract and that bound to the putative receptor must be the authentic radioligand. If the material bound to the putative receptor is not the hormone, it must be established that the active hormone molecule in the target tissue is a metabolite of the circulating hormone. If the material in the extract is largely metabolite but unmetabolized ligand is bound, the ligand binder may in fact be the receptor. However, because experimental characterization of the receptor requires equilibrium binding conditions with a defined concentration of ligand, it may not be possible to demonstrate that the binder is in fact the receptor or to determine the properties of the receptor from a target tissue of high metabolic flux.

(4) Radioligand bound to receptor must be reliably distinguished from the remainder of the radioactivity in the assay. Radioactivity associated with macromolecular material must be clearly separated from free ligand in solution such that cross-contamination does not lead to inaccurate quantitation of receptor. In addition the ligand bound to the receptor, termed specific binding, must be separated from ligand associated with other macromolecules, termed non-specific binding. It is generally believed that non-specific binding of ligand is due to weak and transitory association with many proteins in the extract. The standard way of measuring specific binding to the receptor is by competition experiments with non-radioactive ligand. The rationale for this procedure is that there are a fixed number of receptors in a given sample. If the specific activity of the ligand in the assay is reduced by the addition of non-labeled ligand, the amount of radioactivity bound to the receptor will be reduced accordingly. In contrast, there are an unlimited number of non-specific association or binding sites relative to the solubility limit of the ligand. If the concentration of non-radioactive ligand is increased, binding sites with weaker affinity are recruited. The result is more total ligand molecules bound non-specifically with no reduction in non-specifically bound radioactivity. As a general rule unlabeled ligand in 100-fold excess of the concentration required to saturate the receptors will fully compete with physiologically effective concentrations of ligand reducing binding of radiolabeled ligand to receptor to zero. Any residual radioactivity in the macromolecular fraction is attributed to non-specific binding and subtracted from the total binding in the absence of unlabeled competitor to determine ligand bound to receptor.

Accurate characterization of the receptor requires that non-specific binding be less than 5–10% of the total radioactivity associated with the macromolecular fraction. Experimental analysis of hormone–receptor interaction requires comparing the relative binding of radioligand to receptor in a defined set of conditions, e.g. at various times after addition of hormone. This in turn requires accurate determination of the level of non-specific association. The level of non-specific binding can usually be determined to an accuracy of $\pm$ 10–20%. If the level of non-specific binding is 5%, for every 100 cpm in the macromolecular fraction 5 cpm $\pm$ 0.5–1 cpm are non-specifically bound. The remaining 94–96 cpm are specifically bound, an error of about 2% in determining the level of specific binding. If the level of non-specific binding is 50%, for every 100 cpm in the macromolecular fraction 50 cpm $\pm$ 5–10 cpm are non-specifically bound. The remaining 40–60 cpm are specifically bound, an error of about 20% in determining the level of specific binding. With this range of variability in the data, detecting differences in receptor occupancy in an experimental series would be difficult.

Increasing the specific activity of the ligand and thus the levels of both specifically and non-specifically bound radioactivity at a given ligand concentration does little to alleviate this problem. If the postulated basis of non-specific binding is correct, lowering the ligand concentration should reduce the level of non-specific binding. However, as will be discussed below, it is necessary to study hormone–receptor interactions at biologically active ligand concentrations in order to have significant amounts of ligand bound to receptor (high receptor occupancy). In general, the only solution to the problem of non-specific binding is to do enough replicates of the experiment for statistical analysis or to use a higher-affinity ligand which would permit the assays to be carried out at a lower ligand concentration with corresponding lower levels of non-specific binding.

The active hormone in *Drosophila*, 20-hydroxyecdysone, is of limited usefulness in characterizing the ecdysteroid receptor. Radiolabeled 20-hydroxyecdysone has been biosynthesized from ecdysone with a specific activity of 50–60 Ci $mmol^{-1}$. However, for *Drosophila melanogaster* imaginal discs at hormone concentrations in the physiological range there are high and variable levels of non-specific binding comprising 50% of the total retained radioactivity (Yund, M. and Fristrom, J., 1975; Yund, M., 1979a). Furthermore, the 20-hydroxyecdysone–receptor complex has a high dissociation rate resulting in a half-life of 15–20 min at 25° and 6–8 h at 0–4°. Most of the work on *Drosophila* ecdysteroid receptors has employed a high specific activity form of the ecdysteroid ponasterone A,, first synthesized by King (Yund, M. *et al.*, 1978). Ponasterone A is one of the few known ecdysteroids with a higher biological activity than 20-hydroxyecdysone. This higher biological activity corresponds to a higher affinity for the receptor and a more stable ligand–receptor complex at 0–4°. At ligand concentrations which induce biological response, corresponding to 50–95% receptor occupancy, non-specific binding is negligible.

3.2 Biochemical criteria for identifying receptor

Once assay conditions have been determined to measure reliably specific, competitive binding of hormone to macromolecular components in extract from a target tissue, experiments can be carried out to determine whether this binding does in fact correspond to the hormone receptor, and then to characterize the receptor. This is not a trivial requirement. Other molecules which bind the hormone with moderate to high affinity may be present in the sample of target tissue. Such binding may be fortuitous or may have a physiological function independent of the hormone response under investigation. For example, enzymes involved in synthesis or degradation of ecdysteroids might bind ecdysteroids with reasonable affinity. Embryonic vitellins bind ecdysteroids (Lagueux, M. *et al.*, 1981; Hoffmann, J. and Lagueux, M., vol. 1). This binding is probably important in regulation of ecdysteroid titer during embryogenesis. There is some evidence for ecdysteroid binding in hemolymph of *Locusta migratoria* (Feyereisen, R., 1980). Circulating carrier proteins for other hormones are believed to regulate the concentration of hormone, protect the hormone from degradation or facilitate uptake by target tissues (Goodman, W. and Chang, E., this volume). If such molecules exist for ecdysteroids, they might contaminate the target tissue. However, in most cases the details of the interaction of such molecules with ecdysteroids should be distinguishable from those of the receptor mediating the hormone response.

The criteria for demonstrating that a high-affinity hormone binder from a target tissue is indeed the receptor are somewhat open-ended. The ultimate proof would be for the purified putative receptor to mediate correctly a hormone response such as modulation of transcription of specific genes in a reconstituted system *in vitro*. This has yet to be accomplished in any system, probably due more to our ignorance of transcription and other molecular events under hormonal control than to misidentification of cellular receptors. There are certain properties and permitted parameters for kinetics and binding affinities which a receptor must have to mediate hormone response in a particular tissue. When a list of these criteria has been satisfied, it is generally accepted that the molecular species in question is the *bona fide* receptor.

(1) The receptor should be present in all target cells. It is generally the case, but not a logical necessity, that non-target tissues contain little if any receptor-like molecules.

(2) The receptor can mediate hormone response only if it can detect and respond to hormone titer changes in the physiological range. This puts specific constraints upon the kinetic parameters of hormone–receptor binding. The receptor will be most sensitive to changes in hormone titer at hormone concentrations within one order of magnitude of the equilibrium dissociation constant (K_D) of hormone–receptor binding. At concentrations of hormone equal to the K_D of binding, roughly 50% of the receptors would be occupied by hormone, i.e. 50% of receptors are converted to hormone–receptor complex. At one order of magnitude below this concentration, only about 10% of receptors are occupied. At a hormone concentration one order of magnitude above the K_D of binding about 90% of the receptors are occupied. Increasing hormone concentration by yet another order of magnitude raises the receptor occupancy to only 99%. Thus a hormone receptor is a much more effective and sensitive transducer of the hormone signal when the K_D of binding to hormone is of the same order of magnitude as the physiological concentrations of hormone mediating the response. One would expect that evolution would lead to the coordination of hormone titer and receptor affinity. This has proven to be the case, within experimental error in *in vitro* assay systems, for all hormones examined.

(3) The receptor should have a low capacity for binding hormone. This follows from the expectation that there are a relatively small number of receptors per cell. Saturation of binding should be at hormone concentrations about one or two orders of magnitude above the equilibrium dissociation constant (K_D) of the hormone–receptor binding.

(4) The relative binding affinities of receptor for the hormone and hormone analogs should reflect the relative effectiveness of these molecules in inducing the hormone response. As was discussed above, for a molecule to induce hormone response at a given hormone concentration it must have an affinity for the receptor (K_D for formation of hormone–receptor complex) within an order of magnitude of the effective hormone concentration. A poorer inducer of hormone response would be expected to have lower affinity for the receptor. The simplest hypothesis consistent with present evidence is that the formation of the hormone–receptor complex initiates the response. If this is the case, an exact correspondence between effectiveness in inducing hormone response and strength of binding to receptor would be observed. The minimum requirement is that the receptor has a strong affinity for all active hormone analogs. If hormone–receptor complex formation is not the limiting factor in response, analogs with relatively high affinity for receptor might be poor inducers because of failure to act effectively at another step in hormone response.

(5) The kinetics of hormone–receptor interaction must be consistent with the biological and physiological conditions of hormone response. For example, a putative receptor which requires 30 min to reach its maximum hormone binding is not a good candidate to mediate a hormone effect such as a change in membrane permeability which reaches a maximum in 5 min. Another factor to be considered is the association and dissociation rates which together determine the K_D. The rate at which the receptor binds and releases hormone must be consistent with the physiology and changing hormone titer in the organism at the time of the hormone response. For example, if the dissociation rate of the hormone–receptor complex were such that the complex had a half-life of 24 h at physiological temperature, this receptor system could not easily detect and respond to hormone pulses 8–12 h apart.

In summary, binding of hormone to a macromolecular fraction of a target tissue is not sufficient to identify a receptor. Extensive biochemical characterization of the putative receptor is required. The properties of the putative receptor must conform to the known biology and physiology of the hormone and the target tissue, as well as to the properties inferred from generalizations about molecular mechanisms.

4 ECDYSTEROID RECEPTORS

4.1 Early work

The first observations suggesting that steroid hormones play a role in gene regulation were made in insects. Clever, U. and Karlson, P. (1960) observed a correlation between elevated ecdysteroid titer and transcriptional activity in polytene chromosomes as indicated by changes in chromosome puffing (see Wigglesworth, V., this volume). Early attempts to identify ecdysteroid binding in insects were reported for hemolymph from *Drosophila hydei* (Emmerich, H., 1970a; Butterworth, F. and Berendes, H., 1974). *Pyrrhocoris apterus* (Emmerich, H. 1970b), and *Calliphora erythrocephala* (Thamer, G. and Karlson, P., 1972), for salivary glands from *D. hydei* (Emmerich, H., 1972), for tissue culture cells from *Drosophila melanogaster* (Best-Belpomme, M. and Courgeon, A., 1975) and for *D. hydei* larvae (Butterworth, F. and Berendes, H., 1974). All these experiments used the only ligands available at the time, ecdysteroids (ecdysone and 20-hydroxyecdysone) with relatively low affinity for the receptor which were radiolabeled to a low specific activity. Because the number of total counts bound was very low, results were difficult to reproduce and to interpret (see Butterworth, F. and Berendes, H., 1974). Under these conditions it was not possible to reproducibly demonstrate specific, competitive binding, much less to determine the properties of a hormone binder. The synthesis of [^{3}H]ponasterone A with a specific activity of 122 Ci mmol^{-1} was a major step forward, making possible the demonstration of intracellular proteins having binding properties consistent with ecdysteroid receptors (Yund, M. *et al.*, 1978; Maroy, P. *et al.*, 1978).

4.2 *Drosophila* as a model system

The identification of ecdysteroid receptors and the elucidation of some of their physical and kinetic properties has depended upon both the synthesis of an appropriate radioligand and the selection of a favorable system in which to study hormone action. An appropriate system should have a quantifiable and biologically significant response to the hormone, which can be used to establish kinetic and molecular criteria for hormone–receptor interaction. A receptor mediating the response must have molecular properties consistent with the characteristics of the response. There should be little or no metabolism of the hormone. The tissue should be available in quantities sufficient for biochemical work. An ideal system would offer the opportunity of extending the work beyond the characterization of the binding phenomenon.

Because a major interest in studying steroid hormone receptors is their role in modulation of gene transcription, it is fitting that most of the work on insect ecdysteroid receptors has been done in a genetically well-characterized organism, *Drosophila melanogaster*. The resources of classical genetics and the existence of large numbers of biochemical and morphological mutants coupled with the availability of DNA clones and *in situ* hybridization to polytene chromosomes make *D. melanogaster* a good choice for the study of ecdysteroid action. Imaginal discs and the K_c line of tissue culture cells have so far provided the most information on the properties of ecdysteroid receptors. Salivary glands have been used to describe some of the physical properties of an ecdysteroid binding molecule.

4.3 Imaginal discs

Imaginal discs, small packets of cells found in the hemocoel of the *Drosophila* larva, are precursors of most of the external structures of the adult fly (Oberlander, vol. 2). The increase in ecdysteroid titer at the end of the third larval instar induces disc evagination, a complex series of biochemical and morphological changes by which discs form their respective adult structures (see Doctor and Fristrom, vol. 2). Imaginal discs *in vitro* undergo evagination upon exposure to appropriate concentrations of ecdysteroids (Fristrom, J. *et al.*, 1977; Mandaron, P. *et al.*, 1977), but do not metabolize ecdysteroids (Chihara, C. *et al.*, 1972). A technique for mass isolation of imaginal discs, first developed by Fristrom, J. and Mitchell, H. (1965), yields, with subsequent improvements, about 2 million discs per day (5×10^{10} cells or 4 g, wet weight, of tissue; Eugene, O. *et al.*, 1979) from 1 kg of larvae. The *in vitro* evagination response of imaginal discs provides an extraordinary opportunity for the study of a developmental program that is under hormonal control.

Identification of ecdysteroid receptors mediating the induction of adult differentiation in imaginal discs depended on a correlation of the properties of the hormone response and the observed properties of hormone binding. A bioassay for ecdysteroids based on the evagination response was used to determine the relative biological activities of many ecdysteroids (Chihara, C. *et al.*, 1972; Fristrom, J. and Yund, M., 1976). The dose–response data

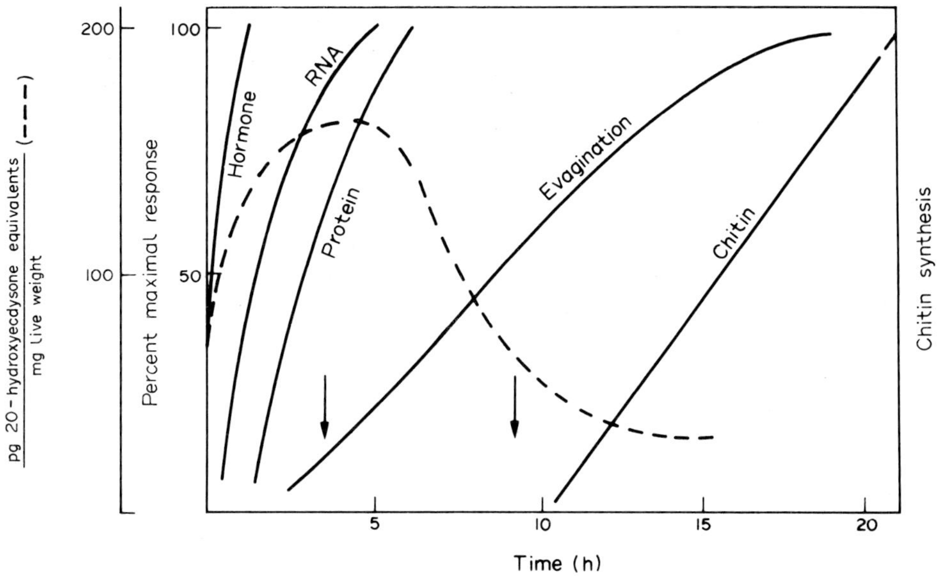

FIG. 2. The *in vitro* schedule of hormone-induced events in imaginal discs. The dotted line indicates the *in vivo* ecdysteroid titers as determined by Hodgetts, R. *et al.*, (1977). See text and the review of Doctor and Fristrom (vol. 2), for discussion of experimental conditions and review of the original data.

established molecular criteria for identifying the ecdysteroid receptor of imaginal discs and served as a standard of comparison for dose–response measurements in cell lines (O'Connor, J. *et al.*, 1980; Cherbas, P. *et al.*, 1980; O'Connor, vol. 8). Table 1 shows the concentration inducing half evagination response in imaginal discs *in vitro* for three selected ecdysteroids; the prohormone ecdysone, the active hormone 20-hydroxyecdysone and the phytoecdysteroid ponasterone A. The hormone receptor mediating the response should have corresponding affinities for these compounds.

Some of the characteristics of hormone response observed *in vitro* are summarized in Figure 2 (for review see Doctor and Fristrom, vol. 2). Hormone at concentrations sufficient to induce evagination is added at 0 time. Events observed, in chronological sequence, are specific, competible binding of hormone in the nucleus, a 2–4-fold increase in RNA synthesis (rRNA and poly A^+RNA), a 2-fold

Table 1: Drosophila *ecdysteroid receptors: analog specificity*

	Ecdysone	20-Hydroxyecdysone	Ponasterone A
Imaginal discs			
Biological activity[a]	1.5×10^{-5} M	4.2×10^{-8} M	1×10^{-9} M
Apparent K_D — cytosol[b]	7.2×10^{-5}	5.2×10^{-8}	2×10^{-9}
Apparent K_d — nuclear extract[c]	—	6.0×10^{-8}	2×10^{-9}
Embryos			
Apparent K_d — nuclear extract[d]	5.0×10^{-5}	2.0×10^{-7}	2×10^{-9}
K_c cells			
Biological activity[e]	3.3×10^{-5}	2.0×10^{-8}	1.6×10^{-9}
Apparent K_D — cytosol[f]	3.6×10^{-5}	2.4×10^{-7}	3.4×10^{-9}
Apparent K_D — nuclear extract[f]	2.7×10^{-5}	2.0×10^{-7}	4.2×10^{-9}

[a] Concentration inducing half evagination (M; Yund, M. *et al.*, 1978)
[b] Relative to 3×10^{-9} M [^{3}H]ponasterone A (M; Yund, M. *et al.*, 1978)
[c] Relative to 3×10^{-9} M [^{3}H]ponasterone A (M; Yund, M., unpublished)
[d] Relative to 3×10^{-9} M [^{3}H]ponasterone A (M; Osterbur, D. and Yund, M., 1982)
[e] Concentration inducing half morphogenetic response (M; Cherbas, L. *et al.*, 1980)
[f] Relative to 5×10^{-10} M [^{3}H]ponasterone A (M; Sage, B. *et al.*, 1982)

increase in general protein synthesis, evagination, and the synthesis of components of the pupal cuticle —chitin and pupal cuticle proteins. Analysis of the RNA and proteins made in the presence and absence of hormone reveals hormone-dependent differences in the synthesis of specific RNAs and nuclear and cytoplasmic proteins. The first arrow on Fig. 2 indicates the time at which the evagination response becomes resistant to RNA synthesis inhibition and independent of the presence of hormone. Induction of chitin synthesis at the time indicated in Fig. 2 requires the withdrawal of hormone at 5–6 h. On this hormonal regimen, the hormone-induced production of chitin becomes independent of RNA synthesis inhibition at the time indicated by the second arrow. The dotted line indicates the approximate corresponding *in vivo* 20-hydroxyecdysone titer (Hodgetts, R. *et al.*, 1977; see Riddiford, vol. 8).

From Figure 2 it is apparent that the hormone binding observed is located in the nucleus at a time consistent with mediation of the change in RNA synthesis. The hormone dependence of chitin synthesis and evagination indicate that hormone must be present for 5 h and subsequently decline in titer more than an order of magnitude to permit both evagination and cuticle synthesis to occur in the observed *in vivo* and *in vitro* time-frames. Similar hormone requirements have been found for the puffing sequence in salivary gland polytene chromosomes at this same developmental stage (see Fristrom, J. and Yund, M., 1980; Richards, vol. 2). From these observations a rapid turnover or dissociation of hormone–receptor complex would be predicted.

The ecdysteroid binding activity in imaginal discs has been investigated over several years using [^{3}H]20-hydroxyecdysone and [^{3}H]ponasterone A (Yund, M. and Fristrom, J., 1975; Yund, M. *et al.*, 1978). Early work on ecdysteroid action in imaginal discs using [^{3}H]20-hydroxyecdysone of low specific activity (3–6 Ci mmol^{-1}) demonstrated specific, competible hormone binding to intact imaginal discs. This specific binding was temperature-dependent, had analog specificity corresponding to biological activity, and had an equilibrium dissociation constant approximately equal to the hormone concentration required for induction of morphogenesis *in vitro* (Yund, M. and Fristrom, J., 1975). However, high levels of non-specific binding made further investigation of ecdysteroid action by this means impractical. The synthesis of [^{3}H]ponasterone A of high specific activity allowed the reliable detection of specific binding at lower hormone concentrations with a corresponding reduction in non-specific binding. [^{3}H]Ponasterone A has been used to identify and characterize the ecdysteroid receptor from imaginal discs (Yund, M. *et al.*, 1978).

Results from competition experiments to determine the analog specificity of *Drosophila* imaginal disc receptors for the three selected ecdysteroids are shown in Fig. 3. In these experiments aliquots of a single receptor preparation were incubated with radiolabeled ligand at a concentration equal to the K_D of binding of the ligand and receptor in the presence of varying concentrations of non-radioactive competitor ecdysteroid. Incubation was carried out under conditions allowing hormone–receptor binding to reach equilibrium, and the amount of radioactive ligand bound was determined. The amount of competing ecdysteroid required to displace 50% of the [^{3}H]ponasterone A binding at this concentration should be equivalent to the K_D for the competitor. Ecdysone has the highest apparent K_D of the three compounds tested and, therefore, has the lowest relative affinity for the receptor; ponasterone A has the smallest K_D and highest affinity for receptor. The apparent K_Ds of all three ecdysteroids are in good agreement with the concentrations of ecdysteroid necessary to induce evagination *in vitro* (Table 1).

To measure the association rate constant of the hormone–receptor complex, aliquots of a receptor preparation and radioligand in concentrations equal to the K_D of receptor binding are incubated for varying times after addition of hormone, and the specific, competitive binding to the receptor determined. The association rate constant (k_a) is calculated from the linear portion of the association curve (Figure 4; Table 2). The value of the association rate constant is an indication of the affinity of the ligand for receptor: the larger the k_a, the higher the affinity. The dissociation rate is measured as a turnover time of the hormone–receptor complex. The incubation of radiolabeled ligand with receptor is allowed to reach equilibrium (maximal binding). Then excess unlabeled ligand is added. As the complex dissociates, the ligand which rebinds is not

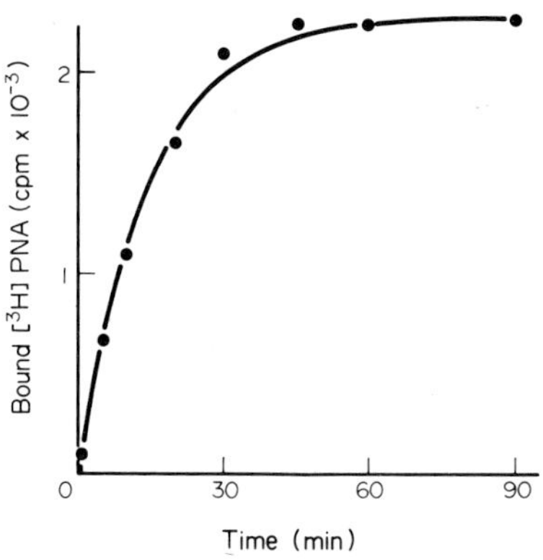

FIG. 3. Analog specificity of the ecdysteroid receptor from imaginal discs. Aliquots of imaginal disc cytosol were incubated with 3×10^{-9} M [^{3}H]ponasterone A and various concentrations of ecdysone analogs under equilibrium binding conditions. The amount of [^{3}H]ponasterone A bound relative to binding in a control aliquot with no competitor added is shown for the analog concentrations tested. The analogs are (■) ponasterone A, (●) 20-hydroxyecdysone and (▲) ecdysone (Yund, M. *et al.*, 1978).

radioactive. The dissociation rate constant (k_d) is calculated from the decrease in radioactivity associated with receptor over time (Figure 5). The kinetic rate constants for imaginal discs at 25° are shown in Table 2. The values correspond to a half-life of binding of approximately 20 min. The equilibrium association constant (K_A) and equilibrium dissociation constant (K_D) are calculated from the rate constants ($K_A = k_a/k_d = 1/K_D$). The K_D, in molar units, should be equivalent to the effective physiological concentration of hormone.

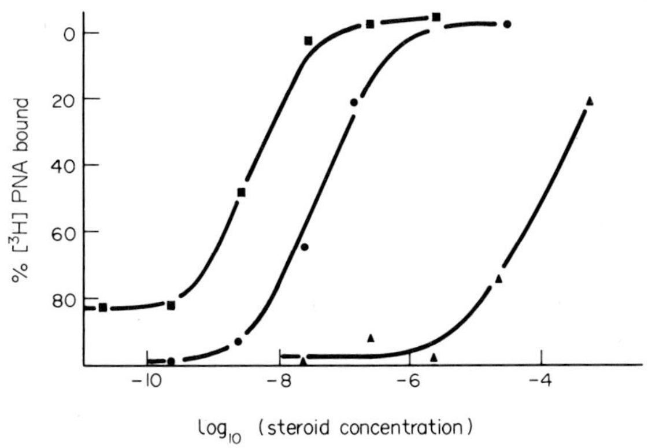

FIG. 4. Association of [^{3}H]ponasterone A with the ecdysteroid receptor from imaginal discs. A 0.3 M KCl extract prepared as described for Fig. 1 was incubated with 3×10^{-9} M [^{3}H]ponasterone A at 25°. Aliquots were collected at the times indicated and the amount of specifically bound [^{3}H]ponasterone A determined for each time point.

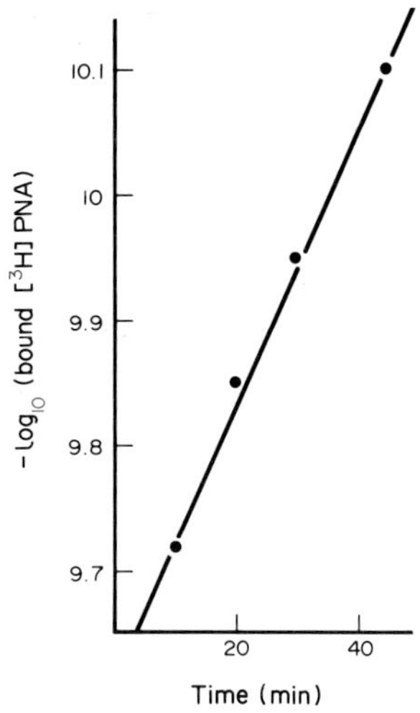

FIG. 5. Dissociation of [^{3}H]ponasterone A from the ecdysteroid receptor from imaginal discs. A 0.3 M KCl extract of imaginal disc nuclei prepared as described in Fig. 1 was incubated with 3×10^{-9} M [^{3}H]ponasterone A at 25° under conditions allowing equilibrium binding. An excess of unlabeled competitor, 2×10^{-7} M 20-hydroxyecdysone, was added at time 0. Aliquots of extract were sampled at the time indicated and the amount of [^{3}H]ponasterone A specifically bound was determined.

The K_Ds for receptor calculated from the kinetic rate constants are shown in Table 2 (for further discussion see Rodbard, D. 1973).

The K_D can also be determined from a Scatchard plot (Scatchard, G. 1949). The amount of hormone bound at equilibrium to a constant amount of receptor is determined over a range of hormone concentrations. For each concentration, bound and free hormone are determined and data plotted as in Figure 6. A straight line indicates one class of binding site is detectable over the concentration range examined; the K_D of binding is determined from the slope of the line. The number of binding sites can be determined from the x intercept. Scatchard analysis of binding from imaginal discs reveals only one class of binding sites in the concentration range tested. The K_D value estimated from the Scatchard data is 3.8×10^{-9} M. This is in good agreement with the K_D estimated from kinetic experiments (Table 2; Yund, M. *et al.*, 1978). This observation implies that the morphogenic response is induced at a hormone concentration at which approximately half of the receptors are occupied by hormone. From Scatchard data it is estimated that there are 1000 ± 300 binding sites per cell in the imaginal discs.

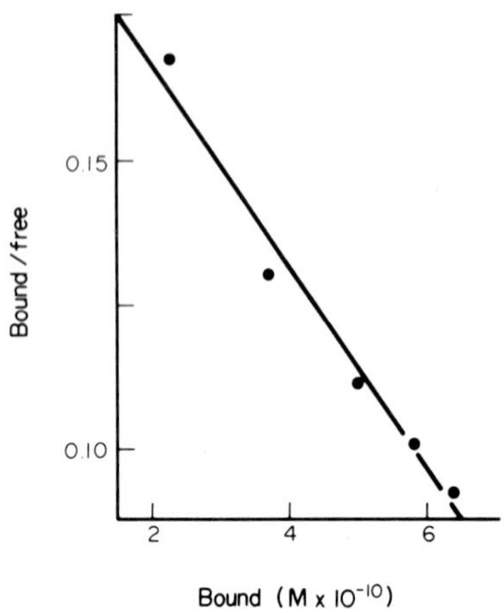

FIG. 6. Scatchard analysis of imaginal disc binding. Aliquots of a 0.3 M KCl extract of imaginal disc nuclei prepared as described in Fig. 1 were incubated with increasing concentrations of [^{3}H]ponasterone A under conditions permitting equilibrium binding. Specifically bound hormone for each concentration is plotted according to the method of Scatchard, G. (1949).

Analysis of these experiments indicates that the ecdysteroid binder detected in imaginal discs is the receptor mediating the RNA and protein synthetic responses, evagination and cuticle production. Specific, competible binding is located in the nucleus before the first synthetic changes are observed. The half-life of the hormone–receptor complex (15–30 min) is consistent with mediation of the evagination response (5 h of hormone exposure required) and dissociation of the complex on subsequent reduction in ecdysteroid titer to permit the initiation of cuticle production (cf. Figure 2). Furthermore, comparison of the analog specificity of biological response and hormone–receptor binding indicates that the biological activity of an ecdysteroid is a direct consequence of its affinity for the hormone receptor (Table 1).

Preliminary characterization of imaginal disc ecdysteroid receptor in crude cellular extract has been accomplished. In size exclusion chromatography on Sephacryl-200, specifically bound ligand from cytoplasmic and nuclear fractions ran near the exclusion limit in a low resolution area corresponding to an apparent molecular weight greater than 200,000 daltons (Yund, M. *et al.*, 1978). Similar molecular weight estimates were obtained on HPLC size exclusion columns in several buffers (Yund, M. and King, D., unpublished observations). In a 5–20% sucrose gradient containing 0.1 M KCl the ecdysteroid–receptor complex sedimented with an S value of 5.95. In an 0.3 M KCl gradient, the complex sedimented at 4.3 S (Table 3). This variance may result from weak electrostatic interaction causing aggregation of receptor with itself or other proteins, or from salt-dependent conformational changes in the receptor itself. As it is not known whether the functional form of the receptor is a monomer, a dimer, or associated specifically with other proteins *in vivo*, it is not possible to determine which, if any, of these forms detected in crude extracts

Table 2: Drosophila *ecdysteroid receptors: kinetic and equilibrium constants*

	k_a (M^{-1} min^{-1})	k_d (min^{-1})	K_D (kinetic) (M)	K_D (Scatchard) (M)
Imaginal discs				
Whole cells[a]	1.5×10^5	3×10^{-2}	2×10^{-7}	—
Cytosol[b,c]	1.2×10^7	3.6×10^{-2}	3×10^{-9}	3.3×10^{-9}
Nuclear extract[b]	6.6×10^{-6}[d]	3.2×10^{-2}[d]	4.5×10^{-9}[d]	3.8×10^{-9}[c]
Embryos				
Nuclear extract[b,e]	—	—	—	6×10^{-9}
K_c cells				
Cytosol[f]	—	2×10^{-2}[g]	—	3.4×10^{-9}[h]
Nuclear extract[f]	—	1×10^{-2}[g]	—	4.2×10^{-9}[h]

[a] [^{3}H]20-hydroxyecdysone, 3×10^{-8} M, 25° (Yund, M. and Fristrom, J., 1975)
[b] [^{3}H]ponasterone A, 3×10^{-9} M, 25°
[c] Yund, M. *et al.*, 1978
[d] Yund, M., unpublished
[e] Osterbur, D. and Yund, M., 1982
[f] [^{3}H]ponasterone A, 5×10^{-10} M, 22°
[g] Sage, B. *et al.*, 1982
[h] Calculated from half-life of 40 min and 60 min (Sage, B. *et al.*, 1982)

corresponds to the "true" receptor. Purification of receptor and knowledge of the molecular mode of action of ecdysteroids are necessary for an accurate assessment.

One surprising result from the study of the ecdysteroid receptors in imaginal discs was the cellular location of the unoccupied receptor. Vertebrate steroid receptors are thought to move from the cytoplasm to the nucleus upon exposure to hormone. Approximately 95% of the unoccupied cellular ecdysteroid receptor was found in the nuclear fraction of imaginal discs (Yund, M. *et al.*, 1978). Hill, R. *et al.* (1982) have demonstrated that the receptor co-purifies with chromatin. As three cell divisions occur in imaginal discs during the third larval instar, only 10–20% of the imaginal disc cells have been exposed to previous high hormone titers during one, or both, of the two larval molts. Whether this previous exposure to high hormone levels or low levels present during the intermolt period prior to isolation is sufficient to cause and sustain nuclear location of the receptor or whether the presence of receptor in the nucleus is independent of hormone binding cannot be determined at this time.

4.4 Tissue culture cells

Typically *Drosophila* cell lines are established from dissociated embryos ranging from 6 to 12 h in age (see e.g. Echalier, G. and Ohanessian, A. 1969). As the *in vitro* characteristics and the ecdysteroid-induced phenotypic effects of the cell lines do not correspond to the known biological response of any tissue *in vivo*, the cell type and developmental state of these cell lines is unknown. Nevertheless, several features of tissue culture cells make cell lines an attractive system for the study of ecdysteroid action. The K_c cell line shows little or no metabolism of hormone (Maroy, P. *et al.*, 1978). *Drosophila* tissue culture cells are relatively easy to maintain and culture in quantities sufficient for biochemical study. The ability to manipulate the environment of cultured cells, and the ease with which they can be selected for traits such as non-responsiveness to hormone (Stevens, B. *et al.*, 1980; Ireland, R. *et al.*, 1982), allows work on cell lines to be extended beyond the characterization of binding phenomenon. Homogeneous cell populations can be obtained using available cloning techniques (Schneider, I. and Blumenthal, A. 1978). This, together with the capability of inducing fusion in these cells (Becker, J. 1972; Rizki, R. *et al.*, 1975), opens up the possibility of somatic cell genetics. Using these techniques it might be possible to assign the locus or loci for the ecdysteroid receptor to a particular chromosome arm or to map the position of genes necessary for the biochemical and morphological responses to the hormone.

Many permanent cell lines of *Drosophila melanogaster* respond to ecdysteroids in a characteristic fashion. These responses include regulation of specific enzyme activity (Best-Belpomme, M. *et al.*, 1978; Cherbas, P. *et al.*, 1977), changes in early protein synthetic patterns (Savakis, G. *et al.*, 1980; Ireland, R. and Berger, E., 1982), and effects on cell proliferation (Courgeon, A.-M., 1972; Cherbas, P. *et al.*, 1980) and aggregation (Courgeon, A.-M., 1972). One response that appears to be common to many of the cell lines is a morphological change in the cells, involving extension of long, branched processes. Cherbas, L. *et al.* (1980) established a method for quantifying the response and determined the analog specificity of sixty ecdysteroids. The concentrations of ecdysone, 20-hydroxyecdysone and ponasterone A required for induction of the morphological response in K_c cells are in good agreement with those inducing the morphological response of imaginal discs (Table 1). The similarity of the binding kinetics of ecdysteroid receptors isolated from K_c cells (Maroy, P. *et al.*, 1978) and from imaginal discs (Table 2) argues that the morphological response is mediated by the same hormone–receptor complex. Subcloned lines selected for non-responsiveness to hormone have little or no detectable specific hormone binding (Stevens, B. *et al.*, 1980; Ireland, R. *et al.*, 1982).

Scatchard analysis of the binding of K_c cells reveals the presence of only one class of binding site in the concentration range tested. The K_D observed is the same as that found for imaginal discs (Table 2; Maroy, P. *et al.*, 1978). Scatchard analysis of ponasterone A binding reveals approximately 1800 binding sites per cell (Beckers, C. *et al.*, 1980). This is comparable to the estimate of approximately 1300 per K_c cell from photoaffinity labeled material (Schaltmann, K. and Pongs, O., 1982) and approximately 1000 per cell for imaginal discs.

However, the amount of receptor in K_c cells is not constant. After the receptor-mediated hormone response the cells enter a quiescent state in which they cease cell division for 100 h. Upon returning to the normal cell cycle these cells are not responsive to additional hormone. This has been shown to be due to a down-regulation of receptor levels after the initial response. This lowered level of receptor persists up to 6 months, after which the cell lines regain the characteristic hormone response (Stevens, B. *et al.*, 1980). Clearly this involves regulation of functional receptor, not the loss of the receptor gene from the cell line.

Ecdysteroid receptor from a crude extract of K_c cells does not display a salt-concentration-dependent change in S values (Table 3). However, the nuclear receptor from fresh or frozen cells runs at 6S, while the cytoplasmic receptor from fresh cells runs at 6S and that from frozen cells at 4.2S (Sage, B. *et al.*, 1982). Photoaffinity cross-linking of [^{3}H]20-hydroxyecdysone to a molecule in the cytoplasm of K_c cells followed by sucrose density centrifugation gives three sedimentation values, 4S, 6S and 9S (Table 3; Schaltmann, K. and Pongs, O., 1982). Sedimentation of cross-linked material from the nuclear fraction of these cells gives only the 9S value, leading to speculation that this 9S value represents the chromatin-binding form of the ecdysteroid-binding protein. The sedimentation properties of cross-linked material were not tested at different salt concentrations.

Time-dependent translocation of bound radioactive hormone from the cytosol to the nuclear fraction has been observed (Maroy, P. *et al.*, 1978). Recent work has found receptors with identical kinetic properties in the cytosol and nuclei of cells grown for many generations in the absence of hormone (Sage, B. *et al.*, 1980). Such hormonally naive cells contain about 1200 receptors per cell in the cytosol and 800 receptors per cell in the nucleus. After hormone exposure translocation of bound hormone through the cytosol to the nuclear compartment occurs (Maroy, P. *et al.*, 1978). Although specifically bound hormone appears in the nucleus, there is no change in the number of receptors in the nuclear and cytoplasmic compartments. This suggests an equilibrium of receptors between the two compartments involving a rebinding of hormone to nuclear receptor or a one for one exchange of hormone–receptor complex and unoccupied receptor. Schaltmann, K. and Pongs, O. (1982) report that photoaffinity-labeled K_c cell receptors translocate from the cytoplasm to the nucleus. The reason for the discrepancy in results from photoaffinity labeling and reversible radioligand binding is not clear at this time (for discussion, see Bonner, J. 1982).

Table 3: Sedimentation coefficients of hormone–receptor complexes

	Nuclear	Cytoplasmic
Imaginal discs[a]	6.0[b], 4.3[c]	—
Embryos[a]	6.0[b], 4.3[c]	—
K_c cells[d]	6.3[e,f]	6.3[e], 4.2[f]
K_c cells[g]	9	4, 6, 7

[a] Osterbur, D. and Yund, M. (1982)
[b] 0.3 M KCl gradient
[c] 0.1 M KCl gradient
[d] Sage, B. *et al.* (1982)
[e] 0–0.2 M KCl gradient, fresh material
[f] 0–0.2 M KCl gradient, frozen–thawed material
[g] Schaltmann, K. and Pongs, O. (1982)

4.5 Embryos

Ecdysteroids regulate cellular and developmental activities throughout the life of the organism. During embryogenesis ecdysteroid-dependent events such as cuticle synthesis occur (Hoffmann, J. and Lagueux, M., vol. 1). Because gram quantities of *Drosophila melanogaster* embryos are readily obtainable, embryos are an attractive source material for an ecdysteroid receptor for study and comparative analysis. We have developed procedures for preparation of an extract of the nuclear fraction of dechlorionated embryos which does not metabolize the hormone (Osterbur, D. and Yund, M., unpublished results). This extract contains ecdysteroid binding proteins with properties very similar to those of the imaginal disc receptor. Because embryos have no known ecdysteroid response which can be manipulated and quantified, descriptive information relating ecdysteroids to the response is not presently obtainable from embryos. Thus, it is not possible to correlate biological response to hormone with properties of the ecdysteroid binder, which is required to demonstrate conclusively that the embryonic ecdysteroid binder is a functional embryonic ecdysteroid receptor.

Scatchard analysis of the binding protein from the nuclear fraction of *Drosophila* embryos reveals only one class of binding site in the concentration range tested. The K_D estimated from the Scatchard data for the binding of [^{3}H]ponasterone A is $6 \pm 2.9 \times 10^{-9}$ M (Osterbur, D. and Yund, M., 1982). This is well within the range of values obtained from both K_c cells and imaginal discs.

Preliminary results in competition experiments using 20-hydroxyecdysone and ecdysone reveal that the embryonic binding molecule has the same relative ordering of affinities for these two compounds as do imaginal discs and tissue culture cells: 20-hydroxyecdysone binds with higher affinity than ecdysone. The K_D values derived from these competition experiments are very similar to those for cells and imaginal discs (Table 1; Osterbur, D. and Yund, M., 1982).

The S values obtained for the embryonic binding molecule are dependent on salt concentration. The embryonic ecdysteroid binding molecule sediments in a broad band in a 0.1 M KCl sucrose density gradient, with a peak of sedimentation at 6.0S. In a 0.3 M KCl sucrose density gradient the sedimentation band is much sharper, peaking at 4.3S. These results are almost identical to the results obtained with the imaginal disc receptor (Table 3; Osterbur, D. and Yund, M., 1982).

Embryos that have been collected over 3 h, then aged for various times, have been used to explore the subcellular location of the ecdysteroid binder. In 0–3 h embryos approximately 45% of the binding activity is found in the nucleus. In 3–6 h embryos 74% of the binding activity is found in the nuclear compartment. This is well before the embryonic ecdysteroid titer peak revealed by radioimmunoassay (Richards, G., 1981). In 6–9 and 9–12 h embryos the amount of binding in the nucleus becomes constant at approximately 85% of the total binding activity (Osterbur, D. and Yund, M., unpublished observations). These data indicate that some portion of the unoccupied embryonic ecdysteroid binding molecules is located in the nucleus.

4.6 Salivary glands

The hormone-dependent changes in the puffing pattern of the polytene chromosomes from the salivary glands of higher Dipterans are a classic and well-studied response to a steroid hormone (Ashburner, M. *et al.*, 1974; Ashburner, M. and Berendes, H., 1978; Richards, G., 1980; vol. 2). These polytene chromosomes are an attractive system for the study of receptor–genome interaction. The ability to manipulate the genetic complement of regions of the chromosomes through the use of duplications and deletions, and the dramatic visualization of this manipulation, provide a good means for the study of ecdysteroid effects on gene activity (Walker, V. and Ashburner, M., 1981). The combination of cytological and biochemical techniques provides a direct means of studying the hormone–receptor complex interaction with the genome.

In a very interesting set of experiments, Gronemeyer, H. and Pongs, O. (1980) have studied the location of photo-cross-linked, endogenous hormone along the salivary gland chromosomes by indirect immunofluoresence using an anti-ecdysteroid antibody. These studies showed the photoaffinity-labeled material bound to the polytene chromosomes at sites which puff in response to the hormone *in vivo*. This suggests a direct interaction with the sites showing a primary response to the hormone. It is interesting that one of the intermolt puffs which regresses upon exposure to 20-hydroxyecdysone, 68C, fluoresces as brightly as do the chromosomal sites which puff in response to the hormone. This observation supports the hypothesis that the hormone has negative as well as positive regulatory functions (Ashburner, M. *et al.*, 1974).

Photoaffinity labeling is a relatively new technique for the study of ecdysteroid binding proteins. It has great potential for use in purifying such binding proteins, as well as for the study of the interaction of the hormone–receptor complex with the genome. The molecular weight estimate for the photoaffinity molecule from salivary glands, 130,000 is identical to that derived for the molecule from K_c cells using SDS polyacrylamide gel electrophoresis. However, the values from sucrose density gradients differ (Table 3).

5 SUMMARY

Sufficient data have been accumulated on the

properties of ecdysteroid binders in *Drosophila* imaginal discs and the *Drosophila* K_c cell line to conclude that these are the hormone receptors mediating the ecdysteroid response. The information of the putative embryonic receptor, although not conclusive, is compelling. Because the observations on salivary glands are entirely with cross-linked material, the kinds of information obtainable by hormone-binding studies are not available. However, the observed S values and chromosome localization are suggestive. The kinetic properties and analog specificity of the imaginal disc, K_c cell and embryonic receptors are not significantly different. The ecdysteroid binding sites of the receptors from these three tissues are equivalent if not identical. This is not sufficient to prove that these receptors are the same molecule (gene product). The part of the molecule which recognizes and binds the hormone might be very similar if not identical, while the portion interacting with other parts of the cell might be different in different tissues, or multiple functional forms of the receptor might exist in a given tissue. The latter has been established for the A and B forms of the progesterone receptor in chick oviduct (Schrader, W. *et al.*, 1981). In the case of the *Drosophila* ecdysteroid receptor, different tissues, different extracts and binding protocols and different gradient procedures were employed. Variation in measurements in crude extracts is the typical situation for steroid hormone receptors (Chamness, G. and McGuire, W., 1972). Differing S values in unpurified material cannot be interpreted as evidence for multiple forms of the receptor. However, multiple forms of the receptor remain a possibility. It is notable that, in contrast to the array of diversified steroid hormones in vertebrates, in *Drosophila* many diverse processes throughout the life cycle are controlled by the one hormone, 20-hydroxyecdysone. This implies that the specificity of the hormone response must lie in tissue-specific molecular differences, perhaps involving different receptors.

Information on the regulation of ecdysteroid receptors is limited. In K_c cells and in imaginal discs the major portion of the unoccupied receptor is found in the nuclear fraction. In neither tissue does the binding of hormone and induction of the hormone response alter the distribution of receptors between the nuclear and cytoplasmic compartments (Hill, R. *et al.*, 1982; Sage, B. *et al.*, 1982). In K_c cells it is possible to follow the post-hormone events in culture. After the hormone induced 100–120 h G_2 arrest of the cell cycle, the cells are refractory to additional hormone. The amount of receptor in the cell is correspondingly reduced and remains low for months. This is suggestive of a hormone-induced differentiation event. The G_2 arrest time of 4–5 days followed by hormone refractoriness would fit well into a developmental time of 10–14 days for *Drosophila*.

The mechanism of ecdysteroid action as revealed by this analysis of ecdysteroid receptor (see chapters 6 to 8, this volume for additional review of ecdysteroid action) is basically the same as the general mechanism established for vertebrate steroid hormone action (see Yund, M., 1979b). In both instances a primary event appears to be interaction of hormone–receptor complex with chromatin, bringing about by some as yet unknown mechanism changes in the transcription of specific genes. The apparent basic evolutionary conservatism in molecules and molecular processes, together with the constant ratio of receptor to DNA in organisms with genome sizes differing by more than an order of magnitude, strongly suggests a common mechanism of hormone regulation of transcription in vertebrates and in *Drosophila*. Extension of this line of reasoning would predict that the function of ecdysteroid receptors would be similar across the order Insecta.

There are two notable differences in the properties of ecdysteroid and vertebrate receptors. Upon tissue fractionation in hypotonic buffer vertebrate steroid receptors are commonly found in the cytoplasmic fraction. The cytoplasmic hormone–receptor complex translocates to the nucleus at 25–37°, but not at 0–4°. Unoccupied ecdysteroid receptors are found in the nuclear fraction. The significance, if any, of the apparent differences in cellular distribution of vertebrate and ecdysteroid receptors cannot be determined at this time. Precise intracellular location of receptor in frozen tissue sections with anti-receptor antibody and fuller understanding of the mechanism of hormone action are necessary to resolve this question. Another conspicuous difference from the vertebrates is the lower K_D of binding of 20-hydroxyecdysone to its receptor and the corresponding higher *in vivo* hormone titer. It is suggested that the relatively rapid developmental program and shorter life cycle of *Drosophila* has required a rapidly turning over hormone–receptor complex resulting in a low K_D of binding. The observation that a crustacean, a crayfish, *Orconectes limosus*, which has a much longer life cycle than *Drosophila*, has ecdysteroid titers two orders of magnitude lower than *Drosophila* and corresponding higher receptor affinities for ecdysteroids (Londershausen, M. and Spindler, K., 1981), tends to support this conclusion.

ACKNOWLEDGEMENTS

We thank B. Sage and J. D. O'Connor for sharing unpublished information, D. King for critical reading of the manuscript, and K. Fechtel and S. Alexander for assistance in preparation of the manuscript and figures. The unpublished work from our laboratory was supported in part by USPHS grant GM-23349 to M. A. Yund and USPHS grant GM-19937 to J. W. Fristrom.

REFERENCES

ASHBURNER, M. and BERENDES, H. D. (1978). Puffing of polytene chromosomes. In *Genetics and Biology of Drosophila*. Edited by M. Ashburner and T. R. F. Wright. Pages 316–395. Academic Press, San Francisco.

ASHBURNER, M., CHIHARA, C., MELTZER, P. and RICHARDS, G. (1973). Temporal control of puffing activity in polytene chromosomes. *Cold Spring Harbor Symp. Quant. Biol. 38*, 655–662.

BECKER, J. L. (1972). Fusions *in vitro* de cellules de *Drosophila melanogaster* induites par la Concanavaline A. *C.R. Acad. Sci. Paris D. 275*, 2969–2972.

BECKERS, C., MAROY, P., DENNIS, R. and O'CONNOR, J. D. (1980). The uptake and release of ponasterone A by the K_C cell line of *Drosophila melanogaster*. *Mol. Cell. Endocr. 17*, 51–59.

BEST-BELPOMME, M. and COURGEON, A.-M. (1975). Presence ou absence de recepteurs saturables de l'ecdysterone dans des clones sensibles ou resistants de *Drosophila melanogaster* in culture *in vitro*. *C.R. Acad. Sci. Paris D 280*, 1397–1400.

BEST-BELPOMME, M., COURGEON, A.-M. and RAMBACH, A. (1978). β-galactosidase is induced by hormone in *Drosophila melanogaster* cell cultures. *Proc. Nat. Acad. Sci. USA 75*, 6102–6106.

BONNER, J. J. (1982). An assessment of the ecdysteroid receptor of *Drosophila*. *Cell 30*, 7–8.

BUTTERWORTH, F. M. and BERENDES, H. D. (1974). Ecdysone-binding proteins in haemolymph and tissues of *Drosophila hydei* larvae. *J. Insect Physiol. 20*, 2195–2204.

CHAMNESS, G. C. and MCGUIRE, W. L. (1972). Estrogen receptor in the rat uterus. Physiological forms and artifacts. *Biochemistry 11*, 2466–2472.

CHERBAS, L., YONGE, C. D., CHERBAS, P. and WILLIAMS, C. M. (1980). The morphological response of K_C-H cells to ecdysteroids: Hormonal specificity. *Wilhelm Roux's Archiv. 189*, 1–15.

CHERBAS, P., CHERBAS, L., DEMETRI, G., MANTEUFFEL-CYMBOROWSKA, M., SAVAKIS, C., YONGE, C. D. and WILLIAMS, C. M. (1980). Ecdysteroid hormone effects on a *Drosophila* cell line. In *Gene Regulation by Steroid Hormones*. Edited by A. K. Roy and J. H. Clark. Pages 278–308. Springer, New York.

CHERBAS, P., CHERBAS, L. and WILLIAMS, C. M. (1977). Induction of acetylcholinesterase activity by β-ecdysone in a *Drosophila* cell line. *Science 197*, 275–277.

CHIHARA, C. J., PETRI, W. H., FRISTROM, J. W. and KING, D. S. (1972). The assay of ecdysones and juvenile hormones on *Drosophila* imaginal discs *in vitro*. *J. Insect Physiol. 18*, 1115–1123.

CLEVER, U. and KARLSON, P. (1960). Induktion von Puff-Voranderungen in den Speicheldrusenchromosomen von *Chironomus tentans* durch Ecdysone. *Exp. Cell Res. 20*, 623–626.

COURGEON, A.-M. (1972). Action of insect hormones at the cellular level. *Exp. Cell Res. 74*, 327–336.

ECHALIER, G. and OHANESSIAN, A. (1970). *In vitro* culture of *Drosophila melanogaster* embryonic cells. *In Vitro 6*, 162–172.

EMMERICH, H. (1970a). Ecdysonbindende proteinfraktionen in den speicheldrusen von *Drosophila hydei*. *Z. vergl. Physiol. 68*, 385–402.

EMMERICH, H. (1970b). Uber die Hamolymphproteine von *Pyrrhocoris apterus* und uber die Bindung von Ecdyson durch Hamolymphproteine. *J. Insect Physiol. 16*, 725–747.

EMMERICH, H. (1972). Ecdysone binding proteins in nuclei and chromatin from *Drosophila* salivary glands. *Gen. Comp. Endocr. 19*, 543–551.

EUGENE, O., YUND, M. A. and FRISTROM, J. W. (1979). Preparative isolation and short-term organ culture of imaginal discs of *Drosophila melanogaster*. *Tissue Culture Association Manual 5*, 1055–1062.

FEYEREISEN, R. (1980). Ecdysteroid-protein interactions: Specificity of the ecdysteroid binding protein from *Locusta migratoria* hemolymph. In *Progress in Ecdysone Research*. Edited by J. A. Hoffmann. Pages 325–334. Elsevier/North Holland, Amsterdam.

FRASER, C. M. and VENTER, J. C. (1980). Monoclonal antibodies to β-andrenergic receptors: Use in purification and molecular characterization of β receptors. *Proc. Nat. Acad. Sci. USA 77*, 7034–7038.

FRISTROM, J. W., DOCTOR, J., FRISTROM, D. K., LOGAN, W. R. and SILVERT, D. J. (1982). The formation of the pupal cuticle by *Drosophila* imaginal discs *in vitro*. *Devel. Biol. 91*, 337–350.

FRISTROM, J. W., FRISTROM, D. K., FEKETE, E. and KUNIYUKI, A. H. (1977). The mechanism of evagination of imaginal discs of *Drosophila melanogaster*. *Am. Zool. 17*, 671–684.

FRISTROM, J. W. and MITCHELL, H. K. (1965). The preparative isolation of imaginal discs from larvae of *Drosophila melanogaster*. *J. Cell Biol. 27*, 445–448.

FRISTROM, J. W. and YUND, M. A. (1976). Characteristics of the action of ecdysones on *Drosophila* imaginal discs cultured *in vitro*. In *Invertebrate Tissue Culture–Research Applications*. Edited by K. Maramorosh. Pages 161–178. Academic Press, New York.

FRISTROM, J. W. and YUND, M. A. (1980). A comparative analysis of ecdysteroid action in larval and imaginal tissues of *Drosophila melanogaster*. In *Progress in Ecdysone Research*. Edited by J. A. Hoffman. *Developments in Endocrinology*. Vol. 7, pages 349–362. Elsevier, North Holland.

GRONEMEYER, H. and PONGS, O. (1980). Localization of ecdysterone on polytene chromosomes of *Drosophila melanogaster*. *Proc. Nat. Acad. Sci. USA 77*, 2108–2112.

HILL, R. J., WATT, F., YUND, M. A. and FRISTROM, J. W. (1982). The effect of 20-hydroxyecdysone on synthesis of chromosomal and cytosol proteins in imaginal discs. *Devel. Biol. 90*, 340–351.

HODGETTS, R. B., SAGE, B. and O'CONNOR, J. D. (1977). Ecdysone titers during postembryonic development of *Drosophila melanogaster*. *Devel. Biol. 60*, 310–317.

IRELAND, R. C. and BERGER, E. M. (1982). Synthesis of low molecular weight heat shock peptides stimulated by ecdysterone in a cultured *Drosophila* cell line. *Proc. Nat. Acad. Sci. USA 79*, 855–859.

IRELAND, R. C., BERGER, E., SIROTKIN, K., YUND, M. A., OSTERBUR, D. and FRISTROM, J. (1982). Ecdysterone induces the transcription of four heat shock genes in *Drosophila* S_3 cells and imaginal discs. *Devel. Biol. 93*, 489–507.

LAGUEUX, M., HARRY, P. and HOFFMANN, J. A. (1981). Ecdysteroids are bound to vitellin in newly laid eggs of *Locusta*. *Mol. Cell. Endocr. 24*, 325–338.

LONDERSHAUSEN, M. and SPINDLER, K.-D. (1981). Characterization of cytoplasmic ecdysteroid receptors in the hypodermis of the crayfish *Orconectes limosus*. *Mol. Cell Endocr. 24*, 253–265.

MANDARON, P., GUILLORMAT, C. and SENGEL, P. (1977). *In vitro* development of *Drosophila* imaginal discs: Hormonal control and mechanism of evagination. *Am. Zool. 17*, 661–670.

MAROY, P., DENNIS, R., BECKERS, C., SAGE, B. A. and O'CONNOR, J. D. (1978). Demonstration of an ecdysteroid receptor in a cultured cell line of *Drosophila melanogaster*. *Proc. Nat. Acad. Sci. USA 75*, 6035–6038.

O'CONNOR, J. D., MAROY, P., BECKERS, C., DENNIS, R., ALVAREZ, C. M. and SAGE, B. A. (1980). Ecdysteroid receptors in cultured *Drosophila* cells. In *Gene Regulation by Steroid Hormones*. Edited by A. K. Roy and J. H. Clark. Pages 263–277. Springer, New York.

OSTERBUR, D. L. and YUND, M. A. (1982). Ecdysteroid binding activity in embryos of *Drosophila melanogaster*. *J. Cell. Biochem. 20*, 277–282.

RAIKOW, R. and FRISTROM, J. W. (1971). Effects of β-ecdysone on RNA metabolism of imaginal discs of *Drosophila melanogaster*. *J. Insect Physiol. 17*, 1599–1614.

RICHARDS, G. (1980). Ecdysteroids and puffing in *Drosophila melanogaster*. In *Progress in Ecdysone Research*. Edited by J. A. Hoffmann. Pages 363–378. Elsevier/North Holland, Amsterdam.

RICHARDS, G. (1981). The radioimmune assay of ecdysteroid titers in *Drosophila melanogaster*. *Mol. Cell. Endocr. 21*, 181–197.

RIZKI, R. M., RIZKI, T. M. and ANDREWS, C. A. (1975). *Drosophila* cell fusion induced by wheat germ agglutinin. *J. Cell Sci. 18*, 113–121.

RODBARD, D. (1973). Mathematics of hormone-receptor interactions. *Adv. Exp. Med. Biol. 36*, 289–326.

SAGE, B. A., TANIS, M. A. and O'CONNOR, J. D. (1982). Characterization of ecdysteroid receptors in cytosol and naive nuclear preparations of *Drosophila* K_c cells. *J. Biol. Chem. 257*, 6373–6379.

SAVAKIS, C., DEMETRI, G. and CHERBAS, P. (1980). Ecdysteroid-inducible polypeptides in a *Drosophila* cell line. *Cell 22*, 665–674.

SCATCHARD, G. (1949). The attraction of proteins for small molecules and ions. *Ann. N.Y. Acad. Sci. 51*, 660–672.

SCHALTMANN, K. and PONGS, O. (1982). Identification and characterization of the ecdysteroid receptor in *Drosophila melanogaster* by photoaffinity labeling. *Proc. Nat. Acad. Sci. USA 79*, 6–10.

SCHNEIDER, I. and BLUMENTHAL, A. B. (1978). *Drosophila* cell and tissue culture. In *The Genetics and Biology of Drosophila*. Edited by M. Ashburner and T. R. F. Wright. Vol. 2a, pages 265–316. Academic Press, San Francisco.

SCHRADER, W. T., BIRNBAUMER, M. E., HUGHES, M. R., WEIGEL, N. L., GRODY, W. W. and O'MALLEY, B. W. (1981). Studies on the structure and function of the chicken progesterone receptor. *Rec. Prog. Hormone Res. 37*, 583–633.

STEVENS, B., ALVAREZ, C. M., BOHMAN, R. and O'CONNOR, J. D. (1980). An ecdysteroid-induced alteration in the cell cycle of cultured *Drosophila* cells. *Cell 22*, 675–682.

THAMER, G. and KARLSON, P. (1972). Nachweis der proteinbindung von ecdysone bei der schmeissfliege meigen *Calliphora erythrocephala*. *Z. Naturf. 27b*, 1191–1195.

WALKER, V. K. and ASHBURNER, M. (1981). The control of ecdysterone-regulated puffs in *Drosophila* salivary glands. *Cell 26*, 269–277.

YUND, M. A. (1979a). Specific binding of 20-hydroxyecdysone to nuclei of imaginal discs of *Drosophila melanogaster*. *Mol. Cell Endocr. 14*, 19–35.

YUND, M. A. (1979b). Ecdysteroid action in imaginal discs of *Drosophila melanogaster*. In *Development of Responsiveness to Steroid Hormones*. Edited by A. M. Kaye and M. Kaye. *Adv. Biosci.* Vol. 25, pages 285–302. Pergamon Press, New York.

YUND, M. A. and FRISTROM, J. W. (1975). Uptake and binding of β-ecdysone in imaginal discs of *Drosophila melanogaster*. *Devel. Biol. 43*, 287–298.

YUND, M. A., KING, D. S. and FRISTROM, J. W. (1978). Ecdysteroid receptors in imaginal discs of *Drosophila melanogaster*. *Proc. Nat. Acad. Sci. USA 75*, 6039–6043.

14 Juvenile Hormone Cellular and Hemolymph Binding Proteins

WALTER G. GOODMAN

University of Wisconsin, Madison, Wisconsin, USA

and

ERNEST S. CHANG

University of California, Davis, California, USA

1 MODES OF TRANSPORT IN THE CIRCULATORY SYSTEM

In insects as well as vertebrates, the transport of trace amounts of biologically active substances from the site of synthesis to the peripheral target tissue is accomplished by the circulatory system. Within this system, the mode of transport for these molecules depends upon the physicochemical nature of the molecule and to some extent its biological function.

Transport of most biologically active polar compounds falls into two categories. The most simple involves molecules which form non-specific, transient associations with proteins or other macromolecules in the circulatory system. In general, this mode of conveyance is associated with molecules which are required in episodic bursts rather than steady-state levels. Polar substances may also be transported via specific plasma transport proteins. In the bound state, these biologically active compounds are dispersed throughout the circulatory system, where they are readily available for maintenance of cell viability. The iron–transferrin complex (Palmour, R. and Sutton, H. 1971) and the vitamin B_{12}–transcobalamin complex (Burger, R. *et al.*, 1975) are examples of this mode of transport.

Another class of biologically important substances, the lipophilic hormones and vitamins, are generally associated with particular plasma proteins (Natelson, S. and Natelson, E. 1980). Their reduced solubility or surface-active nature in plasma may not allow the unbound molecule to develop sufficient titers at peripheral sites. The reversible binding of the lipophilic ligand to a given plasma protein then allows it to be dispersed in an aqueous environment. The role of these proteins in transporting hormones from distant synthetic sites to target tissue is essential in the integration of biochemical, physiological and environmental events. Thus our attention will focus first on this important class of binding proteins.

2 GENERAL ROLES OF HORMONE BINDING PROTEINS IN THE CIRCULATORY SYSTEM

The hormone binding proteins of the circulatory system are involved not only in the mediation of hormone dispersion to distant target tissue but also in the conservation of hormone titers in the plasma at the proper physiological levels. These proteins exert their regulatory effect on circulating hormone levels by: (1) facilitating the passage of hydrophobic hormones from the site of synthesis into the circulatory system; (2) protecting the hormone from enzymatic attack; (3) preventing the rapid loss of hormone due to excretion; (4) reducing non-specific binding; and (5) serving as a reservoir from which the hormone can be transferred to the target site (Westphal, U. 1980).

In addition to their regulatory effects, the hormone binding proteins sequester their respective ligands from areas in which the hormone may have a deleterious effect (Diamond, M. *et al.*, 1969). This hypothesis is especially attractive during the postembryonic development of insects when juvenile hormone (JH) must be rapidly cleared to permit metamorphosis (Gilbert, L. *et al.*, 1978, 1980a,b; Riddiford, L. and Truman, J. 1978; DeKort, C. and Granger, N. 1981). By serving as a scavenger, the protein may facilitate the sequestration and subsequent enzymatic degradation of the hormone (Goodman, W. and Gilbert, L. 1978; Sparks, T. *et al.*, 1979).

There is good evidence that certain ligands are recognized by target tissue only when complexed to their respective binding macromolecules. Specific cell surface receptors have been reported for transport molecules such as transcobalamin (Youngdahl-Turner, P. *et al.*, 1979), transferrin (Glass, J. *et al.*, 1978; Karin, M. and Mintz, B. 1981; Trowbridge, I. and Omary, M. 1981), and vitamin A binding protein (Heller, J. 1975; Rask, L. and Petersen, P. 1976; Chen, C. and Heller, J. 1977; Bhat, M. and Cama, H. 1979). The most elegant demonstration of this interaction has been observed in cholesterol transport. Despite its high solubility in membranes, cholesterol must under physiological conditions be bound to a plasma low-density lipoprotein to facilitate internalization (Brown, M. *et al.*, 1981).

3 CHARACTERISTICS OF HORMONE BINDING PROTEINS

From a general viewpoint, three basic properties of the ligand must be considered in the protein–hormone interaction: electrostatic charge, polarity of the molecule, and the spatial–chemical configuration (Westphal, U. 1971). Since the JHs are not charged, electrostatic effects may be dismissed. The second property of the ligand, polarity, is of central importance to the JH–protein interaction (see section 4.2b). The polarity of a molecule may be gauged by the number and type of electron-attracting "polar" groups. Polarity may be reduced either by decreasing the number of these groups or by introducing electron-repelling "non-polar" groups. Not only must the polarity be correct but spatial organization of the ligand must conform to the three-dimensional characteristics of the binding site. The addition or deletion of key substituents, or the modification of the geometrical or stereochemical configuration, will result in marked differences in the binding affinity.

In reversible, non-covalent associations, the binding forces involved are primarily hydrophobic interaction and hydrogen bonding. The number and variety of hydrogen bonds formed depends on the chemical characteristics of the hormone and the protein binding site. In the case of the JHs, the epoxide and the carbomethoxy group can act as hydrogen acceptors (see section 4.2(b) and Chapter 10, this volume). When electron-repelling groups are abundant both in the ligand and the binding site, hydrophobic interactions become a major factor in binding affinity. In the case of JHs this interaction would occur primarily between the alkyl side-chains of the hormone and the aromatic and aliphatic side-chains of the amino acids found in the binding site.

The hormone binding proteins can be conveniently divided into two classes based on their affinity for, and capacity to bind, their respective ligands. An association constant (K_a) of $10^6\,M^{-1}$ is considered the arbitrary dividing line between high-affinity ($>10^6\,M^{-1}$) and low-affinity ($<10^6\,M^{-1}$) binding proteins. The two categories of binding proteins are also distinguished by the fact that low-affinity proteins have multiple binding sites while high-affinity binders usually have only one site.

The association constants for the high-affinity, low-capacity binding proteins normally encountered in the circulatory system range from $10^7\,M^{-1}$ to $10^9\,M^{-1}$ while those of low-affinity, high-capacity proteins range between $10^4\,M^{-1}$ and $10^6\,M^{-1}$ (Westphal, U. 1980). In a number of cases an inverse relationship exists between the affinity of the protein and its concentration in the circulatory system, i.e., the low-affinity binder occurs at levels 100 to 1000 times higher than those of the high-affinity, low-capacity transport protein. As a consequence, the hormone is usually distributed between both types of macromolecules; however, the role of the low-affinity binding protein is often overlooked when considering the bound or unbound status of a hormone *in vivo* (Westphal, U. 1980).

The two classes of proteins can also be distinguished by their dynamics of binding. The low-affinity, high-capacity binding proteins are generally capable of binding structurally different ligands and contain more than one binding site per molecule for any given ligand. The binding sites for this low-affinity association are considered hydrophobic pockets which lack the rigid three-dimensional conformation necessary to confer specificity. The number of sites filled will therefore vary dramatically, depending upon the polarity of the ligand (Spector, A. 1975; Westphal, U. 1980). These lower-affinity binding proteins are for the most part either lipoproteins or albumin in the vertebrate and lipoproteins in the insect (Whitmore, E. and Gilbert, L. 1972; Gilbert, L. and Chino, H. 1974; Kramer, K. *et al.*, 1974; Gilbert, L. *et al.*, 1976; Law, J. 1980).

In contrast to the low-affinity binders, the high-affinity, low capacity binding proteins have only one site per molecule. The physicochemical nature of the specific binding site, a mixture of hydrophobic and hydrophilic binding forces, results in a three-dimensional pocket into which only those ligands displaying mirror images or relatively close approximations can fit. Although the binding sites have been intensively studied, our information about size and conformation comes from indirect evidence obtained by probing the site with various steric and structural homologs and analogs. The recent success of crystallizing an important steroid hormone transport protein, corticosteroid binding globulin, should pave the way for subsequent X-ray diffraction analysis and offer detailed insight on the hormone-binding domain and its environment (McPherson, A. *et al.*, 1980).

In summary, the hormone–protein complex is held together by non-covalent binding forces which reflect the affinity and specificity of the binder. These forces are responsible for the reversibility of hormone–protein interactions, which are governed by binding equilibria in accordance with the law of mass action.

4 HEMOLYMPH JUVENILE HORMONE BINDING PROTEINS

As the structure and physical properties of the JHs became known, investigators began a search for JH transport molecules. They reasoned that the compatibility of JH with organic solvents precluded high solubility in the aqueous environment of the hemolymph. Although the JHs are relatively insoluble (30 μM JH I in Tris buffer, pH 8.3; Kramer, K. *et al.*, 1976a), their normal physiological concentrations fall far below this value (Chapters 10–12, this volume). However, other evidence justified the need

Table 1: Major hemolymph juvenile hormone binding proteins in various insect species[a]

Species	Stage	Molecular Weight ($\times 10^{-3}$ daltons)	Association Constant (K_a) ($M^{-1} \times 10^{-6}$, 4°)	Reference
Gomphocerus rufus	adult	220	20	Hartmann, R. (1978)
Locusta migratoria	adult	220		Emmerich, H. and Hartmann, R. (1973)
Locusta migratoria	adult		12	Peter, M. *et al.* (1979); Gellissen, G. cited in Hartmann, R. (1978)
Leucophaea maderae[b]	adult		45	Koeppe, J. *et al.* (1981)
Oncopeltus fasciatus	nymph	100	< 0.1	Bassi, S. *et al.* (1977)
Leptinotarsa decemlineata	adult	> 200	< 0.1	Kramer, K. and DeKort, C. (1978)
Tenebrio molitor[c]	larva			Trautmann, K. (1972)
Tenebrio obscurans	larva	> 100		Kramer, K. *et al.* (1976a)
Tenebroides mauritanicus	larva	30		Kramer, K. *et al.* (1976a)
Drosophila hydei	larva	44	10	Klages, G. and Emmerich, H. (1979)
Hyalophora cecropia	pupa	200		Whitmore, E. and Gilbert, L. (1972)
Hyalophora gloveri	pupa	200		Whitmore, E. and Gilbert, L. (1972)
Antheraea polyphemus	pupa			Whitmore, E. and Gilbert, L. (1972)
Manduca sexta	larva	28	10	Kramer, K. *et al.* (1976b); Goodman, W. *et al.* (1978a)
Manduca quinquemaculata	larva	30	11	Kramer, K. and Childs, C. (1977)
Sphinx chersis	larva	30		Kramer, K. and Childs, C. (1977)
Hyloicus chersis	larva	30		Kramer, K. *et al.* (1976b)
Estigmene acraea	larva	30		Mumby, S. and Hammock, B. cited in Kramer, K. *et al.* (1976a)
Trichoplusia ni	larva	19	2	Hammock, B. *et al.* (1977); Sparks, T. and Hammock, B. (1979)
Spodoptera exigua	larva	30		Mumby, S. and Hammock, B. cited in Kramer, K. *et al.* (1976a)
Heliothis virescens	larva	30		Mumby, S. and Hammock, B. cited in Kramer, K. *et al.* (1976a)
Agrotis ipsilon	larva	25		Goodman, W. G. (unpublished observations)
Anagasta kuehniella	larva	30	18	Kramer, K. and Childs, C. (1977)
Caudra cautella	larva	30	70	Kramer, K. and Childs, C. (1977)
Plodia interpunctella	larva	25	76	Ferkovich, S. *et al.* (1975)
Galleria mellonella	larva	25		Rudnicka, M. *et al.* (1979)
Ostrinia nubilalis	larva	25		Goodman, W. G. (unpublished observations)
Diatraea crambidoides	larva	30		Turunen, S. and Chippendale, G. (1981)
Diatraea grandiosella	larva	28–30	3	Turunen, S. and Chippendale, G. (1979, 1981)
Paramyelois transitella	larva	30	4	Kramer, K. and Childs, C. (1977)
Pectinophora gossypiella	larva	30		Mumby, S. and Hammock, B. cited in Kramer, K. *et al.* (1976a)

[a] Modified from Goodman, W. (1983).
[b] JH III used as ligand.
[c] Only radiolabeled analogs tested.

for a protein-mediated delivery system. First, the titer of JH at the target cell must be relatively constant for most of each larval instar. Second, the hormone is surface-active and will adsorb non-specifically to various tissues, thereby reducing the effective concentration in the hemolymph (Akamatsu, Y. *et al.*, 1975; Law, J. 1980). Finally, in a number of insects circulating JH is metabolized by hemolymph esterases (see DeKort, C. and Granger, N. 1981; Chapter 14, this volume). These problems are mitigated by circulating hormone binding proteins (Goodman, W. 1983). Since the early 1970s the search for these proteins has been fruitful (see Table 1), yet the picture is far from complete.

4.1 Juvenile hormone binding lipoproteins

In many insects the lipoproteins represent a

predominant class of hemolymph proteins. Their association with, and transport of, non-polar compounds such as cholesterol, diglycerides and phospholipids suggests that this class of proteins might also transport the JHs. In their studies on *Hyalophora cecropia* hemolymph lipoproteins, Whitmore, E. and Gilbert, L. (1972) demonstrated that JH was bound to certain lipoproteins. Characterization of the lipoproteins by sodium bromide density gradient centrifugation revealed that the major binding lipoprotein was a high-density lipoprotein with a specific density of 1.15–1.17 $g\,ml^{-1}$ (Gilbert, L. *et al.*, 1976). Following this discovery, several investigators reported that lipoprotein–hormone associations represented the major transport mode for JH in a number of species (see Table 1).

With two exceptions in the order Orthoptera, the lipoproteins in all species studied have been characterized by their low affinity for the hormones ($K_a \leqslant 1 \times 10^5\ M^{-1}$) and the ability to bind more than one molecule of hormones per molecule of protein. The proteins display little specificity and bind most non-polar substances. In general, the molecular weights of these proteins exceed 100,000 daltons.

Since a large body of information has accumulated on the transport of JH in *Manduca sexta* hemolymph, the lipoproteins from this insect may serve as a model for the low-affinity, high-capacity juvenile hormone binding proteins. In the larval hemolymph of *Manduca*, six anodally migrating lipoproteins have been detected by electrophoretic separation and lipid crimson staining (Whitmore, E. and Gilbert, L. 1974). One of these proteins, a high-density lipoprotein, will bind JH I both *in vivo* and *in vitro* (Gilbert, L. *et al.*, 1976). Preliminary specificity studies concluded that this high-density lipoprotein (HDL) will bind JH I, JH I acid, various fatty acids and cholesterol. Examination of the binding data by Scatchard plot analysis (Scatchard, G. 1949) revealed that the association constant was $< 10^5\ M^{-1}$ for JH I.

Pattniak, N. *et al.* (1979) characterized an HDL with the same density from the hemolymph of *Manduca*. This protein is composed of 61% protein, 2% carbohydrate and 37% lipid, of which 14% is phospholipid and 23% is neutral lipid. It consists of two subunits, one of 2.85×10^5 daltons and the other of 8.1×10^4 daltons, which have a combined molecular weight of 6×10^5 daltons. Trypsin treatment of the protein indicated that its structural organization differs from that found in the HDLs of mammalian plasma. The mammalian HDLs are composed of a non-polar lipid core covered by a monolayer of phospholipid and cholesterol (Chapman, J. 1980) while the outer surface is composed of polypeptide chains interspersed between phospholipid head groups. In contrast, the smaller peptide chain in the *Manduca* HDL is buried within the molecule (no tryptic fragments recovered), while the larger polypeptide chain is in contact with the surface of the molecule (tryptic fragments recovered) (Pattniak, N. *et al.*, 1979). This structural difference may impose rigidity on the *Manduca* HDL, favoring the rapid exchange of lipophilic ligands such as diglycerides and JH (see chapter 5, volume 10).

A significant departure from the typical low-affinity and high-capacity characteristics of insect lipoproteins has been reported for JH binding hemolymph lipoproteins of two orthopteran species. In the adult grasshopper, *Gomphocerus rufus*, a high-affinity lipoprotein for JH I has been tentatively identified. Preliminary binding studies indicate the protein has an association constant of $2 \times 10^7\ M^{-1}$, which is comparable to that of the well-studied hemolymph proteins in Lepidoptera (Hartmann, R. 1978). Another orthopteran, *Locusta migratoria*, also displays a high-molecular weight JH binding lipoprotein with a relatively high affinity ($1.2 \times 10^7\ M^{-1}$) (Gellissen as cited in Hartmann, R. 1978). Some confusion has resulted from an earlier report that the hemolymph binding protein of *Locusta* was an unstable low-affinity lipoprotein (Emmerich, H. and Hartmann, R. 1973) but re-examination by gel filtration chromatography disclosed that a diglyceride binding protein and the JH binding protein co-migrate as a single peak. The *Locusta* JH binding protein is unique in its ability to discriminate between the JH III enantiomers 10*R* and 10*S*; however, it will not recognize the enantiomers of JH I (Peter, M. *et al.*, 1979). This stereochemical discrimination implies a unique binding site which has been described for only one other JH binding protein, the *Manduca sexta* JH-specific binding protein (Schooley, D. *et al.*, 1978). It will be of considerable interest to compare the homolog specificity of the *Locusta* hormone binding protein with that of the well-studied *Manduca* JH binding protein.

Although certain insect species utilize lipoproteins as the sole means of JH transport, in others the lipoproteins play an auxiliary role. When present in the hemolymph, high-affinity, low-molecular weight binding proteins probably contribute more to the overall regulation of JH titers than lipoproteins of low affinity and specificity. On the other hand, the relatively high concentration of lipoproteins as compared to the JH-specific binding proteins promotes their importance as carriers of JH in the hemolymph. Perhaps our understanding of the complementary role of lipoproteins and JH-specific binding proteins in JH transport and conservation will become clearer as more species from different orders are studied.

4.2 Juvenile hormone-specific binding proteins

In the hemolymph of certain insects the JHs associate not only with lipoproteins but with low molecular weight proteins which bind and transport the hormone specifically. These proteins were first observed in the hemolymph of *Manduca sexta* (Goodman, W. and Gilbert, L. 1974; Kramer, K. *et al.*, 1974). *In vivo* and *in vitro* incubation of [^{3}H]JH I with hemolymph indicated that the radiolabeled hormone was complexed to a single low molecular weight protein and that the primary metabolites of JH, i.e. JH acid, JH acid-diol, and JH diol (Chapter 13, this volume), did not bind. To test the hypothesis that endogenous JH associated with this protein *in vivo*, the low molecular weight JH-specific protein was rapidly purified from hemolymph using preparative slab gel electrophoresis. The area containing the JH binding protein (JHBP) was homogenized in organic solvent and the extract bioassayed (Gilbert, L. *et al.*, 1976; Goodman, W. *et al.*, 1978a). The organic solvent extract displayed considerable bioassay activity, and it was concluded that a hemolymph binding protein specific for JH can and does convey JH through the circulatory system.

4.2.1 Purification and characterization

At the present time only the JHBP from *Manduca* has been purified to apparent homogeneity (Kramer, K. *et al.*, 1976b; Goodman, W. *et al.*, 1978a; Peterson, R. *et al.*, 1982). The *Manduca* JHBP has been isolated from the hemolymph of both late fourth and mid-fifth instar larvae. Although the JHBPs from these two sources share many physical characteristics, they differ in composition and concentration. Qualitative and quantitative changes in hemolymph proteins at the different developmental periods affected the choice of purification methods, as did the paucity of starting material and the low concentration of binding protein in the hemolymph (20–40 μg ml^{-1}). As a consequence, two independent conventional purification schemes were developed. Kramer, K. *et al.* (1976b) employed gel filtration, ion exchange chromatography, and isoelectricfocusing, while Goodman, W. *et al.* (1978a) utilized gel filtration, ion exchange and hydroxylapatite chromatography, and preparative gel electrophoresis. A modification of the method presented by Kramer, K. *et al.* (1976b) has recently been presented which utilized SP-Sephadex as a final step (Peterson, R. *et al.*, 1982). Homogeneity of the final product was determined by sodium dodecyl sulfate gel electrophoresis, analytical gel electrophoresis (Kramer, K. *et al.*, 1976b; Goodman, W. *et al.*, 1978a; Peterson, R. *et al.*, 1982), immunoelectrophoresis (Goodman, W. *et al.*, 1978a), and analytical centrifugation (Kramer, K. *et al.*, 1976b).

A significant new development in the field of juvenile hormone binding protein purification is the use of affinity chromatography. Using JH II as the immobilized ligand, an affinity resin was recently synthesized which can rapidly and efficiently purify the hormone binding protein from *Manduca* hemolymph (Goodman, W. and Goodman, H. 1981). Since this procedure is new, a brief description of the synthesis and utilization of the matrix is in order. To a commercially prepared inert resin, Sepharose 4B, was attached a spacer arm of six carbons ending in a terminal amino group. JH II was converted to its acid by base hydrolysis and purified. The hormone acid was then incubated with the resin, *N*-hydroxysuccinimide, water-soluble carbodiimide and a catalyst, dimethylaminopyridine. Excess reagents were removed by extensive washing and the resin prepared for use. The hormone-binding eluate from gel filtration was stripped of JH, loaded onto the affinity column and exhaustively washed with a borate–glycerol buffer until no absorbance could be detected at 280 nm. The binding protein was then eluted from the column by the addition of 10 μM JH I (see Fig. 1). The binding protein represents the major constituent (85%) of the eluate, with enrichment better than 150-fold and a recovery of > 75%. Although this procedure does not remove all contaminants, the protein can be purified to homogeneity by chromatofocusing (W. Goodman, unpublished observations).

The utilization of this form of chromatography should greatly simplify the isolation of juvenile hormone associated proteins. The success of conventional purification techniques depends upon the composition of the preparation, requiring an entirely new procedure for each protein examined. Use of an affinity column is virtually independent of such considerations. A single protocol will allow one to search for various types of hormone-interacting proteins such as intracellular receptors, biosynthetic enzymes, catabolic enzymes and antisera preparations for immunological determinations of JH.

The physical characteristics of the JHBP from the hemolymph of *Manduca sexta* are presented in Table 2. The molecular weight of the protein from *Manduca* is similar to those of other lepidopteran species but whether this reflects similarities in the primary sequence and binding specificity is as yet undetermined. The relatively high association constant and single binding site of this macromolecule categorize it as a high-affinity, low-capacity binding protein. The association constant does not change dramatically with temperature (Peterson, R. 1977), reflecting the predominance of hydrophobic forces in the binding site. Sequence studies which were

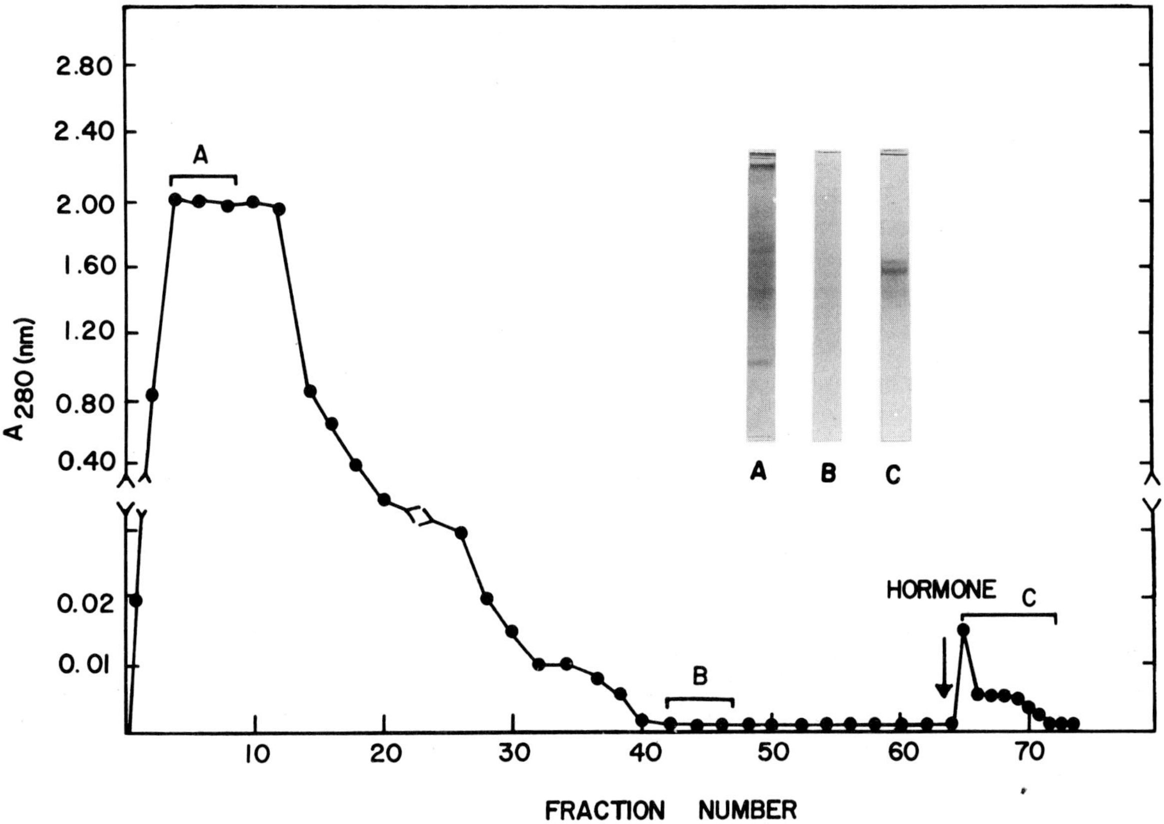

FIG. 1. Elution profile from an affinity column with JH II as the immobilized ligand. Hormone denotes the point at which JH I (10 μM) was introduced to elute the *Manduca* JHBP from the column. The starting material represented the JHBP fraction from preparative gel filtration chromatography of fourth instar *Manduca* hemolymph. The inset represents polyacrylamide gel electrophoretic profiles of: (**A**) fractions 3–7 from the affinity column; (**B**) fractions 43–47; (**C**) fractions 67–73 which represent the binding protein fraction (modified from Goodman, W. and Goodman, H. 1981).

recently performed on the amino terminus of the protein (Peterson, R. *et al.*, 1982) confirmed that the protein is a single polypeptide chain (Kramer, K. *et al.*; 1976b, Goodman, W. *et al.*, 1978a). The sequence is:

[1]Asp–Gln–Gly–Ala–Leu–Phe–Glu–Pro–Cys–Ser–Thr–Gln–Asp–Ile–Ala–Cys–Leu–Ser–?–Ala[20]

According to Peterson, R. *et al.* (1982), the cysteine residues are probably involved in covalent disulfide bonds which would restrict motion at the *N*-terminus.

A comparison of the physical characteristics of *Manduca* JHBP with the vertebrate plasma hormone binding proteins indicates that the insect protein is smaller and has lower affinity constants for each of its ligands (Natelson, S. and Natelson, E. 1980). These properties hold true for the entire spectrum of JH-specific insect binding proteins, with the exception of the specific lipoproteins of the orthopterans (see Table 1). The vertebrate plasma hormone binding proteins often contain > 20% carbohydrate. The carbohydrate presumably does not affect binding; rather, it serves as a glycocalyx, or inert shell, to limit hepatic degradation (Van Baelen, H. and Mannaerts, G. 1974). Although the *Manduca* JHBP has not been completely characterized, the amount of carbohydrate attached can be no more than a few percent. The comparatively slow recirculation of hemolymph through the fat body may preclude the need for this type of protection.

4.2.2 Binding site specificity

The correlation of biological response to binding activity provides information on the dimension and conformation of the binding site as well as insight into the forces involved in binding. The homologs and certain of their isomers have been examined for their ability to competitively displace radiolabeled JH I. The racemic homologs of JH 0, JH I, JH II, and JH III demonstrate different affinities for the binding protein based on their hydrophobicity (Goodman, W. *et al.*, 1976; Kramer, K. *et al.*, 1976a). The relative binding affinity decreases with increasing polarity in the order JH 0 > JH I > JH II > JH III over a 40-fold range. These findings are

Table 2: Some properties of the Manduca sexta *juvenile hormone binding protein*[a]

Molecular weight	
SDS gel electrophoresis	28,000
Gel filtration	29,000
Sedimentation equilibrium	28,000
$S_{20,w}(s)$	2.2
pI	5.0
Association constant (4°)	$1 \times 10^7\ M^{-1}$
Binding sites per molecule	1.0
Subunits	1

[a] From Goodman, W. (1983)

an extension of the polarity rule, which in large measure governs the interactions between steroid hormones and their serum binding proteins (Westphal, U. 1971).

A wide range of compounds structurally related to JH have been tested for displacement activity, including the metabolites of JH, fatty acids, and a number of potent insect growth regulators (Table 3). Only those compounds with an isoprenoid carbon skeleton similar in length to that of JH are effective. Although decreasing the polarity of the molecule enhances the activity, some substitutions which reduce polarity will inhibit binding. For example, methyl farnesenate, an analog of JH III which lacks the epoxide group, is more non-polar yet does not bind (Peterson, R. 1977). Apparently, hydrophilic groups such as the epoxide and carboxylic ester must be conserved for significant hormone–protein interactions to occur.

Using the homologs, geometrical and enantiomeric isomers, and analogs of JH as probes, one can construct a rough map of the binding domain of JHBP. The most comprehensive studies have focused on the racemic geometrical isomers of JH I (Goodman, W. *et al.*, 1978b) and the geometrical and stereochemical isomers of JH III (Peterson, R. *et al.*, 1977; Schooley, D. *et al.*, 1978). The preferential binding of the 2*E* isomer over the 2*Z* isomer suggests that hydrophobic interaction with the protein is enhanced by the methyl of the ester group. The decrease in binding activity of the 2*Z* isomer is probably due not to steric hindrance but to the introduction of the electrophilic carbomethoxy group into a hydrophobic region in the binding site. This contention is borne out by the inability to bind JH acid, which lacks the methyl group and contains a very polar carboxylate anion. The hydrophobic

Table 3: Specificity of the Manduca sexta *juvenile hormone binding protein*[a]

Compound	Ratio of association constants	Compound	Ratio of association constants
1. Homologs[b]		4. JH III Geometrical isomers[e]	
JH 0	3.36	2*E*, 6*E*	0.290
JH I	1.00	2*E*, 6*Z*	0.030
JH II	0.35	2*Z*, 6*E*	0.014
JH III	0.08	2*Z*, 6*Z*	0.007
2. Metabolites[b,c]	—	5. JH III Stereoisomers[f]	
JH I acid	—	JH III 10*R*	6.32
JH III acid	—	JH III 10*R*,*S*	1.00
JH III diol	—	JH III 10*S*	0.20
3. JH I Geometrical isomers[d]		6. Analogs, fatty acids and others[b,c]	—
2*E*, 6*E*, 10*cis*	1.00	ZR 512	—
2*E*, 6*E*, 10*trans*	0.64	ZR 515	—
2*E*, 6*Z*, 10*trans*	0.27	R20458	—
2*E*, 6*Z*, 10*cis*	0.12	palmitic acid	—
2*Z*, 6*E*, 10*cis*	0.08	oleic acid	—
2*Z*, 6*E*, 10*trans*	0.03	linolenic acid	—
2*Z*, 6*Z*, 10*cis*	0.02	methyl epoxy stearate	—
2*Z*, 6*Z*, 10*trans*	0.01	methyl epoxy palmitate	—
		Precocene II	—
		2*E*, 6*E*, Methyl farnesenate	—

[a] Since each group had a different ratio (R = unbound/bound in the absence of unlabeled hormone) and separation of bound and unbound was performed differently, the different groups cannot be compared directly. Dashes signify lack of displacement at the highest concentration tested.
[b] Goodman, W. *et al.*, 1976.
[c] Kramer, K. *et al.*, 1976a.
[d] Goodman, W. *et al.*, 1978b.
[e] Peterson, R. *et al.*, 1977 (calculation of relative affinities different than other groups).
[f] Schooley, D. *et al.*, 1978.

domain of this region is to a certain extent flexible, since addition of electron-repelling groups (up to three methylenes) increases the binding activity. Modification of the C-1 terminus of epoxy farnesenate with a methyl, ethyl, propyl or allyl ester enhances binding activity while introduction of more bulky substituents such as propargyl, butyl or benzyl esters reduces binding activity but does not entirely eliminate it (Peterson, R. 1977). The loss of activity due to bulky esters indicates that the binding domain has definite limits. The ester function can be replaced by an amide with little loss of binding activity since the bulk and polarity are similar (Peterson, R. 1977; Goodman, W. and Goodman, H. 1981). The addition of an extra carbon at the 3′ position significantly enhances relative binding activity, which suggests that lengthening the alkyl side chain brings it in closer contact with electron-repelling groups in the binding site.

The effect of substituting the 6*Z* for the 6*E* configuration leads to a marked reduction in binding activity which may best be explained by the altered intramolecular distances between functional groups. Assuming that the hormone binds to the protein in an extended conformation, changing the geometry will shorten the distance between the epoxide and the carbomethoxy function from ~1.4 to ~1.1 nm. Competition studies strongly suggest that the protein has distinct recognition sites for both substituents (see above). Since the epoxide and the carbomethoxy functions cannot fit the recognition site simultaneously, the 6*Z* isomer lacks the proper dimensions for optimal binding.

Changing the natural 10*cis* to the 10*trans* configuration of the epoxide diminishes the binding activity appreciably. Altering the geometry around the epoxide reverses the position of the ethyl (C-12,13) and methyl (C-11′) groups. This repositioning can interfere with hydrophobic interactions in two ways. Addition of a methylene group may impose steric hindrance, while loss of a methylene group may widen the distance to the amino acid

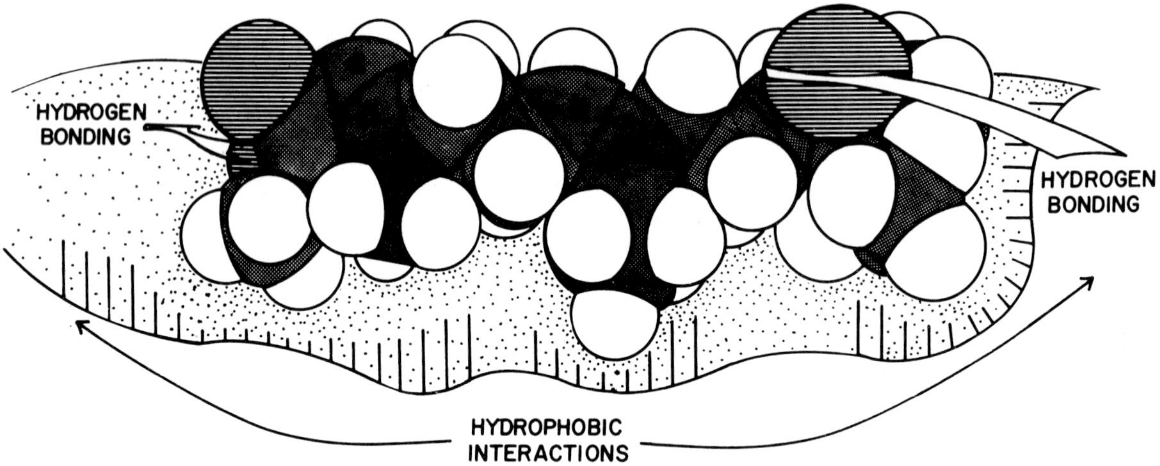

FIG. 2. Space-filling model of the naturally occurring JH I isomer and schematic interpretation of the *Manduca* JHBP binding site. The postulated "hydrophobic face" is directed into the binding groove. The molecule is represented in the most probable conformation.

residues of the binding site. Substitution of the 10*trans* for the 10*cis* also affects hydrophilic interactions that may occur between the epoxide and the binding site. Hydrogen bonding between the oxygen and an electron acceptor in the recognition site may be inferred from binding studies with JH analogs. Introduction of a double bond or a diol at C-10,11 effectively blocks any binding. The specificity of the hydrogen bond formed is demonstrated by the loss of binding when the naturally occurring enantiomer of JH III 10*R* is replaced with its antipode 10*S* or with a racemic mixture (Schooley, D. *et al.*, 1978).

From the evidence presented the binding site can be envisioned as a hydrophobic groove or pocket on the surface of the binding protein. The primary interactions between the hormone and the binding site occur along the alkyl side-chains and the methyl of the ester group, which together constitute a distinct hydrophobic "face" (Fig. 2). The aromatic and aliphatic side-chains of the amino acids in the binding site impart specificity to interactions with this "face". Hydrogen bonding which involves electrophilic groups in the binding domain anchors the ligand at the epoxide and the carbomethoxy group, with the distance between these substituents fixed.

4.2.3 BIOSYNTHESIS AND TITERS

Immunochemical studies demonstrated that JHBP is localized in the hemolymph, fat body and carcass (cuticle with attached epidermal cells and muscle). No JHBP was detected in corpora allata homogenates or medium in which corpora allata were cultured (Nowock, J. *et al.*, 1975). Synthesis of JHBP by the fat body, suggested by the relatively high concentrations found there, was confirmed by immunochemical and autoradiographic methods. Kinetic studies *in vitro* revealed that the newly synthesized protein is released rather than stored in the fat body. The fat body has now been shown to synthesize JH-specific binding protein in several other insects, including the Indian meal moth *Plodia interpunctella* (Ferkovich, S. *et al.*, 1977) and the grasshopper *Gomphocerus rufus* (Hartmann, R. 1978). In *Diatraea grandiosella*, an insect closely related to *Plodia*, investigators were unable to localize the JH-specific binding protein in the fat body or detect its release into culture medium (Turunen, S. and Chippendale, G. 1979).

In certain circumstances, limiting the concentration of JHBP could greatly influence the circulating titer of JH. Should the JH titer exceed equilibrium saturation levels, unbound hormone would be metabolized by hemolymph esterases (see Chapter 13, this volume). In fact, there is sufficient binding protein to complex with virtually all of the hormone at any given point during the fourth instar. Using multiple equilibrium dialysis and immunoprecipitation, Goodman, W. and Gilbert, L. (1978) studied the levels of JHBP in precisely timed first-gate fourth instar *Manduca sexta* larvae. They showed that binding activity peaks just after ecdysis, decreases during the first 24 h and remains low until hour 40, when ecdysteroid concentrations begin to climb (Bollenbacher, W. *et al.*, 1981). The JHBP concentration then increases rapidly until head capsule slippage at hour 55. The levels then begin to decline until ecdysis to the fifth instar at hour 72, when they have dropped to levels slightly lower than those found early in the fourth instar (see Fig. 3). A direct correlation between relative hemolymph JH titers (Fain, M. and Riddiford, L. 1975) and the level of JHBP was noted only at the beginning (0–36 h) of the fourth and fifth instars. During the rest of the fourth instar, JHBP and JH concentrations were inversely related. Based on the JHBP titers in *Trichoplusia ni*, Wing, K. *et al.* (1981) speculate that a similar relationship may emerge; however, identification and quantification of JH in this insect have yet to be performed.

The fluctuations in hormone and binding protein titers throughout the fourth instar raise questions about how JHBP levels are controlled. Since the hormone and the binding protein titers are directly related early in the instar and inversely related during the latter half of the instar, a hormone-related control mechanism is difficult to envision.

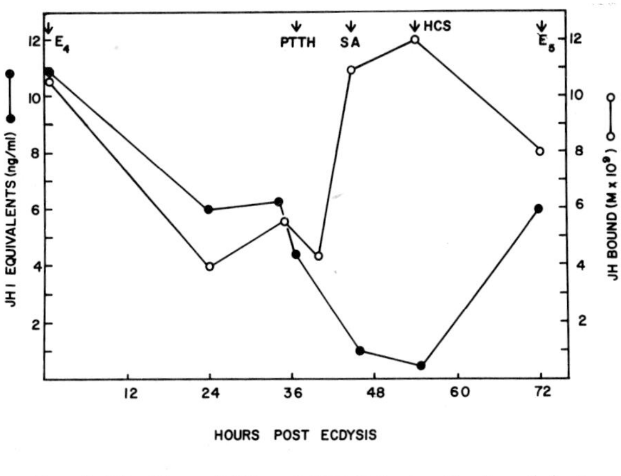

FIG. 3. The titers of JH and JHBP during the fourth instar (first gate) of *Manduca sexta*. JH titers were determined by bioassay (Fain, M. and Riddiford, L. 1975) for second gate-animals and expressed in JH I equivalents. These data were normalized to the first gate by correlating physiological markers such as ecdysis (E), body independent of the head (PTTH), spiracle apolysis (SA), and head capsule slippage (HCS).

On the other hand, the binding protein levels generally follow the fluctuations in overall hemolymph protein concentration. Until more sophisticated methods are developed for measuring and manipulating JH and JHBP titers, control mechanisms will remain obscure.

4.2.4 Functions

Although the most obvious function of the specific JHBP is the transport of hormone, more important is its role in protection of hormone from hemolymph esterases (Chapter 12, this volume). The esterases hydrolyze the methyl ester of JH to yield the biologically inactive JH acid. In *Manduca sexta* two types of esterases have been identified (Weirich, G. *et al.*, 1973). They can be distinguished *in vitro* by their sensitivity to organophosphate inhibitors and their appearance in the hemolymph during specific periods of post-embryonic development. Esterases sensitive to diisopropyl fluorophosphate (DFP), termed general esterases, can efficiently hydrolyze 1-naphthyl acetate and unbound JH, but are limited in their ability to hydrolyze JH bound to the binding proteins (Hammock, B. *et al.*, 1975; Sanburg, L. *et al.*, 1975a,b). During the last larval instar, a population of DFP-insenstitive esterases arise which are capable of hydrolyzing JH in either the bound or unbound state (see DeKort, C. and Granger, N. 1981); these are termed JH-specific esterases. While general esterases are present throughout development, the appearance of JH-specific esterases is linked to those critical periods when the hormone concentration must be minimized.

Another facet of the protective function is the selective conservation of the less polar homologs (Gilbert, L. *et al.*, 1978, 1980a). In those species whose corpora allata synthesize more than one JH homolog, the relative avidities for JHBP will affect the titers found in the hemolymph. Granger, N. *et al.* (1982) have demonstrated that the larval corpora allata of *Manduca* produce a significant amount of JH III *in vitro*, yet the hemolymph contains reduced amounts during the corresponding time period (Peter, M. *et al.*, 1976). Thus, even though the more polar homologs may be produced in significant amounts, the predominance of the less polar homologs in the hemolymph may reflect not only the synthetic output of the corpora allata but selective protection by the binding protein.

It has been proposed that in addition to its protective and transport functions the binding protein may act as a hemolymph "sink" to facilitate the removal of hormone from the corpora allata (Gilbert, L. *et al.*, 1976, 1978; Goodman, W. and Gilbert, L. 1978). In the case of albumin, a non-specific vertebrate steroid binding protein, Ewing, L. *et al.* (1976) have demonstrated that it establishes a gradient to remove hormone from the active gland *in vitro*.

A related premise is that JHBP may scavenge hormone from areas where it has been non-specifically bound, mobilizing it into the hemolymph where it can be metabolized by JH-specific esterases or excreted. This concept is attractive when one considers the highly surface-active nature of the hormone and its propensity to adsorb to non-specific sites such as blood cells and basement and plasma membranes (Akamatsu, Y. *et al.*, 1975; Law, J. 1980). Incubating fat body or epidermis with JH and JHBP *in vitro* enhances the recovery of the hormone from the medium (Nowock, J. *et al.*, 1976; Mitsui, T. *et al.*, 1979), presumably because non-specific adherence to tissue and surfaces in the culture vessel is circumvented. In the insects studied, a large pool of ligand-free protein is available to sequester JH in this manner. For sequestration to be necessary, however, there must be sufficient titers of unbound hormone adhering to peripheral tissues to elicit a biological response. Since virtually all of the circulating hormone is bound (Akamatsu, Y. *et al.*, 1975; Gilbert, L. *et al.*, 1978; Goodman, W. and Gilbert, L. 1978), the levels of unbound hormone remaining would be quite low and could be efficiently metabolized by hemolymph and epidermal esterases (Mitsui, T. *et al.*, 1979).

For those lipophilic hormones which must enter the target cell in order to elicit a biological response, internalization may occur by several different routes. The most widely accepted hypothesis assumes the simple diffusion of unbound hormone across the plasma membrane into the cytoplasm (Gorski, J. and Gannon, F. 1976). In cases other than simple diffusion, processes which involve the plasma binding proteins have been postulated. One model proposes that both hormone and binding protein are recognized at the cell surface and internalized as a complex (Rosenthal, H. *et al.*, 1974). Another model suggests that after the ligand–protein complex binds to the surface of the target cell (Heller, J. 1975) the ligand is then transferred to specific plasma membrane receptors (Maraini, G. *et al.*, 1977).

The involvement of JHBP in hormone recognition and/or transfer at the target site is supported by thermodynamic considerations, such as equilibrium constants, and dynamic parameters, such as rates of association and dissociation. Calculations based on the JHBP concentration (1 μM) (Goodman, W. *et al.*, 1978a) and maximal hormone concentration (10 nM JH I) (Peter, M. *et al.*, 1976) during the fourth instar of *Manduca sexta* and empirically derived values for the association constant ($K_a = 1 \times 10^7\ M^{-1}$) (Goodman, W. *et al.*, 1978a), indicate that virtually all of the hormone is bound to the specific binding protein throughout the instar. Thus, probability would favor recognition of bound rather than unbound hormone at the target cell. An examination of the rates of association (k_a or k_{on}) and dissociation (k_d or k_{off}) can likewise provide important information on the stability of the hormone–protein complex. In vertebrates the steroid hormone–serum binding protein complex is short-lived, with the half-time values ranging from milliseconds to a few seconds at physiological temperatures (Westphal, U. 1978). Virtually all of the data accumulated to date on the interaction of the JHs and the specific binding protein has been obtained using equilibrium or near–equilibrium conditions. However, it has been reported that the half-time of dissociation for the JH–JHBP complex is 15 min (Sanburg, L. *et al.*, 1975a). The slow dissociation rate lends credence to the possibility that a hormone–protein complex may be involved in recognition at the target cell.

The concept that JH and JHBP are recognized as a complex is further supported by experiments using a morphological assay in which JH blocks cuticle deposition induced by molting hormones. When imaginal discs were incubated *in vitro* in the presence of very high levels of JH I (100 μM), little inhibition of cuticle deposition was observed; when low concentrations of JHBP (4–8 nM) were added to the medium, a large synergistic effect was recorded (Sanburg, L. *et al.*, 1975a).

Further evidence for the role of JHBP in eliciting biological activity is derived from a novel experiment using the grasshopper, *Gomphocerus rufus* (Hartmann, R. 1978). When antiserum against the JHBP was injected into adult females, egg maturation was abolished. Since the involvement of JHBP in hormone activity is complex, it is difficult to dissect which function was disrupted by the antiserum. Transport, although impaired by the loss of the protein, could be partially restored by other hemolymph lipoproteins. It may be argued that the loss of egg maturation is directly linked to the loss of protective functions; however, other adult orthopteran species have been shown to compensate for titer loss such as unilateral allatectomy by doubling the output of the remaining gland (Stay, B. and Tobe, S. 1978). Through a feedback mechanism, the circulating JH titer is restored to control levels. Although the loss of binding protein may precipitate the decline in egg maturation in a variety of ways, it is possible that the net result is the loss of recognition at the target cell.

Other investigators have presented evidence that JH is recognized in the unbound state. The observation that the specific binding proteins do not bind the JH analogs or mimics has led to the conclusion that the binding proteins play no part in the recognition or internalization of JH (Nowock, J. *et al.*, 1976; Wyatt, G. and Pan, M. 1978). Since the most potent analogs are usually less polar than JH and display similar physicochemical characteristics, they are most likely transported via lipoproteins as described by Trautmann, K. (1972) and Turunen, S. and Chippendale, G. (1981). Once delivered to the peripheral tissue, the analog could dissociate and enter the cell in the unbound state. Conversely, the analog may remain bound and enter the cell via endocytosis of the lipoprotein (Brown, M. *et al.*, 1981). Within the cell, the analog may then follow the same pathway prescribed for the naturally occurring hormone. Since the naturally occurring hormones and the analogs are neither transported (Goodman, W. *et al.*, 1976; Kramer, K. *et al.*, 1976a) nor metabolized (Hammock, B. and Quistad, G. 1976) in the same manner, the result of analog studies may not accurately reflect the mode of action of the naturally occurring hormones (Chihara, C. *et al.*, 1972; Slade, M. and Wilkinson, C. 1974).

Using epidermal explants from larval *Manduca sexta* Riddiford, L. (1976) demonstrated that unbound JH I could inhibit the commitment to pupation. These data imply that the binding is not involved in recognition and internalization. However, the concentration of hormone necessary to maintain the uncommitted state is at least an order of magnitude higher than that observed *in vivo* (6 nM for early fourth instar and 1.7 nM for early fifth instar) (Peter, M. *et al.*, 1976). It could easily be argued that the high levels of hormone employed (>100 nM) will enter the cell non-specifically with or without the protein. This problem was re-examined in the presence of JHBP to demonstrate that the protein had little or no effect on the commitment to pupation of the explants (Mitsui, T. *et al.*, 1979). Calculation based on the concentrations of JH and JHBP used in these experiments and an affinity constant of 1×10^7 M^{-1} (Goodman, W. *et al.*, 1978a) indicates that less than 33% of the hormone is bound. Since the amount bound never approaches what is observed physiologically, assessment of these experiments remains inconclusive.

An analogous *in vitro* system in mammalian cells may shed some light on the above experiments. Despite its lipoidal nature, uncomplexed cholesterol at physiological concentrations cannot enter the cell unaided. The transfer process requires the presence of a specific serum transport protein, low-density lipoprotein, which binds and transports cholesterol across the plasma membrane via a receptor–endocytosis translocation mechanism (Brown, M. *et al.*, 1981). On the other hand, at non-physiological concentrations cholesterol will enter the cell to promote its effect with or without the lipoprotein (Goldstein, J. *et al.*, 1979).

It is clear that not enough is known about the dynamics of the insect endocrine system to eliminate any of the proposed hypotheses. The contradictory observations reported indicate more effort and new approaches are required to resolve the problems of recognition and internalization. One model that deserves particular attention is the vitamin A–retinol binding protein (RBP) complex in mammalian plasma (Goodman, D. 1975). While the RBP complex and the JHBP complex have a number of features in common, one is particularly relevant: the dynamics of binding are such that virtually all of the ligand circulates in the bound state.

In two very different cell types, the plasma RBP binds to a specific cell surface receptor where retinol is subsequently transferred to a cytoplasmic binding protein (Heller, J. 1975; Rask, L. and Peterson, P. 1976). This model has been elaborated by the discovery of a retinol binding macromolecule in the plasma membrane of one type of target cell (Maraini, G. *et al.*, 1977; Ottonello, S. and Maraini, G. 1981). This protein is distinctly different from either the plasma RBP or the cytoplasmic binding protein (Bashor, M. and Chytil, F. 1975) and may mediate the process of internalization. Although the retinol–RBP scheme for hormone recognition and internalization invites comparison with the JH–JHBP system, it is but one of several intriguing models by which the hormone may enter the target cell.

5 JUVENILE HORMONE RECEPTORS

We shall now turn our attention to the intracellular macromolecules that bind JH. The vertebrate "receptorologists" have outlined the following basic criteria for a cytoplasmic receptor: (1) affinity for the ligand should be high, generally exceeding that of the binding protein in blood by an order of magnitude; (2) the concentration of the receptor should be higher in target than in non-target tissue; (3) the binding site(s) should be saturable; (4) certain structurally related analogs should be capable of inhibiting binding of the ligand (Clark, J. and Peck, E. 1979). Although the cytosolic and nuclear JH binding proteins have not yet been characterized sufficiently, for convenience the term receptor will be used synonymously with intracellular binding protein.

5.1 Selection of a system for the study of hormone receptors

Classically, endocrine studies have been performed by observing the hormone-induced response of whole insects or organs. The use of such an experimental system is advantageous in that the response is often well-characterized and easily interpreted. It is, however, difficult to resolve whether the observed phenomenon is the direct result of the hormone added or rather the final expression of a chain of events. Moreover, the vast majority of receptor research requires monitoring of exogenously applied radiolabeled hormone. The presence of endogenous hormone or the unknown effects of previous endocrine conditioning may complicate these experiments.

As an alternative to whole insects or organs, insect cell lines may be used as model systems. Since the initial report of an insect cell line (Grace, T. 1962), a large number of other lines have subsequently been reported (Hink, W. 1980; chapter 17, volume 10). These insect cell lines have been of immense value in the study of insect endocrinology (for reviews see O'Connor, J. and Chang, E. 1981; Cherbas, L. and Cherbas, P. 1981; chapter 3, volume 8). Their use provides several distinct advantages. First, the cells can be cloned to provide genetically homogeneous populations of cells (Richard-Mollard, C. and Ohanessian, A. 1977). The cells can also be adapted to growth in copious amounts in suspension culture (Lengyel, J. *et al.*, 1975). In the case of cell lines obtained from *Drosophila,* the extensive genetic information available for this species permits the mechanism of control for various physiological processes to be examined at the molecular level (Sang, J. 1981). Lastly, the absence of hormones exogenous to the cell line but endogenous in whole tissue simplifies the design of experiments on hormone uptake and utilization. Certain cell lines which cannot metabolize exogenous hormones allow the investigator to manipulate an essentially "closed system".

Attempts to investigate JH receptors in whole tissue or organs have been successful in several instances. These have included the epidermis from *Manduca* (Riddiford, L. and Mitsui, T. 1978; chapter 2, volume 8) and *Drosophila hydei* (Klages, G. *et al.*, 1980); fat body (Engelmann, F. 1980, 1981) and ovary (Koeppe, J. *et al.*, 1981; chapter 6, volume 8) from the vitellogenic females of the cockroach *Leucophaea maderae*; and the fat body from the locust *Locusta migratoria* (Roberts, P. and Wyatt, G. 1980). JH receptors from an insect cell line have been characterized using the K_c line of *Drosophila melanogaster* (Chang, E. *et al.*, 1980).

5.2 "Characterization" of the receptors

Unlike the hemolymph JH binding proteins, the JH

receptors have yet to be fully characterized. The paucity of tissue, especially in whole insect preparations, and the instability of the receptor in homogenates have made progress slow. The widespread use of radiolabeled hormone to monitor receptor–ligand binding has been hampered by the low specific activity of commercially available hormone, its instability in complex preparations and the non-covalent nature of the interaction between the radiolabeled hormone and receptor. The advent of a photoaffinity-labeled analog of JH may present a solution to some of these problems (Chapter 6, volume 8; Prestwich and Chang, unpublished observations).

Despite these problems, certain physical properties of the receptors have been established. On the basis of velocity sedimentation centrifugation through sucrose gradients (Fig. 4) and molecular sieve chromatography (Fig. 5), an approximate molecular weight of 80,000 Daltons was obtained for the JH receptor from the K_c cell line (Chang, E. *et al.,* 1980).

Kinetic and equilibrium binding data are helpful in determining if an intracellular hormone binding macromolecule is indeed a receptor. By analysis using the Scatchard plot transform (Scatchard, G. 1949), the equilibrium dissociation constant (K_d)

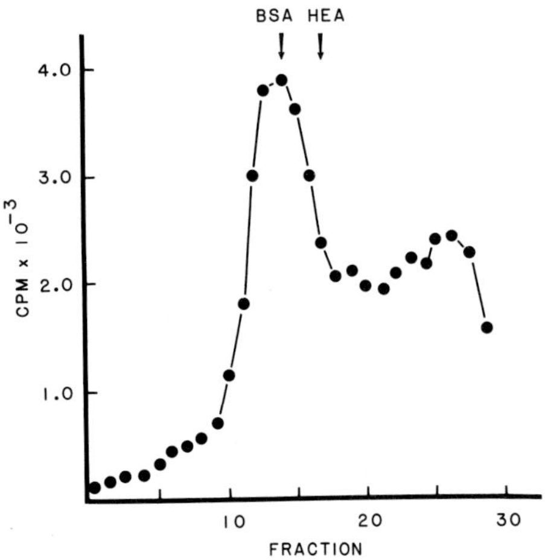

FIG. 4. Velocity sedimentation centrifugation of K_c cell cytosol through a preformed sucrose gradient (10-40%). Cytosol was preincubated with [^{3}H]-JH I for 20 min at 22° followed by 20 min at 4°. The sample (1.3 ml) was layered onto the sucrose and centrifuged at 65,000 rpm for 150 min at 4° (Sorvall TV-865 rotor). Standard proteins were bovine serum albumin (BSA) and hen egg albumin (HEA) (modified from Chang, E. *et al.,* 1980).

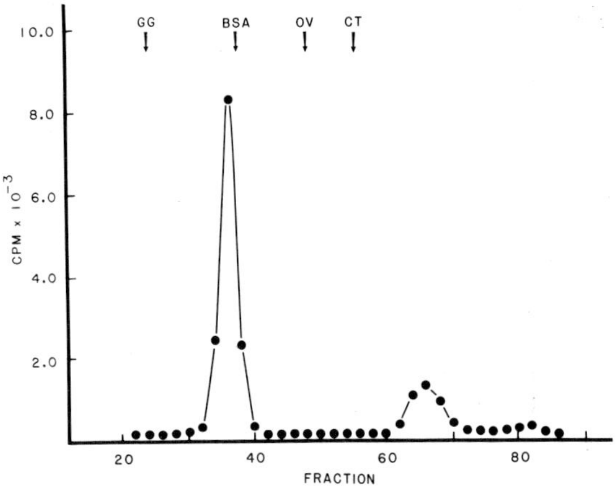

FIG. 5. Sephadex G-100 (superfine, 65 × 1.5 cm) chromatography of K_c cell cytosol (3.0 ml) incubated with [^{3}H]-JH I (20 nM) for 30 min at 21°. Standard proteins were gamma globulin (GG), bovine serum albumin (BSA), ovalbumin (OV), and chymotrypsinogen (CT).

has been determined for several cytoplasmic JH binding macromolecules (see Table 4). These values, ranging from 2.0 nM to 1.0 μM, may be compared to those of the hemolymph binding proteins as an index of relative affinity. In several cases, Scatchard plot analysis has revealed multiple binding components in the cytosol preparation. Interpretation of such phenomena is actually more complex than has been treated in the literature (for a discussion, see Norby, J. *et al.,* 1980). The kinetic rate constants of dissociation and association which are related to the equilibrium dissociation constant by the equation $K_d = k_d/k_a$ have been determined for the JH receptor in *Drosophila hydei* ($k_d = 6.5 \times 10^{-4}\,\mathrm{min}^{-1}$, $k_a = 2 \times 10^6\,\mathrm{M}^{-1}\,\mathrm{min}^{-1}$) and *Drosophila melanogaster* ($k_d = 1.3 \times 10^{-2}\,\mathrm{min}^{-1}$, $k_a = 1.3 \times 10^6\,\mathrm{M}^{-1}\,\mathrm{min}^{-1}$) (Figs. 7 and 8). The equilibrium dissociation constants obtained using these values are in good agreement with those calculated from the Scatchard plots. More importantly, the dissociation rate constant provides information on the temporal stability of the hormone–receptor complex. As a rule, the half-time of dissociation for receptors is rather long when compared to that of the hormone binding proteins of the hemolymph or serum. The slow dissociation rate of the receptor facilitates cellular uptake and retention, thereby increasing the local concentration of the hormone in the target cell (Westphal, U., 1978; Clark, J. and Peck, E., 1979).

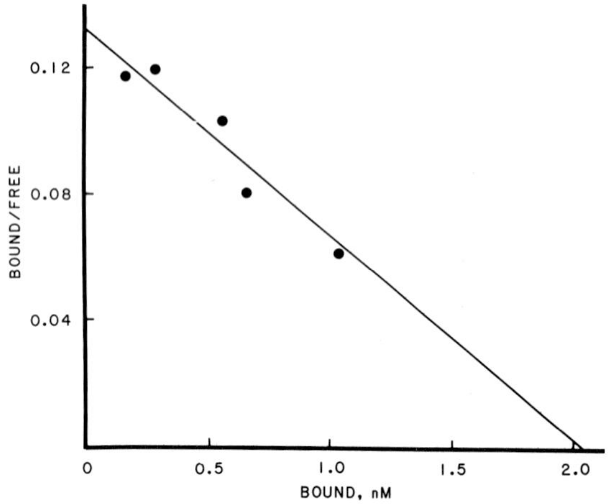

FIG. 6. Scatchard analysis of increasing amounts of radiolabeled hormone added to a constant amount of K_c cell cytosol. [^{3}H]-JH I (2–20 nM) was incubated with 12.5 μl of cytosol in a total volume of 100 μl for 3 h at 23°. The bound hormone was separated from the unbound by means of a dextran-coated charcoal assay. A K_d value is determined from the slope of the line. The line was drawn by the method of least squares (modified from Chang, E. *et al.*, 1980).

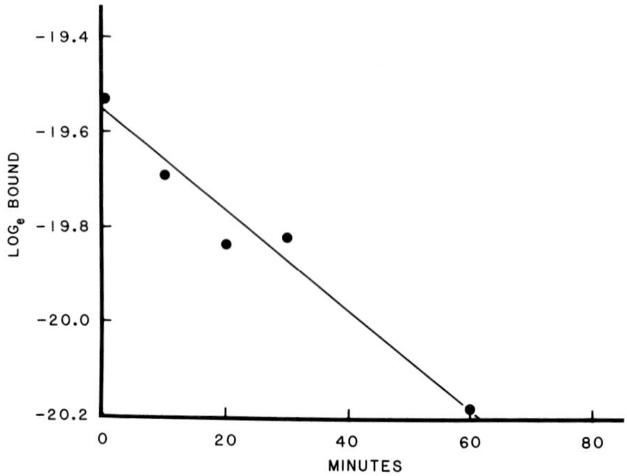

FIG. 7. Rate of dissociation of [^{3}H]-JH I from the cytoplasmic receptor from K_c cells. Radiolabeled hormone (10 nM) was incubated with a 50% solution of cytosol and buffer (10 mM Tris, 5 mM $MgCl_2$, 150 mM KCl, pH 6.9 at 22°) for 3 h and then excess unlabeled JH I was added. At the indicated times, aliquots were removed and assayed for bound hormone by means of a dextran-coated charcoal assay (modified from O'Connor, J. and Chang, E. 1981).

In the *Drosophila* K_c cell line, the cytoplasmic receptor does not bind the acid, diol or acid–diol metabolites of JH. When radiolabeled hormone bound by the receptor is extracted and analyzed by high-performance liquid chromatography, greater than 95% of the radiolabel co-chromatographs with authentic JH standards. In *Drosophila hydei* simple competition studies (Klages, G. *et al.*, 1980) between racemic JH I and the naturally occurring 10*R*,11*S* enantiomer indicated that the 10*R*,11*S* isomer binds preferentially. The receptor discriminates between JH I and JH III, with JH I displaying slightly higher displacement activity. The JH analog methyl farnesenate, which lacks biological activity, shows very poor displacement, as does the potent insect growth regulator ZR515. Specificity may be related to the prevalence of a given form of the hormone in a particular insect. JH receptors in the ovary of the roach *Leucophaea* bind to JH III, the predominant homolog, with the highest affinity. The preferred order of binding is JH III > JH I > JH 0 > ZR515 > JH III acid (Koeppe, J. *et al.*, 1981). In addition, the JH receptor from the *Drosophila* K_c cell line appears to be distinct from the ecdysteroid receptor (Maroy, P. *et al.*, 1978) since the former does not bind ecdysteroids (Chang, E. *et al.*, 1980; Engelmann, F., 1980) and the latter does not bind JH (Chang, E. unpublished observations).

Table 4: Characteristics of the cellular juvenile hormone receptors

Source	Size	Dissociation constant (K_d) (M × 10^9 at 4°)	Binding sites	Reference
Locusta migratoria				
Fat body	12–13S	2.0		Roberts, P. and Wyatt, G. (1980)
Leucophaea maderae				
Fat body		2.5	4.9×10^8 mg^{-1} tissue	Engelmann, F. (1980, 1981)
Ovaries		20.0		Koeppe, J. *et al.* (1981)
Drosophila hydei		3.6	4.6×10^{12} mg^{-1} tissue	Klages, G. *et al.* (1980)
Drosophila melanogaster	80,000 daltons	16.0	1.2×10^9 mg^{-1} cytosol	Chang, E. *et al.* (1980)

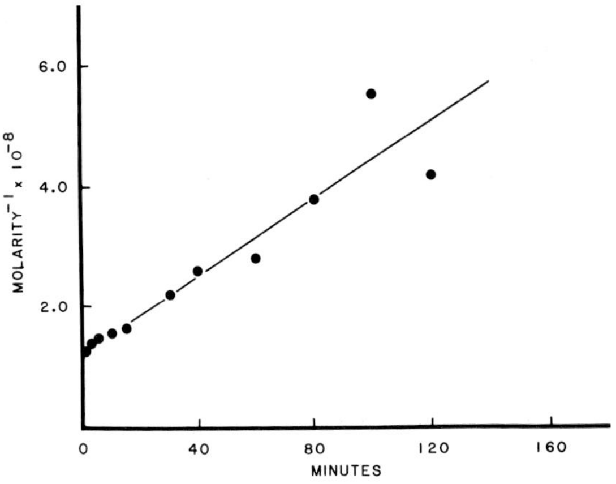

FIG. 8. Rate of association of [^{3}H]-JH I with the cytoplasmic receptor from K_c cells. Radiolabeled hormone (10 nM) was incubated with a 50% solution of cytosol and buffer (10 mM Tris, 5 mM $MgCl_2$, 150 mM KCl, pH 6.9 at 22°) containing 10 μM DFP. Parallel incubations were also carried out in which excess unlabeled JH I (10 μM) was added. Binding was determined by means of a dextran-coated charcoal assay (modified from O'Connor, J. and Chang, E. 1981).

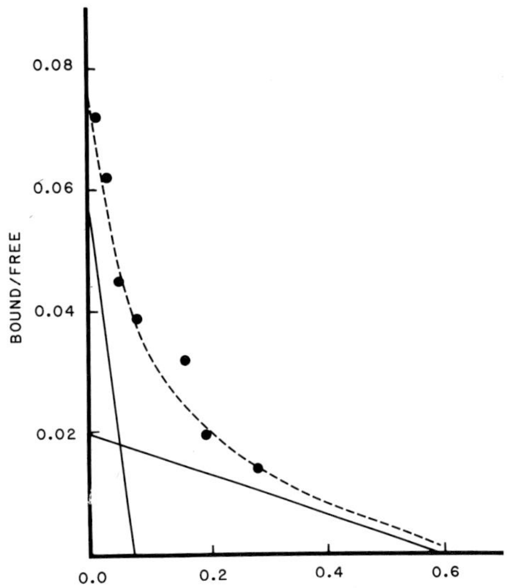

FIG. 9. Scatchard analysis of the unlabeled JH bound to K_c cell nuclei as a function of the amount of hormone added to whole cells. [^{3}H]-JH III (0.2–20 nM) was incubated with 1×10^8 cells ml^{-1} in 10 ml for 30 min. K_d values were determined from correction of the resulting data (Rosenthal, H. 1967) (Chang, E. and Bruce, M. unpublished observations).

5.3 Nuclear juvenile hormone receptors

The specificity and affinity of the cytoplasmic proteins which bind JH imply that they act as receptors *in vivo*. Very little evidence, however, has accumulated in regards to their role in mediating JH action. Experiments by Riddiford, L. and Mitsui, T. (1978) indicate that nuclear binding of JH may be involved in the expression of the hormone's effects. Using epidermal tissue from *Manduca*, they found nuclear JH receptors with a K_d of 2×10^{-8} M prior to pupal commitment. Evidence for nuclear receptors has also been obtained from the *Drosophila* K_c cell line. Nuclei isolated from cells that had never been exposed to exogenous JH did not bind radiolabeled hormone. Therefore it was necessary to incubate intact cells with labeled JH prior to nuclear isolation. Since the cells metabolize JH rapidly (O'Connor, J. and Chang, E. 1981), the esterase inhibitor 1-naphthyl-*N*-propyl carbamate was added to the incubation medium. Nuclei were then isolated and washed with Triton X-100. Determination of the radioactivity bound resulted in a K_d value of 1.2 nM by Scatchard analysis (Fig. 9) (Chang, E. and Bruce, M. unpublished observations). The radioactive material bound was shown to be identical with JH and could be displaced by excess unlabeled hormone (Fig. 10).

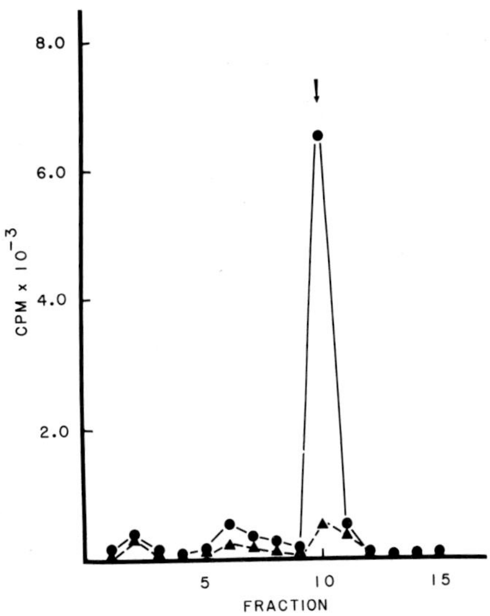

FIG. 10. Thin-layer chromatogram of [^{3}H]-JH III and its metabolites extracted from K_c cell nuclei incubated in the absence (circles) or presence (triangles) of a 1000-fold excess of unlabeled JH III. Samples were extracted with ethyl acetate and analyzed by thin-layer chromatography. Arrow represents the location of an authentic external JH III standard (Chang, E. and Bruce, M. unpublished observations).

These data indicate that JH may influence gene expression at the transcriptional level (Chapters 2 and 3, volume 8). The observed effects of JH upon incorporation of radiolabeled uridine into RNA tend to support this hypothesis (Congote, L. *et al.*, 1969; Chihara, C. and Fristrom, J. 1973; Gavin, J. and Williamson, J. 1978). The ability of JH to inhibit certain ecdysteroid-inducible puffs in *Drosophila* salivary gland chromosomes (Richards, G. 1978) also implies that JH acts at the transcriptional level. The data are insufficient to determine if the nuclear JH receptors are localized on either the chromatin or the nuclear membrane. Although JH has been proposed by different authors to act at both sites (Ashburner, M. and Cherbas, P. 1976; Kroeger, H. 1977), current research supports chromatin as the site of action.

ACKNOWLEDGMENTS

Some of the previously unpublished research presented in this chapter was generously supported by a grant from the American Cancer Society (CD 12) to E.S.C. A portion of the research was supported by grants from the College of Agricultural and Life Sciences and the Graduate School of the University of Wisconsin to W.G.G. We wish to thank our colleague H. L. Goodman for her excellent editorial effort, and Ms. Roxanne Moermond and Ms. Jean Slavin for their typing of the manuscript. This is in part a contribution from the College of Agricultural and Life Sciences of the University of Wisconsin. Literature search was completed in March of 1982.

REFERENCES

Akamatsu, Y., Dunn, P. E., Kezdy, F. J., Kramer, K. J., Law, J. H., Reibstein, D. and Sanburg, L. L. (1975). Biochemical aspects of juvenile hormone action in insects. In *Control Mechanisms in Development*, pp.123–149. Edited by R. H. Meintz and E. Davis. Plenum Press, New York.

Ashburner, M and Cherbas, P. (1976). The control of puffing by ions — the Kroeger hypothesis: a critical review. *Molec. Cell. Endocr. 5*, 89–107.

Bashor, M. and Chytil, F. (1975). Cellular retinol-binding protein. *Biochem. Biophys. Acta 411*, 87–96.

Bassi, S. D., Goodman, W., Altenhofen, C. and Gilbert, L. I. (1977). The binding of exogenous juvenile hormone by the haemolymph of *Oncopeltus fasciatus*. *Insect Biochem. 7*, 309–312.

Bhat, M. K. and Cama, H. R. (1979). Gonadal cell surface receptor for plasma retinol-binding protein. A method for its radioassay and studies on its level during spermatogenesis. *Biochem. Biophys. Acta 587*, 273–281.

Bollenbacher, W. E., Smith, S. L., Goodman, W. and Gilbert, L. I. (1981). Ecdysteroid titer during larval–pupal–adult development of the tobacco hornworm, *Manduca sexta*. *Gen. Comp. Endocr. 44*, 302–306.

Brown, M. S., Kovanen, P. T. and Goldstein, J. L. (1981). Regulation of plasma cholesterol by lipoprotein receptors. *Science 212*, 628–635.

Burger, R. L., Mehlman, C. S. and Allen, R. H. (1975). Human plasma R-type vitamin B_{12} binding proteins: isolation and characterization of transcobalamin I, transcobalamin III, and the normal granulocyte vitamin B_{12} binding protein. *J. Biol. Chem. 250*, 7700–7713.

Chang, E. S., Coudron, T. A., Bruce, M. J., Sage, B. A., O'Connor, J. D. and Law, J. H. (1980). Juvenile hormone-binding protein from the cytosol of *Drosophila* K_c cells. *Proc. Nat. Acad. Sci. 77*, 4657–4661.

Chapman, J. (1980). Animal lipoproteins: chemistry, structure, and comparative aspects. *J. Lipid Res. 21*, 789–853.

Chen, C.-C. and Heller, J. (1977). Uptake of retinol and retinoic acid from serum retinol-binding protein by retinal pigment epithelial cells. *J. Biol. Chem. 252*, 5216–5221.

Cherbas, L. and Cherbas, P. (1981). The effects of ecdysteroid hormones on *Drosophila melanogaster* cell line. In *Advances in Cell Culture*, Vol. 1, pp.91–124. Edited by K. Maramorosch. Academic Press, New York.

Chihara, C., Petri, W., Fristrom, J. and King, D. S. (1972). The assay of ecdysones and juvenile hormones on *Drosophila* imaginal discs *in vitro*. *J. Insect Physiol. 18*, 1115–1124.

Chihara, C. and Fristrom, J. (1973). Effects and interactions of juvenile hormone and β-ecdysone on *Drosophila* imaginal discs cultured *in vitro*. *Devel. Biol. 35*, 36–46.

Clark, J. H. and Peck, E. J. (1979). *Female Sex Steroids*. Springer, New York.

Congote, L. F., Sekeris, C. E. and Karlson, P. (1969). On the mechanism of hormone action. XIII. Stimulation effects of ecdysone, juvenile hormone and ions on RNA synthesis in fat body cell nuclei from *Calliphora erythrocephala* isolated by a filtration technique. *Exp. Cell Res. 56*, 338–346.

deKort, C. A. D. and Granger, N. A. (1981). Regulation of the juvenile hormone titer. *Ann. Rev. Ent. 26*, 1–28.

Diamond, M., Rust, N. and Westphal, U. (1969). High-affinity binding of testosterone and cortisol in normal and androgen treated guinea pigs during various reproductive stages; relationship to masculinization. *Endocr. 84*, 1143–1151.

Emmerich, H. and Hartmann, R. (1973). A carrier lipoprotein for juvenile hormone in the haemolymph of *Locusta migratoria*. *J. Insect Physiol. 19*, 1663–1675.

Engelmann, F. (1980). Endocrine control of vitellogenin synthesis. In *Insect Biology in the Future*, pp.311–324. Edited by M. Locke and D. S. Smith. Academic Press, New York.

Engelmann, F. (1981). Heterogeneity of juvenile hormone binding compounds in fat bodies of a cockroach. *Molec. Cell. Endocr. 24*, 103–112.

Ewing, L., Chubb, C. and Robarie, B. (1976). Macromolecules, steroid binding and testosterone secretion by rabbit testis. *Nature 264*, 84–86.

Fain, M. J. and Riddiford, L. M. (1975). Juvenile hormone titers in the hemolymph during late larval development of the tobacco hornworm, *Manduca sexta*. *Biol. Bull. 149*, 506–521.

Ferkovich, S. M., Silhacek, D. L. and Rutter, R. R. (1975). Juvenile hormone binding proteins in the haemolymph of the Indian meal moth. *Insect Biochem. 5*, 141–150.

Ferkovich, S., Oberlander, H. and Rutter, R. (1977). Release of a juvenile hormone binding protein by fat body of the Indian meal moth, *Plodia interpunctella in vitro*. *J. Insect Physiol. 23*, 297–302.

Gavin, J. A. and Williamson, J. H. (1978). Effects of alpha-amanitin and juvenile hormone analogue (ZR515) on labelling of RNA and protein in adult *Drosophila*. *J. Insect Physiol. 24*, 413–416.

Gilbert, L. I. and Chino, H. (1974). Transport of lipids in insects. *J. Lipid Res. 15*, 339–456.

GILBERT, L. I., GOODMAN, W. and NOWOCK, J. (1976). The possible roles of binding proteins in juvenile hormone metabolism and action. In *Biosynthesis, Metabolism and Cellular Actions,* pp.413–434. C.N.R.S., Paris.

GILBERT, L. I., GOODMAN, W. and GRANGER, N. A. (1978). Regulation of juvenile hormone titer in Lepidoptera. In *Comparative Endocrinology,* pp.471–486. Edited by P. Gillard and H. Boer. Elsevier/North-Holland Biomedical, New York.

GILBERT, L. I., BOLLENBACHER, W. E., GOODMAN, W., SMITH, S. L., AGUI, N., GRANGER, N. and SEDLAK, B. J. (1980a). Hormones controlling insect metamorphosis. *Rec. Prog. Hormone Res. 36,* 401–449.

GILBERT, L. I., BOLLENBACHER, W. E. and GRANGER, N. A. (1980b). Insect endocrinology: regulation of endocrine glands, hormone titer, and hormone metabolism. *Ann. Rev. Physiol. 42,* 493–510.

GLASS, J., NUNEZ, M. T., FISCHER, S. S. and ROBINSON, S. H. (1978). Transferrin receptors, iron transport and ferritin metabolism in Friend erythroleukemia cells. *Biochem. Biophys. Acta 542,* 154–162.

GOLDSTEIN, J. L., HELGESON, J. A. S. and BROWN, M. S. (1979). Inhibition of cholesterol synthesis with compactin renders growth of cultured cells dependent on the low density lipoprotein receptor. *J. Biol. Chem. 254,* 5403–5409.

GOODMAN, D. S. (1975). Retinol-binding protein, prealbumin, and vitamin A transport. In *Progress in Clinical and Biological Research,* Vol. 5, pp.313–330. Edited by G. Jamieson and T. Greenwalt. Alan Liss Publications, New York.

GOODMAN, W. G. (1983). Hemolymph transport of ecdysteroids and juvenile hormones. In *Insect Endocrinology,* Vol. 2, pp. 49–57. Edited by R. H. Downer and H. Laufer. Alan Liss Pub., New York.

GOODMAN, W. and GILBERT, L. I. (1974). Hemolymph protein binding of juvenile hormone in *Manduca sexta. Amer. Zool. 14,* 1289.

GOODMAN, W. and GILBERT, L. I. (1978). The hemolymph titer of juvenile hormone binding protein and binding sites during the fourth larval instar of *Manduca sexta. Gen. Comp. Endocr. 35,* 27–34.

GOODMAN, W. G. and GOODMAN, H. L. (1981). Development of affinity chromatography for juvenile hormone binding proteins. In *Juvenile Hormone Biochemistry,* pp.365–374. Edited by G. E. Pratt and G. T. Pratt. Elsevier/North-Holland Biomedical, New York.

GOODMAN, W., BOLLENBACHER, W. E., ZVENKO, H. L. and GILBERT, L. I. (1976). A competitive binding protein assay for juvenile hormone. In *The Juvenile Hormones,* pp.75–95. Edited by L. I. Gilbert. Plenum Press, New York.

GOODMAN, W., O'HERN, P. A., ZAUGG, R. H. and GILBERT, L. I. (1978a). Purification and characterization of a juvenile hormone binding protein from the hemolymph of the fourth instar tobacco hornworm, *Manduca sexta. Molec. Cell. Endocr. 11,* 225–242.

GOODMAN, W., SCHOOLEY, D. A. and GILBERT, L. I. (1978b). Specificity of the juvenile hormone binding protein: the geometrical isomers of juvenile hormone I. *Proc. Nat. Acad. Sci. 75,* 185–189.

GORSKI, J. and GANNON, F. (1976). Current models of steroid hormone action: a critique. *Ann. Rev. Physiol. 38,* 425–450.

GRACE, T. D. C. (1962). Establishment of four strains of cells from insect tissues grown *in vitro. Nature 195,* 788–789.

GRANGER, N. A., NIEMIEC, S.M., GILBERT, L. I. and BOLLENBACHER, W. E., (1982). JH III biosynthesis by larval corpora allata of *Manduca sexta. J. Insect Physiol. 28,* 385–391.

HAMMOCK, B. D. and QUISTAD, G. B. (1976). The degradative metabolism of juvenoids by insects. In *The Juvenile Hormones,* pp. 374–393. Edited by L. I. Gilbert. Plenum Press, New York.

HAMMOCK, B. D., NOWOCK, J., GOODMAN, W., STAMOUDIS, V. and GILBERT, L. I. (1975). The influence of hemolymph-binding protein on juvenile hormone stability and distribution in *Manduca sexta* fat body and imaginal discs *in vitro. Molec. Cell. Endocr. 3,* 167–184.

HAMMOCK, B. D., SPARKS, T. and MUMBY, S. (1977). Selective inhibition of JH esterases from cockroach hemolymph. *Pestic. Biochem. Physiol. 7,* 517–530.

HARTMANN, R. (1978). The juvenile hormone-carrier in the haemolymph of the acrine grasshopper *Gomphocerus rufus* L.: blocking of the juvenile hormone's action by means of repeated injections of an antibody to the carrier. *Wilhelm Roux's Archiv. 184,* 301–324.

HELLER, J. (1975). Interactions of plasma retinol-binding protein with its receptor. *J. Biol. Chem. 250,* 3613–3619.

HINK, W. F. (1980). The 1979 compilation of invertebrate cell lines and culture media. In *Invertebrate Systems In Vitro,* pp.553–578. Edited by E. Kurstak, K. Maramorosch, and A. Dubendorfer. Elsevier/North-Holland Biomedical, New York.

KARIN, M. and MINTZ, B. (1981). Receptor-mediated endocytosis of transferrin in developmentally totipotent mouse teratocarcinoma stem cells. *J. Biol. Chem. 256,* 3245–3252.

KLAGES, G. and EMMERICH, H. (1979). Juvenile hormone binding proteins in the haemolymph of third instar larvae of *Drosophila hydei. Insect Biochem. 9,* 23–30.

KLAGES, G., EMMERICH, H. and PETER, M. G. (1980). High-affinity binding sites for juvenile hormone I in the larval integument of *Drosophila hydei. Nature 286,* 282–285.

KOEPPE, J. K., KOVALICK, G. E. and LAPOINTE, M. C. (1981). Juvenile hormone interactions with ovarian tissue in *Leucophae maderae.* In *Juvenile Hormone Biochemistry,* pp. 215–213. Edited by G. E. Pratt and G. T. Brooks. Elsevier/North-Holland Biomedical Press, New York.

KRAMER, K. J. and CHILDS, C. N. (1977). Interaction of juvenile hormone with carrier proteins and hydrolases from insect haemolymph. *Insect Biochem. 7,* 397–403.

KRAMER, K. J., DUNN, P. E., PETERSON, R. C. and LAW, J. H. (1976a). Interaction of juvenile hormone with binding proteins in insect hemolymph. In *The Juvenile Hormones.* Edited by L. I. Gilbert, pp. 327–341. Plenum Press, New York.

KRAMER, K. J., DUNN, P. E., PETERSON, R. C., SEBALLOS, H. and LAW, J. H. (1976b). Purification and characterization of the carrier protein for juvenile hormone from the hemolymph of the tobacco hornworm *Manduca sexta* Johannson (Lepidoptera: Sphingidae). *J. Biol. Chem. 251,* 4979–4985.

KRAMER, K. J., SANBURG, L. L., KEZDY, F. and LAW, J. H. (1974). The juvenile hormone binding protein in the hemolymph of *Manduca sexta* (Johannson) (Lepidoptera: Sphingidae). *Proc. Nat. Acad. Sci. 71,* 493–497.

KRAMER, S. and DEKORT, C. A. D. (1978). Juvenile hormone carrier lipoproteins in the hemolymph of the Colorado potato beetle, *Leptinotarsa decemlineata. Insect Biochem. 8,* 87–92.

KROEGER, H. (1977). The control of puffing by ions: a reply. *Mol. Cell. Endocr. 7,* 105–110.

LAW, J. H. (1980). Lipid–protein interactions in insects. In *Insect Biology in the Future,* pp.295–309. Edited by M. Locke and D. S. Smith. Academic Press, New York.

LENGYEL, J., SPRADLING, A. and PENMAN, S. (1975). Methods with insect cells in suspension culture. II. *Drosophila melanogaster.* In *Methods in Cell Biology,* vol 10, pp.195–208. Edited by D. M. Prescott. Academic Press, New York.

MARAINI, G., OTTONELLO, S., GOZZOLI, F. and MERLI, A. (1977). Identification of a membrane protein binding the retinol in retinal pigment epithelium. *Nature 265,* 68–69.

MAROY, P., DENNIS, R., BECKERS, C., SAGE, B. A. and O'CONNOR, J. D. (1978). Demonstration of an ecdysteroid receptor in a cultured cell line of *Drosophila melanogaster. Proc. Nat. Acad. Sci. 75,* 635–638.

MCPHERSON, A., MICKELSON, K. E. and WESTPHAL, U. (1980). Crystallization of corticosteroid binding globulin and α_1-acid glycoprotein. *J. Steroid Biochem. 13,* 991–992.

MITSUI, T., RIDDIFORD, L. M. and BELLAMY, G. (1979). Metabolism of juvenile hormone by the epidermis of the tobacco hornworm, *Manduca sexta. Insect Biochem. 9,* 637–643.

NATELSON, S. and NATELSON, E. (1980). *Principles of Applied Clinical Chemistry,* vol. 3, pp.429–501. Plenum Press, New York.

NORBY, J. G., OTTOLENGHI, P. and JENSEN, J. (1980). Scatchard plot: common misinterpretation of binding experiments. *Analyt. Biochem. 102,* 318–320.

NOWOCK, J., GOODMAN, W., BOLLENBACHER, W. E. and GILBERT, L. I. (1975). Synthesis of juvenile hormone binding proteins by the fat body of *Manduca sexta. Gen. Comp. Endocr. 27,* 230–239.

NOWOCK, J., HAMMOCK, B. D. and GILBERT, L. I. (1976). The binding protein as a modulator of juvenile hormone stability and uptake. In *The Juvenile Hormones,* pp.354–373. Edited by L. I. Gilbert. Plenum Press, New York.

O'CONNOR, J. D. and CHANG, E. S. (1981). Cell lines as a model for the study of metamorphosis. In *Metamorphosis: A Problem in Developmental Biology,* 2nd edition, pp.241–261. Edited by L. I. Gilbert and E. Frieden. Plenum Press, New York.

OTTONELLO, S. and MARAINI, G. (1981). Retinol uptake by isolated retinal pigment epithelium plasma membranes. *Exp. Eye Res. 32,* 69–75.

PALMOUR, R. M. and SUTTON, H. E. (1971). Vertebrate transferrins: molecular weights, chemical composition and iron binding studies. *Biochemistry 10,* 4026–4032.

PATTNIAK, N. M., MUNDALL, E. C., TRAMBUSTI, B. G., LAW, J. H. and KEZDY, F. J. (1979). Isolation and characterization of a larval lipoprotein from the hemolymph of *Manduca sexta. Comp. Biochem. Physiol. 6B,* 469–476.

PETER, M. G., DAHM, K. H. and RÖLLER, H. (1976). The juvenile hormones in blood of larvae and adults of *Manduca sexta* (Joh.). *Z. Naturf. 31c,* 129–131.

PETER, M. G., GUNAWAN, S., GELLISSEN, G. and EMMERICH, H. (1979). Differences in hydrolysis and binding of homologous juvenile hormones in *Locusta migratoria* hemolymph. *Z. Naturf. 34,* 588–598.

PETERSON, R. C. (1977). Juvenile hormone carrier protein from *Manduca sexta*: an improved purification, determination of binding specificity and selected modification. Ph.D. thesis, University of Chicago.

PETERSON, R. C., DUNN, P. E., SEBALLOS, H. L., BARBEAU, B. K., KEIM, P. S., RILEY, C. T., HEINRIKSON, R. L. and LAW, J. H. (1982). Juvenile hormone carrier protein of *Manduca sexta* haemolymph. Improved purification procedure: protein modification studies and sequence of the amino terminus of the protein. *Insect Biochem. 12,* 643–650.

PETERSON, R. C., REICH, M., DUNN, P. E., LAW, J. H. and KATZENELLENBOGEN, J. (1977). Binding specificity of the juvenile hormone carrier protein from the hemolymph of the tobacco hornworm *Manduca sexta* (Johannson) (Lepidoptera: Sphingidae). *Biochemistry 16,* 2305–2311.

RASK, L. and PETERSON, P. A. (1976). *In vitro* uptake of vitamin A from the retinol-binding plasma protein to mucosal epithelial cells from the monkey's small intestine. *J. Biol. Chem. 251,* 6360–6366.

RICHARD-MOLLARD, C. and OHANESSIAN, A. (1977). Méthode de clonage des cellules de Drosophile. Sensibilité aux rayons X de plusieurs lignées cellulaires. *Wilhelm Roux's Archives 181,* 135–149.

RICHARDS, G. (1978). Sequential gene activation by ecdysone in polytene chromosomes of *Drosophila melanogaster.* VI. Inhibition by juvenile hormones. *Devel. Biol. 66,* 32–42.

RIDDIFORD, L. M. (1976). Hormonal control of insect epidermal cell commitment *in vitro. Nature 259,* 115–117.

RIDDIFORD, L. M. and MITSUI, T. (1978). Loss of cellular receptors for juvenile hormone during the change in commitment of the epidermis of the tobacco hornworm, *Manduca sexta.* In *Comparative Endocrinology,* p.519. Edited by P. J. Gaillard and H. Boer. Elsevier, Amsterdam.

RIDDIFORD, L. M. and TRUMAN, J. W. (1978). Biochemistry of insect hormones and insect growth regulators. In *Biochemistry of Insects,* pp.308–355. Edited by M. Rockstein. Academic Press, New York.

ROBERTS, P. E. and WYATT, G. R. (1980). Juvenile hormone binding proteins from fat bodies and from hemolymph of *Locusta migratoria. Amer. Zool. 20,* 935.

ROSENTHAL, H. E. (1967). A graphic method for the determination and presentation of binding parameters in a complex system. *Analyt. Biochem. 20,* 525–532.

ROSENTHAL, H. E., PAUL, M. and SANDBERG, A. A. (1974). Transcortin: a corticosteroid-binding protein of plasma. XII. Immunologic studies on transcortin in guinea pig tissues. *J. Steroid Biochem. 5,* 219–225.

RUDNICKA, M., SEHNAL, F., JAROLIM, V. and KOCHMAN, M. (1979). Hydrolysis and binding of the juvenile hormone in the haemolymph of *Galleria mellonella. Insect Biochem. 9,* 569–575.

SANBURG, L. L., KRAMER, K. J., KEZDY, F. J., LAW, J. H. and OBERLANDER, H. (1975a). Role of juvenile hormone esterases and carrier proteins in insect development. *Nature 253,* 266–267.

SANBURG, L. L., KRAMER, K. J., KEZDY, F. J. and LAW, J. H. (1975b). Juvenile hormone-specific esterases in the hemolymph of the tobacco hornworm, *Manduca sexta. J. Insect Physiol. 21,* 873–888.

SANG, J. H. (1981). *Drosophila* cells and cell lines. In *Advances in Cell Culture,* vol. 1, pp.125–182. Edited by K. Maramorosch. Academic Press, New York.

SCATCHARD, G. (1949). The attractions of proteins for small molecules and ions. *Ann. N.Y. Acad. Sci. 51,* 660–672.

SCHOOLEY, D. A., BERGOT, J., GOODMAN, W. and GILBERT, L. I. (1978). Synthesis of both optical isomers of insect juvenile hormone III and their affinity for the JH-specific binding protein of *Manduca sexta. Biochem. Biophys. Res. Commun. 81,* 743–749.

SLADE, M. and WILKINSON, C. (1974). Juvenile hormone analogs: A possible case of mistaken identity? *Science 181,* 672–674.

SPARKS, T. and HAMMOCK, B. D. (1979). A comparison of the induced and naturally occurring juvenile hormone esterases from last instar larvae of *Trichoplusia ni. Insect Biochem. 9,* 411–421.

SPARKS, T. C., WILLIS, W. S., SHOREY, H. H. and HAMMOCK, B. D. (1979). Haemolymph juvenile hormone esterase activity in synchronous last instar larvae of the cabbage looper, *Trichoplusia ni. J. Insect Physiol. 25,* 125–132.

SPECTOR, A. A. (1975). Fatty acid binding to plasma albumin. *J. Lipid Res. 16,* 165–174.

STAY, B. and TOBE, S. S. (1978). Control of juvenile hormone biosynthesis during the reproductive cycle of a viviparous cockroach. II. Effects of unilateral allatectomy, implantation of supernumerary corpora allata and ovariectomy. *Gen. Comp. Endocr. 34,* 276–286.

TRAUTMANN, K. (1972). *In vitro* Studium dur Trägerproteine von [^{3}H]-markierten juvenilhormonwirksamen Verbindungen in der Hamolymphe von *Tenebrio molitor* L. larvae. *Z. Naturf. 27b,* 263–273.

TROWBRIDGE, I. S. and OMARY, M. B. (1981). Human cell surface glycoprotein related to cell proliferation is the receptor for transferrin. *Proc. Nat. Acad. Sci. 78,* 3039–3043.

TURUNEN, S. and CHIPPENDALE, G. M. (1979). Possible function of juvenile hormone-dependent protein in larval insect diapause. *Nature 280,* 836–838.

TURUNEN, S. and CHIPPENDALE, G. M. (1981). Binding of juvenile hormone, methoprene and hydroprene to haemolymph proteins of larvae of the Southwestern corn borer, *Diatraea grandiosella. Insect Biochem. 11,* 429–435.

VAN BAELAN, H. and MANNAERTS, G. (1974). The role of sialic acid in determining the survival of transcortin in the circulation. *Archiv. Biochem. Biophys. 163,* 53–56.

WEIRICH, G., WREN, J. and SIDDALL, J. B. (1973). Developmental changes of the juvenile hormone esterase activity in haemolymph of the tobacco hornworm, *Manduca sexta. Insect Biochem. 3,* 397–407.

WESTPHAL, U. (1971). *Steroid–Protein Interaction.* Springer, New York.

WESTPHAL, U. (1978). Steroid-binding serum globulins. In *Receptors and Hormones,* pp.443–472. Edited by B. O'Malley and L. Birnbaumer. Academic Press, New York.

WESTPHAL, U. (1980). How are steroids transported in the blood before they enter target cells? In *Steroid Receptors and Hormone Dependent Neoplasia,* pp.1–17. Edited by J. Wittliff and O. Dapunt. Masson, New York.

WHITMORE, E. and GILBERT, L. I. (1972). Haemolymph lipoprotein transport of juvenile hormone. *J. Insect Physiol. 18,* 1153–1167.

WHITMORE, E. and GILBERT, L. I. (1974). Haemolymph proteins and lipoproteins in Lepidoptera — A comparative electrophoretic study. *Comp. Biochem. Physiol. 47B,* 63–78.

WING, K. D., SPARKS, T. C., LOVELL, V. M., LEVINSON, S. O. and HAMMOCK, B. D. (1981). The distribution of juvenile hormone esterase and its interrelationship with other proteins influencing juvenile hormone metabolism in the cabbage looper *Trichoplusia ni. Insect Biochem. 11,* 473–485.

WYATT, G. R. and PAN, M. L. (1978). Insect plasma proteins. *Ann. Rev. Biochem. 47,* 779–817.

YOUNGDAHL-TURNER, P., MELLMAN, I. S., ALLEN, R. H. and ROSENBERG, L. E. (1979). Protein mediated vitamin uptake. Adsorptive endocytosis of the transcobalamin II-cobalamin complex by cultured human fibroblasts. *Exp. Cell Res. 118,* 127–134.

15 Vertebrate Hormones in Insects

KARL J. KRAMER

U.S. Department of Agriculture and Kansas State University, Manhattan, Kansas, USA

1 INTRODUCTION

As our knowledge of insect biochemistry expands, the attention of biological, chemical and entomological scientists is directed more and more to problems of biochemical control and regulation. Accordingly, the interest in the chemical messengers, the hormones, which play such an important role in the regulation of metabolism and development, assumes a greater significance in the field of insect biochemistry.

The insect in the course of its evolution has developed both neural and hormonal systems to enable it to cope with the complexities of its environment, providing for homeostasis, integration, differentiation and growth. Some of these systems are truly unique and are found only in insect species while others are also found in other groups of organisms. This chapter is concerned with substances found in insects that are chemically or immunochemically similar to some of the hormones found in vertebrates.

There are three major classes of vertebrate hormones. The lowest molecular weight group is derived from two amino acids: tyrosine, which is the precursor of catecholamines and thyroid hormones, and tryptophan, which is converted to serotonin and melatonin. Another group of hormones is derived from fatty acids or cholesterol including the adrenocortical and gonadal steroids and also the prostaglandins. The largest group of hormones is the peptide–protein hormones. These vary in size from less than 10 to several hundred amino acids. Most of the subject-matter of this review concerns hormones in the latter class, the peptide hormones, but the prostaglandins are also covered. Information on other classes can be found elsewhere in this series.

2 METHODS OF DETECTION

A real breakthrough in hormone research occurred a few years ago when it became clear that radioimmunological and immunocytochemical procedures could be employed to detect materials and cell-types in invertebrates that had been first identified in vertebrates. These techniques have revolutionized our view of the gastroenteropancreatic neuroendocrine system and its evolution. Consequently, comparative endocrinology is making rapid progress, and it is difficult to make a comprehensive review which does not soon become outdated. Notwithstanding, the following is an effort to summarize the metastable state of our knowledge in this field. Other recent reviews include Frontali, N. and Gainer, H. (1977) and Haynes, L. (1980).

Using bioassay, radioimmunological and immunohistochemical techniques, a number of laboratories around the world have shown that vertebrate peptide-like substances are present in neuroendocrine and/or gastrointestinal tissues of very primitive organisms in the invertebrate group of animals. Even unicellular organisms such as bacteria, fungi, yeast, and protozoans appear to contain such substances. These peptides are indeed ancient molecules and must have appeared very early in evolution.

The bulk of the evidence for vertebrate hormone-like substances in insects is immunochemical. There are a number of peptides present in invertebrates that share immunochemical similarities with vertebrate peptides. The function of the invertebrate peptides is not known with certainty because there are many fewer biological data available and practically no chemical data. Only in the case of an insulin-like peptide (Kramer, K. *et al.*, 1982) and a pancreatic polypeptide (Duve, H. *et al.*, 1982) have any chemical compositional data been reported from an insect species. Thus, the true extent to which an insect peptide factor resembles a vertebrate peptide remains obscure. More chemical information such as isolation, characterization and molecular sequencing is needed.

The cells of origin of many of these invertebrate peptides are known and it seems likely that they arise in neuroectodermal tissue. For many years a neat conceptual barrier has been built by vertebrate physiologists between the gut and the brain. It was believed that peptides unique to the nervous system carried information from the brain to the endocrine glands from which peptide hormones were released into the blood. Until recently cells of the neural crest could not be implicated as ancestors or antecedents of peptide hormone producing cells. Now parallel immunocytochemical studies on mouse embryo pancreas have demonstrated that catecholamine-synthesizing enzymes (found only in cells of neuroectodermal origin) are present in cells that also contain glucagon and insulin (Teitelman, G. *et al.*, 1981, 1982). Thus, two of the enteropancreatic endocrine cell-types have a neuroectodermal origin. It is interesting to note that some of the early reports of these enteropancreatic peptides being present in neuroendocrine tissue dealt with insects (Tager, H. *et al.*, 1975, 1976; Kramer, K. *et al.*, 1977b).

In the past few years many of the peptide hormones have been detected in peripheral and central nervous systems. Now there are more than 20 "neuroactive" peptides found in various regions of the vertebrate nervous system. Many are well known, such as growth hormone, insulin, glucagon, and prolactin. Others are less familiar, such as bombesin, first found in frog skin, which stimulates the secretion of digestive enzymes and also regulates body temperature, and substance P, which apparently carries a pain signal. No-one knows the function of these peptides in the nervous system. Perhaps they are neurotransmitters or neuromodulators. For example, insulin increases acetylcholinesterase activity in rat brain (Catalan, R. *et al.*, 1981). In insects a large number of peptides occur in the neurosecretory system and may function similarly.

As stated earlier immunoassay can be used for the study of insect peptides that are chemically similar to vertebrate hormones. A radioimmunoassay (RIA) procedure already developed for previously well-characterized peptides from vertebrates (usually mammals) can be applied to the detection of related peptides within the insect. Currently available RIA systems for detection of peptide hormones frequently employ antisera raised against human, porcine, bovine or other mammalian peptides. Some of these include insulin, glucagon, gastrin, human growth hormones, renin, somatostatin, and pancreatic polypeptide. The use of these systems for the study of peptide hormones in species other than those from which the immunogen was obtained requires consideration of the question of species specificity of the immune reaction. Although commercial immunoassay kits are validated for their native immunogen, they may or may not behave properly in assays for closely related molecules found in other species.

The amino acid sequences of peptides from different species have diverged during the course of evolution. Their divergence has frequently occurred through single-base mutations and has generally been conservative, resulting in no significant alteration in the biological function of many peptides. Species differences in a hormone are therefore likely to occur in regions of the molecule not directly involved in its biological action. For example, insulin or a closely related molecule is native to *Escherichia coli*, *Tetrahymena pyriformis*, *Neurospora crassa*, *Aspergillus fumigatus* (Le Roith, D. *et al.*, 1980, 1981b) and also several invertebrate animal species (see Table 1). The antigenic site(s) of the peptide, however, is probably more highly mutable and one cannot predict whether a heterologous RIA system (one developed with one animal species and utilized in another) will work.

For most peptide families, sequence, structure and immunologic activity have been conserved, at least to a fractional extent. For example, examination of sequence data shows that the structures of the vertebrate hypoglycemic hormone, insulin, and the hyperglycemic hormone, glucagon, have been relatively well preserved during evolution (Dayhoff, M. 1972; Falkmer, S. *et al.*, 1973; Falkmer, S. and Ostberg, Y. 1977). The estimated rates of mutational acceptance from insulin and glucagon are 4 and 1.5 accepted point mutations per 100 residues per 100 million years, respectively (McLaughlin, P. and Dayhoff, M. 1972). Thus, it is predicted that on the average an insulin-like peptide will change in sequence by only 40% during the 900 million years since the divergence of insects and mammals and that the glucagon-like peptide has changed by only 15%. It is therefore not surprising that a heterogeneous population of antibodies directed toward a vertebrate hormone cross-react with a non-vertebrate hormone, even though perhaps at a much reduced potency. Antibodies raised against porcine or bovine glucagons react strongly with avian, piscine, other mammalian and invertebrate glucagon-like peptides.

Nevertheless, peptide immunology, in general, is not easily predictable. Immunochemical identity among peptides may require more than the identity of primary structure of a molecule, since insulins from a number of species reported to have identical amino acid sequences (Smith, L. 1966) are indistinguishable immunochemically with some, but not all, antisera (Berson, S. and Yalow, R. 1961, 1966). For example, guinea pig insulin has almost full biological potency but less than 10^{-3} the immunochemical potency of other mammalian insulins in immunoassay systems employing antibodies developed in man or guinea pig to bovine or porcine insulin (Berson, S. and Yalow, R. 1966). Insulins from fish and many lower vertebrates are even less reactive in these systems (Yalow, R. and Berson, S. 1964). Yet these antibodies cross-react with insulin-like peptides in invertebrates, unicellular eukaryotes and bacteria (Tager, H. *et al.*, 1976; Falkmer, S. and Ostberg, Y. 1977; Plisetskaya, E. *et al.*, 1978; Maier, V. *et al.*, 1978; Duve, H. *et al.*, 1979; Le Roith, D. *et al.*, 1980, 1981a, 1981b).

Although the use of heterologous RIA systems has proved to be extremely useful in the identification and localization of known peptides in many tissues, failure to observe sought-for immunoreactivity in extracts does not rule out the presence of peptides of similar structure and biologic function when the immunogen used for production of the antibodies, and the extract for which the potency is sought, are from different species. Even when there is sufficient immunologic similarity between species to permit the use of a heterologous system for assay, accurate quantification depends on the availability of appropriate standards. As with biologic assays, it may be necessary to define units in terms of tissue equivalents, since accumulation of sufficient quantities of peptide for development of standard units may be practically impossible. As a precautionary measure, a proteolytic enzyme inhibitor such as di-isopropyl

Table 1: Occurrence of insulin-like peptides in insects and other invertebrates

Species	Tissue source	Assay	Quantity*	Reference
Insects				
Apis mellifera, honeybee	adult gut	RIA†	200	Ishay, J. *et al.*, 1976
	larval midgut	RIA	200	Ishay, J. *et al.*, 1976
	whole body	RIA	200	Maier, V. *et al.*, 1978
	royal jelly	glucose oxidation in rat adipose tissue	200	Dixit, P. and Patel, N. 1964
	worker jelly	glucose oxidation in rat adipose tissue	200	Dixit, P. and Patel, N. 1964
	royal jelly	RIA	15	Kramer, K. *et al.*, 1977b
	royal jelly	hypotrehalosemia in *M. sexta* larvae	+	Kramer, K. *et al.*, 1977b
	royal jelly	RIA	0.025	Kramer, K. *et al.*, 1982
	hemolymph	RIA	4	Moreau, R. *et al.*, 1981
	midgut	RIA	200, $MW_{app} = 1.8 \times 10^4$	
	brain	IC	+	Maier, V. *et al.*, 1981
Bombyx mori, silkmoth	brain, corpora cardiaca, corpora allata	IC	+	Yui, R. *et al.*, 1980
Calliphora vomitoria, blow fly	brain	RIA, membrane receptor,	3 pg per head	Duve, H. *et al.*, 1979
	6–8 median neuro-secretory cells, hemolymph	lipogenesis and hypotrehalosemia, IC	+	Duve, H. and Thorpe, A. 1979, Duve, H. *et al.*, 1981a, 1981b
Drosophila melanogaster, fruit fly	whole body	hypoglycemia in mice	2	Meneses, P. and Ortiz, M. 1975
	hemolymph	RIA	1000	Seecof, R. and Dewhurst, S. 1974
	head, whole body	RIA, glucose oxidation in rat adipocytes	2	Le Roith, D. *et al.*, 1980, 1981a
Eristalis aeneus, hoverfly	brain	IC	+	El-Salhy, M. *et al.*, 1980
Hymenopteran‡ species	gut	RIA	+	Ishay, J. *et al.*, 1976
Locusta migratoria, locust	corpus cardiacum	RIA, hypolipaemia	500 ng per gland	Orchard, I. and Loughton, B. 1980
Manduca sexta, tobacco hornworm	hemolymph	RIA, amino acid analysis	1 pg ml^{-1} 10	Kramer, K. *et al.*, 1980 Kramer, K. *et al.*, 1982
	corpus cardiacum/corpus allatum complexes	RIA	1 ng $gland^{-1}$	Tager, H. *et al.*, 1976
	corpus cardiacum/corpus allatum complexes	hypotrehalosemia in *M. sexta* larvae	+	Tager, H. *et al.*, 1976
	brain	IC	+	El-Salhy, M. *et al.*, 1982, 1983
Periplaneta americana, cockroach	brain	RIA	+	Tager, H. *et al.*, 1976
Plodia interpunctella, Indian meal moth	head	RIA	0.2 ng $head^{-1}$	Tager, H. *et al.*, 1976
Other invertebrates				
Anodonta cygnea, mollusc	intestinal epithelium, hemolymph	RIA, IC	+	Plisetskaya, E. *et al.*, 1978
Asterias rubens, starfish	pyloric caeca	glycogen deposition in mouse tissue	> 1000	Wilson, W. and Falkmer, S. 1965
Branchiostoma lanceolatum, cephalochordate	gut epithelium	IC	+	Reinecke, M. 1981

Table 1: Occurrence of insulin-like peptides in insects and other invertebrates—Continued

Species	Tissue source	Assay	Quantity*	Reference
Buccinum undatum, whelk	acid–alcohol extracts of esophagus, stomach, intestine	glycogen deposition in mouse tissue. Activity neutralized by insulin antiserum	728 476 168	Davidson, J. *et al.*, 1971
Carcinus maenas, crab	hepatopancreas extracted by acid–alcohol	glycogen deposition in mouse tissue	540	Davidson, J. *et al.*, 1971
Ciona intestinalis, sea squirt	digestive tract extracted by acid–alcohol	glycogen deposition in mouse tissue. Activity neutralized by insulin antiserum	228	Davidson, J. *et al.*, 1971
Eledone cirrosa octopus	acid–alcohol extract of intestine, cecum	glycogen deposition in mouse tissue. Activity neutralized by insulin antiserum	3840 64	Davidson, J. *et al.*, 1971
Helix pomatia, snail	protein gland and hepatopancreas	glucose oxidation in epididymal fat pad. Activity neutralized by insulin antiserum	+	Amman, I. *et al.*, 1967
Homarus americanus, American lobster	digestive gland	glucose incorporation into glycogen	+	Sanders, B. 1980
Mya arenaria, clam	whole body extracted using acid–alcohol	blood sugar levels of rabbit	+	Collip, J. 1923
Mytilus edulis, mollusc	gut	IC	+	Fritsch, H. *et al.*, 1976 Fritsch, H. and Sprang, R. 1977
Ostrea edulis, oyster	gel-filtered acid–alcohol extract of hepatopancreas	glycogen deposition in rat and oyster tissue	+	de Martinez, N. *et al.*, 1965
Pecten maximum, scallop	acid–alcohol extract of esophagus	glycogen deposition in mouse tissue	396	Davidson, J. *et al.*, 1971
Pisaster ochraceous, starfish	pyloric caeca extracted by acid–alcohol	glycogen deposition in mouse tissue. Activity neutralized by antiserum to bovine insulin	288	Wilson, W. and Falkmer, S. 1965
Strophocheilus oblongus, snail	hepatopancreas and gut	RIA	+	Marques, M. and Falkmer, S. 1976
Styela clava, ascidian	esophageal epithelium	IC	+	Bevis, P. and Thorndyke, M. 1978
Unio pictorium, mollusc	intestinal epithelium, hemolymph	RIA, IC	+	Plisetskaya, E. *et al.*, 1978

*Units are ng vertebrate insulin equivalents per g tissue or ml of fluid unless otherwise noted. If not quantitated, positive assay denoted by +.
†RIA and IC are used as abbreviations for radioimmunoassay and immunocytochemistry, respectively.
‡A total of ten different hymenopteran species were examined including *Apis mellifera*, *Vespa orientalis*, *Vespa crabro*, *Paravespula vulgaris*, *Paravespula germanica*, *Dolichovespula media*, *Dolichovespula saxonica*, *Polyrachis simplex*, *Chalicodoma sicula*, and *Chrysis* sp.

phosphorofluoridate, phenylmethanesulfonyl fluoride, soybean trypsin inhibitor or trasylol (Zyznar, E. 1981) should be incorporated into the immunoassay because peptides and proteins are likely to be damaged by proteases that may be present in crude extracts.

The insects utilized most commonly to date for peptide hormone surveys are from four orders: Diptera, Hymenoptera, Lepidoptera, and Orthoptera. These include three fly species (*Calliphora, Drosophila, Eristalis*), honeybee (*Apis*), two moth species (*Manduca* and *Bombyx*) and the locust

(*Locusta*). Most other orders still need to be examined for the presence of vertebrate hormone-like substances. A number of invertebrates other than insects have been tested for vertebrate hormone-like materials and these are included in this survey for comparative purposes.

3 INSULIN

Insulin is a two-chain polypeptide of molecular weight 6000 held together by disulfide linkages (Larner, J. and Haynes, R. 1975a). Its action is very complex and deficiency leads to hyperglycemia, reduction in the rate of glucose transport across certain cell membranes, reduction in the activity of glycogen synthase, increase in the activity of pyruvate dehydrogenase, increase in rate of conversion of protein to glucose, hyperlipemia, ketonemia and azoturia.

In 1975, T. Normann suggested that an insulin-like activity resulting from decapitation of the blow fly, *Calliphora erythrocephala*, was due to a lack of hypotrehalosemic hormone of cephalic origin. The cephalic neuroendocrine gland complex, the corpus cardiacum–corpus allatum (CC–CA), was a likely release site for this hormone. Other investigators have also demonstrated hypertrehalosemia in other Diptera after selective removal of the corpus cardiacum of the blow fly, *Phormia regina* (Chen, A. and Friedman, S. 1977) or of the CC–CA in the house fly, *Musca domestica* (Lui, T. 1973). About the same time, an insulin-like peptide was detected in neuroendocrine tissues and hemolymph from other dipteran species (Seecof, R. and Dewhurst, S. 1974; Duve, H. *et al.*, 1979; Duve, H. and Thorpe, A. 1979; El-Salhy, M. *et al.*, 1980). Those results suggested that the dipteran hypotrehalosemic hormone and the insulin-like peptide were one and the same. Now similar evidence is available from lepidopteran, hymenopteran, orthopteran and other dipteran species (Table 1). All of the above evidence points to an insulin-like peptide functioning in insect carbohydrate metabolism. A recent report also indicated that a locust insulin-like peptide exhibits hypolipaemic activity as well (Orchard, I. and Loughton, B. 1980).

Insulin-like peptides have been identified by immunological, immunohistochemical, and biological assays in 18 species of Insecta (Table 1) taken from Hymenoptera (Dixit, P. and Patel, N. 1964; Ishay, J. *et al.*, 1976; Yui, R. *et al.*, 1980; Moreau, R. *et al.*, 1981), Orthoptera (Tager, H. *et al.*, 1976; Orchard, I. and Loughton, B. 1980), Diptera (Seecof, R. and Dewhurst, S. 1974; Meneses, P. and Ortiz, M. 1975; Duve, H. and Thorpe, A. 1979; Duve, H. *et al.*, 1979; Le Roith, D. *et al.*, 1980; Kramer, K. *et al.*, 1980). The amino acid composition of an insulin-like peptide isolated from an insect has recently been reported (Kramer, K. *et al.*, 1982). The material called "insect insulin" is similar to mammalian insulin in its solubility in acid, elution profile on Bio-Gel P-10 and P-30 (molecular size), cross-reactivity with guinea pig anti-insulin immunoglobulin, biological activity, and amino acid composition (Table 2).

Table 2: Amino acid compositions of Manduca *immunoreactive insulin and porcine insulin (Kramer, K.* et al., *1982)*

Amino acid	*Manduca*	Porcine
Aspartic acid	3.1	3
Threonine + glycine	8.0*	6.0
Serine	6.8	3
Glutamic acid	6.2	7
Proline	n.d.*	(1)
Alanine	2.0†	2
Valine	2.9	3
Methionine	n.d.*	0
Isoleucine	2.1	2
Leucine	1.8	6
Tyrosine	2.8	4
Phenylalanine	1.0	3
Histidine	2.1	2
Lysine	0.9	1
Arginine	1.1	1
Tryptophan	n.d.*	0
Cysteine	n.d.*	(6)
Total	40.8	43(50)
Ratio, basic : acidic	0.44	0.40

*Threonine and glycine not resolved and proline not detected by HPLC analysis. Methionine, tryptophan and cysteine not stable to acid hydrolysis. For comparison to porcine peptide, all residues except the above were utilized.

†Averaged values from two analyses. Integer values were normalized to 2.0 residues of alanine.

Insulin-like peptides have been semiquantified in whole body (Meneses, P. and Ortiz, M. 1975), heads (Le Roith, D. *et al.*, 1980, 1981a) and hemolymph (Seecof, R. and Dewhurst, S. 1974) from *Drosophila melanogaster*; whole body (Maier, V. *et al.*, 1978), gut tissue (Ishay, J. *et al.*, 1976), royal jelly (Dixit, P. and Patel, N. 1964; Kramer, K. *et al.*, 1977b;

Kramer, K. *et al.*, 1982) and worker jelly (Dixit, P. and Patel, N. 1964) from *Apis mellifera*; gut tissue from several hymenopteran species (Ishay, J. *et al.*, 1976; hemolymph and brain–corpus cardiacum–corpus allatum complex from *Manduca sexta* (Tager, H. *et al.*, 1976; Kramer, K. *et al.*, 1980); head from *Plodia interpunctella* (Tager, H. *et al.*, 1976); corpus cardiacum–corpus allatum from *Periplaneta americana* (Tager, H. *et al.*, 1976); corpus cardiacum from *Locusta migratoria* (Orchard, I. and Loughton, B. 1980); and brain from *Calliphora vomitoria* (Duve, H. *et al.*, 1979). Levels detected range from picograms to more than micrograms per gram of tissue or millilitre of hemolymph. The peptide has been localized immunohistochemically in the median neurosecretory cells and nerve fibers from them reaching to the corpus cardiacum and corpus allatum of *Bombyx mori* (Yui, R. *et al.*, 1980), and cells in the pars intercerebralis of *Eristalis aeneus* (El-Salhy, M. *et al.*, 1980) and *Manduca sexta* (El-Salhy, M. *et al.*, 1982).

The presence of insulin-like material has been confirmed in 17 other invertebrate species (for review see Falkmer, S. *et al.*, 1973, 1981) and also several unicellular eukaryotes (a ciliated protozoan and two unicellular fungi, Le Roith, D. *et al.*, 1980) and a bacterium (Le Roith, D. *et al.*, 1981b). Those results suggest that a peptide hormone can be highly conserved over extremely long periods of evolution (perhaps 900 million years) or that late recombination events can occur involving extramurally introduced genes.

The specific localization of insulin-like peptides in brain tissue and hemolymph of insects is somewhat reminiscent of the vertebrate case where insulin has been found not only in pancreas, blood, and gastrointestinal tract, but also in brain (Havrankova, J. *et al.*, 1978; Eng, J. and Yalow, R. 1980; Dorn, A. *et al.*, 1982). The identification of insulin-like and other vertebrate hormone-like peptides in tissues of organisms as diverse as insects, mammals and bacteria further suggests that the peptides and their anatomical distribution arose during the evolution of prokaryotes. Whether the presence of insulin in certain tissues such as brain was due to its

Table 3: Effects of vertebrate insulin on lower life forms

Species	Tissue	Effect	Quantity	Reference
Acetabularia mediteranea, algae	whole cell	plasma membrane binding and cellular internalization	30 nM	Legros, F. *et al.*, 1975
Agrius cinglulata, sweetpotato hornworm	fat body	triglyceride, diglyceride and fatty acid released	10 μM	Chang, F. 1974
Balanus nubilis, barnacle	muscle	sugar transport stimulated	μM	Baker, P. and Carruthers, A. 1980
Danaus plexippus, monarch butterfly	fat body	triglyceride, diglyceride and fatty acid released	10 μM	Chang, F. 1974
Drosophila melanogaster, fruit fly	embryonic cell line	differentiation enhanced, effect abolished when insulin-antibody included	μM	Seecof, R. and Dewhurst, S. 1974
	embryonic cell lines	growth stimulated	0.4 μM	Mosna, G. and Barigozzi, C. 1976
	imaginal disk	growth stimulated	3 nM	Davis, K. and Shearn, A. 1977
	imaginal disk	molting hormone induced, evagination stimulated	nM	Martin, P. and Shearn, A. 1980
	four cell lines of embryonic origin	replaced serum requirement in supporting cell growth	1 mg ml^{-1}	Mosna, G. 1981
Polycelis nigra, planarian	whole cell	glucose uptake stimulated	2 μM	Csaba, G. and Kadar, M. 1978
Tetrahymena pyriformis, protozoan	whole cell	glucose uptake stimulated	40 nM	Csaba, G. and Lantos, T. 1976
	whole cell	cAMP level increased	mM	Csaba, G. and Nagy, S. 1976
Vespa orientalis, oriental hornet	larval hemolymph	glucose uptake stimulated	0.1 mM	Ishay, J. 1975

Table 4: Occurrence of glucagon-like peptides in insects and other lower life forms

Species	Tissue source	Assay	Quantity	Reference
Apis mellifera, honeybee	whole body	RIA of gel-filtered extract	200 ng g^{-1}	Maier, V. *et al.*, 1978
	brain	IC	+	Maier, V. *et al.*, 1981
Astacus fluviatilis, starfish	intestinal gland	RIA	+, $MW_{app} = 6 \times 10^3$	Maier, V. *et al.*, 1975
Branchiostoma lanceolatum, cephalochordate	whole body	RIA	+	Assan, R. *et al.*, 1969
	gut	IC	+	Van Noorden, S. and Pearse, A. *et al.*, 1976; Reinecke, M. 1981
Cancer pagurus, crab	hepatopancreas	RIA	+	Assan, R. *et al.*, 1969
Cerastoderma edule, mollusc	digestive system	IC	+	Banks, I. *et al.*, 1980
Cynthia papillosa, tunicate	whole body	RIA	+	Assan, R. *et al.*, 1969
Eristalis aeneus, hoverfly	brain	IC	+	El-Salhy, M. *et al.*, 1980
Helix pomatia, snail	hepatopancreas	RIA	+	Assan, R. *et al.*, 1969
Hydra attenuata, hydra	nervous system	IC	1 ng g^{-1}	Grimmelikhuijzen, C. and Holst, J. 1981
Lymnaea stagnalis, pond snail	cerebral, parietal and visceral ganglia	IC	+	Schot, L. *et al.*, 1981
Manduca sexta, tobacco hornworm	brain, corpus cardiacum, corpus allatum	RIA of gel-filtered extracts glycogenolysis	1 ng $gland^{-1}$, $MW_{app} = 3 \times 10^3$	Tager, H. *et al.*, 1975, 1976
	hemolymph	RIA of gel-filtered extracts	nM, $MW_{app} = 3 \times 10^3$	Kramer, K. *et al.*, 1980
	midgut	RIA of gel-filtered extracts	200 pg g^{-1}, $MW_{app} = 1.5 \times 10^4$	Tager, H. and Kramer, K. 1980
	brain	IC	+	El-Salhy, M. *et al.*, 1982, 1983
Patella caerulae, limpet	hepatopancreas	RIA	+	Assan, R. *et al.*, 1969
Periplaneta americana, American cockroach	corpus cardiacum–corpus allatum	RIA of gel-filtered extracts	1 pg $head^{-1}$	Tager, H. *et al.*, 1976
	midgut caeca	IC	+	Iwanaga, T. *et al.*, 1981
Plodia interpunctella, Indian meal moth	head	RIA of gel-filtered extracts	10 pg $head^{-1}$	Tager, H. *et al.*, 1976

production *in situ* or its uptake and sequestration from plasma remains open to question (Eng, J. and Yalow, R. 1980, 1981; Rosenzweig, J. *et al.*, 1980).

Vertebrate insulin can affect lower life-forms physiologically in a number of ways (Table 3). These include growth stimulation, lipid mobilization, sugar uptake, and cellular internalization. Apparently insulin-like peptides from highly diverse species are cross-reactive biologically. A noteworthy effect of vertebrate insulin on insect tissues is that it stimulated growth of *Drosophila* imaginal disks and several cell lines in culture (Seecof, R. and

Dewhurst, S. 1974; Mosna, G. and Barigozzi, C. 1976; Davis, K. and Shearn, A. 1977). It was even used to completely replace serum in fruit fly embryonic cell cultures *in vitro* (Mosna, G. 1981). The main role of serum in both mammalian and insect cell culture media is believed to be to supply various hormones. Apparently the *Drosophila* cells have an absolute requirement for only insulin.

4 GLUCAGON

Glucagon is a single-chain polypeptide with a molecular weight of nearly 3500 (Larner, J. and Haynes, R. 1975b). While insulin serves as a hormone of fuel storage, glucagon serves as a hormone of fuel mobilization. The known actions of glucagon appear to result from stimulation of cyclic AMP synthesis (chapter 8, volume 8). Glucagon shares amino acid sequences with secretin, vasoactive intestinal peptide and glucose-dependent insulin releasing peptide, and together they make up a hormone family (Dockray, G. 1979). Several reviews on the phylogeny and ontogeny of glucagon production have appeared previously (Falkmer, S. and Marques, M. 1972; Falkmer, S. and Van Noorden, S. 1982).

In 1944 a glucagon-like activity was first demonstrated in invertebrates when aqueous extracts of crab eyestalks produced hyperglycemia (Abramowitz, A. *et al.*, 1944). The responsible factor was never characterized chemically but it was localized to the sinus gland.

Table 5: Occurrence of somatostatin-like peptides in insects and other lower life forms

Species	Tissue source	Assay	Quantity	Reference
Acheta domesticus, house cricket	hemolymph, brain, Malpighian tubules	RIA	+	Patton, R. and Kuo, C. 1977
Branchiostoma lanceolatum, cephalochordate	gut epithelium	IC	+	Reinecke, M. 1981
Ciona intestinalis, sea squirt	esophagus, stomach	RIA	$100\ \text{ng g}^{-1}$	Falkmer, S. *et al.*, 1977
	gastrointestinal tract	IC	+	Fritsch, H. *et al.*, 1978
	perikarya of cerebral ganglion and medulla	IC	+	Fritsch, H. *et al.*, 1979
Dugesia lugubris, planarian	cerebral, ganglion, nerve trunks	IC	+	Schilt, J. *et al.*, 1981
Eristalis aeneus, hoverfly	brain, ventral ganglia	IC	+	El-Salhy, M. *et al.*, 1980
Hydra attenuata, hydra	nervous system	IC	$800\ \text{pmole g}^{-1}$	Grimmelihuijzen, C. and Holst, J. 1981
Locusta migratoria, locust	median neurosecretory cells of brain, corpus cardiacum	IC	+	Doerr-Schott, J. *et al.*, 1978
Lymnaea stagnalis, pond snail	cerebral ganglion	IC	+	Schot, L. *et al.*, 1981
Manduca sexta, tobacco hornworm	brain	IC	+	El-Salhy, M. *et al.*, 1982, 1983
Panulirus interruptus, spiny lobster	eyestalk	IC	+	Mancillas, J. *et al.*, 1981
Periplaneta americana, cockroach	midgut	IC	+	Iwanaga, T. *et al.*, 1981
Porcellio dilatatus, isopod	protocerebrum, tritocerebrum, subesophageal ganglion, ventral nerve cord ganglia	IC	+	Martin, G. and Dubois, M. 1981

Table 6: Occurrence of gastrin/cholecystokinin/caerulein-like peptides in insects and other lower life forms

Species	Tissue source	Assay	Quantity	Reference
Aplysia californica, sea hare	gut and blood	RIA	5 ng g^{-1}	Straus, E. *et al.*, 1975
Bombyx mori, silkworm	medial neurons in brain and prothoracic ganglion	IC	+	Yui, R. *et al.*, 1980
Branchiostoma lanceolatum, cephalochordate	gut	IC	+	Van Noorden, S. and Pearse, A. 1976; Reinecke, M. 1981
Calliphora erythrocephala, blow fly	medial neurosecretory cells of pars intercerebralis	IC	+	Duve, H. and Thorpe, A. 1981c
	brain	RIA	4 ng g^{-1}, two components purified by immunoaffinity adsorption and gel filtration	Dockray, G. *et al.*, 1981
Ciona intestinalis, sea squirt	gut	IC	+	Fritsch, H. *et al.*, 1978
Eristalis aeneus, hoverfly	brain and fused ventral ganglia	IC	+	El-Salhy, M. *et al.*, 1980
Hydra attenuata, hydra and *Tealia felina*, sea anemone	nervous system	IC	5 fmole g^{-1}	Grimmelikhuijzen, C. *et al.*, 1980
Lumbricus terrestris, earthworm	intestine	RIA	0.1–1 ng g^{-1}	Rzasa, P. *et al.*, 1982
Lymnaea stagnalis, pond snail	all ganglia	IC	+	Schot, L. *et al.*, 1981
Manduca sexta, tobacco hornworm	brain–corpus cardiacum–corpus allatum complex	RIA of gel filtered extract, IC	10 ng g^{-1}	Kramer, K. *et al.*, 1977a El-Salhy, M. *et al.*, 1982, 1983
Otala lactea, land snail	blood and gastrointestinal tract	RIA	2 ng g^{-1}	Straus, E. *et al.*, 1975
Styela clava, sea squirt	cerebral vesicle	IC	+	Thorndyke, M. 1981

Twenty-five years later the chemical nature of a similar factor in insects was suggested (chapter 4, volume 8). Experimental approaches implicating the corpus cardiacum and other components of the neurosecretory system in the storage and secretion of a proteinaceous hypertrehalosemic hormone included ablation of the endocrine tissue by surgery (Normann, T. and Duve, H. 1969; Vejbjerg, K. and Normann, T. 1974), stimulation of tissue electrically or mechanically (Normann, T. and Duve, H. 1969; Vejbjerg, K. and Normann, T. 1974) and injection of tissue extracts into test animals (Steele, J. 1963; Brown, B. 1965; Natalizi, G. and Frontali, N. 1966; Keeley, L. and Friedman, S. 1967; Mordue, W. and Goldsworthy, G. 1969; Goldsworthy, G. and Mordue, W. 1974; Tager, H. *et al.*, 1976). Typically, injection of gland extracts resulted not only in hypertrehalosemia, but also in reduction of the glycogen content of the fat body (Goldsworthy, G. 1970; Wiens, A. and Gilbert, L. 1967; Tager, H. *et al.*, 1976; Hanaoka, K. and Takahashi, S. 1976), indicating a similarity of action of this factor (trehalagon) in insects to that of glucagon in vertebrates. Administration of bovine/porcine glucagon to insects evoked hyperglycemia and hypertrehalosemia (Bounias, M. 1979).

Glucagon-like peptides have now been detected by immunochemical or biological techniques in 14 species of invertebrates of which five are insects (Table 4). Tissue sources included whole body (200 $\mu g\ g^{-1}$), head (10 $pg\ g^{-1}$), gut (200 $pg\ g^{-1}$), hepatopancreas, brain (1 $mg\ g^{-1}$), ganglia and hemolymph (nM). There appear to be multiple forms of the peptide with apparent molecular weights ranging from 3×10^3 to 15×10^3 depending on the tissue used for extraction. A partially purified extract of a glucagon-like peptide ($MW_{app} = 3 \times 10^3$) from the corpus cardiacum of the tobacco hornworm, *Manduca sexta*, promoted glycogenolysis in the fat body from the same species.

One source of a glucagon-like peptide in the hoverfly, *Eristalis aeneus*, was a group of cells present in the protocerebral hemisphere of the brain (El-Salhy, M. *et al.*, 1980). The immunoreactive peptide may then be transferred to the ring gland for storage. In the pond snail, *Lymnaea stagnalis* (Schot, L. *et al.*, 1981) cells containing a glucagon-like substance were observed in the lateral lobes of the cerebral, parietal and visceral ganglia. A large glucagon-like peptide has been detected in the midgut of the tobacco hornworm (Tager, H. and Kramer, K. 1980). This peptide ($MW_{app} = 1.6 \times 10^4$) cross-reacted with an antiserum for glucagon-like peptides of apparent molecular weight $> 3.5 \times 10^3$ but did not react with two other "pancreas-specific peptide" ($MW_{app} = 3.5 \times 10^3$) antiglucagon sera. Thus, a high molecular weight glucagon-like component predominated in the midgut of *M. sexta*, whereas a low molecular weight form dominated in neurosecretory tissue and hemolymph (Tager, H. *et al.*, 1975, 1976; Kramer, K. *et al.*, 1980). Midgut cells of the cockroach, *Periplaneta americana*, exhibited glicentin-like immunoreactivity (Iwanaga, T. *et al.*, 1981).

As is the case for the mammalian intestinal peptide glicentin or enteroglucagon (Faloona, G. and Unger, R. 1974), the physiological function of the midgut glucagon-like peptide in the hornworm is unknown. It is likely that the insect peptide of molecular weight 16,000, like its mammalian counterpart, represents a precursor of the lower molecular weight hormone and that its structure is very similar to that of the hormone with extensions both at its carboxyl and amino-termini (Moody, A. *et al.*, 1977; Tager, H. and Markese, J. 1979). Although the midgut peptide has not yet been isolated for further characterization, physical and immunological similarities between glucagon-related peptides from the gut of *Manduca sexta* and the intestines of higher vertebrates exist.

Recently, immunochemical and immunocytochemical techniques have been used to identify and characterize glucagon-related peptides in the central nervous system of the rat and human (Loren, I. *et al.*, 1979; Tager, H. *et al.*, 1980; Sanders, D. *et al.*, 1981). The results suggested that glucagon-containing peptides that have undergone the intestinal type of post-translational modifications are present in neuronal cells of the rat hypothalamus. Multiple glucagon-like peptides were also evident in partially purified extracts of insect nervous tissue (Tager, H. *et al.*, 1976). The identification of glucagon-like peptides in nervous tissues of species as diverse as invertebrates and

Table 7: Occurrence of pancreatic polypeptide-like peptides in insects and other lower life forms

Species	Tissue source	Assay	Reference
Achatina fulica, giant African snail	gut and nervous system	IC	Van Noorden, S. *et al.*, 1980
Bombyx mori, silkmoth	cells at medial surface of adult brain lobe and axons	IC	Yui, R. *et al.*, 1980
Branchiostoma lanceolatum, cephalochordate	gut epithelium	IC	Reinecke, M. 1981
Calliphora erythrocephala, blow fly	adult brain	RIA, IC, purification and amino acid composition	Duve, H. and Thorpe, A. 1980; Duve, H. *et al.*, 1981, 1982
Eristalis aeneus, hoverfly	cells in fused ventral ganglia and protocerebrum of larval brain	IC	El-Salhy, M. *et al.*, 1980; El-Salhy, M. 1981
Lumbricus terrestris, earthworm	ventral nerve cord and ganglia	IC	Sundler, F. *et al.*, 1977
Lymnaea stagnalis, pond snail	all ganglia but pleural and buccal	IC	Schot, L. *et al.*, 1981
Manduca sexta, tobacco hornworm	brain	IC	El-Salhy, M. *et al.*, 1982, 1983
Periplaneta americana, cockroach	midgut	IC	Iwanaga, T. *et al.*, 1981

Table 8: Occurrence of substance P-like peptides in insects and other lower life forms

Species	Tissue source	Assay	Reference
Achatina fulica, giant African snail	nervous system	IC	Van Noorden, S. *et al.*, 1980
Ciona intestinalis, sea squirt	neural ganglia	IC	Fritsch, H. *et al.*, 1979
Eristalis aeneus, hoverfly	brain	IC	El-Salhy, M. *et al.*, 1980
Helix pomatia, snail	ganglia	bioassay	Gubitz, H. 1966
Hydra attenuata, hydra	nervous system	IC	Taban, C. and Cathieni, M. 1979
	nervous system	1 fmole g^{-1}	Grimmelikhuijzen, C. *et al.*, 1981a
Lymnaea stagnalis, pond snail	cerebral and pedal ganglia	IC	Schot, L. *et al.*, 1981
Manduca sexta, tobacco hornworm	pars intercerebralis	IC	El-Salhy, M. *et al.*, 1982, 1983
Panulirus interruptus, spiny lobster	optic neuropile	IC	Mancillas, J. *et al.*, 1981
Procambarus clarkii, crayfish	eyestalk	IC	Mancillas, J. *et al.*, 1981
Tealia felina, sea anemone	whole body	1 fmole g^{-1}	Grimmelikhuijzen, C. *et al.*, 1980

Table 9: Occurrence of vasopressin-like peptides in insects and other lower life forms

Species	Tissue source	Assay	Amount	Reference
Achatina fulica, giant African snail	intestinal ganglia	IC for arg–vasopressin	+	Van Noorden, S. *et al.*, 1980
Acheta domesticus, cricket	brain and subesophageal ganglia	RIA	pg $gland^{-1}$	Strambi, C. *et al.*, 1978, 1979
Blabera fusca, cockroach	nervous system	RIA, IC	+	Strambi, C. *et al.*, 1979
Cephalopod	neurohemal organ	IC	+	Martin, R. *et al.*, 1980
Clitumnus extradentatus	ganglia	IC	+	Remy, C. *et al.*, 1977
Gryllus domesticus	nervous system	RIA, IC	+	Strambi, C. *et al.*, 1979
Locusta migratoria, locust	subesophageal ganglion, central nervous system	IC	+	Remy, C. *et al.*, 1979; Remy, C. and Girardie, J. 1980
	subesophageal ganglion, rectum, hemolymph, Malpighian tubules	RIA	pg mg $protein^{-1}$	Proux, J. and Rougon-Rapuzzi, G. 1980
Lymnaea stagnalis, pond snail	buccal, cerebral, pedal, gastric, pharyngeal, pallial and intestinal ganglia	IC	+	Schot, L. *et al.*, 1981
Octopus vulgaris, octopus	vena cava neurosecretory gland	IC	+	Martin, R. *et al.*, 1980
Periplaneta americana, American cockroach	nervous system	RIA, IC	+	Strambi, C. *et al.*, 1980
Polistes gallicus	nervous system	RIA, IC	+	Strambi, C. *et al.*, 1979

mammals suggested that the anatomical distribution of these peptides arose remarkably early during the evolution of animals.

Biological assays have shown that there are multiple hyperglycemic factors present in the cerebral neuroendocrine system of insects. Besides a glucagon-like hyperglycemic peptide in several insects, other factors such as adipokinetic hormone in locusts also exhibit carbohydrate-elevating activity (Loughton, B. and Orchard, J. 1981; this volume; Steele, J. E. (1983), volume 8). Activation of locust fat body phosphorylase achieved with a corpus cardiacum extract was due partly to adipokinetic hormone and to a second factor, perhaps a glucagon-like one (Gade, G. 1981). Multiple factors were present in the tobacco hornworm also. Tager, H. *et al.* (1975, 1976) reported that a trypsin-sensitive glucagon-like substance decreased the glycogen content of fat body, while Ziegler, R. (1979) described a glycogen phosphorylase activating factor which is not trypsin-sensitive. A rather high molecular weight hyperglycemic peptide was purified from sinus glands of the crayfish; it has a minimal size of $\sim 7 \times 10^3$ (Keller, R. 1981). Von Norstrand, M. *et al.*, (1980) have detected a peptide in the cephalic endocrine system of the honeybee *Apis mellifera* that exhibits red pigment-concentrating activity in shrimp and hyperglycemic activity in the cockroach. The latter factor has not been chemically characterized.

5 SOMATOSTATIN

In vertebrates somatostatin is a tetradecapeptide found in the brain and digestive tract which exhibits multiple physiological effects including negative control of hormone release from pancreas, gastric antrum, duodenum and pituitary, inhibition of enzyme and electrolyte secretion in pancreas and stomach, blockage of water and chloride secretion in colon, and increased absorption of sodium chloride across ileal brush border (Vale, W. *et al.*, 1977). These effects have been tentatively explained in terms of competition for receptor occupancy, inhibition of cyclic AMP production, or selective activation of phosphoprotein phosphatase.

Somatostatin-like peptides have been detected in five species of insects and seven species of other invertebrates (Table 5). They have been localized primarily in nervous and digestive tissues by immunological techniques. In both the hoverfly (*Eristalis aeneus*) and locust (*Locusta migratoria*) somatostatin immunoactive cells are widely distributed, being found in the brain (medial neurosecretory cells of the pars intercerebralis), nervous part of corpus cardiacum, and ventral ganglia from which bundles of immunoactive nerve fibers extend in various directions (Doerr-Schott, J. *et al.*, 1978; El-Salhy, M. *et al.*, 1980). Quantitative measurements were obtained with a deuterostomian invertebrate, *Ciona intestinalis*, where mg per g wet weight levels were detected in the esophagus and

Table 10: Occurrence of neurophysin-like proteins in insects and other lower life forms

Species	Tissue source	Assay	Comment	Reference
Achatina fulica, giant African snail	ganglia	IC	–	Van Noorden, S. *et al.*, 1980
Acheta domesticus, cricket	brain, rectum, Malphighian tubule and hemolymph	IC	–	Girardie, J. and Remy, C. 1981
Clitumnus extradendatus	subesophageal ganglion	IC	–	Remy, C. *et al.*, 1977
Dugesia lugubris, planarian	cerebral ganglion and nerve trunks	IC	–	Schilt, J. *et al.*, 1981
Locusta migratoria, locust	subesophageal ganglion	IC	–	Remy, C. *et al.*, 1979
	subesophgeal ganglion	RIA	acidic protein, $MW_{app} = 10^4$	Camier, M. *et al.*, 1980
	brain and corpus cardiacum	–	a.a. composition, $MW_{app} = 1.1 \times 10^4$	Friedel, T. *et al.*, 1980

Table 11: Occurrence of enkephalin-like and endorphin-like peptides in insects and other life forms

Species	Tissue source	Assay	Amount	Reference
Enkephalin				
Achatina fulica, giant African snail	alimentary canal	IC	+	Van Noorden, S. *et al.*, 1980
Cephalopod	neurohemal organ	IC	metenkephalin +	Martin, R. *et al.*, 1979
Eristalis aeneus, hoverfly	brain, ventral ganglia, nerve fibers	IC	+	El-Salhy, M. *et al.*, 1980
Hirudinea, leech	midbody ganglion	IC	+	Zipser, B. 1980
Locusta migratoria, locust	met-enkephalin found in optic lobes, cerebral ganglion, subesophageal ganglion, corpus cardiacum, Malpighian tubules, mandibular muscles and oviduct; leu-enkephalin found in optic lobes, cerebral ganglion, subesophageal ganglion, and corpus cardiacum	RIA of gel-filtered extracts	~ 1 pg mg protein^{-1}, $MW_{app} = 7\text{–}10 \times 10^2$	Gros, C. *et al.*, 1978
	met-enkephalin found in four neurons in lateral protocerebrum of brain	IC	+	Remy, C. and Dubois, M. 1981
Lumbricus terrestris, earthworm	nervous system	IC	+	Alumets, J. *et al.*, 1979
Lymnaea stagnalis, pond snail	met-enkephalin found in cerebral and pedal ganglia	IC	+	Schot, L. *et al.*, 1981
Manduca sexta, tobacco hornworm	brain	IC	+	El-Salhy, M. *et al.*, 1982, 1983
Panulirus interruptus, spiny lobster	eyestalk and retina	IC	+	Mancillas, J. *et al.*, 1981
Procambarus clarkii, crayfish	eyestalk	IC	+	Mancillas, J. *et al.*, 1981
Endorphin				
Bombyx mori, silkworm	subesophageal ganglion	IC	+	Remy, C. *et al.*, 1979
Dendrobaena subrubicunda, earthworm	subesophageal ganglion	IC	+	Remy, D. and Dubois, M. 1979
Eristalis aeneus, hoverfly	pars intercerebralis, calyx	IC	+	El-Salhy, M. *et al.*, 1980
Manduca sexta, tobacco hornworm	brain	IC	+	El-Salhy, M. *et al.*, 1982, 1983
Thaumetyroea pityocampa	subesophageal ganglion	IC	+	Remy, C. *et al.*, 1978

stomach (Falkmer, S. *et al.*, 1977). The peptide occurs in the cerebral ganglion and medulla of this species also (Fritsch, H. *et al.*, 1979).

Somatostatin appears to be a phytogenetically old hormone since it was detected in animals belonging to thc protostomian evolutionary line (Arthropoda, Mollusca, Annelida) (Falkmer, S. *et al.*, 1978; Falkmer, S. 1979; Van Noorden, S. and Falkmer, S. 1980). Its function in all of these invertebrates is presently unknown.

6 GASTRIN/CHOLECYSTOKININ/CAERULEIN

Gastrin is a heptadeca- or larger peptide in vertebrates that causes an increase in the secretion of acid-containing gastric juices. Nearly all of the biological activity is found in the carboxyl-terminal tetrapeptide. Cholecystokinin is a larger peptide that causes the gall bladder to contract. Its carboxyl-terminal pentapeptide is identical to that of gastrin. The carboxyl-terminal tetrapeptide is a potent releaser of insulin and other pancreatic islet cell hormones and also decreases food intake in rats and monkeys (Rehfeld, J. *et al.*, 1980). Caerulein, a decapeptide isolated from amphibian skin, shares the identical carboxyl-terminal sequence of five amino acids. Antibody raised against any of these peptides will probably cross-react with the other peptides.

Gastrin/cholecystokinin/caerulein-like peptides have been detected in four species of insects (silkworm, *Bombyx mori*; hornworm, *Manduca sexta*; hoverfly, *Eristalis aeneus*; and blow fly, *Calliphora vomitoria*) and nine species of other invertebrates including snails, earthworms and several sea animals (Table 6). Immunoreactivity was demonstrated in brain neurons and prothoracic ganglia in insects. There is also immunochemical evidence for the presence of a gastrin-like peptide in gut tissues and hemolymph from several invertebrate species. Both gut and nervous tissues contain low mg per g wet weight levels. In two mollusc species (Straus, E. *et al.*, 1975) and one insect (Kramer, K. *et al.*, 1977a), the immunoreactive peptide(s) was similar to vertebrate gastrin in size and susceptibility to proteolytic enzymes. In some other invertebrates no gastrin-like material was detected in nervous tissues (see references to Table 6).

Results from using a number of antisera directed against different parts of either gastrin or cholecystokinin suggest that the insect peptide is homologous at the carboxyl-terminal region to the vertebrate counterpart. From blow fly brain, three components were partially purified that were similar to the C-terminal fragments of cholecystokinin and gastrin (Dockray, G. *et al.*, 1981). Porcine cholecystokinin stimulated gastric release of acid phosphatase and other proteins in the ascidian *Styela clava* (Bevis, P. and Thorndyke, M. 1981). Pentagastrin did not induce gastric secretion (Thorndyke, M. and Bevis, P. 1980). This same species exhibited antigastrin immunoreactivity in endocrine cells of the gut (Bevis, P. and Thorndyke, M. 1979).

7 PANCREATIC POLYPEPTIDE

The physiological role of pancreatic polypeptide which is found in both pancreatic and neural tissues in vertebrates is equivocal. It is thought to be involved in the regulation of some peptide hormone titers because it inhibits insulin release in the rat but does not affect glucagon levels (Murphy, W. *et al.*, 1981). Human, porcine, ovine, bovine and avian pancreatic polypeptides have been purified and their primary structures determined (Floyd, J. *et al.*, 1977). Mammalian polypeptides differ by not more than two or three amino acids while avian and bovine have homology at only 16 out of 36 amino acid residues. Pancreatic polypeptide immunoreactive material has been shown by immunocytochemistry to occur in the brain of mammals (Loren, I. *et al.*, 1979) which implies that it is one of the series of gastroenteropancreatic/brain peptides.

Recent immunocytochemical studies have shown that pancreatic polypeptide-like immunoreactive material is present in the brain of four insect species while one species, *Periplaneta americana*, exhibits immunoreactivity in midgut (Table 7). Four other species of invertebrates exhibited positive immunoreactivity in nervous tissue and one, the giant African snail, *Achatina fulica*, also showed pancreatic polypeptide-like immunoreactivity in gut tissue. Localization in nervous tissue appeared to be varied. For example, immunoreactive cells were found in *Bombyx mori* located near the medial surface

Table 12: Some miscellaneous vertebrate peptide-like substances found in insects and other life forms

Peptide	Species	Source	Assay	Vertebrate activity	Reference
Adrenocorticotropin (ACTH)	*Ciona intestinalis*, sea squirt	ganglia	RIA	stimulates adrenocortical hormone synthesis	Georges, D. and Dubois, M. *et al.*, 1979
Adrenocorticotropin	*Dugesia lugubris*, earthworm	neurons	RIA		Schilt, J. *et al.*, 1981
Adrenocorticotropin	*Lymnaea stagnalis*, pond snail	neurons	RIA		Boer, H. *et al.*, 1979
Bombesin	*Achatina fulica*, giant African snail	neurons	IC	pressor activity	Van Noorden, S. 1980
Bombesin	*Hydra attenuata*, hydra	nervous system	IC, RIA, 0.2 pmole g^{-1}		Grimmelikhuijzen, C. *et al.*, 1981c
Calcitonin	*Ciona intestinalis*, sea squirt	neural ganglia, alimentary tract	IC	inhibition of bone resorption of calcium	Fritsch, H. *et al.*, 1979, 1980
Calcitonin	*Lymnaea stagnalis*, snail	cerebral and pedal ganglia	IC		Schot, L. *et al.*, 1981
Calcitonin	*Manduca sexta*, hornworm	brain	IC		El-Salhy, M. *et al.*, 1982, 1983
Calcitonin	*Styela clava*, ascidian	pharynx	IC		Thorndyke, M. and Probert, L. 1979
Glucose-dependent insulinotrophic peptide (GIP)	*Lymnaea stagnalis*, snail	cerebral, parietal and visceral ganglia	IC	stimulates insulin secretion	Schot, L. *et al.*, 1981
Glucose-dependent insulinotrophic peptide	*Manduca sexta*, hornworm	brain	IC		El-Salhy, M. *et al.*, 1982, 1983
Luteinizing hormone-releasing hormone (LHRH)	*Cionia intestinalis*, sea squirt	nervous system	IC	stimulates release of luteinizing hormone	Georges, D. and Dubois, M. 1980
Melanotropin	cephalopod	neurohemal organ	IC	regulates pigmentation	Martin, R. *et al.*, 1980
Melanotropin	*Lymnaea stagnalis*, snail	cerebral and pedal ganglia	IC		Schot, L. *et al.*, 1981
Neurotensin	several invertebrates	digestive tract	RIA, IC		Reinecke, M. *et al.*, 1980
Neurotensin	bacteria	—	RIA		Bhatnagar, Y. and Carraway, R. 1981
Neurotensin	*Hydra attenuata*, hydra	nervous system	IC, RIA, 300 fmole g^{-1}		Grimmelikhuijzen, C. *et al.*, 1981b
Neurotensin	*Tealia felina*, sea anemone	nervous system	RIA		Grimmelikhuijzen, C. *et al.*, 1980
Oxytocin	*Achatina fulica*, snail	nervous system	IC	smooth muscle stimulant	Van Noorden, S. *et al.*, 1980
Oxytocin	*Lymnaea stagnalis*, snail	CNS	IC		Schot, L. *et al.*, 1981
Secretin	*Achatina fulica*, snail	nerve fibers, plexuses and ganglia	IC	stimulates secretion of pancreatic juice	Van Noorden, S. *et al.*, 1980

Table 12: Some miscellaneous vertebrate peptide-like substances found in insects and other life forms—Continued

Peptide	Species	Source	Assay	Vertebrate activity	Reference
Secretin	*Branchiostoma lanceolatum*	gut epithelium	IC		Reinecke, M. 1981
Secretin	*Cerastoderma edule*, mollusc	digestive system	IC		Banks, I. *et al.*, 1980
Secretin	*Eristalis aeneus*, hoverfly	brain, fused ventral ganglia	IC		El-Salhy, M. *et al.*, 1980
Secretin	*Lymnaea stagnalis*, snail	cerebral gangion	IC		Schot, L. *et al.*, 1981
Secretin	*Styela clava*, ascidian	gastric epithelium	IC		Bevis, P. and Thorndyke, M. 1979
Thyrotropin releasing factor (TRF)	*Lymnaea stagnalis*, snail	nervous system and hemolymph	RIA	stimulates release of thyrotropin	Grimm-Jorgensen, Y. 1978, 1979
Thyrotropin releasing factor	gastropod	circumesophageal ganglia	RIA		Grimm-Jorgensen, Y. 1978; Grimm-Jorgensen, Y. *et al.*, 1975
Vasoactive intestinal polypeptide (VIP)	*Branchiostoma lanceolatum*, cephalochordate	gut epithelium	IC	relaxes smooth muscle	Reinecke, M. 1981
Vasoactive intestinal polypeptide	*Lumbricus terrestris*, earthworm	nervous system	IC		Sundler, F. *et al.*, 1977
Vasoactive intestinal polypeptide	*Lymnaea stagnalis*, snail	cerebral ganglion	IC		Schot, L. *et al.*, 1981
Vasoactive intestinal polypeptide	*Manduca sexta*, hornworm	brain	IC		El-Salhy, M. *et al.*, 1982, 1983
Vasoactive intestinal polypeptide	*Periplaneta americana*, cockroach	midgut	IC		Iwanaga, T. *et al.*, 1981
Vasotocin	*Aplysia californica*, sea hare	nervous system	IC		Lubroviak, K. *et al.*, 1979
Vasotocin	*Lymnaea stagnalis*, snail	pharyngeal, pallial and visceral ganglia	IC		Schot, L. *et al.*, 1981

of the brain lobe, anterior to the medial neurosecretory cell group, while pancreatic polypeptide-immunoreactive cells occurred only in the fused ventral ganglia and protocerebrum of the hoverfly.

Using 5 kg of *Calliphora* adult heads, a pancreatic polypeptide-like material was partially purified whose physical behavior and immunochemical cross-reactivity are identical with those of the bovine hormone (Duve, H. *et al.*, 1981a). This neuropeptide has now been purified and its amino acid composition is homologous to vertebrate pancreatic polypeptide (Duve, H. *et al.*, 1982). Apparently, the material in the brain of the blow fly has close structural similarity to the mammalian hormone. No immunoreactivity was detected in blow fly hemolymph.

8 SUBSTANCE P

Substance P is an undecapeptide first extracted from brain and intestine, and more recently found in neural retinal tissue in vertebrate species (Chang, M. *et al.*, 1971). It is structurally homologous with eledoisin and physalaemin, two powerful hypotensive agents. To date, substance P does not have a

Table 13: Occurrence of prostaglandins and prostaglandin synthetase in insects and other lower life forms

Species	Source	Assay	Identification	Reference
Acheta domesticus, house cricket	gonads	formed from arachidonic acid	—	Destephano, D. *et al.*, 1974, 1976
Acheta domesticus, house cricket	testis, seminal vesicles, spermatophores	RIA	PGE_2, 5–500 ng g^{-1}	Destephano, D. and Brady, U. 1977
Bombyx mori, silk moth	gonads	TLC, UV, fluorescence, enzyme assay, contraction of smooth muscle	PGE_1, PGE_2, $PGF_{1\alpha}$, $PGF_{2\alpha}$	Setty, B. and Ramaiah, T. 1979
Musca domestica, house fly	gonads	biosynthesis from arachidonic acid catalyzed by microsomal enzyme	PGE_2	Wakayama, E. *et al.*, 1980
Homarus, lobster; *Lumbricus*, earthworm; *Mytilis*, mussel; *Cyanea*, jellyfish; *Anthoplexaura*, horny coral	whole animal, gills, stomach	biosynthesis from eicostrienoic acid	PGE_1	Christ, E. and VanDorp, D. 1972
Teleogryllus commodus, Australian field cricket	spermatheca	TLC, HPLC	PGE_2	Loher, W. *et al.*, 1981

clear function, being both a hypotensive, sialogogic, and contractile factor. It may act as a neurotransmitter, sending a pain message along sensory nerve fibers.

Substance P has been detected in only two insects, the hoverfly and the hornworm (Table 8). It was found in the brain, located in a region posterior to the corpora pedunculata. No nerve fibers were found to contain immunoreactive substance P. Seven other invertebrate species contained the peptide in their nervous system, including the sea squirt, hydra, lobster, crayfish and three kinds of snails.

9 VASOPRESSIN

Vasopressin is a cyclic nonapeptide present in vertebrate neurohypophyseal extracts (Brazeau, P. 1975). It is present in the posterior pituitary of all mammals in one of either two forms: arginine vasopressin or lysine vasopressin. These peptides act as antidiuretics, probably by altering plasma osmotic pressure. Arginine vasopressin has also been found to facilitate long-term memory (Bohus, B. *et al.*, 1978).

Six species of insects contain a vasopressin-like peptide (Table 9). Both diuretic and antidiuretic functions have been demonstrated (Strambi, C. *et al.*, 1978; Proux, J. and Rougon-Rapuzzi, G. 1980). In the locust vasopressin immunoreactivity occurred in brain, ventral nerve cord, Malpighian tubules and hemolymph. The peptide was apparently synthesized in the subesophageal ganglion, transported axonally to the brain and nerve cord, released into hemolymph, and degraded in the rectum and Malpighian tubules. Five other invertebrate species exhibit vasopressin-like immunoreactivity in the nervous system.

10 NEUROPHYSIN

Neurophysins are rather large neurosecretory proteins found in the hypothalamus of vertebrates that may be involved in the transport and storage of certain hormones. Three forms have been isolated from vertebrate tissues.

The locust and cricket exhibit neurophysin-like immunoreactivity in nervous tissue, hemolymph and rectal tissue (Table 10). In *Locusta*, a vasopressin-like peptide is associated with a neurophysin-like carrier protein in the subesophageal ganglia (Remy, C. *et al.*, 1979). The physical and chemical properties of insect neurophysin are analogous to those of the vertebrate neurophysins (Camier, M. *et al.*, 1980).

The nervous tissues of three other invertebrate species also contain neurophysin-like proteins.

11 ENKEPHALIN AND ENDORPHIN

Enkephalin and endorphin are pain-killing neurotransmitter peptides found in the brain and intestinal nerves that inhibit the contraction of smooth muscle in mammals (Guillermin, C. 1976). Two different forms of enkephalin have been identified: leu- and met-enkephalins (Hughes, J. 1975). Either or both peptides have been found in invertebrates, including the locust, fly, hornworm, snail, lobster, leech, earthworm and cephalopod (Table 11).

Certain molluscs exhibit immunoreactive enkaphalin in nerves of the posterior part of the alimentary canal and also visceral, pedal and cerebral ganglia. Insects possess immunoreactivity in the pars intercerebralis, corpora pedunculata, fused ventral ganglia, calyx cell, optic lobe, cerebral ganglia, subesophageal ganglia, corpora cardiaca, Malphigian tubules, oviduct and mandibular muscle. Endorphin immunoreactive cells are observed in the pars intercerebralis, calyx, or subesophageal ganglion of the earthworm, silkworm, hoverfly or hornworm (Table 11).

12 MISCELLANEOUS VERTEBRATE PEPTIDES

Approximately twelve other vertebrate peptide-like substances including adrenocorticotropin, bombesin, calcitonin, glucose-dependent insulinotrophic peptide, luteinizing hormone-releasing hormone, melanotropin, neurotensin, oxytocin, secretin, thyrotropin releasing factor, vasoactive intestinal polypeptide and vasotocin have been identified in several invertebrate species (Table 12). Most of these were not insect species and, in a majority of cases, tissue sources were derived from a portion of the nervous system. A few species exhibited immunoreactivity in the digestive tract or hemolymph.

13 PROSTAGLANDINS

Prostaglandins are 20-carbon unsaturated carboxylic acids with a cyclopentane ring (Douglas, W. 1975). There are several classes of these lipid hormones which are distinguished by substituents of the cyclopentane ring and the number of double bonds in the side-chains. Prostaglandins have been detected in almost every tissue and fluid of the vertebrate body.

Prostaglandins identified in invertebrates include PGE_1, PGE_2, $PGF_{1\alpha}$ and $PGF_{2\alpha}$ (Table 13). These are the most abundant derivatives which are found in vertebrates. In the house cricket, nearly mg per g levels were found in the testis (Destephano, D. and Brady, U. 1977). *In vitro* synthesis of prostaglandins from arachidonic acid or eicostrienoic acid has been demonstrated in crickets, flies and various other invertebrates. In the cricket *Teleogryllus commodus* the prostaglandin-synthesizing enzyme complex is transferred from the male spermatophore into the female (Loher, W. *et al.*, 1981). PGE_2 is then synthesized in the female's spermatheca from arachidonic acid. Oviposition is eventually stimulated by PGE_2.

14 CONCLUDING REMARKS

From the preceding, it can be seen that numerous vertebrate hormone-like substances occur in invertebrates, especially in elements of the central nervous system (Table 14). Most of the evidence is based on results obtained with immunocytochemistry and radioimmunoassay, with the use of antisera to vertebrate neuropeptides as well as gut peptides. All of these biologically active peptides are phylogenetically ancient substances that have a wide distribution in the animal kingdom. Morphological and functional studies suggest that they act both as neurohormones and as neurotransmitters/neuromodulators in a great number of animal species. Seven of these peptides have been found in hemolymph and either nervous or gut tissue from insects, suggesting that these may indeed be hormones as defined classically (Table 14). Five are found in all three types of tissues (gastrin/cholecystokinin/caerulein, glucagon, insulin, neurophysin and vasopressin).

On the basis of these findings it may be expected that the reverse is also true, that peptides originally thought present only in invertebrates are

Table 14: General tissue distribution of vertebrate hormone-like peptides in invertebrates

Nervous tissue	Gut tissue	Hemolymph
adrenocoticotropin	calcitonin	gastrin/cholecystokinin/caerulein
bombesin	gastrin/cholecystokinin/caerulein	glucagon
calcitonin	glucagon	insulin
endorphin	insulin	neurophysin
enkephalin	neurophysin	thyrotropin releasing factor
gastrin/cholecystokinin/caerulein	neurotensin	vasopressin
glucagon	secretin	
glucose-dependent insulinotrophic peptide	pancreatic polypeptide	
insulin	somatostatin	
luteinizing hormone-releasing hormone	vasoactive intestinal polypeptide	
melanotropin	vasopressin	
neurophysin		
oxytocin		
pancreatic polypeptide		
secretin		
somatostatin		
substance P		
thyrotropin releasing factor		
vasoactive intestinal polypeptide		
vasopressin		
vasotocin		

endogenous to vertebrate tissues. So far two such substances have been reported. A neurohormone (head differentiation factor) from *Hydra attenuata* and *Anthopleura elegantissima* (sea anemone) was detected in human and rat brain and rat intestine (Schaller, H. *et al.*, 1977; Bodenmuller, H. and Schaller, H. 1981; Schaller, H. and Bodenmuller, H. 1981). The CNS of fish, *Poecilia latipinna*, and several species of vertebrates contains a molluscan tetrapeptide (cardioexcitatory peptide) which is also present in insects (*Leptinotarsa decemlineata*, *Periplaneta americana*, *Locusta migratoria* and *Pieris brassicae*) (Price, D. and Greenberg, M. 1977; Boer, H. *et al.*, 1980; Dockray, G. *et al.*, 1981; Weber, E. *et al.*, 1981). Certainly in the future more peptides common to vertebrates and invertebrates will be discovered and characterized, and our understanding of the nature, physiological role and metabolism of these subtances will be broadened.

Finally, a word about how such studies might relate to insect control. The presence of vertebrate hormone-like substances in insects suggests that pharmacological agents which affect hormone action in vertebrates may function in insects. Some vertebrate hyperglycemic and hypoglycemic agents including alloxan, streptozotocin, chlorpropamide, tolbutamide, phenformin and neutral red have been screened for activity against insects (Kramer, K. *et al.*, 1978, 1979). Although not as effective as they were in vertebrates, those chemicals were nevertheless active as effectors of carbohydrate metabolism in insects. Just as pharmaceutical chemists have synthesized drugs with metabolic activity optimized for vertebrate animals, pesticide chemists could develop similar compounds (insect pharmacology) that may have potential for use as insect growth regulators. Many chemicals have already been synthesized and tested by pharmaceutical companies, but most have been set aside because of unfavorable results with vertebrate animals. Interest in these materials should be renewed and they, and also new compounds, should be screened for peptide hormone-like activity in insects.

ACKNOWLEDGEMENTS

I am grateful to Dr. Peter Dunn, Purdue University,

and Dr. Sture Falkmer, University of Lund, Sweden for critical comments. This paper formed contribution No. 82–299–j, Department of Biochemistry, Kansas Agricultural Experiment Station, Manhattan, Kansas 66506. Cooperative investigation between AR, USDA and the Kansas Agricultural Experiment Station.

REFERENCES

ABRAMOWITZ, A. A., HISAW, F. L. and PAPANDREA, D. (1944). The occurrence of a diabetogenic factor in the eyestalks of crustaceans. *Biol. Bull. 86*, 1–5.

ALUMETS, J., HAKANSON, R., SUNDLER, F. and THORELL, J. (1979). Neuronal localization of immunoreactive enkephalin and β-endorphin in the earthworm. *Nature (London) 279*, 805–806.

AMMAN, I., MELANI, F. and GROSCHEL-STEWART, U. (1967). Insulinaktivität bei Schnecken (*Helix promatia*). In *Die Pathogenese des Diabetes mellitus. Die endokrine Regulation des Fettstoffwechsels*, pp.96–98. Edited by E. Klein. Springer-Verlag, Berlin.

ASSAN, R., TCHOLROUTSKY, G. and ROSSELIN, G. (1969). Caractérisation radioimmunologique de glucagon dans les tissus digestifs de diverses espèces animales. *Pathol. Biol. 17*, 747–749.

BAKER, P. F. and CARRUTHERS, A. (1980). Insulin stimulates sugar transport in giant muscle fibres of the barnacle. *Nature (London) 286*, 276–279.

BANKS, I., SLOAN, J. M. and BUCHANAN, K. (1980). Glucagon-like (GLI) and secretin-like immunoreactivity (SLI) within the digestive system of the bivalve mollusc *Cerastoderma edule*. *Regul. Pept. 1*: Suppl. 1, S7.

BARKER, J. L., IFSHIN, M. and GAINER, H. (1975). Studies in bursting pacemaker potential activity in molluscan neurons. III. Effects of hormones. *Brain Res. 84*, 501–510.

BERSON, S. A. and YALOW, R. S. (1961). Immunochemical distinction between insulin with identical amino acid sequences. *Nature 191*, 1392–1393.

BERSON, S. A. and YALOW, R. S. (1966), Insulin in blood and insulin antibodies. *Amer. J. Med. 40*, 676–690.

BEVIS, P. J. R. and THORNDYKE, M. C. (1978). Endocrine cells in the esophagus of the ascidian *Styela clava*. A cytochemical and immunofluorescence study. *Cell Tiss. Res. 187*, 153–158.

BEVIS, P. J. R. and THORNDYKE, M. C. (1979). A cytochemical and immunofluorescence study of endocrine cells in the gut of the ascidian *Styela clava*. *Cell Tiss. Res. 199*, 139–144.

BEVIS, P. J. R. and THORNDYKE, M. C. (1981). Stimulation of gastric enzyme secretion by porcine cholecystokinin in the ascidian *Styela clava*. *Gen. Comp. Endocr. 45*, 458–464.

BHATNAGER, Y. M. and CARRAWAY, R. E. (1981). Bacterial peptides with C-terminal similarities to bovine neurotensin. *Peptides 2*, 51–59.

BODENMULLER, H. and SCHALLER, H. C. (1981). Conserved amino acid sequence of a neuropeptide, the head activator, from coelenterates to humans. *Nature 293*, 679–680.

BOER, H. H., SCHOT, L. P. C., ROUBOS, E. W., MAAT, A., LODDER, J. C., REICHELT, D. and SWAAB, D. F. (1979). ACTH-like immunoreactivity in two electronically coupled giant neurons in the sand snail *Lymnaea stagnalis*. *Cell Tiss. Res. 202*, 231–240.

BOER, H. H., SCHOT, L. P. C., VEENSTRA, J. A. and REICHELT, D. (1980). Immunocytochemical identification of neural elements in the central nervous system of a snail, some insects, a fish, and a mammal with an antiserum to the molluscan cardio-excitatory tetrapeptide FMRF-amide. *Cell Tiss. Res. 213*, 21–27.

BOHUS, B., KOVACS, G. L. and DE WIED, D. (1978). Oxytoxin, vasopressin and memory: opposite effects on consolidation and retrieval processes. *Brain Res. 157*, 414–417.

BOUNIAS, M. (1979). Relations doses/effets de l'insuline et du glucagon sur la glycémie de l'abeille. *C.R. Soc. Biol. 173*, 1031–1036.

BRAZEAU, P. (1975). Agents affecting the renal conservation of water. In *The Pharmacological Basis of Therapeutics*, pp.848–859. Edited by L. Goodman and A. Gilman. Macmillan, New York.

BROWN, B. E. (1965). Pharmacologically active constituents of the cockroach corpus cardiacum: resolution and some characteristics. *Gen. Comp. Endocr. 5*, 387–401.

CAMIER, M., GIRARDIE, J., REMY, C., GIRARDIE, A. and COHEN, P. (1980). Identification of immunoreactive neurophysin-like proteins in the central nervous system of an insect, *Locusta migratoria*. *Biochem. Biophys. Res. Commun. 93*, 792–796.

CATALAN, R. E., MARTINEZ, A. M., MATA, F. and ARAGONES, M. D. (1981). Effect of insulin on acetylcholinesterase activity. *Biochem. Biophys. Res. Commun. 101*, 1216–1220.

CHANG, F. (1974). Effects of vertebrate adipokinetic hormones on the rate of *in vitro* lipid release in insects. *Comp. Biochem. Physiol. 49B*, 567–578.

CHANG, M. M., LEEMAN, S. E., and NIALL, H. D. (1971). Amino acid sequence of substance P. *Nature New Biol. 232*, 86–88.

CHEN, A. C. and FRIEDMAN, S. (1977). Hormonal regulation of trehalose metabolism in the blow fly, *Phormia regina*: interaction between hypertrehalosemic and hypotrehalosemic hormones. *J. Insect. Physiol. 23*, 1223–1232.

CHRIST, E. J. and VANDORP, D. A. (1972). Comparative aspects of prostaglandin biosynthesis in animal tissues. *Biochim. Biophys. Acta 270*, 537–545.

COLLIP, J. B. (1923). The demonstration of an insulin-like substance in tissues of the clam, *Mya arenaria*. *J. Biol. Chem. 55*, 39–44.

CSABA, G. and KADAR, M. (1978). Effects of epinephrine, glucagon, and insulin on glucose metabolism of *Planaria*. *Endokrinologie 71*, 113–115.

CSABA, G. and LANTOS, T. (1976). Effects of insulin on the glucose uptake of Protozoa. *Experientia 31*, 1097–1098.

CSABA, G. and NAGY, S. U. (1976). Effect of vertebrate hormones on cyclic AMP level in *Tetrahymena*. *Acta Biol. Med. Germ. 35*, 1399–1401.

DAVIDSON, J. K., FALKMER, S., MEHROTRA, B. K., AND WILSON, S. (1971). Insulin assays and light microscopical studies of digestive organs in Protostomian and Deuterostomian species and in Coelenterates. *Gen. Comp. Endocr. 17*, 388–401.

DAVIS, K. T. and SHEARN, A. (1977). *In vitro* growth of imaginal disks from *Drosophila melanogaster*. *Science 196*, 438–440.

DAYHOFF, M. O. (1972). Hormones, active peptides and toxins. In *Atlas of Protein Sequence and Structures*. 5. National Biomedical Research Foundation. Silver Springs, Maryland. Vol. 5, D-173–D-228.

DE MARTINEZ, N. R., GARCIA, M. C., SALAS, M. and CANDELA, J. L. R. (1965). Protein with insulin-like activity isolated from oyster (*Ostrea edulis*) hepatopancreas. *Gen. Comp. Endocr. 20*, 305–311.

DESTEPHANO, D. B. and BRADY, U. E. (1977). Prostaglandin and prostaglandin synthetase in the cricket, *Acheta domesticus*. *J. Insect. Physiol. 23*, 905–911.

DESTEPHANO, D. B., BRADY, U. E. and LOVINS, R. C. (1974). Synthesis of prostaglandin by reproductive tissue of the male house cricket, *Acheta domesticus*. *Prostaglandins 6*, 71–79.

DESTEPHANO, D. B., BRADY, U. E. and WOODALL, L. B. (1976). Partial characterization of prostaglandin synthetase in the reproductive tract of the male house cricket, *Acheta domesticus*. *Prostaglandins 11*, 261–273.

DIXIT, P. K. and PATEL, N. G. (1964). Insulin-like activity in larval foods of the honeybee. *J. Insect Physiol. 202*, 189–190.

DOCKRAY, G. J. (1979). Evolutionary relationships of the gut hormones. *Fed. Proc. 38*, 2295–2301.

DOCKRAY, G. J., DUVE, H. and THORPE, A. (1981). Immunochemical characterization of gastrin/cholecystokinin-like peptides in the brain of the blow fly, *Calliphora vomitoria*. *Gen. Comp. Endocr. 45*, 491–496.

DOCKRAY, G. J., VAILLANT, C. and WILLIAMS, R. G. (1981). New vertebrate brain-gut peptide related to a molluscan neuropeptide and an opioid peptide. *Nature 293*, 656–657.

DOERER-SCHOTT, J., JOLY, L. and DUBOIS, M. P. (1978). Cells reacting with a somatostatin antiserum in the pars intercerebralis of an insect, *Locusta migratoria*. *C.R. Acad. Sci. 286*, 93–95.

DORN, A., BERSTEIN, H. G., RINNE, A., HAHN, H. and ZIEGLER, M. (1982). Insulin-like immunoreactivity in the human brain. *Histochemistry 74*, 293–300.

DOUGLAS, W. W. (1975). Prostaglandins. In *The Pharmacological Basis of Therapeutics*, pp.640–647. Edited by L. Goodman and A. Gilman. Macmillan, New York.

DUVE, H. and THORPE, A. (1979). Immunofluorescent localization of insulin-like material in the median neurosecretory cells of the blow fly, *Calliphora vomitoria* (Diptera). *Cell. Tiss. Res. 200,* 200–191.

DUVE, H. and THORPE, A. (1980). Localisation of pancreatic polypeptide (PP)-like immunoreactive material in neurones of the brain of the blow fly, *Calliphora erythrocephala* (Diptera). *Cell. Tiss. Res. 210,* 101–109.

DUVE, H., THORPE, A., NEVILLE, R. and LAZARUS, N. R. (1981a). Isolation and partial characterization of pancreatic polypeptide-like material in the brain of the blow fly, *Calliphora vomitoria. Biochem. J. 197,* 767–770.

DUVE, H. and THORPE, A. (1981c). Gastrin/cholecystokinin (CCK)-like immunoreactive neurons in the brain of the blow fly, *Calliphora erythrocephala* (Diptera). *Gen. Comp. Endocr. 43,* 381–391.

DUVE, H., THORPE, A. and LAZARUS, N. R. (1979). The isolation of material displaying insulin-like immunological and biological activity from the brain of the blow fly, *Calliphora vomitoria. Biochem. J. 184,* 221–227.

DUVE, H., THORPE, A., LAZARUS, N. R. and LOWRY, P. J. (1982). A neuropeptide of the blow fly *Calliphora vomitoria* with an amino acid composition homologous with vertebrate pancreatic polypeptide. *Biochem. J. 201,* 429–432.

EL-SALHY, M. (1981). Immunohistochemical localization of pancreatic polypeptide (PP) in the brain of the larval instar of the hoverfly, *Eristalis aeneus* (Diptera). *Experientia 37,* 1009–1010.

EL-SALHY, M., FALKMER, S. and KRAMER, K. J. (1982). The brain of the tobacco hornworm, *Manduca sexta,* as a neuroendocrine organ. A clue to the evolution of the diffuse neuroendocrine system. Abstr. 4th Int. Symp. on Gastrointestinal Hormones, Stockholm.

EL-SALHY, M., GRIMELIUS, L., FALKMER, S., ABOU-ELELLA, R. and WILANDER, E. (1980). Immunohistochemical evidence of gastroenteropancreatic neurohormonal peptides of vertebrate type in the nervous system of a dipteran insect, the hoverfly, *Eristalis aeneus. Regulatory Peptides 1,* 187–204.

ENG, J. and YALOW, R. S. (1980). Insulin recoverable from tissues. *Diabetes 29,* 105–109.

ENG, J. and YALOW, R. S. (1981). Evidence against extrapancreatic insulin synthesis. *Proc. Nat. Acad. Sci. USA 78,* 4576–4578.

FALKMER, S. (1979). Immunocytochemical studies of the evolution of islet hormones. *J. Histochem. Cytochem. 27,* 1281–1282.

FALKMER, S. and MARQUES, M. (1972). Phylogeny and ontogeny of glucagon production. In *Glucagon: Molecular Physiology, Clinical and Therapeutic Implications,* pp.343–361. Edited by P. J. Lefebvre and R. H. Unger. Pergamon Press, Oxford.

FALKMER, S. and OSTBERG, Y. (1976). Production of islet hormones in invertebrates, cyclostomes and primitive erathostomes. In *The Evolution of Pancreatic Islets.* Edited by T. Grillo, L. Leibson and A. Epple. Pergamon Press, Oxford.

FALKMER, S. and OSTBERG, Y. (1977). Comparative morphology of pancreatic islets in animals. In *The Diabetic Pancreas.* Edited by B. Volk and K. Wellmann. Plenum Press, New York.

FALKMER, S., CARRAWAY, R. E., EL-SALHY, M., EMDIN, S. O., GRIMELIUS, L., REHFELD, J. F., REINECKE, M. and SCHWARTZ, T. F. (1981). Phylogeny of the gastroenteropancreatic neuroendocrine system. A review with special reference to the occurrence of CCK-like and neurotensin-like polypeptides in lower vertebrates and invertebrates. *UCLA Forum Med. Sci. 23,* 21–42.

FALKMER, S. and VAN NOORDEN, S. (1982). Ontogeny and phylogeny of the glucagon cell. In *Handbuch der Experimentallen Pharmakologie.* Edited by G. V. R. Born, A. Farah, H. Herken and A. D. Welch. Volume *Glucagon.* Edited by P. J. Lefebvre. Springer-Verlag, Heidelberg (in press).

FALKMER, S. and VAN NOORDEN, S. (1982b). Ontogeny and phylogeny of the glucagon cell. In *Handbuch der Experimentellen Pharmakologie. Glucagon.* Edited by G. V. Born, A. Farah, H. Herken and A. D. Welch. Springer-Verlag (in press).

FALKMER, S., EMDIN, S., HAVU, N., LUNDGREN, G., MARQUES, M., OSTBERG, Y., STEINER, D. F. and THOMAS, N. W. (1973). Insulin in invertebrates and cyclostomes. *Amer. Zool. 13,* 625.

FALKMER, S., ELDE, R. P., HELLERSTROM, C., PETERSON, B., EFENDIC, S., FOHLMAN, J. and SILJEVALL, J. (1977). Some phylogenetical aspects in the occurrence of somatostatin in the gastro-entero-pancreatic endocrine system. A histological and immunocytochemical study, combined with quantitative radioimmunological assays of tissue extracts. *Arch. Histol. Jap. 40,* 99–117.

FALKMER, S., ELDE, R. P., HELLERSTROM, C. and PETERSON, B. (1978). Phylogenetic aspects of somatostatin in the gastroenteropancreatic (GEP) endocrine system. *Metabolism 27,* 1193–1196.

FALOONA, G. and UNGER, R. (1974). Biological and immunological activity of pancreatic glucagon and enteric glucagon-like immunoreactivity. *Israel J. Med. Sci. 10,* 1324–1331.

FLOYD, J. C., FAJANS, S. S., PEK, S. and CHANCE. R. E. (1977). Pancreatic polypeptide. *Rec. Prog. Horm. Res. 33,* 519–570.

FOX, R. M. and FOX, J. W. (1964). Ectodermal structures. In *Introduction to Comparative Entomology,* p.238. Reinhold, New York.

FRIEDEL, T., LOUGHTON, B. G. and ANDREW, R. D. (1980). A neurosecretory protein from *Locusta migratoria. Gen. Comp. Endocrinol. 41,* 487–498.

FRITSCH, H. A. R. and SPRANG, R. (1977). On the ultrastructure of polypeptide hormone-producing cells in the gut of the ascidian, *Ciona intestinalis* L. and the bivalve *Mytilus edulis* L. *Cell Tiss. Res. 177,* 407–414.

FRITSCH, H. A. R., VAN NOORDEN, S. and PEARSE, A. G. E. (1976). Cytochemical and immunofluorescence investigations on insulin-like producing cells in the intestine of *Mytilus edulis* L. (Bivalvia). *Cell Tiss. Res. 165,* 365–369.

FRITSCH, H. A. R., VAN NOORDEN, S. and PEARSE, A. G. E. (1978). Localization of somatostatin- and gastrin-like immunoreactivity in the gastrointestinal tract of *Ciona intestinalis* L. *Cell Tiss. Res. 186,* 181–185.

FRITSCH, H. A. R., VAN NOORDEN, S. and PEARSE, A. G. E. (1979). Localization of somatostatin, substance P, and calcitonin-like activity in the neural ganglia of *Ciona intestinalis. Cell Tiss. Res. 202,* 263–274.

FRITSCH, H. A. R., VAN NOORDEN, S. and PEARSE, A. G. E. (1980). Calcitonin-like immunocytochemical staining in the alimentary tract of *Ciona intestinalis* L. *Cell Tiss. Res. 205,* 439–444.

FRONTALI, N. and GAINER, H. (1977). Peptides in invertebrate nervous system. In *Peptides in Neurobiology,* pp.259–293. Edited by H. Gainer. Plenum, New York.

GADE, G. (1981). Activation of fat body glycogen phosphorylase in *Locusta migratoria* by corpus cardiacum extract and synthetic adipokinetic hormone. *J. Insect Physiol. 27,* 155–161.

GEORGES, D. and DUBOIS, M. P. (1979). Immunological evidence for an ACTH-like antigen in the nervous ganglion of a protochordate *Ciona intestinalis* (Ascidiacea). *Arch. Anat. Mic. Morph. Exp. 68,* 121–125.

GEORGES, D. and DUBOIS, M. P. (1980). Mise en évidence par une technique d'immunofluorescence d'un antigène de type LH-RH dans le système nerveux de *Ciona intestinalis* (Tunicier: Ascidiae). *C.R. Acad. Sci. Paris 290,* 29–31.

GIRARDIE, J. and REMY, C. (1981). Particularités histocytologiques de 2 cellules neurosecrétorices à vasopressine et neurophysine-like du criquet migrateur. *J. Physiol. Paris 77,* 61–65.

GOLDSWORTHY, G. J. (1970). The action of hyperglycaemic factor from the corpus cardiacum of *Locusta migratoria* on glycogen phosphorylase. *Gen. Comp. Endocr. 14,* 78–85.

GOLDSWORTHY, G. J. and MORDUE, W. (1974). Neurosecretory hormones in insects. *J. Endocr. 60,* 529–558.

GRIMM-JORGENSEN, Y. (1978). Immunoreactive thyrotropin releasing factor in a gastropod. Distribution in the central nervous system and haemolymph of *Lymnaea stagnalis. Gen. Comp. Endocr. 35,* 387–390.

GRIMM-JORGENSEN, Y. (1979). Effect of thyrotropin releasing factor on body weight of the pond snail *Lymnaea stagnalis. J. Exp. Zool. 208,* 169–176.

GRIMM-JORGENSEN, Y., MCKELVY, J. F. and JACKSON, I. M. D. (1975). Immunoreactive thyrotropin releasing factor in gastropod circumoesophageal ganglia. *Nature (London) 254,* 620.

GRIMMELIKHUIJZEN, C. J. P. and HOLST, J. J. (1981). Unpublished observations cited in Falkmer and Van Noorden, 1982.

GRIMMELIKHUIJZEN, C. J. P., SUNDLER, F. and REHFELD, J. F. (1980). Gastrin/CCK-like immunoreactivity in the nervous system of coelenterates. *Histochemistry 69,* 61–68.

GRIMMELIKHUIJZEN, C. J. P., BALFE, A., EMSON, P. C., POWELL, D. and SUNDLER, F. (1981a). Substance P-like immunoreactivity in the nervous system of Hydra. *Histochemistry 71*, 325–333.

GRIMMELIKHUIJZEN, C. J. P., CARRAWAY, R. E., ROKAENS, A. and SUNDLER, F. (1981b). Neurotesin-like immunoreactivity in the nervous system of *Hydra*. *Histochemistry 72*, 199–209.

GRIMMELIKHUIJZEN, C. J. P., DOCKRAY, G. J. and YANAIHARA, N. (1981c). Bombesin-like immunoreactivity in the nervous system of *Hydra*. *Histochemistry 73*, 171–180.

GROS, C., LAFON-CAZAL, M. and DRAY, F. (1978). Presénce de substances immunoréactivement apparentées aux enkephalines chez un insecte, *Locusta migratoria*. *C.R. Acad. Sci. Paris 287*, 647–650.

GUBITZ, H. (1966). Substanzen mit P-Wirkung aus verschiedenen nervosen Geweben und ihr Abbau mit Fermenten aus dem Nervensystem verschiedener Tiere. *Z. Biol. 115*, 194–214.

GUILLERMIN, C. R. (1976). Peptides from brain with potent opiate activity. *C.R. Acad. Sci. Paris 282*, 783–789.

HANAOKA, K. and TAKAHASHI, S. Y. (1976). Effect of a hyperglycemic factor on hemolymph trehalose and fat body carbohydrates in the American cockroach. *Insect Biochem. 6*, 621–625.

HAVRANKOVA, J., SCHWECHEL, D., ROTH, J. and BROWNSTEIN, M. (1978). Identification of insulin in rat brain. *Proc. Nat. Acad. Sci. USA 75*, 5737–5741.

HAYNES, L. W. (1980). Peptide neuroregulators in invertebrates. *Prog. Neurobiol. 15*, 205–245.

HEMMINGSEN, A. M. (1924). The action of insulin in the frog and some invertebrates. *Skand. Arch. Physiol. 46*, 56–63.

HUGHES, J. (1975). Identification of two related pentapeptides from the brain with potent opiate agonist activity. *Nature (London) 258*, 577–579.

ISHAY, J. (1975). Glucose levels in *Vespa orientalis*: the effect of starvation. *Comp. Biochem. Physiol. 52A*, 91–96.

ISHAY, J., GITTER, S., GALUN, R., DORON, M. and LARON, Z. (1976). The presence of insulin in and some effects of exogenous insulin on hymenopteran tissues and body fluids. *Comp. Biochem. Physiol. 54A*, 203–206.

IWANAGA, T., FUJITA, T., NISHIITSUTSUJI-UWO, J. and ENDO, Y. (1981). Immunohistochemical demonstration of PP-, somatostatin-, enteroglucagon- and VIP-like immunoreactivities in the cockroach midgut. *Biomed. Res. 2*, 202–207.

KEELEY, L. L. and FRIEDMAN, S. (1967).Corpus cardiacum as a metabolic regulator in *Blaberas discoidalis*. 1. Long term effects of cardiatectomy in whole body and tissue respiration and in trophic metabolism. *Gen. Comp. Endocrinol. 8*, 129–134.

KELLER, R. (1981). Purification and amino acid composition of the hyperglycemic neurohormone from the sinus gland of *Orconectes limosus* and comparison with the hormone from *Carcinus maenas*. *J. Comp. Physiol. 141*, 445–450.

KRAMER, K. J., CHILDS, C. N. and SPEIRS, R. D. (1979). Effect of neutral red on carbohydrate levels in the tobacco hornworm, *Manduca sexta* (L.) (Lepidoptera: Sphingidae). *Comp. Biochem. Physiol. 64C*, 229–230.

KRAMER, K. J., CHILDS, C. N., SPEIRS, R. D. and JACOBS, R. M. (1982). Purification of insulin-like peptides from insect hemolymph and royal jelly. *Insect Biochem. 12*, 91–98.

KRAMER, K. J., JACOBS, R. M., SPEIRS, R. D. and HENDRICKS, L. H. (1978). Effect of vertebrate hypoglycemic and β-cell cytotoxic agents in insects. *Comp. Biochem. Physiol. 61C*, 95–97.

KRAMER, K. J., SPEIRS, R. D. and CHILDS, C. N. (1977a). Immunochemical evidence for a gastrin-like peptide in insect neuroendocrine system. *Gen. Comp. Endocr. 32*, 423–426.

KRAMER, K. J., TAGER, H. S., CHILDS, C. N. and SPEIRS, R. D. (1977b). Insulin-like hypoglycemic and immunological activities in honeybee royal jelly. *J. Insect Physiol. 23*, 293–295.

KRAMER, K. J., TAGER, H. S. and CHILDS, C. N. (1980). Insulin-like and glucagon-like peptides in insect hemolymph. *Insect Biochem. 10*, 179–182.

LARNER, J. and HAYNES, R. C. (1975a). Insulin and oral hypoglycemic drugs. In *The Pharmacological Basis of Therapeutics*, pp.1507–1520. Edited by L. Goodman and A. Gilman. Macmillan, New York.

LARNER, J. and HAYNES, R. C. (1975b). Glucagon. In *The Pharmacological Basis of Therapeutics*, pp.1527–1529. Edited by L. Goodman and A. Gilman. Macmillan, New York.

LEGROS, F., UYTDENHOFF, P., DUMONT, I., HANSON, B., JEANMART, J., MASSANT, B. and CONRARD, V. (1975). Specific binding of insulin to the unicellular alga *Acetabularia mediterranea*. *Protoplasma 86*, 119–134.

LE ROITH, D., LESNICK, M. A. and ROTH, J. (1981a). Insulin in insects and annelids. *Diabetes 30*, 70–76.

LE ROITH, D., SHILOACH, J., ROTH, J. and LESNICK, M. A. (1980). Evolutionary origin of vertebrate hormones: substances similar to mammalian insulins are native to unicellular eukaryotes. *Proc. Nat. Acad. Sci. USA 77*, 6184–6188.

LE ROITH, D., SHILOACH, J., ROTH, J., and LESNICK, M. A. (1981b). Insulin or a closely related molecule is native to *Escherichia coli*. *J. Biol. Chem. 256*, 6533–6536.

LE ROITH, D., SHILOACH, J., ROTH, J. and LESNICK, M. A. (1981c). Unpublished. Quoted in Le Roith *et al.*, 1981b.

LOHER, W., GANJIAN, I., KUBO, I., STANLEY-SAMUELSON, D. and TOBE, S. S. (1981). Prostaglandins: their role in egg-laying of the cricket *Teleogryllus commodus*. *Proc. Nat. Acad. Sci. USA 78*, 7835–7838.

LOREN, I., ALUMETS, J., HAKANSON, R., SUNDLER, F. and THORELL, J. (1979). Gut-type glucagon immunoreactivity in nerves of the rat brain. *Histochemistry 61*, 335–341.

LOREN, I., ALUMETS, J., HAKANSON, R., SUNDLER, F. and THORELL, J. (1979). Pancreatic polypeptide immunoreactivity in mammalian brain. *Cell Tiss. Res. 200*, 179–186.

LOUGHTON, B. G. and ORCHARD, I. (1981). The nature of the hyperglycemic factor from the glandular lobe of the corpus cardiacum of *Locusta migratoria*. *J. Insect Physiol. 27*, 383–385.

LUBROVIAK, K., MOORE, G., THORNHILL, J. A. and LEDERIS, K. (1979). The vertebrate neuropeptide hormone, arginine vasotocin, is present in the nervous system of the invertebrate *Aplysia californica*. *Fed. Proc. 38*, 1745.

LUI, T. P. (1973). The effect of allatectomy on blood trehalose in the female housefly, *Musca domestica* L. *Comp. Biochem. Physiol. 46A*, 109–113.

MAIER, V., GRUBE, D., STEINER, R., GREISCHEL, A. and PFEIFFER, E. (1981). Glucagon and insulin in the brain of the honeybee. Absts. Internat. Sympos. "Brain–Gut Axis: The New Frontier", p.119. Florence, 29 June–1 July.

MAIER, V., KRODER, A., KELLER, R. and PFEIFFER, E. F. (1975). Glucagon-like activity (GLI) in the intestine of *Porcus domesticus* and *Astacus fluviatilis*. *Acta Endocr. Suppl. 193*, 41.

MAIER, V., WITZNICK, G., KELLER, R. and PFEIFFER, E. F. (1978). Insulin-like and glucagon-like immunoreactivities in the honeybee (*Apis mellifera*). *Acta Endocr. 87*, 69–70.

MANCILLAS, J. R., MCGINTY, J. F., SELVERSTON, A. I., KARTEN, H. and BLOOM, F. E. (1981). Immunocytochemical localization of enkephalin and substance P in retina and eyestalk neurons of lobster. *Nature 293*, 676–678.

MARQUES, M. and FALKMER, S. (1976). Effects of mammalian insulin on blood glucose level, glucose tolerance, and glycogen content of musculature and hepatopancreas in a gastropod mollusk, *Strophocheilus oblongus*. *Gen. Comp. Endocr. 31*, 217–223.

MARTIN, G. and DUBOIS, M. P. (1981). A somatostatin-like antigen in the nervous system of an isopod *Porcellio dilatatus* Brandt. *Gen. Comp. Endocr. 45*, 125–130.

MARTIN, P. and SHEARN, A. (1980). Development of *Drosophila* imaginal discs *in vitro*: effects of ecdysone concentration and insulin. *J. Exp. Zool. 211*, 291–301.

MARTIN, R., FROSCH, D., WEBER, E. and VOIGT, K. H. (1979). Met-enkephalin-like immunoreactivity in a cephalopod neurohemal organ. *Neurosci. Lett. 15*, 253–257.

MARTIN, R., FROSCH, D. and VOIGT, K. H. (1980). Immunocytochemical evidence for melanotropin- and vasopressin-like material in a cephalopod neurohemal organ. *Gen. Comp. Endocr. 42*, 235–243.

MCLAUGHLIN, P. J. and DAYHOFF, M. O. (1972). Evolution of species and proteins: a time scale. In *Atlas of Protein Sequence and Structures*. National Biomedical Research Foundation, Silversprings, Maryland. Vol. 5, 47–52.

MENESES, P. and ORTIZ, M. (1975). A protein extract from *Drosophila melanogaster* with insulin-like activity. *Comp. Biochem. Physiol. 51A*, 483–485.

MOODY, A. J., JACOBSEN, H., SUNDBY, F., FRANDSEN, E. R., BAETENS, D. and ORCI, L. (1977). Gastrointestinal glucagon-like immunoreactivity. In *Glucagon: its Role in Physiology and Clinical Medicine*, pp.129–138. Edited by P. P. Foa, J. S. Bajaj and N. L. Foa. Springer, New York.

MORDUE, W. and GOLDSWORTHY, G. J. (1969). The physiological effects of corpus cardiacum extracts in locusts. *Gen. Comp. Endocr. 12*, 360–369.

MOREAU, R., RAOELISON, C. and SUTTER, B. C. J. (1981). An intestinal insulin-like molecule in *Apis mellifica* L. (Hymenoptera). *Comp. Biochem. Physiol. 69A*, 79–83.

MOSNA, G. (1981). Insulin can completely replace serum in *Drosophila melanogaster* cell culture *in vitro*. *Experientia 37*, 466–467.

MOSNA, G. and BARIGOZZI, C. (1976). Stimulation of growth by insulin in *Drosophila* embryonic cells *in vitro*. *Experientia 32*, 855–856.

MURPHY, W. A., FRIES, J. L., MAYERS, C. A. and COY, D. H. (1981). Human pancreatic polypeptide inhibits insulin release in the rat. *Biochem. Biophys, Res. Commun. 101*, 189–193.

NATALIZI, G. M. and FRONTALI, N. (1966). Purification of insect hyperglycemic and heart accelerating hormones. *J. Insect Physiol. 12*, 1279–1287.

NORMANN, T. C. (1975). Neurosecretory cells in insect brain and production of hypoglycemic hormone. *Nature 254*, 259–261.

NORMANN, T. C. and DUVE, H. (1969). Experimentally induced release of a neurohormone influencing hemolymph trehalose level in *Calliphora erythrocephala*. *Gen. Comp. Endocr. 12*, 449–459.

ORCHARD, I. and LOUGHTON, B. G. (1980). A hypolipaemic factor from the corpus cardiacum of locusts. *Nature, London 286*, 494–496.

PATTON, R. L. and KUO, C. K. (1977). The distribution of the STH-like protein in insects. *Insect Biochem. 7*, 487–489.

PEARSE, A. G. E. (1976). Peptides in brain and intestine. *Nature, London 262*, 92–94.

PLISETSKAYA, E., KAZAKOV, V. K., SOLBITSKAYA, L. and LEIBSON, L. G. (1978). Insulin-producing cells in the gut of freshwater bivalve molluscs *Anodenta cygnea* and *Unio pictorum* and the role of insulin in the regulation of their carbohydrate metabolism. *Gen. Comp. Endocr. 35*, 133–145.

PRICE, D. A. and GREENBERG, M. J. (1977). Structure of a molluscan cardioexcitatory neuropeptide. *Science 197*, 670–671.

PROUX, J. and ROUGON-RAPUZZI, G. (1980). Evidence for vasopressin-like molecules in migratory locust. Radioimmunological measurements in different tissues: Correlation with various states of hydration. *Gen. Comp. Endocr. 43*, 378–383.

PROUX, J., ROUGON-RAPUZZI, G. and REMY, C. (1981). Influence de la pars intercerebralis et des cellules neurosecrétrices sous-ocellaires médianes sur la teneur en "vassopressine-like" dans la claire nerveuse ventrale et l'hémolymphe du criquet migrateur. Etude radioimmunologique et immunohistologique. *J. Physiol. Paris 77*, 66–67.

REHFELD, J. F., LARSSON, L. I., GOLTERMANN, N. R., SCHWARTZ, T. W., HOLST, J. J., JENSEN, S. L. and MORLEY, J. S. (1980). Neural regulation of pancreatic hormone secretion by the C-terminal tetrapeptide of CCK. *Nature 284*, 33–38.

REINECKE, M. (1981). Immunohistochemical localization of polypeptide hormones in the endocrine cells of the digestive tract of *Branchiostoma lanceolatum*. *Cell Tiss. Res. 219*, 445–456.

REINECKE, M., CARRAWAY, R. E., FALKMER, S., FEURLE, G. E., and FORSSMANN, W. G. (1980). Occurrence of neurotensin immunoreactive cells in the digestive tract of lower vertebrates and deuterostomian invertebrates. A correlated immunohistochemical and radioimmunological study. *Cell Tiss. Res. 212*, 173–178.

REMY, D. and DUBOIS, M. P. (1979). Localisation par immunofluorescence de peptides analogues à l'α-endorphine dans les ganglions infra-oesophagiens du lombricide *Dendrobaena subrubicunda* Eisen. *Experientia 35*, 137–138.

REMY, C. and DUBOIS, M. P. (1981). Immunohistochemical evidence of methionine enkephalin-like material in the brain of the migratory locust. *Cell Tiss. Res. 218*, 271–278.

REMY, C. and GIRARDIE, J. (1980). Anatomical organization of two vasopressin–neurophysin-like neurosecretory cells throughout the central nervous system of the migratory locust. *Gen. Comp. Endocr. 40*, 27–35.

REMY, C., GIRARDIE, J. and DUBOIS, M. P. (1977). Exploration immunocytologique des ganglions cérébroides at sous-oesophagiens du plasme *Clitumnus extradentatus*: existence d'une neurosécrétion apparentée à la vasopressin–neurophysine. *C.R. Acad. Sci. 285*, 1495–1497.

REMY, C., GIRARDIE, J. and DUBOIS, M. P. (1978). Présence dans le ganglion sous-oesophagien de la chénille processionaire du pin (*Thaumetopoea pityocampa* Schiff) de cellules révélées en immunofluorescence par un articorps anti-α-endorphine. *C.R. Acad. Sci. 286*, 651–653.

REMY, C., GIRARDIE, J., and DUBOIS, M. P. (1979). Vertebrate neuropeptide-like substances in the suboesophageal ganglion of two insects: *Locusta migratoria* R. and F. (Orthoptera) and *Bombyx mori* L. (Lepidoptera): immunocytochemical investigation. *Gen. Comp. Endocr. 37*, 93–100.

ROSENZWEIG, J. L., HAVRANKOVA, J., LESNICK, M. A., BROWNSTEIN, M. and ROTH, J. (1980). Insulin is ubiquitous in extrapancreatic tissues of rats and humans. *Proc. Nat. Acad. Sci. USA 77*, 572–576.

RZASA, P., KALOUSTIAN, K. U. and PROKOP, E. K. (1982). Immunochemical evidence for a gastrin-like peptide in the intestinal tissues of the earthworm *Lumbricus terrestris*. *Comp. Biochem. Physiol. 71A*, 613–615.

SANDERS, B. M. (1980). Insulin-like activity and immunoreactivity in *Homarus americanus*. *Amer. Zool. 20*, 933.

SANDERS, D. J., ZAHEDI-ASL, S., MARR, A. P., PERRY, E. K. and PERRY, R. H. (1981). Glucagon in human brain: distribution in controls and patients with alzheiner disease. Abstr. Internat. Sympos. "Brain–Gut Axis: The New Frontier". Florence, p.160. 29 June–1 July.

SCHALLER, H. C. and BODENMULLER, H. (1981). Isolation and amino acid sequence of a morphogenetic peptide from Hydra. *Proc. Nat. Acad. Sci. USA 78*, 7000–7004.

SCHALLER, H. C., FLICK, U. and DARAI, G. (1977). A neurohormone from *Hydra* is present in brain and intestine of rat embryos. *J. Neurochem. 29*, 393–394.

SCHILT, J., RICHOUX, J. P. and DUBOIS, M. P. (1981). Demonstration of peptides immunologically related to vertebrate neurohormones in *Dugesia lugubris* (Turbellaria: Tricladida). *Gen. Comp. Endocr. 43*, 331–335.

SCHOT, L. P. C., BOER, H. H., SWAAB, D. F. and VAN NOORDEN, S. (1981). Immunocytochemical demonstration of peptidergic neurons in the central nervous system of the pond snail *Lymnaea stagnalis* with antisera raised to biologically active peptides of vertebrates. *Cell Tiss. Res. 216*, 273–291.

SEECOF, R. C. and DEWHURST, S. (1974). Insulin is a *Drosophila* hormone and acts to enhance the differentiation of embryonic *Drosophila* cells. *Cell Diff. 3*, 63–70.

SETTY, B. N. Y. and RAMAIAH, T. R. (1979). Isolation and identification of prostaglandins from the reproductive organs of male silkmoth, *Bombyx mori* L. *Insect Biochem. 9*, 613–617.

SMITH, L. F. (1966). Species variation in the amino acid sequence of insulin. *Amer. J. Med. 40*, 662–666.

STEELE, J. E. (1963). The site of action of insect hyperglycemic hormone. *Gen. Comp. Endocr. 3*, 46–52.

STRAMBI, C., STRAMBI, A., CUPO, A., ROUGON-RAPUZZI, G. and MARTIN, N. (1978). Etude des taux d'une substance apparentée à la vasopressine dans le système nerveux de grillons soumis à différentes conditions hypgromètriques. *C.R. Acad. Sci. Paris 287*, 1227–1230.

STRAMBI, C., ROUGON-RAPUZZI, G., CUPO, A., MARTIN, N. and STRAMI, A. (1979). Mise en évidence immunocytologique d'un compose apparenté à la vasopressine dans le système nerveux du grillon *Acheta domesticus*. *C.R. Acad. Sci. D. 288*, 131–133.

STRAMBI, C., STRAMBI, A., ROUGON-RAPUZZI, G., CUPO, A. and MARTIN, N. (1979). Radioimmunological and immunocytological study of a hormone related to vasopressin in the insect. *J. Physiol. (Paris) 75*, 63.

STRAUS, E., YALOW, R. S. and GAINER, H. (1975). Molluscan gastrin concentration and molecular forms. *Science 190*, 687–689.

SUNDLER, F., HAKANSON, R., ALUMETS, J. and WALLES, B. (1977). Neuronal localization of pancreatic polypeptide (PP) and vasoactive intestinal peptide (VIP) in the earthworm (*Lumbricus terrestris*). *Brain Res. Bull. 2*, 61–65.

TABAN, C. H. and CATHIENI, M. (1979). Localization of substance P-like immunoreactivity in *Hydra*. *Experientia 35*, 811–812.

TAGER, H. S. and KRAMER, K. J. (1980). Insect glucagon-like peptides: evidence for high molecular weight form in midgut from *Manduca sexta* (L.). *Insect Biochem. 10,* 617–619.

TAGER, H. S. and MARKESE, J. (1979). Intestinal and pancreatic glucagon-like peptides: evidence for identity of higher molecular weight forms. *J. Biol. Chem. 254,* 2229–2233.

TAGER, H., HOHEMBOKEN, M., MARKESE, J. and DINERSTEIN, R. J. (1980). Identification and localization of glucagon-related peptides in rat brain. *Proc. Nat. Acad. Sci. USA 77,* 6229–6233.

TAGER, H. S., MARKESE, J., KRAMER, K. J. and SPEIRS, R. D. (1975). Glucagon-like immunoreactivity in insect corpus cardiacum. *Nature (London) 254,* 707–708.

TAGER, H. S., MARKESE, J., KRAMER, K. J., SPEIRS, R. D. and CHILDS, C. N. (1976). Glucagon-like and insulin-like hormones of the insect neurosecretory system. *Biochem. J. 156,* 515–520.

TEITELMAN, G., JOH, T. H. and REIS, D. J. (1981). Transformation of catecholaminergic precursors into glucagon (A) cells in mouse embryonic pancreas. *Proc. Nat. Acad. Sci. USA 78,* 5225–5229.

TEITELMAN, G., JOH, T. H. and REIS, D. J. (1983). *Neuroscience 10,* 907–922.

THORNDYKE, M. C. (1981). Peptide systems in protochordates. I. The brain–gut axis. *Gen. Comp. Endocr. 45*

THORNDYKE, M. C. and BEVIS, P. J. R. (1980). Paracrine control of gastric secretion in protochordates. *Regul. Pept. 1,* 115.

THORNDYKE, M. C. and PROBERT, L. (1979). Calcitonin-like cells in the pharynx of the ascidian *Styela clava. Cell Tiss. Res. 203,* 301–309.

VALE, W., RIVIER, C. and BROWN, M. (1977). Regulatory peptides of the hypothalamus. *Ann. Rev. Physiol. 29,* 473–527.

VAN NOORDEN, S. and FALKMER, S. (1980). Gut–islet endocrinology — some evolutionary aspects. *Invest. Cell Pathol. 3,* 21–35.

VAN NOORDEN, S. and PEARSE, A. G. E. (1976). The localization of immunoreactivity to insulin, glucagon and gastrin in the gut of *Amphioxus lanceolatus.* In *The Evolution of Pancreatic Islets,* pp.163–178. Edited by T. Grillo, L. Leibson and A. Epple. Pergamon Press, Oxford.

VAN NOORDEN, S., FRITSCH, H. A. R., GRILLO, T. A. I., POLAK, J. M. and PEARSE, A. G. E. (1980). Immunocytochemical staining for vertebrate peptides in the nervous system of a gastropod mollusc. *Gen. Comp. Endocr. 40,* 375–376.

VAN NORSTRAND, M., CARLSEN, J., JOSEFSSON, L. and HERMAN, W. S. (1980). Studies on a peptide with red pigment-concentrating and hyperglycemic activity from the cephalic endocrine system of the honeybee, *Apis mellifera. Gen. Comp. Endocr. 42,* 526–533.

VEJBJERG, K. and NORMANN, T. C. (1974). Secretion of hyperglycemic hormone from the corpus cardiacum of flying blow flies, *Calliphora erythrocephala. J. Insect Physiol. 20,* 1189–1192.

WAKAYAMA, E. J., DILLWITH, J. W. and BLOMQUIST, G. J. (1980). *In vitro* biosynthesis of prostaglandins in the reproductive tissues of the male housefly, *Musca domestica* L. *Amer. Zool. 20,* 904.

WEBER, E., EVANS, C. J., SAMUELSSON, S. J. and BARCHAS, J. V. (1981). Novel peptide neuronal system in rat brain and pituitary. *Science 214,* 1248–1251.

WENIG, K. and JOACHIM, J. (1936). Der Einfluss des Insulins auf der Lymphzuckenspiegel bei der Seidenraupe (*Bombyx mori*). *Biochem. Z. 285,* 98–100.

WIENS, A. W. and GILBERT, L. I. (1967). Regulation of carbohydrate mobilization and utilization in *Leucophaea maderae. J. Insect Physiol. 13,* 779–794.

WILSON, W. and FALKMER, S. (1965). Starfish insulin. *Can. J. Biochem. 43,* 1615–1624.

YALOW, R. S. and BERSON, S. A. (1964). Reaction of fish insulins with human insulin antisera. *N. Engl. J. Med. 270,* 1171–1181.

YUI, R., FUJITA, T. and ITO, S. (1980). Insulin-, gastrin-, pancreatic polypeptide-like immunoreactive neurons in the brain of the silkworm, *Bombyx mori. Biomed. Res. 1,* 42–46.

ZIEGLER, R. (1979). Hyperglycemic factor from the corpora cardiacum of *Manduca sexta* (L.). *Gen. Comp. Endocr. 29,* 350–357.

ZIPSER, B. (1980). Identification of specific leech neurones immunoreactive to enkephalin. *Nature 283,* 857–858.

ZYZNAR, E. S. (1981). A rationale for the application of trasylol as a protease inhibitor in radioimmunoassay. *Life Sci. 28,* 1861–1866.

ADDENDUM

Several papers have recently appeared that describe the identification and/or localization of vertebrate hormone-like peptides in invertebrates, including:

BENEDECZKY, I., KISS, J. Z. and SOMOGYI, P. (1982). Light and electron microscopic localization of substance P-like immunoreactivity in the cerebral ganglion of the locust with a monoclonal antibody. *Histochem. 75,* 123–131.

CHAMBERLAIN, S. C. and ENGBRETSON, G. A. (1982). Neuropeptide immunoreactivity in *Limulus.* I. Substance P-like immunoreactivity in the lateral eye and protocerebrum. *J. Comp. Neurol. 208,* 304–315.

DUVE, H. and THORPE, A. (1982). Vertebrate-type brain, gut peptides in the brain of the blow fly *Calliphora vomitoria*: immunocytochemical localization of α-endorphin-like peptide. *Gen. Comp. Endocrinol. 46,* 371.

DUVE, H. and THORPE, A. (1982). The distribution of pancreatic polypeptide in the nervous system and gut of the blow fly, *Calliphora vomitoria* (Diptera). *Cell Tiss. Res. 227,* 67–77.

ENGELHARDT, R. P., DHAINAUT-COURTOIS, N. and TRAMU, G. (1982). Immunohistochemical demonstration of a CCK-like peptide in the nervous system of a marine annelid worm, *Nereis diversicolor O. F. Müller. Cell Tiss. Res. 227,* 401–411.

EL-SALHY, M., FALKMER, S., KRAMER, K. J. and SPEIRS, R. D. (1983). Immunohistochemical investigations of neuropeptides in the brain, corpora cardiaca and corpora allata of an adult lepidopteran insect, *Manduca sexta* L. *Cell Tiss. Res. 232,* 295–317.

ENDO, Y., IWANAGA, T., FUJITA, T. and NISHIITSUTSUJI-UWO, J. (1982). Localization of pancreatic polypeptide (PP)-like immunoreactivity in the central and visceral nervous systems of the cockroach *Periplaneta. Cell Tiss. Res. 227,* 1–9.

FRITSCH, H. A. R., VAN NOORDEN, S. and PEARSE, A. G. E. (1982). Gastrointestinal and neurohormonal peptides in the alimentary tract and cerebral complex of *Ciona intestinalis. Cell Tiss. Res. 223,* 369–402.

FUJITA, T., YUI, R., IWANAGA, T., NISHIITSUTSUJI-UWO, J., ENDO, Y. and YANAIHARA, N. (1981). Evolutionary aspects of "Brain–gut peptides": an immunohistochemical study. *Peptides 2*: suppl. 2, 123–131.

GRIMM-JORGENSEN, Y. (1983). Immunoreactive somatostatin in two pulmonate gastropods. *Gen. Comp. Endocr. 49,* 108–114.

HANSEN, B. L., HANSEN, G. N. and SCHARRER, B. (1982). Immunoreactive material resembling vertebrate neuropeptides in the corpus cardiacum and corpus allatum of the insect *Leucophaea maderae. Cell Tiss. Res. 225,* 319–329.

MURTAUGH, M. P. and DENLINGER, D. L. (1982). Prostaglandins E and F $_{2\alpha}$ in the house cricket and other insects. *Insect Biochem. 12,* 599–603.

MUNDAL, S., GHOSH, B. and CHOUDHURI, D. K. (1982). Insulin and lycopodium extract acting as a hypoglycemic agent in *Lohita grandis* Gray (Heteroptera: Pyrrhocoridae: Insecta). *Curr. Science 51,* 51–52.

MOREAU, R., GOURDOUX, L., LEQUELLEC, Y. and DUTRIEU, J. (1982). Endocrine control of hemolymph carbohydrates in *Locusta migratoria*: comparison between effects of two endogenous hormonal extracts and effects of insulin and glucagon. *Comp. Biochem. Physiol. 73A,* 669–673.

OSBORNE, N. N., CUELLO, A. C. and DOCKRAY, G. J. (1982). Substance P and cholecystokinin-like peptides in *Helix* neurone and cholecystokinin and serotonin in a giant neurone. *Science 216,* 409–411.

OSBORNE, N. N. and DOCKRAY, G. J. (1982). Bombesin-like immunoreactivity in specific neurones of the snail *Helix aspersa* and an example of the coexistence of substance P and serotonin in an invertebrate neurone. *Neurochem. Internat. 4,* 175–180.

OSBORNE, N. N., PATEL, S. and DOCKRAY, G. (1982). Immunohistochemical demonstration of peptides, serotonin and dopamine-β-hydroxylase-like material in the nervous system of the leech *Hirudo medicinalis. Histochem. 75,* 573–583.

PESTARINO, M. (1982). Occurrence of different secretin-like cells in the digestive tract of the ascidian *Styela plicata* (Urochordata, Ascidiacea). *Cell Tiss. Res. 226,* 231–235.

PROUX, J., ROUGON, G. and CUPO, A. (1982). Enhancement of excretion across locust Malphigian tubules by a diuretic vasopressin-like hormone. *Gen. Comp. Endocrinol. 47,* 449–457.

RAABE, M. (1982). Insect Neurohormones. Plenum Press, New York. 300pp.

SHEN, L. P., HOBART, P., CRAWFORD, R., PICTET, R. and RUTTER, W. J. (1982). Endocrine pancreas specific genes in vertebrates and invertebrates. *J. Supramol. Struc. Cellular Biochem. 1*: suppl. 5, 402.

STEFANO, G. B. and SCHARRER, B. (1981). High affinity binding of an enkephalin analog in the cerebral ganglion of the insect *Leucophaea maderae* (Blattaria). *Brain Res. 225*, 107–14.

TELLER, J. K., ROSINSKI, G., PILC, L., KASPRZYL, A. and LESICKI, A. (1983). The presence of insulin-like hormone in heads and midguts of *Tenebrio molitor* L. (Coleoptera) larvae. *Comp. Biochem. Physiol. 74A*, 463–465.

VENTURINI, G., CAROLEI, A., PALLADINI, G., MARGOTTA, U. and LAURO, M. G. (1983). Radioimmunological and immunocytochemical demonstration of met-enkephalin in Planaria. *Comp. Biochem. Physiol. 74C*, 23–25.

Species Index

Author Index

Subject Index